W9-DDQ-652

Optimum Pipe Size Selection

OPTIMUM PIPE SIZE SELECTION

By

Claude B. Nolte, Ph.D.
Technical Consultant
Newport Beach, California

First Edition
1978

Trans Tech Publications
Clausthal, Germany

This Edition
1979

Gulf Publishing Company
Book Division
Houston, London, Paris, Tokyo

ACKNOWLEDGEMENTS

The assistance and cooperation of the following persons and companies is gratefully acknowledged.

William J. Beckett
Julia Bishop
Chemical Engineering Magazine
Clark Brothers Company
Helen Cooper
Crane Company
Fluor Engineers & Constructors
Thomas T. Gill
Sheila Hansburger
E.R. McCartney
Arthur G. McKee & Company
Ryle Miller

PREFACE

To many of us who witnessed the use as well as used the original methods which now appear in Nolte's books, it is a great satisfaction to see the material now available for engineers and designers. This is the first time, to my knowledge, that all of the formulae, tables and charts have been brought together under one cover. By including the background for the original data with the development of its use for the present pipe sizing method, the book becomes a textbook for the neophyte as well as a cook book for the experienced engineer and designer.

Nolte has injected much original thought into the work allowing him to identify and use the typical pipe concept for a simplified approach to "Least Annual Cost" and "Pressure Drop Available" for determining sizing for economic piping systems. Inclusion of a means to adjust for pipe materials and labor costs easily, has increased the applicability of this system. Numerous standard and original nomographs are included to expedite the use of the methods outlined.

Nolte is to be congratulated for revitalizing some economic approaches which may have, unfortunately, been sidelined and for contributing fresh thought to a very necessary function of plant engineering to make it possible to obtain consistent results.

May 1978 — Chester S. Beard

Library of Congress Catalog Number 78-75079

ISBN 0-87201-650-1

This edition published 1979 by Gulf Publishing Co., Houston, Texas.

TABLE OF CONTENTS

LIST OF NOMOGRAPHS

INTRODUCTION

Commonly the Introduction to a book may be passed over or given only cursory attention. In this instance, important information is presented so the reader is urged to peruse this section of the book with care.

The message to be given here may seem unrelated to Optimum Pipe Size Selection because the data will deal mainly with the correct selection of the pressure drop to be taken across control valves within the piping loop. As will be shown, there is a distinct and important correlation between piping friction, equipment pressure loss, and the pressure drop to be allotted to the control valve. This is offered as a unique concept, published by the author years ago, and updated here with additional data.

Successful plant operation depends upon the establishment of a proper balance between the sum of the equipment piping pressure losses, and the design pressure drop assigned to the control valve, at the normal rate of flow for the system.

The importance of this factor often is not recognized. Rather, an arbitrary allowance of 10 psi to 15 psi is made for the control valve. In some applications, the friction in the pipe, heat exchangers, etc., does not entail more than 20 to 25 psi. In such a case, provided that the variation in the rate of flow is modest, this balance of the pressure losses will result in satisfactory valve performance.

However, when an engineer is faced with the selection of the proper pressure loss to be allowed for the valve when such simple conditions do not prevail, it then becomes essential to consider the factors which influence the flexibility of the control valve. These factors are: the probable variations in the flow rate (during normal, and even abnormal conditions), the initial loading of the valve, and the percentage of the total system friction losses which are alloted to the control valve.

A control valve is essentially a variable resistance placed in the system to maintain a specific rate of flow against varying head losses or to hold a specified pressure differential with varying flow rates. Generally an automatic control valve is not required in a system where variations in flow or pressure drop are unimportant.

One can set the flow rate for a lawn sprinkler, for example, by an adjustment of the faucet. If the pressure in the system changes and the rate of flow increases or lessens, the effect will be unimportant. On the other hand, the feed rate to a butane splitter must be kept extra constant. In process plants, some loops require more rigorous control than others. The system is designed for a certain throughput, but as a consequence of variations, for example, in the gasoline content of the gas or crude stock composition, the actual flow rates in various pipes may be larger or smaller than design. Prediction of the precise magnitude of the variations may be difficult. However, using the methods to be described, great precision is not required to obtain proper control. Approximations of variations in flow rates will suffice to give satisfactory performance.

Conferences with process engineers generally will reveal whether the probable plant variations will produce flow rates which are greater or lesser than the design conditions and whether the degree of difference will be in the order of 10 percent or 30 percent, or more.

Examples will help to clarify the picture. In a gasoline extraction plant, consider the flow of hot, lean oil from the stripper. Normally, this will not be subject to significant variation in flow rate because the lean oil flow to the absorber is set by a flow controller, and the system pressure is set. Consequently, the pressure drop to be absorbed by the control valve regulating the flow through the heat exchangers to the lean-oil surge tank need only be a fraction of the friction in the exchangers and the pipelines, and good control will still be attained. By contrast, the back-pressure regulator on the stripper overhead accumulator will be required to accomodate wide variations in flow. Therefore, it must be designed to provide a large percent of the total friction loss to maintain control.

Heat exchanger and pipe friction must be known with a fair degree of accuracy. It is to be realized that the most important source of friction is in the equipment rather than in the pipe; hence, one should obtain best possible estimates from the manufacturer.

The challenge of correct control valve selection may be reduced to the basic determination of how much pressure drop must be alloted to the control valve, in comparison to the friction in the pipes and equipment, at the design flow rate, after considering the probable flow rate variations. If the allottment to the valve is too small, minor variations in flow rate, or system pressure differences, may result in the valve shifting to the wide open or nearly closed position even though at the design flow rate the valve would be at normal opening. In either case, one is in danger of losing control! If too much pressure loss is assigned to the control valve, power will be wasted. Therefore, proper control valve size selection can influence plant operation and economics.

"Valve loading" designates the ratio of the actual flow rate through the valve to the maximum capacity of the valve (wide open) at a specific pressure drop. For a valve capable of passing 100 gpm with a pressure drop of 10 psi, if the valve were to be passing 60 gpm with the 10 psi drop, the valve loading would be 0.6.

Control valves normally are to be selected so that the loading at design conditions will be in the range of 0.20 to 0.85, with the final values depending upon the magnitude of process lags and the probable variations in rate of flow, and whether these will be likely to be above or below the normal flow rate. Ideally a valve should be able to control with a loading from essentially 1.0 down to about 0.05. At the practical level, such a wide range usually is not possible. The change in the flow rate for a given percent of valve-stem travel often is not constant over the entire range of the valve, certainly not for the most practical plug styles. This means that the valve's sensitivity will not be in balance with the sensitivity of the controller when the valve loading deviates greatly during normal variations in plant operations. For the operation to be satisfactory, generally it is recognized that the over-all sensitivity of the controller and the valve must be in balance with the particular system which is being controlled. If the system has extensive time lags, the balance between the system and the controller-valve behavior readily can be upset should changes occur in the controller-valve sensitivity balance. Hence, in making the valve selection, care is to be exercised to minimize the possibility of introducing sensitivity changes by requiring the valve to work over too broad a range. A valve may be, in theory, able to operate over a wide range, but in the practical sense probably cannot do so without altering the system response.

Time lags represent the interval of time which will elapse between a variation in flow resulting from a change in valve position and the appearance of, and recognition of, the influence of that change at the sensing apparatus of the control instrument.

When the system time lag is small, or where periodic variation in the rate of flow induced by a non-balanced system will not be detrimental, control valve loading from 0.95 to 0.10, possibly 0.05 may well be acceptable. As a general rule, upper limits of 0.85 are recommended, while the lower limits lie in the 0.2 to 0.3 range, depending upon the type of control and the resultant possible process lags. Table I-1 gives more detailed information.

Additional consideration of the significance of "valve loading" is appropriate to insure that the material which follows will be clearly understood. Valve loading is the ratio of the actual flow to the maximum capacity of the valve at the pressure drop available for the design conditions. If the pressure drop across the valve alters as the flow rate changes, then the maximum capacity, which is to be used to determine valve loading, also will change. *This is a point which can be overlooked.*

If a valve has a maximum capacity of 100 gpm at design pressure drop and the actual flow is 60 gpm, it may be assumed that the valve has a "spare capacity" of 40 gpm. That is not necessarily correct because the maximum capacity will remain 100 gpm *only* if the pressure drop does not change. In practice, the pressure drop across the valve upon a change in flow rate *must change to some* degree; the question is: how much.

If the friction in the system, exclusive of the amount across the valve, is small compared to that assigned to the valve at design conditions, then the amount of variation in the pressure drop across the valve with a process change will be small. For practical purposes the maximum capacity of the valve will be independent of changes in the rate of flow. Consequently, the change in loading will be essentially proportional to the change in flow rate.

To demonstrate this, assume the pressure drop across a valve at the design flow rate of 60 gpm is 10 psi while the pipe friction is 0.25 psi and the capacity of the valve with a 10 psi drop is 100 gpm. The valve loading would be 0.6. A reduction of the flow rate to 30 gpm would result in a decrease of valve loading to 0.297. For practical purposes, that is in proportion to the change in flow rate. This occurs because the pressure drop across the valve would increase by a mere 0.18 psi, insignificant compared to the initial 10 psi.

Usually, in practical applications, the pressure drop through the equipment is equal to or larger than the valve friction at design conditions. Hence, the change in valve loading will be *greater* than the change in flow rate — by quite a margin.

In the system described, consider the effect of including an exchanger, thereby making the pipeline and exchanger friction 10 psi rather than 0.25 psi as before when only a pipeline was present. All other factors remain as before. With a decrease of flow to 39 gpm, the pipe and heat exchanger pressure loss would decrease from 10 psi to about 2.5. The total system pressure drop must remain at 20 psi. With only 2.5 psi taken by the pipe and exchanger, the rest must be absorbed by the valve. This will have a definite influence upon the valve's maximum capacity under the new conditions. The maximum capacity will become 135 gpm instead of 100. The increase is in proportion to the square root of the change in pressure drop. With the actual flow being 30 gpm, the valve loading becomes 0.23. Under conditions involving large process time lags, such a loading *could* place the valve at the threshold of cycling, leading to process instability.

Consider the effect of an *increase* in the flow rate using the same example. Assume the rate to be increased by 20%, from 60 to 72 gpm. The pipe and exchanger friction will increase from 10 psi to 14.4 psi, leaving only 5.6 psi for the valve. The maximum capacity of the valve will drop from 100 to 75 gpm. With a flow of 72 gpm, the valve loading will become 0.96. For practical purposes, the valve can be considered wide open and the process out of control. This with a mere 20% increase in flow rate.

As the amount of pressure drop outside the control valve is increased, the effects become more pronounced. Consider the same example, except allow the pipe and exchanger friction to be 40 psi under design conditions while retaining the valve drop at 10 psi. Again, assume the flow rate to be cut in half, from 60 gpm to 30. The pressure drop in the pipe and heat exchanger will decrease from 40 to 10 psi requiring the valve to absorb the extra 30 psi. At the new level of 40 psi, the valve capacity will jump from 100 gpm to 200 and the valve loading will diminish to 0.15. This level of loading will be too low for good control, with the *possible* exception of level regulation.

With the same arrangement, consider the effect of an increase of flow rate by merely 10% to 66 gpm. The pressure drop in the pipe and exchanger will rise to 48.4 psi. With a total available system pressure change of only 50 psi, the drop for the valve will plummet from 10 to 1.6 psi. The maximum capacity for the valve under these conditions is only 40 gpm, a rate far below the desired 66. In this situation, the valve would be wide open with a change in flow rate of a mere 3 gpm, or an increase of but 5%!

Plant engineers and operators have employed methods of compensating for the difficulties introduced by the incorrect pressure loss distribution between valves and equipment. These consist of either absorbing some extra pressure loss in a pinched block, if the valve is operating too close to its seat, or supplementing the flow through a control valve which is operating nearly wide open by opening the bypass to some degree. These can be used to attempt to offset a poor design. However, the limitations on the flexibility of the installation become even more undesirable.

In the example just given, if the control valve were to be required to handle half the design flow rate for some temporary condition, the valve could be forced to operate at or near the original design loading by closing a block valve until it absorbed the extra 30 psi. However, with a small increase in flow rate, the control valve would fly open and lose control. To retain the control of the process, the block valve setting would require readjustment each time the flow rate altered even a small amount. A pinched block valve may temporarily stop a valve and controller from hunting but at the expense of system flexibility. With the use of that approach, a control system becomes semi-automatic.

The cracked bypass introduces even more drastic limitations to good control. With the control valve already wide open and the requirement of an additional increase in flow to handle some special condition, the action of opening the bypass will reduce the pressure loss across the control valve thereby reducing its "grasp" of the control.

In an earlier example, it was noted that an increase of 10% above design flow rate was not possible with the valve initially absorbing 20% of the pressure drop. The valve could only pass 40 gpm at the available pressure drop of 1.6 psi. The opened bypass would pass the extra 26 gpm. This flow would still pass through the bypass even if the control valve were to close completely. Thus, the opened bypass reduces control system sensitivity.

This is the way a measure of reduction in control system sensitivity was introduced into on-off controllers in years past. One faces the danger of losing control of the process by wandering, as compared to hunting, as would be the case with the pinched block valve. To avoid this, controller sensitivity, or the amount of bypassing, would need to be altered each time the process flow underwent a change.

The cracked bypass influence has *additional* disadvantages. The proportion of friction allotted to the control valve would be reduced from the original 20% to 3.3%. With a small decrease in flow, about 3 to 4 gpm, the valve would be nearly closed. With the control valve shut tight, the flow through the bypass could be close to the original design rate of 60 gpm.

In short, even though these methods can be used in emergency situations, and may be appropriate *at those times,* they are by no means to be thought of as substitutes for proper control valve and equipment pressure drop balance. If control valves are correctly integrated into the flow system during plant design, these expedients should not be required during plant operation with the possible exception of some extreme emergency.

The step by step procedures to be followed in establishing the correct pressure drop balance between control valves and equipment, if one were to use this time consuming approach, would be as follows:

1. A normal pressure drop through the valve would be assumed. From a valve sizing formula or chart, and the known normal flow rate, the valve loading would be computed.

2. After establishing the probable values of the abnormal flow rates (greater and lesser than normal) the equipment and pipe friction would be computed at these rates. From this the new pressure drop conditions across the valve would be established and the new maximum rates of flow determined from sizing charts or formulae.

3. The valve loading at the abnormal conditions would be calculated.

4. Study of these figures would reveal whether or not the appropriate pressure drop for the valve was chosen in step 1. If not, another approximation of the pressure drop through the valve would be made and the process repeated until the resulting upper and lower levels of valve loading matched the needs of the process.

Clearly, such a program would be too complex and time consuming. All the steps may be combined into one equation which enables the engineer to accomplish the various operations in one step. The derivation follows:

Let:

D_v = Design pressure drop in the valve, psi
D_f = New pressure drop in valve at new flow rate, psi
D_t = Total pressure drop in system
D_p = Total pressure drop in pipe and other equipment
C = Valve coefficient
Q_1 = Design flow rate
Q_2 = New flow rate
Q_m = Maximum capacity of the valve at design pressure drop
L_1 = Initial valve loading, fractional
L_2 = Final valve loading, fractional
R = Change in flow rate
F = Fraction of friction in valve

By definition:

$$Q_m = C\sqrt{D_v} \quad (1\text{-}1)$$

$$Q_1 = CL_1\sqrt{D_v} \quad (1\text{-}2)$$

$$Q_2 = CL_2\sqrt{D_f} \quad (1\text{-}3)$$

$$Q_2 = RL_1 \cdot Q_m \quad (1\text{-}4)$$

Substituting Equation (1-3) in Equation (1-4),

$$CL_2\sqrt{D_f} = RL_1 \cdot Q_m \quad (1\text{-}5)$$

Substitute Equation (1-1) and rearranging:

$$\frac{L_2}{L_1} = \frac{R\sqrt{D_v}}{\sqrt{D_f}} \qquad (1\text{-}6)$$

As will be shown in Chapter 3, in Equation (3-8), the pressure loss in piping is not proportional to the square of the flow rate. The exponent is 1.84 which means that the change in the pipe and the equipment will be proportional to the flow rate to the 1.84 power. Therefore, the pipe and equipment friction at the new flow rate becomes $R^{1.84}D_p$.

By definition,

$$D_f = D_p - D_p^{1.84} + D_v \qquad (1\text{-}7)$$

and

$$D_p = D_t - D_v \qquad (1\text{-}8)$$

Hence by substitution of Equation (1-8) into Equation (1-7):

$$D_f = (D_t - D_v) - (D_t - D_v)\,R^{1.84} + D_v \qquad (1\text{-}9)$$

$$\frac{D_f}{D_t} = (1 - F) - (1 - F)\,R^{1.84} + F \qquad (1\text{-}10)$$

Simplify:

$$D_f = D_t\,[R^{1.84}(F - 1) + 1] \qquad (1\text{-}11)$$

Substituting Equation (1-11) in Equation (1-7):

$$\frac{L_2}{L_1} = \frac{R\sqrt{D_v}}{\{D_t\,[R^{1.84}(F - 1) + 1]\}^{0.5}} \qquad (1\text{-}12)$$

By definition:

$$D_v = D_t\,F \qquad (1\text{-}13)$$

Substituting in Equation (1-12)

$$\frac{L_2}{L_1} = \frac{R\sqrt{F}}{[R^{1.84}(F - 1) + 1]^{0.5}} \qquad (1\text{-}14)$$

Equation (1-14) gives the change in valve loading in terms of the change in flow rate and the fraction of total friction allotted to the control valve. The presence of R in two forms complicates the determination of R, the change in flow rate, when that is being sought. However, Graph 1* based upon a modification of Equation (1-14), permits ready solution of all control valve calculations.

* All nomographs are in Appendix A.

Recommended values for valve loading under different types of control are given in Table I-1.

TABLE I-1
Valve Loading for Different Types of Control

Type of Control	Upper Loading	Lower Loading
Flow Control	80 %	40 %
Temperature Control	80 %	35 %
Pressure Reducing	80 %	30 %
Back Pressure	85 %	25 %
Level Control	90 %	20 %

Those forms of control, in which process lags may well be larger, require the narrower band of loading. Flow control with an orifice-type sensor calls for the narrowest loading range. This is required because of the great non-linearity of the common orifice-actuated flow sensing mechanism. With a larger change in flow, the sensitivity of the valve and controller will become unbalanced. If a flow sensor is of the type where the signal is linear with flow, then such control falls in the same range as pressure reduction.

A common misapplication of control valves can well be mentioned. The author has observed this error sufficiently often in the course of consulting work to justify a clarification and explanation of the nature of this situation.

The need arises to control the output from a centrifugal pump. The proper method is to place the control valve in the discharge pipe from the pump. The misapplication occurs when the control valve is used as a continuous bypass between the discharge and the suction. Such an approach generally is used because of concern that problems could arise if the discharge from the pump were to be fully shut off under some plant conditions. Certainly, under specific circumstances that *might* be true.

In some applications, for example boiler feedwater pumps, an emergency-condition bypass to suction during low steaming rates is essential to protect the pump. That is a wholly different situation than is being discussed. The approach to which the objection is being placed involves *continuous* bypassing to the pump suction to control the final flow rate away from the pump, which will *guarantee* undesirable conditions, *permanently!*

Up to three problems will result. (A) The net positive suction head required by the pump will be increased. If allowance is not made for this in the pump selection and installation, loss of suction or rough running can occur. (B) The power consumed by the pump will increase significantly, and this increase is *continuous,* producing a permanent wastage of energy. (C) In high pressure applications, continuous bypassing to the suction can increase installation costs for a heavy duty control valve and can result in increased maintenance costs as well.

The bypass to the suction makes the pump operate far out on the performance curve at all times. The pump probably would have been selected for the normal flow required for the process. However, with the bypass present and in continuous use, the flow out of the pump will increase as far as it can go. The result will be an increase in net positive suction head required and an increase in power consumed.

One example of misapplication occurred at the feed pump for a butediene plant. The contracting engineering company had installed a bypass control valve around the pump. The valve was actuated by a flow controller. The pump took suction from a number of feedstock tanks in the farm. When a tank was drawn down to about one third full, the pump would lose suction. The plant management had planned to relocate the pump in a pit to increase the head at a projected cost of many thousands of dollars. When investigation turned up the blunder, the bypass valve was replaced with another in the discharge line. The problem was eliminated at a cost of a few hundred dollars with the added benefit of a power saving as well.

In an underground water flooding project for a West Coast oil field, very high pressure pumps were injecting water into an oil reservoir to prevent land subsidence. The rate of injection was regulated by a controller. The designer of the plant had placed the control valve as a continuous bypass from discharge to suction. Because of the very high pressure drop, the control valve seats required frequent replacement. Consulting assistance was sought in an effort to choose a valve construction which would reduce the frequency of the costly replacements. Instead, the recommendation was made to place a control valve in the discharge pipe and abandon the bypass arrangement. Not only were costly valve repairs eliminated, but the power bill was cut by $ 12,000 annually as well!

With the use of Graph 1, the engineer may make a rapid and accurate determination of the correct percentage of fluid pressure loss which must be assigned to the valve to ensure proper control over the flow variations which may be anticipated. In a similar manner, if the percent of total friction assigned to the valve is known and with the allowable change in loading known from the process, the change in flow which may be tolerated without loss of control may be established with ease.

These concepts will play an important role in the selection of the optimum pipe size as will be unfolded in the Chapters to come.

Figure I-1

Gas Cycling Plant in California

CHAPTER 1

FUNDAMENTALS OF OPTIMUM PIPE SIZE SELECTION

For decades, chemical engineers have used various rules of thumb for selecting the size of pipe in continuous process plants. Often these methods result in sizes which are not the correct selection for the operating conditions. This causes the plant to be less efficient to operate or more costly to erect. With the emphasis on conservation of power and material resources, a rational method for proper pipe size selection which results in the least annual cost with highest performance becomes essential.

The methods presented in this book, as opposed to commonly employed rules, can result in saving of 4 to 18 % in investment or operating costs, which translates into $ 400,000 to $ 1,800,000 for a $ 10,000,000 facility. Even before our awareness of the energy shortage, that would be a nice economic plum.

The state of the art in pipe selection was brought out in an article printed recently in a prominent engineering magazine (36)*. The theory of pipe size selection, in the classical sense, provides ample formulae for determining pressure losses, the effects of fittings in pipelines and so on. Difficulty arises when one realizes that if the textbook approach were to be used, before the sizes can be selected, one must know the length of pipe runs; but those depend upon other facets of plant design, which in turn are influenced by the size of the pipe! To follow this method would require an iterative program involving successive design approximations wherein pipe sizing and other plant elements would be jockeyed toward the final configuration.

Clearly, that mode of design is rarely practical. Therefore, the engineer falls back upon the use of pipe sizing methods which "worked before". Generally these are "educated estimates". The article (36) included two charts intended to supply "confidence factors" as guides for the less experienced chemical engineer. As so commonly happens when one uses such rules of thumb, for certain plant conditions (not adequately defined in the article, by the way), the pipe sizes recommended for *modest* flow rates *could* be acceptable. However, for higher rates of flow the sizes recommended would result in operating costs which would be 190 % of the optimum value! It would appear that valid experience data for small flows were extrapolated to larger flows with the inevitably disastrous consequences. This rule of thumb is but *one of many,* often handed down through a generation of engineers, which, when put to the test of producing accurate answers over a spread of operating conditions, simply does not match up.

In engineering design, the successful rule would be based upon years of trial and error under conditions where the parameters which influence pipe size selection: pressure, temperature, amortization rate, pipe material costs and power costs have remained relatively the same through those years. Examples of such industries could be water works and manufactured gas. With those conditions established, determining a correct size requires experimenting, making mistakes, and correcting them over and over again. Eventually, a "good" rule can be created. Used exclusively for the environs within which it was evolved, the rule could give accurate and reliable data.

* Numbers in parantheses refer to References at the end of the book.

Unfortunately, when a "good" rule, one which has proven its worth through the years, is applied *outside the boundaries of its creation,* problems can arise. The engineer using the rule may not *know* its boundaries and hence emerge with pipe sizes which are far from optimum.

To demonstrate the magnitude of the errors in pipe selection which can occur, consider the effects of using a "good rule" when the parameters, which were fixed in the environment in which it was created, are changed. The amount of the changes will be kept within normal plant variations. Assume a liquid is being pumped at a pressure of 120 psi. The pipe is made of carbon steel. The amortization rate is for 7 years. Power cost is $ 0.0218 kwhr. Table 1-1 gives the percent of change in the optimum size from that established for the original conditions.

TABLE 1-1
Change in Optimum Size with Changes in Parameters

Value of Parameter	Pipe Size (% of original)
2500 psi flanges needed	75
Pipe used 10% of time	68
Pipe used 50% of time	89
Pipe material is 316 stainless steel	83
Amortization in 2.5 years	84
Cost of power is halved	89
Cost of power is doubled	112
Amortization in 15 years	114
These influences are multiplicative:	
2500 psi flanges and 316 s.s.	62
Amortization 15 years and power cost doubled	128

Such errors, significant though they are, can be dwarfed by those which develop when the source of the rule and the nature of the application do not match. Returning to the article mentioned earlier, for larger flow rates the size which this approach would advise, using "confidence factors" as the basis, would be 63% of the optimum size. Initial cost would be less, of course, but the excessive power costs, which go on year after year would be 868% higher than they should be! A review of a recent plant design (made by a major contractor) revealed pipe sizes from 150% to 200% larger than needed, with corresponding excessive initial costs from 185% to 280%.

Returning to the influences of common parameters, as listed in Table 1-1, the trend of technology will increase the quantity of variables the engineer must face in plant design. These changes in processes — and economics — increase the difficulty of developing "newer" rules for advancing engineering requirements; most engineers have neither the time nor the funds to gain the needed experience.

It is a fact that plants with incorrectly sized pipes may function acceptably with no one being aware that money and materials have been, or are being, wasted. That is part of the challenge; because a plant can still make product if the pipes are a couple of sizes too small or too large; sloppy design *can* get by! It is more likely that the undersized line will be detected because it may interfere with the process. For the oversized pipe, that is not so likely, although a 1$^1/_2$ inch control valve in a 4 or 6 inch line *should* alert the designer of a mistake. If the valve is not undersized, then the pipe *must* be too large, however, this *is* a common occurrence which goes unrecognized!

However, the engineer need not rely upon the rule of thumb approach. The alternative is a rational method based upon fundamentals of fluid dynamics laced with sound economics. Such a system is provided in this book with all the tools needed to select the optimum pipe size for all situations the engineer will encounter today and in the future. The optimum size means that one which will provide the lowest annual total cost consistent with the designated operating conditions and performance requirements. This system has been proved through thirty years of usage by design engineers working in widely diversified industries.

The procedures were developed by the author for a major engineering contractor. The assignment came during World War II when newer processes were being introduced into the petrochemical industry which included parameters and operating conditions of a wholly different character than had been common experience in earlier years. Furthermore, people who had training in pipe fluid dynamics were practically unobtainable. To handle these challenges, a pipe size selection system was created which enabled the inexperienced engineer, accurately and swiftly to select the pipe size which would be consistent from one service to the next and from one project to the next. This approach gave important results in 1944. The engineering time needed to select pipe sizes that were neither too large *nor too small* was cut by 75% of that used formerly. Tons of excess steel in oversize piping were eliminated.

Today, the system can be even more useful, and appropriate, when one considers widely varying pressures, temperatures and the use of special piping materials with far higher costs than steel, and the need to conserve energy.

The methods offered in this book can be used to size pipes without final and detailed knowledge of the plant plot plan, or of the exact length of the lines, or the details of the fittings in them. The methods may be applied to process plants, power plants, refineries, gasoline plants, LNG and LPG facilities, air conditioning systems, coal or oil gasification plants as well as ship-board piping.

The system provides speed and accuracy in pipe size selection through the use of nomographs to obtain the answers from the mathematical models which are at the heart of the approach. After establishing the parameters of the flowing material and deciding which of the three modes of pipe size selection applies, the correct size can be found in 10 to 30 seconds. What is more important, different engineers will arrive at the same selection! Use of this book will put the independent consultants, the 2 to 3 men design firms, and the largest contracting concerns on the same basis. All can select with equal skill the optimum pipe size, now and in the future, regardless of prior knowledge and experience with a specific process, exotic pipe materials, pressure levels, or geographic location.

For all their simplicity and speed, the methods take into consideration the influences of all the significant variables which alter pipe size selection.

The optimum pipe size is controlled by one of three modes of selection: the **LEAST ANNUAL COST** (LAC), **PRESSURE DROP AVAILABLE** (PDA) or **VELOCITY ALLOWABLE** (VA). All pipelines, with the exception of those sized for strength or some similar arbitrary reason, fall into one or more of these categories. In most plants, the first two comprise the larger proportion of the total.

LEAST ANNUAL COST

Least Annual Cost (LAC) applies when a fluid, liquid or gas, (or a combination of both) is being set into motion by a pump, compressor or blower. The approach is to balance the costs of operation with amortized cost of construction to provide a size which results in the lowest annual charge for the pipe system. Under the same flow rate, the amortized cost for a larger pipe is greater than for a smaller, while the annual operating costs are less for the larger. With the optimum size, the sum of the two charges are at the minimum. The mathematical models upon which this mode is based, explained in detail in Chapter 2, do not require knowledge of the length of the pipe nor the fittings used.

Engineers have been inclined to avoid selection based upon economic consideration. They have had concerns about the time consumed and the difficulty of obtaining the required data. These past challenges have been eliminated by using the approach given in this book.

When a control valve is used in the piping beyond a pump, compressor or blower, an objection may be raised to the concept that the pipe size should be based upon economics. Should not this be controlled by the Pressure Drop Available mode? Speaking practically, the pressure loss across the valve should only be enough to maintain control. Therefore this pressure loss is analogous to that occurring in conjunction with other equipment, heat exchangers, for example, which does not influence the size of pipe. A complete analysis of the applicable principles is given in the Introduction. LAC applies to steam supply and exhaust pipe lines for some classes of steam turbines.

PRESSURE DROP AVAILABLE

Pressure Drop Available (PDA) applies when a small, or even a large, pressure loss may be (or must be) absorbed by the pipe. A control valve may or may not be present. Except for very special cases, this would not be used in conjunction with a pumped or compressed fluid. According to conventional calculation procedures, this mode requires at least an approximation of the length of the pipe and the quantity and nature of the fittings in the system before the selection can be made. Such is not the case for most conditions using the approach given in this book.

On some projects PDA may cover more forms of pipe classes than does LAC. It can apply under some conditions to the suction piping for pumps or compressors. It *may* apply when a control valve is used for regulation. A sub-group of this mode applies when gas or liquid pressure is being reduced substantially. Another variation applies to the lines conducting steam to and from reciprocating pumps and to heating equipment. PDA can apply, also, to certain aspects of two-phase flow applications.

VELOCITY ALLOWABLE

Velocity Allowable (VA) can be the logical factor used in pipe size selection in several circumstances. The purpose may be either to keep the velocity below some upper limit or to keep it above some vital minimum value. Often, two-phase flow is present when VA is the controlling aspect.

During two-phase flow one may need to make certain that the velocity is high enough to insure entrainment of liquid drops in the gaseous carrier, or of particles in a liquid or gaseous carrier.

Conversely, the requirement during two-phase flow may be to make sure the velocity does not exceed that which could cause erosion of the pipe as the result of droplet (or solid particle) impingement or, indeed, the rupture of the pipe from the impact of liquid slugs on elbows.

Other maximum velocity limits relate to avoidance of vortices at the outlet of vessels and prevention of bubble entrainment under similar circumstances.

NOZZLE OR EQUIPMENT CONNECTION SIZE

The sizes of the connection on pumps, vessels, exchangers and similar equipment need not necessarily match the connecting pipe size. Certainly, decisions based upon the manufacturing costs may well result in pump connections which are smaller than the optimum pipe size. In similar manner, if a column *could* be used in another service in the future, the manufacturer may elect to use a nozzle size larger than the current application requires. Consequently, one does better to select the correct size for the service and swage to match the connection on the apparatus.

This brief overview of the principles of Optimum Pipe Size Selection provides the basis upon which the remainder of this book is constructed. The Chapters which follow will lay out the development of the mathematical models and their application to practical aspects of day to day decisions encountered in the selection of pipe sizes for all forms of process facilities.

The Appendices contain additional information which may be of assistance to the chemical engineer. Included is a section devoted to properties of various hydrocarbons, etc.

In the organization of a book of this nature, which may be used by engineers with varying degrees of fluid dynamics background, compromises must be made. The skilled person could find a plan, which elaborated upon facets of a typical problem, needlessly detailed and therefore undesirable. On the other hand, the person who requires data which had been omitted in the interests of simplicity, could be dissatisfied in equal measure.

The book has been organized to present the development of the three modes of selection in their logical sequence. Where fluid dynamics concepts apply, they are introduced with minimal description of their derivation. However, for those who wish to review, and those who may need to attain basic familiarity with the principles, complete information is provided in succeeding Chapters. A reference is given to the location for more data when a subject which is not given full explanation is first introduced.

Figure 1-1

Gas Recycling Plant

CHAPTER 2

LEAST ANNUAL COST

INTRODUCTION

A rational mode of selecting pipe sizes requires that it be capable of dealing with the variety of conditions encountered in modern process facilities. Under some conditions, the pressure drop available may be the controlling factor. Other instances may require that velocity, a maximum or a minimum, will govern. In many cases, the deciding factor will be the lowest yearly cost, including the cost of the pipe itself plus the costs of pushing the fluid through it, be that fluid a gas or a liquid (or a mixture of both). The correct size should be such that the relationship of the *total* operating costs to investment will be the most favorable.

When a liquid is flowing under the influence of gravity, or when the pressure on a gas or liquid is being reduced, the cost of operation becomes essentially zero — only maintenance remains because no input of energy is required. The optimum size then becomes the smallest, the least expensive, which will permit the flow to take place with no undesirable side effects. This size then requires the lowest amortized capital cost consistent with proper performance. However when the fluid is being pumped, compressed or blown, the additional cost for this purpose must be considered.

The total annual cost for an economically selected pipe, of a given size and schedule, will be the same regardless of the nature of the fluid, assuming the flow rate, in either case, to be the same percent of the economic maximum. The cost is the same if the fluid is water, gasoline or natural gas, assuming the power costs are the same, of course. Consider a 6 inch pipe handling the economic maximum for three fluids with densities of 62.4, 6.24 and 0.624 pounds per cubic foot. The total annual costs, per foot of pipe, are the same for all three, namely $ 8.20.

EARLIER HISTORY

Heretofore, the detailed analysis, which the economic approach has seemed to require, has rendered the method impractical, except, perhaps, in cases of cross country pipelines or instances of very expensive alloy piping. A significant step was taken in 1937 by GENEREAUX of E. I. DuPont de Nemours & Co. (1). He prepared a mathematical model which included the significant parameters which must be considered for an economic analysis. The "fixed" ones, those which would be essentially invariable for a given project, were combined into one constant which could be changed by the user when needed. This was a worthwhile simplification. Unfortunately, when the formula, and a nomograph based upon it, were reprinted in a prominent handbook for the chemical engineer, an error was made (not present in the original article) in the description of the flow units used. The pipe sizes one would obtain using the erroneous data would be too large. Consequently, the "economic" approach to pipe size selection received poor publicity and was not widely accepted.

A similar approach was presented in 1940 by JOHNSON and MAKER (2), who also recognized the importance of a rational approach to pipe selection. They wrote of the need to abandon the ideas of fixed limitations upon pressure drop or velocity in the process of making a choice of pipe size. The concepts in their paper were developed in a rigorous fashion. Unfortunately, in their care, they gave equal value to all eventualities, without really calling to the attention of the readers that many parameters they emphasized simply did not require consideration by the engineer, but could be lumped into an overall factor.

As they presented their formulae, individual knowledge for each pipe, calling for information on a number of cost items, was required. Similarly, the nomographs required the determination of the friction factor. As a result of the complexities in their method, it was not widely used, which is unfortunate, for the approach is valid even though the method is time-consuming. Here again can be seen the value in the GENEREAUX approach which bypassed these difficulties.

The GENEREAUX approach has a sound basis and can be adapted easily to units familiar to the engineer. The fundamental premise states that the combined annual costs of operating a pipeline should be so proportioned that the total is made as low as practical.

LEAST ANNUAL COST PREMISES

On the surface, this concept may appear to require extensive knowledge which is not obtainable at the time the pipe size generally is selected. Without knowing the exact length of the pipe, as well as the number of elbows and fittings, how can the capital cost, let alone the operating cost be determined? Without such knowledge, does not the whole premise fall apart?

To establish the least cost, one need not examine the entire pipe, rather one can consider the amortised capital cost plus the operating costs on *ONE FOOT* of the pipe. If the costs of a section can be made minimal, the cost for the entire pipe also will be minimal. This concept, mentioned (and maybe originated) by GENEREAUX, cuts the Gordian Knot of complexity and renders the problem soluble with ease. Within certain limitations, which are readily established, one need not be concerned about the length of the pipe nor the fittings in it. *For a given plant,* the least costly size will be essentially a function of flow rate, fluid density, piping material and pressure rating. There are other variables, to be sure. However, for the most part, these tend to maintain the same ratios of value or size, one to the other. These include the cost of pipe compared to the cost of power, or compared to construction costs. Such relationships can change with time; thus a periodic review is practical. It is worthy of note that over the last thirty years, even though costs for specific elements have risen considerably, the offsetting items similarly have moved up. The net result calls for little change in the optimum pipe size. This is true even though the costs for power yet have not risen quite as much as have those for labor and pipe.

Actually, this does make sense. If a plant that was well designed ten years ago were to be out of balance, economy-wise, today, a nearly insurmountable stumbling block would be thrown into the engineer's path. However, in the *future* these ratios of costs *may not* stay the same. The rapid increases in the cost of power might someday be enough to call for larger pipe sizes than are now the economic choice. Therefore, the system used must include provision for incorporation, with ease, of the influences of such possibilities.

Figure 2-1

Gasoline Plant

From one plant to another, certainly, some of the variables may be different. The amortization rate, the hours of operation per year, the cost of power, all these may well have other values. However, after the influences any such elements may have are introduced at the start of the project, no further attention need be given to them.

There is an important exception to the concept that costs retain about the same ratio in the upward movement. When the materials are purchased in one economic sphere and the plant is erected in another locale by local labor, the ratio of costs will be different from those which apply when the purchase and construction are in the same economic strata, even if not in the same geographical vicinity. As a consequence, the optimum pipe size *may* well be different for the same flow rate, etc. On the other hand, in this situation, the costs of power may be lower than those in the country from which the materials were purchased. That could offset the change in pipe size dictated by the lower erection labor cost. Allowance for such conditions can be made with ease, and the optimum pipe size readily selected.

BASIC MATHEMATICAL MODEL FOR LAC

A mathematical model for economic pipe size selection, which is independent of the size of the pipe, offers the most simplicity in usage. This approach eliminates the need to adjust the factor(s) in the equation(s) for various pipe sizes. GENEREAUX demonstrated that for most metallic pipe systems, the costs of the pipe, valves, fittings, erection,. etc., are proportional to the 1.5 power of the pipe diameter. This holds today as has been shown by the studies of MARSHALL and BRANDT (4). This relation permits the introduction of the cost of any size pipe system to be related to the cost of a specific size of pipe. GENEREAUX chose 1 inch as the reference. However, current practice shows that a better choice is 2 inch because reliable large purchase cost data are more readily obtained for this size rather than the 1 inch.

The cost of one foot any size pipe of a given material can be closely approximated in terms of the cost of 2 inch pipe multiplied by the diameter of the pipe raised to the 1.5 power.*

$$C_P = 0.353\,XD^{1.5} \qquad (2\text{-}1)$$

Where:
C_P = Cost of pipe size D, dollars per foot
X = Cost of 2 inch pipe, dollars per foot
The amortized capital cost of one foot of the pipe installation can be expressed as:

$$C_P = 0.353\,(a + b)\,(F + 1)XD^{1.5} \qquad (2\text{-}2)$$

Where:
a = Amortization rate, reciprocal of years, fractional
b = Maintenance, fractional
F = Factor expressing the cost of valves, fittings, welding, supports, erection, etc., as a multiplier relative to the costs of the pipe alone. If the costs of erection and fittings were to be 675% of pipe cost, F would be 6.75
X = Cost of one foot of 2 inch pipe of a given schedule and material, dollars per foot
D = Pipe internal diameter, inches

* The 1.5 exponent applies for metallic pipe materials and for some lined pipes. It does not apply to plastic pipe.

The change in the amortized capital costs with size for one foot of pipe is shown by Figure 2-2.

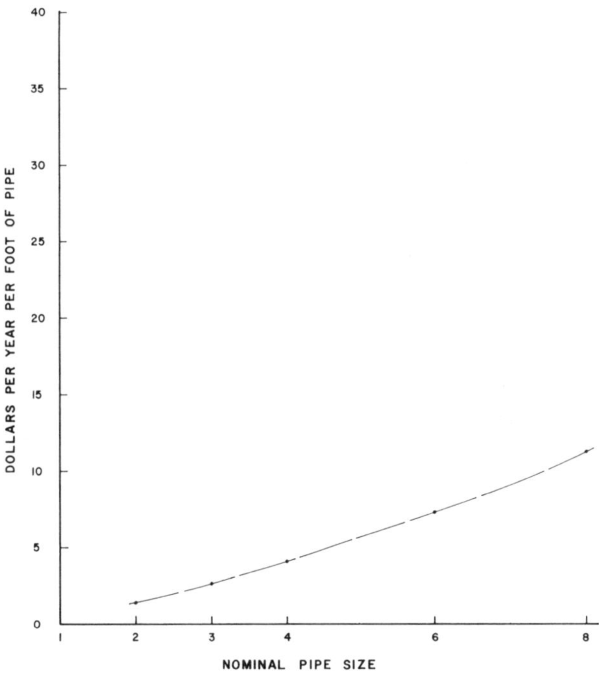

Figure 2-2
Amortized Capital Costs for One Foot of Pipe

The second factor in the total annual cost of a pipe installation is the expense incidental to moving the fluid through it. This does *not* refer to the cost of raising the pressure of a gas or liquid to satisfy a process requirement nor to bucking a static head of a column of liquid nor to accelerating a fluid column. It applies solely to the *frictional* losses encountered in transporting the fluid through the pipe.

For this approach, the cost of moving the fluid may be considered to be the product of flow volume and frictional pressure loss. The hourly consumption of energy may be expressed as:

$$1000\,(W/\varrho)\,(144\,\Delta P_1) \quad \text{(foot-pounds, force)} \qquad (2\text{-}3)$$

Converting to more convenient units, this becomes:

$$\frac{1000\,W\,(144\,\Delta P_1)}{2{,}654{,}200\,E\varrho} \quad \text{(kwhr)} \qquad (2\text{-}4)$$

In dollars per year this converts to:

$$C_f = \frac{0.0542\ W\ \Delta P_1\ YK}{E\varrho} \qquad (2\text{-}5)$$

Where:

W = *Thousands* of pounds of flowing material per hour

ΔP_1 = Pressure drop in pounds per square inch, per *foot* of pipe

Y = Hours of operation per year

K = Cost of electrical energy in \$/kwhr (The cost of other forms of energy may be employed in the working equations.)

E = Efficiency of the pump and driver, fractional

ϱ = Density of the fluid, pounds per cubic foot

To make a workable mathematical model, the pipe size must be expressible in terms of flow rate and fluid characteristics. This requires a mathematical expression for pressure drop, in pounds per square inch per foot of pipe, in terms of the smallest number of variables. As will be discussed in detail in Chapter 3, pressure loss, in the turbulent region, can be expressed by the FANNING equation as modified by GENEREAUX.

$$\Delta P_1 = 0.1325\ W^{1.84}\ \mu^{0.16}/\varrho D^{4.84} \qquad (2\text{-}6)$$

Where:

ΔP_1 = Pressure drop in pounds per square inch, per foot of pipe

W = Flow rate of material, *thousands* of pounds per hour

μ = Viscosity of the flowing fluid, centipoise

ϱ = Fluid density, pounds per cubic foot

D = Pipe inside diameter, inches

Substituting Equation (2-6) for ΔP_1 in Equation (2-5);

$$C_f = 2.84 \times 10^6\ W^{2.84}\ \mu^{0.16}\ YK/D^{4.84}\ \varrho^2\ E \qquad (2\text{-}7)$$

Where in addition to the above legend:

Y = Hours of operation per year

K = Cost of power, \$/kwhr

E = Efficiency of the pump and driver, fractional

Figure 2-3 illustrates the variation in the annual operating costs for one foot of pipe system with various sizes.

The total annual cost of one foot of the pipe is the sum of Equations (2-7) and (2-2):

$$C_t = 0.353\ (a + b)\ (F + 1) \times D^{1.5} + \frac{2.84 \times 10^6\ W^{2.84}\ \mu^{0.16}\ YK}{D^{4.84}\ \varrho^2\ E} \qquad (2\text{-}8)$$

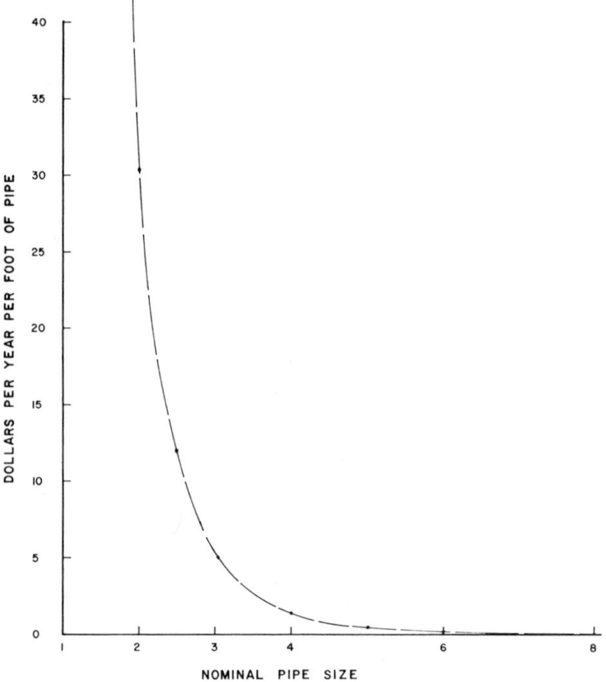

Figure 2-3
Annual Cost of Operating One Foot of Pipe

Using Equation (2-8) to calculate the costs for various sizes of pipe handling the same rate of flow of water in a pump driver system, reveals interesting data. Note the slow change in the amortized capital costs compared to the operational costs.

TABLE 2-1
Ratio of Total Operating Costs to Pipe Size

Pipe Size (inches)	C_p (\$)	C_F (\$)	C_T (\$)	Ratio of Total Costs
1	0.50	1013.	1013.	194
1.5	0.92	142.	142.	27
2	1.41	35.40	37.	7.0
2.5	1.98	12.00	14.	2.7
3	2.60	5.00	7.60	1.45
4	3.98	1.24	5.22	1.00
6	7.32	0.17	7.49	1.43
8	11.27	0.04	11.31	2.17

From the total cost column, it can be seen that the optimum size is a 4 inch with the lowest figure of \$ 5.22 per year. Note that the choice of either a 3 inch or a 6 inch would result in annual costs substantially greater than with the optimum size, by 43 to 45 %. This illustrates the importance of choosing the *correct size*. Making the pipe one size larger, "to be on the safe side", may well insure the plant will "work", but at what a cost!

Figure 2-4 illustrates the total annual cost of a piping system.

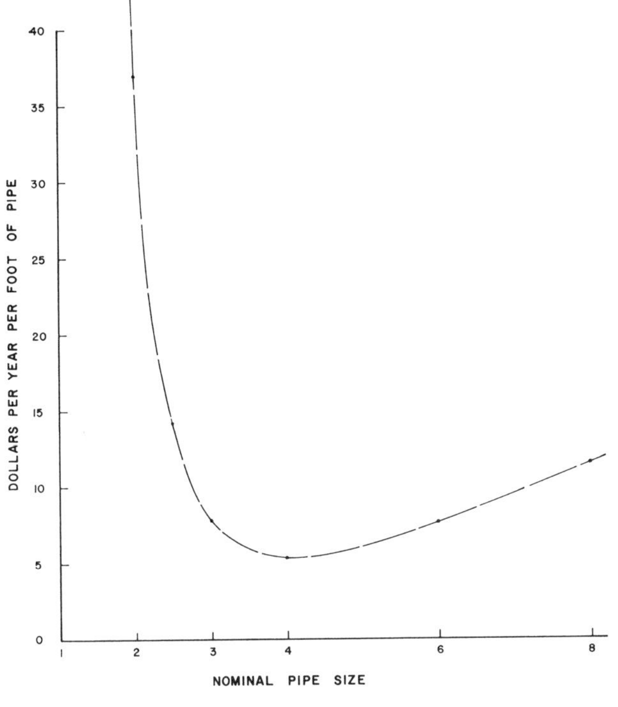

Figure 2-4

Total Annual Costs of One Foot of Pipe

Clearly, a process, such as was used for demonstration to provide the data for Table 2-1, would be costly, and impractical in most cases. It would, however, give an accurate pipe size. There is a simple way to obtain the optimum size with one calculation; an equation which defines the lowest cost pipe can be obtained by differentiating Equation (2-8) and setting the result equal to zero. This gives:

$$D^{4.84+1.5} = \frac{0.0348\, W^{2.84}\, \mu^{0.16}\, YK}{0.53\,(a+b)\,(F+1)\, XE\, \varrho^{2}} \qquad (2\text{-}9)$$

By reducing the equation to the first power of D, the result is:

$$D = \frac{W^{0.448}\,\mu^{0.025}}{\varrho^{0.316}} \left[\frac{0.0657\, YK}{(a+b)\,(F+1)\, XE}\right]^{0.158} \qquad (2\text{-}10)$$

Where:

D = Economic pipe internal diameter, inches
W = Flow rate, *thousands* of pounds per hour
μ = Viscosity of the flowing fluid, centipoise
ϱ = Density of the flowing fluid, pounds per cubic foot
Y = Hours of operation per year
K = Cost of power, \$/kwhr
a = Amortization rate, reciprocal of years, fractional
b = Maintenance, fractional
F = Factor for cost of fittings, valves and erection compared to bare pipe cost
X = Cost of one foot of 2 inch schedule 40 carbon steel pipe, \$/foot
E = Efficiency of pump (or compressor) and driver, fractional

The parameters within the bracket are those which tend to remain constant for a given plant. Should a variation occur, hours of operation of a specific pipe, for example, this can be recognized, and a simple correction applied. Furthermore, because the exponent applied to the parameters within the bracket is 0.158, large changes in the values within the bracket may occur before the influence will be sufficient to alter the pipe size. This was demonstrated by Table 1-1, which illustrates the magnitude of the change in the optimum size with variations in the parameters within the brackets.

APPLICATION TO COMMERCIAL PIPE SIZES

Equation (2-10) yields an exact pipe size, in inches and decimals, which would be the optimum selection for the designated flow conditions. The engineer does not have available to him an infinite variety of sizes but rather must choose from definite commercial ones.

If the optimum size lies a bit above a commercial size, one might presume that the correct approach would be to employ the next larger size. To do so, however, increases the amortized cost of the pipe, while it is true, decreasing the operating cost. For a limited increase in flow rate above the optimum one as calculated from Equation (2-10), for a commercial size, the total cost may well be lower than going to the next larger size. The greater power expense will be offset with a lower amortized capital cost.

To illustrate the point, consider the costs for a 4 inch and a 6 inch pipe conducting water with various flow rates. The amortized capital cost for the 4 inch would be \$ 3.80 per foot. For the 6 inch, it would be \$ 7.09. At a flow rate of 250 gpm, shown by Equation (2-10) to be the economic rate for a pipe with inside diameter of 4.025 inch, the annual operating cost would be \$ 0.80, for a total of \$ 4.60. The operating cost for a 6 inch pipe at the same flow would be \$ 0.11, far below the 4 inch. However, the *total* annual cost for the 6 inch would be \$ 7.20, quite a bit more than for the 4 inch.

Therefore the 6 inch would not be the least expensive choice at, for example, 265 gpm. Table 2-2 shows the change in costs as the flow rate is increased in the two sizes.

The size of the factor for 4 inch pipe suggests the potential desirability of the use of 5 inch rather than the common practice of jumping to 6 inch. The 5 inch

TABLE 2-2
Total Costs per Foot of Pipe for Water Flow

	Dollars per Year 4 inch				Dollars Per Year 6 Inch		
gpm	C_P	C_F	C_T		C_P	C_F	C_T
250	3.80	0.80	4.60		7.09	0.11	7.20
330	3.80	1.69	5.49		7.09	0.23	7.32
380	3.80	2.54	6.34		7.09	0.35	7.44
405	3.80	3.05	6.84		7.09	0.42	7.51
430	3.80	3.63	7.43		7.09	0.50	7.59
440*	3.80	3.81	7.61	*crossover	7.09	0.52	7.61
445	3.80	3.94	7.74		7.09	0.54	7.63
450	3.80	4.13	7.93		7.09	0.57	7.66

At the "crossover" flow rate of 440 gpm, the costs for both sizes are equal. For flows above the 440, the 6 inch would be the best selection. At lower rates, the 4 inch would be the optimum choice. The percent of increase above the calculated rate depends upon the spread between diameters of commercial pipe. For 2 inch versus 3 inch, the ratio is 1.68. For 3 inch versus 4 inch it is 1.44. For 8 inch versus 10 inch, it 1.35. As the difference in diameter from one size to the next narrows, the spread ratio becomes smaller.

The amount of flow beyond that calculated by Equation (2-10), which can be handled and still have the lowest annual cost can be expressed thus:

$$W_{max} = W_1 \times \left(\frac{D_L}{D_S}\right)^{1.33} \qquad (2\text{-}11)$$

Where:

W_{max} = Maximum flow allowable

W_1 = Calculated economic flow for the *nearest smaller commercial pipe size*

D_L = Next larger commercial inside diameter above that indicated by Equation (2-10)

D_S = Next smaller commercial inside diameter indicated

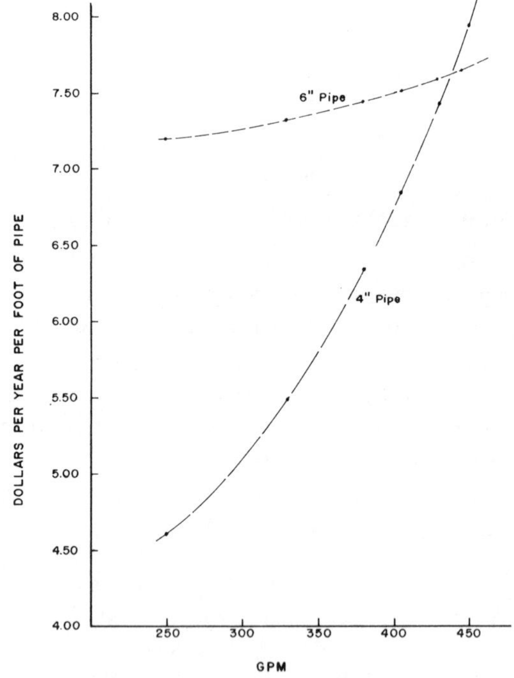

Figure 2-5

Total Annual Costs for 4 inch and 6 inch Pipe with Various Flow Rates of Water

Table 2-3 gives values of the factor $(D_L/D_S)^{1.33}$ for commonly used pipe sizes. These data apply (with trivial differences) for all schedules of pipe.

TABLE 2-3
Values of $(D_L/D_S)^{1.33}$

Size	Factor	Size	Factor	Size	Factor
0.5	1.45	3	1.44	14	1.19
0.75	1.38	4	1.71	16	1.17
1.00	1.77	6	1.44	18	1.16
1.5	1.39	8	1.35	20	1.28
2.0	1.27	10	1.26	24	1.39
2.5	1.33	12	1.13	30	1.28

would consume less power in friction loss than the 4 inch, hence would be a more economic size.

On the various nomographs for economic pipe selection, an adjustment has been made to reflect another way of allowing for the spread between commercial size pipes. The point of change from one pipe size to the next has been altered according to Equation (2-12). This diameter is the one wherein the crossover from the smaller size to the larger occurs. This is analogous to the diameter of the economic size which would match the crossover flow rate shown in Table 2-2.

Where:

$$D_c = D_L^{0.60}\, D_S^{0.40} \qquad (2\text{-}12)$$

D_c = Inside diameter representing the proper crossover between two adjacent commerical pipe sizes

D_L = Larger commercial pipe size, inside diameter, indicated by Equation (2-10)

D_S = Smaller commercial pipe size, inside diameter

All the nomographs for the economic selection automatically provide this adjustment. If one is using any of the equations given in this Chapter for selecting the economic size, the calculated pipe size is compared to the values of D_c provided in Table 2-4. If the calculated size is *greater* than the nearest commercial pipe inside diameter but less than D_c, that commercial size is the optimum. If the calculated size is greater than D_c, the next larger pipe size is the proper choice.

EXAMPLE 2-1:

From Equation (2-10), the economic size is found to be 8.51 inch. From Table 2-4 the D_c for 8 inch is 9.16. Use 8 inch pipe. If the economic size calculated had been 9.25 inch, which is greater than 9.16, 10 inch pipe would be the correct selection.

INFLUENCE OF FITTINGS AND VALVES

Just as the capital cost of a pipe line must include the additional costs for fittings and valves, (excluding control valves), etc., the operating cost, which is the result of the power consumed to overcome friction developed during the movement of the fluid through it, will be increased by the extra friction which occurs in such fittings and valves. As will be discussed in detail in Chapter 6, this effect can be expressed as "equivalent length." That is to say, an elbow of a given size produces the same friction (resistance to flow) as a specific number of feet of straight pipe. Similarly, a tee produces a resistance equivalent to some other length of pipe, and so forth.

TABLE 2-4
Values of D_c for Various Pipe Sizes and Schedules

Size	D_c		
	Sched. 40	Sched. 80	Sched. 160
1.5	1.87	1.75	1.54
2	2.31	2.16	1.94
2.5	2.80	2.65	2.41
3	3.62	3.43	3.09
4	5.15	4.89	4.41
6	7.16	6.71	6.11
	Sched. 40	Sched. 100	Sched. 160
8	9.16	8.50	7.78
10	11.14	10.25	9.44
12	12.65	11.69	10.75
14	14.23	13.19	12.14
16	16.10	14.97	13.76
	0.25 inch wall		
10	11.42		
12	13.00		
14	14.67		
16	16.67	Sched. 100	
18	18.78	16.72	
20	21.82	19.45	
24	26.94		
30	32.95		

Figure 2-6

Complex Piping

TYPICAL PIPE LINE ADJUSTMENT

Statistical studies of plant pipe systems reveal that a "typical pipe line" will have an average length and will contain a selection of fittings which are sufficiently constant to be used in pipe size computations with safety.

This typical pipeline is made up of 93 feet of 6 inch pipe with 1.6 cocks, 10.2 turns, 2.1 tees, 5.9 flanges and 32.6 welds.

It follows that a "typical pipe" of a given size will have an equivalent length (which is to be considered in determining the flow friction) that is longer by quite a bit than the 93 feet of pipe alone. The larger the pipe, the greater will be the equivalent length, as a percent of the actual pipe length. Consequently, Equation (2-10), to present a factual answer, requires an additional element which is a function of pipe size.

The afore mentioned study revealed patterns indicating that the equivalent length can be described by the following mathematical model:

$$L + L_e = CL^j \qquad (2\text{-}13)$$

Where:
L = Actual pipe length, feet
L_e = Equivalent pipe length for fittings, etc.
C = Constant depending on pipe size
j = Exponent depending on pipe size
Values of C and j are given in Table 2-5.

TABLE 2-5
Values of *C* and *j* for Equation (2-13)

Size	C	j
2	1.94	0.91
3	2.45	0.88
4	3.06	0.85
6	6.81	0.74
8	10.76	0.68

Incorporating Equation (2-13) into Equation (2-10) yields the fundamental model for economic pipe size selection:

$$D = \frac{W^{0.479}\,\mu^{0.027}}{\varrho^{0.337}}\left[\frac{0.0657\,Y\,K}{(a+b)\,(F+1)\,E\,X}\right]^{0.169} \qquad (2\text{-}14)$$

Where:
D = Inside diameter of economic pipe size, inches
W = Flow rate, *thousands* of pounds per hour
μ = Viscosity of the flowing fluid, centipoise
ϱ = Density of the flowing fluid, pounds per cubic foot
Y = Hours of operation per year
K = Cost of power, \$/kwhr
a = Amortization rate, reciprocal of years, fractional
b = Maintenance, fractional
F = Factor for cost of fittings, valves and erection compared to bare pipe cost, fractional
X = Cost of one foot of 2 inch schedule 40, carbon steel pipe, \$/foot
E = Efficiency of pump (or compressor) and driver, fractional

VALUES OF VARIABLES USED

The values of the variables used herein are as follows:

Y = 7880 hours per year, normal operation
K = Cost of electric power, an average around the United States, as of the end of 1975, \$0.0218 per kwhr
a = Amortization in seven years, 0.143
b = Maintenance 0.01
F = 6.75
X = \$1.32, as of the end of 1975
E = Depends upon the nature of the apparatus, refer to specific equation(s) or nomograph(s)

INFLUENCE OF VARIATION OF PARAMETERS

It is worthy of emphasis that the variables within the brackets usually will not change from the basic values listed above. Under some circumstances, however, they may be different. Therefore, a clear understanding of their significance is essential. Similarly, because Equation (2-14) forms the working basis for all following formulae for selecting the economic pipe size, exploration of its elements is useful to insure correct usage.

It will be recalled the value of n in Equation (2-1) had been established as lying between 1.4 and 1.6 by *GENEREAUX* and that the average value of 1.5 has been built into the economic size equations. For some piping materials, the exponent is not 1.5, and, therefore, the equations (and the corresponding nomographs) require adjustments when such materials are used. The following tables define those materials where the exponent is and is not within the 1.4 to 1.6 band.

TABLE 2-6
Size Exponent in the Band 1.4 to 1.6

Carbon steel	Glass-lined steel
304 S. S.	Rubber-lined steel
316 S. S.	Saran-lined steel
Nickel	Polypropylene-lined steel
Alloy 20	Penton-lined steel
Monel	Kynar-lined steel
Inconel 600	TFE-lined steel
Aluminium	FEP-Teflon-lined steel
Glass	Zirconium 1 inch to 3 inch
	Titanium 1 inch to 3 inch

For the materials listed in Table 2-7, the exponent has a value from 0.6 to 0.8. Therefore, to use nomographs based upon all economic pipe size equations, (2-14), and those which follow, adjustments must be made. The information is provided in Chapter 7, Table 7-11.

TABLE 2-7
Size Exponent in the Band 0.60 to 0.80

Hastelloy B
Glass-fiber-reinforced plastic
Titanium-lined steel
Zirconium-lined steel
Tantalum-lined steel

INFLUENCE OF VISCOSITY UPON ECONOMIC SIZE

The small size of the exponent for viscosity, 0.027, means that the size of the pipe will not change significantly, unless the viscosity is widely different from that of water, i. e. one centipoise. For the average liquid application, the value of the factor will be essential unity. A tenfold change in viscosity will alter the pipe size by 6 % and normally need not be considered. On the other hand, if the viscosity is quite high, above 200 centistokes, the flow character may no longer be turbulent (the conditions under which Equation (2-14) applies), but rather laminar.* Another mathematical model will apply, as will be discussed later in this Chapter.

* Chapter 3 covers the difference between turbulent and laminar flow and how to determine which form is present.

INFLUENCE OF BRACKETED VALUES UPON ECONOMIC SIZE

The elements of Equation (2-14) within the brackets constitute the cost facet of the economic selection method. The success of the system is established by the choice of the parameters required fully to account for influences on the pipe size. Included are the economic and operational variations which can be encountered in different industries and different processes as well as different economic climates, from time to time, in this country and others.

To the parameters in the working equations and the nomographs have been assigned practical and probable values which will be applicable for many, if not virtually all, circumstances the engineer will encounter. However, to recognize those times when other values should be used, the basis for selection of the values and their significance should be understood.

The exponent applied to the bracketed values, 0.169, diminishes the influence of the values within it to a significant extent. A change of 34 % is required before the pipe size will be altered by 8 %. Consequently *precise* knowledge about the bracketed parameters is not necessary. Naturally, the best available data should be used.

When the values applicable require alteration, factors are supplied, as discussed in detail in Chapter 7, which permit corrections to be made easily and swiftly.

The hours of operation per year were chosen as 7880 which equals 90 %. This value is considered proper to allow for common maintenance and occasional equipment break-down.

The efficiency of the mechanical apparatus, pump or compressor, plus prime mover, varies with the particular device. For electrically driven pumps the value may be high, 55 %, whereas for gas engine driven compressors, it would be lower, 23 %. The values chosen are defined with each equation and nomograph.

Power costs do vary from one part of the country (and the world) to another, and as of 1975, they have been rising. However, with the small exponent, substantial changes can take place before the influence would be detected in the pipe size. The value used is an average prevailing in the United States during 1975, namely, $ 0.0218 per kwhr.

Consideration of the form of power can be important. That derived from a company-owned plant (whether steam or gas engine driven) can change the power cost depending upon the fuel used and other considerations. If power is in the form of gas (or fuel oil), whether it is purchased or derived from company-owned sources can be significant.

Finally, the type of power and driver can influence the level of the efficiency of the driver and mover.

The amortization rate chosen, 7 years, may not apply to all projects. For some, 20 years may be appropriate, or 3 for another.

The allowance for maintenance has been set at 1 %. This could be low for some plants. However, except in *very* unusual circumstances, the influence of variations in the frequency and extent of maintenance will be insignificant.

The cost of pipe deserves special consideration. It is the price paid for one foot of 2 inch pipe, schedule 40, carbon steel, in the quantity applicable to the amount of the average pipe size (commonly 6 inch) which will be used in the plant. Correction factors are furnished for each nomograph (and each equation, if they are used) which adjust for the difference in cost of higher schedule pipe and for pipe materials other than carbon steel. Consequently, cost data are needed only for the size and type designated as the basis, 2 inch schedule 40, carbon steel.

Should one prefer to use the actual value of the pipe, rather than depend upon the relationships between carbon steel costs and other materials, (for example, if a correction for the particular pipe material is not provided in Chapter 7), the appropriate equation for the type of service would be used and the cost of one foot of the specific pipe would be introduced, with no other correction applied.

For conditions of operation where heavier pipe than schedule 40 and stronger fittings than 300 lb ASA are used, separate nomographs are furnished for ease of usage.

The parameter F, the relationship between the cost of bare pipe and the total pipe cost, deserves special discussion. This approach for establishing the total pipe system cost, using the cost of 2 inch pipe as the starting point, may appear to be dubious. However, extensive studies have shown this *can* be trusted. Compensations do occur, when the method would appear to be in doubt, which make the approach quite valid.

An example can assist in establishing acceptance of these ideas. An important influence upon the cost of a given length of pipe can be the presence of a valve. Adding one to a "typical" pipe system will increase its capital cost by about 17 %. However, the equivalent length, which influences the operating cost, will also rise about 9 %. When the ratio between the two is adjusted by the exponent of 0.169, the change in the economic pipe size is a mere 1 %. Adding *six* valves will reduce the economic size by 5 %. That *might* be enough to justify moving to the next smaller size if the line were loaded to full capacity without the valves.

Valves introduce the major amount of added cost to a pipe system. When *their* influence is not major, one can see that the concept of using F as a basis for establishing the cost in the economic balance makes sense.

This is further borne out when one realizes that the influence of an elbow or a tee on the cost of operating the pipe is even more closely matched to the amortized capital cost.

Figure 2-7

Simple Field Piping

The value for F for longer lines is smaller than that which applies to short lines. On the other hand, the lessened influence on the equivalent length results in a shift such that they nearly balance each other.

Rules of thumb have contended that long and simple lines should be larger than shorter, more complex, ones because the latter cost more per foot. While that is true, the *operating* costs for the short line increase a bit more *rapidly*. Hence, the actual adjustment which would be appropriate is the opposite of the "rule". The field pipe line of optimum size can handle a bit more than the plant line. For 1000 feet compared to the "average" 93 feet pipeline within a plant, the inter-plant pipe can pass 8 to 10% more depending upon pipe size.

PRACTICAL UNITS

Equation (2-14) does not lend itself to application with the common chemical engineering terms used in the U. S. For liquids, gpm (U. S. gallons per minute) and specific gravity, have been chosen. For gases, thousands of standard cubic feet per day, mscf/d, is the rate parameter along with specific gravity, temperature, pressure and deviation from the perfect gas law. For steam, the units are thousands of pounds per hour and pressure.

LIQUID ECONOMIC PIPE SIZES

For the mathematical model for liquids being pumped within a plant, the source of power is assumed to be commercial electricity at average industrial rate. For steam or gas-engine driven pumps, correction factors are furnished with the nomographs.

In its base form, Equation (2-14) becomes the starting point.

$$D = \frac{W^{0.479}\,\mu^{0.027}}{\varrho^{0.337}}\left[\frac{0.0657\,Y\,K}{(a+b)\,(F+1)\,E\,X}\right]^{0.169} \tag{2-14}$$

The values of the variables within the brackets are as follows:

Y = 7880 hours per year, normal operation
K = Cost of electric power, an average around the United States in 1975, $0.0218 per kwhr
a = Amortization in seven years, 0.143
b = 0.01
F = 6.75
E = 0.55
X = $1.32

The bracketed value to the 0.169 power becomes 1.55.

$$W = \frac{60 \times 8.35 \times Q \times S}{1000} \tag{2-15}$$

$$\varrho = 62.4\,S \tag{2-16}$$

Equation (2-14) for liquid turbulent flow converts to:

$$D = 0.276\,Q^{0.479}\,S^{0.142}\,\mu^{0.027} \tag{2-17}$$

Where:
D = Economic internal pipe diameter, inches*
Q = Rate of flow, U.S. gallons per minute at flowing temperature
S = Specific gravity at flowing temperature, water at 60° F equals 1.0
μ = Viscosity at flowing temperature, centipoise

Equation (2-17) defines the economic pipe size for liquids when commercial electric power is used. For steam or gas engine drive, the constant is 0.265 rather than 0.276.

For most liquids, the viscosity value to the 0.027 power will be virtually unity and can therefore be ignored without noticeable effect. However, if the viscosity were to be 0.15 centipoise, and no allowance were made, the pipe size would be too large by 5%. If the viscosity were to be 6 centipoise, the pipe would be small by 5%. For 35 centipoise the size would be small by 10%.

For the sake of ease in use, the nomographs for the liquid economic size do not include a viscosity term, but a correction means is provided.

It must be recognized that Equation (2-17) applies *only* within the turbulent region. See Chapter 3 for discussion of turbulent and laminar flow conditions. If the viscosity is above 200 centipoise, Reynolds Number must be determined using Graph 31, based upon Equation (3-5), and the nature of the flow is established, see Chapter 3. If the flow should not be turbulent, Equation (2-36), to follow, must be used.

For metric (S I) units the equation for the economic pipe size, in centimeters, for liquid service, becomes:

$$D_{si} = 1.717\,Q_{si}^{0.479}\,S_{si}^{0.142}\,\mu_{si}^{0.027} \tag{2-17 M}$$

---

* If flange rating is higher than 600 lb ASA, correct size by F_d, see Chapter 7.

Where:
D_{si} = Economic internal pipe diameter, centimeters*
Q_{si} = Rate of flow, cubic meters per hour
S_{si} = Specific gravity, kg/liter, at flowing temperature, water at 4° C = 1.00
μ_{si} = Viscosity, kg/m sec

Equation (2-17 M) with the factor of 1.717 is based upon the values of K and X as defined earlier in this Chapter. For use where these values may not apply, as for example, in Europe, the local costs are to be converted to the U. S. equivalent and the value of the bracketed parameters, as altered by the exponent, 0.169, is to be calculated and divided by 1.55. The value of 1.717 is to be multiplied by that ratio.

Neither Equation (2-17), nor Equation (2-17 M), has been adjusted to allow for the spread in commercial pipe sizes. Use Table 2-4 when pipe size is determined using the equations. That does not apply when one of the nomographs associated with Equation (2-17) is used.

EXAMPLES OF ECONOMIC LIQUID PIPE SIZE SELECTION

Having developed the formula for the economic size selection for liquids, consideration may now be given to the levels of pressure drop and velocity which will be found in pipe so selected. Table 1-1 lists the changes in pipe sizes which will accompany different values of the parameters which do influence pipe size selection, but which mainly are ignored in the common methods of pipe selection. Assume that a liquid is being pumped through 3 inch schedule 40 steel pipe at 150 gpm. If the flange rating were to be altered to ASA rating 2500 psi, and the cost of power were one half the normal used in the derivation of the equations, then for the same flow, the optimum pipe size would be the original size multiplied by the factors listed in Table 1-1, as follows:

$$3.068 \times 0.89 \times .75 = 2.048.$$

For contrast, if the amortization rate were to be 15 years rather than 7, and if the power costs were double the rate used herein, the optimum size would be larger than the original:

$$3.068 \times 1.14 \times 1.12 = 3.917.$$

* See note for Equation (2-17).

Consider the pressure drops and velocities which would result, using these two pipe diameters, even though they are not standard.

TABLE 2-8
Examples of Liquid Economic Pipe Size Selection

Size (inches)	Pressure Drop (psi/100)	Velocity (ft/sec)
2.048	19	15
3.068	2.6	6.5
3.917	0.81	4.0

These figures show with clarity how using rules of thumb can result in ambiguous answers! Yet, if applied to a specific industry for which they were developed, the results could be correct. The third example listed could well fit a waterworks installation with the "rules" being not over 1 psi drop per hundred feet and not over 5 feet per second for velocity. Use these blindly in another application, and the answers would be incorrect.

GAS ECONOMIC PIPE SIZE

For gas-driven reciprocating or centrifugal compressors, derivation of the economic pipe size selection equation commences with an adaption of the fundamental Equation (2-14). Where the source of power is gas rather than electricity, the constant within the brackets will have a different value, 0.000553 rather than 0.0657.

$$D = \frac{W^{0.479}\,\mu^{0.027}}{\varrho^{0.337}} \left[\frac{0.000553\,Y\,K_t}{(a+b)\,(F+1)\,E\,E_g\,X}\right]^{0.169} \tag{2-14 A}$$

Where:
D = Inside pipe diameter, inches
W = Rate of flow, *thousands* of pounds per hour
μ = Viscosity at flow temperature, centipoise
ϱ = Density at flow conditions, pounds per cubic foot
Y = Hours of operation per year
K_t = Cost of natural gas, dollars per Therm (1,000,000 BTU)
a = Amortization rate, fractional, reciprocal of years
b = Maintenance, fractional
F = Relative cost of fittings and erection compared to cost of pipe, fractional
E = Efficiency of compressor, fractional
E_g = Efficiency of engine, fractional
X = Cost of one foot of two inch, schedule 40 carbon steel pipe, dollars

Using the following values of the parameters within the brackets:

$Y = 7880$
$K = \$0.70$ per Therm
$a = 0.143$ per seven year rate
$b = 0.01$
$F = 6.75$
$E = 0.90$
$E_g = 0.366$
$X = \$1.32$

The equation may then be simplified to:

$$D = 1.35\ W^{0.479}\ \mu^{0.027}/\varrho^{0.337} \qquad (2\text{-}14\ B)$$

For ease of usage, Equation (2-14 B) may be converted into engineering units:

$$\text{mcf/d} = 9105.6\ W/28.97\ G \qquad (2\text{-}18)$$

$$\varrho = 2.696\ G\ P/T\ Z \qquad (2\text{-}19)$$

Then the working equation for gas becomes:

$$D = 0.0615\ M^{0.479}\ G^{0.142}\ \mu^{0.027}\ \left[\frac{TZ}{P}\right]^{0.337} \qquad (2\text{-}20)$$

Where:

D = Economic internal pipe diameter, inches*
M = *Thousands* of standard cubic feet per day at 60° F and 14.69 psia
G = Specify gravity of gas, air at 14.69 psia at 60° F equals 1.00
T = Gas flowing temperature, degrees Rankine
μ = Gas viscosity at flowing temperature, centipoise
Z = Deviation from perfect gas law, fractional
P = Pressure, pounds per square inch, absolute

For electrically driven compressors using K at 0.0218 \$/kwhr, the constant becomes 0.0666 rather than 0.0615.

Equation (2-20 M) below with the factor of 0.418 is based upon the values of K and X as defined earlier. For use where these values may not apply, for example, in Europe, the local costs are to be converted into the U. S. equivalent, and the value of the bracketed parameters, as altered by the exponent, 0.169, is to be calculated and divided by 1.35. Then 0.418 is to be multiplied by that ratio, and the result is to be used in the equation in place of 0.418.

* If flange rating is higher than 600 ASA, correct the size by F_d in Chapter 7.

For metric (SI) units the equation for the economic pipe size in inches, for gas service becomes:

$$D_{si} = 0.418\ M_{si}^{0.479}\ G_{si}^{0.142}\ \mu_{si}^{0.027}\ \left[\frac{K'Z}{P_{si}}\right]^{0.337} \qquad (2\text{-}20\ M)$$

Where:

D_{si} = Economic internal pipe diameter, centimeters
M_{si} = Flow rate, *thousands* of Nm^3/hr (0° C – 760 mm Hg)
G_{si} = Specific gravity of gas, kg/liter
μ_{si} = Gas viscosity at flowing temperature, kg/m sec
K' = Gas flowing temperature, degrees Kelvin
P_{si} = Pressure, kilo Pascals

For electrically driven compressors using K as 0.0218 \$/kwhr, the constant becomes 0.444 rather than 0.411.

Neither Equation (2-20) nor (2-20 M) has been adjusted to allow for the spread in commercial pipe sizes. Use Table 2-4 when pipe size is determined using the equations. That does not apply when one of the nomographs associated with Equation (2-20) is used.

STEAM ECONOMIC PIPE SIZE — RESIDUAL STEAM USED FOR HEATING

The economic steam pipe size mathematical model considers the attainment of minimum annual cost using a different approach. Economic steam pipe size selection mainly applies for piping to and from steam turbines. In other cases, as detailed in Chapter 11, the method is other than the economic balance.

In process plants, steam turbines may fall into two categories: the heat in exhaust steam is used for other process purposes, or it is not. Equations for both conditions will be developed.

In the first case, not all the heat value has been extracted from the steam after it leaves the turbine. In the second case, most of the available energy has been consumed, which means that the entire cost of the steam is considered in the development of an economic model.

The cost of a steam pipe is considered to be made up of the following:

A. The cost of the pipe, installation, valves, fittings, supports and insulation.

B. The cost of heat loss.

C. The cost of the pressure losses due to flow friction.

The annual cost of owning the pipe is expressed by Equation (2-2).

$$C_p = 0.353\,(a + b)\,(F + 1)\,X\,D^{1.5} \quad (2\text{-}2)$$

Where:

a = Amortization rate, reciprocal of years, fractional

b = Maintenance, fractional

F = Relative cost of fittings, valves and erection compared to cost of pipe, fractional

X = Cost of one foot of 2 inch schedule 40 carbon steel pipe, dollars

D = Pipe diameter, inches

The heat loss for pipe insulated in normal manner can be expressed as:

$$H = 11.61\,D^{0.69}\,\Delta T^{0.25} \quad (2\text{-}21)$$

Where:

H = Btu lost per hour per foot of pipe

D = Pipe diameter, inches

ΔT = Temperature difference, degrees F, between steam at operating pressure and air

Allowing the value of heat to be \$ 0.70 per million Btu, with an operating time of 7880 hours per year and with a boiler efficiency of 75 %, the annual cost of heat loss can be expressed thus:

$$C_h = 0.0854\,D^{0.69}\,\Delta T^{0.25} \quad (2\text{-}22)$$

The cost of the pressure lost due to pipe friction is considered to be the additional steam rate induced in a turbine by a small reduction in the differential pressure across the steam turbine. The steam rate can be expressed as:

$$\log R = 1.30/\log\left(\frac{P}{P_B}\right)^{0.283} \quad (2\text{-}23)$$

Where:

R = Steam rate, pounds per horsepower hour

P = Supply pressure, psia

P_B = Exhaust pressure, psia

The extra steam per hour resulting from the reduced differential pressure across the turbine is:

$$W = \frac{R \text{ x Horsepower}}{\text{Efficiency of Turbine}} \quad (2\text{-}24)$$

Combining these yields the annual cost of pressure drop:

$$C_f = \frac{1.05 \text{ x } 10^{-4}\,W^{2.84}\,\mu^{0.16}\,P_B^{0.60}\,Y\,K_t\,i}{D^{4.84}\,(P - P_B)\,1.67\,_{10}c} \quad (2\text{-}25)$$

$$c = \frac{1.30}{\left[\log\frac{P}{P_B}\right]^{0.283}}$$

Where:

C_f = Cost of friction in the pipe

W = Rate of flow of steam, *thousands* of pounds per hour

μ = Viscosity of steam, centipoise

P_B = Exhaust pressure of steam, psia

P = Supply pressure of steam, psia

Y = Hours of operation per year

K_t = Cost of fuel

i = Fraction of energy value remaining in exhaust steam

D = Pipe diameter

The total cost for the pipe line per year is the sum of the three elements: the cost of the pipe, plus the cost of the heat loss and the cost of the pressure drop. When differentiated, set equal to zero and simplified, two equations result: One for supply lines to turbines and one for the exhaust lines.

For steam supply lines where exhaust steam is used for heating:

$$D_s = \frac{11.40\,P_B^{0.103}\,W^{0.543}}{P^{0.495}}\left[\frac{0.000553\,Y\,K_t\,i}{(a+b)\,(F+1)\,E_g\,X}\right]^{0.191} \quad (2\text{-}26)$$

For steam exhaust lines from turbines where exhaust steam is used for heating:

$$D_e = \frac{13.44\,W^{0.543}}{P^{0.294}\,P_B^{0.134}}\left[\frac{0.000553\,Y\,K_t\,i}{(a+b)\,(F+1)\,E_g\,X}\right]^{0.191} \quad (2\text{-}27)$$

Where:

D_s = Inside diameter of supply pipe, inches

D_e = Inside diameter of exhaust pipe, inches

P_B = Exhaust steam pressure, psia

P = Supply steam pressure, psia

W = Rate of flow, *thousands* of pounds per hour

Y = Hours of operation per year

K_t = Cost of fuel, dollars per million Btu (Therm)

i = Fraction of energy value remaining in exhaust steam

a = Amortization rate, reciprocal of years, fractional

b = Maintenance, fractional
F = Relative cost of fitting and valves plus erection, compared to the cost of pipe, fractional
E_g = Boiler efficiency, fractional
X = Cost of one foot of two inch, schedule 40, carbon steel pipe, dollars

In arriving at these equations, the viscosity influence, which is small because of the low value of the exponent, has been incorporated into the pressure parameter. As with the other economic equations, the influence of the effective length of fittings and valves has been included.

WORKING ECONOMIC STEAM EQUATIONS — RESIDUAL HEAT USED

Working equations are obtained by introducing practical values for the elements within the brackets. The values used are as follows:

Where:
Y = 7880 hours per year, (90%)
K_t = \$0.70 per Therm (million Btu)
i = 0.50
a = Amortization 0.143, 7 years
b = Maintenance, one percent, 0.01
F = 6.75
E_g = 0.85
X = \$1.32

The working equation for steam *supply* lines to turbines when the residual heat in the steam *is* used for process heating is as follows:

$$D_s = 11.7\, P_B^{0.103}\, W^{0.543}/P^{0.495*} \quad (2\text{-}28)$$

The working equation for steam *exhaust* lines from turbines when the residual heat in the steam *is* used for process heating is as follows:

$$D_e = 13.8\, W^{0.543}/P^{0.294}\, P_B^{0.134*} \quad (2\text{-}29)$$

For metric (SI) units, the equations for the economic pipe sizes, in centimeters, for steam to and from the turbines, when the residual heat *is* used for heating, become:

$$D_s = 84.5\, P_{Bsi}^{0.103}\, W_{si}^{0.543}/P_{si}^{0.495} \quad (2\text{-}28\ M)$$

for supply, and

$$D_e = 111\, W_{si}^{0.543}/P_{si}^{0.294}\, P_{Bsi}^{0.134} \quad (2\text{-}29\ M)$$

for exhaust.

Where:
D = Economic internal pipe diameter, centimeters
P_{Bsi} = Steam exhaust pressure, kilo Pascals
W_{si} = Flow rate of steam, thousands of kilograms per hour
P_{si} = Steam supply pressure, kilo Pascals

Equations (2-28 M) and (2-29 M), with factors of 84.5 and 111 are based upon the values of K and X as defined earlier. For use where these values may not be applicable, for example in Europe, the local costs are to be converted into the U.S. equivalent, and the value(s) of the bracketed parameters, as altered by the exponent, 0.191, is to be calculated and divided by 1.026. Then the appropriate factor, 84.5 or 111, is to be multiplied by the resulting ratio, and that figure is to be used in the equation in place of 84.5 or 111 as the case may be.

The Equations (2-28), (2-29), (2-28 M) and (2-29 M) have not been adjusted for the spread in commercial pipe sizes.

When pipe size is determined using the equations, use Table 2-4. When Graphs 46 or 47 associated with Equations (2-28) and (2-29) are used, this is automatically provided.

Equations (2-28) and (2-29), or associated nomographs, along with the metric equations, are to be used *only* where the residual energy (latent heat) will be used for heating. The energy alloted to pipe friction has been adjusted to allow for the average value of the heating capacity remaining in the exhaust steam. When the residue steam is not used for heating, the correct selection of the pipe size will be made with a different equation, as will be covered later in this Chapter.

The size of the supply pipe will be influenced by the level of the *downstream* pressure as well as the upstream. This results from the premise that the cost of friction is determined by the change in steam rate for the turbine as the pressure differential across it alters. As the back pressure is increased, meaning that there will be less drop across the turbine, a smaller pressure loss is compatible with the least annual cost.

* See foot note for Equation (2-17).

EXAMPLES OF ECONOMIC PIPE SIZE SELECTION FOR STEAM — RESIDUAL HEAT USED

Table 2-9 demonstrates typical pipe sizes for the same flow with different downstream pressures, and the pressure drops associated with the pipe size.

TABLE 2-9
Economic Pipe Size and Pressure Drop for 10,000 Pounds per Hour of Steam

(Residual Heat Being Used)

P (psia)	P_B (psia)	D_S (inch)	D_E (inch)	Pressure Loss Supply	psi per 100 Feet Exhaust
315	40	3.54	5.42	1.55	0.20
315	75	3.78	4.99	1.12	0.29
315	100	3.89	4.80	0.97	0.36
215	40	4.28	6.06	0.88	0.16
215	75	4.56	5.58	0.65	0.25

In Table 2-2, the costs per foot per year for the same rate of flow of water through different pipe sizes was given. In the flow of steam, a slightly different pattern appears when the cost of heat loss requires consideration. Table 2-10 illustrates the changes in annual cost associated with a flow of 64,000 pounds per hour of steam to a turbine at 315 psia, when the exhaust pressure is 40 psia, and where various sizes of pipe are considered. Please note that non-standard pipe sizes are treated as if they could be obtained at costs commensurate with standard pipe sizes for the purpose of making the comparison, only.

TABLE 2-10
Costs per Foot per Year for Various Steam Pipe Sizes

(Residual Heat Being Used)

Size (inch)	C_P ($)	C_H ($)	C_F ($)	C_T ($)	Cost Ratio
8	12.51	1.55	11.12	25.18	1.11
9	14.93	1.68	6.28	22.89	1.01
9.5	16.19	1.74	4.84	22.77	1.00
10	17.49	1.81	3.77	23.07	1.01
11	20.17	1.93	2.38	24.48	1.08
12	22.99	2.05	1.54	26.58	1.17

The calculated optimum size would be a non-standard 9.5 inch. The choice would lie between 8 inch and 10 inch. From Table 2-4 the crossover from 8 inch to 10 inch pipe occurs at a calculated diameter of 9.16. Therefore, the optimum size would be 10 inch rather than 8 inch. The values shown on Table 2-10 lead to the same decision. If one were to follow a common practice of "playing the safe side" by going to one size larger, 12 inch, the costs would be greater than if one dropped back one size to 8 inches!

Consider the influence upon plant costs of just one pipe of typical equivalent length using this example. Consider that the 10 inch is the correct size. The typical length of a 10 inch pipe is 183 feet. Even though the typical lengths of 8 inch and 12 inch are different from that, 183 feet would represent the correct figure for this study because the origin and terminus of the pipe will not alter in this case with a change in size.

Should 8 inch be used rather that 10 inch, over the 7 year amortization period, the costs would be $ 2,700 higher. If 12 inch were used the costs would be $ 4,500 higher. Proper selection of pipe size really can offer the opportunity to make substantial cost savings in plant construction and operation.

STEAM ECONOMIC PIPE SIZE — RESIDUAL STEAM NOT USED FOR HEATING

When the exhaust steam is not used for process heating, the economic pipe size is established via a different concept. All the energy used in the steam generation is now assignable to the economic pipe selection rather than a portion, as was the case in the prior situation. The boiler is treated as a "compressor" driving the steam through the piping. The cost of the energy used in steam generation is based upon fuel(s) rather than electricity. The cost of heat loss through pipe insulation must be considered. The influence of steam viscosity is included within the density parameter. With these adjustments to Equation (2-14), the result is:

$$D = 0.90 \frac{W^{0.479}}{\varrho^{0.337}} \left[\frac{0.000553\, Y\, K_t}{(a+b)(F+1)\, E_g\, H_L\, X} \right]^{0.169} \tag{2-30}$$

Where:

D = Inside diameter of pipe, inches
W = Rate of flow, *thousands* of pounds per hour
ϱ = Density of steam, pounds per cubic foot
Y = Hours of operation per year
K_t = Cost of fuel, dollars per Therm (million Btu)

a = Amortization rate, reciprocal of years, fractional
b = Maintenance, fractional
F = Relative cost of fittings and valves plus erection, compared to the cost of pipe, fractional
E_g = Boiler efficiency, fractional
H_L = Factor for heat loss, fractional
X = Cost of one foot of two inch, schedule 40, carbon steel pipe, dollars

A working equation is obtained by introducing practical values for the elements within the brackets. The values used are as follows:

Y = 7880 hours per year, (90%)
K_t = \$0.70 per Therm (million Btu)
a = Amortization in 7 years, 0.143
b = Maintenance, 1%, 0.01
F = 6.75
E_g = 0.85
H_L = 1.08
X = \$1.32

The value of the Therm at \$ 0.70 can be considered appropriate for gas in the oil fields and for refineries and petrochemical plants. On the other hand power plants, or other concerns purchasing gas from pipeline companies, generally pay a higher price. As of the fall of 1974, this is fixed at \$ 1.25 per Therm. In Chapter 7, factors for correcting equations or nomographs for different fuel costs are provided.

With the values given above, the equation for the economic pipe size for steam when the exhaust is not used for heating is:

$$D = 1.02\, W^{0.479}/\varrho^{0.337*} \qquad (2\text{-}31)$$

This applies to supply and exhaust lines to and from turbines as well as for headers. It generally is not applicable to piping to heating equipment. The approach for this is covered in Chapter 3.

For metric (SI) units, the equation for the economic size when the exhaust is not used for heating, is:

$$D_{si} = 8.51\, W_{si}^{0.479}/\varrho_{si}^{0.337*} \qquad (2\text{-}31\ M)$$

Where:
D_{si} = Economic internal pipe diameter, centimeters
W_{si} = Flow rate of steam, *thousands* of kilograms per hour
ϱ_{si} = Steam density, kilograms per cubic meter

Equation (2-31 M) with the factor 8.51 is based upon the values of K and X as defined earlier. For use where these values may not be applicable, for example in Europe, the local costs are to be converted into the U.S. equivalent and the value of the bracketed parameters, as altered by the exponent, 0.169, is to be calculated and divided by 1.136. Then 8.51 is to be multiplied by that ratio and to be used in its place in the equation.

The Equations (2-31) and (2-31 M) have not been adjusted for the spread in commercial pipe sizes. When the pipe size is determined by using either of the equations, use Table 2-4. When any of the nomographs associated with Equation (2-31) are used, the allowance is automatically included.

Figure 2-8

Vacuum Unit

ECONOMIC PIPE SIZE SELECTION FOR LAMINAR FLOW

Thus far, consideration of economic pipe size has been based upon turbulent flow conditions. When highly viscous liquids are being handled, the flow pattern may be laminar*. Then the foregoing formu-

* If flange rating is above 600 ASA, correct size using F_d, correction for flange rating, see Chapter 7.

* For discussion of the difference between laminar and turbulent flow, see Chapter 3.

lae are not applicable. Whenever the viscosity is 200 centipoise, or higher, the type of flow must be established by calculating Reynolds Number, see Chapter 3. If Reynolds Number is 2320 or below, the flow is laminar and a different formula is needed.

The equation for the economic pipe size during laminar flow is developed by substituting the formula for pressure drop during laminar flow. The development of this equation is given in Chapter 3.

The equation for pressure drop is:

$$\Delta P_1 = \frac{0.000274\, Q\, \mu}{D^4} \tag{2-32}$$

Then the cost of friction becomes:

$$C_f = \frac{0.119 \times 10^{-6}\, Q^2\, \mu\, Y\, K}{D^4\, E} \tag{2-33}$$

The cost of pipe is as given in Equation

$$C_p = 0.353 \times D^{1.5} \tag{2-1}$$

The amortized cost of one foot of pipe is as given in Equation

$$C_p = 0.353\,(a+b)\,(F+1) \times D^{1.5} \tag{2-2}$$

Following the previously given method as described, the economic pipe size for laminar flow is:

$$D = 0.173\, Q^{0.307}\, \mu^{0.182} \left[\frac{0.0139\, Y\, K}{(a+b)\,(F+1)\, X\, E}\right]^{0.182} \tag{2-34}$$

Where:

ΔP_1 = Pressure drop during laminar flow, psi per foot
Q = Rate of flow, U.S. gallons per minute
μ = Flowing viscosity, centipoise
D = Inside diameter of pipe, inches
C_f = Cost of pumping against pipe friction, dollars per foot of pipe per year
C_p = Cost of pipe, of size D, dollars per foot
Y = Hours of operation per year
K = Cost of electrical energy, $/kwhr
X = Cost of one foot of 2 inch, schedule 40, carbon steel pipe, dollars per foot
E = Efficiency of pump and driver, fractional
a = Amortization rate, reciprocal of years, fractional
b = Maintenance, fractional
F = Factor for cost of fittings, valves and erection, compared to bare pipe

As was done for turbulent flow equations, the influence of the pipe equivalent length, compared to the straight pipe length, must be incorporated into Equation (2-34) by combining it with Equation (2-13). The result is the formula:

$$D = 0.173\, Q^{0.393}\, \mu^{0.197} \left[\frac{0.00656\, Y\, K}{(a+b)\,(F+1)\, E\, X}\right]^{0.197} \tag{2-35}$$

The values for the variables with the brackets used in the working equation, below, have been selected as follows:

Y = 7880 hours per year, (90%)
K = Cost of electrical power, an average around the United States in 1975, 0.0218 $/kwhr
a = Seven years, 0.143
b = 0.01 (one percent)
F = 6.75
E = 0.55
X = $1.32, as of the end of 1974

Incorporating these values gives a working equation:

$$D = 0.182\, Q^{0.393}\, \mu^{0.197} \tag{2-36}$$

The equation applies solely when the flow is laminar. When it is used, Reynolds Number must be below 2320, see Chapter 3.

For metric (SI) units, the equation for the economic pipe size for laminar flow conditions becomes:

$$D_{si} = 3.23\, Q_{si}^{0.393}\, \mu_{si}^{0.197} \tag{36 M}$$

Where:

D_{si} = Economic internal pipe diameter, centimeters
Q_{si} = Flow rate of liquid, cubic meters per hour
μ_{si} = Flowing viscosity, kg/m sec

Neither Equation (2-36), nor Equation (2-36 M) have been adjusted for the spread of commercial pipe sizes. Use Table 2-4 to select the correct size.

Table 2-11 illustrates the optimum flow rates and the associated pressure drop and velocities for several pipe sizes for two viscosities

Notice that as the viscosity increases, the optimum flow rate diminishes, but the total annual cost remains constant for a given pipe size. However, the *cost* per *gallon* of liquid pumped per year rises sharply as the viscosity increases.

TABLE 2-11

Optimum Flow Rates, Pressure Drops for Two Viscosities

Size (inch)	Viscosity (cp)	Flow Rate (gpm)	ΔP_{100} (psi)	Velocity (ft/sec)	Cost per foot per year ($)
4	300	249	7.8	6.3	6.16
4	1000	137	14.0	3.5	6.16
6	300	590	3.6	6.6	9.79
6	1000	324	6.6	3.6	9.79
8	300	1112	2.3	7.1	14.02
8	1000	609	4.1	3.9	14.02

TABLE 2-12
Economic Pipe Size Selection-Reduction

Pipe Size (inches)	Minimum Length in Feet of Reduced Size (Welded) Pipe Required to Pay for Reduction						
	1/4″ Wall Thickness (Sch. 20 in 12″ Size) (Sch. 30 in 8″ and 10″ Size)		3/8″ Wall Thickness (Sch. 30 in 8″, 10″ and 12″ Size)		1/2″ Wall Thickness (Sch. 40 in 8″, 10″ and 12″ Size)		1/4″ Wall Thickness (Sch. 20 in 12″ Size) (Sch. 30 in 8″ and 10″ Size) Field Fab. Eccent. Reducer
	Concentric Reducers	Eccentric Reducers	Concentric Reducers	Eccentric Reducers	Concentric Reducers	Eccentric Reducers	Eccentric Reducers (Only)
14 x 8	48.6	72.2	18.7	28.0	14.3	21.1	45.0
14 x 10	223.0	324.0	25.6	37.3	16.7	24.8	182.0
14 x 12	1525.0	2180.0	52.0	73.6	37.8	52.8	1180.0
16 x 10	48.5	73.4	24.8	37.5	22.5	34.0	33.2
16 x 12	105.0	156.0	46.2	68.0	34.0	49.5	65.0
16 x 14	116.0	170.2	116.0	168.7	90.0	129.0	67.8
18 x 12	80.9	122.0	42.0	62.9	31.0	46.0	47.2
18 x 14	79.2	117.0	67.8	100.0	52.5	76.0	41.1
18 x 16	192.7	283.0	139.2	204.0	110.6	158.0	95.8
20 x 14	64.9	96.4	52.0	77.1	47.5	58.6	42.0
20 x 16	104.0	153.5	74.0	108.5	57.9	83.0	57.1
20 x 18	222.0	325.0	154.5	224.0	118.5	183.0	116.0
24 x 16	70.2	106.6	52.1	78.7	63.7	51.7	36.0
24 x 18	82.8	124.5	62.2	93.2	40.0	59.3	36.7
24 x 20	105.0	152.0	78	116.0	49.0	70.5	43.3

Comparison with Table 2-2 illustrates the differences in annual cost per year for the same size pipe with turbulent versus laminar flow conditions. The variation is not as great as might be assumed, considering the change in viscosity present.

ECONOMICS OF PIPE SIZE REDUCTION

When the process calls for a flow to be split between two or more pipes after an initial run in one pipe, a decision must be made whether or not it is less costly to reduce the pipe size beyond the branch(es). Factors which will influence this decision will be the sizes of the pipes and the length of the run(s) beyond the side branch. The cost of continuing the larger pipe for the full run must be balanced against the lesser cost of the small pipe plus the added cost of the reducer and its installation. Tables 2-12 and 2-13 provide the results of extensive studies of these relative costs and show the minimum smaller run of pipe for a specific size reduction which will justify the use of the smaller pipe.

It will be seen that typically, the smaller the sizes of pipe under consideration, the shorter will be the length of the smaller pipe required to justify the reduction in size. For example, using schedule 40 pipe and assuming concentric reducers, dropping from 3 inch to 2 inch will be the economic approach if the length of the 2 inch will be more than 13.5 feet. If the sizes are 10 inch and 6 inch, the length of the smaller size must be at least 16.5 feet. With 20 inch and 16 inch, the minimum run of the 16 inch would be 104 feet.

The minimum run of the small pipe will depend upon the percent of reduction as well as the size of the pipe. As the difference in size is lessened, the length of smaller pipe which is required to justify the reduction increases. This occurs because the dif-

TABLE 2-13
Economic Pipe Size Selection-Reduction

Pipe	Minimum Length in Feet of Reduced Size (Welded) Pipe Required to Pay for Reduction					
	Schedule 40 Pipe		Schedule 80 Pipe		Schedule 160 Pipe	
Size (inches)	Concentric Reducers	Eccentric Reducers	Concentric Reducers	Eccentric Reducers	Concentric Reducers	Eccentric Reducers
2 x 1	17.4	24.9	15.5	20.8	11.1	15.7
2 x $1^1/_4$	31.1	39.5	25.8	33.6	27.1	37.5
2 x $1^1/_2$	54.2	69.0	38.0	49.4	56.5	65.2
3 x $1^1/_2$	10.3	13.7	10.3	14.5	—	10.2
3 x 2	13.5	17.0	14.2	18.5	9.9	13.1
3 x $2^1/_2$	30.7	38.2	29.0	36.8	19.2	25.1
4 x 2	9.1	12.0	9.4	12.8	5.7	7.8
4 x $2^1/_2$	10.6	18.4	13.9	18.4	8.5	10.8
4 x 3	27.0	33.9	26.2	34.0	13.0	17.5
6 x $2^1/_2$	10.1	—	8.4	—	—	—
6 x 3	10.3	14.1	9.6	13.4	—	—
6 x 4	14.5	19.4	12.7	17.1	8.5	11.5
8 x 3	10.2	—	9.3	—	—	—
8 x 4	11.4	16.0	10.1	14.6	—	—
8 x 6	24.6	32.5	18.6	26.0	—	—
10 x 4	11.4	—	10.1	—	—	—
10 x 6	16.2	22.5	15.0	21.7	—	—
10 x 8	32.7	43.6	33.7	47.4	—	—
12 x 6	12.5	—	13.8	—	—	—
12 x 8	17.7	25.1	23.7	34.5	—	—
12 x 10	26.5	36.0	49.5	68.9	—	—

ference in cost becomes less as the sizes approach each other. For example, a 12 inch versus 6 inch requires 12.5 feet whereas a 12 inch versus 10 inch only becomes practical with 26.5 feet.

During this study, consideration was given to the economics of block valve sizes used in conjunction with control valves. If the control valve is two pipe sizes smaller than the line in which it is installed, the block valves should be intermediate in size. The block valves should be line size if the control valve is one size smaller than the pipe in which it is installed.

CONTROL VALVES IN ECONOMIC SIZE PIPE LINES

The presence of a control valve in a pipe line may be thought to eliminate the use of the Least Annual Cost concept in selecting the pipe size. As is discussed in the Introduction, only enough pressure loss should be taken through the valve to maintain control, and no more. Therefore, the Least Annual Cost *still* applies. The friction through the valve, like pressure drop through an exchanger, establishes an initial cost base-line which is higher than would be the case if the valve were not present. That does not negate the concept of striving for the total cost to be minimized.

SUMMARY OF LEAST ANNUAL COST CONCEPTS

From the content of this Chapter, it has been demonstrated that the selection of pipe size based upon the plan of aiming for the Least Annual Cost, over the anticipated life of the plant or facility, can be used as a workable and practical approach. The engineer needs only to know the barest fundamentals about the project to establish whether or not the working equations, or the nomographs based upon them, contain rational values for the cost determining aspects of the economic selection process. If the match is good, as generally will be the case, the engineer may proceed, using the data defining the flowing fluid and the flow rate to select the size.

If the project calls for different costs for power than the "typical" used herein, or if the amortization rate is not seven years, to mention a couple of possibilities, at the start of the work, a basic correction is determined for all pipes to be chosen. This will be applied to all selections according to the Least Annual Cost method.

CHAPTER 3

SELECTION ACCORDING TO PRESSURE DROP AVAILABLE

INTRODUCTION

Many pipe lines in process plants should be chosen according to the pressure drop which is available or which must be dissipated. In the conventional approach, the engineer is expected to make an estimate of the length of the pipe and the valves and fittings which will be used in the final configuration. Such estimates are as good, or as poor, as the experience and the proclivities of the person doing the work. Two engineers probably would arrive at different values for the estimate.

The statistical studies mentioned earlier in Chapter 2 can be applied to this form of pipe size selection, as well as to the Least Annual Cost. With accuracy quite ample for at least 90% of the pipes sized by this approach, the statistical information shows that, for a given size of pipe, in the typical plant, the length of straight pipe and the equivalent length* due to fittings, valves, etc., can be expressed via a mathematical model. When this equation is combined with pressure drop formulae, one can quickly select a pipe size which will fall within the range of total pressure drop established as being correct for the system.

The pressure loss which is applicable may be controlled by the process alone, or it may be established according to the principles presented in the Introduction. The methods which the engineer may use to establish the allowable total pressure drop will be covered later.

* See Chapter 6 for definition and derivation of equivalent length.

Because the control valve pressure drop and piping pressure drop *are* intertwined, there can be an advantage to having pipe size calculations and control valve selection made by the same engineer.

DERIVATION OF BASIC PRESSURE DROP FORMULAE

The formulae used to determine pressure loss in piping are fundamental to fluid dynamics. With an assumption that engineers who are familiar with these disciplines may welcome a review, a brief recapitulation of the derivation of the appropriate equations will be presented before moving on to the derivation of the combination formulae for "typical" plant piping system selection.

A common and basic expression for the pressure drop occuring during the flow of fluids under turbulent conditions was proposed by CHEZY in 1775.

$$h = \frac{flV^2}{2dg} \tag{3-1}$$

Where:

h = Head loss in feet of fluid
f = Friction factor, related to Reynolds Number
l = Pipe length, feet
V = Velocity of fluid flow, feet per second
d = Pipe internal diameter, feet
g = Acceleration due to gravity

This equation also has been credited to DARCY, to EYTELWEIN, to WEISBACH and to J. T. FANNING. We shall use the reference to FANNING hereafter.

Essentially, it states that the head loss is the product of the velocity head (See Chapter 4) times a coefficient (friction factor).

Equation (3-1) can be expressed in terms which are more convenient for the engineer by introducing pressure drop in pounds per square inch, pipe dimensions in inches and flow rates in gpm or mcf/d.

For liquids the equation becomes:

$$psi/100 = 1.35 f\, S\, Q^2/D^5 \qquad (3\text{-}2)$$

For gases:

$$psi/100 = 1.26 \times 10^{-3} f\, G\, M^2\, T\, Z\, /\, P\, D^5 \qquad (3\text{-}3)$$

Where:

psi/100 = Pressure drop, pounds per square inch per one hundred feet of pipe
f = Friction factor
S = Specific gravity at flowing temperature, water at 60° F equals 1.00
Q = Rate of flow, U.S. gallons per minute at flow temperature
D = Pipe inside diameter, inches
G = Specific gravity, air at 14.69 psia and 60° F equals 1.00
M = Rate of flow, *thousands* of standard cubic feet per day
T = Temperature, degrees Rankine (degrees F + 460)
Z = Deviation from perfect gas law, fractional
P = Pressure, pounds per square inch, absolute

In metric (SI) units, these equations become:

For liquids

$$dP_{si} = 6442\, Q_{si}^2\, f\, S_{si}/D_{si}^5 \qquad (3\text{-}2\ M)$$

For gases

$$dP_{si} = 3.06 \times 10^7\, M_{si}^2\, G_{si} f K' Z/P_{si}\, D_{si}^5 \qquad (3\text{-}3\ M)$$

Where:

dP_{si} = Pressure drop, kilograms per square centimeter, per thousand meters of pipe
Q_{si} = Rate of flow, cubic meters per hour, at flow temperature
S_{si} = Specific gravity of liquid at flowing temperature, kilograms per liter, water at 4° C is 1.00
f = Friction factor
M_{si} = Rate of flow of gas, *thousands* of normal cubic meters per hour
G_{si} = Specific gravity of gas, air at one atmosphere and 4° C equals 1.00
K' = Temperature, degrees Kelvin
Z = Deviation from perfect gas law, fractional
P_{si} = Pressure, kilo Pascals
D_{si} = Pipe inside diameter, centimeters

Figure 3-1 Pipe Rack

REYNOLDS NUMBER

The friction factor is related to the dimensionless group called Reynolds Number (5), the surface roughness in the pipe, and to the pipe size, as well. Figure 3-2, sometimes called STANTON'S Diagram (6), illustrates the relationships.

Care should be taken in the use of flow equations involving the friction factor f. The values obtained for f from various sources may not be consistent with the flow equation being used. The value for the most commonly used version is four or eight times those used in other references. Mixing the equations and values for f from different sources can lead to gross errors.

The values of f used herein will match the majority usage but will be different (by the factor of four) from one common reference, PERRY'S Chemical Engineers Handbook (7).

Reynolds Number is an expression of the ratio of the inertial forces to the viscous forces in a flow system. When the value of the number is low, the flow is laminar. As the value increases, through increased flow rate or by other means, the flow begins to exhibit turbulence. The lowest value at which this can occur was established by SCHILLER (11) as 2320. As the number increases, the degree of turbu-

FIGURE 3-2

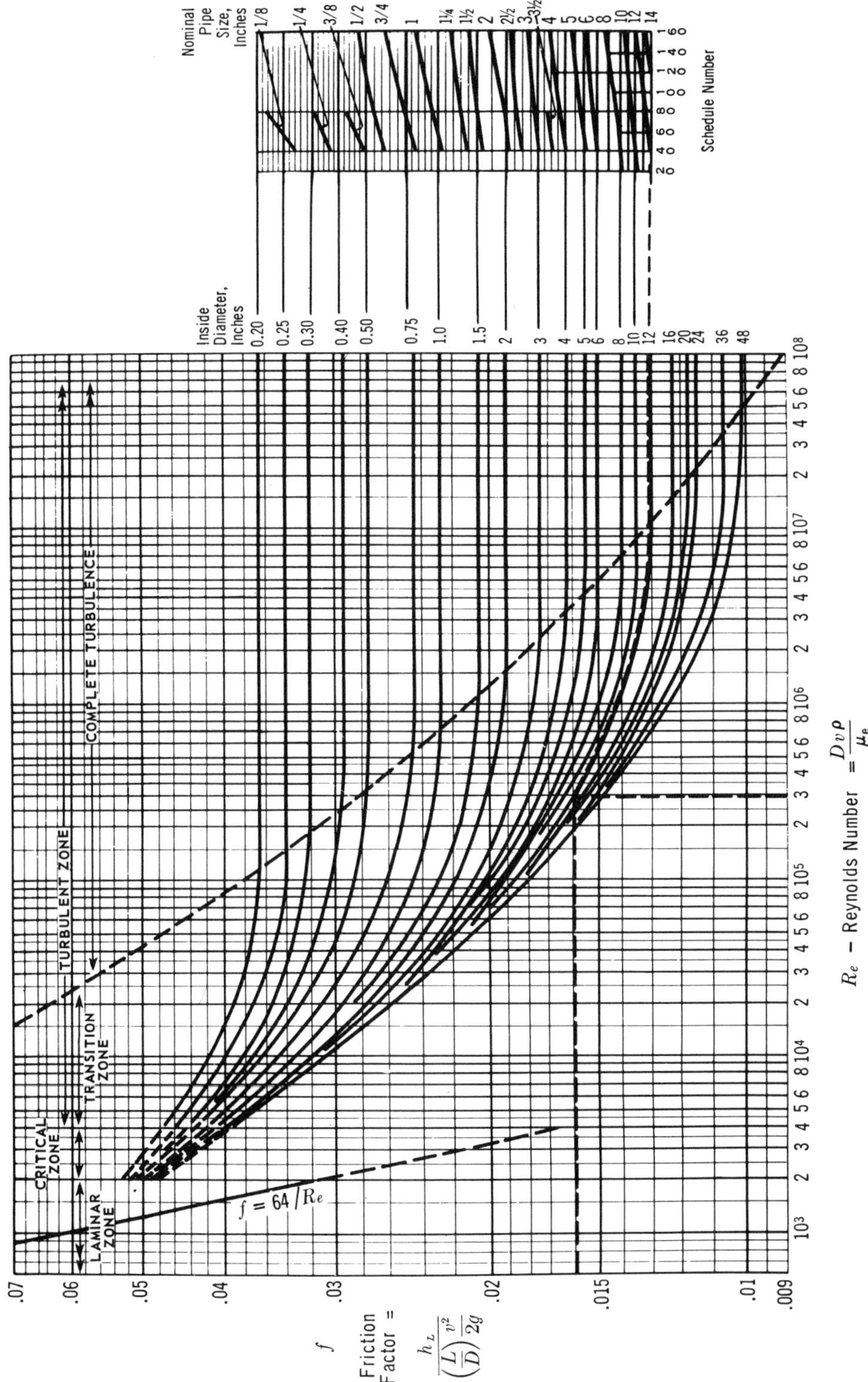

Problem: Determine the friction factor for 12-inch Schedule 40 pipe at a flow having a Reynolds number of 300,000.

Solution: The friction factor (f) equals 0.016.

FRICTION FACTORS FOR CLEAN COMMERCIAL STEEL PIPE

(From TP-410, FLOW OF FLUIDS, Courtesy Crane Company)

(Derived from L. F. Moody, Trans. A.S.M.E. Nov. 1944)

lence grows. The best explanation available, so far, suggests that in any flow there are present small disturbances. When the flow is laminar and therefore Reynolds Number is small, the viscous forces dominate and the disturbances are dampened out quickly. When the flow is turbulent and the inertial forces are dominant, the disturbances are augmented.

As Figure 3-2 shows, the friction factor, at Reynolds Numbers below the critical 2320, is a simple exponential value of the Number, while at higher values, the relation is more complex. In the laminar flow region, the HAGEN-POISEUILLE Law* applies, which states that the loss in pressure through a pipe is inversely proportional to the fourth power of the pipe diameter.

At values of Reynolds Number which are smaller for small pipes and larger for large pipes, the degree of turbulence reaches a maximum, and thereafter the value of f becomes independent of the Reynolds Number. The velocities at which this occurs are mainly above those encountered in process work where fluids generally are being pumped or compressed. Where two-phase flow exists or where fluids are being "blown-down", the Reyonolds Number may be large enough for stabilization of the friction factor to occur.

One common form of Reynolds Number is:

$$N_R = \frac{\varrho L V}{\mu'} \tag{3-4}$$

Where:
N_R = Reynolds Number
ϱ = Density, pounds per cubic foot
L = Linear dimension of the conduit, feet
V = Velocity of flow, feet per second
μ' = Viscosity, slugs per foot-second

This may be converted to variables more common in engineering to give the two working equations for liquids and for gases.

$$N_R = \frac{3157\,Q S}{D \mu} \tag{3-5}$$

for liquids, and

$$N_R = \frac{20.0\,M G}{D \mu} \tag{3-6}$$

for gases.

Where:
Q = Rate of flow, U.S. gallons per minute at flowing temperature
S = Specific gravity at flowing temperature, water at 60° F equals 1.00
D = Pipe inside diameter, inches
μ = Viscosity, centipoise, at flow temperature
M = Rate of flow, *thousands* of standard cubic feet per day
G = Specific gravity, air at 14.69 psia and 60° F is 1.00

In metric (SI) units these equations become:

$$N_R = 35.3\,Q_{si}\,S_{si} \,/\, D_{si}\,\mu_{si} \tag{3-5 M}$$

for liquids, and

$$N_R = 45.5\,M_{si}\,G_{si} \,/\, D_{si}\,\mu_{si} \tag{3-6 M}$$

for gases.

Where:
N_R = Reynolds Number
Q_{si} = Rate of flow, cubic meters per hour, at flow temperature
S_{si} = Specific gravity of liquid at flowing temperature, kilograms per liter, water at 4° C is 1.00
D_{si} = Pipe inside diameter, centimeters
μ_{si} = Viscosity, kilograms per meter seconds, at flowing temperature
M_{si} = Rate of flow of gas, *thousands* of normal cubic meters per hour
G_{si} = Specific gravity of gas, air at one atmosphere and 4° C equals 1.00

THE GENEREAUX SIMPLIFICATION

From Figure 3-2, STANTON'S Diagram, one can see that the relation between Reynolds Number and pipe size is not a simple one. However, if an expression could be written which would satisfactorily define f as an expotential function of N_R, such a compromise would simplify calculations of pressure drop. Various investigators have proposed means to do this. The one offered by GENEREAUX (1), which can be expressed thus,

$$f = 0.14/N_R^{0.16} \tag{3-7}$$

fits the requirements of the methods used herein. It produces "safe-side" values compatible with the flow rates normally encountered in process plants. The dashed line on Figure 3-2 illustrates the comparison with the true values of f with the compromise produced by Equation (3-7). Notice that at the higher values of N_R, the approximation tends to follow the larger pipe size values, and at the lower, the smaller pipe size values. This conforms with the "typical"

* Based upon their research in 1839—1840.

conditions encountered in process plants where larger values of N_R usually are associated with bigger pipe and vice versa. Hence, even though at first glance it might seem that Equation (3-7) could miss a good deal of the ground, this is not likely in common practice. However, where the best accuracy is required, the Equation (3-7) would not be used.

By combining Equation (3-7) with Equation (3-1), a general expression for pressure drop for fluids, in the turbulent region, is obtained.

$$\text{psi}/100 = \frac{13.25\, W^{1.84}\, \mu^{0.16}}{\varrho\, D^{4.84}} \quad (3\text{-}8)$$

By substituting terms which more commonly suit the process engineer, the result for liquids and for gases becomes:

$$\text{psi}/100 = \frac{0.0595\, Q^{1.84}\, \mu^{0.16}\, S^{0.84}}{D^{4.84}} \quad (3\text{-}9)$$

for liquids

$$\text{psi}/100 = \frac{1.27 \times 10^{-4}\, M^{1.84}\, G^{0.84}\, \mu^{0.16}\, Z\, T}{P\, D^{4.84}} \quad (3\text{-}10)$$

for gases.

Where:

psi/100 = Pressure drop, pounds per square inch per one hundred feet
W = Rate of flow, *thousands* of pounds per hour
μ = Viscosity, centipoise
ϱ = Density, pounds per cubic foot
D = Pipe internal diameter, inches
Q = Rate of flow, U.S. gallons per minute at flowing temperature
S = Specific gravity, at flowing temperature, water at 60° F equals 1.00
M = Rate of flow, *thousands* of standard cubic feet per day
G = Specific gravity, air at 14.69 psia and 60° F equals 1.00
Z = Deviation from perfect gas law, fractional
T = Temperature, degrees Rankine (460 + F)
P = Pressure, pound per square inch, absolute

For most liquid calculations, the viscosity term in the equation becomes essentially unity. However, appropriate nomographs based upon this equation provide means to include viscosity influences with ease.

In metric (SI) units, these equations become:

$$dP_{si} = 5.77\, Q_{si}^{1.84}\, \mu_{si}^{0.16}\, S_{si}^{0.84} / D_{si}^{4.84} \quad (3\text{-}9\text{ M})$$

for liquids,

and

$$dP_{si} = 2.72 \times 10^5\, M_{si}^{1.84}\, G_{si}^{0.84}\, \mu_{si}^{0.16}\, Z K' / P_{si}\, D_{si}^{4.84}$$

for gases. (3-10 M)

Where:

dP_{si} = Pressure drop, kilograms per square centimeter, per thousand meters of pipe
Q_{si} = Rate of liquid flow, cubic meters per hour, at flow temperature
μ_{si} = Viscosity, kilograms per meter seconds, at flow temperature
S_{si} = Specific gravity of liquid at flowing temperature, kilograms per liter, water at 4° C is 1.00
D_{si} = Pipe inside diameter, centimeters
M_{si} = Rate of flow of gas, *thousands* of normal cubic meters per hour
G_{si} = Specific gravity of gas, air at one atmosphere and 4° C equals 1.00
Z = Deviation from perfect gas law, fractional
K' = Temperature, degrees Kelvin
P_{si} = Pressure, kilo Pascals

If the pressure loss in a stretch of pipe exceeds 10% of the initial pressure, the calculation must be limited to a shorter portion of pipe; initial pressure minus the pressure loss through the shorter section (not over 10% of the original) is used to set the pressure for a subsequent calculation. This iterative process is continued for successive sections to the end of the pipe line.

FURTHER SIMPLIFICATION FOR GAS FLOW

The actual values for the viscosities of gases may be difficult to obtain. For the determination of pipe size in process plants, this problem may be overcome by relating the gas viscosity to other properties, such as molecular weight, which *are* readily obtainable. In conjunction with temperature, otherwise required in gas pressure drop calculations, accomodation can be made simultaneously for the influence upon viscosity caused by temperature. One must also provide for the effects of pressure upon gas viscosity. These can be enough to alter pressure drop by plus 25 to 30%, in extreme cases which can occur when the pressure is five times critical and the temperature is slightly above the critical value.

All these influences can be incorporated into a variation of Equation (3-10) that simplifies its use for paraffin hydrocarbon gases. For these gases, viscosity may be expressed as a function of molecular weight. Specific gravity also is a function of that parameter. By graphical solution it can be shown that:

$$G^{0.84}\,\mu^{0.16} = 0.0335\,\overline{MW}^{0.788} \qquad (3\text{-}11)$$

at a constant temperature.

Where:
G = Specific gravity, air at 14.69 psia at 60° F equals 1.00
μ = Viscosity, centipoise
$\overline{MW}$ = Molecular weight

The influence of temperature upon the viscosity of the common gases can be expressed as:

$$\mu_T = \mu_b \left[\frac{T}{520}\right]^{0.87} \qquad (3\text{-}12)$$

Where:
μ_b = Viscosity at 60° F, centipoise
μ_T = Viscosity at some temperature, T, centipoise
T = Temperature of the gas, degrees Rankine (°F+460)

Combining Equations (3-11) and (3-12) yields an expression for the effect of the viscosity of the gas with a given molecular weight, corrected for the actual temperature.

$$\mu^{0.16} = 0.0145\,T^{0.134}\,\overline{MW}^{0.788} \qquad (3\text{-}13)$$

The effects of pressure on the viscosity of gases was correlated by COMINGS and EGLY (8) in terms of the reduced pressure and reduced temperature, see Figure 6-4. This form of correlation is based upon the principle of corresponding states. Two substances should have similar properties at corresponding conditions with reference to some basic properties such as the critical pressure and critical temperature (9).

The laws of CHARLES and BOYLE, when combined, yield the perfect or ideal gas law, which states that for ideal gases, the product of the pressure and volume, divided by the absolute temperature, is a constant. While this relation, for practical purposes, may apply to many gases at moderate pressure and temperature, the statement is not wholely accurate. For certain process calculations, corrections must be applied. The deviation from the ideal gas law is designated as Z, a ratio.

$$PV = ZCT \qquad (3\text{-}14)$$

Where:
P = Pressure
V = Volume
Z = Deviation factor, fractional
C = Constant dependent upon the molecular weight of the gas
T = Temperature, absolute

The deviation from the perfect gas law, Z, may be expressed through the use of reduced pressure and reduced temperature. This is one of the more common applications of the principle of corresponding states. Figure 3-4 illustrates the relationship between the reduced pressure and the reduced temperature, upon the increase, or decrease, of the density of a gas, at various pressure and temperature conditions, when compared to the theoretical density according to the ideal gas law. (See Chapter 6 for information on the critical conditions).

Under some conditions, the ideal gas law may suffice for engineering calculations. However, as the critical conditions for a gas are approached, the actual density of a gas may be as much as four times the value which the gas law would predict. Under the conditions encountered in most process plants, the deviation will be greatest when temperatures are near atmospheric and when pressures are near 500 psi.

The reduced pressure is the actual pressure divided by the absolute critical pressure:

$$P_r = P/P_c \qquad (3\text{-}15)$$

The reduced temperature is the actual temperature, degrees Rankine, divided by the critical temperature, degrees Rankine:

$$T_r = T/T_c \qquad (3\text{-}16)$$

Where:
P_r = Reduced pressure, fractional
T_r = Reduced temperature, fractional
P = Pressure, pounds per square inch, absolute
T = Temperature, degrees Rankine (°F + 460)
P_c = Critical pressure of the gas, psia
T_c = Critical temperature of the gas, degrees Rankine

For mixtures of gases, mole-average critical constants, called the pseudo-critical pressure and the pseudo-critical temperature, are used.*

* For a discussion of the nature of, and determination of, the critical and pseudo-critical conditions, see Chapter 6.

* More data on critical pressure and temperature are given in Chapter 6.

FIGURE 3-3

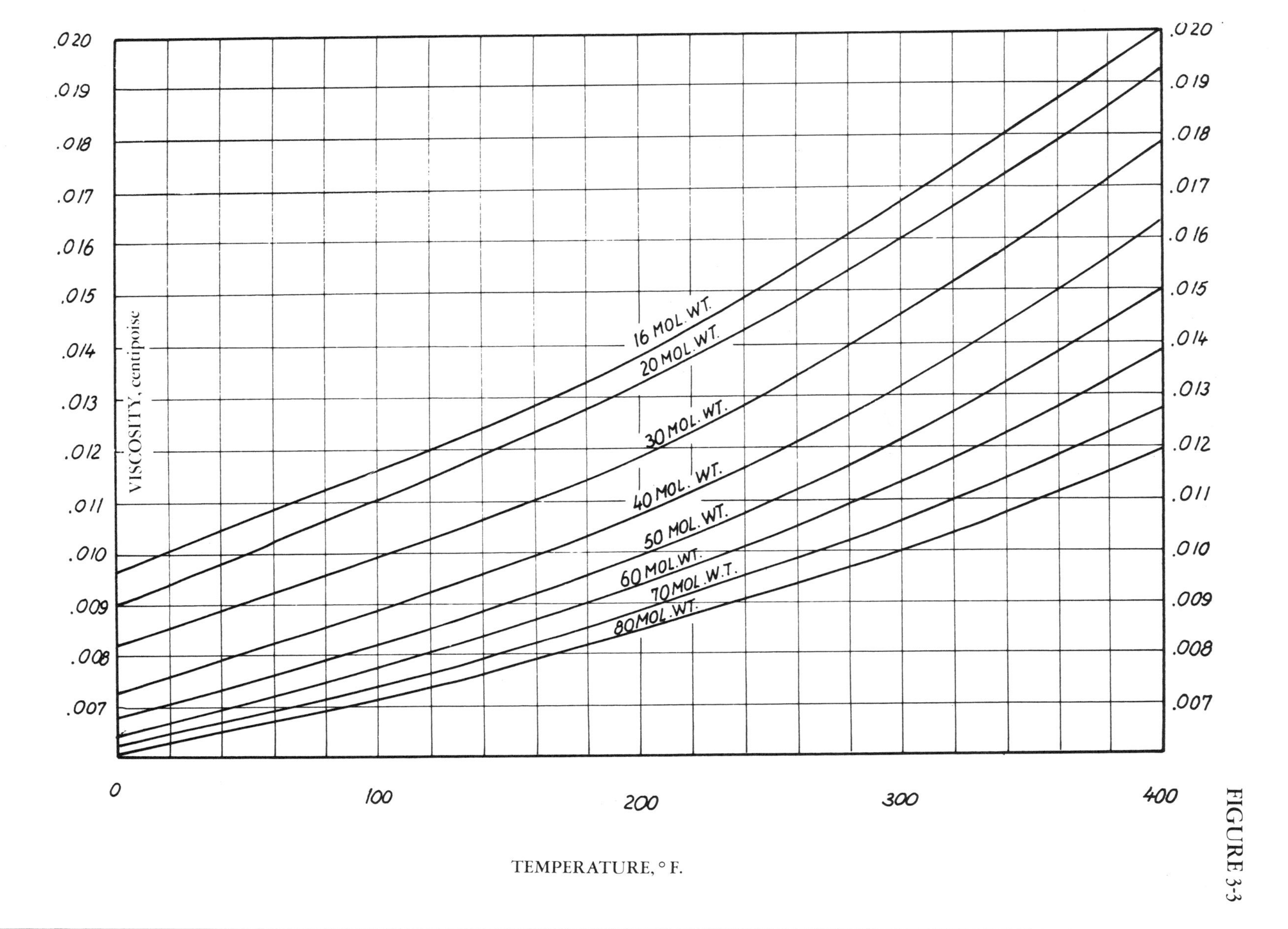

VISCOSITY OF HYDROCARBON GASES

As shown by Z in Equation (3-10), deviations from the perfect gas must be included in pressure drop calculations whenever Z is significantly different from unity. The same method may be used for the influence of pressure upon gas viscosity.

The combined effect, which may *only be used with Equation* (3-17) (see below), is presented in Figure 3-5, showing the relation, Ψ, between the combination of Z and the elevated pressure value of viscosity versus reduced pressure and temperature.

$$\Psi = Z \left(\frac{\mu_P}{\mu_T} \right)^{0.16} \quad (3\text{-}17)$$

Where:
Z = Deviation from ideal gas law = PV/RT
μ_P = Viscosity of gas at flowing temperature and pressure
μ_T = Viscosity of the gas at the flowing temperature and atmospheric pressure

Equation (3-10) expressing the pressure drop for gases may now be further simplified by the substitution of Ψ for both the Z and viscosity terms to provide the gas equation in its most condensed form.

$$\text{psi}/100 = \frac{1.68\text{x}10^{-6}\, M^{1.84}\, \overline{MW}^{0.784}\, T^{1.15}\, \Psi}{P\, D^{4.84}} \quad (3\text{-}18)$$

Where:
psi/100 = pressure drop, pounds per square inch per one hundred feet of pipe
M = Rate of flow, *thousands* of standard cubic feet per day
$\overline{MW}$ = Molecular weight
T = Temperature, degrees Rankine (°F+460)
Ψ = Factor combining deviation for the gas law and the relation between molecular weight and viscosity, see Equation (3-17)

This can be used for common paraffin hydrocarbon gases when flow is turbulent. It may be used for other gases when the highest accuracy is not required. For calculations requiring the highest accuracy, Equation (3-10) is preferred.

However, as an example of the commonly existing, close agreement between Equation (3-18) and the unmodified FANNING formula, Equation (3-3), consider the following situation. The pressure drop through a 4 inch pipe, one hundred feet long, carrying 2.5 million standard cubic feet per day of air at 115 psia and 60 °F is 0.695 psi by Equation (3-3) versus 0.678 psi by Equation (3-18). The difference of 2.5 % is insignificant compared to the uncertainty which may exist in the other variables involved; the rate of flow may, in itself, be an approximation!

The gas pressure drop Graphs 39, 40 and 41 are based upon Equation (3-18). In the absence of *reliable* viscosity data, Equation (3-18), or the nomographs based upon it, may still be the best compromise. Extensive tables are given in Appendix B for the viscosity of various gases. Chapter 6 covers means for estimating gas viscosities when accurate data are not available.

In metric (SI) units, the condensed pressure drop equation for gas becomes:

$$dP_{si} = 1909\, M_{si}^{1.84}\, \overline{MW}^{0.788}\, K'^{1.34}\, \Psi / P_{si}\, D_{si}^{4.84} \quad (3\text{-}18\text{ M})$$

Where:
dP_{si} = Pressure drop, kilograms per square centimeter, per thousand meters of pipe
M_{si} = Rate of flow of gas, *thousands* of normal cubic meters per hour
$\overline{MW}$ = Molecular weight of the gas
K' = Temperature, degrees Kelvin
Ψ = Factor combining deviation from the ideal gas law and the relation between molecular weight and viscosity, see Equation (3–17)
P_{si} = Pressure, kilo Pascals
D_{si} = Pipe inside diamter, centimeters

MODIFIED AND STANDARD FANNING EQUATIONS FOR STEAM

The pressure drop for steam flow can best be determined using an adaption of the modified FANNING Equation (3-8). For saturated steam, the pressure may be used to express the viscosity and the density, over the range of pressure from 27 inches Hg to 500 psig with adequate accuracy. When the closest approximation of pressure drop is needed, the FANNING equation including the calculated value for friction factor, based upon Reynolds Number, is the best choice.

For *saturated* steam over the range of 27 inches Hg to 500 psig, the modified FANNING formula is:

$$\text{psi}/100 = 2198\, W^{1.84} / D^{4.84}\, P^{0.935} \quad (3\text{-}19)$$

FIGURE 3-4

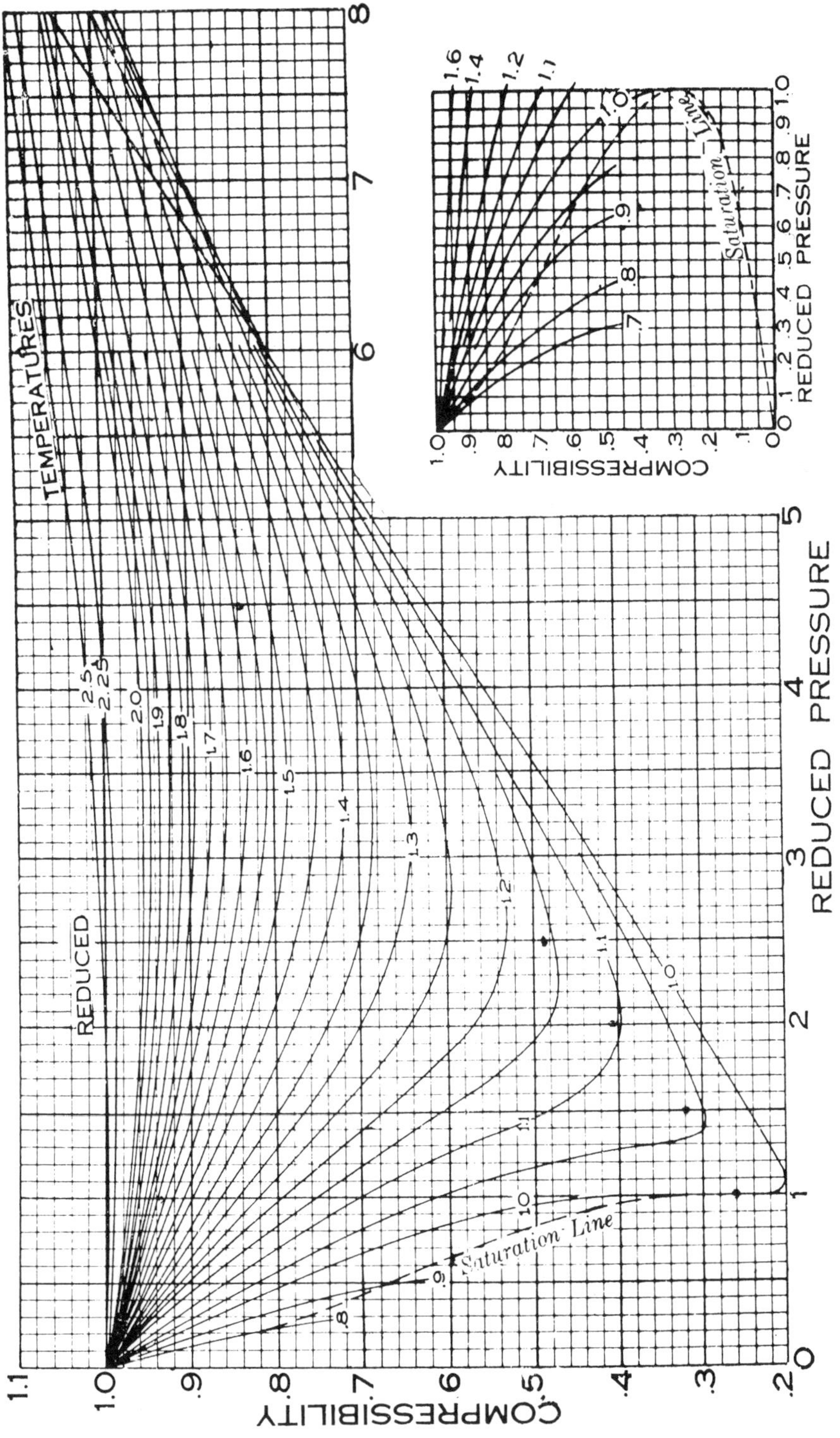

DEVIATION FROM IDEAL GAS LAWS, Z

Z VERSUS T_R AND P_R

Where:

psi/100 = Pressure drop, pounds per square inch per hundred feet of pipe

W = Steam flow rate, *thousands* of pounds per hour

D = Pipe inside diameter, inches

P = Steam pressure, pounds per square inch, absolute

Graphs 48 and 49 are based upon Equation (3-19) and include a means for correcting for superheat.

The unmodified FANNING equation for steam is:

$$\text{psi}/100 = 337.6\, W^2 f / \varrho D^5 \qquad (3\text{-}20)$$

Reynolds Number for steam is expressed as:

$$N_R = 6310\, W / D\mu \qquad (3\text{-}21)$$

Where, in addition to the legend above:

f = Friction factor, fractional

ϱ = Density of steam, pounds per cubic foot

μ = Viscosity of steam, centipoise

N_R = Reynolds Number, fractional

In metric (SI) units, the modified FANNING equation for saturated steam is:

$$dP_{si} = 1.20 \times 10^7\, W_{si}^{1.84} / P_{si}^{0.935}\, D^{4.84} \qquad (3\text{-}19\ \text{M})$$

The unmodified FANNING equation is:

$$dP_{si} = 5.53 \times 10^6\, W_{si}^2 f / D^5\, \varrho_{si} \qquad (3\text{-}20\ \text{M})$$

Reynolds Number is:

$$N_R = 35.3\, W_{si} / D_{si}\, \mu_{si} \qquad (3\text{-}21\ \text{M})$$

Where:

dP_{si} = Pressure drop, kilograms per square centimeter per thousand meters of pipe

W_{si} = Steam flow rate, *thousands* of kilograms per hour

D_{si} = Pipe inside diameter, centimeters

P_{si} = Steam pressure, kilo Pascals

f = Friction factor

ϱ_{si} = Density of steam, kilograms per cubic meter

μ_{si} = Viscosity of steam, kilograms per meter second

N_R = Reynolds Number, fractional

WILLIAMS AND HAZEN FORMULA FOR WATER FLOW

In addition to the FANNING equations, other formulae have been developed for determining the pressure drop of specific fluids. While not all will be discussed in this book, one equation for water does deserve mention. This is the approach developed by WILLIAMS and HAZEN in 1920 (10). Engineers in water works had recognized that cast iron and steel pipes carrying water would, with the passage of time, exhibit increasing pressure losses. This was due to the influence of corrosion and/or scale which would form on the inside walls. In effect, viewed from the concepts of the FANNING approach, this would signal an increase in the value of f for older pipe compared to new.

The WILLIAMS and HAZEN formula is:

$$\text{psi}/100 = \frac{452\, Q^{1.85}}{C^{1.85}\, D^{4.86}} \qquad (3\text{-}22)$$

Where:

psi/100 = Pressure drop, pounds per square inch per 100 feet of pipe

Q = Rate of flow, U.S. gallons per minute.

C = Pipe coefficient, see below

D = Pipe internal diameter, inches

The coefficient C is essentially a friction factor. Table 3-1 gives values for various pipes with the influence of age in water service.

The equation is applicable only to water at normal atmospheric temperatures.

TABLE 3-1
Table of C Values for WILLIAMS and HAZEN Formula

Constant	Type of pipe
140	New smooth steel pipe
	New smooth brass pipe or tubing
	Glass tubing
130	Copper tubing
	Ordinary brass pipe
	New cast iron
125	Old steel pipe
120	Wood stave pipe
	Concrete pipe
	New wrought iron pipe
	Four to six years old cast iron pipe
110	Ten to twelve years old cast iron pipe
	Vitrified pipe
	Spiral riveted steel, flow with lap
100	Spiral riveted steel, flow against lap
	Thirteen to twenty years old cast iron pipe
90	Twenty-six to thirty years old cast iron pipe
60	Corrugated steel pipe

FIGURE 3-5

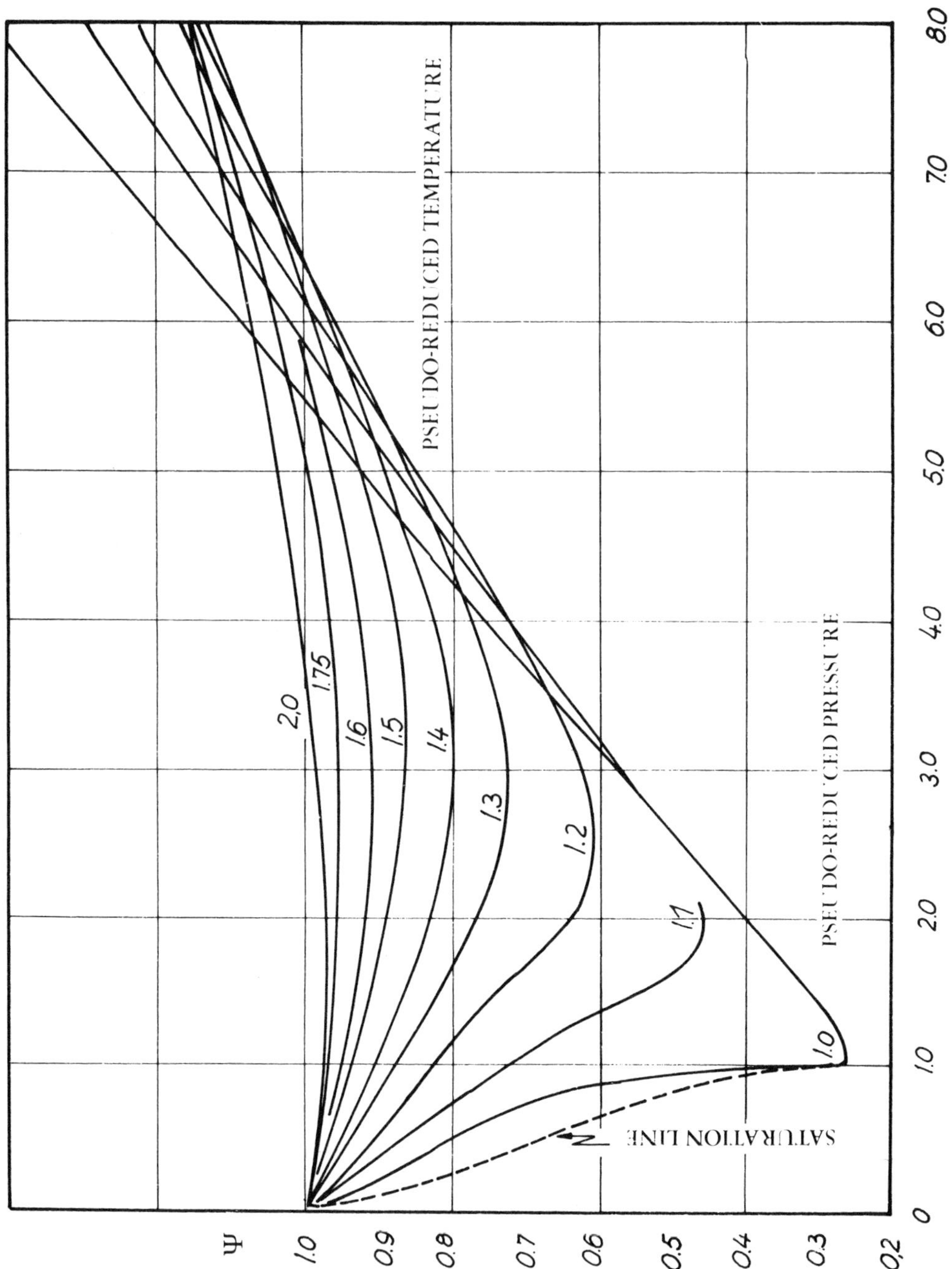

Ψ VERSUS T_R AND P_R

(Sources: G. G. Brown, Comings & Egly, Sage & Lacey)

The similarity between the modified FANNING Formula, Equation (3-9), is worth noting. The rate of flow exponents are 1.85 for WILLIAMS and HAZEN versus 1.84 for Equation (3-9). The diameter exponent is 4.86 compared to 4.84. Hence the same pragmatic approximations were made by WILLIAMS and HAZEN in 1920 and GENEREAUX in 1937, even though they were, in all probability, conceived relative to different fluids.

Comparing the WILLIAMS and HAZEN formula with Equation (3-9) one finds that, for new steel pipe, the WILLIAMS and HAZEN coefficient would be 126 for water at 60° F. This is less than Table 3-1 designates which gives an indication of the "safety factor" in Equation (3-9). The 126 value is very close to the 125 for old steel pipe!.

The pressure drop calculated by the WILLIAMS and HAZEN method of using C of 142 in a 6 inch pipe carrying 600 gpm of water at 60° F is the same as that calculated using Equation (3-2), which does *not* have the GENEREAUX simplification, namely, 1.01 psi per 100 feet.

Under the same flow conditions, the value of the pressure drop as determined using the modified FANNING Equation, (3-9), is 1.09 psi per 100 feet of pipe. Here again can be seen the nature and size of the safety factor built into the modified FANNING equation.

When the most *accurate results* are required, the modified FANNING is not the best choice. However for *most* plant work, it suffices quite well. Considering the rapidity with which pipe calculaitons can be made using the nomographs based upon it, the distinct advantages of this method become obvious.

PRESSURE LOSS IN A "TYPICAL" PROCESS PLANT PIPE SYSTEM

At the beginning of this Chapter, it was stated that considerations similar to those applied in Chapter 2 regarding the Least Annual Cost method of pipe size selection, could be used in selection by the Pressure Drop Available approach. With a series of condensed equations established for calculating pressure drop for various fluids, the means of combining them with the applicable effective length of "typical" process pipe lines is the next step toward developing the simplified formulae for pipe size selection when Pressure Drop Available is the basis.

In Chapter 6, the statistical studies of the composition of "typical" pipe lines in process plants are explored.

Equation (3-22 A) defines the sum of the actual pipe length and the equivalent length for "typical" process pipes.

$$L_T = 17.3\, D^{1.41} \qquad \text{(3-22 A)}$$

Where:
L_T = Actual pipe length plus equivalent pipe length
D = Inside diameter of pipe, inches

The product of $L_T/100$ and the pressure drop per hundred feet, derived from the appropriate formulae for the fluid under consideration, yields the total pressure drop for the "typical" pipeline handling that material.

For process plant design, a more practical concept involves developing equations, suggested above, to yield the pipe size required to keep the total pressure loss in a "typical" line within some designated limit. This limit may be set by the process requirements, the allowable pressure drop for the pipe compared to a control valve, as described in the Introduction, or it may be an arbitrary value. Whatever the source, the limit can be accommodated through the equations to follow.

Combining the various equations developed thus far in this Chapter with the equation for the total length of a "typical" pipe provides the formulae desired for the pipe size selection wherein the Pressure Drop Available mode is applicable, and where the flow is turbulent.

WORKING EQUATIONS FOR PRESSURE DROP AVAILABLE MODE FOR "TYPICAL" LINE

For all **fluids** incorporating the modified FANNING equation:

$$D = 1.274\, W^{0.536}\, \mu^{0.047}/(\Delta P_T)^{0.29}\, \varrho^{0.29} \qquad \text{(3-23)}$$

For **liquids** incorporating the modified FANNING equations:

$$D = 0.263\, Q^{0.536}\, \mu^{0.047}\, S^{0.250}/\Delta P_T^{0.29} \qquad \text{(3-24)}$$

For **gases** incorporating the modified FANNING equation:

$$D = 0.0441\, M^{0.536}\, \mu^{0.047}\, G^{0.250} \left(\frac{ZT}{\Delta P_T P} \right)^{0.292} \qquad \text{(3-25)}$$

For **gases** with the viscosity combined into molecular weight:

$$D = 0.0125\, M^{0.536}\, \overline{MW}^{0.230}\, T^{0.34} \left(\frac{\Psi}{\Delta P_T P}\right)^{0.292} \quad (3\text{-}26)$$

For **steam,** incorporating the modified FANNING equation:

$$D = 5.65\, W^{0.536} / \Delta P_T^{0.292}\, P^{0.273} \quad (3\text{-}27)$$

Where:

D = Pipe inside diameter, inches
W = Rate of flow, *thousands* of pounds per hour
μ = Viscosity at flow temperature, centipoise
ΔP_T = Pressure drop in the entire line, pounds per square inch
ϱ = Fluid density at flow conditions, pounds per cubic foot
P = Gas pressure, pounds per square inch, absolute
Q = Rate of flow of liquid at flow temperature, U.S. gallons per minute
S = Liquid specific gravity, at flow temperature, water at 60° F is 1.00
G = Gas specific gravity, air at 60° F and 14.69 psia equals 1.00
M = Gas rate of flow, *thousands* of standard cubic feet per day at 60° F and 14.69 psia
$\overline{MW}$ = Molecular weight
T = Temperature, degrees Rankine (°F+46°)
Ψ = Factor combining deviation from the gas law and the relation between molecular weight and viscosity, see Equation (3-17).

For metric (SI) units, these equations become:
For **liquids:**

$$D_{si} = 3.60\, Q_{si}^{0.536}\, \mu_{si}^{0.047}\, S_{si}^{0.250} / \Delta P_{si}^{0.292} \quad (3\text{-}24\text{ M})$$

For **gases** with viscosity to be introduced:

$$D_{si} = 21.7\, M_{si}^{0.536}\, \mu_{si}^{0.047}\, G_{si}^{0.250} \left(\frac{Z\,K'}{\Delta P_{si}\, P_{si}}\right)^{0.292} \quad (3\text{-}25\text{ M})$$

For **gases** with viscosity included in the equation via molecular weight:

$$D_{si} = 113\, M_{si}^{0.536}\, \overline{MW}^{0.230}\, K'^{0.34} \left(\frac{\Psi}{\Delta P_{si}\, P_{si}}\right)^{0.292} \quad (3\text{-}26\text{ M})$$

For **steam,** incorporating the modified FANNING equation:

$$D_{si} = 65.3\, W_{si}^{0.536} / \Delta P_{si}^{0.292}\, P_{si}^{0.273} \quad (3\text{-}27\text{ M})$$

Where:

D_{si} = Pipe internal diameter, centimeters
Q_{si} = Rate of liquid flow, cubic meters per hour, at flow temperature
μ_{si} = Viscosity, kilograms per meter second, at flow temperature
S_{si} = Specific gravity of liquid at flowing temperature, kilograms per liter, water at 4° C is 1.00
ΔP_{si} = Pressure drop in the total line, kilo Pascals
M_{si} = Rate of flow of gas, *thousands* of normal cubic meters per hour
G_{si} = Specific gravity of the gas, air at one atmosphere and 4° C equals 1.00
Z = Deviation from ideal gas law, fractional
K' = Temperature, degrees Kelvin
P_{si} = Pressure, kilo Pascals
$\overline{MW}$ = Molecular weight
Ψ = Factor combining deviation from the ideal gas law and the relation between molecular weight and viscosity, see Equation (3-17)
W = Rate of steam flow, *thousands* of kilograms per hour

PRACTICAL ASPECTS OF SIMPLIFIED PDA MODE

At the first consideration of this simplified PDA mode of pipe size selection, objections may be raised, based largely upon earlier approaches to pipe size selection which required estimates to be made of the length of pipe and the number and style of fittings, values, etc. For a typical pipe system, however, the statistical data is likely to be more reliable, and certainly more consistent, than will be "guestimates" usually employed.

Certainly, the engineer must be mindful of the existence, from time to time, of possible situations where the length of straight pipe will exceed the "typical" line as considered here. This can occur when product is being conducted to storage, but ordinarily the Least Annual Cost mode may apply for that circumstance. If a true PDA line runs to another unit, then consideration of the total straight pipe may well be justified.

With the spread between diameters in commercial pipe, it develops, however, that considerable latitude is available to the engineer without being overly concerned about the abnormal pipe system. From pipe sizes 1$^1/_2$ inch to 8 inch, the actual length plus equivalent length can vary by 200% and still the total pressure loss will be within tolerance. For pipe sizes above 8 inch, the variation can be 100%.

In those circumstances where the pipe is thought not to match the typical, this method will *still* provide a rapid means to obtain a first approximation (which often may turn out to be the correct size).

Figure 3-6
Absorption Oil Cooler Piping

CHAPTER 4

VELOCITY ALLOWABLE

In traditional pipe size selection, velocity has been one of the common parameters considered. Velocity determination is independent of the length of the pipe as well as the number and style of fittings used. Thus, in the absence of the statistical data on pipe line configurations, presented herein, the velocity approach would permit pipe size selection without knowing details about the pipe system. This assumes that the velocity criteria employed would allow for the variations in pressure rating, material to labor ratios, etc., and that pipes so selected would meet the true needs of the system. Such approaches *could* be established. The traditional methods were not developed in this manner.

There are those conditions where velocity, either above some minimum, or below a maximum, will be the optimum selection. Commonly, these conditions may occur in multiphase flow rather than single phase. Before the former is considered in Chapter 5, a brief review of the principles and formulae which apply to determination of velocity in single phase flow will be considered.

VELOCITY CONDITIONS

In all conditions of fluid flow, the velocity of the flowing material is *not* a constant value over the cross-section of the pipe. Where the velocity is below the critical Reynolds Number ($N_R = 2320$), and hence where the flow is laminar, the resistance to flow is due almost entirely to viscosity. There is little motion between the walls of the pipe and the fluid, so "friction", in the sense of "rubbing" of the fluid on the pipe essentially does not occur. Velocities range from nearly zero at the walls to much larger values at the center of the pipe. Therefore, the flow profile has a sharp, "bullet", shape.

During turbulent flow, the pattern of movement is more complex; one cannot set up a simple description of what is happening. As a whole, the pattern of the velocity profile is less predictable. The greater "mixing" taking place in turbulent flow blunts the "point" of the pattern characteristic of laminar conditions.

In all the formulae for velocity of flow, the *average* velocity (which may also be called the bulk velocity) is being implied even though rarely so defined.

FORMULAE FOR VELOCITY

As a general statement, the equation for the velocity of flow may be given as:

$$V = \text{Cubic feet per second/pipe area.} \qquad (4\text{-}1)$$

For liquids, this may be altered to convenient units to yield

$$V = 0.409\, Q/D^2 \qquad (4\text{-}2)$$

or

$$V = 0.01193\, B/D^2 \qquad (4\text{-}3)$$

Where:

V = Velocity (average), feet per second
Q = Rate of flow, U.S. gallons per minute at flow temperature
D = Pipe inside diameter, inches
B = Rate of flow, barrels (42 U.S. gallons) per day at flow temperature

For gases, with convenient units:

$$V = 0.06\ MTZ/PD^2 \qquad (4\text{-}4)$$

Where:

M = Rate of flow, *thousands* of standard cubic feet per day (14.69 psia and 60° F)

T = Temperature of flowing gas, degrees Rankine (°F + 460)

Z = Deviation from perfect gas law, fractional

P = Gas pressure, pounds per square inch, absolute

For steam flow, the velocity formula is:

$$V = 51\ W/D^2 \varrho \qquad (4\text{-}5)$$

Where, in addition to the above:

W = Rate of steam flow, *thousands* of pounds per hour

ϱ = Steam density, pounds per cubic foot

For metric (SI) units, the formulae become:
For liquids,

$$V_{si} = 0.219\ Q_{si}/D_{si}^2 \qquad (4\text{-}2\ M)$$

For gases,

$$V_{si} = 1310\ M_{si}\ K'Z/P_{si}\ D_{si}^2 \qquad (4\text{-}4\ M)$$

For steam,

$$V_{si} = 1601\ W_{si}/D_{si}^2\ \varrho_{si} \qquad (4\text{-}5\ M)$$

Where:

V_{si} = Average velocity, meters per second

Q_{si} = Rate of liquid flow, cubic meters per hour, at the flow temperature

D_{si} = Pipe inside diameter, centimeters

M_{si} = Gas rate of flow, *thousands* of normal cubic meters per hour

K' = Flowing temperature, degrees Kelvin

Z = Deviation from perfect gas law, fractional

P_{si} = Gas flowing pressure, kilo Pascals

W_{si} = Steam flow rate, *thousands* of kilograms per hour

ϱ_{si} = Steam density, kilograms per cubic meter

TERMINAL VELOCITIES

In the computation of gas or steam velocity, the pressure will be diminishing as the substance travels the length of the pipe. Generally, this influence upon the calculated velocity will be trivial. However, if the pressure change should be more than 10% of the original absolute pressure, the pressure used should be selected according to the significance of the final result. If the initial velocity is important, the initial pressure is used. If the final velocity is the controlling factor, the ending pressure is chosen.

MAXIMAL VELOCITIES

When gas (or steam) at an elevated pressure is being discharged through a pipe, the maximum velocity which it can attain is the velocity of sound in the specific gas at the conditions present at the end of the pipe. For a short pipe, a valve or an orifice, the terminal acoustic velocity is considered to occur when the final absolute pressure is 50% of the initial absolute pressure. A further reduction of the downstream pressure will not produce an increase in flow rate nor in the terminal velocity.

While this rule of thumb will suffice for certain approximations, exacting design calculations, where the pipe length is substantial, will call for a more accurate method. Exploration of how the acoustic velocity becomes the limiting factor and the methods used to determine the maximum flow under blowdown or similar conditions is indicated.

When the pressure in a gas diminishes as the substance passes along the length of the pipe, the energy used to overcome the frictional resistance is converted into heat. This results in an increase of the intrinsic energy of the gas. This may be expressed thus:

$$H = I_2 - I_1 = \frac{1}{J} \int_{v_1}^{v_2} P dv \qquad (4\text{-}6)$$

Where:

H = Quantity of heat added

I_1 = Initial intrinsic energy of the gas

I_2 = Final intrinsic energy of the gas

J = Factor used to convert work energy to heat energy

v_1 = Specific volume, initial

v_2 = Specific volume, final

P = Pressure, absolute

c_v = Specific heat at constant volume

c_p = Specific heat at constant pressure

T_1 = Temperature, initial

T_2 = Temperature, final

This equation applies whether or not the heat is added at constant volume. Therefore,

$$I_2 - I_1 = c_v (T_2 - T_1) \qquad (4\text{-}7)$$

Thus the change in intrinsic energy is proportional to the change in temperature. When heat is added (from friction) while the pressure is constant, Equation (4-6) then becomes

$$c_p (T_2 - T_1) = c_v (T_2 - T_1) + \frac{P (v_2 - v_1)}{J} \qquad (4\text{-}8)$$

From this it can be determined that

$$c_p - c_v = R/J \qquad (4\text{-}9)$$

Referring to Equation (4-7) and substituting from (4-9) the value of R/J, and allowing $k = c_p/c_v$, it will be seen that

$$dI = \frac{d\,(Pv)}{J\,(k-1)} \qquad (4\text{-}10)$$

Thus it can be recognized that the ratio of the specific heats of gases at constant volume and constant pressure requires consideration in exploring the behavior of gases during conditions of maximum velocity.

To explore further the reasons behind the concept that acoustic velocity is the limitation of the rate of gas flow, additional thermo-dynamic considerations deserve discussion. DODGE (12) states that the equation for determining the maximum flow rate for a given gas under specific conditions can be written in simplified form as:

$$W = K\left[r^{\frac{2}{k}} - r^{\frac{k+1}{k}}\right] \qquad (4\text{-}11)$$

Where:
W = Weight flow rate
K = Constant which is assumed independent of r or W
r = Ratio of the initial to final pressures, P_2/P_1
k = Ratio of specific heats

This equation shows that the flow will be zero under several conditions: when r equals one, when P_2 equals zero, or when r equals zero. The rate of flow will increase as the pressure changes, but one could not expect that this would increase indefinitely, unless one has an "infinite" flow rate. Knowing that the rate *does* increase up to a given maximum, if the rate of change of W with respect to r is determined by differentiating Equation (4-14) and setting the result equal to zero, the maximum rate of change can be found.

$$r_c = \left(\frac{2}{1+k}\right)^{\frac{k}{k-1}} \qquad (4\text{-}12)$$

This value for r_c may be substituted into the equation for an isentropic expansion of an ideal gas:

$$V_2 = \sqrt{\frac{2g\,k\,P_1\,v_1}{k-1}\left[1 - r^{\frac{k-1}{k}}\right]} \qquad (4\text{-}13)$$

and after elimination of $P_1\,v_1$, by the adiabatic–expansion equation,

$$P_1\,v_1 = P_2\,v_2\,(r)^{\frac{1-k}{k}} \qquad (4\text{-}14)$$

the final result is:

$$V_2 = (gk\,P_2\,v_2)^{1/2} \qquad (4\text{-}15)$$

This is the velocity of sound in a given gas. Therefore, the velocity of the gas under free expansion will increase until the gas reaches the acoustic velocity. There can be no further increase. The maximum velocity level will occur at the terminus of the conduit.

For another approach to this behavior of gases, DODGE and THOMPSON offer an alternate explanation which arrives at the same result (13) (pages 392 to 411).

The conditions in a pipe when the flow is allowed to proceed *toward* the maximum velocity may not progress to the acoustic velocity. When the gas may be *approaching* acoustic levels without attaining them, DODGE and THOMPSON (13) show that the behavior may be described as follows:

When the critical velocity has been reached, the gas will be at a pressure, P_0, and have a specific volume, v_0.

The conditions are described by:

$$P_0/P_1 = N_M^2\left\{\left(\frac{k-1}{k+1}\right)\left[1 + \frac{2}{(k-1)\,N_M^2}\right]\right\}^{\frac{1}{2}} \qquad (4\text{-}16)$$

The ratio between the values of velocity and specific volume of the gas at any point in the pipe is constant. Therefore, the ratio between the final and initial velocities is expressed by the term within the brackets in this equation. In the interior of an insulated pipe, the conditions may be expressed as:

$$fl/D = -\frac{(k+1)}{2k}\log_e\frac{V_2^2}{V_1^2} + \frac{1}{k}\left(\frac{1}{N_M^2} + \frac{k-1}{2}\right)\left(1 - \frac{V_2^2}{V_1^2}\right) \qquad (4\text{-}17)$$

From Equation (4-17) one may now calculate values of fl/D for a value of N_M and for a series of values of V_2/V_1. With the friction factor known, the length of the pipe corresponding to a specific velocity may be found.

The ratio of the pressure at any point in the pipe, to the initial pressure, can be found from:

$$\frac{P_1}{P_2} = \frac{V_2}{V_1}\left[1 + \frac{(k-1)\,N_M^2}{2}\left(1 - \frac{V_2^2}{V_1^2}\right)\right] \qquad (4\text{-}18)$$

Where:

V_1 = Gas velocity, initial
V_2 = Gas velocity, final
k = Ratio of specific heats, c_v/c_p
N_M = Mach's Number, actual velocity divided by the acoustical velocity
f = Friction factor
l = Length of the pipe
D = Pipe inside diameter

A convenient solution for problems of this type can be made using a graph depicting a family of curves for a number of conditions. Such a family of curves is shown in Figure 4-1. The diagonal line shows values of Mach's Number at the *beginning* of the pipe for those conditions where the velocity at the *terminus* is acoustic. The curved lines show the values of Mach's Number at the *beginning* of the pipe when the terminal velocity is *less* than acoustic. The diagonal line is used when the ratio of the pressures at the start and terminus are lower than shown on the ordinate, for a given fl/D value. The curves are used when the ratio is greater than shown on the ordinate for a given fl/D.

Figure 4-1 applies to gases with a k of 1.32. For other values of k, the pressure ratio on the ordinate must be corrected by the values of Table 4-1.

TABLE 4-1

Correction Factors for Values of k (Figure 4-1)
(Multiply Ordinate Value by Factor to Obtain True r_c)

k	Factor	k	Factor
1.001	1.112	1.40	0.974
1.05	1.106	1.45	0.958
1.10	1.079	1.50	0.945
1.15	1.057	1.55	0.930
1.20	1.040	1.60	0.915
1.25	1.024	1.65	0.904
1.30	1.003	1.70	0.891
1.35	0.991	1.75	0.876
		1.80	0.865

The earlier approximation of the critical pressure ratio of 50% for gases may now be re-examined, and a more exact method may be explored from the material just presented.

The 50% "rule" applies when fl/D is small; for an orifice, valve or short pipe, and for one value of k. As shown by Equations (4-12) and (4-18), and in Table 4-2, this approximation can lead to errors when applied without caution.

Equation (4-12) depicts the relation between the critical pressure ratio, r_c, and the ratio of the specific heats, k, (also called the "adiabatic expansion coefficient"). Values for r_c for values of k are given in Table 4-2.

TABLE 4-2

Values of Critical Pressure Ratio Versus k

k	r_c	k	r_c
1.005	0.605	1.40	0.528
1.01	0.603	1.45	0.519
1.05	0.600	1.50	0.512
1.10	0.585	1.55	0.504
1.15	0.573	1.60	0.496
1.20	0.564	1.65	0.490
1.25	0.555	1.70	0.483
1.30	0.544	1.75	0.475
1.35	0.537	1.80	0.469

APPROXIMATION OF K

Computations by EDMISTER (14) show that for hydrocarbon gases one may calculate the value of k when no other data are available. In the case of mixed gases, the arithmetic molal average of the C_p for the mixture may be used in the formula:

$$k = \frac{\overline{MW}\,C_p}{(\overline{MW}\,C_p - 1.987)} \qquad (4\text{-}19)$$

Where:

$\overline{MW}$ = Molecular weight of the gas (28.96 x G may be used for a mixture in the absence of other information)
C_p = Specific heat of the gas at constant pressure, at the gas temperature

It is to be noted that the value of C_p, and hence k, decreases with increasing temperature. For exacting calculations, when the gas temperature may change during the transit through the pipe, use of a mean value lying between the two temperature extremes may be advisable. To illustrate the magnitude of the

FIGURE 4-1

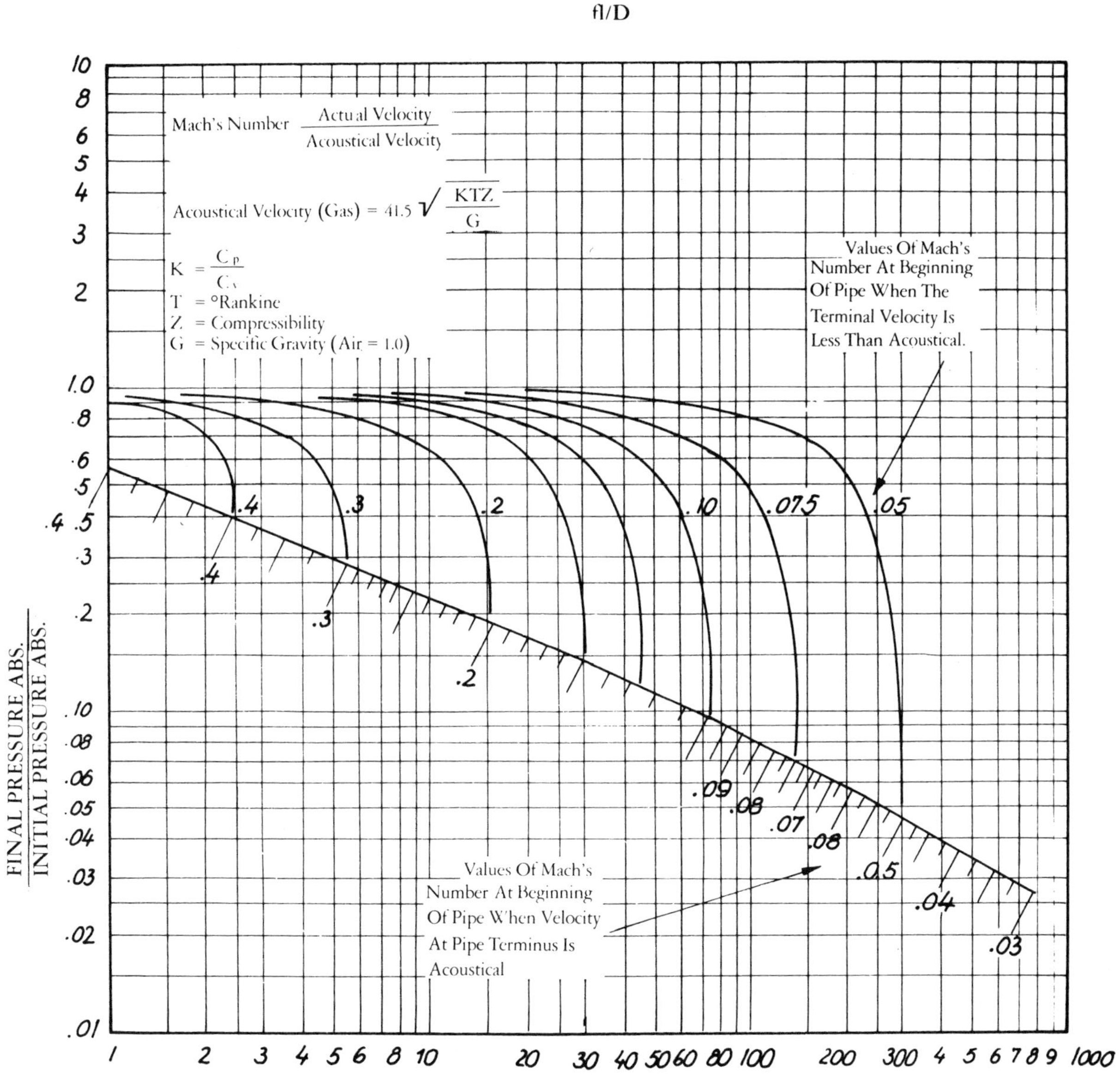

Mach's Number $= \frac{\text{Actual Velocity}}{\text{Acoustical Velocity}}$

Acoustical Velocity (Gas) $= 41.5\sqrt{\frac{KTZ}{G}}$

$K = \frac{C_p}{C_v}$
T = °Rankine
Z = Compressibility
G = Specific Gravity (Air = 1.0)

Values Of Mach's Number At Beginning Of Pipe When The Terminal Velocity Is Less Than Acoustical.

.4 .3 .2 .10 .075 .05

.4 .3 .2 .09 .08 .07 .06 .05 .04 .03

Values Of Mach's Number At Beginning Of Pipe When Velocity At Pipe Terminus Is Acoustical

FINAL PRESSURE ABS. / INITIAL PRESSURE ABS.

10 8 6 5 4 3 2 1.0 .8 .6 .5 .4 .3 .2 .10 .08 .06 .05 .04 .03 .02 .01

1 2 3 4 5 6 8 10 20 30 40 50 60 80 100 200 300 4 5 6 7 8 9 1000

fl/D

D = Pipe I.D., in.
L = Pipe Length, ft
f = Coefficient of Fricition x 12

Pipe Size, in.	f
1	0.28
2	0.24
3	0.22
4	0.19
6	0.18
10	0.17
20	0.14

fl/D VERSUS PRESSURE RATIO

(Gas flow at High Rates through Long Lines or at Acoustical Velocities) difference, Table 4-3 gives k values for methane at various temperatures.

TABLE 4-3

Values of k for Methane at Various Temperatures

Temperature, °F	50	100	150	200	250	300
k	1.311	1.299	1.285	1.272	1.263	1.241

Equation (4-19) shows the value of k will be larger for gases with lower molecular weights and lower for those with larger molecular weight. For example, for monatomic gases, such as helium, argon, and mercury k is about 1.65. For diatomic gases such as hydrogen, carbon monoxide, nitrogen, it is about 1.40. For the tri-, tetra-, and pentatomic, such as carbon dioxide, methane and the like, it is approximately 1.30.

VELOCITY HEAD

Velocity head is one aspect of fluid mechanics which often may be ignored with impunity but which, in certain instances, can be of great importance in engineering calculations. This is the kinetic energy, taken from the total system energy, which is required to bring the fluid to its working velocity.

Various forms of energy are present in fluids in motion. The principle of the conservation of energy requires that the total energy of a system remain constant. That is to say, the total of the kinetic, potential, pressure and frictional energies is constant even though the values of the individual portions may change. This concept, as applied to fluids, was propounded by DANIEL BERNOUILLI in 1738. A modern version, which includes the frictional effects, can be expressed as:

$$\frac{d(V^2)}{2g} + dz + v dP + dE_f = \text{Constant} \quad (4\text{-}20)$$

Where:
V = Velocity
z = Elevation of the fluid with respect to some reference
P = Pressure
E_f = Work done against pipe friction
v = Specific volume

The first term is the kinetic energy, the energy of motion, commonly called, in fluid dynamics, "velocity head". This may be expressed as:

$$h = V^2/2g \quad (4\text{-}21)$$

Where:
h = Height of a fluid column equal to the kinetic energy
V = Velocity of flow
g = Acceleration due to gravity

Using terms related to *pressure* can be more useful to the engineer. For liquids, Equation (4-21) becomes:

$$P_e = 0.00672\ V^2 S \quad (4\text{-}22)$$

While for gases it is:

$$P_e = 2.92 \times 10^{-4}\ PGV^2/TZ \quad (4\text{-}23)$$

Where:
P_e = Equivalent pressure, pounds per square inch
V = Velocity of flow, feet per second
S = Liquid specific gravity, water at 60° F equals 1.00
G = Gas specific gravity, air at 14.69 psia and 60° F equals = 1.00
P = Pressure, pounds per square inch, absolute
T = Temperature, degrees Rankine (460 + ° F)
Z = Deviation from true gas law, fractional

For metric (SI) units, velocity head in terms of pressure becomes, for liquids:

$$P_{esi} = 0.50\ V_{si}^2\ S_{si} \quad (4\text{-}22\ M)$$

And for gases:

$$P_{esi} = 1.75 \times 10^{-3}\ P_{si}\ G_{si}\ V_{si}^2/K'Z \quad (4\text{-}23\ M)$$

Where:
P_{esi} = Equivalent pressure, kilo Pascals
V_{si} = Velocity of flow, meters per second
S_{si} = Liquid specific gravity, kilograms per cubic decimeter
P_{si} = Pressure, kilo Pascals
G_{si} = Gas specific gravity, kilograms per cubic decimeter
K' = Temperature, degrees Kelvin
Z = Deviation from perfect gas law, fractional

The energy converted to velocity head is reversible (except for small frictional or approach losses) so that if the flow ceases, the pressure will again rise to virtually the original level.

In gas streams, at reasonable pressure levels, the influence of velocity head generally can be ignored. At very high pressures, this factor could require consideration. In liquid streams, because the density is far greater than for most gases, velocity head is more likely to be of a value large enough to warrant a check in systems where the available driving force is limited.

PRESSURE FROM A DIRECTIONAL CHANGE (IMPACT PRESSURE)

When a fluid is flowing through an elbow or a tee, the change in direction of the flow stream will produce a pressure upon the surface of that fitting which induces the change in direction. DODGE and THOMPSON (13) (pp. 120-121) present an analysis of the development of the forces, generated under these conditions which can be transfered into pressure, with the following result:

$$P_t = 2\,[P + (0.0026\ V^2\,\varrho)] \qquad (4\text{-}24)$$

Where:
P_t = Impact pressure plus the static pressure, pounds per square inch
P = Static pressure, pounds per square inch, gauge
V = Fluid velocity, feet per second
ϱ = Fluid density, pounds per cubic foot

The pressures generated at high velocity of flow may be sufficiently large to cause pipe rupture. As an example, using Equation (4-24), it can be shown that at the bull of a tee with water flowing at 50 feet per second in a pipe with a static pressure of 100 psi, the total pressure would rise to 1007 psi. Air with a density of 0.45 pounds per cubic foot flowing at the acoustic velocity would develop a total pressure of 3500 psi at the bull of a tee.

Therefore, under conditions where velocities may become quite high, the total pressure should be checked to be *sure* the pressure rating of the selected pipe will not be exceeded. Of course, if the pipe is completely straight, which is generally not true in most process plants, this velocity limitation would not apply.

In metric (*SI*) units:

$$P_{tsi} = 13.8\,(0.145\ P_{si} + 0.056\ V_{si}^2\,\varrho_{si}) \qquad (4\text{-}24\ M)$$

Where:
P_{tsi} = Impact pressure plus static pressure, kilo Pascals
P_{si} = System pressure, kilo Pascals
V_{si} = Velocity of flow, meters per second
ϱ_{si} = Fluid density, kilograms per cubic decimeter

SUMMARY

The theory and formulae given in this Çhapter will be further expanded in Chapter 5 for dealing with multiphase flow conditions wherein the main criteria for pipe size selection, based upon velocity, will apply.

Figure 4-2
Butadiene Plant

CHAPTER 5

TWO-PHASE FLOW

LIQUID-GAS FLOW

In fluid dynamics, the most complex calculations will be encountered when the engineer is dealing with the simultaneous flow of two phases. Not only are the formulae more complex, the results are less reliable. Under good conditions, uncertainties up to plus/minus 25% may occur. When the conditions are poor, the potential errors may rise to plus/minus 50%.

Two-phase flow may incorporate various combinations: liquid and gas, liquid and solid, gas and solid. There is another possible combination, two immiscible liquids. Consideration of that form of two-phase flow is beyond the scope of this book.

CONTROLLING CRITERIA

During two-phase flow, the Optimum Pipe Size Selection may be made using one of two controlling factors: Pressure Drop Available or Velocity Allowable.

The PDA may be established according to the principles laid down in the INTRODUCTION, or it may be established by the requirements of the process.

The VA will depend upon whether the second phase must be entrained in the first, or whether the requirement is to avoid erosion or rupture of elbows or pipe which could be caused by a two-phase velocity which is too high. These limitations are set forth later in this Chapter.

LIQUID-GAS TWO-PHASE CALCULATIONS

The technical literature contains thousands of publications on the subject of liquid-gas two-phase flow investigations. This subject has been a favorite for engineering dissertations. Hence, many studies, which while learned and searching, may cover facets of two-phase flow rich in unusual problem areas that may not deal with the situations the design engineer commonly encounters. An attempt will be made here to focus attention upon the conditions which are more common and to provide guidance in avoiding those which may offer greater levels of uncertainty.

In most plants the common form of two-phase flow will be liquid and gas. Frequently this is the combination of steam and condensate being discharged from steam traps or continuous drainers. Similar conditions will be found in the flashing of natural gasoline and other hydrocarbon mixtures, the effluent from fired heaters, as well as in the flashing of refrigerants such as ammonia or Freons.

Under such conditions, conversion of a part of the liquid into gas takes place quickly and the ratio of gas to liquid, rapidly established, remains relatively constant. Under such conditions, fluid dynamic calculations, while more complex than for single phase applications, are not greatly so.

Other conditions may be encountered where the fraction of gas versus the fraction of liquid is not constant. Rather, the ratio of gas to liquid changes as the flow progresses. This can occur during the passage of liquid being vaporized through a heat exchanger or fired heater, or as the effluent from such equipment rises vertically upward, whereby the pressure on the system is reduced and hence vaporization increases. This is especially significant at low pressure or vacuum conditions. Another such situation can occur if liquid at the bubble point undergoes a gradual pressure reduction, with or without an accompanying temperature increase. Then the calculations become complex. Not only must one contend with the energy losses associated with the frictional effects and acceleration, but also those connected with the formation of the interfaces between the liquid and the gas. These interfaces require energy for their generation, also for the increase in surface area, together with the effects of movement through the pipe. According to DEGANCE and ATHERTON (16), the lack of consideration of the first two items accounts for the errors which can be encountered in simplified two-phase calculations.

As a liquid, which is in motion in a horizontal pipe, gradually undergoes vaporization, bubbles appear as the first change. As the vaporization progresses, and assuming the flow velocity is modest, the two phases may separate with liquid at the bottom and gas at the top. With sufficient gas velocity, the surface of the liquid is kicked into waves. With still greater velocity, the flow may become annular, with the liquid clinging to the walls and the gas (perhaps with a liquid mist entrained) traveling at a higher velocity through the central core. These modes are DUKLER'S segregated flows (15).

MODES OF TWO-PHASE FLOW

During intermittent flow, plugs of gas travel along the upper surface of the pipe, separated by plugs of liquid with action like a percolator, except in the horizontal. Slugs of liquid are followed by sections of gas.

DUKLER'S distributive flow may appear as bubble flow, where bubbles of many sizes have collected at the top of the pipe and are carried along with the liquid.

With more vapor and higher velocity, mist flow occurs when droplets of various sizes are being carried in suspension in the gas stream. While one can layout the types of flow and their general characteristics, the awkward fact remains that the means for predicting the nature of the flow present in any given situation tends to be highly speculative. Various experimentors have prepared flow regime maps (18), (19), and (20). Probably each definition applied with reasonable accuracy for the conditions under which it was established. However, another user may easily overstep the applicable bounds without realizing he has done so.

An early attempt was made by BAKER, based upon data developed from tests on the flow of air and water (18). While he did break down, systematically, two-phase flow into the various regimes, attempts by others to use his approach have resulted in large errors. This lead KNOWLES (19) to develop another correlation, which DEGANCE and ATHERTON (16) believe to be more reliable. However, KNOWLES' approach does not define all the flow regimes. Furthermore, there is confusion over the formulae set forth originally compared to those in the redefined version. This results in uncertainty when one attempts to apply them to the regime maps in the first publication.

PRACTICAL CONSIDERATIONS

One can best fall back upon pragmatic data related, in so far as that be possible, to engineering experience. The following *guidelines*, as applied to air flow near atmospheric pressure in *horizontal* pipes, can be useful. See also computations for ENTRAINMENT, farther in this Chapter.

Bubble or froth flow: Bubbles of gas are dispersed throughout the liquid. Superficial liquid velocities are from 5 to 15 feet per second while gas superficial velocities are from 1 to 10 feet per second.*

Plug flow: Alternate plugs of liquid and gas move along the upper portion of the pipe cross-section. Superficial velocities are less than 2 feet per second for the liquid, and less than 3 feet per second for the gas.

Stratified flow: The liquid flows smoothly along the bottom of the pipe cross-section and the gas

* Superficial velocity is the value attained if the phase referenced is assumed to occupy the entire pipe cross-sectional area without considering the other phase at all.

travels above it. Superficial velocities are less than 0.5 feet per second for the liquid, and from 2 to 10 feet per second for the gas.

Wavy flow: The interface between the gas and the liquid has waves traveling along it. Superficial velocities are less than 1 foot per second for the liquid and up to 15 feet per second for the gas.

Slug flow: As the wave form develops and intermittently is picked up and carried along with the gas, a slug forms which travels along the pipe at a higher velocity than the superficial for the liquid. The impact of such slugs as the flow rounds a bend can cause vibration and even pipe rupture. Generally, slug flow *may* occur when gas superficial velocity lies between 15 and 20 feet per second.

Annular flow: The liquid flows as a layer around the pipe inner surface, with the gas passing through the hollow core. Some liquid will be entrained in the gas as a spray. The superficial velocity of the gas will be above 20 feet per second.

Spray flow: This is an advanced phase of annular flow in which virtually all the liquid is entrained as fine droplets. This occurs at velocities well above 20 feet per second.

DUKLER has pointed out that for practical purposes, there are three modes of two-phase liquid-gas flow which can occur in a piping system: segregated, intermittent and distributed.

Segregated flow, present when a pipe is running partially full, really need not be treated as two-phase. Equations provided for full running pipes may be used for partially full conditions. The equivalent diameter which matches the hydraulic characteristics of the actual channel is determined from the hydraulic radius. This is the cross-sectional flowing area of the liquid divided by the wetted perimeter.

$$R_h = A_l / P_w \tag{5-1}$$

Where:
R_h = Hydraulic radius
A_l = Cross-sectional area of the flowing liquid
P_w = Wetted perimeter of the pipe

The equivalent diameter is four times the hydraulic radius.

Corrections for the velocity and carrying capacity for pipes running partially full of liquid are given in Table 5-1.

TABLE 5-1

Carrying Capacity of Partially Full Pipes

% Full	Velocity (% of full)	Carrying Capacity (% of full)
100	100	100
95	111	106.3
90	115	107.3
80	116	98
70	114	84
60	108	67
50	100	50
40	88	33
30	72	19
25	65	14
20	56	9
10	36	3

The increase in velocity below 100% full and down to 50% can be recognized as the effect of lessened friction on the upper surface of the liquid in contact with gas, as compared to the pipe wall surface.

Returning to DUKLER'S modes of two-phase flow, the next is intermittent. This constitutes the condition where calculations are complex and the least precise. The flow conditions are unsteady. Within plant areas, therefore, they are to be avoided when possible because their very irregular character can lead to erratic plant behavior. SIMPSON (17) indicates that the Froude Number* may be useful in predicting the presence of two-phase (gas-liquid) slug flow when the piping is vertical. Slug flow in this condition can lead to pressure pulsations and vibration.

When bubbles are present and the pipe diameter is larger than one inch, and if the liquid viscosity is less than 100 centipoise, Figure 5-1 is applicable.

Froude Number for the gas phase is:

$$N_{Frg} = \frac{0.608\, V_g}{\sqrt{D}} \sqrt{\frac{\varrho_g}{\varrho_l - \varrho_g}} \tag{5-2}$$

Froude Number for the liquid phase is:

$$N_{Frl} = \frac{0.108\, V_l}{\sqrt{D}} \sqrt{\frac{\varrho_l}{\varrho_l - \varrho_g}} \tag{5-3}$$

* Froude Number is a dimensionless constant related to the ratio of the inertial to the gravitational forces.

Where:
N_{Frg} = Froude Number, gas
N_{Frl} = Froude Number, liquid
V_g = Superficial of the gas, feet per second
V_l = Superficial of the liquid, feet per second
D = Pipe diameter, inches
ϱ_g = Gas density, pounds per cubic foot
ϱ_l = Liquid density, pounds per cubic foot

Metric equivalents for the equations in this Chapter are given at the end.

VERTICAL DOWNWARD TWO-PHASE FLOW

When two-phase flow is downward, bubbles can develop which tend to flow counter-current to the liquid. The result (under liquid flow velocities the magnitude of which are dependent upon the elements in Froude Number) can be a suspended bubble which may partially block the channel and interfere with flow. SIMPSON (17) has studied these phenomena and KELLY (18) also has contributed valuable data.

At Froude Number for liquid below 0.31 to 0.58, bubbles tend to remain trapped in the pipe or rise against the liquid flow, whereas, above these values, the bubbles, swept along with the liquid, will be carried out of the pipe. Under the conditions where the bubble is suspended, pulsation and vibration can be experienced.

The magnitude of the range of Froude Numbers over which the transition can occur is an indication of the uncertainties associated with the measurements required in the experimental work. The designer will do well to stay comfortably away from the doubtful region.

When two phases, gas and liquid, are present in a vessel and the liquid is being withdrawn, the velocities at the outlet require consideration to avoid the simultaneous withdrawal of gas. Such a condition can have profoundly disturbing effects upon the process and the equipment, for example, vapor entering a pump along with the liquid.

When the flow at the outlet is rotational, vortices can form. A cross installed in the outlet is an effective means to prevent vortex formation.

During non-rotational flow from a vessel the liquid, in effect, forms a circular weir around the outlet. When the liquid Froude Number is less than about 0.3 to 0.55 or when the ratio of the height of the liquid to the diameter of the pipe (in the same units) is more than 0.25, the vapor is not likely to be drawn into the pipe along with the liquid. However, at higher values of Froude Number, vapor will be entrained in the liquid unless the depth of the liquid is ample. The equation given by HARLEMAN (21), gives Froude Numbers for the liquid which will prevent gas entrainment during vertical downflow.

$$N_{Frl} = 3.24\,(h/d)^{2.5} \qquad (5\text{-}4)$$

Where:
N_{Frl} = Froude Number, liquid, see Equation (5-3)
h = Depth of liquid in vessel above outlet, feet
d = Diameter of outlet, feet

PERCENTAGE OF THE TWO PHASES

The pressure drop and the velocity determinations, which are better termed approximations in gas-liquid two-phase flow, require knowledge of the relative amounts of the two phases present in the pipe. Where hot condensate is flashing while being transferred from a higher to a lower pressure, the determination of the percent flashed is expressed by the difference in enthalpy between the unflashed and the flashed condensate divided by the heat of vaporization.

$$F = \left(\frac{E_1 - E_2}{H}\right) 100 \qquad (5\text{-}5)$$

Where:
F = Percent of liquid flashed by weight
E_1 = Enthalpy of the unflashed liquid
E_2 = Enthalpy of the flashed liquid
H = Heat of vaporization after flashing

The equation applies to all pure liquid compounds. For mixtures of compounds, equilibrium calculations, the method for which is beyond the scope of this book, must be used to determine the liquid-vapor ratios after flashing.

For steam condensate flash determination, Graph 2 is provided. For other *pure* compounds, the enthalpies and the heats of vaporization may be determined from Graphs 4 and 5, and then entered into Equation (5-5).

FIGURE 5-1

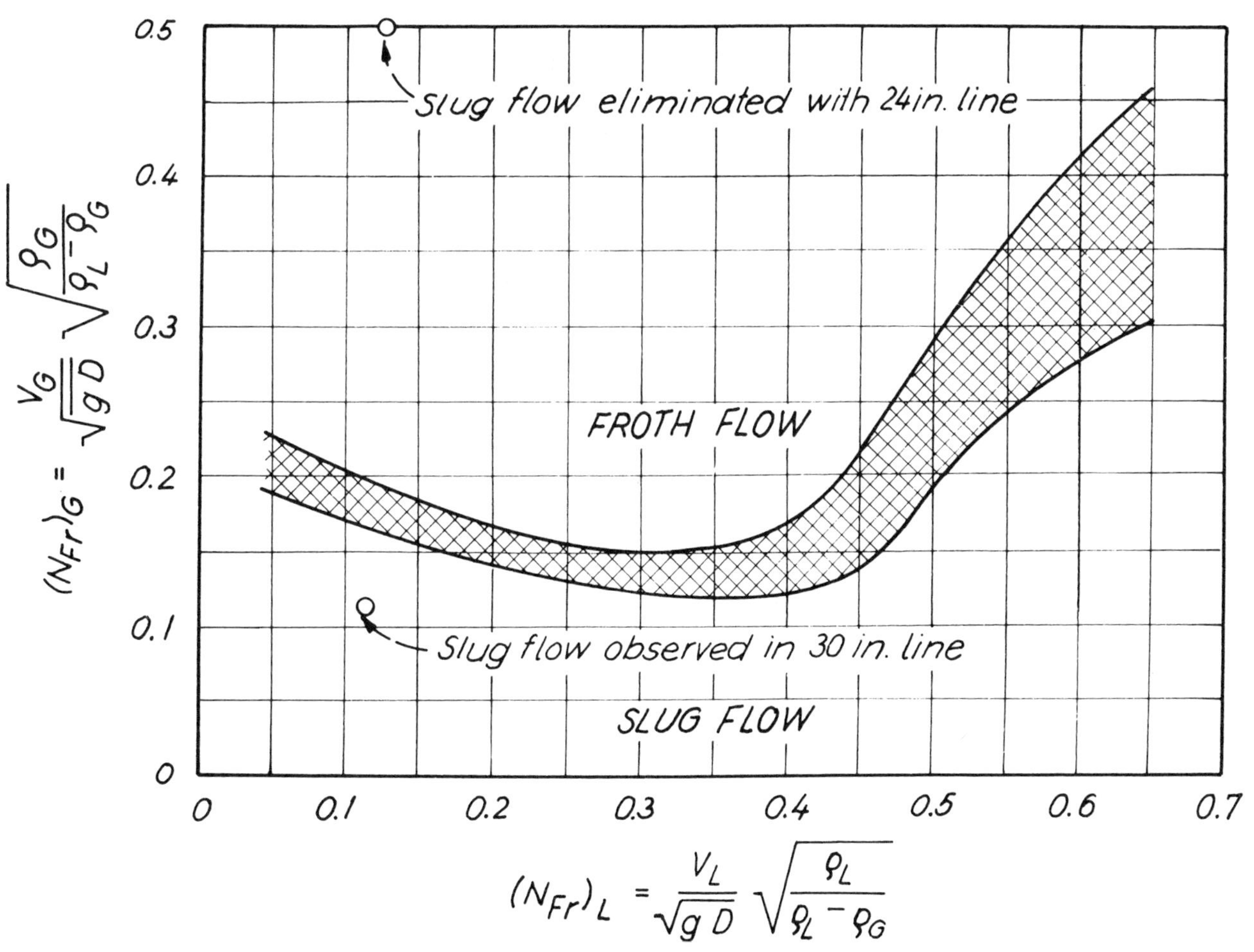

FROUDE NUMBERS FOR SLUG VERSUS FROTH FLOW

(Chemical Engineering, June 17, 1968)

Where the two-phase flow is the result of relatively complete flashing upon a pressure reduction, the modified FANNING Equation (3-8) can be used to determine pressure drop.

$$psi/100 = \frac{13.25\ W^{1.84}\ \mu^{0.16}}{\varrho_{av}\ D^{4.84}} \qquad (3\text{-}8)$$

Where:

psi/100 = Pressure drop, pounds per square inch per 100 feet of pipe

W = Rate of flow, *thousands* of pounds per hour of the combined liquid and vapor

μ = Viscosity at flow temperature, centipoise

ϱ_{av} = Mean density (by weight percent), pounds per cubic foot

The last term, mean density, is determined by:

$$\varrho_{av} = \frac{501\ S}{[0.08\ (100 - F)] + \left(\frac{1.86\ F\ S\ T\ Z}{G\ P}\right)} \qquad (5\text{-}6)$$

Where:

ϱ_{av} = Mean density of the mixture, pounds per cubic foot

S = Specific gravity of liquid before flashing, at flowing temperature, water at 60° F equals 1.00

F = Percent of liquid flashed by weight

T = Final temperature of mixture, degrees Rankine (°F+460)

Z = Deviation from perfect gas laws, fractional

G = Specific gravity of gas, air at 14.69 psia and 60° F equals 1.00

P = Final pressure, pounds per square inch, absolute

The various flow regimes in two-phase liquid-gas flow monitor the selection of the viscosity to be used in the formulae, which is an important consideration in pressure drop calculations. For spray flow, the most common form which the engineer will encounter, the viscosity of the *gaseous* phase is used. For the modes in which the gas is present as bubbles in the liquid, the *liquid* viscosity is employed, as is the case with plug flow and slug flow. Stratified and wavy flow call for a mean value between the liquid and gas viscosities.

This explains some of the difficulty which has been experienced in attempts to develop methods of calculation of pressure losses in systems which start as liquid and end as a mixture of liquid and gas. At the start of the vaporization process, the liquid viscosity monitors the friction. As the regimes progress through the several stages, changes take place. For one section the influence may fluctuate from one to the other. Finally, assuming there is enough vaporization occuring, the gas viscosity influence takes over.

LIQUID-GAS FLOW PRESSURE DROP

Experience has shown that Equation (3-8) will provide adequate approximations of the pressure drop in two-phase turbulent flow systems where the degree of flash is substantially completed at the beginning of the pipe. Should the pressure drop so calculated be more than ten percent of the initial pressure, the pipe length must be segregated into smaller increments and the pressure drop over each of these computed. The final pressure of each segment becomes the initial pressure for the next. The pressure drop increments are summed for the total drop. The amount of flashing should be reconsidered, based upon the lowered pressure in each segment, for there may be some influence.

This approach has been verified for steam condensate systems, such as would exist down-stream of steam traps, by RUSKAN (22) using computer studies compared with other methods presented in the literature. The differences in approximated pressure drops were found to be insignificant while the calculation time was greatly diminished.

When the flashing will be progressive, as in vaporization in heater tubes or in a long field discharge line from oil and gas separators, to mention two examples, Equation (3-8) cannot be expected to yield meaningful data. For such conditions, more sophisticated methods are required. The correlation prepared by LOCKHART and MARTINELLI (23), developed in 1949, has been generally accepted, probably more for its simplicity rather than for its accuracy when compared with later works, such as that of DUKLER (24). MARTINELLI expects calculated pressure drop to be within ± 50% whereas DUKLER expects ± 25%. When the engineer has access to a computer, the more elegant approach offered by DUKLER can be the better selection. However, it is to be noted that MARTINELLI'S correlations have been used for many years with acceptable success.

LIQUID-GAS FLOW VELOCITY

The general equation for velocity, which can be used for two-phase liquid-gas flow, is:

$$V = 183.3\ \overline{CFS}/D^2 \qquad (5\text{-}7)$$

Where:
V = Average flow velocity, feet per second
$\overline{CFS}$ = Flow rate, cubic feet per second
D = Pipe inside diameter, inches

The rate of flow for the combined phases of any liquid after flashing can be found from:

$$\overline{CFS} = \frac{Q}{3600}\left[0.08\,(100 - F) + \frac{1.86\,F\,S\,T\,Z}{G\,P}\right] \qquad (5\text{-}8)$$

Where:

$\overline{CFS}$ = Average flow rate of mixture, cubic feet per second

Q = Liquid flow rate before flashing, U.S. gallons per minute at flow temperature

F = Percent of liquid flashed, by weight

S = Specific gravity of liquid before flashing, at flow temperature, water at 60° F equals 1.00

T = Final temperature of mixture, degrees Rankine (°F+460)

Z = Deviation from perfect gas law, fractional

G = Specific gravity of the gas phase, air at 14.69 psia and 60° F equals 1.00

P = Final pressure of the mixture, pounds per square inch, absolute

For steam-condensate mixtures, the average density of the mixture can be obtained from Graph 3. The flow rate in cubic feet per second can be found from:

$$\overline{CFS} = W/3.6\,\varrho_{av} \qquad (5\text{-}9)$$

Where:

$\overline{CFS}$ = Average rate of flow of the steam-condensate mixture, cubic feet per second

W = Combined steam and condensate flow rate, *thousands* of pounds per hour

ϱ_{av} = Average density of the steam-condensate mixture, pounds per cubic foot

ENTRAINMENT

At velocities determined by the gas density, liquid specific gravity and the size of the droplets, liquids will be entrained in an upward flowing gas stream. When pipe size selection is being made for a liquid-gas two phase system, the velocity must be adequate to insure entrainment does occur. If the velocity is below that value, the liquid flow will be intermittent, as in a perculator. From a practical standpoint, the same size of pipe can be used in the horizontal section of the run as is required for the vertical portion.

STOKES investigated the behavior of droplets (and solid particles) suspended in gas. PERRY'S Chemical Engineers Handbook (7) gives comprehensive information on his work. A simplified presentation, suitable for pipe size selection, is shown in Figure 5-3. This is derived from values found in the field and verified by STOKES' data.

The vertical scale is the flowing *gas* density, *exclusive* of the presence of the liquid. The horizontal scale is the *total* velocity of the mixture (not the superficial velocity). Figure 5-3 gives safe-side data for entrainment of droplets one millimeter in diameter within the range of viscosities commonly encountered in plant design.

Figure 5-2

Piping Around Columns

EROSION VELOCITY

When droplets, entrained in a gas stream, impinge upon a surface, such as the wall of an elbow or a part of a valve, with sufficient velocity, the metal will be work-hardened, becoming more brittle. In time, the bombardment can cause erosion of the metal. Consequently, in all two-phase flow applications, care *must* be taken to avoid exceeding critical values of velocity which depend upon the average specific gravity of the flowing mixture and upon the size of the pipe.

Figure 5-4 gives the velocities for long radius elbows where erosion can be expected to commence. If the velocity must be kept higher for some reason, use of gradual pipe bends will avoid difficulty.

IMPACT VELOCITY

When slug flow is present, or possible, consideration must be given to the possibility of vibration, noise or even pipe rupture caused by the potentially destructive impact of liquid slugs upon incident surfaces of elbows, fittings or valves. The equation given in Chapter 4, for the pressure developed by any fluid which impinges upon a surface not parallel to the flow stream, also can be employed to calculate the pressure generated by slugs of liquid.

$$P_t = 2\,(P + 0.0026\, V^2 \varrho) \qquad (4\text{-}24)$$

Where:

P_t = Impact pressure plus the static pressure, pounds per square inch
P = Static pressure, pounds per square inch
V = Gas superficial velocity, feet per second
ϱ = Liquid density, pounds per cubic foot

The shock generated by the impact of a slug of liquid traveling at high velocity can cause transient effects which *may* by higher than Equation (4-24) will predict. Because of this uncertainty, an ample safety factor for pipe strength is advisable. Twice the usual is a safe practice.

SUMMARY

As a closing consideration for two-phase, liquid-gas flow calculations, the engineer will do well to be aware that these, more than in any other phase of fluid dynamics, are at best approximations. If the process or other engineering aspects of the situation could be adversely influenced by poor approximations of pressure loss, one would be well advised to use more than one approach and average the results. In addition, however, whenever practical, play it on the safe side, which can mean using a size larger as insurance, if velocity or pressure loss must be kept down, or a size smaller if the need is to have enough velocity.

TWO-PHASE LIQUID-SOLID FLOW

PARTICLE SIZE AND VELOCITY

One of the most important considerations in liquid-solids flow is the size of the particles in the mixture. Very fine materials form non-settling, homogeneous slurries, which behave like viscous muds. They show a uniform distribution of the particles in the liquid. Such homogeneous slurries are normally to be found where particle size is small and with low specific gravity and relatively high solids concentration. Slurries made up of coarse and dense particles are characterized by the tendency of the materials to settle out of the liquid. Such heterogeneous slurries have flow behaviour which is dependent upon the velocity. At very high velocity, above 65 feet per second, there will be a suspension with virtually no settling. From 25 to 65 feet per second there will be a heterogeneous suspension wherein the larger particles will begin to concentrate near the bottom of the pipe. From 10 feet per second to 25, heterogeneous settling (saltation) occurs. Larger particles will drop out and then be picked up again, moving along the bottom of the pipe. The velocity range for this behavior is narrow; thus small changes in composition or velocity can upset the balance. Below 10 feet per second, moving bed flow occurs.

Clearly, for heterogeneous conditions, flow should be fast enough to avoid the moving bed mode.

Table 5-2 lists particle size versus solid specific gravity. For particle sizes above the value shown, the slurry behavior will tend toward heterogeneous. At lower values, the behavior will tend toward homogeneous.

TABLE 5-2

Particle Size Versus Specific Gravity

Specific Gravity	1.0	2.0	3.0	4.0	5.0
Particle Size (Microns)	1000	300	150	110	100

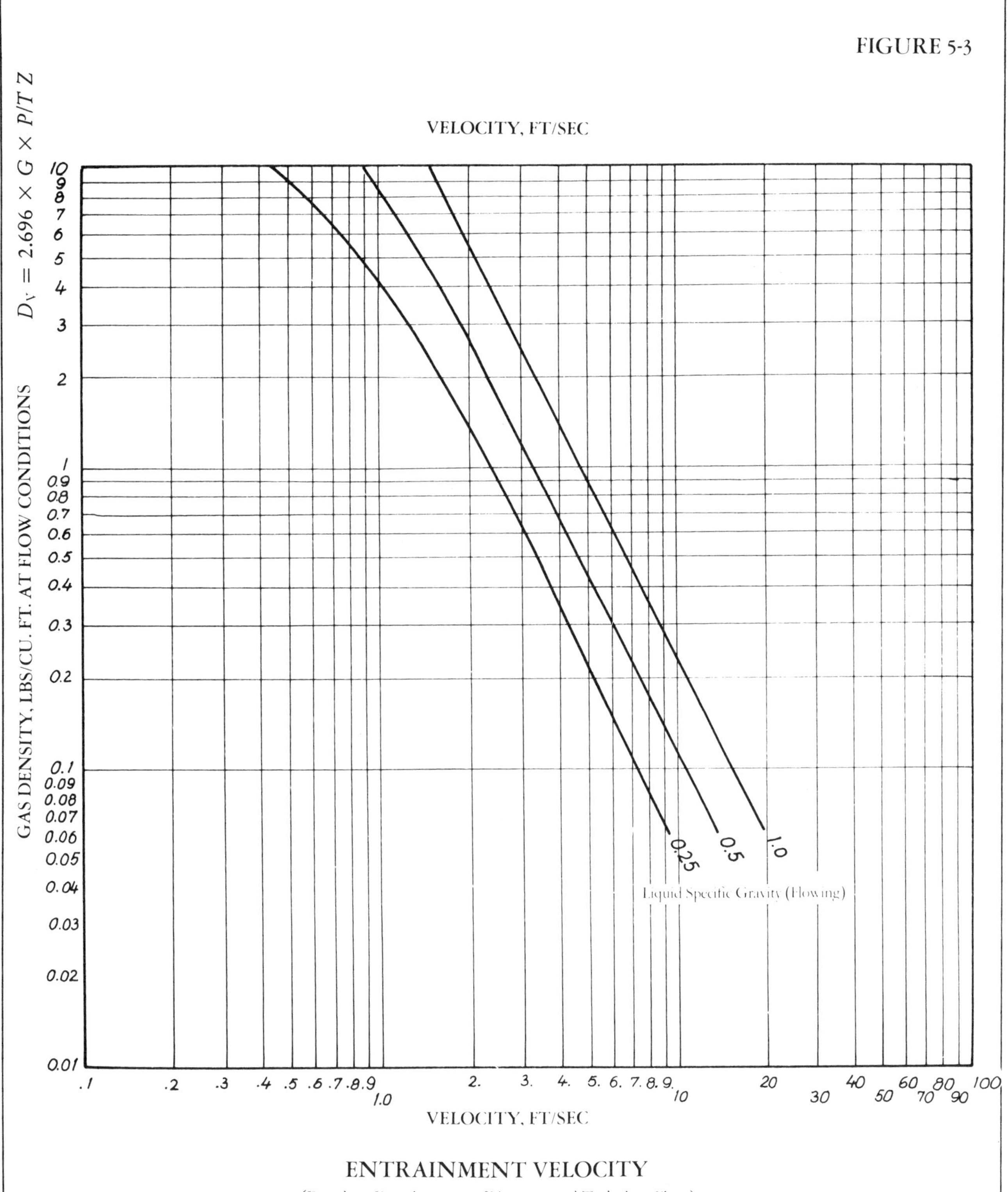

ENTRAINMENT VELOCITY

(Based on Droplets 1 mm Diameter and Turbulent Flow)

$$V = 0.611\sqrt{\frac{D_L - D_V}{D_V}}$$

(Checked against data in Perry: "Chem. Eng. Handbook" 2nd Ed., page 1852)

V = Velocity, ft/sec

D_L = Liquid Density, lbs/cu. ft.

D_V = Flowing Gas Density, lbs/cu. ft.

T = Absolute Temperature

P = Absolute Pressure

Z = Compressibility

G = Gas Specific Gravity (Air = 1.00)

PRESSURE DROP

The friction losses during homogeneous slurry flow can be calculated using the FANNING Equation (3-2) which incorporates the friction factor. Reynolds Number can be found from Equation (3-5).

The economic pipe size can be found from Equation (2-17). The specific gravity will be that of the average of the mixture. The viscosity is to be that of the liquid.

For heterogeneous slurries, the friction losses may be determined by adding the friction losses due to the homogeneous component to those which are present in the heterogeneous portion. The losses in the heterogeneous portion are best computed by the equation proposed by DURAND (34).

$$i_m = i \left\{ 1 + 81\, C_V \left[\frac{d\,(S_s - S_l)}{V_m^2\, C_D^{0.5}\, S_l} \right]^{1.5} \right\} \qquad (5\text{-}10)$$

Where:
i_m = Total friction loss of the mixture
i = Friction loss due to the vehicle liquid
C_V = Volumetric concentration of solids
d = Pipe diameter (feet)
S_s = Specific gravity of the solids
S_l = Specific gravity of the liquid
V_m = Velocity of the slurry, feet per second
C_D = Particle drag coefficient

DEPOSITION VELOCITY

The limiting velocity of deposition is given by FADDICK (35) as:

$$V_m/[2gd\,(S-1)]^{0.5} = 0.255 + 1.88\, d' \qquad (5\text{-}11)$$

Where:
V_m = Velocity of mixture, feet per second
g = Gravitational constant, feet per second per second
d' = Pipe inside diameter, feet
S = Solids specific gravity
d' = Mean particle size, mm

ZANDI and GOVATOS (36) suggest a means to predict the onset of bedload deposition as follows:

$$\frac{V_m^2 \sqrt{C_D}}{C_V D\,(s-1)} = 11 \qquad (5\text{-}12)$$

It is suggested that reference be made to the original articles for more detailed discussion of the various data given above regarding Equations (5-10) through (5-12).

TWO-PHASE SOLID-GAS FLOW

CONTROLLING FACTORS

As was discussed for liquid-solid two-phase flow, the modes of transport can vary greatly, depending upon the gas-solid weight ratio, the density of the gas, the density of the particles, their size and shape. The range can be from an essentially homogeneous mixture, which behaves nearly like a more dense gas, to the other extreme where the particles are transported by saltation, alternately moving and depositing in a manner similar to the movement of sand dunes.

TRANSPORT VELOCITY

Data for the velocity required to transport the different materials varies considerably from one source to another. One can assume this depends upon such variables as the drag coefficient for the substances, which is not always included in the material specifications. Certainly a statement such as, "sand, 150 pounds per cubic foot, 100 mesh" does not fully define its aerodynamic characteristics; these *do* influence the minimum velocity needed to transport the sand. Also, there can be a difference in the needed velocity depending upon whether the flow is horizontal or vertically upward.

Table 5-3 offers a list of minimum transport velocities in feet per second for a variety of materials. In the absence of other data, these can be used as starting point.

Transport velocity is the criterion which determines the pipe size selection. When possible, obtain recommended transport velocity from the system designer. The velocity resulting from a selection based upon the economic method might not be high enough to insure proper material movement.

PRESSURE DROP

The pressure loss in the piping can be expressed as a factor applied to the pressure drop calculated upon the basis of the flow of the transport gas only, using

FIGURE 5-4

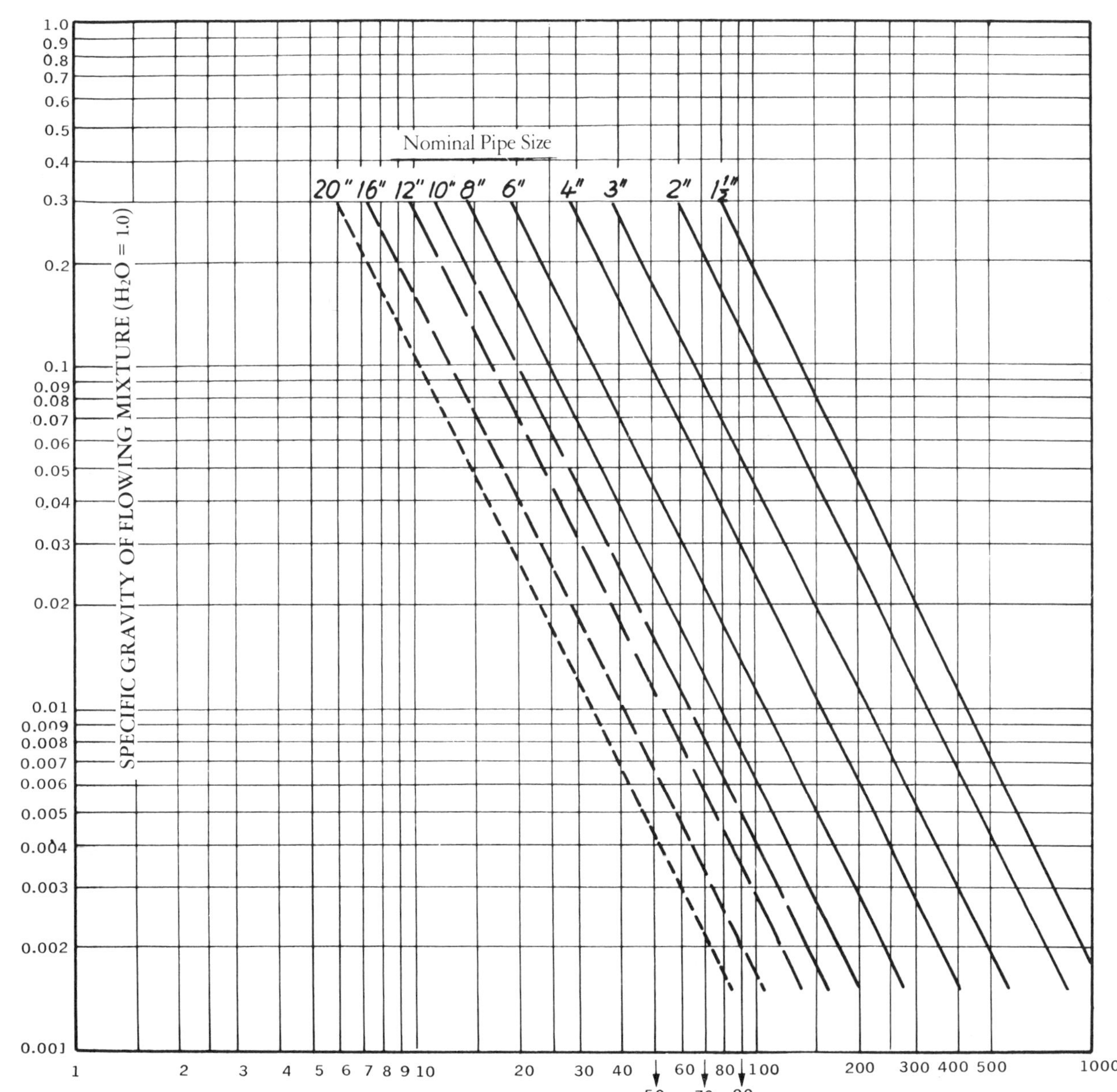

EROSION VELOCITY

(where erosion will be expected in L.R. Schedule 40 elbows)

TABLE 5-3
Minimum Pick-Up Velocities with Air at Atmospheric Pressure as The Transport Medium

Material	Density	Size	Velocity (ft/sec)	Material	Density	Size	Velocity (ft/sec)
Alum	50	100-8 M	85	Malt	32	0.25 inch	70
Alumina	60	100-8 M	85	Oats	26	0.25 inch	70
Bentonite	50	-100 M	85	Perlite	5		65
Cellulose acetate	10		65	Plastic pellets	35	0.1 inch	70
Charcoal	25	0.5 inch	65	Plastic powder	25	100 M	60
Clay	50	pulver.	85	T.P.S.			85
Coffee beans	42	0.25 inch	70	Rice		0.1 inch	70
Coke			75	Salt	80	100 M	85
Corn	45	0.25 inch	70	Salt Cake	85		85
Corn grits	40		70	Soap Chips	15	0.25 inch	65
Diatomaceous earth	10		65	Soda ash	35	100 M	70
Flour	40		60	Soda ash	55		75
Grain, spent	30	0.1 to 0.5 inch	65	Starch			65
Lime, pebble	53	0.5 inch	85	Sugar			85
Limestone	80	pulver.	85	Wheat	48	0.25 inch	70
Lime hydrate	25	100 M	75	Wood chips	25	variable	85

Equation (3-10). The factor is a function of velocity (superficial), the ratio of material weight flow rate to the gas weight flow rate, expressed thus:

$$F_1 = \frac{R_c}{K_1} + 1 \qquad (5\text{-}13)$$

Where:
F_1 = Gas pressure drop
R_c = Pounds of material per second / pounds of air per second
K_1 = Velocity factor

TABLE 5-4

Velocity Factor K_1 as Function of Gas Velocity

Velocity, ft/sec	K_1
35	1.15
50	2.14
65	3.11
85 +	3.5

The superficial velocity of the gas can be determined by Equation (4-4), using the assumption that only the gas is flowing in the pipe.

TWO-PHASE FLOW METRICATION

Froude Number for gases

$$N_{Frg} = \frac{31.8\, V_{gsi}}{\sqrt{D_{si}}} \sqrt{\frac{\varrho_{gsi}}{\varrho_{lsi} - \varrho_{gsi}}} \qquad (5\text{-}2\text{ M})$$

Froude Number for liquids

$$N_{Frl} = \frac{31.8\, V_{lsi}}{\sqrt{D_{si}}} \sqrt{\frac{\varrho_{lsi}}{\varrho_{lsi} - \varrho_{gsi}}} \qquad (5\text{-}3\text{ M})$$

HARLEMAN equation to prevent gas entrainment during vertical down-flow of vapor

$$N_{Frl} = 3.24 \left(\frac{h_{si}}{d_{si}}\right)^{2.5} \qquad (5\text{-}4\text{ M})$$

General equation to determine flashing of condensate

$$F = 100\,(E_1 - E_2)\,/\,H \qquad (5\text{-}5\text{ M})$$

The mean density of a two-phase mixture

$$\varrho_{avsi} = \frac{8029\, S_{si}}{\left[0.08\,(100 - F) + \left(\dfrac{23.1\, F\, S_{si}\, K/Z}{G_{si}\, P_{si}}\right)\right]}$$

Velocity during two-phase flow (5-6 M)

$$V_{si} = \frac{353 \text{ m}^3/\text{hr}}{D_{si}^2} \qquad (5\text{-}7 \text{ M})$$

The rate of flow for combined phases of any liquid after flashing

$$\text{m}^3/\text{hr} = 0.125\, Q_{si}$$

$$\left[0.08\,(100 - F) + \left(\frac{23.1\, F\, S_{si}\, K'Z}{G_{si}\, P_{si}} \right) \right] \qquad (5\text{-}8 \text{ M})$$

For steam-condensate mixtures, the flow rate can be determined from

$$\text{m}^3/\text{hr} = 1000\, W_{si}/\varrho_{avsi} \qquad (5\text{-}9 \text{ M})$$

Terminology for equations (5-2 M) through (5-9 M) is as follows:

ϱ_{gsi} = Density of the gas phase, kilograms per liter
N_{Frg} = Froude Number for gases
V_{gsi} = Velocity of the gas, meters per second
ϱ_{lsi} = Density of the liquid phase, kilograms per liter
D_{si} = Pipe inside diameter, millimeters
N_{Frl} = Froude Number for liquid
h_{si} = Depth of liquid in vessel above outlet, meters
d_{si} = Diameter of outlet, meters
F = Percent of liquid flashed, by weight
E_1 = Enthalpy of the unflashed liquid
E_2 = Enthalpy of the flashed liquid
H = Heat of vaporization, after flashing
ϱ_{avsi} = Mean density of the mixture, kilograms per cubic meter
S_{si} = Specific gravity of the liquid before flashing, at flow temperature kilograms/per cubic meter
K' = Final temperature of the mixture, degrees, Kelvin
Z = Deviation from the perfect gas law, fractional
G_{si} = Specific gravity of the gas phase kilograms per liter
P_{si} = Final pressure, kilo Pascals
V_{si} = Average velocity of flowing mixture, meters per second
m³/hr = Rate of flow of mixture, cubic meters per hour
W_{si} = Combined steam and condensate flow rate, kilograms per hour

CHAPTER 6

FLUID MECHANICS PARAMETERS

EQUIVALENT LENGTH OF FITTINGS, ETC.

In the previous Chapters, formulae have been given for the calculation of pressure losses as fluids flow through pipes. Clearly, the usual process piping system is made up of more than only straight pipe sections. These may range from a gradual bend to the abrupt alteration of the flow path within a globe valve. The statistical frequency of the existence of various fittings and valves in a typical process plant piping system is discussed in Chapter 3. The relation between pipe size and equivalent length of valves and fittings is provided through Equations (2-13) and (3-22 A). However, discussion of the nature of equivalent length and its use when required in fluid dynamics calculations deserves more consideration for those cases where the statistical approach is not feasible.

As explained in Chapter 4, an alteration in the velocity of a fluid will produce a corresponding change in the system pressure. An increase of velocity is achieved at the expense of the system pressure.

When fluids encounter the configurations inevitably associated with valves and fittings, increases in velocity do occur with a consequent diminution of the system pressure. Furthermore, the frictional resistance to the flow is increased, due, among other factors, to greater surface roughness, protruding structures and the impingement of the fluid upon the conduit walls during a change in flow direction.

Considering the variety of valves and fittings offered by manufacturers, the task of developing pressure loss equations for each item and each size (as well as make) renders this approach impossible. Fortunately, tests have shown that one may relate the pressure losses to the velocity head with the result that the pressure loss becomes proportional to the ratio L/D, the length of a pipe to its diameter. Theoretically, the relation should include the friction factor, f, as well, but for most engineering calculations, this may be considered as constant.

L/D then becomes proportional to the length of straight pipe which offers the same resistance to flow as would the particular fitting or valve under the same flow conditions. Assuming the geometry of the valve or fitting to remain dynamically similar from one size to the next, which is sometimes (but not always) true, the same ratio could apply for all sizes of the flow element.

Because valve and fitting design must take into consideration good manufacturing economics, fluid flow similarity often cannot be maintained over broad ranges of sizes. Therefore, the L/D ratio for a small plug valve may be different from that for a large one. However, this still results in a great simplification in calculation because the equations used to determine pressure loss in straight pipe can be used with fittings, etc., by merely adding to the length of straight pipe, the "equivalent length" assignable to the fitting. The pressure loss is determined by multiplying the pipe length plus the equivalent length by the pressure loss per unit length. The influence upon pressure drop of other forms of flow disturbance may be determined using the equivalent length approach.

The following Table lists the factor which when

multiplied by the nominal pipe size in *inches* yields the length of pipe in *feet* which would generate pressure loss equivalent to the fitting or other flow disturbance listed.

TABLE 6-1

Factors for Equivalent Lengths of Various Piping Elements

Element	Factor
Gate valve	1.0
Plug cock, 2 to 4 inch	1.0
Welding elbow, $R/D = 1.5$	1.2
Entrance loss	1.2
Long radius screwed elbow	1.4
Welding elbow, $R/D = 1.0$	1.7
Run of tee	1.7
Standard screwed elbow	2.4
Exit loss	2.4
Borda entrance	3.7
Welding tee through side	3.7
Screwed tee through side	4.8
Check valves, 6 inch and up	5.0
Plug cocks 6 to 12 inch	5.0
Angle valve, open	14.0
Globe valve, open	27.0

The equivalent length of standard swages is given in Table 6-2. Note that the equivalent length may be based upon the pressure loss calculated either for the larger or smaller pipe size.

For *estimating* the equivalent length of the bends, tees, entrance and exit losses which will occur in different common piping systems, the following factors may be used. The pipe size is multiplied by the factor. The equivalent length thus derived is added to the estimated length of the straight pipe to be used.

TABLE 6-2

Equivalent Length of Standard Swages

Size	Equivalent Length in Feet Based on Pressure Drop in Larger Size Line	Size	Equivalent Length in Feet Based on Pressure Drop in Smaller Size Line
$1\times1^1/_2$	9.6	$1\times1^1/_2$	1.3
$1^1/_2\times2$	3.0	$1^1/_2\times2$	0.90
$2\times2^1/_2$	1.7	$2\times2^1/_2$	0.72
$2^1/_2\times3$	3.8	$2^1/_2\times3$	1.2
3×4	8.1	3×4	2.4
4×6	38.0	4×6	5.2
6×8	10.8	6×8	2.9
8×10	13.5	8×10	4.6
10×12	8.2	10×12	3.9
12×14	2.3	12×14	1.4
14×16	5.9	14×16	3.1
16×18	4.6	16×18	2.6
18×20	4.5	18×20	2.7
20×24	18.	20×24	7.3
1×2	58.0	1×2	2.2
$1^1/_2\times2^1/_2$	16.0	$1^1/_2\times2^1/_2$	1.9
2×3	18.0	2×3	2.5
$2^1/_2\times4$	38.0	$2^1/_2\times4$	3.7
3×6	182	3×6	7.5
4×8	204	4×8	9.3
6×10	121	6×10	10.3
8×12	70	8×12	10.7
10×14	28	10×14	8.5
12×16	21.0	12×16	6.9
14×18	29.0	14×18	8.8
16×20	26.0	16×20	8.8
18×24	58	18×24	15

EQUIVALENT LENGTH OF BENDS

The flow of liquids or gases in bends has been studied by various investigators including BEIJ (25). Rotation of the fluid at right angles to the direction of flow sets up an auxilliary flow path which is superimposed upon the main one. This rotation appears to result from the combined action of centrifugal force and the frictional resistance on the conduit wall. The pressure loss in a bend is the sum of the influence of the pipe length plus secondary losses, the tangential loss and the curvature influence. All can be expressed as an equivalent length of straight pipe. Figure 6-1

TABLE 6-3

Equivalent Length of Typical Assemblages

Service	Factor
Overhead vapor line to one condenser located adjacent to column	10.7
Overhead vapor line to condenser bank located adjacent to column	13.4
Reboiler liquid line	9.4
Reboiler vapor line	6.0

FIGURE 6-1
CHART FOR RESISTANCE OF 90 DEGREE BENDS

The chart at the right shows the resistance of 90 degree bends to the flow of fluids in terms of equivalent lengths of straight pipe.

Resistance of bends greater than 90 degrees is found using the formula:

$$\frac{L}{D} = R_t + (n - 1)\left(R_l + \frac{R_b}{2}\right)$$

n = total number of 90° bends in coil
R_t = total resistance due to one 90° bend, in L/D
R_l = resistance due to length of one 90° bend, in L/D
R_b = bend resistance due to one 90° bend, in L/D

Problem: Determine the equivalent lengths in pipe diameters of a 90 degree bend and a 270 degree bend having a relative radius of 12.

Solution: Referring to the "Total Resistance" curve, the equivalent length for a 90 degree bend is **34.5** pipe diameters.

The equivalent length of a 270 degree bend is:

$L/D = 34.5 + (3 - 1)[18.7 + (15.8 \div 2)]$
$L/D =$ **87.7** pipe diameters

Note: This loss is less than the sum of losses through three 90 degree bends separated by tangents.

From *Pressure Losses for Fluid Flow in 90 Degree Pipe Bends* by K. H. Beij. Courtesy of Journal of Research of National Bureau of Standards, Vol. 21, July, 1938.

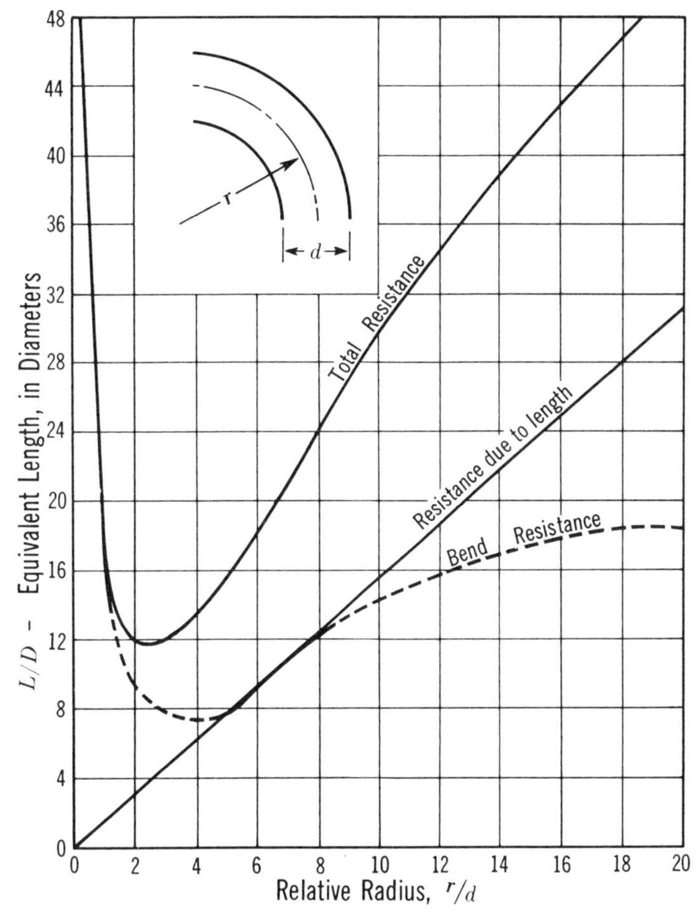

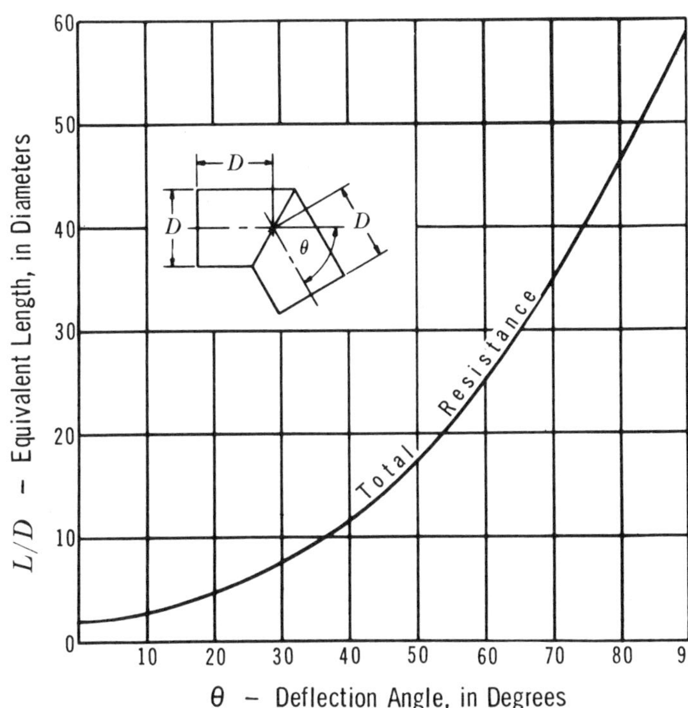

FIGURE 6-2
CHART FOR RESISTANCE OF MITER BENDS

(From TP-410, FLOW OF FLUIDS, Courtesy Crane Company)

Problem: Determine the equivalent length in pipe diameters of a 40 degree miter bend.

Solution: Referring to the "Total Resistance" curve in the chart, the equivalent length is **12** pipe diameters.

shows both the equivalent length and the total losses for a single ninety degree bend, and the equivalent length for the secondary losses for each *subsequent* ninety degree turn after the *initial* one. The equivalent length of a coil (where there are no tangential sections between ninety degree turns) can be expressed as:

$$L_e = l_t + (n-l)\, l_s \qquad (6\text{-}1)$$

Where:
L_e = Equivalent length of continuous coil, feet
l_t, l_s from Figure 6-1
n = Number of ninety degree bends in the coil

The resistance of mitered bends is shown in Figure 6-2. The factor given multiplied by the pipe size in inches gives the total equivalent length in feet.

Figure 6-2A

Mitered Bends

Graph 71 and the associated Tables B-14 to B-16 supply additional data for determining equivalent lengths of various piping elements.

FLOWING SPECIFIC GRAVITY FOR LIQUIDS

The various equations dealing with liquids given in the prior Chapters specify that both the specific gravity and the rate of flow are to be considered at the flowing temperature. Both are influenced by the effect of temperature upon the density of the liquid. The degree of change for a given temperature increase or decrease will depend upon the chemical composition of the liquid.

The composition establishes the critical temperature. At the critical temperature, the density of the liquid is at its lowest value, twelve to eighteen pounds per cubic foot. Between 60° F (16° C) and the critical temperature (T_c) the density changes at an ever increasing rate, altering very rapidly as the critical region is approached.

For water, the critical temperature is 707° F (375° C), whereas for normal butane it is 307° F (153° C). Consequently, for a temperature change from 60° F (16° C) to 100° F (38° C), the change in specific gravity for the butane will be considerably more than for the water, 3.9% versus 0.9%. Hence, one must take care when using charts showing changes in liquid specific gravity versus temperatures. Those applicable, for example, to hydrocarbons, may not be accurate when used for other liquids. However, these limitations may not be called out.

Figure 6-3 is a good mean for most hydrocarbons in the range of temperatures shown.

For higher temperatures and for other materials, alternate methods should be employed. If the critical temperature is known, or can be estimated, Graph 6 provides the means to determine the specific gravity of a liquid at any temperature. The specific gravity at the base temperature, usually 60° F or 4° C, must be known. Recognize, however, when the actual temperature is close to the critical, calculated values of the specific gravity, regardless of the method used to determine them, are subject to considerable potential errors.

VISCOSITY

The viscosity of a substance is a measure of its internal resistance to flow. In liquids it is an indicator of the lubricative qualities. In the centimeter-gram-second system, absolute viscosity in poise can be defined as the number of dynes required to move a plane of one square centimeter, at a distance of one centimeter from another plane of the same area, through a distance of one centimeter in one second. The commonly used unit is one hundredth of a poise, the centipoise. The viscosity of water at 60° F (16° C) is essentially one centipoise.

FIGURE 6-3

SPECIFIC GRAVITY - TEMPERATURE RELATION SHIP FOR PETROLEUM OILS

(Reproduced by permission from the Oil and Gas Journal)
(From TP-410, FLOW OF FLUIDS, Courtesy Crane Company)

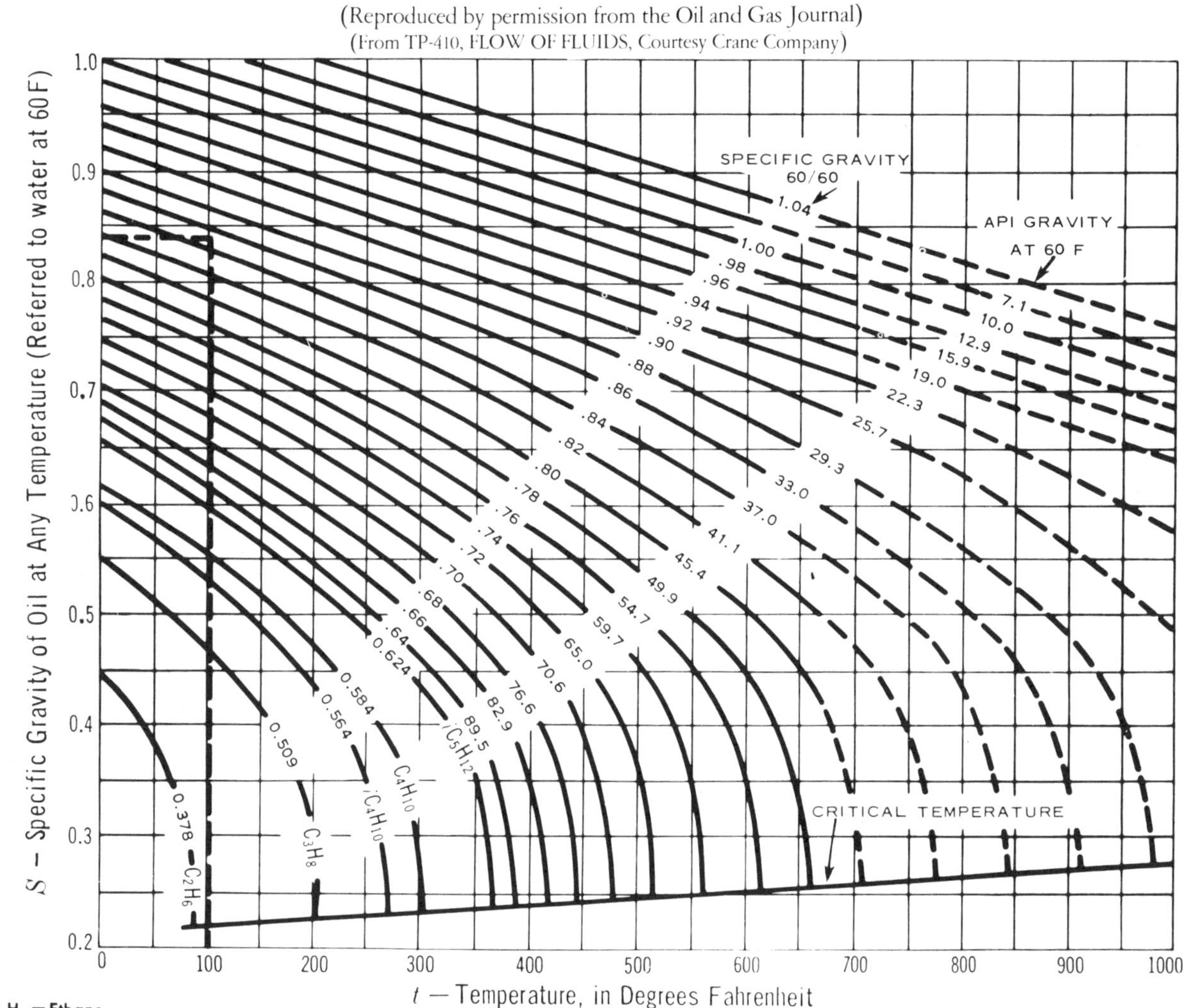

C_2H_6 = Ethane
C_3H_8 = Propane iC_4H_{10} = Isobutane
C_4H_{10} = Butane iC_5H_{12} = Isopentane

Example: The specific gravity of an oil at 60 F is 0.85 The specific gravity at 100 F = 0.83

To find the weight density of a petroleum oil at its flowing temperature when the specific gravity at 60 F/60 F is known, multiply the specific gravity of the oil at flowing temperature (see chart above) by 62.4, the density of water at 60 F.

WEIGHT DENSITY AND SPECIFIC GRAVITY* OF VARIOUS LIQUIDS

Liquid	Temp. t Deg. Fahr.	Weight Density ρ Lbs. per Cu. Ft.	Specific Gravity S
Acetone	60	49.4	0.792
Ammonia, Saturated	10	40.9	...
Benzene	32	56.1	...
Brine, 10% Ca Cl	32	68.05	...
Brine, 10% Na Cl	32	67.24	...
Bunkers C Fuel Max.	60	63.25	1.014
Carbon Disulphide	32	80.6	...
Distillate	60	52.99	0.850
Fuel 3 Max.	60	56.02	0.898
Fuel 5 Min.	60	60.23	0.966
Fuel 5 Max.	60	61.92	0.993
Fuel 6 Min.	60	61.92	0.993
Gasoline	60	46.81	0.751
Gasoline, Natural	60	42.42	0.680
Kerosene	60	50.85	0.815
M. C. Residuum	60	58.32	0.935

Liquid	Temp. t Deg. Fahr.	Weight Density ρ Lbs. per Cu. Ft.	Specific Gravity S
Mercury	20	849.74	...
Mercury	40	848.03	...
Mercury	60	846.32	13.570
Mercury	80	844.62	...
Mercury	100	842.93	...
Milk	...	†	...
Olive Oil	59	57.3	0.919
Pentane	59	38.9	0.624
SAE 10 Lube‡	60	54.64	0.876
SAE 30 Lube‡	60	56.02	0.898
SAE 70 Lube‡	60	57.12	0.916
Salt Creek Crude	60	52.56	0.843
32.6° API Crude	60	53.77	0.862
35.6° API Crude	60	52.81	0.847
40° API Crude	60	51.45	0.825
48° API Crude	60	49.16	0.788

*Liquid at 60 F referred to water at 60 F.

†Milk has a weight density of 64.2 to 64.6.

‡100 Viscosity Index.

Values in the table at the left were taken from *Smithsonian Physical Tables*, Mark's *Engineers' Handbook*, and [12]Nelson's *Petroleum Refinery Engineering*.

VISCOSITY MEASUREMENT

The more commonly used viscometers do not measure the force as defined above, but rather measure the time required for a given sample to flow through the apparatus. Two versions are in common use in the United States, **Saybolt Universal** and **Saybolt Furol.** Measurement must be made at a constant temperature. Common temperatures used are 100, 130 or 210° F for Universal and 122 or 210° F for Furol. The latter is used for liquids with relatively higher viscosities.

Other methods for viscosity measurement, using the flow approach, include: Redwood Admiralty, Degrees Engler, Redwood Standard and Barbey.

These forms of viscosity measurement depend not only upon the viscosity of the liquid, but also upon the weight of the head of liquid above the orifice which generates the force pushing the liquid through it. Hence, the rate of flow through a device, and the consequent measure of viscosity, is influenced by the density of the liquid. The viscosity so measured is related to the *kinematic* viscosity in the centimeter-gram-second system, called the stoke, or in more commonly used magnitude, the centistoke. The relation between the centistoke and the centipoise is given as follows:

Centistokes × Specific Gravity = Centipoise.

Measurements with the various *flow* forms of viscometer, as mentioned above, may be translated to centistokes (and then to centipoise using the specific gravity) by reference to Graph 7.

VISCOSITY AT FLOW TEMPERATURE

For fluid dynamics calculations the required viscosity of a liquids is at the *flowing* temperature. Similar to density, the influence of temperature upon viscosity is dependent upon chemical composition. Graph 8 shows the viscosities for straight-run petroleum fractions and some pure compounds at various temperatures. While these data are not absolutely exact, they are sufficiently accurate for fluid dynamics calculations. The viscosities are expressed in *centistokes;* whereas the equations employ *centipoise.* Conversion to the latter is done by multiplying centistokes by the specific gravity at the flowing temperature.

Graph 9 gives data for straight run fractions but is based upon the average boiling point and the U.O.P. Characterization Factor, *K*, explained later on.

Graph 10 applies to a variety of crude oils.

Additional data on viscosities of various liquids at different temperatures can be found in PERRY (7), pages 3-199 to 3-201. Also consult the International Critical Tables. References are given in Table 3-262 in PERRY (7).

CHANGES IN VISCOSITY WITH HIGH PRESSURE

The influence of pressure upon liquid viscosity generally is not significant until the pressure exceeds 600 psi. At that level the effect upon most oils will be about the same as a temperature change of 2° F (1.1° C). Liquids with a complex molecular structure will undergo a greater change in viscosity than will the more simple forms. For example, the change in viscosity of ethyl alcohol, for a given alteration in pressure, will be ten times that of mercury. ANDRADE's approach (26), provides a method for approximating the effect of pressure on liquid viscosity.

Appendix B presents viscosity data for a variety of liquids under various temperatures, and in some cases, at various pressures.

VISCOSITIES OF LIQUID MIXTURES

To establish the viscosity of mixtures of miscible liquids, the empirical KENDALL-MONROE method can be useful (27). For immiscible liquids, the viscosity of the blend may be calculated using the process developed by TAYLOR (28). This assumes that the diameter of the droplets in the dispersion is less than 0.03 mm.

GAS VISCOSITY

Appendix B contains viscosity data for gases under different conditions of temperature and pressure. Additional information may be found in PERRY (7), pages 3-196 to 3-198.

The viscosity of a paraffin hydrocarbon gas at a temperature different from that of the reference may be calculated, with accuracy sufficient for flow calculations, from the relationship given by PERRY (29):

$$\mu_T = \mu_{60^\circ} \left[\frac{T}{520}\right]^{0.87} \tag{6-2}$$

Where:

μ_T = Gas viscosity at flow temperature, centipoise
$\mu\ 60^\circ$ = Gas viscosity at 60° F, centipoise
T = Flow temperature, degrees Rankine (°F+460)

To determine the viscosity of other gases at a temperature different from the reference, the following values may be substituted for the exponent, 0.87, in Equation (6-2).

TABLE 6-4

Alternate Exponents for Equation (6-2)

Gas	Exponent
Air	0.77
Ammonia	0.98
Argon	0.82
Carbon Dioxide	0.94
Carbon Monoxide	0.76
Chlorine	1.0
Helium	0.69
Hydrogen	0.70
HCl	1.03
Methane	0.87
Neon	0.66
Nitric Oxide	0.78
Nitrogen	0.77
Nitrous Oxide	0.89
Oxygen	0.81

For gases other than those covered by the Table above, the method of HISCHFELDER, given in PERRY (7), pp. 3-229 and 3-230, provides information with 3 to 15% accuracy.

Corrections for the influence of pressure can be made using Figure 6-4 which relates the viscosity change to the reduced pressure and the reduced temperature. Additional methods are given in PERRY (7) page 3-230.

CHARACTERIZATION FACTOR "K"

From time to time, the engineer will be requested to provide preliminary size information when complete data about the fluid(s) to be handled are not available. Even though ideally he should not be expected to make estimates about the character of the fluids, this situation may arise and require *some* means of coping with it.

For hydrocarbons, the following procedure has been a valuable aid. The degree of paraffinicity of mixtures of hydrocarbon compounds is expressed as a function of the "average boiling temperature". This refers to the boiling point of the partially evaporated hydrocarbon mixture when one half of the molal constituents in the original sample have been evaporated. This temperature, in conjunction with the specific gravity of the initial liquid, can be used to determine the "characterization factor", K, as proposed by WATSON and NELSON (30), and revised by SMITH and WATSON (33).

$$K = T_b^{0.33}/S \tag{6-3}$$

Where:

K = Characterization Factor (degree of paraffinicity)*
T_b = Molal average boiling point from a standard distillation, degrees Rankine (°F+460)
S = Initial sample specific gravity at 60° F, water at 60° F = 1.00

Typical values for hydrocarbons are as follows (32).

TABLE 6-5

Characterization Factors

Material	Characterization Factor
Paraffin hydrocarbons	12.7
Paraffin-base oils	12.5
Mixed-base oils	11.8
Cracked materials	10.5
Asphalt-base oils	10.2
Benzene	9.8

Appendix B contains extensive additional values of K for various petroleum crudes and fractions from around the world, as well as for pure compounds.

* Not the same as the K referenced as the equilibrium constant.

Earlier reference has been made to the critical constants for pure compounds. These are the pressure and temperature at which the distinction between gaseous and liquid phases disappear and one phase, only, is present. KAY (31) introduced the concept that pseudo-critical constants could be assigned to mixtures of compounds, thereby permitting the prediction of properties of such mixtures. The exhaustive studies of SMITH and WATSON (33) of the critical properties of mixtures provided a series of graphs from which both the true and the pseudo-critical pressures and temperatures of hydrocarbon mixtures may be estimated.

These data have been incorporated into Graphs 13 to 15.

Graph 13 solves Equation (6-3) relating *K*, specific gravity and molal average boiling temperature.

Graph 14 provides the relation between specific gravity, *K* and the molecular weight of the mixture.

Graph 15 presents the relation between *K*, the initial specific gravity and the critical conditions, both pressure and temperature. In the case of mixtures of hydrocarbons, this would be pseudocritical rather than critical conditions.

If one knows the value of any two of the parameters, the remaining ones can be found from the Graphs. In the absence of exact values, one may be able, using Table 6-4 or the data in Appendix B, to select an appropriate *K* Factor for hydrocarbons which, with the specific gravity (or API gravity), will permit the values of other parameters to be estimated with sufficient accuracy for pipe size calculations.

As an example, determination of the pseudocritical conditions for a material can permit a better approximation of the specific gravity at elevated temperature or an estimation of the viscosity. This process is discussed in more detail in Chapter 7.

FIGURE 6-4

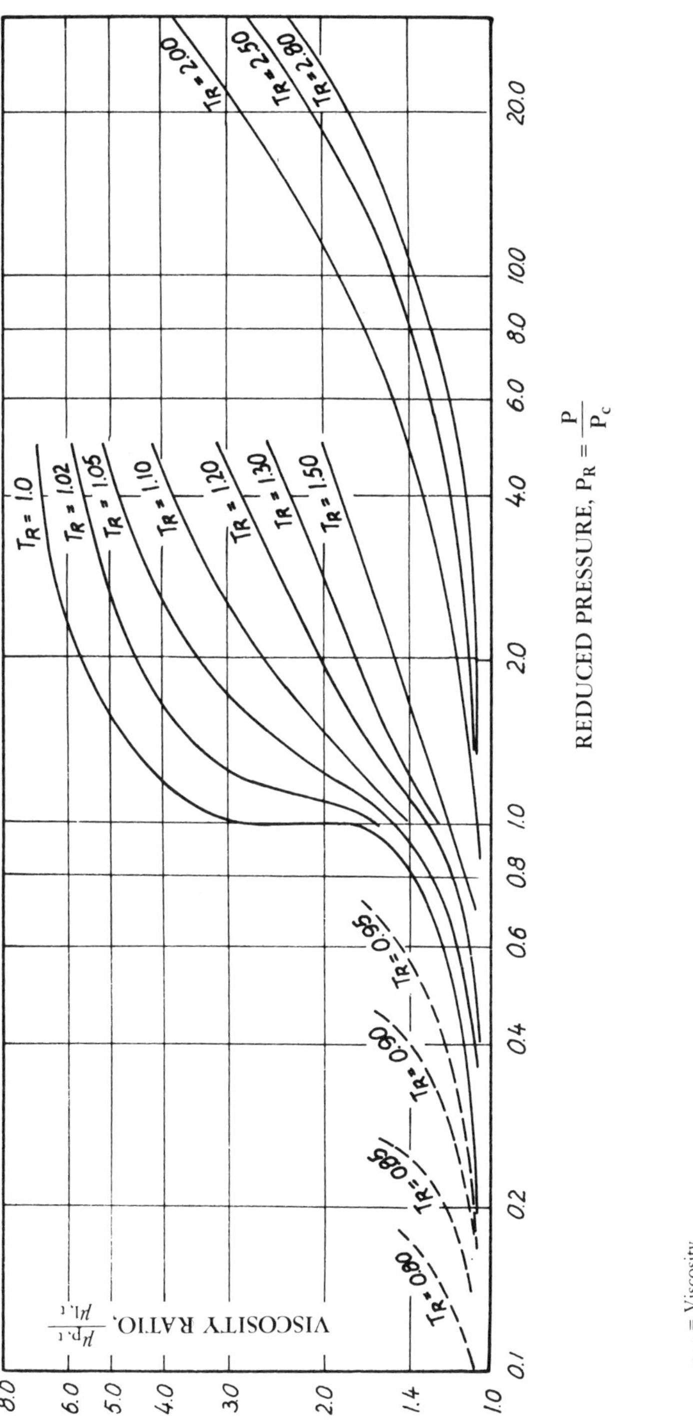

$\mu_{P,t}$ = Viscosity

$\mu_{1,t}$ = Viscosity at given Temperature but at low Pressure

P = Static Pressure

P_c = Critical Pressure

T = Absolute Temperature

T_c = Critical Temperature, absolute

$P_R = \frac{P}{P_c}$ $\quad T_R = \frac{T}{T_c}$

EFFECT OF PRESSURE ON GAS VISCOSITY

FIGURE 6-5

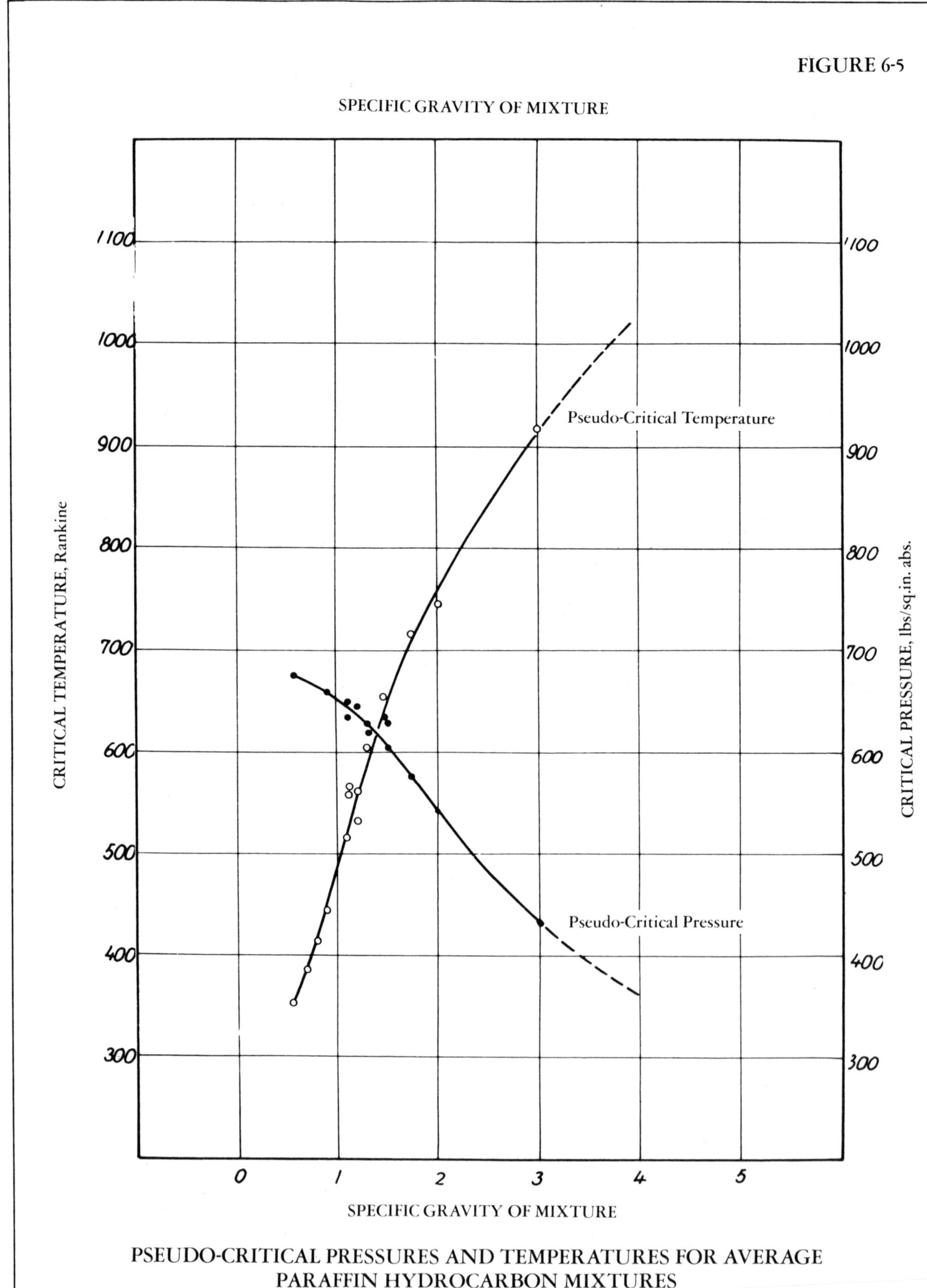

PSEUDO-CRITICAL PRESSURES AND TEMPERATURES FOR AVERAGE PARAFFIN HYDROCARBON MIXTURES

CHAPTER 7

PREPARATIONS

THE FLOW SHEET AND PLOT PLAN

In Chapter 1 the various modes of pipe size selection, according to the Optimum Process, are described. The first step in a pipe size selection project is an examination of the flow sheet for the purpose of establishing the mode which applies to each pipe, while at the same time, becoming familiar with the process, equipment, control valves, etc. The pipes may be coded according to the mode to make another evaluation unnecessary as the work of the size selection proceeds. Review of the plot plan will assist the engineer to recognize which pipes, if any, may have unusually long runs and require special care in PDA modes. Look carefully for those pipes in which two-phase flow may be encountered. These do take special precautions, as is mentioned in earlier Chapters.

Data on pump characteristics are useful, both as to centrifugal and reciprocating types. Collect the information on steam rates for turbines, and steam consumption for reciprocating pumps, if available. If not available, reasonable estimating data are provided later in this Chapter. Collect steam rates and pressures in steam heated apparatus; exchangers, reboilers, etc. Check specifications of equipment wherein pressure losses of flowing fluids will occur; exchangers, filters, fired heaters, condensors, etc.

Ascertain if the process has special requirements as to piping pressure losses. For example, a vacuum still overhead vapor line pressure loss is far more stringent than would apply for a deethanizer column. Where are the condensors, in the cooling tower or adjacent to the column?

What can be the effects of process upsets on the rates of flow of various streams? The Introduction explains the significance of this form of information whenever control valves are a part of the pipe system.

Consideration is to be given to the relief valve discharge piping systems, both liquid and gas. What valves are likely to blow at the same time under various plant abnormalities. These data are essential for the correct relief header size selection. One should not *necessarily* assume that all relief valves will blow at the same time. Granted they might in some plants and under some conditions. However, making that assumption without careful checks will result in larger piping than needed. On the other side of the coin, if that *is* likely to occur, the piping *must be large enough to handle it,* regardless of greater costs.

A study of the type(s) of steam traps to be used plus determination of their characteristics is important. Look at the influences of start-up on the amounts of condensate which must be handled and at what pressure difference. This will influence the amount of flashing and, possibly, the correct downstream piping sizes selection. Consider in similar fashion, the continuous drainers, if any. Such a careful study of the data provided on the flow sheet, or other source of flow data, can profitably be made early in the project. Identifying missing information at the start will give more time for collecting the missing data, particularly if the source is not readily available.

Look for pressure, temperature, specific gravity, viscosity (where needed), ratio of liquid to vapor in two-phase conditions, and flowing specific gravities (for liquids).

Each pipe on the flow sheet should be identified as to fitting rating and pipe material, if other than 300 ASA and carbon steel.

PIPE SIZE CALCULATION SHEET

When pipe size selection is begun, some type of form for defining the flow information and recording the pipe size selection mode and method is appropriate. Forms used by one company are shown in Figures 7-1 to 7-3. Some companies prefer to have a single sheet per pipe line to permit more detailed data to be collected.

LIQUID SPECIFIC GRAVITY

The specific gravity of liquids may be shown only for 60° F, not at the flowing temperature. The value at the *operating temperature* is required. Similarly, the rate of flow may be given in gallons at 60° F, rather than at the flowing temperature. When this occurs corrections are to be made using the data given in Chapter 6.

The method can vary according to the nature of the liquid. For the ordinary hydrocarbons, Figure 6-3 can be used. Enter the chart with the specific gravity at 60° F along the right hand diagonal lines. Follow the line to the intersection of the flowing temperature on the essentially vertical fan of lines. Move downward and to the left along the second grid of diagonal lines to read the specific gravity at the flowing temperature on the left side of the chart.

EXAMPLE: A crude oil fraction with a specific gravity of 0.86 at 60° F is flowing at 500° F, at 250 gpm, measured at 60° F. What is the hot specific gravity and flow rate? Performing the procedure just described, the answer is 0.68. The hot flow rate will be the cold rate multiplied by the ratio of the cold to the hot specific gravities. Hence, the answer is: $250 \times 0.86 / 0.68 = 316$ gpm at 500° F.

If the temperature lies beyond the limits given on Figure 6-3, base the determination upon the method using critical (or pseudocritical) temperature. Having established (or approximated) the critical temperature, the change of the specific gravity at the flowing temperature may be found from Graph 6.

EXAMPLE 7-1:

A paraffin base oil with a specific gravity of 0.80 at 60° F is flowing at the rate of 200 gpm (cold) at 800° F. What are the actual gravity and flow rate?

SOLUTION:

Knowing that the oil is paraffinic, the Characterization Factor can be found from Table 6-4, Chapter 6, to be 12.5. Using Graph 15, the critical temperature of 1300° R is found by laying a straight edge to intersect 12.5 on the *K* scale and 0.80 on the specific gravity scale. The absolute critical temperature in degrees Rankine is read on the left hand scale. Using Graph 6, a straight edge is laid upon 1300° absolute on the center scale and 800° F on the left one. The ratio of the specific gravities, 0.62, is read on the right-hand scale.

The actual flow rate is the given rate divided by the gravity ratio: $200/0.62 = 323$ gpm

The actual gravity is the given value multiplied by the gravity ratio: $0.8 \times 0.62 = 0.50$

The method for water requires that the specific gravity at the flowing temperature be found from Table 7-1, and the rate of flow calculated by dividing the cold flow rate by the flowing specific gravity.

LIQUID VISCOSITIES

When viscosity data are not available for liquids, the information contained in Appendix B should be consulted. If that does not contain the desired data, and the material is a pure compound or a mixture thereof, or a *straight run* hydrocarbon fraction, Graph 8 may be helpful. For pure compounds, lay a straight edge from the temperature through the index for the compound and read the viscosity (in *centistokes*) on the right-hand scale. If a mixture of two pure compounds is present, an average may be struck between the indices for the two.

Scales are given for three classes of crude fractions. Choose the appropriate scale and lay the straight edge from the temperature through the specific gravity which matches the fraction.

The kinematic viscosity obtained, multiplied by the actual specific gravity, gives the absolute viscosity in centipoise.

FIGURE 7-1

Rev				
By				
Date				

PUMP AND HYDRAULIC SYSTEM CALCULATION

Project No.________ Made By ________ Date ________ Pump Tag No.________

Client________ Checked By ________ Date ________

Unit________ Pump Service ________

SYSTEM SKETCH

FLUID PROPERTIES

Fluid ________

Temp. ________ °F VP @ T ________ psia

SG @ T ________ Visc @ T ________ cp

FLOW RATES AND LINE SIZING

Flow Data Source ________

Line Section	1	2	3	4	5
From					
To					
Flow, lb/hr					
Temp, °F					
SG					
Visc., CP					
GPM					
Line Size					
Velocity					
psi/100 ft					
Equiv L					
Total △P					

Form No. 566-1-7-74

SAMPLE CALCULATION SHEET

FIGURE 7-2

PRESSURE BALANCE

Note: Design Pressure Drops based on Flow Rates ________ • Normal.

SUCTION SIDE		Branch From ______ @ Normal Flow	Branch From ______ @ Design Flow	Branch From ______ @ Normal Flow	Branch From ______ @ Design Flow
Starting Pressure	psig	•			
Elevation	ft/psi	/	/	/	/
Line Loss	psi				
	psi				
	psi				
	psi				
	psi				
SUCTION PRESSURE	PSIG				

DISCHARGE SIDE		Branch To ______ @ Normal Flow	Branch To ______ @ Design Flow	Branch To ______ @ Normal Flow	Branch To ______ @ Design Flow	Branch To ______ @ Normal Flow	Branch To ______ @ Design Flow
End Pressure	psig						
Elevation	ft/psi	/	/	/	/	/	/
Line Loss	psi						
	psi						
	psi						
	psi						
	psi						
	psi						
Orifice Plate	psi						
Control Valve	psi						
DISCHARGE PRESSURE	PSIG						

NPSH AVAILABLE

Boiling Liquids	Non-Boiling
Starting Elev ______ ft	Vap. Press. ______ psia
Pump Cntr Elev ______ ft	Suction ______ psig =
Line Loss ______ psi = ______ ft	______ psia
NPSH Avail = ______ ft	NPSH Avail = ______ psi
	• 2.31 / sg = ______ ft

Pump $\triangle$P = ______ − ______ = ______ psi

Pump $\triangle$H = ______ psi • 2.31 / ______ sg = ______ ft.

BHP = ______ GPM • ______ psi / (1714• ______ η_p = ______ HP

Power = ______ BHP • 0.7457 / ______ η_m = ______ kw

CONTROL VALVE SIZING

Flow Rate, GPM						
Specific Gravity						
Viscosity, cp						
$\triangle$P Available						
Calculated Cv						
Est Cont Valve Size						

DESIGN PRESSURE

Suction Vessel Max WP ______ psig

\+ Max Elevation = ______ ft = ______ psig

\+ Shutoff Head = ______ • ______ = ______ psi

Design Pressure = ______ psig

Form No. 566-8-74

SAMPLE CALCULATION SHEET

FIGURE 7-3

PIPING PRESSURE DROP SUMMARY

LINE NO.	P&I NO.	DESCRIPTION		I.D. INCHES	FLUID	PHASE	PRESS. PSIG	TEMP. °F	FLOW RATE LBS./HR.	@ OPER. COND.			VEL-OCITY FPS	ΔP/100' PSI	EQUIV. LENGTH FT.	TOTAL LINE ΔP PSI	REV. NO.
		FROM	TO							S.G. or (LBS/CF)	VISC. CP	GPM or (CFS)					

NOTES:

CLIENT:	CONTRACT NO.	MADE BY:	REV. NO.	SH. NO.
PROJECT:		DATE:		

SAMPLE CALCULATION SHEET

TABLE 7-1
Specific Gravity and Viscosity of Water

° F	lbs/ft³	lbs/gal	Sp. Gr.	Absolute Viscosity cp
32	62.42	8.344	1.0008	1.79
35	62.42	8.344	1.0008	1.69
40	62.42	8.344	1.0008	1.55
45	62.42	8.344	1.0008	1.42
50	62.41	8.343	1.0006	1.31
55	62.39	8.341	1.0004	1.21
60	62.37	8.338	1.0000	1.12
65	62.34	8.334	0.9995	1.04
68	62.32	8.331	0.9992	1.00
70	62.30	8.329	0.9989	0.97
75	62.27	8.324	0.9983	0.92
80	62.20	8.317	0.9976	0.86
85	62.17	8.311	0.9967	0.81
90	62.12	8.304	0.9959	0.77
95	62.06	8.296	0.9950	0.72
100	62.00	8.288	0.9940	0.68
105	61.93	8.279	0.9929	0.65
110	61.86	8.270	0.9918	0.62
115	61.79	8.260	0.9907	0.58
120	61.72	8.250	0.9895	0.56
125	61.64	8.239	0.9882	0.53
130	61.56	8.229	0.9869	0.51
135	61.47	8.217	0.9855	0.49
140	61.38	8.206	0.9842	0.47
145	61.29	8.194	0.9827	0.45
150	61.20	8.181	0.9812	0.43
155	61.10	8.168	0.9796	0.41
160	61.00	8.155	0.9781	0.40
165	60.90	8.141	0.9764	0.39
170	60.80	8.128	0.9748	0.38
175	60.69	8.113	0.9731	0.36
180	60.58	8.098	0.9713	0.35
185	60.47	8.084	0.9655	0.34
190	60.36	8.068	0.9677	0.33
195	60.24	8.053	0.9659	0.31
200	60.13	8.037	0.9640	0.30
205	60.00	8.021	0.9620	0.29
210	59.89	8.004	0.9600	0.29
215	59.75	7.987	0.9580	0.28
220	59.62	7.970	0.9559	0.27
225	59.50	7.953	0.9539	0.26
230	59.37	7.936	0.9519	0.25
235	59.24	7.918	0.9497	0.25
240	59.11	7.902	0.9477	0.24
245	58.98	7.884	0.9456	0.24
250	58.84	7.866	0.9434	0.23
255	58.70	7.847	0.9412	0.22
260	58.56	7.828	0.9389	0.22
265	58.42	7.809	0.9366	0.21
270	58.27	7.789	0.9343	0.21
275	58.12	7.769	0.9318	0.20
280	57.96	7.748	0.9292	0.19
285	57.80	7.727	0.9268	0.19
290	57.65	7.707	0.9243	0.19
295	57.49	7.685	0.9217	0.19
300	57.33	7.664	0.9192	0.18
342	55.87	7.469	0.896	0.158
392	53.97	7.215	0.865	0.136
435	52.16	6.973	0.836	0.123
482	49.90	6.671	0.800	0.113
525	47.62	6.366	0.763	0.104
572	44.46	5.944	0.713	0.095
617	40.98	5.478	0.657	0.084
662	35.86	4.794	0.575	0.071
680	33.01	4.413	0.530	0.063
698	28.20	3.770	0.452	0.053
705	22.59	3.020	0.362	0.038

EXAMPLE 7-2:

A mixture of 30% butane and 70% pentane has a specific gravity of 0.617 at 60° F. What is the viscosity at 120° F?

SOLUTION:

Laying a straight edge from 120° F on the left scale through an estimated point 30% of the distance from butane to pentane yields, on the right-hand scale, a viscosity of 0.27 centistokes. From Table 7-2, the critical temperatures of butane and pentane are found to be 766° R and 840° R. Multiply 766 × 0.3 and add to 840 × 0.7 yields a pseudo-critical temperature of 818° R for the mixture. From Graph 6, as described above, the ratio of the specific gravities is found to be 0.94. The given specific gravity multiplied by this ratio yields the actual specific gravity which is needed to convert the kinematic viscosity from Graph 8 into centipoise. 0.617 × 0.94 = = 0.58, the specific gravity at 120° F; 0.27 × 0.58 = = 0.17, the viscosity in centipoise at 120° F.

TABLE 7-2
Physical Properties of Hydrocarbons and Miscellaneous Compounds

Compound	Symbol	Molecular Formula	Molecular Weight	Boiling Point, °F	Characterization Factor	Critical Temp. ° Rankine	Crit. Pres. psia	Latent Heat at Normal Boiling Point	N-Value C_p/C_v	Liquid Sp. Gr.	Gallons per Mol
Paraffin Series											
Methane	C_1	CH_4	16	—258.5	23.7	344	673	245	1.306	.3	6.40
Ethane	C_2	C_2H_6	30	—128.2	18.5	549	708	211	1.189	.374	9.64
Propane	C_3	C_3H_8	44	— 43.8	14.7	666	617	183	1.128	.508	10.41
Isobutane	C_{4i}	C_4H_{10}	58	10.9	13.8	733	530	158	1.088	.563	12.38
n-Butane	C_4	C_4H_{10}	58	31.1	13.5	766	551	166	1.090	.584	11.94
Isopentane	C_{5i}	C_5H_{12}	72	82.2	13.1	829	482	146	1.066	.625	13.84
n-Pentane	C_5	C_5H_{12}	72	97.0	13.0	846	485	153	1.070	.631	13.71
n-Hexane	C_6	C_6H_{14}	86	155.7	12.8	914	434	146	1.057	.664	15.57
n-Heptane	C_7	C_7H_{16}	100	209.1	12.8	972	397	138	1.045	.688	17.47
n-Octane	C_8	C_8H_{18}	114	258.1	12.8	1024	361	131	1.040	.707	19.38
n-Nonane	C_9	C_9H_{20}	128	303.3	12.8	1073	337	125	1.034	.722	21.31
n-Decane	C_{10}	$C_{10}H_{22}$	142	345.2	12.8	1115	312	120	1.028	.734	23.25
Olefin Series											
Ethene	C_{2-}	C_2H_4	28	—154.7	16.4	510	748	208	1.258	.4	8.50
Propene	C_{3-}	C_3H_6	42	— 54.0	14.2	658	668	189	1.180	.5	10.02
Butene-1	$C_{4\text{-}1}$	C_4H_8	56	20.4	13.1	751	619	174	1.145	.600	11.23
Butene-2	$C_{4\text{-}2}$	C_4H_8	56	33.7	12.9	771	619	183	—	.615	10.96
Isobutene	C_{4i-}	C_4H_8	56	19.4	13.0	753	580	167	—	.603	11.17
Pentene	C_{5-}	C_5H_{10}	70	86.2	12.7	854	595	148	—	.646	13.00
Diolefin Series											
1,3-Butadiene	—	C_4H_6	54	23.5	12.8	786	—	194	—	.614	10.58
Aromatic Series											
Benzene	—	C_6H_6	78	176.2	9.75	1011	700	170	—	.882	10.63
Toluene	—	C_7H_8	92	231.1	10.2	1069	611	156	—	.870	12.72
o-Xylene	—	C_8H_{10}	106	291.2	10.3	1137	542	149	—	.883	14.44
m-Xylene	—	C_8H_{10}	106	282.2	10.5	1114	526	147	—	.867	14.71
p-Xylene	—	C_8H_{10}	106	281.1	10.5	1111	515	146	—	.864	14.76
Cumene	—	C_9H_{12}	120	306.1	10.6	1145	473	138	—	.863	16.67
Miscellaneous											
Phenol	—	C_6H_{60}	94	358.6	—	1246	889	—	—	1.081	10.44
Hydrogen	—	H_2	2	—422.9	—	60	188	194	1.41	.071	3.38
Nitrogen	—	N_2	28	—320.4	—	227	492	86	1.404	.808	4.16
Carbon Dioxide	—	CO_2	44	—109.3	—	548	1073	—	1.304	1.22	4.33
Hydrogen Sulfide	—	H_2S	34	— 76.5	—	673	1306	236	1.320	.790	5.17

EXAMPLE 7-3:

A California fraction with a specific gravity of 0.76 relative to 60° F is at 350° F. What is the viscosity in centipoise?

SOLUTION:

On Graph 8 lay a straight edge from 350 on the left scale through 0.75 on the California Fractions scale which yields a kinematic viscosity of 0.45 centistokes. From Figure 6-3, the specific gravity at 350° F is found to be 0.6. Hence, the viscosity in centipoise will be the product of the kinematic viscosity and the actual specific gravity: $0.6 \times 0.45 = = 0.27$ centipoise.

For crude oil viscosities, Graph 10, which provides data for a number of typical crude oils, may be used in the manner just described.

Sometimes the viscosities of a liquid at two temperatures may be known, but the value for another temperature is required. Graph 11 is used. Draw a *faint* pencil line between the first temperature and appropriate viscosity in *centistokes* and the second temperature and the matching viscosity. The intersection of the two lines is a turning point to determine the viscosity at any other temperature. The viscosity as given may require conversion to centistokes.

EXAMPLE 7-4:

A product has a viscosity of 400 Saybolt Second Universal at 100° F and 60 Saybolt Universal at 210° F. What will be the kinematic viscosity at 500° F, at 60° F?

SOLUTION:

Using Graph 7, convert the two SSU values into centistokes. 400 SSU equals 90 centistokes, and 60 SSU equals 10 centistokes. Using Graph 11, establish an intersection of two lines connecting 100° F with 90 centistokes and 210° F with 10 centistokes. Using the intersection as a turning point, it is found that the viscosity at 500° F is 1.35 centistokes, and at 60° F it is 450 centistokes.

PIPE NUMBERING

In establishing the method used to number the pipes in a plant, allowance should be made for ready identification of the pipe for size selection purposes. A variety of schemes have been propounded and used satisfactorily. The one presented herein has proved practical from the viewpoint of the various departments in a contracting engineering organization. Accordingly it is offered as *one* workable approach.

1. Continue the same number from one piece of apparatus to another, but not beyond. Note paragraph 13 for an exception. The line number may, when necessary, originate at one piece of equipment and terminate at another line. A control valve is not considered a piece of apparatus and hence the number continues beyond it unless the flange rating on either side is different, as noted in paragraph 2.

2. Change the pipe number when the classification (pressure rating of the fittings) changes.

3. Number around apparatus in essentially the following manner:
   - A. Feed
   - B. Overhead to condenser
   - C. Condenser to accumulator
   - D. Gas from accumulator
   - E. Water draw-off
   - F. Pump suction
   - G. Pump discharge
   - H. Reflux
   - I. Overhead condensate out
   - J. Reboiler liquid
   - K. Reboiler vapor
   - L. Bottoms out of reboiler
   - M. Heating medium in, unless steam or other utility
   - N. Heating medium out, unless steam or other utility

4. Funnel Drains
   - A. One number from vessel to valve
   - B. Entire funnel system, including accumulator header branches and funnels, has one number.

5. Pressure Drains

Pipe from apparatus to last valve ahead of drain header takes one number. Pipe from last valve takes drain header number.

6. Spare Apparatus

Where a common spare is used for two items, the pipe from the spare to both other pipes is a separate number from either of the other.

Where the spare is for one piece of equipment only, the same line number is used for both connections.

7. One pipe number may include two or more sizes but not more than one pipe and fitting rating.

8. Use one number for each steam, water, air, sealing oil, and other utility main supply headers, provided the pipe fitting classification remains constant. Give separate numbers to branches, except as noted in Paragraph 6.

9. If a number is cancelled, do not reuse.

10. Sample, control, instrument or bleed connections on a pipe take its number.

11. Bypasses: The number of the upstream main pipe continues through the branch, the valve and companion flange. The number of the downstream branch is the same as the pipe it ties into.

12. With multiple equipment of the same kind, the header takes one number; the branches take separate numbers.

13. For series or parallel exchangers, condensers, dryers, etc., the number of the incoming pipe continues through the manifolding. The number of the outgoing pipe starts at the last exchanger, etc. or at the termination of the manifold.

14. Use one number to steam trap and another from it. When trap is draining a pipe, use its number to the trap.

15. In writing the description of a pipe system, when possible avoid reference to another pipe number when stating origin or terminus.

ADJUSTMENTS FOR THE ECONOMIC PIPE SELECTION FORMULAE AND NOMOGRAPHS

Certain elements within the brackets in the economic formulae can change within a job, at times, as well as from job to job, as discussed in Chapter 3.

Please note that such adjustments to the equations and the nomographs are not *always* needed. Actually, the opposite is more likely to occur. However, circumstances *may* arise when the "typical" values used in the equations and nomographs will be too far removed from the actual (more than 30% higher or lower) for the results to fall within plus or minus 5% of the optimum pipe size. Then the engineer must employ the adjustment factors.

The conditions may call for one or more corrections. In a domestic plant where most pipe material is carbon steel but some portions require 304 stainless steel, the factor for stainless lines would be determined and only used thereafter where needed. On the other hand, if a job is located in an area where power costs are high, all pipes chosen for that plant might call for one correction for that parameter plus additional correction(s) for special pipe material(s). One does *not* compute a correction for each pipe!

Rather, when needed, a factor is computed for a group of pipes and/or all the pipes in a plant.

It should be further noted that the adjustments for conditions different than those used as "typical" are multiplicative. If more than one parameter is changed, the product of the various factors will be the final one to be used, whether for the equations or the nomographs. In considering the need for using adjustments, keep in mind that some may increase the pipe size while others will decrease it. One may well find that the two or more factors may cancel out, and no alteration from the size computed with the equations or nomographs "as is", will result.

Earlier, the notation was made that until the parameter under consideration changed by more than 30% from the "typical", no adjustment would be needed. Where more than one element differs from the "norm", the same guide applies; only when the product of *all* changes exceeds 30% increase, or decrease, in the value within the bracket, before being raised to the specific exponent, need one consider determination and use of the adjustment factor.

Down through the years, experience has shown, within the United States, values which depend upon the economy: power costs, labor costs, and material costs, for the most part, move together, with the total result staying within the 30% band. Therefore, because one aspect offsets another, the economic pipe size has remained the same for a given rate of flow over the past 30 years. However, this has not been the case for foreign projects. Nor does it imply that one may ignore the amortization rate or the percentage of usage. These can and do change, and should be considered when approaching any new project.

In summary, decrease in hours of operation or power costs will *reduce* the optimum pipe size. A decrease in the cost of pipe, field assembly labor rate, the amortization rate, or the efficiency of the pump and driver will *increase* the pipe size.

The parameters which could be different are the following:

a) The cost of power
b) The hours of operation (percent of useage)
c) The cost of the pipe system, per foot
d) The rate of amortization of the plant

Where the *formulae* are being used, the correction(s) would be applied to the pipe size obtained from the basic equation. When the *nomographs* are used, the mode calls for the *rate of flow to be adjusted before* it is used with the graph. The resulting size of pipe determined *requires no further correction.* Consequently, two factors are presented herein. Care must be taken to employ the one which matches the method of computation: an equation or a nomograph. F_d refers to the correction of the pipe diameter, while F_r refers to the rate of flow adjustment.

Both the formulae for calculating the two forms of correction are given below together with tables of values of the factors for the parameters which more commonly may be encountered in practice.

HOURS OF OPERATION

The "typical" hours of operation are considered to be 7880 per year which is 90% usage. To correct for other levels of operation when equations are used:

$$F_d = 0.428\ Y_w^{0.169} \tag{7-1}$$

Where:
F_d = Correction factor for pipe size
Y_w = Hours of use of the pipe system, *per week*

Table 7-3 gives values of the correction when the formulae are used for different number of hours of operation per week.

TABLE 7-3

Values of F_d for Different Hours of Operation per Week

(Multiply the pipe size obtained from formulae by F_d)

Hours of Operation, per week	10	20	40	60	75	100	120
Value of F_d	0.63	0.71	0.80	0.86	0.89	0.93	0.96

For the correction to be applied to the rate of flow when the nomographs are used:

$$F_r = 0.170\ Y_w^{0.353} \tag{7-2}$$

Where:
F_r = Correction factor for the flow rate
Y_w = Hours of use of pipe system, *per week*

Table 7-4 gives values of the correction to be used when nomographs are used for various hours of operation per week.

TABLE 7-4
Values of F_r for Different Hours of Operation per Week

(Multiply the previously corrected flow rate by F_r before using in the nomograph)

Hours of Operation, per week	10	20	40	60	75	100	120
Value of F_r	0.38	0.49	0.63	0.72	0.78	0.86	0.92

POWER COSTS

The "typical" value for electric power used in the derivation of the various economic formulae is 0.0218 $/kwhr. When the actual cost is other than that, the factor to be applied to the size pipe determined by the formulae is:

$$F_d = 1.909\ K_e^{0.169} \tag{7-3}$$

Where:
K_e = Electric power costs, dollars per kwhr

Table 7-5 gives values of the correction when the formulae are used which apply when the cost of electric power is other than the "typical".

To adjust the flow rate before entering the nomographs, the correction is:

$$F_r = 3.86\ K_e^{0.353} \tag{7-4}$$

Table 7-6 gives value of the correction to be used, when the nomographs are employed, and when the cost of electric power is other than "typical".

The equations for economic size selection, where gas engine or steam drives are used, rather than commercial electric power, employs the cost of the Therm (one million Btu). The "typical" cost used was $0.70. If the actual cost is other than that, and the equations for pipe size selection are used, the correction is:

$$F_d = 1.062\ K_g^{0.169} \tag{7-5}$$

Where:
K_g = Cost of fuel, dollars per Therm

TABLE 7-5
Values of F_d for Different Electric Power Costs
(Multiply the pipe size obtained from the formula by F_d)

Power cost, $/kwhr	0.007	0.01	0.015	0.0218	0.025	0.030	0.035	0.040
Values of F_d	0.83	0.88	0.94	1.00	1.025	1.057	1.08	1.11

TABLE 7-6
Values of F_r for Different Electric Power Costs
(Multiply the previously corrected flow rate by F_r before using in the nomograph)

Power cost, $/kwhr	0.007	0.01	0.015	0.0218	0.025	0.030	0.035	0.040
Values of F_r	0.67	0.76	0.88	1.00	1.05	1.12	1.18	1.24

TABLE 7-7
Values of F_d for Different Fuel Costs
(Multiply the pipe size obtained from the formula by F_d)

Fuel cost, $ per Therm	0.20	0.30	0.45	0.70	1.00	1.50	2.00	2.50
Values of F_d	0.81	0.87	0.93	1.00	1.06	1.14	1.19	1.24

TABLE 7-8
Values of F_r for Different Fuel Costs

Fuel cost, $ per Therm	0.20	0.30	0.45	0.70	1.00	1.50	2.00	2.50
Values of F_r	0.64	0.74	0.86	1.00	1.13	1.31	1.45	1.57

TABLE 7-9
Values of F_d for Different Amortization Rates
(Multiply the pipe size obtained from the formula by F_d)

Amortization, years	2	3	5	7	10	15	20	25
Values of F_r	0.82	0.87	0.95	1.00	1.05	1.12	1.17	1.21

TABLE 7-10
Values of F_r for Different Amortization Rates

Amortization rate, years	2	3	5	7	10	15	20	25
Values of F_r	0.65	0.75	0.89	1.00	1.12	1.28	1.39	1.48

TABLE 7-10 A
Values of F_r for Values of Z

Z	0.30	0.40	0.50	0.60	0.70	0.80	1.00	1.10
F_r	0.43	0.53	0.61	0.70	0.78	0.86	1.00	1.07

TABLE 7-11
Values of F_d for Different Pipe Costs

Cost of 2 inch pipe, $	0.50	0.65	1.00	1.32	1.75	2.50	3.50	5.00
F_d	1.18	1.12	1.05	1.00	0.95	0.89	0.85	0.80

When the formulae are used for pipe size selection, Table 7-7 gives the correction to be used when the value of gas (or other fuel) is other than $0.70 per Therm.

When the nomographs are to be used, the flow rate is to be multiplied before using in the nomograph by:

$$F_r = 1.134\ K_g^{0.353} \qquad (7\text{-}6)$$

Where:
F_g = Cost of fuel, dollars per Therm

When the nomographs are used for pipe size selection, Table 7-8 gives values of F_r by which the flow rate is to be multiplied before entering the nomograph, when the value of fuel is other than the "typical", $0.70 per Therm.

AMORTIZATION RATE

To adjust the formulae for amortization rate (other than the seven years considered "typical"), use:

$$F_d = 0.728 / (a + 0.01)^{0.169} \qquad (7\text{-}7)$$

Where:
a = Amortization rate, fractional, reciprocal of years

When the formulae are used for pipe size selection, Table 7-9 gives the values of the correction to be used when the amortization rate is other than the "typical" seven years.

To adjust the flow rate to be used in the nomographs (for other than the seven years considered "typical"), multiply the actual rate by:

$$F_r = 0.515 / (a + 0.01)^{0.353} \qquad (7\text{-}8)$$

Where:
a = Amortization rate, fractional, reciprocal of years

When the nomographs are used for the pipe size selection, Table 7-10 gives values of F_d by which the flow rate is to be multiplied before entering the nomograph, when the amortization rate is other than the "typical" seven years.

PIPE AND LABOR COSTS

The equations and nomographs for selecting pipe using the LAC mode are based upon the 1975 typical cost for two inch, carbon steel, schedule 40 pipe, of $1.32 per foot. If the actual cost of pipe should deviate from this figure by 30%, plus or minus, an adjustment should be applied.

When materials other than carbon steel are used, an adjustment should be used.

Please note that corrections for pipe schedule other than 40 are provided in the adjustment for the fitting rating (see Table 7-14).

In the preparation of the equations and nomographs, for establishment of typical costs for a pipe system, norms were chosen for labor rates and productivity for the erection of the pipe system. In parts of the world, or the United States, the actual rates which apply may be greater, or less, than these norms which are $13.00 per hour at an efficiency of 80%. So long as the actuals are within 30%, plus or minus, no adjustment is to be made.

In plants being erected in countries other than the United States, where materials may be purchased in one economic sphere, where the typical pipe cost may differ from $13.00 per foot, and erected in another economic sphere with the assembly being done by local workers with labor rates and productivity different from the norm, an adjustment may be needed.

DEVIATION FROM THE PERFECT GAS LAW

When the Optimum Pipe Size is being selected for gas, using the nomographs, under conditions of flow where the deviation, Z, is significant, a correction to the flow rate, F_r, must be applied to the nominal flow rate before entry is made into a nomograph. When Equation (2-17) is used, Z may be entered into the formula direct; hence values of F_d when Z is significant are not needed.

PIPE COSTS DIFFER FROM NORM

When the actual pipe cost differs by more than 30% from $1.32 per foot for two inch, schedule 40, carbon steel, Table 7-11 provides the corrections, F_d, to be applied to the pipe size derived from the equations.

When the nomographs are used, Table 7-12 gives the corrections, F_r, to be applied to the flow rate before entering into the nomograph.

TABLE 7-12
Values of F_r for Different Pipe Costs

Cost of 2 inch pipe, $	0.50	0.65	1.00	1.32	1.75	2.50	3.50	5.00
F_r	1.41	1.28	1.12	1.00	0.91	0.79	0.71	0.63

TABLE 7-13
Adjustments for Economic Size Equations or Nomographs for Special Pipe Materials

Pipe Material		F_d	F_r	c	d	e
Carbon steel, schedule 40		1.00	1.00	1.86	0.600	1.00
304 Stainless steel, schedule 40		0.86	0.72	2.94	0.753	4.43
304 Stainless steel, schedule 5		0.91	0.82	3.73	0.671	1.99
316 Stainless steel, schedule 40		0.83	0.68	2.71	0.767	6.41
316 Stainless steel, schedule 5		0.88	0.77	3.68	0.750	2.52
Nickel, schedule 40		0.77	0.57	2.91	0.818	11.04
Nickel, schedule 5		0.82	0.66	4.24	0.764	4.62
Alloy 20, schedule 40		0.79	0.62	2.18	0.764	11.32
Alloy 20, schedule 5		0.86	0.73	2.81	0.736	4.48
Monel, schedule 40		0.79	0.61	2.87	0.821	8.61
Monel, schedule 5		0.85	0.71	3.90	0.764	3.60
Inconel 600, schedule 40		0.74	0.53	3.48	0.821	12.14
Inconel 600, schedule 5		0.79	0.61	2.98	0.764	8.73
Hastelloy B, schedule 40	Note 1	0.68	0.45	1.68	0.869	43.45
Hastelloy B, schedule 5	Note 1	0.75	0.55	1.69	0.807	22.44
Aluminum, schedule 40		0.93	0.87	2.67	0.713	1.79
Glass		0.89	0.78	1.77	0.835	4.50
F.R.P. (glass fiber reinforced plastic)	Note 1	0.98	0.99	2.13	0.229	2.87
Glass lined steel, schedule 40		0.86	0.74	1.41	0.459	10.65
Rubber lined steel, schedule 40		0.91	0.81	1.96	0.422	5.38
Titanium lined steel, schedule 40	Note 1	0.76	0.55	2.13	0.422	18.59
Zirconium lined steel, schedule 40	Note 1	0.74	0.53	1.97	0.422	24.37
Tantalum lined steel, schedule 40	Note 1	0.64	0.40	1.34	0.422	86.63
Saran lined steel*		0.94	0.87	1.92	0.422	4.11
Penton lined steel**		0.88	0.76	1.65	0.422	8.46
Polypropylene lined steel		0.93	0.86	3.00	0.422	4.39
Kynar lined steel***		0.87	0.75	1.61	0.422	9.21
TFE lined steel****		0.83	0.67	1.73	0.422	11.60
FEP lined steel*****		0.85	0.71	1.73	0.422	10.15

* Polyvinylidene chloride
** Chlorinated polyether
*** Polyvinylidene fluoride
**** Teflon, tetrafluoroethylene
***** Teflon, fluorinated ethylene-propylene
(Based upon cost data assembled by MARSHALL and BRANDT (4).
Note 1: Sizes one inch to three inch, only.

PIPE MATERIAL IS OTHER THAN CARBON STEEL

The values of F_d and F_r, to be used with the equations or nomographs, respectively, for a variety of piping materials, are given in Table 7-13. This Table also provides other parameters, *c*, *d*, and *e* which will be discussed in following paragraphs.

LABOR AND MATERIAL COSTS DEVIATE FROM THE NORM

When both facets must be considered, the parameters also given in Table 7-13 are used in the following formulae.

When the several LAC equations are used for pipe size selection, use:

$$F_d = 1.53 / (cXe + dL_r / E)^{0.169} \qquad (7\text{-}9)$$

When nomographs are used to select the pipe size, use:

$$F_r = 2.42 / (cXe + dL_r/E)^{0.353} \qquad (7\text{-}10)$$

Where:

X = Cost in dollars of one foot of two inch schedule 40 carbon steel pipe delivered to the plant site

L_r = Field labor rate, dollars per hour including direct labor costs, fringe benefits, contractor's overhead and profit, in the country where the plant is being erected

E = Labor efficiency, fractional

c = Factor from Table 7-13. This is the typical ratio of total pipe system material cost to pipe cost excluding control valves and associated block and bypass valves, fractional

d = Factor from Table 7-13. This is the typical hours of labor to erect one foot of two inch pipe, fractional

e = Factor from Table 7-13. This is the typical ratio of cost of actual pipe material used to cost of carbon steel schedule 40, fractional

The possibility exists that the cost of power, as well as the labor rate, may both be less in a foreign country. The influence of one may totally or partially offset the influence of the other. Consequently, under such conditions, the optimum pipe size may be altered but little, if at all, from the size as established by the use of the equations or nomographs without corrections being applied.

INFLUENCE OF FLANGE RATING

When flange ratings other than 150 ASA to 600 ASA are being used, and the equations are employed to select the LAC sizes, corrections to these sizes are to be made using the values of F_d given in Table 7-14. When nomographs are used, separate graphs are furnished for the different flange ratings.

TABLE 7-14

Values of F_d for Various Flange Ratings
(Multiply pipe size obtained from formula by F_d)

Flange rating	900	1500	2500
F_d	0.93	0.83	0.76

VERY HIGH OPERATING PRESSURES

Nomographs are supplied for economic selection for the several standard pressure ratings up to 2500 psi. For greater pressures than those suitable for 2500 psi fittings, the applicable *equation* should be used and the size calculated be multiplied by F_d according to Table 7-15.

TABLE 7-15

Values of F_d for Very High Operating Pressures
(Multiply pipe size obtained from formula by F_d)

Working pressure, psi	10,000	15,000	20,000	25,000
Values of F_d	0.71	0.68	0.66	0.64

INFLUENCE OF MULTIPLE CORRECTIONS

The various forms of F_d and F_r are to be used in sequence when more than one variant is required to define the operating conditions. For either factor, the final result is obtained by successive multiplication. *Note* that when several variables are required, and each acts to reduce the pipe size, the reduction in economic pipe size could, in *unusual* cases, result in the velocity of rupture being approached, in which case the *latter limitation* would control rather than the economic based upon LAC.

DATA REQUIRED FOR STEAM DRIVEN RECIPROCATING PUMPS

Before the size pipe required to furnish steam to a reciprocating pump can be established, the steam consumption and the steam chest pressure are required. The engineer making the pipe size selection may not always have these data in the project specifications.

The steam consumption of a reciprocating pump depends upon the hydraulic horsepower it must deliver, upon the diameter of the steam cylinder, the steam chest pressure and the steam exhaust pressure.

The steam chest pressure is *not* the pressure in the steam supply main. Rather it is the pressure required to be acting upon the steam cylinder, which is established by the ratio of the areas of the steam and liquid cylinders at the head difference generated.

The hydraulic horsepower is a function of the flow rate being delivered by the pump and the increase in liquid pressure which the pump is producing, as well as the pump type. Graph 16 provides a ready means to determine the hydraulic horsepower. The pressure scale on the right side refers to the INCREASE in pressure provided by the pump.

EXAMPLE 7-5:
A steam driven 10 x 9 reciprocating duplex pump takes suction at 50 psig and discharges liquid at 160 psi with a flow rate of 250 gpm. What is the hydraulic horsepower?

SOLUTION:
The increase in pressure is 160 minus 50 equals 110 psi. Connect 110 on the right scale with 250 on the left and read the answer of 7 hhp on the center scale.

Before the steam consumption can be determined, the steam chest pressure must be established using Graph 17. If the exhaust pressure is atmospheric, Graph 18 provides the steam consumption direct. If a higher exhaust pressure exists, a Back Pressure Factor must be obtained from Graph 19. This is used in place of the steam supply pressure in Graph 18.

EXAMPLE 7-6:
For the pump described in the example above, what will be the steam consumption if the exhaust is to atmosphere? With a 40 psig exhaust?

SOLUTION:
Using Graph 17, align 9 on the "*L*" scale with 10 on the "*S*" scale to locate 4.7 on the turning scale "*A*". Align this with 110 on the "*H*" scale and read the steam chest pressure of 100 psig on the "*C*" *scale*.

For the atmospheric exhaust steam pressure, using Graph 18, align 7 hhp on scale "*H*" with 100 psig on scale "*R*" to locate turning point on scale "*A*" of 2.82. Align this with 10 on the "*D*" scale and read steam consumption as 750 pounds per hour on the "Duplex" scale.

When the *exhaust pressure* is greater than atmospheric, the *steam chest pressure* will be increased by the gauge pressure of the exhaust steam.

SOLUTION:
With the 40 psig exhaust, the steam chest pressure becomes 140 rather than 100. Using Graph 19, align 140 on the left scale with 40 on the right and read a Back Pressure Factor of 16.5 on the center scale. Enter this into Graph 18 on the "*R*" scale and align with 7 on the "*H*" scale to read a value of 3.5 on the "*A*" scale. Align this with 10 on the "*D*" scale and read steam consumption of 1300 pounds per hour on the duplex section of the "Q" scale.

The steam chest pressure influences the size of the pipe supplying steam to the pump as well as being part of the determination of the steam flow rate. In the example just presented, the piping requirements will be quite different for the two cases, not only because the higher exhaust pressure calls for more steam. The higher steam chest pressure means that less pressure difference between the steam header and the chest will be available. Assuming the header pressure to be 150 psig, with the 40 psig exhaust, 10 psi would be the available pressure to be consumed by the piping and the steam control valve. With the atmospheric exhaust, the pressure drop available would be 50 psi.

The pattern of the discharge of liquid from a reciprocating pump is far from a continuous stream. In all fluid flow calculations, steam supply or liquid discharge, allowance must be made for this phenomenon. This applies to selection of the pipe size whether based upon economic selection, pressure drop available or allowable velocity.

The duplex pump flow pattern is essentially 130% of the nominal flow rate. For simplex pumps the factor is 160%. In calculating pressure drop, velocity, or the economic size, the mean flow rate must be increased by this ratio before being used in the formulae, or corresponding nomographs.

DATA REQUIRED FOR STEAM TURBINES

Normally, the steam consumption of turbines will be furnished by the supplier. However, there may be times when the engineer selecting pipe sizes may be required to make estimates of these data. Graph 22 provides this information with adequate accuracy for *pipe size selection* purposes; not necessarily for others. The theoretical steam rate for the size turbines to be found in process plants can be obtained by aligning the supply pressure on the left scale with the exhaust pressure on the right. The theoretical rate in pounds per hydraulic horsepower per hour is read on the center scale. The grid yields the turbine efficiency in terms of the supply pressure and the turbine horsepower. The actual steam rate will be the theoretical divided by the efficiency, fractional.

EXAMPLE 7-7:

What is the working steam rate of a 100 HP turbine with a supply pressure of 200 psig and an exhaust of 15″ mercury vacuum? What will it be if the exhaust pressure is 5 psig?

SOLUTION:

Using Graph 22, align 200 on the left scale with 15″ on the right and read the theoretical rate as 10.75 pounds per horsepower per hour. From the grid find the efficiency to be 31 %. The actual flow will be:

$$100 \times 10.75 / 0.31 = 3467 \text{ pounds per hour.}$$

For the higher back pressure, again using Graph 22, align 200 on the left with 5 psig on the right and obtain the theoretical steam rate of 14.5 pounds per horsepower per hour. The actual flow will be:

$$100 \times 14.5 / 0.31 = 4677 \text{ pounds per hour}$$

In contrast to the reciprocating steam pumps previously discussed, the pressure of the steam at the turbine inlet should be considered at essentially the header pressure. While some engineers may control the flow from a steam turbine driven centrifugal pump by throttling the steam flow to the driver, this practice is not recommended for usual applications. Turbines incorporate a throttling governor to prevent runaway speeds, of course. Smaller sizes may employ a governor to control speed rather than using nozzle regulation. The point being made here is the avoidance of a steam control valve *outside* the turbine casing in the steam supply pipe. The far better approach to flow or other control is to provide a control valve in the pump discharge.

DATA REQUIRED FOR RECIPROCATING COMPRESSORS

The pattern of the flow of gases from reciprocating compressors is, like that of reciprocating liquid pumps, quite irregular. However in contrast to the latter, the flow pattern is much more complex and can not be defined as a simple percentage. Rather, determination requires consideration of various parameters to arrive at the specific flow pattern for a particular application.

At the suction side as well as on the discharge end, gas flow does not commence as soon as the piston begins a stroke. Incoming flow will be inhibited until the residual gas in the cylinder has fallen to the level in the suction pipe. "Clearance pockets" provided to enable the compressing capacity of a constant speed compressor to match the load, provide "dead space" not swept by the piston. The gas within that space will expand during the suction stroke. The greater the clearance volume, the smaller will be the percentage of the stroke during which gas can enter the cylinder.

In a matching manner, the discharge from the cylinder on the compression stroke will be delayed when a greater clearance volume is used. The gas in the pocket must be recompressed before discharge can commence. To present an absurd case to explain the point, with sufficient clearance volume (the amount would depend upon the compression ratio) no gas would be discharged; the pressure in the cylinder would reach the discharge line pressure only when the stroke was complete. Therefore, no flow at all would occur.

The flow pattern in the branch pipes from a compressor will depend upon: the clearance, the compression ratio, and the number and type of cylinders.

Figure 7-4 provides the relation between the compression ratio, the percentage of clearance and the volumetric efficiency during the suction stroke. Figure 7-5 relates the same parameters during the discharge stroke.

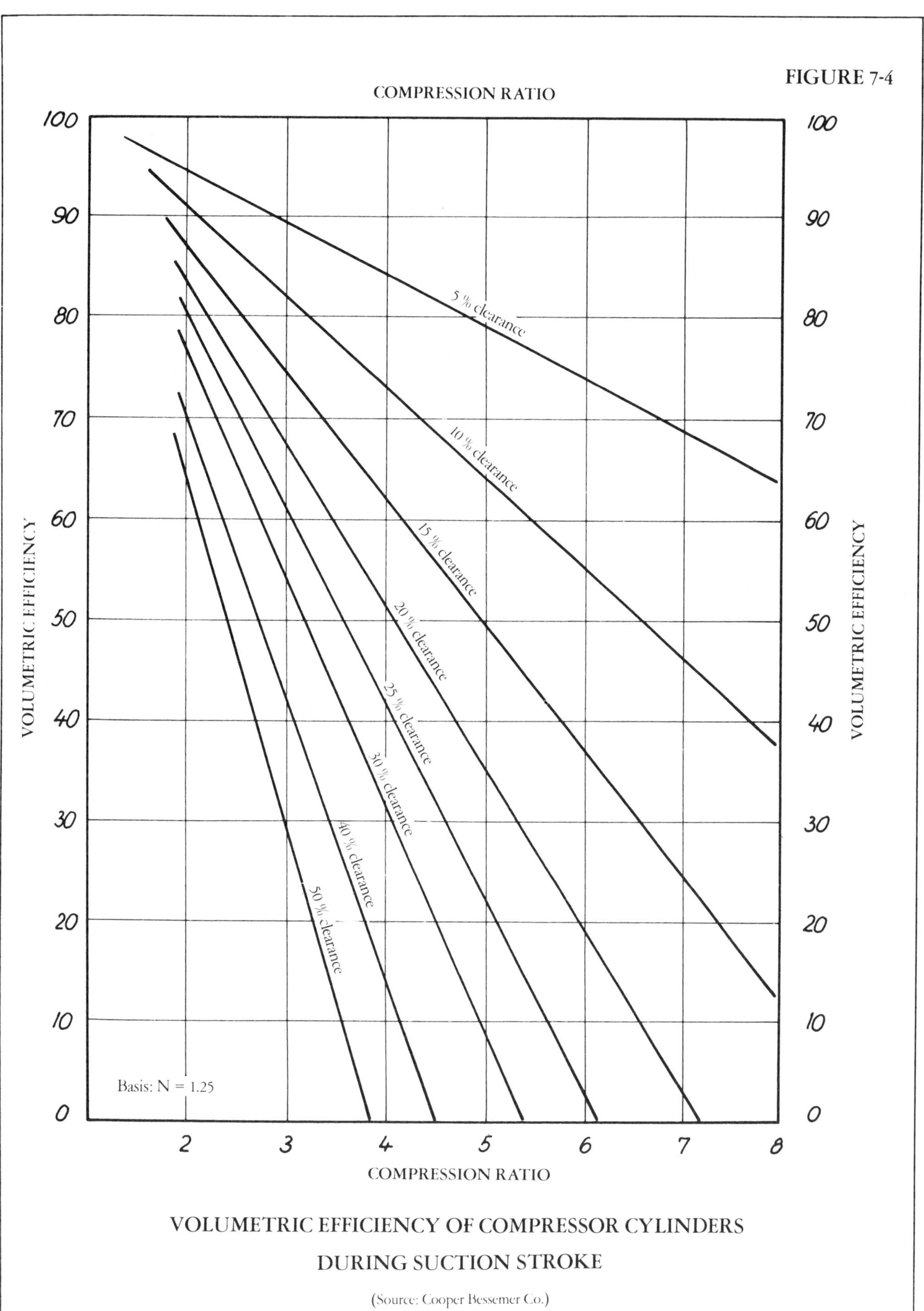

VOLUMETRIC EFFICIENCY OF COMPRESSOR CYLINDERS DURING SUCTION STROKE

(Source: Cooper Bessemer Co.)

FIGURE 7-5

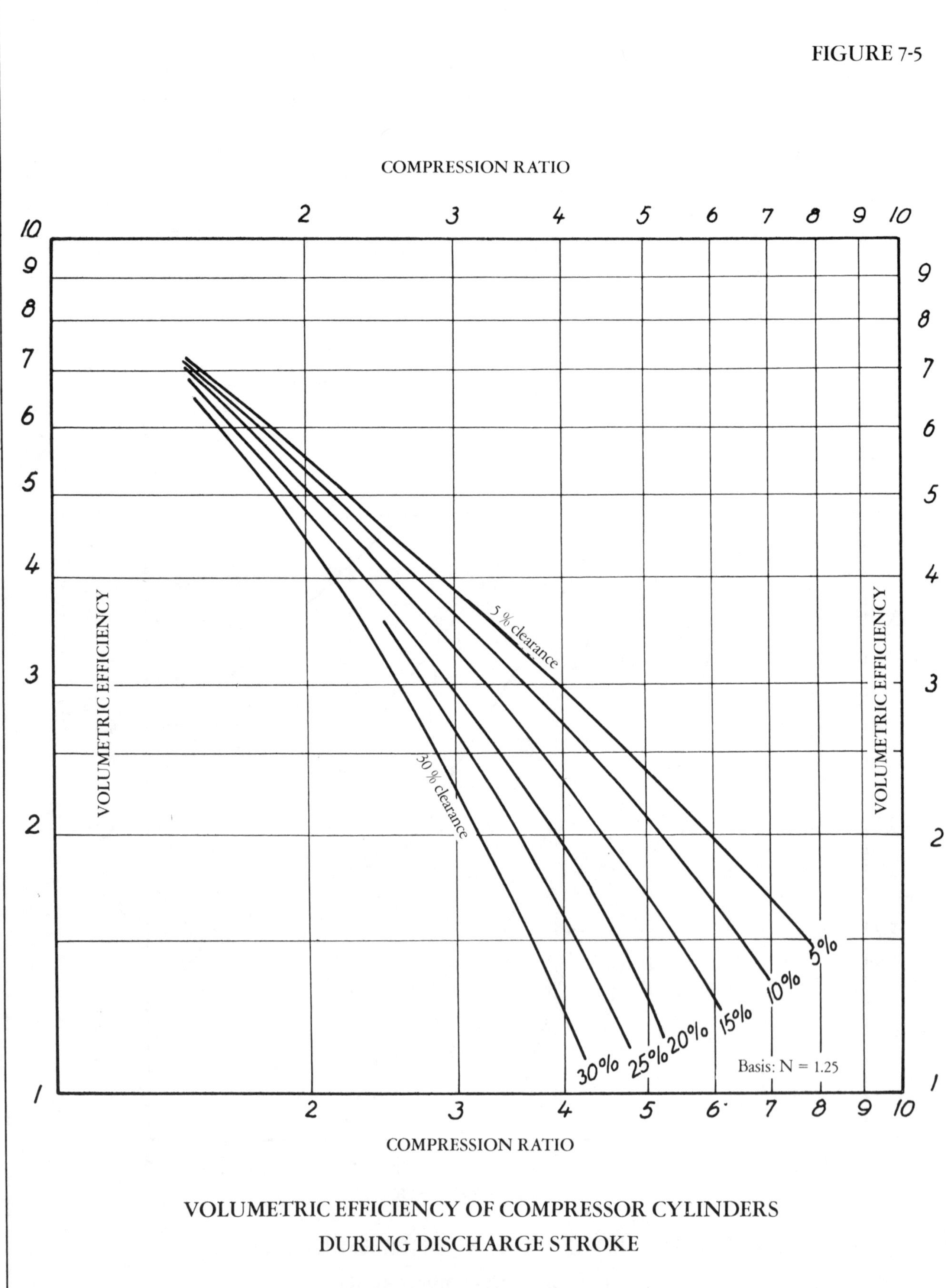

VOLUMETRIC EFFICIENCY OF COMPRESSOR CYLINDERS DURING DISCHARGE STROKE

FIGURE 7-6

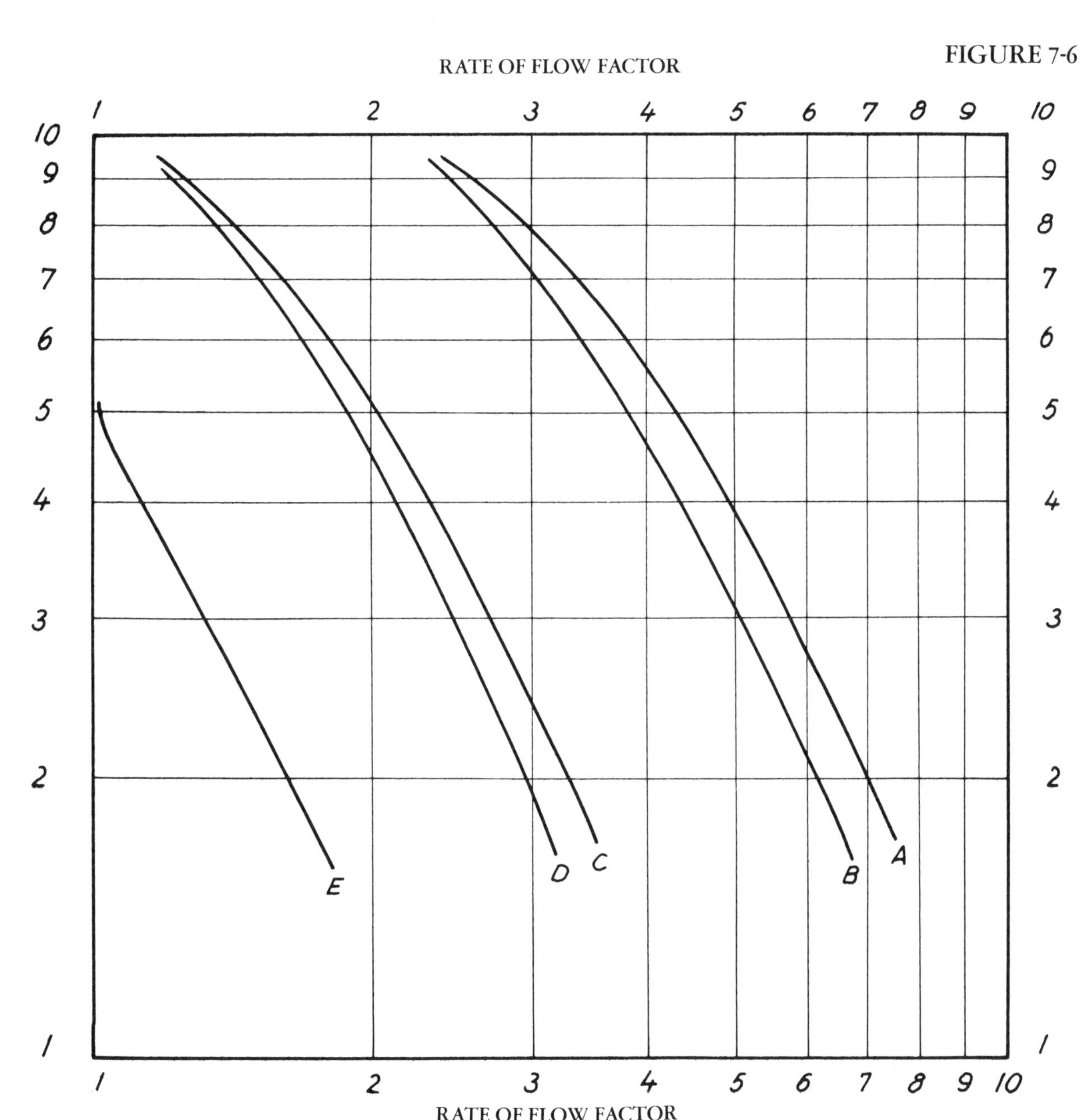

RATE OF FLOW FACTORS FOR SIZING PIPE TO AND FROM INDIVIDUAL COMPRESSORS

Curve A – Forward Stroke of Single Acting Cylinder
Curve B – Return Stroke of Single Acting Cylinder
Curve C – One Double Acting Cylinder
Curve D – Two Double Acting Cylinders
Curve E – Three or Four Double Acting Cylinders

(Source: Cooper Bessemer Co.)

FIGURE 7-7

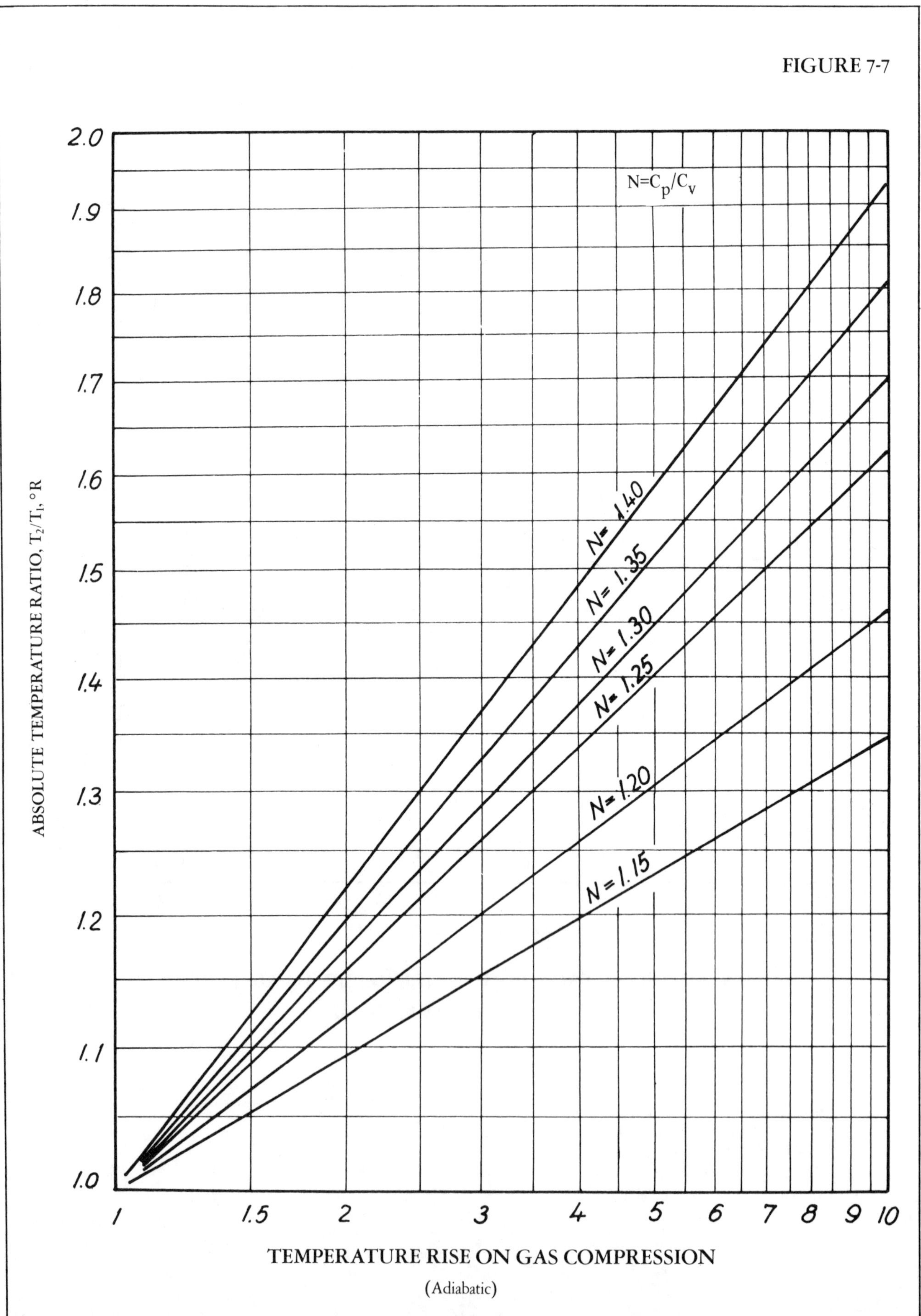

TEMPERATURE RISE ON GAS COMPRESSION

(Adiabatic)

Figure 7-6 presents the momentary flow ratio (rate of flow factor) as a function of the volumetric efficiency and the type of cylinder system. The rate of flow factor is applied to the nominal flow rate before determining the optimum pipe size by any of the methods, whether formulae or nomographs are employed. Similarly, the nominal flow rate is to be corrected by that factor before calculations of pressure drop or velocity are made.

If the process data should not include the discharge temperature of the effluent gas from a compressor, this information can be obtained from Figure 7-7.

EXAMPLE 7-8:

A two cylinder, double acting compressor takes suction at 300 psig and discharges at 1000 psig. The clearance is 15 %. What are the rate of flow factors for suction and discharge lines?

SOLUTION:

The compression ratio is 1015 divided by 315 which is 3.22. Enter Figure 7-4 at the bottom with 3.22. With a clearance of 15 %, the volumetric efficiency is 71.5 % during suction.

For the discharge condition, enter compression ratio into the bottom scale of Figure 7-5. The efficiency will be 30.5 %.

Enter Figure 7-6 on the vertical scale with 71.5 % efficiency for the suction and read flow factor of 1.5 on the *D* curve. Enter 30.5 % on vertical scale for discharge and read 2.45 on the *D* scale.

CHAPTER 8

PIPE SIZE SELECTION FOR PUMP OR COMPRESSOR DISCHARGE

The method of selecting the discharge piping for a compressor or pump falls in the category of LEAST ANNUAL COST (LAC), as introduced in Chapter 1 and amplified in Chapter 2.

Whenever a liquid or a gas is being driven by a pump or a compressor through the piping and *equipment,* the optimum pipe size is that which results in the least annual total of amortized construction costs plus annual operating costs of the system.

An objection may be raised to using this approach when a control valve is included. Is not the valve provided to consume the excess pressure generated by the pump or compressor beyond that needed to drive the fluid through the pipe and other equipment? As outlined in detail in the Introduction, the *correct* selection of the control valve does not result in the consumption of *excess* pressure. Rather the properly selected valve provides *only* that amount of pressure drop needed to maintain control of the process under expected parameter variations. Therefore, a balance exists between pressure losses in the piping and in the valve which validates the use of the selection on the basis of LAC.

A few exceptions to this definition can exist. If the discharge piping from a pump or compressor includes a side stream whereby a *portion* of the fluid is being delivered to a *lower* pressure than the main stream, that pipe size would not be selected by the LAC approach. Rather, it would be considered under Pressure Drop Available in Chapter 10.

If the fluid is transported to a vessel, heat exchanger, or other process apparatus, by a pump or compressor, and then moves on to other equipment operating at essentially the same pressure, the pipe size beyond the apparatus also would be selected by the LAC method. However, that would not apply if the fluid were being dropped to a much lower pressure beyond the vessel, etc. Then PDA would be used.

If a fluid is being processed at a natural higher pressure and, when the processing is completed, it is reduced to a lower pressure to go to storage or another process, the pipes handling the material *within the processing* area are treated as LAC, beyond as PDA. The reasoning is that if too much of the natural pressure were to be consumed using PDA in the process selection, the pressure reduction would require a pump or compressor to restore the pressure. Hence, LAC is appropriate. In other words, use LAC, even if no pump or compressor is included, provided the system would require one to be used if excess pressure were to be consumed by the use of smaller pipe.

LAC PIPE SIZES FOR LIQUIDS

For centrifugal or gear pumps, or when the pipe size being selected is sufficiently far downstream from reciprocating pumps so the flow rate is steady (and in the turbulent region), the pipe size is to be chosen by using Graphs 25 through 28. These are constructed using the typical values for the bracketed parameters (see Chapters 2 and 3), and assume electric drives with turbulent flow. If the bracketed values for a plant are different from the typical, tables described in Chapter 7, and repeated with each nomograph, provide the factors by which the actual flow rates are to be multiplied before entry into the nomographs. Nomographs are supplied for each pipe schedule and flange rating.

If formulae are to be used for LAC selection, the pipe size is determined by using Equation (2-17) as developed in Chapter 2.

$$D_c = 0.276\,Q^{0.479}\,S^{0.142}\,\mu^{0.025} \qquad (2\text{-}17)$$

Where:

D_c = Economic internal pipe diameter, inches, at the crossover point for commercial pipe sizes, see Table 2-4*

Q = Rate of flow, U.S. gallons per minute at the flowing temperature

S = Specific gravity at the flowing temperature, water at 60° F equals 1.00

μ = Viscosity at the flowing temperature, centipoise

The factor 0.276 applies when commercial electric power is used to drive the pump. For steam or gas-engine drivers, the constant is 0.310. See Chapter 2 for details.

Because the exponents associated with specific gravity and viscosity are small, significant changes in these parameters can occur before the influence upon the economic diameter will justify another pipe size selection. Therefore, the following tables of pipe sizes for liquids can provide a quick method for LAC selection for common fluids being pumped using electric power. For steam or gas engine drivers, the correction for the flow rate, F_r, is 1.25.

The selections obtained from these tables are based upon the typical values of the variables within the economic brackets in Equation (2-17) for liquid pipe line selection. Should the actual values for a project differ from the typical, the corrections to be applied to the flow rate before entry into the *nomographs* likewise are applied to the actual flow rate before using the tables. Values of F_r for different conditions are given in Chapter 7.

To use the tables for metric (SI) units, multiply the flow rate in cubic meters by 4.403 to convert to gpm. The nominal pipe size selected will be in *inches*.

The formula used to derive the LAC flow rate for different pipe sizes is a variation of Equation (2-17), rewritten to define optimum flow rate in terms of pipe size, specific gravity and viscosity.

$$Q = 14.7\,D_c^{2.088}/S^{0.296}\,\mu^{0.052} \qquad (8\text{-}1)$$

Where:

Q = Economic flow rate, U.S. gallons per minute, at the flow temperature

D_c = Pipe inside diameter, inches, representing the crossover between two commercial pipe sizes, see Equation (2-12) and Table 2-4*

S = Specific gravity, at flow temperature, water at 60° F equals 1.00

μ = Viscosity, at flow temperature, centipoise

The left column applies to liquids with specific gravities from 1.55 to 1.02. The right column applies to liquids with specific gravities from 0.98 to 0.65.

For Table 8-3, the constant in Equation (8-1) is 12.6. For Table 8-4, it is 9.97. For Table 8-5 it is 8.29.

For metric (SI) units, Equation (8-1) becomes,

$$Q_{si} = 15.9\,D_{csi}^{2.088}/S_{si}^{0.296}\,\mu_{si}^{0.052} \qquad (8\text{-}1\text{ M})$$

Where:

Q_{si} = Cubic meters per hour, at flow temperature

D_{csi} = Pipe inside diameter, centimeters, representing the crossover between two commercial pipe sizes, see Equation (2-12)*

S_{si} = Specific gravity, kilograms per cubic meter at flow temperature

μ_{si} = Viscosity at flow temperature, kilograms per meter second

Figure 8-1

Reciprocating Pump Piping

* If flange rating is above 600 ASA, correct the calculated size by F_d for flange rating, see Chapter 7.

EXAMPLE 8-1:

In a plant with electric drivers, and where the typical bracketed economic parameters apply, a centrifugal pump is discharging 800 gpm through schedule 40 carbon steel pipe with 300 ASA flange rating.

SOLUTION:

From Table 8-1, using the center column, a 6 inch will carry 895 gpm; use 6 inch.

Graph 25, LAC LIQUID SIZE, 600 ASA, could be used. Connect 800 on the "*Q*" Scale with 1.0 on the "*G*" Scale. Intersection on the "*D*" Scale lies between 4 and 6, so use 6 inch.

EXAMPLE 8-2:

Same conditions as in EXAMPLE 8-1, except the driver is a gas engine with the "typical" cost for gas, $0.70 per Therm.

SOLUTION:

Correction to the flow rate, 0.80, noted at the beginning of this Chapter, is applicable because the drive is a gas engine

$$800 / 0.80 = 1000.$$

From Table 8-1, 8 inch will carry 1500 gpm; use 8 inch.

TABLE 8-1
LAC Pipe Sizes for Various Flow Rates of Common Liquids
(Schedule 40 pipe with 150 lb ASA to 600 lb ASA Flange Rating)

Size (inches)	Liquid Sp. Gr. ave. 1.28, Visc. ave. 1.47 (gpm)	Water (gpm)	Liquid Sp. Gr. ave. 0.82, Visc. ave. 0.66 (gpm)
1.5	50	54	59
2	77	84	92
2.5	115	126	136
3	196	216	234
4	410	450	488
6	815	895	970
8	1365	1500	1625
10	2055	2255	2445
12	2680	2940	3185
14	3425	3760	4075
16	4435	4865	5175

* **NOT** the commercial dimensions. Also corrected for flange rating if above 600 lb ASA. See Chapter 7.

TABLE 8-2
LAC Pipe Sizes for Various Flow Rates of Common Liquids
(0.25 inch pipe wall, carbon steel, with 150 lb ASA or 300 lb ASA Flange Rating)

Size (inches)	Liquid Sp. Gr. ave. 1.28, Visc. ave. 1.47 (gpm)	Water (gpm)	Liquid Sp. Gr. ave. 0.82, Visc. ave. 0.66 (gpm)
10	2165	2375	2575
12	2835	3115	3375
14	3650	4010	4345
16	4765	5230	5670
18	6115	6710	7275
20	8365	9180	9950
24	12985	14255	15455
30	19775	21700	23530
36		28940	
40		39280	

TABLE 8-3
LAC Pipe Sizes for Various Flow Rates of Common Liquids
(Schedule 80 carbon steel pipe with 900 lb ASA Flange Rating)

Size (inches)	Liquid Sp. Gr. ave. 1.28, Visc. ave. 1.47 (gpm)	Water (gpm)	Liquid Sp. Gr. ave. 0.82, Visc. ave. 0.66 (gpm)
1.5	37	41	44
2	57	63	68
2.5	88	96	105
3	151	165	180
4	316	345	376
6	610	670	725
(Schedule 100 carbon steel pipe with 900 lb ASA Flange Rating)			
8	1000	1100	1190
10	1480	1625	1760
12	1950	2140	2320
14	2505	2750	2980
16	3265	3580	3885
18	4110	4515	4890
20	5640	6190	6710

TABLE 8-4
LAC Pipe Sizes for Various Flow Rates of Common Liquids
(Schedule 160 carbon steel pipe with 1500 lb ASA Flange Rating)

Size (inches)	Water (gpm)	Liquid Sp. Gr. ave. 0.82, Visc. ave. 0.66 (gpm)
1.5	25	27
2	40	43
2.5	63	68
3	105	114
4	221	240
6	436	473
8	723	784
10	1083	1175
12	1420	1540
14	1830	1985
16	2380	2580

TABLE 8-5
LAC Pipe Sizes for Various Flow Rates of Common Liquids
(Schedule 160 carbon steel pipe with 2500 lb ASA Flange Rating)

Size (inches)	Water (gpm)	Liquid Sp. Gr. ave. 0.82, Visc. ave. 0.66 (gpm)
1.5	20	22
2	33	36
2.5	52	56
3	87	95
4	184	199
6	363	393
8	600	650
10	900	975
12	1180	1280
14	1520	1650
16	1975	2140

EXAMPLE 8-3:

In a plant with electric drive, a loading line is used 20 hours per week to handle a liquid with 0.80 specific gravity, and 0.60 centipoise viscosity. "Typical" bracketed economic parameters apply.

SOLUTION:

From Table 7-4, the correction to the flow rate for the reduced operation time is 0.49.

$$800 \times 0.49 = 392$$

From Table 8-1, the right hand column, a 4 inch line will carry 488 gpm, use 4 inch.

EXAMPLE 8-4:

An electric drive pump, in a location where cost of power is $0.01/kwhr, water at 50 psi is being pumped at 800 gpm, the temperature is 65° F. Determine LAC pipe size.

SOLUTION:

From Table 7-6 the value of F_r is 0.76 when power cost is $0.01/kwhr.

$$800 \times 0.76 = 608$$

From Table 8-1, center column, a 6 inch can carry 895, a 4 inch, 450. Use the 6 inch.

EXAMPLE 8-5:

Acidic water is being pumped at 800 gpm through Schedule 5 316 stainless steel pipe using an electric drive in a plant where bracketed costs are "typical". What is the LAC size?

SOLUTION:

From Table 7-13, F_r for schedule 5 316 S.S. pipe is 0.77.

$$800 \times 0.77 = 616$$

From Table 8-1, 6 inch can carry 895 and 4 inch 450. Use 6 inch.

EXAMPLE 8-6:

A high pressure system requiring schedule 80 carbon steel pipe with 900 ASA flanges is carrying 110 gpm of liquid with specific gravity of 1.18 and viscosity of 1.3. The drive is steam. What is the LAC size?

SOLUTION:

From beginning of this chapter, the correction to the flow rate, F_r, for steam drive, is 1.25.

$$110 \times 1.25 = 138$$

From Table 8-3, left hand column, the capacity of a 2.5 inch pipe is 88 gpm and 3 inch is 151. Use 3 inch.

EXAMPLE 8-7:

The reflux line of a splitter column handles 500 gpm of hydrocarbon liquid at 450 psi, specific gravity is 0.374, and viscosity 0.12 centipoise at 100° F. The following values for the bracketed parameters apply to the entire operation. Amortization is seven years, electric power is 0.16$/kwhr. Usage is 150 hours per week (90%). Pipe cost is $1.35 per foot for 2 inch schedule 40 carbon steel pipe. Selection is being made by formulae. What is the optimum size?

SOLUTION:

A reflux line will be a pump discharge; thus the optimum size will be according to LAC. Equation (2-17), the formula for LAC for liquids under turbulent flow is used, with suitable corrections to be applied for deviation of the bracketed parameters.

Equation (2-17):

$$D = 0.276\, Q^{0.479}\, S^{0.142}\, \mu^{0.027}$$

Where:

D = Economic internal pipe diameter, inches
Q = Rate of flow, U.S. gpm at flowing temperature
S = Specific gravity at flowing temperature, water at 60° F equals 1.00
μ = Viscosity at flowing temperature, centipoise

$$D = 0.276 \times 500^{0.479} \times 0.374^{0.142} \times 0.12^{0.027} = 4.49$$

For this plant, the value of F_d will differ slightly from the typical because the power cost, $0.016, is less than that used in developing Equation (2-17). From Table 7-5 the correction, F_d, for 0.015$/kwhr is 0.94. Use 0.95 for F_d. All other values for the bracketed parameters are close enough to the typical that no further corrections are justified.

$$0.95 \times 4.49 = 4.23$$

The calculated diameter of 4.23 inches is compared to the values of D_c for schedule 40 (adequate for 450 psi at 100° F) as shown on Table 2-4. 3 inch D_c is 3.62, too small. Use 4 inch having a D_c of 5.15, larger than the calculated 4.23.

EXAMPLE 8-8:

A plant is being constructed in the Middle East. Materials are being purchased in Europe and the delivered cost of 2 inch schedule 40 carbon steel pipe is $1.05 per foot. Labor cost at the site is $3.15 per hour. Labor efficiency is 70%. The cost of gas is $0.45 per Therm. Drivers will be steam turbine or gas engine. Amortization is to be 12 years. Pipe size selection will be done using the nomographs. What will be the overall correction factor, F_r, to be applied to the flow rates for which sizes are to be calculated?

SOLUTION:

Starting with Equation (7-10), and using Table 7-13, calculate the labor and material correction.

$$F_r = 2.42/(cXe + dL_r/E)^{0.353} \qquad (7\text{-}10)$$

Where:

c = Factor from Table 7-13
X = Delivered cost of one foot of 2 inch schedule 40 carbon steel pipe, dollars
e = Factor from Table 7-13
d = Factor from Table 7-13
L_r = Field labor rate, dollars per hour, including, fringe benefits, overhead and profit
E = Labor efficiency, fractional

c = 1.86 from Table 7-13, carbon steel, schedule 40
X = $ 1.05, given
e = 1.00, from Table 7-13
d = 0.600, from Table 7-13
L_r = $3.15, given
E = 0.70 given

Then: $F_r = 2.42/(1.86 \times 1.05 \times 1.00 + 0.600 \times 3.15/0.70)^{0.353}$

$F_r = 1.41$

The correction for fuel cost obtained from Table 7-8 is 0.86.

The correction for steam or gas engine drivers is 0.92 as defined at the beginning of this chapter.

The correction for 12 year amortization, from Table 7-9, is 1.07.

The overall factor, F_r, by which *all* flow rates will be multiplied before entry into the nomographs will be the product of the series determined:

$$F_r = 1.41 \times 0.86 \times 0.92 \times 1.07 = 1.19$$

DISCUSSION:

This will result in the optimum pipe size being 9% larger than a selection made using the typical values of economic parameters used throughout this book.

If such sizes were used, the amortized cost of the plant would be distinctly higher than if the optimum sizes, using the applicable parameters, were employed.

The difference in the total annual operating costs of a pipe line, when the correct size selection has been made using actual bracketed parameter data versus the assumed "typical" values used in the book, can be shown by an adaptation of Equation (2-8), the formula for the total annual cost of one foot of pipe.

$$C_r = (F_d^{1.5} + 1/F_d^{4.84})/2 \quad (8\text{-}2)$$

Where:

C_r = Cost ratio, fractional
F_d = Correction to the economic pipe diameter, per Equation (7-9)

For the Example 8-8, above, the cost ratio would be Equation (7-9):

$$F_d = 1.53 / (cXe + dL_r/E)^{0.169}$$

Where:

F_d = Correction to the calculated LAC pipe size
c = Factor from Table 7-13
X = Cost in dollars per foot of 2 inch schedule 40 carbon steel pipe delivered to the plant site
e = Factor from Table 7-13
d = Factor from Table 7-13
L_r = Field labor rate, dollars per hour, at plant site
E = Labor efficiency, fractional
$F_d = 1.53/(1.86 \times 1.05 \times 1.00 + 0.60 \times 3.15/0.7)^{0.169}$
$F_d = 1.18$

Equation (8-2):

$$C_r = (1.18^{1.5} + 1 / 1.18^{4.84}) / 2 = 0.86$$

Thus it may be seen that the *correct* size, even though *larger* than the size which would be incorrectly obtained using Equation (2-17) without the required correction, will have a total annual cost which is 14% *less*.

In Example 8-8, the increase in size required by the lowered cost of the pipe installation, partially is offset by the lowered fuel value. Hence, the correct economic selection would result in about 75% of the sizes being one larger than if the straight book values were used. Note, however, the difference in size, in some cases, can be more than shown in this example.

EXAMPLE 8-9:

A plant is being erected in Alaska. Materials are purchased in Northern California where the cost of 2 inch schedule 40 carbon steel pipe is \$1.32 per foot, but by the time it is shipped to the job site, the delivered cost is \$1.75. The labor rate is \$20 per hour and the efficiency of labor is 70%. The value of fuel is \$0.40 per Therm. Amortization is 7 years. What is the overall value of F_d to be used to adjust the pipe diameter determined by using LAC equations?

SOLUTION:

The correction for deviation of labor and material costs is calculated using Equation (7-9).

Equation (7-9):

$$F_d = 1.53 / (cXe + dL_r / E)^{0.169}$$

Where:
F_d = Correction to the calculated LAC pipe size
c = Factor from Table 7-13
X = Cost in dollars per foot of 2 inch schedule 40 carbon steel pipe delivered to the plant site
e = Factor from Table 7-13
d = Factor from Table 7-13
L_r = Field labor rate, dollars per hour, at plant site
E = Labor efficiency, fractional

Substituting actual values:
c = 1.86, Table 7-13
X = \$1.75, given
e = 1.00, Table 7-13
d = 0.60, Table 7-13
L_r = \$20, given
E = 0.70, given
$F_d = 1.53 / (1.86 \times 1.75 \times 1.00 + 0.60 \times 20 / 0.70)^{0.169}$
$F_d = 0.92$

From Table 7-7, the effect of fuel cost, F_d, is 0.91.

Multiply 0.92 by 0.91 to obtain 0.84, total F_d, the factor to be applied to the calculated pipe diameter. Thus, the optimum size would be 16% smaller than that which would be selected if typical values in this book were used. If the latter sizes *were* to be used, the amortised total cost for the pipe would be 19% higher than with the optimum size, using the applicable parameters.

Examples 8-8 and 8-9 demonstrate the economic value of using the actual working parameters for installations which lie outside the average economic sphere.

LAC PIPE SIZES FOR RECIPROCATING PUMP DISCHARGE

The pattern of discharge from reciprocating pumps is periodic; thus the average flow and the peak flow are different. For a simplex pump, the mean peak flow is 1.62 times the nominal discharge. For a duplex, it is 1.28 times the nominal.

In the selection of pipe sizes on the *LAC basis,* to allow for the irregularity of flow from such pumps, the nominal flow rate must be increased by 1.18 for simplex and 1.09 for duplex pumps before entering the flow rate into *nomographs* or *Tables 8-1* to *8-5.*

When using the *economic Equation* (2-17), the nominal flow is increased by the discharge mean flow factors, 1.28 for the duplex and 1.62 for the simplex.

EXAMPLE 8-10:

The reflux line for a tower operating at 75 psig is fed by a reciprocating simplex steam pump. At the flow temperature of 100° F, the flow rate is 184 gpm and the specific gravity is 0.92. Choose the LAC pipe size.

SOLUTION:

With no special mention, one can assume the economic parameters follow the typical used in this book. The pressure and temperature call for schedule 40 pipe and low flange rating.

The correction to the flow rate for a simplex pump, if one is using the LAC Tables, is 1.18.

$$1.18 \times 184 = 217$$

From Table 8-1 right column, 3 inch pipe will carry 234 gpm and a 4 inch 488, use 3 inch.

EXAMPLE 8-11:

A duplex hot oil pump is discharging 125 gpm of a California fraction at 620° F, with a hot specific gravity of 0.67, against a pressure of 900 psig. The pipe is schedule 80 with 900 ASA flange rating. Pipe size is to be selected using equations rather than nomographs.

SOLUTION:

The discharge of a pump calls for LAC treatment.
From Graph 9, APPROXIMATE VISCOSITIES OF STRAIGHT RUN FRACTIONS, find kinematic viscosity to be 0.63. Multiply by 0.67, the hot specific gravity, to obtain the viscosity in centipoise.

$$0.67 \times 0.63 = 0.42$$

The rate of flow for a reciprocating pump must be corrected, as stated earlier in this chapter, by 1.28.

$$1.28 \times 125 = 160$$

LAC selection is made using:

Equation (2-17):

$$D_c = 0.276\, Q^{0.479}\, S^{0.142}\, \mu^{0.027}$$

Where:

D_c = Economic internal pipe diameter, inches, at the cross-over point for commercial pipe sizes, Table 2-4

Q = Rate of flow, U.S. gpm at flow temperature

S = Specific gravity at the flowing temperature, water at 60° F equals 1.00

μ = Viscosity at the flowing temperature, centipoise

$$D_c = 0.276 \times 160^{0.479} \times 0.67^{0.142} \times 0.42^{0.027}$$

$$D_c = 2.90$$

Correct D_c by F_d for the 900 ASA flange rating, Table 7-14,

$$0.93 \times 2.90 = 2.70$$

D_c is the crossover diameter for the economic size, in inches, for commercial pipe sizes. From Table 2-4, 2.5 inch schedule 80 pipe has a D_c equal to 2.65. Three inch D_c is 3.43. Use 3 inch pipe.

LAC PIPE SIZES FOR GASES

In a system wherein the gas pressure has been developed by compression, all pipe sizes would be chosen by the LAC method. This would apply where ever the gas is to be conducted: to coolers, scrubbers, absorption towers, treaters, etc.

In the direct discharge pipe branches from reciprocating compressors, the flow will be irregular and an accounting for this must be made, as noted below. After several gas streams from reciprocating equipment have been blended, the flow will be smooth and a correction for pulsative flow will not be needed.

In the discharge branches from centrifugal compressors, the flow will also be smooth and require no special handling.

LAC PIPE SIZES FOR COMPRESSOR DISCHARGE

For centrifugal compressors, wherein the flow rate is essentially steady, the LAC pipe size is chosen using Graphs 35 through 38.

These are constructed using the typical values for the bracketed parameters in the formulae and assume gas powered drivers. If the bracketed economic values for a plant are different from the typical, tables in Chapter 7 provide the factors by which the actual flow rates are to be multiplied before entry into the several nomographs, which are provided for different pipe schedules an associated flange ratings.

If formulae are to be used for LAC selection, the pipe size is determined by using Equation (2-20) as developed in Chapter 2.

Equation (2-20):

$$D_c = 0.0615\ M^{0.479}\ G^{0.142}\ \mu^{0.027}\ \left(\frac{TZ}{P}\right)^{0.337}$$

Where:

D_c = Economic internal pipe diameter, inches, at the crossover point for commercial pipe sizes, see Table 2-4
M = *Thousands* of standard cubic feet per day at 60° F and 14.69 psia
G = Specific gravity of gas, air at 14.69 psia and 60° F equals 1.00
T = Gas flowing temperature, degrees Rankine (°F+460)
μ = Gas viscosity at flowing temperature, centipoise
Z = Deviation from perfect gas law, fractional
P = Pressure, pounds per square inch, absolute

For electrically driven compressors, using the value of electric energy at 0.0218 $/kwhr, the constant becomes 0.0666 rather than 0.0615.

EXAMPLE 8-12:

In a domestic plant located in California, 20,000,000 standard cubic feet per day of gas with a specific gravity of 0.71, temperature of 215 °F and 2000 psig are being handled downstream of centrifugal compressors. Pipe schedule is 160 and flange rating is 1500 ASA. What is the optimum pipe size?

SOLUTION:

Because the gas is being compressed, LAC is the proper mode. With a centrifugal compressor, the flow will be steady.

Using the nomograph approach, choose Graph 37, LAC GAS SIZE, 1500 ASA, to match the pipe schedule and the flange rating. This is based upon gas with a specific gravity of 0.70, thus no gravity correction is needed. Align 20,000 on the "Q" Scale with 215 on the "T" Scale and locate a turning point of 1.7 in the "A" Scale. Align 1.7 on the "A" Scale with 2000 on the "P" Scale and locate pipe size on Scale "D" as lying between 4 and 6. Use 6 inch.

EXAMPLE 8-13:

In a gasoline plant in Texas, gas from the well head is being stripped of gasoline components. Downstream of the inlet scrubber, the gas goes to the stripper column. The flow is 32,000 mcf/d at 600 psig and 100° F with a specific gravity of 1.25. The stripped gas will be delivered to a commercial pipe line. What is the proper size?

SOLUTION:

The pressure and temperature call for schedule 40 pipe and 600 ASA flange rating. Because the plant is in the U.S. and nothing is noted about amortization, etc., consider that no special economic parameters apply.

Gas in a commercial pipe line is, or will be, compressed. Hence, LAC applies, even though no actual compressing is involved in direct fashion.

Graph 35 applies for LAC gas selection for schedule 40 pipe and 600 ASA flange rating. The specific gravity differs from the norm used in the graph, 0.70. Thus a correction to the flow rate is needed. From the scale in the lower right corner of Graph 35, read the correction as 1.17.

$$32{,}000 \times 1.17 = 37{,}440$$

The pressure and temperature conditions offer the possibility that the deviation from the perfect gas law, Z, could be significant. See Chapter 3. From Figure 6-5 find for a gas of specific gravity 1.25 that T_c is 538° R, and F_c is 705 psia. The reduced temperature, T_r, is found by dividing the actual temperature (degrees Rankine) by the T_c and the reduced pressure, P_r, is found by dividing the actual absolute pressure by the P_c.

$T_r = (100 + 460) / 538 = 1.04$
$P_r = 615/705 = 0.87$

Using Figure 3-4, find Z is 0.66; this is a large deviation from the perfect gas law and must be considered. From Table 7-10 A find F_r is 0.75. This is applied to 37,440 to yield 28,080. Align this value on Scale "Q" with 100 on Scale "*T*" to obtain the turning point Scale "*A*". Align this with 600 on Scale "*P*" to find the size lies between 4 and 6 on Scale "*D*". Select 6 inch.

If the influence of Z had been overlooked, the pipe selected would have been too large, 8 inch rather than 6 inch.

Alternately, the solution to Example 8-13 could be found using

Equation (2-20):

$$D_c = 0.0615\ M^{0.479}\ G^{0.142}\ \mu^{0.027}\ \left(\frac{TZ}{P}\right)^{0.337}$$

Where:

D_c = Economic internal pipe diameter, inches, at the cross-over point for commercial pipe sizes, see Table 2-4

M = *Thousands* of standard cubic feet per day at 60° F and 14.69 psia

G = Specific gravity of gas, air at 14.69 psia and 60° F equals 1.00

μ = Gas viscosity at flowing temperature, centipoise

T = Temperature of gas, degrees Rankine (°F+460)

Z = Deviation from perfect gas law, fractional

P = Gas pressure, pounds per square inch, absolute

The deviation from the perfect gas law, Z, is found as described above.

From Figure 3-3 obtain viscosity for the gas as 0.0115

$$D_c = 0.0615 \times 32{,}000^{0.479} \times 1.250^{.142} \times 0.0115^{0.027} \times \frac{(560 \times 0.66)^{0.337}}{615}$$

$D_c = 6.82$

From the Table 2-4, for schedule 40 pipe, 6 inch has a D_c of 7.16 which is larger than the calculated value. 4 inch has a D_c of 5.15, less than the calculated value. Hence the correct size is 6 inch.

LAC PIPE SIZES FOR RECIPROCATING COMPRESSORS

The pattern of gas flow discharge from a reciprocating compressor is discontinuous. The degree of irregularity, influenced by various factors, is discussed in Chapter 7. The rate of flow factor is a measure of the flow pattern. This factor is used to correct the nominal gas flow rate before entry into the LAC pipe size nomographs.

The number of cylinders served by the pipe will effect the rate of flow factor. For the pipe between a volume bottle and one cylinder, the rate factor is determined using curve "*C*". The pipe between two cylinders and one bottle requires the use of curve "*D*". Curve "*E*" applies when three or more cylinders discharge through one pipe to a bottle.

The suggested capacity for volume bottles for pressures up to 500 psi is 5 cylinder volumes. Between 500 and 1500 psi, it is 12 cylinder volumes. From 1500 to 4000 psi, 18 volumes.

If volume bottles are not used, the discharge header size should be two pipe sizes larger than the LAC size to avoid impairment of compressor efficiency. Header size is *not reduced* as branches are taken from it to the machines.

Pipe beyond discharge headers will not require correction for the rate of flow factor.

If the discharge from one or two cylinders is piped direct to a cooler, vessel, or column, etc., the rate of flow factor must be applied before the pipe size is selected.

EXAMPLE 8-14:

In a gasoline plant being erected in West Texas, 4,250 mcf/d of a gas with a specific gravity of 0.62, at 120 psia and 75° F is being compressed to 360 psia with a gas driven, two cylinder, double acting compressor. The clearance is 11 %. Pipe size selection is to be made using a nomograph. What should be the size of the discharge line?

SOLUTION:

From Figure 7-5, using the compression ratio of 3, and the clearance (11 %), read efficiency of 35.5 %.

Using Figure 7-6, enter efficiency of 35.5 %, extend to curve "*D*" for two cylinder double acting, and read 2.25 as the Rate of Flow Factor.

The discharge temperature is not given. Using Figure 7-7, enter compression ratio of 3 and read temperature ratio as 1.25. Suction temperature is given as 75° F which, in absolute units, is

$$75 + 460 = 535° \text{ R}$$

$$535 \times 1.25 = 669$$

Converting to degrees F, this becomes 209°.

The nominal rate of flow is given as 4,250 mcf/d. This is multiplied by the Rate of Flow Factor of 2.25 to yield 9,563.

The parameters within the brackets for the LAC equation lie within the typical range.

At the pressure and temperature under consideration, schedule 40 pipe with 300 ASA flanges would apply. Therefore, use Graph 35, LAC GAS SIZE, to 600 ASA.

The specific gravitiy of the gas, being less than 0.7, the basis for Graph 35, the specific gravity correction will be used. For 0.62, the factor is 0.965. The previously corrected flow rate, 9,563, multiplied by 0.965 is 9,228.

Align 9,228 on "Q" Scale on Graph 35 with 209 on the "*T*" Scale, locate 2.48 on Scale "*A*". Align that value with *gauge* pressure of 345 on the "*P*" Scale. Read pipe size between 4 and 6 on the "*D*" Scale. Correct size is 6 inch.

Figure 8-2

Compressor Piping

CHAPTER 9

PIPE SIZE SELECTION FOR PUMP OR COMPRESSOR SUCTIONS

INTRODUCTION

The methods of selection for compressor or pump suction piping may be LAC, PDA, or VA depending upon operating conditions. The details of these methods are discussed in Chapters 1 through 4. As with discharge piping, see Chapter 8, the requirements are different for reciprocating versus centrifugal apparatus. With the former, the instantaneous rate of flow, larger by varying degrees than the average, is of prime importance in pipe size selection. For pump suctions, the volatility and/or amount of gas saturation strongly influences the correct size.

PUMP SUCTIONS

For cold, non-gas-saturated liquids, where ample head exits, such as ten feet of liquid, the LAC approach, using Table 8-1, is applicable. However, the operating conditions for most pump suctions are *less favorable* and call for larger pipe sizes than LAC would provide.

For many applications, it is possible that gas or vapor may be released from the liquid as the pressure on the system is reduced, whether caused by the effects of velocity head (Chapter 4), or by piping pressure losses. Pipe size can be selected according to PDA, allowing for the NPSH* requirement of the pump.

However, extensive experience has provided a practical method of determination of the carrying capacities of various pipe sizes for suction purposes. Table 9-1 provides the capacity for various pipe sizes for cold liquids that are non-gas-saturated and where the normal liquid head is available.

Separate columns are supplied for reciprocating pumps to allow for the instantaneous flow rate which is greater than the nominal by factors given in Chapter 8.

Figure 9-1
Pump Suction Piping

Equilibrium liquids are at the boiling point or at the condensing point. For these, the suction pipe size selection must be more conservative because a reduction in pressure through pipe friction can cause the liquid to vaporize, producing small bubbles in the stream. To a lesser extent, this can occur in gas saturated cold liquids, such as cooling tower water.

* Net positive suction head.

TABLE 9-1
Pump Suction Sizes
(Non-Equilibrium, Non-Gas-Saturated, Cold Liquids)

	Sched. 40 to 12 inch Sched. 10 for larger			Sched. 80 to 6 inch Sched. 100 for larger		
Size (inches)	Centrifugal (gpm)	Duplex (gpm)	Simplex (gpm)	Centrifugal (gpm)	Duplex (gpm)	Simplex (gpm)
24	7750	6300	4800			
20	5260	4150	3400	4100	3200	2550
18	4150	3400	2550	3300	2650	2150
16	3250	2550	2000	2500	1950	1550
14	2300	1850	1450	1900	1500	1175
12	1900	1500	1175	1500	1175	925
10	1200	940	710	1000	780	640
8	720	570	440	620	480	375
6	375	300	240	340	270	225
4	155	120	94	140	110	84
3	74	58	46	65	51	40
2.5	42	34	27	37	29	23
2	24	19	15	22	17	14
1.5	14	11		12		

Pump manufacturers must be aware of the possibility of bubble formation through pressure reduction at the eye of a centrifugal pump. Because of differences in pump inlet design, the amount of velocity increase, and therefore the pressure reduction caused by it, will vary.

The specifications of a given pump will include the designation of the NPSH (net positive suction head) required for satisfactory pump operation. NPSH simply means the pressure, expressed in *feet of liquid* at the inlet center line, required at the pump suction, *above* the equilibrium or vapor pressure of the liquid *at the pumping temperature*, to insure proper pump performance. If the head is less than called for, due either to insufficient depth of liquid ahead of the pump, or due to excessive pressure loss, or velocity increase, in the pipe, serious consequences can result. As the liquid begins to bubble (boil), cavitation will develop. The bubbles of vapor will collapse as the velocity pattern within the pump decreases. When this occurs against an impellor blade or other surface, metal deformation may occur (similar to shot peaning) accompanied by loud sounds and vibration. The effect of cavitation, in addition to causing pump damage, seriously can reduce the pump capacity. It must be avoided.

For gas saturated liquids, or those near the boiling or condensing point, experience has established correct suction pipe sizes as expressed in Tables 9-2 and 9-3. These assume the use of normal practices as to depth of liquid above the suction pipe inlet, keeping the piping short and using minimum elbows and NO more than *one* run through the bull of a tee.

For pipe sizes larger or smaller than those listed in the tables, establish the velocity from Graph 33, VELOCITY FOR LIQUIDS, for the closest two pipe sizes, note the pattern of change, and select the logical velocity for the size desired. The flow rate can be found by choosing a size with the required velocity employing Graph 33.

EXAMPLE 9-1:

Cold water is being drawn from a storage tank, vented to the air, located 20 feet above grade. The flow rate is 650 gpm. The centrifugal pump is electrically driven. Select the suction pipe size.

SOLUTION:

Under the conditions given, ample head, cold water that is not air saturated, the selection would

TABLE 9-2
Cooling Tower Suction Pipes
(Schedule 40 to 12 inch, schedule 10 larger)

Size (inch)	Rate (gpm)	Size (inch)	Rate (gpm)	Size (inch)	Rate (gpm)
30	9000	16	2010	8	455
24	5200	14	1500	6	250
20	3500	12	1150	4	100
18	2750	10	750	3	51

TABLE 9-3
Pump Suction Pipe Sizes for Equilibrium Liquids

	Sched. 40 to 12 inch Sched. 10 larger			Sched. 80 to 6 inch Sched. 100 larger		
Size (inch)	Centrifugal (gpm)	Duplex (gpm)	Simplex (gpm)	Centrifugal (gpm)	Duplex (gpm)	Simplex (gpm)
30	7000	5400	4350	3200		
24	4650	3700	2900	2100	2500	1950
20	2800	2200	1725	1650	1650	1300
18	2200	1750	1375	1275	1300	1025
16	1625	1275	1000	920	1000	780
14	1200	975	750	740	720	565
12	950	750	590	510	580	460
10	625	500	390	300	400	330
8	375	295	225	170	235	185
6	200	155	125	70	135	108
4	84	66	52	36	54	46
3	45	35	28	22	28	22
2.5	27	22	17	15	17	13
2	17	12	10		12	
1.5	11					

be via LAC. With the driver specified as electric and no other special values for standard parameters, Table 8-1 is the simplest (and correct) method. A 4 inch can handle 450 gpm and a 6 inch 895. Use the 6 inch.

EXAMPLE 9-2:

A cooling tower suction line carries 1100 gpm. Select the correct size.

SOLUTION:

Table 9-2 specifies cooling tower suction pipe sizes. A 12 inch can carry 1150 gpm. Use 12 inch.

EXAMPLE 9-3:

A reciprocating simplex pump is handling 35 gpm of liquid hydrocarbon reflux, with specific gravity of 0.58. The rate of flow and specific gravity are given at 60° F but the flow temperature is 120° F. Select the correct suction pipe size.

SOLUTION:

The flow rate must be corrected for the expansion of the liquid at the flow temperature. From Figure 6-3 determine that the flowing specific gravity is 0.54. The flowing rate becomes

$$0.58/0.54 \times 35 = 37.6$$

Reflux will be at or near equilibrium condition, therefore, Table 9-3 will apply. A 4 inch will carry 52 gpm to a simplex pump. A 3 inch will carry 28. Use the 4 inch.

Observe that if the pump had been centrifugal, the capacity of a 3 inch, 45 gpm, would have been ample. The pulsating character of the flow into the simplex pump requires the use of the larger pipe.

COMPRESSOR SUCTION PIPING — RECIPROCATING

The selection of reciprocating compressor piping depends upon the number of cylinders being served by the pipe, the compression ratio and the compressor effficiency.

A compressor cylinder does not intake gas during the entire suction stroke. Therefore, the momentary average rate of flow will be higher than the nominal. When one cylinder, only, is served by a pipe, the momentary average rate of flow, which must be used in the selection, can be from 2.5 to 3 times the mean flow rate. The method for determining the momentary average flow rate is given in Chapter 7, using Figure 7-6.

The number of cylinders served by the pipe affects the rate factor. The size pipe from a volume bottle leading to one double acting cylinder is determined from Curve "*C*" to establish the rate factor. If two cylinders take suction from a common volume bottle, the pipe connecting the bottle and the header is selected using Curve "*D*" to find the factor. If three or four cylinders are involved, Curve "*E*" is used.

Pipe serving several machines would have a factor of 1.00. Thus the pipe to a compressor battery would require no correction.

Curves "*A*" and "*B*" are for single acting cylinder machines.

If volume bottles are used, the header should be selected by LAC, allowing for the rate of flow factor. No reduction in size is made following the take-offs for each machine.

The capacity for volume bottles for pressure up to 500 psi is 7 cylinder volumes on the suction. For 500 to 1500 psi, use 12 cylinder volumes. For pressures up to 4000 psi, use 18 volumes.

If volume bottles are not used, the header size should be two larger than the LAC.

Except as otherwise noted, suction piping, with due correction for the rate of flow factor, is made using LAC methods.

EXAMPLE 9-4:

A battery of compressors delivers gas of specific gravity 0.73 at the rate of 19,200 mcf/d. The suction pressure is 175 psig; the gas is at 75° F. Value of fuel and related parameters are normal, volume bottles are not used. The rate of flow factor is 2.3. Determine the pipe size to the battery, the header size, and the size of the eight suction branches.

SOLUTION:

For the pipe to the battery, the flow factor is 1.00. The mode of selection would be LAC. At the pressure and temperature given, Graph 35, LAC GAS SIZE 600 lb ASA, would apply.

The specific gravity of 0.73 is so close to 0.70, the basis for the Graph, that no gravity correction is needed. Align 19,200 on the "Q" Scale with 75 on the "*T*" Scale to locate 1.87 on the "*A*" Scale. Align this with 175 on the "*P*" Scale to locate pipe size on the "*D*" Scale. The size falls between 6 and 8 on the "*D*" Scale. Use 8 inch for the line to the header.

With no volume bottles being used, the suction header should be two sizes larger than LAC so use 12 inch.

For the branches, the flow rate is divided by 8 and multiplied by 2.3, the rate of flow factor, to yield 5,520. Using Graph 35 as just described, determine the branch sizes to be 4 inch for each.

If this should be less than the suction nozzle size, consultation with the manufacturer may be warranted. The larger size may well be due to manufacturing simplification or it may represent the pipe size need of the machine at the specified load.

CHAPTER 10

PIPE SIZE SELECTION FOR PRESSURE DROP ALLOWABLE

Chapter 3 discusses the derivation of pressure drop formulae and their usage. This Chapter will cover the practical applications of these concepts. Examples will be provided, as well, of computations of pressure drop which will occur in pipe size selections made by modes other than PDA.

Fundamentally, this mode of selection is based upon the consumption of a specified pressure loss through a particular length of pipe. However, as discussed in Chapter 3, typical pipe systems comprise predictable equivalent lengths, resulting from the influences of the fittings as well as the actual pipe length, so satisfactory selection of pipe size can be made *without the need for exact information about the fittings in each pipe.*

Except in rare cases, PDA is *not* the mode of pipe size selection to be used for discharge piping for *compressors* or *pumps.* For these, LAC is the correct mode.

Examples of circumstances where pressure drop available *is* the correct approach may help to clarify when PDA applies.

EXAMPLE 10-1:

1100 gpm of water at 65° F are to flow from a tank with a liquid surface 15 feet above the level in the cooling tower basin. There are five bends and a control valve in the system. The control valve absorbs 4.3 psi. The length of pipe is 125 feet. Determine the correct pipe size.

SOLUTION:

The available pressure drop (from the head of water) in psi is

$$15 \times 0.433 = 6.5 \text{ psi}$$

The pressure loss through the control valve is to be deducted, so the amount remaining for drop through the pipe is

$$6.5 - 4.3 = 2.2$$

The equivalent length of the fittings will be substantial, perhaps as much as the pipe. Assume the actual length plus equivalent length is 250 feet

$$2.2 \times 100/250 = 0.88$$

Therefore, the pressure loss per *hundred feet* of pipe must be less than 0.88 psi.

Using Graph 29, PRESSURE LOSS FOR WATER, align 1100 on the rate Scale with 0.88 on the C-125 Scale; you will find the size is a bit over 8 inch on the diameter scale. The pressure drop per 100 feet for 10 inch pipe is 0.34 psi/100. Try 10 inch.

From Table 6-1 determine that the factor for equivalent length of a long radius welding elbow is 1.2, for an entrance it is also 1.2, while for an exit, it is 2.4. Thus the equivalent length of the fittings is computed as:

Elbows: 1.2 x 5 x 10	=	60
Entrance: 1.2 x 10	=	12
Exit: 2.4 x 10	=	24
Total:		96 feet

The total length (pipe plus fittings) would then be

$$125 + 96 = 221$$

The pressure loss in the 10 inch pipe would be

$$2.21 \times 0.34 = 0.75 \text{ psi}$$

This is less than the amount available so try 8 inch pipe.

The equivalent length for the fittings is:

Elbows: 1.2 x 5 x 8	=	48
Entrance: 1.2 x 8	=	9.6
Exit: 2.4 x 8	=	19.2
Total:		76.8 feet

The sum of the equivalent length and the actual pipe is

$$125 + 76.8 = 201.8 \text{ feet}$$

Using Graph 29, align 1100 on the rate Scale with 8 inch on the "*D*" Scale to find pressure loss per hundred feet is 1.01. The total pressure drop in the pipe and fittings will be

$$2.018 \times 1.01 = 2.04$$

versus the 2.2 available. Therefore, the 8 inch is the correct size.

EXAMPLE 10-2:

Pentane is being conducted to a low pressure treater, from which it goes to storage. The specific gravity at 60° F is 0.63. The flowing temperature is 90° F. The pressure in the cooler is 35 psi and in the treater is 25 psi. The pressure drop through the liquid level control valve at the cooler is 8.0 psi. Determine the proper pipe size when the flow rate is 167 gpm.

SOLUTION:

The flow rate must be corrected for the actual temperature. Use Figure 6-3 to find the specific gravity at 90° F is 0.615. The flow at actual temperature is

$$167 \times 0.63/0.615 = 171$$

(When the flowing temperature is substantially above atmospheric, the correction to the flow rate will become appreciable compared to this example.)

The kinematic viscosity at the flow temperature is found from Graph 8. Align 90 on the temperature Scale with pentane to read 0.35 centistokes. Multiply by the specific gravity at 90° F, 0.615, to obtain absolute viscosity in centipoise

$$0.615 \times 0.35 = 0.22$$

The pressure drop assigned to the pipe is the pressure in the cooler less the pressure in the treater, less the pressure drop in the control valve

$$35 - 25 - 8 = 2 \text{ psi}$$

The pipe size is found by using Graphs 51, SIMPLIFIED PDA—WATER, and 52, SPECIFIC GRAVITY—VISCOSITY CORRECTION. First determine from Graph 52 the influence of the viscosity and specific gravity of the liquid. Align 0.22 on the "*V*" Scale with 0.615 on the "*S*" Scale to find the value of "C" as 0.70. This indicates that the actual liquid is equal in fluid flow characteristics to 0.70 times that of water. Align the 0.70 on the "C" Scale with 171 on the "Q" Scale to locate the turning point on the "*A*" Scale. Rotate about the turning point to 1.0 on the "C" Scale and read the *corrected flow rate* on the "Q" Scale as 118.

Transfer 118 to the "Q" Scale on Graph 51, and align with 2 (the pressure drop assigned to the pipe) on the "*F*" Scale. Read the correct pipe size on "*D*" Scale as 2.78 inches. Use 3 inch.

EXAMPLE 10-3:

What will be the pressure drop per 100 feet in the 3 inch pipe selected for Example 10-2, above?

SOLUTION:

Using Graph 29, PRESSURE LOSS FOR WATER, align 171 on the "Q" Scale with 3 inch schedule 40 pipe on the "*D*" Scale to read 3.5 on the C-125 Scale. Using Graph 30, VISCOSITY—SPECIFIC GRAVITY CORRECTION, align 0.21, the viscosity, on the "V" Scale with 0.615 on the "S" Scale to read 0.475 on the "F" Scale. Align 0.475 on the "F" Scale with (the value found from Graph 29). Rotate about the value of 5.14 on the "A" Scale to 1.0 on the "F" Scale to read 1.65 on the "P" Scale which is the pressure drop per 100 feet in the 3 inch line.

EXAMPLE 10-4:

For the pipe system in Example 10-3, what will be the total pressure drop if the system consists of 41 feet of pipe, 10 long radius welding elbows, one cock and two tees having the flow with the run?

SOLUTION:

From Table 6-1 select the equivalent length factors for the various fittings and multiply by the pipe size and the quantity:

Elbows: 10 x 1.2 x 3	=	36
Tees: 2 x 1.7 x 3	=	10.2
Cock: 1.0 x 3	=	3
Total:		49.2 feet

Add the straight pipe

$$41 + 49.2 = 90.2 \text{ feet}$$

The product of the number of 100s of feet of pipe (equivalent total) and the pressure drop per 100 feet yields the total pressure loss in the pipe

$$0.902 \times 1.65 = 1.49 \text{ psi}$$

It will be noted that the actual estimated pressure drop is well below the allowable value, 2.0, defined in Example 10-2, above.

If the design conditions for Example 10-2 had called for the pentane to be sent to storage, rather than to a treater, wherein the distance would require 400 feet of pipe rather than the typical encountered in a plant area, the use of Graphs 51, SIMPLIFIED PDA—WATER, and 52, SPECIFIC GRAVITY—VISCOSITY CORRECTION, would not have been proper. Rather, the equivalent length of the total system, including typical fittings, could be estimated from Equation (2-13) and the allowable pressure drop per hundred feet estimated. Using the information from Graphs 51 and 52 as a starting point for the pipe size, determine the probable drop in the total line using Graphs 29, PRESSURE DROP—WATER, and 30, SPECIFIC GRAVITY—VISCOSITY CORRECTION, as described in Example 10-3, above, to find the pressure drop per hundred feet. An iterative process could be required to determine the proper size.

It is worth noting that if such a system were being considered, the working conditions *probably* would have provided more available pressure drop in the pipe. As much as 12 psi might be available rather than 2, as specified in Example 10-2. To explore this effect, another example will be considered.

EXAMPLE 10-5:

Pentane at the flow rate in Example 10-2 is being sent to storage from the cooler. The pressure in the tank may be up to 15 psi. Using Graphs 51, SIMPLIFIED PDA-WATER, and 52, SPECIFIC GRAVITY—VISCOSITY CORRECTION, determine an approximate pipe size and check to see if this will be correct if the length of straight pipe is 405 feet with 10 elbows, two cocks and two tees, one via the bull and one via the run.

SOLUTION:

The pressure drop available in the pipe system will be

$$35 - 15 - 8 = 12 \text{ psi}$$

From Graphs 51 and 52 (used as described in Example 10-2, above), the pipe size is estimated at 2 inch. The pressure drop per 100 feet, from Graphs 29 and 30, as used in Example 10-3, above, is 6.8 psi.

Using Equation (2-13), determine the approximate total equivalent length for a typical 2 inch line with 405 feet of straight pipe, as 458 feet.

Equation (2-13):

$$L + L_e = CL^j$$
$$L + L_e = 1.94\, L^{0.91}\,*$$

Where:

L = Actual pipe length, feet

L_e = Equivalent pipe length of fittings, feet

$$L + L_e = 1.94 \times 405^{0.91} = 458$$

For 3 inch, the total equivalent length is calculated to be 457 feet, as follows

* Values of c and j from Table 2-5.

Elbow: 1.2 x 10 x 3	= 36
Tee, run 1.7 x 1 x 3	= 5.1
Tee, bull: 3.7 x 1 x 3	= 11.1
	52.2
Pipe:	405
Total:	457.2 feet

Hence, for the 2 inch pipe, the maximum allowable pressure loss per 100 feet, which would generate the required flow would be

$$12/4.57 = 2.63$$

The calculated pressure drop per 100 feet in the 2 inch at a flow of 171 gpm is 6.8 psi/100, far *above* the 2.6 available. Therefore, the 2 inch is too small.

The pressure drop in the 3 inch pipe, as calculated above, in Example 10-3, is 1.65 psi/100. Hence, the 3 inch is correct. The total pressure loss in the piping system, exclusive of the control valve, would be

$$1.65 \times 4.57 = 7.5 \text{ psi}$$

This is below the allowed 12 psi so the 3 inch is satisfactory.

EXAMPLE 10-6:

Hydrocarbon gases, of specific gravity 0.774, at a pressure of 430 psi and temperature 90° F, flow to a reabsorber at the rate of 4,673 mcf/d. The pressure in the reabsorber is 427 psi. The system contains 8 elbows and 183 feet of straight pipe. Determine the suitable pipe size.

DISCUSSION:

In such a process situation one can assume that compression is involved either upstream or downstream. Consequently, one could expect that the proper mode of selection would be LAC. However, when a process requirement *is specified,* to be certain the pressure in the reabsorber is no less than the process specifies, PDA should be used. For this example, straight PDA as well as the simplified approach will be used for comparison. Additionally, LAC will be tried as an experiment.

SOLUTION A:

Apply regular pressure drop computation. *Assume* the total equivalent length to be 220 feet. With 3 psi available, the pressure drop per 100 feet is 1.36. Because the gas is other than air, the calculated allowable drop must be adjusted using Graph 41 correction for Graphs 39 and 40. The correction factor is 0.74. The adjusted value of the allowable pressure drop to be used in Graph 40, pressure drop for gas, 100 to 6000 psi, is:

$$1.36 / 0.74 = 1.84 \text{ psi/100 feet}$$

Using Graph 40, align 4.673 (million) on the "Q" Scale with the average pressure between the two vessels, 428.5 psi, on the "*P*" Scale to locate 4.0 on Scale "*A*". Align 4.0 with 1.84 on the "*F*" Scale to find the size lies between 3 and 4 on the "*D*" Scale. Assume 4 is correct but check. Using Graphs 40 and 41 determine pressure drop per 100 feet to be 0.39 psi/100.

From Table 6-1, establish that the factor for equivalent length of an elbow is 1.2, for an entrance, 1.2 and for an exit, 2.4. Calculate the equivalent length of 4 inch.

Elbows: 8 x 1.2 x 4	= 38.4
Entrance: 1.2 x 4	= 4.8
Exit: 2.4 x 4	= 9.6
Total:	52.8 feet

Add the straight pipe to obtain 236 feet.
Total pressure drop in 4 inch pipe will be

$$0.39 \times 236/100 = 0.92 \text{ psi}$$

This being less than the allowed 3 psi, the 4 inch is the correct size.

SOLUTION B:

Use the LAC approach, with Graph 35, LAC GAS PIPE SIZE, 600 lb ASA.

The specific gravity being 0.774, the correction to the rate of flow is found on the "specific gravity correction" scale in the lower right corner to be 1.03. The corrected rate to be used with Graph 35 is

$$1.03 \times 4{,}673 = 4{,}810$$

Align 4.81 on the "Q" Scale with 90 on the "*T*" Scale and locate 3.05 on the "*A*" Scale. Align 3.05 on "*A*" with 428.5 on the "*P*" Scale to find pipe size on the "*D*" Scale lies between 3 and 4. Hence use 4 inch.

Thus one can observe that either conventional PDA pressure drop calculation or LAC provides the same answer *for this example.* This may not always be true.

DISCUSSION:

According to Equation (3-22 A), the total equivalent length of a typical 4 inch pipe system will be 122 feet. This is far less than the specified amount of straight pipe. Therefore, the simplified PDA approach, using Graphs 53, SIMPLIFIED PDA GAS, and 54, SPECIFIC GRAVITY TEMPERATURE CORRECTION, probably would not give correct results, *but* could be used for a first approximation, rather than the LAC method, as demonstrated above. This is shown in the next solution.

SOLUTION C:

Using Graphs 53 and 54, the simplified PDA size selection would be 3 inch. Calculating the pressure drop for this size, using Graphs 40 and 41, find the pressure drop in 3 inch is 1.45 psi/100. For the estimated total equivalent length of 240 feet, when the straight pipe is 183 feet, according to Equation (2-13), the total pressure drop would be 3.5 psi, well above the allowable 3 psi. Therefore, 4 inch is the correct size.

FUEL GAS PIPE SIZES

Fuel gas supply may be taken from a compressor suction. Even though the gas is being moved by compression, LAC does not apply if the pressure is being reduced to serve the engines. Rather, PDA is correct. The same can apply to gas feeding the burners of a boiler.

EXAMPLE 10-7:

Fuel gas, specific gravity 0.60, at a pressure of 60 psi, supplies the compressor house at the rate of 2,000 mcf/d. At the compressor fuel header, the pressure is dropped to 30 psi. From the compressor building layout, the straight pipe length appears to be no more than 70 feet.

SOLUTION:

Under the conditions outlined, 10% of the pressure reduction may be assigned to the pipe, that is 3 psi. Using Graph 54, SPECIFIC GRAVITY VISCOSITY CORRECTION, align 60 on the temperature scale with 0.60 on the specific gravity scale to locate correction factor of 0.66. Multiply the flow rate in millions of cubic feet by the factor

$$2.0 \times 0.66 = 1.32$$

Using Graph 53, SIMPLIFIED PDA—GAS, align 60 on the "*P*" Scale with 3 on the "*F*" Scale to determine turning point on the "*A*" Scale.

Rotate about the turning point to align 1.32 on the "*M*" Scale and find pipe size on the "*D*" Scale is 1.96. Use 2 inch.

DISCUSSION:

The close approach of the calculated size to the used size may appear to be risky. Recall, the pressure loss through the regulator is considerable. It can accomodate minor excess pressure loss in the pipe leading to the regulator. Therefore, the 2 inch size is a satisfactory selection.

OVERHEAD VAPOR PIPE SIZES

Overhead vapor lines from fractionating towers to condensers can represent a critical usage of PDA. The pressure drop normally should not exceed 0.5 psi/100 feet for systems from 15 psi to 500 psi. However, the limiting pressure loss is *that specified on the process flow sheet.* For vacuum systems, the allowable loss per 100 feet may be less than 0.5 psi, TOTAL.

The length of the straight pipe, which will likely be required, is to be estimated from the plot plan. The overall factors for fittings, etc., shown on Table 6-3 can be used in the absence of actual information. Graphs 39 or 40, PRESSURE DROP FOR GAS, along with 41, VISCOSITY—SPECIFIC GRAVITY—TEMPERATURE CORRECTION, are used to calculate the pressure loss per 100 feet. For a starter, a first approximation may be made using Graph 35, LAC GAS SIZE, to 600 lb ASA.

If the calculated size is larger than the column nozzle, prompt, written notification should be given to the Project Engineer!

When the overhead condenser is placed directly above the accumulator, the pipe size connecting them must be generous. The liquid should be flowing as a hollow tube around the walls of the pipe, allowing a channel for vapor flow inside. If not so arranged, a liquid column may form and, surging up and down, set up pressure fluctuations in the fractionator.

The Froude Number, discussed in Chapter 5, can assist in selecting the adequate size.

A rule, which is generously safe, calls for the return line to be one pipe size smaller than the overhead vapor line. However, if *less* than *half* of the vapor is condensed, the return line should be the *same* size as the vapor line.

When the condenser is placed in the coil shed of the cooling tower, the *overhead* line is sized as stated above. However, two conditions may exist for the *return* line to the accumulator. If complete condensation will occur, and a control valve is used in the vapor line, the combined pressure loss in the vapor and return lines should be no more than 10% of the pressure drop across the control valve. If no control valve is used, the pressure drop in the return line should be made essentially the same as in the vapor line.

When incomplete condensation is anticipated, and if vertical upward travel of the mixture will be required, the two-phase return line must be sized to provide ample velocity for droplet entrainment. See Chapter 5 for more details of this calculation.

REBOILER LINES

Reboiler lines, both vapor and liquid, must be sized with care because a *limited pressure differential* exits to cause flow. This is the difference in the liquid heads in the reboiler and the column. In the absence of definite information to the contrary, allow 4.5 inches of liquid head for the vapor line and the same for the liquid line. This applies to kettle-type reboilers. For thermosyphon style, more head may be available, but the return line must be calculated using two-phase flow methods, see Chapters 5 and 17.

COOLER AND REHEATER LINES

Cooler and reheater lines, on absorption towers or stills, are selected according to PDA, following somewhat the same methods as just described for reboiler pipes. A limited liquid head exists to drive the fluid(s) through the pipe and the apparatus. This head is produced by the two columns of fluid in the lines to and from the cooler or reheater.

For coolers, the difference in the specific gravities in the two pipes may not be significant and, in establishing the head, use the flowing specific gravity and the *difference* in *elevation* of the nozzles to determine the pressure drop available. The total, less the losses through the cooler, can be used to overcome fluid frictional losses in the piping.

For reheaters, extra care may be appropriate if some vaporization should occur in the heater, which will call for the pressure loss in the return piping to be calculated using two-phase flow concepts as set forth in Chapters 5 and 17.

The liquid head, which is the source of pressure driving the liquid (and possibly gases as well), through the reheater and the piping, is determined by using the *flowing specific gravity* in the supply line and the *difference* in the *elevation* of the outlet and inlet nozzles on the tower or column. The pressure loss available for the piping, the total head as calculated above, less the loss through the reheater, is to be divided essentially equally between the supply and return piping if two-phase conditions are present in the latter.

CHAPTER 11

STEAM PIPING FOR HEATING AND RECIPROCATING PUMPS

These pipes are selected by PDA, whether for steam to heaters or to drive reciprocating pumps. In most cases, a control valve regulates the steam flow.

HEATING PIPING

For steam heating applications, the pressure loss through the shell side of the apparatus will be minimal. Therefore, the procedure given in the Introduction, for establishing the ratio of the pressure drop through the piping as compared to that through the valve, need not be used. The general rule of assigning 10% of the pressure drop across the valve to the pressure drop through the piping is satisfactory.

The pressure drop through the valve is established by the temperature desired in the heater, which fixes the downstream pressure. The upstream is that of the steam supply system.

EXAMPLE 11-1:

The temperature in a reheater requires the shell steam pressure to be 115 psig. The steam supply pressure is 150 psig. The steam flow is 2800 pounds per hour. Select the pipe size beyond the control valve, and determine the pressure loss per hundred feet and the velocity.

SOLUTION:

The pressure drop across the valve will be 90% of 35 psi, or 31.5. This leaves 3.5 psi for the drop through the pipe. Use Graph 55, SIMPLIFIED PDA-STEAM, to select the size for the "typical" pipe. Align 3.5 on the "*F*" Scale with 115 on the "*P*" Scale and locate the turning point on the "*A*" Scale. Align the latter with 2.8 on the "*W*" Scale and find the size on the "*D*" Scale is between 2 inch and 1.5 inch, so select 2 inch.

Using Graph 49, PRESSURE DROP FOR STEAM, 5 PSIG TO 500 PSIG, align 115 on the "*P*" Scale with 2.8 on the "Q" Scale to find turning point at 3.15 on the "*A*" Scale. Align 3.15 with internal pipe size, 2.067 inches on the "*D*" Scale and read pressure drop on the "*F*" Scale as 4.6 psi per 100 feet.

Using Graph 50, VELOCITY FOR STEAM, align 115 on the "*P*" Scale with 2.8 on the "Q" Scale to locate turning point on "*A*" Scale. Align this with 2 inch schedule 40 mark on the "*D*" Scale to read velocity of 6300 feet per minute on the "*V*" Scale.

RECIPROCATING PUMP PIPING

For pipe size selection for steam to reciprocating pumps, refer to Chapter 7, DATA REQUIRED FOR RECIPROCATING PUMPS, to fix the steam chest pressure. The pressure loss in the pipe will be 10% of the difference between the header pressure and the steam chest pressure. The remainder is assigned to the control valve.

EXAMPLE 11-2:

The steam header pressure is 250 psig. A simplex reciprocating pump requires an average rate of 3500 pounds per hour of steam at 100 psig in the chest. Determine the pipe sizes to the control valve and beyond.

SOLUTION:

The total pressure loss in the pipe and the valve will be

$$250 - 100 = 150 \text{ psi}$$

To the pipe is assigned 10%, or 15 psi. Because of the considerable difference in steam pressures before and beyond the control valve, the two pipes probably will not be the same size. Therefore, one half of the total pressure loss will be assigned to each portion. The momentary flow rate for a simplex pump is 160% of the average which also applies to the steam flow, as well. Therefore the rate used to select the size will be

$$3500 \times 1.6 = 5{,}600$$

The sizes will be established using Graph 55, SIMPLIFIED PDA, STEAM.

For the upstream portion, align 7.5 on the "*F*" Scale with 250 on the "*P*" Scale to find the turning point on the "*A*" Scale. Align this with 5.6 on the "*W*" Scale to find the size on the "*D*" Scale lies between 1.5 inch and 2.0 inch. Use 2.0 inch.

For the downstream portion, also using Graph 55, align 7.5 on the "*F*" Scale with 100 on the "*P*" Scale to locate the turning point on the "*A*" Scale. Align the latter with 5.6 on the "*W*" Scale to find the pipe on the "*D*" Scale lies between 2.0 inch and 2.5 inch. Use 2.5 inch.

The size selection for the exhaust steam from the pump also is made according to PDA, using the same pressure loss assigned to the supply pipe. Graph 55 is used.

EXAMPLE 11-3:

The simplex pump described in Example 11-2 is exhausting to a header with 15 psig pressure. Select the pipe size.

SOLUTION:

The allowable pressure loss will be 7.5 psi, as before. The momentary flow rate for pipe size selection will be 5600 pounds per hour, as previously established. Using Graph 55, align 7.5 on the "*F*" Scale with 15 on the "*P*" Scale to locate the turning point on the "*A*" Scale. Align this with 5.6 on the "*W*" Scale to find the pipe size on the "*D*" Scale lies between 3 and 4 inch. Use 4 inch.

CHAPTER 12

STEAM PIPING FOR TURBINES

The headers and branch lines for steam turbines, both supply and exhaust, are selected according to the LAC approach. In Chapter 2, the theory and development of the concepts and formulae are given. The pipe size selection method will depend upon whether or not the residual heat in the steam will, or will not, be used for heating. Table 12-1 illustrates the degree of difference in pipe size for the two conditions.

As demonstrated by Table 2-9, under the condition where the heat in the residue steam will be used, the *supply* and *exhaust* pipe sizes will be *different* when the *ratio* of the pressures to and from the turbine *changes.* Such is not the case when the heat is *not* used. Therefore, the relationship between the steam pipe size and the use, or not, of the heat, is *not* solely a function of the steam supply (or exhaust) *pressure.* Rather, it depends upon the ratio of the two pressures, as is shown by Table 12-1. For 100 psig supply (with atmospheric exhaust) the size is larger when the heat is used. For the 300 psig supply condition (and atmospheric exhaust), the reverse is true. Under conditions where the supply would be 100 psig but the exhaust was to be a higher pressure, the correct size, when the heat was to be used, would be smaller than where it would not.

For pipe size selection via nomographs, the following are used:

LAC EXHAUST STEAM *USED* FOR HEATING — Graph 46

LAC EXHAUST STEAM *NOT USED* FOR HEATING — Graph 47

PRESSURE DROP FOR STEAM, VAC. TO 5 PSIG — Graph 48

PRESSURE DROP FOR STEAM, 5 PSIG TO 500 PSIG — Graph 49

VELOCITY OF STEAM — Graph 50

EXAMPLE 12-1:

Saturated steam at 400 psig is supplied to a turbine at the rate of 35,000 pounds per hour. The exhaust pressure is 65 psig. Select the correct supply pipe size and the correct exhaust pipe size.

SOLUTION:

With the exhaust pressure at the given level, the steam will be used for some other purpose, probably heating. Equations (2-28) and (2-29) apply, or if preferred, Graph 46, LAC EXHAUST STEAM *USED* FOR HEATING, may be employed.

Equation (2-28):

$$D_s = 11.7\ P_B^{0.103}\ W^{0.543} / P^{0.495}$$

Where:

D_s = Supply pipe internal diameter, inches
P_B = Steam back pressure, pounds per square inch, absolute
W = Steam flow rate, *thousands* of pounds per hour
P = Steam supply pressure, pounds per square inch, absolute

TABLE 12-1
Economic Pipe Size for 10,000 Pounds per Hour of Saturated Steam

Supply (psig)	Exhaust (psig)	PIPE SIZE: Exhaust heat used, Supply (inches)	PIPE SIZE: Exhaust heat used, Exhaust (inches)	PIPE SIZE: Exhaust heat not used, Supply (inches)	PIPE SIZE: Exhaust heat not used, Exhaust (inches)
300	0	3.13	6.17	3.50	9.34
100	0	5.16	8.31	4.86	9.34

For the supply pipe size, use

$$D_s = 11.7 \text{ x } (65+14.7)^{0.103} \text{ x } 35^{0.543}/(400+14.7)^{0.495}$$

$$D_s = 11.7 \text{ x } 1.57 \text{ x } 6.89/19.76 = 6.41$$

From Table 2-4, the crossover diameters for various commercial pipe sizes in schedule 40 weight are compared with the calculated size of 6.41 inches. For 4 inch, the value is 5.15, for 6 inch it is 7.16. Therefore, use 6 inch.

Equation (2-29):

$$D_e = 13.8\ W^{0.543}/P^{0.294}\ P_B^{0.134}$$

will be used for the exhaust pipe size.

$$D_e = 13.8 \text{ x } 35^{0.543}/(400+14.7)^{0.294} \text{ x } (65+14.7)^{0.134}$$

$$D_e = 13.8 \text{ x } 6.89/5.88 \text{ x } 1.80 = 8.97$$

From Table 2-4, the cross-over diameters for various commercial pipe sizes in schedule 40 weight are compared with the calculated size of 8.97 inches. The value for 8 inch is 9.16, above the calculated figure, so use 8 inch.

SOLUTION USING GRAPH 46

For the supply pipe size, align 400 on the "*P*" Scale with 65 on the "*B*" Scale to locate 2.7 on the turning Scale "*A*". Align 2.7 with 35 on the "Q" Scale to find size lies below 6 inch and above 4 inch. Use 6 inch.

For the exhaust line, align 400 on the "*E*" Scale with 65 on the "*B*" Scale to locate 4.95 on the "*A*" Scale. Align 4.95 with 35 on the "Q" Scale to find size lies below an 8 inch and above a 6 inch. Use the 8 inch.

As the supply and exhaust pressures approach each other, the sizes become closer together, the supply size increasing and the exhaust size decreasing.

EXAMPLE 12-2:

Conditions in Example 12-1 apply except the exhaust pressure is 15 psig rather than 65 psig. Exhaust steam used for heating. Select the supply and exhaust pipe sizes.

SOLUTION:

For the supply pipe, use Equation (2-28), (or Graph 46), noted above.

$$D_s = 11.7 \text{ x } (15+14.7)^{0.103} \text{ x } 35^{0.543}/(400+14.7)^{0.495}$$

$$D_s = 11.7 \text{ x } 1.42 \text{ x } 1.89/19.76 = 5.79$$

Comparison with Table 2-4 shows that a 6 inch pipe is still correct, but under some conditions, where the size chosen for the higher back pressure might lie closer to the limiting value of D_c, the next larger pipe size would apply.

For the exhaust pipe, use Equation (2-29), noted above.

$$D_e = 13.8 \text{ x } 35^{0.543}/(400+14.7)^{0.294} \text{ x } (15+14.7)^{0.134}$$

$$D_e = 13.8 \text{ x } 6.89\ /\ 5.88 \text{ x } 1.58 = 10.23$$

Comparison of this value with Table 2-4 indicates that the correct size is 10 inch rather than 8 inch, as determined in the previous example.

When the exhaust pressure is at a level for which the temperature of the residual heat is too low to be significantly useful, the concept of this model is lost and should not be used. Rather, the formula developed for the condition when the residual heat in exhaust steam is *not being used*, is applicable.

When the exhaust steam is *not* to be used for heating, the exhaust pressure will be *atmospheric*, or lower. Consider the variations in pipe size which result for the same flow rate and supply pressure as in the prior examples, but where the exhaust conditions demonstrate that the residual heat is *not* to be used for heating.

EXAMPLE 12-3:

A steam turbine requires 35,000 per hour of steam at 400 psig. The exhaust is to atmosphere. Select the supply and exhaust pipe sizes.

SOLUTION:

Because the exhaust is going to atmosphere, the heat remaining in the steam will not be used. Therefore, Equation (2-31) or Graph 47, LAC EXHAUST STEAM *NOT USED* FOR HEATING, would be employed. For this example, the equation will be used.

Equation (2-31):

$$D = 1.02\, W^{0.479} \,/\, \varrho^{0.337}$$

Where:
D = Pipe inside diameter, inches
W = Steam flow rate, *thousands* of pounds per hour
ϱ = Steam density, pounds per cubic foot

Density of steam is found from Table B-14 to be 0.893

$$D = 1.02 \times 35^{0.479}/0.893^{0.337}$$

$$D = 1.02 \times 5.49/0.96 = 5.83$$

From Table 2-4, the D_c of 6 inch schedule 40 pipe is 7.16, thus the correct size is 6 inch.

For the exhaust pipe, Equation (2-31) shown above, also is used because for LAC selection, when the heat in the exhaust steam is *not* being used, the same formula applies. See Table B-14 for steam density of 0.0373.

$$D = 1.02 \times 35^{0.479}/0.0373^{0.337}$$

$$D = 1.02 \times 5.49/0.33 = 16.97$$

Using Table 2-4, compare the calculated value with the tabulation. The 16 inch pipe has a D_c of 16.67, less than the calculated 16.97. Therefore, use 18 inch.

EXAMPLE 12-4:

A steam turbine requires 35,000 pounds per hour of steam at 400 psig. The exhaust pressure will be 20 inches of mercury vacuum. Select the correct pipe sizes for supply and exhaust.

SOLUTION:

For the supply pipe, the size determined for Example 12-3 will apply because the exhaust pressure does not influence the supply pipe size when the heat in the steam is **not** to be used. The low exhaust pressure demonstrates that this is the case.

For the exhaust pipe, because of the lower pressure, the size will be different from the prior example. When the heat is not used, the same equation is used for both the supply and exhaust pipe size selection. Therefore, use Equation (2-31) (or Graph 47).

Equation (2-31):

$$D = 1.02\, W^{0.479} \,/\, \varrho^{0.337}$$

Where:
D = Pipe inside diameter, inches
W = Steam flow rate, *thousands* of pounds per hour
ϱ = Steam density, pounds per cubic foot

The steam density is found from Table B-14 to be 0.0134.

$$D = 1.02 \times 35^{0.479} \,/\, 0.0134^{0.337}$$

$$D = 1.02 \times 5.49 \,/\, 0.23 = 24.35$$

From Table 2-4, compare the calculated value with the values of D_c to select 24 inch pipe with 0.25 inch wall.

CHAPTER 13

PRESSURE LET – DOWN FOR GAS

INTRODUCTION

When the pressure of gases is being reduced, for example through a control valve, the pipe size selection will be made using the PDA mode. The amount of pressure loss which can be assigned to the pipe system is established according to the principles in the Introduction. If *all* the loss in pressure will be divided between the valve and the piping, and the rate of *flow will not vary significantly,* one may assign 10% to 20% of the total loss to the pipe and the remainder to the valve. However, best practice calls for an analysis of the flow conditions and the determination of the correct portion of the pressure drop to be assigned to the pipe, established by using Graph 1, CONTROL VALVE SELECTION.

NOISE CONSIDERATIONS

When gas at very high pressure is being reduced to a lower one, velocities in the lower pressure portion may become quite high. Consideration must be given to the possibility of excessive noise generation. Experience has shown that a velocity of 425 feet per second in steam at 20 psig did not produce significant noise.

When the velocity approaches the sonic, and the amount of flow is large, the noise level may become higher than OSHA limits. At this date, exact means for predicting noise levels in piping are not well established.

When a control valve is being used to absorb the major share of the pressure loss, *it* will become the most likely source of excessive noise, rather than the piping. Valve manufacturers have developed, and are improving, their techniques for predicting and eliminating excessive noise (39).

RETROGRADE CONDENSATION

While the condition may be encountered but rarely, the engineer must be alert to the possibility, under pressure reduction, of condensation of gas into liquid. In certain natural gas reservoirs, this phenomenon, known as "retrograde condensation", will occur upon reduction of the pressure. For retrograde condensation to be possible, in the initial state of the gas mixture, the *pressure* will be *above critical* and the *temperature below.* (See Chapter 6 for information on the critical pressure and temperature.) As the pressure is reduced, the single phase (gaseous) will shift to two phases, liquid along with gaseous. In a pure substance, complete condensation will take place, with the gas becoming totally liquid. The process is analagous to the reverse of vaporization. With further pressure reduction, the liquid will reach the boiling point and transfer, once again, into the gaseous state.

If retrograde condensation should be a possibility, consideration must be given to the gas-liquid mixture velocity, either to insure entrainment, should that be needed in a pipe carrying the mixture upward, or to avoid erosion through excessive velocity. Pipe size selection must include the methods presented in Chapter 17.

GAS LET-DOWN

In the majority of cases, pressure let-down of gas becomes a special version of the method presented in Chapter 10 for PDA.

EXAMPLE 13-1:

Natural gas with a specific gravity of 1.20 at 2000 psig and at 112° F is being blown down to 1500 psig. The flow rate could be from 25,000,000 cubic feet per day to 1,500,000. The drop through the pressure reducing regulator is 450 psi leaving 50 psi for the pipe. The pipe length is 425 feet upstream of the regulator and 25 feet downstream. The upstream pipe has 6 elbows and two tees through the bull. The downstream has two elbows. Determine the upstream and downstream pipe sizes.

SOLUTION:

The pipe size will be selected using the maximum flow rate. To fulfill the conditions given, and in the absence of knowledge of the size of the pressure reducing valve (which would give a clue to the starting point in selecting the pipe size) some means is needed to choose the first approximation for the correct size. Equation (3-26) PDA for the "typical" line pressure drop available mode, or Graphs 53 and 54, based upon it, are used.

Equation (3-26):

$$D = 0.0125\ M^{0.536}\ \overline{MW}^{0.230}\ T^{0.34} \left(\frac{\Psi}{\Delta P_T P}\right)^{0.292}$$

Where:

D = Pipe inside diameter, inches
M = Gas flow rate, *thousands* of standard cubic feet per day, at 60° F and 14.69 psig
$\overline{MW}$ = Molecular weight
T = Flowing temperature, degrees Rankine (°F+460)
Ψ = Factor combining deviation from ideal gas laws and viscosity, see Equation (3-17)
ΔP_t = Pressure drop in the entire line, pounds per square inch
P = Gas pressure, pounds per square inch, absolute

Equation (3-26) includes Ψ, so a check of its value is made first, to determine if it will be significant. From Figure 6-5, the critical values for the gas are found to be T_c = 550° R and P_c = 638 psia. The pseudo-reduced values are,

$$T_r = (460 + 112) / 555 = 1.03$$
$$P_r = (2000 + 14.7) / 638 = 3.16$$

Referring to Figure 3-5 and introducing 1.03 and 3.16, find Ψ is 0.6. Thus it is seen that the deviation from the perfect gas law is sufficient in this example that it must be considered in the pipe size selection.

$\overline{MW}$ is determined from the specific gravity to be 34.7.
Making the first approximation, using Equation (3-26), above,

$$D = 0.0125 \times 25{,}000^{0.536} \times 34.7^{0.23} \times 572^{0.34} \left(\frac{0.6}{50 \times 2015}\right)^{0.292}$$

$$D = 0.0125 \times 227.7 \times 2.26 \times 8.66 \times 0.029 = 1.66$$

Next, determine the pressure drop using the nearest size pipe which would be 2 inch schedule 80 with an inside diameter of 1.939. See Table B-13.

Using Equation (3-18):

$$\text{psi}/100 = \frac{1.68 \times 10^{-6}\ M^{1.84}\ \overline{MW}^{0.784}\ T^{1.15}\ \Psi}{P\ D^{4.84}}$$

Where:

psi/100 = Pressure loss, pounds per square inch per 100 feet of pipe
M = Flow rate, *thousands* of standard cubic feet per day, at 60° F and 14.69 psig
$\overline{MW}$ = Molecular weight
T = Gas temperature, degrees Rankine (°F + 460)
Ψ = Factor combining deviation from ideal gas laws and viscosity, see Equation (3-17)
P = Gas pressure, pounds per square inch, absolute
D = Pipe inside diameter, inches

Substituting:

$$\text{psi}/100 = \frac{1.68 \times 10^{-6} \times 25{,}000^{1.84} \times 34.7^{0.784} \times 550^{1.15} \times 0.6}{2015 \times 1.939^{4.84}}$$

$$\text{psi}/100 = 57.31$$

This is too high because the *total* allowable pressure drop is to be 50 psi in the entire line. Try the next pipe size larger, 3 inch. The internal diameter is 2.900 inches for schedule 80. The pressure drop will be proportional to the ratio of the diameters to the 4.84 power.

$$57.31 \times (1.939 / 2.900)^{4.84} = 8.17$$

The total equivalent length of the upstream pipe, from Table 6-1, will be,

6 elbows: 6 x 1.2 x 3	=	21.6
2 tees (bull): 2 x 3.7 x 3	=	22.2
Straight pipe:		425
Total:		468.8 feet

Total pressure drop in the upstream pipe will be

$$4.69 \times 8.17 = 38.3 \text{ psi}$$

For the downstream pipe, the pressure drop will be in inverse proportion to the pressures in the same size pipe

$$8.17 \times 2{,}000/1{,}500 = 10.9 \text{ psi}/100$$

The total equivalent length of the downstream section will be

2 elbows: 2 x 1.2 x 3	=	7.2
Straight pipe:	=	25
Total:		32.2 feet

Total pressure drop in the downstream pipe will be

$$0.32 \times 10.9 = 3.5 \text{ psi}$$

The total pressure drop in the pipe system, both segments will be

$$3.5 + 38.3 = 41.8 \text{ psi}$$

This value is less than the allowable 50 psi; therefore, the 3 inch is the correct size, both upstream and down.

If, however, the downstream pressure were a smaller fraction of the upstream than shown in this example, the downstream size might well be larger than the upstream. Commonly, the downstream pipe size will be larger. Therefore, the control valve should be located as close as possible to the final destination of the pipe to permit the smaller size to be used over the greater length, thereby reducing costs.

BLOW-DOWN TO ATMOSPHERIC PRESSURE

When gas is being blown down essentially to atmospheric pressure from a substantially higher pressure, the mode of selection will be according to Velocity Allowable, rather than PDA. The limitation to flow will be attainment of, or approach to, the maximum velocity, i.e., the velocity of sound in the gas at the exit pressure and temperature. The concepts are explained in Chapter 4 and expanded in Chapter 17.

CHAPTER 14

RELIEF VALVE DISCHARGE HEADERS

RATE OF FLOW

In this phase of pipe size selection, in contrast to most, the engineer may well be the person who establishes the flow rates to be considered in making the size selection. An analysis must be made of the probability of the simultaneous discharging of groups of relief valves into various branches of the relief header system. Clearly, in many emergency situations, only one valve will be activated. Under some conditions, two or more may be brought into action. The likelihood of *all* relief valves discharging simultaneously, generally is beyond probability in a total plant, and even in one section, except in some catastrophic emergency.

EMERGENCY CONDITIONS

The opening of one relief valve generally will eliminate the need for activation of valves on equipment located downstream. An exception could be the presence of a fire which could raise the pressures in a series of vessels and other equipment. The possibilities of simultaneous relief being required from several pieces of apparatus are so diverse that no hard and fast rules can be laid down. The engineer must comb the process flow sheet, considering the effects of various emergencies which could arise.

After the possible combinations of affected valves for various emergencies have been collected, the flow rates can be determined by summing the relief capacities of the valves.

BACK-PRESSURE EFFECTS

The type of relief valve will monitor the size of the header required to handle the loads imposed. If the discharge rate of the valve will be reduced by back-pressure in the relief header, a larger sized system must be employed than would be the case if the back-pressure does not influence the discharge rate. The relief valve manufacturer should be consulted for information relative to back-pressure effects.

The size of the relief header system may be reduced, if local conditions will permit supplementary relief valves, of some sort, on the header at strategic points to reduce the pressure in a serious emergency, through venting to atmosphere.

When possible, relief headers should be kept as short as practical. This approach can mean more blow-down tanks and/or flares or other means to limit air pollution. Thus, a balance is needed between costs of fewer and larger headers and the costs of extra numbers of associated apparatus.

LIQUID AND GAS RELIEF HEADERS

Separate header systems should be used for liquid and gas relief unless there is *no possibility* of simultaneous relief of both fluids into the header. Determining the pressure drop of two-phase systems is, at best, a good, and, at worst, a poor approximation. The engineer should not risk plant safety by attempting to reduce redundancy in the relief system by combining the two in the face of possible simultaneous usage.

CALCULATION METHODS

In calculating the sizes of the branches and the main lines of the relief system for gases, the PDA mode may apply when loads are moderate. However, VA often will be the proper selection, using the approach to, or attainment of, acoustic velocity, as discussed in Chapters 4 and 17.

For liquid relief header systems, the method of pipe size determination will be PDA, discussed in Chapters 3, and 10.

CHAPTER 15

STEAM TRAP PIPING

INTRODUCTION

The size of the piping, especially downstream, if too small, can adversely effect the performance of steam traps. On the other hand, making the downstream piping very large, for safety, can raise plant construction costs. The nature and frequency of the discharge of condensate from the trap system served by one header will depend upon the characteristics of the various traps being used. Start-up loads may be quite different than those during normal operations. Except for continuous drainers, the discharge rate for the trap *will not* be the average, but will depend upon a variety of factors which include the trap type, size, and operating conditions, including the downstream pressure.

DETERMINATION OF THE FLOW PATTERN

For bucket, disc, or thermostatic traps, the flow rate of condensate is established by the capacity of the trap, as specified by the manufacturer, at the particular pressure difference. With the correct selection, the discharge *will not be continuous.* For example, if the condensate rate were to be 2000 pounds per hour, the maximum capacity of the correct trap would be about 6000 pounds per hour. Therefore, it would be discharging at an *average rate* of 2000 pounds per hour operating for 33% of the time. The rate used for pipe size selection must be the *trap capacity;* in this case, about 6000 pounds per hour.

An exception may occur if the start-up load will be substantially greater than the normal flow. Then, start-up load could be the controlling factor, thereby calling for larger pipes.

For continuous drainers, by contrast, the pipe size selection is made according to the condensate flow rate from the apparatus being served, which will be less than the *capacity* of the drainer.

In downstream header size selection, the intermittent flow character of individual traps will tend to be smoothed out as the units which supply it grow in number. Beyond 12 units, one may assume the flow to be the *average rate* of *all* traps. From 6 to 11 units, assume 1.5 times the average, while for 3 to 5, use double the average.

A particular trap in a bank of 3 to 5 units being served by one header may be in critical service. Now and then, the random pattern of discharge could result in *all* traps being activated at the same time. Then build-up of downstream pressure might reduce the discharge capacity of all traps, *including* the *critical one,* for as long as the pattern would hold. This condition is not likely to continue beyond a few dumps. However, to remove the possibility of process interruption, the header, *in this instance,* should be made large enough to handle the flow which would result if all traps dumping into it were operating at the same moment.

If the traps selected should have capacities more than the suggested 2 to 3 times the average condensate flow rate, the recommendations given must be adjusted accordingly. To give an example, if the traps will discharge 4 to 6 times the required condensate flow, the safe number at which the average flow could be used would increase to 20 traps.

The same considerations are applied as sub-headers feed into larger headers, etc. When the *total* number of traps reaches the quantity given, above, the flow pattern is figured according to the recommended ratios.

PIPE STORAGE CAPACITY

The length of the downstream section, being the larger size, should be as short as possible to reduce piping costs. As a side benefit, the storage capacity of the longer upstream pipe can provide a safety factor. If trap discharge capacity should be diminished temporarily by a pressure build up in the downstream pipe system, the excess condensate could be collecting in the upstream pipe awaiting the time when the pressure would return to the normal level. For a critical service condition, as mentioned above, the downstream header need not be increased to avoid process difficulties *if* the upstream pipe has capacity to hold 4 to 6 dumps of condensate.

EXAMPLE 15-1:

A reheater is condensing 12,000 pounds per hour of steam, producing 24 gpm of condensate, at 200 psig. The bucket steam trap discharges to a header with 100 psig pressure. The trap requires 90 psi to discharge 50 gpm of condensate. Using Equation (3-24), the pipe size has been found to be 2 inch. From the manufacturer's specifications, the time for a single dump is 1.2 seconds. What length of pipe is needed to provide capacity for 6 dumps?

SOLUTION:

From data in Table B-12, the volumetric capacity of 2 inch pipe is calculated to be 0.17 gallons per foot. The gallons per dump will be

$$(1.2/60) \times 50 = 1.0$$

The length required will be that amount which will hold 6 dumps, or 6 gallons,

$$1 \times 6/0.17 = 34 \text{ feet}$$

It follows that, if the length of pipe were less than sufficient to accomodate 1 gallon, in this case less than 6 feet, the condensate would be backing up into the equipment which might impair its functioning.

CONDENSATE FORMED IN PIPE

The rate of condensate formation in steam pipes can be determined by using Graph 57, STEAM CONDENSATION ON BARE PIPE. Correctly insulated pipe will condense 15 to 25% of the amount shown for *bare* pipe. The "*T*" Scale provides both the temperature difference between steam and air at 60° F and the corresponding steam pressure. The "*D*" Scale the nominal, as well as actual, size for commercial pipe.

EXAMPLE 15-2:

How much condensate will be developed in a 10 inch 300 psi steam header, 225 feet long, which has been correctly insulated?

SOLUTION:

Using Graph 57, align 300 on the pressure portion of the "*T*" Scale with 300 hundred on the "*L*" Scale to find the turning point on the "*A*" Scale, 6.1. Align this with the bold face 10 on the "*D*" Scale to read 480 pounds per hour of condensate for each 100 feet of *bare* pipe. With insulation, the condensate rate will be 120 pounds per hour. The total will be

$$120 \times 225/100 = 270 \text{ pounds per hour}$$

UPSTREAM PIPE SIZES

The pipe size upstream should be kept small if the heat in the system is being conserved via flash-steam recovery. In any event, the basis for selection is PDA. Because the upstream pipe will be smaller than the downstream, within the limits of other considerations of a practical nature, the trap should be located as close as possible to the destination of the condensate, be that a condensate-steam header, flash tank, or the like. Simple economics is the reasoning: make the larger pipe (downstream) as short as is feasible.

As a workable approach, assign to the piping 10% of the pressure drop from the source of the condensate to the destination. Because the upstream run will be the longer, to it will be assigned 70% of the total pipe pressure drop allowed.

The pressure drop in the upstream piping, selected on this basis, could result in the generation of a *small* quantity of flash steam. If *smaller* sizes than herein recommended were to be used, *significant* flashing *could* occur which might influence trap operation.

EXAMPLE 15-3:

A bucket trap, with a capacity of 18,000 pounds per hour will drain a piece of equipment developing 6000 pounds per hour of condensate. The steam header pressure is 175 psig, the pressure in the equipment is 140 psig. The condensate will drain into a 60 psig system. Determine the correct upstream pipe size.

SOLUTION:

The pressure in the steam header is not involved; only the pressure in the equipment, namely 140 psig. The difference between it and the condensate system at 60 psig, that is, 80 psi, will be absorbed by the trap and the piping. Allow 10% for the pipe, and 70% of that to be assigned to the upstream section,

$$80 \times 0.1 \times 0.7 = 5.6 \text{ psi}$$

Assuming the entire 5.6 psi will be absorbed by the upstream pipe, some flash steam will be developed by the pressure reduction. From Graph 2, PERCENT STEAM FLASHED FROM HOT CONDENSATE, the amount may be found. Align 140 on the "*H*" Scale with 134, (140-6), on the "*L*" Scale to determine the flash is 0.5%. This is so small as to be relatively insignificant; however, to be rigorous, it will be considered in this example.

From Table B-14, find the saturation temperature for steam at 140 psig is 360° F. From Table 7-1, find the specific gravity of water at 360° F is 0.92 and the viscosity is 0.13 centipoise.

Determine the density of the mixture of water with a trace of steam as

$$62.4 \times 0.92 \times 0.995 = 57.1 \text{ pounds per cubic feet}$$

The practical mode of pipe size selection is via Simplified PDA, see Chapter 3, using Equation (3-23)

$$D = 1.274\, W^{0.536}\, \mu^{0.047} / (\Delta P_T)^{0.29}\, \varrho^{0.29}$$

Where:

D = Pipe inside diameter, inches

W = Condensate rate, *thousands* at pounds per hour

μ = Viscosity at flow temperature, centipoise

ΔP_T = Pressure drop in the entire line, pounds per square inch

ϱ = Density of the fluid mixture, pounds per cubic feet

Using the applicable values

$$D = 1.27 \times 18^{0.536} \times 0.13^{0.047} / 5.6^{0.29} \times 57.1^{0.29}$$

$$D = 1.02 \text{ inches}$$

From Table B-12, find that 1 inch schedule 40 pipe has an inside diameter of 1.049, slightly larger than the calculated value, so consider using 1 inch.

In using Equation (3-23) one assumes the length of the pipe will be the "typical". For 1 inch size that would be, from Equation (3-22 A), 18 feet. If the distance from the condensate source to the destination, by the plot plan, will be 15 feet or more, the 1 inch size would be too small because the required size is close to the actual diameter of 1 inch pipe. Under conditions where the distance would be greater than 15 feet, the next size larger should be used. If on the other hand, the distance were to be 12 feet or less, the one inch would be proper.

Consider the effect of ignoring the 0.5% flash. The difference lies in the ratio of the two densities to the 0.29 power. The result is an increase of 0.5% in the pressure drop and 0.15% in the diameter, truly insignificant.

DOWNSTREAM PIPE SIZES

The downstream piping carries a mixture of water (condensate) and flash steam, requiring consideration of two-phase flow, as presented in Chapter 5 and expanded in Chapter 17. For a first approximation, the Simplified PDA approach is used, via Equation (3-23).

The downstream pipe should be as short as practical, as explained above. The pressure drop assignable to the downstream pipe is to be 30% of the total for the trap piping. The entire drop in the piping is to be 10% of the pressure difference between the source of the condensate and its destination. The remainder of the pressure loss is assigned to the trap. In the selection, consideration is to be given to erosion, impact and entrainment velocities.

EXAMPLE 15-4:

Groupings of traps discharge into a common condensate return pipe line. The distance from the most remote trap to the header will be about 200 feet. Two traps, at the far end of the header, dump in at the same location. One has a capacity of 20,000 pounds per hour with the upstream pressure being 200 psig and the header pressure at 100 psig. The average rate is 7000 pounds per hour. The second operates at 180 psig, and has a capacity of 10,000 pounds per hour into the 100 psig header. The average rate is 5000 pounds per hour. Downstream about 100 feet, 6 more traps, with capacities of 15,000 pounds per hour each, operating at 170 psig, (average rate is 7000 pounds per hour, each), will dump into the common header. Determine the correct sizes for the initial and final sections of the header.

SOLUTION:

The initial section, with two traps dumping into it, will require determination of the flow pattern, as discussed earlier in this Chapter. The number of traps involved at the initial end is two; thus, there will be no reduction in the effects, of the intermittent character of the flow. Therefore, the basis for selection will be upon the sum of the individual trap capacities, 30,000 pounds per hour, because both *could* be discharging at the same time.

The amount of flashing will be different for each because the initial trap pressures are different, 200 psig and 180 psig. Use Graph 3, SPECIFIC GRAVITY OF MIXTURE FROM FLASHED CONDENSATE, to find the specific gravity of the water-steam mixture from each trap.

Align 200 on the "*H*" Scale with 100 on the "*L*" Scale to find the specific gravity is 0.06. Likewise find the specific gravity for the other trap operating at 180 psig to be 0.07. The specific gravity of the mix will be

$$0.06 \times 20/30 = 0.04$$
$$0.07 \times 10/30 = 0.023$$
$$0.040 + 0.023 = 0.063$$

The density will be

$$0.063 \times 62.4 = 3.93 \text{ pounds per cubic foot.}$$

With the density far lower than that of water, the flow pattern likely will be homogeneous so the applicable viscosity will be that of steam at 100 psig. See Figure B-2 to find the viscosity is 0.017 centipoise.

The pressure drop in the downstream section will be 30% of the 10% of the total pressure difference, 2.1 psi.

Use Equation (3-23), SIMPLIFIED PDA, for the first approximation.

Equation (3-23):

$$D = 1.27\, W^{0.536}\, \mu^{0.047} \,/\, (\Delta P_T)^{0.29}\, \varrho^{0.29}$$

Where:
- D = Pipe inside diameter, inches
- W = Thousands of pounds per hour of condensate and steam
- μ = Steam viscosity, centipoise
- ΔP_T = Pressure drop in the entire line, pounds per square inch
- ϱ = Density of the fluid mixture, pounds per cubic foot

$$D = 1.27 \times 30^{0.536} \times 0.017^{0.47} \times / \; 2.1^{0.29} \times 3.93^{0.29}$$

$$D = 3.54 \text{ inches}$$

The calculated size is below the 4 inch schedule 40 inside diameter of 4.025, as shown on Table B-12, so 4 inch is to be used.

For the second section, the pattern of flow will be different from the first portion because more traps will be dumping into it, 8 instead of 2. As discussed earlier in this Chapter, the flow rate to use would be 1.5 times the overall *average*.

$$7000 + 5000 + (6 \times 7000) = 54{,}000$$
$$54{,}000 \times 1.5 = 81{,}000$$

For the second section, the pressure drop available will be decreased by that already taken up in the first portion, essentially one half. Thus the allowable for the second section will be 1.1 psi.

The average density, calculated as above, is 4.65, and the average specific gravity is 0.075.

Again use Equation (3-23) to determine the first approximation.

$$D = 1.27 \times 81^{0.536} \times 0.019^{0.047} \,/\, 1.1^{0.29} \times 4.65^{0.29}$$

$$D = 6.92 \text{ inches}$$

The nearest standard size, 8 inch, is larger than the calculated size and is the proper selection.

For both sections, a check is needed for the various velocities which could influence the size: erosion, impact and entrainment.

The velocity of the mixture is determined using

Equation (4-2)

$$V = 0.409\, Q \,/\, D^2$$

Where:
V = Mean velocity, feet per second
Q = Rate of flow of the mixture, gallons per minute
D = Diameter of the pipe, inches

For the initial 4 inch section, the flow rate is obtained from the specific gravity of the mixture.

$$30{,}000 \,/\, 501 \times 0.063 = 950 \text{ gpm}$$

The average velocity is, using Equation (4-2), as given above

$$V = 0.409 \times 950 \,/\, 4.026^2 = 24 \text{ feet per second.}$$

For the final 8 inch section, the flow rate is

$$81{,}000 \,/\, 501 \times 0.075 = 2156 \text{ gpm}$$

The average velocity is

$$V = 0.409 \times 2156 \,/\, 7.94^2 = 14 \text{ feet per second.}$$

The **erosion** velocity is determined from Figure 5-4 to be 62 feet per second in the 4 inch and 28 feet per second in the 8 inch. These are well above actual velocities, so there is no concern.

The **entrainment** velocity is determined from Figure 5-3, using steam density as 0.26 pounds per cubic foot, from Table B-14, to be 9.5 feet per second. The actual velocities in either section are well above that, so there is no cause to question the size selections for this parameter.

The **impact** velocity pressure is calculated from Equation (4-24):

$$P_t = 2\,[P + (0.0026\, V^2 \rho)]$$

Where:
P_t = Impact pressure *plus* the *static pressure,* pounds per square inch
P = Static pressure, pounds per square inch
V = Fluid mixture velocity, feet per second
ρ = Mixture density, pounds per cubic foot

$$P_t = 2\,[115 + (0.0026 \times 24^2 \times 4.65)] = 244 \text{ psig}$$

That value is a fraction of the safe working pressure of schedule 40 pipe whether 4 inch or 8 inch. Thus, these two selections are correct in all regards.

CONTINUOUS DRAINERS

The flow through continuous drainers will be steady, rather than intermittent, in contrast to the case with bucket traps, disc traps or thermostatic traps. Therefore, in selecting the pipe sizes, both upstream and downstream, the allowances specified in the earlier examples to care for the higher momentary flow character of intermittent traps are not used. Thus, the size selection will be based upon the rate of condensation of the steam in the apparatus, NOT upon the *capacity* of the drainer.

In the prior EXAMPLE 15-1, if a drainer were being used, the condensate flow rate introduced in Equation (3-23) would have been 6000 pounds per hour rather than 18,000, reducing the calculated diameter to essentially half that for the bucket trap, 0.54 inches versus 1.02 inches.

START-UP CONDENSATE LOADS

At the start-up of a plant, when equipment is cold, the condensate loads will be different from those during operation, often much larger. Also, the pressures may be other than those which will occur during normal operation. Depending upon the time allowed for start-up, as well as the load factors, consideration *may be* required of these parameters in selecting the trap piping sizes. In doing so, however, included in such consideration must be the cost factor of the larger pipe sizes which this approach would entail. Opting for the smaller sizes and taking more time for start-up, in some cases, could be the wiser choice.

BATCH OPERATION CONDENSATE PIPING

In selecting the pipe sizes of condensate systems for batch operation, the initial condensate load will be the controlling rate, in most cases.

CHAPTER 16

TWO-PHASE FLOW APPLICATIONS

INTRODUCTION

The fundamental concepts covering two-phase (liquid-gas) flow are developed in Chapter 5. Unless the user is truly familiar with these principles, a review is warranted before pipe selections are made.

For most applications the engineer will encounter, the proportions of the liquid and gas in two-phase mixtures essentially will be established at the commencement of the containing pipe. Examples are: the effluent from a fired heater, or from a heat exchanger, equilibrium liquid flashing through a valve.

When possible, which *may* not be practical for the first example, make the pipe containing the two-phase flow as short as is possible. If the mixture is being conducted to a vessel, consider placing the valve *on* the vessel, or as close as can be arranged. To insure that these requirements are met, special instructions must be transmitted to the piping layout crew chief. Otherwise, all the careful planning can be for naught; so communicate!

The calculation of the pressure loss through a fired heater, where vaporization is taking place, ordinarily is the responsibility of the manufacturer and hence would not be handled by the pipe size selection engineer. Should he be obliged to make such calculations, refer to the methods developed by LOCKHART and MARTINELLI (23) or DUKLER (24).

The effluent from one, or more, oil field traps may be conducted for long distances downstream of the control valve. Or, if the valve is placed at the end of the pipe adjacent to the destination but the elevation of the pipe is increased along the way, sufficient loss in pressure may occur to initiate flashing.

In such a system, located in a desert region, the temperature of the conducted liquid may be increased by solar heat thereby bringing about vaporization. For these, or similar applications, wherein the liquid-to-gas ratio will be undergoing a gradual change, the methods of MARTINELLI and LOCKHART (23) or those of DUKLER (24) would be the better choice.

For the more common applications, mentioned earlier, the various formulae or nomographs presented in this book will provide the appropriate size. Depending upon the nature of the application, LAC, PDA or VA may be the proper selection.

PRESSURE LET-DOWN FOR LIQUIDS

When fluids at equilibrium, other than hot water, are being subjected to pressure reduction, with the consequence of liquid vaporization (flashing), the determination of the downstream pipe size, while related to the approach used in Chapter 15, generally will be more complex. In some cases it will be far more so.

For one consideration, the degree of vaporization is by no means as easily determined as is the case with flashing condensate. Rather than using a simple alignment chart, equilibrium calculations, beyond the scope of this book, are required. Thus, for the pipe size selection, the *process engineer* must provide this essential data to the piping engineer.

The viscosity of the mixture is not as readily established. Rather than dealing with a single compound, as water or steam, the applicable viscosity, whether of the liquid phase or vapor phase system, requires that an approximation be made.

Fundamentally, the approach to the pipe selection is via PDA but with consideration of VA required in some cases.

In determining the allowable pressure loss in the piping, the concepts set forth in the Introduction are to be used. These provide the means to establish the ratio between the pressure drop in the pipe compared to the pressure loss in the control valve.

The upstream segment is to be the greatest portion of the total pipe length. This is done for two reasons. (1) The upstream size will be smaller and, therefore, should be the longest segment. (2) The selection of the upstream pipe is simple; whereas the selection of the downstream size is more complex and fraught with the possibility of potential errors. Thus, keeping it short reduces the likelyhood of significant deviation between the calculated pressure drop and the actual.

UPSTREAM PIPE SIZE

A good first approximation of the correct pipe size can be obtained by choosing the next pipe size larger than the port size of the control valve.

The pressure drop is determined using the modified FANNING Formula with Graphs 29, PRESSURE LOSS FOR WATER, and 30, VISCOSITY-GRAVITY CORRECTION, or Equation (3-9), and the estimated equivalent length. The resulting total pressure drop is compared with the allowable. If it consumes up to 70% of the total, that size could be correct, depending upon the results of the calculation for the downstream segment.

As an alternate approach, Graphs 51, SIMPLIFIED PDA-WATER, and 52, SPECIFIC GRAVITY-VISCOSITY CORRECTION, or Equation (3-24) may be used for a trial determination of the pipe size. If the upstream pipe segment is estimated to be short, or the pipe size calculated lies midway between commercial pipe sizes, this method will provide a safe size. Otherwise, the total pressure drop should be calculated, as explained in the preceding paragraph.

In determining the pressure losses in the upstream piping, be aware of the possibility that, with substantial pressure loss in the upstream segment, some vapor may be generated by flashing. If the amount is small, less than 1 to 2%, the influence likely will be negligible. If greater than 3%, effects should be determined using Graphs 29, PRESSURE LOSS FOR WATER, and 30, VISCOSITY-SPECIFIC GRAVITY CORRECTION, or Equation (3-9), with the flow rate and the specific gravity adjusted for the effects of the vaporization.

The viscosity to be used, normally, under upstream vaporization conditions, will be that of the liquid. Should the vaporization be enough to cause the mixture to attain entrainment velocity (see Chapter 4) wherein the gaseous viscosity would be used in the two-phase flow calculation, this would demonstrate the pipe size chosen is too small.

DOWNSTREAM PIPE SIZE

When the pressure on an equilibrium liquid is reduced beyond a restriction, such as a control valve, nearly instant flashing will occur with a portion of the liquid becoming a gas. The mechanism is described in Chapter 5. Because of the considerable increase in the volume of the gas — liquid mixture, the velocity and pressure loss, which would be experienced if the same pipe size as used upstream were to be employed, would be excessive. Therefore, downstream piping for such conditions, virtually without exception, will be larger, possibly more than one pipe size larger.

The vapor volume and the liquid volume must be known. Normally, determining these is the responsibility of the chemical engineer. If these data cannot be otherwise obtained, and the material being flashed is a *pure compound*, a good approximation of the percentage of flashing can be established using Graphs 4, ENTHALPY, and 5, HEAT OF VAPORIZATION, and entering the resulting data into Equation (5-5).

The downstream pipe length must be kept as short as practical to reduce costs.

If a section will require the mixture to rise vertically, the velocity in this portion must be sufficient to entrain the droplets. Certainly, low velocities, which could permit slug flow, must be avoided, see Froude Number, Equations (5-2) and (5-3), and Figure 5-1, Chapter 5.

Unless there is *no possibility* of vertical upward flow of the two-phase mixture in the pipe, the first approximation of the correct size will be based upon a velocity sufficient to insure entrainment. See Figure 5-3 to establish the lowest velocity allowable. The rate of flow of the mixture in CFS is found from Equation (5-8) and the velocity from Equation (5-7). The velocity selected must be below that at which erosion of pipe elbows will occur; see Figure (5-4).

Should the entrainment velocity lie above that of erosion, unlikely but possible, the pipe must be made up of gradual bends rather than commercial elbows. If this should occur, the engineer must notify the piping layout crew chief of this requirement.

For a final check, in a critical application where pressure drop available is limited, the engineer should obtain an isometric sketch of the pipe segment so that actual lengths and fittings used may be considered in the calculation of the total pressure drop. The pressure drop per 100 feet is determined using Equation (3-8). The flow rate, *W*, is the combined rate for the liquid and the gas. The density is the mean value of the mixture as determined from Equation (5-6). Assuming the velocity is above entrainment, the viscosity will be that of the gas phase.

EXAMPLE 16-1:

Oil from a 1500 psig absorber flashes through a liquid level control valve. The flow entering the valve is 421 gpm with a specific gravity of 0.72. Leaving the valve the flow is 353 gpm with specific gravity of 0.79 at 105° F and 425 psig, plus 3,170 mcfd of gas with specific gravity of 0.69. Select the correct sizes above and below the control valve.

SOLUTION:

Schedule 80 pipe will be used to the valve. To avoid vortex formation at the vessel outlet, the velocity should not exceed 7 feet per second up to the first elbow. Using Graph 33, VELOCITY FOR LIQUIDS, align 421 on the "Q" Scale with 7 on the "V" Scale. On the "D" Scale, find the size lies between 4 inch and 6 inch. Use 6 inch.

Beyond the first turn, the size may be reduced. The first approximation of the pipe size is found from the Simplified PDA Graphs 51 and 52 or Equation (3-24)

$$D = 0.263\, Q^{0.536}\, \mu^{0.047}\, S^{0.25} / \Delta P_T^{0.29}$$

Where:

Q = Rate of flow of liquid at flow temperature, U.S. gallons per minute

μ = Viscosity of liquid at flow temperature, centipoise

S = Specific gravity of liquid, water at 60° F equals 1.00

ΔP_T = Pressure drop in the entire pipe, pounds per square inch

The viscosity of the liquid is found from Graph 8, APPROXIMATE VISCOSITIES OF STRAIGHT-RUN FRACTIONS. Align 105 on the temperature Scale with 0.72 on the California Fractions Scale to read 0.71 on the centistoke Scale. Convert to centipoise by multiplying by the flowing specific gravity, 0.72, to yield 0.51 centipoise.

The allowable pressure drop in the pipe is found as follows:

$$(1500 - 435) \times 0.7 \times 0.10 = 75.$$

This is the difference in the initial and final pressures, with 10% alloted to the valve and 70% of the remainder for the upstream pipe segment.

Equation (3-24):

$$D = 0.263 \times 421^{0.536} \times 0.51^{0.047} \times 0.72^{0.250} \ / \ 75^{0.29}$$

$$D = 1.71 \text{ inches}$$

The velocity in the nearest schedule 80 2 inch pipe size would be found using Graph 33, VELOCITY FOR LIQUIDS. Align 421 on the "Q" Scale with 1.939 (From Table B-13), on the "D" Scale to read the velocity on the "V" Scale as 46 feet per second.

There is a possibility of some flashing in a pipe with this velocity. Hence, the erosion velocity must be explored using Figure 5-4. Use the liquid specific gravity to estimate that erosion could occur at 40 feet per second. Thus, the 2 inch pipe is too small.

From Graph 33, VELOCITY FOR LIQUIDS, the velocity in a 2.5 inch pipe would be 30 feet per second, and the erosion velocity would be 32 feet per second; still not safe. Using 3 inch, the velocity would drop to 20 feet per second, well below the 25 feet per second for the erosion velocity, so 3 inch is the correct selection.

For the downstream section, erosion is likely to be the limiting factor. Therefore, start with a determination of the flowing cubic feet per second, leading up to the velocity calculation in various estimated sizes. The two-phase rate of flow is found from

Equation (5-8):

$$\overline{CFS} = Q/3600\left[0.08\,(100 - F) + \frac{1.86\,F\,S\,T\,Z}{GP}\right]$$

Where:

$\overline{CFS}$ = Average flow rate of mixture, cubic feet per second

Q = Liquid flow rate before flashing, U.S. gallons per minute at flow temperature
F = Percent of liquid flashed, by weight
S = Specific gravity of liquid before flashing at flow temperature, water at 60° F equals 1.0
T = Final temperature of mixture, degrees Rankine (°F+460)
Z = Deviation from perfect gas law, fractional
G = Specific gravity of the gas phase, air at 14.69 psia and 60° F equals 1.00
P = Final pressure of the mixture, pounds per square inch, absolute

The percent flashed by weight is the ratio of the flow rates corrected for the difference in specific gravity, before and after flashing.

$$100 \times [(421 \times 0.72) - (353 \times 0.79) / 421 \times 0.72] =$$

$$= 8\,\%.$$

The deviation from true gas law, Z, is next found.

The pseudo-critical temperature and pseudo-critical pressure are found for a gas with specific gravity of 0.69, from Figure 6-5, as 390° R and 670 psia, respectively. The ratio of the actual temperature to the T_c is

$$565/390 = 1.45$$

The ratio of the pressure to the P_c is

$$440/670 = 0.66$$

From Figure 3-4, the deviation, Z, is found to be 0.93.

Entering the values into Equation (5-8)

$$\overline{CFS} = 421/3600$$

$$\left[0.08\,(100 - 8) + \frac{1.86 \times 8.0 \times 0.72 \times 565 \times 0.93}{0.69 \times 440}\right]$$

$$\overline{CFS} = 3.03$$

Next, the erosion velocity is checked for likely pipe sizes, such as 4 inch and 6 inch, using Figure 5-4. This calls for the specific gravity (relative to water) of the mixture. The density of the mixture is found from

Equation (5-6):

$$\varrho_{av} = \frac{501\,S}{[0.08\,(100 - F)] + \dfrac{(1.86\,F\,S\,T\,Z)}{GP}}$$

Where:
ϱ_{av} = Density of the two-phase mixture, pounds per cubic foot
S = Specific gravity of the liquid before flashing, water at 60° F = 1.00
F = Percent liquid flashed by weight, fractional
T = Final temperature, degrees Rankine (°F+460)
Z = Deviation from ideal gas law, fractional
G = Specific gravity of the flashed gas, air at 14.69 psia and 60° F equals 1.00
P = Final pressure, pounds per square inch, absolute

Introducing the actual values

$$\varrho_{av} = \frac{501 \times 0.72}{[0.08\,(100 - 8.0)] + \dfrac{(1.86 \times 8.0 \times 0.71 \times 565 \times 0.93)}{0.69 \times 440}}$$

$$\varrho_{av} = 14.1$$

Divide the density by 62.4 (the density of water) to obtain the specific gravity of the mix, relative to water

$$14.1/62.4 = 0.23$$

From Figure 5-4, find the erosion velocity for 4 inch pipe to be 34 feet per second, for 6 inch to be 22 feet per second.

Determine the average velocity of the mixture from

Equation (5-7):

$$V = 183.3\,\overline{CFS}/D^2$$

Where:
V = Velocity of flow, feet per second
$\overline{CFS}$ = Rate of flow, cubic feet per second
D = Pipe inside diameter, inches

Introducing values for 4 inch pipe

$$V = 183.3 \times 3.03 / 4.026^2 = 34$$

This is the same as the erosion velocity for the 4 inch size. Therefore, unless the pipe will be a straight run from the valve to the destination, 4 inch is too small. The velocity for 6 inch is found to be 15 feet per second. Compared to the erosion velocity of 22 feet per second for 6 inch, this would be safe. Therefore, use the 6 inch.

The entrainment velocity must be checked, unless there will not be a section of the pipe where vertical rise of the mixture will occur, a rather unlikely, but possible, condition.

This is determined from Figure 5-3, which calls for the density of the gas phase alone, not of the mixture. This is found using

Equation (2-19):

$$\varrho = 2.696\, G\, P \,/\, T\, Z$$

Where:

ϱ = Gas density, pounds per cubic foot

G = Gas specific gravity, air at 14.69 psia and 60° F is 1.00

P = Pressure, pounds per square inch, absolute

T = Gas temperature, degrees Rankine (°F+460)

Z = Deviation from true gas law, fractional

Introducing the actual values

$$\varrho = 2.696 \times 0.69 \times 440 \,/\, 565 \times 0.93$$

$$\varrho = 1.56$$

From Figure 5-3, find that the entrainment velocity is 3.2 feet per second, so the 6 inch selection is correct.

If the length of pipe will be significant, the pressure drop per hundred feet is to be found using Equation (3-8).

With the velocity so far beyond the level for entrainment, the impact velocity need not be considered because slug flow cannot occur.

LAC TWO-PHASE APPLICATIONS

When a liquid is pumped through a heat exchanger or fired heater, and vaporization will occur, the outlet piping size normally will be based upon the LEAST ANNUAL COST mode, as developed in Chapter 2.

For the first approximation, determine the pipe size using Equation (2-14). The rate of flow, W, will be the combined rate of both phases. The density will be obtained from Equation (5-6). The viscosity will be assumed to be that of the gaseous phase. Having selected the tentative pipe size, the velocity is determined using Equation (5-7). The flow rate in CFS is found from Equation (5-8). Unless there is no possibility of vertical upward flow of the mixture, the velocity must be at least that of entrainment, see Figure 5-3, and below that of erosion. See Figure 5-4.

If the pipe size so selected provides a velocity between the two limits, the size is correct.

If the velocity is below that of entrainment, and vertical upward flow will be present (normally the case) the size of the pipe must be reduced to provide entrainment, even though the pressure loss resulting will be higher than is proper for the LAC concept. In this case, operation must take precedent over economics.

If the velocity obtained by either approach lies above that of erosion, unlikely, but possible, the pipe segment must be made up with gradual pipe bends rather than commercial elbows. This information must be communicated to the piping layout Crew Chief.

REFRIGERATION PIPE SIZE SELECTION

In a refrigeration system, the gaseous refrigerant is compressed and cooled before expansion. Just as the segments of pipe up to the point of the expansion valve would be chosen by the LAC mode, so would those beyond. Granted, pressure is being dissipated across the expansion valve; but, downstream of it, the compressor is handling the gas. Therefore, even in the two-phase flow portion of the piping, LAC selection applies.

Notice, however, that the operating time of the system normally will be quite a bit less than 100%. Therefore, the appropriate factors for reduced operating time should be used, see Table 7-3, in establishing the LAC sizes for all portions of the piping.

CHAPTER 17

SPECIAL VELOCITY CONSIDERATIONS

INTRODUCTION

In Chapter 5, the formulae, and their development as related to the selection of pipe sizes based upon velocity, either minimum or maximum, are presented. Even though this basis is not used frequently in the approach developed in this book, there are conditions where its use is required. The circumstances will include the need for entrainment, the avoidance of erosion, limitations for impact rupture of the pipe and the maximum flow attainable through a pipe.

ENTRAINMENT

When *bubbles* of *gas* are carried in a *liquid*, entrainment will be maintained unless the liquid velocity is quite low, slower than those usually employed in practical plant design. Thus velocity consideration is not critical in this mode.

On the other hand, when liquid droplets are being carried by a gas, if the required velocity is not maintained, the liquid will settle to the bottom of the pipe to produce intermittent flow, such as *plug, stratified, wavy*, or *slug*, see Chapter 5.

The velocity, at which droplet entrainment will occur, depends upon the size and specific gravity of the droplet together with the density of the gas. The lower the gas density, the greater will be the entrainment velocity. Thus, low pressure systems call for higher velocities to insure droplet transport.

Figure 5-3 provides a safe basis for etablishing the minimum velocity required to insure entrainment of 1 mm diameter drops.

Observe that the density scale is not related to the average density of the mixture but only is based upon the density of the *gas* itself.

EXAMPLE 17-1:

Determine the entrainment velocity for a mixture of air, at 60° F and 15 psig, with water.

SOLUTION:

The density of the air is determined using

Equation (2-19):

$$\varrho = 2.696\, G\, P \,/\, T\, Z$$

Where:

ϱ = Gas density, pounds per cubic foot
G = Specific gravity, air at 14.69 psia and 60° F equals 1.00
P = Pressure, pounds per square inch, absolute
T = Temperature, degrees Rankine (°F+460)
Z = Deviation from ideal gas law, fractional

Introducing the actual values

$$\varrho = 2.696 \times 1.0 \times 29.7 \,/\, (60 + 460) \times 1.00$$

$$\varrho = 0.154$$

From Figure 5-3, find the minimum velocity for entrainment is 12.4 fps.

To demonstrate the influence of pressure (and thus density) of the gas upon the entrainment velocity, consider the same problem with a higher pressure.

EXAMPLE 17-2:

Determine the entrainment velocity for a mixture of air at 60° F and 1500 psig with water.

SOLUTION:

Using Equation (2-19), as above, the density is calculated as 7.85 pounds per cubic foot. From Figure 5-3, the entrainment velocity is found to be 1.6 fps, compared to 12.4.

The influence of liquid density makes a difference, as well. Consider the effect on the last example of specific gravity of the liquid of 0.5, rather than that of water.

EXAMPLE 17-3:

Determine the entrainment velocity for air at 1500 psig when it is carrying liquid with a specific gravity of 0.5.

SOLUTION:

From Equation (2-19) as above, find the gas density to be 7.85. From Figure 5-3, the entrainment velocity is found to be 1.05 fps, compared to 1.6 fps.

IMPACT VELOCITY

When liquid and gas move through a piping system in alternate slugs, the force generated on an elbow, valve or tee from the hammering of the liquid slugs can produce a pressure sufficient to rupture the pipe if the velocity of flow is high. The degree of influence depends upon the curvature in the elbow or other impediment to free flow. The sharper the curve, the greater will be the effect of the impact.

Equation (4-24) defines the total pressure developed which is dependent upon the square of the gas *superficial* velocity and the density of the liquid. Though not included in the formula, the mass of liquid in a section moving through the pipe also can influence the effect upon the pipe. Therefore, the equation is an approximation and one does well to allow an ample factor of safety. In checking for possible rupture, half the nominal pipe pressure rating should be used.

EXAMPLE 17-4:

The rate of flow of a liquid with specific gravity of 0.85 is such that the velocity is 36 fps. The pipe is schedule 40, and the static pressure is 505 psig. Determine the total pressure (impact plus static) in the pipe. The safe working stress in the pipe is 16,000 psi. Determine if 6 inch pipe will be safe.

SOLUTION:

Equation (4-24):

$$P_t = 2\,(P + 0.0026\,V^2\,\varrho)$$

Where:

P_t = Impact pressure plus the static pressure, pounds per square inch
P = Static pressure, pounds per square inch, gauge
V = Velocity, feet per second
ϱ = Liquid density, pounds per cubic foot

Substituting values

$$P_t = 2\,(505 + 0.0026 \times 36^2 \times 62.4 \times 0.85)$$

$$P_t = 1367 \text{ psi}$$

From Table B-12 find the thickness of 6 inch schedule 40 pipe is 0.28 inches. Next refer to Graph 63, STRENGTH OF HOLLOW CYLINDERS which is based upon a stress of 10,000 psi versus the 16,000 in this example. Therefore, the pressure to be used on Graph 63 is determined by multiplying the actual pressure by 1.6, yielding 854. Align 854 on the "*P*" Scale with the outside diameter of 6 inch pipe, as shown on "*D*" Scale, to find the required wall thickness of 0.27 on the "*A*" Scale. If the temperature had been substantially above 100° F, the thickness, as shown on "*A*" Scale, would require correction by aligning 0.27 on the "*A*" Scale with the actual temperature on the "*T*" Scale to locate the corrected value on the "*B*" Scale.

The actual pipe thickness is above the required 0.27 inch wall so the 6 inch schedule 40 pipe can be used.

The effect of impact can be reduced by using fabricated bends and avoiding tees if the velocity must be above the impact limit.

EROSION VELOCITY

When liquid and gas will flow together at velocities which will insure entrainment of the liquid, if the velocity is excessive, the impingement of the droplets on the surfaces of elbows, tees an valves may cause erosion of the metal.

The level of velocity at which erosion will take place in long radius elbows made of carbon steel pipe is shown in Figure 5-4. Pipe made of more resistant steel can be subjected to higher velocity, in proportion to the allowable stress, compared to that of carbon steel. Another method to avoid erosion, when the velocity must be higher than that shown as allowable by Figure 5-4, is to fabricate the pipe system using very long radius bends rather than commercial welding elbows.

In using Figure 5-4, recognize that the vertical scale calls for the specific gravity of the *mixture*, not of the liquid nor of the gas alone. Equation (5-6) is used to determine the density of the mixture, from which the specific gravity (relative to water) can be established.

EXAMPLE 17-5:

The specific gravity of a mixture of gas and oil is 0.10. Determine the erosion velocity for 6 inch pipe.

SOLUTION:

Given the specific gravity and the pipe size, refer to Figure 5-4 to find the allowable velocity is 34 fps.

TERMINAL VELOCITY

The reasons which explain how it is possible that gas will reach a maximum velocity, which cannot be exceeded, are discussed in Chapter 4 under TERMINAL VELOCITIES. Under blow-down conditions or similar maximum flow applications, this phenomenon will control the passage of the gas through the pipe. Generally, for very short conduits, the rule that a final pressure less than half the initial pressure will not increase the flow rate, can be used. However, when the pipe system is long, relative to the diameter, more specialized calculations must be used.

Equation (4-18) provides the basis for the determination when the ratio of the length to the diameter is large. However, the graphical solution using Figure 4-1 is far less complicated.

EXAMPLE 17-6:

Methane is being blown down through a 1 inch pipe, 1200 feet long, from 270 psig and 60° F to atmosphere. Determine the probable maximum flow rate.

SOLUTION:

Methane is indicated in Chapter 4 to have a k value of 1.3, and thus Figure 4-1 can be used without correction. The value of fL/D is calculated to be 240. From Figure 4-1, the ratio of the initial to final pressure is found to be 0.052.

Reference to Figure 4-1 reveals that the velocity at the end of the pipe will be acoustic because the values of the pressure ratio and fL/D intersect on the diagonal line at the value of Mach's number of 0.056. This means that the velocity of the gas will be 5.6% of the velocity of sound, in that gas, at the *beginning* of the pipe. The velocity of sound in the gas is found using Equation (4-15 A), shown on Figure 4-1.

Equation (4-15 A):

$$V_a = 41.5\,(k\,T\,Z\,/\,G)^{0.5}$$

Where:

V_a = Acoustic velocity in gas at flowing conditions
k = Ratio of specific heats, C_p/C_v
T = Flowing temperature, degree Rankine (°F+460)
Z = Deviation from ideal gas law fractional
G = Specific gravity of gas, air at 14.69 psia and 60° F equals 1.00

Substituting values

$$V_a = 41.5\,(1.3 \times 520 \times 1.0\,/\,0.55)^{0.5}$$

$$V_a = 1455 \text{ fps}$$

The Mach's number multiplied by V_a yields the velocity at the beginning of the pipe

$$1455 \times 0.056 = 81.5 \text{ fps}$$

The rate of flow to produce the initial velocity can be found from Graph 42, VELOCITY FOR GAS, or Equation (4-4). Align 270 on the "P" Scale with 1.049 (pipe internal diameter) on the "D" Scale to locate 2.88 on the "A" Scale. Align 2.88 with 81.5 on the "V" Scale to read 819 thousand standard cubic feet per day.

If the intersection of the values of the pressure ratio and fL/D were to lie *below* the diagonal line, there could be *no additional flow*, beyond that which would result from the pressure ratio at the intersection of the value of fL/D and the diagonal line.

EXAMPLE 17-7:

In the prior example, consider the starting pressure to be 500 psig. Determine the maximum rate of flow.

SOLUTION:

The pressure ratio is now

$$14.7 / 514.7 = 0.029$$

The intersection of the value of fL/D, 240, (as calculated above) with 0.029 lies *below* the diagonal line on Figure 4-1. This indicates that pressure at the end of the pipe will be higher than 14.7. It will be the value established by the pressure ratio compatible with the intersection of the value of fL/D, 240, with the diagonal line, namely 0.052. The final pressure at the end of the pipe will be

$$0.052 \times 514.7 = 26.8 \text{ psia}$$

Because the gas is free to flow at its maximum rate, one might assume the pressure at the terminus would be 14.7; actually it will be almost double that. Also, if the gas were being discharged into a vessel at 26.8 psia rather than to air at 14.7, the rate of flow *would not be altered.*

In this example, because the parameters which govern the acoustic velocity have not changed from Example 17-6, above, the value will remain 1455 fps, and the initial velocity also will remain at 81.5 fps. However, at a *higher pressure,* the *amount* of gas which will generate that initial velocity will be larger. Using Graph 42, VELOCITY FOR GAS, as described above, the flow rate is 1479 thousand standard cubic feet per day.

When the initial pressure and fL/D are such that the terminal velocity will be *less* than acoustic, Figure 4-1 will still provide the solution.

EXAMPLE 17-8:

Methane is being blown down through a 1 inch pipe 60 feet long from an initial pressure of 27.7 psia at a temperature of 160° F to atmospheric pressure. Determine the probable maximum flow rate.

SOLUTION:

The value of fL/D is 12 and the intersection of it and the diagonal line on Figure 4-1 lies at a minimum pressure ratio of 0.21. However, the assumed pressure ratio is

$$14.7 / 27.7 = 0.53$$

The intersection of 0.53 and 12 lies at a Mach's Number of 0.2, indicating that the initial velocity will be 20% of the acoustic.

Because the temperature is higher than in Example 17-6, the acoustic velocity via Equation (4-15 A), above, will not be the same as in the prior two examples.

$$V_a = 41.5\,(1.3 \times 620 \times 1.0 / 0.55)^{0.5}$$

$$V_a = 1589 \text{ fps}$$

Thus, the initial velocity will be

$$1589 \times .2 = 318 \text{ fps}$$

Graph 42, VELOCITY FOR GAS, will be used to find the flow rate which will produce the *initial* velocity of 318 fps. The temperature of the gas is not 60° F, the basis for Graph 42. Notice that a scale for temperature (and Z) correction is placed at the bottom. If the nomograph is used to determine the velocity for a given temperature with a specified flow rate, the latter is adjusted by the temperature factor before entry into the nomograph. The same applies if Z deviates from 1.0.

In this instance, the reverse approach will be used. A flow rate with gas at 60° F will be found from the nomograph and that rate will be adjusted to match the flowing temperature of 160° F.

Transform the pressure in psia to gauge

$$27.7 - 14.7 = 13$$

Align 13 on the "*P*" Scale with 1.049 on the "*D*" Scale to find the turning point on the "*A*" Scale as 1.87. Align 1.87 with 318 on the "*V*" Scale to read 310 on the "Q" Scale.

This value is the flow rate in standard cubic feet, at a flowing temperature of 60° F which would produce the velocity of 318 fps. However, because the actual temperature is greater than 60°, a lesser quantity of gas will be required to produce the velocity.

The horizontal scale at the bottom of Graph 42 provides the factor by which the calculated quantity is to be *divided* to yield the actual flow rate

$$310 / 1.19 = 260 \text{ mcf/d}$$

If the downstream pressure were to be increased, the flow rate *would* be reduced. This may be seen from the fact that the intersection of the values of fl/D and the *actual* pressure ratio lie above the diagonal line.

If the pipe terminated in a vessel operating at a vacuum, the flow would increase as the final pressure was decreased. However, that would apply only until the pressure ratio reached 0.21, when the pressure in the vessel was reduced to 5.82 psia, or 8.88 psi below zero. Any further increase in the amount of vacuum would bring the theoretical intersection below the diagonal line, representing a physically impossible state.

CHAPTER 18

COMPARATIVE CAPACITIES OF PIPES

On occasion, the engineer may be required to determine fluid dynamic relations falling outside the normal flow problems, such as the pipe size which would be equivalent to the carrying capacity of an annulus or a pair of lines in parallel. A group of nomographs are provided to handle such problems.

EQUIVALENT PIPE SIZE FOR AN ANNULUS

Graph 58 provides the basis for determining the equivalent diameter of pipe which corresponds, hydraulically, to a given annular channel. If the pressure loss through an annulus is to be determined, Graph 58 provides the means to select the diameter of pipe which matches the annulus. Flow calculations, using nomographs or formulae provided elsewhere in this book, may then be applied to the equivalent pipe in place of the annulus.

To use Graph 58, the ratio of the outer (smaller) diameter to the inner (larger) diameter is calculated. This value is aligned on the right-hand scale with the outer diameter of the annulus on the left-hand scale. The equivalent circular diameter is read on the center scale.

EXAMPLE 18-1:

Water flows through a heating system composed of 4 inch and 2 inch schedule 40 steel pipe at the rate of 155 gpm. Determine the pressure drop per 100 feet and the velocity.

SOLUTION:

From Table B-12 find the inside diameter of the 4 inch is 4.026, and the outside diameter of 2 inch is 2.375.

The ratio of the diameters is

$$2.375/4.026 = 0.59$$

Align 0.59 on the ratio scale with 4 on the outer diameter scale on Graph 58 to find the equivalent diameter to be 2.8 inches.

Using Graph 29, PRESSURE LOSS FOR WATER, align 155 on the rate Scale with 2.8 on the internal pipe diameter Scale to find the pressure drop on the pressure loss Scales. If the pipe is new, use C-125. If old, use C-100. Assume new, so pressure drop per 100 feet is 4.5 psi.

Using Graph 33, VELOCITY FOR LIQUIDS, align 155 on the rate Scale with 2.8 on the pipe diameter Scale to read velocity as 8.3 fps.

COMPARABLE CAPACITIES

When the need arises to divide a flow into more than one pipe, Graph 61 is used. This could occur when a series of branches are to be split off from a main line. It may also be used if the flows from a number of pipes are to be collected into one line.

EXAMPLE 18-2:

Determine the pipe size required to carry the flows from 8 pipes 2 inches in diameter.

SOLUTION:

Align 8 on the ratio Scale with 2 on the right Scale to read the required diameter as 4.5 inch.

COMPLEX PIPE LINES, EQUIVALENT LENGTH

Graph 59 is used to determine the equivalent length of one size which matches a given length of some other size pipe.

LOOPED LINES

When pipes are run in parallel, or looped, use Graph 60 to select combinations which will have the best possible match of sizes. If the sides of the loop are not of the same length, use Graph 59 to determine a size which has the same characteristics as when the length matches that of the other leg.

EXAMPLE 18-3:

A pair of pipes, one of 12 inch and the other of a size to be selected, handle the output of an 18 inch and then will deliver fluid to another 18 inch. The length of the 12 inch is 750 feet, and the length of the size to be determined is 110 feet. Determine the unknown size.

SOLUTION:

From Graph 60, determine that the parallel line, *assuming* it to be *750* feet long also, would be 15.5 inch. Using Graph 59, align 15.5 on left scale with 750 on right scale to locate 10.5 on the turning scale. Align 10.5 with 110 on the right scale to read the required diameter for 110 feet, 10.8 inches.

HEADER DESIGN

When a series of side take-off lines are fed by a common pipe or header, it may be important, in some applications but not all, for the discharge from all branches to be the same. If modes of control are placed in the branches, this requirement may not be present. If, on the other hand, no control is possible, as, for example, in a gas burner or in a system supplying gas to a group of compressors, then inherent in the design should be the provision for equal flow through each portion.

It is to be recognized that two opposing influences are at work in a header. As the fluid is drawn off through each branch, the velocity in the header will be decreasing. With such a decrease, there will be an *increase* in the static pressure as the result of the loss of velocity head, see VELOCITY HEAD, Chapter 5. At the same time, the flow through the pipe will produce a pressure *loss* due to the friction of the flow.

The amount of fluid delivered through each take-off depends upon the pressure at the particular take-off location. Thus a balance between these two influences is desired, thereby keeping the pressure constant.

Experience has shown that the ratio of the summed areas of outlets to the area of the pipe is a controlling criterion. This ratio should be kept in the vicinity of 5 to 70 for the most efficient distribution. The size header resulting from the larger value will be several pipe sizes larger than the size selected by LAC and, hence, would be used only where the very best distribution is required, for example on a gas burner.

SIMPSON (17) describes a rigorous method for header selection which may be justified in some instances.

Figure 18-1

Dehydrogenation Unit

The velocities associated with LAC design are higher than desirable for good distribution in a header. If header size is not increased over the LAC, the *first outlet port* should be no less than 5 header pipe diameters *beyond* the inlet.

CHAPTER 19

METHODS OF APPROACH

The challenge in pipe size selection becomes the determination of the basis upon which the selection is to be made. As familiarity with the work is gained, the recognition of the correct approach becomes almost automatic for the more common applications, and the analysis of the unusual lines become simpler. To repeat, there are three possible classifications for line size selection.

a. There is a given pressure that may be consumed.

b. The fluid is being pumped or compressed, and a pipe size yielding the minimum yearly operating cost should be selected; this cost includes the cost of installation (amortized over some period) and the cost of pumping or compressing against the pipe line friction.

c. There is a minimum or maximum velocity limitation.

d. Actually, a fourth classification exists, a selection made for structural strength of pipe or any other arbitrary reason.

All pipes within a plant can be classified as one of the above types, although more than one check may be required to determine which one is limiting.

It should be understood the limitations set forth in this book are not to be considered inflexible. As experience is gained, engineering judgment and common sense will indicate those places where the recommendations must be tempered; although frequently the alteration actually will be the more accurate recognition of the correct limiting factors.

To present a summary for a general guide, a number of examples of each of the types, as well as a resumé of general limiting conditions, follows:

A. LINES LIMITED BY PRESSURE DROP AVAILABLE

1. Overhead vapor lines from fractionating columns are ordinarily limited to a pressure drop of 0.5 psi/100. If the line to the condenser is very long or if exceptionally close cuts are required, this may be reduced to 0.5 psi total. If in doubt consult the Process Engineer.

2. The pipes between a still partial condenser accumulator and final condenser have pressure drop limitations expressed on the flow sheet.

3. Kettle-type reboiler circuits are limited to a drop of 9 inches head of liquid for both lines.

4. Thermo-syphon reboilers have a head limitation equal to the liquid head from the bottom of the oil outlet nozzle on the vessel to the top of the exchanger, minus the liquid-vapor head (average density) from the top of the exchanger to the bottom of the liquid vapor return nozzle on the column. (See also velocity limitation). Reheaters and inter-coolers are similar unless a pump is placed in the circuit.

5. Any system not involving a pump or compressor and in which flow is induced by a difference in pressure between the two vessels is limited by that pressure differential.

a. When no control valves are involved, the actual differential between the two vessels may be consumed. Consider static head in liquid flow which may increase or decrease the pressure drop available for line friction.

b. When control valves are used, determine the differential that may be consumed in pipe line, versus the control valve, by the method given in the Introduction. The pressure drop through exchangers, mixers, preheaters, etc. is not to be considered as pipe line friction. Such pressure drops, as well as static head that must be bucked, should be subtracted from the differential between the vessels to obtain the amount available for control valve and pipe line. Ordinarily, if there is no liquid vaporization or extreme expansion of gases beyond the valve, the line size should be from one to two pipe sizes larger than the control valve size.

6. There are a few special conditions in which a pump discharge line should be selected on a pressure drop basis. One is for the high pressure lean oil going to a low pressure absorber with a control valve in the line. In this case, the pipe from the pump to the high pressure absorber would be an LAC selection, while the branch to the low pressure absorber would be selected on the basis of pressure drop available, taking up to 20% in the line and the remainder in the valve.

   Overhead product lines from the reflux pump discharge are similar.

7. When a line from a vessel to a control valve is short and is used as a meter run, the size of the line is fixed by the requirements of the meter and orifice.

8. Vent lines from accumulators are handled like type 5-b above. The same is true if high pressure absorber residue gas is blown down for fuel gas.

B. PIPES LIMITED BY LAC SELECTION

1. In general this includes the suction and discharge headers for a compressor and the discharge lines from a pump. Branch lines to and from compressors should be selected as explained in Chapters 9 and 18; however the manufacturer of the machine may specify these sizes.

2. The presence of a control valve in a pump or compressor discharge does not rule the line out as an LAC size classification. The pressure drop across the valve will be acting as a governing agent only; thus, the size selection still will be LAC, and ordinarily, the motor valve will be one size smaller. Of course, if a side stream is taken from the main discharge line and it passes through the regulator to a considerably lower pressure, the pipe is then classed as PDA.

C. PIPES LIMITED BY VELOCITY CONDITIONS (VA)

1. Pump suction lines are the most common selections of this type.

2. The liquid outlet lines from any vessel in which liquid and vapor are in equilibrium (absorbers, reboilers, stills, accumulators, flash-tanks) should be sized to limit velocity to 7 feet per second until one bend has been made in the piping, to prevent a vortex from drawing vapor as well as liquid into the pipe.

3. The feed to fractionating columns should not enter the vessel at a rate exceeding 10 feet per second, 5 feet is better. This applies to two-phase as well as liquid flow. Up to the last elbow into the vessel, this limitation does not hold. The purpose of reducing the entrance velocity is to avoid sweeping the liquid seal off the tray and thus destroying its efficiency.

4. Sufficient velocity to insure entrainment should be provided for:

   a. Liquid-vapor feed to fractionating columns, up to the last elbow (see above).

   b. The liquid-vapor return line of a thermosyphon reboiler.

   c. The liquid-vapor line from a reflux condenser located in the cooling tower when the accumulator entrance connection is higher than the condenser outlet.

5. Lines carrying liquid and vapor at low pressures sometimes may be limited by erosion velocity rather than pressure drop; when high pressure drops are taken in the line this should be checked. Consider, especially, the first elbow after a motor valve when flashing occurs.

6. The velocity limitation for two-phase flow at pressures over 400 psi, such as rich oil lines from high pressure absorbers to vent tanks, etc., is determined by the impact limitation, see Chapter 17.

7. Noise in steam lines where pressure drop is not limiting may be undesirable. So far, little is known about it, but reports indicate that at 425 feet per second at 20 psi, the noise level was moderate.

D. PIPES SELECTED ON AN ARBITRARY BASIS

1. The line from a shell and tube reflux condenser mounted above the accumulator is arbitrarily made one size smaller than the overhead line.

2. Lines to storage or small lines placed on piperacks may be arbitrarily limited to 1.5 inch or 2 inch for strength or rigidity.

CHAPTER 20

CONCLUSION

The methods of pipe size selection presented herein may seem more comprehensive than required for the average project. However, this book is for more than the average job — it is intended to be used for virtually all the challenges the engineer will encounter, over the years, as processes, materials and locale all undergo change, even radical change. By becoming familiar with the workings of the various modes of pipe size selection and emphasing those which apply to his needs, the engineer will develop an ease of use that will facilitate the applications of these principles.

At the beginning, more time may be required, but only until one establishes familiarity with the commonly used nomographs. It is to be recalled that these will speed the selection of pipe sizes significantly and the use of them is urged. Presently, the first benefit of the use of Optimum Pipe Size Selection will emerge — reduction in engineering time to make the selections. Experience has shown this can be reduced to *less* than one-half percent of the total plant costs, on the average.

The second benefit lies in the certainty of the correct size for all applications. No longer does one need to ponder which rule of thumb to use. The *correct size* can be chosen swiftly and accurately.

The third benefit is the reduction of plant costs. According to Mr. Ryle Miller, an editor for Chemical Engineering Magazine, such savings can amount to between 4 and 18% of the plant costs. With a construction backlog now on the books of the engineering organizations which is well over $3,000,000,000, as of the writing of this book (1975), such potential savings are astonishing figures of $120,000,000 to $540,000,000! This assumes that all such projects would be designed according to the principles laid down herein; and one must admit that is a bit optimistic! Nonetheless, the message can be read with clarity; the use of Optimum Pipe Size Selection techniques can save manpower, speed up project completion, reduce materials consumption and save power — all of which adds up to saving money!

References

1. GENERAUX: Chem. & Met. Eng., 44 (5) 241—248 (1937)
2. JOHNSON & MAKER: Proc. Tenth Mid Year Meeting, American Petroleum Institute, Sec III, 7—23, (1940)
3. MENDEL, O.: Chem. Eng. 255—256, June 17, 1968
4. MARSHALL, S. & BRANDT, J.: Chem. Eng. 94—106, Oct. 28, 1974
5. REYNOLDS, O.: Phil. Trans. Royal Soc., London, 1883
6. STANTON, T. E.: "Friction", Chaps II & IV. Longmans Green & Co., London 1932
7. PERRY, et al.: Chemical Engineer's Handbook, 4th Edition, McGraw-Hill, N. Y. 1950
8. COMINGS & EGLY: Industrial Eng. Chem., 32, 714—718, (1940)
9. BROWN, SOUDERS & SMITH: Ind. Eng. Chem., 24, 513, (1932)
10. WILLIAMS & HAZEN: "Flow of Water in Pipes", John Wiley & Sons, Inc., New York, 1920
11. SCHILLER, L.: Forschungsarbeiten Ver. deut. Ing., Vol 248 p. 16
12. DODGE, B. F.: Chemical Engineering Thermodynamics, McGraw-Hill, Inc., New York, 1944
13. DODGE, R. A. & THOMPSON, M. J.: Fluid Mechanics, McGraw-Hill, Inc., New York, 1937
14. EDMISTER, W. C.: Ind. Eng. Chem., 32, 373, (1940)
15. DUKLER, A. E.: 11th Advanced Seminar — Two-Phase Liquid Gas Flows, AIChE Meeting, New York, Nov. 1967
16. DeGANCE & ATHERTON: Phase Equilibria, Flow Regimes, Energy Loss: Chem. Eng., April 20, 1970, page 158
17. SIMPSON, L. L.: Chemical Engineering, page 207, June 17, 1968
18. BABER, O.: Experiences with Two-Phase Pipelines, Paper for Canadian Natural Gas Proc. Assn., Calgary, Sept. 15, 1960
19. KNOWLES, C. R.: M. S. Thesis, Univ. Texas, Austin, 1965
20. WICKS & DUKLER: AIChE Journal, 6, 463—468, 1960
21. HARDEMAN, D., et al.: 8th Congress Inter. Assoc. Hydraulic Research, Aug. 1950
22. RUSKAN, R. P.: Chem. Eng., Vol 82, No 27, pp 101—103
23. LOCKHARDT, R. & MARTINELLI, R.: Chem. Eng. Progress, Jan 1949, pp 39—48
24. DUKLER, A., et al.: AIChE Journal, Jan 1964, pp 44—51
25. BEIJ, K. H.: Journal of Research of the National Bureau of Standards, Vol 21 July, 1938
26. ANDRADE: Endeavor, 13, 117 (July 1954)
27. KENDALL & MONROE: J. Am. Chem. Soc., 39, 1787 (1917)
28. TAYLOR: Proc. Royal Soc. (London), 138 A, 41 (1932)
29. PERRY, et al.: Chemical Engineer's Handbook, 2nd Edition, McGraw-Hill, New York, 1941
30. WATSON, K. M. & NELSON, E. F.: Ind. Eng. Chem. 25, 880—887 (1932)
31. KAY, W. B.: Ind. Eng. Chem., 28, 1014—1019, (1936)
32. NELSON, W. L.: Petroleum Refinery Engineering, McGraw-Hill New York, 1936, page 104
33. SMITH, R. L. & WATSON, K. M.: Ind. Eng. Chem., 29, 1408—1414, (1937)
34. DURAND, R. & CONDOLIOUS, E.: "Experimental Investigations on the Transport of Solids in Pipes", Journal d'Hydraulique, Société Hydrotechnique de France, Grenoble, June 1952
35. FADDICK, R. R.: "A Mineral Slurry Data Book", U.S. Bureau of Mines, March 31, 1972
36. ZANDI, I. & GOVATOS, T.: "Heterogeneous Flow of Solids in Pipelines", Journal of the Hydraulic Div., ASCE, Vol 92, No HY3, Proc. Paper 5244, May 1967, pp 145—159
37. CHIMES, A. R.: Chem. Eng., Aug 5 1974, pp 118—120
38. WASP, E. et al.: Slurry Pipeline Transportation, Trans Tech Publications, Clausthal (Germany), 1977
39. CHEREMISINOFF, N. P.: Estimating Noise from Control Valves, Pollution Engineering, Vol 9, No. 6, pp 48—50, June 1977

APPENDIX A
NOMOGRAPHS

A nomograph (or alignment chart) is a graphical method for rapid and accurate solution of mathematical equations. Because of the character of formulae used in fluid flow calculations, involving exponentials of a decimal nature, nomographs are particularly useful. Even with the common-place availability of computers, the use of the nomograph *still* can be a time saver. In less time than it takes to prepare and punch the information for the computer, the engineer, with the aid of the nomograph, can establish the correct pipe size.

Granted, when special two-phase flow calculations must be made using the methods of DUKLER or MARTINELLI, the computer approach is by far the better. For routine size establishment, however, such is not the case.

In the development of a nomograph, in the interests of having the scales as large as possible and to avoid undue complexity, the designer will select for the scales those parameters which most commonly will vary. Those elements which less commonly will change are set aside to be entered into the problem by the application of correctional factors. These are applied to the value of a specific parameter before it is used in the nomograph.

The use of such factors applies mainly to the LAC nomographs, of which there are a number in this volume. To speed the use, the more commonly used factors are reproduced on the facing page of each LAC graph along with references to the locations in the text of any other related materials.

Examples of the usage of the various nomographs are provided in the Chapters relating to the specific formulae for which they are designed.

For nomographs with more than three scales, a key to the usage is provided to indicate the order in which the scales are to be used.

EXAMPLE A-1:

The key reads, "*PDA, AQF*". *P* and *D* and Q are known. What is *F*?

SOLUTION:

Align *P* with *D* to obtain *A*. Align *A* with Q to read *F*. Alternately, if *D*, Q, and *F* are known; to obtain *P*, first align Q with *F* to find *A*. Then align *A* with *D* to find *P*. At least 3 of the 4 variables must be known to establish the value of the fourth. If, however, only two are known, one can quickly chart a series of possible answers permiting a rapid solution via comparison with other data.

In the use of the nomographs, a transparent straight edge works better than a ruler. Marking the page with a pencil line between two values will result in the nomograph becoming very cluttered. A better method is to place a blunt but pointed object on the turning scale and rotate the straight edge about it. A blunted needle set into a wooden holder makes a good tool.

GRAPH 1

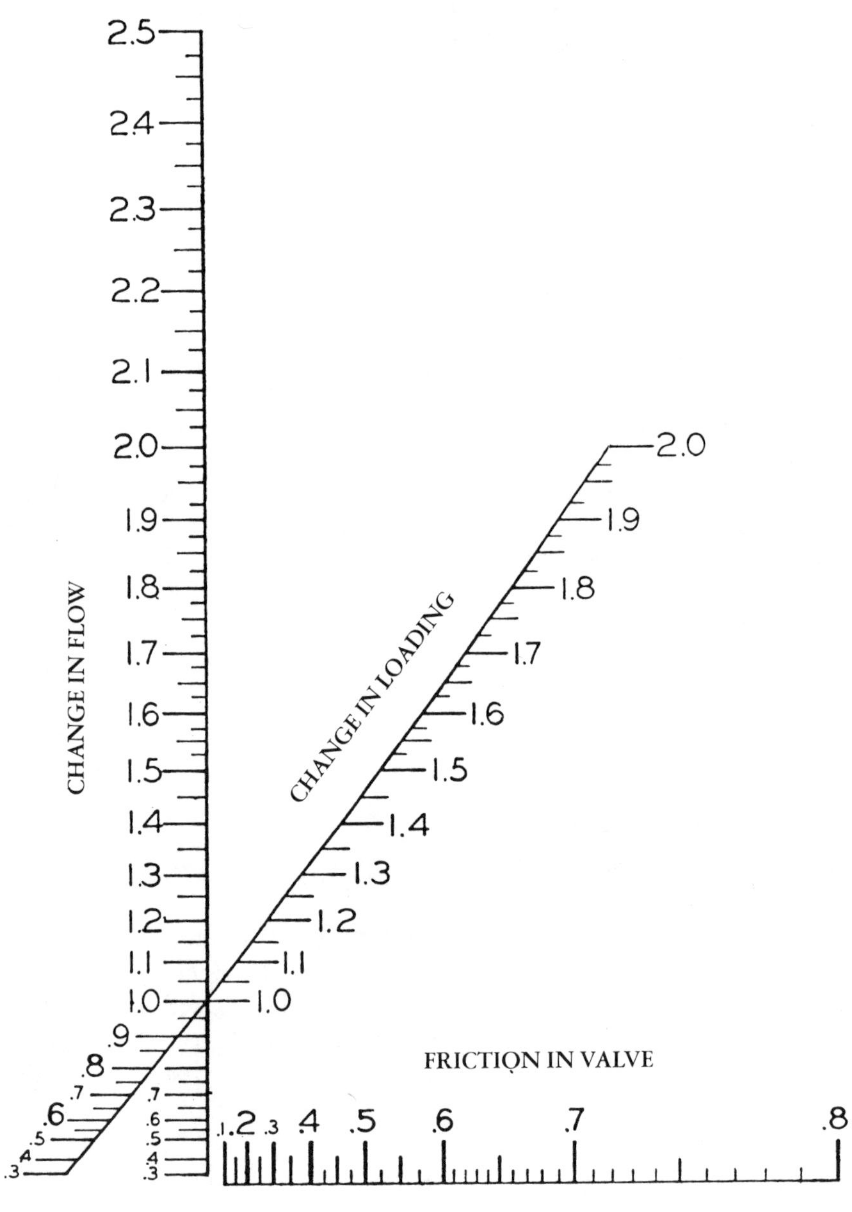

CONTROL VALVE SELECTION

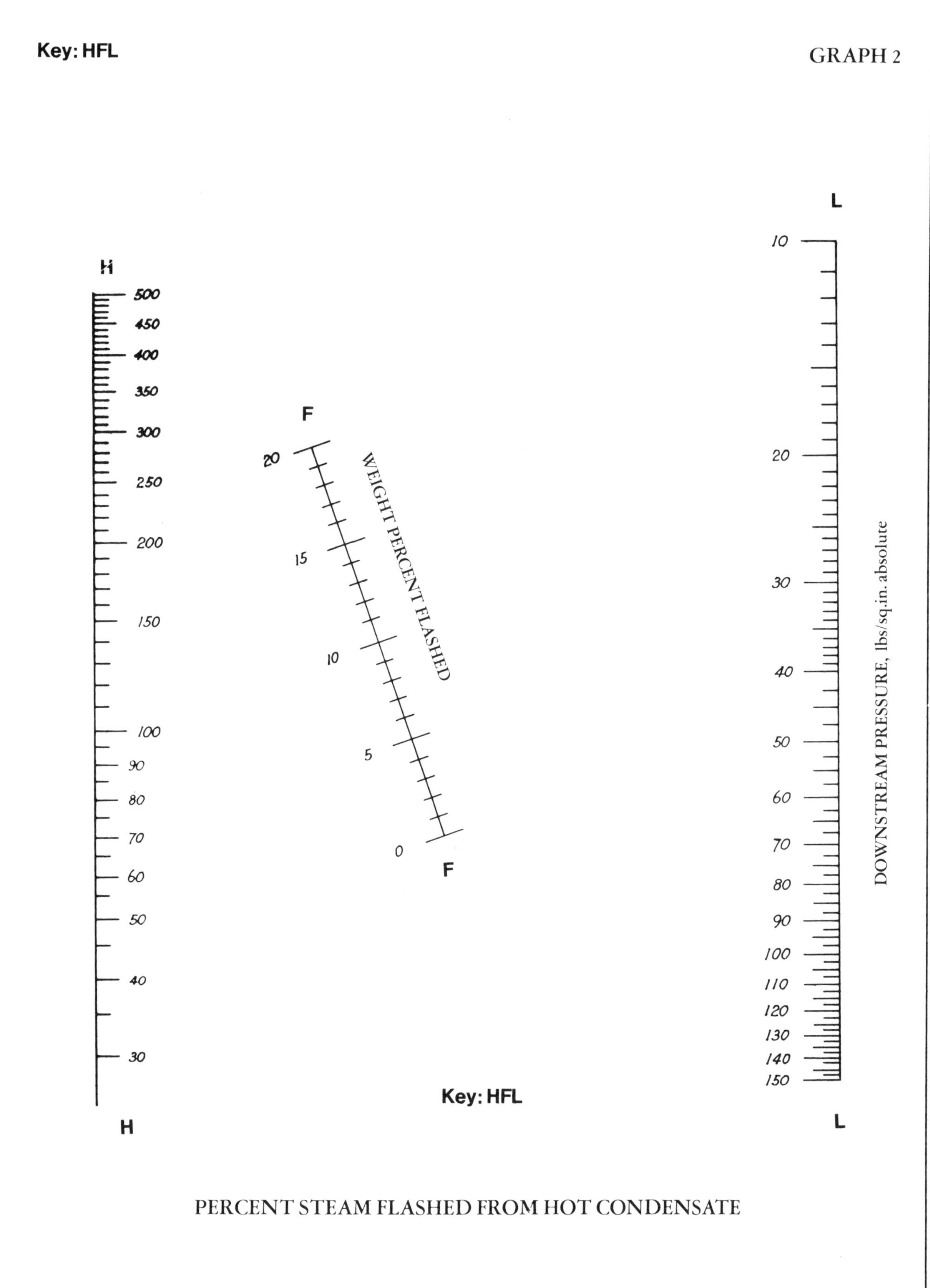

PERCENT STEAM FLASHED FROM HOT CONDENSATE

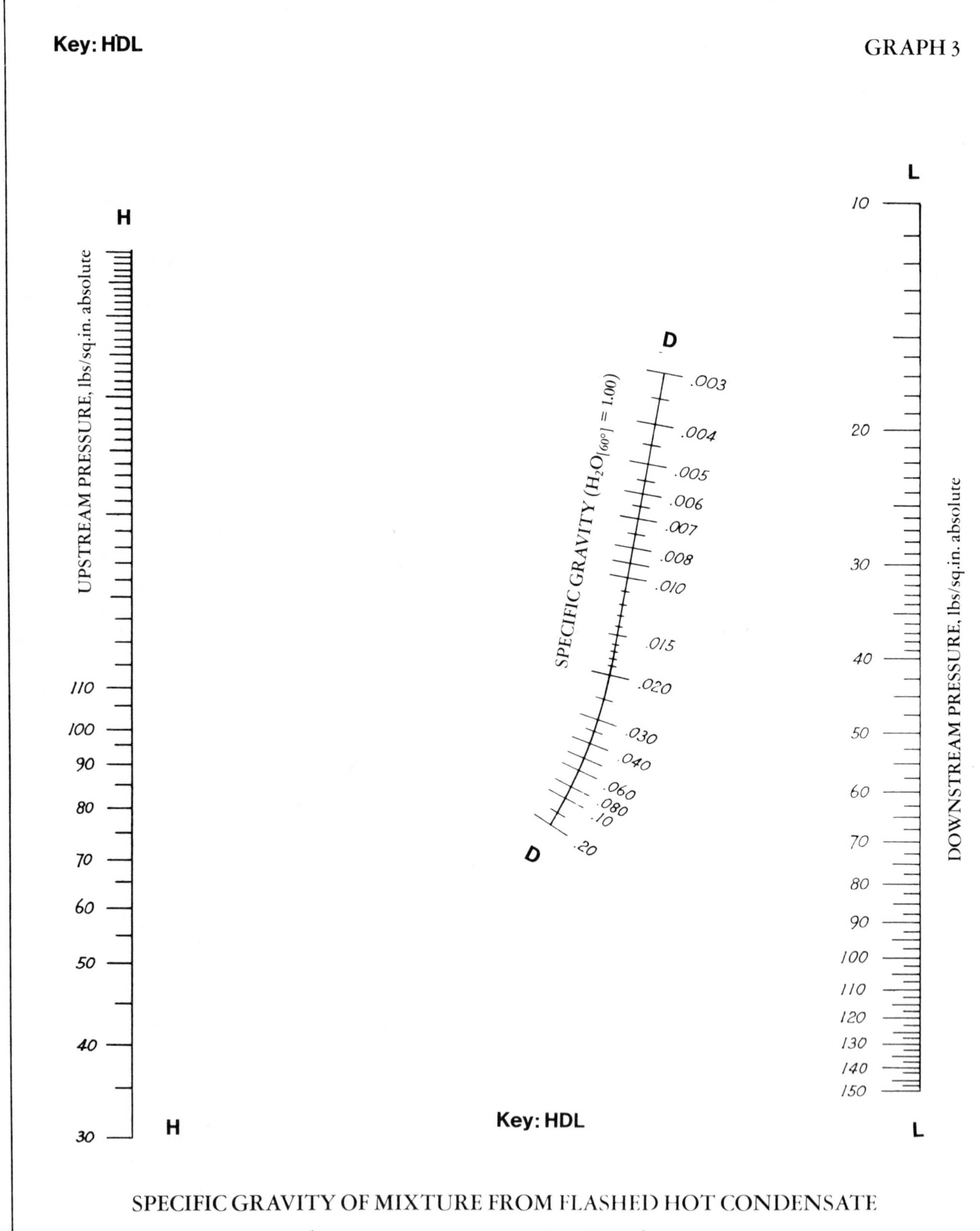

SPECIFIC GRAVITY OF MIXTURE FROM FLASHED HOT CONDENSATE

(Density equals specific gravity multiplied by 62.4)

GRAPH 4

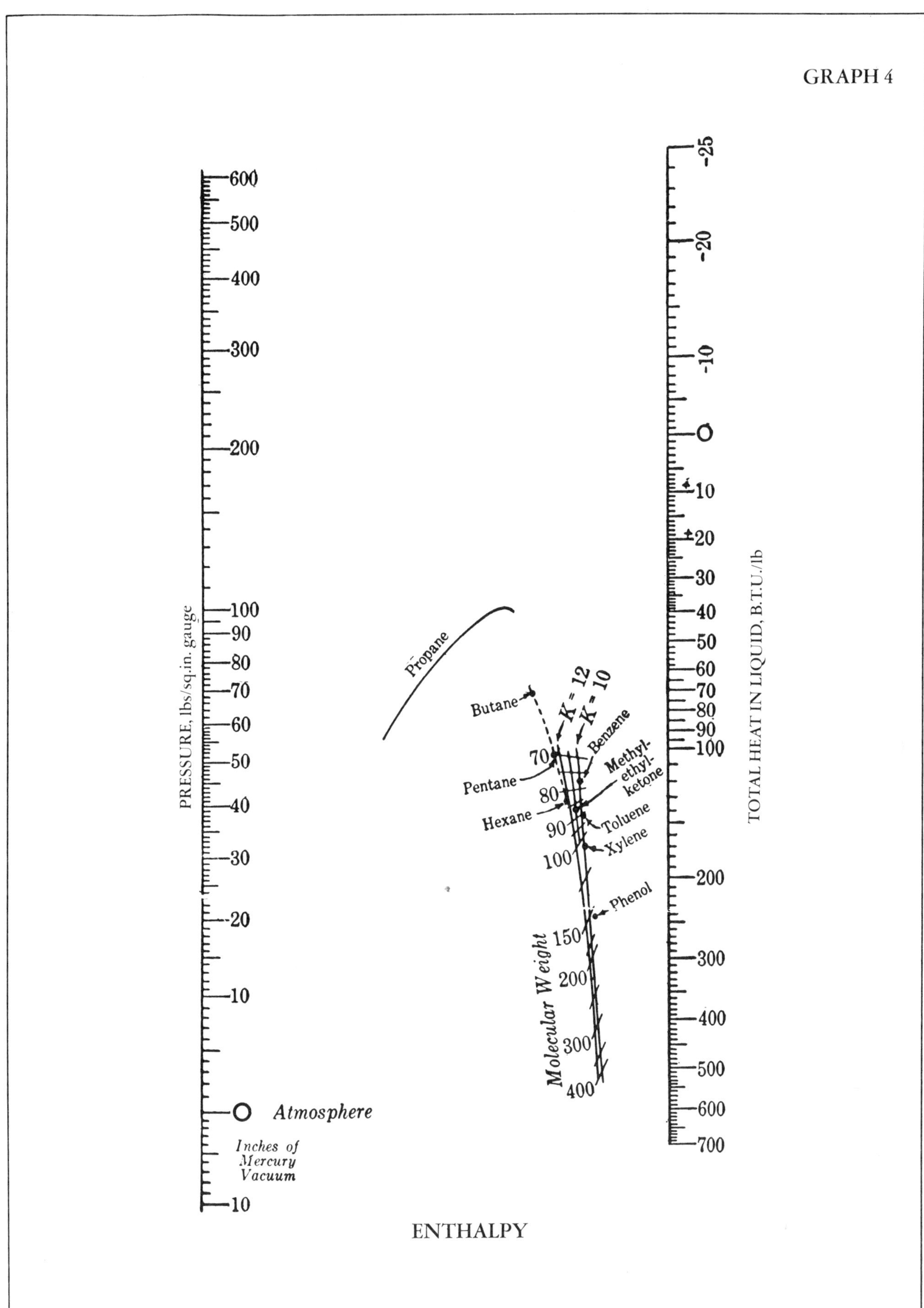
PRESSURE, lbs/sq.in. gauge
600
500
400
300
200
100
90
80
70
60
50
40
30
20
10
O Atmosphere
Inches of Mercury Vacuum
10
Propane
Butane
K = 12
K = 10
Benzene
Methyl-ethyl-ketone
Pentane
Hexane
Toluene
Xylene
Phenol
70
80
90
100
150
200
300
400
Molecular Weight
TOTAL HEAT IN LIQUID, B.T.U./lb
-25
-20
-10
O
10
20
30
40
50
60
70
80
90
100
200
300
400
500
600
700

ENTHALPY

GRAPH 5

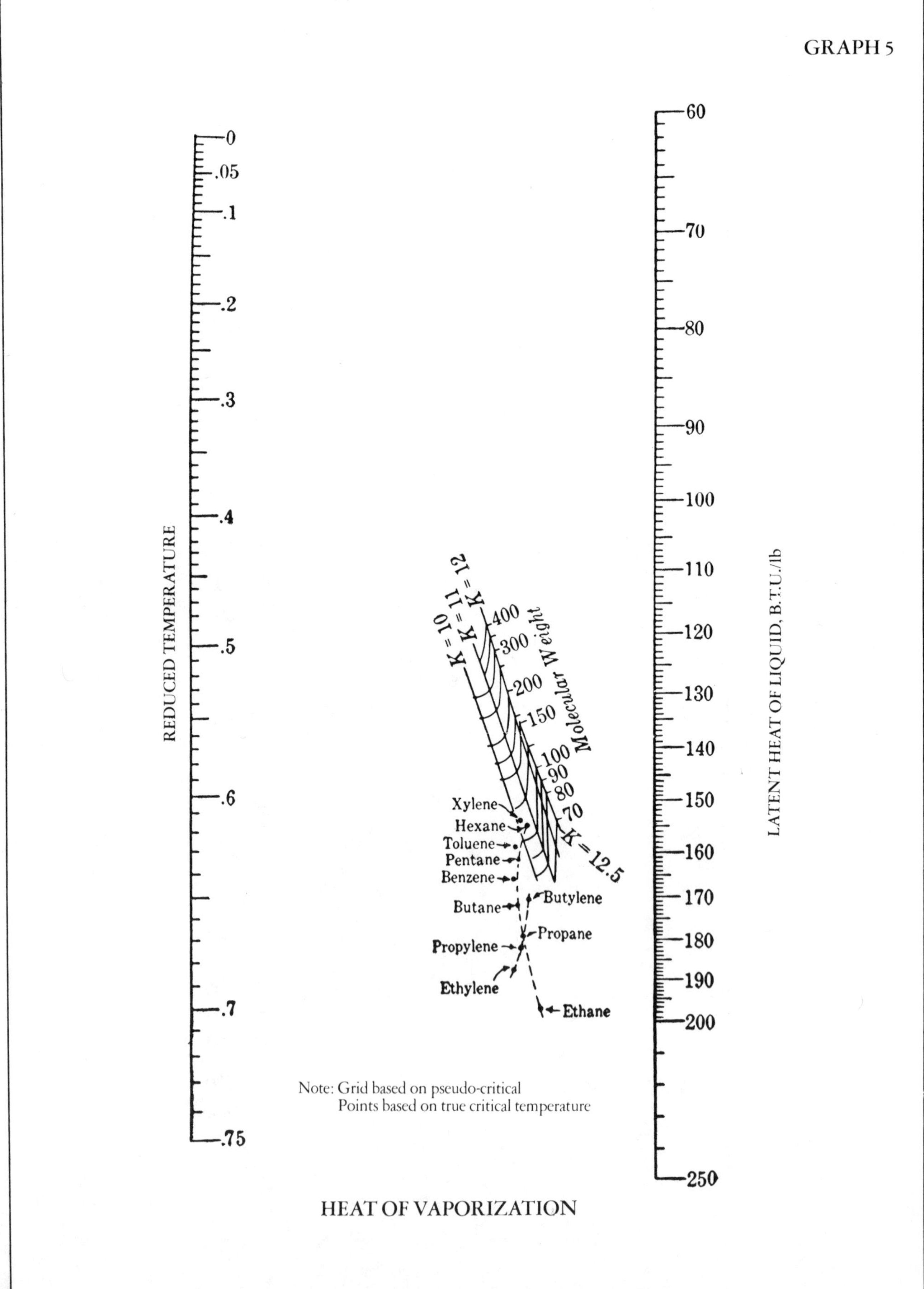

HEAT OF VAPORIZATION

GRAPH 6

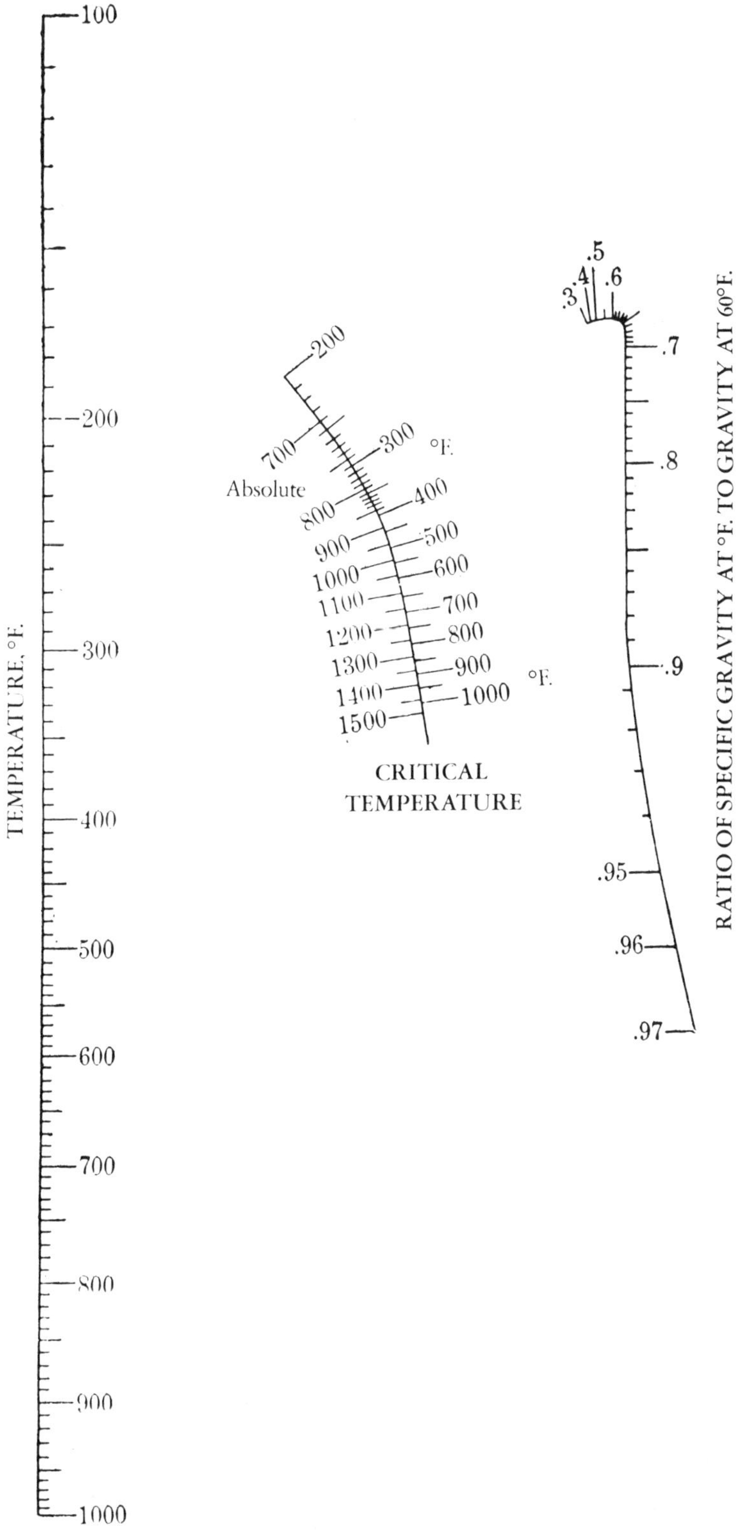

TEMPERATURE, °F.

100
200
300
400
500
600
700
800
900
1000

CRITICAL TEMPERATURE

°F.: 200, 300, 400, 500, 600, 700, 800, 900, 1000 °F.

Absolute: 700, 800, 900, 1000, 1100, 1200, 1300, 1400, 1500

RATIO OF SPECIFIC GRAVITY AT °F. TO GRAVITY AT 60°F.

.3 .4 .5 .6 .7 .8 .9 .95 .96 .97

THERMAL EXPANSION OF HYDROCARBONS

GRAPH 7

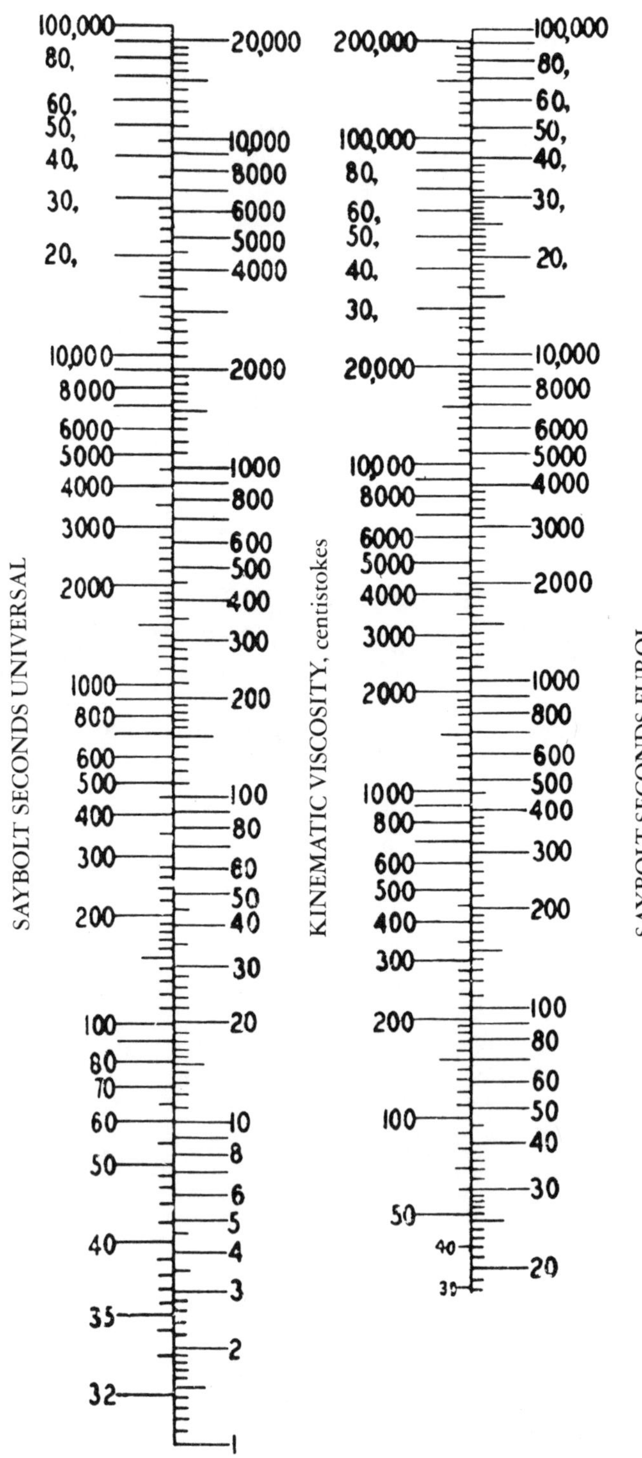

VISCOSITY CONVERSIONS

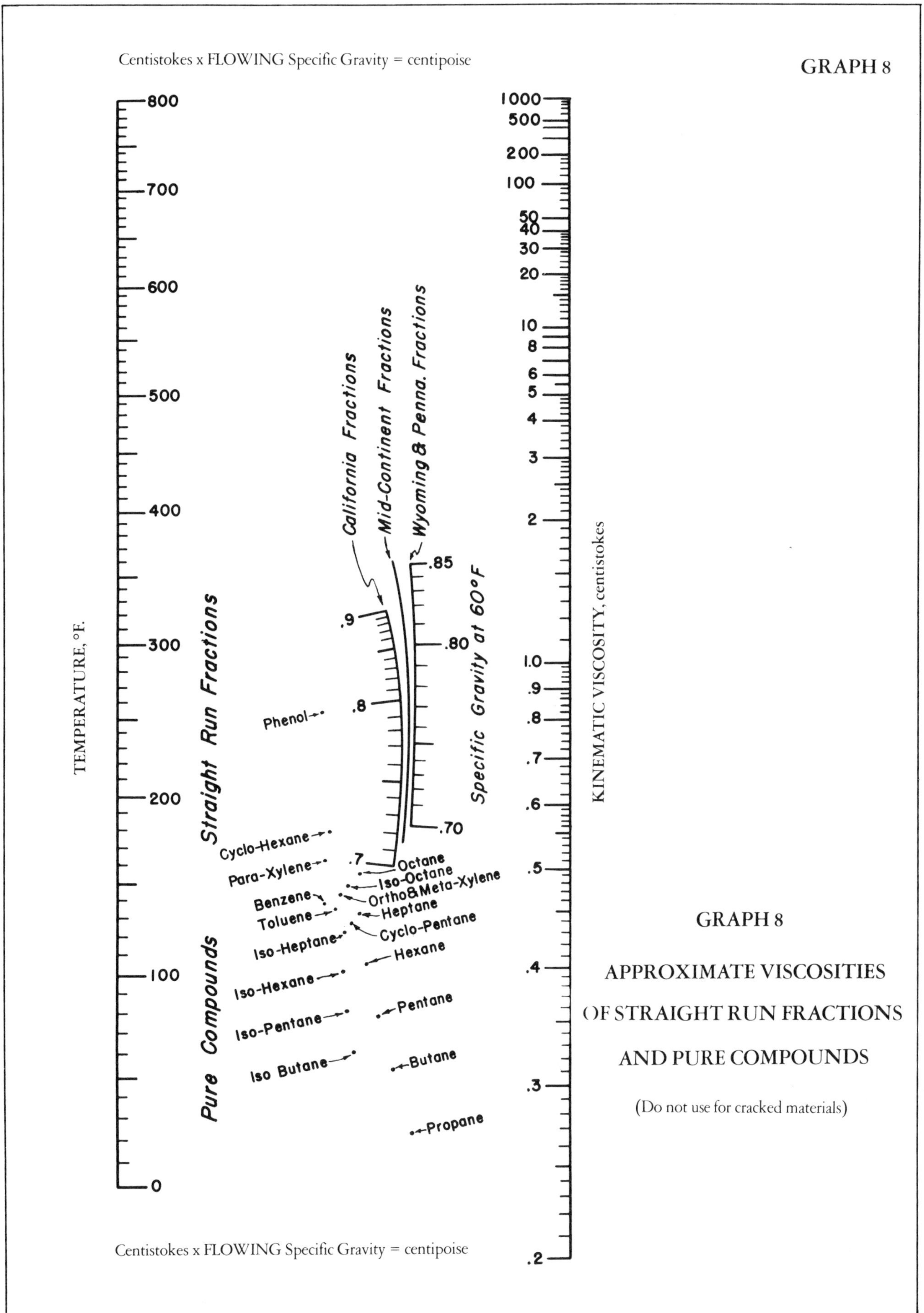

GRAPH 8

APPROXIMATE VISCOSITIES OF STRAIGHT RUN FRACTIONS AND PURE COMPOUNDS

(Do not use for cracked materials)

GRAPH 9

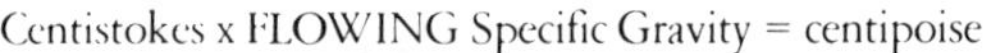

Centistokes x FLOWING Specific Gravity = centipoise

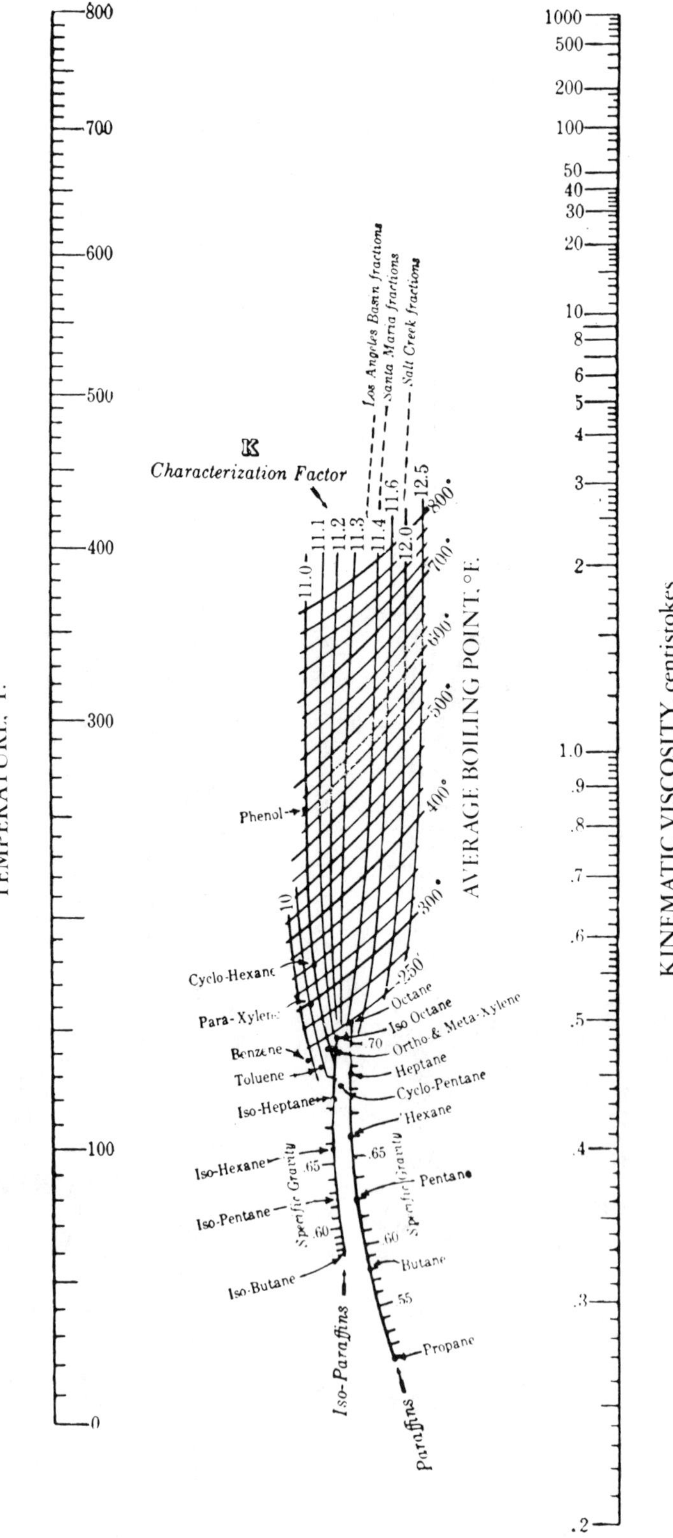

APPROXIMATE VISCOSITIES OF STRAIGHT RUN FRACTIONS

(Based upon K and average boiling point)

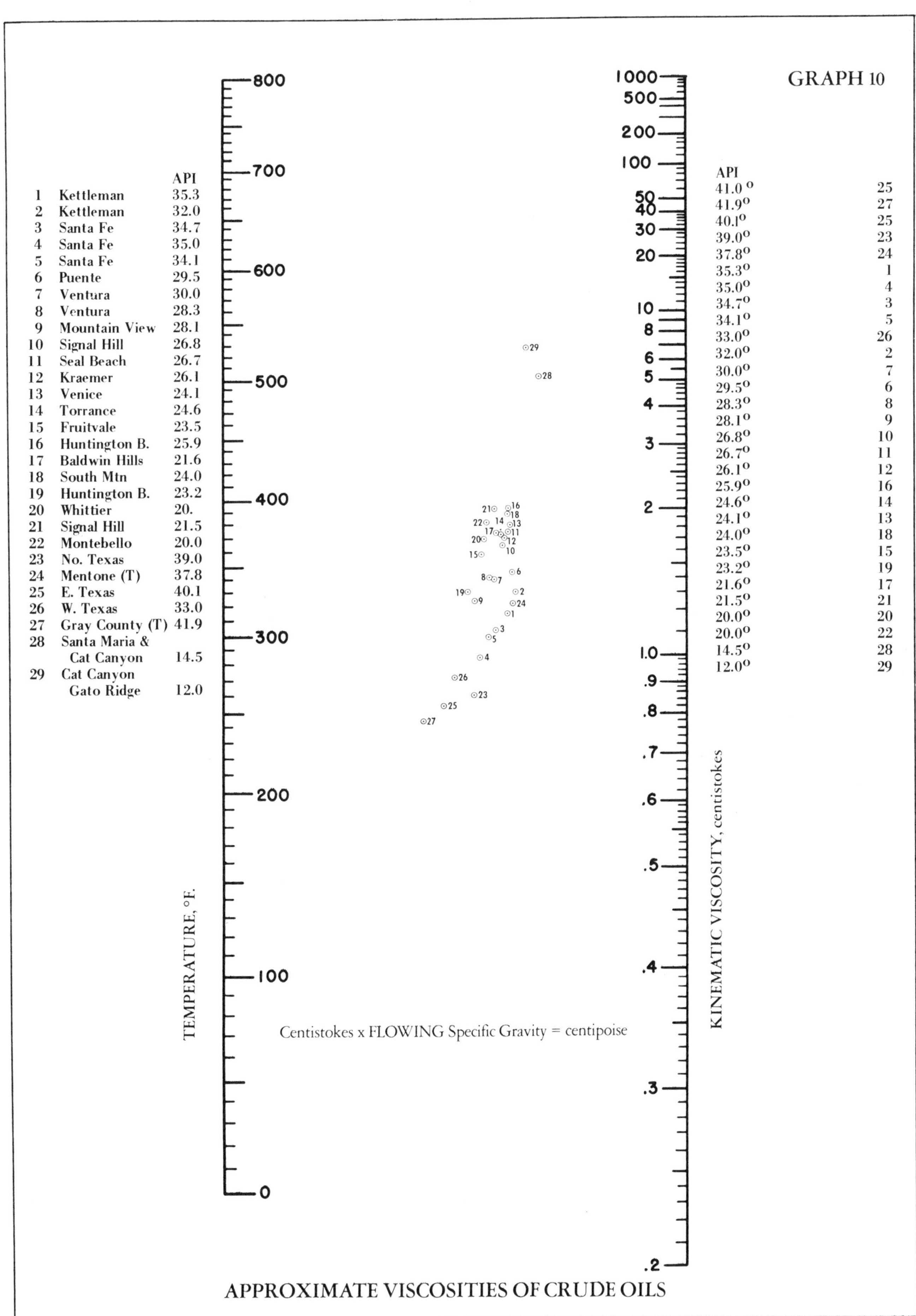

APPROXIMATE VISCOSITIES OF CRUDE OILS

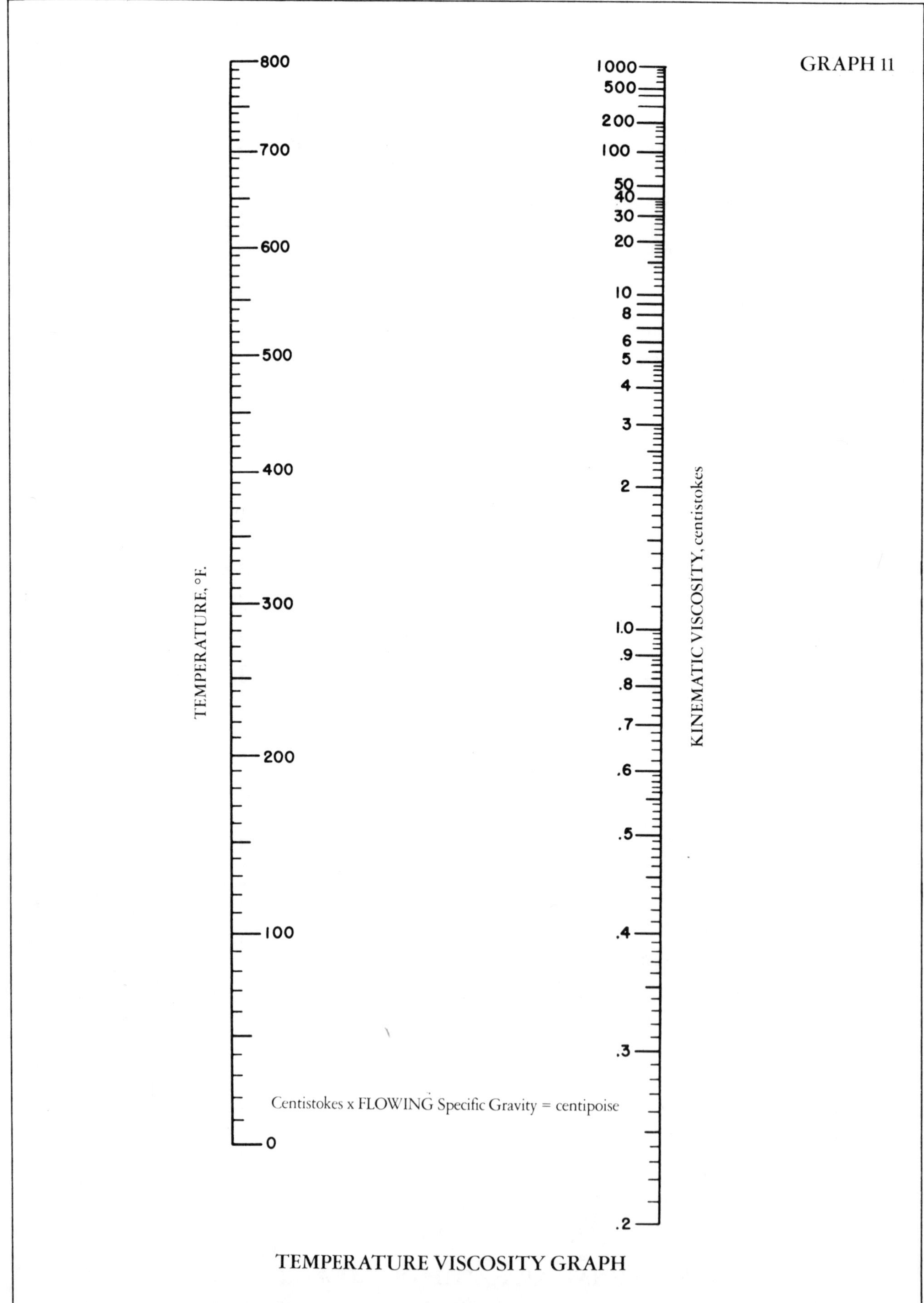

TEMPERATURE VISCOSITY GRAPH

GRAPH 12

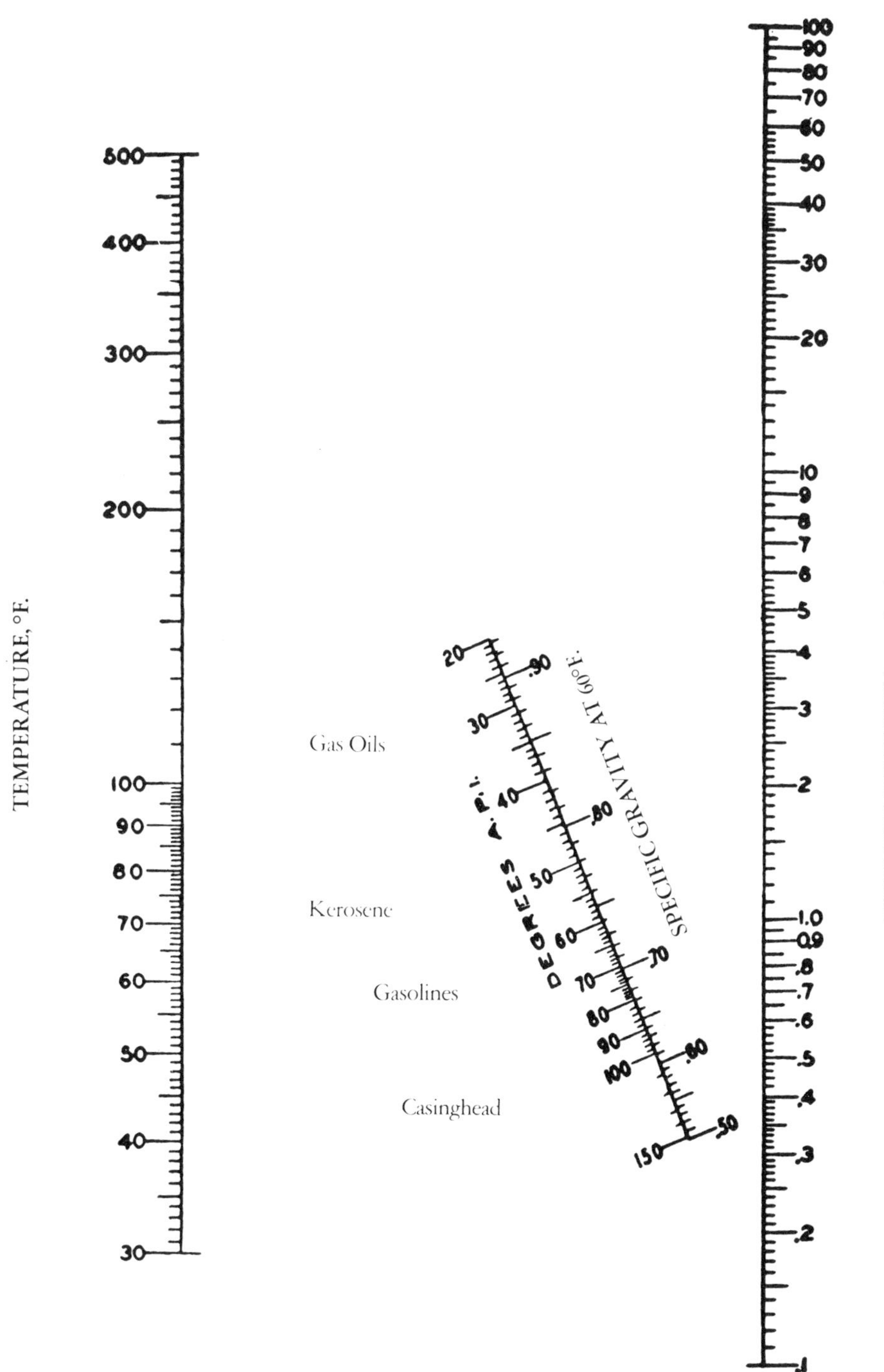

APPROXIMATE VISCOSITIES OF VARIOUS HYDROCARBONS

GRAPH 13

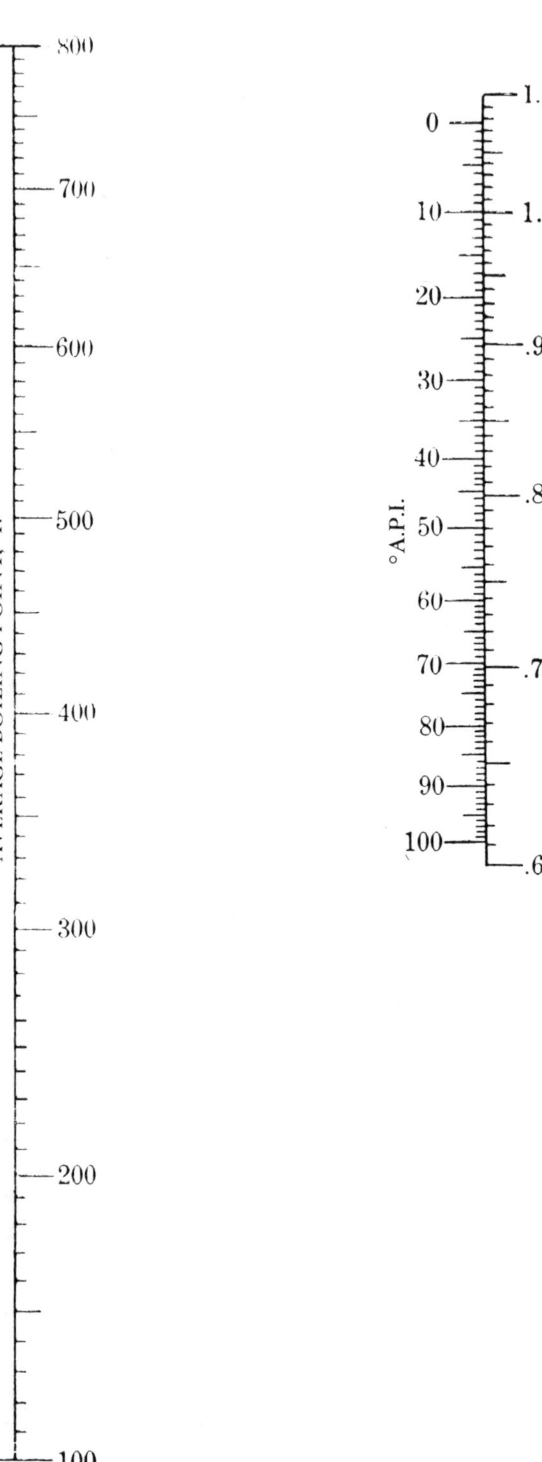

CHARACTERIZATION FACTOR, SPECIFIC GRAVITY AND AVERAGE BOILING POINT

GRAPH 14

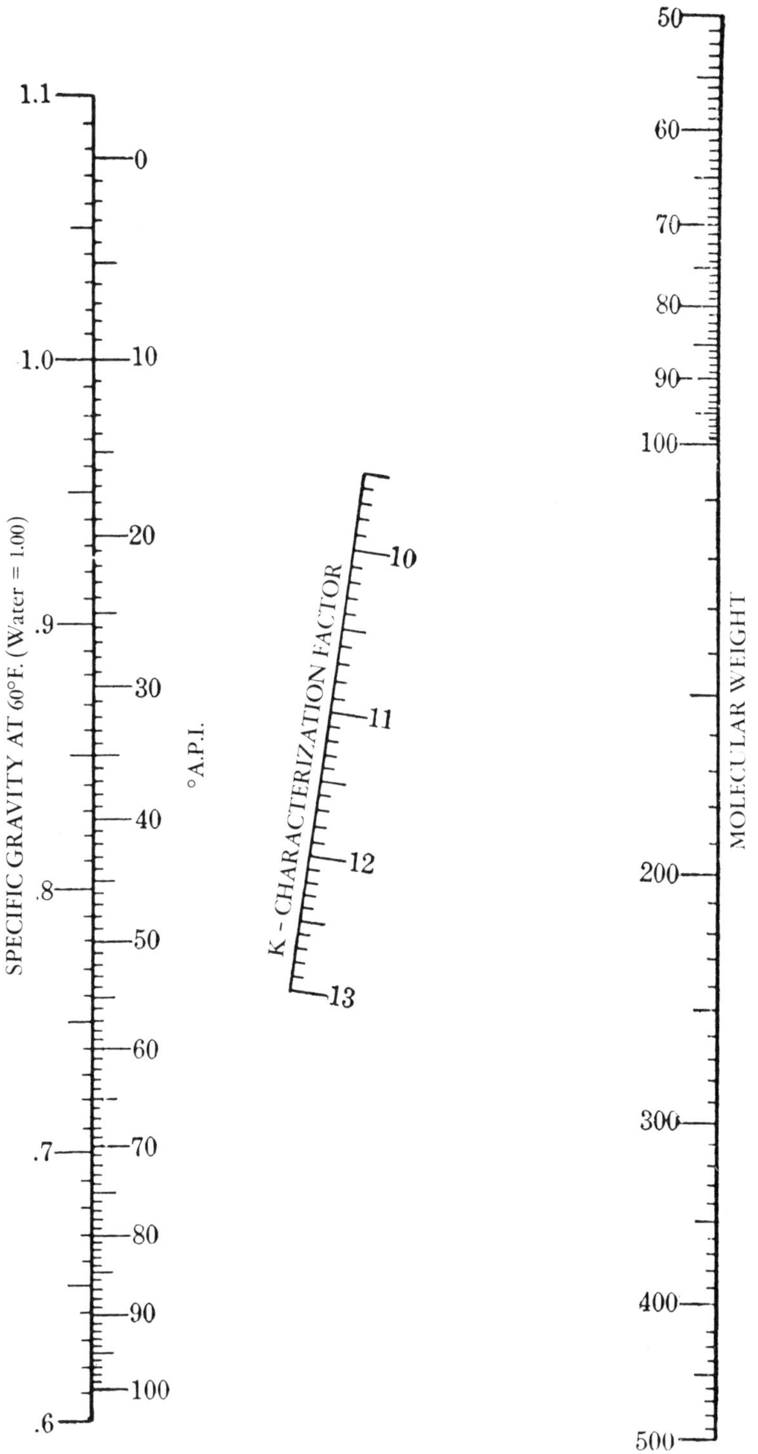

CHARACTERIZATION FACTOR, SPECIFIC GRAVITY AND MOLECULAR WEIGHT

GRAPH 15

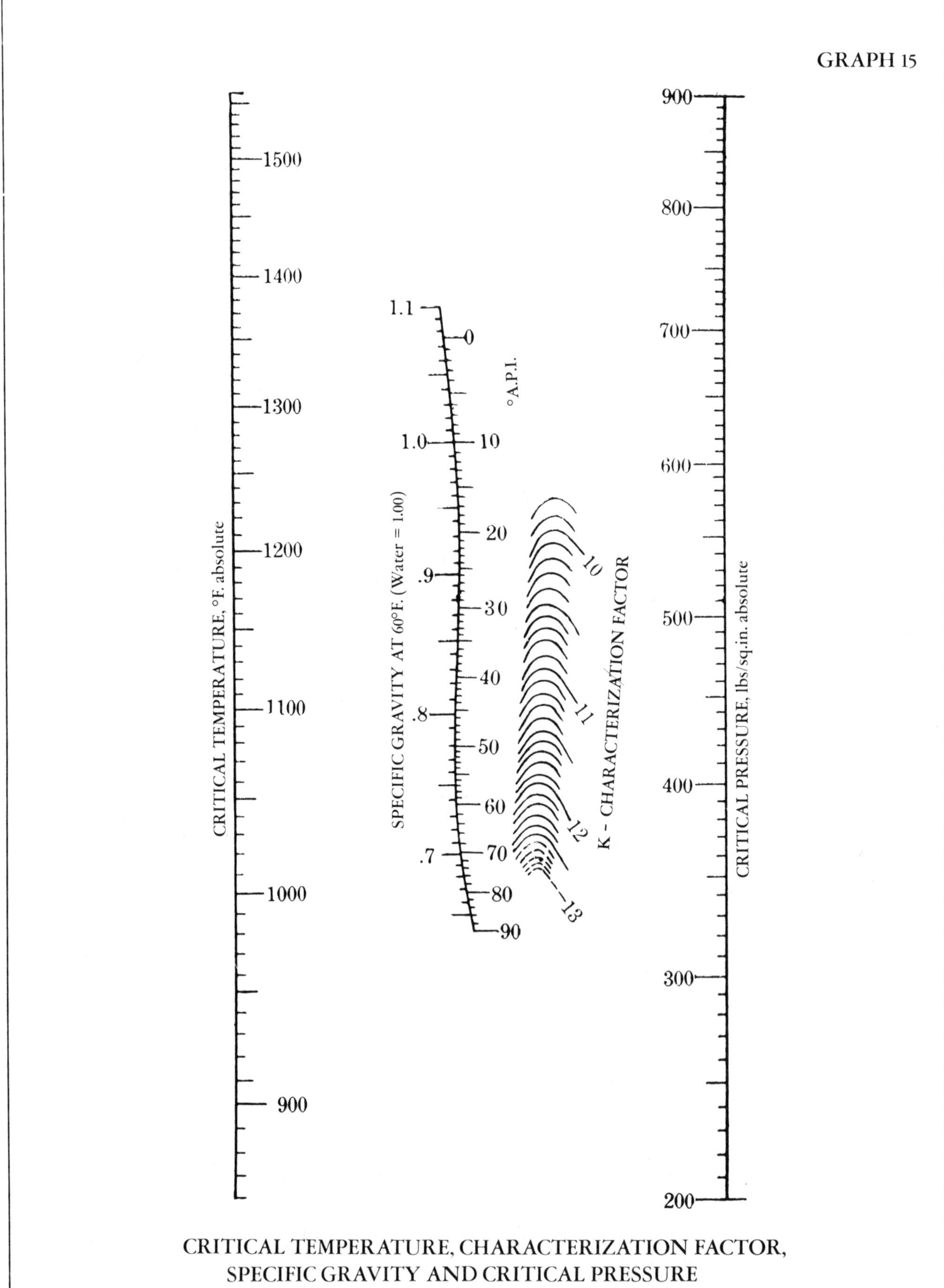

CRITICAL TEMPERATURE, CHARACTERIZATION FACTOR, SPECIFIC GRAVITY AND CRITICAL PRESSURE

GRAPH 16

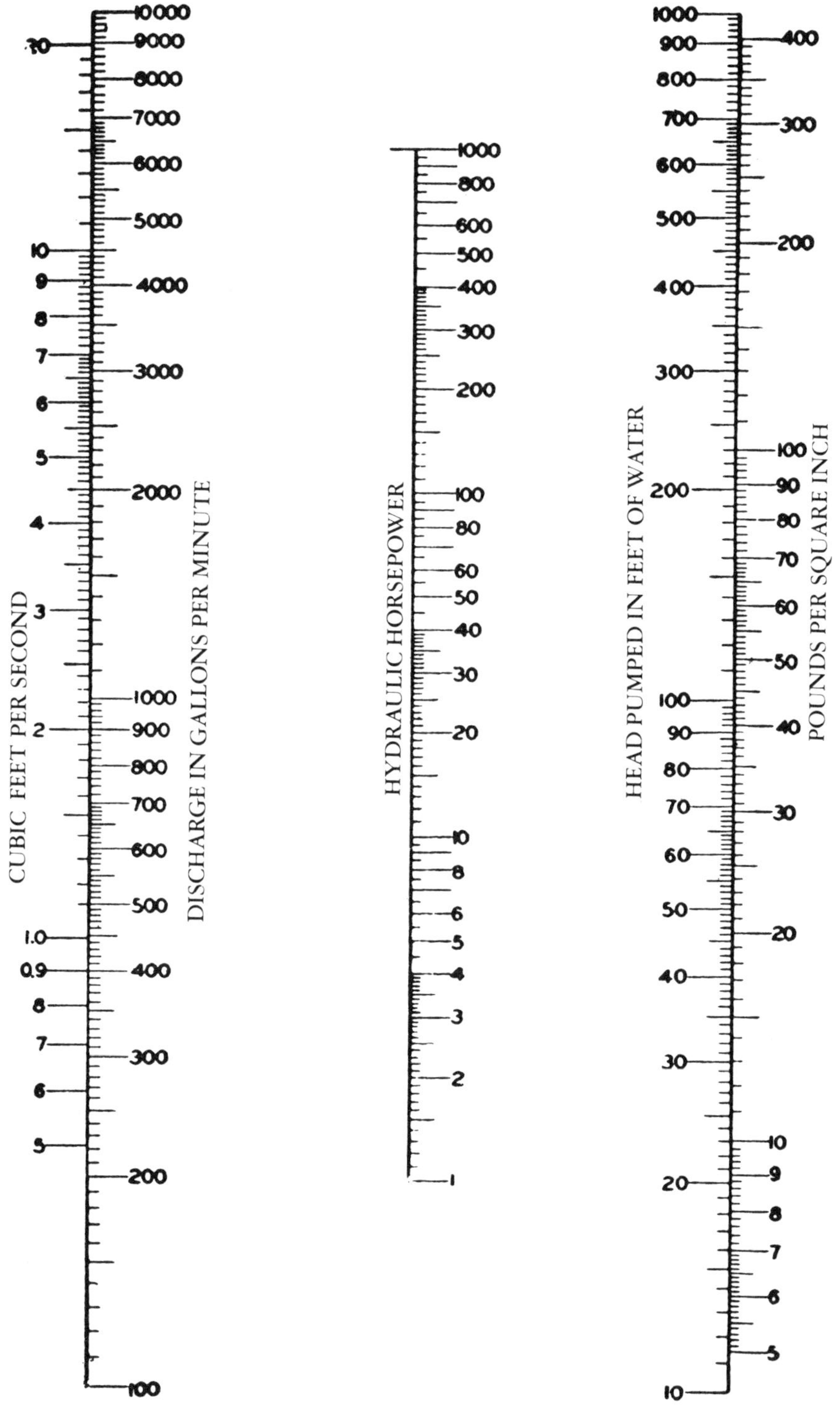

CUBIC FEET PER SECOND
20 10 9 8 7 6 5 4 3 2 1.0 0.9 8 7 6 5

DISCHARGE IN GALLONS PER MINUTE
10000 9000 8000 7000 6000 5000 4000 3000 2000 1000 900 800 700 600 500 400 300 200 100

HYDRAULIC HORSEPOWER
1000 800 600 500 400 300 200 100 80 60 50 40 30 20 10 8 6 5 4 3 2 1

HEAD PUMPED IN FEET OF WATER
1000 900 800 700 600 500 400 300 200 100 90 80 70 60 50 40 30 20 10

POUNDS PER SQUARE INCH
400 300 200 100 90 80 70 60 50 40 30 20 10 9 8 7 6 5

HYDRAULIC HORSEPOWER

Key: LAS
HAC

GRAPH 17

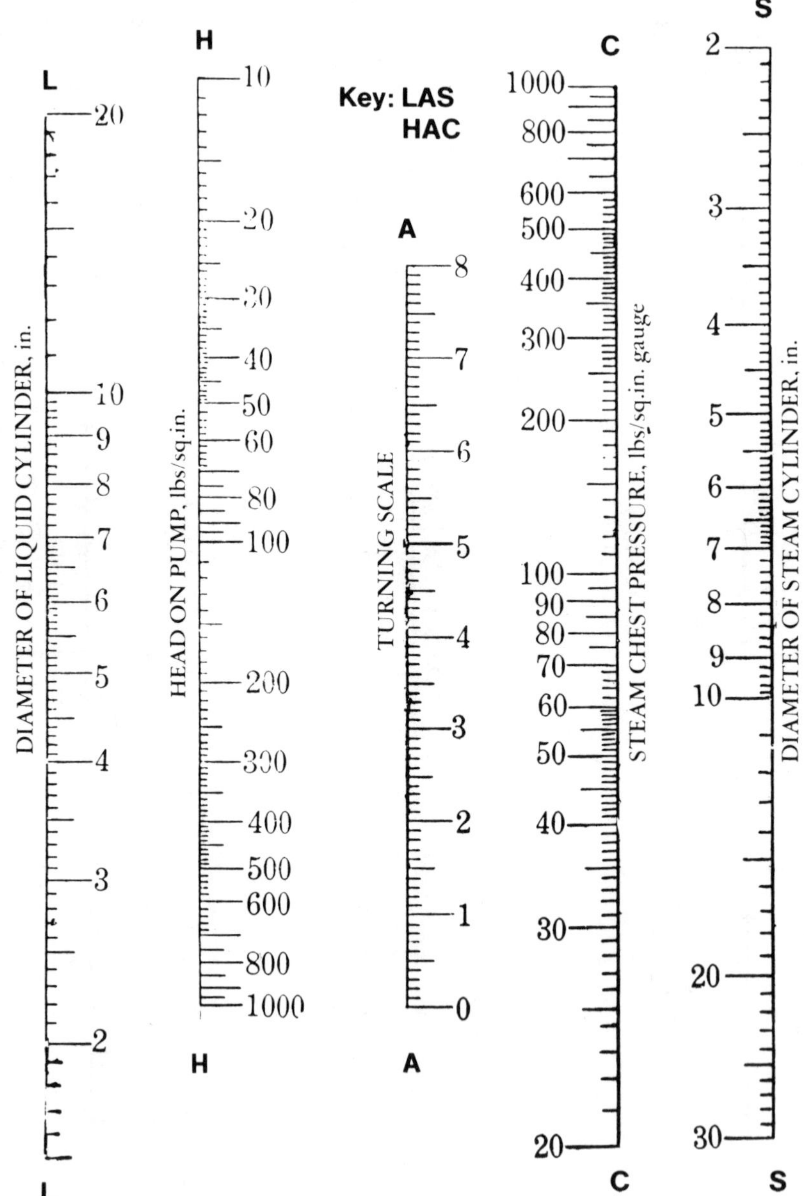

STEAM CHEST PRESSURE
(Add exhaust pressure)

GRAPH 18

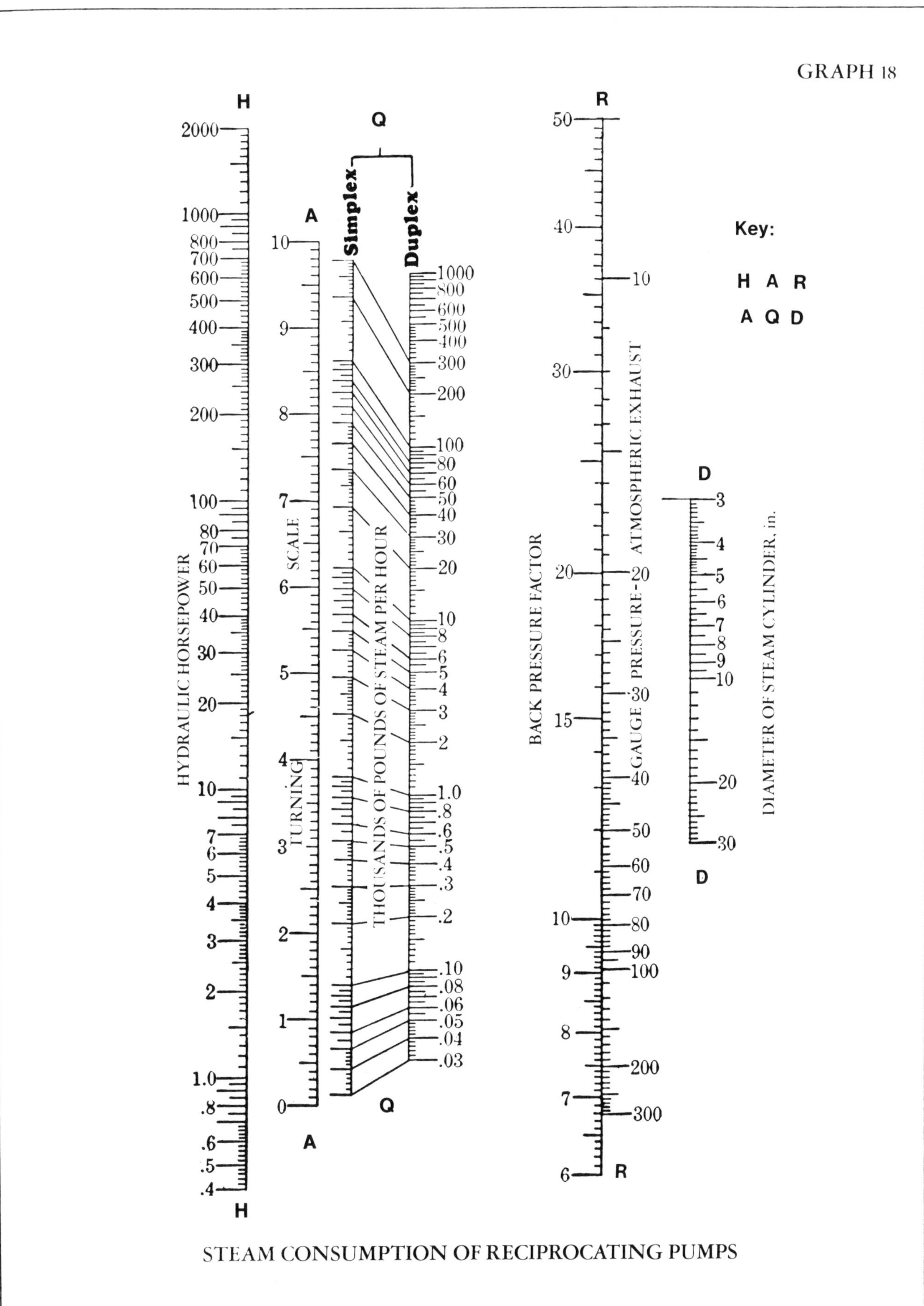

STEAM CONSUMPTION OF RECIPROCATING PUMPS

GRAPH 19

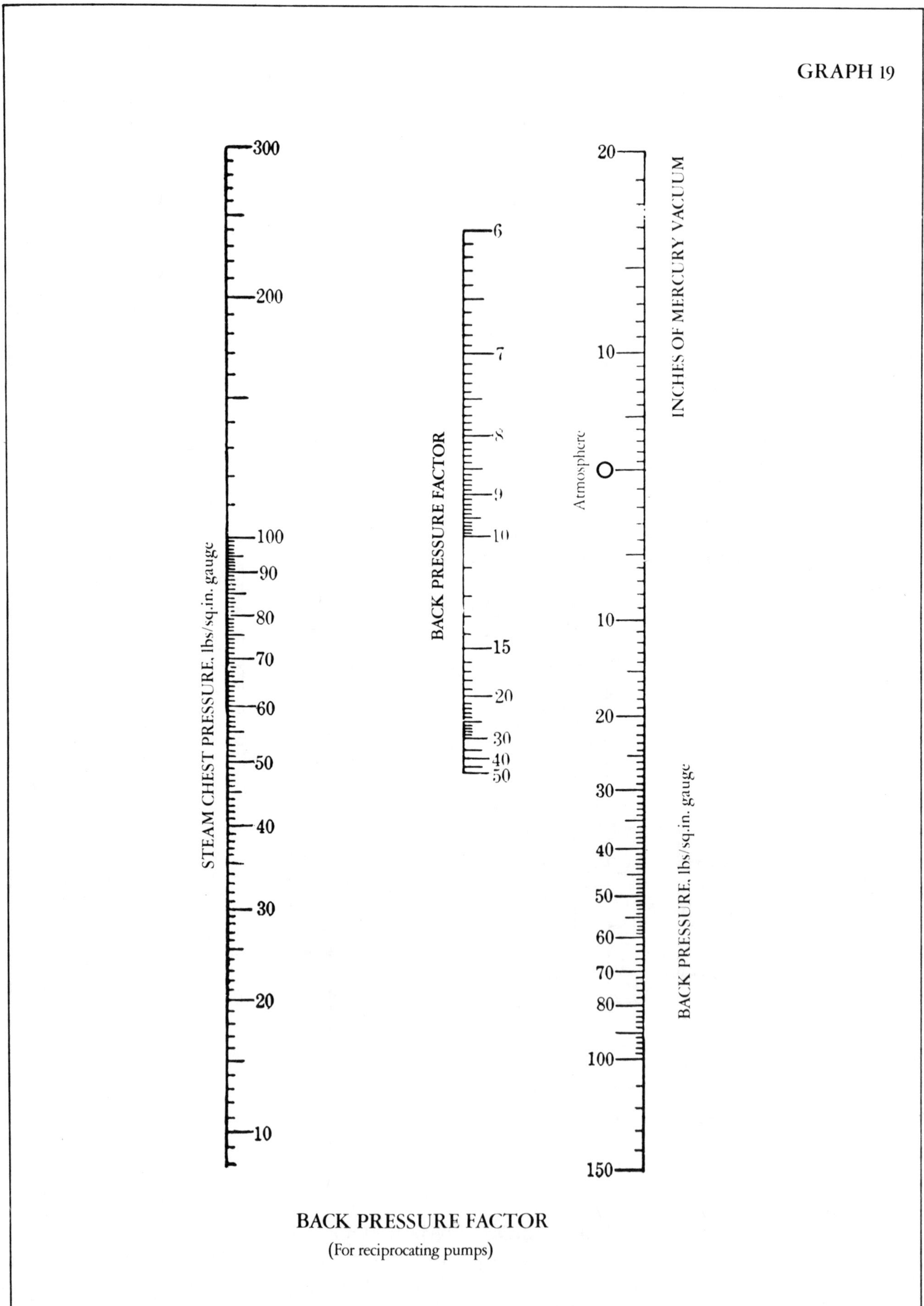

BACK PRESSURE FACTOR

(For reciprocating pumps)

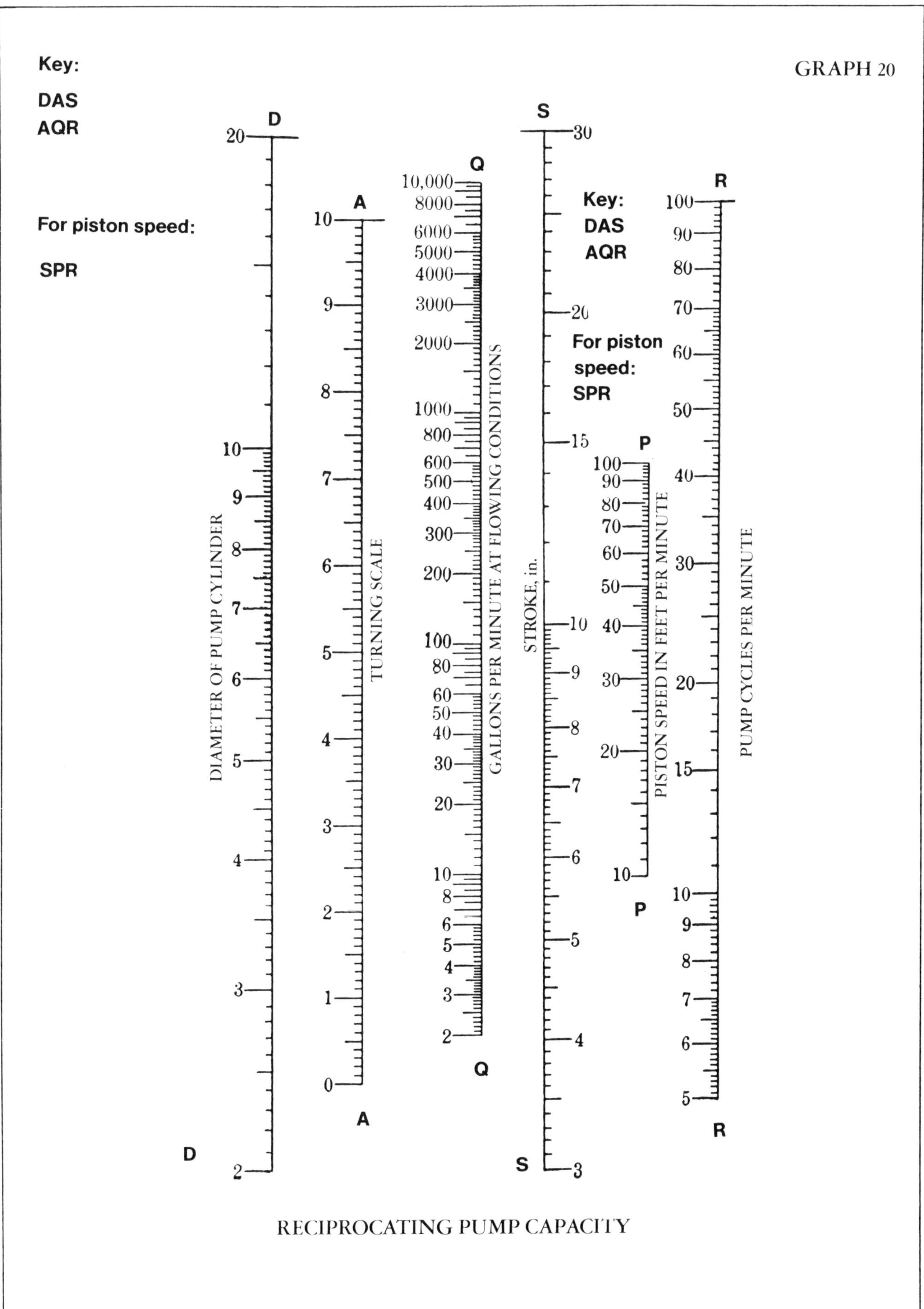

RECIPROCATING PUMP CAPACITY

GRAPH 21

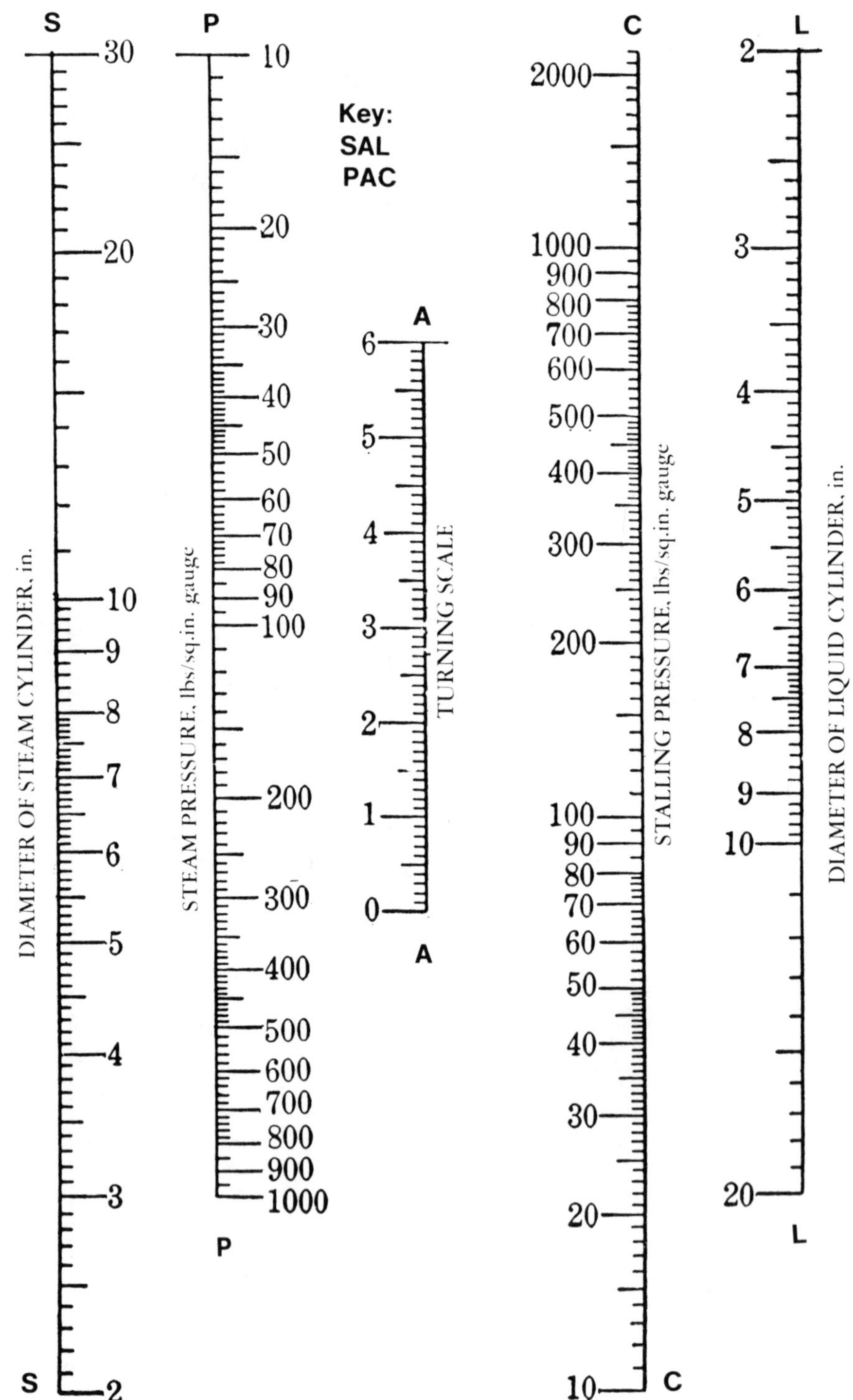

STALLING PRESSURE OF RECIPROCATING PUMPS

(Add pump section pressure)

GRAPH 22

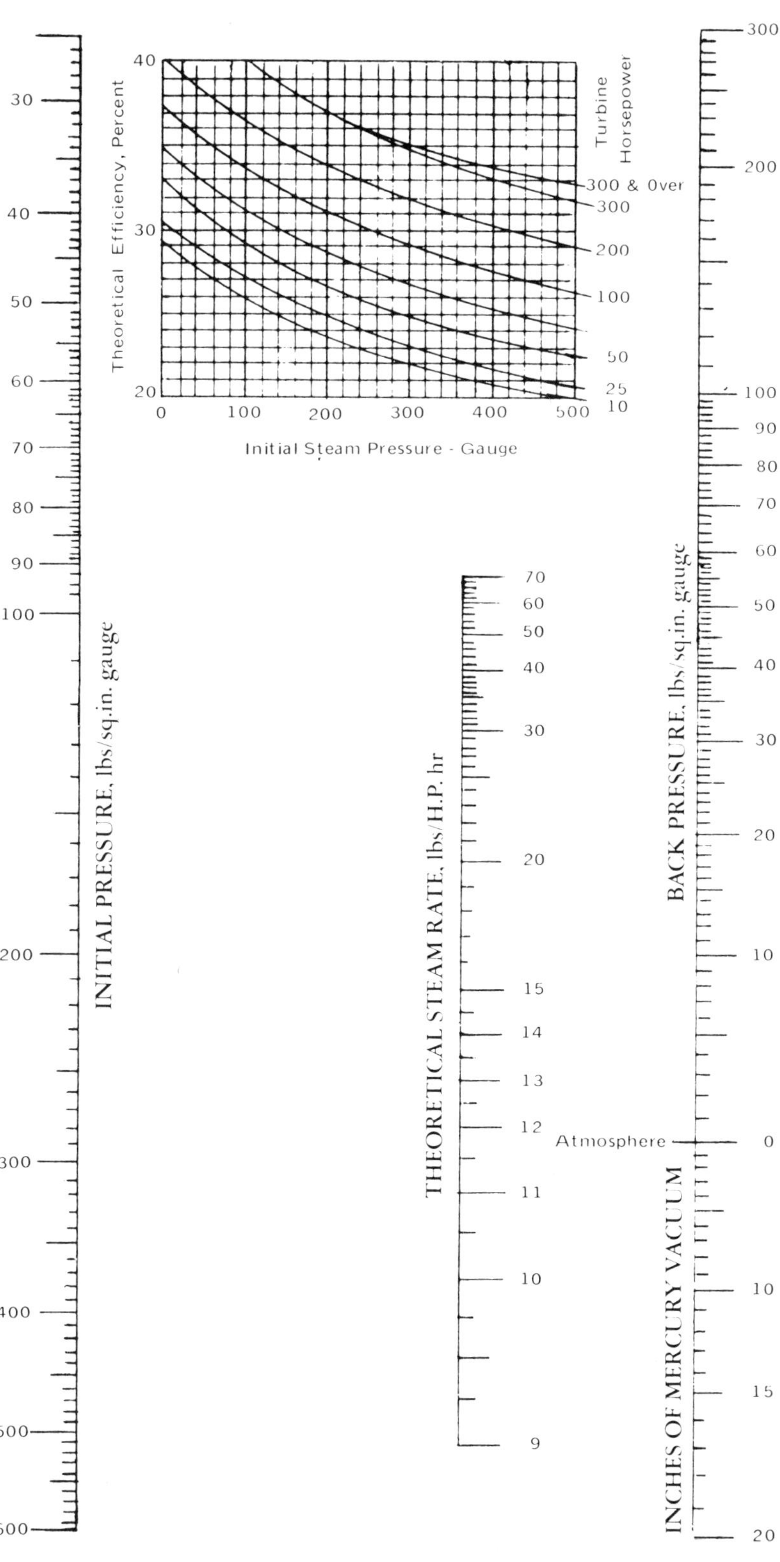

TURBINE STEAM RATES

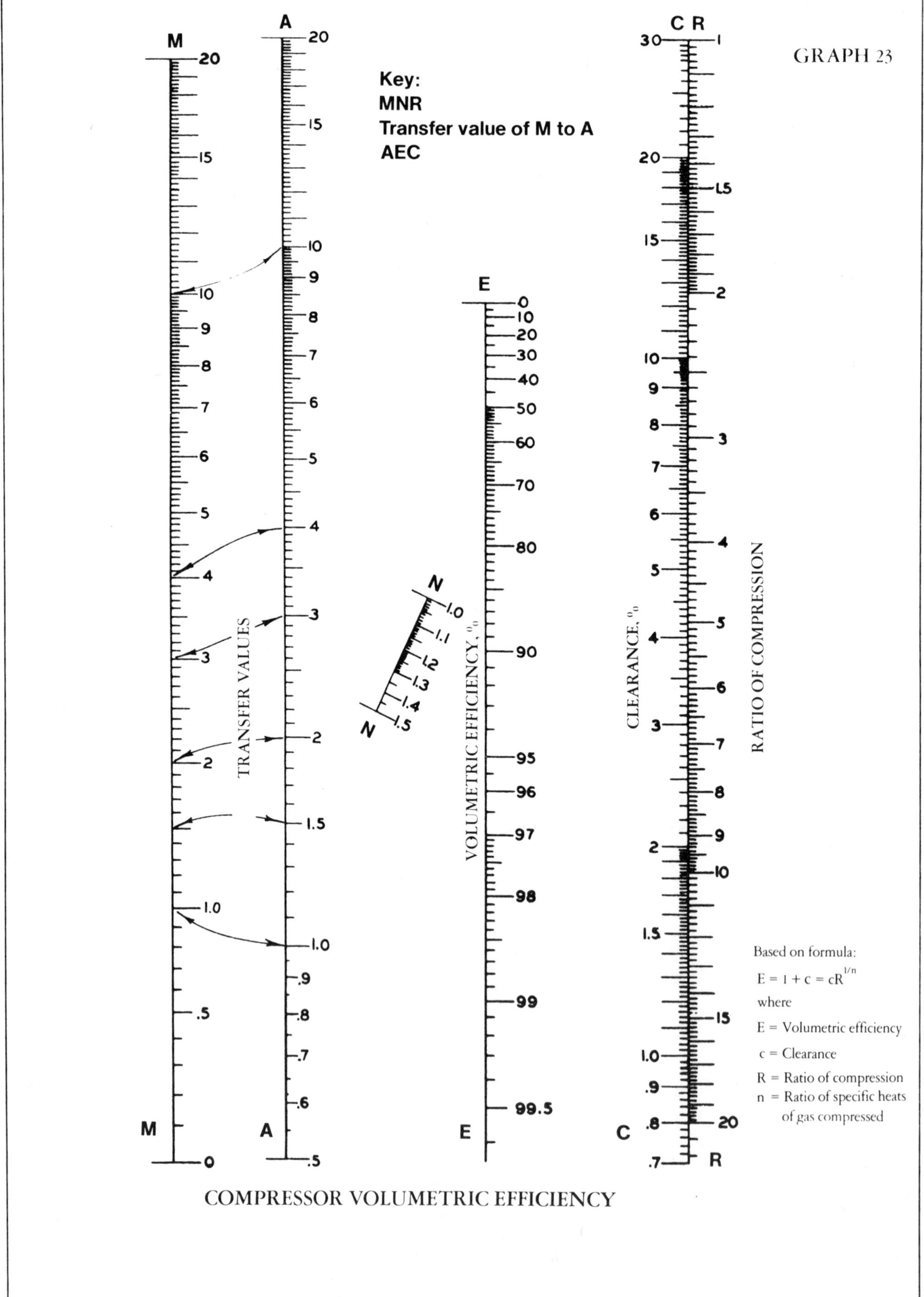

COMPRESSOR VOLUMETRIC EFFICIENCY

GRAPH 24

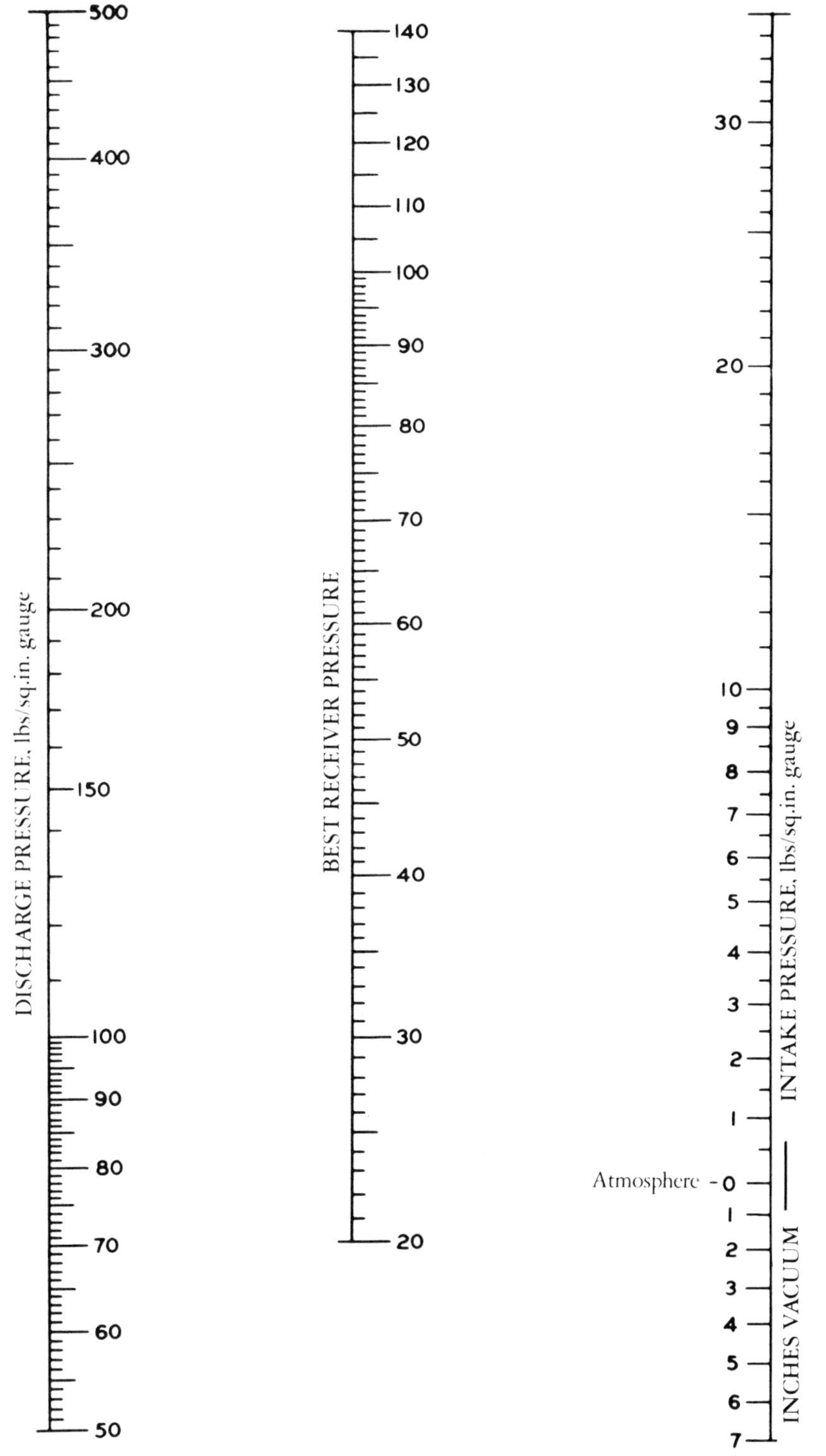

DISCHARGE PRESSURE, lbs/sq.in. gauge

500
400
300
200
150
100
90
80
70
60
50

BEST RECEIVER PRESSURE

140
130
120
110
100
90
80
70
60
50
40
30
20

INTAKE PRESSURE, lbs/sq.in. gauge

30
20
10
9
8
7
6
5
4
3
2
1
Atmosphere - 0

INCHES VACUUM

1
2
3
4
5
6
7

BEST RECEIVER PRESSURE

Two Stage Compression

Summary of Factors F_r for Graph 25

Multiply given flow rate, as previously corrected, if needed, by the factor or factors before entering rate in the nomograph.

Centrifugal Pumps

Type of Drive	Electric motor	Internal Combustion	Steam Turbine
F_r	1.00	1.25	1.25

Reciprocating Steam Pumps

Type	Simplex	Duplex
F_r	1.18	1.09

Hours of Operation

Per week	10	20	40	60	75	100	120
F_r	0.38	0.49	0.63	0.72	0.78	0.86	0.92

Electric Power Costs*

$ per kwhr	0.007	0.01	0.015	0.0218	0.025	0.030	0.035	0.04
F_r	0.67	0.76	0.88	1.00	1.05	1.12	1.18	1.24

Fuel Costs*

$ per therm	0.20	0.30	0.45	0.70	1.00	1.50	2.00	2.50
F_r	0.64	0.74	0.85	1.00	1.13	1.31	1.45	1.57

Pipe Costs 2 Inch Schedule 40 Carbon Steel

$ per foot	0.50	0.65	1.00	1.32	1.75	2.50	3.50	5.00
F_r	1.41	1.28	1.12	1.00	0.91	0.79	0.71	0.63

Amortization

Years	2	3	5	7	10	15	20	25
F_r	0.65	0.75	0.89	1.00	1.12	1.28	1.39	1.48

*Do not use both.
Factors are multiplicative.
See also Table 7-13 for factors for materials other than carbon steel.
See Equation (7-10) when labor and material costs deviate from norm.

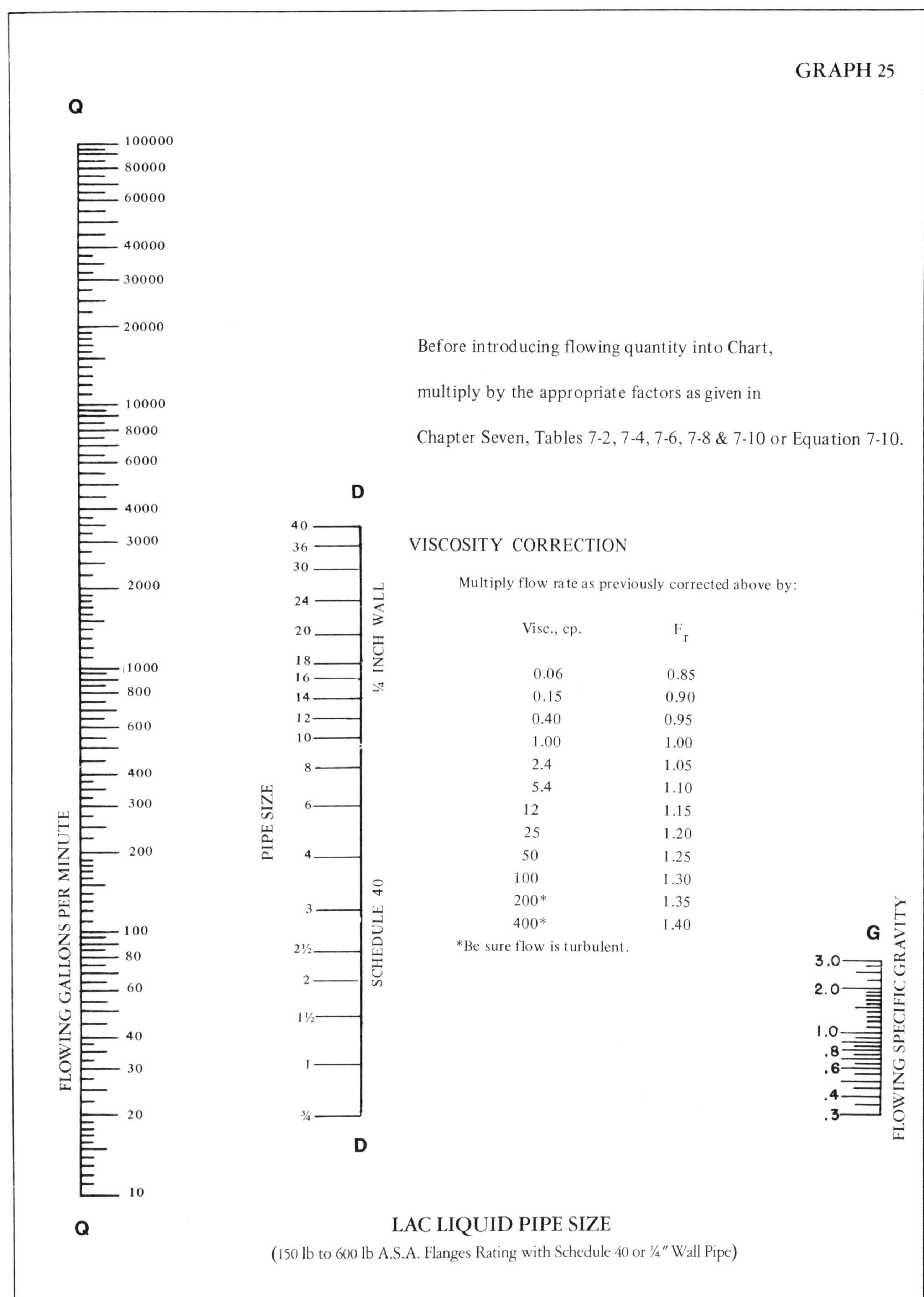

Before introducing flowing quantity into Chart, multiply by the appropriate factors as given in Chapter Seven, Tables 7-2, 7-4, 7-6, 7-8 & 7-10 or Equation 7-10.

VISCOSITY CORRECTION

Multiply flow rate as previously corrected above by:

Visc., cp.	F_r
0.06	0.85
0.15	0.90
0.40	0.95
1.00	1.00
2.4	1.05
5.4	1.10
12	1.15
25	1.20
50	1.25
100	1.30
200*	1.35
400*	1.40

*Be sure flow is turbulent.

LAC LIQUID PIPE SIZE

(150 lb to 600 lb A.S.A. Flanges Rating with Schedule 40 or ¼" Wall Pipe)

Summary of Factors F_r for Graph 26

Multiply given flow rate, as previously corrected, if needed, by the factor or factors before entering rate in the nomograph.

Centrifugal Pumps

Type of Drive	Electric motor	Internal Combustion	Steam Turbine
F_r	1.00	1.25	1.25

Reciprocating steam pumps

Type	Simplex	Duplex
F_r	1.18	1.09

Hours of Operation

Per week	10	20	40	60	75	100	120
F_r	0.38	0.49	0.63	0.72	0.78	0.86	0.92

Electric Power Costs*

$ per kwhr	0.007	0.01	0.015	0.0218	0.025	0.030	0.035	0.04
F_r	0.67	0.76	0.88	1.00	1.05	1.12	1.18	1.24

Fuel Costs*

$ per therm	0.20	0.30	0.45	0.70	1.00	1.50	2.00	2.50
F_r	0.64	0.74	0.85	1.00	1.13	1.31	1.45	1.57

Pipe Costs 2 Inch Schedule 40 Carbon Steel

$ per foot	0.50	0.65	1.00	1.32	1.75	2.50	3.50	5.00
F_r	1.41	1.28	1.12	1.00	0.91	0.79	0.71	0.63

Amortization

Years	2	3	5	7	10	15	20	25
F_r	0.65	0.75	0.89	1.00	1.12	1.28	1.39	1.48

*Do not use both.
Factors are multiplicative.
See also Table 7-13 for factors for materials other than carbon steel.
See Equation (7-10) when labor and material costs deviate from norm.

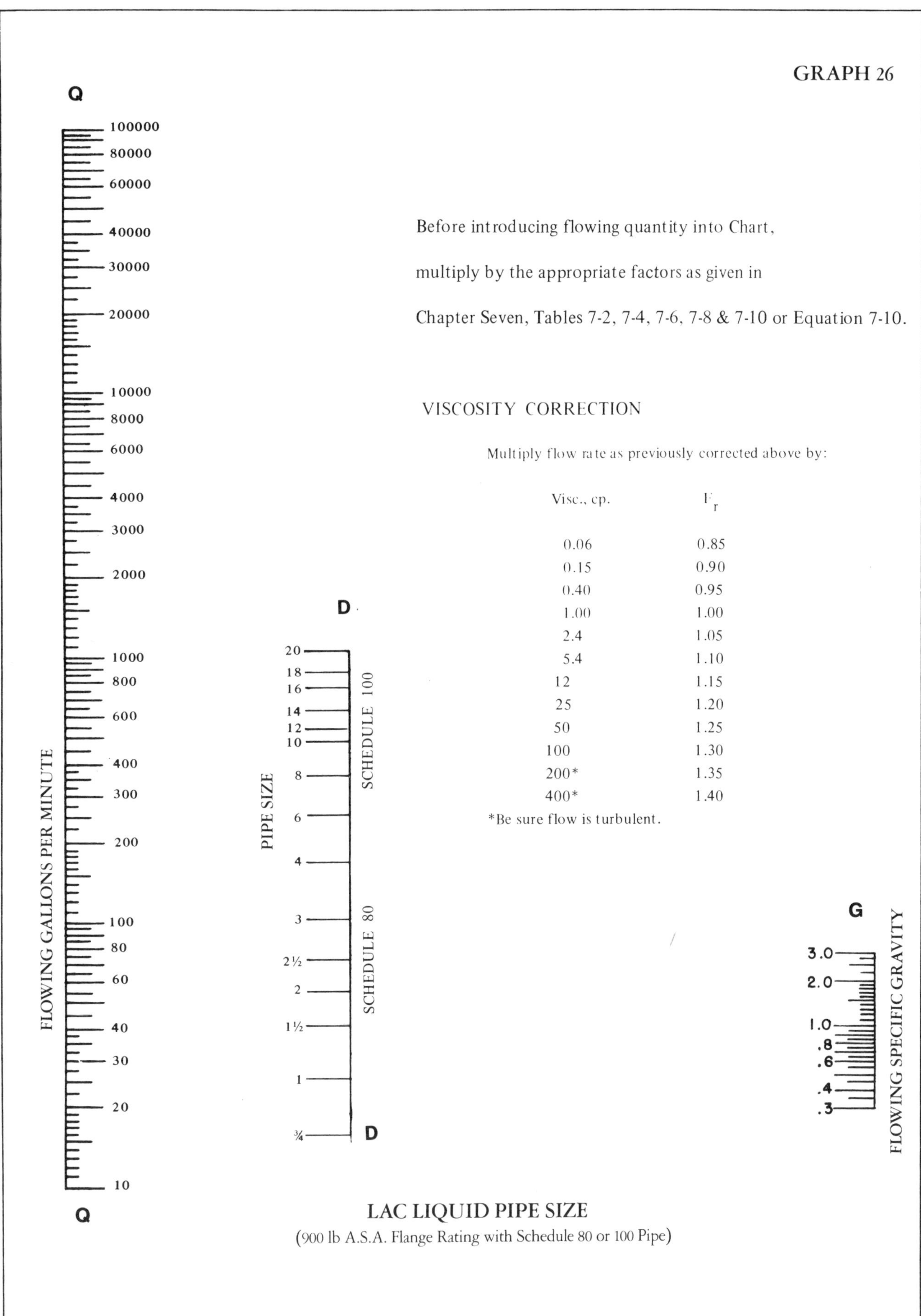

Before introducing flowing quantity into Chart, multiply by the appropriate factors as given in Chapter Seven, Tables 7-2, 7-4, 7-6, 7-8 & 7-10 or Equation 7-10.

VISCOSITY CORRECTION

Multiply flow rate as previously corrected above by:

Visc., cp.	F_r
0.06	0.85
0.15	0.90
0.40	0.95
1.00	1.00
2.4	1.05
5.4	1.10
12	1.15
25	1.20
50	1.25
100	1.30
200*	1.35
400*	1.40

*Be sure flow is turbulent.

LAC LIQUID PIPE SIZE
(900 lb A.S.A. Flange Rating with Schedule 80 or 100 Pipe)

Summary of Factors F_r for Graph 27

Multiply given flow rate, as previously corrected, if needed, by the factor or factors before entering rate in the nomograph.

Centrifugal Pumps

Type of Drive	Electric motor	Internal Combustion	Steam Turbine
F_r	1.00	1.25	1.25

Reciprocating Steam Pumps

Type	Simplex	Duplex
F_r	1.18	1.09

Hours of Operation

Per week	10	20	40	60	75	100	120
F_r	0.38	0.49	0.63	0.72	0.78	0.86	0.92

Electric Power Costs*

$ per kwhr	0.007	0.01	0.015	0.0218	0.025	0.030	0.035	0.04
F_r	0.67	0.76	0.88	1.00	1.05	1.12	1.18	1.24

Fuel Costs*

$ per therm	0.20	0.30	0.45	0.70	1.00	1.50	2.00	2.50
F_r	0.64	0.74	0.85	1.00	1.13	1.31	1.45	1.57

Pipe Costs 2 Inch Schedule 40 Carbon Steel

$ per foot	0.50	0.65	1.00	1.32	1.75	2.50	3.50	5.00
F_r	1.41	1.28	1.12	1.00	0.91	0.79	0.71	0.63

Amortization

Years	2	3	5	7	10	15	20	25
F_r	0.65	0.75	0.89	1.00	1.12	1.28	1.39	1.48

*Do not use both.
Factors are multiplicative.
See also Table 7-13 for factors for materials other than carbon steel.
See Equation (7-10) when labor and material costs deviate from norm.

GRAPH 27

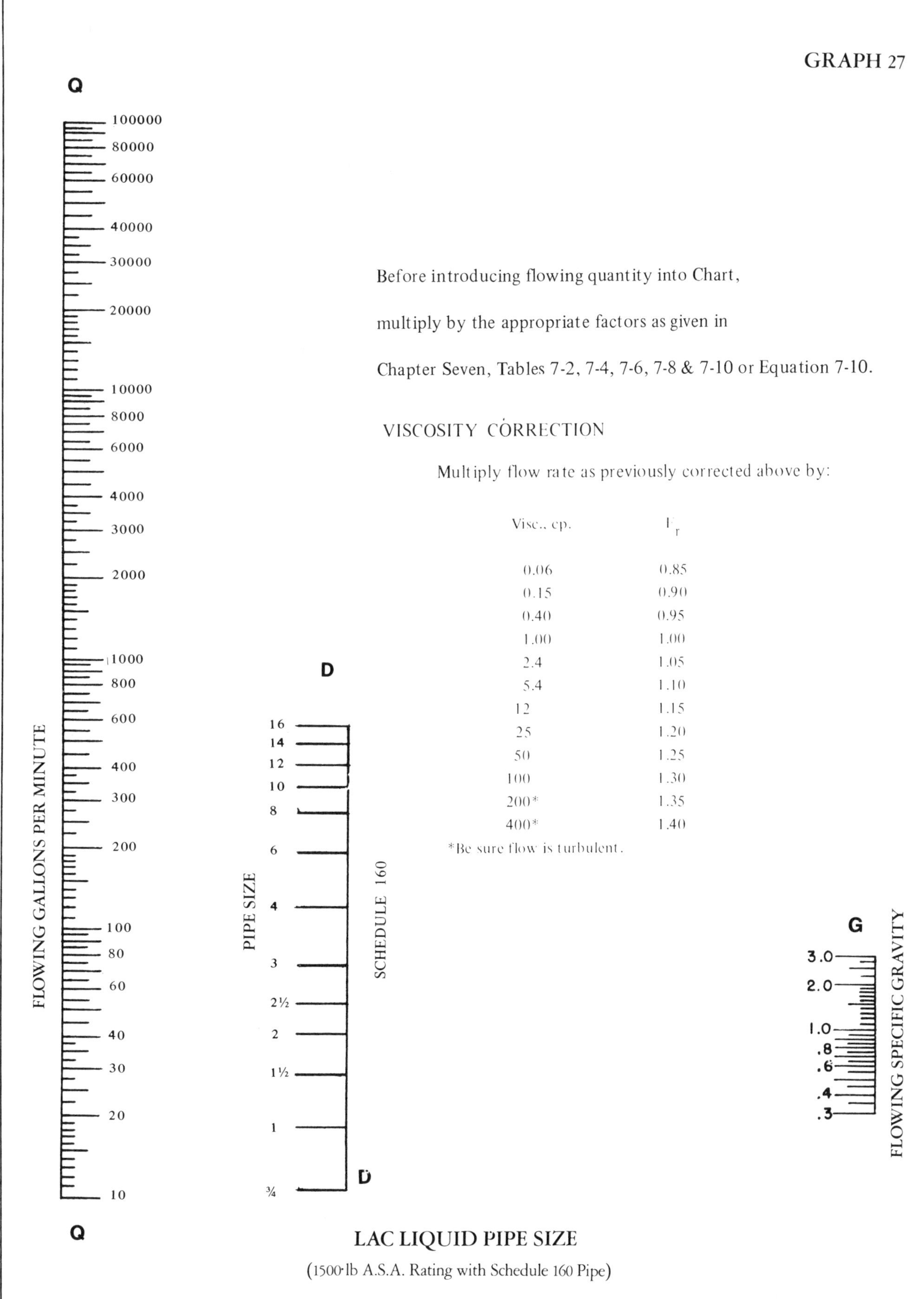

Before introducing flowing quantity into Chart, multiply by the appropriate factors as given in Chapter Seven, Tables 7-2, 7-4, 7-6, 7-8 & 7-10 or Equation 7-10.

VISCOSITY CORRECTION

Multiply flow rate as previously corrected above by:

Visc., cp.	F_r
0.06	0.85
0.15	0.90
0.40	0.95
1.00	1.00
2.4	1.05
5.4	1.10
12	1.15
25	1.20
50	1.25
100	1.30
200*	1.35
400*	1.40

*Be sure flow is turbulent.

LAC LIQUID PIPE SIZE

(1500-lb A.S.A. Rating with Schedule 160 Pipe)

Summary of Factors F_r for Graph 28

Multiply given flow rate, as previously corrected, if needed, by the factor or factors before entering rate in the nomograph.

Centrifugal Pumps

Type of Drive	Electric motor	Internal Combustion	Steam Turbine
F_r	1.00	1.25	1.25

Reciprocating Steam Pumps

Type	Simplex	Duplex
F_r	1.18	1.09

Hours of Operation

Per week	10	20	40	60	75	100	120
F_r	0.38	0.49	0.63	0.72	0.78	0.86	0.92

Electric Power Costs*

$ per kwhr	0.007	0.01	0.015	0.0218	0.025	0.030	0.035	0.04
F_r	0.67	0.76	0.88	1.00	1.05	1.12	1.18	1.24

Fuel Costs*

$ per therm	0.20	0.30	0.45	0.70	1.00	1.50	2.00	2.50
F_r	0.64	0.74	0.85	1.00	1.13	1.31	1.45	1.57

Pipe Costs 2 Inch Schedule 40 Carbon Steel

$ per foot	0.50	0.65	1.00	1.32	1.75	2.50	3.50	5.00
F_r	1.41	1.28	1.12	1.00	0.91	0.79	0.71	0.63

Amortization

Years	2	3	5	7	10	15	20	25
F_r	0.65	0.75	0.89	1.00	1.12	1.28	1.39	1.48

*Do not use both.
Factors are multiplicative.
See also Table 7-13 for factors for materials other than carbon steel.
See Equation (7-10) when labor and material costs deviate from norm.

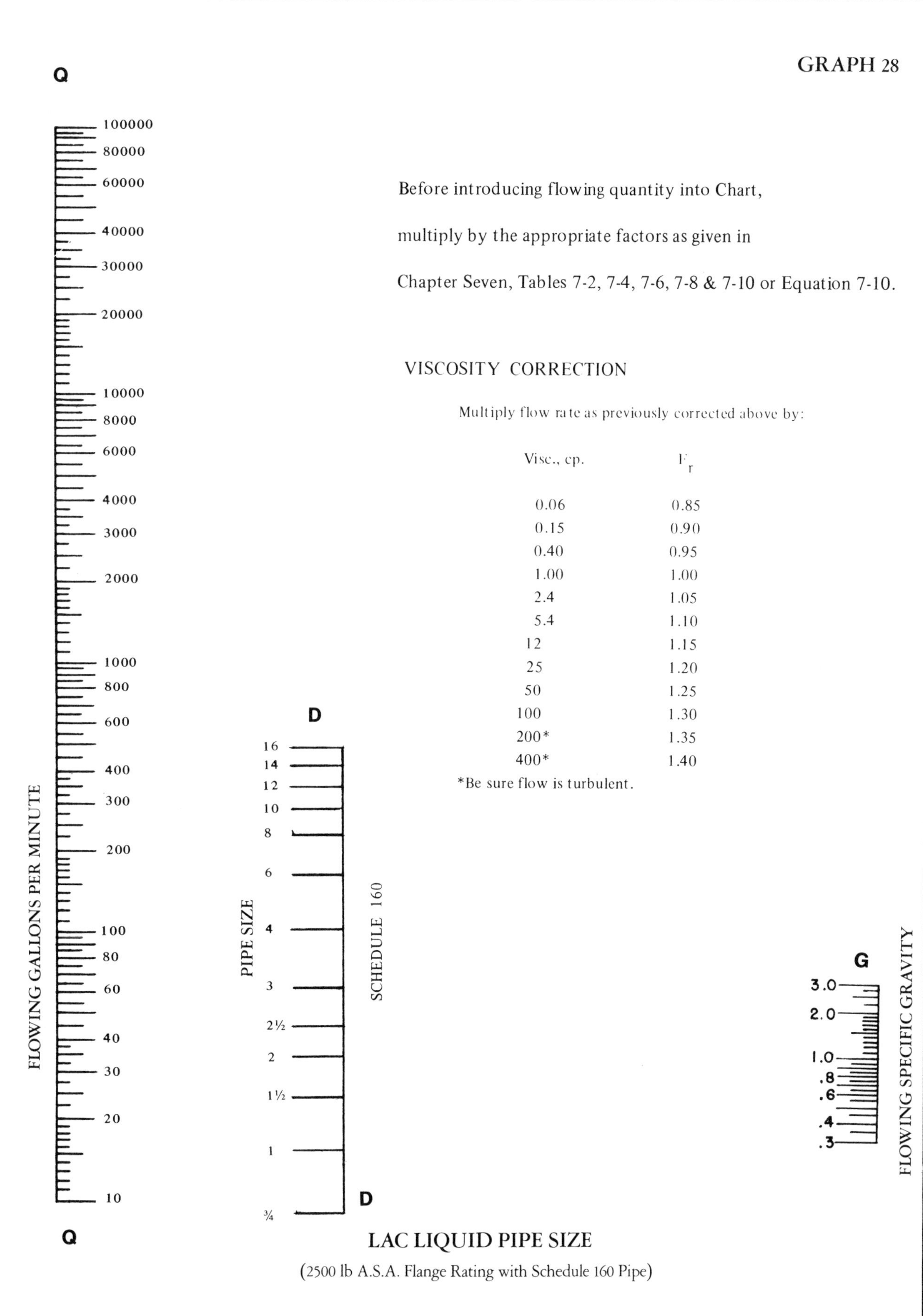

Before introducing flowing quantity into Chart, multiply by the appropriate factors as given in Chapter Seven, Tables 7-2, 7-4, 7-6, 7-8 & 7-10 or Equation 7-10.

VISCOSITY CORRECTION

Multiply flow rate as previously corrected above by:

Visc., cp.	F_r
0.06	0.85
0.15	0.90
0.40	0.95
1.00	1.00
2.4	1.05
5.4	1.10
12	1.15
25	1.20
50	1.25
100	1.30
200*	1.35
400*	1.40

*Be sure flow is turbulent.

LAC LIQUID PIPE SIZE

(2500 lb A.S.A. Flange Rating with Schedule 160 Pipe)

GRAPH 29

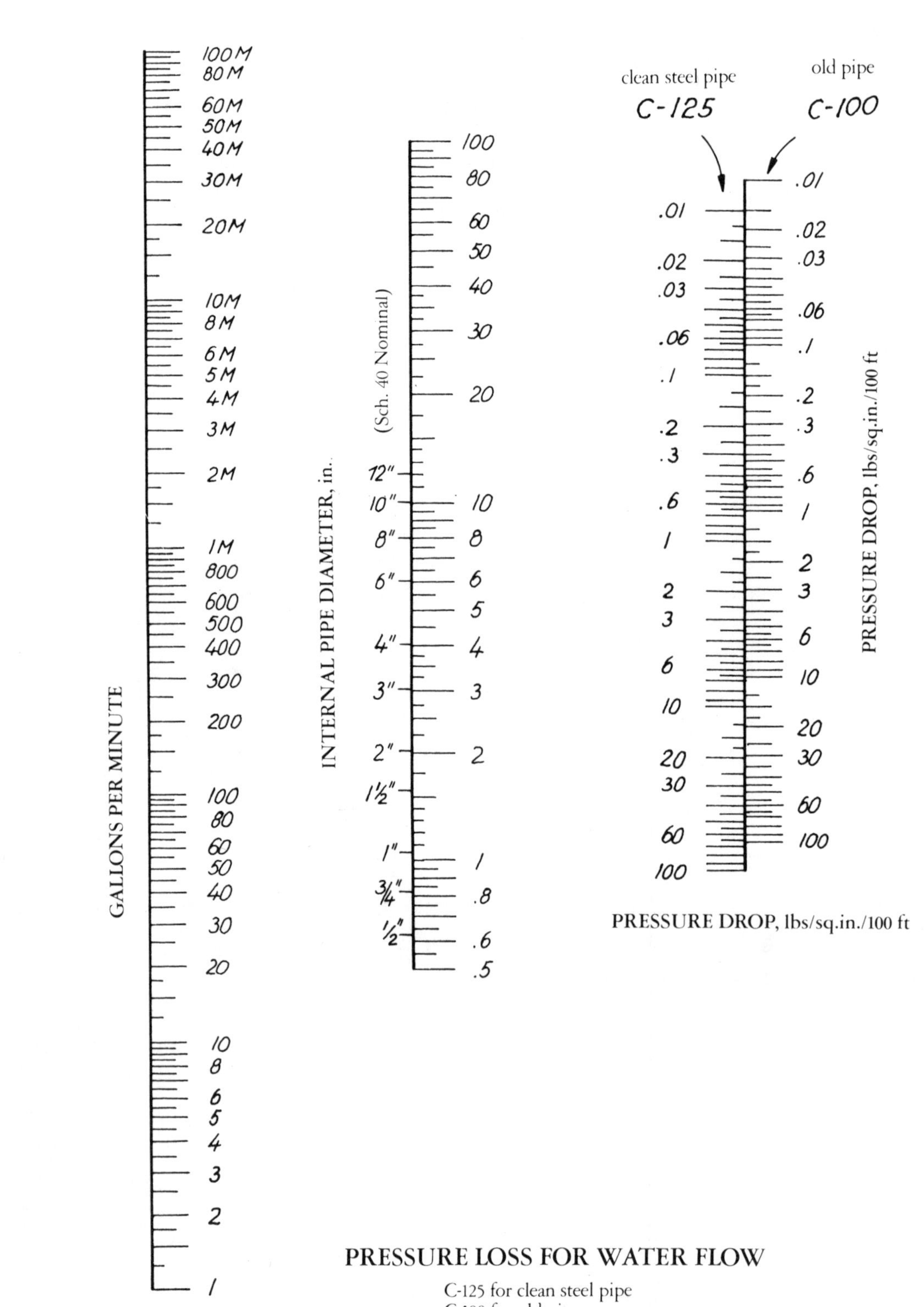

PRESSURE LOSS FOR WATER FLOW

C-125 for clean steel pipe
C-100 for old pipe

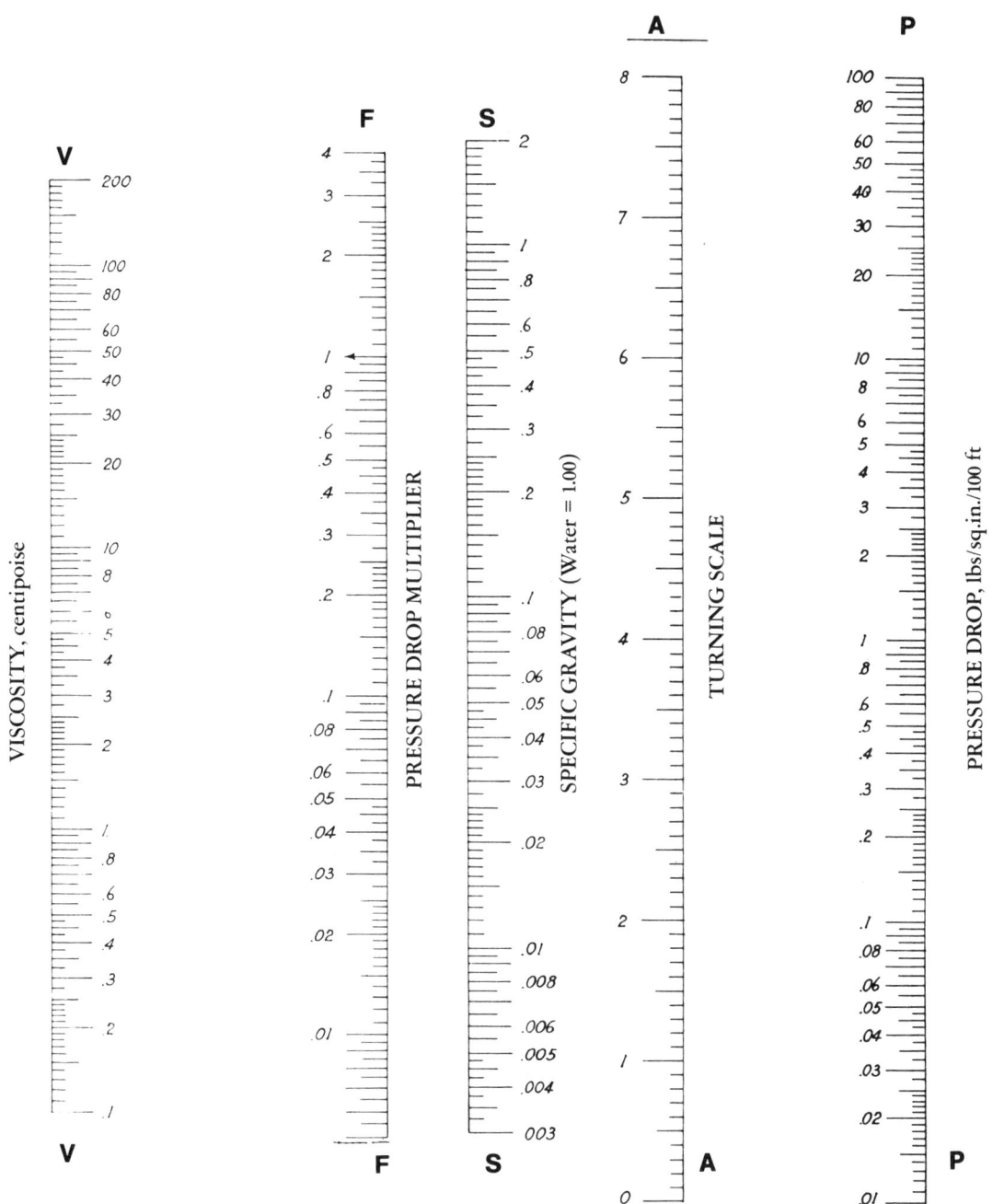

V — VISCOSITY, centipoise: 200, 100, 80, 60, 50, 40, 30, 20, 10, 8, 6, 5, 4, 3, 2, 1, .8, .6, .5, .4, .3, .2, .1

F — PRESSURE DROP MULTIPLIER: 4, 3, 2, 1, .8, .6, .5, .4, .3, .2, .1, .08, .06, .05, .04, .03, .02, .01

S — SPECIFIC GRAVITY (Water = 1.00): 2, 1, .8, .6, .5, .4, .3, .2, .1, .08, .06, .05, .04, .03, .02, .01, .008, .006, .005, .004, 003

A — TURNING SCALE: 8, 7, 6, 5, 4, 3, 2, 1, 0

P — PRESSURE DROP, lbs/sq.in./100 ft: 100, 80, 60, 50, 40, 30, 20, 10, 8, 6, 5, 4, 3, 2, 1, .8, .6, .5, .4, .3, .2, .1, .08, .06, .05, .04, .03, .02, .01

Locate F from VFS. Locate A from FAP using for P value obtained on Graph 29 where C = 125. Rotate about A to F = 1 and read corrected △ P on P.

VISCOSITY - SPECIFIC GRAVITY CORRECTION FOR LIQUID PRESSURE LOSS

GRAPH 31

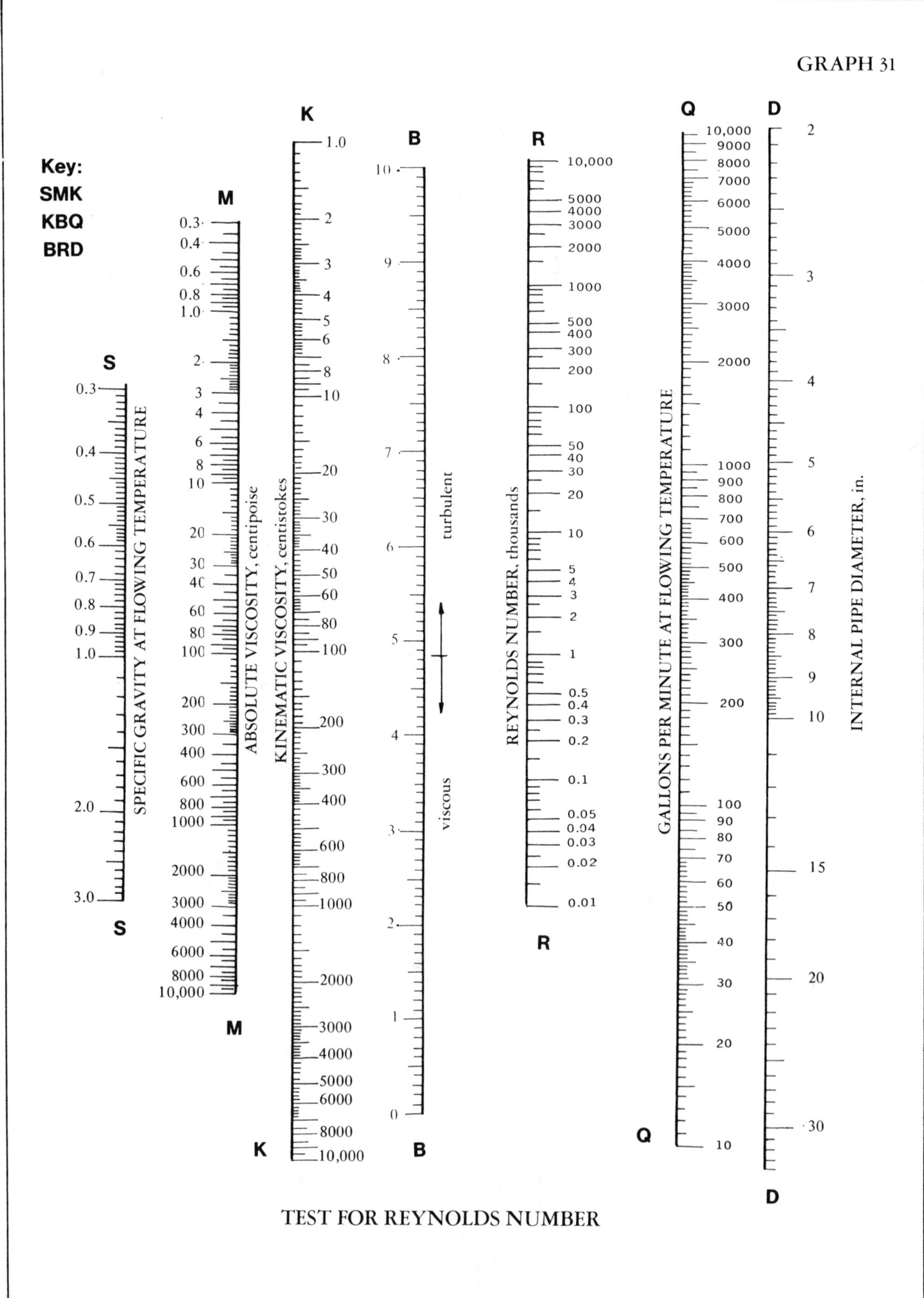

TEST FOR REYNOLDS NUMBER

GRAPH 32

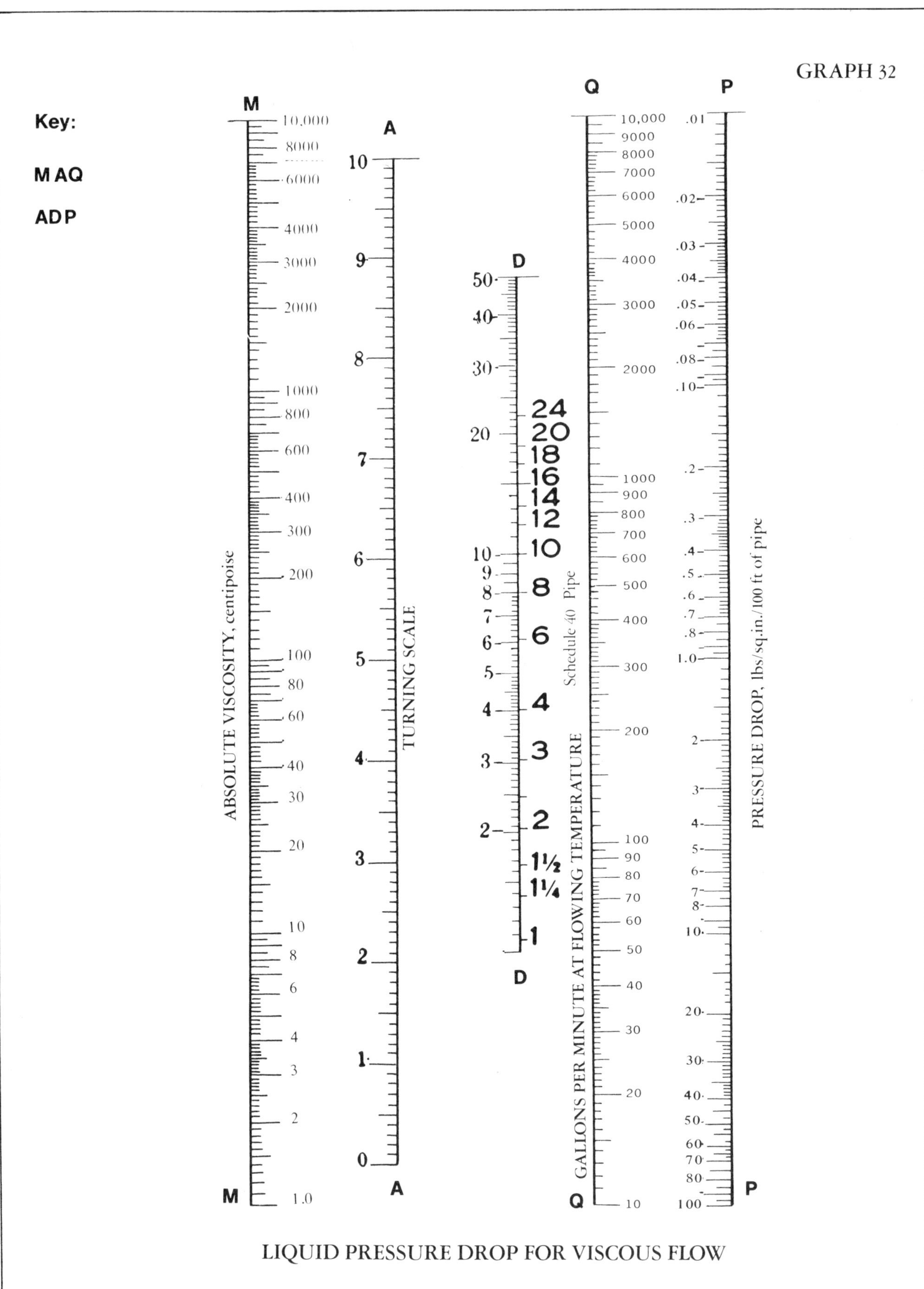

LIQUID PRESSURE DROP FOR VISCOUS FLOW

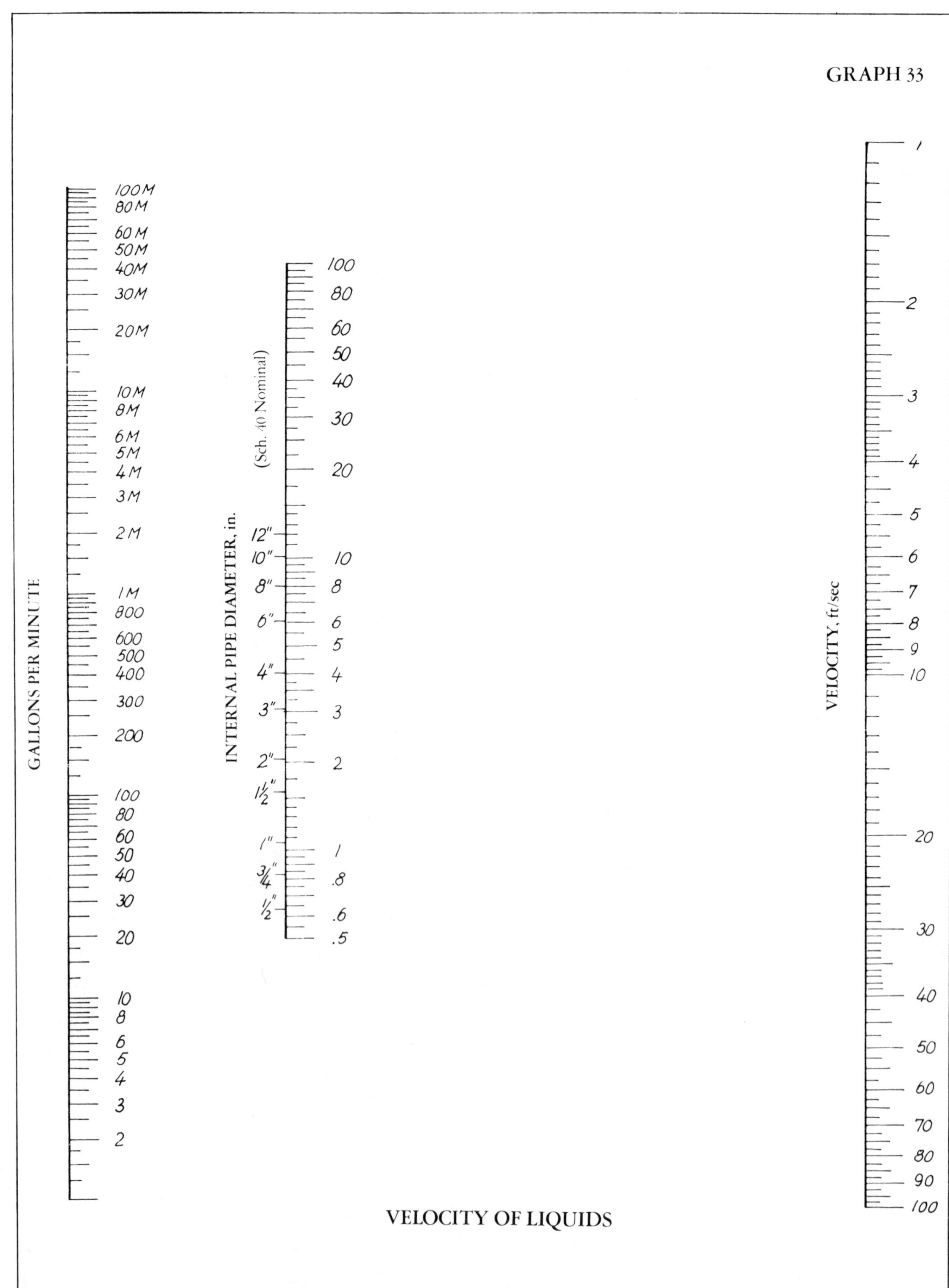

VELOCITY OF LIQUIDS

GRAPH 34

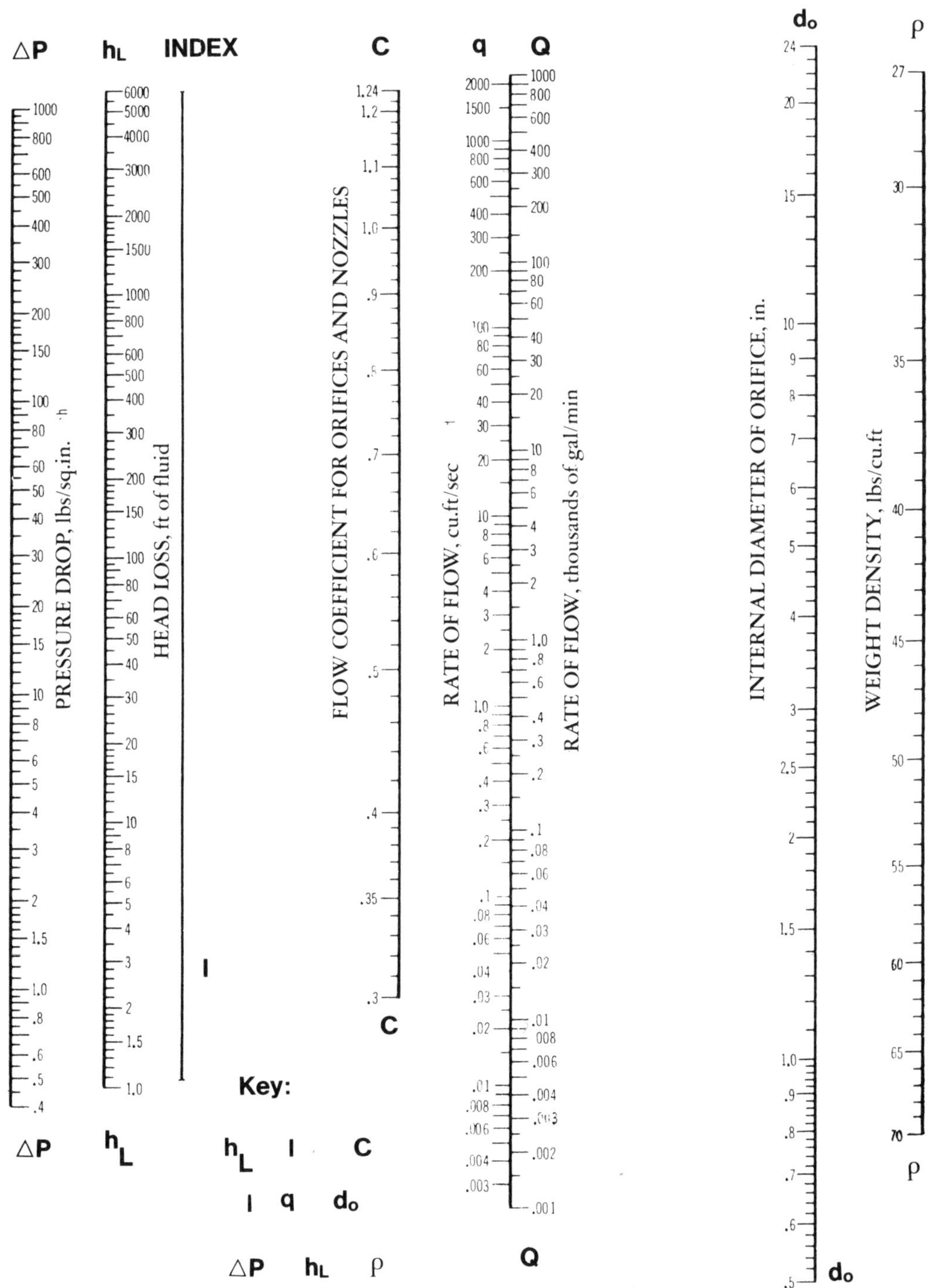

FLOW OF LIQUIDS THROUGH NOZZLES AND ORIFICES

(From TP-410, FLOW OF FLUIDS, Courtesy Crane Company)

Summary of Factors F_r for Graph 35

Multiply given flow rate, as previously corrected, if needed, by the factor or factors before entering rate in the nomograph.

Hours of Operation

Per week	10	20	40	60	75	100	120
F_r	0.38	0.49	0.63	0.72	0.78	0.86	0.92

Fuel Costs*

\$ per therm	0.20	0.30	0.45	0.70	1.00	1.50	2.00	2.50
F_r	0.64	0.74	0.85	1.00	1.13	1.31	1.45	1.57

Electric Power Costs*

(Including correction for electric drive)

\$ per kwhr	0.007	0.01	0.015	0.0218	0.025	0.030	0.035	0.04
F_r	0.57	0.64	0.74	0.85	0.89	0.95	1.00	1.05

Deviation from Perfect Gas Laws

Z	0.30	0.40	0.50	0.60	0.70	0.80	1.00	1.10
F_r	0.43	0.53	0.61	0.70	0.78	0.86	1.00	1.07

Pipe Costs 2 Inch Schedule 40 Carbon Steel

\$ per foot	0.50	0.65	1.00	1.32	1.75	2.50	3.50	5.00
F_r	1.41	1.28	1.12	1.00	0.91	0.79	0.71	0.63

Amortization

Years	2	3	5	7	10	15	20	25
F_r	0.65	0.75	0.89	1.00	1.12	1.28	1.39	1.48

*Do not use both.
Factors are multiplicative.
See also Table 7-13 for factors for materials other than carbon steel.
See Equation (7-10) when labor and material costs deviate from norm.

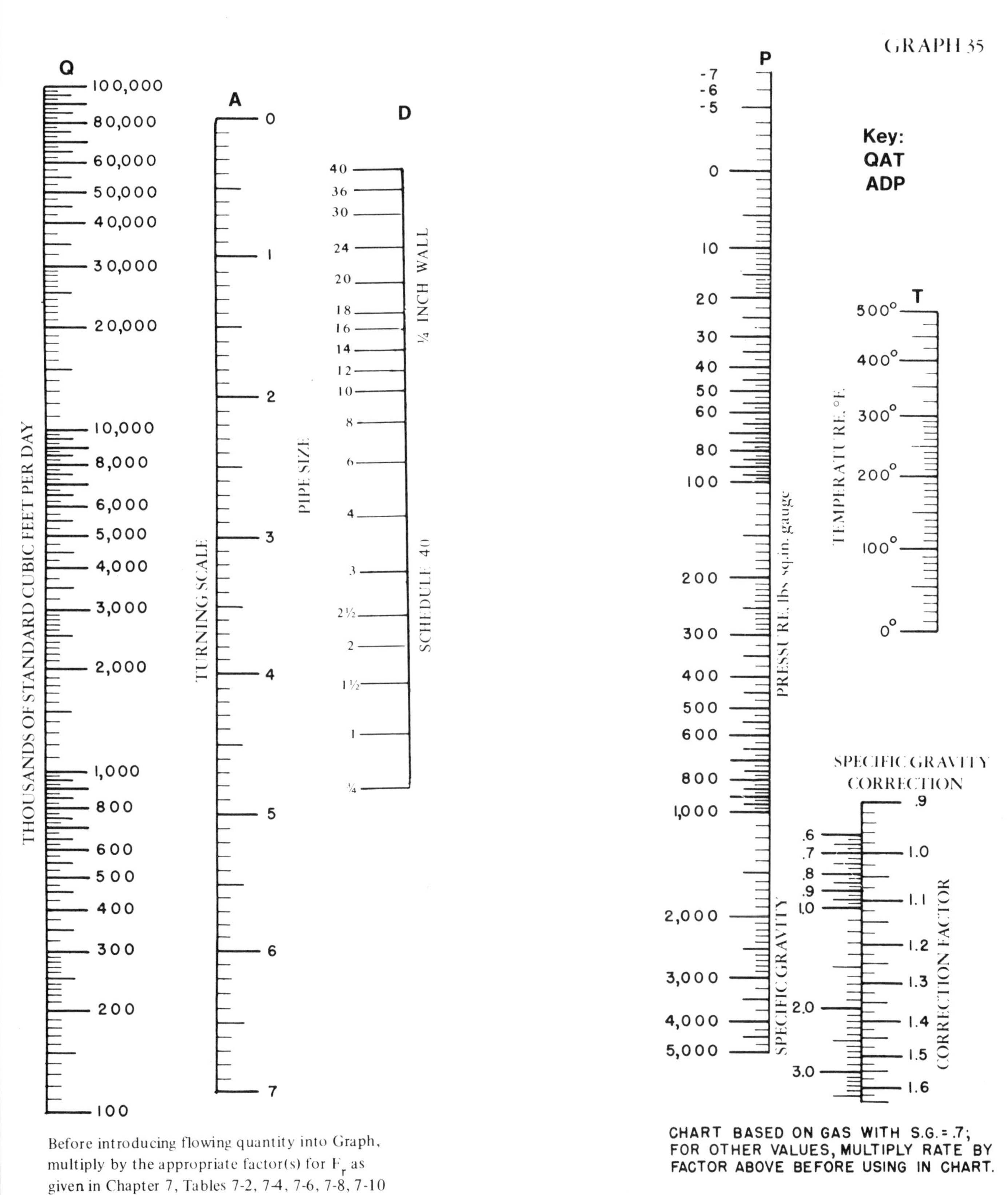

Before introducing flowing quantity into Graph, multiply by the appropriate factor(s) for F_r as given in Chapter 7, Tables 7-2, 7-4, 7-6, 7-8, 7-10 or Equation 7-10

LAC GAS PIPE SIZE

(150 to 600 lb A.S.A. Flange Rating with Schedule 40 or ¼" Wall Pipe)

Summary of Factors F_r for Graph 36

Multiply given flow rate, as previously corrected, if needed, by the factor or factors before entering rate in the nomograph.

Hours of Operation

Per week	10	20	40	60	75	100	120
F_r	0.38	0.49	0.63	0.72	0.78	0.86	0.92

Fuel Costs*

$ per therm	0.20	0.30	0.45	0.70	1.00	1.50	2.00	2.50
F_r	0.64	0.74	0.85	1.00	1.13	1.31	1.45	1.57

Electric Power Costs*
(Including correction for electric drive)

$ per kwhr	0.007	0.01	0.015	0.0218	0.025	0.030	0.035	0.04
F_r	0.57	0.64	0.74	0.85	0.89	0.95	1.00	1.05

Deviation from Perfect Gas Laws

Z	0.30	0.40	0.50	0.60	0.70	0.80	1.00	1.10
F_r	0.43	0.53	0.61	0.70	0.78	0.86	1.00	1.07

Pipe Costs 2 Inch Schedule 40 Carbon Steel

$ per foot	0.50	0.65	1.00	1.32	1.75	2.50	3.50	5.00
F_r	1.41	1.28	1.12	1.00	0.91	0.79	0.71	0.63

Amortization

Years	2	3	5	7	10	15	20	25
F_r	0.65	0.75	0.89	1.00	1.12	1.28	1.39	1.48

*Do not use both.
Factors are multiplicative.
See also Table 7-13 for factors for materials other than carbon steel.
See Equation (7-10) when labor and material costs deviate from norm.

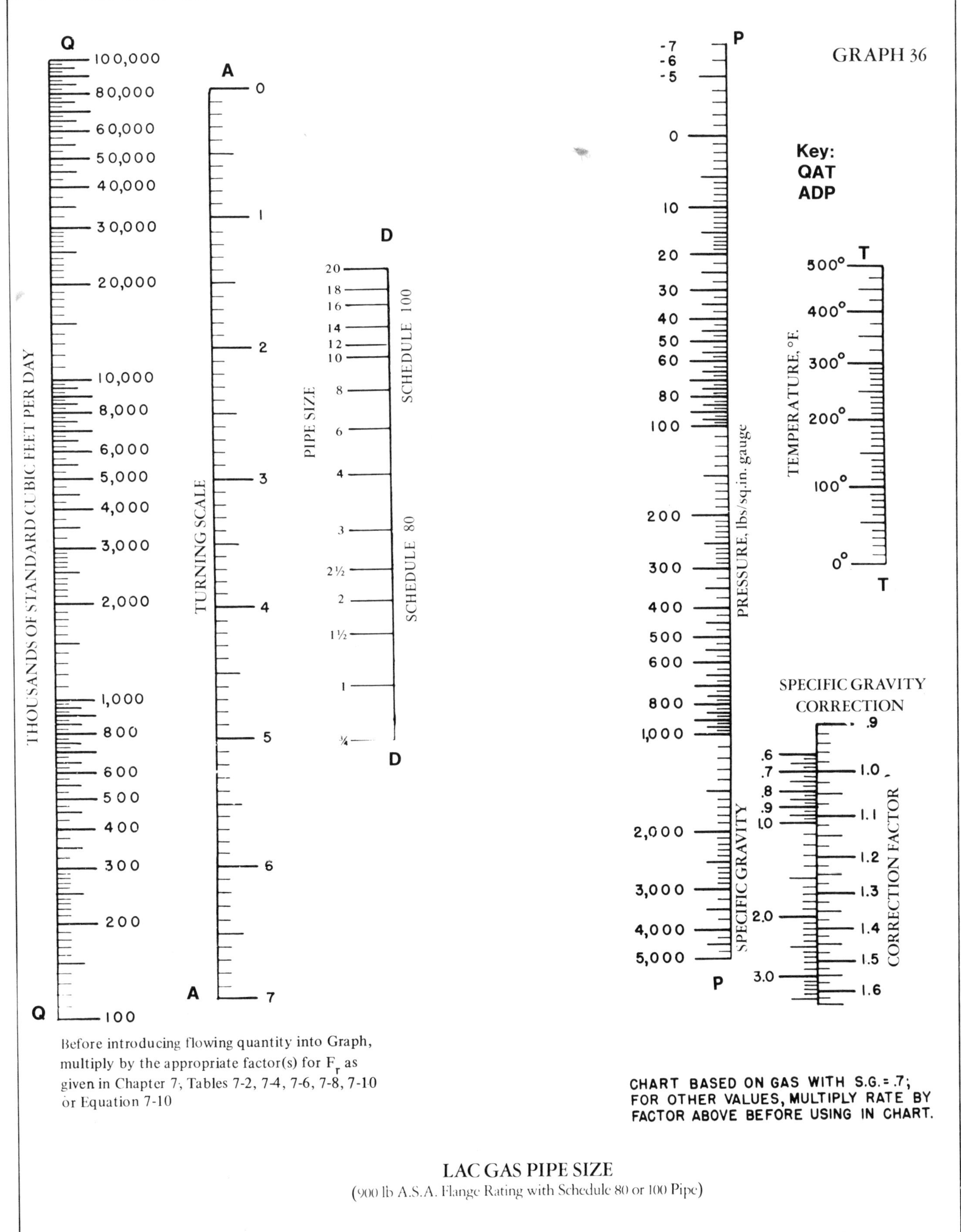

Before introducing flowing quantity into Graph, multiply by the appropriate factor(s) for F_T as given in Chapter 7, Tables 7-2, 7-4, 7-6, 7-8, 7-10 or Equation 7-10

CHART BASED ON GAS WITH S.G.= .7; FOR OTHER VALUES, MULTIPLY RATE BY FACTOR ABOVE BEFORE USING IN CHART.

LAC GAS PIPE SIZE

(900 lb A.S.A. Flange Rating with Schedule 80 or 100 Pipe)

Summary of Factors F_r for Graph 37

Multiply given flow rate, as previously corrected, if needed, by the factor or factors before entering rate in the nomograph.

Hours of Operation

Per week	10	20	40	60	75	100	120
F_r	0.38	0.49	0.63	0.72	0.78	0.86	0.92

Fuel Costs*

\$ per therm	0.20	0.30	0.45	0.70	1.00	1.50	2.00	2.50
F_r	0.64	0.74	0.85	1.00	1.13	1.31	1.45	1.57

Electric Power Costs*
(Including correction for electric drive)

\$ per kwhr	0.007	0.01	0.015	0.0218	0.025	0.030	0.035	0.04
F_r	0.57	0.64	0.74	0.85	0.89	0.95	1.00	1.05

Deviation from Perfect Gas Laws

Z	0.30	0.40	0.50	0.60	0.70	0.80	1.00	1.10
F_r	0.43	0.53	0.61	0.70	0.78	0.86	1.00	1.07

Pipe Costs 2 Inch Schedule 40 Carbon Steel

\$ per foot	0.50	0.65	1.00	1.32	1.75	2.50	3.50	5.00
F_r	1.41	1.28	1.12	1.00	0.91	0.79	0.71	0.63

Amortization

Years	2	3	5	7	10	15	20	25
F_r	0.65	0.75	0.89	1.00	1.12	1.28	1.39	1.48

*Do not use both.
Factors are multiplicative.
See also Table 7-13 for factors for materials other than carbon steel.
See Equation (7-10) when labor and material costs deviate from norm.

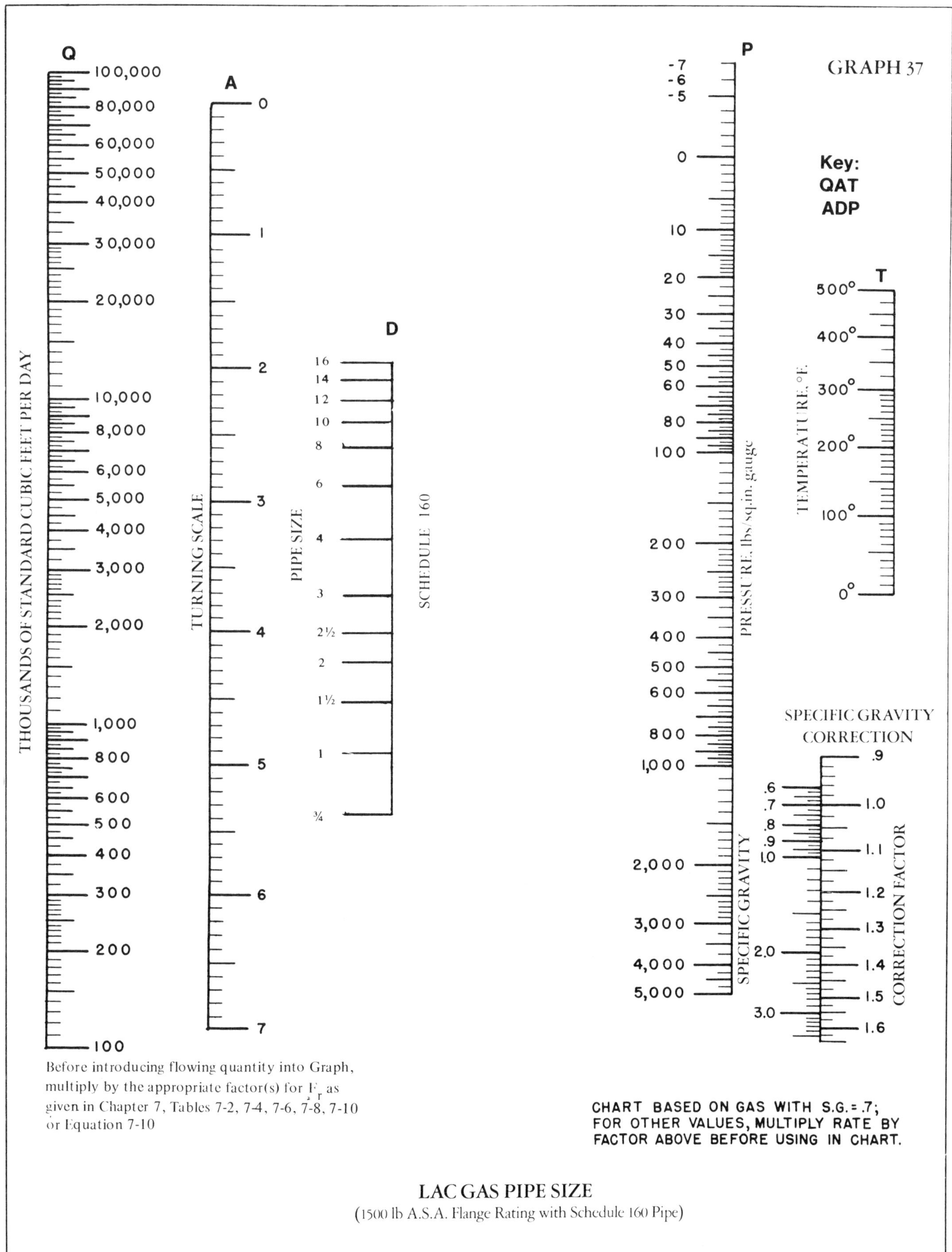

Before introducing flowing quantity into Graph, multiply by the appropriate factor(s) for F_r as given in Chapter 7, Tables 7-2, 7-4, 7-6, 7-8, 7-10 or Equation 7-10

LAC GAS PIPE SIZE

(1500 lb A.S.A. Flange Rating with Schedule 160 Pipe)

Summary of Factors F_r for Graph 38

Multiply given flow rate, as previously corrected, if needed, by the factor or factors before entering rate in the nomograph.

Hours of Operation

Per week	10	20	40	60	75	100	120
F_r	0.38	0.49	0.63	0.72	0.78	0.86	0.92

Fuel Costs*

$ per therm	0.20	0.30	0.45	0.70	1.00	1.50	2.00	2.50
F_r	0.64	0.74	0.85	1.00	1.13	1.31	1.45	1.57

Electric Power Costs*
(Including correction for electric drive)

$ per kwhr	0.007	0.01	0.015	0.0218	0.025	0.030	0.035	0.04
F_r	0.57	0.64	0.74	0.85	0.89	0.95	1.00	1.05

Deviation from Perfect Gas Laws

Z	0.30	0.40	0.50	0.60	0.70	0.80	1.00	1.10
F_r	0.43	0.53	0.61	0.70	0.78	0.86	1.00	1.07

Pipe Costs 2 Inch Schedule 40 Carbon Steel

$ per foot	0.50	0.65	1.00	1.32	1.75	2.50	3.50	5.00
F_r	1.41	1.28	1.12	1.00	0.91	0.79	0.71	0.63

Amortization

Years	2	3	5	7	10	15	20	25
F_r	0.65	0.75	0.89	1.00	1.12	1.28	1.39	1.48

*Do not use both.
Factors are multiplicative.
See also Table 7-13 for factors for materials other than carbon steel.
See Equation (7-10) when labor and material costs deviate from norm.

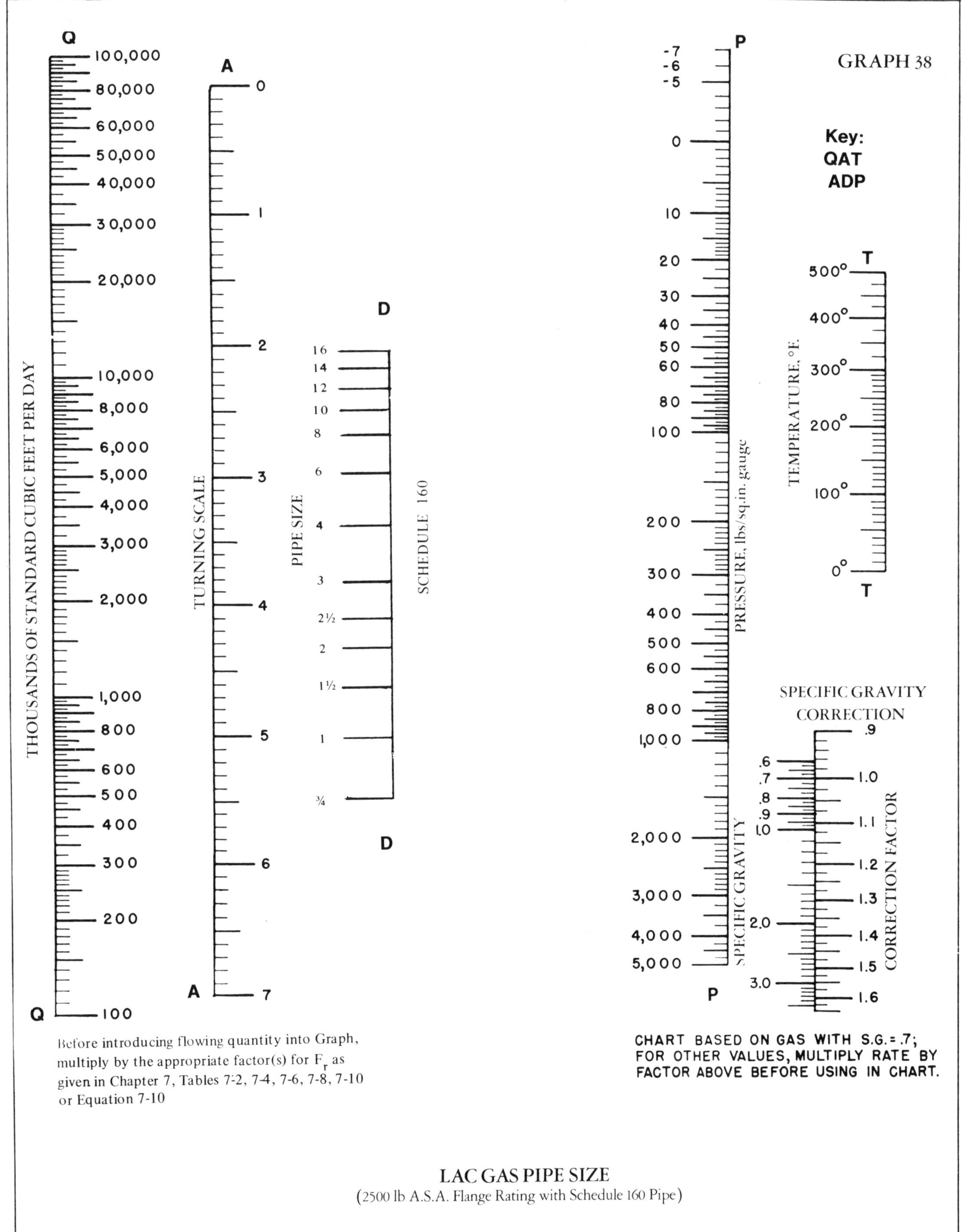

Before introducing flowing quantity into Graph, multiply by the appropriate factor(s) for F_T as given in Chapter 7, Tables 7-2, 7-4, 7-6, 7-8, 7-10 or Equation 7-10

LAC GAS PIPE SIZE

(2500 lb A.S.A. Flange Rating with Schedule 160 Pipe)

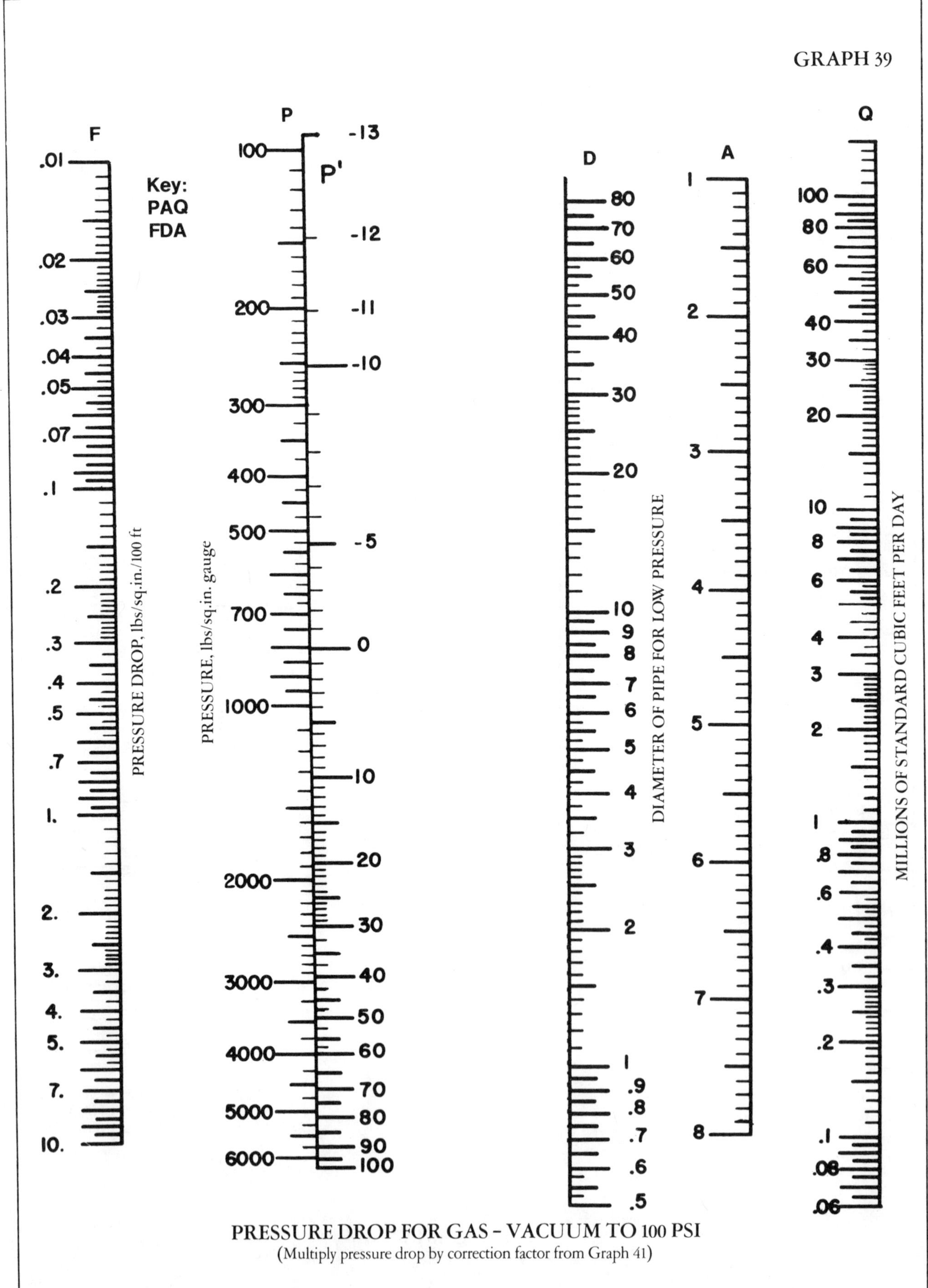

PRESSURE DROP FOR GAS – VACUUM TO 100 PSI
(Multiply pressure drop by correction factor from Graph 41)

GRAPH 40

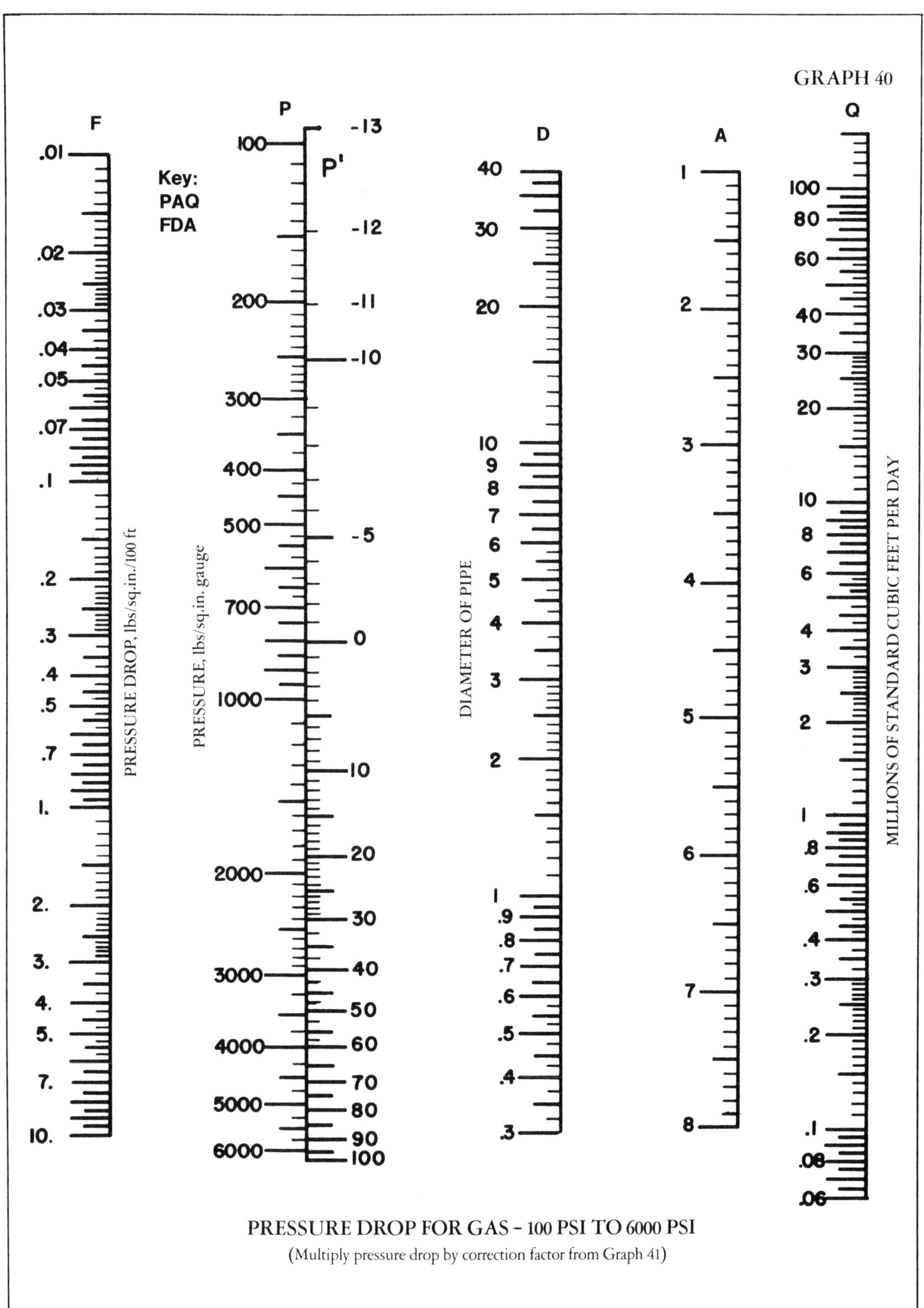

PRESSURE DROP FOR GAS – 100 PSI TO 6000 PSI

(Multiply pressure drop by correction factor from Graph 41)

GRAPH 41

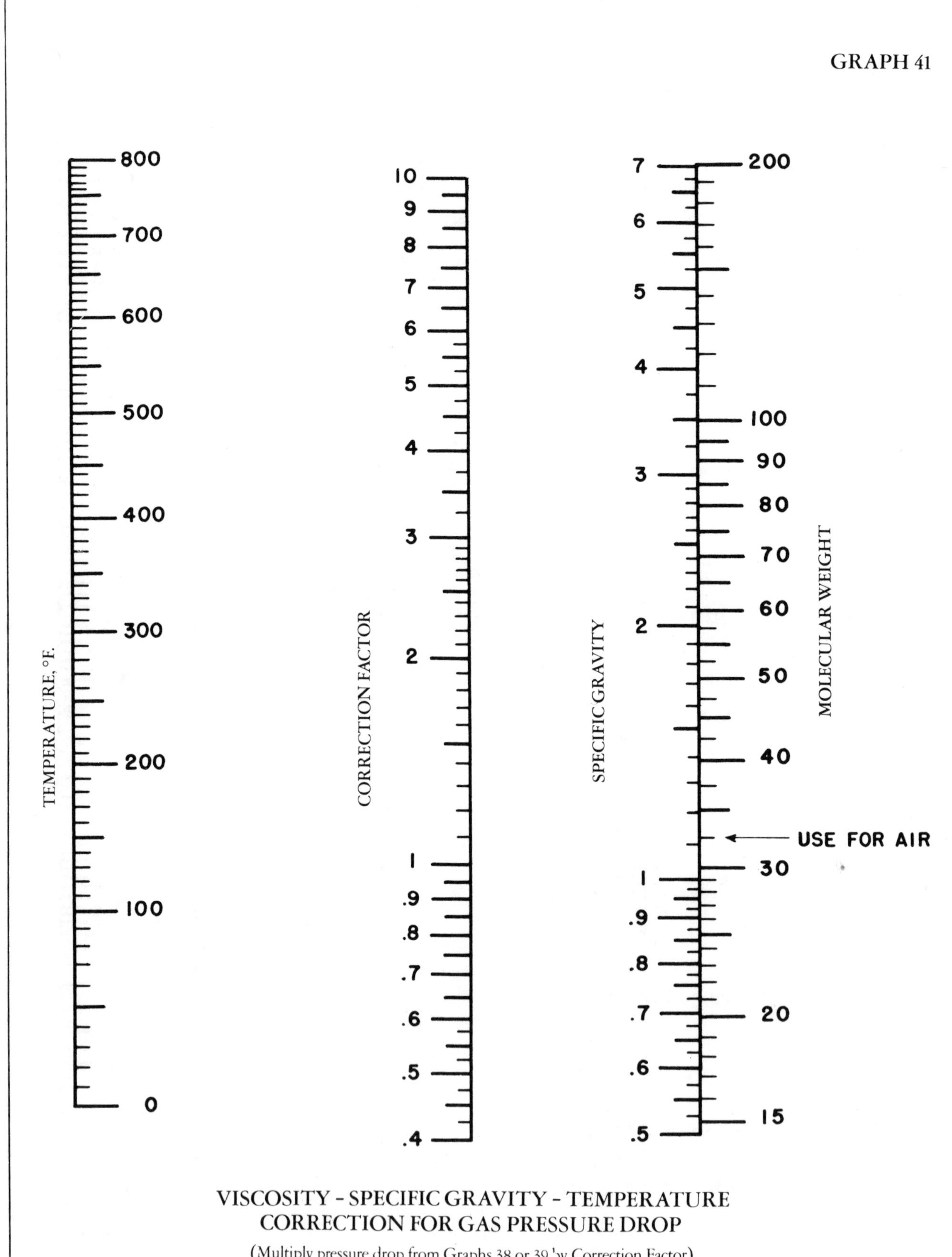

VISCOSITY - SPECIFIC GRAVITY - TEMPERATURE CORRECTION FOR GAS PRESSURE DROP

(Multiply pressure drop from Graphs 38 or 39 by Correction Factor)

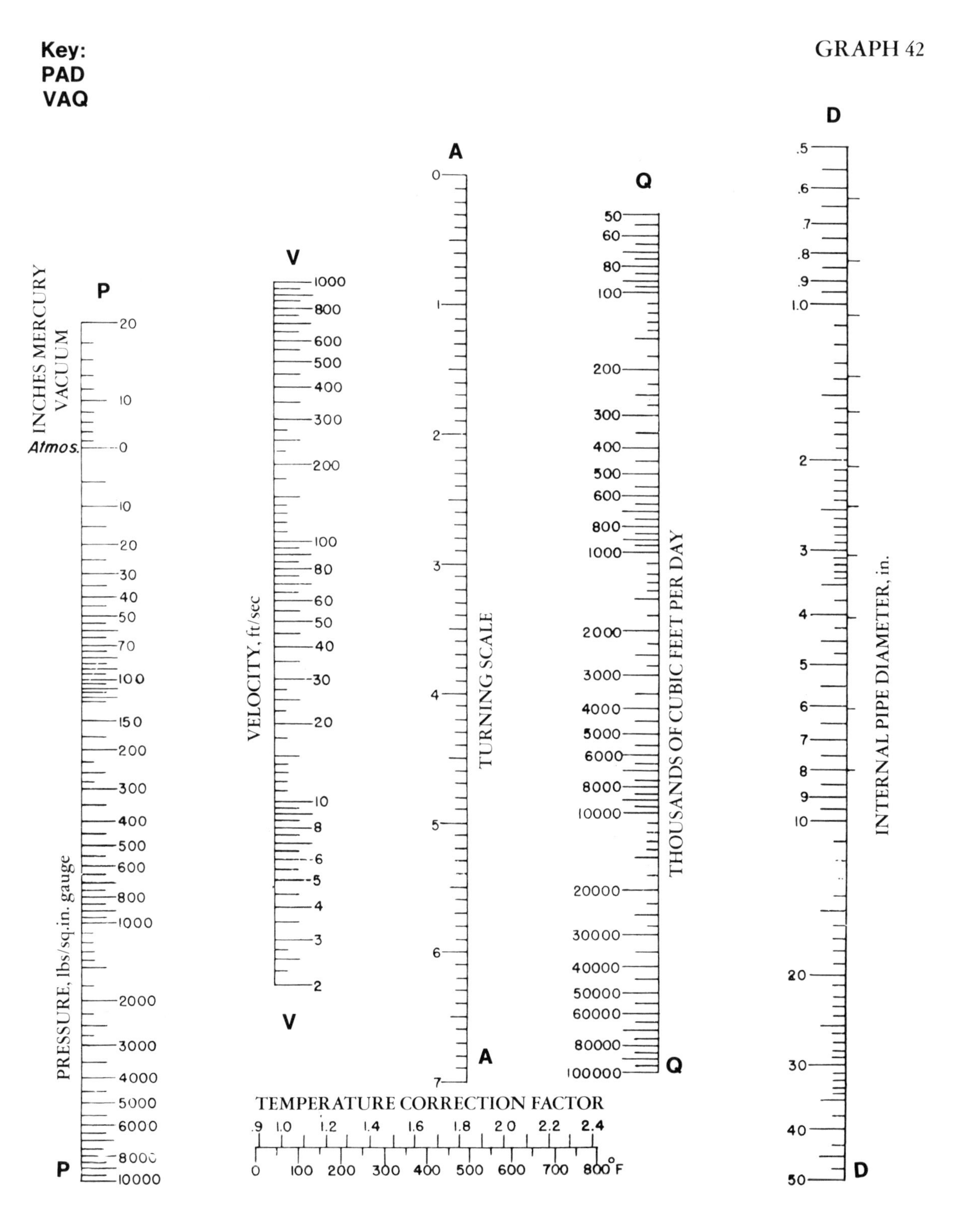

VELOCITY FOR GAS

Note: When temperature is not 60°F. and/or Z is not 1.0 multiply Q by Z and by Temperature Correction Factor before using chart.

GRAPH 43

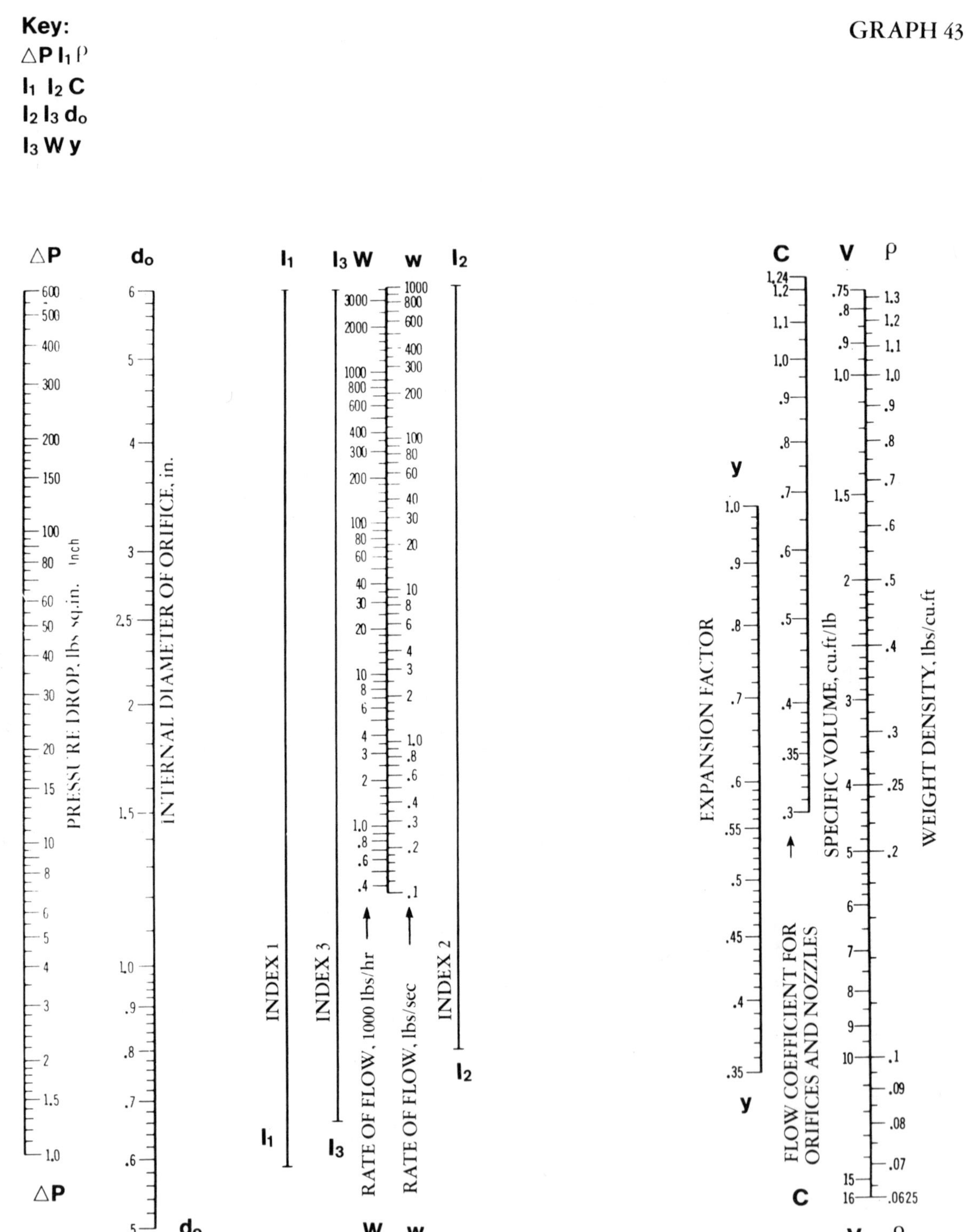

FLOW OF COMPRESSIBLE FLUIDS THROUGH NOZZLES AND ORIFICES

(From TP-410, FLOW OF FLUIDS, Courtesy Crane Company)

GRAPH 44

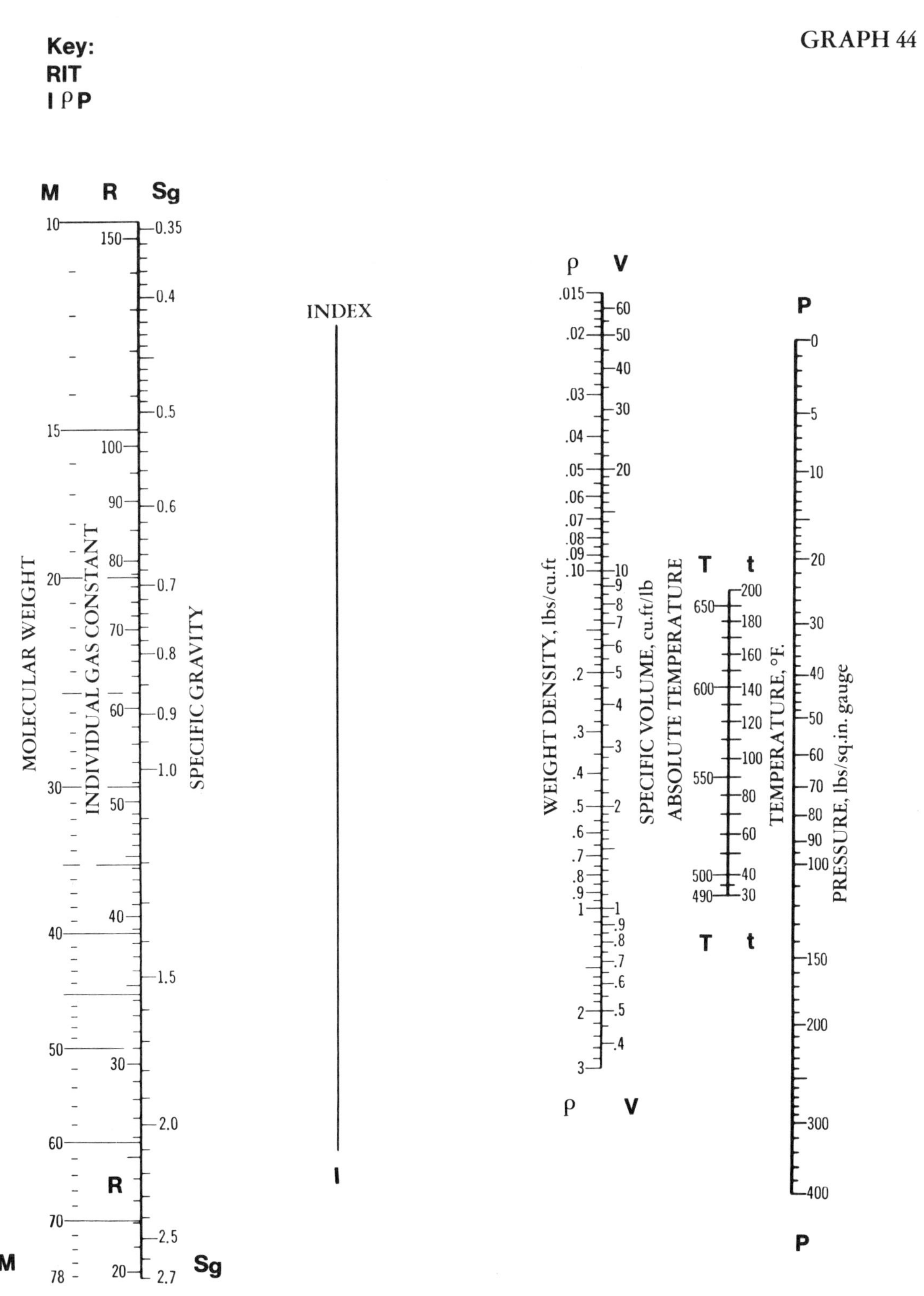

WEIGHT DENSITY AND SPECIFIC VOLUME OF GASES AND VAPORS

(From TP-410, FLOW OF FLUIDS, Courtesy Crane Company)

GRAPH 45

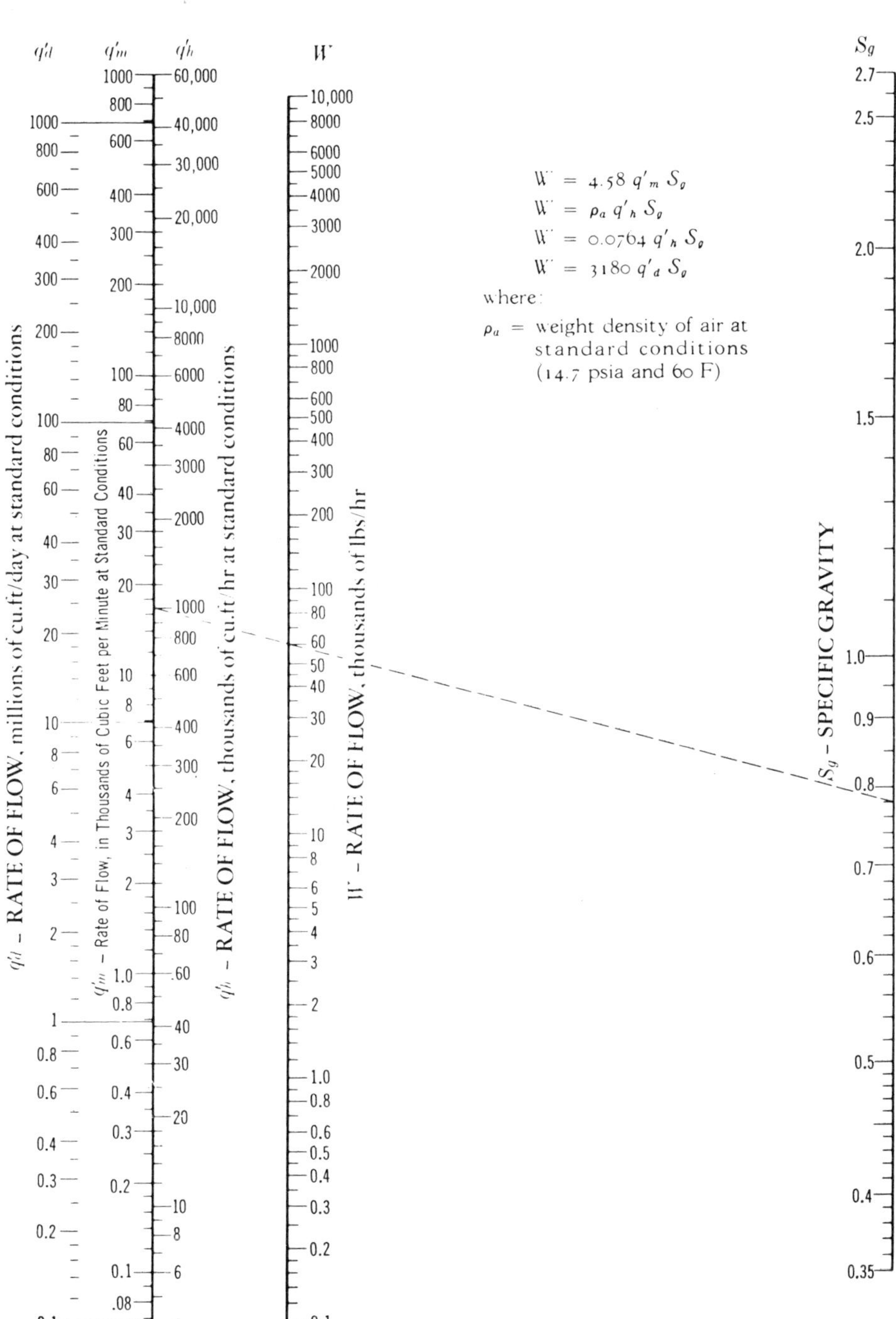

EQUIVALENT VOLUME AND WEIGHT FLOW RATES OF COMPRESSIBLE FLUIDS

(From TP-410, FLOW OF FLUIDS, Courtesy Crane Company)

Summary of Factors F_r for Graph 46

Multiply given flow rate, as previously corrected, if needed, by the factor or factors before entering rate in the nomograph.

Hours of Operation

Per week	10	20	40	60	75	100	120
F_r	0.38	0.49	0.63	0.72	0.78	0.86	0.92

Fuel Costs

\$ per therm	0.20	0.30	0.45	0.70	1.00	1.50	2.00	2.50
F_r	0.64	0.74	0.85	1.00	1.13	1.31	1.45	1.57

Pipe Costs 2 Inch Schedule 40 Carbon Steel

\$ per foot	0.50	0.65	1.00	1.32	1.75	2.50	3.50	5.00
F_r	1.41	1.28	1.12	1.00	0.91	0.79	0.71	0.63

Amortization

Years	2	3	5	7	10	15	20	25
F_r	0.65	0.75	0.89	1.00	1.12	1.28	1.39	1.48

Factors are multiplicative.
See Equation (7-10) when labor costs deviate from norm.

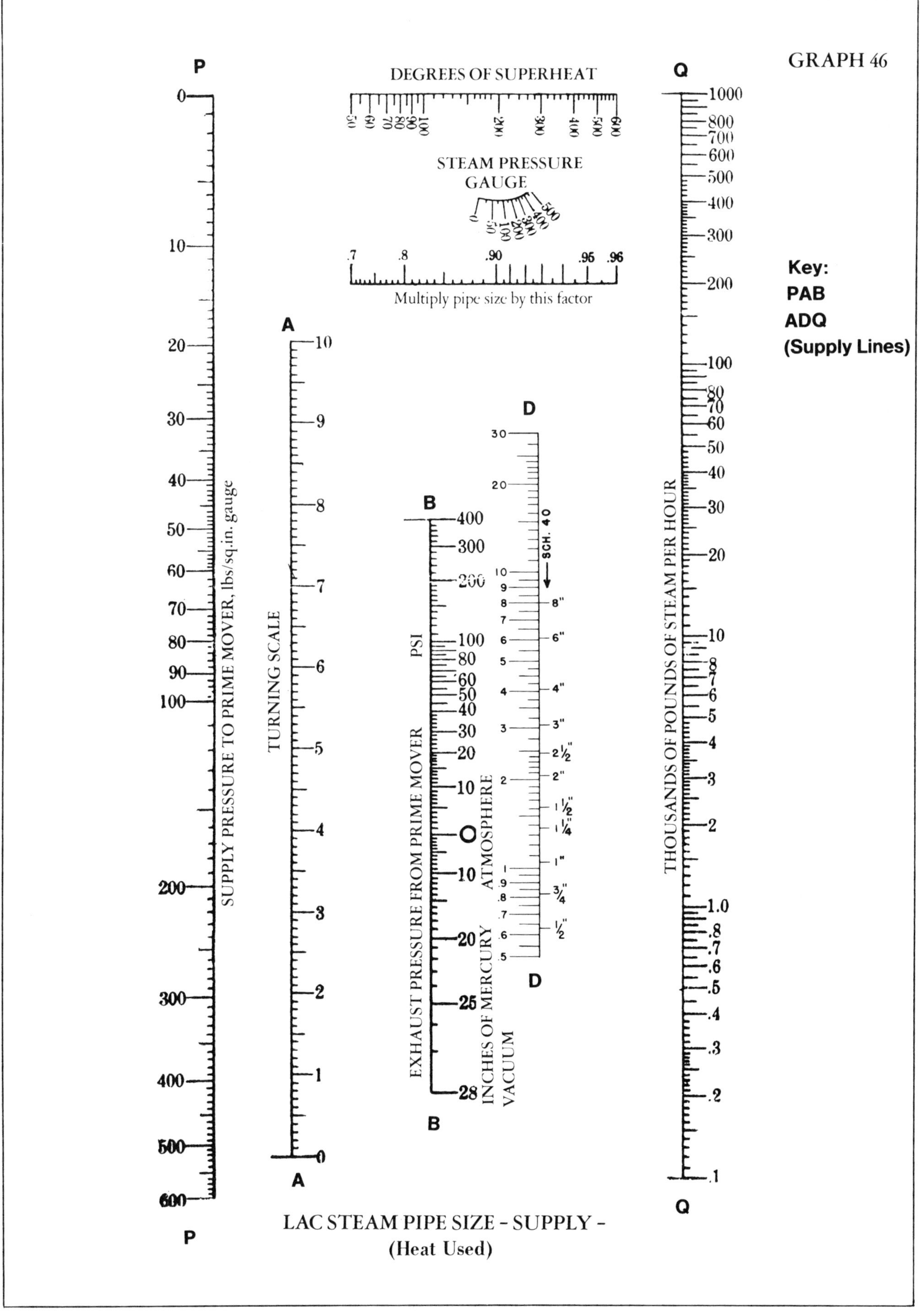

LAC STEAM PIPE SIZE – SUPPLY –
(Heat Used)

Summary of Factors F_r for Graph 46 A

Multiply given flow rate, as previously corrected, if needed, by the factor or factors before entering rate in the nomograph.

Hours of Operation

Per week	10	20	40	60	75	100	120
F_r	0.38	0.49	0.63	0.72	0.78	0.86	0.92

Fuel Costs

$ per therm	0.20	0.30	0.45	0.70	1.00	1.50	2.00	2.50
F_r	0.64	0.74	0.85	1.00	1.13	1.31	1.45	1.57

Pipe Costs 2 Inch Schedule 40 Carbon Steel

$ per foot	0.50	0.65	1.00	1.32	1.75	2.50	3.50	5.00
F_r	1.41	1.28	1.12	1.00	0.91	0.79	0.71	0.63

Amortization

Years	2	3	5	7	10	15	20	25
F_r	0.65	0.75	0.89	1.00	1.12	1.28	1.39	1.48

Factors are multiplicative.
See Equation (7-10) when labor costs deviate from norm.

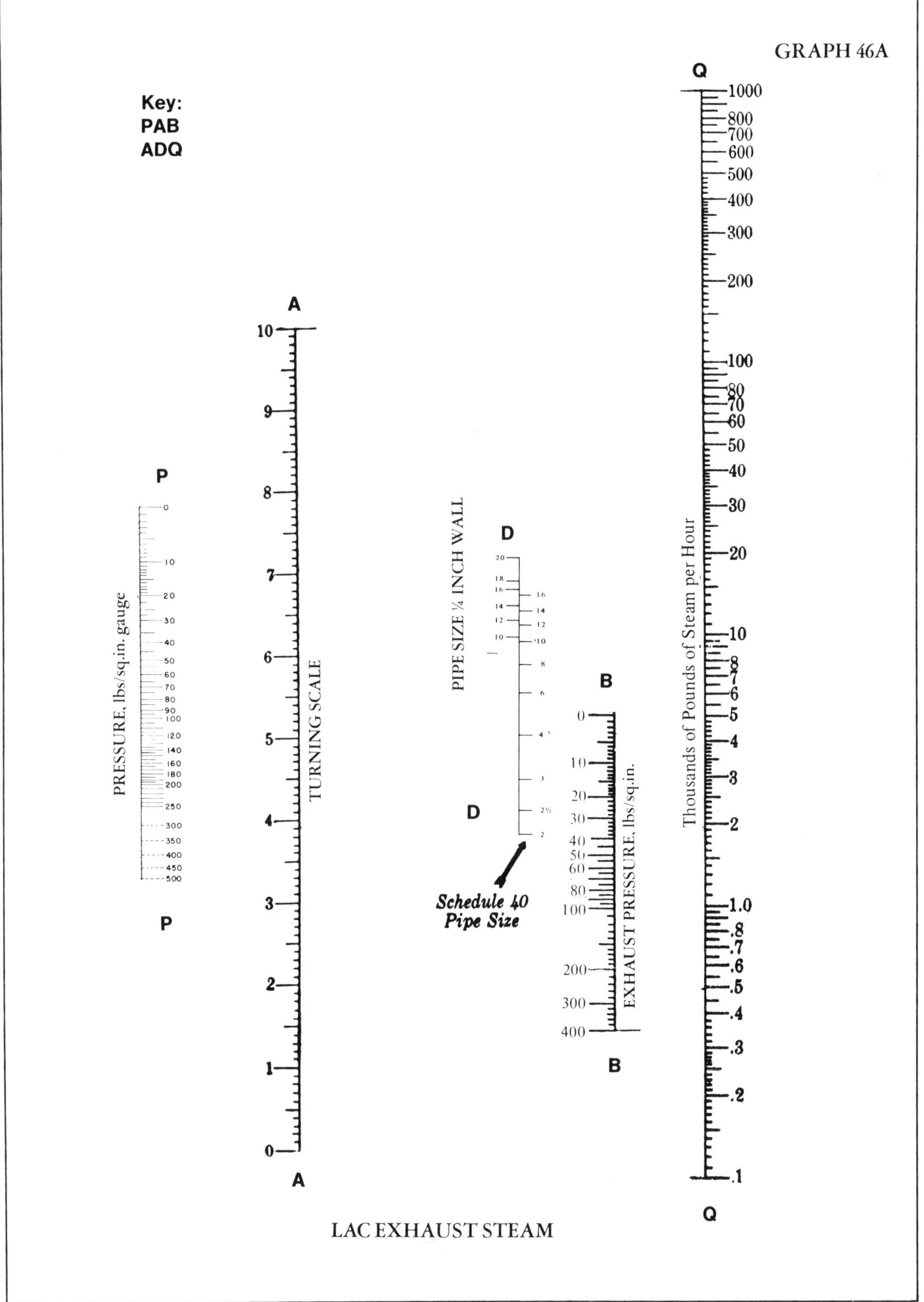
GRAPH 46A
Key:
PAB
ADQ
P
PRESSURE, lbs/sq.in. gauge
A
TURNING SCALE
D
PIPE SIZE ¼ INCH WALL
Schedule 40
Pipe Size
B
EXHAUST PRESSURE, lbs/sq.in.
Q
Thousands of Pounds of Steam per Hour
LAC EXHAUST STEAM

Summary of Factors F_r for Graph 47

Multiply given flow rate, as previously corrected, if needed, by the factor or factors before entering rate in the nomograph.

Hours of Operation

Per week	10	20	40	60	75	100	120
F_r	0.38	0.49	0.63	0.72	0.78	0.86	0.92

Fuel Costs

$ per therm	0.20	0.30	0.45	0.70	1.00	1.50	2.00	2.50
F_r	0.64	0.74	0.85	1.00	1.13	1.31	1.45	1.57

Pipe Costs 2 Inch Schedule 40 Carbon Steel

$ per foot	0.50	0.65	1.00	1.32	1.75	2.50	3.50	5.00
F_r	1.41	1.28	1.12	1.00	0.91	0.79	0.71	0.63

Amortization

Years	2	3	5	7	10	15	20	25
F_r	0.65	0.75	0.89	1.00	1.12	1.28	1.39	1.48

Factors are multiplicative.
See Equation (7-10) when labor costs deviate from norm.

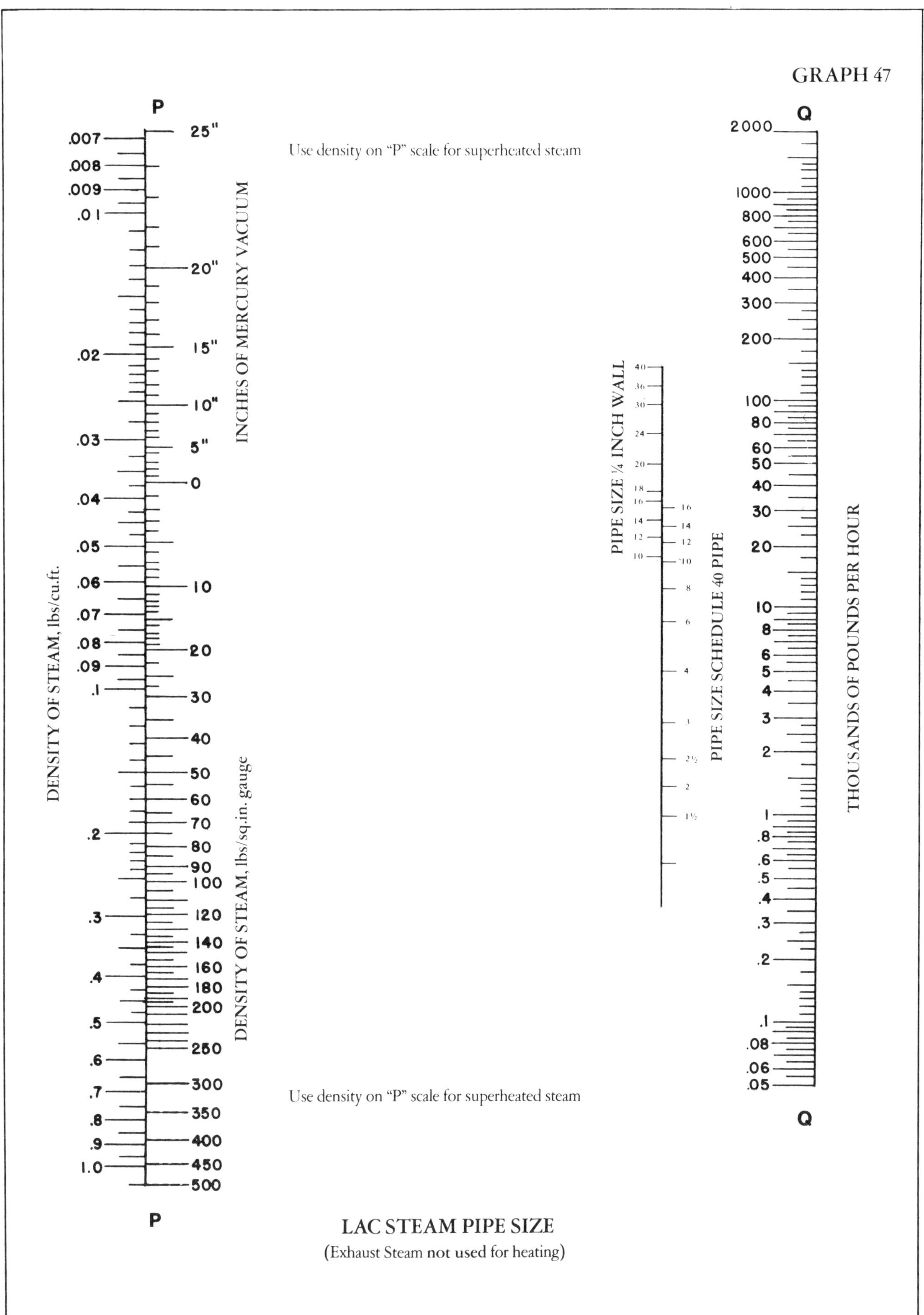

LAC STEAM PIPE SIZE

(Exhaust Steam not used for heating)

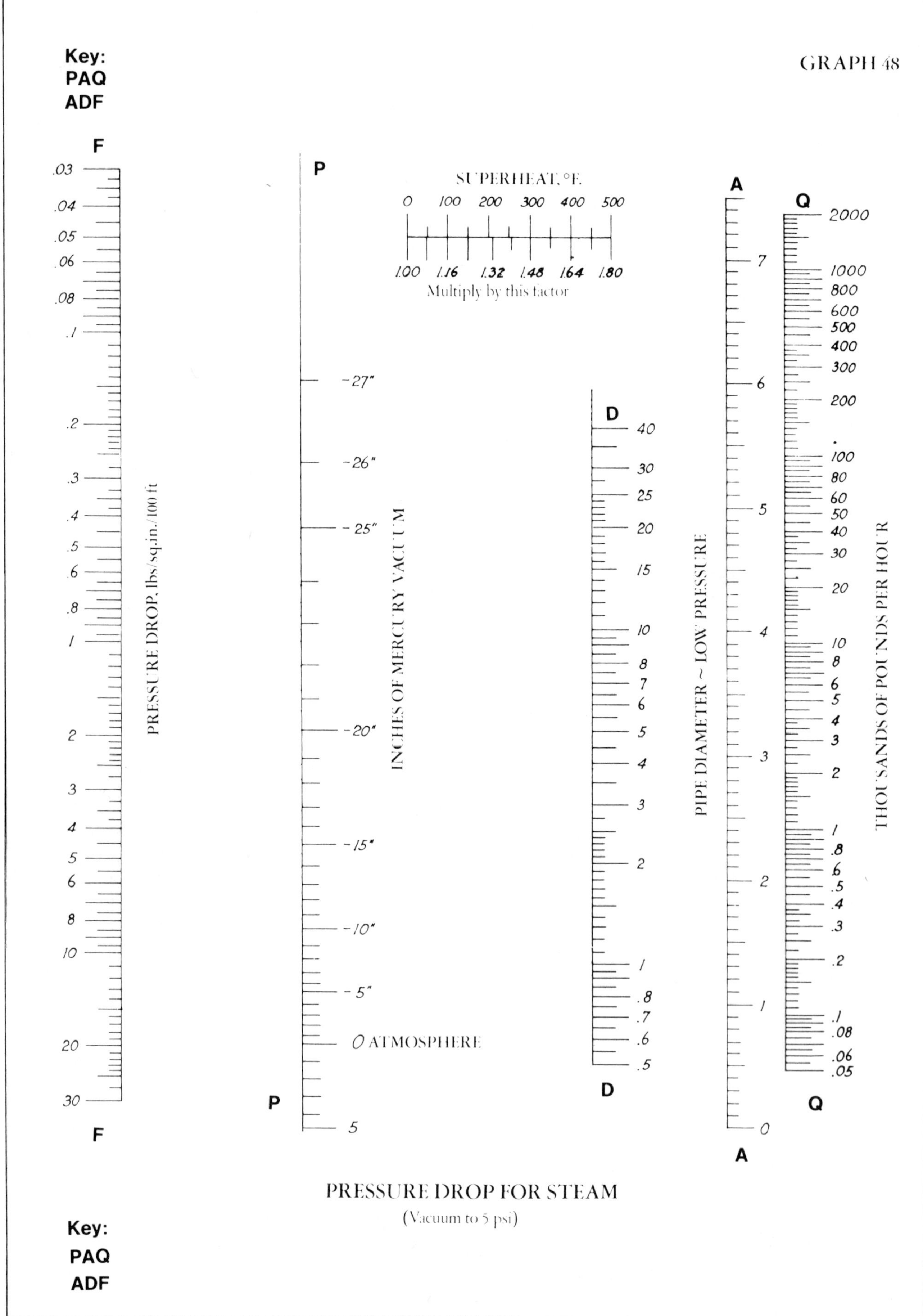
Key:
PAQ
ADF
GRAPH 48
F
P
SUPERHEAT, °F.
0 100 200 300 400 500
1.00 1.16 1.32 1.48 1.64 1.80
Multiply by this factor
A
Q
.03 .04 .05 .06 .08 .1 .2 .3 .4 .5 .6 .8 1 2 3 4 5 6 8 10 20 30
PRESSURE DROP, lbs/sq.in./100 ft
-27" -26" -25" -20" -15" -10" -5" 0 ATMOSPHERE 5
INCHES OF MERCURY VACUUM
D
40 30 25 20 15 10 8 7 6 5 4 3 2 1 .8 .7 .6 .5
PIPE DIAMETER ~ LOW PRESSURE
7 6 5 4 3 2 1 0
2000 1000 800 600 500 400 300 200 100 80 60 50 40 30 20 10 8 6 5 4 3 2 1 .8 .6 .5 .4 .3 .2 .1 .08 .06 .05
THOUSANDS OF POUNDS PER HOUR
PRESSURE DROP FOR STEAM
(Vacuum to 5 psi)
Key:
PAQ
ADF

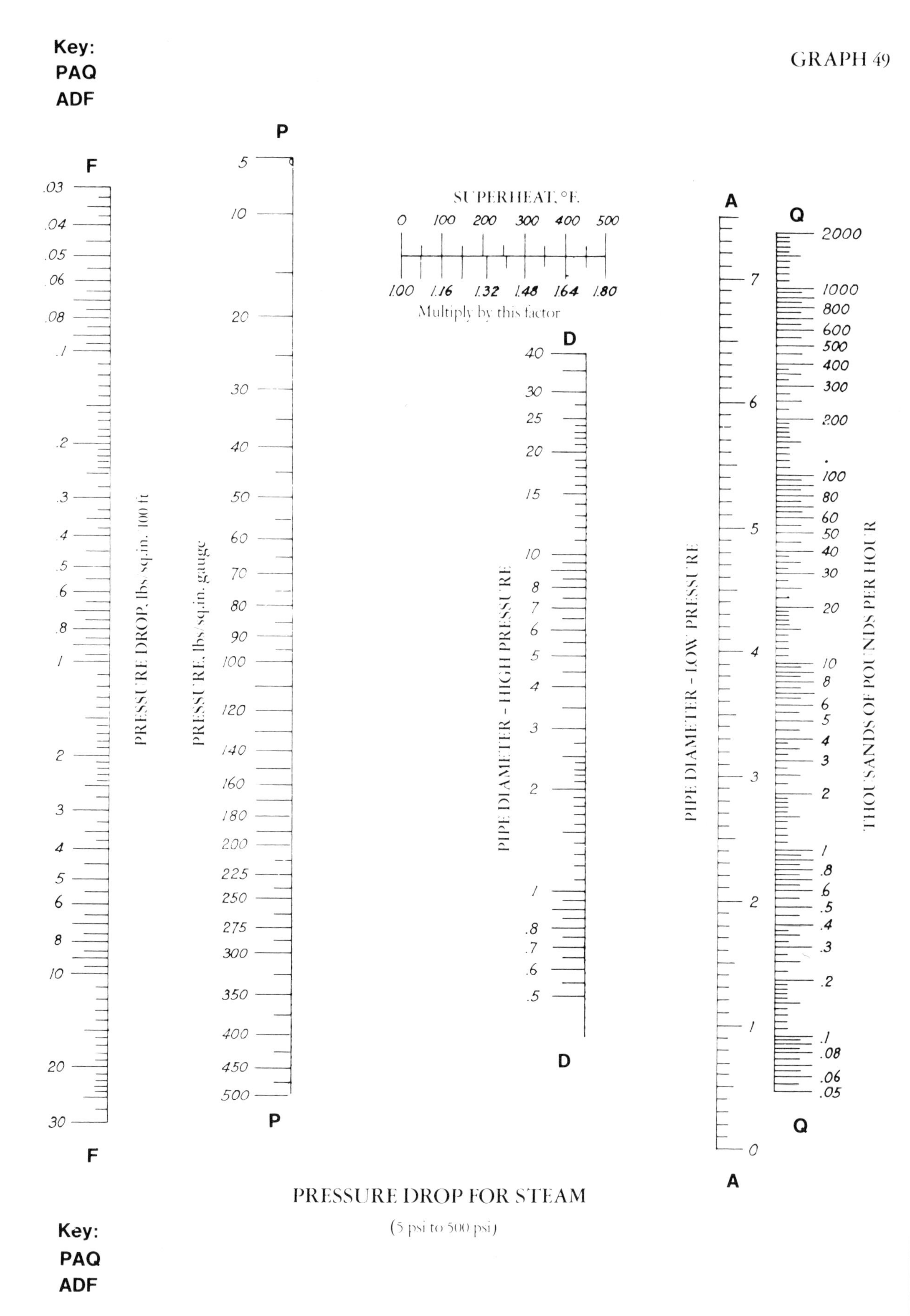

PRESSURE DROP FOR STEAM

(5 psi to 500 psi)

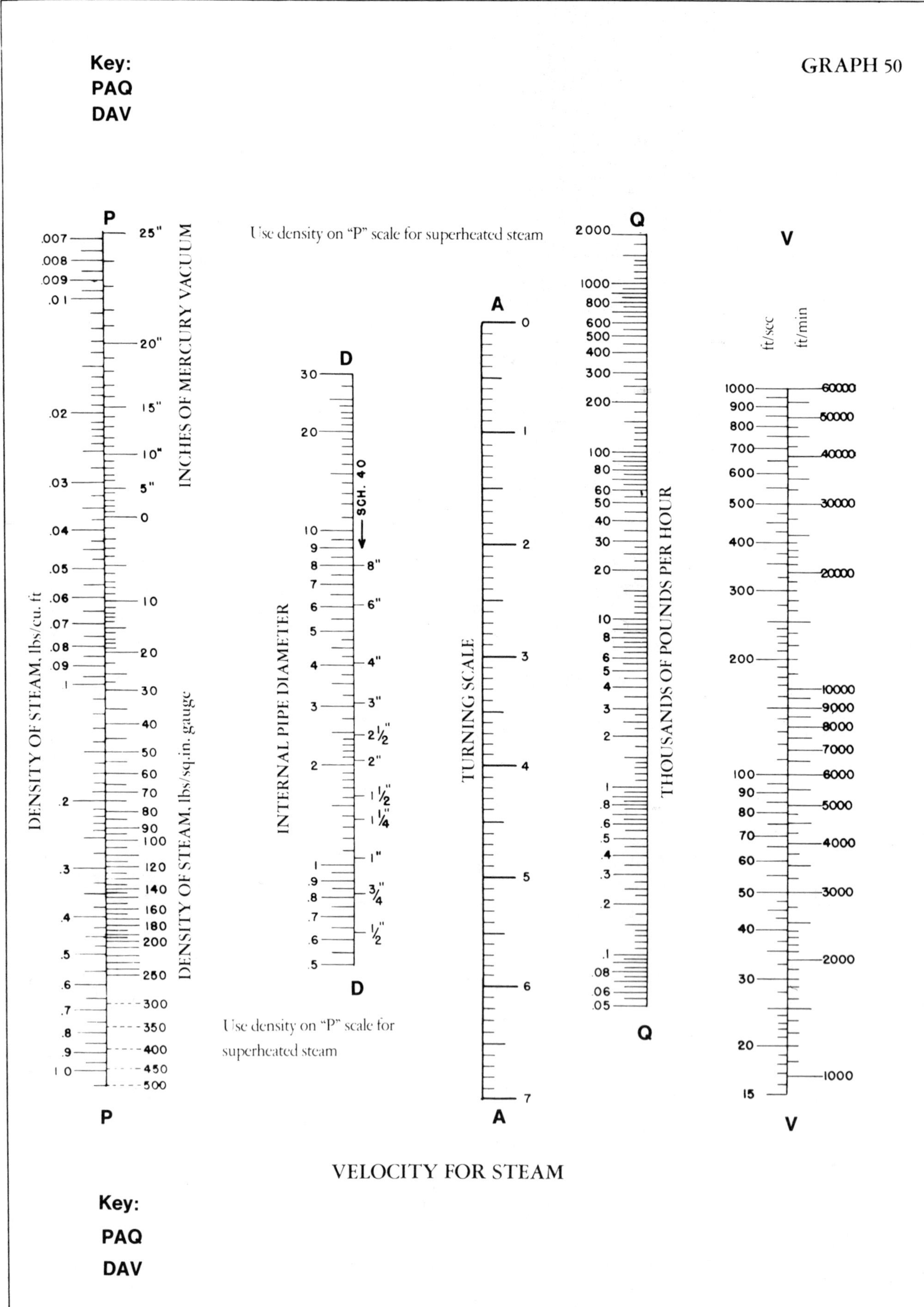
Key:
PAQ
DAV
GRAPH 50
P
INCHES OF MERCURY VACUUM
DENSITY OF STEAM, lbs/cu. ft
DENSITY OF STEAM, lbs/sq.in. gauge
Use density on "P" scale for superheated steam
D
SCH. 40
INTERNAL PIPE DIAMETER
A
TURNING SCALE
Q
THOUSANDS OF POUNDS PER HOUR
V
ft/sec
ft/min
Use density on "P" scale for superheated steam
VELOCITY FOR STEAM
Key:
PAQ
DAV

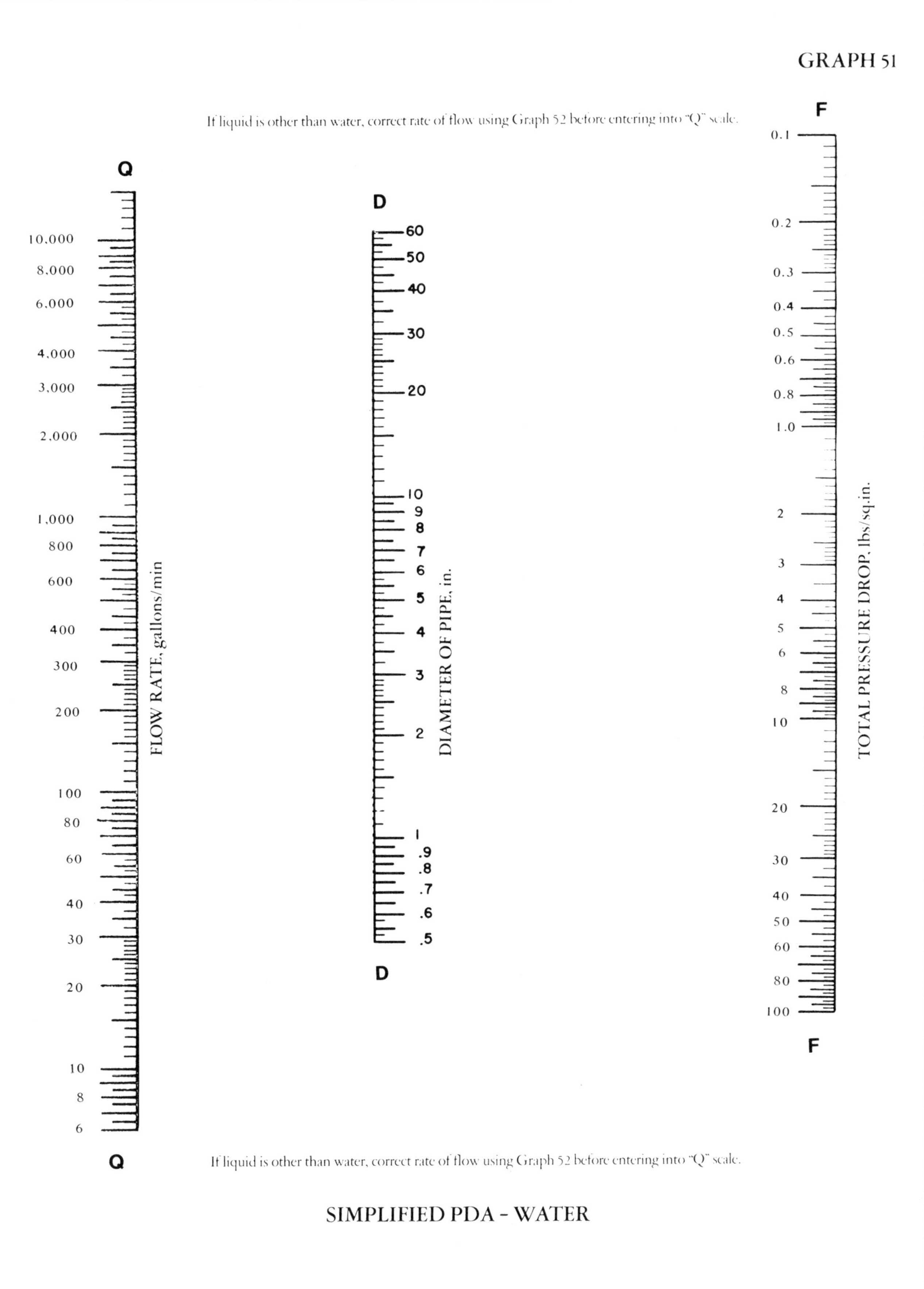

SIMPLIFIED PDA - WATER

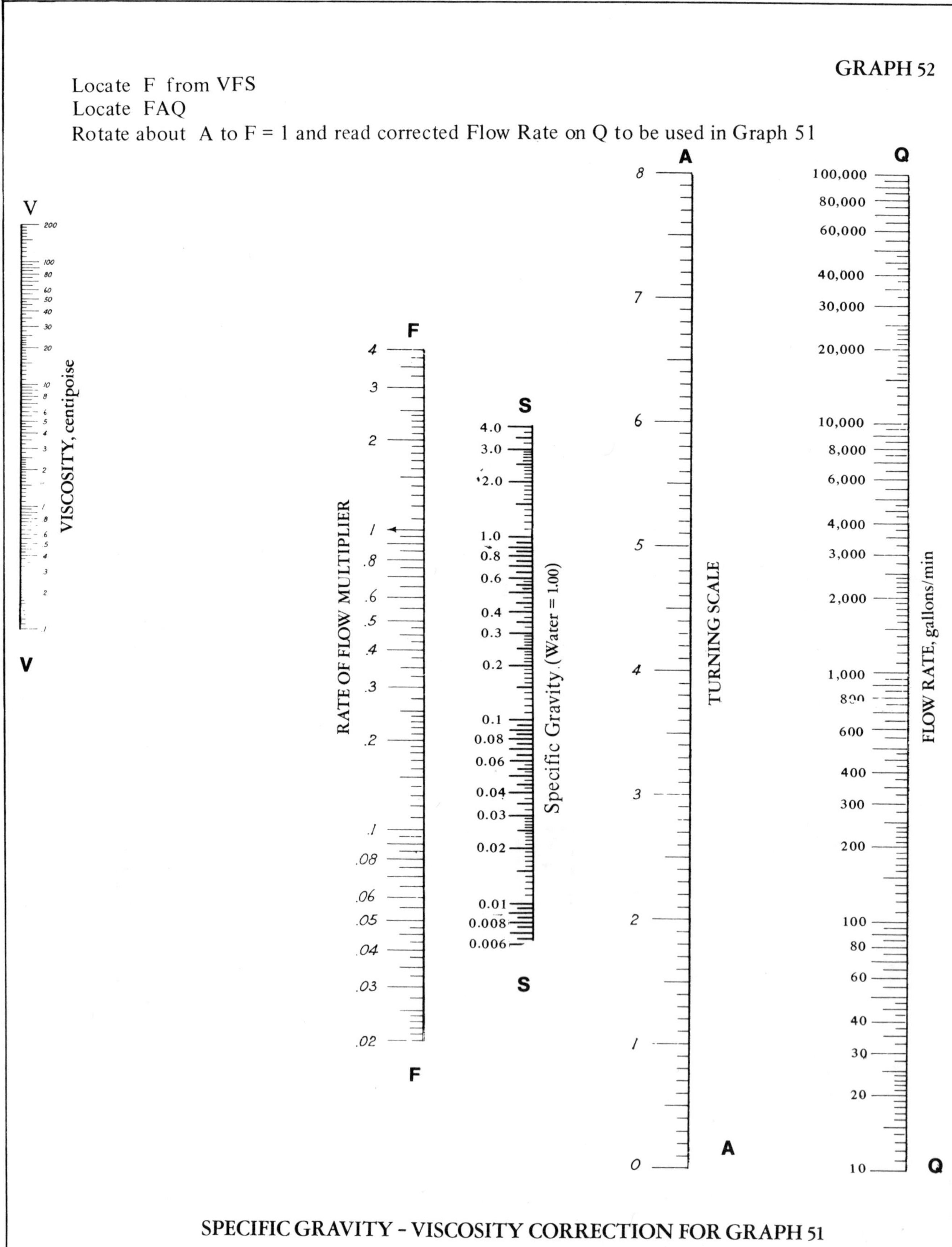

SPECIFIC GRAVITY - VISCOSITY CORRECTION FOR GRAPH 51

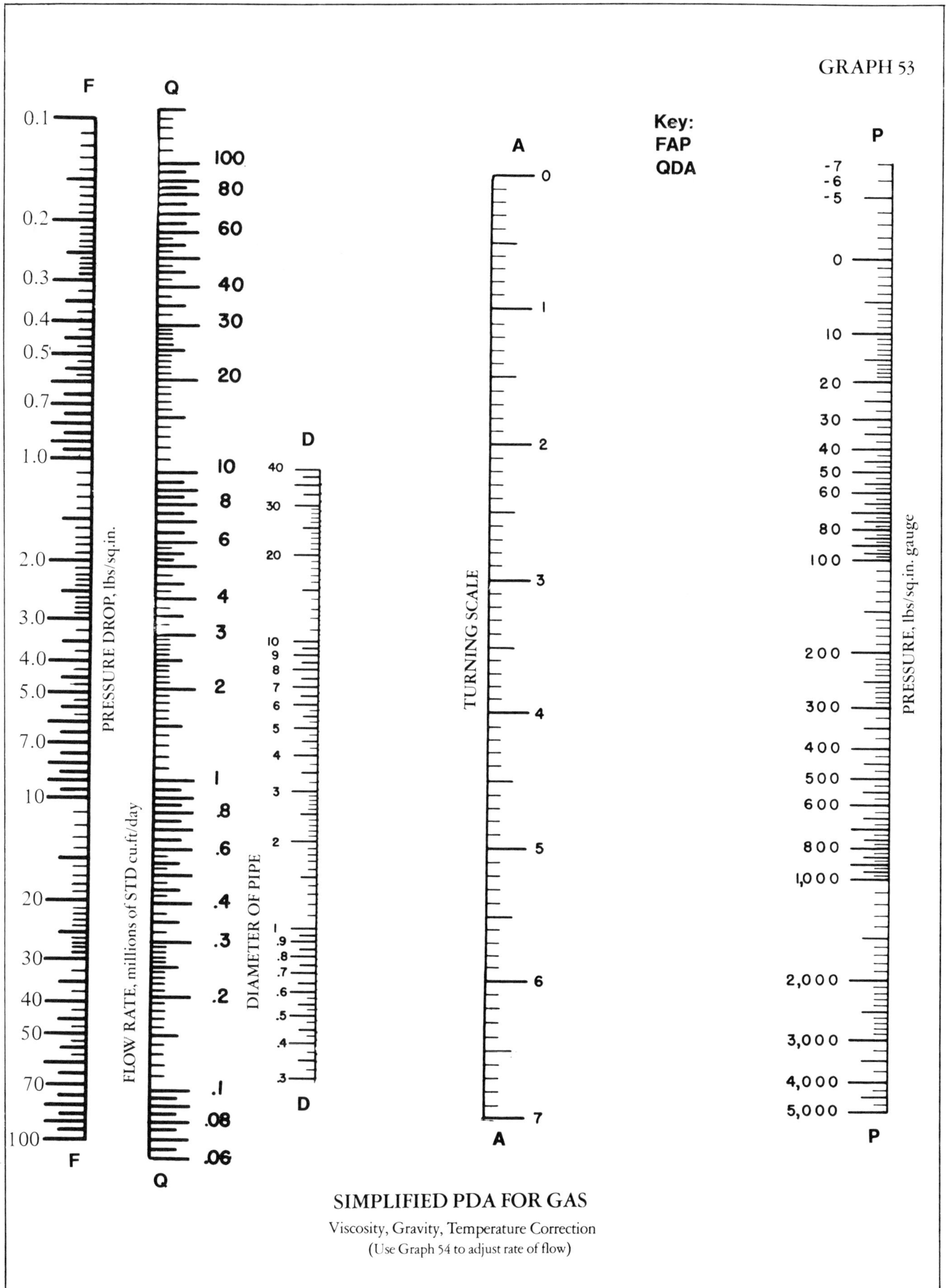

SIMPLIFIED PDA FOR GAS

Viscosity, Gravity, Temperature Correction

(Use Graph 54 to adjust rate of flow)

GRAPH 54

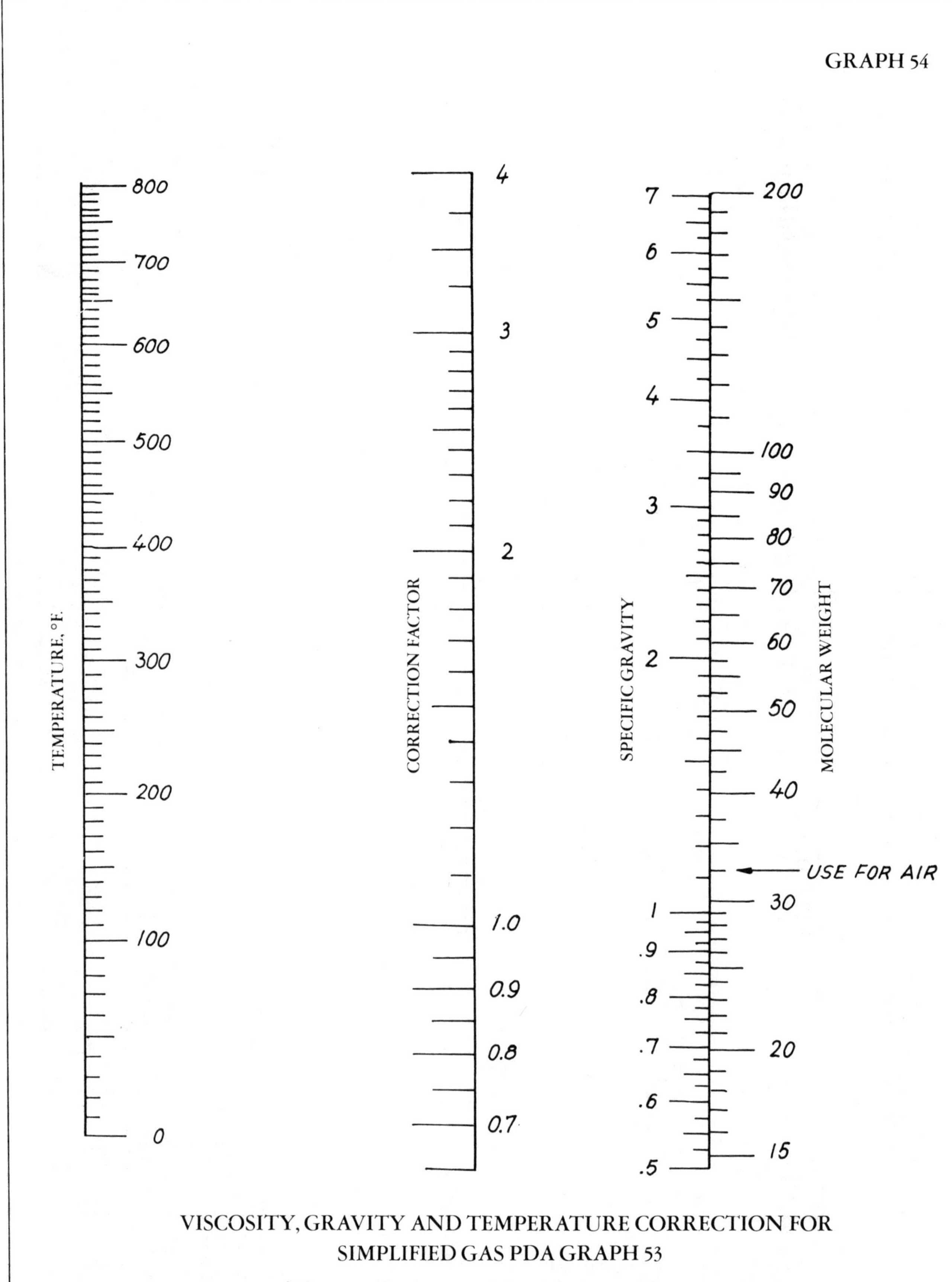

VISCOSITY, GRAVITY AND TEMPERATURE CORRECTION FOR SIMPLIFIED GAS PDA GRAPH 53

(Adjust rate of flow by correction factor before entering into Graph 53)

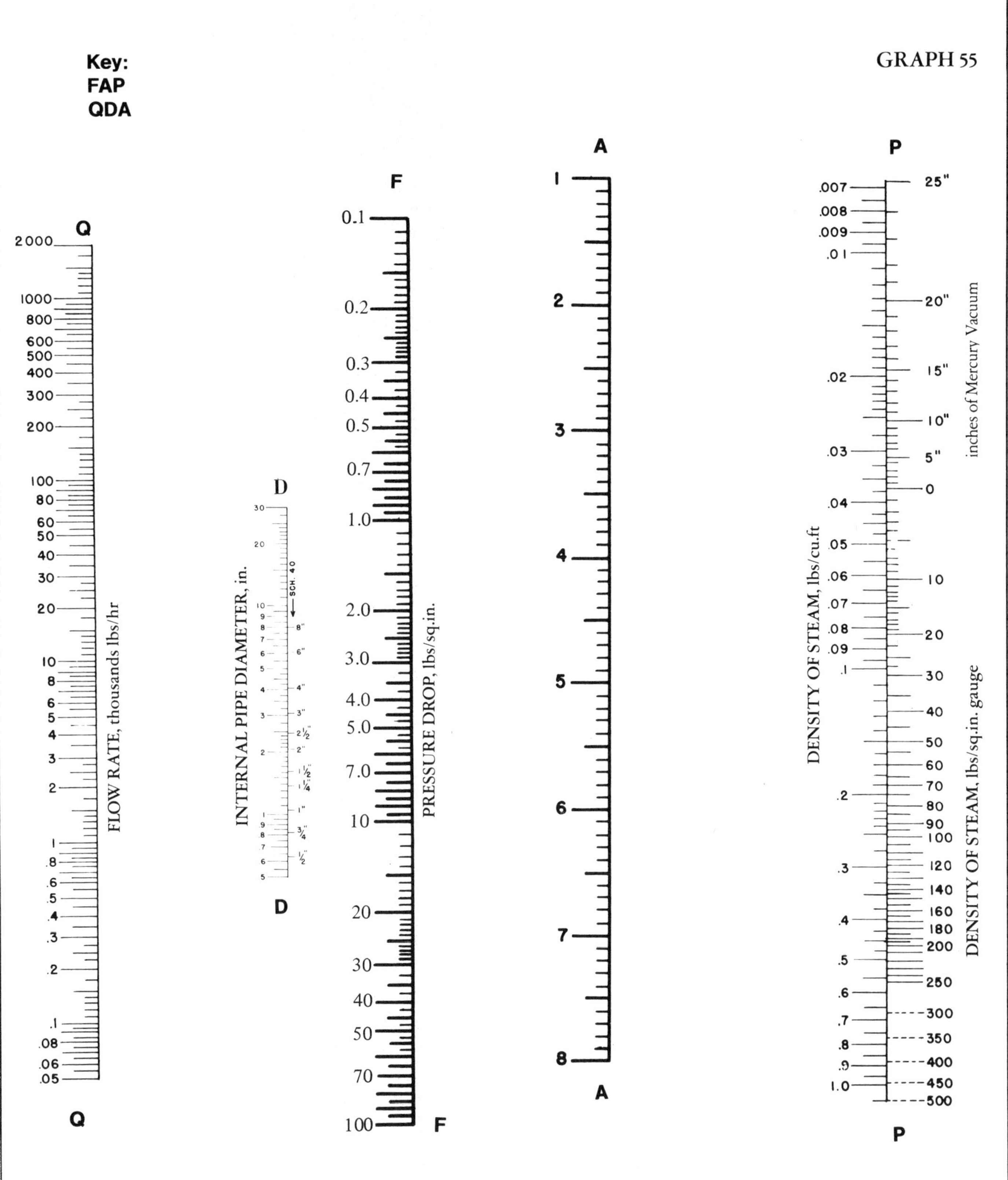

SIMPLIFIED PDA FOR STEAM

For superheated steam use density on the "P" scale

GRAPH 56

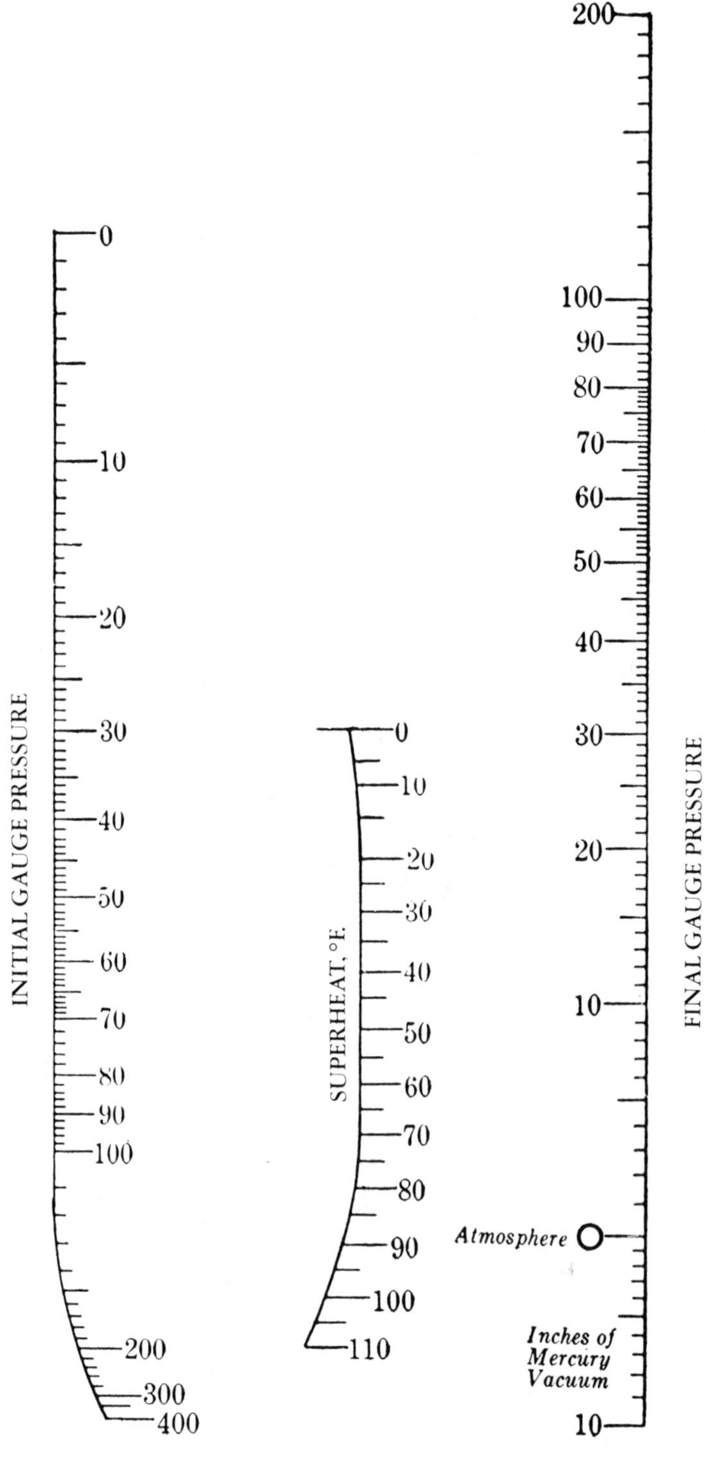

TEMPERATURE OF THROTTLED STEAM

GRAPH 57

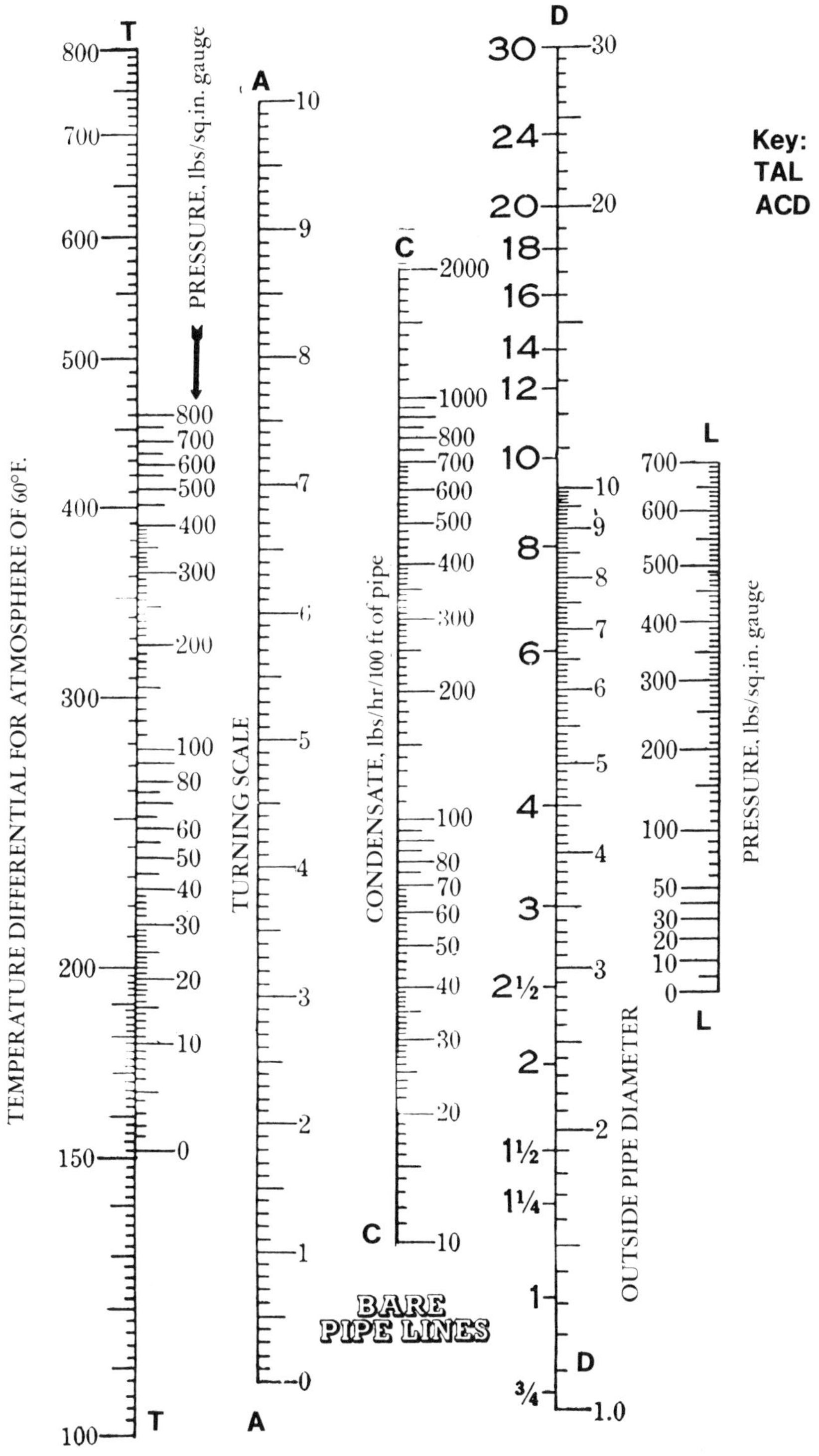

T
TEMPERATURE DIFFERENTIAL FOR ATMOSPHERE OF 60°F.
800 700 600 500 400 300 200 150 100
PRESSURE, lbs/sq.in. gauge
800 700 600 500 400 300 200 100 80 60 50 40 30 20 10 0
T

A
TURNING SCALE
10 9 8 7 6 5 4 3 2 1 0
A

C
CONDENSATE, lbs/hr/100 ft of pipe
2000 1000 800 700 600 500 400 300 200 100 80 70 60 50 40 30 20 10
C

D
OUTSIDE PIPE DIAMETER
30 24 20 18 16 14 12 10 8 6 4 3 2½ 2 1½ 1¼ 1 ¾
30 20 10 9 8 7 6 5 4 3 2 1.0
D

Key:
TAL
ACD

L
PRESSURE, lbs/sq.in. gauge
700 600 500 400 300 200 100 50 30 20 10 0
L

BARE PIPE LINES

STEAM CONDENSATION ON BARE PIPES

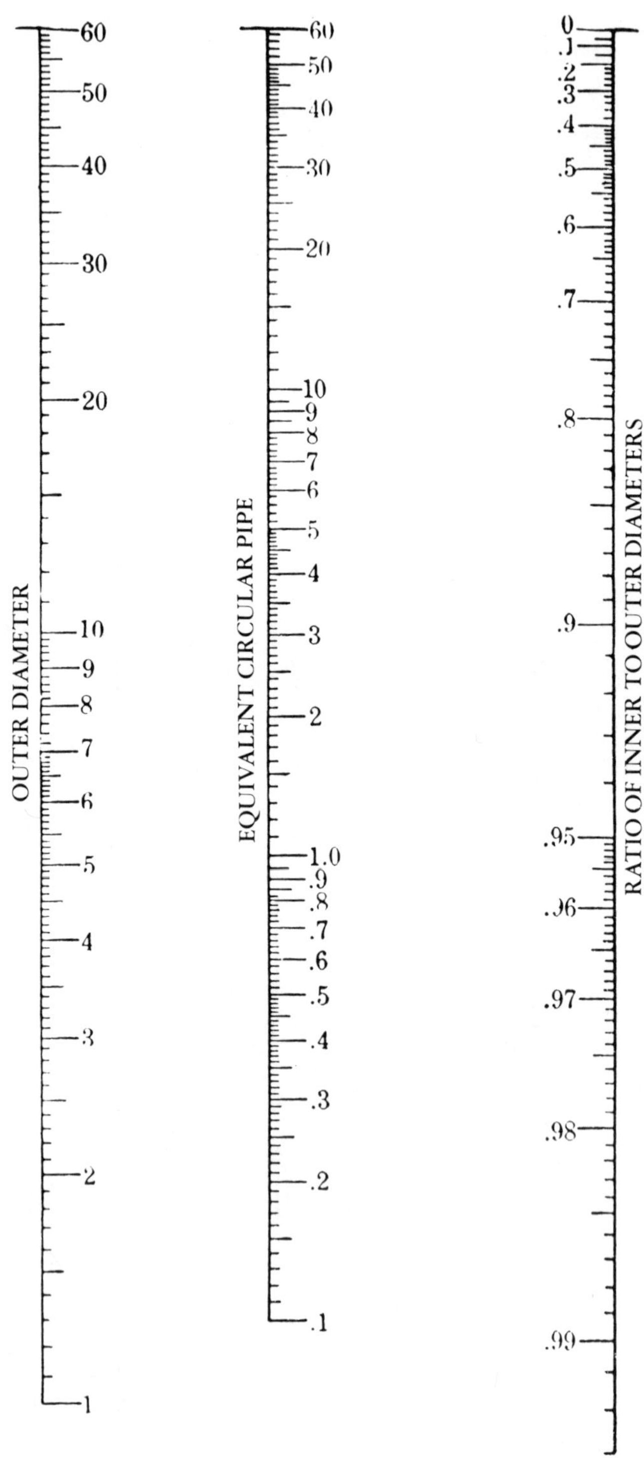

EQUIVALENT DIAMETER OF ANNULUS BETWEEN TWO PIPES

INTERNAL DIAMETER OF PIPE, in.

20" O.D. — 20; 16" O.D. — 15; 12"; 10" — 10; 9; 8" — 8; 7; 6" — 6; 5; 4" — 4; 3" — 3; 2½"; 2" — 2

TURNING SCALE

13; 12; 11; 10; 9; 8; 7; 6; 5; 4; 3; 2; 1

LENGTH, ft

100; 150; 200; 300; 400; 500; 600; 700; 800; 900; 1,000; 1,500; 2,000; 3,000; 4,000; 5,000; 6,000; 7,000; 8,000; 9,000; 10,000; 15,000; 20,000

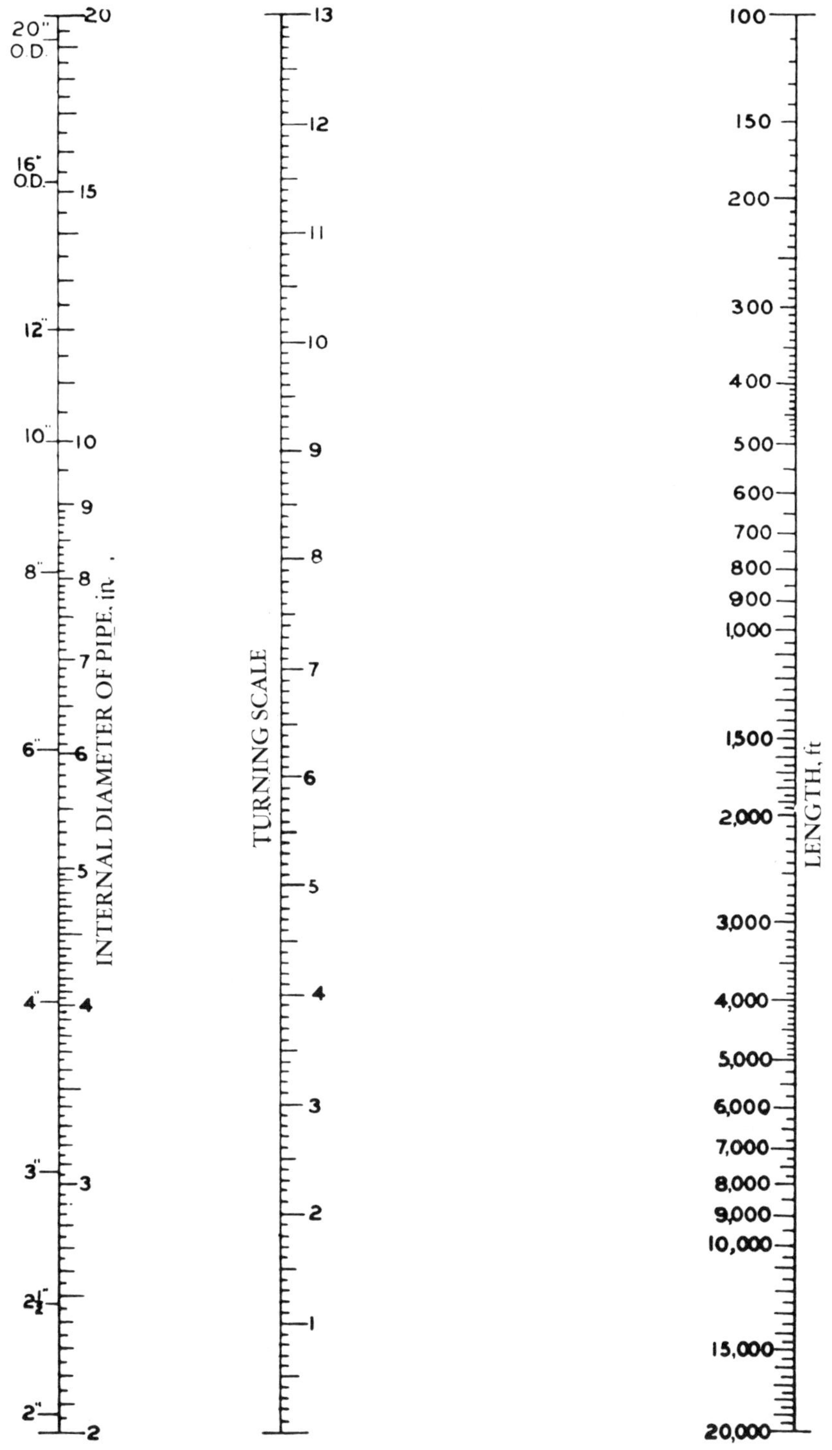

COMPARABLE LENGTHS OF VARIOUS PIPE SIZES

(Draw line from pipe size to length and pivot around intersection on turning scale to secure size and length with equal pressure drop)

GRAPH 60

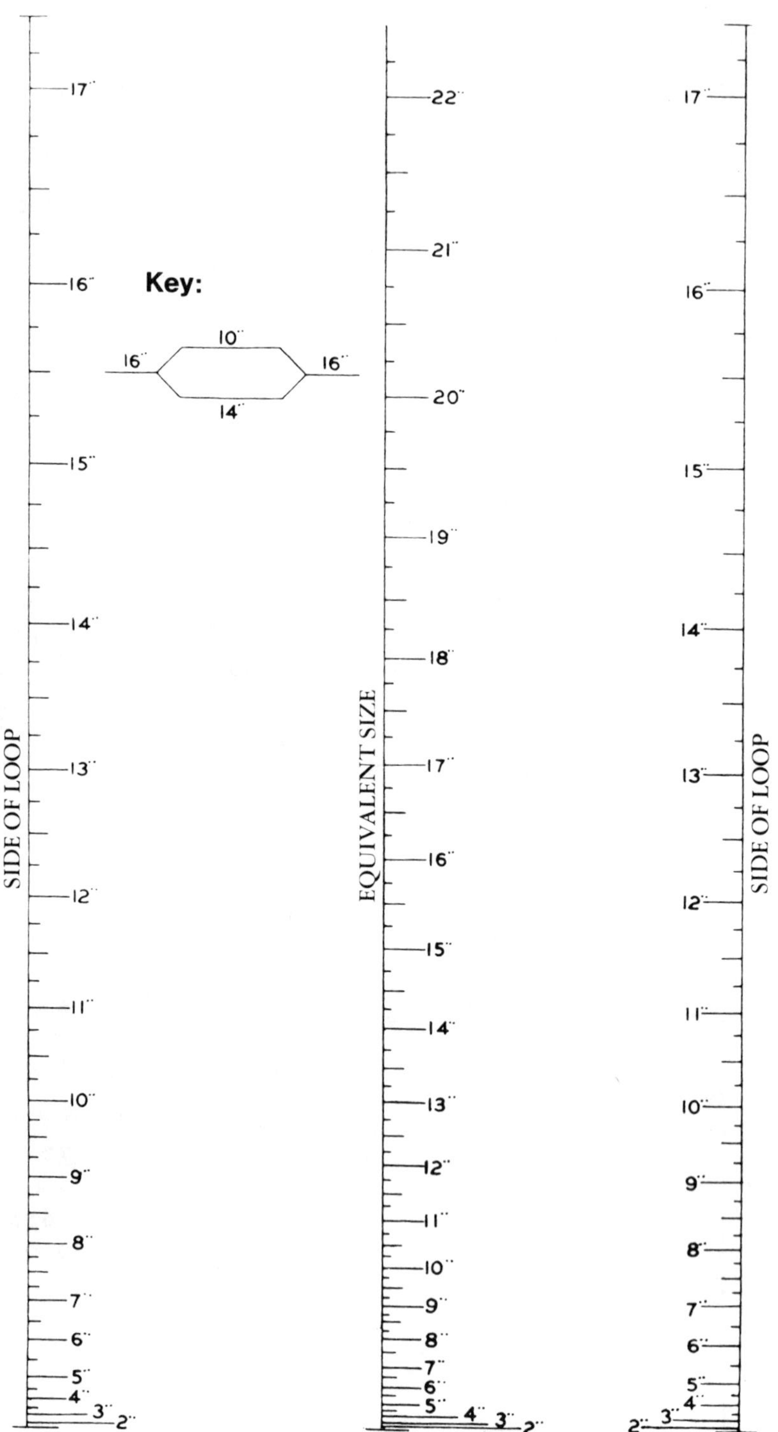

EQUIVALENT SIZE OF LOOPED PIPES

(Sides of loop must be of same length. If they are not, use equivalent length chart)

GRAPH 61

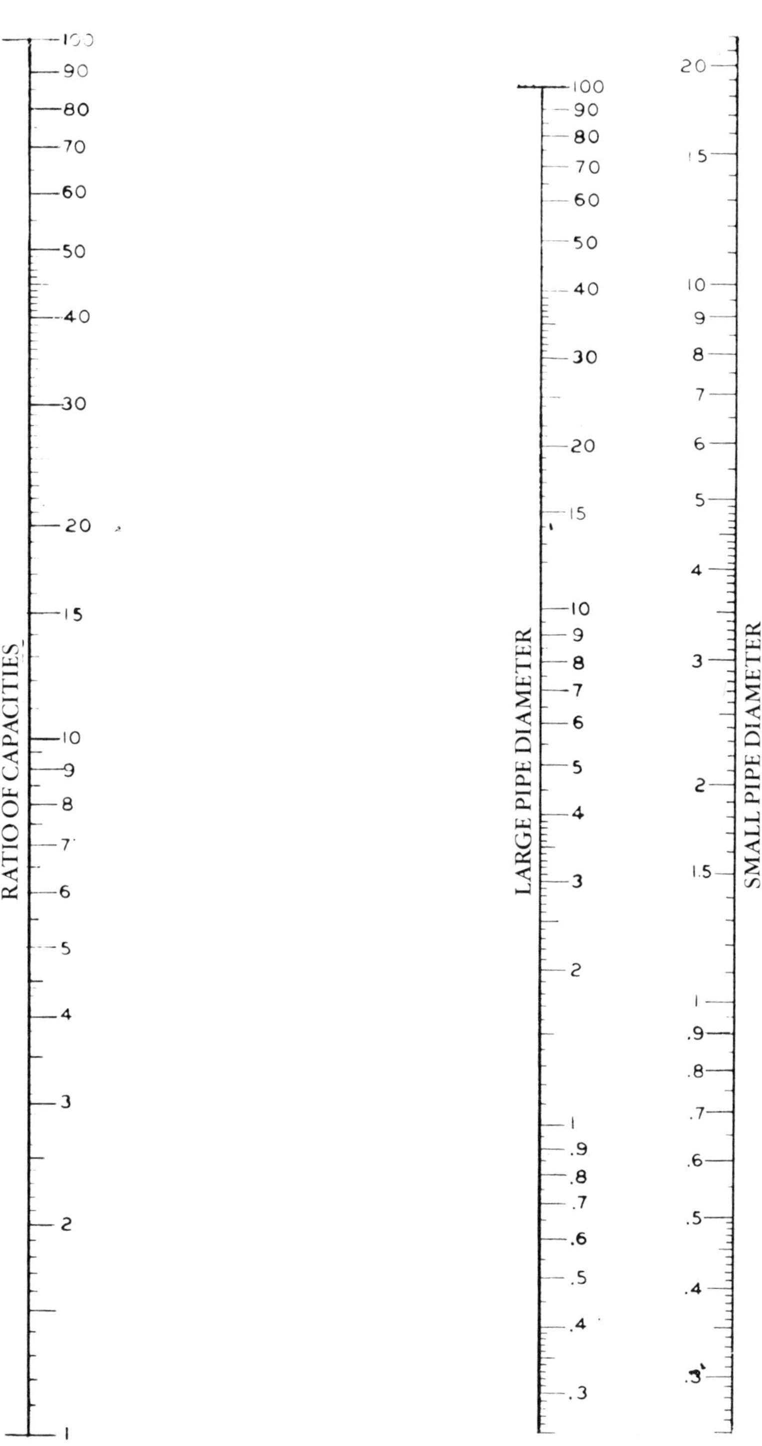

COMPARATIVE CAPACITIES OF PIPES

GRAPH 62

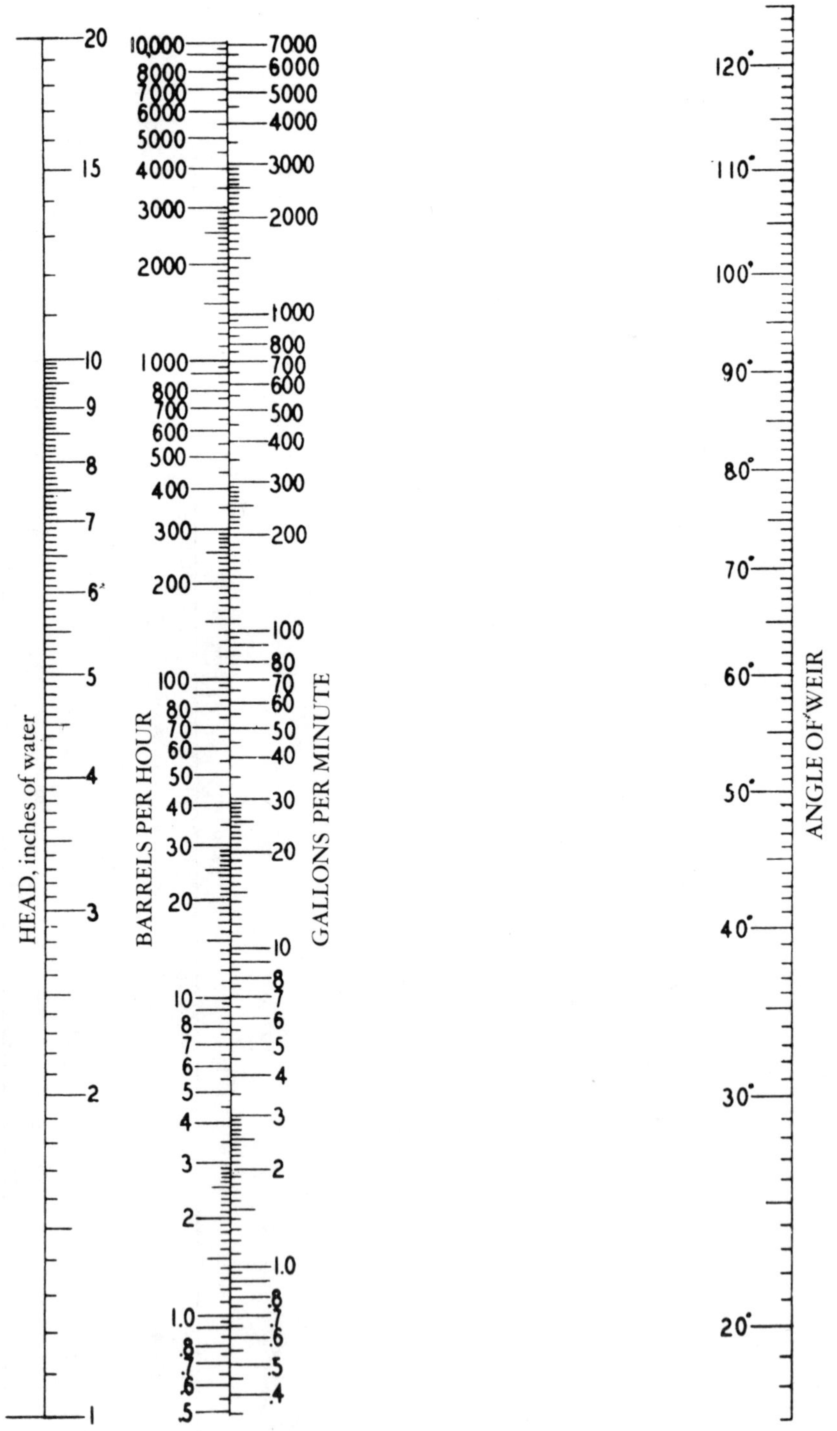

HEAD, inches of water

BARRELS PER HOUR

GALLONS PER MINUTE

ANGLE OF WEIR

CAPACITY OF SHARP EDGED V-NOTCH WEIRS

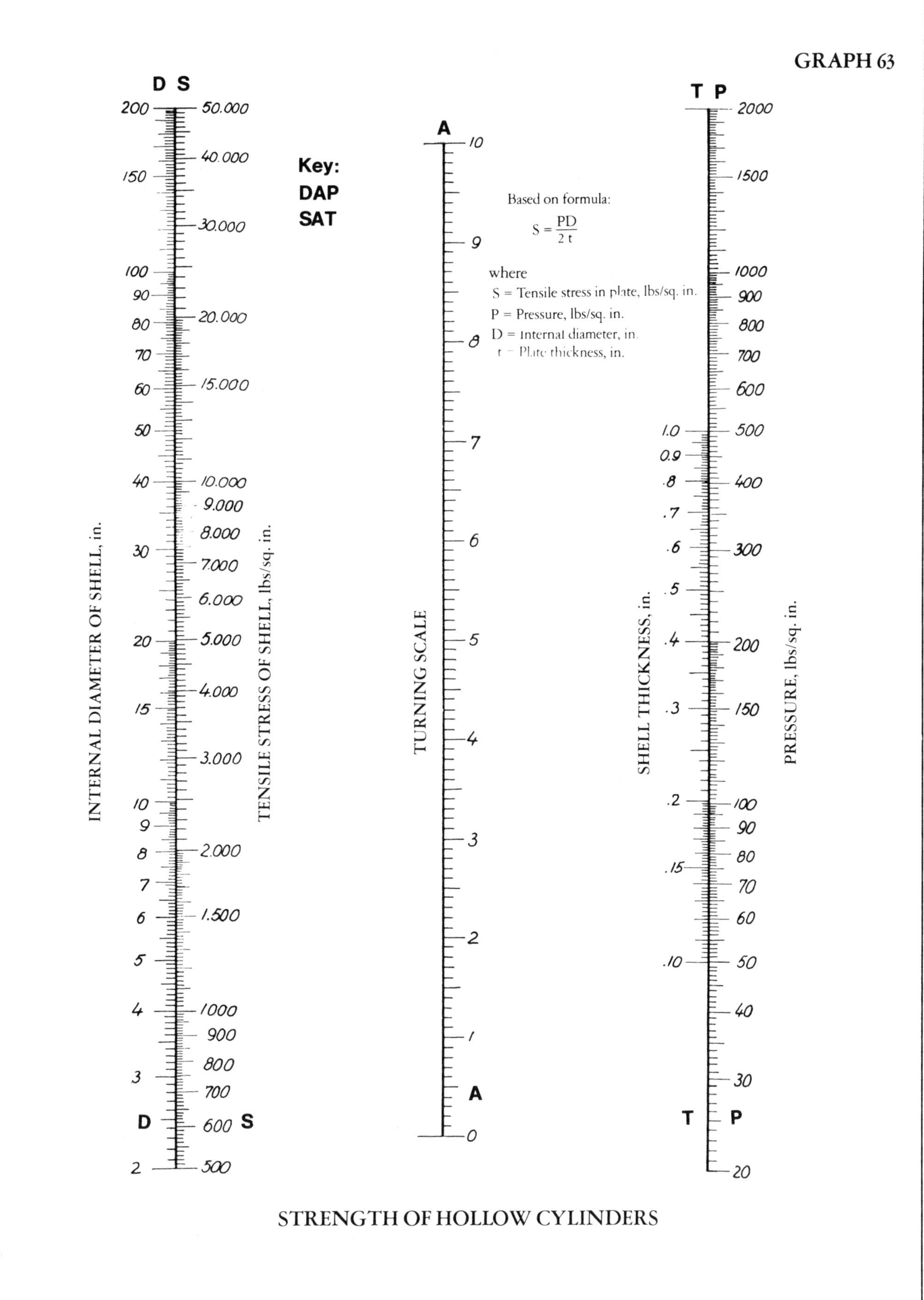

STRENGTH OF HOLLOW CYLINDERS

GRAPH 64

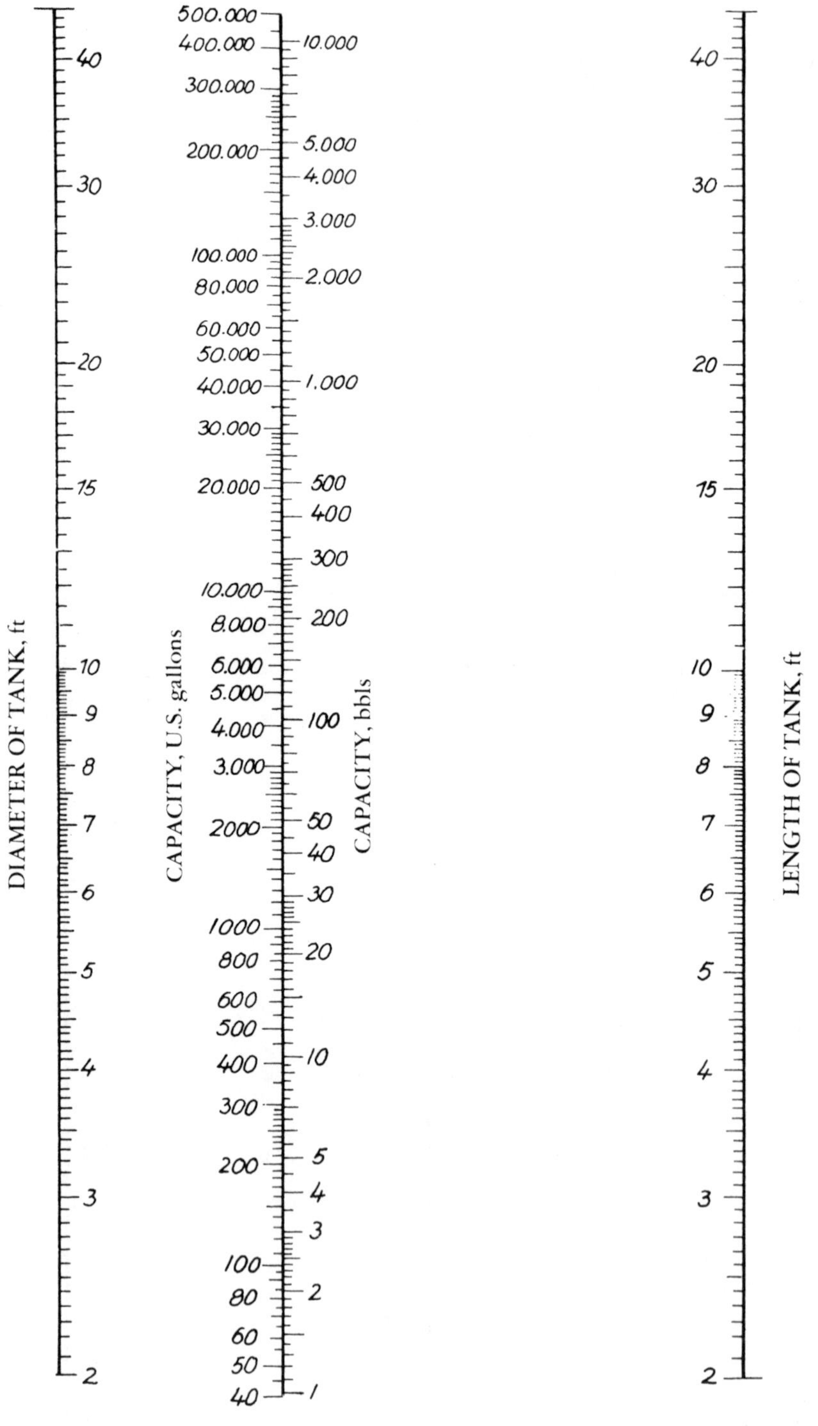

DIAMETER OF TANK, ft

CAPACITY, U.S. gallons

CAPACITY, bbls

LENGTH OF TANK, ft

CAPACITY OF HORIZONTAL CYLINDRICAL TANKS

GRAPH 65

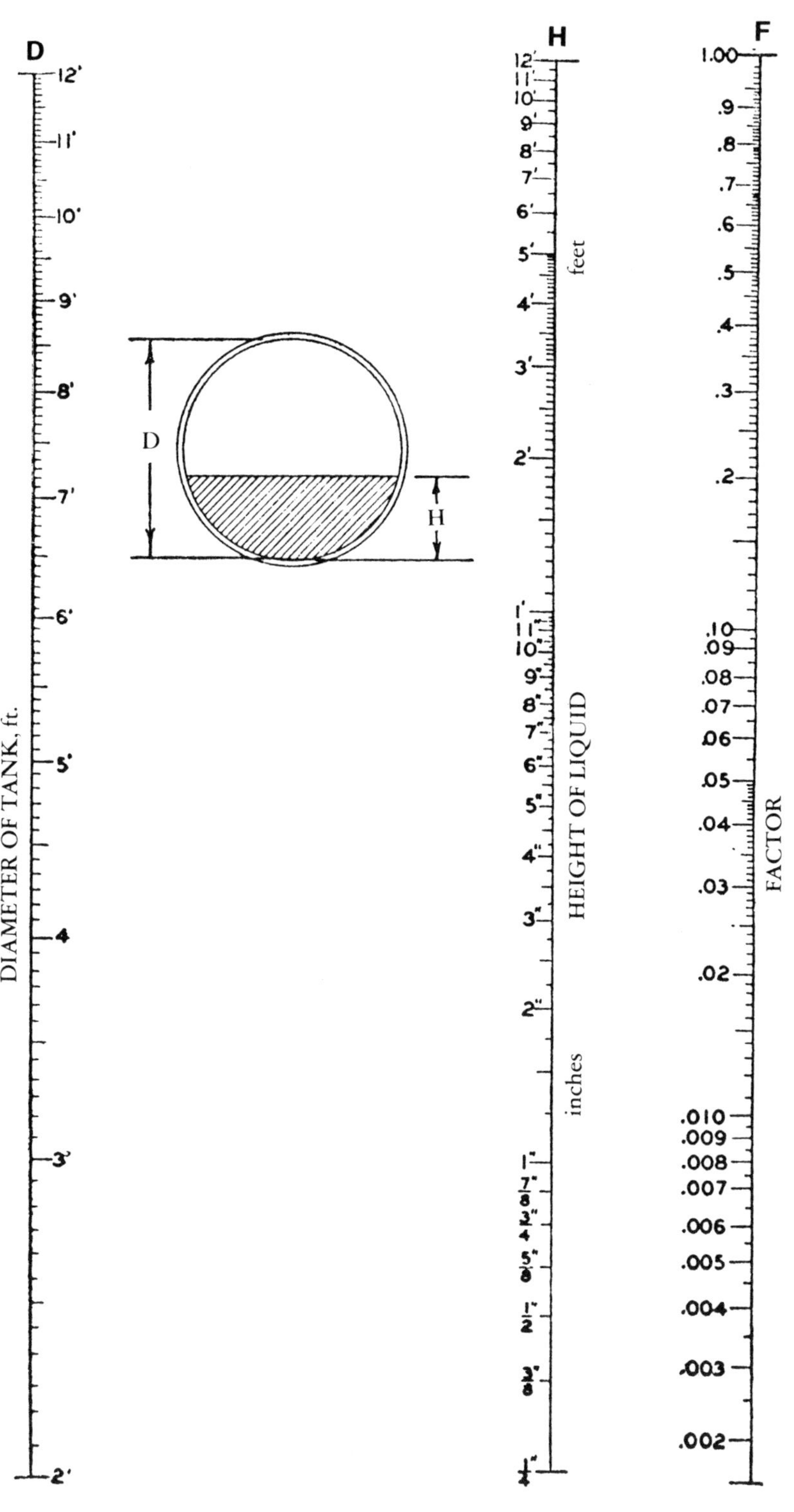

CAPACITY OF HORIZONTAL TANKS

GRAPH 66

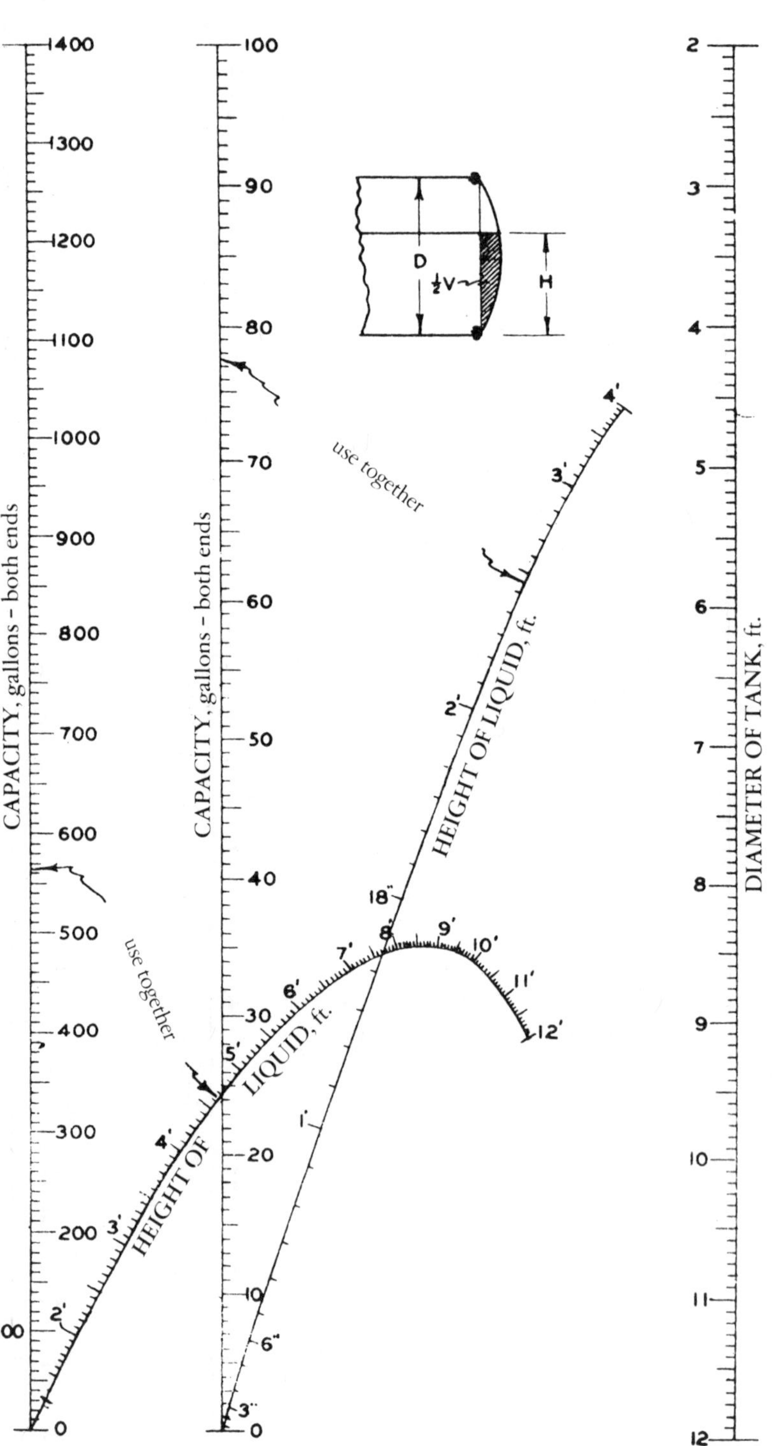

CAPACITY OF END BULGES IN HORIZONTAL TANKS

GRAPH 67

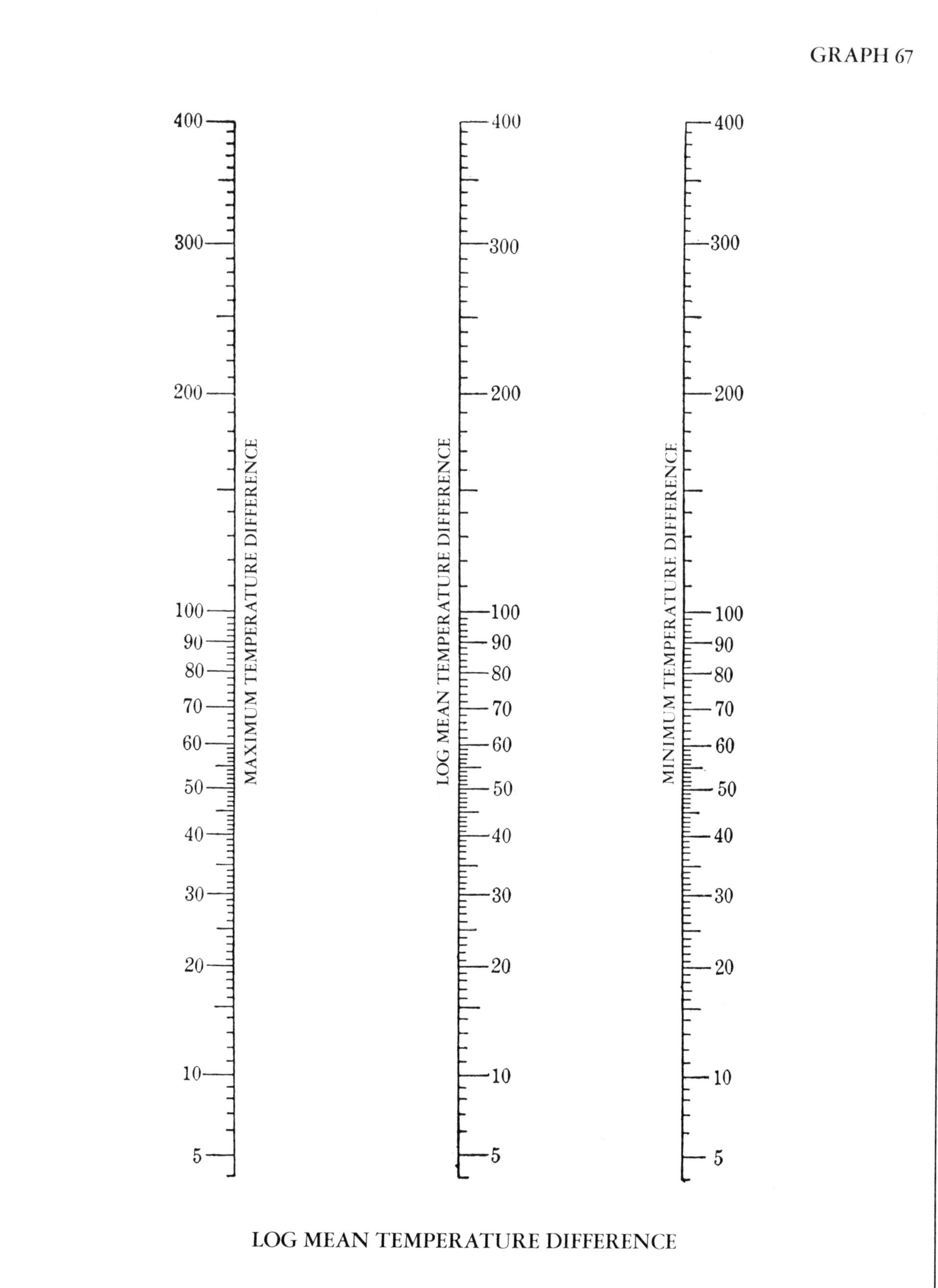

LOG MEAN TEMPERATURE DIFFERENCE

GRAPH 68

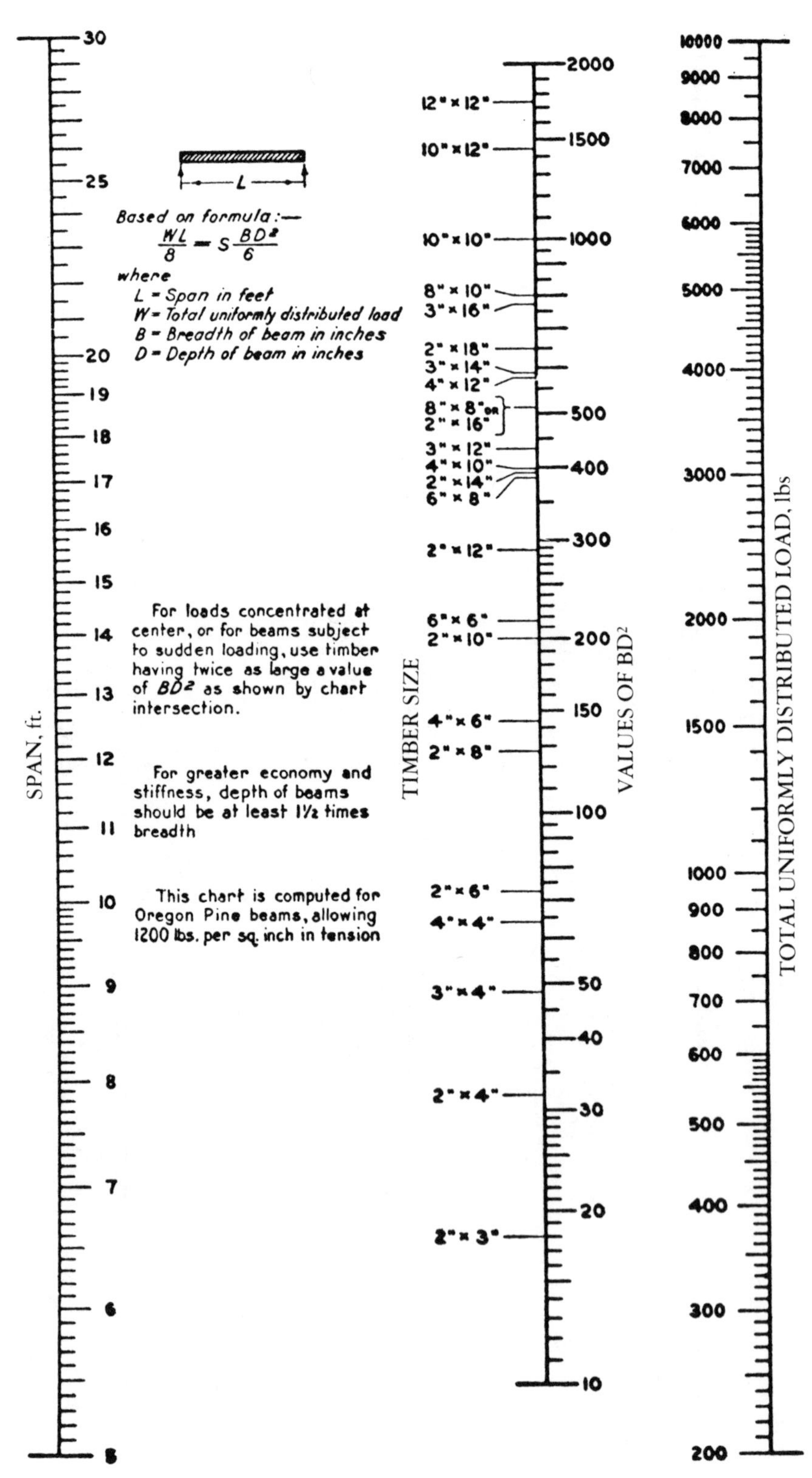
SPAN, ft.
30
25
20
19
18
17
16
15
14
13
12
11
10
9
8
7
6
5
Based on formula:—
$\frac{WL}{8} = S\frac{BD^2}{6}$
where
L = Span in feet
W = Total uniformly distributed load
B = Breadth of beam in inches
D = Depth of beam in inches
For loads concentrated at center, or for beams subject to sudden loading, use timber having twice as large a value of BD^2 as shown by chart intersection.
For greater economy and stiffness, depth of beams should be at least 1½ times breadth
This chart is computed for Oregon Pine beams, allowing 1200 lbs. per sq. inch in tension
TIMBER SIZE
12" × 12"
10" × 12"
10" × 10"
8" × 10"
3" × 16"
2" × 18"
3" × 14"
4" × 12"
8" × 8" or 2" × 16"
3" × 12"
4" × 10"
2" × 14"
6" × 8"
2" × 12"
6" × 6"
2" × 10"
4" × 6"
2" × 8"
2" × 6"
4" × 4"
3" × 4"
2" × 4"
2" × 3"
VALUES OF BD^2
2000
1500
1000
500
400
300
200
150
100
50
40
30
20
10
TOTAL UNIFORMLY DISTRIBUTED LOAD, lbs
10000
9000
8000
7000
6000
5000
4000
3000
2000
1500
1000
900
800
700
600
500
400
300
200

STRENGTH OF WOODEN BEAMS

GRAPH 69

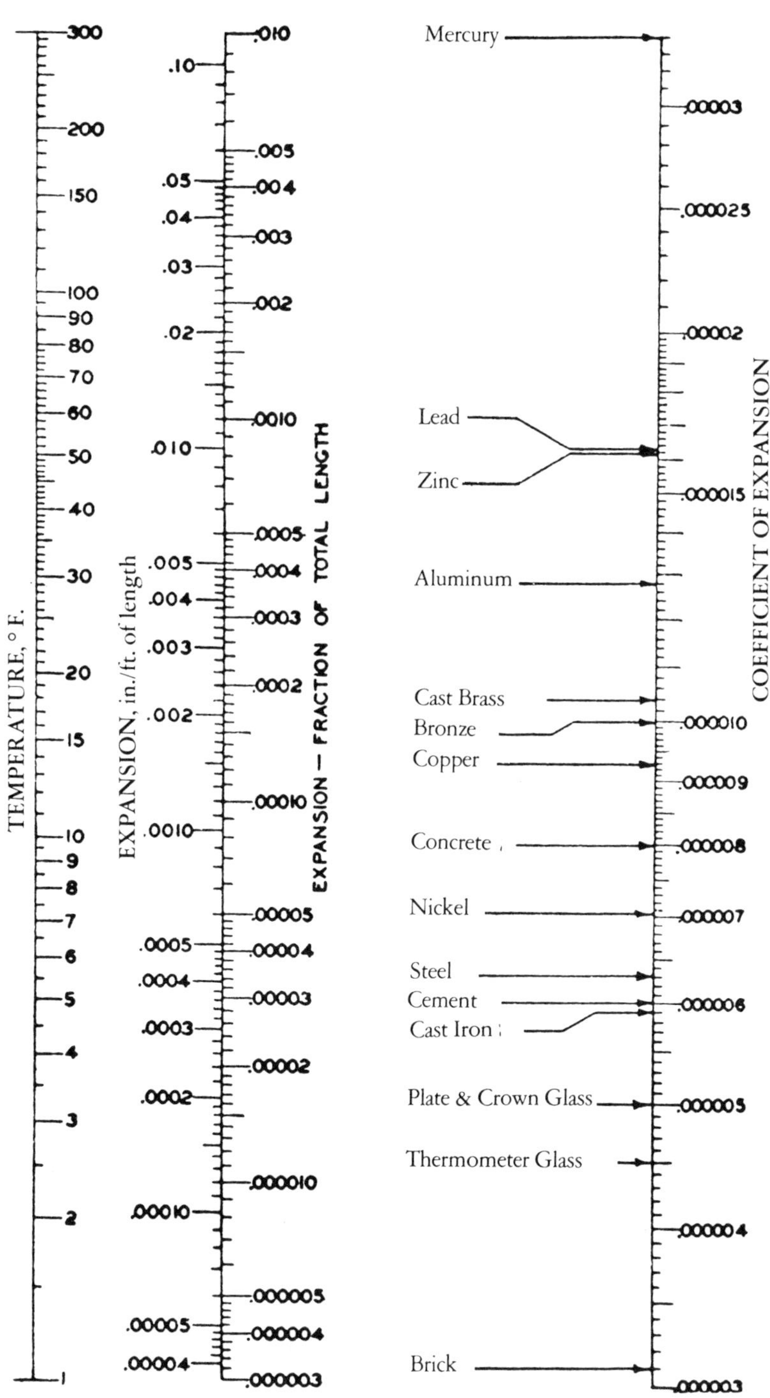

THERMAL EXPANSION OF METALS AND OTHER MATERIALS

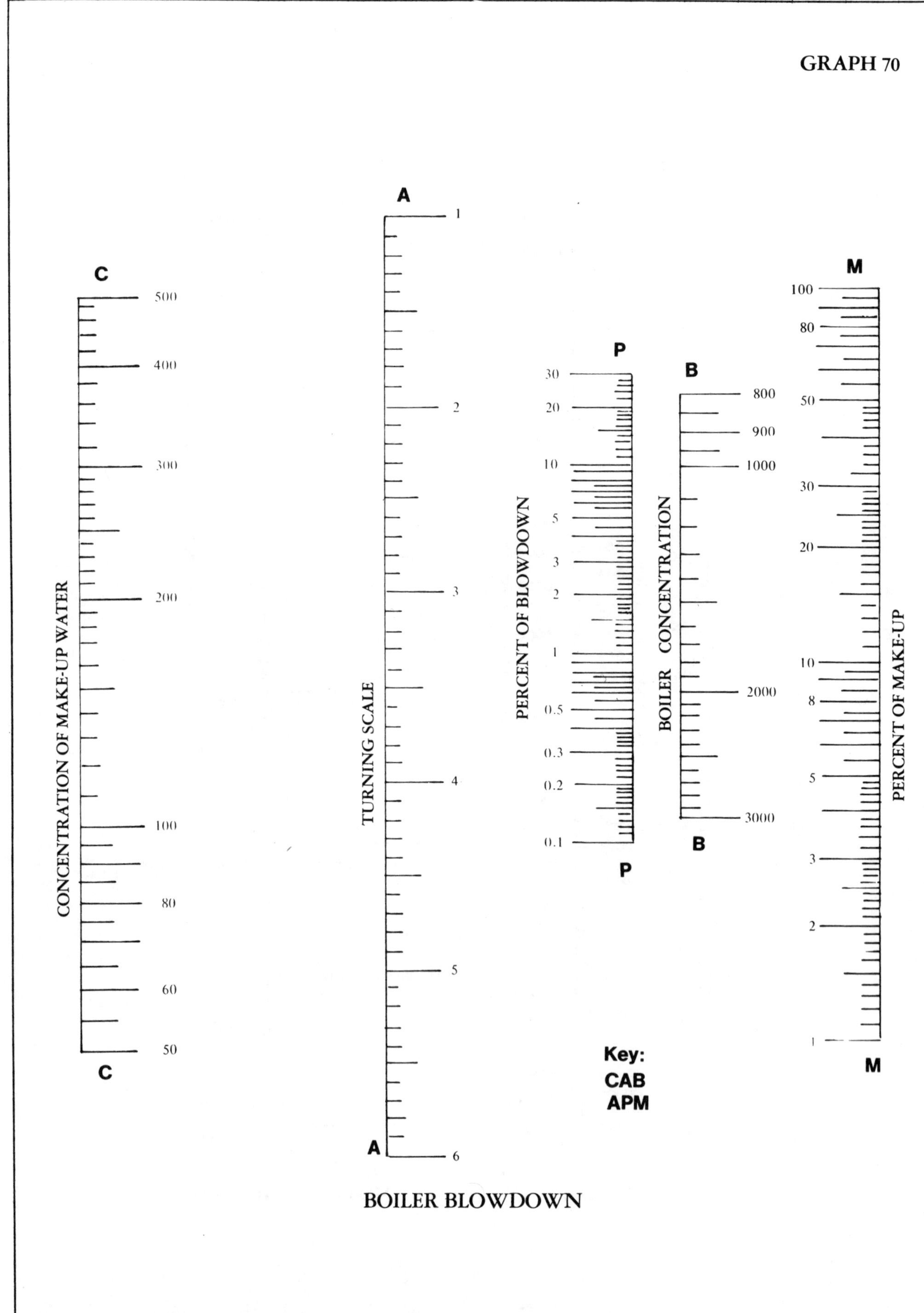

BOILER BLOWDOWN

GRAPH 71

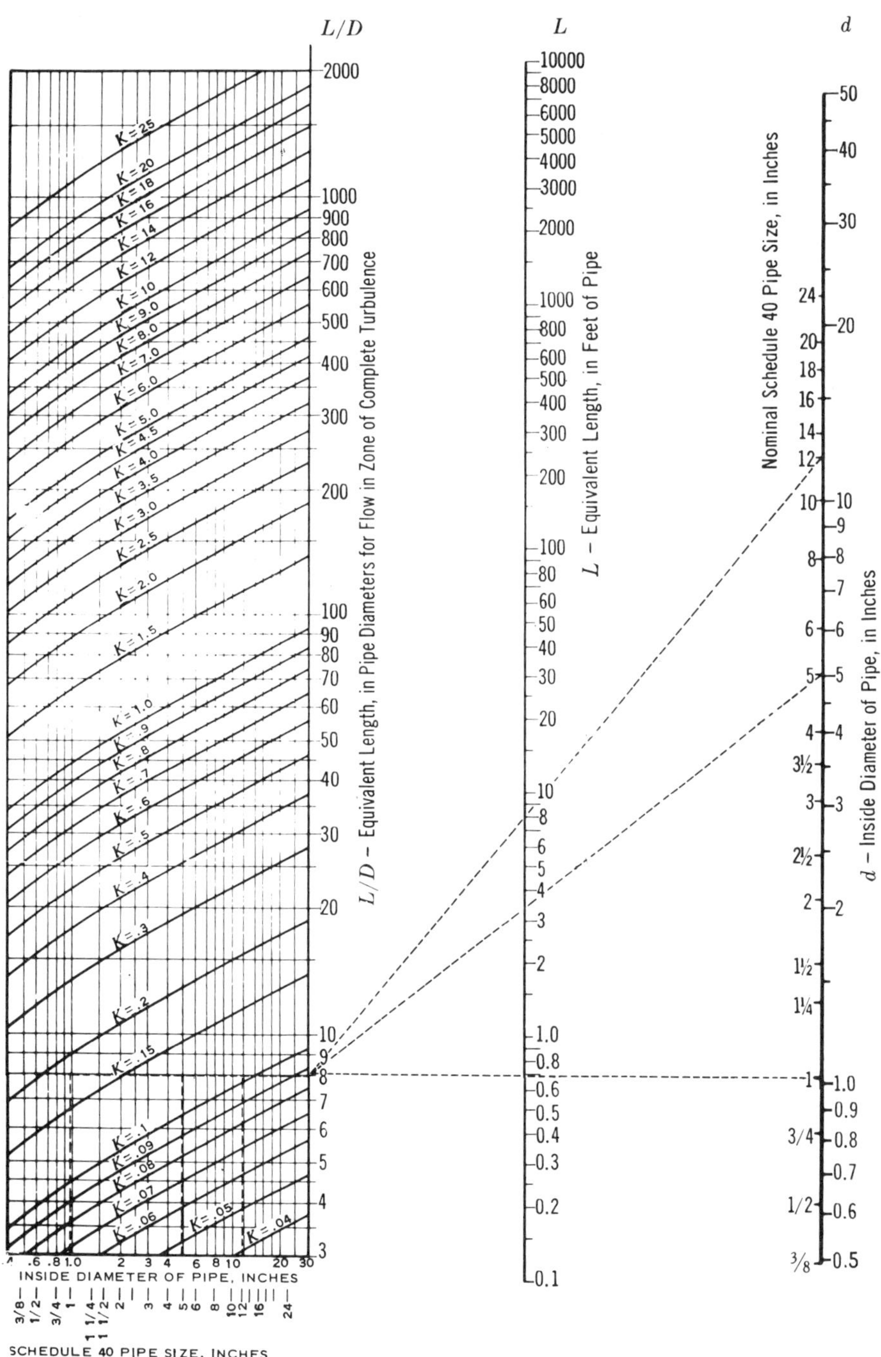

Problem: Find the equivalent length in pipe diameters and feet of Schedule 40 clean commercial steel pipe, and the resistance factor *K* for 1, 5, and 12-inch fully-opened gate valves with flow in zone of complete turbulence.

Solution

Valve Size	1″	5″	12″	*Refer to*
Equivalent length, pipe diameters	8	8	8	Figure B-16
Equivalent length, feet of Sched. 40 pipe	0.7	3.4	7.9	Dotted lines on chart.
Resist. factor *K*, based on Sched. 40 pipe	0.18	0.12	0.10	

(From TP-410, FLOW OF FLUIDS, Courtesy Crane Company)

EQUIVALENT LENGTHS L AND L/D AND RESISTANCE COEFFICIENT K

See Figures B-15 to B-17 for values of K

APPENDIX B
SUPPLEMENTARY TECHNICAL DATA

A collection of additional data useful in fluid dynamics calculations is provided. These include information on fluid viscosities, values of the U.O.P. Characterization Factor, data on specific gravity and density of fluids, dimensions of standard pipe, and the like.

FIGURE B-1

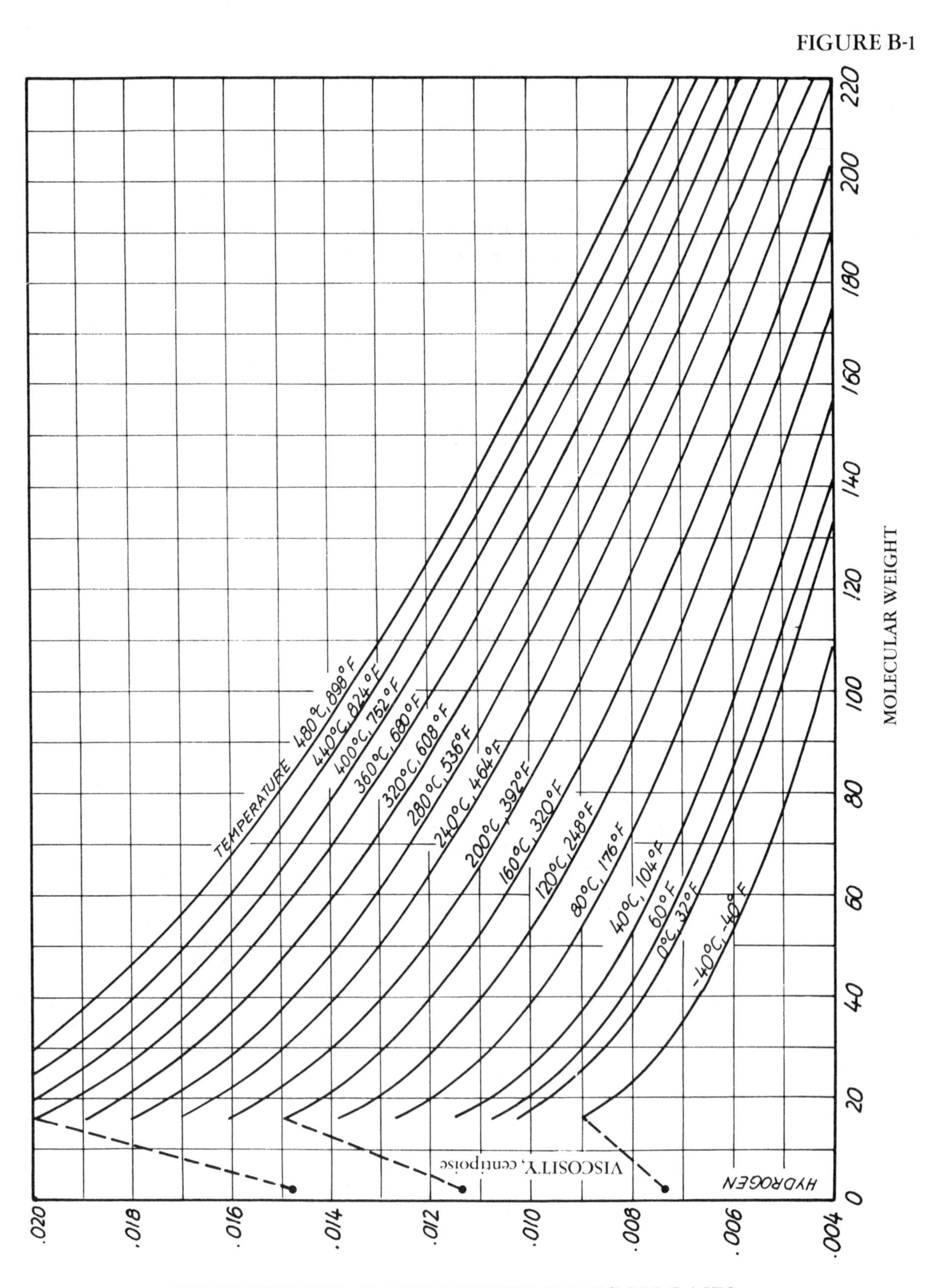

VISCOSITY OF PARAFFINIC HYDROCARBON GASES

AT ATMOSPHERIC PRESSURE

FIGURE B-2

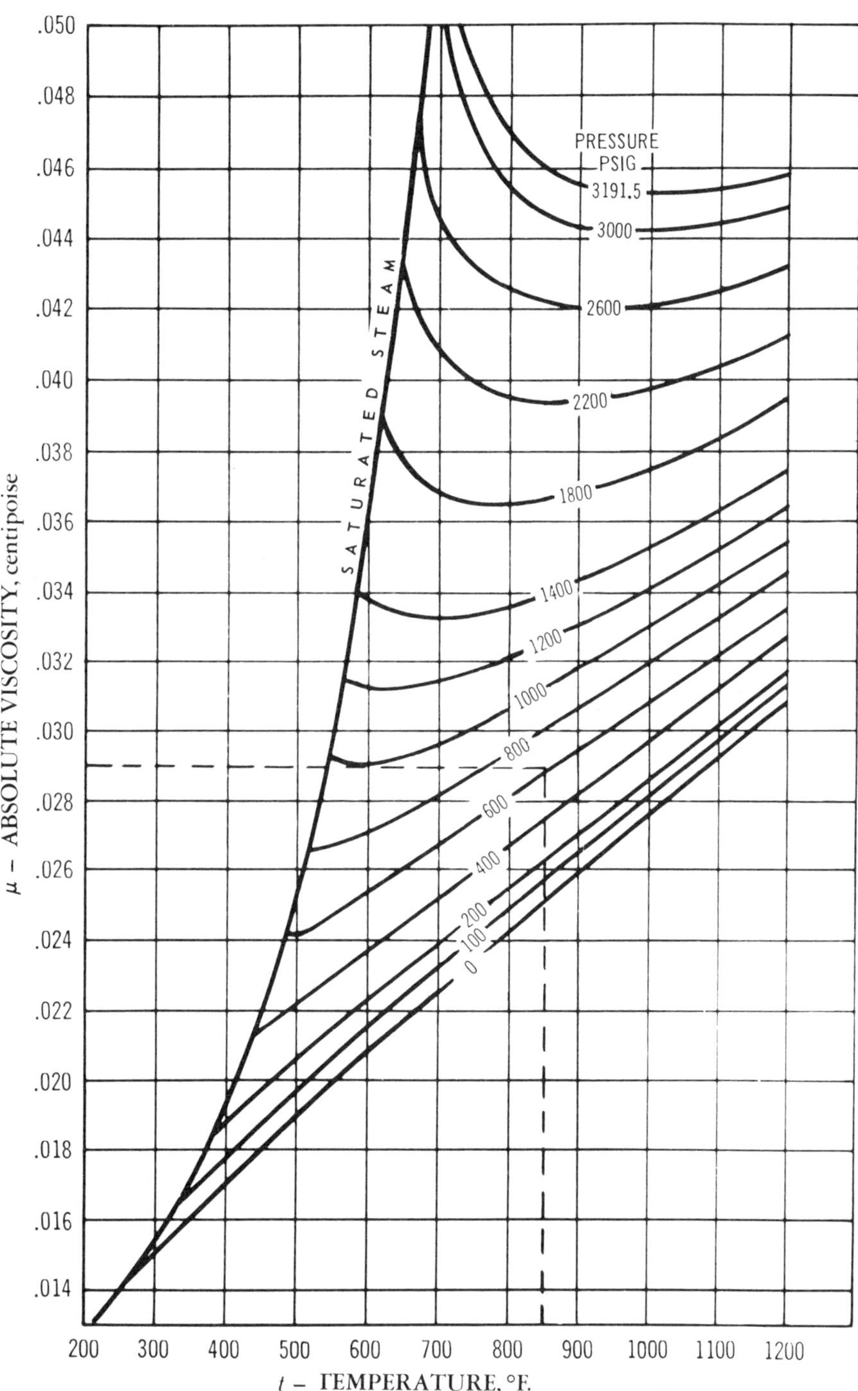

Example: Viscosity of 600 psig, 850 °F. steam is 0.029 centipoise

(Adapted from: Philip J. Potter: Steam Power Plants, Copyright 1949, The Ronald Press Company)

VISCOSITY OF STEAM

(From TP-410, FLOW OF FLUIDS, Courtesy Crane Company)

FIGURE B-3

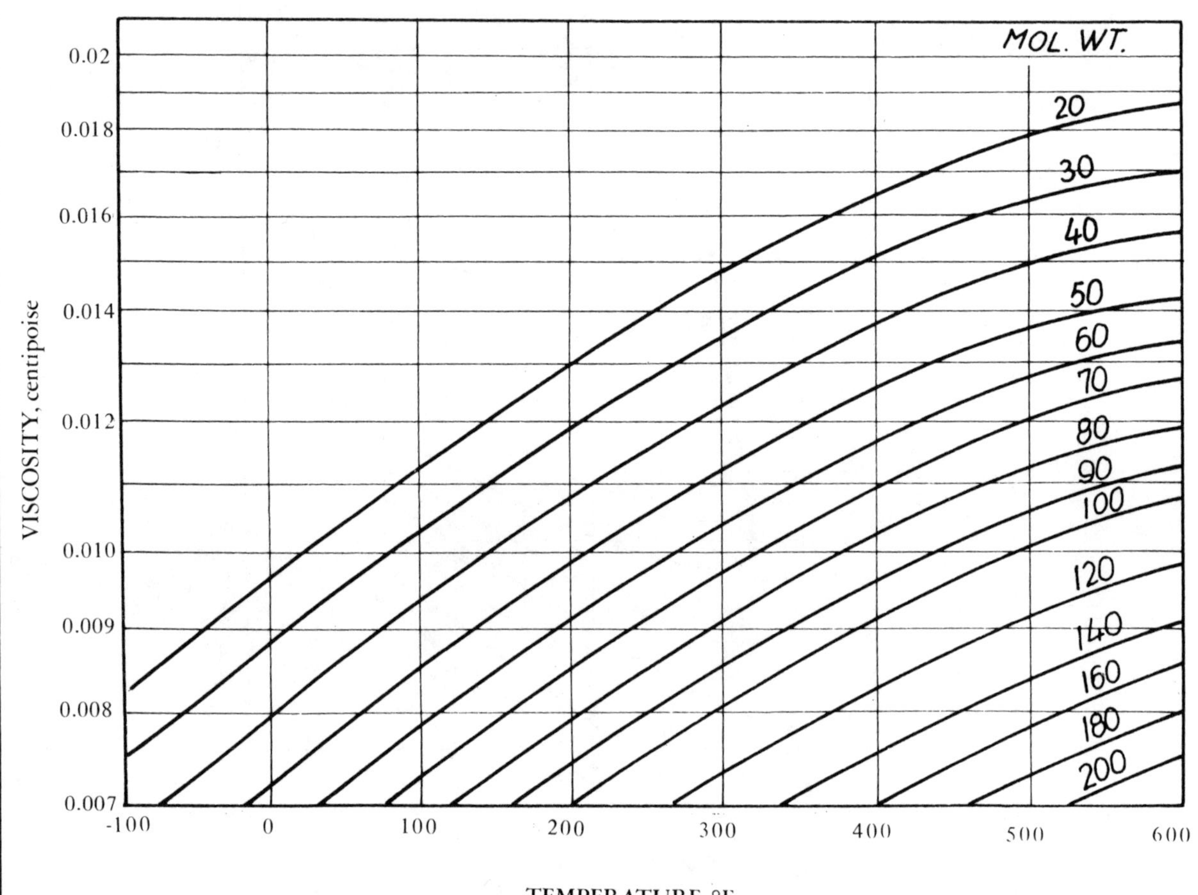

VISCOSITY - TEMPERATURE RELATION FOR PARAFFINIC HYDROCARBON GASES AT 760 mm PRESSURE

FIGURE B-4

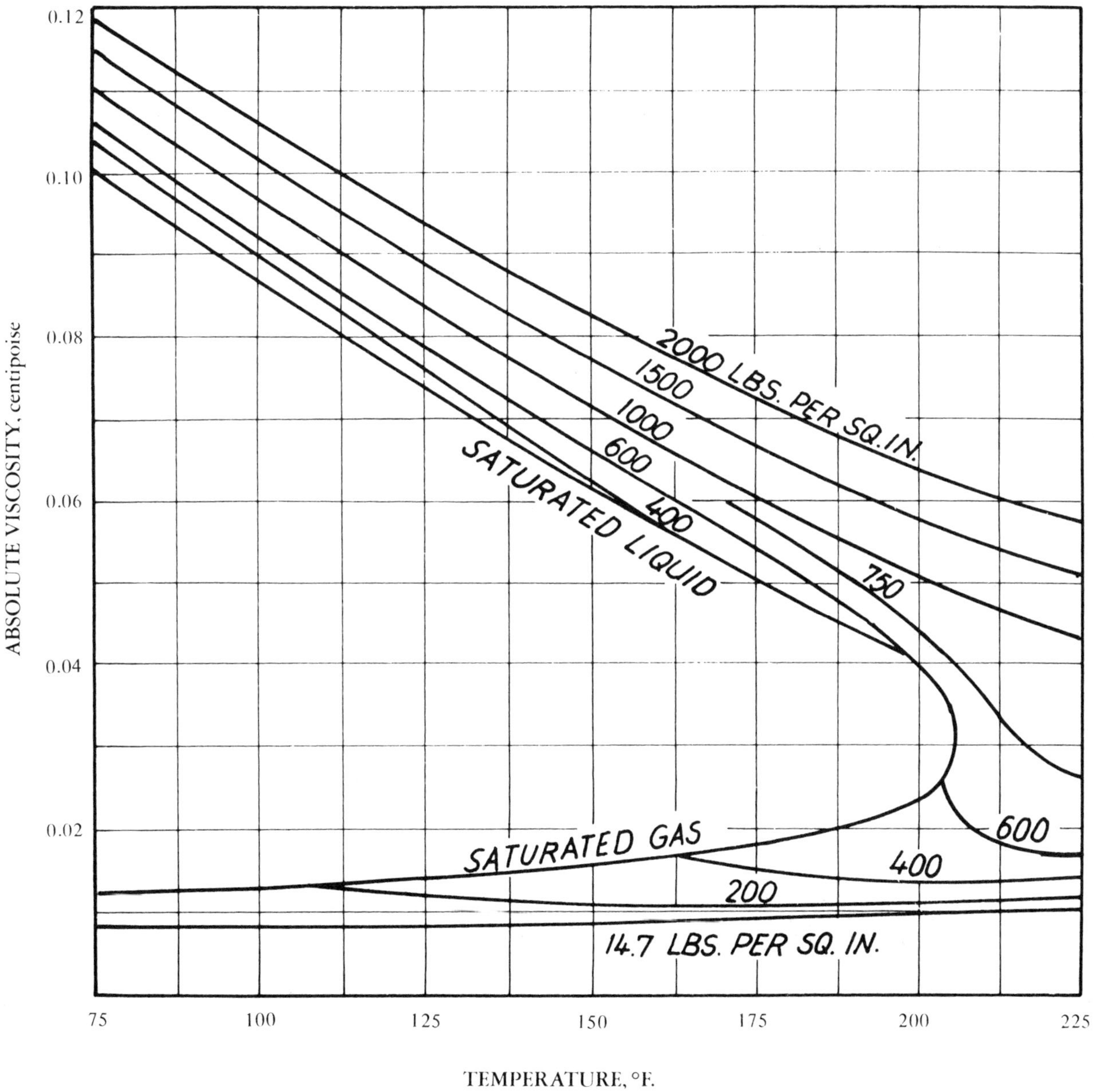

VISCOSITY – TEMPERATURE RELATION FOR PROPANE

FIGURE B-5

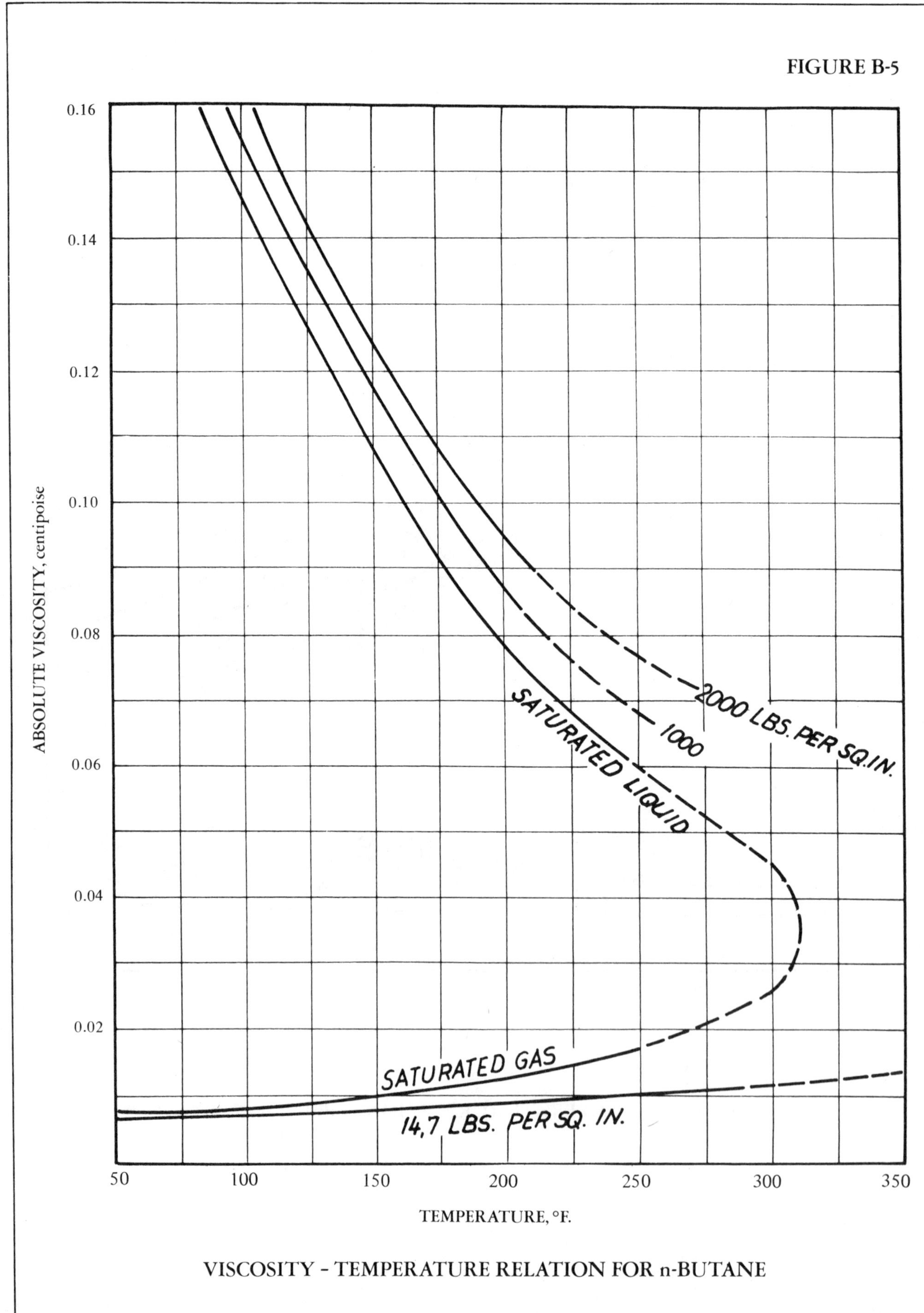

VISCOSITY - TEMPERATURE RELATION FOR n-BUTANE

FIGURE B-6

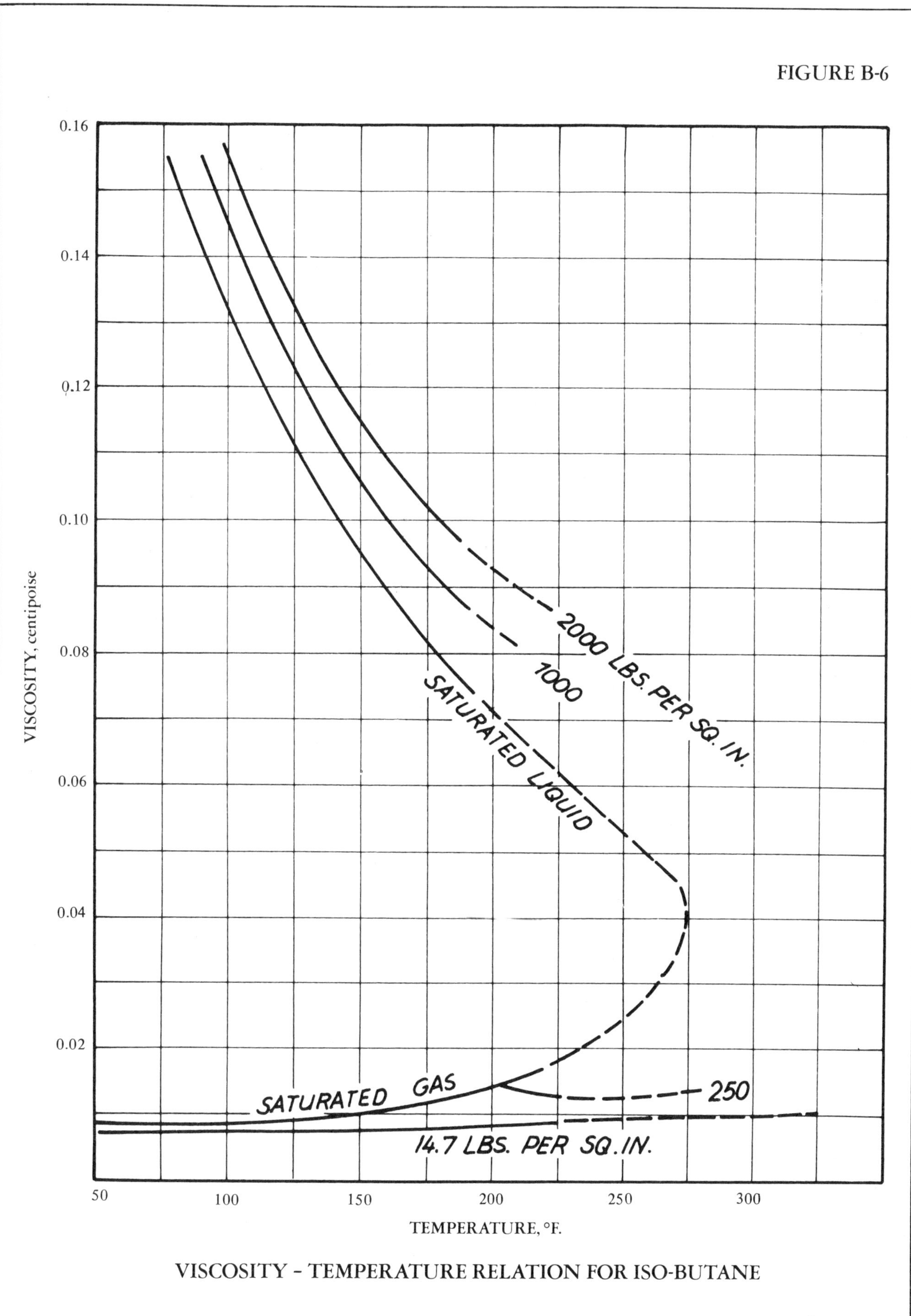

VISCOSITY - TEMPERATURE RELATION FOR ISO-BUTANE

FIGURE B-7

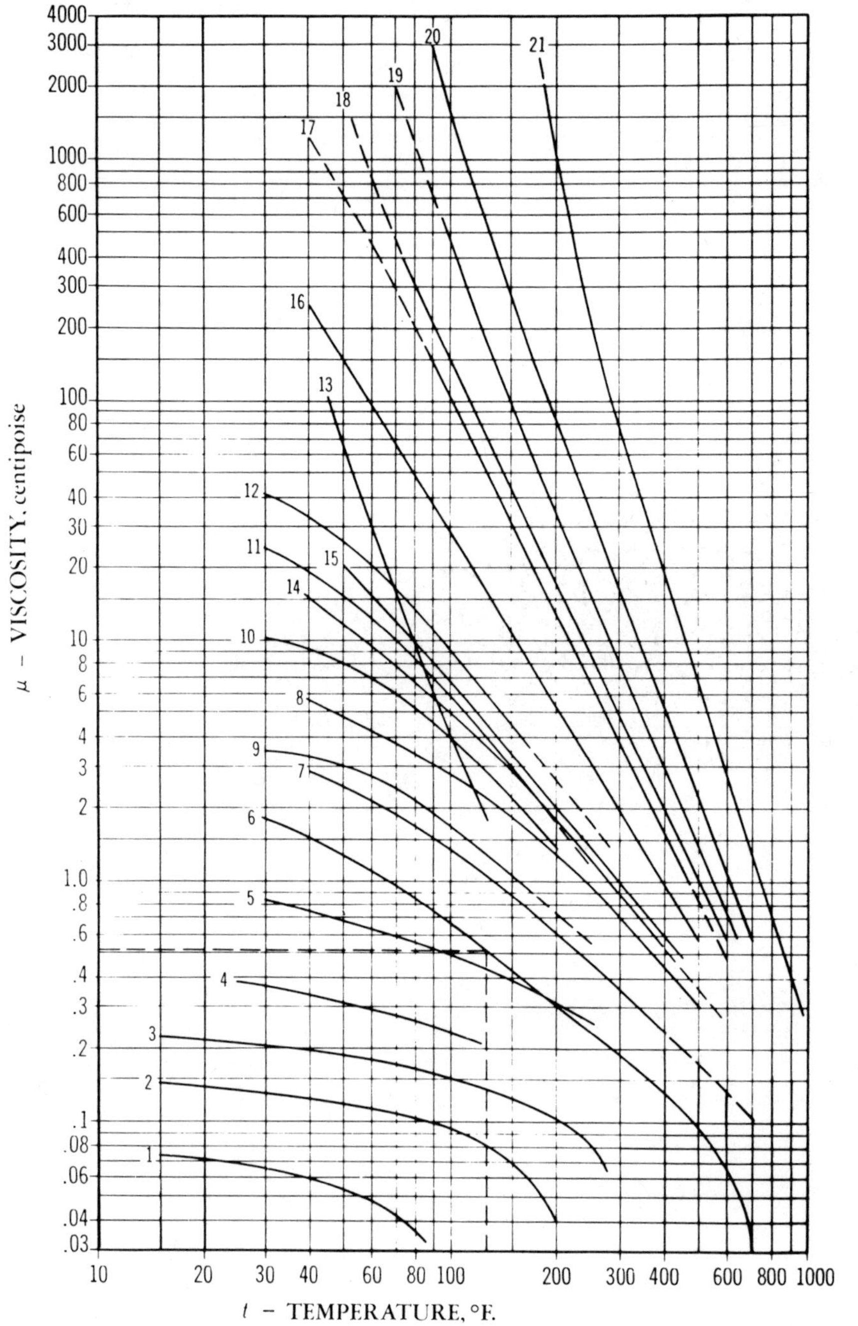

1. Ethane (C_2H_6)
2. Propane (C_3H_8)
3. Butane (C_4H_{10})
4. Natural Gasoline
5. Gasoline
6. Water
7. Kerosene
8. Distillate
9. 48 Deg. API Crude
10. 40 Deg. API Crude
11. 35.6 Deg. API Crude
12. 32.6 Deg. API Crude
13. Salt Creek Crude
14. Fuel 3 (Max.)
15. Fuel 5 (Min.)
16. SAE 10 Lube (100 V.I.)
17. SAE 30 Lube (100 V.I.)
18. Fuel 5 (Max.) or Fuel 6 (Min.)
19. SAE 70 Lube (100 V.I.)
20. Bunker C Fuel (Max.) and M.C. Residuum
21. Asphalt

Example: The viscosity of water at 125°F is 0.52 centipoise (Curve No. 6).

VISCOSITY OF WATER AND LIQUID PETROLEUM PRODUCTS

(From TP-410, FLOW OF FLUIDS, Courtesy Crane Company)

(Data extracted in part by permission from the Oil and Gas Journal)

FIGURE B-8

(From TP-410, FLOW OF FLUIDS, Courtesy Crane Company)

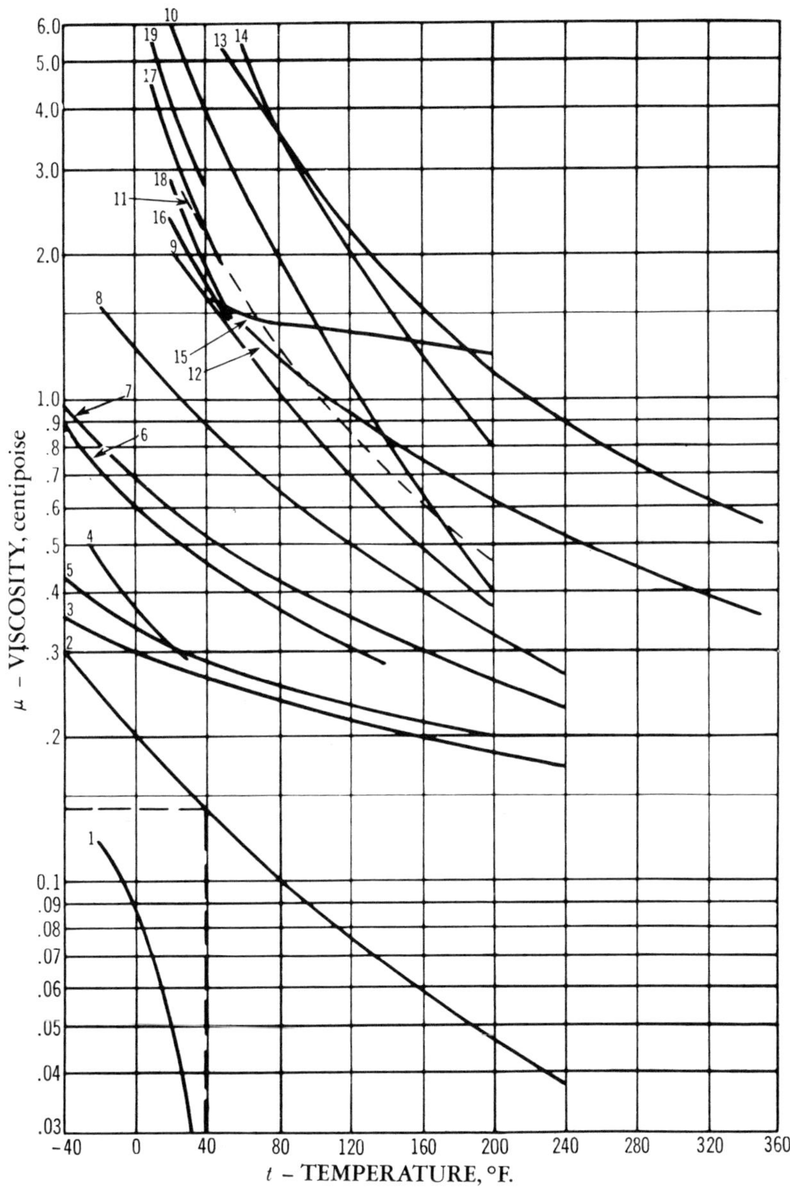

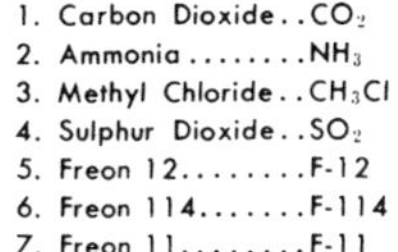
1. Carbon Dioxide . . CO_2
2. Ammonia . . . . NH_3
3. Methyl Chloride . . CH_3Cl
4. Sulphur Dioxide . . SO_2
5. Freon 12 . . . . F-12
6. Freon 114 . . . . F-114
7. Freon 11 . . . . F-11
8. Freon 113 . . . . F-113
9. Ethyl Alcohol
10. Isopropyl Alcohol
11. 20% Sulphuric Acid . . . . . . 20% H_2SO_4
12. Dowtherm E
13. Dowtherm A
14. 20% Sodium Hydroxide . . 20% NaOH
15. Mercury
16. 10% Sodium Chloride Brine . . . 10% NaCl
17. 20% Sodium Chloride Brine . . . 20% NaCl
18. 10% Calcium Chloride Brine . . 10% $CaCl_2$
19. 20% Calcium Chloride Brine . . 20% $CaCl_2$

Example: The viscosity of ammonia at 40 F is 0.14 centipoise.

VISCOSITY OF VARIOUS LIQUIDS

FIGURE B-9

The curves for hydrocarbon vapors and natural gases in the chart at the upper right are taken from Maxwell[15]; the curves for all other gases in the chart are based upon Sutherland's formula, as follows:

$$\mu = \mu_0 \left(\frac{0.555\ T_0 + C}{0.555\ T + C}\right)\left(\frac{T}{T_0}\right)^{3/2}$$

where:

μ = viscosity, in centipoise at temperature T.

μ_0 = viscosity, in centipoise at temperature T_0.

T = absolute temperature, in degrees Rankine (460 + deg. F) for which viscosity is desired.

T_0 = absolute temperature, in degrees Rankine, for which viscosity is known.

C = Sutherland's constant.

Note: The variation of viscosity with pressure is small for most gases. For gases given on this page, the correction of viscosity for pressure is less than 10 per cent for pressures up to 500 pounds per square inch.

Fluid	Approximate Values of "C"
O_2	127
Air	120
N_2	111
CO_2	240
CO	118
SO_2	416
NH_3	370
H_2	72

Upper chart example: The viscosity of sulphur dioxide gas (SO_2) at 200 F is 0.016 centipoise.

Lower chart example: The viscosity of carbon dioxide gas (CO_2) at about 80 F is 0.015 centipoise.

Viscosity of Various Gases

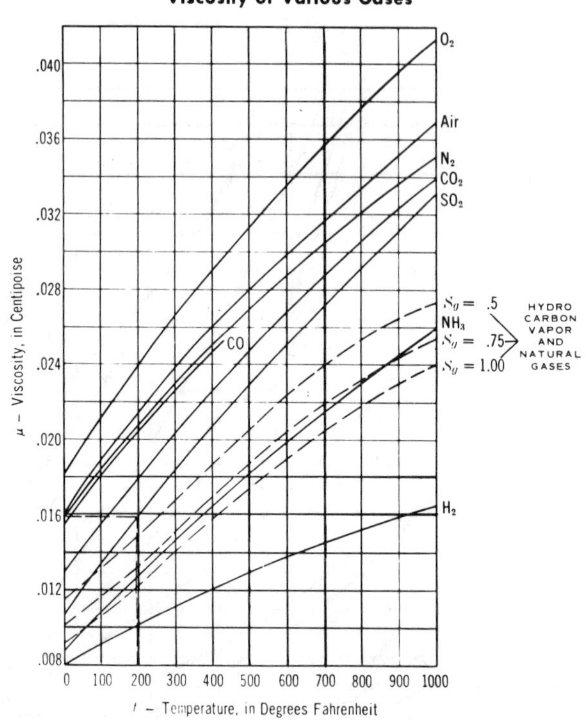

Viscosity of Refrigerant Vapors[11]
(saturated and superheated vapors)

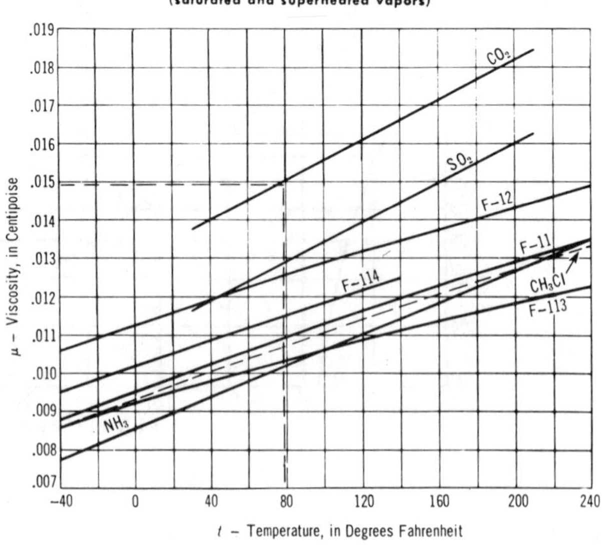

VISCOSITY OF GASES AND VAPORS

(From TP-410, FLOW OF FLUIDS, Courtesy Crane Company)

FIGURE B-10

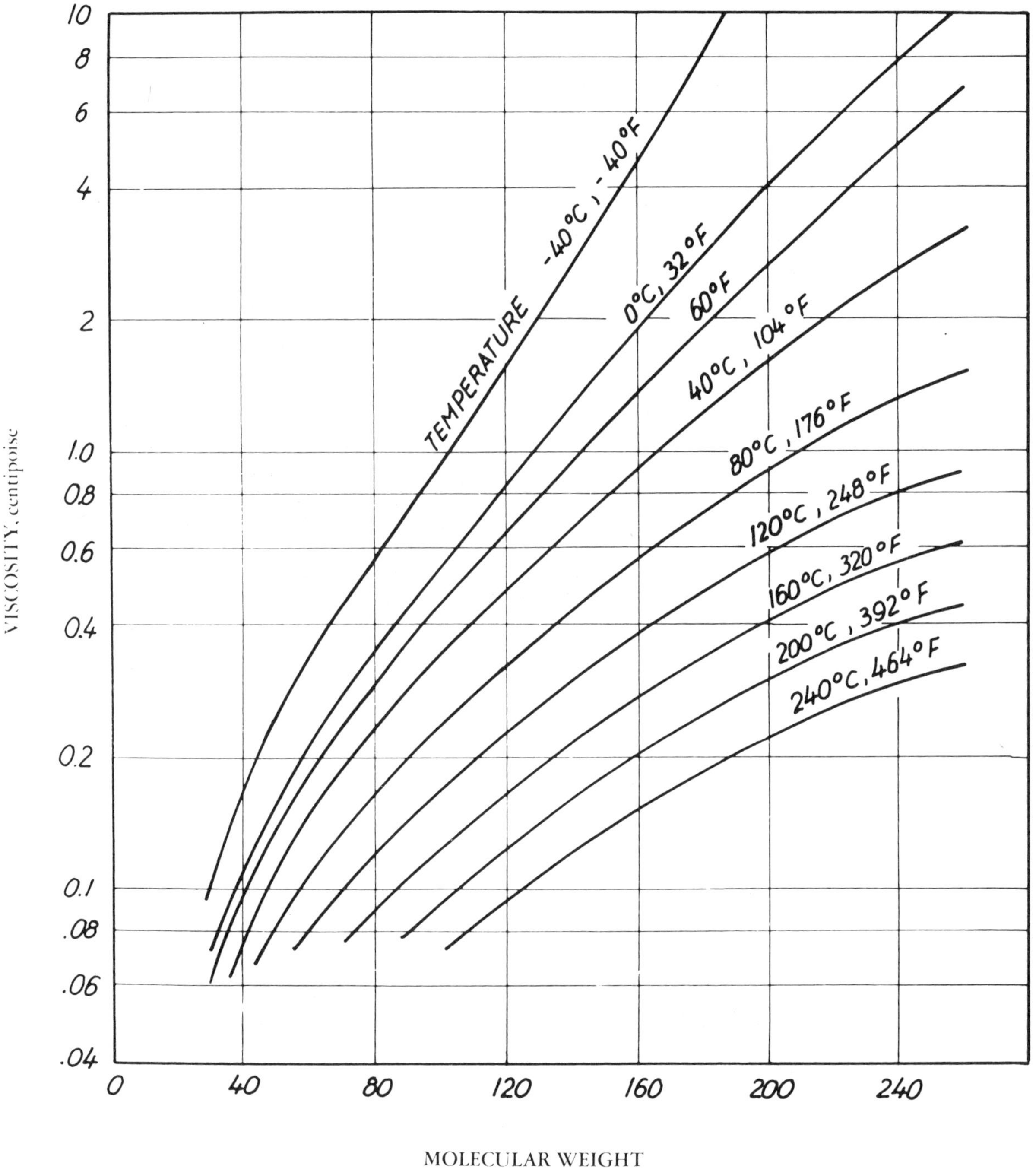

VISCOSITY OF PARAFFINIC HYDROCARBON LIQUIDS AT ATMOSPHERIC PRESSURE

FIGURE B-11

(From TP-410, FLOW OF FLUIDS, Courtesy Crane Company)

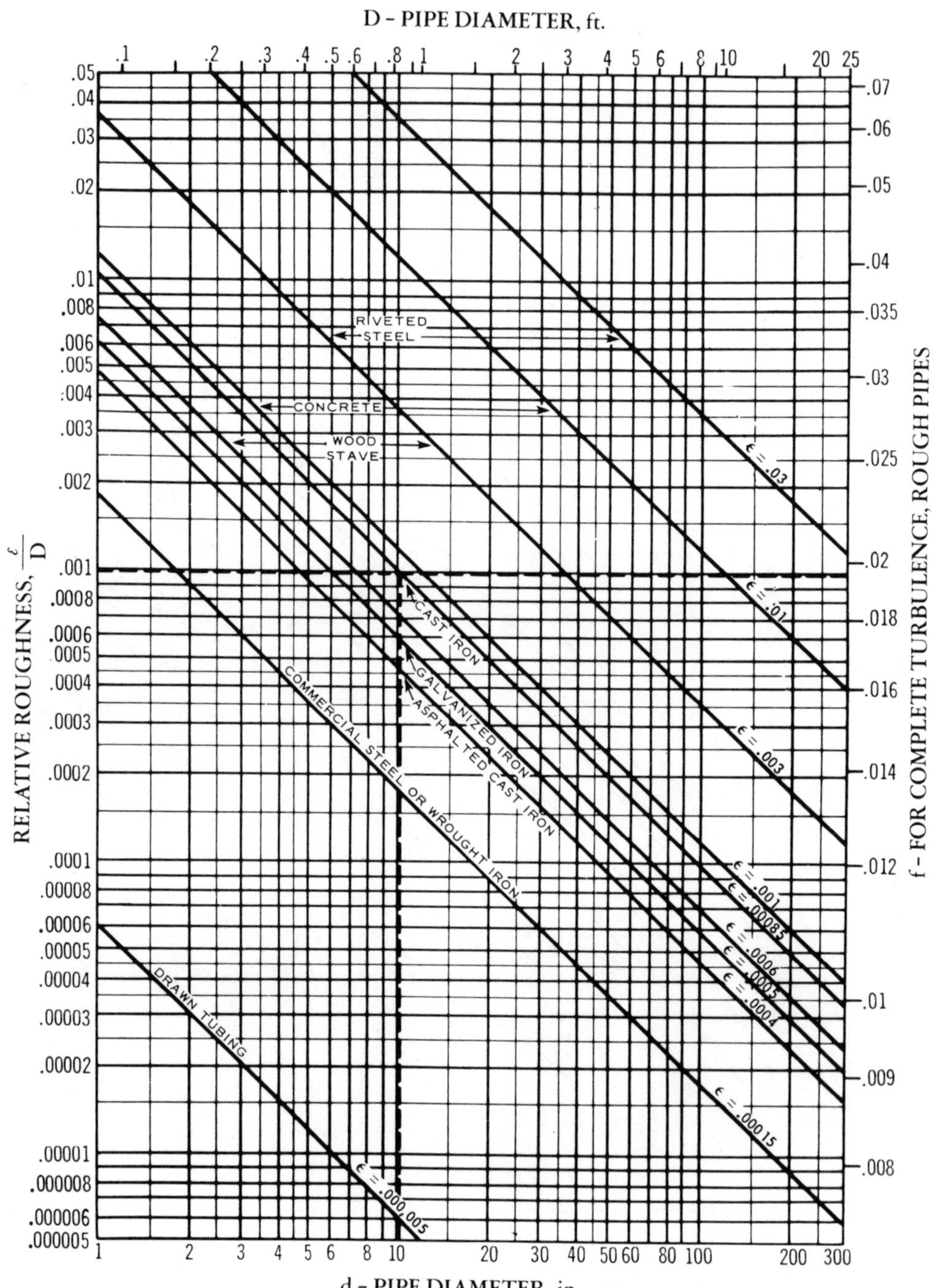

(Data extracted from "Friction Factors for Pipe Flow" by L. F. Moody with permission of the publisher, The American Society of Mechanical Engineers)

Problem: Determine absolute and relative roughness, and friction factor, for fully turbulent flow in 10-inch cast iron pipe (I.D. = 10.16″).

Solution: Absolute roughness (ε) = 0.00085; relative roughness (ε/D) = 0.001; friction factor at fully turbulent flow (f) = 0.0196.

RELATIVE ROUGHNESS OF PIPE MATERIALS AND FRICTION FACTORS FOR COMPLETE TURBULENCE

(From TP-410, FLOW OF FLUIDS, Courtesy Crane Company)

FIGURE B-12

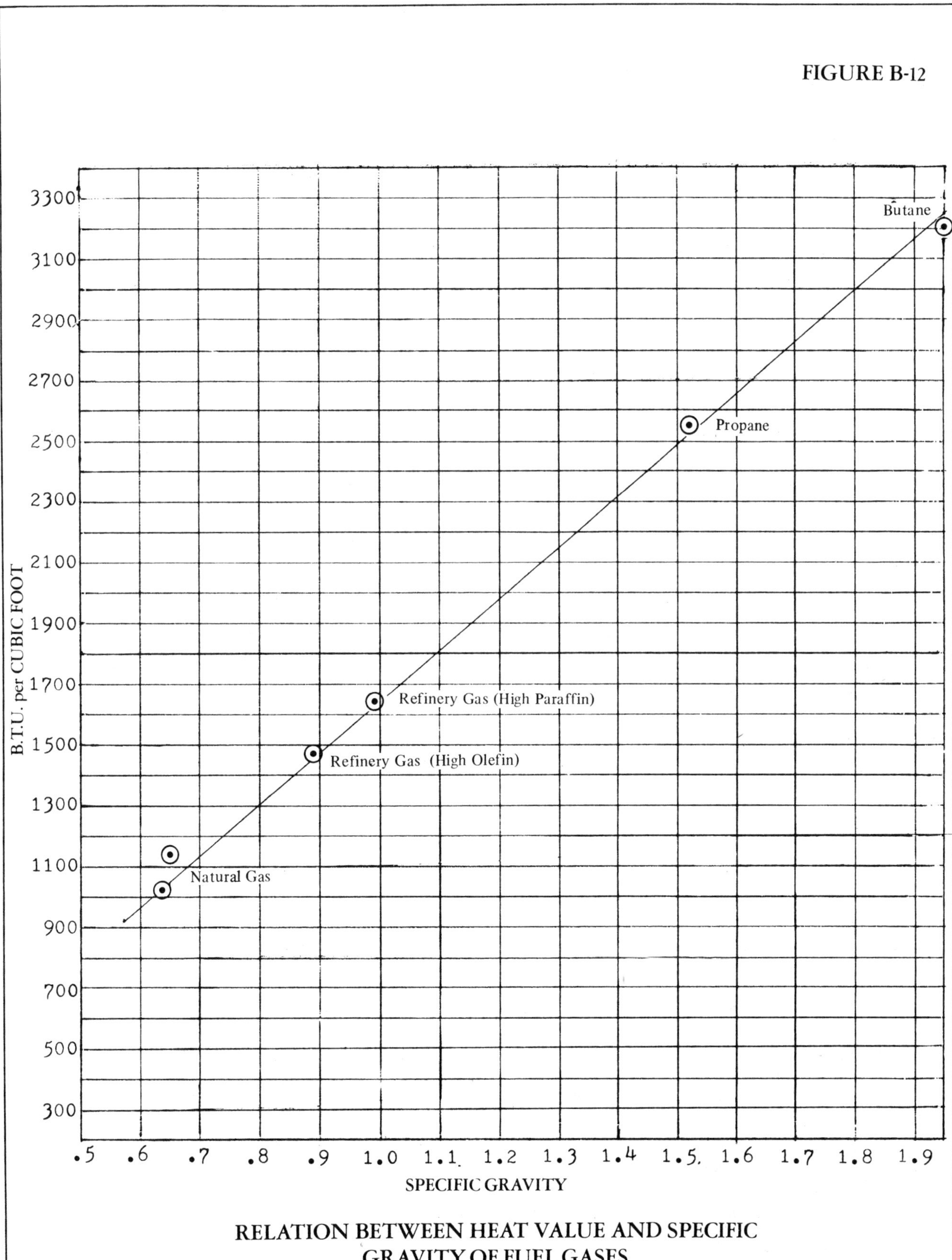

RELATION BETWEEN HEAT VALUE AND SPECIFIC GRAVITY OF FUEL GASES

FIGURE B-13

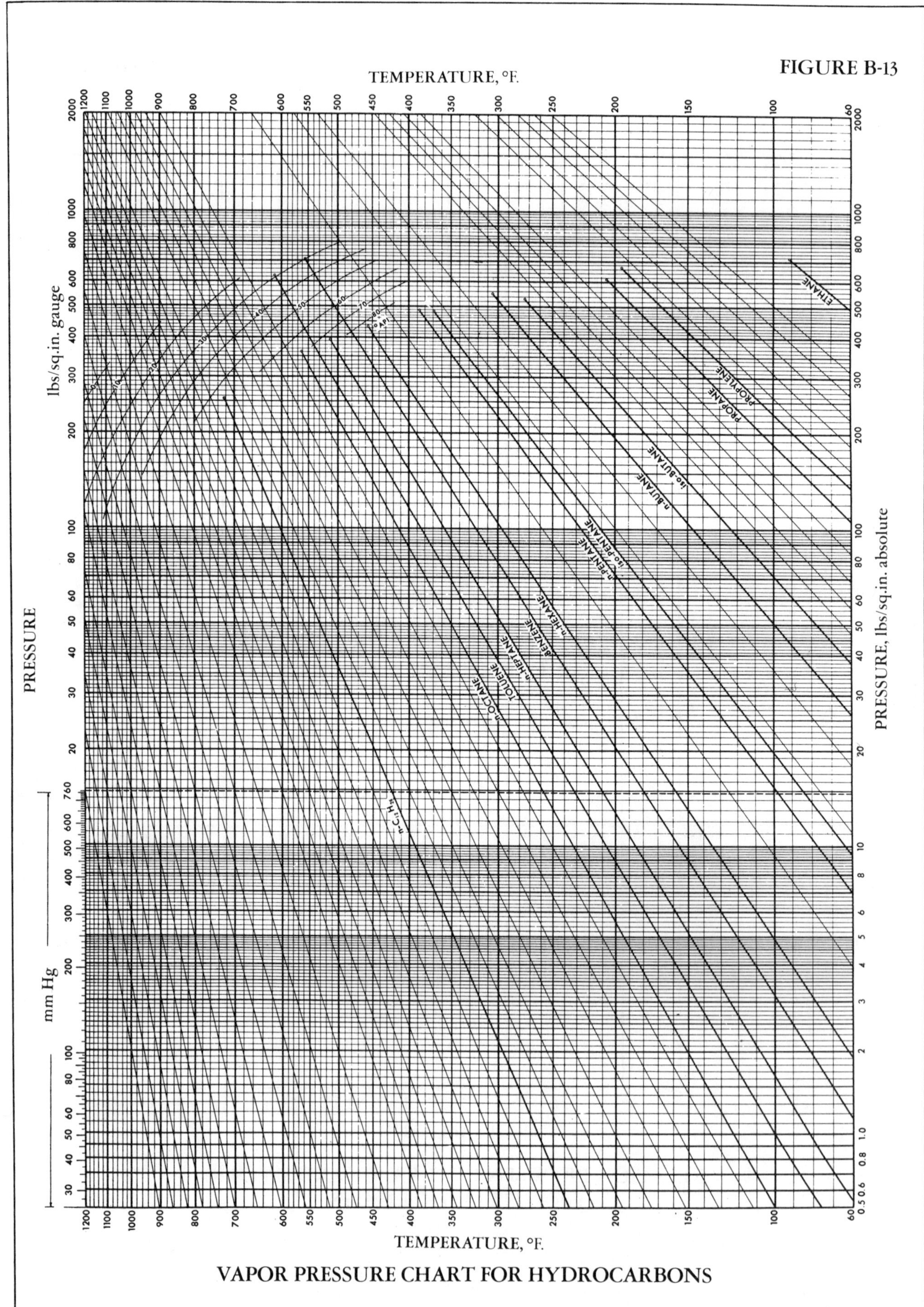

VAPOR PRESSURE CHART FOR HYDROCARBONS

FIGURE B-14

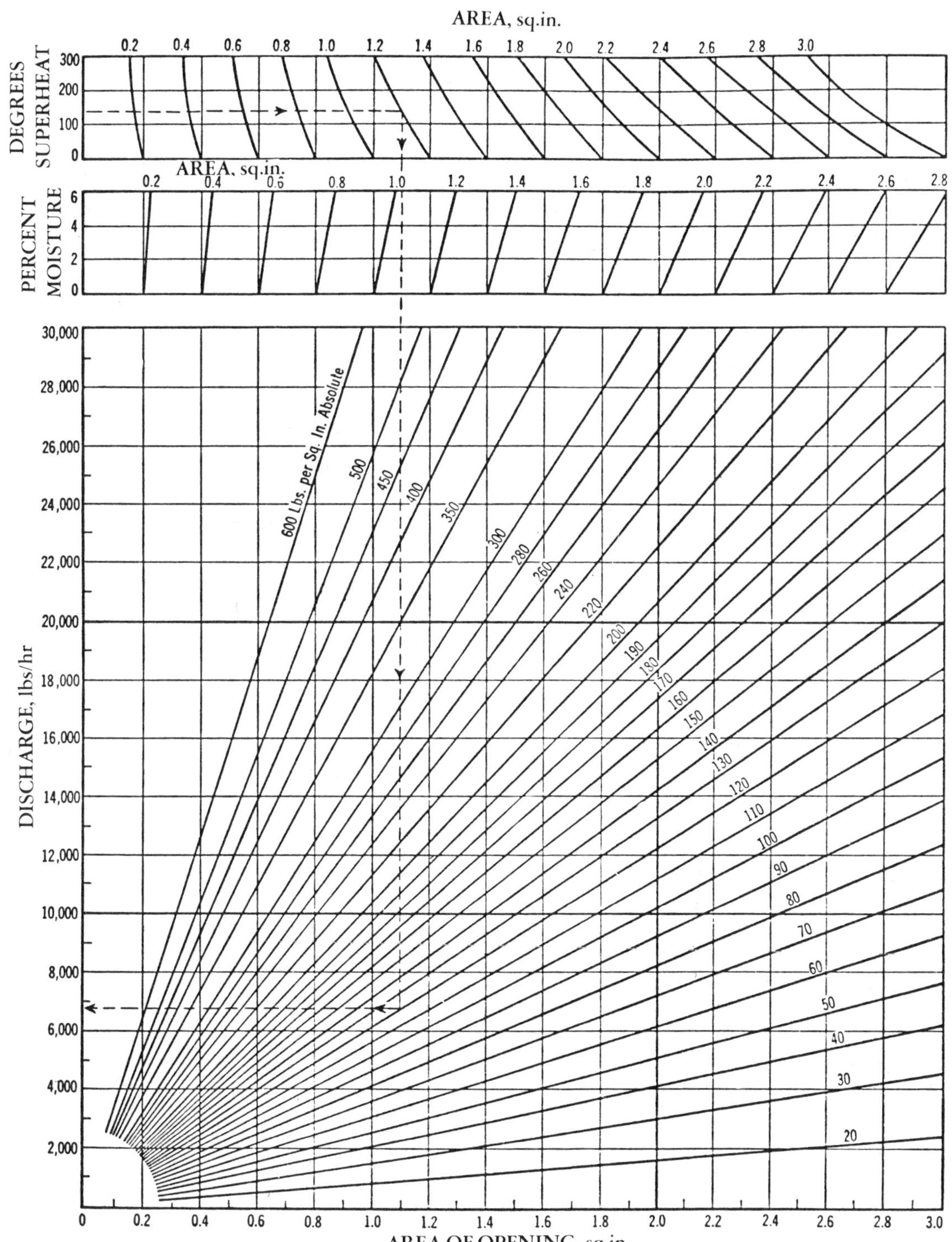

DISCHARGE OF STEAM TO ATMOSPHERE THROUGH ORIFICES, NOZZLES, AND SHORT TUBES

(From TP 410, FLOW OF FLUIDS, Courtesy Crane Company)

FIGURE B-15

GATE VALVES
Wedge Disc, Double Disc, or Plug Type

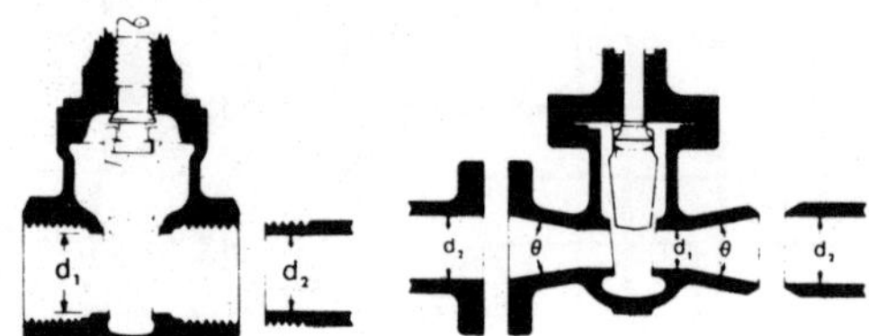

If: $\beta = 1$, $\theta = 0$ . . . . $K_1 = 8\,f_T$
$\beta < 1$ and $\theta \leqq 45°$ . . . . K_2 = Formula 5
$\beta < 1$ and $\theta > 45° \leqq 180°$ . . . K_2 = Formula 6

SWING CHECK VALVES

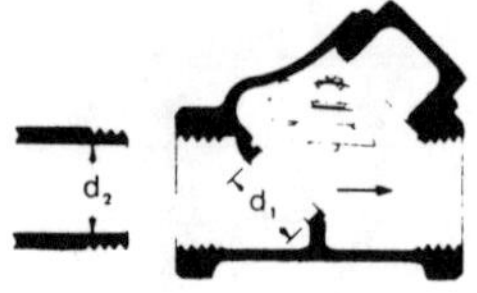

$K = 100\,f_T$

Minimum pipe velocity (fps) for full disc lift $= 35\sqrt{\bar{V}}$

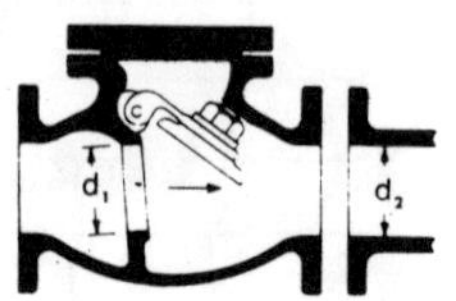

$K = 50\,f_T$

Minimum pipe velocity (fps) for full disc lift $= 48\sqrt{\bar{V}}$

GLOBE AND ANGLE VALVES

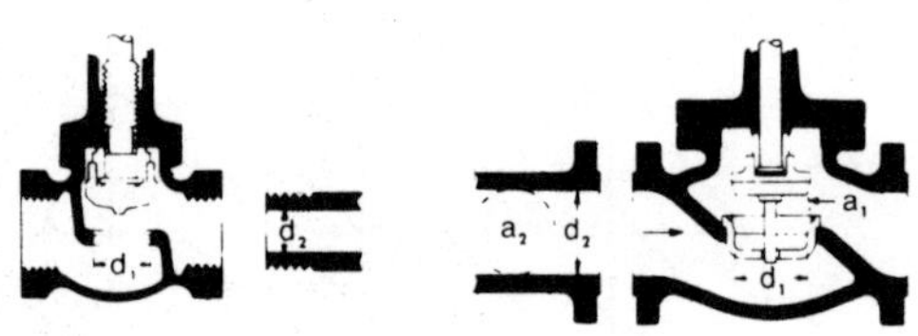

If: $\beta = 1 \ldots K_1 = 340\,f_T$

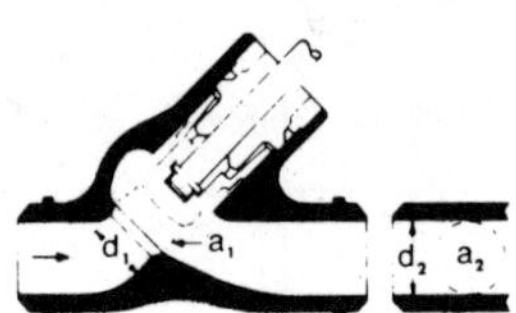

If: $\beta = 1 \ldots K_1 = 55\,f_T$

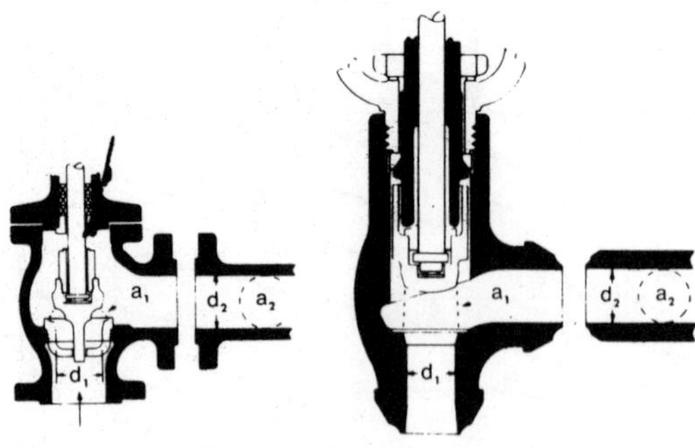

If: $\beta = 1 \ldots K_1 = 150\,f_T$ If: $\beta = 1 \ldots K_1 = 55\,f_T$

All globe and angle valves, whether reduced seat or throttled,

If: $\beta < 1 \ldots K_2$ = Formula 7

LIFT CHECK VALVES

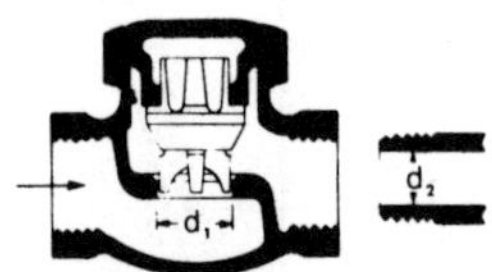

If: $\beta = 1 \ldots K_1 = 600\,f_T$
$\beta < 1 \ldots K_2$ = Formula 7

Minimum pipe velocity (fps) for full disc lift $= 40\,\beta^2\sqrt{\bar{V}}$

If: $\beta = 1 \ldots K_1 = 55\,f_T$
$\beta < 1 \ldots K_2$ = Formula 7

Minimum pipe velocity (fps) for full disc lift $= 140\,\beta^2\sqrt{\bar{V}}$

TILTING DISC CHECK VALVES

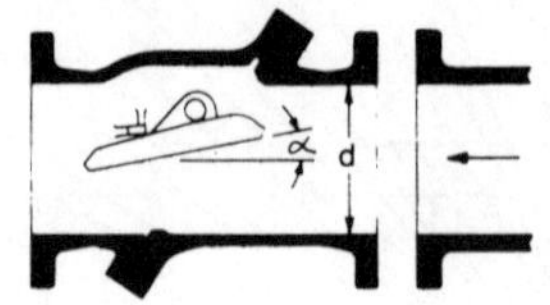

	$\alpha = 5°$	$\alpha = 15°$
Sizes 2 to 8" . . . K =	$40\,f_T$	$120\,f_T$
Sizes 10 to 14" . . . K =	$30\,f_T$	$90\,f_T$
Sizes 16 to 48" . . . K =	$20\,f_T$	$60\,f_T$
Minimum pipe velocity (fps) for full disc lift =	$80\sqrt{\bar{V}}$	$30\sqrt{\bar{V}}$

REPRESENTATIVE RESISTANCE COEFFICIENTS (K) FOR VALVES AND FITTINGS

(For Graph 71)

(From TP-410, FLOW OF FLUIDS, Courtesy Crane Company)

FIGURE B-16

STOP-CHECK VALVES (Globe and Angle Types)

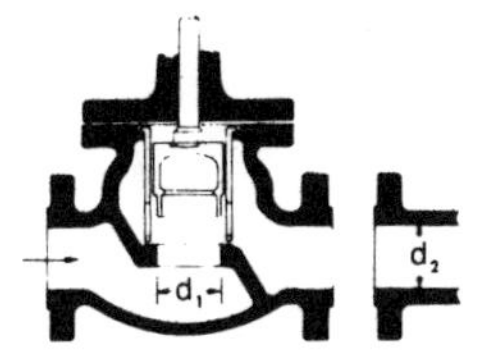

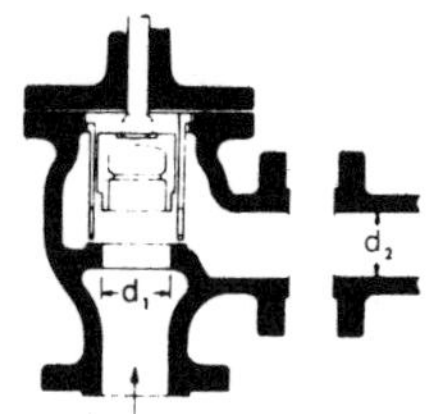

If:	If:
$\beta = 1 \ldots K_1 = 400\, f_T$	$\beta = 1 \ldots K_1 = 200\, f_T$
$\beta < 1 \ldots K_2 =$ Formula 7	$\beta < 1 \ldots K_2 =$ Formula 7
Minimum pipe velocity for full disc lift $= 55\, \beta^2 \sqrt{\bar{V}}$	Minimum pipe velocity for full disc lift $= 75\, \beta^2 \sqrt{\bar{V}}$

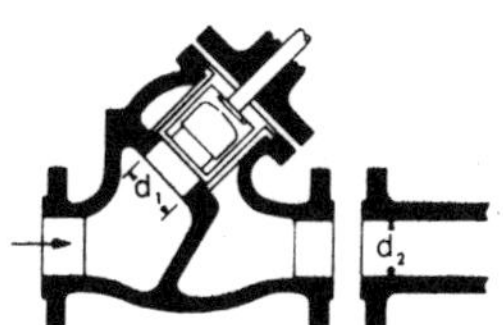

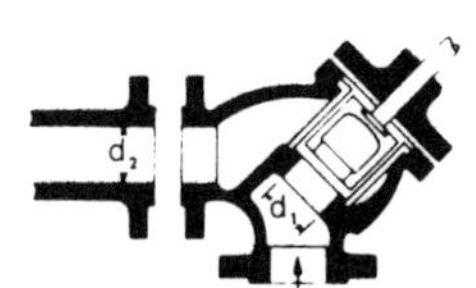

If:	If:
$\beta = 1 \ldots K_1 = 350\, f_T$	$\beta = 1 \ldots K_1 = 300\, f_T$
$\beta < 1 \ldots K_2 =$ Formula 7	$\beta < 1 \ldots K_2 =$ Formula 7

Minimum pipe velocity (fps) for full disc lift $= 60\, \beta^2 \sqrt{\bar{V}}$

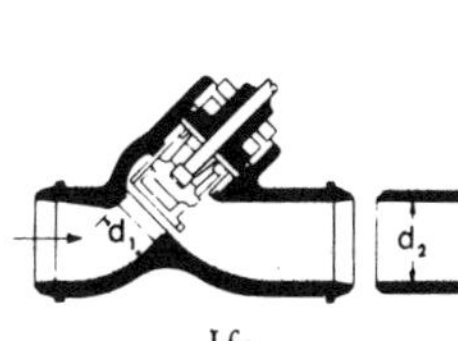

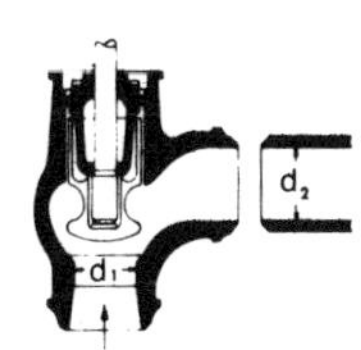

If:	If:
$\beta = 1 \ldots K_1 = 55\, f_T$	$\beta = 1 \ldots K_1 = 55\, f_T$
$\beta < 1 \ldots K_2 =$ Formula 7	$\beta < 1 \ldots K_2 =$ Formula 7

Minimum pipe velocity (fps) for full disc lift $= 140\, \beta^2 \sqrt{\bar{V}}$

FOOT VALVES WITH STRAINER

Poppet Disc	Hinged Disc
	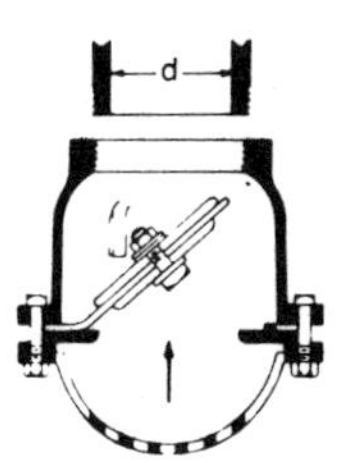
$K = 420 f_T$	$K = 75 f_T$
Minimum pipe velocity (fps) for full disc lift $= 15 \sqrt{\bar{V}}$	Minimum pipe velocity (fps) for full disc lift $= 35 \sqrt{\bar{V}}$

BALL VALVES

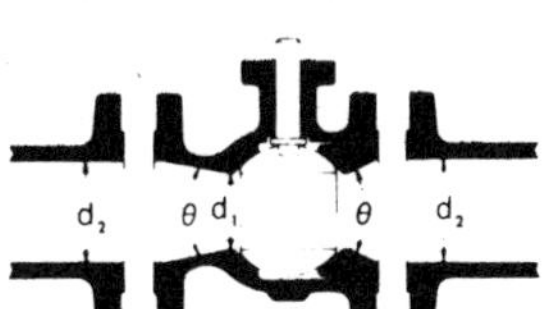

If:
$\beta = 1,\ \theta = 0 \ldots\ldots\ldots\ldots K_1 = 3\, f_T$
$\beta < 1$ and $\theta \leqq 45° \ldots\ldots\ldots K_2 =$ Formula 5
$\beta < 1$ and $\theta > 45° \leqq 180° \ldots K_2 =$ Formula 6

BUTTERFLY VALVES

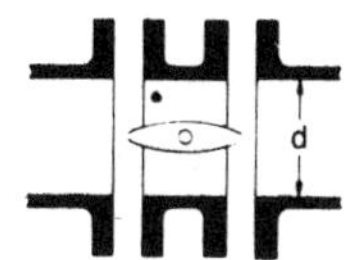

Sizes 2 to 8″ . . . $K = 45\, f_T$
Sizes 10 to 14″ . . . $K = 35\, f_T$
Sizes 16 to 24″ . . . $K = 25\, f_T$

REPRESENTATIVE RESISTANCE COEFFICIENTS (K) FOR VALVES AND FITTINGS

(For Graph 71)

(From TP-410, FLOW OF FLUIDS, Courtesy Crane Company)

FIGURE B-17

PLUG VALVES AND COCKS

Straight-Way

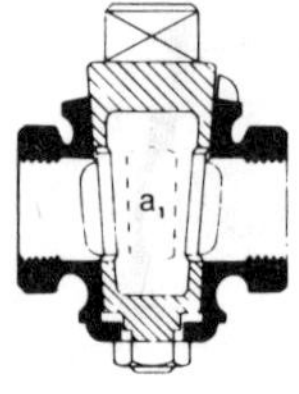

3-Way

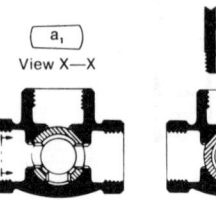

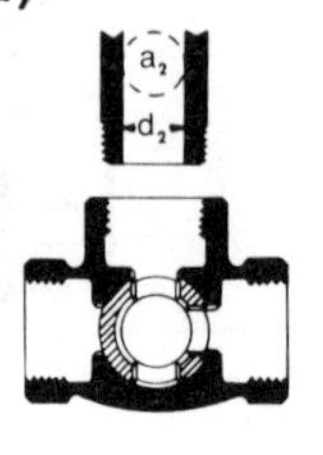

If: $\beta = 1$, $K_1 = 18\,f_T$

If: $\beta = 1$, $K_1 = 30\,f_T$

If: $\beta = 1$, $K_1 = 90\,f_T$

If: $\beta < 1 \ldots K_2 =$ Formula 6

STANDARD ELBOWS

90°

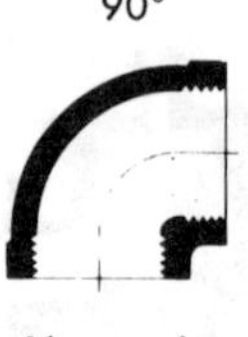

$K = 30\,f_T$

45°

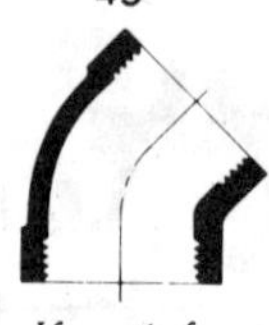

$K = 16\,f_T$

STANDARD TEES

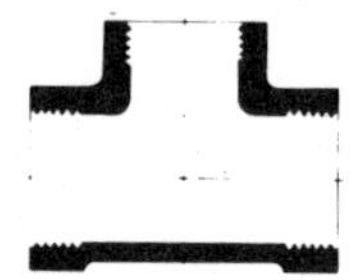

Flow thru run....... $K = 20\,f_T$
Flow thru branch.... $K = 60\,f_T$

MITRE BENDS

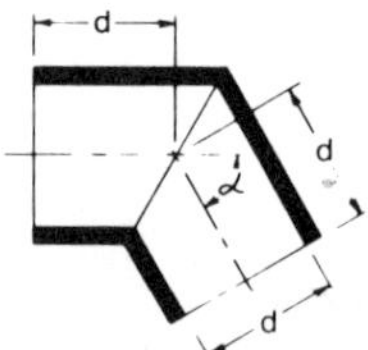

α	K
0°	2 f_T
15°	4 f_T
30°	8 f_T
45°	15 f_T
60°	25 f_T
75°	40 f_T
90°	60 f_T

90° PIPE BENDS AND
FLANGED OR BUTT-WELDING 90° ELBOWS

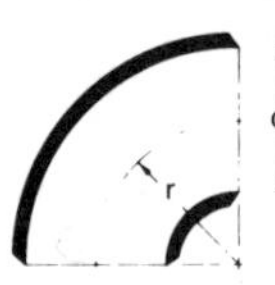

r/d	K	r/d	K
1	20 f_T	10	30 f_T
2	12 f_T	12	34 f_T
3	12 f_T	14	38 f_T
4	14 f_T	16	42 f_T
6	17 f_T	18	46 f_T
8	24 f_T	20	50 f_T

The resistance coefficient, K_B, for pipe bends other than 90° may be determined as follows:

$$K_B = (n - 1)\left(0.25\ \pi\ f_T \frac{r}{d} + 0.5\ K\right) + K$$

n = number of 90° bends
K = resistance coefficient for one 90° bend (per table)

PIPE ENTRANCE

Inward Projecting

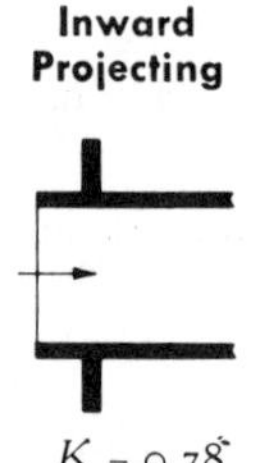

$K = 0.78$

r/d	K
0.00*	0.5
0.02	0.28
0.04	0.24
0.06	0.15
0.10	0.09
0.15 & up	0.04

*Sharp-edged

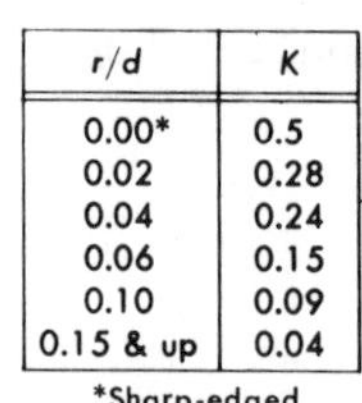

Flush

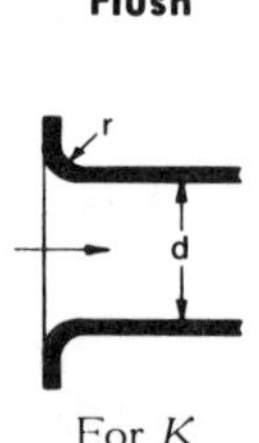

For K, see table

CLOSE PATTERN RETURN BENDS

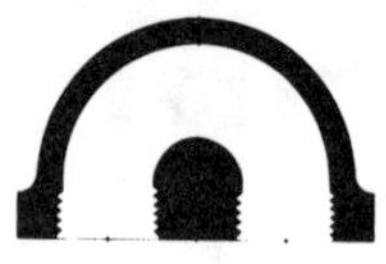

$K = 50\,f_T$

PIPE EXIT

Projecting　Sharp-Edged　Rounded

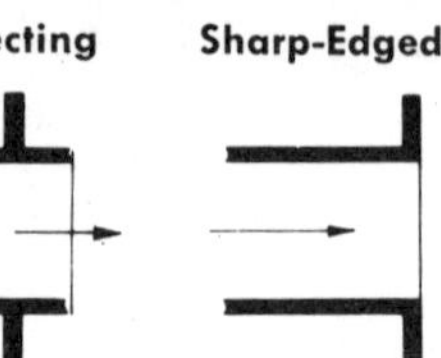

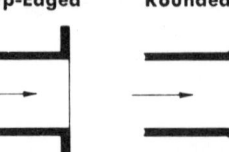

$K = 1.0$　$K = 1.0$　$K = 1.0$

REPRESENTATIVE RESISTANCE COEFFICIENTS (K) FOR VALVES AND FITTINGS

(For Graph 71)

(From TP-410, FLOW OF FLUIDS, Courtesy Crane Company)

VISCOSITIES OF HYDROCARBONS
Tables B-1 to B-6

CHARACTERIZATION FACTORS FOR CRUDES
Tables B-7 to B-10

DATA FOR COMMERCIAL WROUGHT IRON PIPE
Tables B-11 to B-13

TABLE B-1
Viscosities of the Normal Paraffinic Liquids

Hydrocarbon	Empirical Formula	Mol. Wt.	Viscosity in Centipoises at					Density			Kinematic Viscosity in Centistokes at				
			0° C (32° F)	20° C (68° F)	50° C (122° F)	80° C (176° F)	100° C (212° F)	d_4^0 (32° F)	d_4^{50} (122° F)	d_4^{100} (212° F)	0° C (32° F)	20° C (68° F)	50° C (122° F)	80° C (176° F)	100° C (212° F)
n-Butane	C_4H_{10}	58	0.208	0.172	0.137	(0.114)	(0.102)	0.6010	0.5370	0.4730	0.3445	0.299	0.2551	(0.2285)	(0.2155)
n-Pentane	C_5H_{12}	72	0.282	0.231	0.1805	(0.147)	(0.130)	0.6454	0.5892	0.5514	0.437	0.3685	0.3065	(0.2580)	(0.2355)
n-Hexane	C_6H_{14}	86	0.398	0.319	0.2415	0.1915	(0.1675)	0.6760	0.6370	0.5980	0.5890	0.483	0.379	0.312	(0.280)
n-Heptane	C_7H_{16}	100	0.520	0.411	0.305	0.2335	0.1975	0.7000	0.6590	0.6180	0.743	0.601	0.4625	0.368	0.320
n-Octane	C_8H_{18}	114	0.7015	0.5395	0.3845	0.2885	0.243	0.7185	0.6782	0.6379	0.9765	0.7685	0.567	0.441	0.381
n-Nonane	C_9H_{20}	128	0.969	0.709	0.487	0.360	0.3055	0.7325	0.6930	0.6540	1.325	0.988	0.718	0.5375	0.467
n-Decane	$C_{10}H_{22}$	142	1.298	0.907	0.6015	0.4525	0.357	0.7450	0.7060	0.6690	1.742	1.243	0.852	0.6615	0.5335
n-Undecane	$C_{11}H_{24}$	156	1.725	1.168	0.7465	0.5265	0.435	0.7550	0.7175	0.6810	2.285	1.576	1.040	0.7555	0.6390
n-Dodecane	$C_{12}H_{26}$	170	2.239	1.466	0.9015	0.618	0.501	0.7644	0.7286	0.6928	2.929	1.954	1.237	0.874	0.725
n-Tetradecane	$C_{14}H_{30}$	198	—	2.178	1.259	0.832	0.661	0.7780	0.7440	0.7075	—	2.851	1.692	1.152	0.9345
n-Hexadecane	$C_{16}H_{34}$	226	6.194	3.516	1.858	1.172	0.927	0.7891	0.7544	0.7197	7.848	4.535	2.462	1.597	1.288
n-Octadecane	$C_{18}H_{38}$	254	—	4.529	2.344	1.368	1.023	0.7960	0.7630	0.7280	—	5.769	3.072	1.844	1.406
n-Dotriacontane	$C_{32}H_{66}$	450	—	—	—	5.945	4.09	0.8240	0.7936	0.7632	—	—	—	7.668	5.359

TABLE B-2
Viscosities of Iso-Paraffinic Liquids

Hydrocarbon	Empirical Formula	Mol. Wt.	Viscosity in Centipoises at					Density			Kinematic Viscosity in Centistokes at				
			0° C (32° F)	20° C (68° F)	50° C (122° F)	80° C (176° F)	100° C (212° F)	d_4^0 (32° F)	d_4^{50} (122° F)	d_4^{100} (212° F)	0° C (32° F)	20° C (68° F)	50° C (122° F)	80° C (176° F)	100° C (212° F)
2-Methylpentane	C_6H_{14}	86	0.3735	0.2985	0.226	0.1795	—	0.6711	0.6284	0.5857	0.5565	0.4565	0.3596	0.2980	—
3-Methylpentane	C_6H_{14}	86	0.394	0.322	0.2465	—	—	0.6819	0.6389	0.5959	0.5780	0.4845	0.3860	—	—
2:3-Dimethylbutane	C_6H_{14}	86	0.4945	0.3835	—	—	—	0.6794	0.6354	0.5914	0.7276	0.5795	—	—	—
2:2-Dimethylbutane	C_6H_{14}	86	0.477	0.3735	—	—	—	0.6677	0.6284	0.5831	0.7142	0.5750	—	—	—
2-Methylhexane	C_7H_{16}	100	0.4765	0.381	0.2825	0.217	0.1815	0.6960	0.6533	0.6106	0.6845	0.5610	0.4325	0.3455	0.2975
2:2:3-Trimethylbutane	C_7H_{16}	100	0.8015	0.5915	—	—	—	0.7068	0.6648	0.6228	1.134	0.8575	—	—	—
2:2:4-Trimethylpentane	C_8H_{18}	114	0.647	0.5025	0.359	0.269	0.227	0.7081	0.6680	0.6279	0.9135	0.7260	0.5375	0.4180	0.3615
2:7-Dimethyloctane	$C_{10}H_{22}$	142	1.047	0.8205	0.5755	0.4285	0.3595	0.7390	0.7005	0.6621	1.417	1.132	0.8215	0.6325	0.5430
2:11-Dimethyldodecane	$C_{14}H_{30}$	198	2.630	1.531	0.851	0.555	0.4365	0.7820	0.7484	0.7130	3.363	1.991	1.137	0.7630	0.6120
2-Methylheptadecane	$C_{18}H_{38}$	254	7.638	3.963	1.919	1.143	0.875	0.7946	0.7593	0.7243	9.574	5.078	2.527	1.548	1.208
3:12-Diethyltetradecane	$C_{18}H_{38}$	254	9.010	4.700	2.355	1.437	1.122	0.8059	0.7722	0.7385	11.18	5.932	3.049	1.911	1.519
3-Ethyloctadecane	$C_{20}H_{42}$	282	11.61	5.598	2.553	1.466	1.089	0.8091	0.7750	0.7421	14.35	7.039	3.294	1.941	1.468
2:12-Dimethyleicosane	$C_{22}H_{46}$	310	—	6.457	2.818	1.514	0.9955	—	—	—	—	—	—	—	—
2:19-Dimethyleicosane	$C_{22}H_{46}$	—	—	—	—	—	—	0.8083	0.7757	0.7431	—	8.119	3.633	2.002	1.340
4-Propylnonadecane	$C_{22}H_{46}$	310	18.90	8.807	3.827	2.148	1.600	0.8109	0.7771	0.7440	23.31	11.05	4.925	2.835	2.150
2-Methyltricosane	$C_{24}H_{50}$	338	—	13.24	4.550	2.280	1.629	0.8146	0.7827	0.7508	—	16.51	5.813	2.986	2.169
5-Butyleicosane	$C_{24}H_{50}$	338	—	12.91	5.060	2.654	1.950	0.8150	0.7815	0.7498	—	16.12	6.474	3.478	2.600
7:12-Dimethyl-n-dihexyl-(9:10)-octadecane	$C_{32}H_{66}$	450	—	35.81	11.32	5.129	—	—	—	—	—	—	—	—	9.281
7:16-Dimethyl-9:11:12:14-tetra-n-hexyldocosane	$C_{48}H_{98}$	678	—	227.5	39.9	13.12	(7.762)	0.9014	0.8689	0.8364	—	256.1	45.92	15.45	—

TABLE B-3
Viscosities of the Acyclic Olefin Liquids

Hydrocarbon	Empirical Formula	Mol. Wt.	Viscosities in Centipoises at					Density			Kinematic Viscosity in Centistokes at				
			0° C (32° F)	20° C (68° F)	50° C (122° F)	80° C (176° F)	100° C (212° F)	d_4^0 (32° F)	d_4^{50} (122° F)	d_4^{100} (212° F)	0° C (32° F)	20° C (68° F)	50° C (122° F)	80° C (176° F)	100° C (212° F)
Trimethylethylene	C_5H_{10}	70	0.2535	0.211	(0.1665)	(0.138)	(0.124)	0.6850	0.6365	—	0.3701	0.3171	0.2616	—	—
Isoprene	C_5H_8	68	0.2605	0.216	(0.171)	(0.141)	(0.125)	0.6912	0.6437	—	0.3769	0.3219	0.2657	—	—
Diallyl	C_6H_{10}	82	0.3405	0.273	0.208	(0.166)	(0.145)	0.7074	0.6580	—	0.4814	0.3971	0.3161	0.2640	—
2:2:3-Trimethylbutene-3	C_7H_{14}	98	0.625	0.475	(0.334)	—	—	0.7235	0.6793	—	0.8640	0.7630	0.4916	—	—
2:4:4-Trimethylpentene (1 and 2)	C_8H_{16}	112	0.6965	0.5345	0.380	0.2825	0.2365	0.7330	0.6913	0.6496	0.9501	0.7461	0.5496	0.4239	0.3641
n-Dodecene-1	$C_{12}H_{24}$	168	2.004	1.346	0.851	0.589	0.483	0.7744	0.7384	0.7024	0.587	1.771	1.153	0.8217	0.6877
2:11-Dimethyl-dodecadiene	$C_{14}H_{26}$	194	3.095	1.738	1.012	0.695	0.566	0.8220	0.7869	0.7509	3.765	2.155	1.286	0.9080	0.7537
n-Hexadecene-1	$C_{16}H_{32}$	224	5.408	3.177	1.718	1.096	0.871	0.7968	0.7615	0.7262	6.787	4.060	2.256	1.480	1.199
2-Methylheptadecene-2	$C_{18}H_{36}$	252	7.143	3.837	1.875	1.127	0.875	0.8089	0.7761	0.7449	8.831	4.823	2.415	1.486	1.174
3:12-Diethyltetra-decanediene-2, 12	$C_{18}H_{34}$	250	7.218	3.987	2.125	1.368	1.090	0.8249	0.7929	0.7609	8.752	4.909	2.681	1.768	1.433
3-Ethyloctadecene-2	$C_{20}H_{40}$	280	10.54	5.272	2.535	1.466	1.086	0.8181	0.7844	0.7511	12.88	6.557	3.232	1.919	1.445
2:19-Dimethyleicosadiene	$C_{22}H_{42}$	306	7.834	3.972	1.905	1.122	0.851	0.8311	0.7976	0.7630	9.426	4.857	2.389	1.445	1.115
4-Propylnonadecene-3	$C_{22}H_{44}$	308	17.16	8.248	3.738	2.100	1.575	0.8182	0.7858	0.7512	20.97	10.25	4.756	2.745	2.097
2-Methyltricosene-2	$C_{24}H_{48}$	336	—	(16.90)	5.303	2.485	1.726	0.8334	0.7994	0.7656	—	20.61	6.634	3.190	2.254
4-Butyleicosene-4	$C_{24}H_{48}$	336	24.63	10.94	4.708	2.543	1.855	—	—	—	—	—	—	—	—
5-Butyleicosene-4	$C_{24}H_{28}$	—	—	—	—	—	—	0.8205	0.7889	0.7566	30.01	13.54	5.968	3.306	2.452
Tricaprylene	$C_{32}H_{64}$	336	—	25.94	8.67	3.90	—	0.8505	0.8175	0.7845	—	30.98	10.60	4.889	—
7-Methyl-8-β-octyltetra-cosylene (7, 8 and 8, 9)		448	(55—46)	20.99	7.015	3.069	(1.950)	0.8604	0.8206	0.7876	64.46	24.77	8.549	3.832	2.476

TABLE B-4
Viscosities of the Monocyclic Naphthenic Liquids

Hydrocarbon	Empirical Formula	Mol. Wt.	Viscosities in Centipoises at					Density			Kinematic Viscosity in Centistokes at				
			0° C (32° F)	20° C (68° F)	50° C (122° F)	80° C (176° F)	100° C (212° F)	d^0	d^0	d^0	0° C (32° F)	20° C (68° F)	50° C (122° F)	80° C (176° F)	100° C (212° F)
cycloPentane	C_5H_{10}	70	0.556	0.4375	0.3245	(0.2525)	(0.2190)	0.7621	0.7211	0.6801	0.7295	0.5865	0.4500	0.3625	0.3220
Methylcyclopentane	C_6H_{12}	84	0.647	0.501	0.3615	0.274	(0.2345)	0.7676	0.7226	0.6776	0.8430	0.6680	0.5000	0.3940	0.3460
Ethylcyclopentane	C_7H_{14}	98	0.7415	0.569	0.411	0.3125	0.2685	0.7859	0.7449	0.7039	0.9435	0.7395	0.5520	0.4340	0.3815
1:3-Dimethylcyclopentane	C_7H_{14}	98	0.622	0.508	0.3665	0.2785	0.2380	0.7639	0.7199	0.6759	0.8145	0.6810	0.5080	0.4015	0.3605
n-Propylcyclopentane	C_8H_{16}	112	0.9035	0.6935	0.491	0.3665	0.312	0.7950	0.7490	0.7030	1.163	0.8940	0.6555	0.5080	0.4440
n-Butylcyclopentane	C_9H_{18}	126	1.202	0.8915	0.6095	0.4455	0.3715	0.8003	0.7613	0.7223	1.502	1.136	0.8005	0.6040	0.5145
cycloHexane	C_6H_{12}	84	1.455	0.9865	0.6095	0.413	0.332	0.7966	0.7506	0.7046	1.827	1.268	0.8120	0.5715	0.4710
Methylcyclohexane	C_7H_{14}	98	0.991	0.728	0.4945	0.3665	0.3095	0.7860	0.7447	0.7034	1.261	0.9460	0.6640	0.5095	0.4400
Ethylcyclohexane	C_8H_{16}	112	1.161	0.8375	0.5715	0.417	0.350	0.8044	0.7621	0.7198	1.443	1.064	0.7500	0.5660	0.4865
o-Dimethylcyclohexane	C_8H_{16}	112	1.321	0.942	0.625	0.4465	0.369	0.8031	0.7638	0.7245	1.644	1.196	0.8185	0.6035	0.5095
m-Dimethylcyclohexane	C_8H_{16}	112	0.912	0.6855	0.4795	0.3555	0.299	0.7882	0.7485	0.7088	1.157	0.8875	0.6405	0.4905	0.4220
p-Dimethylcyclohexane	C_8H_{16}	112	1.081	0.776	0.527	0.3855	0.3245	0.7887	0.7487	0.7087	1.371	1.004	0.7040	0.5320	0.4580
n-Propylcyclohexane	C_9H_{18}	126	1.403	0.993	0.6605	0.4755	0.3945	0.8078	0.7705	0.7332	1.737	1.252	0.8570	0.6355	0.5380
1:3:5-Trimethylcyclohexane	C_9H_{18}	126	0.891	0.676	0.474	0.354	0.3005	0.7880	0.7505	0.7130	1.131	0.8745	0.6315	0.4865	0.4215
n-Butylcyclohexane	$C_{10}H_{20}$	140	1.910	1.300	0.828	0.5795	0.473	0.8151	0.7787	0.7423	2.343	1.624	1.063	0.7660	0.6370
isoButylcyclohexane	$C_{10}H_{20}$	140	1.722	1.225	0.7905	0.552	0.4465	0.8102	0.7737	0.7372	2.125	1.540	1.022	0.7340	0.6055
								20° C		80° C					
Cetylcyclohexane	$C_{22}H_{42}$	292	—	14.95	5.82	3.12	—	0.8287	0.8086	0.7885	—	18.04	7.197	3.958	—
Octadecylcyclohexane (11)	$C_{24}H_{48}$	336	47.63	19.52	7.467	3.745	2.599	0.8431	0.8118	0.7805	56.5	23.5	9.2	4.72	3.33
(l-Butyloctadecyl)cyclohexane (12)	$C_{28}H_{56}$	392	95.50	32.99	10.59	4.838	3.210	0.8450	0.8146	0.7830	113.0	39.5	13.0	6.08	4.10
Docosylcyclohexane (13)	$C_{28}H_{56}$	392	75.48	30.09	11.04	5.355	3.643	0.8479	0.8174	0.7869	89.0	36.0	13.5	6.7	4.63
(l-Butyldocosyl)cyclohexane (14)	$C_{38}H_{64}$	448	104.7	47.89	15.05	6.682	4.363	0.8518	0.8225	0.7932	123.0	57.0	18.3	8.3	5.5

Note: Numbers following names of hydrocarbons refer to those in Mikeska's paper.

Reference: J.I.P.T., Vol. 24, 1938, pp. 321—337.

TABLE B-5
Viscosities of Polycyclic Hydrocarbon Liquids

Hydrocarbon	Empirical Formula	Mol. Wt.	Viscosity in Centipoise at					d^0_4	d^{50}_4	d^{100}_4	Kinematic Viscosity in Centistokes at				
			0° C (32° F)	20° C (68° F)	50° C (122° F)	80° C (176° F)	100° C (212° F)	(32° F)	(122° F)	(212° F)	0° C (32° F)	20° C (68° F)	50° C (122° F)	80° C (176° F)	100° C (212° F)
Indene	C_9H_8	116	2.938	1.816	1.052	0.7345	0.5945	1.0152	0.9692	0.9232	2.895	1.820	1.085	0.7800	0.644
Hydrindene	C_9H_{10}	118	2.239	1.493	0.9185	0.6635	0.542	0.9813	0.9378	0.8942	2.280	1.550	0.9790	0.7280	0.6060
Naphthalene	$C_{10}H_8$	128	—	(2.483)	(1.422)	0.964	0.761	1.0373	0.9991	0.9609	—	2.430	1.423	0.9875	0.7920
1:2:3:4-Tetrahydro-naphthalene	$C_{10}H_{12}$	132	3.606	2.234	1.256	0.867	0.698	0.9866	0.9469	0.9072	3.655	2.300	1.325	0.9395	0.7695
Decahydronaphthalene	$C_{10}H_{18}$	138	4.169	2.630	1.466	0.991	0.7815	0.8982	0.8609	0.8236	4.640	2.980	1.705	1.180	0.9490
Dicyclopentyl	$C_{10}H_{18}$	138	1.923	1.403	0.9185	0.6775	0.560	0.8761	0.8389	0.8017	2.195	1.630	1.095	0.8295	0.6985
a-Methylnaphthalene	$C_{11}H_{10}$	142	6.109	3.334	1.679	1.102	0.867	1.0347	0.9979	0.9611	5.900	3.270	1.680	1.130	0.9020
β-Methylnaphthalene	$C_{11}H_{10}$	142	—	(2.564)	1.361	0.910	0.721	1.0562	0.9960	0.9357	—	2.485	1.365	0.9480	0.7705
Phenylcyclopentane	$C_{11}H_{14}$	146	2.667	1.778	1.086	0.785	0.647	0.9618	0.9228	0.8838	2.775	1.880	1.175	0.8730	0.7320
cycloHexylcyclopentane	$C_{11}H_{20}$	152	4.645	2.723	1.469	0.991	0.780	0.8907	0.8535	0.8183	5.215	3.110	1.720	1.190	0.9555
Diphenyl	$C_{12}H_{10}$	154	—	(3.733)	(1.959)	1.253	0.9595	1.0502	1.0096	0.9690	—	3.610	1.940	1.270	0.9900
Phenylcyclohexane	$C_{12}H_{16}$	160	3.681	2.377	1.380	0.955	0.762	0.9591	0.9183	0.8775	3.840	2.520	1.505	1.070	0.8685
Dicyclohexyl	$C_{12}H_{22}$	166	7.413	4.027	1.995	1.279	0.982	0.8988	0.8638	0.8288	8.245	4.550	2.310	1.520	1.185
Dibenzyl	$C_{14}H_{14}$	182	—	(4.624)	2.198	1.387	1.064	1.0027	0.9655	0.9283	—	4.680	2.275	1.470	1.145
Dicyclohexylethane	$C_{14}H_{26}$	194	17.99	8.128	3.266	1.901	1.409	0.8893	0.8560	0.8227	20.23	9.280	3.815	2.275	1.710
Hexylnaphthalene (26)	$C_{16}H_{20}$	212	23.93	8.84	3.28	1.73	1.26	0.9730	0.9420	0.9110	24.6	9.20	3.48	1.87	1.38
Dihydrodiethyl-anthracene	$C_{18}H_{20}$	236	—	157.0	17.30	5.176	—	1.0864	0.9949	0.9634	—	154.8	17.39	5.305	—
Misobutylnaphthalene	$C_{18}H_{24}$	240	—	47.21	8.511	3.420	(2.249)	0.9424	0.9104	0.8784	—	50.78	9.350	3.835	2.560
1:2:3:4-Tetrahydro-1:4-diisobutylnaphthalene	$C_{18}H_{28}$	244	—	40.30	8.630	3.420	—	0.9259	0.8939	0.8619	—	44.15	9.655	3.910	—
Dihexylnaphthalene (27)	$C_{22}H_{32}$	296	155.3	42.0	11.16	4.63	2.95	0.9477	0.9152	0.8847	164	45.0	12.2	5.16	3.33
Dihydrodiisoamyl-anthracene	$C_{24}H_{32}$	320	—	4920	184.5	23.66	—	0.9868	0.9499	0.9130	—	5062	194.2	25.60	—
1:1-Diphenyl-hexadecene-1	$C_{28}H_{40}$	376	164.9	46.8	13.1	5.56	3.70	0.9316	0.8991	0.8666	177.0	50.9	14.6	6.32	4.27
1:1-Diphenylhexadecane	$C_{28}H_{42}$	378	166.3	47.4	13.1	5.55	3.68	0.9265	0.8940	0.8615	179.6	51.9	14.6	6.34	4.27
Trihexylnaphthalene (28)	$C_{28}H_{44}$	380	305.8	75.4	18.18	6.97	4.32	0.9267	0.8952	0.8637	330	82.5	20.3	7.96	5.0
Trihexylnaphthalene (29)	$C_{28}H_{44}$	380	434.3	100.2	22.65	8.33	5.04	0.9240	0.8920	0.8600	470	110	25.4	9.54	5.86
Octadecylnaphthalene (15)	$C_{28}H_{44}$	380	124.5	42.5	13.35	5.96	3.92	0.9223	0.8897	0.8571	135	46.7	15.0	6.85	4.57
Octadecyltetrahydro-naphthalene (31)	$C_{28}H_{48}$	384	124.1	42.6	13.55	6.08	4.02	0.8989	0.8691	0.8393	138	48.0	15.6	7.14	4.79
1:1-Dicyclohexyl-hexadecane	$C_{28}H_{54}$	390	275.4	70.8	17.3	6.86	4.35	0.8921	0.8596	0.8271	308.7	80.5	20.2	8.16	5.13

TABLE B-5 (continued)

Hydrocarbon	Empirical Formula	Mol. Wt.	Viscosity in Centipoise at 0° C (32° F)	20° C (68° F)	50° C (122° F)	80° C (176° F)	100° C (212° F)	d_4^0 (32° F)	d_4^{50} (122° F)	d_4^{100} (212° F)	Kinematic Viscosity in Centistokes at 0° C (32° F)	20° C (68° F)	50° C (122° F)	80° C (176° F)	100° C (212° F)
Octadecyldecahydro-naphthalene (37)	$C_{28}H_{54}$	390	154.6	52.0	16.35	7.21	4.70	0.8785	0.8475	0.8168	176	60.0	19.3	8.7	5.76
5:14-Diphenyloctadeca-diene-5, 13 (50)	$C_{30}H_{42}$	402	254	64.4	15.6	6.18	3.98	0.9461	0.9151	0.8841	269	69.0	17.0	6.9	4.5
1:1-Diphenyl-octadecene-1 (47)	$C_{30}H_{44}$	404	211.1	59.45	16.08	6.53	4.14	0.9465	0.9140	0.8815	223	63.7	17.6	7.3	4.7
Dihydro-di-β-octyl-anthracene	$C_{30}H_{44}$	404	—	1476	115.1	(24.75)	—	0.9646	0.9331	0.9016	—	1550	123.4	27.10	—
1-Phenyl-2-benzyl-heptadecene-2	$C_{30}H_{44}$	404	430.5	78.5	17.06	6.70	4.28	0.9367	0.9042	0.8717	459.6	85.0	18.9	7.57	4.92
1:1-Diphenyl-octadecane (48)	$C_{30}H_{46}$	406	195.0	55.75	15.2	6.31	4.02	0.9420	0.9100	0.8780	207	60	16.7	7.08	4.58
5:14-Diphenyl-octadecane (51)	$C_{30}H_{46}$	406	336	77.2	17.7	6.71	4.18	0.9206	0.8896	0.8586	365	85	19.9	7.7	4.87
Octadecyldiphenyl (40)	$C_{30}H_{46}$	406	182	57.2	16.8	7.15	4.61	0.9278	0.8968	0.8658	196	62.5	18.7	8.14	5.33
1-Phenyl-2-benzyl-heptadecane	$C_{30}H_{46}$	406	380.2	81.5	18.4	6.92	4.35	0.9296	0.8969	0.8642	409.0	88.9	20.5	7.89	5.03
5:14-Dicyclohexyl-octadecane (52)								—	—	—	690	142	28.9	10.4	6.3
1-cycloHexyl-2-hexa-hydrobenzyl-heptadecane	$C_{30}H_{58}$	418	242.3	64.0	16.3	6.52	4.17	0.8990	0.8665	0.8340	269.5	72.25	18.8	7.70	5.00
(1-Butyloctadecen-1-yl) naphthalene (18)	$C_{32}H_{50}$	434	233.6	65.6	17.5	7.08	4.44	0.9236	0.8926	0.8616	253	72	19.6	8.1	5.15
Docosylnaphthalene (16)	$C_{32}H_{52}$	436	217	69.3	20.3	8.54	5.45	0.9121	0.8311	0.8501	238	77	23.0	9.9	6.41
(1-Butyloctadecyl) naphthalene (17)	$C_{32}H_{52}$	436	269.5	74.8	19.6	7.78	4.49	0.9132	0.8827	0.8522	295	83	22.2	9.0	5.27
(1-Butyloctadecen-1-yl) tetrahydronaphthalene (34)	$C_{32}H_{54}$	438	411.8	103.6	25.2	9.51	5.83	0.9053	0.8748	0.8443	455	116	28.8	11.1	6.9
Docosyltetrahydro-naphthalene (32)	$C_{32}H_{56}$	440	214	68.9	20.7	8.76	5.63	0.8959	0.8649	0.8339	239	78	23.9	10.35	6.75
(1-Butyloctadecyl) tetrahydronaphthalene (33)	$C_{32}H_{56}$	440	511.2	119.3	26.8	9.64	5.73	0.8970	0.8650	0.8330	570	135	31	11.4	6.88

TABLE B-5 (continued)

Hydrocarbon	Empirical Formula	Mol. Wt.	Viscosity in Centipoise at 0° C (32° F)	20° C (68° F)	50° C (122° F)	80° C (176° F)	100° C (212° F)	d_4^0 (32° F)	d_4^{50} (122° F)	d_4^{100} (212° F)	Kinematic Viscosity in Centistokes at 0° C (32° F)	20° C (68° F)	50° C (122° F)	80° C (176° F)	100° C (212° F)
Docosyldecahydronaphthalene (38)	$C_{32}H_{62}$	446	260	81.4	23.6	9.78	6.20	0.8783	0.8473	0.8163	296	94	27.8	11.8	7.6
1-Butyloctadecyldecahydronaphthalene (39)	$C_{32}H_{62}$	446	416.6	109.8	26.55	10.01	6.07	0.8770	0.8460	0.8150	475	127	31.4	12.1	7.45
1-Phenyl-1-naphthyloctadecene-1 (49)	$C_{34}H_{46}$	454	1502	260.6	44.8	13.87	7.77	0.9628	0.9333	0.9038	1560	274	48	15.1	8.6
(1-Butyloctadecen-1-yl) diphenyl (41)	$C_{34}H_{52}$	460	463	117	27.9	10.4	6.30	0.9258	0.8948	0.8638	500	128	31.2	11.9	7.3
(1-Butyloctadecyl) diphenyl (42)	$C_{34}H_{54}$	462	530.3	129.0	29.5	10.64	6.37	0.9145	0.8836	0.8527	580	143	33.4	12.3	7.47
Docosyldiphenyl (43)	$C_{34}H_{54}$	462	248.7	77.2	21.9	9.15	5.84	0.9210	0.8900	0.8590	270	85	24.6	10.5	6.8
(2-Butyleicosyl)naphthalene (21)	$C_{34}H_{56}$	464	306	85.7	22.8	9.00	5.63	0.9149	0.8839	0.8529	335	95	25.8	10.4	6.6
(1-Butyloctadecyl)diphenyl (46)	$C_{34}H_{66}$	474	865.8	183.6	28.85	12.55	7.27	0.8835	0.8505	0.8175	980	211	33.9	15.1	8.90
(1-Ethyl-2-butyleicosen-1-yl)naphthalene (22)	$C_{35}H_{58}$	478	597	140.3	29.9	10.8	6.50	0.9115	0.8805	0.8495	655	156	33.9	12.5	7.65
(1-Butyldocosen-1-yl) naphthalene (20)	$C_{36}H_{58}$	490	247.2	74.4	20.65	8.48	5.35	0.8988	0.8673	0.8358	275	84	23.8	10.0	6.4
(1-Butyldocosyl)naphthalene (19)	$C_{36}H_{60}$	492	278.3	80.7	21.8	8.77	5.51	0.8947	0.8637	0.8327	311	91.5	25.2	10.4	6.62
(1-Ethyl-2-butyleicosyl) naphthalene (25)	$C_{36}H_{60}$	492	696.5	155.2	32.8	11.4	6.68	0.9045	0.8735	0.8425	770	174	37.5	13.3	7.95
(1-Butyldocosen-1-yl) tetrahydronaphthalene (35)	$C_{36}H_{62}$	494	554.6	137.6	31.3	11.6	6.99	0.8945	0.8635	0.8325	620	156	36.2	13.7	8.4
(1-Butyldocosyl)tetrahydronaphthalene (36)	$C_{36}H_{64}$	496	594.8	146.9	33.8	12.35	7.37	0.8919	0.8604	0.8289	667	167	39.3	14.7	8.90
(1-Butyldocosen-1-yl) diphenyl (44)	$C_{38}H_{60}$	516	607	149.5	35.4	12.9	7.76	0.9345	0.9035	0.8725	650	162	39.2	14.6	8.9
(1-Butyldocosyl)diphenyl (45)	$C_{38}H_{62}$	518	704.2	170.3	38.05	13.6	7.97	0.9086	0.8782	0.8478	775	190	43.3	15.8	9.4
(1:2-Dibutyleicosen-1-yl)naphthalene (24)	$C_{38}H_{62}$	518	633.4	146.4	32.3	11.4	6.79	0.9051	0.8741	0.8431	700	164	37	13.3	8.05
(1:2-Dibutyleicosyl) naphthalene (23)	$C_{38}H_{64}$	520	789.3	170.9	35.05	12.0	7.03	0.9073	0.8763	0.8453	870	191	40	14.0	8.32

TABLE B-6
Viscosities of the Monocyclic Aromatic Liquids

Hydrocarbon	Empirical Formula	Mol. Wt.	0° C (32° F)	20° C (68° F)	50° C (122° F)	80° C (176° F)	100° C (212° F)	d_4^{0} (32° F)	d_4^{50} (122° F)	d_4^{100} (212° F)	0° C (32° F)	20° C (68° F)	50° C (122° F)	80° C (176° F)	100° C (212° F)
Benzene	C_6H_6	78	0.9015	0.6515	0.4365	0.317	0.2655	0.8952	0.8532	0.8112	1.007	0.7415	0.5115	0.383	0.3275
Toluene	C_7H_8	92	0.764	0.5835	0.417	0.3135	0.2665	0.8816	0.8406	0.7996	0.8665	0.6745	0.4960	0.3840	0.3334
Ethylbenzene	C_8H_{10}	106	0.877	0.668	0.4785	0.3595	0.305	0.8854	0.8414	0.7974	0.9905	0.7970	0.5685	0.4410	0.3825
o-Xylene	C_8H_{10}	106	1.102	0.913	0.556	0.411	0.3445	0.8972	0.8572	0.8172	1.228	0.9225	0.6485	0.4935	0.4215
m-Xylene	C_8H_{10}	106	0.8015	0.6165	0.4445	0.3375	0.2885	0.8811	0.8384	0.7957	0.9095	0.7140	0.5305	0.4150	0.3625
p-Xylene	C_8H_{10}	106	0.8455	0.6425	0.458	0.3435	0.291	0.8783	0.8350	0.7917	0.9630	0.7460	0.5485	0.4245	0.3675
n-Propylbenzene	C_9H_{12}	120	1.178	0.959	0.589	0.4315	0.362	0.8786	0.8390	0.7995	1.341	0.9955	0.7020	0.5290	0.4530
n-Butylbenzene	$C_{10}H_{14}$	134	1.478	1.050	0.7015	0.5045	0.420	0.8824	0.8434	0.8044	1.674	1.212	0.8320	0.6150	0.5220
Cetylbenzene	$C_{22}H_{36}$	300	—	10.90	4.75	2.68	—	0.8730	0.8397	0.8064	—	12.68	5.656	3.269	—
Octadecylbenzene (1)	$C_{24}H_{42}$	330	29.0	13.25	5.70	3.08	2.22	0.8696	0.8384	0.8052	33.0	15.45	6.8	3.76	2.76
(1-Butyloctadecen-1-yl) benzene (5)	$C_{28}H_{48}$	384	68.45	25.5	8.77	4.17	2.82	0.8830	0.8520	0.8210	77.5	29.3	10.3	5.0	3.43
Docosylbenzene (2)	$C_{28}H_{50}$	386	70.70	25.5	8.60	4.05	2.93	0.8674	0.8351	0.8028	81.5	29.8	10.3	4.96	3.40
(1-Butyloctadecyl)benzene (3)	$C_{28}H_{50}$	386	78.35	28.15	9.40	4.39	2.95	0.8706	0.8394	0.8081	90.0	32.8	11.2	5.35	3.65
(1-Butyldocosen-1-yl) benzene (6)	$C_{32}H_{56}$	440	43.55	23.25	11.21	6.44	4.77	0.8729	0.8434	0.8142	49.9	27.0	13.3	7.8	5.86
(1-Butyldocosyl)benzene (4)	$C_{32}H_{58}$	442	138.2	45.40	13.90	6.05	3.95	0.8692	0.8372	0.8058	159.0	53.0	16.6	7.4	4.9
Dioctadecylbenzene (7)	$C_{42}H_{78}$	582	186.4	71.85	24.70	11.31	7.51	0.8669	0.8371	0.8072	215.0	84.0	29.5	13.8	9.3
(1-Butyloctadecyl) (1-butyloctadecen-1-yl) benzene (9)	$C_{50}H_{92}$	692	536.9	156.4	41.90	16.26	9.92	0.8660	0.8380	0.2100	620.0	183.0	50.0	19.8	12.25
Di-(1-butyloctadecyl) benzene (10)	$C_{50}H_{94}$	694	546.5	153.0	39.95	15.15	9.22	0.8620	0.8320	0.8020	634.0	180.0	48.0	18.6	11.5
Trioctadecylbenzene (8)	$C_{60}H_{114}$	834	581.3	178.8	48.60	19.04	11.53	0.8676	0.8338	0.8090	670.0	209.0	58.0	23.2	14.25

TABLE B-7
Characterization Factor for Crudes

Source	°API	K
MIDDLE EAST		
Kuwait reduced crude, after 615° F.E.P.	16.1	11.6
Arabian reduced crude, after 625° F.E.P.	18.3	11.7
Arabian topped crude after 400° E.P.	24.5	11.8
Flashed Arabian reduced crude after 500 I.B.P.	12.8	11.5
Aghajari topped crude after 400° E.P.	23.9	11.8
Wafra topped crude after 400° E.P.	18.7	11.7
Seria topped crude after 400° E.P.	28.9	11.6
CHINA		
Kansu topped crude after 400° E.P.	26.8	12.2
BURMA		
Nanorkatiya topped crude after 400° E.P.	24.9	11.7
Nanorkatiya reduced crude after 650° F	20.9	11.8
ITALY		
Gela reduced crude after 720 E.P.	1.8	11.1
Gela topped crude after 400 E.P.	8.7	11.0
MEXICO		
Esequiel ordonez reduced crude after 640 E.P.	9.1	11.4
Panuco reduced crude, 550 I.B.P.	6.6	11.4
Visb. Panuco bottoms 258 I.B.P.	9.6	11.0
Tenextepec topped crude after 400 E.P.	13.4	11.5
Sharmex topped crude after 400 E.P.	13.9	11.5
Aceite topped crude after 400 E.P.	20.2	11.7
SOUTH AMERICA		
Bahia reduced crude after 670° F	30.9	12.6
Bahia topped crude after 580 I.B.P.	33.4	12.7
Lagumar topped crude after 400 E.P.	22.6	11.8
Nova Olinda topped crude after 400 E.P.	33.8	12.3
La Rosa topped crude after 400 E.P.	18.6	11.7
Aqua Caliente reduced crude after 700 E.P.	24.6	12.2
Morey topped crude after 400 E.P.	15.8	11.6
Magallanes reduced crude after 525 E.P.	32.6	12.3
Candeias topped crude after 400 E.P.	27.6	12.4
Itaparica topped crude after 400 E.P.	31.2	12.5
Lake Maricaibo Lana #6 topped crude 420 I.B.P.	25.3	11.9
Oficia and Lagonar asphalt	8.1	11.6
Tigre asphalt	7.2	11.5
CANADA		
Conrad reduced crude after 630 E.P.	13.9	11.4
Conrad topped crude after 400 E.P.	18.9	11.5
Princess topped crude after 400 E.P.	22.1	11.8
Redwater topped crude after 400 E.P.	23.9	11.8
Taber topped crude after 640 E.P.	13.7	11.5
Stettler topped crude after 400 E.P.	20.0	11.6
Virden topped crude after 400 E.P.	23.1	11.7
Leduc reduced crude after 560 E.P.	23.6	11.8
Roseray topped crude after 500 E.P.	18.1	11.6
Smiley topped crude after 400 E.P.	21.6	11.7
Fosterton reduced crude after 640 E.P.	11.2	11.4

TABLE B-7
Characterization Factor for Crudes

Source	°API	K
Fosterton reduced crude after 400 E.P.	16.1	11.5
Fort William blended reduced crude after 640 E.P.	14.4	11.5
Midale reduced crude after 530 E.P.	15.7	11.5
Midale reduced crude after 645 E.P.	14.0	11.4
Glenevis topped crude after 400 E.P.	14.0	11.5
Pembina reduced crude after 750 E.P.	18.7	11.8
Coleville reduced crude after 640 E.P.	8.7	11.4
ILLINOIS		
Aurora reduced crude after 620 E.P.	20.4	11.9
Kenland reduced crude after 640 E.P.	20.6	11.9
Matoon reduced crude after 640 E.P.	21.4	11.9
Pana reduced crude after 640 E.P.	20.7	11.9
MONTANA — WYOMING		
Mule Creek reduced crude after 700 E.P.	31.2	12.5
Light Worland reduced crude after 620° F	18.8	11.8
Cutbank reduced crude after 640 E.P.	20.0	11.7
Gage reduced crude after 600 E.P.	23.8	11.9
Ragged Point reduced crude after 600 E.P.	20.3	11.8
Light Elk Basin reduced crude after 700 E.P.	17.3	11.5
Mel Stone reduced crude after 600 E.P.	20.8	11.9
Sumutra reduced crude after 600 E.P.	21.1	11.9
Hawn reduced crude after 650 E.P.	21.7	11.8
Sun Phillips Dynneson red. cr. after 640 E.P.	27.4	12.2
Poplar reduced crude after 650 E.P.	24.0	12.0
Byron topped crude after 400 E.P.	18.9	11.7
Dry Creek reduced crude after 550 E.P.	23.2	11.9
Kevin Sunburst reduced crude after 552 I.B.P.	21.0	11.7
ROCKY MOUNTAIN		
Scotts Bluff reduced crude after 550 E.P.	26.4	12.0
Dener Julesburg asphalt	8.8	11.5
Wilson Creek reduced crude after 650 E.P.	17.6	11.7
KANSAS — OKLAHOMA		
Phillipsburg reduced crude	21.2	12.0
Lindsborg reduced crude after 436 E.P.	22.8	12.0
Midland reduced crude after 572 I.B.P.	22.9	11.9
Shallow Water reduced crude	17.3	11.7
Ben Franklin reduced crude after 494 I.B.P.	24.8	12.0
Dawnport reduced crude after dist.	27.2	12.1
Konalty reduced crude after 600 E.P.	25.2	12.1
Maysville reduced crude after 600 E.P.	26.4	12.2
MICHIGAN		
Edmore reduced crude after 525 E.P.	31.2	12.3
Fork reduced crude after 525 E.P.	31.9	12.2
CALIFORNIA		
Lombardi reduced crude after 525 E.P.	9.6	11.4
Edna topped crude after 400 E.P.	15.0	11.5
Placerita topped crude after 280 I.B.P.	15.0	11.4
Kern County Blend reduced crude after 680 E.P.	15.5	11.6
Kettleman topped crude after 400 E.P.	25.2	11.8

TABLE B-7
Characterization Factor for Crudes

Source	°API	K
TEXAS		
Upton reduced crude after 650 E.P.	16.7	11.6
Toborg reduced crude after 650 E.P.	12.7	11.4
Sharon Ridge reduced crude after 720 E.P.	11.2	11.4
Scurry reduced crude after 650 E.P.	23.3	11.6
Opelika reduced crude after 650 E.P.	29.9	11.3
East Texas reduced crude after 650 E.P.	21.5	11.9
Sprayberry topped crude after 400 E.P.	27.7	12.1
Wheeler topped crude after 400 E.P.	28.8	12.1
Quito topped crude after 400 E.P.	29.0	12.0
Somerset topped crude after 400 E.P.	21.0	11.8
Seeligson reduced crude after 525 E.P.	28.0	11.8
Placedo topped crude after 400 E.P.	22.8	11.3
Slaughter topped crude after 400 E.P.	21.2	11.7
NEW MEXICO		
San Juan Basin reduced crude after 600 E.P.	25.6	12.0
Drickey Queen topped crude after 400 E.P.	25.4	12.0
Hospan topped crude after 400 E.P.	26.1	12.0
Hobbs topped crude after 400 E.P.	22.5	11.7
Barber reduced crude after 650 E.P.	12.8	11.5
LOUISIANA		
Point Alahache reduced crude after 640 E.P.	23.6	12.0
Cox Bay reduced crude after 640 E.P.	20.2	11.8
Ostrela reduced crude after 640 E.P.	20.7	11.8
Ostrela asphalt	8.6	11.6
Lake Washington reduced crude after 525 E.P.	19.1	11.8
Delta Farms reduced crude after 520 E.P.	25.7	11.9

TABLE B-8
Characterization Factor for Crude Fractions

SOURCE	White Products (K)	Gas Oils and Heavier (K)
Pennsylvania	12.1 to 12.6	12.1 to 12.6
Rodessa	12.1 to 12.6	12.1 to 12.6
Panhandle	11.9 to 12.2	12.1 to 12.6
Mid Continent	11.9 to 12.2	11.9 to 12.2
Kuwait	11.9 to 12.6	11.7 to 12.2
Iraq	11.9 to 12.2	11.7 to 12.2
Iranian	11.9 to 12.2	11.7 to 12.2
East Texas	11.7 to 12.0	11.9 to 12.2
South Louisiana	11.7 to 12.0	11.9 to 12.2
Jusepin	11.7 to 12.0	11.7 to 12.0
West Texas	11.7 to 12.0	11.7 to 12.0
Tia Juana	11.7 to 12.0	11.5 to 11.8
Colombian	11.5 to 11.8	11.5 to 11.8
Lagunillas	11.3 to 11.6	11.3 to 11.6

TABLE B-9
Characterization Factors of Hydrocarbon Types of Approximately the Same Boiling Points

Name	Type	B.P.°F	K
Benzene	Aromatic	176	9.73
Cyclohexene	Cyclo Olefin	181	10.58
Cyclohexane	Cyclo Paraffin	177	10.98
3-Hexyne	Acetylene	179	11.86
2-Me-1,5-Hexadiene	Diolefin	191	11.97
2,3,3-Trimeth.-1-butene	Mono Olefin	172	11.98
2,4-Dimeth. Pentane	Paraffin	177	12.7

TABLE B-10
Characterization Factor for Various Hydrocarbons

Source	K	°API
Kuwait Vacuum Gas Oil	11.8	23.0
Kuwait Vacuum Gas Oil, hydro treated	12.2	29.1
Iranian Vac. Gas Oil	11.85	24.0
Oklahoma Vac. Gas Oil	11.8	29.1
Arabian Vac. Gas Oil	12.0	30.0
Michigan Vac. Gas Oil	12.1	29.5
Iranian Vac. Gas Oil	11.6	20.0
Nigerian Vac. Gas Oil	11.7	20.8
West Texas Vac. Gas Oil	11.9	22.1
Louisiana Vac. Gas Oil	12.1	28.0
Wyoming Vac. Gas Oil	12.1	28.8
Taikei Vac. Gas Oil	12.6	35.0
Thermally Cracked Gasoline	11.7	
Cracking Unit Recycle	10.4	
Cracked Residuum (Bunker C)	10.0 to 11.0	

TABLE B-11

Physical Data for Commercial Wrought Iron Pipe (Based on ANSI B36.10 Wall Thickness) (From TP-410, FLOW OF FLUIDS, Courtesy of Crane Co.)

	Nominal Pipe Size	Outside Diameter	Thickness	Inside Diameter		Inside Diameter Functions (In Inches)				Transverse Internal Area	
				d	D	d^2	d^3	d^4	d^5	a	A
	Inches	Inches	Inches	Inches	Feet					Sq. In.	Sq. Ft.
Schedule 10	14	14	0.250	13.5	1.125	182.25	2460.4	33215.	448400.	143.14	0.994
	16	16	0.250	15.5	1.291	240.25	3723.9	57720.	894660.	188.69	1.310
	18	18	0.250	17.5	1.4583	306.25	5359.4	93789.	1641309.	240.53	1.670
	20	20	0.250	19.5	1.625	380.25	7414.9	144590.	2819500.	298.65	2.074
	24	24	0.250	23.5	1.958	552.25	12977.	304980.	7167030.	433.74	3.012
	30	30	0.312	29.376	2.448	862.95	25350.	744288.	21864218.	677.76	4.707
Schedule 20	8	8.625	0.250	8.125	0.6771	66.02	536.38	4359.3	35409.	51.85	0.3601
	10	10.75	0.250	10.25	0.8542	105.06	1076.9	11038.	113141.	82.52	0.5731
	12	12.75	0.250	12.25	1.021	150.06	1838.3	22518.	275855.	117.86	0.8185
	14	14	0.312	13.376	1.111	178.92	2393.2	32012.	428185.	140.52	0.9758
	16	16	0.312	15.376	1.281	236.42	3635.2	55894.	859442.	185.69	1.290
	18	18	0.312	17.376	1.448	301.92	5246.3	91156.	1583978.	237.13	1.647
	20	20	0.375	19.250	1.604	370.56	7133.3	137317.	2643352.	291.04	2.021
	24	24	0.375	23.25	1.937	540.56	12568.	292205.	6793832.	424.56	2.948
	30	30	0.500	29.00	2.417	841.0	24389.	707281.	20511149.	660.52	4.587
Schedule 30	8	8.625	0.277	8.071	0.6726	65.14	525.75	4243.2	34248.	51.16	0.3553
	10	10.75	0.307	10.136	0.8447	102.74	1041.4	10555.	106987.	80.69	0.5603
	12	12.75	0.330	12.09	1.0075	146.17	1767.2	21366.	258304.	114.80	0.7972
	14	14	0.375	13.25	1.1042	175.56	2326.2	30821.	408394.	137.88	0.9575
	16	16	0.375	15.25	1.2708	232.56	3546.6	54084.	824801.	182.65	1.268
	18	18	0.438	17.124	1.4270	293.23	5021.3	85984.	1472397.	230.30	1.599
	20	20	0.500	19.00	1.5833	361.00	6859.0	130321.	2476099.	283.53	1.969
	24	24	0.562	22.876	1.9063	523.31	11971.	273853.	6264703.	411.00	2.854
	30	30	0.625	28.75	2.3958	826.56	23764.	683201.	19642160.	649.18	4.508
Schedule 40	⅛	0.405	0.068	0.269	0.0224	0.0724	0.0195	0.005242	0.00141	0.057	0.00040
	¼	0.540	0.088	0.364	0.0303	0.1325	0.0482	0.01756	0.00639	0.104	0.00072
	⅜	0.675	0.091	0.493	0.0411	0.2430	0.1198	0.05905	0.02912	0.191	0.00133
	½	0.840	0.109	0.622	0.0518	0.3869	0.2406	0.1497	0.09310	0.304	0.00211
	¾	1.050	0.113	0.824	0.0687	0.679	0.5595	0.4610	0.3799	0.533	0.00371
	1	1.315	0.133	1.049	0.0874	1.100	1.154	1.210	1.270	0.864	0.00600
	1¼	1.660	0.140	1.380	0.1150	1.904	2.628	3.625	5.005	1.495	0.01040
	1½	1.900	0.145	1.610	0.1342	2.592	4.173	6.718	10.82	2.036	0.01414
	2	2.375	0.154	2.067	0.1722	4.272	8.831	18.250	37.72	3.355	0.02330
	2½	2.875	0.203	2.469	0.2057	6.096	15.051	37.161	91.75	4.788	0.03322
	3	3.500	0.216	3.068	0.2557	9.413	28.878	88.605	271.8	7.393	0.05130
	3½	4.000	0.226	3.548	0.2957	12.59	44.663	158.51	562.2	9.886	0.06870
	4	4.500	0.237	4.026	0.3355	16.21	65.256	262.76	1058.	12.730	0.08840
	5	5.563	0.258	5.047	0.4206	25.47	128.56	648.72	3275.	20.006	0.1390
	6	6.625	0.280	6.065	0.5054	36.78	223.10	1352.8	8206.	28.891	0.2006
	8	8.625	0.322	7.981	0.6651	63.70	508.36	4057.7	32380.	50.027	0.3474
	10	10.75	0.365	10.02	0.8350	100.4	1006.0	10080.	101000.	78.855	0.5475
	12	12.75	0.406	11.938	0.9965	142.5	1701.3	20306.	242470.	111.93	0.7773
	14	14.0	0.438	13.124	1.0937	172.24	2260.5	29666.	389340.	135.28	0.9394
	16	16.0	0.500	15.000	1.250	225.0	3375.0	50625.	759375.	176.72	1.2272
	18	18.0	0.562	16.876	1.4063	284.8	4806.3	81111.	1368820.	223.68	1.5533
	20	20.0	0.593	18.814	1.5678	354.0	6659.5	125320.	2357244.	278.00	1.9305
	24	24.0	0.687	22.626	1.8855	511.9	11583.	262040.	5929784.	402.07	2.7921
Schedule 60	8	8.625	0.406	7.813	0.6511	61.04	476.93	3725.9	29113.	47.94	0.3329
	10	10.75	0.500	9.750	0.8125	95.06	926.86	9036.4	88110.	74.66	0.5185
	12	12.75	0.562	11.626	0.9688	135.16	1571.4	18268.	212399.	106.16	0.7372
	14	14.0	0.593	12.814	1.0678	164.20	2104.0	26962.	345480.	128.96	0.8956
	16	16.0	0.656	14.688	1.2240	215.74	3168.8	46544.	683618.	169.44	1.1766
	18	18.0	0.750	16.500	1.3750	272.25	4492.1	74120.	1222982.	213.83	1.4849
	20	20.0	0.812	18.376	1.5313	337.68	6205.2	114028.	2095342.	265.21	1.8417
	24	24.0	0.968	22.064	1.8387	486.82	10741.	236994.	5229036.	382.35	2.6552
Schedule 80	⅛	0.405	0.095	0.215	0.0179	0.0462	0.00994	0.002134	0.000459	0.036	0.00025
	¼	0.540	0.119	0.302	0.0252	0.0912	0.0275	0.008317	0.002513	0.072	0.00050
	⅜	0.675	0.126	0.423	0.0353	0.1789	0.0757	0.03200	0.01354	0.141	0.00098
	½	0.840	0.147	0.546	0.0455	0.2981	0.1628	0.08886	0.04852	0.234	0.00163
	¾	1.050	0.154	0.742	0.0618	0.5506	0.4085	0.3032	0.2249	0.433	0.00300
	1	1.315	0.179	0.957	0.0797	0.9158	0.8765	0.8387	0.8027	0.719	0.00499
	1¼	1.660	0.191	1.278	0.1065	1.633	2.087	2.6667	3.409	1.283	0.00891

From TP-410, FLOW OF FLUIDS, Courtesy of Crane Co

Commercial Wrought Steel Pipe Data

(Based on ANSI B36.10 Wall Thicknesses)

TABLE B-12

Physical Data for Commercial Wrought Iron Pipe (Based on ANSI B36.10 Wall Thickness) (From TP-410, FLOW OF FLUIDS, Courtesy of Crane Co.)

Schedule	Nominal Pipe Size	Outside Diameter	Thickness	Inside Diameter		Inside Diameter Functions (In Inches)				Transverse Internal Area	
				d	D	d^2	d^3	d^4	d^5	a	A
	Inches	Inches	Inches	Inches	Feet					Sq. In.	Sq. Ft.
Schedule 80—cont.	1½	1.900	0.200	1.500	0.1250	2.250	3.375	5.062	7.594	1.767	0.01225
	2	2.375	0.218	1.939	0.1616	3.760	7.290	14.136	27.41	2.953	0.02050
	2½	2.875	0.276	2.323	0.1936	5.396	12.536	29.117	67.64	4.238	0.02942
	3	3.5	0.300	2.900	0.2417	8.410	24.389	70.728	205.1	6.605	0.04587
	3½	4.0	0.318	3.364	0.2803	11.32	38.069	128.14	430.8	8.888	0.06170
	4	4.5	0.337	3.826	0.3188	14.64	56.006	214.33	819.8	11.497	0.07986
	5	5.563	0.375	4.813	0.4011	23.16	111.49	536.38	2583.	18.194	0.1263
	6	6.625	0.432	5.761	0.4801	33.19	191.20	1101.6	6346.	26.067	0.1810
	8	8.625	0.500	7.625	0.6354	58.14	443.32	3380.3	25775.	45.663	0.3171
	10	10.75	0.593	9.564	0.7970	91.47	874.82	8366.8	80020.	71.84	0.4989
	12	12.75	0.687	11.376	0.9480	129.41	1472.2	16747.	190523.	101.64	0.7058
	14	14.0	0.750	12.500	1.0417	156.25	1953.1	24414.	305176.	122.72	0.8522
	16	16.0	0.843	14.314	1.1928	204.89	2932.8	41980.	600904.	160.92	1.1175
	18	18.0	0.937	16.126	1.3438	260.05	4193.5	67626.	1090518.	204.24	1.4183
	20	20.0	1.031	17.938	1.4948	321.77	5771.9	103536.	1857248.	252.72	1.7550
	24	24.0	1.218	21.564	1.7970	465.01	10027.	216234.	4662798.	365.22	2.5362
Schedule 100	8	8.625	0.593	7.439	0.6199	55.34	411.66	3062.	22781.	43.46	0.3018
	10	10.75	0.718	9.314	0.7762	86.75	807.99	7526.	69357.	68.13	0.4732
	12	12.75	0.843	11.064	0.9220	122.41	1354.4	14985.	165791.	96.14	0.6677
	14	14.0	0.937	12.126	1.0105	147.04	1783.0	21621.	262173.	115.49	0.8020
	16	16.0	1.031	13.938	1.1615	194.27	2707.7	37740.	526020.	152.58	1.0596
	18	18.0	1.156	15.688	1.3057	246.11	3861.0	60572.	950250.	193.30	1.3423
	20	20.0	1.281	17.438	1.4532	304.08	5302.6	92467.	1612438.	238.83	1.6585
	24	24.0	1.531	20.938	1.7448	438.40	9179.2	192195.	4024179.	344.32	2.3911
Schedule 120	4	4.50	0.438	3.624	0.302	13.133	47.595	172.49	625.1	10.315	0.07163
	5	5.563	0.500	4.563	0.3802	20.82	95.006	433.5	1978.	16.35	0.1136
	6	6.625	0.562	5.501	0.4584	30.26	166.47	915.7	5037.	23.77	0.1650
	8	8.625	0.718	7.189	0.5991	51.68	371.54	2671.	19202.	40.59	0.2819
	10	10.75	0.843	9.064	0.7553	82.16	744.66	6750.	61179.	64.53	0.4481
	12	12.75	1.000	10.750	0.8959	115.56	1242.3	13355.	143563.	90.76	0.6303
	14	14.0	1.093	11.814	0.9845	139.57	1648.9	19480.	230137.	109.62	0.7612
	16	16.0	1.218	13.564	1.1303	183.98	2495.5	33849.	459133.	144.50	1.0035
	18	18.0	1.375	15.250	1.2708	232.56	3546.6	54086.	824804.	182.66	1.2684
	20	20.0	1.500	17.000	1.4166	289.00	4913.0	83521.	1419857.	226.98	1.5762
	24	24.0	1.812	20.376	1.6980	415.18	8459.7	172375.	3512313.	326.08	2.2645
Schedule 140	8	8.625	0.812	7.001	0.5834	49.01	343.15	2402.	16819.	38.50	0.2673
	10	10.75	1.000	8.750	0.7292	76.56	669.92	5862.	51291.	60.13	0.4176
	12	12.75	1.125	10.500	0.8750	110.25	1157.6	12155.	127628.	86.59	0.6013
	14	14.0	1.250	11.500	0.9583	132.25	1520.9	17490.	201136.	103.87	0.7213
	16	16.0	1.438	13.124	1.0937	172.24	2260.5	29666.	389340.	135.28	0.9394
	18	18.0	1.562	14.876	1.2396	221.30	3292.0	48972.	728502.	173.80	1.2070
	20	20.0	1.750	16.5	1.3750	272.25	4492.1	74120.	1222981.	213.82	1.4849
	24	24.0	2.062	19.876	1.6563	395.06	7852.1	156069.	3102022.	310.28	2.1547
Schedule 160	½	0.840	0.187	0.466	0.0388	0.2172	0.1012	0.04716	0.02197	0.1706	0.00118
	¾	1.050	0.218	0.614	0.0512	0.3770	0.2315	0.1421	0.08726	0.2961	0.00206
	1	1.315	0.250	0.815	0.0679	0.6642	0.5413	0.4412	0.3596	0.5217	0.00362
	1¼	1.660	0.250	1.160	0.0966	1.346	1.561	1.811	2.100	1.057	0.00734
	1½	1.900	0.281	1.338	0.1115	1.790	2.395	3.205	4.288	1.406	0.00976
	2	2.375	0.343	1.689	0.1407	2.853	4.818	8.138	13.74	2.241	0.01556
	2½	2.875	0.375	2.125	0.1771	4.516	9.596	20.39	43.33	3.546	0.02463
	3	3.50	0.438	2.624	0.2187	6.885	18.067	47.41	124.4	5.408	0.03755
	4	4.50	0.531	3.438	0.2865	11.82	40.637	139.7	480.3	9.283	0.06447
	5	5.563	0.625	4.313	0.3594	18.60	80.230	346.0	1492.	14.61	0.1015
	6	6.625	0.718	5.189	0.4324	26.93	139.72	725.0	3762.	21.15	0.1469
	8	8.625	0.906	6.813	0.5677	46.42	316.24	2155.	14679.	36.46	0.2532
	10	10.75	1.125	8.500	0.7083	72.25	614.12	5220.	44371.	56.75	0.3941
	12	12.75	1.312	10.126	0.8438	102.54	1038.3	10514.	106461.	80.53	0.5592
	14	14.0	1.406	11.188	0.9323	125.17	1400.4	15668.	175292.	98.31	0.6827
	16	16.0	1.593	12.814	1.0678	164.20	2104.0	26961.	345482.	128.96	0.8956
	18	18.0	1.781	14.438	1.2032	208.45	3009.7	43454.	627387.	163.72	1.1369
	20	20.0	1.968	16.064	1.3387	258.05	4145.3	66590.	1069715.	202.67	1.4074
	24	24.0	2.343	19.314	1.6095	373.03	7204.7	139152.	2687582.	292.98	2.0346

From TP 410, FLOW OF FLUIDS, Courtesy Crane Company

Commercial Wrought Steel Pipe Data

(Based on ANSI B36.10 Wall Thicknesses)

TABLE B-13

Physical Data for Commercial Wrought Iron Pipe (Based on ANSI B36.10 Wall Thickness) (From TP-410, FLOW OF FLUIDS, Courtesy of Crane Co.)

Nominal Pipe Size	Outside Diameter	Thickness	Inside Diameter		Inside Diameter Functions (In Inches)				Transverse Internal Area	
			d	D	d^2	d^3	d^4	d^5	a	A
Inches	Inches	Inches	Inches	Feet					Sq. In.	Sq. Ft.
Standard Wall Pipe										
1/8	0.405	0.068	0.269	0.0224	0.0724	0.0195	0.00524	0.00141	0.057	0.00040
1/4	0.540	0.088	0.364	0.0303	0.1325	0.0482	0.01756	0.00639	0.104	0.00072
3/8	0.675	0.091	0.493	0.0411	0.2430	0.1198	0.05905	0.02912	0.191	0.00133
1/2	0.840	0.109	0.622	0.0518	0.3869	0.2406	0.1497	0.0931	0.304	0.00211
3/4	1.050	0.113	0.824	0.0687	0.679	0.5595	0.4610	0.3799	0.533	0.00371
1	1.315	0.133	1.049	0.0874	1.100	1.154	1.210	1.270	0.864	0.00600
1 1/4	1.660	0.140	1.380	0.1150	1.904	2.628	3.625	5.005	1.495	0.01040
1 1/2	1.900	0.145	1.610	0.1342	2.592	4.173	6.718	10.82	2.036	0.01414
2	2.375	0.154	2.067	0.1722	4.272	8.831	18.250	37.72	3.355	0.02330
2 1/2	2.875	0.203	2.469	0.2057	6.096	15.051	37.161	91.75	4.788	0.03322
3	3.500	0.216	3.068	0.2557	9.413	28.878	88.605	271.8	7.393	0.05130
3 1/2	4.000	0.226	3.548	0.2957	12.59	44.663	158.51	562.2	9.886	0.06870
4	4.500	0.237	4.026	0.3355	16.21	65.256	262.76	1058.	12.730	0.08840
5	5.563	0.258	5.047	0.4206	25.47	128.56	648.72	3275.	20.006	0.1390
6	6.625	0.280	6.065	0.5054	36.78	223.10	1352.8	8206.	28.891	0.2006
8	8.625	0.277	8.071	0.6725	65.14	525.75	4243.0	34248.	51.161	0.3553
	8.625S	0.322	7.981	0.6651	63.70	508.36	4057.7	32380.	50.027	0.3474
10	10.75	0.279	10.192	0.8493	103.88	1058.7	10789.	109876.	81.585	0.5666
	10.75	0.307	10.136	0.8446	102.74	1041.4	10555.	106987.	80.691	0.5604
	10.75S	0.365	10.020	0.8350	100.4	1006.0	10080.	101000.	78.855	0.5475
12	12.75	0.330	12.090	1.0075	146.17	1767.2	21366.	258300.	114.80	0.7972
	12.75S	0.375	12.000	1.000	144.0	1728.0	20736.	248800.	113.10	0.7854
Extra Strong Pipe										
1/8	0.405	0.095	0.215	0.0179	0.0462	0.00994	0.002134	0.000459	0.036	0.00025
1/4	0.540	0.119	0.302	0.0252	0.0912	0.0275	0.008317	0.002513	0.072	0.00050
3/8	0.675	0.126	0.423	0.0353	0.1789	0.0757	0.03201	0.01354	0.141	0.00098
1/2	0.840	0.147	0.546	0.0455	0.2981	0.1628	0.08886	0.04852	0.234	0.00163
3/4	1.050	0.154	0.742	0.0618	0.5506	0.4085	0.3032	0.2249	0.433	0.00300
1	1.315	0.179	0.957	0.0797	0.9158	0.8765	0.8387	0.8027	0.719	0.00499
1 1/4	1.660	0.191	1.278	0.1065	1.633	2.087	2.6667	3.409	1.283	0.00891
1 1/2	1.900	0.200	1.500	0.1250	2.250	3.375	5.062	7.594	1.767	0.01225
2	2.375	0.218	1.939	0.1616	3.760	7.290	14.136	27.41	2.953	0.02050
2 1/2	2.875	0.276	2.323	0.1936	5.396	12.536	29.117	67.64	4.238	0.02942
3	3.500	0.300	2.900	0.2417	8.410	24.389	70.728	205.1	6.605	0.04587
3 1/2	4.000	0.318	3.364	0.2803	11.32	38.069	128.14	430.8	8.888	0.06170
4	4.500	0.337	3.826	0.3188	14.64	56.006	214.33	819.8	11.497	0.07986
5	5.563	0.375	4.813	0.4011	23.16	111.49	536.6	2583.	18.194	0.1263
6	6.625	0.432	5.761	0.4801	33.19	191.20	1101.6	6346.	26.067	0.1810
8	8.625	0.500	7.625	0.6354	58.14	443.32	3380.3	25775.	45.663	0.3171
10	10.75	0.500	9.750	0.8125	95.06	926.86	9036.4	88110.	74.662	0.5185
12	12.75	0.500	11.750	0.9792	138.1	1622.2	19072.	223970.	108.434	0.7528
Double Extra Strong Pipe										
1/2	0.840	0.294	0.252	0.0210	0.0635	0.0160	0.004032	0.00102	0.050	0.00035
3/4	1.050	0.308	0.434	0.0362	0.1884	0.0817	0.03549	0.01540	0.148	0.00103
1	1.315	0.358	0.599	0.0499	0.3588	0.2149	0.1287	0.07711	0.282	0.00196
1 1/4	1.660	0.382	0.896	0.0747	0.8028	0.7193	0.6445	0.5775	0.630	0.00438
1 1/2	1.900	0.400	1.100	0.0917	1.210	1.331	1.4641	1.611	0.950	0.00660
2	2.375	0.436	1.503	0.1252	2.259	3.395	5.1031	7.670	1.774	0.01232
2 1/2	2.875	0.552	1.771	0.1476	3.136	5.554	9.8345	17.42	2.464	0.01710
3	3.500	0.600	2.300	0.1917	5.290	12.167	27.984	64.36	4.155	0.02885
3 1/2	4.000	0.636	2.728	0.2273	7.442	20.302	55.383	151.1	5.845	0.04059
4	4.500	0.674	3.152	0.2627	9.935	31.315	98.704	311.1	7.803	0.05419
5	5.563	0.750	4.063	0.3386	16.51	67.072	272.58	1107.	12.966	0.09006
6	6.625	0.864	4.897	0.4081	23.98	117.43	575.04	2816.	18.835	0.1308
8	8.625	0.875	6.875	0.5729	47.27	324.95	2234.4	15360.	37.122	0.2578

From TP 410, FLOW OF FLUIDS, Courtesy Crane Company

Commercial Wrought Steel Pipe Data

(Based on ANSI B36.10 Wall Thicknesses)

TABLE B-14
Properties of Saturated Steam

Gauge Pressure psig	Temperature °F	Density #/cu. ft.	Heat of Liquid BTU per lb.	Latent heat of Vaporization BTU per lb.	Enthalpy BTU per lb.	Gauge Pressure psig	Temperature °F	Density #/cu. ft.	Heat of Liquid BTU per lb.	Latent heat of Vaporization BTU per lb.	Enthalpy BTU per lb.
0	212	0.0373	180	970	1150	200	388	0.468	362	837	1199
5	227	0.049	194	960	1154	205	390	0.479	364	835	1199
10	239	0.060	208	952	1160	210	392	0.489	366	833	1199
15	250	0.072	217	946	1163	215	394	0.500	368	832	1200
20	259	0.083	226	940	1166	220	395	0.510	370	830	1200
25	267	0.094	234	935	1169	225	397	0.521	372	828	1200
30	274	0.105	242	929	1171	230	399	0.532	373	827	1200
35	280	0.116	248	925	1173	235	401	0.543	375	826	1201
40	287	0.127	255	920	1175	240	402	0.554	377	824	1201
45	292	0.138	261	916	1177	245	404	0.565	379	822	1201
50	298	0.149	266	912	1178	250	406	0.576	381	820	1201
55	303	0.160	271	908	1179	255	407	0.587	383	819	1202
60	307	0.171	277	904	1181	260	409	0.598	385	817	1202
65	312	0.181	281	901	1182	265	411	0.605	386	816	1202
70	316	0.192	285	898	1183	270	412	0.615	388	814	1202
75	320	0.203	289	895	1184	275	414	0.625	390	812	1202
80	324	0.214	293	892	1185	280	416	0.636	391	811	1202
85	328	0.225	297	889	1186	285	417	0.647	393	809	1202
90	331	0.235	301	886	1187	290	419	0.657	395	808	1203
95	335	0.246	304	884	1188	295	420	0.668	397	806	1203
100	338	0.256	308	881	1189	300	422	0.679	398	805	1203
105	341	0.267	311	879	1190	305	423	0.690	400	803	1203
110	344	0.278	314	876	1190	310	425	0.700	401	802	1203
115	347	0.288	317	874	1191	315	426	0.711	402	801	1203
120	350	0.299	321	871	1192	320	428	0.721	404	799	1203
125	353	0.310	323	869	1192	325	429	0.732	406	797	1203
130	355	0.320	327	866	1193	330	430	0.743	407	796	1203
135	358	0.331	329	864	1193	335	432	0.754	409	795	1204
140	361	0.341	332	862	1194	340	433	0.764	411	793	1204
145	363	0.352	334	860	1194	345	434	0.775	413	791	1204
150	366	0.362	338	857	1195	350	436	0.786	414	790	1204
155	368	0.373	340	855	1195	355	437	0.796	416	788	1204
160	371	0.383	343	853	1196	360	438	0.807	417	787	1204
165	373	0.394	345	851	1196	365	440	0.818	418	786	1204
170	375	0.404	348	849	1197	370	441	0.828	419	785	1204
175	377	0.415	350	847	1197	375	442	0.839	421	783	1204
180	379	0.425	352	845	1197	380	443	0.849	422	782	1204
185	382	0.436	354	843	1197	385	444	0.860	424	780	1204
190	384	0.446	357	841	1198	390	445	0.871	425	779	1204
195	386	0.457	359	839	1198	395	447	0.882	426	778	1204

TABLE B-14
Properties of Saturated Steam

Gauge Pressure psig	Temperature °F	Density #/cu. ft.	Heat of Liquid BTU per lb.	Latent heat of Vaporization BTU per lb.	Enthalpy BTU per lb.	Gauge Pressure psig	Temperature °F	Density #/cu. ft.	Heat of Liquid BTU per lb.	Latent heat of Vaporization BTU per lb.	Enthalpy BTU per lb.
400	448	0.893	427	777	1204	0	212	0.0373	180	970	1150
405	449	0.903	428	776	1204	1	211	0.0362	177	972	1149
410	450	0.914	430	774	1204	2	209	0.0350	176	973	1149
415	451	0.925	431	773	1204	3	207	0.0338	174	974	1148
420	453	0.936	432	772	1204	4	205	0.0327	172	975	1147
425	454	0.947	434	770	1204	5	203	0.0315	171	976	1147
430	455	0.957	435	769	1204	6	201	0.0303	169	977	1146
435	456	0.961	436	768	1204	7	199	0.0292	167	978	1145
440	457	0.979	438	766	1204	8	197	0.0280	166	980	1144
445	458	0.990	439	765	1204	9	195	0.0268	162	981	1143
450	460	1.000	440	764	1204	10	193	0.0256	160	983	1143
455	461	1.011	442	762	1204	11	190	0.0244	158	984	1142
460	462	1.022	443	761	1204	12	187	0.0232	155	986	1141
465	463	1.033	445	759	1204	13	185	0.0220	153	987	1140
470	464	1.044	446	758	1204	14	182	0.0207	149	989	1138
475	465	1.055	447	757	1204	15	179	0.0196	146	991	1137
480	466	1.066	448	756	1204	16	176	0.0183	143	993	1136
485	467	1.077	449	755	1204	17	172	0.0171	140	995	1135
490	468	1.088	450	754	1204	18	169	0.0159	137	997	1134
495	469	1.099	451	753	1204	19	166	0.0146	133	999	1132
500	470	1.100	453	751	1204	20	161	0.0134	129	1001	1130
505	471	1.121	454	750	1204	21	157	0.0121	125	1004	1129
510	472	1.132	455	749	1204	22	152	0.0109	120	1007	1127
515	473	1.143	456	748	1204	23	147	0.0095	114	1010	1124
520	474	1.154	457	747	1204	24	141	0.0082	109	1013	1122
525	475	1.166	458	745	1203	25	134	0.0069	102	1017	1119
530	476	1.176	459	744	1203	26	126	0.0056	94	1021	1115
535	477	1.187	460	743	1203	27	115	0.0043	84	1027	1111
540	478	1.198	461	742	1203	28	101	0.0020	70	1035	1105
545	479	1.209	463	740	1203	29	79	0.0017	47	1048	1095
550	480	1.220	464	739	1203	30	—				
555	481	1.231	465	738	1203						
560	482	1.242	466	737	1203						
565	483	1.254	467	736	1203						
570	483	1.265	468	735	1203						
575	484	1.276	468	734	1202						
580	485	1.287	469	733	1202						
585	486	1.299	470	732	1202						
590	487	1.309	471	731	1202						
595	488	1.320	472	730	1202						
600	489	1.332	473	729	1202						

TABLE B-15
Capacity Constants for Partially Filled, Horizontal Cylindrical Tanks

Portion of Diam. at Liq. Level	Portion of Full Capacity	Portion of Diam. at Liq. Level	Portion of Full Capacity	Portion of Diam. at Liq. Level	Portion of Full Capacity	Portion of Diam. at Liq. Level	Portion of Full Capacity
0.01	0.0017	0.26	0.2066	0.51	0.5128	0.76	0.8155
0.02	0.0047	0.27	0.2179	0.52	0.5255	0.77	0.8263
0.03	0.0087	0.28	0.2292	0.53	0.5383	0.78	0.8369
0.04	0.0134	0.29	0.2407	0.54	0.5510	0.79	0.8474
0.05	0.0187	0.30	0.2523	0.55	0.5636	0.80	0.8576
0.06	0.0245	0.31	0.2640	0.56	0.5763	0.81	0.8677
0.07	0.0308	0.32	0.2759	0.57	0.5889	0.82	0.8776
0.08	0.0375	0.33	0.2878	0.58	0.6014	0.83	0.8873
0.09	0.0446	0.34	0.2998	0.59	0.6140	0.84	0.8967
0.10	0.0520	0.35	0.3119	0.60	0.6264	0.85	0.9059
0.11	0.0599	0.36	0.3241	0.61	0.6389	0.86	0.9149
0.12	0.0680	0.37	0.3364	0.62	0.6513	0.87	0.9236
0.13	0.0764	0.38	0.3487	0.63	0.6636	0.88	0.9320
0.14	0.0851	0.39	0.3611	0.64	0.6759	0.89	0.9401
0.15	0.0941	0.40	0.3736	0.65	0.6881	0.90	0.9480
0.16	0.1033	0.41	0.3860	0.66	0.7002	0.91	0.9554
0.17	0.1127	0.42	0.3986	0.67	0.7122	0.92	0.9625
0.18	0.1224	0.43	0.4111	0.68	0.7241	0.93	0.9692
0.19	0.1323	0.44	0.4237	0.69	0.7360	0.94	0.9755
0.20	0.1424	0.45	0.4364	0.70	0.7477	0.95	0.9813
0.21	0.1526	0.46	0.4490	0.71	0.7593	0.96	0.9866
0.22	0.1631	0.47	0.4617	0.72	0.7708	0.97	0.9913
0.23	0.1737	0.48	0.4745	0.73	0.7821	0.98	0.9952
0.24	0.1845	0.49	0.4872	0.74	0.7934	0.99	0.9983
0.25	0.1955	0.50	0.5000	0.75	0.8045	1.00	1.000

TABLE B-16
Capacity Constants for Partially Filled, Elliptical Heads

Fraction of Full Depth	Fraction of Full Capacity	Fraction of Full Depth	Fraction of Full Capacity	Fraction of Full Depth	Fraction of Full Capacity	Fraction of Full Depth	Fraction of Full Capacity
0.01	0.0003	0.26	0.168	0.51	0.517	0.76	0.853
0.02	0.001	0.27	0.179	0.52	0.530	0.77	0.865
0.03	0.003	0.28	0.190	0.53	0.545	0.78	0.875
0.04	0.005	0.29	0.202	0.54	0.560	0.79	0.885
0.05	0.007	0.30	0.216	0.55	0.574	0.80	0.896
0.06	0.011	0.31	0.228	0.56	0.590	0.81	0.905
0.07	0.014	0.32	0.241	0.57	0.604	0.82	0.913
0.08	0.019	0.33	0.256	0.58	0.619	0.83	0.922
0.09	0.023	0.34	0.270	0.59	0.633	0.84	0.930
0.10	0.028	0.35	0.282	0.60	0.648	0.85	0.939
0.11	0.033	0.36	0.297	0.61	0.662	0.86	0.945
0.12	0.039	0.37	0.310	0.62	0.674	0.87	0.953
0.13	0.046	0.38	0.324	0.63	0.688	0.88	0.960
0.14	0.052	0.39	0.338	0.64	0.702	0.89	0.966
0.15	0.060	0.40	0.352	0.65	0.715	0.90	0.972
0.16	0.069	0.41	0.368	0.66	0.729	0.91	0.977
0.17	0.077	0.42	0.382	0.67	0.742	0.92	0.982
0.18	0.086	0.43	0.398	0.68	0.756	0.93	0.986
0.19	0.095	0.44	0.412	0.69	0.769	0.94	0.989
0.20	0.104	0.45	0.428	0.70	0.784	0.95	0.993
0.21	0.114	0.46	0.441	0.71	0.797	0.96	0.995
0.22	0.124	0.47	0.457	0.72	0.808	0.97	0.997
0.23	0.135	0.48	0.471	0.73	0.820	0.98	0.998
0.24	0.146	0.49	0.487	0.74	0.831	0.99	0.999
0.25	0.156	0.50	0.500	0.75	0.844	1.00	1.00

APPENDIX C
INTERNATIONAL SYSTEM OF UNITS (SI)
AND
CONVERSIONS FROM U.S. UNITS

The International System of Units (SI), which has been accepted by The American National Standards Institute, has adopted the following basic and derived units of measurement.

BASIC SI UNITS

UNIT	NAME OF UNIT	SYMBOL
Length	Meter	m
Mass	Kilogram	kg
Time	Second	s
Electric current	Ampere	A
Temperature	Kelvin	K
Luminous intensity	Candela	cd

In addition to the six basic units, SI includes derived units which make it a more complete and coherent system of units, suitable for measurements in engineering. Some of the derived units have been given non-definitive names which carry specific definitions.

SOME DERIVED SI UNITS (With Non-Definitive Names)

UNIT	NAME OF UNIT	SYMBOL
Force	Newton	$N = kg\ m/s^2$
Work, Energy, Quantity of heat	Joule	$J = Nm$
Power	Watt	$W = J/s$
Electrical potential	Volt	$V = W/A$

Other derived units carry no special names and must always be expressed in terms of units from which they were derived.

SOME DERIVED SI UNITS (Without Special Names)

UNIT	SI UNIT	SYMBOL
Area	Square Meter	m^2
Volume	Cubic Meter	m^3
Density (mass density)	Kilogram per Cubic Meter	kg/m^3
Pressure	Newton per Square Meter	N/m^2

MULTIPLES AND SUB-MULTIPLES

The following multiples and sub-multiples of the SI Units are formed by means of the prefixes below:

Factor by Which the Unit is Multiplied	Prefix	Symbol
10^{12}	tera	T
10^{9}	giga	G
10^{6}	mega	M
10^{3}	kilo	k
10^{2}	hecto	h
10	deca	da
10^{-1}	deci	d
10^{-2}	centi	c
10^{-3}	milli	m
10^{-6}	micro	μ
10^{-9}	nano	n
10^{-12}	pico	p
10^{-15}	femto	f
10^{-18}	atto	a

The symbol of a prefix is considered to be combined with the unit symbol to which it is directly attached, forming with it a new unit symbol which can be raised to a positive or negative power and which can be combined with other unit symbols to form symbols for compound units.

Examples:

$1\ \mathrm{cm}^3 = (10^{-2}\mathrm{m})^3 = 10^{-6}\mathrm{m}^3$

$1\ \mu\mathrm{s}^{-1} = (10^{-6}\mathrm{s})^{-1} = 10^{6}\mathrm{s}^{-1}$

$1\ \mathrm{mm}^2/\mathrm{s} = (10^{-3}\mathrm{m})^2/\mathrm{s} = 10^{-6}\mathrm{m}^2/\mathrm{s}$

Compound prefixes should not be used; for example, write nm (nanometer) instead of (m m).

CONVERSION FACTORS

The following tables provide conversion from the Foot, Pound, Second (FPS) System to the International System of Units (SI), formerly known as the MKSA System. The tables may be read either from left to right or vice versa, depending upon what is to be converted.

The notation is in accordance with ASTM Designation: E380-16.

Reading From Left To Right:

UNITS "A"	MULTIPLIED BY	OR	DIVIDED BY	EQUALS UNITS "B"

Reading From Right To Left:

EQUALS UNITS "A"	DIVIDED BY	OR	MULTIPLIED BY	UNITS "B"

LINEAR MEASURES

ft (U.S.)	12	8.333×10^{-2}	inch
	3.048×10^{-1}	3.2808	m — meters
	1.8939×10^{-4}	$5.280 \times 10^{+3}$	miles (U.S. statute)
inch (U.S.)	2.5400×10^{1}	3.9370×10^{-2}	mm — millimeters
miles (U.S. statute)	1.6094	6.214×10^{-1}	km — kilometers

VOLUME AND VOLUME FLOW

cu ft (U.S.)	7.4805	1.337×10^{-1}	gallon (U.S.)
	2.8317×10^{-2}	3.5314×10^{1}	m^3 — cubic meters
	1.781×10^{-1}	5.6148	barrels (U.S.)
std cu ft (60° F, 1 atm)	2.679×10^{-2}	3.7326×10^{1}	nm^3 — normal cu meters (0° C, 760 mm Hg.)
gallon (U.S.)	2.38×10^{-2}	4.2×10^{1}	barrels
	3.7854×10^{-3}	2.6417×10^{2}	m^3 — cubic meters
	3.7854	2.6417×10^{-1}	l — liters
gallon (U.S.)	8.327×10^{-1}	1.2010	gallon (Imperial)
barrels (U.S.)	1.590×10^{-1}	6.290	m^3 — cubic meters
	1.590×10^{2}	6.290×10^{-3}	l — liters
ft^3/sec	$1.0194 \times 10^{+2}$	9.810×10^{-3}	m^3/hr — cubic meters/hour
	4.488×10^{2}	2.228×10^{-3}	U.S. gpm
ft^3/min	1.6990	5.886×10^{-1}	m^3/hr — cubic meters/hour
ft^3/hr	1.247×10^{-1}	8.020	gpm
mscf/day (60° F — atm)	1.1163	8.958×10^{-1}	Nm^3/hr — normal m^3/hr (0° C — 760 mm Hg)

NOTE: 4° C = 39.2° F, 4° C is a common cgs reference temperature

Reading From Left To Right:

UNITS "A"	MULTIPLIED BY	OR	DIVIDED BY	EQUALS UNITS "B"

Reading From Right To Left:

EQUALS UNITS "A"	DIVIDED BY	OR	MULTIPLIED BY	UNITS "B"

VOLUME AND VOLUME FLOW (Continued)

macf/day	1.1799	8.4753×10^{-1}	m^3/hr — cubic meters/hour
	5.1948	1.925×10^{-1}	gpm
U.S. gpm	2.271×10^{-1}	4.4033	m^3/hr — cubic meters/hour
	3.4286×10^{1}	2.9166×10^{-2}	b/d — barrels/day
b/d — barrels/day	6.6250×10^{-3}	1.5094×10^{2}	m^3/hr — cubic meters/hour

MASS FLOW AND MASS VELOCITY

lb/hr (lb av)	1.0714×10^{-2}	9.3333×10^{1}	long tons/day (FPS System)
	1.2000×10^{-2}	8.3333×10^{1}	short tons/day (U.S.)
	1.0886×10^{-2}	9.1859×10^{1}	metric tons/day
	4.5359×10^{-1}	2.20462	kg/hr — kilograms/hour
lb mole/hr	4.5359×10^{-1}	2.20462	k mol/hr
kg/hr	2.3621×10^{-2}	4.2335×10^{1}	long tons/day (FPS System)
	2.6455×10^{-2}	3.7799×10^{1}	short tons/day (U.S.)
	2.4000×10^{-2}	4.1667×10^{1}	metric tons/day
long tons/day * (FPS System)	1.120	8.929×10^{-1}	short tons/day
	1.016	9.842×10^{-1}	metric tons/day
short tons/day * (FPS System)	9.072×10^{-1}	1.1023	metric tons/day
lb mass/hr ft^2	4.882	2.0482×10^{-1}	kg/m^2hr
lb mass/hr ft	1.488	6.720×10^{-1}	kg/m hr

PRESSURE

lb/in^2	6.8046×10^{-2}	1.4696×10^{1}	atm — atmosphere
	7.0307×10^{-2}	1.4223×10^{1}	kg/cm^2 — kilograms/sq.cm.
	6.8946×10^{1}	1.4504×10^{-2}	millibar
	2.036	4.9116×10^{-1}	inches Hg (32° F)
	2.042	4.8971×10^{-1}	in Hg (60° F)
	5.1714×10^{1}	1.9337×10^{-2}	mm Hg or Torr (32° F) (millimeters of mercury)
	2.7681×10^{1}	3.6127×10^{-2}	inches H_2O (39.2° F)
	2.3067	4.335×10^{-1}	ft of H_2O (39.2° F)
	7.0309×10^{-1}	1.4223	m of H_2O — m. water col. (39.2° F)
	6.894	0.14505	kilo pascals

* 1 long ton = 2240 lbs, 1 short ton = 2000 lbs.

Reading From Left To Right:

UNITS "A"	MULTIPLIED BY	OR	DIVIDED BY	EQUALS UNITS "B"

Reading From Right To Left:

EQUALS UNITS "A"	DIVIDED BY	OR	MULTIPLIED BY	UNITS "B"

PRESSURE (Continued)

lb/ft^2	4.8824	2.048×10^{-1}	kg/m^2 — kilograms sq.m. mm of H_2O — mm water col.
kg/cm^2	9.678×10^{-1}	1.0332	atm — atmosphere
	9.8066×10^{2}	1.0197×10^{-3}	millibar
	2.8959×10^{1}	3.4532×10^{-2}	inches Hg (60° F)
	7.3556×10^{2}	1.3595×10^{-3}	mm Hg or Torr (32° F)
	3.9370×10^{2}	2.540×10^{-3}	inches H_2O (39.2° F)
	3.2808×10^{1}	3.0479×10^{-2}	ft of H_2O (39.2° F)
	10.0	1.0×10^{-1}	m of H_2O — m water col.
atm	1.0131×10^{3}	9.869×10^{-4}	millibar
	2.9921×10^{1}	3.3421×10^{-2}	inches Hg (32° F)
	7.600×10^{2}	1.3158×10^{-3}	mm Hg or Torr (32° F)
	3.3900×10^{1}	2.950×10^{-2}	ft of H_2O (39.2° F)
	1.0333×10^{1}	9.678×10^{-2}	m of H_2O — m water col. (39.2° F)
inches Hg (32° F)	1.3595×10^{1}	7.3554×10^{-2}	inches H_2O (39.2° F)
	1.1330	8.8265×10^{-1}	ft of H_2O (39.2° F)
	3.4533×10^{-1}	2.8958	m of H_2O — m water col. (39.2° F)
mm Hg or Torr (32° F)	5.3526×10^{-1}	1.8683	inches H_2O (39.2° F)
	4.4605×10^{-2}	2.2419×10^{1}	ft of H_2O (39.2° F)
	1.3596×10^{-2}	7.3554×10^{1}	m of H_2O — m water col. (39.2° F)

NOTE: In Europe — One physical atm = 1.0332 kg/cm^2 = 14.696 psi = 760 mm Hg.
One technical atm = 1.0000 kg/cm^2 = 14.223 psi = 735.559 mm Hg.

DENSITY, SPECIFIC GRAVITY, CONCENTRATION
PRESSURE DROP PER UNIT LINE LENGTH

lb/ft^3	1.601846×10^{-2}	6.2428×10^{1}	kg/dm^3
	1.601846×10^{1}	6.2428×10^{-2}	kg/m^3
	1.600342	6.2487×10^{1}	spec. grav. t/H_2O at 60° F
	1.60189×10^{-2}	6.2427×10^{1}	spec. grav. t/H_2O at 4° C kg/liter
	1.3368×10^{-1}	7.4805	lb/U.S. gallon
	5.6146	1.7811×10^{-1}	lb/barrel (U.S.)
lb/U.S. gallon	1.1983×10^{-1}	8.3454	kg/dm^3
lb/barrel (U.S.)	2.8530×10^{-3}	3.5051×10^{2}	kg/dm^3
spec. grav. t/H_2O at 4° C	1.0	1.0	kg/liter
spec. grav. t/H_2O at 60° F	9.99034×10^{-1}	1.000967	kg/liter spec. grav. t/H_2O at 4° C

Reading From Left To Right:

UNITS "A"	MULTIPLIED BY	OR	DIVIDED BY	EQUALS UNITS "B"

Reading From Right To Left:

EQUALS UNITS "A"	DIVIDED BY	OR	MULTIPLIED BY	UNITS "B"

DENSITY, SPECIFIC GRAVITY, CONCENTRATION PRESSURE DROP PER UNIT LINE LENGTH (Continued)

		for solids	
spec. grav. 60° F/H_2O at 60° F	0.9991	1.0009	spec. grav. 15° C/H_2O at 4° C*
		for liquids	
	0.9996	1.0004	
spec. grav. t/4° C	0.999973	1.000027	kg/dm^3
spec. grav. t/60° F	0.999023	1.000977	kg/dm^3
grains/100 scf	2.4188×10^1	4.13428×10^{-2}	milligram/Nm^3
grains/100 scf	$\frac{5.4238 \times 10^2}{MWgas}$	1.8437 x 10-3 x MWgas	PPM (milligrams/kg)
			ft of H_2O/100 ft (39.2° F)
psi/100 ft	2.3067	4.335×10^{-1}	m of H_2O/100 m
			kg/cm^2/1000 m

DENSITY OF GASES

$$\text{DENSITY} = \frac{\text{Mol. weight x pressure}}{\text{constant x temp. x compress. factor}}$$

density	constant	$\frac{1}{\text{constant}}$				pressure	temperature x Compressibility Factor
lb/ft^3	10.731	9.3188×10^{-2}	x	MW	x	psia	— °Rz
lb/ft^3	7.3020×10^{-1}	1.36945	x	MW	x	atm abs	— °Rz
kg/cm^3	8.4788×10^{-2}	1.1794×10^1	x	MW	x	kg/cm^2 abs	— °Kz

VISCOSITY

cp-centipoise = 0.01 gm/cm-sec	1.0×10^{-3}	1.0×10^3	kg/m sec
	3.6000	2.7778×10^{-1}	kg/m hr
	6.7197×10^{-4}	1.4882×10^3	lb/ft sec
	2.4191	4.1338×10^{-1}	lb/ft hr
	2.08854×10^{-5}	4.788026×10^4	lb mass-sec/ft^2
	7.51878×10^{-1}	1.33001×10^1	lb mass-hr/ft^2
cs-centistoke = 0.01 cm^2/sec	1.0×10^{-6}	1.0×10^6	m^2/sec
	3.6×10^{-3}	2.7778×10^2	m^2/hr
	1.0764×10^{-5}	$9.2903 \times 10^{+4}$	ft^2/sec
	3.8750×10^{-2}	2.5807×10^1	ftf^2/hr

* Depends on coefficient of thermal expansion.

Reading From Left To Right:

UNITS "A"	MULTIPLIED BY	OR	DIVIDED BY	EQUALS UNITS "B"

Reading From Right To Left:

EQUALS UNITS "A"	DIVIDED BY	OR	MULTIPLIED BY	UNITS "B"

HEAT AND THERMAL PROPERTIES*

BTU	2.5200×10^{-1}	3.9683	Kcal
BTU/lb	5.556×10^{-1}	1.8000	Kcal/kg
BTU/scf	9.4060	1.0632×10^{-1}	Kcal/Nm^3
BTU/acf	8.89992	1.1237×10^{-1}	Kcal/m^3
BTU/1b °F	1.0000	1.0000	Kcal/kg °C
BTU/hr °F	1.4882	6.7195×10^{-1}	Kcal/m hr °C
BTU in/hr ft^2 °F	1.2402×10^{-1}	8.0632	Kcal/m hr °C
BTU/hr ft^2	2.7125	3.6866×10^{-1}	Kcal/m^2 hr
BTU/hr ft^2 °F	4.8825	2.0481×10^{-1}	Kcal/m^2 hr °C

*The above values are based on the I. T. Calorie (International Steam Table Calorie).

scf: at 60° F and 1 atm (29.92″ Hg), dry
Nm^3: at 0° C and 760 mm Hg, dry
acf: actual cubic feet

WORK AND POWER

BTU (I. T.)	7.7817×10^{2}	1.2851×10^{-3}	ft — lb
	2.9307×10^{-4}	3.4121×10^{3}	kw-hr — kilowatt hour
kw-hr	2.6552×10^{6}	3.7662×10^{-7}	ft-lb (force)
	3.6710×10^{5}	2.7241×10^{-6}	kg m
kcal	1.1630×10^{-3}	8.5985×10^{2}	kw-hr
	4.2693×10^{2}	2.3423×10^{-3}	kg m
ft-lbs/sec	1.3826×10^{-1}	7.2327	kg m/sec
hp (FPS System)	5.50×10^{2}	1.81818×10^{-3}	ft-lb/sec
	1.0139	9.8632×10^{-1}	ps or cv or pk (metric horsepower)
	7.457×10^{-1}	1.3410	kw — kilowatt
hp (Metric)	7.500×10^{1}	1.3333×10^{-2}	kg m/sec
	7.355×10^{-1}	1.3596	kw
kw	1.00×10^{3}	1×10^{-3}	Newton meter/sec or watt
	1.0197×10^{2}	9.8068×10^{-3}	kg m/sec
	7.3753×10^{2}	1.3558×10^{-3}	ft-lb/sec
BTU/hr (I. T.)	3.9301×10^{-4}	2.544×10^{3}	hp (FPS System)
	3.9846×10^{-4}	2.510×10^{3}	metric horsepower
	2.9307×10^{-4}	3.412×10^{3}	kw
kcal/hr	1.5596×10^{-3}	6.4119×10^{2}	hp (FPS System)
	1.5812×10^{-3}	6.3243×10^{2}	metric horsepower
	1.1630×10^{-3}	8.5985×10^{2}	kw

Reading From Left To Right:

UNITS "A"	MULTIPLIED BY	OR	DIVIDED BY	EQUALS UNITS "B"

Reading From Right To Left:

EQUALS UNITS "A"	DIVIDED BY	OR	MULTIPLIED BY	UNITS "B"

MOLAL VOLUMES — MOLAL GAS CONSTANTS

lb mole	3.7946×10^{2}	2.6353×10^{-3}	scf (60° F; 1 atm)
lb mole	3.5902×10^{2}	2.7854×10^{-3}	cf (32° F; 1 atm)
lb mole	10.167	9.836×10^{-2}	Nm^3
lb mole/hr	9.107	1.0980×10^{-1}	mscf/day
K mol	8.3660×10^{2}	1.1953×10^{-3}	scf (60° F; 1 atm)
K mol	7.9150×10^{2}	1.2633×10^{-3}	ft^3 (32° F; 1 atm)
K mol	2.2414×10^{1}	4.4616×10^{-2}	Nm^3
K mol/hr	2.0078×10^{1}	4.980×10^{-2}	mscf/day

pv = ZRT; Pressure x Molal Volume = Molal Gas Constant x Temp.

pressure	molal volume	$\frac{1}{\text{constant}}$	constant	temp.
psia	ft^3/lb mole	9.3188×10^{-2}	10.731	°Rankine
atm abs	ft^3/lb mole	1.36949	7.032×10^{-1}	°Rankine
atm abs	m^3/k mol	1.2187×10^{1}	8.2056×10^{-2}	°Kelvin
kg/cm^2 abs	m^3/k mol	1.1795×10^{1}	8.4783×10^{-2}	°Kelvin

Note:
°Kelvin = (°Centigrade + 273°)
°Rankine = (°Fahrenheit + 460°)

VOLUME FLOW TO MASS FLOW RELATIONSHIPS

$$\text{Volume Flow} = \frac{\text{Mass Flow}}{\text{Constant x Density,}} \text{ for Vapors}$$

$$\text{Volume Flow} = \frac{\text{Mass Flow}}{\text{Constant x Spec. Grav.,}} \text{ for Liquids}$$

volume flow	constant	$\frac{1}{\text{constant}}$		mass flow		density
gpm	5.007×10^{2}	2.0×10^{-3}	x	lb/hr	—	spec. grav.
gpm	8.021	1.25×10^{-1}	x	lb/hr	—	lb/ft^3
mcf/day	2.601×10^{3}	3.84×10^{-4}	x	lb/hr	—	spec. grav.
mcf/day	4.17×10^{1}	0.024	x	lb/hr	—	lb/ft^3
Barrels/sec	1.262×10^{6}	7.924×10^{-7}	x	lb/hr	—	spec. grav.
BPSD	1.461×10^{1}	6.86×10^{-2}	x	lb/hr	—	spec. grav.
liters/min	1.323×10^{2}	7.56×10^{-3}	x	lb/hr	—	spec. grav.
liters/min	6.0×10^{1}	1.67×10^{-2}	x	kg/hr	—	spec. grav.
m^3/hr	2.2046×10^{3}	4.536×10^{-4}	x	lb/hr	—	spec. grav.
m^3/hr	3.531×10^{1}	2.83×10^{-2}	x	lb/hr	—	lb/ft^3
m^3/hr	1000	10^{-3}	x	kg/hr	—	spec. grav.

Reading From Left To Right:

UNITS "A"	MULTIPLIED BY	OR	DIVIDED BY	EQUALS UNITS "B"

Reading From Right To Left:

EQUALS UNITS "A"	DIVIDED BY	OR	MULTIPLIED BY	UNITS "B"

$$\text{Volume Flow} = \frac{\text{Mass Flow x Temp.}}{\text{Press.}} \text{x Compr. Factor x Constant}$$

volume flow	$\frac{1}{\text{constant}}$	constant		mass flow		temp.		pressure
ft^3/hr	9.3188×10^{-2}	10.731	x	lb mole/hr	x	°R x Z	—	psia
	1.3695	7.302×10^{-1}	x	lb mole/hr	x	°R x Z	—	atm abs
Macf/day	3.8828	2.5754×10^{-1}	x	lb mole/hr	x	°R x Z	—	psia
	5.7060×10^{1}	1.7525×10^{-2}	x	lb mole/hr	x	°R x Z	—	atm abs
m^3/hr	1.1795×10^{1}	8.4783×10^{-2}	x	k mol/hr	x	°K x Z	—	kg/cm^2 abs

Note: acf = actual cubic feet

HYDRAULIC POWER TO FLOW — HEAD RELATIONSHIPS

$$N = \frac{\text{Mass Flow x Head}}{\text{Constant}} = \frac{\text{Mass Flow x Diff. Press.}}{\text{Constant x Density}}$$

$$N = \frac{\text{Vol. Flow x Diff. Press.}}{\text{Constant}} = \frac{\text{Vol. Flow x Head x Density}}{\text{Constant}}$$

power	constant	$\frac{1}{\text{constant}}$	flow		head diff. press.		density
hp (FPS System)	$3.954 \times 10^{+3}$	2.53×10^{-4}	gpm	x	ft of liq.	x	spec. gr.*
	2.47×10^{5}	4.05×10^{-6}	gpm	x	ft of liq.	x	lb/ft^3
	1.714×10^{3}	5.83×10^{-4}	gpm	x	psi		—
	5.88×10^{4}	1.7×10^{-5}	barrels/day	x	psi		—
	1.98×10^{6}	5.05×10^{-7}	lb/hr	x	ft of liq.		—
	8.58×10^{5}	1.165×10^{-6}	lb/hr	x	psi	—	spec. grav
	1.375×10^{4}	7.2727×10^{-5}	lb/hr	x	psi	—	lb/ft^3
kw	5.302×10^{3}	1.88×10^{-4}	gpm	x	ft of liq.	x	spec. gr.*
	3.31×10^{5}	3.021×10^{-6}	gpm	x	ft of liq.	x	lb/ft^3
	2.299×10^{3}	4.35×10^{-4}	gpm	x	psi	x	—
	7.885×10^{4}	1.27×10^{-5}	barrels/day	x	psi		—
	2.655×10^{6}	3.77×10^{-7}	lb/hr	x	ft of liq.		
	1.15×10^{6}	8.69×10^{-7}	lb/hr	x	psi	—	spec. gr.
	1.844×10^{4}	5.42×10^{-5}	lb/hr	x	psi	—	lb/ft^3

* ft. or meters of liquid x spec. grav. = ft. or meters of water column.

Reading From Left To Right:

UNITS "A"	MULTIPLIED BY	OR	DIVIDED BY	EQUALS UNITS "B"

Reading From Right To Left:

EQUALS UNITS "A"	DIVIDED BY	OR	MULTIPLIED BY	UNITS "B"

HYDRAULIC POWER TO FLOW — HEAD RELATIONSHIPS (Continued)

power	constant	$\frac{1}{\text{constant}}$	flow		head diff. press.		density
	3.67×10^{2}	2.72×10^{-3}	m^3/hr	x	m liq. col.	x	spec. gr.*
	3.67×10^{1}	2.72×10^{-2}	m^3/hr	x	kg/cm^2		—
	3.67×10^{5}	2.72×10^{-6}	kg/hr	x	m liq. col.		—
	3.67×10^{4}	2.72×10^{-5}	kg/hr	x	kg/cm^2	—	spec. gr.
kw	8.810×10^{3}	1.135×10^{-4}	metric tons/day	x	m liq. col.		—
	8.810×10^{2}	1.135×10^{-3}	metric tons/day	x	kg/cm^2	—	spec. gr.
	2.7×10^{2}	3.70×10^{-3}	m^3/hr	x	m liq. col.	x	spec. gr.*
	2.7×10^{1}	3.7×10^{-2}	m^3/hr	x	kg/cm^2		—
metric	2.70×10^{5}	3.70×10^{-6}	kg/hr	x	m liq. col.		—
horse	2.70×10^{4}	3.7×10^{-5}	kg/hr	x	kg/cm^2	—	spec. gr.
power	6.480×10^{3}	1.54×10^{-4}	metric tons/day	x	m liq. col.		—
	6.48×10^{2}	1.54×10^{-3}	metric tons/day	x	kg/cm^2	—	spec. gr.

* ft. or meters of liquid x spec. grav. = ft. or meters of water column.

Index

D

E

F

G

H

I

J

K

L

M

U

V

W

X,Y

S0-ASK-421

U.S. AIR FORCE

A COMPLETE HISTORY

U.S. AIR FORCE

A COMPLETE HISTORY

Lieutenant Colonel Dik Alan Daso, USAF (Ret)
Curator of Modern Military Aircraft
National Air and Space Museum

THE AIR FORCE HISTORICAL FOUNDATION
Smithsonian National Air and Space Museum, Washington, D.C.

BEAUX ARTS EDITIONS

THE AIR FORCE HISTORICAL FOUNDATION

Page 2: *U.S. Air Force commands air and space for American military operations. (USAF)*

The Air Force Historical Foundation, established in 1953, is an independent, 501 (c)(3) nonprofit organization dedicated to the preservation, perpetuation, and publication of the history and traditions of American military aviation. The Foundation emphasizes the history and traditions of the United States Air Force, its predecessor organizations dating to the Aeronautical Division of the United States Army Signal Corps in 1907, and the men and women whose lives and dreams were devoted to flight.

Past presidents of the Foundation read like a Who's Who of air power pioneers and leaders and include General Carl A. Spaatz, General Hoyt S. Vandenberg, Major General Benjamin D. Foulois, General Curtis E. LeMay, and General Bernard A. Schriever. The logo of the Foundation, a Wright aircraft designated Signal Corps No. 1, was the first aircraft accepted by the United States Army in 1909.

The Foundation's activities include the quarterly air power journal *Air Power History*; a series of biographies of aviation greats; an annual awards program to recognize outstanding United States Air Force Academy and Air Force Reserve Officers Training Corps cadets and active duty USAF and Royal Air Force officers; and biennial symposia on diverse aspects of aerospace power of special historical interest and significance.

Individuals may learn more about the Air Force Historical Foundation and become a Foundation member at www.afhistoricalfoundation.org.

The Air Force Historical Foundation
Post Office Box 790
Clinton, Maryland 20735-0790
www.afhistoricalfoundation.org
e-mail: execdir@afhistoricalfoundation.org
(301) 736-1959; fax: (301) 981-3574

Beaux Arts Editions
Published by Universe Publishing
A Division of Rizzoli International Publications, Inc.
300 Park Avenue South
New York, NY 10010
www.rizzoliusa.com

Project Editor: Melissa C. Payne
Series Editor: James O. Muschett
Designer: Charles J. Ziga
Copy Editor: Deborah T. Zindell

ISBN-13: 978-0-88363-453-0

Printed in China

All photography and illustrations courtesy U.S. Air Force unless otherwise credited.

CONTENTS

PREFACE

For the Members of the World's Finest Air Force—
Past, Present, and Future

This book is made up of three different elements arranged in chronological order. The first element is a chronicle of events and achievements of individuals and organizations that forged the independent United States Air Force in 1947, and then continued the rich traditions already established as a branch within the U.S. Army into the twenty-first century. The second element is a collection of short essays written by more than thirty scholars—experts in specific air and space history fields—which expand upon specific events, aircraft, technologies, or people throughout each chapter. These essays follow related chronological entries and consist of two pages placed face to face with a slightly different background design so that they might be easily identified. The third element is the introductory narrative, which is designed to provide a broad contextual background for each chapter. Throughout the book are laced more than one thousand illustrations including works of art, photographs, and artifact images. It is hoped that these illustrations will help the reader to better visualize and appreciate both the technology of flight and the vast range of emotions surrounding flying activities—in the air and on the ground.

This volume makes every attempt to include elements of aerospace technology as well as the human side of the Air Force experience. This "Complete History" is intended to be a thorough but not an all-inclusive work. While no book about the U.S. Air Force can include a detailed history of every unit, weapon system, or Airman throughout time, this three-dimensional approach provides a summary that highlights pivotal events and influential individuals that have shaped America's primary air arm. Surprisingly, the history of the United States Air Force begins well before the Wright Flyer skimmed across the sands of Kitty Hawk in 1903.

In only one century, the invention of the airplane has revolutionized the way daily lives are lived. People travel routinely not only from coast to coast but from continent to continent. Mail delivery that once took a week or more to travel from New York to California today takes only hours. Countless lives have been saved by swift evacuation of injured by medical helicopters or medical aircraft that are equipped to treat victims even while flying to distant hospitals. A world once separated by distance has over the last century become a world separated only by time. Overall, it is a much smaller place.

In that same century, the airplane has changed warfighting from traditional linear ground battles between determined surface combatants, to fluid three-dimensional combat zones where land armies and surface naval fleets that do not enjoy the protection of a superior air force are never truly safe from attack—from land, sea, or air forces.

The U.S. Air Force is unique among the services because of its special relationship to the U.S. Army. Until 1947, America's "Air Force" was part of America's Army. Although attempts to separate the two had been initiated before the First World War, there were those Airmen who did not support such a split. Little money was available, especially during the Great Depression, and creating a new administrative structure was time consuming and costly. As the development of aviation technology allowed a more significant role for airplanes in war, the responsibility to employ the new aerial weapon with skill, knowledge, and courage also grew.

As long-range strategic bombers like the B-17 and B-24 left the drawing board and became reality, airpower employment became a specialty that only trained Airmen could handle—both in the air and on the ground. In essence, the creation of the independent U.S. Air Force on September 18, 1947, was a direct result of the creation of a specialty field—strategic, long-range, offensive airpower employment. Strategic bombardment was so drastically beyond the capabilities of traditional land and naval forces that establishing an independent military branch to accomplish that mission was a natural evolution. Forged in global combat during the Second World War, America's Air Force was ready for the independence that had been sought by Billy Mitchell, Frank Andrews, and Hugh Knerr, and achieved by Hap Arnold, George Kenney, Tooey Spaatz, and Jimmy Doolittle.

This book is a chronicle of the dynamic changes in aviation technology, the significant contributions made by aviators, and the moments when Airmen and their machines met and shaped the history of America's Air Force.

—Lieutenant Colonel Dik A. Daso, USAF (Ret)

TAKING FLIGHT

1783–1914

TAKING FLIGHT

1783–1914

The legend of Icarus and Daedalus has long stood as a metaphor for manned flight. Daedalus, imprisoned on the Isle of Crete for committing murder, had, while in captivity, fathered a son named Icarus. Eventually, Daedalus and Icarus were imprisoned in the Labyrinth, a maze that Daedalus had designed to keep the Minotaur from devouring humans. Daedalus devised a way to escape from the Labyrinth by fashioning wings from feathers and wax. After affixing the wings to Icarus' back, Daedalus warned his son not to fly too close to the ocean and not too near the sun so that the wings would not become wet nor would they melt from the sun's heat. Icarus, young and full of life, was thrilled with the freedom the wings had provided him. He soared and looped and then flew too high where the sun melted the binding wax. He fell into the sea while Daedalus successfully escaped to Sicily.

Even though dreams of flying predate the myth of Daedalus and Icarus, the inherent element of technology and the thrill and the danger of soaring above landlocked humans apparent in that tale still grips us today.

Mankind first left the earth in balloons filled with heated air in 1783. Soon after, hydrogen gas was used to lift early aeronauts skyward. Almost immediately military applications for balloons became obvious. In Europe, balloonists reconnoitered enemy positions as early as 1794.

By the 1860s, balloons had developed into a viable tool for observing enemy positions and troop movements. During the American Civil War, both Confederate and Union Armies used balloons for observation. The Union expanded their balloon force to a total of seven balloons under the direction of Thaddeus Lowe. Lowe's balloons delivered reconnaissance during several battles between 1861 and 1863, including intelligence gathered at the Battle of Fair Oaks that turned the tide of the fighting and assured a Union victory. When General George McClellan was relieved of command in 1863, enthusiasm for the weapon waned. Lowe, not a very good financial manager, resigned from the balloon corps in 1863, and the corps was disbanded shortly after his departure.

Yet balloons were not completely predictable in their performance. They were at the mercy of prevailing winds, very hard to conceal, and easy targets for other air machines during the Great War.

It was the invention of the heavier-than-air, self-propelled, controllable "aeroplane" by two American bicycle makers in 1903 that truly began the march toward the realization of the potential for manned flight.

The real miracle of the Wrights' aircraft was not initially apparent. For the first few years, they could only fly a straight path and for just a few short minutes. It was only after the brothers had perfected the ability to turn their airplane in flight that their achievement became a world-altering invention. Control in three dimensions made all the difference. It was the ability to control all axes of orientation from the pilot's seat that made possible all that developed in aviation technology after 1903.

These early years of flight were experimental, thrilling, dangerous, and transformational. Mankind had left the surface of the earth and floated skyward like a great winged creature. The invention of the airplane fulfilled the dreams of past centuries and would inspire those looking skyward for generations to come.

Pages 8–9: *Large crowds gather to witness the flight trials of America's first military airplane—the Wright Military Flyer. ("The Wright Brothers at Fort Myer—July 30, 1909," John T. McCoy, U.S. Air Force Art Collection)*

Above: *On December 17, 1903, the Wright Flyer took flight at Kitty Hawk, North Carolina, changing the world forever. (NASM)*

Right: *This photograph that includes Lieutenant Benjamin Foulois (center), Wilbur Wright (holding flag), and Orville Wright (right) was one of the images that provided inspiration for John McCoy's painting on the previous spread. (USAF)*

TAKING FLIGHT

1783–1914

"Not within a thousand years would man ever fly."
—Wilbur Wright, 1901

1783

19 September
Joseph and Etienne Montgolfier, sons of a paper manufacturer in Annoy, France, send the first living creatures aloft in a hot air balloon. The sheep, rooster, and duck return to earth safely after only a few minutes in the air over the palace at Versailles, France.

15 October
Frenchman François Pilâtre de Rozier becomes the first human to ascend in a balloon—the world's first aeronaut. Designed by the Montgolfier brothers, the balloon is filled with heated air created by burning straw and wool upon a fire grate mounted beneath the silk balloon. Rozier is able to remain airborne for more than four minutes at the full 80-foot extension of the tether.

21 November
The first hot air balloon free ascent is successfully accomplished from the grounds of the Royal Palace of La Muette (present-day Paris) when de Rozier and the Marquis d'Arlandes persuade King Louis to allow them to make the flight. The balloon flies across the Seine River and then over the city of Paris, covering 25 miles. The flight is witnessed by many, including American Benjamin Franklin.

1 December
The first free flight in a hydrogen balloon is made by Jacques Alexandre César Charles and M. Robert. They ascend and then drift a distance of 27 miles from Les Tuileries, Paris, to Nesles.

1784

4 June
The first free flight in a hot air balloon by a woman is made by Madame Elisabeth Thible at Lyon, France.

24 June
The first American to fly aloft on a tethered balloon is a 13-year-old Baltimore youth by the name of Edward Warren. He is launched on his adventure by Peter Carnes, a lawyer and tavern keeper from Bladensburg, Maryland.

Left: *The Montgolfier balloon first carried man aloft in 1783. (NASM Art Collection)*

Opposite: *Union troops use hydrogen generators to fill one of Thaddeus S. C. Lowe's observation balloons used during the Battle of Fair Oaks. (USAF)*

1785

7 January
J. P. F. Blanchard, a Frenchman, and Dr. John Jefferies, an American loyalist living in England since the War for American Independence, make the first flight across the English Channel.

1793

9 January
Frenchman Jean Pierre Blanchard makes the first balloon ascension in the United States from the Walnut Street Prison in Philadelphia, Pennsylvania (then the nation's capital), to Gloucester County, New Jersey. This balloon is filled not with hot air, but with hydrogen gas. The flight lasts just over 45 minutes.

1819

2 August
Charles Guille becomes the first man to parachute jump from a balloon in the United States. The jump is made from an altitude of approximately 8,000 feet over Long Island, New York.

1861

18 June
American aeronaut Thaddeus S. C. Lowe sends the first ever aerial telegraph message from his balloon named *Enterprise.*

24 September
Balloonist Thaddeus Lowe ascends to an altitude of 1,000 feet across the Potomac River, near Washington, D.C., and helps to aim Union fire at Confederate troops.

1 October
The U.S. Army forms a Balloon Corps for the first time, consisting of five balloons and 50 men.

1862

16 April
Thaddeus S. C. Lowe, assigned to Union general Fitz-John Porter's division, ascends in a hydrogen balloon to reconnoiter Confederate positions. When his tether breaks, his craft begins to drift over enemy lines where he takes advantage and accomplishes detailed reconnaissance of the situation. Lowe allows the balloon to continue to ascend in hopes that prevailing winds might blow him back over to the Union side of the lines. As he ascends the winds shift and once back in friendly Union territory, he releases some of his hydrogen and lands safely among friends. Lowe continued with near-daily ascents through the months of May and June in and around Richmond, Virginia.

31 May–1 June
During the Battle of Fair Oaks (Seven Pines) a hydrogen balloon, the *Intrepid,* is used by the Union to observe enemy troop movements.

THE FIRST MILITARY AIRMEN

Tom Crouch

A great blue and gilt balloon, standing seventy feet tall, carried Pilâtre de Rozier and François Laurent into the air on the afternoon of November 21, 1783. For the first time in history, human beings left the ground in free flight. That evening, Benjamin Franklin, the leader of the American diplomatic mission in Paris, described the event in a letter to his friend, Sir Joseph Banks, the president of the Royal Society. The new invention, he suggested, would have considerable military reconnaissance value.

Franklin's prophecy began to come true just ten years later, when Louis-Bernard Guyton Morveau, scientific advisor to the Committee of Public Safety, suggested that the armies of revolutionary France be equipped with tethered observation balloons. Scientists Charles Coutelle and Nicholas Conté constructed the balloon *L'Entreprenant* with government funding and provided a demonstration that led French officials to create the 1st Compagnie d'Aérostiers, the world's first military aviation unit, on April 2, 1794.

The Corps d'Aérostiers grew to include three balloons by 1796 and saw service on several fronts. A unit participating in the Egyptian campaign of 1798 saw little action and lost their equipment in the Battle of Aboukir Bay. The record failed to impress Napoleon, who disbanded balloon operations following the return of the armies to France in 1799.

The notion of military observation balloons had begun to intrigue Americans as early as 1840, when the War Department considered making night ascents to locate Seminole Indian campfires in the Everglades. Six years later, John Wise, the most prominent of the antebellum American balloonists, suggested using a large balloon to bomb a Mexican fortress into submission.

When the Civil War broke out in the spring of 1861, a number of aeronauts who had built reputations as aerial showmen before the war stepped forward with proposals to form an observation balloon unit. One of these early volunteers, John La Mountain, operated a balloon in support of U.S. Army units stationed at Fort Monroe, near Hampton Roads, Virginia.

It was Thaddeus Sobieski Constantine Lowe, however, who succeeded in organizing a larger balloon corps. A 29-year-old New Hampshire man who had made national headlines in April 1861 with a 900-mile balloon voyage from Cincinnati, Ohio, to Unionville, South Carolina, Lowe arrived in Washington, D.C., two months later with his balloon, *Enterprise.* Smithsonian Secretary Joseph Henry introduced him to the Secretary of War and President Lincoln, and then arranged for him to make a demonstration flight from the spot where the National Air and Space Museum now stands. Lowe carried a telegrapher aloft with him and sent a message to the White House describing the view of the military camps circling Washington, "nearly fifty miles in diameter." By the end of June, he was making tethered observation flights from advanced federal positions near Falls Church.

Opposite: *Thaddeus Lowe ascends in the* Intrepid *to reconnoiter an ongoing Civil War battle. (USAF)*

Right: *Thaddeus Lowe makes a tethered observation flight near Washington, D.C. (*Lt. Lowe's Balloon Tethered at 1,000 Feet Above the York River in 1862 *by John McCoy, USAF Art Collection)*

action during the fighting on the Virginia Peninsula in the spring of 1862. Lowe made significant contributions to the survival, if not the victory, of Union forces at the Battle of Fair Oaks, one of the famous Seven Days engagements of that campaign.

As a civilian contractor, Lowe sometimes had difficulty arranging transportation and logistical support. As a result, the aeronauts missed the Battle of Antietam in September 1862. Members of the corps saw action during the Battle of Fredericksburg that December, and at Chancellorsville in early May 1863. Convinced that Army officials failed to appreciate the potential of his corps, Lowe resigned shortly thereafter when his request for increased funding was refused. The Union balloons continued to be operated by members of Lowe's corps for some months more, but would never again see action.

Later, the limited use of observation balloons during the Campaign in Cuba in 1898 did little to bolster military confidence in aerial operations. A balloon became trapped in the trees on July 1, 1898, drawing Spanish artillery fire down on U.S. troops who would attack Kettle and San Juan Hills later that day. While balloon activities were once again reduced following the Spanish-American War, the surviving aeronautical unit served as the mechanism for the introduction of both the powered airship and the airplane to the U.S. Army only a decade later.

During World War I, both the Allied and Central Powers employed sausage-shaped observation balloons tethered close to the front, spotting artillery and keeping an eye on movement behind the enemy lines. The intrepid airmen dangling beneath the hydrogen-filled balloons were equipped with parachutes to give them at least a chance of escaping the flames when enemy fighter planes attacked. More than a century after the first military aeronauts ventured aloft, the observation balloon had finally come into its own at the very moment when it was being replaced by photographic reconnaissance conducted by airplanes.

Left: *Otto Lilienthal died in an 1896 glider crash. Here, he pilots his 1895 glider without incident. (NASM)*

1867

16 April
Wilbur Wright is born near Millville, Indiana, to Reverend Milton Wright and the former Susan Catherine Koerner.

1871

19 August
Orville Wright is born in Dayton, Ohio, to Reverend Milton Wright and the former Susan Catherine Koerner.

1878

24 December
Charles DeForest Chandler is born in Cleveland, Ohio. A military balloonist, Chandler becomes the commander of the first Army aviation school in College Park, Maryland. In 1907, Chandler becomes the first Officer in Charge of the Aeronautical Division of the Army Signal Corps.

1879

9 December
Benjamin Delahauf Foulois is born in Washington, Connecticut. He becomes one of the first three military pilots taught to fly by the Wright Brothers.

1886

25 June
Henry Harley Arnold is born near Philadelphia, Pennsylvania. He enters the Military Academy at West Point in 1903 and later becomes one of the first military pilots in history.

1891

Samuel Pierpont Langley, an American physicist and the third secretary of the Smithsonian Institution, begins his study of heavier-than-air craft by constructing models that include steam-powered engines for propulsion. He publishes a book, *Experiments in Aerodynamics,* summarizing the results of his early model tests.

1896

6 May
Samuel P. Langley successfully flies Aerodrome No. 5, the first pilotless, engine-driven, heavier-than-air craft. He launches the plane by using a catapult mounted atop a houseboat anchored near Quantico, Virginia, where it flies approximately 3,300 feet and then 2,300 feet on consecutive attempts.

16 May
Orville and Wilbur Wright begin selling a bicycle of their own design—the Wright Special.

10 August
Manned glider pilot Otto Lilienthal dies after he crashes in an invention he created two days before. Previously, Lilienthal had successfully glided in biplane and monoplane gliders.

14 December
James Harold Doolittle is born in Alameda, California. "Jimmy," as he is known to all Americans, leads the famous raid against Tokyo on April 18, 1942, and rises to three-star rank and commander of the Eighth Air Force during World War II.

1897

12–14 August
Clément Ader attempts to fly a twin-engine airplane but is unsuccessful.

14 October
Clément Ader attempts flight once again, but fails again. His funding is withdrawn.

1899

30 May
Wilbur Wright sends a letter to the Smithsonian Institution seeking information about aeronautical matters.

June–December
The Wrights obtain information from a variety of sources concerning aeronautical experiments and national weather data.

1900

13 May
Wilbur Wright contacts Octave Chanute, author of *Progress in Flying Machines* (1894), an early technically oriented aeronautical book.

October
The Wrights fly their first glider on the sands of Kitty Hawk, North Carolina. During these test flights, stability and control are examined. The craft is flown as both a glider and as a manned aircraft.

1901

26–27 June
The Wright Brothers meet Octave Chanute when he visits Dayton, the home of the Wrights' bicycle shop.

9 July
The Wright 1900 Glider is smashed by near 100-mile per-hour winds at Kitty Hawk.

27 July
The Wright 1901 Glider flies for the first time.

1902

January
Wilbur Wright sends technical data from the brothers' wind tunnel experiments and photos of the device used to measure forces created in the wind tunnel to Octave Chanute.

Right: *The Wrights gathered valuable flight data by flying kites and gliders, like this 1902 manned glider flown on the sand dunes at Kitty Hawk, North Carolina. (NASM)*

Above: *Samuel Pierpont Langley's failed flying machine, the Aerodrome A, is prepared for launch. (NASM)*

Below: *Charles M. Manley piloted the attempts to launch the Aerodrome from the top of a houseboat afloat in the Potomac River near Washington, D.C. (NASM)*

19 September
Having returned to Kitty Hawk, the Wrights begin flights using a new, larger glider. The performance of the 1902 Glider had been calculated during previous experiments and proved to be accurate in flight.

December
Construction of the Wright Flyer four-cylinder engine begins. Experiments with propellers continue.

1903

12 February
Wrights test their airplane motor for the first time. It breaks the following day during further tests.

23 March
The Wright brothers file their first airplane patent in the U.S. It is based upon the Wright 1902 Glider successfully flown at Kitty Hawk the year before.

8 August
Samuel Pierpont Langley, the third secretary of the Smithsonian Institution, flies an exact-scale gasoline model miniature—one-quarter-scale—of his Aerodrome A, which he intends to launch from a houseboat on the Potomac River. This model includes an improved engine from an earlier model that flew two years before.

25 September
Orville and Wilbur Wright arrive at Kitty Hawk. The next day they begin repairs on the 1902 house and construct a new shed in which to build the 1903 Wright Flyer.

7 October
Samuel P. Langley's full-sized, 48-foot-wingspan Aerodrome is launched unsuccessfully by catapult built on a houseboat in the Potomac River. The Aerodrome is piloted by Charles M. Manly, who also built the 52-horsepower engine that powers the craft.

9 October
Wrights begin assembling the 1903 Flyer. Although it is essentially completed by November 5, cracks in the propeller shafts force delays until new ones can be shipped to Kitty Hawk from Dayton, Ohio.

8 December
Samuel P. Langley's airplane—the Aerodrome A—is catastrophically wrecked when it again unsuccessfully launches from its houseboat-based catapult. Charles M. Manley, again the pilot, is rescued from the icy Potomac River without injury after becoming trapped beneath the broken craft. No further attempts to fly the craft are ever made.

14 December
Wilbur Wright makes an unsuccessful attempt to fly the 1903 aircraft. The aircraft stalls immediately after takeoff, settling to the ground 105 feet from the launching point. Minor damage is quickly repaired. The flight is not considered either sustained or controlled.

17 December
At 10:35 a.m. Orville Wright makes the first controlled, sustained, power-driven free flight in a heavier-than-air machine. The flight lasts 12 seconds and covers 120 feet. Three additional flights are made during the day. The second flight, piloted by Wilbur, lasts about 12 seconds and covers 175 feet. The third, piloted by Orville, lasts 15 seconds and covers more than 200 feet. The fourth and final flight made by the 1903 Wright Flyer is piloted by Wilbur and lasts 59 seconds and traverses 852 feet. Shortly after the fourth flight, a gust of wind destroys the 1903 Flyer. It is further disassembled and shipped back to Dayton for storage.

Above, left: *Orville Wright, by Efrem Melik. (NASM Art Collection)*

Above, right: *Wilbur Wright, by Efrem Melik (NASM Art Collection)*

THE WRIGHT BROTHERS: FIRST WINGS

Peter L. Jakab

Wilbur and Orville Wright launched a new era in the history of humankind with their creation of the first successful, powered, heavier-than-air flying machine. The Wrights began serious experimentation in aeronautics in 1899, first flew a powered airplane on December 17, 1903, and perfected their craft by 1905. In this short period, with remarkable originality, they defined the essential elements of the problem, conceived creative technical solutions, and built practical mechanical design tools and components that resulted in a viable aircraft. They did much more than simply coax a machine off the ground. They established the fundamental principles of aircraft design that are still in place today.

Why Wilbur and Orville? On the surface, the fact that the Wrights did invent a successful airplane quickly and with little assistance would suggest that "sheer genius" had to have been at the core of their achievement. Probing deeper, however, it becomes apparent that there were a number of specific research techniques, innate conceptual skills, and personality traits that came together in a unique way that largely explain why these two people invented the airplane.

First and foremost, the Wright brothers' approach to mechanical flight was grounded in strict engineering techniques. Indeed, they not only invented the airplane, but they invented the practice of aeronautical engineering in the process. In particular, they pioneered techniques for using a wind tunnel in aircraft design that, in rudimentary terms, are still used today.

Merged with this basic engineering perspective and its associated practices were a number of conceptual capabilities and approaches present in the Wrights' method. Among the most important was their capacity for developing conceptual models of a problem that could then be transformed into practical hardware. The brothers' considerable ability for turning abstract ideas into workable machinery reveals itself over and over again in their aeronautical work.

The Wrights reinforced these innate talents with several sound approaches to technological innovation. They developed a series of gliders and powered airplanes that were based on a single, evolving basic design, modifying only a few factors at a time. Though seemingly an obvious approach, many of the Wrights' contemporaries jumped from one radical design to another.

Wilbur and Orville also understood that an airplane was not just one invention, but numerous inventions all working in concert to produce a workable flying machine. The Wrights' unwavering attention to the complete technological system of mechanical flight, including the pilot, was crucial to their success.

Although the Wrights solved a myriad of problems that made up the creation of a successful airplane, the method of control they devised was foremost. Unlike their predecessors, who attempted to control their craft by shifting bodyweight to alter the center of gravity, the Wrights chose to control their aircraft aerodynamically. For lateral control, they warped, or twisted, the wings such that one side was presented to the wind at a higher angle than the other, which generated differing amounts of lift on either side of the airplane. The pilot could easily and deftly balance or turn the airplane by manipulating the wings in this manner. (Modern airplanes use movable surfaces called ailerons to achieve this, but the principle is the same.) For climb and descent, a movable elevator was used to control the movement of the center of the lifting pressure relative to a fixed center of gravity of the airplane. If the center of

pressure (CP) was ahead of the center of gravity (CG), the airplane would climb, while if the CP was behind the CG, the airplane would descend. These techniques were ground-breaking because they enabled the pilot to effectively control the airplane in three-dimensional space, and because they were aerodynamically based, they did not limit the size and weight of the craft, as pilot weight-shifting did. All airplanes continue to use this basic system of control developed by the Wright brothers.

The Wrights' propeller design was also a standout of original thinking. They reasoned if a wing moving horizontally produces a vertical lift force, a similar airfoil-shaped structure turned on its side and spun to create the flow of air over the surface would produce a horizontal thrust; in essence, they envisioned a rotary wing. All aircraft propellers since are based on this revolutionary concept. It was one of the most creative aspects of their entire airplane.

The true significance of the Wright brothers' airplane was not merely that it was the first to fly, but that it embodied all the critical elements of every subsequent airplane. Far more than a technical curiosity, it was a seminal design that could evolve and be developed into what we have today. This is the legacy of the Wright brothers' aeronautical work.

Opposite: *Wilbur (far) and Orville (near) flying their 1901 Wright glider as a kite. (NASM)*

Above: *The Wright brothers, Orville (right) and Wilbur (left), flew together in the same airplane only once. On May 25, 1910, over Huffman Prairie near Dayton, Ohio, the men that had invented the airplane in 1903 shared the exhilaration of flight seated side-by-side. The 8' by 10' original painting hangs at Carillon Historical Park, Dayton, Ohio, and is dedicated to Dr. Tom Crouch, National Air and Space Museum. ("The Bishop's Boys," by Dean Mosher, courtesy of the artist)*

Left: *Thomas Baldwin pilots this early Signal Corps dirigible. (USAF)*

Opposite: *Aeronaut Charles Levée drifts away from the U.S. Military Academy at West Point in the 25-foot L'Alouette balloon on a frigid February morning in 1905. (Robert and Kathleen Arnold Collection)*

1904

April–May
The Wright Brothers build a stronger airplane including a more powerful engine that produces 18 horsepower, a 30 percent increase over their 1903 Flyer.

26 May
The Wrights make the first flight with their second aircraft. Although the original Wright Flyer made only four flights, this 1904 model would fly more than 100.

30 July
Modifications of the 1904 Wright Flyer are completed. The blade width of the propellers is increased and the radiator and gas tank are moved aft to improve stability.

3 August
The first successful circuit flight of a U.S. dirigible (navigable balloon) is flown near Oakland, California, piloted by Captain Thomas S. Baldwin.

13 August
Wilbur Wright flies 1,340 feet in just over 32 seconds shattering his record from the fourth flight at Kitty Hawk.

7 September
The Wrights use the first catapult launching system—a weight and derrick device—to assist in obtaining takeoff speed. Resembling an oil derrick, the device is used until wheels are added to the aircraft.

20 September
Until now, the Wrights only flew in straight lines. On this date, Wilbur makes one complete circle in the 1904 Flyer.

25 October
Captain Thomas S. Baldwin's airship, the *California Arrow*, flies in a circle during the St. Louis Exposition. The airship, piloted by Roy Knabenshue, becomes the first controllable, motor-driven lighter-than-air craft to fly a circular flight path in the United States.

9 November
Wilbur Wright flies for more than five minutes—the first to do so—and covers nearly three miles over Huffman Prairie near Dayton, Ohio.

1905

18 January
The Wright brothers conduct initial talks with the U.S. Government to negotiate a sale of the airplane to the Army. No agreements are reached.

11 February
L'Alouette, a 25-foot balloon piloted by Charles Levée, is launched from the siege battery at the Military Academy, West Point, New York. The launch is witnessed by Cadet Henry H. Arnold (class '07), who would later rise to command the Army Air Forces during World War II.

23 June
The 1905 Wright Flyer—also known as Wright Flyer III—makes its first flight at Huffman Prairie near Dayton, Ohio. This is the first fully controllable aircraft and is capable of turning, banking, and remaining airborne for up to 30 minutes.

5 October
Capping a series of long-duration flights, Orville Wright flies the Wright Flyer III 29 laps around their Huffman Prairie Flying Field, remaining airborne for nearly 40 minutes. The flight totals more than 24 miles, setting a world record for both distance and time aloft. The average speed attained on this flight is approximately 38 miles per hour.

9 October
In a letter to the U.S. Board of Ordnance and Fortification, the Wrights describe their airplane and offer the machine for sale.

27 October
The U.S. Board of Ordnance and Fortification turns down an offer made by the Wright Brothers to supply airplanes for observation purposes. They misunderstand the offer as a request for funds to conduct research on the aircraft.

30 November
The Aero Club of America is established.

1906

27 February
Samuel Pierpont Langley dies in Aiken, South Carolina.

2 March
Robert H. Goddard speculates that atomic energy might be used for space travel. His thoughts are preserved within the pages of his *Green Notebook.*

22 May
On their third attempt, the Wrights are issued their first U.S. Government patent on their flying machine—a three-axial, airplane-control system.

11 August
Mrs. C. J. Miller becomes the first American woman airship passenger. Her husband, Major Miller, pilots the craft.

September
Glenn H. Curtiss and Captain Thomas Baldwin, a U.S. Army balloonist, meet with the Wright brothers for the first time at their Dayton workshop.

30 September
Major H. B. Hersey and Lieutenant Frank Lahm win the first Gordon Bennett Balloon Race while piloting an Army balloon. They fly 648 kilometers from Paris to Searborough, England.

1907

4 June
Army Corporal Edward Ward is detailed to learn about balloon manufacturing techniques. Ward is the first noncommissioned officer in the Army's new balloon division.

14 June
Henry Harley Arnold, later known as Hap, graduates in the West Point class of 1907. Arnold would rise to the rank of five-star general during World War II.

1 August
Captain Charles DeForest Chandler takes command of the newly established Aeronautical Division, U.S. Army Signal Corps. The division is the first responsible for balloons and heavier-than-air craft. This division becomes the forerunner of the U.S. Air Force. The first military airplane will not be acquired by the division until August 1909.

1 October
The Aerial Experiment Association is formed by Dr. Alexander Graham Bell after a suggestion to do so by his wife. Along with Dr. Bell, association members are F. W. Baldwin, J. A. D. McCurdy, Glenn H. Curtiss, and Lieutenant Thomas Selfridge.

17 October
Signal Corps Balloon No. 10 piloted by Captain Charles DeForest Chandler and J. C. McCoy wins the Lahm Cup for ballooning. They fly nearly 475 miles in a little more than 20 hours from St. Louis, Missouri, to Walton, West Virginia.

7 November
The Army Signal Corps is given $25,000 to procure an airship by the Board of Ordnance and Fortification.

30 November
The first airplane company in America is formed by aviator Glenn H. Curtiss. The main factory is located near Hammondsport, New York.

5 December
Wilbur Wright offers the U.S. Government the opportunity to purchase an airplane that will carry two individuals for the price of $25,000. The Board of Ordnance and Fortification asks the Signal Corps to submit detailed specifications for such a purchase.

6 December
Towed by a motor boat, Lieutenant Thomas Selfridge soars above Bras d'Or Lake, Nova Scotia, in Cygnet I, a kite built by Dr. Alexander Graham Bell. Selfridge is aloft for seven minutes.

16 December
The Chief Signal Officer calls for bids for procurement of an airship (lighter-than-air).

23 December
The U.S. Army Signal Corps solicits bids for a heavier-than-air flying machine due to the Board of Ordnance and Fortification on February 1, 1908. The formal specification (Signal Corps Specification 486) is issued by Brigadier General James Allen, Chief Signal Officer.

Above: *Captain Charles Chandler ready to take off in a Wright aircraft at College Park, Maryland. (NASM)*

Below: *Cadet Henry Harley Arnold, U.S. Military Academy class of 1907. (Robert and Kathleen Arnold Collection)*

1908

28 January
Lieutenant Frank P. Lahm, accompanied by Henry W. Alden and J. G. Obermeier, fly a balloon from Canton, Ohio, to Oil City, Pennsylvania. The 100-mile trip takes two hours and 20 minutes.

8 February
The U.S. War Department accepts the Wrights' bid of $25,000 to furnish one heavier-than-air flying machine within 200 days. The machine is to carry two men, fly at a speed of 40 miles per hour, and weigh approximately 1,200 pounds. The formal contract is signed soon after specifying delivery of the airplane by August 28. Additional bids by J. F. Scott and A. M. Herring were also approved by the Secretary of War, but these men never produced any planes.

24 February
The U.S. Army contracts with Captain Thomas S. Baldwin to build a lighter-than-air ship for $6,750.

12 March
The Red Wing aircraft flies for the first time at Lake Keuka, Hammondsport, New York—320 feet on its first attempt. It is designed by Lieutenant Thomas Selfridge for the Aerial Experiment Association and is flown on this occasion by F. W. Baldwin.

4 April
Robert H. Goddard uses the term "jet propulsion" to describe a possible method of achieving space flight. In his notes he describes a crude combustion chamber and propulsion nozzle.

11 April
Lieutenant Frank P. Lahm takes over command of the Aeronautical Division, U.S. Army Signal Corps, Office of the Chief Signal Officer.

22 April
Captain Charles DeForest Chandler pilots a balloon in which Theodore Roosevelt, Jr., the President's son, and Captain Fitzhugh Lee, the President's military aide de camp, are airborne four hours and 30 minutes during an ascent that terminates in Delaware City.

Above: *Brigadier General James Allen, Chief Signal Officer. (US Army)*

Right: *Traditionally, the horse cavalry provided mobility and reconnaissance for the Army. Frank Lahm introduces his steed to the new military machine that would eventually replace the horse. (USAF)*

Left: *Lieutenant Thomas Selfridge became the first military man to fly solo in an aircraft. The White Wing featured hinged ailerons instead of the "wing warping" technique used on Wright aircraft. (NASM)*

Below: *Glenn Curtiss pilots his* June Bug *near Hammondsport, New York. Curtiss and the Wrights will argue over aircraft patents for many years. (NASM)*

6–14 May
The Wrights fly the 1905 Flyer for the first time after modifications made so that the pilot and passenger could sit upright rather than lie prone as all earlier models required. On the 14th, the first passenger flight occurs when Wilbur Wright takes one of his employees, Charles Furnas, into the sky as he prepares to meet the U.S. Army requirements to carry two men. These flights occur at Kill Devil Hill, the site of their first flight in 1903.

19 May
Lieutenant Thomas E. Selfridge becomes the first American military man to solo in an airplane. He flies an aircraft designed by F. W. Baldwin known as the White Wing, near Hammondsport, New York, future home of the Curtiss Aircraft Factory. The White Wing design features hinged ailerons rather than the Wright design, which uses "wing warping" for directional control.

31 May
Glenn Curtiss announces that his aircraft company located in Hammondsport, New York, is prepared to produce and deliver aircraft at the cost of $5,000 for delivery in 60 days or less.

21 June
Glenn Curtiss flies his first airplane, the *June Bug*, making him the third aviation pioneer in America. This is the third airplane built under the auspices of Dr. Bell's Aerial Experiment Association. Curtiss flies three hops during the day.

4 July
Glenn H. Curtiss pilots his *June Bug* airplane for more than one mile near Hammondsport, New York. Averaging 39 miles per hour on the one minute, 42-and-one-half-second flight, Curtiss wins the Scientific American trophy. This flight is the first "official" test flight of an aircraft made in America. Measurements

Right: *Orville Wright draws a sizeable crowd while orbiting Fort Myer during aircraft acceptance tests in 1908. (NASM)*

Below: *Lieutenant Thomas Selfridge is killed when a propeller on the 1908 Flyer breaks, severing important wing support wires while airborne. Orville, the pilot, suffers a broken hip but survives the disaster. (USAF)*

were made by a representative of the Federation Aeronatique Internationale (F.A.I.), the organization that still certifies aviation records.

17 July
In Kissimmee, Florida, a municipal ordinance is passed regulating aircraft use within city limits. This ordinance is the world's first legislation regulating aviation activities.

23 July
Thomas Baldwin delivers the Army's first airship for trials at Fort Myer, Virginia, in fulfillment of his government contract. Included in the deal is the hydrogen production plant needed to fill the airship.

4 August
Captain Baldwin's dirigible begins flight tests at Fort Myer, Virginia.

22 August
The Army accepts Captain Baldwin's dirigible (renamed Signal Corps Dirigible No. 1) after successful flight tests at Fort Myer, Virginia. It cost $6,750 and carries two men and a payload of 450 pounds.

3 September
Orville Wright flies at Fort Myer, Virginia, becoming familiar with the testing area where the U.S. Army trials will be held.

9 September
Breaking all endurance records, Orville Wright orbits Fort Myer for 57 minutes and 25 seconds at an altitude greater than 100 feet.

17 September
Lieutenant Thomas Etholen Selfridge is killed when the aircraft piloted by Orville Wright crashes at Fort Myer from an altitude of approximately 75 feet. One of the propellers splits and cuts one of the support wires for the aircraft rudder. Orville is critically injured in the crash and spends the next six months recuperating from a broken hip. Selfridge is buried at Arlington Cemetery four days later.

1909

The first edition of *Jane's All the World's Airships* is published at Exeter, England. Fred T. Jane is the book's editor. He changes the title of the book to *Jane's All the World's Aircraft* the next year.

12 January
A French company, General Aerial Navigation, pays the Wright Brothers $100,000 for rights to their patents as well as an investment interest in their airplane company.

31 January
The publication *New York World* offers a prize of $10,000 for any flight made from New York City to Albany, New York, during the upcoming Hudson-Fulton celebration to be held that fall.

28 March
Wilbur Wright leaves for Rome, Italy, to perform flight demonstrations with a duplicate of the Military Flyer just delivered to the Army, fulfilling the original $30,000 contract agreement.

15–25 April
Wilbur Wright pilots the first aircraft to take motion pictures from the air. The feat occurs in Centocelle, Italy, in a Wright biplane. The moving pictures are shown in a theater in Dayton, Ohio, on April 19.

10 June
President William Howard Taft presents special Aero Club of America Medals to Orville and Wilbur Wright in ceremonies at the White House.

26 June
Glenn Curtiss flies over New York City for the first time during exhibition flights at the Morris Park aerodrome.

29 June
Having returned to Fort Myer, Orville Wright makes practice flights in preparation for the resumption of Army trials that were canceled after last September's fatal flying accident.

25 July
Frenchman Louis Charles-Joseph Blériot crosses the English Channel in an aircraft of his own design, winning a £ 1,000 prize offered by the *London Daily Mail* to the first aviator able to cross the English Channel in either direction. He takes off from Les Barrages, France, and lands at Dover, England.

27 July
With Lieutenant Frank P. Lahm, a military balloonist, aboard as the Army observer, Orville Wright meets and exceeds the Army Signal Corps specifications for the purchase of a Wright aircraft. The importance of the test is compounded as it is flown before a crowd of 10,000, including the President of the United States, William Howard Taft.

Above: *Louis Blériot and his aircraft, the Blériot XI monoplane, in which he became the first aviator to cross the English Channel in July 1909. (NASM)*

Below: *Onlookers watch as Louis Blériot begins his flight from France across the English Channel. (NASM)*

30 July
With Lieutenant Benjamin D. Foulois aboard as the Army observer, Orville Wright averages just over 42 miles per hour while flying a cross-country speed test required by Army specifications. The additional two miles per hour achieved by the aircraft above the required 40 miles per hour requirement earns the Wrights an additional $5,000 bonus, making the total cost of the first military aircraft $30,000.

2 August
General James Allen, Chief Signal Officer of the Army, formally approves the purchase of the Wright airplane after the Army observers report is formally filed. The airplane, a Wright Model A biplane, is later named *Miss Columbia*.

23 August
Glenn Curtiss becomes the first American to hold the world speed record when he flies just over 43 miles per hour in his Rheims Racer. He accomplishes this feat at the world's first major air meet held in Rheims, France.

25 August
College Park, Maryland, becomes the first airfield used under a lease agreement by the Army Signal Corps.

7 September
Frenchman Eugéne Lefebvre is killed when his Wright Model A crashes. He becomes the first pilot to die in a powered aircraft while at the controls of the craft.

4 October
Wilbur Wright flies from Governor's Island, New York, out past the Statue of Liberty, up the Hudson River to Grant's Tomb, and then returns to Governor's Island. He wins the $10,000 prize offered by the *New York World*. As a safety precaution, a canoe is attached to the center of the aircraft to help keep it afloat in the event that a water landing is necessary. The canoe is painted bright red.

7 October
Glenn Curtiss becomes the first American to hold the F.A.I. pilot certificate. He is issued Aero Club of France Certificate No. 2.

Above: *"Lieutenant Frank Lahm—First Serviceman to Fly," by Richard Green. (USAF Art Collection)*

Below: *Orville Wright successfully retraces the path of early Hudson River explorers in order to win a $10,000 prize. To prevent sinking in case of an accident, a bright red canoe is affixed to the plane's midsection. (Library of Congress)*

1909 WRIGHT MILITARY FLYER

Peter L. Jakab

The 1909 Wright Military Flyer is the world's first military airplane. In 1908, the U.S. Army Signal Corps advertised for bids for a two-seat observation aircraft. The general requirements were as follows: that it be designed to be easily assembled and disassembled so that an Army wagon could transport it; that it would be able to carry two people with a combined weight of 350 pounds, and sufficient fuel for 125 miles; and that it would be able to reach a speed of at least 40 miles per hour in still air. (This speed performance would be calculated during a two-lap test flight over a five-mile course, with and against the wind.) It must demonstrate the ability to remain in the air for at least one hour without landing, and then land without causing any damage that would prevent it from immediately starting another flight. It should be able to ascend in any sort of country in which the Signal Corps might need it in field service and be able to land without requiring a specially prepared spot; be able to land safely in case of accident to the propelling machinery; and be simple enough to permit someone to become proficient in its operation within a reasonable amount of time.

The purchase price was set at $25,000 with 10 percent added for each full mile per hour of speed over the required 40 miles per hour and 10 percent deducted for each full mile per hour under 40 miles per hour.

Wilbur and Orville Wright were invited to submit a bid. They constructed a two-place, wire-braced biplane with a 30–40 horsepower Wright vertical four-cylinder engine driving two wooden propellers, similar to the aircraft Wilbur had been demonstrating in Europe in 1908. This airplane made its first flight at Fort Myer, Virginia, on September 3, 1908. Several days of very successful and increasingly ambitious flights followed. Orville Wright set new duration records day after day, including a 70-minute flight on September 11. He also made two flights with a passenger.

On September 17, however, tragedy occurred. At 5:14 p.m., Orville took off with Lieutenant Thomas E. Selfridge, the Army's observer, as his passenger. The airplane had circled the field four-and-a-half times when a propeller blade split. The aircraft, then at 150 feet, safely glided to 75 feet, when it then plunged to earth. Orville was severely injured, including a broken hip. Lieutenant Selfridge died of his injuries a few hours later without regaining consciousness. Selfridge was the first person killed in a powered airplane accident. The aircraft was destroyed.

On June 3, 1909, the Wrights returned to Fort Myer with a new airplane to complete the trials begun in 1908. (Wilbur had been flying in Europe the previous year and had thus been absent from Fort Myer in 1908.) The engine was the same as in the earlier aircraft, but the 1909 model had a smaller wing area and modifications to the rudder and the wire bracing. Lieutenant Frank P. Lahm and Lieutenant Benjamin D. Foulois, future Army pilots, were the Wrights' passengers. On July 27, with Lahm, Orville made a record flight of one hour, 12 minutes, and 40 seconds, covering approximately 40 miles. This satisfied the Army's endurance and passenger carrying requirements. To establish the speed of the airplane, a course was set up from Fort Myer to Shooter's Hill in Alexandria, Virginia, a distance of five miles. After waiting several days for optimum wind conditions, Orville and Foulois made the ten-mile round trip on July 30. The outbound lap speed was 37.7 miles per hour and the return lap was 47.4 miles per hour, giving an average speed of 42.5 miles per hour. For the two miles per hour over the required 40, the Wrights earned an additional $5,000, making the final sale price of the airplane $30,000.

Upon taking possession of the Wright Military Flyer in 1910, the War Department renamed the aircraft "Signal Corps No. 1." The Army then conducted flight training at nearby College Park, Maryland, and at Fort Sam Houston in San Antonio, Texas. Various modifications were made to the Wright Military Flyer during this period. The most significant was the addition of wheels to the landing gear. (The original design took off from a dolly traveling down a launching rail with a catapult, and landed on skids.)

Early in 1911, the Signal Corps placed an order with the Wrights for two of their new Wright Model B airplanes. In addition, the War Department proposed shipment of the original 1909 Army airplane to the Wright Company factory in Dayton, Ohio, to have it rebuilt with Model B controls and other improvements. The Wright Company quoted a price of $2,000 for the upgrade, but advised against it because of the many design improvements that had been made during the intervening two years. The manager of the Wright Company, Frank Russell, learned that the Smithsonian Institution was interested in the first Army airplane and would welcome its donation to the national museum. The War Department agreed and approved the transfer on May 4, 1911. The Wright Military Flyer remains on public display at the Smithsonian in Washington, D.C., just across the Potomac River from Fort Myer, where it first flew in 1909.

Opposite: *On July 27, 1909, Frank P. Lahm and Orville Wright prepare to take off in an effort to satisfy the one-hour endurance requirement established by the Army. (NASM)*

Below: *The 1909 Wright Military Flyer now hangs in the Smithsonian Institution, National Air and Space Museum, in downtown Washington, D.C. (NASM)*

Left: *Lieutenant Foulois took a few flying lessons from the Wrights, but taught himself how to fly by trial and error. For a time, Foulois was the only pilot in the Army. (NASM)*

23 October
Lieutenant Foulois receives his first flight instruction from Wilbur Wright at College Park, Maryland.

26 October
Wilbur Wright instructs two U.S. Army officers, 1st Lieutenant Frederick E. Humphries and 2nd Lieutenant Frank P. Lahm, who both solo a Wright aircraft at College Park for the first time on this day. Humphries soloes first.

27 October
Wilbur Wright takes aloft the first woman airplane passenger flown in America. Mrs. Ralph H. Van Deman enjoys a four-minute hop around the airfield at College Park, Maryland.

1910

10 January
The first air meet in America is held in California. The Aero Club of California hosts the event at Dominguez Field near Los Angeles. It is during this air meet that a young James H. "Jimmy" Doolittle sees airplanes fly for the very first time, inspiring him to become an Army pilot himself.

19 January
Lieutenant Paul Beck, Army Signal Corps, acts as the Army's first aerial bombardier when he drops three two-pound sandbags from a Farman biplane piloted by Louis Paulhan, attempting to hit a target on the ground during the Los Angeles Flying Meet.

10 February
Wilbur and Orville Wright receive the first-ever Langley Medal from the Smithsonian Institution in Washington, D.C. The medal is awarded for meritorious investigation in connection with aeronautical science and its application to aviation, and is presented by the Chief Justice of the United States, Melville W. Fuller. The Chief Justice also acts as the Chancellor of the Smithsonian Board of Regents.

15 February
Signal Corps aviators move flying operations from College Park to San Antonio, Texas, looking for better flying weather. Although clouds are fewer, winds are often very gusty, and flying operations are limited.

2 March
Lieutenant Benjamin Foulois flies solo for the first time near San Antonio at Fort Sam Houston on four flights totaling nearly one hour in the air. He accomplishes this feat by corresponding with the Wrights, who send him flight instruction by mail. On the fourth flight, the gasoline feed pipe breaks and forces Foulois to make an emergency landing. A hard landing cracks several parts of the structure and grounds the plane for 10 days. At the time, Foulois is the only pilot in the Aeronautical Division of the Signal Corps.

Left: *Bishop Milton Wright offers a prayer while his sons Wilbur (left) and Orville (with mustache) bow in respect. (NASM)*

Below: *Glenn Curtiss accompanied by his companion. Note the steering wheel rather than the stick used by the Wrights. (NASM)*

19 March
The Wrights open a flying school near Montgomery, Alabama, the current site of Maxwell Air Force Base.

5 May
At the Mt. Weather Observatory in Virginia, a kite soars to an unbelievable altitude of 23,800 feet. This establishes a new world altitude record for kites.

6 May
Professor David Todd of Amherst College ascends to 5,000 feet in the balloon *Massachusetts*, and uses a telescope to make observations of Halley's Comet. He makes four sketches of the comet. The balloon is piloted by Bostonian Charles J. Glidden. Todd's wife Mabel is also a passenger on the flight.

25 May
Orville Wright takes his father, 82-year-old Bishop Milton Wright, for a seven-minute airplane ride. Orville and Wilbur fly for the first and only time together in an airplane. Orville is the pilot.

29 May
Glenn Curtiss flies from Albany to New York City in two hours and 50 minutes. He wins the Scientific AmericanTrophy—for the third consecutive time for this accomplishment—and the trophy is retired to Curtiss permanently.

29 June
The first Wright Model B aircraft is completed.

30 June
Glenn Curtiss drops practice bombs on a target shaped like a battleship floating on Lake Keuka, New York. It is the same lake where Curtiss once tested a propeller-driven wind-wagon, a tri-bladed ice boat, across the frozen surface while testing the design of airship propellers.

BENJAMIN DELAHAUF FOULOIS

Benjamin Foulois was born in Connecticut on December 9, 1879. He was educated in public schools until he began an apprenticeship with his father, who was a plumber. In 1898, he enlisted in the First U.S. Volunteer Engineers and served in Puerto Rico until 1899. He was mustered out of the volunteers and then reenlisted in the Regular Army during campaigns in the Philippines during 1899. He was commissioned a 2nd Lieutenant in 1901 and continued exploring and mapping on Mindanao.

During the early 1900s, Foulois attended the Army's Infantry-Cavalry School at Fort Leavenworth, Kansas, participated in operations with the Army of Cuban Pacification, and then returned to Leavenworth to attend Signal School. He was then assigned to the Office of the Chief Signal Officer in Washington, D.C.

By 1909 Foulois had piloted the Army's first dirigible, and was one of the first Army officers introduced to flying by Wilbur and Orville Wright. He was Orville's passenger during the final Fort Myer test flight that established a world speed, altitude, and endurance record and met the Army specification for a two-person aircraft that flew at least 40 miles per hour.

From 1909 until 1911, Lieutenant Foulois was the only active pilot in the U.S. Army. Having taken a few basic flights with the Wrights, Foulois moved the Army's only aircraft—Signal Aeroplane No. 1—to San Antonio, Texas, where he taught himself to fly by trial and error. He wrote letters to the Wrights explaining his flight experiences—mostly crashes—and they wrote back with suggestions on how to improve his flight skills.

In 1914, Captain Foulois organized and took command of the First Aero Squadron, which later moved to Fort Sam Houston, Texas, by air. This redeployment marked the first cross-country flight by an entire flying unit. The squadron, flying Curtiss JN-2 biplanes, participated in the Mexican Punitive Expedition from March to August in 1916, under overall command of General John J. Pershing. Although the air mission to Mexico was generally considered a failure, the experience gained in planning for logistics support and operating from austere locations was valuable later in Foulois' career.

During World War I, Major Foulois headed the Joint Army and Navy Technical Committee. In that post he crafted the funding bill for aviation as America entered the war. His detailed proposal earned rapid congressional approval in the amount of $640 million, an unprecedented amount for those days. It was an achievement that brought Foulois particular satisfaction later in his life.

Foulois was temporarily promoted to Brigadier General as his responsibilities swelled. He served in a variety of aviation posts in Europe, beginning as the Chief of Air Service, American Expeditionary Forces, and ending as the Assistant Chief of the Air Service, Service of Supply. Foulois also assisted in drafting relevant aviation clauses incorporated into the Treaty of Versailles.

After a tour as air attaché to several European countries, Lieutenant Colonel Foulois (now permanent rank) returned to Fort Leavenworth and attended the Army Command and General Staff School. After a

brief assignment as commander of Mitchel Field, New York, Foulois was appointed Assistant Chief of the Air Corps and promoted to brigadier general in 1927. As assistant chief, Foulois led the May 1931 Air Corps Coast Defense Exercises, which included the participation of almost every airplane in the Air Corps' inventory. The success of these trials earned him both the Mackay Trophy for that year as well as a promotion to Chief of the Air Corps and another star.

Major General Foulois held the distinction of being the first Chief of the Air Corps who was actually a military aviator. While Air Corps Chief, Foulois pressed for more solid doctrinal studies and tasked his assistant, Brigadier General Oscar "Tubby" Westover, to expand the curriculum at the Air Corps Tactical School and plan for the establishment of an independent operational organization that later became the General Headquarters Air Force (GHQ Air Force). Foulois pushed hard for dedicated research and development activities, some that led directly to the development of the XB-15 and B-17 programs. During all this, he flew. Foulois logged more flight hours than almost anyone in the Air Corps.

In 1934, when contract problems resulted in the cancellation of air mail service, Foulois volunteered his Air Corps to fill the void. The resulting operations tarnished Foulois' career despite his continued emphasis on safety in the air and caution in mission accomplishment. That spring, 66 crashes resulted in the loss of twelve airmen and indicated that night and poor weather training was seriously lacking. Eventually, this shortcoming was rectified with increased funding and better instrument training for military pilots.

After the so-called "Mail Fiasco" had ended, Foulois suffered continual harassment from Congress and the press about contracting issues and questions of Air Corps roles and responsibilities. By December 1935, Foulois was tired and ready to retire. He took his last O-38 flight on December 25, 1935—from Washington to Kitty Hawk and back—and retired the following week after 37 years of service. From 1956 to 1965 he served as president of the Air Force Historical Foundation.

Foulois enjoyed a peaceful retirement in New Jersey until his wife became ill in 1959. He moved to Andrews Air Force Base so that he could be near her until she died in 1961. Benjamin Foulois died on April 25, 1967, and was buried in his hometown of Washington, Connecticut.

Opposite: *Major General Benjamin D. Foulois. (USAF)*

Above: *Lieutenant Benjamin Foulois over Fort Sam Houston, Texas. ("Gallant Beginning," by Keith Ferris, USAF Art Collection)*

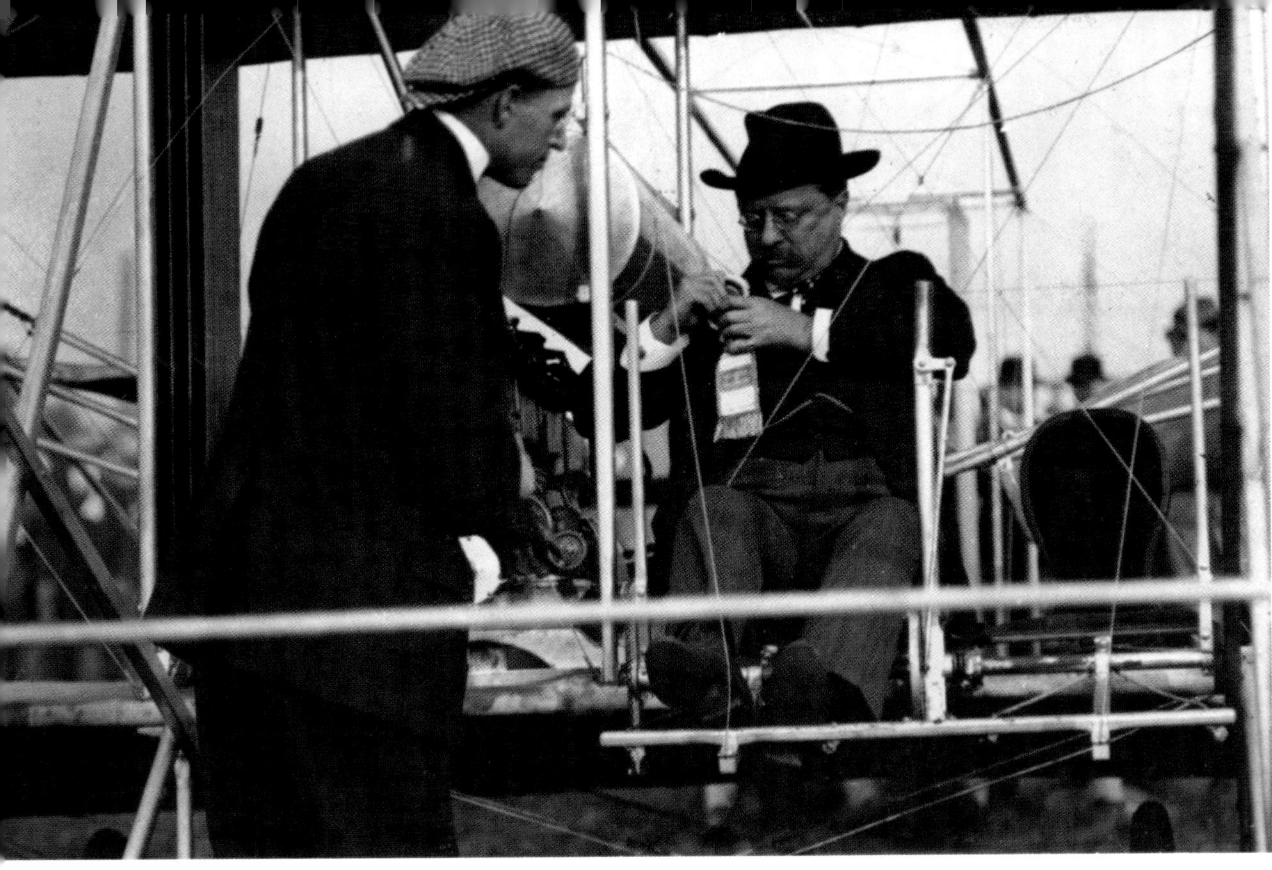

Left: *Former President Teddy Roosevelt is strapped in by his pilot Arch Hoxsey. Roosevelt is the first Chief Executive to ever fly in an airplane. (NASM)*

1 July
Captain Charles DeForest Chandler is replaced as the commander of the Aeronautical Division by Captain Arthur S. Cowan. In 1911, Cowan would select a young infantry officer, Lieutenant Henry H. Arnold, for flight training.

9 July
Walter Brookins reaches 6,175 feet altitude in a Wright biplane while flying near Atlantic City, New Jersey. He is the first pilot to fly higher than one mile in an airplane. For this achievement Brookins wins $5,000 offered by the Atlantic City Aero Club for the first to break the "mile-high" barrier.

8 August
Signal Corps No. 1 is modified by Corporal Glen Madole when he adds wheels to the airplane skids. This important change allows the aircraft to take off without the use of the launching rail or cumbersome "derrick" catapult launching tower. The pilot, Lieutenant Foulois, notes that the shock absorbing springs on the newly installed wheels are not strong enough for the forces of landing and has Madole replace them the following week.

20 August
As a passenger in a Curtiss biplane piloted by Charles F. Willard, a Curtiss employee, over Sheepshead Bay Track, New York, Army Lieutenant Jacob Fickel becomes the first U.S. military man to shoot a firearm from an airplane—a .30-caliber Springfield rifle.

27 August
Flying in a Curtiss biplane at Sheepshead Bay, New York, James A. D. McCurdy transmits and receives radio signals with a ground station. The signals are generated by an H. M. Horten radio set.

31 August
Glenn H. Curtiss flies from Euclid Beach, Ohio, to Cedar Point, a distance of 65 miles across Lake Erie. He wins a $5,000 prize for the flight offered by the *Cleveland Press* newspaper.

2 September
Blanche Scott solos in a Curtiss pusher airplane, becoming the first American woman to do so. She accomplishes this feat at the Curtiss factory airfield in Hammondsport, New York.

11 October
Former President Theodore Roosevelt becomes the first chief executive to fly. He takes a short hop as a passenger on a Wright biplane piloted by Arch Hoxsey at Kinloch Park near St. Louis, Missouri.

7 November
Accomplishing the first air cargo mission, Philip O. Parmalee carries a bolt of silk from Dayton to Columbus, Ohio, flying a Wright B-10 aircraft.

23 November
Aeronautics pioneer Octave Chanute dies. A naturalized U.S. citizen born in Paris, Chanute was a gifted engineer and designed bridges and the stockyards in Chicago and Kansas City. He mentored the Wright brothers early in their studies of aeronautics and is generally credited as the first aviation historian.

Right: *Philip Parmalee (right) and Lieutenant Foulois (left) prepare to fly an observation flight along the Mexican border, where they will practice air-to-ground communication. (NASM)*

Below: *Philip Parmalee uses tape to attach an instrument to a support strut of his flyer. Cockpit instrumentation was nonexistent in the early days. (NASM)*

1911

15 January
Lieutenant Myron S. Crissy, flying as a passenger in a Wright biplane piloted by Philip O. Parmalee, drops a 36-pound live bomb that he designed on a target from an altitude of 1,500 feet.

17 January
The Glenn H. Curtiss aviation school is established at on North Island, San Diego, California, opening four days later. This airfield becomes the Signal Corps Aviation School.

21 January
Army Lieutenant Paul W. Beck sends telegraphic signals to the ground over Selfridge Field, Michigan. He becomes the first Army aviator to communicate by radio from an airplane. Beck built the transmitter himself and sends the signal from a Wright biplane piloted by Philip O. Parmalee at 100 feet altitude and one-and-one-half-mile range from the receiver.

26 January
Glenn H. Curtiss accomplishes the first American seaplane flight at San Diego Bay, California. The plane is a Curtiss biplane with pontoons and is called a "hydro-airplane."

30 January
J. A. D. McCurdy flies a Curtiss "hydro-airplane" from Key West, Florida, to a landing some 10 miles from Havana, Cuba, where he is picked up by a Navy torpedo boat. His flight is the longest over-water flight up to that date.

7 February
Flying an Antoinette aircraft from North Island, California, to a border location near Tijuana, Mexico, Harry S. Harkness delivers a military message to deployed forces there. The trip covers 21 miles in 25 minutes.

27 February
Along the Mexican border near Laredo, Texas, Lieutenant Benjamin Foulois and his instructor pilot, Philip O. Parmalee from the Wright Company, use a Wright B Flyer to demonstrate the potential of the airplane used in coordination with ground troops.

3 March
U.S. military aviation (Signal Service of the Army) receives its first ever congressional appropriation not to exceed $125,000 ($25,000 available immediately) for the purchase, maintenance, operation, and repair of aircraft for the 1912 fiscal year.

Lieutenant Benjamin Foulois and civilian pilot Philip Parmalee fly from Laredo, Texas, to Eagle Pass in an effort to demonstrate the potential for uses of airplanes

Left: *Early College Park aviators (back row, l to r): Captain Frederick B. Hennessy, Lieutenant Henry H. Arnold, Lieutenant Roy C. Kirtland, Captain Frank M. Kennedy, Lieutenant Samuel H. McCleary, Lieutenant Harold Geiger; (front row, l to r): Lieutenant Lewis C. Rockwell, Lieutenant Thomas DeWitt Milling. (Robert and Kathleen Arnold Collection)*

in war. They drop messages and receive radio signals during the mission.

17 March
A Curtiss D pusher-type airplane sporting a tricycle landing gear is demonstrated to the U.S. Army. It is accepted and later becomes Army Signal Corps Aeroplane No. 2.

11 April
College Park becomes the first permanent Army flight school after Lieutenant Arnold and Lieutenant Milling arrive from their initial flight training in Dayton, Ohio. The Signal Corps requests that four permanent hangars be constructed to house the school and the aircraft.

21 April
Lieutenant Henry H. Arnold and Lieutenant Thomas DeWitt Milling receive orders to proceed to Dayton, Ohio, for flying lessons at the Wright Flying School.

27 April
The Curtiss IV Model D and the Wright Type B are accepted as the Army's second and third aircraft at Fort Sam Houston, Texas. Both are pusher-type aircraft.

4 May
The Wright Military Flyer—Signal Corps Aeroplane No. 1—is earmarked for donation to the Smithsonian Institution when the War Department approves the transfer.

7 May
U.S. Army Lieutenant Arnold, Lieutenant Milling, and Navy Lieutenant John Rodgers arrive in Dayton, Ohio, to begin a course of instruction in flying from the Wright Flying School. Arnold and Milling complete their training six days later, becoming the first Army pilots trained at the Wright School.

10 May
Lieutenant George E. M. Kelly is killed when the Curtiss Model D pusher-type aircraft he is flying crashes violently into the ground at Fort Sam Houston, Texas. He becomes the first student pilot to die while flying his own aircraft. In 1917, Kelly Field, located southwest of San Antonio, Texas, is named after him.

20 June
Captain Charles DeForest Chandler assumes command of the Aeronautical Division for the second time, replacing Captain Arthur Cowan, the man who replaced him in 1910.

3 July
The Signal Corps Aviation School is established at College Park, Maryland. Captain Charles DeForest Chandler is appointed to be the school's commander while Lieutenants Arnold and Milling become the school's only instructors. Both finished flight school just weeks before in Dayton at the Wright Flying School.

2 August
Harriet Quimby earns her flying license and becomes the first female pilot in America. She receives F.A.I. Certificate No. 37.

4 August
Harriet Quimby becomes the first woman to fly at night.

14–25 August
Harry N. Atwood flies from St. Louis to New York, a distance of 1,155 miles, accomplishing the longest cross-country flight flown to date.

4 September
Lieutenant Milling becomes the first Army pilot to fly at night after completing the 160 mile tri-state air race. He lands his plane by the light of gasoline flares, which illuminate the landing zone near Boston.

26 September
Lieutenant Milling and two passengers set the world endurance record for three people of one hour 54 minutes near Nassau Boulevard, New York. For this flight, Milling wins the Rodman Wannamaker Trophy.

30 September
Lieutenant Henry H. Arnold flies as a "stunt pilot" for scenes in the movie *The Military Air Scout*. The filming follows the conclusion of the Aero Club of America air meet at Nassau Boulevard, New York. Legend has it that the film director began to call Arnold "Hap," because of his perpetual smile. The fact is he did not begin using that nickname until later in his adult life, after the death of his mother in 1931.

10 October
Lieutenant Milling tests a bomb sight and a device to release bombs devised by Riley E. Scott, who accompanies Milling on the flight to verify the function of the mechanism. Milling becomes the first Army pilot to drop bombs from a military aircraft.

20 October
The original Wright Military Flyer is delivered to the Smithsonian Institution to be put on display. The aircraft now hangs in the "Early Flight" gallery in the National Air and Space Museum on the National Mall in downtown Washington, D.C.

22 October
During the Italo-Turkish War, the first use of the airplane in combat occurs when an Italian reconnaissance Blériot aircraft is sent to accomplish surveillance of Turkish forces near Azizia. The first bombs are dropped by Italian planes on November 1.

26 October
The 1909 Wright Military Flyer (Signal Corps No. 1) is exhibited at the Smithsonian following restoration at the Dayton factory after use in San Antonio by Lieutenant Foulois.

28 November
The Signal Corps Aviation School pack up their aircraft and move to Barnes Farm, an airfield near Augusta, Georgia, in search of better winter flying weather. Instead, Augusta sees the worst blizzard to hit the region in fifteen years. Tents collapse under the weight of the snow and as the snow begins to melt, deep flood waters further impair flight operations until the spring.

17 December
Cameraman E. R. Shaw makes the first aerial motion picture film in America during a flight near Beaumont, Texas.

Below: *"Miss Harriet Quimby," by Flohri. (NASM Poster Collection)*

1912

20 January
Wilbur Wright, seeking feedback to be used in the construction of a follow-on airplane, visits the Signal Corps detachment in Augusta, Georgia. He spends two days in conference with Captain Chandler, Lieutenant Arnold, and Lieutenant Roy C. Kirtland.

25 January
Lieutenant Henry H. Arnold establishes the Army altitude record while flying a Wright aircraft at Augusta, Georgia. The flight lasts almost one hour and achieves the height of 4,674 feet.

27 January
Clarence H. Mackay (Máck'ee) establishes the award of a trophy in his name, which is to be awarded annually. Mackay stipulates that the trophy competition should be held under specific annual rules or may be awarded by the War Department for the most meritorious military flight of the year. Today, the Mackay Trophy is awarded for the single most meritorious Air Force flight of the year.

23 February
The "Military Aviator" rating is established in War Department Bulletin (number 2).

20 March
The Wrights test a new six-cylinder aircraft engine that will greatly improve available horsepower when mounted on later Wright airplanes.

21 March
Lieutenant Frank P. Lahm makes the first American military flight at an overseas location when he flies the Wright Model B at Fort William McKinley, Philippine Islands. Lahm had opened the Air School ten days before with two students, Lieutenant Moss L. Love and Corporal Vernon L. Burge. Corporal Burge becomes the first enlisted man taught to fly and qualifies on June 14.

15 April
The Army Signal Corps orders its first tractor-type airplane from the Burgess Company.

16 April
Aviatrix Harriet Quimby, America's first female pilot, becomes the first woman to cross the English Channel piloting an aircraft—a Blériot XI monoplane. Headlines for the day, however, are focused upon the tragic sinking of the ocean liner *Titanic.*

6 May
Three Signal Corps airplanes fly from College Park, Maryland, to Chevy Chase, Maryland, in the first ever multi-plane cross-country mission.

17 May
Arthur L. "Owl" Welsh receives instruction from Orville Wright on the operation of the new Wright C aircraft. Over the next two years, several airmen will die at the controls of that model of aircraft. Welsh perished in the Wright C less than one month after these lessons.

30 May
Wilbur Wright, the world's first pilot, dies of typhoid fever in Dayton, Ohio, at the age of 45.

5 June
Lieutenant Colonel C. B. Winder completes flight training at the Army Aviation School near Augusta, Georgia, and becomes the first National Guard pilot returning to serve in the Ohio National Guard.

7 June
While flying as a passenger with Lieutenant Milling, Captain Charles DeForest Chandler becomes the first aviator to fire a machine gun (Lewis low-recoil weapon) from the air. Despite promising results, the Army had not actually accepted the Lewis Gun for general use, and the order for ten such weapons is never filled.

Lieutenant John P. Kelly is assigned to the Signal Corps at College Park, Maryland. Kelly is the first air medical officer (Flight Surgeon) in the Signal Corps.

11 June
While conducting the official Army acceptance tests of the Wright C aircraft pilot Lieutenant Leighton W. Hazelhurst and Wright employee "Owl" Welsh, are killed during the speed tests. Lieutenant "Tommy" Milling observes the accident and describes it as a stall followed by a violent nose-first crash.

14 June
Corporal Vernon Burge becomes the first enlisted man to qualify as a pilot when he completes the flying training course at the Army Air School in the Philippine Islands.

5 July
Pursuant to the establishment of the Military Aviator rating in February, Captain Chandler, Lieutenant Milling. and Lieutenant Arnold become the first Army pilots to qualify for the rating. To qualify, the pilot is required to climb to an altitude of 2,500 feet (recorded by an onboard barograph), fly for five minutes in wind conditions of 15 miles per hour (measured by anemometer near the ground), carry a passenger to an altitude of 500 feet, and then perform a landing, touching down within 150 feet of a predetermined landing point, with the engine shut off prior to touching the ground. Additionally, the pilot has to execute a volplane (engine off glide) from an altitude of 500 feet and come to rest within 300 feet of a predetermined location. The final requirement is to navigate a cross-country flight greater than 20 miles at an average altitude of 1,500 feet and report upon ground features or other special matters of observation. These initial requirements are modified in 1913.

29 July–1 August
Lieutenant Foulois carries out a number of airborne radio tests in a Wright B Flyer. Transmissions were successfully received from distances as great as ten miles.

10–17 August
Lieutenant Milling and Lieutenant Foulois take part in Army ground maneuvers near Bridgeport, Connecticut. This marks the first participation of Army aircraft with Army ground forces during a training exercise.

Opposite: *The Wright military aircraft could be disassembled and transported in a train car. Here, the College Park contingent prepares to deploy their aircraft to Augusta, Georgia, in search of better flying weather. (Robert and Kathleen Arnold Collection)*

Right: *Harriet Quimby climbs into her Blériot XI monoplane and then becomes the first woman to cross the English Channel on April 16, 1912. (NASM)*

THE WRIGHT PROPELLERS

Jeremy R. Kinney

On the morning of December 17, 1903, the world's first successful aerial propellers took to the air on the Wright Flyer. A propeller is a series of twisted airfoils, or blades, that convert the energy supplied by its power source through helical motion. The blades, connected to a central hub, strike the air at a certain angle, called pitch, and generate thrust by creating an area of high pressure behind the propeller, which pushes the airplane forward. The idea of the aerial propeller dates back to the helical screw concept of the fifteenth-century inventor-philosopher Leonardo da Vinci. By the late nineteenth century, experimenters rejected power transmission devices such as flapping wings, oars, and paddle wheels in favor of the propeller.

Wilbur and Orville Wright used their "mind's eye," or non-verbal thinking, to envision the airplane as a synergistic technical system based on the four forces of flight—aerodynamics, control, propulsion, and structures—during the process of creating their historic flyer. Within that framework, the Wrights theorized, designed, and constructed the world's first practical fixed-pitch aerial propeller and the aerodynamic theory to calculate its performance.

Wilbur believed that studying the design of ship propellers would help them design their aeronautical propeller, but they quickly realized that no formal theories for such designs existed. It was left to them to establish such theoretical work. After many heated engineering discussions, Wilbur and Orville realized that a propeller was simply a rotating, twisted wing moving in a helical path. As a result, the brothers used airfoil data calculated from their wind tunnel to design blades able to convert the energy of their twelve-horsepower engine into thrust. Using a drawknife and hatchets, they carved and shaped the propellers from two-ply spruce wood, covered them in linen, and sealed them with aluminum powder mixed in varnish. The two original Wright propellers, once connected to the engine via a chain-and-sprocket transmission system, were 66 percent efficient—which was enough to get the flyer off the ground at Kitty Hawk.

Wilbur and Orville went on to improve the design of their propellers. Their 1908 and 1909 propellers featured a swept-back leading edge, which the Wrights described as "bent end," a configuration that generated approximately 75–80 percent efficiency. Such unprecedented performance, surprisingly, was not realized again until the 1930s. The pioneering knowledge the Wrights gained during their process of invention was lost on successive generations until the rise of professional aeronautical engineering during World War I.

The Wright propellers did possess limitations related to their configuration and construction materials. They were fixed-pitch in configuration, which was simple in operation, efficient for one operating regime, and gave adequate performance for aircraft that operated at low altitudes. Wilbur and Orville made them from wood because the material was strong, light, and easy to fabricate. For propellers to be efficient enough to contribute to the overall performance and military mission of the airplane over the course of the twentieth century, engineers needed to develop a metal variable-pitch propeller, which allowed the pitch angle at which each propeller blade rotated through the air to vary according to different flight conditions. Metal construction ensured the

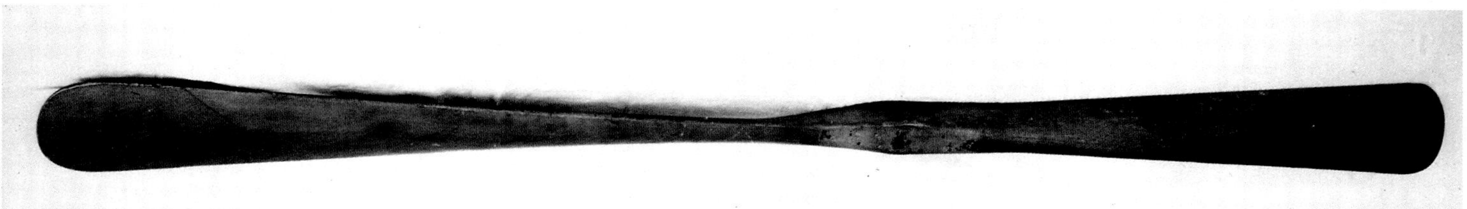

survival of the propeller in extreme environmental conditions, but it was much more difficult to hand-work than wood. The Wrights never experimented with variable-pitch and they rejected metal as a propeller construction material because their fixed-pitch wood propellers delivered exceptional performance for their designs.

Orville traveled to Fort Myer, Virginia, near Washington, D.C., in September 1908 to demonstrate a Wright airplane to the American military. Tragedy struck when one of the bent-end propellers failed in flight. In the ensuing crash, Orville was seriously injured and his passenger, Lieutenant Thomas E. Selfridge, was killed, becoming the first passenger fatality in an airplane accident. Orville would recuperate and return a year later with the Military Flyer, which became the American military's first powered flying machine.

Despite those setbacks, the importance of the Wrights' invention of the aerial propeller cannot be underestimated. The Wrights' twisted, rotating wing propeller was a significant step above the flat plates of previous propeller experimenters. The combination of the aerial propeller and the reciprocating internal combustion piston engine, perfected by Wilbur and Orville, was the main form of propulsion for the first 50 years of heavier-than-air flight.

Opposite: *An early Wright wind tunnel used to gather data for the 1903 Flyer. (NASM)*

Above: *One of the original Wright Flyer propellers. (NASM)*

Left: *A rear view of a later model Wright aircraft. (NASM)*

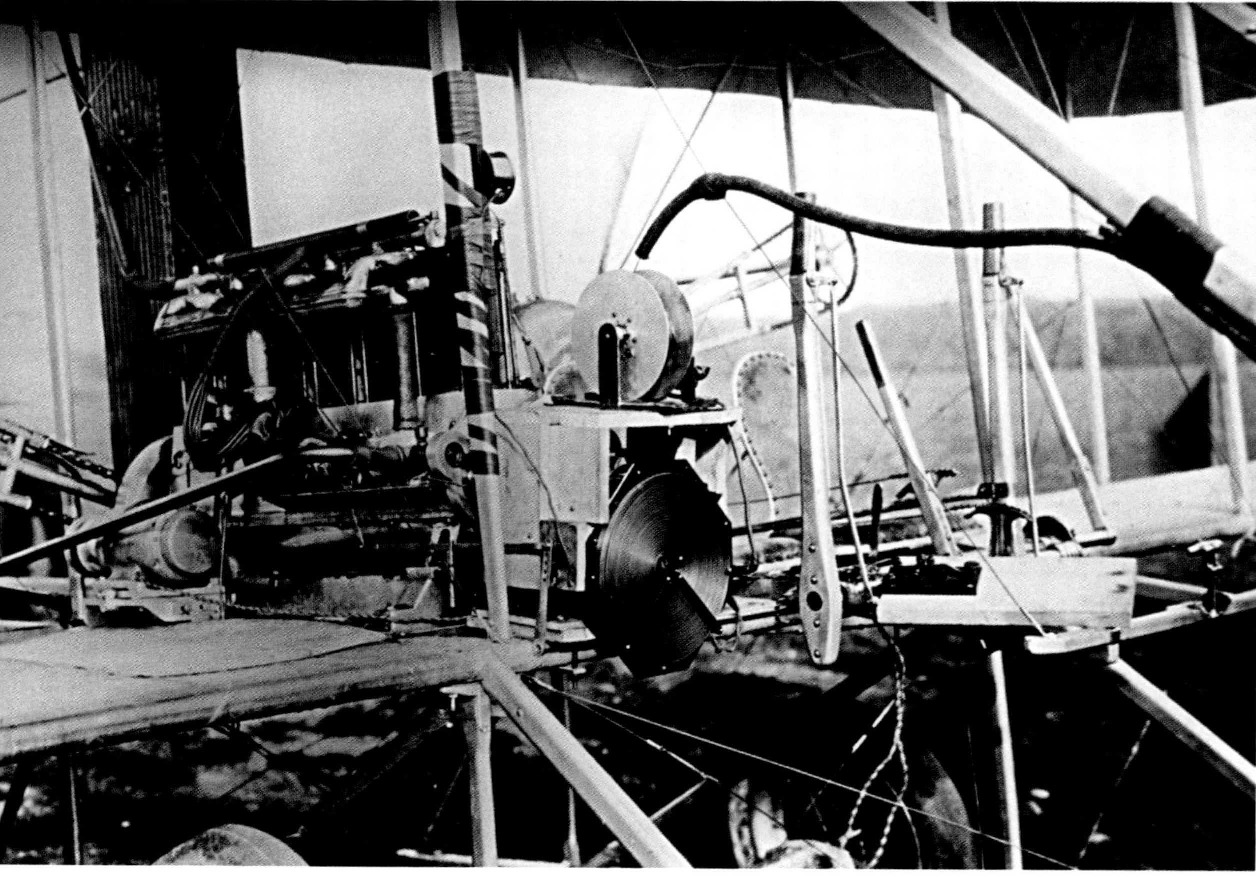

Left: *During flight experiments at Fort Riley, Kansas, radio signals were transmitted from the aircraft to a ground receiver. The long wire transmitter is seen here, mounted on the aircraft. (Robert and Kathleen Arnold Collection)*

28 September
Multiple fatalities occur for the first time over College Park as Lieutenant Lewis C. Rockwell, piloting Signal Corps No. 4 (a Wright B Flyer), crashes while attempting a gliding landing. Along with Rockwell, his passenger, Corporal Frank S. Scott, is killed. Scott is the first enlisted man to die in an airplane crash.

1 October
Lieutenant Arnold and his passenger Lieutenant Alfred L. P. Sands escape death by only a few feet as the Wright aircraft Arnold is piloting suffers a series of accelerated stalls and becomes uncontrollable. Miraculously, Arnold rights the airplane just in time. Thankful to be alive, Arnold grounds himself for the next four years.

9 October
The first competition for the Mackay Trophy takes place near College Park, Maryland. A triangular reconnaissance course is to be flown and a troop concentration located during the flight. Lieutenants Arnold and Milling begin the contest, but Milling is forced to withdraw after becoming ill. Lieutenant Arnold completes the task and is awarded the first Mackay Trophy. Arnold describes the trophy's size by mentioning that about four gallons of beer might easily be poured into the large, silver, goblet-shaped trophy.

5–13 November
During military exercises at Fort Riley, Kansas, artillery spotting utilizing direct communication with ground forces (radio, drop cards, and smoke signals) is successfully practiced for the first time by Lieutenant Henry H. Arnold and Follett Bradley, his observer.

27 November
Three Curtiss-F two-seat flying boats are purchased by the Army Signal Corps.

8 December
The Signal Corps Aviation School at North Island, San Diego, California, is established. The first to arrive are those who have been trained to fly Curtiss aircraft, also known as the "Curtiss Contingent."

1913

16 January
Dr. Thaddeus S. C. Lowe, Civil War balloonist, dies in Pasadena, California. He is 81 years old.

11 February
Representative James Hay of West Virginia introduces House Resolution 28728—the first bill advocating a separate aviation corps. The bill is defeated.

Right: *Robert Goddard's patent drawings for his 1914 "Rocket Apparatus." (NASM)*

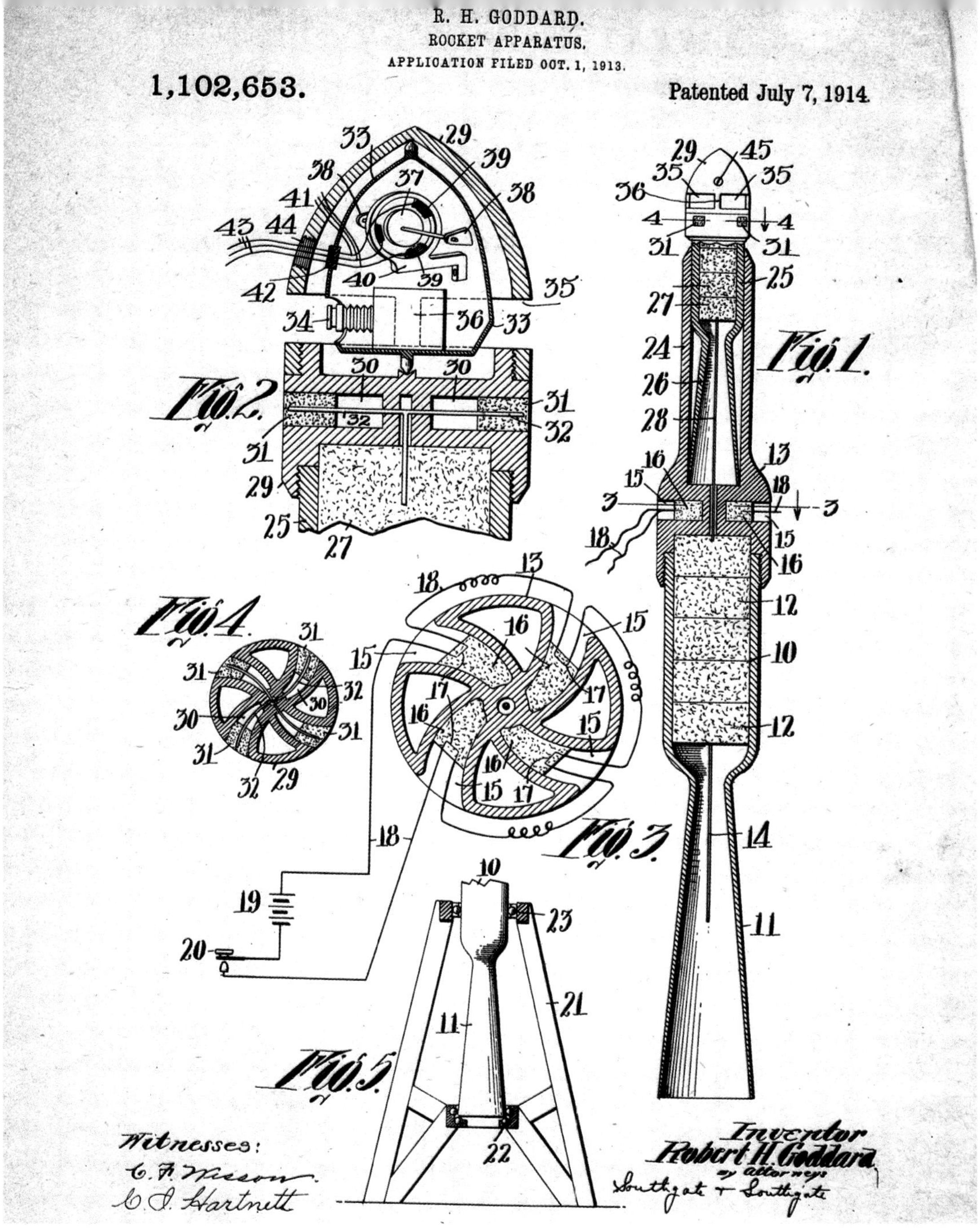

13 February
The Board of Regents of the Smithsonian Institution appoints a committee to advise the Langley Field Aerodynamic Project.

17 February
The first autopilot device is tested on an Army aircraft. Called a gyrostabilizer by the inventor, Elmer Sperry, the device is tested and improved and a variation of the gyrostabilizer is even used on an unmanned aircraft tested by the Army in 1918 for use during World War I.

2 March
Flight pay of 35 percent above base pay is authorized by Congress for those volunteering for aviation duties.

5 March
The 1st Provisional Aero Squadron is established to support General Pershing's Punitive Expedition into Mexico. The unit is based at Texas City, Texas.

31 March
Lieutenant W. C. Sherman makes the first aerial map from an aircraft. Sherman rides as a passenger with Lieutenant Tommy Milling from San Antonio to Texas City, Texas, drawing as they fly.

27 May
The War Department publishes General Order No. 39, which states that all qualified military aviators will receive a Military Aviator's Certificate and a badge. At that time, 24 officers are qualified.

28 May
Lieutenant Tommy Milling and Lieutenant W. C. Sherman set the two-man endurance and distance record when they fly 220 miles from Texas City to San Antonio, Texas. The flight takes four hours and 22 minutes.

30 May
The Massachusetts Institute of Technology (MIT) establishes an aerodynamics course under the guidance of Navy officer Jerome C. Hunsaker.

12 June
The first Curtiss tractor-type aircraft (Signal Corps No. 21) is accepted by the Army Signal Corps.

1 June
In the skies over Los Angeles, California, Georgia "Tiny" Broadwick becomes the first woman to make a parachute jump in the U.S. She is 18 years old and leaps from 1,000 feet. She jumps from an aircraft piloted by Glenn L. Martin over Griffith Field.

10 September
Major Samuel Reber assumes command of the Aeronautical Division of the Signal Corps. He will hold the position after the division name is changed to the Aviation Section until 1916.

1 October
Dr. Robert H. Goddard files the paperwork for his first patent for a "Rocket Apparatus." Patent number 1,102,653 is granted on July 7, 1914.

4 December
The War Department publishes General Order No. 75, which establishes the standard organization for an Aero Squadron.

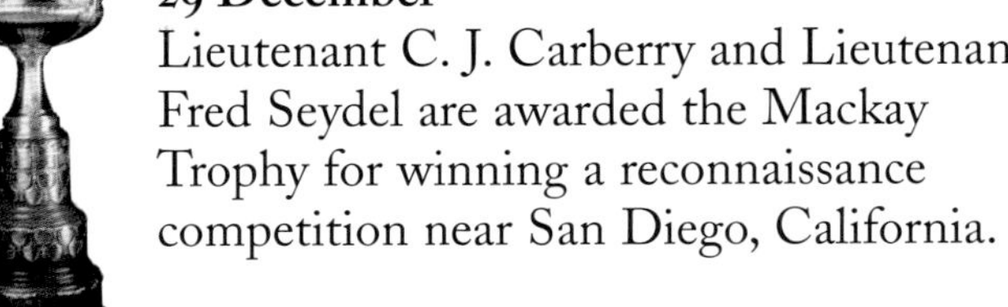

29 December
Lieutenant C. J. Carberry and Lieutenant Fred Seydel are awarded the Mackay Trophy for winning a reconnaissance competition near San Diego, California.

31 December
Orville Wright demonstrates an automatic aircraft stabilizer at Dayton, Ohio. He receives the 1913 Collier Trophy for his efforts.

1914

5 February
Lieutenant Joseph C. Morrow, Jr., becomes the last "Military Aviator" qualified under the original rules for the rating. He passed his test on December 27, 1913.

24 February
In a meeting of the Signal Corps Aviation School in San Diego, California, all pusher-type aircraft are condemned as too dangerous to fly. A large number of accidents in 1913 and 1914, many resulting in fatalities while flying the Wright C aircraft, are behind the outcry.

28 May
After borrowing the Langley Aerodrome from the Smithsonian Institution, Glenn H. Curtiss, having made dozens of modifications to the aircraft (including the addition of pontoons), successfully flies for five seconds from a lake near Hammondsport, New York. The distance covered is 150 feet.

24 June
The Signal Corps Aviation School at San Diego receives a Curtiss J (Signal Corps No. 29).

28 June
Archduke Franz Ferdinand, heir to the Austro-Hungarian throne, is assassinated by a member of a Serbian extremist group known as the Black Hand. This act is the catalyst for the outbreak of World War I.

7 July
Dr. Robert H. Goddard receives a patent for a multi-stage step rocket design.

14 July
Dr. Goddard is awarded a second patent for a liquid-fueled gun rocket. Goddard is awarded 214 patents between 1914 and 1956.

Opposite: *Major Samuel Reber assumed command of the Aeronautical Division of the Signal Corps in late 1913. (USAF)*

Right: *Glenn Curtiss flies a highly modified Langley Aerodrome float plane. He had borrowed the artifact from the Smithsonian Institution in an attempt to validate Langley's earlier aeronautical work. (NASM)*

18 July
The Aviation Section of the U.S. Army Signal Corps is established, replacing the former Aeronautical Division. Lieutenant Colonel Samuel Reber, head of the Aeronautical Division since last October, becomes the first commander of the Aviation Section, for which 60 officers and students plus 260 enlisted men are authorized. Reber is later relieved of command in the wake of the court-martial of Lieutenant Colonel Lewis E. Goodier, who was reprimanded for inciting military aviators to bring charges against commanders at the San Diego school.

28 July
World War I begins in Europe when Austria-Hungary declares war on Serbia. Mobilization of European armies begins to gain momentum.

1 August
Germany declares war on Russia, setting in motion a chain of declarations of war that pave the way for the engagement of armies in Europe.

1–16 December
Lieutenant Herbert Arthur Dargue and Lieutenant Joseph O. Mauborgne successfully demonstrate two-way radio communications during a series of tests near Fort William McKinley, Philippines. They are flying a Burgess-Wright biplane.

8 December
Lawrence B. Sperry is awarded the Collier Trophy for his experiments in gyroscopic stabilization of aircraft. Sperry demonstrates the device for a committee of the Aero Club of America.

11 December
Lieutenant H. A. Dargue pilots a Burgess-Wright aircraft with communications specialist J. O. Mauborgne on board. They receive radio signals from over ten miles away for the first time in Army aviation history. Five days later the same duo will both receive and transmit radio telegraphy from the same aircraft flying in the Philippine Islands.

23 December
Captain T. F. Dodd and Lieutenant S. W. Fitzgerald are awarded the Mackay Trophy by winning a reconnaissance competition.

THE UNITED STATES AIR FORCE ART PROGRAM

Jeffery S. Underwood

Left: *"2nd Lieutenant H. H. Arnold Flies the Wright B Flyer at College Park, Maryland-1911," by John McCoy. (USAF Art Collection)*

Opposite: *"Curtiss JN-4 Jenny-1917," by L. Mahan. (USAF Art Collection)*

Long before Homer's epic poetry recounted the war at Troy or Herodotus wrote the history of the Persian Wars, artists recorded military achievements through painting and sculpture. Depictions of weapons on cave walls, soldiers on Egyptian tombs, enameled bricks showing Persian archers, Greek warriors on vases, or rows of clay soldiers in ancient Chinese burial sites often provide the only information about otherwise unrecorded military struggles. Moreover, these artworks remind us of the human element of war in a way that the written word never can.

By the nineteenth century, military painting had become a major art form, perhaps best demonstrated by David's paintings of the French Revolution and the Napoleonic Wars. The advent of photography, newspaper war correspondents, and mass communications made it possible to illustrate military actions quickly around the world, but military art continued to capture the human element of warfare in its own unique way.

When the United States entered World War I, the U.S. Army already had a collection of military art, and it took an interest in producing an official documentation of the war. Eight artists received commissions as captains and went to Europe to record the efforts of the American Expeditionary Forces, but the program ended after the Armistice. During World War II, the Army sent both military and civilian artists overseas to record the war, but Congress soon withdrew the funding. Therefore, the Army decided to support civilian artists the same way as war correspondents. In addition, the U.S. Navy created a similar art program to document its war effort.

After the reorganization of America's military in 1947, the Army transferred 800 military aviation works of art to the newly established U.S. Air Force. Over time, the Air Force added paintings by notable French and British artists as well as German art that had been captured during World War II. General Curtis LeMay initiated the senior leader portrait program during his tenure as Chief of Staff. Together, these artworks formed the foundation of the Air Force Art Collection.

To aid recruiting, the Air Force sponsored 30 cartoonists on a tour of Air Force installations in 1951, and the following year they sponsored artists from the Society of Illustrators of New York. Pleased with the results of these tours, Air Force officials created the USAF Art Program and formally invited the Society of Illustrators of New York to participate. The program officially sponsored trips for civilian members of the Society to Air Force installations around the world. Eventually, the Societies of Illustrators of Los Angeles and of San Francisco, the Midwest Air Force Artists, the Southwest Society of Air Force Artists, and numerous independent artists joined the program.

Often, the artists who make officially sponsored tours donate their artwork to the Air Force Art Collection. These works portray the people, equipment, activities, facilities, and other historically significant subjects related to the Air Force and its predecessor organizations, and the societies review the work of their members before offering them as gifts.

The Air Force Art Program Office manages the Air Force Art Collection for the Secretary of the Air Force. This office acquires, maintains, controls, and exhibits the collection, assists the professional artists participating in the program, and handles the donated artwork in accordance with Air Force regulations. It also establishes the appraisal guidelines for art in the collection and for newly donated art.

Since the number of works in the collection is small, they may be exhibited only in officially designated areas, such as offices or reception areas in higher headquarters, official residences of senior Department of Defense and Department of the Air Force personnel, the National Air and Space Museum, the Air Force Academy, or the National Museum of the United States Air Force. The Air Force Art Program does not loan artwork in the collection indefinitely to private individuals or private organizations, but it will provide film negatives or transparencies for reproduction or for publication in a book or research paper on a case-by-case basis. The public often sees artwork from the Air Force Art Collection in public art shows and as the cover art for numerous Office of Air Force History publications.

AIR SERVICE DURING WORLD WAR I

1915–1918

12
32501
10

AIR SERVICE DURING WORLD WAR I

1915–1918

In 1914, as Europe plunged into a deadly conflict known at the time as the Great War, President Woodrow Wilson made a pledge to keep America out of the conflict. He succeeded until German U-boats unleashed unrestricted attacks upon ships bound for Europe, and Germany attempted to bring Mexico into the war through a secret communiqué. The message was intercepted and decoded and came to be known as the Zimmerman Telegram. The imminent threat to American commerce and lives at sea could no longer be ignored, and on April 6, 1917, America declared war against Germany.

America was totally unprepared to fight a foreign war in 1917. Little had been done in the way of production, training for combat was inadequate, and manpower was lacking. In April 1917, the Aeronautics Division of the Signal Corps, manned by 131 officers and 1,087 enlisted men, was comprised of seven squadrons that flew 55 aircraft, most of these obsolete. Essentially, American combat airpower did not exist. A meager force of men and machines made up the Air Service in 1914.

When military actions against Pancho Villa in Mexico began in 1916, the Army's JN-2 Jenny aircraft were not up to the task. Only two of the original eight aircraft assigned to the expedition into Mexico survived the operation.

America had no aircraft industry to speak of, and no way to build one in time to contribute aircraft to the war effort. In the end, it was the British-designed de Havilland DH-4 that was built in American factories by the thousands. Yet, only a few hundred ever made it to the war zone. However, American ingenuity and production capability excelled in one area: engine production. One thing that automobile manufacturers could do well was build engines. During the war years, more than 32,000 engines had been built; 15,500 of those were the famous Liberty Engine, a 12-cylinder, 400-horsepower machine that would be used for the next two decades to power American military aircraft. The Liberty Engine was America's single independent technological contribution to the war effort.

The invention of the internal combustion engine in 1859, which was greatly improved during the following four decades, became the catalyst for the mechanization of maneuver warfare—a radical transformation. The European battlefield became an entertaining and quizzical mix of horse-drawn caissons and motorized ambulances. Trenches laced the European landscape while wood and wire biplanes flew overhead reconnoitering enemy positions, dropping bombs on entrenched ground troops, and dogfighting with their airborne adversaries. Later, motorized tanks appeared on the same battlefields as horse cavalry.

When America did enter the war, the early pioneer airmen rose rapidly to positions of authority, some achieving hero status, including Billy Mitchell, Eddie Rickenbacker, Raoul Lufbery, Benny Foulois, Tooey Spaatz, Henry H. Arnold, and Frank Luke.

For the most part, the "war to end all wars" was a European affair. Although the image of the exquisitely dressed heroic aviator seemed a stark contrast to the mud-covered infantry in the trenches, there was little glory in the consequences of the war. France lost nearly an entire generation of its men while economies and societies were totally destroyed.

But the American Expeditionary Force's contribution in the last eight months of the conflict turned the terrible tide of the stalemate into an Allied victory by November 11, 1918. Above the quagmire, Americans shot down more than 750 enemy aircraft, dropped more than 130 tons of bombs, while flying approximately 35,000 hours during the war.
In operations near the end of the war, Americans commanded and participated in the first mass aerial attacks in history. Brigadier General William Billy Mitchell's air armada, which numbered nearly 1,500 airplanes, was vital to Allied success at St. Mihiel and the Meuse-Argonne.

On the ground, more than 50,000 American soldiers died on the fields of battle. Away from the battle zone more than 60,000 perished in non-theater operations. After the war, as if the world had not suffered enough, a great flu pandemic erupted, taking the lives of an estimated 25–50 million people worldwide. Soldiers, sailors, and airmen returning from Europe were unceremoniously quarantined upon the ships that had just carried them home. Families that had endured the trials of a world war now suffered agony and death at the hands of an incurable virus. In history, even the Black Death (1347–1351) was less calamitous than the influenza pandemic of 1918–1919. An estimated one-half million Americans died of the Spanish Flu during the pandemic, and fully one-fifth of the entire U.S. population was infected at some point during the year.

After the fighting had ended and the Treaty of Versailles had been signed, a tenuous peace prevailed while the world's militaries demobilized and went home. Wartime strength of the Air Service shrank from 190,000 in November 1918 to 81,000 only two months later. By June 1919, the Air Service consisted of a skeleton 5,500 officers and 21,500 enlisted. Appropriations for air dropped from $460 million during fiscal year 1919 to a meager $25 million the next. Training ceased, production slammed to a halt, huge stockpiles of wartime machinery were sold, and airfields and depots were abandoned. America's military was gutted from wingtip to wingtip.

Despite rapid demobilization, it was apparent that airplanes had become permanent features of mechanized warfare of the future. Yet it would take two decades for America's air arm to develop into something more than a second-rate air force.

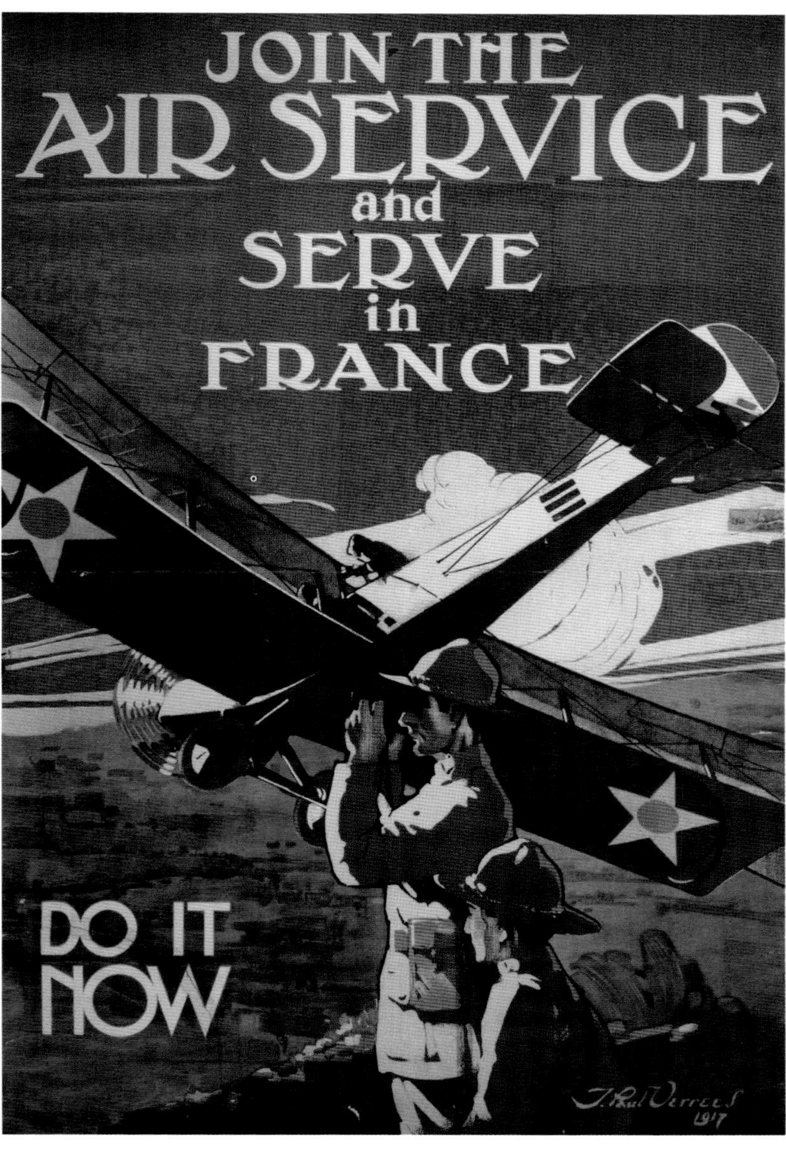

Pages 50–51: *This painting depicts a squadron of American-built DH-4s of the 11th Day Bombardment Squadron returning from a mission. They are attacked by German Fokkers, but Spads of the 94th Pursuit Group commanded by Captain Eddie Rickenbacker rush into the fray. ("First American-Built Battle Planes See Action-1918," by John McCoy, USAF Art Collection)*

Above: *This early World War I recruiting poster emphasizes the military role of observation, artillery spotting, and reconnaissance rather than bombing and pursuit. (NASM Poster Collection)*

AIR SERVICE DURING WORLD WAR I

1915–1918

"I guess we considered ourselves a different breed of cat, right in the beginning. We flew through the air and the other people walked on the ground; it was as simple as that."
—Carl A. "Tooey" Spaatz

1915

Albert Einstein publishes his general theory of relativity. Einstein will later play an important role in the development of nuclear theory in the 1930s.

15 January
Lieutenant J. C. Carberry and Lieutenant Arthur C. Christie set the official American two-man altitude record when they fly a tractor-type OXX Curtiss 100 to an altitude of 11,690 feet. The mission lasts for one hour and 13 minutes.

Lieutenant B. Q. Jones, piloting a Martin tractor biplane at San Diego, California, sets a new official American one-man flight endurance record of eight hours 53 minutes. For the flight he is awarded the Mackay Trophy.

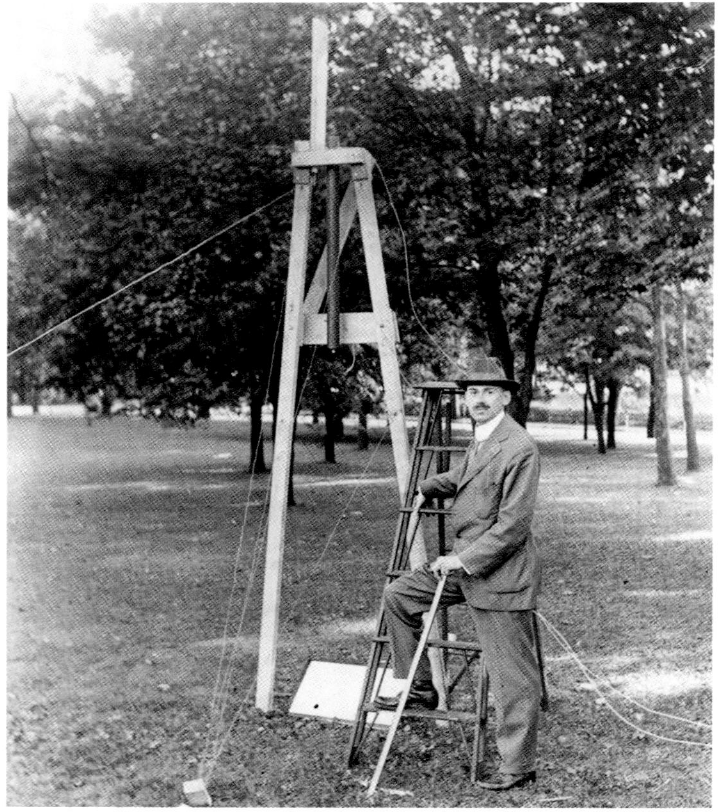

19 February
Robert H. Goddard begins experimentation with solid fuel rockets. Goddard measures the efficiency of gunpowder-propelled Coston signal rockets, which he later modifies with smokeless powder, steel tube construction, and an exhaust nozzle.

3 March
The Advisory Committee for Aeronautics is established by congressional approval as part of the Naval Appropriations Act. Later renamed to include the word "National" in its title, the NACA would lead the integration of theoretical aeronautical science to practical applications for the next four decades. The NACA is the first-ever government organization created to support aviation research and development. At the first meeting on April 23, Chief Signal Officer Brigadier General George P. Scriven is named as the organization's first chairman, while secretary of the Smithsonian Institution, Dr. Charles D. Walcott, was elected the first chairman of the NACA Executive Committee.

1 April
French aviator, Lieutenant Roland Garros, shoots down an enemy aircraft with a machine gun that fires directly through the arc of his aircraft propeller. Not being synchronized with prop rotations, metal deflectors are affixed to the propeller blade to deflect bullets away from the blade.

7 May
The Ocean Liner *Lusitania* is torpedoed off the coast of Ireland. Nearly 1,200 are killed in the attack; 128 are Americans.

Left: *Dr. Robert H. Goddard seen with his 1915 rocket experiments. (NASM)*

Opposite, bottom: *Anthony Fokker improved upon the Garros system and invented a synchronized system that linked engine rotation to the machine-gun firing mechanism. This allowed bullets to be fired through the propeller arc without fear of striking the blade. (NASM)*

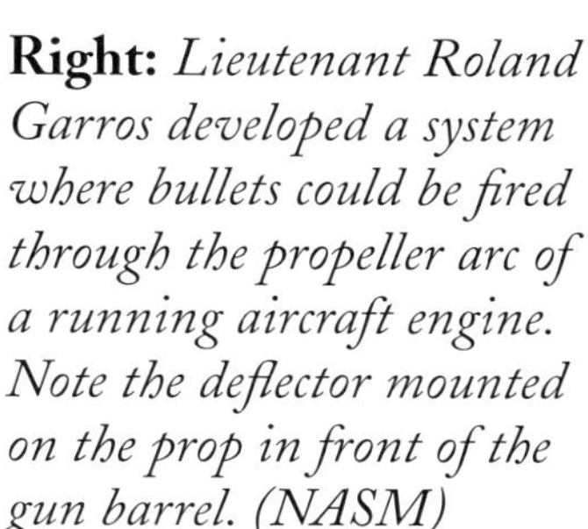

Right: *Lieutenant Roland Garros developed a system where bullets could be fired through the propeller arc of a running aircraft engine. Note the deflector mounted on the prop in front of the gun barrel. (NASM)*

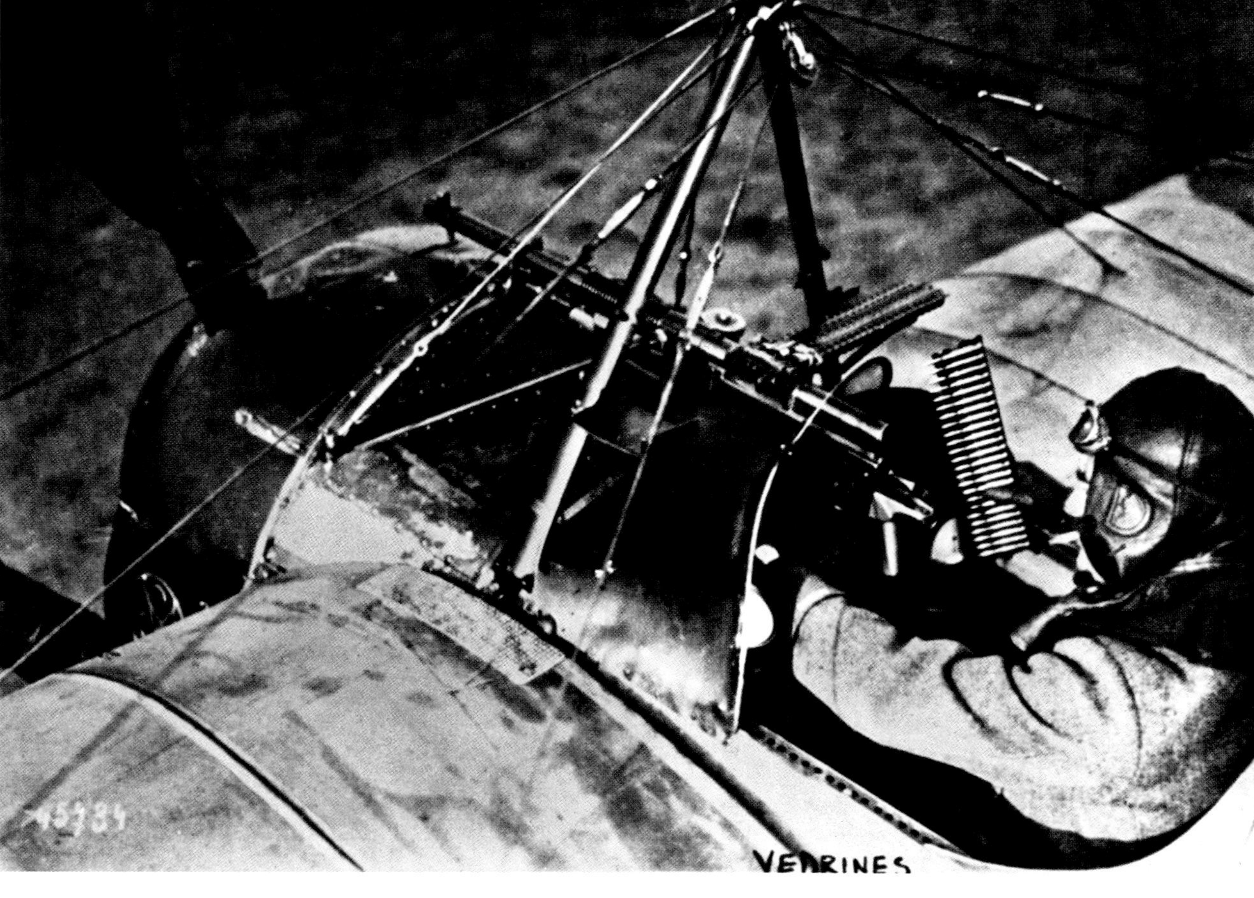

20 May
The first mass-production aircraft accepted by the U.S. Army, the Curtiss JN-2, is selected for use by the 1st Aero Squadron.

22 June
Pilot Jack Vilas takes Wisconsin State Forester E. M. Griffith for a flight to observe a raging fire, the first aerial forest fire patrol flown.

1 July
Anthony Fokker's synchronized machine gun, which fires through the propeller arc without striking the blade, is used by a German pilot to shoot down an Allied aircraft.

3 July
Lieutenant B. Q. Jones becomes the first Army pilot to perform a loop in an aircraft near San Diego, California.

13 October
Orville Wright sells the Wright Company to a group of investors.

15 October
William E. Boeing begins flight school in California. He will rise to be the president of one of the largest and most productive aircraft companies in history after becoming the president of Pacific Aero Products. He changes the company name to Boeing Airplane Company in 1917.

1 November
The first aviation National Guard unit, the Aviation Detachment of the 1st Battalion of the Signal Corps, is organized in New York. Captain Raynal Cawthorne Bolling, the unit organizer, is named its first commander.

9 December
Written by Jerome C. Hunsaker and E. B. Wilson, NACA Report No. 1, *Report of Behavior of Aeroplanes in Gusts*, is published.

11 December
Four Portuguese Army officers become the first foreign students to enter a U.S. flying training program when they report to the Signal Corps Aviation School in San Diego, California.

12 December
Grover C. Loening designs an airplane constructed of an all-steel frame and covered in fabric. The plane, built by the Sturtevant Aeroplane Company, flies successfully for the first time.

19 December
Six Curtiss JN-3 aircraft fly 439 miles from Fort Sill, Oklahoma, to Fort Sam Houston, Texas. This marks the first time a squadron, the 1st Aero Squadron, flies cross-country as a unit.

1916

5 January
The 1st Company of the 2d Aero Squadron departs San Francisco for the Philippine Islands. This is the first time that a complete aviation unit is deployed for duty outside of the United States.

9 March
Francisco "Pancho" Villa, leading a mounted force of more than 1,000 men, executes several raids on American border settlements in New Mexico. Although U.S. cavalry chase him into Mexico, they are unable to track him down.

15 March
The 1st Aero Squadron, under the command of Captain Foulois, joins General John J. Pershing during the search for Pancho Villa in Mexico. With only a handful of underpowered Curtiss JN-2 Jenny aircraft (modified as JN-3s), Foulois and his small force arrive in Columbus, New Mexico, the site of one of the hostile raids, and begin flying reconnaissance missions in support of Pershing's forces.

21 March
Escadrille Americaine is formed in France after authorization from the French government is given. Volunteer American pilots fill the unit, which eventually changes its name to the now famous *Lafayette Escadrille*.

2 April
Colonel William Glassford arrives at Rockwell Field to assume command of the Signal Corps Aviation School.

3 April
Captain Billy Mitchell reports to the Chief Signal Officer and then takes command of the Aeronautics Division.

7 April
While attempting to deliver messages to the American consul in Chihuahua City, Mexico, Captain Foulois and Lieutenant Bert Dargue are fired upon by locals who were, in theory, also trying to capture Pancho Villa.

Above: *Lieutenant Benny Foulois (left) and General John J. Pershing pursued Pancho Villa into Mexico in 1916. (USAF)*

Left: *Insignia for the Lafayette Escadrille, 1916. (USAF)*

Right: *U. S. Air Service JN-2 (modified) in Mexico, summer 1916. (USAF)*

Below: *Major General George O. Squier was Chief of the Aviation Section of the Signal Corps from May 20, 1916, until February 19, 1917. ("Major General George O. Squier," by Bjorne Egeli, USAF Art Collection)*

20 May
Lieutenant Colonel George O. Squier takes command of the Aviation Section of the Signal Corps.

3 June
Congress passes the National Defense Act of June 3, 1916. Until this date, the Aviation Section had been limited to 60 officers and 260 enlisted men.

8 June
The National Advisory Committee for Aeronautics (NACA) calls its first open meeting. Representatives of industry and the government are invited to attend.

18 June
Lafayette Escadrille pilot H. Clyde Balsey becomes the first American aviator shot down during World War I. He survives the incident, which occurs near Verdun, France.

23 June
Lafayette Escadrille member Victor Emmanuel Chapman is shot and killed while flying a Nieuport over enemy lines near Verdun. Chapman is the first American aviator killed during the Great War.

1 July
The British Army begins the Battle of the Somme by sending 100,000 infantry troops across no-man's-land into the sites of deeply entrenched German machine guns. The British attack is cataclysmic—approximately 57,000 casualties result from the futile frontal assault.

13 July
The First Aero Company, New York National Guard, is called to federal service during the border crisis with Mexico. Although they do not deploy, this is the first time that an air unit of the National Guard is called to duty by federal authorities.

7 August
The Wright Company and the Glenn L. Martin Company merge. The result is the creation of the Wright Martin Aircraft Corporation.

29 August
The NACA receives an appropriation of $82,500 as part of the Navy appropriations bill for 1917. A large piece—$69,000—is used for construction of the aeronautical laboratory at Langley Field, Virginia. The Signal Corps receives more than $14 million from the Army appropriations bill to be used for military aeronautics.

DE HAVILLAND DH-4

Walter J. Boyne

Left: *The durable DH-4 was powered by the Liberty-12 engine, which had been produced by the thousands during World War I. The DH-4 was a favorite among "Barnstormers" and also used as a training aircraft. (*DH-4 Flying the Mail in the 1920s *by John McCoy, USAFAP).*

Opposite: *After World War I, DH-4s filled many aviation roles. This one carried mail across the country. (USAF)*

The de Havilland DH-4 was the only airplane built in the United States to serve in combat over the Western Front during World War I. Despite its relatively good performance, it was much maligned, gaining an unfair reputation as a "flaming coffin." Besides its wartime role, it is important as the first combat plane to be mass-manufactured in the United States, with 4,356 being built. It also pioneered the use of the famous Liberty V-12 engine. After the war, DH-4's, often rebuilt and modified, served for many years in a wide variety of military and civil duties.

The handsome, if somewhat angular, biplane began life in Great Britain as the Aircraft Company Ltd. (Airco) D.H. 4. Designed by Geoffrey de Havilland, it first flew in mid-August 1916, powered by a B.H.P. engine of 200 horsepower. Intended as a reconnaissance/bomber aircraft, the D.H. 4 was fast for the time at 113 mph. Most production aircraft received the 275-hp Rolls-Royce Eagle engine, which brought top speed to over 120 mph, faster than almost every German fighter. The key to both the D.H. 4's performance and versatility was its combination of clean lines, generous wing area, high aspect ratio wings, and the excellent Rolls Royce engine.

As the United States had neither combat aircraft nor an industry to make them, a decision was made to purchase foreign designs and manufacture them in America. Even though the D.H. 4 had been in combat since 1917, its performance justified its adoption for American use, and the Dayton-Wright, Fisher, and Standard companies were selected to build it.

The speed with which American industry transformed the D.H. 4 into the DH-4 was remarkable. A contract with Dayton-Wright was signed on August 14, 1917. Drawings (which used the English measuring system) and a sample aircraft were sent from Great Britain. The first American aircraft, although virtually hand-built and powered by the equally new Liberty engine, made its first flight only two months later on October 29, 1917.

In order to build the DH-4, the United States also had to create sources for aircraft instruments, tanks, radiators, fabric, dope, varnish, fittings, guns, tires, and every other element.

For publicity purposes, the first war-zone-bound DH-4 left the Dayton factory by dint of enormous effort, on the night of December 31, 1917. However, there were serious delays in delivery and production, and the first four DH-4s did not arrive in France until May 6, 1918. Early in August, the 135th Aero Squadron flew the first mission ever by American-built planes, powered by American-built engines, and flown by American pilots. Brigadier General Benjamin D. Foulois was in the lead plane, which was flown by 1st Lieutenant A. Blair Thaw.

Initially, and oddly, the DH-4s were assigned as fighters to serve as top cover for SPADs engaged in ground attack work. The DH-4 quickly proved unsuitable for this task, and as more arrived in France, they were given reconnaissance and bombing missions. American pilots were not entirely pleased by the DH-4 and many made unfavorable comparisons with the Breguet 14 and Salmson 2A2 aircraft with which they were familiar. In the DH-4, a pressurized fuel tank separated the pilot and observer, making communication difficult. The aircraft soon gained a reputation for catching fire, but postwar analysis revealed that DH-4s were shot down in flames at the same rate as other observation types. Some 696 were delivered to France, with a maximum of 196 in service at the front at any time. They were used both by the Army Air Service and the Navy-Marine Northern Bombing Group.

Many wartime aircraft were modified to be DH-4B's by relocating the now unpressurized fuselage fuel tank and squeezing the pilot and observer seats closer together. The landing gear was moved forward to avoid nosing over. In the early postwar years, Boeing and Fokker constructed gas-welded steel tube fuselages to create the DH-4M1 and 2 series, respectively. These served as trainers and liaison planes for years, the last ones being retired in 1932.

The Post Office Department relied on DH-4s as its standard equipment until 1927, when they were turned back to the Army and subsequently used by the Forest Service.

The DH-4 taught the Air Service and American industry how to procure and produce a modern, if already semiobsolescent warplane. Just as with the Curtiss P-40 in World War II, the DH-4 delivered adequate performance, and was available in quantity when warplanes were vitally needed.

Left: *A JN-4 "Jenny" turns to final approach during a training mission. ("Kelly Field, Texas, 1925," by Bill Fleming, USAF Art Collection)*

Opposite, top: *This war bonds poster warned of the possibility of aerial attack upon the Statue of Liberty by enemy aircraft, a scare tactic to help fill wartime coffers. (NASM Poster Collection)*

Opposite, bottom: *Lawrence B. Sperry developed early gyroscopic stabilizers used in aircraft and unmanned drones. (NASM)*

2 September
Flying in two airplanes, Lieutenant William A. Robertson and Corporal Albert D. Smith exchange radiotelegraph messages with Lieutenant Henry A. Dargue and Captain Clarence C. Culver over North Island, California. The aircraft are about two miles apart during the experiment.

13 September
In San Diego at the Signal Corps Aviation Station, an aeronautics course for field officers is established.

27 September
Charles D. Wolcott, secretary of the Smithsonian Institution, authorizes a grant of $5,000 to Robert Goddard to develop a rocket capable of investigations into weather and other phenomena.

18 November
Captain Raynal C. Bolling leads seven JN-4 aircraft form the 1st Aero Company on the first National Guard cross-country flight. The seven-plane formation flies from Mineola, New York, to Princeton, New Jersey, and then returns.

23 November
The NACA recommends purchasing a large plot of land north of Hampton, Virginia, to be used as an airfield by Army and Navy aircraft. Named Langley Field, this location will eventually house the first NACA laboratory facilities.

20 December
The Army Balloon School is established at Fort Omaha, Nebraska.

22 December
Lawrence B. Sperry files a patent request for an "aerial torpedo," basically an unmanned airplane. The design for the weapon includes gyroscopic stabilizers to keep the aircraft level in flight and an engine revolution counter that automatically shuts off the fuel supply after a certain number of revolutions have been reached. Early experiments with this technology take place in an aircraft (sometimes with Lawrence Sperry onboard to monitor the system). Later, the Sperrys would apply this technology to the U.S. Army's Liberty Eagle project. The Liberty Eagle (also known as the Kettering Bug and the Flying Bomb) was a small, pilotless biplane that incorporated the Sperry gyro and revolution counter components. Sperry's patent, number 1,418,605, is finally granted in 1922.

30 December
Near Hampton, Virginia, the U.S. Army establishes a new aviation school. It would be named Langley Field the next year. Today, Langley AFB remains the oldest Air Force base still active.

1917

3 January
Germany announces that they will unleash their underwater forces in a campaign of unrestricted submarine warfare in the Atlantic Ocean. The campaign begins on the first day of February.

9 January
Captain Henry H. Arnold leaves the Aviation School at San Diego to go to Panama where he will assume command of the 7th Aero Squadron. His duties will include finding a location to deploy his squadron, working with the Navy to protect the canal, and then standing up the squadron which, as of this date, had no airplanes.

3 February
The United States severs diplomatic relations with Germany.

19 February
Lieutenant Colonel John B. Bennet takes command of the Aviation Section of the Signal Corps. Lieutenant Colonel Squier is promoted to Chief Signal Officer.

24 February
Great Britain, whose intelligence forces had intercepted and deciphered a cable from Germany to Mexico, releases the contents of the Zimmerman Telegram to officials of the United States.

THAT LIBERTY SHALL NOT
PERISH FROM THE EARTH
BUY LIBERTY BONDS
FOURTH LIBERTY LOAN

28 February
In experiments at North Island, voice communication is established from an airplane to the ground transmitted by a radiotelephone. This experiment marks the beginning of the quest to establish easily understandable communications between aircraft and individuals on the ground.

8 March
Count Ferdinand von Zeppelin, the inventor of the rigid airship, dies at age 78.

13 March
Chief Signal Officer Brigadier General George O. Squier approves the establishment of the Intelligence subdivision. This is the birth of Army Air Intelligence.

20 March
President Wilson's cabinet unanimously votes in favor of war with Germany.

2 April
President Woodrow Wilson asks Congress for a declaration of war against the Central Powers. He says, "The world must be made safe for democracy."

Left: *An Allied airfield in Europe. ("Issoudun, August 1918," by Jules Andre Smith, U.S. Army Art Collection)*

6 April
The U.S. declares war on the Imperial German Government and enters the Great War. Manning the Aviation Section of the Signal Corps are 35 pilots and 1,987 enlisted men, assigned to fly 55 training aircraft.

30 April
Major Billy Mitchell flies over the trenches of no-man's-land. He is the first American Army officer to fly into hostile territory during the war.

16 May
To assist in directing the massive industrial buildup of America's aircraft industry, President Wilson establishes the Aircraft Production Board. The board is chaired by Howard Coffin, who brings with him many years of experience with the Hudson Motor Car Company.

23 May
In a cable to the American government, French Premier Alexander Ribot asks the U.S. to send pilots and planes to assist in the war against the Central Powers. The Ribot Cable specifies a need for 5,000 pilots, 4,500 aircraft, and 50,000 airplane mechanics, as well as a request for 16,500 engines and planes to be produced and delivered from January to June in 1918. Major Benjamin Foulois is charged to transform the requests in the Ribot Telegram into a detailed plan. Included in Foulois' work are estimations of student pilot requirements, numbers and locations of training fields, selection of aircraft manufacturers, and program budgeting.

26 May
The Aviation Section of the Signal Corps gains staff recognition when Major T. F. Dodd is assigned to the staff of the commander in chief of the American Expeditionary Forces (AEF) as the aviation officer.

28 May
The Selective Service Act, passed in Congress, eventually results in the drafting of more than 3 million men for the war effort.

29 May
In a room in the Willard Hotel in Washington, D.C., Colonel Edwin Deeds gathers together several engine designers and they begin to conceptualize the "Standardized Aircraft Engine" for the war. The engine finally produced as a result of this initial meeting is the 12-cylinder Liberty, capable of producing 400 horsepower. When production finally ends, more than 15,000 Liberty engines, of various configurations, have been built.

2 June
The Army's Aviation Section is reorganized and renamed—Airplane Division, Army Signal Corps.

5 June
The first American aviation unit, the 1st Aeronautical Detachment, arrives in France.

16 June
Nearly 100 skilled civilian airplane mechanics are the first group of aviation personnel to be deployed overseas during World War I. They sail to England,

where they are to study French and British aircraft production and then return to the U.S. to assist in the buildup of American aircraft factories.

17 June
In an effort to qualify and quantify aircraft production in the U.S., a team of investigators led by Harvard Law School graduate Major Raynal Bolling is sent by the Aircraft Production Board to Europe to evaluate their aircraft production programs and the operational use and success of airpower in the war.

30 June
Lieutenant Colonel Billy Mitchell becomes Aviation Officer for the American Expeditionary Force (AEF). He takes the place of Major T. F. Dodd.

3 July
American soldiers of the American Expeditionary Forces (AEF) begin arriving in France.

4 July
The first 8-cylinder Liberty engine arrives at the Bureau of Standards in Washington, D.C, for testing. Building the first engine from design to assembly takes only six weeks.

20 July
Scott Field, named after the first enlisted man killed in an aircraft, is designated by the War Department to be built at a site near Shiloh Valley Township, Illinois. Corporal Frank S. Scott, who died on September 28, 1912, is the only enlisted man to be honored by having an air base named after him. His pilot, Lieutenant Lewis Rockwell, was also killed as a result of the crash.

23 July
Major Benjamin Foulois is named officer-in-charge of the Airplane Division, Army Signal Corps.

24 July
In response to the Ribot Telegram of May 24, and Major Foulois' efforts to meet the requirements, the largest congressional appropriation ever awarded to Army aviation is authorized. The $640 million appropriation is far beyond any other. Additionally, the Aviation Section is authorized to expand to nearly 10,000 officers and more than 87,000 enlisted troops.

27 July
A British-built D.H. 4 aircraft arrives in the United States to serve as the model for American de Havilland DH-4 production. The American version of the DH-4 would be powered by the 12-cylinder Liberty engine. About 4,500 are built in the U.S. before the end of the war, but only a few hundred make it to Europe as combat fighters.

28 July
The 29th Provisional Construction Squadron (later the 400th CS) arrives at Liverpool Harbor. It is the first American Aero Squadron to arrive in the service of the AEF.

13 August
Major Ralph Royce commands the 1st Aero Squadron when they set sail for Europe. The 1st Aero Squadron is the first flying unit to report for duty in the AEF.

Right: *The British-designed de Havilland DH-4 was the only U. S.-manufactured aircraft to fly during World War I. It was powered by an American-designed and -built Liberty engine. (USAF)*

EDGAR S. GORRELL

George Williams

Mention strategic planning to most students of airpower, and the Great War seldom enters the conversation. If pressed, enthusiasts and experts may recall the colorful achievements of William "Billy" Mitchell and Benjamin "Benny" Foulois and their cooperative endeavors with their counterparts in the British, French, and Italian air services during 1917–1918. The role played by Colonel Edgar S. Gorrell, U.S. Air Service, will either be slighted or overlooked entirely.

Gorrell has received little credit for his contributions to military aviation and, thus, has never emerged as an influential figure in American airpower. Not an entirely original thinker, the energetic Gorrell nonetheless typified the best tradition of the American military by his pragmatic approach to military airpower theory.

A 1912 graduate of the Military Academy at West Point, and a qualified military aviator when the United States declared war on Germany in April 1917, Gorrell deployed with the advance group of the American Expeditionary Forces (AEF) on June 16, as a part of what was known as the Bolling Mission. Arriving in Paris in July 1917, well ahead of the bulk of the AEF, he served as head of the U.S. Air Service Technical Section and was responsible for investigating requirements and implementing policy and programs to jump-start the nascent American bombing program.

Unsurprisingly, French and British air officers, including Major General Hugh M. Trenchard, head of the RFC/RAF, wielded considerable influence upon the youthful, hard-charging Gorrell.

An English liaison officer, Major Lord Tiverton, later second Earl of Halsbury, was particularly influential in determining American bombardment doctrine. At the time, however, his contributions were not acknowledged. As part of his duties with the British Air Commission in Paris in September 1917 Tiverton had formulated a comprehensive plan for the aerial bombardment of the German homeland.

A few months later, in late November 1917, Gorrell submitted his own proposal for a massive Allied bombardment campaign against Germany through the American chain of command. Gorrell's Plan, lauded as an inspired, original example of American thought and the cornerstone of much of subsequent strategic airpower doctrine, drew heavily upon Tiverton's submission to the Air Board of the British Admiralty written just three months before.

Tiverton's September 1917 bombing proposal for the British had addressed four broad areas: objectives, offensive forces, the need to concentrate bombing effort, and impact upon German civilian morale. As the first practical study produced by any Allied nation of the feasibility of strategic bombardment, Tiverton's detailed proposal for a long-range bombing campaign against German cities set an early standard for offensive airpower doctrine.

Gorrell's scheme for the U.S. Air Service mirrored Tiverton's overall structure, duplicating his sequence of paragraphs, transposing exact phrases and terminology, and accepting his logical arguments and conclusions, virtually word for word. Unsurprisingly, the conclusion to the body of Gorrell's paper echoed almost exactly the urgent tone of Tiverton's final paragraph.

In any case, the bombing scheme was not implemented before the Armistice, largely due to production and training shortfalls. Gorrell wrote a second paper in early 1918, "The Future Role of American Bombardment Aviation," repeating Tiverton's tenets and adding Trenchard's views on bombing's moral effect. Predictably, it met the same fate as its more notable but still obscure predecessor. However, both of Gorrell's proposals survived to be cited, consulted, and accepted during the interwar years at the Air Corps Tactical School.

Edgar Gorrell's skill lay in his drawing together operational lessons from the more experienced Allied air services for American benefit. Thanks largely to Gorrell's energy and realistic initiatives, American airpower thought implicitly benefited from Tiverton's method of systematic analysis and research. Lord Tiverton's seminal approach to airpower theory, preserved in Gorrell's papers, was eventually incorporated into U.S. bombardment doctrine that came to fruition during the Second World War. In the long run, the pragmatic Gorrell has thus enjoyed the attention and qualified acclaim that has thus far eluded Major Lord Tiverton, his English counterpart and staff mentor.

Opposite: *Flying French Breguet bombers, the 96th was the first American group to bomb German installations. ("The 96th Bombardment Squadron-1918," by John McCoy, USAF Art Collection)*

Above: *Colonel Edgar S. Gorrell. (NASM)*

Left: *Colonel Henry H. Arnold stands beside an early-production Liberty Engine. (Robert and Kathleen Arnold Collection)*

21 August
The L. W. F. Engineering Company's "Model F" aircraft flies for the first time. The Model F is the first aircraft powered by a Liberty engine.

23 August
The 1st Reserve Aero Squadron sails from New York headed for France. This unique unit will later be redesignated as the 26th Aero Squadron.

25 August
The 12-cylinder Liberty Engine is ordered into mass production after the test model passes its 50-hour test with flying colors. During the test, the engine produces an average of 300 horsepower.

3 September
Brigadier General William L. Kenly, a Field Artillery officer, takes command of U.S. aviation activities in Europe when he is named as the Chief of the Air Service for the AEF. At the same time, Colonel Billy Mitchell is named as the air commander in the Zone of Advance.

13 September
Having served under General Pershing in Mexico, the 1st Aero Squadron becomes the first aviation unit to serve as part of the American Expeditionary Force. Their primary mission becomes battlefield reconnaissance and artillery spotting. The unit is also credited with 13 air-to-air victories during the war.

21 September
General Tasker H. Bliss is appointed Chief of Staff, United States Army.

16 October
During testing at Langley Field, Virginia, radio-telephone transmissions of 25 miles between two planes and 45 miles between planes and ground stations are achieved.

18 October
The Signal Corps, recognizing the need for a centralized location to carry out aeronautical research and development, establishes McCook Field near Dayton, Ohio, home of the Wright brothers, inventors of the airplane.

The Signal Corps establishes the Aviation Medical Research Board.

23 October
The American infantry fires its first shots in combat from the trenches of France.

29 October
Test pilot Howard Rinehart flies the first American-built, British-designed de Havilland DH-4 during its first flight near Dayton, Ohio.

6 November
In Russia, Vladimir Ulyanov Lenin leads the

Communists during the October Revolution and successfully establishes a Communist government there.

7 November
Eugene Jacques Bullard becomes the first African American pilot to score an aerial victory during World War I. Bullard had left the U.S. to avoid rampant racism in 1912. He joined the French Army and fought at Verdun. Later he retrained as a pilot, flying 20 combat missions and credited with one victory.

21 November
At Cambrai, the first major tank battle in history is fought when British forces break through the previously unbreakable Hindenburg Line. Surprised by their success, the British are unable to capitalize on the initial breakthrough.

27 November
Brigadier General William Kenly is replaced as Chief of the Air Service for the American Expeditionary Force (AEF) in Europe by Brigadier General Foulois.

1918

19 January
The School of Aviation Medicine opens at Hazelhurst Field in Mineola, New York. Under the command of Major William H. Wilmer doctors and nurses are trained in the skills required to treat injured military aviation personnel. Their first experiment is to develop a pressure chamber to simulate conditions at high altitude. The altitude chambers still used today are the descendants of those built in 1918.

20 January
Colonel Billy Mitchell takes command of the air assets as Chief of Air Service, I Army Corps, at Neufchateau, France.

23 January
The first American balloon flight occurs as part of the AEF near Marne, France.

5 February
While flying in a French-piloted aircraft as a gunner, Lieutenant Stephen W. Thompson shoots down an enemy aircraft and becomes the first American in American military history to be credited with an aerial victory. Lieutenant Thompson had been invited aboard a Breguet bomber when the French gunner became ill. While under attack over the town of Saarbrucken, Thompson's defensive machine-gun fire downed an Albatross D. III fighter.

7 February
The Joint Army/Navy Technical Board resolves to standardize instrumentation on Army and Navy planes.

18 February
The 95th Aero Squadron arrives in France and is the first complete American fighter unit in the theater of operations, while the 103rd Pursuit Squadron, AEF, is formed with former members of the Lafayette Escadrille, which began operations at the front under French control.

Below: *Eugene Bullard was the first African American pilot to score an aerial victory in World War I in November 1917. (NASM)*

Following Spread: *Lieutenant Stephen W. Thompson, a gunner on a French-piloted Breguet bomber, shoots down an enemy Albatross and becomes the first American to score an aerial victory. (*The Air Forces First Victory—5 February 1918 *by Eugene Knight, USAFAP)*

Left: *The Nieuport 28 was the first pursuit-type aircraft used by an American unit, under American command, supporting American troops during World War I. This aircraft is on display at the National Air and Space Museum, Steven F. Udvar Hazy Center, near Washington, D.C. (NASM)*

23 February–5 March
The 2d Balloon Company begins combat operations under I Corps command at Toul, France. This is the first occasion that an Army aeronautical unit serves in support of frontline American troops. The 2d Company is the first to see action, but more than 35 balloon companies participate in operations during the rest of the war.

6 March
Elmer and Lawrence Sperry successfully test an unpiloted aircraft near Long Island, New York. The Curtiss/Sperry "Flying Bomb" is constructed from a Curtiss N-9 seaplane and intended for launch from a catapult. Trails using this launch method are not successful. Finally, the "Flying Bomb" is mounted on the top of an automobile and driven into the wind until it lifts off and flies straight ahead for about 1,000 yards and then dives to the ground when the engine is cut off by the revolution counter.

8 March
In pressure room tests at the Signal Corps laboratory in Mineola, New York, a simulated pressure altitude of 34,000 feet is achieved during an early altitude chamber flight lasting 24 minutes.

11 March
The first Distinguished Service Cross awarded to an Army airman is earned when Lieutenant Paul Baer shoots down an enemy aircraft. He is the first pilot flying with an American unit to down an aircraft.

12 March
The first Air Service pilot to be killed on a combat mission, Captain Phelps Collins from the 103d Aero Squadron, dies near Paris when his SPAD XIII crashes after a high-altitude dive.

19 March
The first American operations across enemy lines are flown by the 94th Aero Squadron. The 94th is nicknamed the "Hat-in-the-Ring" Squadron because of its aircraft insignia.

1 April
American Aviation Headquarters opens in Rome, Italy, under an agreement in which Italian pilots will train American pilots.

In Britain, Major General Hugh Trenchard becomes Chief of Staff for the first independent air service in history. The Royal Air Force (RAF) is formed by combining the Royal Flying Corps and the Royal Naval Air Service.

6 April
Lieutenant J. C. McKinney takes night photographs by using magnesium flares to illuminate the ground. His pilot, a civilian, is Norbert Carolin.

14 April
Piloting Nieuport 28s from the 94th Aero Squadron, Lieutenant Alan Winslow and Lieutenant Douglas Campbell shoot down two German fighters in a 10-minute air battle over Toul Airdrome. Campbell is the first American pilot to down an enemy aircraft.

On this same day, Captain Edward V. Rickenbacker flies his first combat mission in France. He will become America's "Ace of Aces" during World War I.

21 April
German ace Manfred von Richthofen, known as the Red Baron, is shot down by Canadian aviator Captain A. Roy Brown. The Red Baron had 80 confirmed kills

Right: *Lieutenant Eddie Rickenbacker and members of the 94th Aero Squadron ("Hat-in-the-Ring") standing next to a French Spad. Several members of the unit had flown with the Lafayette Escadrille early in the war. (USAF)*

Below: *British Major General Hugh Trenchard led the first independent air force in history when the Royal Air Force (RAF) was formed in 1918. (NASM)*

when he himself was finally shot down. Although there remains some dispute over credit for the victory, most evidence points to Captain Brown.

23 April
The long-anticipated first shipment of Liberty engines arrives in Pauillac, France.

29 April
Lieutenant Edward V. Rickenbacker helps shoot down an enemy aircraft and receives a half credit for the kill.

Construction plans for a five-foot wind tunnel are approved by the NACA for construction at the Langley Memorial Aeronautical Laboratory.

7 May
Captain Rickenbacker scores his first kill when he shoots down a German Pfalz aircraft. He is flying a Nieuport 28 with the 94th Aero Squadron.

11 May
The first Liberty-engine-powered, American-built DH-4 aircraft begins service with the American Expeditionary Forces in France. Only a few hundred of the U.S. DH-4 versions are flown in combat.

15 May
U.S. Army pilots establish a permanent airmail route between Washington, D.C., and New York City. It is the first such permanent airmail route sponsored by the government. The mail route is covered by a fleet of seven Curtiss JN-4H Jenny biplanes.

17 May
The first flight of an American-made DH-4 built by the Dayton Wright Company is made in France.

18 May
The 96th Squadron is established at Amanty Airdrome, France. This is the first unit specifically formed to perform daytime bombardment.

19 May
Famed Lafayette Escadrille pilot Raoul G. Lufbery is killed in combat. While in Escadrille service Lufbery shot down 17 aircraft and then transferred to the AEF.

21 May
Major General William L. Kenly takes command of the Division of Military Aeronautics as Army aviation splits off from the Signal Corps and is placed directly

EDWARD V. RICKENBACKER (1890–1973)

Walter J. Boyne

Edward V. Rickenbacker used his native intelligence, strong character, and energy to rise from abject poverty to become successively a prosperous automobile salesman, a top racing car driver, the American Ace of Aces in World War I, a visionary automobile manufacturer and, in his final career, the cost-cutting head of a profitable Eastern Airlines.

His greatest fame came in World War I, when he became the leading American ace with 26 official victories, a total revised downward to 24-1/3 by the USAF Historical Research Center in the 1960s. Initially rebuffed by the U.S. Army as not being suitable officer material, Rickenbacker gave up a racing career where he was earning the modern equivalent of a million dollars per year to volunteer as an enlisted man, serving as a driver on General Pershing's staff. At 27, he was considered "too old" for flying but nonetheless adroitly maneuvered his way into flying school in France. His profanity and bad grammar caused many of his more polished colleagues to consider him to be too rough hewn, but there was no doubting his flying ability and his understanding of the maintenance aspects of flying.

Rickenbacker was with the 94th Aero Squadron when it entered combat in the spring of 1918, flying the fast but fragile Nieuport 28 fighter. His first victory came on April 25, and by May 30 he had shot down five aircraft. During much of this time he flew with a severe mastoid condition, resulting in extreme pain in his jaw and ears, which was aggravated by the sudden changes in altitude encountered in a dog-fight. He was sidelined by the illness, and did not return to combat until September, when he was promoted to captain and assumed command of the 94th. His squadron was now equipped with the more rugged (if less reliable) SPAD XIII, an aircraft suited to his aggressive tactics. In the next two months Rickenbacker scored 21 victories, one of which was not credited until well after the war.

The former racing star proved to be an excellent squadron commander, concerned for the welfare of both his officers and men, and knowledgeable not only of air combat, but also of maintenance and logistics. In 1930, he was awarded the Medal of Honor for a September 30, 1918, action in which he attacked seven enemy aircraft and shot down one fighter and one observation plane.

After the war, Rickenbacker returned to his first love, the automobile, and founded the Rickenbacker Motor Car Company, which used the famous "Hat in the Ring" insignia of the 94th as its identifying symbol. Despite the car's many advanced features, including four-wheel brakes, the company failed in 1927. Typically, Rickenbacker assumed full responsibility for the firm's $250,000 debt, which he subsequently paid off out of his private earnings. In the following years he became president of the Indianapolis Speedway and an executive with General Motors. In 1933, he was asked by GM to take over one of its troubled companies, Eastern Airlines. Rickenbacker ran Eastern as he had run the 94th, with an iron hand but a kindly eye for hardworking employees. He restored the airline to profitability and purchased it in 1938.

In February 1941, he was almost fatally injured in the crash of an Eastern Douglas DST transport on approach to the Atlanta airport. He was still recuperating from his injuries when President Franklin D. Roosevelt asked him to undertake a series of special missions. One of these involved a flight to Australia in October 1942. The B-17 carrying him and his colleagues became lost, ran out of fuel, and crashed in the Pacific Ocean. For the next three weeks, Rickenbacker and seven companions floated in three tiny rafts, all but one managing to survive exhaustion, thirst, and hunger before being rescued. Many of the survivors credited Rickenbacker with maintaining their will to live.

After the war, Rickenbacker resumed active control of Eastern Airlines, but in time he made a number of business decisions which ultimately forced his retirement in 1963. He wrote two very readable memoirs, both with the assistance of ghost-writers. The first was *Fighting the Flying Circus,* written with Laurence LaTourette Driggs, while the second was *Rickenbacker, An Autobiography,* written with Booten Herndon.

Rickenbacker was valorous and exemplified the Air Force core values, but perhaps his greatest contribution to the service was his demonstration that mutual respect and consideration between officers and enlisted troops is of the utmost importance.

Opposite: "*Captain Edward V. Rickenbacker,*" *by Howard Chandler Christy. (NASM Art Collection)*

Below: *America's leading World War I flying ace, Captain "Eddie" Rickenbacker. (USAF)*

Left: *Brigadier General Billy Mitchell greets Chief of the Air Service, Major General Mason M. Patrick, after a flight. (USAF)*

Below: *Brigadier General "Billy" Mitchell and General John J. Pershing worked closely during World War I as American Expeditionary Force (AEF) operations expanded. Mitchell became Pershing's air commander in France. (USAF)*

under the Secretary of War. The Bureau of Aircraft Production is also established and John D. Ryan is named its director the following week. These two divisions form the Army Air Service, which is formally established on May 24.

29 May
Brigadier General Mason Patrick takes command of the Air Service of the American Expeditionary Forces.

12 June
The first American unit to accomplish daylight bombing, the 96th Aero Squadron, attacks hostile targets near Dommary-Baroncourt, France.

6 July
An American-manned, French-built observation balloon of the 2d Balloon Company is shot down by a German Albatross. It is the first U.S. observation balloon shot down during the war.

14 July
Lieutenant Quentin Roosevelt of the 95th Aero Squadron is shot down in his Nieuport 28 over France. The youngest son of former President Theodore Roosevelt is buried near the crash.

20 July
The 148th Pursuit Squadron begins field operations with the Royal Air Force at Capelle Airdrome near Dunkirk.

24 July
At a conference of the Allied commanders, the American Army, under the command of General John J. Pershing, is assigned to attack the German lines at the salient near St. Mihiel. The attack will be the first fought under the American flag on the Western Front. Pershing issues orders which establish his battle commanders, including his Chief of Air Service, Colonel William Mitchell.

Right: *Strategic bombardment was attempted during World War I but yielded limited results. American bomber development began seriously after the war had ended. Aircraft, like this Martin MB-1 bomber, were also used to carry mail after the war. (NASM)*

2 August
Eighteen American-built DH-4 aircraft powered by Liberty engines patrol the battle front during missions flown from Ourches, France. This marks the first combat patrols by American-built planes.

7 August
In preparation for the attack at St. Mihiel, the Air Service commander, Colonel Billy Mitchell, requests intelligence information from the G-2 staff. Among the requested information are airdrome locations, troop concentrations, enemy travel routes, and potential enemy transportation choke points (road intersections, rail stations, and the like).

17 August
The U.S. Army's Martin MB-1 bomber flies for the first time. Although the MB-1 does not see action during the Great War, it becomes the Air Service standard bomber aircraft through the 1920s. Later, the type is modified and used by the Post Office Department.

19 August
In the most detailed description of aircraft procedures issued to date, Colonel Billy Mitchell's office releases Air Service Circular No. 1, which clearly delineates everyday operations for observation, pursuit, and bombardment aircraft. Significant detail is given to the mission and execution of tasks by the "Infantry Airplane." Communication techniques from radio to ground signaling to the engaged infantry take up dozens of pages in the manual. Details covering the pursuit mission and bombardment include descriptions and diagrams of appropriate formations to be used when attacking the enemy in the air or on the ground. Interestingly, "night reconnaissance" is also discussed as a method to establish enemy troop movements the day before battle.

24 August
A number of French aviation units are assigned to serve under the command of the Air Service of the First Army. British bomber forces, however, are not subordinated to Mitchell for the battle. They remain under the command of Major General Hugh Trenchard but do coordinate their bombardment activities with Colonel Mitchell's staff.

28 August
John D. Ryan becomes the first Director of the Air Service, also holding the title of Assistant Secretary of War.

Air Service reports indicate that a total of 1,467 American and French airplanes will be available for the offensive at St. Mihiel as well as 21 Royal Air Force balloons. Cooperation with the Royal Air Force bombers is ongoing as the date for the battle approaches.

3 September
General Pershing lifts a previous restriction upon aerial activity to disguise AEF troop concentrations of the past two weeks around the St. Mihiel salient. Increased reconnaissance flights to secure photographs of the enemy positions begin on a large scale.

7 September
Field Orders No. 9 issued by General Pershing's headquarters provide by-name detailed responsibilities for combat operations and each day's objective for aviation assets. French aviation assets are formally brought under Colonel Mitchell's command. During the following four days, plans to execute the operation continues at the unit levels across the St. Mihiel front.

Practical aerial troop transport is demonstrated when 18 soldiers are airlifted in several aircraft between two bases in Illinois.

11 September
Colonel Billy Mitchell issues orders for the Air Service to be executed the next morning. He orders that the "Air Service will take the offensive at all points with the objective of destroying the enemy's air service, attacking his troops on the ground and protecting our own air and ground troops." Mitchell's orders are issued through his Chief of Staff, Air Service Lieutenant Colonel Thomas DeWitt Milling.

12 September
The first major American air offensive of the war begins at St. Mihiel, France. In this remarkable battle more than 1,480 aircraft participate—the largest force ever assembled. Brigadier General Billy Mitchell is in command of the Allied air forces.

The "Arizona Balloon Buster," Lieutenant Frank Luke, shoots down his first enemy observation balloon. Most combat reports from the morning reflect extremely cloudy and rainy weather for much of the day. Heavy artillery fire is ongoing after the ground attack begins. Reports for the First Pursuit Group indicate that enemy observation balloons are aloft but are immediately reeled back to the ground when attacked by American patrols. The balloons remain grounded for the duration of the day. Limited enemy fighter activity is reported, and anti-aircraft fire is significant only in the immediate vicinity of the St. Mihiel salient.

At the end of the day, following a highly successful offensive where all AEF units reached their objectives, orders for the following day's operations are issued. German forces had been observed retreating from the St. Mihiel salient in large numbers. Pershing's orders are simple. They read, "The attack will continue tomorrow with a view to completing the hostile defeat and gathering the booty." He also orders the Air Service to continue to attack as described in previous Field Orders No. 9.

13 September
American forces of the First Army, AEF, complete the junction between northern and southern attacks. The successful pincer movement traps thousands of German soldiers who are taken prisoner. AEF objectives are achieved across the entire line of attack and reconnaissance forces begin probing the well-defended Hindenburg Line to establish if the Germans will defend or withdraw from that location. With improving weather, the Air Service is ordered to pursue and attack enemy convoys of materiel and troops with reconnoitering positions north of the Hindenburg Line.

Left: *Lieutenant Frank Luke, the "Arizona Balloon Buster," received the Medal of Honor for shooting down 14 enemy balloons and four enemy aircraft before being shot down and killed by enemy troops in combat. (NASM)*

Right: *Captain Eddie"Rickenbacker received the Medal of Honor for engaging seven enemy aircraft and shooting down two of them. The battle is depicted in this painting. ("Rickenbacker's Medal of Honor" by William Marsalko, USAF Art Collection)*

14 September
While AEF ground forces continue to press forward along the Hindenburg Line, enemy aircraft are seen in greater numbers, often defending artillery batteries. Fokker and Pfalz pursuit planes flying in formations numbering between seven and 15 aircraft begin to appear in greater numbers in response to the AEF offensive. Clear skies facilitate more activity along the battle front.

15 September
Major Carl Spatz (later changed to Spaatz) shoots down an enemy Fokker aircraft that is part of a formation of five Fokkers. The AEF takes two days to consolidate gains and replenish supplies in preparation for another major offensive operation—The Meuse Argonne.

18 September
Major R. W. Schroeder sets a world altitude record of 28,899 feet over McCook Field, Dayton, Ohio.

25 September
While on a voluntary patrol near Billy, France, 1st Lieutenant Edward V. Rickenbacker of the U.S. Army Air Corps, 94th Aero Squadron, Air Service attacks seven enemy planes, shooting down two of them, one Fokker and one Halberstadt. For his actions he is awarded the first Medal of Honor for air combat. Because the recommendation for the award was lost during the war, his medal is not awarded until 1930.

25 September
Battle Orders No. 7 is issued by Colonel Billy Mitchell, Chief of the Air Service, First Army, delineating offensive operations during the anticipated Meuse-Argonne campaign. This aerial campaign will continue until the end of hostilities on November 11.

26 September
At dawn, the Air Service takes to the skies as part of a massive Allied offensive at the Meuse-Argonne. The orders directing air operations are simple: "Our air service will take the offensive at all points . . . with the objective of destroying the enemy's air service, attacking his troops on the ground and protecting our own air and ground troops." The air battle will persist until the armistice is signed on November 11, 1918.

29 September
2nd Lieutenant Frank Luke Jr., U.S. Army Air Corps, 27th Aero Squadron, 1st Pursuit Group, Air Service, voluntarily starts patrol on German observation balloons and dies of a chest wound after being forced to land in Near Murvaux, France. Completely surrounded by enemies, he defends himself with his automatic pistol. Lieutenant Luke earns the name "Arizona Balloon Buster" for his success at attacking enemy observation balloons. At the time of his death, Luke had shot down 14 enemy balloons and four airplanes in less than three weeks (September 12–29). Luke is posthumously awarded the Medal of Honor for his actions this day.

Left: *Lieutenants Goettler and Bleckley drop supplies to the "Lost Battalion" before being shot down over France. They received the Medal of Honor for their bravery. ("Airdrop to the Lost Battalion," by Merv Corning, USAF Art Collection)*

Below: *Mechanics assemble a Liberty Eagle in the factory. (USAF)*

2 October
The Liberty Eagle unmanned flying bomb is tested successfully near Dayton, Ohio. It flies for a total of nine seconds, reaching a speed of 42 miles per hour. More successful tests occur later in the month.

6 October
Lieutenant Harold E. Goettler (pilot), U.S. Army Air Corps, 50th Aero Squadron, Air Service, dies immediately after his plane is shot down near Binarville, France, while attempting to drop a package of supplies to a battalion of the Army's 77th Division—the Lost Battalion—which had been cut off by German forces.

Lieutenant Erwin R. Bleckley (observer), U.S. Army Air Corps, 130th Field Artillery, 50th Aero Squadron, Air Service, dies near Binarville, France, after receiving fatal wounds in an attempt to deliver supplies to a battalion of the Army's 77th Division—the Lost Battalion. His plane is brought down by enemy rifle and machine-gun fire from the ground. Both Goettler and Bleckley are posthumously awarded the Medal of Honor.

9 October
More than 350 bombers and pursuit planes drop 32 tons of bombs between La Wavrille and Damvillers, France. It is the largest single concentration of aircraft for one mission to date.

12 October
Aircrews assigned to the 185th Pursuit Squadron accomplish the first night flight operations over France flown by Americans during the war.

30 October
Captain Eddie Rickenbacker shoots down his final enemy aircraft of the war. His total of 26 confirmed victories was revised to 24-1/3 by the Office of Air Force History. Upon his return to the United States, Rickenbacker is awarded the Mackay Trophy for his combat record as America's "Ace of Aces."

6–7 November
Representatives of the Air Service and Ordnance Department gather at Aberdeen proving grounds where Dr. Robert Goddard test fires a number of tube-launched, recoilless, solid-propellant rocket guns for their review.

10 November
The last two American aerial victories of the war are made by airmen of the 94th Aero Squadron (Major Maxwell Kirby) and two crews from the 104th Observation Squadron. The final patrol by the Army Air Service is flown by members of the 3d Aero Squadron later in the day.

11 November
At 11 a.m. the Great War comes to an end. The American Air Service had deployed a total of 45 aircraft squadrons and 37 balloon companies. At war's end the Air Service consisted of 20,568 officers, 175,000 enlisted men, 3,538 aircraft in service in the AEF, and another 4,865 aircraft in service in the U.S.

14 November
Brigadier General Billy Mitchell is appointed Chief of the Air Service, Third Army.

4 December
Major Albert D. Smith commands a flight of four Curtiss JN-4 aircraft, which take off from San Diego, California. The four Jennies take 18 days to complete the Army's first transcontinental flight. They land near Jacksonville, Florida, on December 22. Only one of the four planes, that belonging to Major Smith, completes the trip from coast to coast.

23 December
Major General C. T. Menoher is appointed Director of Air Service. He takes command on January 2, 1919.

Right: *Brigadier General Billy Mitchell as Chief of the Air Service for the Third Army. He flew this Spad XVI aircraft during the Battle of St. Mihiel. (USAF)*

THE LIBERTY EAGLE

Left: *The Liberty Eagle—better known as the Kettering Bug—was produced by the Army in limited numbers in 1918 and 1919. Here, several "Bugs" are lined up next to the launching track during early testing. The project produced the first unmanned aerial weapon for use by the Air Service. (USAF)*

An eager crowd of Army brass was seated in grand stands at a secluded airfield near Dayton. They were to witness the flight test of a new weapon—a tiny, pilotless biplane that had the potential to strike the enemy by air without danger to pilots or soldiers. As the miniature craft lifted into the air, the development team began to smile and those in attendance began to wonder how this new missile could be used in fighting the Great War that was devouring Europe. Just then, the aircraft pivoted off its course, swooped and dove, and then headed straight toward the reviewing stands. Fortunately, it crashed landed a few hundred feet from the invited guests—much to the embarrassment of the once-confident development team.

One of the disheartened developers was Colonel Henry H. Arnold, Deputy Director of the Aviation Division of the Signal Corps. Arnold's task force of civilian scientists and military project managers had produced the Army's first guided missile. It was called the Liberty Eagle but unofficially was referred to as "FB," which stood for flying bomb, and also the "Kettering Bug," after the principal designer.

The unpiloted machine was a beautiful wood-crafted, mini-biplane. Early versions had been crafted from papier mâché. It housed a two-stroke Ford engine and carried a 225-pound warhead. The Bug had no wheels and was launched from a wagon-like contraption, which ran on a long section of portable track. The engine was cranked at one end of the track, which was pointed directly at the intended target. When the engine was fully revved, the mechanical counter was engaged and the Bug was released. When it reached flying speed, it lifted off and flew straight ahead toward the target, climbing to a preset altitude that was controlled by a supersensitive aneroid barometer. When the Bug reached its altitude, the barometer sent signals to small flight controls that were moved

Right: *A small team of engineers and mechanics start the Ford-built engine and prepare the Liberty Eagle for launch. Tests were flown in Dayton, Ohio, and in Florida. Note that the cart with wheels was reused after each launch. (USAF)*

by a system of cranks and a bellows from a player piano for altitude control. A gyro helped maintain the stability of the craft, the barometer helped maintain altitude, but only the design of the wings assured directional stability. Kettering designed the Bug to fly straight ahead until a mechanical counter had sensed a calculated number of engine rotations required to carry the weapon to the intended target. When the preset number of revolutions was reached, a cam fell into place and the wings folded. The Bug pointed to the ground, looking much like a diving falcon swooping down on its prey. Unfortunately, the Bug was not as deadly, nor as fast, as a falcon. After the first ill-fated test, lateral controls were added that rectified the instability problem caused by over-dependence upon the dihedral—the slightly upward angle of the wing—for lateral stability.

On the Liberty Eagle team were Orville Wright; Elmer Sperry—who spearheaded the Navy's aerial torpedo project a few months earlier; Robert Millikan—soon to be president of Caltech and a future Nobel Laureate; and Charles "Boss" Kettering—the primary engineer. Most test flights were accomplished at a remote test field in the Florida panhandle on the wide open sand dunes which existed in that day. As important as the development of the machine was the cultivation of the members of the team, particularly Millikan, who would play a crucial scientific role for the air arm in the 1930s and again during the Second World War. Arnold never forgot his experiences in production, administration, scientific experimentation, or testing. Nor did he forget the men who had helped design and build the missile.

Colonel Arnold was sent to Europe with the device in an effort to persuade Gen. John J. Pershing to use it in the waning days of World War I. Despite several setbacks, Arnold finally made it to the Western Front during November 1918, only a few days before the armistice went into effect. Because the weather was so terrible, however, the Army's secret weapon flew no combat missions. The project, along with most other weapons projects, was terminated shortly thereafter. But the idea of the Liberty Eagle project was not eliminated forever.

In the fall of 1939, Arnold wrote his old friend Charles Kettering, now vice-president of General Motors, wanting to develop glide bombs to be used in the event of war. Arnold wanted a device that could be used by the hundreds that might keep his pilots away from enemy flak barrages. Even though strategic airpower doctrine centered on precision, daylight, and high-altitude bombing, Arnold was willing to sacrifice some level of bombing accuracy to minimize potential danger to his airmen. Initially, Arnold considered reviving the Bug. About 50 were in storage from the First World War and a mass production plan already existed. The idea was finally dropped in 1942 because of the weapon's relatively short 200-mile range.

Today's Air Force depends upon the technology of precision-guided munitions that have their roots in the Liberty Eagle, developed and tested during the First World War.

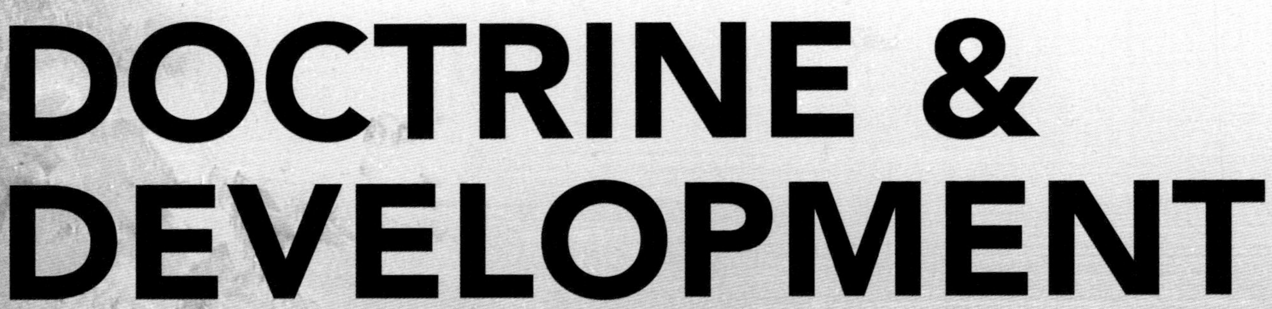

DOCTRINE & DEVELOPMENT

1919–1940

BB
80
80
BB
82
BREADY
91

DOCTRINE & DEVELOPMENT

1919–1940

Immediately following the end of hostilities in November 1918, America's military demobilized. Air Service appropriations plummeted from $460 million to $25 million in the 1919 budget. The Air Service officer corps was gutted from 190,000 to 81,000 by January 1919, and by the end of June had shriveled to 5,500 officers and 21,500 enlisted men. In all, 95 percent of the officers commissioned for the war and nearly all of the wartime enlistees were discharged. Equipment production ceased, machinery was liquidated, airfields and maintenance depots were abandoned, and production contracts were summarily cancelled.

Those aviators who remained in the Air Service began the slow process of reorganizing for peacetime and searching for ways to improve aircraft and define their mission. Many pilots were tasked to generate positive publicity for the Air Service by participating in traveling aerial demonstrations. Tooey Spaatz, who shot down three German aircraft during the war, and Jimmy Doolittle were two of those whose public aerial displays were well received across the nation. But it was the outspoken Brigadier General Billy Mitchell who brought the failings of American military aviation to the forefront during the 1920s.

Mitchell, who had commanded the largest aerial armada in history during the St. Mihiel offensive, took the offensive in the U.S. as an advocate for an independent air arm and for the capabilities and potential of the aerial weapon. He orchestrated bombing demonstrations against naval vessels, which inflamed Navy leadership and fueled early battles between the services over the possible roles for the airplane in both the Army and the Navy. He was openly critical of U.S. Army and Navy leadership and wrote many articles qualifying his arguments. His eventual court-martial for his highly critical campaign immortalized him in Air Force lore as one of the earliest, and certainly the most outspoken, advocates of centralized command of American aviation—as well as a champion of an independent air arm. Mitchell's experiments led to the immediate development of an aircraft designed specifically to fulfill the dream of true strategic bombardment—the NBL-1, at that time the largest airplane in the world.

The XNBL-1, designed by Walter H. Barling in 1923, was the first American airplane built with a wingspan as long as the distance of the Wright brothers first flight. The Barling stood three stories tall and 65 feet long. Six Liberty engines (four tractors and two pushers) provided thrust enough to allow a cruising speed of 100 mph, and the 2,000-gallon fuel capacity allowed the plane to remain airborne for a full 12 hours without bombs. The Barling was capable of carrying one 10,000-pound bomb, then under development, for two hours. The Barling flew for the first time at McCook Field on August 22, 1923, but it performed poorly. When fueled and loaded with simulated bombs, it was so heavy that only two runways in America could withstand its weight. Still, the Barling was not a total loss. Valuable wind tunnel data, parts design, and other aeronautical engineering problems were addressed and solved during the Barling's development. In that way, the Barling influenced the design of the B-17, B-24, and B-29 bombers that became the backbone of America's strategic bombing campaign during World War II.

Strategic bombers must fly tremendous distances. To increase aircraft range for pursuit planes and bombers, Major Henry H. Arnold approved the trials of a new, dangerous, potentially revolutionary advance in aviation operations: mid-air refueling. Nothing more than hoses, ropes, and gas cans provided such capability. Audacity and fearlessness played a larger role in the success of the trials than the machinery involved. On June 27, 1923, 1st Lieutenant Frank Seifert and 1st Lieutenant Virgil Hine achieved two successful contacts in a modified DH-4 aircraft. A second, even more successful, test occurred in August.

Pages 82–83: *"Y1B-17: Birth of the Flying Fortress," by Jane Bready. (USAF Art Collection)*

Right: *The mammoth Barling bomber used a unique set of landing gear "trucks" to handle the enormous weight of the plane. Six Liberty engines propelled the craft. ("Barling Bomber," by Sidney Bradd, USAF Art Collection)*

Aerial spectacles became routine occurrences during the Roaring Twenties. Air Service aircraft spread out along the west coast and began patrolling forested areas, spotting fires before they blazed out of control—an effort that saved thousands of acres of timber and millions of dollars for the lumber industry as well. Doolittle crossed the country by air, from the Florida coast to the Pacific Ocean, in less than 24 hours. In the late 1920s, he would be the first pilot to land an airplane without reference to the outside—flying using instrumentation alone. Special Douglas World Cruiser aircraft were manned by Air Service crews in the first-ever flight around the world. By the end of the decade, the newly renamed Army Air Corps had experimented with aerial refueling to the point where it became possible to remain airborne as long as the mechanical function of the airplane permitted. During the famous flight of the "?" (*Question Mark*), the seven-day airborne odyssey over Pasadena was terminated by mechanical failures rather than a problem receiving fuel, oil, or food.

By the time America succumbed to the pressures of worldwide economic depression in 1929, the Air Corps had developed more capable airplanes. Wood and wire biplanes were eventually replaced by all-metal monoplanes with retractable landing gear and powerful engines. Aircraft loads and combat ranges increased to the point that, by the end of the 1930s, Air Corps B-17s routinely flew missions of 2,000 miles or more.

Publicity for the Air Corps remained an important part of its mission while military budgets were tight. But publicity can be good or bad, as the Air Corps found out in 1934 when they were asked to deliver the U.S. mail via air during a particularly harsh winter. Pilots were killed and planes were lost while trying to carry out what some commanders called an impossible task.

While technical development moved at a deliberate pace, airpower theory and doctrine developed at breakneck speeds. In the early 1930s, when the Air Corps Tactical School (ACTS) moved to Maxwell Field in Montgomery, Alabama, an environment of academic freedom and intellectual effort contributed significantly to how air weapons would be used during wartime. Air theory was to Maxwell Field as technical development was to Wright Field.

Organizational issues also became critical during these years. Based upon the reports of the Drum Board, the Baker Board, and the Howell Commission, as well as the experiences gained from experiments with the Provisional GHQ formed in late 1934, the General Headquarters (GHQ) Air Force was finally created on March 1, 1935. The Drum Board had issued the requirement for 2,320 aircraft, which remained unchanged by the other two committees, largely because of the need for expediency and budget preservation throughout the committee process.

By the dawn of the 1940s, the Army Air Corps had made large aviation technology leaps, had developed doctrine to employ these new air weapons, and established contacts with scientists and industrialists to make expansion of aeronautical research and aircraft production run smoothly when such efforts were needed. Although not an independent service, the foundations of America's Air Force were being assembled and would soon be tested in the crucible of war.

DOCTRINE & DEVELOPMENT

1919–1940

"Progress in aeronautics is being made at such a rapid rate that the only way to keep abreast of other nations is actually to keep abreast, year by year, never falling behind."
—From the 1923 National Advisory Committee for Aeronautics annual report

1919

2 January
Major General Charles T. Menoher assumes command of the U.S. Army Air Service. During the fall of 1917, Menoher held command of the 42d Infantry Division, which was the first American unit to hold sole responsibility for a section of the battle front. The 42d was in direct contact with the enemy for nearly six months in the trenches that weaved across France.

6 January
In preparation for the advent of national air mail flights, four Curtiss JN-4H aircraft fly more than 4,000 miles across the country, taking pictures and selecting airfields that will assist future air mail pilots.

24 January
During a remarkable piece of stunt flying, Lieutenant Temple M. Joyce of the Army Air Service performs 300 consecutive loops over the Allied base at Issoudun, France.

10 March
Brigadier General Billy Mitchell assumes command of Military Aeronautics under the Director of Air Service, replacing Major General W. L. Kenley as director.

19 April
Captain E. F. White and mechanic H. M. Schaefer set an American distance record for a nonstop flight when they fly a DH-4B-Liberty 400 from Chicago to New York in six hours and 50 minutes. They cover 738.6 miles.

28 April
Civilian Leslie L. Irving makes the first successful free-fall parachute jump using a prototype Model AA chute designed by James F. Smith. Irving jumps from 1,500 feet from a U.S.-built de Havilland D-9 over McCook Field, Ohio. Though he fractured his ankle upon landing, the results of the test encourage the U.S. Army to purchase 400 of the Model AA, which were produced by Irving's own company.

16–27 May
A Navy-Curtiss NC-4 Flying Boat and its five-man crew successfully complete the first aerial crossing of the Atlantic Ocean. Their route took them from Newfoundland to Portugal, with one stop in the Azores. Although three NC-4s began the journey, only one successfully completed the trip.

19 May
Master Sergeant Ralph W. Bottriell earns the Distinguished Flying Cross when he successfully test-jumps the Model AA manually operated, free-fall-type backpack parachute. Bottriell developed the parachute, which is deployed by using a D-shaped handle, known to all today as the "D-ring."

Opposite: *Brigadier General Billy Mitchell served as the Director of Military Aeronautics before he fell into disfavor and was exiled to Texas. His 1925 court-martial drew attention to the many problems that existed in military aviation at the time. (NASM)*

Right: *Assembly of the Kettering Bug, also known as the Liberty Eagle, was simple. Here, engineers check each "Bug" individually. Although intended to be used above the trenches of France during World War I, the weapon never made it into combat. (USAF)*

1 June
Under the command of Major Henry H. Arnold, commander of Rockwell Field, California, the first aerial fire patrol of the West Coast begins in response to a request from the District Forester from San Francisco.

28 June
The Treaty of Versailles is signed. The treaty, harsh toward Germany as the aggressor nation in World War I, forbids them an air force as well as levies costly reparations to the victors of the Great War. Such conditions are later responsible for the rise of German nationalism and the Nazi Party.

24 July
Flying a Martin MB-2-Liberty 400 bomber, Army Air Service Lieutenant Colonel R. S. Hartz, Lieutenant E. E. Harmon, and crew begin the first attempted flight around the United States peripheral borders (coastal and border circuit flight). The crew takes off from Bolling Field near Washington, D.C., heads to the north and then follows a counterclockwise route that eventually totals nearly 10,000 air miles and takes a total flight time of almost 115 hours flown over the period of 108 days. The flight is completed on November 9, 1919.

1 September
Lieutenant Lester B. Sweely, Air Service Reserve, drops an external, center-line mounted 300-pound demolition bomb from his DH-4B. This divebombing experiment takes place at the Aberdeen Proving Ground.

26 September
A series of aerial tests of the Army's Liberty Eagle (also Kettering Bug) are initiated near Arcadia, Florida. Of the 14 test launches only five become airborne. The last test flies more than 15 miles and proves to be the longest of all the test launches.

8 October
An Army transcontinental reliability and endurance test begins in New York. A total of 44 planes make the trip from east coast to west coast, 15 make the flight from west to east. Only 10 make the round trip, which is won based upon elapsed time. Lieutenant B. W. Maynard, piloting a DH-4, wins the test and earns the Mackay Trophy for his accomplishment.

27 October
Director of the Air Service, Major General Charles T. Menoher issues a final report rejecting a congressional proposal that would have established an independent air force department. The proposal was introduced by Senator Harry S. New and Representative Charles F. Curry.

30 October
Successful tests are accomplished on the reversible-pitch propeller, which will allow aircraft to use the propeller as a brake during landing to shorten stopping distances. The test is done at McCook Field, Dayton, Ohio.

Left: *Lieutenant Doolittle had this DH-4B modified at McCook Field in preparation for his 1922 cross-country flight. An extra fuel and oil tank was added to the upper wing and the unoccupied cockpit. (NASM)*

1920

25 February
An Air Service School is authorized for establishment at Langley Field, Virginia.

27 February
Major R. W. "Shorty" Schroeder flies his Packard-LePere LUSAC-11 biplane powered by a Liberty 400 engine to a world record altitude of 33,114 feet. The flight takes place over McCook Field, Ohio.

1 April
NACA Technical Report No. 91, *Nomenclature for Aeronautics*, is approved for publication in an effort to standardize terminology used in aeronautical science across America.

4 June
The Army Reorganization Bill of 1920 establishes the Air Service as a combat arm of the Army on par with the cavalry, artillery, and infantry. Flight pay of 50 percent over basic pay is authorized and the "airplane pilot" rating is established. The commander's title is changed from Director to Chief of Air Service, and authorizations are made for 1,514 officers and 16,000 enlisted men on a permanent basis.

8 June
Lieutenant John H. Wilson makes an unofficial world record parachute jump over San Antonio, Texas, from an altitude of 19,861 feet.

11 June
The NACA conducts its first test of the five-foot wind tunnel—Wind Tunnel 1—at Langley Field, Virginia. It is the first test by the NACA conducted by their own staff at their own facilities. Although not remarkable in its design or capability, Wind Tunnel 1 leads NACA scientists and engineers to more advanced wind tunnel concepts.

28 June
Army and Navy officials are encouraged by the National Advisory Committee for Aeronautics to assign air officers to the Massachusetts Institute of Technology to study aeronautical engineering. One of the most famous airmen to do so is Jimmy Doolittle, who receives both a master's and PhD from MIT in the mid-1920s.

1 July
A French-designed Hispano-Suiza engine capable of firing a 37mm cannon shell through the propeller shaft and out the nose of an airplane is built by the Wright Aeronautical Company.

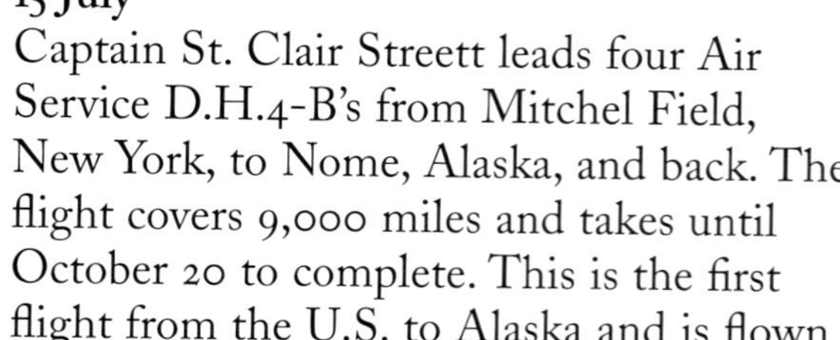

15 July
Captain St. Clair Streett leads four Air Service D.H.4-B's from Mitchel Field, New York, to Nome, Alaska, and back. The flight covers 9,000 miles and takes until October 20 to complete. This is the first flight from the U.S. to Alaska and is flown over largely uncharted and rough mountainous terrain. Streett is awarded the Mackay Trophy for the mission.

11 September
Three airships fly formation using radio communication and direction at Langley Field, Virginia.

1 November
Major Thomas DeWitt Milling, who learned to fly at the Wright Flying School along with Henry H. Arnold, assumes command of the Field Officers School at Langley Field, Virginia. Milling's school eventually evolves into the Air Corps Tactical School and moves to Maxwell Field in Montgomery, Alabama. This school becomes the hotbed of airpower theory and doctrine during the interwar years.

25 November
The first Pulitzer Air Race is won by Lieutenant Corliss C. Moseley piloting a Verville-Packard 600 over Mitchel Field, New York. Moseley covers 132 miles with an average speed of 156.5 miles per hour. One additional requirement for competing aircraft was a landing speed less than 75 miles per hour.

1921

10 January
McCook Field engineers test an advanced engine design that consists of three banks of six cylinders each. The engine produces 700 horsepower.

21–24 February
Air Service Lieutenant William D. Coney makes a solo transcontinental flight from Rockwell Field, California, to Jacksonville, Florida. Total flying time on this flight is 22 hours and 27 minutes. Flights of this kind set the stage for later transcontinental attempts that would take approximately the same amount of flight time, but be flown on the same day.

22–23 February
Jack Knight and E. M. Allison fly the first transcontinental mail delivery, which arrives in New York City from San Francisco after a little over 33 hours. De Havilland DH-4s built in the U.S. make the journey, a significant portion in bad weather, after 14 stops—three days quicker than traditional railroad delivery. Knight pilots the night leg from North Platte, Nebraska, to Chicago, Illinois.

23 March
Lieutenant A. G. Hamilton makes a parachute jump from 23,700 feet over Chanute Field, Illinois.

8 June
Piloted by Lieutenant Harold R. Harris, an Army Air Service D-9-A airplane flies with a pressurized cabin. It is the first flight of this kind and will eventually develop into fully pressurized crew and passenger aircraft during the 1930s.

15 June
Bessie Coleman receives her Federal Aeronautic Internationale (FAI) pilot's license. She is the first African American woman to earn her F.A.I. license.

13–21 July
Brigadier General Billy Mitchell leads the First Provisional Air Brigade flying from Langley Field, Virginia, in tests designed to prove the ability of aircraft to attack and destroy naval vessels. Using

Right: *Billy Mitchell organized bombing trials, using surplus naval vessels as targets. These MB-2s were the backbone of the test force that sent the captured German battleship* Ostfriesland, *once thought unsinkable by U.S. Naval officers, to the bottom. (USAF)*

MASON PATRICK

Robert White

From 1917 to 1927 there was a dramatic change in American views concerning the role of aircraft in national defense. The majority views of Congress, the War Department, and the public in general were transformed in a tumultuous decade that witnessed heated editorial debate, endless legislative battles, interservice rivalries, and the much-publicized court-martial of Billy Mitchell. In the midst of this debate, a quiet, determined individual set the United States Air Service on the road to independence as a separate service and ensured the future of America's nascent commercial aviation. Major General Mason M. Patrick was responsible for building the early foundation upon which the United States Air Force would eventually stand.

Patrick is one of the great unsung American airpower pioneers. An engineering officer by training, this 1886 U.S. Military Academy graduate did not begin his military aviation career until May 1918. It was then, in the midst of America's involvement in World War I, that General John J. Pershing selected Patrick to command the air arm of the American Expeditionary Force. Patrick's no-nonsense approach to running the Air Service brought order to chaos—primarily by mitigating friction between Billy Mitchell and Benny Foulois—that had plagued Pershing's air arm. With Patrick in charge, the AEF Air Service began to provide the support that Pershing desperately required.

After the war, Patrick returned to his engineering role thinking that he had left military aviation behind him for good. But by the autumn of 1921 the Air Service was embroiled in disputes over independence from the Army, suffered from severe fiscal woes, and witnessed continuing personality clashes which degraded its combat efficiency. The newly appointed Chief of Staff, General Pershing, turned to Patrick, whom he knew and trusted, to take charge of the Air Service.

Initially, Patrick struggled to keep the Air Service functioning. He did not seriously consider attempting to gain autonomy—a goal fiercely sought by Billy Mitchell, his deputy. Patrick recognized Mitchell's strengths and weaknesses and used Mitchell's talents where appropriate and constrained his deputy's political and publicist personality when necessary. Patrick even endorsed Mitchell's retention as his deputy in 1924, but, in the end, he could not save Mitchell from himself. Mitchell's outspoken criticism of both military and civilian leadership ultimately resulted in his court-martial and conviction. He resigned in February 1926.

From October 1921 through 1927, it fell to Patrick to orchestrate the behind-the-scenes policies and politics that eventually resulted in the creation of the U.S. Army Air Corps in July 1926. Additionally, he oversaw an impressive five-year aviation procurement program. Patrick was one of the first aviation officers to recognize the potential of airpower during times of both war and peace. He was remarkably successful in gaining support for the three legs of the aviation triangle—military aviation, commercial aviation, and the aviation manufacturing base—and he was responsible for legislation which greatly enhanced each of these. Patrick even found time to earn his pilot's wings—at age 59—still the record for the oldest military member to achieve aviator status.

Although Patrick was trained an engineer, he became the first true leader of American military

Opposite: *Although they had similar ideas about airpower, Mason Patrick (left) chose to work within the Army system, while Billy Mitchell (right) conducted a highly public campaign highlighting the shortcomings of American airpower. (USAF)*

Right: *"Major General Mason Patrick," by Sidney Dickinson. (USAF Art Collection)*

aviation. He was no less a believer in the strategic and independent capabilities of the Air Force than his deputy, Billy Mitchell, but Patrick took a much more practical, gradualist, and ultimately successful approach to solving intractable Air Service problems. While leading the young Air Service, Patrick developed doctrine that explicitly supported the need for autonomy. He believed that if the Air Service was tied by doctrine to ground force missions, there existed no rationale to support autonomy for an independent air arm and there would be no impetus for additional monies to support the infrastructure and mission of such a separate service. Patrick keenly appreciated the critical role that doctrine played in the resource debate, and he judiciously supported efforts that justified an independent air arm. But there were constraints.

Patrick knew the value of air power, but more importantly, he grasped the limitations and capabilities of air power during that particular phase of its development. Patrick built and implemented an air power agenda that included the acquisition of state-of-the-art military aircraft, the rapid development of commercial aviation, Air Service officer professionalization, and critical legislative initiatives that established his ideas in law. Patrick's tenure as chief of the Air Service and Air Corps prepared the way for an independent Air Force.

Left: *Billy Mitchell's bombing tests included experiments with tactics. Among these were smoke-screening and bombardment with phosphorus weapons, which would hinder deck operations. Here, Mitchell's MB-2 gets low to deliver the final blow. ("General Mitchell's Bombers Sink the Ostfriesland—1921," by Robert Lavin, USAF Art Collection)*

Martin MB-2 and Hadley Page bombers, Mitchell's men sink three ships moored approximately 75 miles from the airfield off the Virginia Capes. The captured German battleship, *Ostfriesland*, considered unsinkable by many in the U.S. Navy, is one of the three ships sent to the bottom during the eight days of tests. The other two are the German destroyer *G-102* and the light cruiser *Frankfort*.

29 July
After successfully sinking several ships the week before, Billy Mitchell's aircraft stage a mock attack with more than a dozen Army bombers upon New York City. Attempting to demonstrate the vulnerability of East Coast cities, Mitchell claimed that the attack would have leveled the city. Mitchell's ultimate aim was to prove that the naval forces were no longer capable of providing adequate coastal defense and that such a mission rightfully belonged with the Air Service.

3–4 August
Army Air Service Lieutenant John A. Macready performs the first-ever crop dusting operation while flying a Curtiss JN-6 Jenny. The aircraft is fitted with a large tank filled with insecticide dust, which Macready dispenses over a grove of caterpillar-infested Catalpa trees near Troy, Ohio. Macready spreads the dust from treetop level, and in two days the caterpillars succumb.

10 August
The Navy Bureau of Aeronautics under the command of Admiral William A. Moffett is established. Although the Chief of Naval Operations had approved "bureau" status in February 1920, the announcement is not made public until this date.

18 September
Lieutenant J. A. Macready sets a world altitude record of 34,508 feet while piloting a Packard LePere fighter biplane over McCook Field, Ohio. Macready, who earns the Mackay Trophy, flies a circular flight path during his climb to altitude, which expanded to nearly 70 miles at the top of his ascent.

23 September
The surplus battleship *Alabama* is sent to the bottom of the Chesapeake Bay after being sunk by Air Service bombers with a 2,000-pound bomb during tests.

5 October
Major General Charles T. Menoher assumes command of the Air Service, replacing Major General Mason M. Patrick.

18 October
Brigadier General Billy Mitchell sets a world speed record of nearly 223 miles per hour for a one-kilometer course while piloting a Curtiss R6 at Mount Clemens, Michigan.

12 November
Aerial stunt man Wesley May straps a five-gallon can of gasoline onto his back and steps from the wing of an airborne Lincoln Standard piloted by Frank Hawks, to the wing skid of a JN-4 Jenny flown by Earl S. Daugherty over Long Beach, California. He empties the contents of the can into the Jenny's fuel tank, completing the first-ever air-to-air refueling of an airborne aircraft.

28 November
Ludwig Prandtl, lauded German aerodynamicist, publishes *Application of Modern Hydrodynamics*, as NACA Report No. 116. One of Prandtl's brightest pupils, Theodore von Kármán, will move from Germany to head the Caltech aeronautical laboratory in the 1930s.

1922

20 March
Perhaps as a reaction to Mitchell's 1921 bombing tests, and fearing the loss of control over aviation assets, the U.S. Navy's first commissioned aircraft carrier, the USS *Langley* (CV-1), is constructed by modifying an existing ship, the collier USS *Jupiter*, with a carrier deck.

23 March
Edgar Buckingham of the Bureau of Standards points out in his NACA Report No. 159, *Jet Propulsion for Airplanes,* that at 250 miles per hour, jets would burn four times the fuel of piston engines. But, at higher speeds the fuel efficiency of jets would dramatically increase.

12 June
Air Service Captain A. W. Stevens leaps from 24,200 feet over McCook Field, Ohio, setting a new altitude record. Stevens is carried aloft by a Martin bomber with supercharged engines.

16 June
Lieutenant Clayton L. Bissell initiates a series of night cross-country flights originating at Bolling Field, D.C., and arriving at Langley Field, Virginia. Bissell then returns during the hours of darkness, an effort to demonstrate that night flying can be as routine as day flying.

Henry Adler Berliner makes a successful helicopter flight at College Park, Maryland. The craft lifts 12 feet into the air and hovers. The flight is witnessed by representatives of the U.S. Bureau of Aeronautics.

Below: *The Berliner (Emile) Helicopter hovering over the ground, 1922. (NASM)*

INTERWAR AERONAUTICAL DEVELOPMENT AT MCCOOK AND WRIGHT FIELDS

Jeremy R. Kinney

Left: *In addition to aircraft, other equipment was developed and tested at McCook Field. The Barling Bomber was housed in this enormous, free-standing hangar. New high-volume refueling systems were also developed for the Barling to decrease refueling ground time. (Robert and Kathleen Arnold Collection)*

On February 27, 1920, Major Rudolph "Shorty" Schroeder flew his wood-and-fabric Packard-LePere LUSAC-11 biplane higher than 33,000 feet over Dayton, Ohio. The army test pilot withstood temperatures approaching 70 degrees below zero, oxygen deprivation, and carbon monoxide poisoning from engine exhaust to fly higher than any person ever had before. Almost 20 years later, the crew of the all-metal Lockheed XC-35 pressurized cabin monoplane cruised comfortably through that same sub-stratosphere without the oxygen bottles or warm, heavy clothing that Schroeder necessarily relied upon to pioneer high-altitude flight. These flights, and many others during the interwar period, pushed the limits of human endurance and aeronautical technology at the army's aeronautical research and development facilities at McCook Field from 1917 to 1927 and Wright Field from 1927 to 1941.

After entering World War I in April 1917, the United States committed itself to an ambitious aviation production program supported by an unprecedented $640 million government appropriation (approximately $11 billion in modern currency). The government expected a virtually nonexistent American aviation industry to deliver 22,625 aircraft and 45,625 engines to the Army by July 1918.To assist in that goal, the Army established the Engineering Division of the United States Army Air Service at McCook Field in Dayton, Ohio. The Army named the field after a prominent Ohio family that distinguished itself during the Civil War and chose Dayton due to the central location between the industrial Midwest and Washington, D.C. The facility began operations in December 1917 and employed an impressive cadre of military and civilian engineers tasked with modifying and improving existing military aircraft while also developing new planes.

Paltry budgets, tenuous government-industrial relations, and national enthusiasm for the airplane characterized the early interwar years. By the mid-1920s, the army had outgrown the McCook facility. Above the main hangar, the cautionary message, "THIS FIELD IS SMALL—USE IT ALL," warned pilots of the McCook's close proximity to downtown Dayton. The passage of the Air Corps Act of 1926 resulted in the creation of Materiel Division. To support that division a new facility called Wright Field, in honor of Wilbur and Orville Wright, opened northeast of Dayton with the requisite space to keep pace with the rapid development of military aircraft.

Under the leadership of engineer-officers such as Lieutenant Colonel Jesse G. Vincent (1918), Colonel Thurman H. Bane (1918–1923) Brigadier General William E. Gillmore (1926–1929), the Army's myriad of engineering activities at McCook and Wright fields enabled American military aircraft to fly (and fight) higher, faster, and farther. The innovative aircraft designed or specified by the Army included the multi-role USD-9A "battle plane" (1918); the Verville-Sperry R-3 air racer with retractable landing gear (1922); the de Bothezat helicopter (1922); the Fokker C-2 Bird of Paradise that made the first transpacific flight from California to Hawaii with advanced navigational instruments (1927); and the all-metal Boeing B-9 monoplane bomber of 1932, which spurred on the development of more-advanced designs, primarily the Martin B-10 (1933) and the four-engine Boeing B-17 Flying Fortress heavy bomber (1935).

Research tools such as the McCook 14-inch and Wright five-foot wind tunnels, the propeller whirl test rig, power plant torque stands, and the structures-testing laboratory provided fundamental data and simulated flight conditions to ensure the safety and durability of new equipment. When it was not economically or technologically feasible for private industry to initiate research, the Army started critical development programs on parachutes, variable-pitch propellers, turbosuperchargers, sodium-cooled exhaust valves, engine coolants, and high-octane gasoline. Either directly or by close association, the Army received nine Collier Trophies, the most prestigious award given in American aeronautics each year, for its contributions during the interwar period.

The Army disseminated its pioneering information to the aeronautical community. General information reached a broader audience through technical orders and bulletins such as the multi-edition *Handbook of Instructions for Airplane Designers*. Officer-students assigned to the engineering school took advanced technical knowledge back to operational units. Moreover, many engineers started their careers in the employment of the Army at McCook and Wright fields before going through the "revolving door" to industry to continue their important work.

During the 1920s and 1930s, the slow, fabric-covered, strut-and-wire-braced biplane of 1918 transformed into the modern high-speed, streamlined, cantilever-wing monoplane of 1938. The Army, through its engineering facilities at McCook and Wright fields, nurtured the new aviation industry and its engineering community to provide the fundamental infrastructure from which modern airplanes evolved. As a result, a distinctive, refined, and highly sophisticated Army style of aeronautical engineering emerged that reflected the American aviation industry's emphasis on immediate and practical solutions.

Left: *Aircraft wing strength was tested by the "sandbag" method, which allowed uniform weight to be distributed over the entire surface of the wing. Often the wings were loaded until failure. (Robert and Kathleen Arnold Collection)*

29 June
Lawrence Sperry demonstrates his long-range, radio-guided, aerial torpedo near Mitchel Field, New York. During the test, which ranged from striking targets at 30, 60, and 90 miles from the launch point, a "mother ship" is required within one mile from the unmanned torpedo to guide it using radio signals. Although given a bonus for the success of the tests, neither the Army nor the Navy buys the gizmo. During World War II, the concept will be revived as part of Project Aphrodite—a series of unmanned glide bombs and radio-controlled war-weary aircraft loaded with nitrostarch and directed by radio control from a following "mother ship."

16 August
The Sperry airway lighting system is demonstrated at McCook Field, Ohio. The light is the precursor to the modern airport rotating beacon.

21 August
Lawrence B. Sperry releases the landing wheels from his aircraft while in flight during experiments with skid landings flown in Farmingdale, New York.

4 September
Lieutenant Jimmy Doolittle makes the first transcontinental flight in a highly modified DH-4B-Liberty 400. Doolittle takes off from Pablo Beach, Florida, refuels at Kelly Field in San Antonio, Texas, flying a total of 2,163 miles in less than 24 hours.

14–23 September
Major H. A. Strauss, commanding the non-rigid airship C-2, flies the first transcontinental flight of an airship from Langley Field, Virginia, to the Army Balloon School at Ross Field, Arcadia, California.

29 September
Dr. Robert H. Goddard makes a report to Smithsonian Secretary Charles G. Abbot concerning developments in multiple-charge rockets. Apparently not impressed, Abbot terminates Goddard's funding a few days later.

5 October
Flying a Fokker T2-Liberty 375, Lieutenant J. A. Macready and Lieutenant P. G. Kelly set a flight endurance record of 35 hours and 18 minutes over Rockwell Field, California, earning the Mackay Trophy.

13 October
Lieutenant T. J. Koenig wins the Liberty Engine Builders Trophy at the National Air Races held at Selfridge Field, Michigan. Koenig flies a LePere-Liberty 400 to an average speed of just under 129 miles per hour over a 257-mile course.

14 October
Lieutenant R. L. Maughan wins the Pulitzer Trophy Race flying a Curtiss R-6 with a speed of 206 miles per hour. Of the 24 aircraft in the competition, the top four places are all Curtiss racing designs.

18 October
Brigadier General Billy Mitchell breaks the world absolute speed record, setting a new mark of 224.05 miles per hour over a one-kilometer course. The speed is the average of four passes. Mitchell, flying a Curtiss R-6-D12 Curtiss 375 in the skies over Selfridge Field, Michigan, becomes the first American military pilot to hold the absolute record and also the first record holder certified outside of France.

20 October
Lieutenant Harold R. Harris parachutes from his Loening PW-2A aircraft when its wings collapse in flight over McCook Field, Ohio. Harris is the first military pilot to be saved by the use of a personal parachute in flight. A civilian air mail pilot, C. C. Eversola, had used his personal parachute to safely escape an uncontrollable DH-4 in February 1921.

23 October
The reversible propeller is demonstrated for the first time at Bolling Field, D.C., by the manufacturer, the American Propeller Company.

3–4 November
Lieutenant John A. Macready and Oakley G. Kelly, flying a Fokker T-2 aircraft, set a world distance record of 2,060 miles when they fly from San Diego, California, to Benjamin Harrison, Indiana. Aircraft engine problems force them to stop 800 miles short of their original goal of New York.

8 November
The School of Aviation Medicine is established from the former Air Service Medical Research Laboratory and School for Flight Surgeons.

18 December
The first flight of the first rotorcraft flown by the Army Air Service occurs at McCook Field, Ohio. The vehicle, developed and flown by Russian immigrant George de Bothezat, climbs as high as six feet off the ground and is airborne for one minute and 45 seconds. The second hop is piloted by Major Thurmond H. Bane, who rises to a height of six feet, traverses 300 feet over the ground, and then hovers for about two minutes. Bane becomes the first Army helicopter pilot.

1923

5 January
Experimental cloud seeding is accomplished over McCook Field under the direction of Professor W. D. Bancroft of Cornell University.

6 February
The Army's newest airship, the D-2, flies for the first time at Scott Field, Illinois. The D-2 remains airborne for just over one hour and climbs to an altitude of 1,000 feet during the trial.

7 February
Lieutenant Russell A. Meredith transports a physician from Selfridge Field to help a critically ill man on Beaver Island, Michigan. The winter conditions had hampered ground transportation and the lake was frozen solid. Meredith was awarded the Distinguished Flying Cross for his courage and skill.

1 March
The TC-1, the largest American nonrigid airship, is delivered to the Air Service by Goodyear Tire and Rubber Company.

5 March
An external auxiliary, jettisonable fuel tank is tested on a MB3A at Selfridge Field, Michigan. The tank works as advertised and adds significant range, a total of 400 miles, in flight.

27 March
The Lassiter Board (named for its chairman, Major General William Lassiter) suggests that legislation is needed that would expand the Air Service and reorganize its functions. The most significant of these would allow independent operations at the strategic level beyond the reach of surface forces.

Opposite: *Lieutenant Jimmy Doolittle pauses a moment during the one fuel stop made during his September 4, 1922, cross-country flight. The dice painted on the side of the airplane were often added to aircraft that he flew during his life. On one side of the airplane, a "seven" was shown. On the other, an "eleven" was revealed. (NASM)*

Right: *The de Bothezat helicopter in tethered flight. Major Thurman Bane is piloting during this December 18, 1922, McCook Field flight. (NASM)*

Left: *Refueling in mid-air by Captain Lowell H. Smith and Lieutenant John P. Richter at Rockwell Field, California, during the summer of 1923. They stayed in the air four days using modified DH-4 aircraft. (USAF)*

16–17 April
While practicing for their upcoming transcontinental flight, Lieutenant John A. Macready and Lieutenant Oakley G. Kelly fly a Fokker T-2 for 36 hours and four minutes over a distance of 2,516 miles, setting world records for duration and distance. By carrying 10,800 pounds of supplies, they also set a world record for single engine payload mark.

2–3 May
Lieutenant John A. Macready and Lieutenant Oakley G. Kelly accomplish the first nonstop transcontinental flight in a Fokker T2-Liberty 375 aircraft. The flight from Roosevelt Field, Long Island, New York, to Rockwell Field, San Diego, California, takes 26 hours and 50 minutes and covers more than 2,500 miles. This award makes Macready the first and only three-time recipient of the Mackay Trophy.

14 May
The prototype Curtiss PW-8 pursuit aircraft is accepted by the Army.

17 May
Major Thomas Scott Baldwin dies in Buffalo, New York, at age 69. Baldwin, a pioneer in parachute development, was the first man to jump from a balloon using a parachute for his descent. Baldwin was also the first Air Service balloon pilot and supervised the early Army balloon program.

26 May
Lieutenant H. G. Crocker, flying a DH-4B-Liberty 400, makes a south-to-north transcontinental crossing from Houston, Texas, to Gordon, Ontario, in 11 hours and 55 minutes.

20 June
The Gallaudet CO 1, Liberty 400, all-metal airplane flies for the first time. The plane is designed and built by the Air Corps Engineering Division at McCook Field, Ohio.

27 June
Four Air Service lieutenants flying two DH-4 aircraft over Rockwell Field, California, perform the first aerial refueling between two aircraft by the use of hoses. Two contacts are made during this first flight and a more successful attempt occurs during August. The experiment was ordered by the Rockwell Field commander, Major Henry H. Arnold.

4 July
Army S6 balloon pilot Lieutenant R. S. Olmstead and his aide, Lieutenant J. W. Shoptaw, fly to victory at the National Elimination Balloon Race. They travel from Indianapolis, Indiana, to Marilla, New York, a total distance of nearly 500 miles.

22 August
The XNBL-1 Barling bomber flies for the first time at McCook Field, Ohio, piloted by Lieutenant H. R. Harris. Designed by Walter H. Barling and built by the Witteman Aircraft Corporation, the Barling featured three wings (two with control surfaces), a wingspan of 120 feet (the length of the Wright's first flight), six Liberty engines (four tractors and two

pushers), and stood three stories tall. During the test, the Barling flies for 28 minutes, travels 25 miles, and cruises at 93 miles per hour. When fully loaded with fuel and simulated bombs, the Barling was so heavy that only two runways in the country could support its weight. Only one Barling was ever built.

23 August
The Polikarpov I-1 cantilever, low-wing monoplane flies for the first time. The I-1 is powered by an American Liberty Engine.

27–28 August
Lieutenant Lowell H. Smith and Lieutenant John P. Richter remain aloft for 37 hours and 15 minutes flying a DH-4B-Liberty 400 over Rockwell Field, California, by repeated aerial refueling. Not only do they set a new world refueled duration record, but they also set a distance record of 3,293 miles during the flight.

5 September
Army bombers sink two condemned naval vessels, the *New Jersey* and the *Virginia*, during bombardment tests near Cape Hatteras, North Carolina. General John J. Pershing witnesses the tests.

13 September
Lieutenant J. F. Whitely and Lieutenant H. D. Smith and crew set out on a transcontinental tour flying a Martin-2-Liberty 400. They depart from Langley Field, Virginia, and fly to Rockwell Field, California, on an 8,000-mile journey that takes three months. The mission was accomplished in anticipation of the establishment of a national airway system. They return to Langley on December 14, 1923.

18 September
Army Lieutenant Rex K. Stoner, flying a Sperry M-1 Messenger, successfully accomplishes a mid-air hook-up with an Army D-3 airship. The "landing" is made with a device mounted to the top wing of the Messenger that catches a set of webbing that hangs below the airship. The Messenger, designed to carry dispatches between air and ground commanders, is also the smallest aircraft ever built for the Air Service.

25 October
The Barling Bomber (NBX-1) piloted by H. R. Harris, sets a weight carrying record for aircraft when a 3,000 kilogram load is carried aloft for one hour and 19 minutes and climbs to an altitude of 5,344 feet.

1 November
Robert H. Goddard successfully launches a liquid-oxygen-and-gasoline-powered rocket motor on a test stand. Both the oxygen and gasoline are pumped to the combustion chamber by devices attached to the rocket itself.

13 December
Lawrence B. Sperry, Elmer Sperry's son and daring test pilot, drowns when he crashes his Sperry Messenger during an attempt to cross the English Channel.

Right: *Army Air Service MB-2 pilots who made a successful round trip flight from Langley Field, Virginia, to the Pacific Coast. (NASM)*

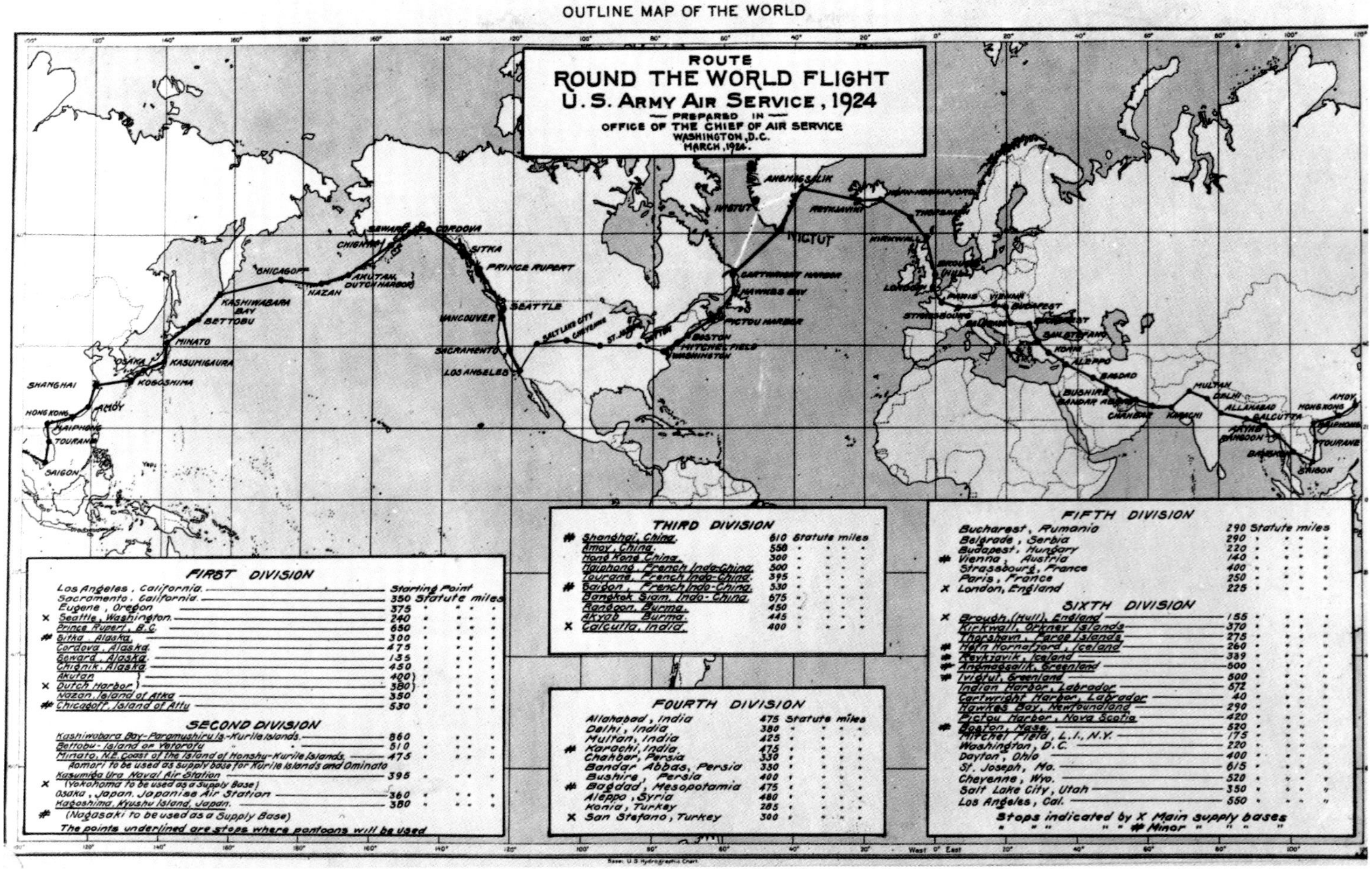

1924

22 February
Lieutenant J. Macready, piloting an XC-05A-Liberty 400, climbs to 38,704 feet, setting an American altitude record.

27 February
Air Service Corporal C. E. Conrad leaps with a parachute from a DH-4B biplane over Kelly Field, Texas, from an altitude of 21,500 feet.

4 March
Two DH-4s and two Martin bombers drop a series of bombs on the Platte River at North Bend, Nebraska, in an attempt to break up the surface ice. Although the operation goes on for six hours, the ice is finally shattered enough to clear the river, averting the potential for heavy flooding.

7 March
Lieutenant E. H. Barksdale and Lieutenant B. Jones fly a DH-4B-Liberty 400 from McCook Field, Ohio, to Mitchel Field, New York, using only cockpit instruments—a distance of 575 miles. Takeoff and landing are made by use of outside references.

6 April
Four Army Air Service planes—Douglas World Cruisers—set out from Seattle, Washington, to circumnavigate the globe.

30 April
The Douglas World Cruiser *Seattle* crashes into an Alaskan mountainside. Miraculously, the crew survives. A second DWC, the *Boston*, lands in rough Atlantic waters and sinks later in the mission. Major Martin and his mechanic Sergeant Harvey take the next 10 days to walk to safety, assisted by supplies found in an old mountain cabin.

2 May
Lieutenant Macready and Lieutenant Albert Stevens establish an unofficial two-man altitude record while flying a LePere aircraft. Stevens takes an aerial photo of Dayton, Ohio, which covers 19 square miles of the town.

19 May
Lieutenant J. Macready, piloting a LePere Liberty 400, sets a new altitude record of 35,239 feet over Dayton.

1 July
Dr. George W. Lewis is appointed Director of

Aeronautical Research of the National Advisory Committee for Aeronautics. He will hold the position for 23 years.

3 August
Douglas World Cruiser *Boston* loses oil pressure and is forced to land in rough North Sea waters and is later destroyed by the pounding waves.

28 September
Two of the original four Douglas World Cruisers, the *Chicago* and the *New Orleans*, land in Seattle after the first successful circumnavigation of the globe. As part of the journey, the World Cruisers make the first transpacific flight and the first westbound Atlantic crossing. This World Flight takes 175 days, covers approximately 27,000 miles, during which just over 371 hours of flight time are accumulated. Lieutenant Lowell Smith and Lieutenant Leslie Arnold crew the *Chicago* while Lieutenant Erik Nelson and Lieutenant John Harding crew the *New Orleans*. For successful completion of the World Flight, the flight officers are awarded the Mackay Trophy.

28 October
In the first-ever cloud-seeding exercise, Air Service planes drop electrified sand into cloud formations over Bolling Field near Washington, D.C. The experiment successfully dissipates much of the clouds.

1925

18 January
The Loening Amphibian is unveiled at Bolling Field after a year of secretive development. The all-metal plane, built for the Army Air Service, is powered by an inverted Liberty 400 engine, and with a crew of three is capable of nonstop flight of 700 miles.

23 May
The first National Advisory Committee for Aeronautics (NACA) Aircraft Engineering Research Conference is held. During the conference, NACA officials promote the transfer of government technology and expertise to industrial firms.

12 June
Daniel Guggenheim awards $500,000 to New York University for the establishment of a permanent School of Aeronautics. Soon after, the Daniel Guggenheim Fund for the Promotion of Aeronautics is formed and expands his philanthropy across the country.

22 August
The Air Service announces the development of a system that is capable of recording each heading that an aircraft maintains in flight. It is the early precursor to the "black box."

12 September
President Calvin Coolidge appoints Dwight W. Morrow to chair a board to advise on U.S. air policy.

Opposite: *Map of the first successful around-the-world flight completed by two of four Douglas World Cruisers in September 1924. (NASM)*

Right: *The Douglas World Cruiser* Chicago, *seen in the wheeled configuration, today resides in the Pioneers of Flight gallery at the National Air and Space Museum in Washington, D.C. (NASM)*

THE FIRST WORLD FLIGHT

Carroll V. Glines

Left: *The Douglas World Cruiser* Chicago *in flight with pontoons used in many water landings during the World Flight. (NASM)*

No one can deny that World War I had a beneficial effect on fixed wing aviation. At first tolerated only as "the eyes of the artillery," airplanes became full-fledged weapons of war with distinct capabilities of their own. Improvements had come rapidly through a few veteran leaders who believed the airplane had great potential for national defense. One of them was Major General Mason M. Patrick, Chief of the Army Air Service. He encouraged his pilots to demonstrate aircraft potential by setting altitude, distance and speed records. He also appointed a committee of officers to determine if a flight around the world was feasible.

The committee suggested some basic requirements. Most important was that five two-place, open-cockpit, single-engine biplanes with interchangeable wheels and floats be built to Air Service specifications by the Douglas Aircraft Company of California. Of the five aircraft, called World Cruisers, one would be for testing and the other four would attempt the flight.

The flight was planned for early spring of 1924, flying westward against the prevailing winds. The flight route would follow the Aleutian Islands before the spring fogs, advance through the Far East ahead of the typhoons and monsoons, then cross the Atlantic before the Arctic weather began. The cooperation of the U.S. Navy, the Coast Guard, and the State Department was considered essential to support the mission.

A group of pilots was authorized to travel in advance along the probable routes over the Pacific and Atlantic oceans to establish supply bases and make preparations for the maintenance and repair of the World Cruisers along the most difficult legs of the flight. Their favorable reports led to final approval for this difficult undertaking. At the same time, there was considerable interest by other countries to make such a flight. British, French, Italian, Portuguese, and Argentine airmen were either already planning or attempting around-the-world flights.

The American flight, led by Major Frederick L. Martin, officially began from Seattle, Washington, on April 6, 1924. Unfortunately, Martin crashed into a mountain near Port Moller, Alaska, the plane a total loss. After a diligent search, the other three planes were ordered to continue along the Aleutians under the leadership of Lieutenant Lowell H. Smith. Meanwhile, Martin and Sergeant Alva L. Harvey, his mechanic, were found cold but unhurt.

Fighting poor weather and with many delays, the three remaining airplanes arrived in Japan, where they were received enthusiastically. They proceeded to China, French Indo-China, and Burma, fighting new perils such as thousands of curious natives threatening to ram the planes with their boats, extreme heat and engine problems. Smith's plane was forced down by engine failure and he became ill but recovered; the flight continued after the engine was replaced. Meanwhile, the pilots of the other nations had to terminate their flights for various reasons, mostly because of inadequate advance preparations.

In Calcutta, the pontoons of the World Cruisers were exchanged for wheels, the engines were replaced, and the planes overhauled for the journey over the scorching deserts of Mesopotamia (now Iraq), Persia (now Iran), and Turkey. The weary but eager fliers made up time on the route to Paris via Bucharest, Belgrade, Budapest, and Vienna. They were received in each city by increasingly larger crowds, banquets, and receptions. They met General Pershing in Paris and the Prince of Wales in London but were anxious to undertake the most dangerous leg of all—the flight across the North Atlantic.

The trio of planes flew to northwestern Scotland to have new engines installed and the wheels replaced with pontoons. U.S. Navy ships were stationed along the route for rescue and weather reporting. Halfway between Scotland and Iceland, Lieutenant Leigh Wade had to ditch his plane in high seas when an engine oil pump failed. Wade and his mechanic, Sergeant, Henry H. Ogden, were rescued by a British fishing trawler and taken aboard a Navy destroyer. The plane was not recovered.

The two remaining planes landed in Labrador safely. The fifth World Cruiser was flown to Nova Scotia where Wade and Ogden rejoined the flight. The three crews proceeded to Seattle, the official starting point, via New York, Washington, D.C., and several western cities. They had succeeded where the crews and planes of the other nations had not. The American Army Air Service pilots had established a benchmark of courage and persistence for others to follow.

Left: *Eight of the original World Flight crewmembers pose by one of the four Douglas World Cruisers in Seattle before the departure: Major Frederick Martin, Sergeant Alva Harvey, Lieutenant Lowell Smith, Sergeant ArthurTurner, Lieutenant Leigh Wade, Sergeant Henry Ogden, Lieutenant Erik Nelson, Lieutenant John Harding. Sergeant Turner became ill and was replaced by Lieutenant Leslie Arnold. (Leigh Wade Collection)*

Left: *This Curtiss R3C-2 was flown by Lieutenant Jimmy Doolittle to win the 1925 Schneider Cup Race. Earlier, Lieutenant Cy Bettis had flown the same plane with wheels (the R3C-1) to victory in the Pulitzer Race. (NASM)*

12 October
Flying a Curtiss R3C-1 racer at Mitchel Field, New York, Lieutenant Cyrus Bettis sets two new world speed records—249.3 mph over a 100-kilometer course and 249 mph over a 200-kilometer course. Bettis shares the Mackay Trophy for his flights.

26 October
Lieutenant Jimmy Doolittle flies the Curtiss R3C-2 floatplane to a Schneider Cup victory near Baltimore Harbor. Doolittle's average speed is just over 230 miles per hour. He shares the Mackay Trophy for the flight.

27 October
Jimmy Doolittle sets a world seaplane speed record of 245.713 miles per hour in the same R3C-2 aircraft that had won the Schneider Cup the day before.

30 November
The Morrow Board (named for its head, Dwight W. Morrow), also known as the President's Aircraft Board, releases a report which examined the possibility of an independent air force. Falling short of recommending an air force with equal stature as the Army and the Navy, the Morrow Board suggests that the Air Service should be renamed as the Air Corps, an assistant secretary of war for air should be established, and also recommends a five-year expansion plan for the air arm.

6 December
Dr. Robert H. Goddard tests a liquid-fuel rocket, which fires for 10 seconds, produces a chamber pressure of 100 pounds, and successfully lifts its own weight.

17 December
Airpower firebrand Billy Mitchell is found guilty of "conduct of a nature to bring discredit upon the military service," after a highly publicized court-martial. Mitchell had accused military leaders of "almost treasonable" administration surrounding the loss of two airship crashes in the preceding months. His rank and pay are suspended for five years, but he chooses to retire from military life instead.

24 December
Pratt and Whitney completes construction of its first aircraft engine. It is the now famous Wasp.

1926

1 January
Henry Reid becomes the Engineer-in-Charge of the National Advisory Committee for Aeronautics Langley Memorial Aeronautical Laboratory. He holds the position for 34 years.

8 January
Lieutenant Orvil Anderson pilots the world's largest semi-rigid airship, the RS-1, on its maiden voyage. Anderson and a crew of eight men attain a speed of 40 miles per hour as they circle Scott Field, Illinois. The RS-1 is 282 feet long, 70 feet in diameter, and displaces 755,500 cubic feet when fully inflated.

Right: *The RS-1 was the largest lighter-than-air, semi-rigid vehicle built during its day. Lieutenant Orvil Anderson and his crew of eight piloted the vehicle on its first flight. (NASM)*

16 January
The Daniel Guggenheim Fund for the Promotion of Aeronautics is established. During its existence the fund would help establish aeronautical departments in seven major universities, provide stipends to entice European aeronautical scientists like Theodore von Kármán to take positions in these universities, and also fund independent aeronautical experimentation like those conducted in 1929 at Mitchel Field by Jimmy Doolittle to examine blind, or "zero-zero," landings.

27 January
Billy Mitchell resigns from the Army. His resignation is effective on February 1.

29 January
Lieutenant J. A. Macready sets an American altitude record of 38,704 feet while flying a XCO5-A over McCook Field, Ohio.

8 March
Dr. Robert H. Goddard tests an oxygen-pressure-fed rocket motor on a static stand at the Clark University.

16 March
Robert H. Goddard successfully launches the world's first liquid-fueled rocket. The launch is made near Auburn, Massachusetts. The flight lasts for 2.5 seconds, accelerates to 60 miles per hour, and achieves the height of 184 feet.

2 July
The Air Corps Act of 1926 becomes law. The name of the Air Service is officially changed to the United States Army Air Corps (AAC) with an authorized strength of 16,650 personnel and 1,800 aircraft. A five-year expansion program is authorized to fill the new strength requirements.

Congress authorizes the establishment of the Distinguished Flying Cross. The medal is made retroactive to American entry into World War I, April 6, 1917.

As part of H.R. 10827, Congress directs that 20 percent of the peacetime pilot force will be comprised of enlisted men.

Major General Mason M. Patrick assumes command of the renamed Army Air Corps.

16 July
F. Trubee Davidson is appointed as the first Assistant Secretary of War for Air.

16 September
Captain Charles Lindbergh (Air Corps Reserve) runs out of gas during a night mail delivery flight in heavy fog. He is forced to jump from the airplane and use his parachute to reach the safety of the ground. Only three weeks later, Lindbergh will resort to the same method to save his life after a similar situation. In all, he jumps four times to safety during his flying career.

Left: *Captain Ira C. Eaker piloted one of the Loening Amphibians on the Pan-American Goodwill Tour. Eaker would rise to the rank of lieutenant general during the Second World War. (USAF)*

7 December
Flying an old de Havilland aircraft, Wright Field flight surgeon Captain Charles T. Buckner is flown to an altitude of 28,000 without any oxygen. He is studying the effects of high altitude on the human body. Buckner's practical experiments are the first intentional studies in aerospace physiology.

21 December
Five Air Corps crews, flying Loening AO-1A Amphibians, set out from Kelly Field in San Antonio, Texas, to begin a Pan-American Goodwill Tour of South and Central America. Major Herbert A. Dargue leads the mission which will cover approximately 20,000 miles, visit 25 countries, and return to Bolling Field, D.C., on May 2, 1927. Unfortunately, one of the airplanes crashes, killing two Air Corps crewmen—Captain Clinton F. Woolsey and Lieutenant John W. Benton. All the Pan-American flyers are awarded the Mackay Trophy for their service. The mission demonstrates the feasibility of trans-hemispheric, long-distance flight.

1927

21 April
Joseph S. Ames is elected chairman of the NACA. Ames is one of the nation's leading physicists and a professor at Johns Hopkins University.

28 April
The Ryan NYP *Spirit of St. Louis* makes its first flight. The *Spirit* contains five fuel tanks that together give the aircraft a range of 4,000 miles.

2 May
Major Bert Dargue and four of the five AO-1A Amphibians land at Bolling Field in Washington, D.C., ending the Pan American goodwill flight.

4 May
Captain H. C. Gray, piloting a free-flying balloon, set an unofficial altitude record of 42,479 feet above Scott Field, Illinois.

20–21 May
Charles A. Lindbergh, flying the Ryan NYP *Spirit of St. Louis*, completes the first solo, nonstop transatlantic flight, which covers 3,590 miles. Lindbergh took off from New York and lands in Paris 33 hours and 32 minutes later. Lindbergh, a captain in the Missouri National Guard's 110th Observation Squadron, is later awarded the Distinguished Flying Cross and a special Congressional Medal of Honor.

25 May
Famed aviator and racer Jimmy Doolittle successfully accomplishes the first outside loop. The maneuver is accomplished after several months of thought and contemplation while recuperating in a D.C. hospital after breaking both ankles in a bar accident in South

America while demonstrating the Curtiss Hawk to interested governments.

22 June
John F. Victory is appointed secretary to the NACA. Victory, assistant to the secretary since 1917, becomes the corporate knowledge for the agency over time. He remains in this post until the 1958 changeover to NASA, and then until his retirement in 1960 as special assistant to T. Keith Glennan, the administrator for NASA.

28–29 June
Lieutenant Lester J. Maitland and his navigator, Lieutenant Albert F. Hegenberger, fly from Oakland, California, to Wheeler Field, Oahu. They fly for 25 hours and 50 minutes, just over 2,400 miles, in the Fokker C2-3-Wright 220 *Bird of Paradise*. In addition to the Mackay Trophy, both airmen are awarded the Distinguished Flying Cross.

20 July
Colonel Charles A. Lindbergh begins an aerial tour of the U.S. in the *Spirit of St. Louis*. The Guggenheim Fund for the Promotion of Aeronautics sponsors the trip which covers 48 states and lasts until October 23d.

12 October
Wright Field is dedicated on several thousand acres of land just to the east of Dayton, Ohio. McCook Field, the air arm's home for research and development for the previous decade, closes.

4 November
In a tragic record-setting flight, Captain Hawthorne C. Gray ascends to 42,470 feet, a record for all aircraft, in an Army hydrogen balloon but dies from lack of oxygen before he can descend to a safe altitude. The record is nullified because of Gray's death.

10 December
Colonel Charles A. Lindbergh visits the U.S. House of Representatives where the House passes a bill awarding him the Congressional Medal of Honor.

14 December
Major General James Fechet assumes command of the Army Air Corps, replacing Mason Patrick. Fechet rose from the enlisted ranks to this command position.

1928

11 January
Commander Marc A. Mitscher, piloting a UO-1, makes the first takeoff and landing on the aircraft carrier USS *Saratoga*. Mitscher is the Air Officer for the carrier and will go on to command the USS *Hornet* during the Tokyo Raid led by Lieutenant Colonel James H. "Jimmy" Doolittle in April 1942.

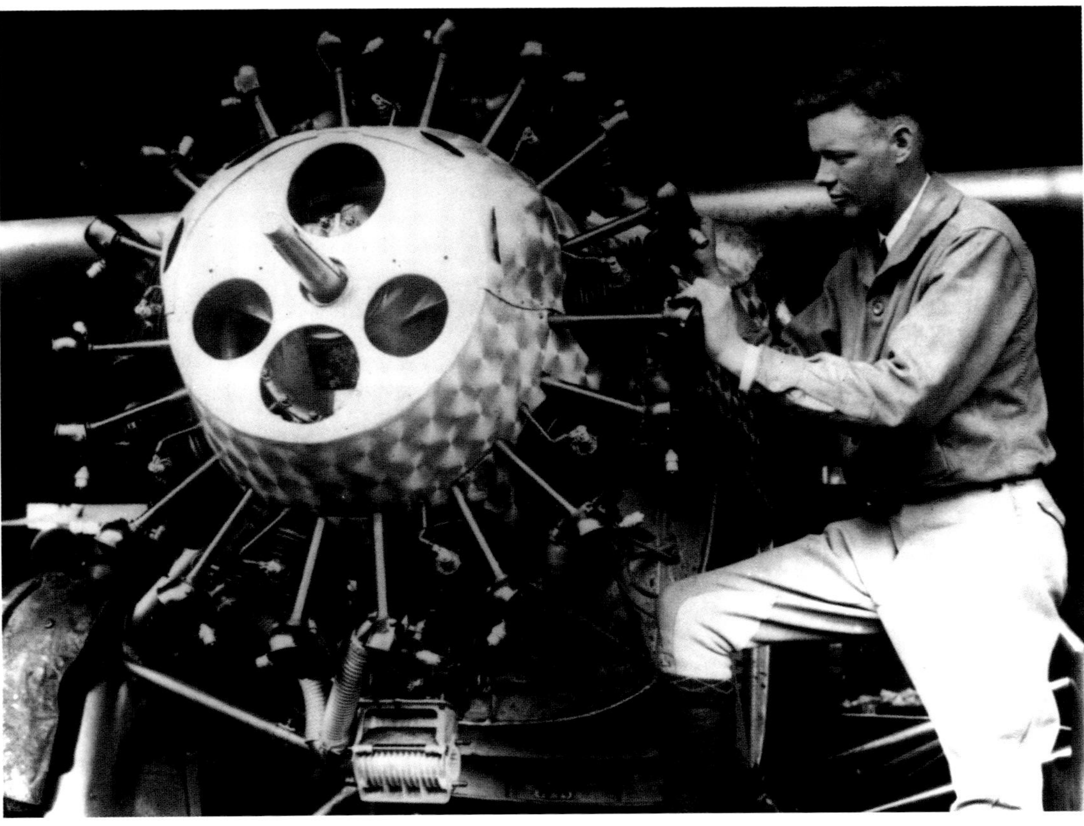

Right: *Captain Charles A. Lindbergh (Missouri National Guard) works on the Wright J-5C engine that powered his transatlantic crossing in the Ryan NYP* The Spirit of St. Louis. *(NASM)*

Left: *Randolph Field circa 1930. (NASM)*

Opposite, clockwise from top left: *Fokker C-2* Question Mark *positioned to receive fuel behind the C-1 refueling aircraft; Record-setting crew: Sergeant Hooe, Lieutenant Quesada, Lieutenant Halverson, Captain Eaker, and Major Spatz; Fokker C-2* Question Mark *in flight; Major Spatz (later Spaatz) grabs the hose for fuel transfer. (All images USAF)*

3 February
Lieutenant H. A. Sutton begins to test the spin characteristics of airplanes at Wright Field. His investigations are critically important to improving aircraft safety. He earns the Mackay Trophy for his flying.

15 February
President Calvin Coolidge authorizes the establishment of a new Army Air Corps training center near San Antonio, Texas. An innovative, circular design will allow each of the squadrons to fly, live, and work near each other in different quadrants of the central housing circle. The headquarters building is cleverly designed to conceal the base water tower.

1–9 March
Lieutenant Burnie R. Dallas and Beckwith Havens, flying a Loening Amphibian, make the first transcontinental amphibious airplane flight. Their total flight time is 32 hours and 45 minutes.

15–21 April
Former Air Corps Lieutenant Carl B. Eielson and Britain Hubert Wilkins make the first east-to-west Arctic crossing from Point Barrow to Green Harbor, Spitzbergen. Flying in a Lockheed Vega outfitted with skis, they endure five days of weather-related delays but finally make the crossing of 2,200 miles in a total of 20 hours and 20 minutes flying time.

12 May
Army Air Corps Reservist Lieutenant Julian S. Dexter completes a two-month mission mapping over 60,000 square miles of the Florida Everglades. Dexter flies for 65 total hours to complete the task.

Lieutenant R. W. Douglas and Lieutenant J. E. Parker, flying in two Boeing PW9-D12 Curtiss aircraft, set a record for single-seat aircraft when they fly from France Field, Panama Canal Zone, to Bolling Field, D.C.

9 June
In a remarkable display of consistent performance in the air, Lieutenant Earle Partridge wins the Army Air Corps aerial gunnery match at Langley Field, Virginia. This victory is his third in a row. Partridge will rise to high command positions during World War II.

15 June
The first aircraft-to-train mail transfer is made when Lieutenant Karl S. Axtater and Lieutenant Edward H. White fly their Air Corps blimp over the moving Central train. They are able to deliver a mail satchel to the clerk on the train, thus completing the operation.

16 June
Superchargers designed to deliver sea-level pressure when at 30,000 feet altitude are successfully completed at Wright Field. A new and improved liquid oxygen system is also tested with excellent results.

30 June
Captain William E. Kepner and Lieutenant William O. Eareckson capture first place in the International Gordon Bennett Balloon Race held in Detroit by soaring for 460 miles. Since this victory is the third consecutive by American pilots, the trophy is permanently retired when awarded after the race.

19 September
The first diesel engine designed for aircraft propulsion is tested at Utica, Michigan. The engine is designed by I. M. Woolson and built by the Packard Motor Car Company.

10 October
Captain St. Claire Streett and his observer and photographer, Captain A. W. Stevens, climb to 37,854 feet, setting an altitude record for aircraft carrying more than one person in flight. The event occurs over Wright Field, Ohio.

11 November
Having crossed over the globe during April, former Air Corps Lieutenant Carl B. Eielson and Britain Hubert Wilkins, flying in the same Lockheed Vega used during that flight, make the first flight over Antarctica.

3 December
An association of pioneer airmen meets in Chicago during an air meet taking place there. The "Early Birds" roster is an impressive group of men including Benny Foulois, Hap Arnold, Glenn Curtiss, and Igor Sikorsky.

19 December
Harold F. Pitcairn makes the first autogiro flight in America at Willow Grove, Pennsylvania.

1929

1–7 January
Taking off on New Years Day, the crew of the Fokker C2-3 Wright 220 *Question Mark* will set an unofficial endurance record of 150 hours and 40 minutes while orbiting over Los Angeles and Pasadena, California. The mission commander is Major Carl "Tooey" Spatz, Captain Ira C. Eaker, 1st Lieutenant Harry Halverson, 2nd Lieutenant Elwood R. Quesada, and Staff Sergeant Roy Hooe. Two Douglas C-1 aircraft refuel the *Question Mark* 37 times during the mission. The C-2 was named *Question Mark* because there was no certain way to know how long the experimental flight would remain airborne.

7 January
The comic strip *Buck Rogers in the 25th Century* debuts in newspapers across the country. Written by Phillip Nowlan and illustrated by Richard Calkins, the series fires the imagination of many youngsters of the day.

9–16 January
The first airplane ferried by the Army Air Corps to a foreign station occurs when Major Paul Beck commands a C-2 Army transport, which flies 3,130 miles from Wright Field to France Field, Panama Canal Zone.

27 February
The Distinguished Flying Cross is awarded to Orville and Wilbur (posthumously) Wright by former Secretary of War Dwight Davis.

14 April
Patent application is made for the "pilot maker" flight trainer by inventor Edward Albert Link. The Link Trainer will become an essential part of every pilot's training before the Second World War begins.

16 May
The Paramount Studios movie *Wings* is awarded the Oscar for best picture of the year. Such an honor is indicative of the nation's attraction to flight and flying.

17 July
Dr. Robert H. Goddard successfully launches a liquid-fueled rocket that carries a camera to take pictures of a thermometer and a barometer mounted in view of the lens. The rocket ascends to 171 feet above the ground near Auburn, Massachusetts.

15 August
Lieutenant Nicholas B. Mamer and Arthur Walker fly nonstop from Spokane, Washington, to the east coast and back. Flying a Buhl Sesquiplane *Spokane Sun God*, the crew covers 7,200 miles on the record setting trip and is refueled in flight a total of 11 times. The airplane is powered by a Wright Whirlwind engine.

24 September
Following 18 months of planning and preparation, Lieutenant Jimmy Doolittle makes the first blind flight from takeoff to landing. Although he had accomplished the flight earlier that day in complete fog, the accomplishment was not official until successfully flown with a flight observer and ground observers as witnesses. Lieutenant Ben Kelsey occupied the front cockpit of the Consolidated NY-2 aircraft while Doolittle was enclosed beneath a cockpit hood that eliminated any outside references for the duration of the flight. Later in life, Doolittle considers this flight his single most important contribution to aviation.

Left: *Lieutenant Jimmy Doolittle landed this Consolidated NY-2 without outside references on September 24, 1929. Doolittle considered that blind landing his most significant contribution to aviation. (NASM)*

Right: *The Curtiss P-1 Hawk was a favorite of many aviators for its power and maneuverability. (USAF)*

29 October
On this Tuesday, one of the worst in Stock Market history, stock values fall an average of 10–12 percent. Poor monetary policy is eventually blamed for the economic crisis. Black Tuesday, as it becomes known, marks the end of the Roaring Twenties and the beginning of the Great Depression in America, which lasts for ten years.

23 November
Robert H. Goddard receives a $50,000 grant from the Daniel Guggenheim Fund for the Promotion of Aeronautics for rocket research.

29 November
Army Captain Ashley McKinley, the trip photographer, accompanies Navy Commander Richard E. Bird, Bernt Balchen, and Harold June during the first flight over the South Pole. Balchen pilots the Ford Trimotor on the flight.

31 December
The Daniel Guggenheim Fund for the Promotion of Aeronautics terminates its programs.

1930

8–29 January
Major Ralph Royce leads a deployment of Curtiss P-1C Hawks from Selfridge Field, Michigan, to Spokane, Washington. The objective of the journey is to train for the harsh conditions that might be experienced in arctic flying. Major Royce is awarded the Mackay Trophy for his mission.

4 April
The American Interplanetary Society is formed in New York, initially composed of 11 men and one woman. Formed to promote spaceflight and rocket experimentation, this organization is the progenitor of today's American Institute of Aeronautics and Astronautics (AIAA).

6 April
Captain Frank Hawks completes a transcontinental flight of 2,860 miles from San Diego to New York, taking 36 hours and 47 minutes while piloting a glider in tow by another aircraft.

8 April
The first Daniel Guggenheim Medal for Aeronautics is presented to Orville Wright.

12 April
A group of 19 P-12 pilots shatter the old formation altitude record of 17,000 feet by climbing while in formation to 30,000 feet over Mather Field, California. Captain Hugh M. Elmendorf of the 95th Pursuit Squadron leads the mission.

21 April
Colonel Charles Lindbergh, flying a Lockheed Sirius monoplane, sets a new transcontinental speed record of 14 hours and 45 minutes. He and his wife, Anne Morrow Lindbergh, cover 2,530 miles flying from Glendale, California, to Roosevelt Field, Long Island, New York.

16 June
Elmer Ambrose Sperry dies in New York City at age 69. Sperry invented the airplane gyrocompass and numerous stabilizer devices for aircraft. He founded several companies during his life and was a consultant to the Navy on the topics of mines, torpedoes, and navigation at sea.

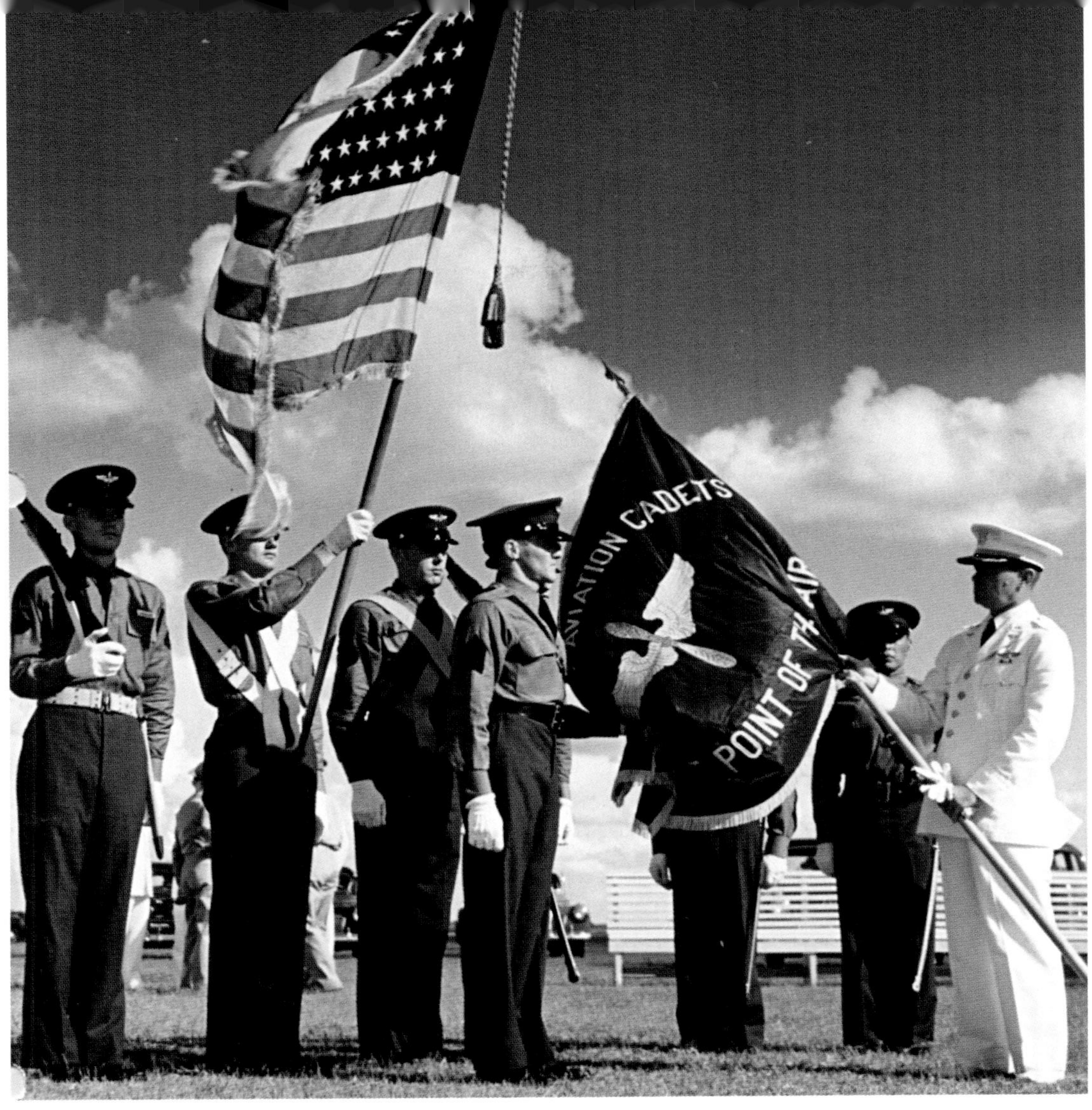

Left: *When Randolph Field opened in 1930, it was envisioned as the West Point of the Air. In this image, Colonel Idwal H. Edwards unfurls the flag during his command in 1941. (USAF)*

Opposite: *Jimmy Doolittle assists ground crew in refueling his Laird Super Solution during the 1931 cross-country Bendix Race. Not only did he win the race but he also made the first coast-to-coast flight in less than 12 hours. (NASM)*

20 June
Randolph Field in San Antonio, Texas, opens for flying. During dedication ceremonies more than 200 aircraft perform a flyover for the 15,000 who came to witness the ceremony. Conducting primary and basic flight training, this base becomes the "West Point of the Air" that pioneer military aviators have envisioned for many years. Today, Randolph Air Force Base is still the hub for Pilot Instructor Training in the Air Force.

23 July
Pioneer aviator and aircraft designer Glenn Hammond Curtiss dies in Buffalo, New York, from complications after an appendectomy. He was 52 years old. Curtiss is buried within sight of his Hammondsport aircraft factory. Curtiss built the *June Bug*, which won the first Scientific American Trophy for the first public flight in America greater than one kilometer.

28 September
Daniel Guggenheim, long-time supporter of aeronautics in America, dies at Sands Point, New York. He was 73 years old.

6 November
Captain Edward V. Rickenbacker is presented with the Medal of Honor by President Herbert Hoover. Rickenbacker was nominated for the award during the Great War but the paperwork was lost.

9–10 November
Captain Roy W. Ammel makes the first solo, nonstop flight between New York and the Panama Canal Zone. Piloting a Lockheed Sirius Blue Flash, Ammel crosses 2,700 miles in 24 hours and 35 minutes.

30 December
Dr. Robert H. Goddard launches his fifth rocket, the first from his new test facility near Roswell, New Mexico. The 11-foot-long rocket soars 2,000 feet into the sky and accelerates to 500 miles per hour.

1931

9 January
The Air Corps takes primary responsibility for U.S. coastal defense after Army Chief of Staff, General Douglas MacArthur, and Chief of Naval Operations, Admiral William Pratt, agree to the assignment.

15 February
The Air Corps flies the first night deployment of aircraft when 19 planes venture from Selfridge Field, Michigan, to Bolling Field, D.C.

9 May
The now famous A-2 leather flying jacket is approved for production.

21–30 May
The Air Corps holds the largest aerial exercise in its history when 1,400 aircrew, manning more than 660 planes, conduct training missions over vast areas of the country. Brigadier General Benjamin D. Foulois, commander of the provisional division formed for the mission, is in charge of a composite force of pursuit, attack, bombardment, observation, and transport aircraft. Foulois is awarded the Mackay Trophy for his command efficiency during this massive aerial exercise.

27 May
The first wind tunnel for full-size aircraft (a full-scale wind tunnel) begins operating at Langley Field under the supervision of the National Advisory Committee for Aeronautics (NACA).

15–31 July
The Air Corps Tactical School moves from Langley Field to Maxwell Field, Montgomery, Alabama. Today, Maxwell Air Force Base remains the intellectual heart of the Air Force.

11 August
Major John F. Curry assumes command of the Air Corps Tactical School newly established at Maxwell Field, Alabama.

4 September
Lieutenant James H. Doolittle flies from Los Angeles to Cleveland, Ohio, in nine hours and 10 minutes. For this he wins the first Bendix Transcontinental Race, flying a Laird racer powered by a 550-horsepower Wasp Junior engine. He refuels and continues on to New York where he successfully completes the first coast-to-coast flight in less than 12 hours when he touches down in Newark, New Jersey, after 11 hours and 16 minutes. Not done yet, he refuels and flies back to Cleveland to refuel and claim his Bendix Trophy, then continues to St. Louis, his home at the time.

18 September
Japanese forces attack Chinese forces in Manchuria, starting hostile activities that will last until 1945.

26 October
The prototype for the de Havilland DH-82A Tiger Moth flies for the first time. Large numbers are built and are used as trainers by civil and military pilots.

20 December
Major General Benny Foulois assumes command of the Army Air Corps, replacing Major General James E. Fechet. Like Fechet, Foulois had enlisted in the Army and then rose to general officer rank.

1932

1–31 January
The 11th Bombardment Squadron stationed at March Field, California, flying Condor bombers, provides relief packages by airdrop to snowbound Navajo and Hopi Indians. Lieutenant Charles H. Howard is the squadron commander for these hazardous humanitarian missions. For their intrepidity during these missions, the 11th squadron receives the Mackay Trophy.

26 January
Pioneer aviator Edward Stinson dies when his Stinson monoplane runs out of fuel over Lake Michigan. During his attempted emergency landing, he strikes a flagpole and dies of his injuries. Three passengers are seriously injured in the accident.

22 February
Early aeronautical pioneer Colonel Thurman H. Bane dies in New York. Bane was instrumental in early aviation technology experimentation, research, and development while commander at McCook Field, Ohio. Projects such as the Barling Bomber, Liberty Eagle, cantilever monoplane, and variable pitch propeller were accomplished under Bane's leadership.

20 March
The Boeing XP-936, later XP-26, flies for the first time, piloted by Boeing test pilot Les Tower. It is the first all-metal monoplane purchased by the Air Corps. The P-26 Peashooter is the last Army pursuit plane to feature an open cockpit, nonretractable landing gear, and external wire bracing for its wings. All of these high-drag features adversely affect the Pea Shooter's performance.

9 May
Captain Albert F. Hegenberger flies the first solo instruments-only flight at Wright Field near Dayton, Ohio, piloting a Consolidated NY-2. He wins the Collier Trophy for this aerial achievement.

20–21 May
Amelia Earhart, flying a Lockheed Vega, becomes the first woman to accomplish a solo, nonstop flight across the Atlantic Ocean. She pilots her Lockheed Vega a distance of 2,026 miles, taking 15 hours and 40 minutes to complete the flight. Earhart wins the Harmon Trophy and other awards for her accomplishment.

24 July
Alberto Santos-Dumont, Brazilian aviation pioneer, dies in Sao Paulo.

31 August
Captain A. W. Stevens and Lieutenant C. D. McAllister fly to an altitude of five miles over Fryeburg, Maine, to photograph a solar eclipse.

3 September
Major James H. Doolittle, U.S. Army Reserve, flies the dangerous Granville GeeBee Racer to a new world speed record for land planes over a three-kilometer course near Cleveland, Ohio. The powerful Wasp engine propels Doolittle at 294 miles per hour.

21 September
The intensity of cosmic rays are measured by spherical devices placed into Condor bombers at March Field, California. These experiments are conducted by Dr. Robert Millikan, a Nobel Laureate from Caltech, under the command of aircrew from Lieutenant Colonel Arnold's 11th Bombardment Squadron.

15 October
The Institute of Aeronautical Sciences is founded by a group interested in aviation as an emerging science. Jerome C. Hunsaker is chosen as the president. This organization will eventually merge with the American Rocket Society and form the American Institute of Aeronautics and Astronautics (AIAA).

12 November
On a farm near Stockton, New Jersey, the American Interplanetary Society (AIS) fires its first rocket in a static test environment. AIS No. 1 burns for approximately 30 seconds while producing 60 pounds of thrust. The AIS changes its name to American Rocket Society (ARS) in 1934.

19 November
Wilbur and Orville Wright are honored with the dedication of a national monument at Kitty Hawk, North Carolina, the location of the first powered, manned flight.

1 December
Wernher von Braun, a 20-year-old college student, begins work for the German Army's secret rocket program.

5 December
Orville Wright, co-inventor of the airplane, receives the first honorary fellowship of the Institute of Aeronautical Sciences.

1933

19 February
Former Chief Signal Officer of the Army Brigadier General James Allen dies. Allen sought to expand aviation-related activities within the Signal Corps and was responsible for the funding and development of early Army aircraft, particularly the Wright Military Flyer.

15-22 July
Famed aviator Wiley Post becomes the first person to fly solo around the world. Piloting a Lockheed Vega *Winnie Mae*, the journey takes seven days, 18 hours, 49 minutes at an average speed of just under 135 miles per hour.

11 October
The Drum Board (after its head Major General Hugh A. Drum) recommends the formation of a General Headquarters Air Force consisting of 1,000 planes to be used independently from surface forces. Secretary of War George H. Dern approves the committee's report.

27 November
The first production Martin B-10 bomber is delivered to the Army Air Corps. A remarkable technological achievement, the all-metal monoplane features an enclosed cockpit, retractable landing gear, internal bomb bay, rotating gun turret, and with its powerful twin-engine configuration, is faster than most fighters of the time.

1934

9 February
All commercial airline contracts to carry the U.S. mail are summarily cancelled by the Roosevelt administration.

19 February
Under presidential order, the Army Air Corps begins delivering the U.S. mail in place of commercial carriers. Even though flying a restricted delivery schedule, poor winter weather and inadequate training result in accidents and fatalities while attempting to accomplish the task.

10–19 March
Air Corps delivery of the U.S. mail is suspended after the loss of nine airmen. Service begins again on the 19th after upgrades to equipment and further reduction of the number of routes and flights.

Opposite: *Boeing P-26 Peashooters in echelon formation. (NASM)*

Below: *Martin YB-10 bombers in close trail formation. (USAF)*

Following spead: *"Doolittle in Gee Bee Racer" by Ferdinand Warren. (NASM Art Collection)*

NR2100
11
7 11
SARA
Wasp
Goodrich Tires

THE AIR MAIL EMERGENCY OF 1934

Carroll V. Glines

Left: *In February 1934, the Army Air Service took over delivery of the U.S. Mail after contracts were cancelled by the President. (Robert and Kathleen Arnold Collection)*

Opposite, above: *A U. S. Army mail plane flies over the mountains in Utah during the winter of 1934. (Robert and Kathleen Arnold Collection)*

Opposite, below: *Lieutenant Colonel Hap Arnold commanded the Western District and flew this specially marked P-12. (USAF Art Collection)*

Air mail service in the United States was pioneered by pilots of the Army Air Service on May 15, 1918, with flights between Washington, D.C., Philadelphia, and New York. During the next seven years the Air Service proved that the mail could be flown from place to place swiftly, reliably, and safely. In 1925, the Post Office Department gradually turned over the task to private civilian contractors. By the following year, contract airlines were shuttling back and forth over 12 air mail routes.

Passengers slowly became a source of additional revenue, and the airline industry began to grow but was still dependent on government air mail contracts. Many small airlines were formed, merged with others, and gradually gave shape to a highly competitive, geographically integrated transportation system. Intense competition led to charges of fraud and collusion in the award of air mail contracts. Congressional investigators alleged that several companies had been favored with contracts without public bidding. President Franklin Roosevelt ordered the cancellation of all air mail contracts on February 9, 1934.

Major General Benjamin D. Foulois, then chief of the Army Air Corps, was casually asked by the Second Assistant Postmaster General if the Air Corps could take over the task. Foulois replied he thought it could in about 10 days if the President wanted them to try. He did not realize that such an informal answer to a casual question would elicit such an import response. He was ordered to take over the air mail routes on February 19, 1934.

Foulois promptly organized a country-wide route structure into three zones and appointed senior flying officers to run them. He made careful plans but knew that only a small number of Army pilots had any extensive night or instrument flying experience and very few of their aircraft were equipped with landing, navigation, or cockpit lights. Only the airliners had new gyro instruments and experience in radio beam flying techniques, which were developed by civilian airlines, not the military aviation branches.

Without instruments or radios, Army pilots began making familiarization flights over their assigned routes without the mail. On the first night of practice runs, there were two fatal crashes, both in bad weather. A few days later, the first pilot crashed with mail aboard and was killed. Again weather was a factor but the lack of instruments and radio equipment was the underlying cause. There were several more fatal accidents caused by the lethal combination of darkness, poor weather, and inexperience. Foulois was publicly criticized for "those air mail murders."

Other problems developed. Funds had not been transferred from the Post Office Department to the Army Air Corps. Enlisted men were living in shacks and existing on food handouts from the local townspeople. Supplies were not available and work had to be done in the open in howling blizzards and drenching rains.

Foulois was ordered to the White House and severely reprimanded by President Roosevelt. Foulois asked for a 10-day stand-down period of operations, which was granted. When operations resumed,

conditions were improved by the introduction of a few properly modified planes and rudimentary training in instrument flying.

In the following weeks, negotiations for new air mail contracts commenced and by the end of April 1934 Foulois was instructed to phase out operations. On June 1, the last bag of mail was flown by an Army Air Corps plane and the books were closed on a tragic but vitally critical chapter in the history of American aviation.

Postmaster General James A. Farley personally praised the Air Corps for their "fine courage and dauntless determination" and predicted that Congress would give more adequate support to the Army. The Air Corps did not receive any other words of tribute, but much had been accomplished under difficult circumstances. Over 13,000 hours of flying time had been logged, and pilots had flown over 1.5 million miles carrying 777,000 pounds of mail.

In the end, it was the loss of airmen and aircraft, not their successes, which were significant. The crashes and deaths had pointed out as nothing else could in peacetime that the United States military air arm was dangerously inadequate. The air mail experience had emphatically drawn attention to the state of neglect into which Air Corps procurement and training had fallen. General Foulois reminisced years later, "Flying the mail was a godsend for the Air Corps and the country. Without it, I'm convinced we would have never recovered from the disaster at Pearl Harbor."

Left: *Lieutenant Colonel Hap Arnold led a flight of 10 Martin B-10 bombers on a round-trip flight from Washington, D.C., to the Territory of Alaska. The B-10 was the most advanced aircraft of its day. (NASM)*

22 May
Captain W. T. Larson is awarded the 1933 Mackay Trophy for his research and development of procedures of aerial frontier defense specifically pertaining to instrument flying and blind takeoffs and landings.

24 March
Major General George O. Squier dies in Washington, D.C. Squier was instrumental in the procurement of the Army's first airplane and acted as Chief of the Air Service during World War I.

1 June
Commercial contracts now renegotiated, the Air Corps is relieved from delivering domestic air mail.

18 June
The Boeing Company decides to spend company money on the design of the Model 299. The money turns out to be well spent, as the result of the company-funded work will be the B-17 Flying Fortress.

28 June
The Boeing Aircraft Company is awarded a contract for the design of the B-17 Flying Fortress.

18 July
The Baker Board (for its head Newton D. Baker, former Secretary of War) issues a report that reaches similar conclusions to the 1933 Drum Board. Both groups suggest the formation of a centrally controlled aerial strike force.

19 July
Lieutenant Colonel Hap Arnold, leading ten Martin B-10s, sets out on a round trip flight from Washington, D.C., to Alaska and back.

28 July
Major W. E. Kepner, Captain A. W. Stevens, and Captain O. A. Anderson ascend to 60,613 feet in an Air Corps National Geographic Stratosphere balloon named *Explorer I*. All are awarded the Distinguished Flying Cross for the mission.

20 August
Lieutenant Colonel Hap Arnold is awarded the Mackay Trophy upon returning to Bolling Field, D.C., after completion of a 7,000-mile trip to Alaska with 10 Martin B-10 bombers. It is the first journey of such distance during which air-to-ground radio contact is maintained for the entire trip. While in Alaska, the B-10s fly aerial mapping missions and mingle with the local population.

Right: *Wiley Post and the Lockheed Vega* Winnie Mae. *Post was the first man to fly solo around the world in 1933. (NASM)*

5 September
Famed aviator Wiley Post climbs his Lockheed Vega *Winnie Mae* to stratospheric altitudes while wearing a pressure suit developed by the B. F. Goodrich Company. This cumbersome suit becomes the first step toward the development of other advances in pressurized flight.

8 November
Captain Edward V. Rickenbacker, Captain Charles W. France, and Silas Moorehouse set a new record for passenger transport when they fly from Los Angeles, California, to Newark, New Jersey, in 12 hours and four minutes.

17 November
Captain Fred C. Nelson wins the Mitchell Trophy Race at Selfridge Field, Michigan. His average speed is 217 miles per hour.

1935

11–12 January
Amelia Earhart makes the first-ever solo flight from Wheeler Field, Hawaii, to Oakland, California, piloting a Lockheed Vega powered by a Pratt and Whitney Wasp engine. The flight takes 18 hours and 16 minutes.

15 January
Army Reserves Major Jimmy Doolittle sets a transcontinental record for passenger aircraft when he and two passengers travel from Los Angeles to New York in 11 hours and 59 minutes, a record broken only three weeks later. The flight also sets a new west-to-east transcontinental record.

1 March
Following the recommendations of the Drum and Baker Boards, the General Headquarters (GHQ) Air Force commanded by Brigadier General Frank M. Andrews is activated at Langley Field, Virginia. It is the first air unit established with the authority to manage tactical aviation independently from the ground commander. It is a compromise but still an important first step toward the establishment of an independent Air Force.

8 March
Dr. Robert H. Goddard launches a pressure-equalized, liquid-propellant rocket from his Roswell, New Mexico, test facility. The rocket features a recovery parachute, soars to 1,000 feet, and accelerates to 700 miles per hour. This launch is Goddard's finest thus far. He surpasses this flight in May, achieving a launch altitude of 7,500 feet when he fires a large, 84-pound rocket.

THE MARTIN B-10 BOMBER

Jeffery S. Underwood

Left: *Martin B-10: the first all-metal, monoplane, retractable gear bomber ever built. (NASM)*

Opposite: *"Martin B-10B, 9th Bomb Squadron," by Walter Jefferies. (USAF Art Collection)*

The first modern bomber delivered in large quantities, the Martin B-10 bomber, incorporated numerous technological advances, and it greatly influenced the development of air power doctrine during the interwar years. Nevertheless, the B-10 remains largely forgotten, overshadowed by the more famous Boeing B-17.

Aircraft design and construction underwent a revolution toward the end of the 1920s as all-metal monoplanes replaced the wood and fabric biplanes. At its new factory near Baltimore, Maryland, the Glenn L. Martin Company took the lead in bomber design with its all-metal Model 123. Built as a private venture, the Model 123's advanced elements included a monocoque fuselage, a streamlined midwing with the twin-engines fitted into the leading edges rather than placed into nacelles, and a retractable undercarriage. To carry its bomb load internally, the aircraft had a deep belly, which gave it a whale-like appearance.

In March 1932, Martin sent its Model 123 to Wright Field, Ohio, for evaluation by the U.S. Army Air Corps (AAC) with the designation XB-907. The bomber reached a remarkable speed of 197 miles per hour, but its wings suffered from a serious flutter problem. AAC engineers at Wright Field who had recently developed a light, but strong, cantilevered wing employing "stressed skin" technology, proposed a solution. This wing's external aluminum skin acted as a load-carrying component, not just a covering. The added strength and rigidity allowed for an eight-foot increase in the airplane's wingspan. In return for helping fix the wing flutter problem, the AAC insisted that Martin replace the nose gunner's open cockpit with a manually operated turret.

The B-10 also had more powerful Wright R-1820-19 Cyclone radial engines, which produced 675 horsepower and were protected by full-engine cowlings for improved streamlining. The modified aircraft, redesignated the XB-907A, achieved a speed of 207 miles per hour—faster than any American fighter in service. For this achievement, the Martin Company received the 1932 Collier Trophy. Thereafter, the revolutionary stressed skin technology used on the Martin bomber became a structural characteristic on all U.S. combat aircraft in World War II.

This fast bomber influenced the formulation of American air power doctrine by convincing AAC planners that fighters could not stop modern bombers from reaching their targets. Furthermore, they concluded that bombers flying in close formation on daylight missions needed no fighter escort because the combined defensive armament of the formation would be enough protection. These beliefs evolved into the daylight strategic bombing doctrine adopted by U.S. Army Air Forces (AAF) in World War II.

In 1933, the AAC purchased the XB-907A from Martin and an additional 48 production aircraft. Martin built the first 14 as service test aircraft designation of YB-10 ("Y" for service test) with Wright R-1820-25 Cyclone radial engines of 675 horsepower, sliding canopies that enclosed the pilot and rear gunner's positions, provisions for a radio operator, and a ventral gun position. Thirty-two more aircraft were delivered as B-12As, differing from the YB-10 by using Pratt and Whitney R-1690-11 Hornet radial engines.

During the summer of 1934, Lieutenant Colonel Hap Arnold, who later commanded the AAF in World War II, successfully led a flight of 10 YB-10s from Washington, D.C., to Fairbanks, Alaska, and back. For this round-trip flight of 7,360 miles, much of it over uncharted territory, Arnold won the 1934 Mackay Trophy. Besides demonstrating the reliability and speed of the B-10, the Alaska flight proved that the AAC could rapidly move its modern aircraft to defend any part of North America.

Pleased with the Martin bomber, the AAC ordered 103 B-10B aircraft, the primary service version. Powered by two Wright R-1820-33 engines of 775 horsepower, the B-10B had a maximum speed of 215 miles per hour, and its armament consisted of three defensive .30-cal. machine guns and capacity to carry 2,200 pounds of bombs. In addition to assignments in the continental United States, B-10s served with AAC units overseas in the Panama Canal Zone and the Philippine Islands. Export versions found their way to Argentina, China, Siam, Turkey, the Dutch East Indies, and one even went to the Soviet Union. Obsolete and replaced by newer B-17s and B-18s by the time World War II started, the remaining B-10s performed secondary duties like towing targets. No B-10s saw combat with the AAF, but a few saw action in China and the Dutch East Indies.

The only surviving B-10, an export version once sold to Argentina, is on display at the National Museum of the United States Air Force at Wright-Patterson AFB, Ohio.

Left: *Doctor Robert H. Goddard setting up his 1935 A-series rocket. (NASM)*

Below: *Army Air Forces North American AT-6C Texans in formation near Foster Field, Texas, during World War II. (NASM)*

Opposite, above: Explorer II *ready for launch, November 1935. (USAF)*

Opposite, below: *Major General Oscar Westover, Chief of the Air Corps (left), and Brigadier General Hap Arnold, his deputy. (Robert and Kathleen Arnold Collection)*

28 March
Dr. Robert H. Goddard launches the first gyroscopically controlled rocket. It soars to 4,800 feet and accelerates to 550 miles per hour.

1 April
The North American NA-16, later the AT-6, flies for the first time. A majority of Army pilots train in the AT-6 before moving on to combat aircraft.

31 May
Hickam Field, located near Fort Kamehameha, Hawaii, is dedicated.

28 July
Under the hand of company test pilot Les Tower, the Boeing Model 299 flies for the first time at Seattle, Washington. Christened the Flying Fortress by interested press writers, the name was formalized before the production model took to the skies. Hap Arnold once said that the B-17 was "airpower you could put your hands on."

15 August
Tragedy strikes as famed satirist Will Rogers and stunt pilot Wiley Post are killed in a take-off crash at Point Barrow, Alaska. They are flying a hybrid Lockheed Orion-Explorer.

20 August
The Boeing Model 299 is flown to Wright Field, Ohio, for tests. Test pilot Leslie R. Tower and crew make the 2,100-mile trip at an average speed of 232 miles per hour.

24 August
Brigadier General Frank M. Andrews, piloting a Martin B-12W bomber modified with pontoon floatation gear, sets three new speed-with-payload seaplane records during flights from Langley Field, Virginia, to Floyd Bennett Field, New York, and back.

13 September
Howard Hughes flies 352 miles per hour in his Hughes Special, setting a new international speed record for land planes.

17 September
The largest nonrigid airship ever constructed in the U.S., the TC-14, makes it first flight from its assembly point, Scott Field, Illinois. The TC-14 is the largest of its type in the world.

30 October
The Boeing Model 299 crashes on takeoff during flight testing. Gust-locking mechanisms were not removed before flight and an uncontrollable flying condition resulted.

11 November
Captain Orvil A. Anderson and Captain Albert W. Stevens climb to 72,395 feet in an enclosed gondola, the *Explorer II*, setting a world altitude record. This balloon ascent is part of a joint National Geographic-Army Air Corps stratospheric research project. The men are airborne for more than eight hours and drift nearly 350 miles during the ascent. They are later awarded the Mackay Trophy and the Hubbard Gold Medal of the National Geographic Society for the flight.

12 December
Army Lieutenant Hugh F. McCaffery and his five-man crew fly from San Juan, Puerto Rico, to Miami, Florida, setting a record for amphibian-type aircraft of 1,033 miles.

17 December
The Douglas DC-3 flies for the first time near Santa Monica, California. The DC-3 will become one of the most successful passenger liners in history as well as one of the most successful military aircraft (C-47 Dakota) ever built—used during three different wars. Originally called the Douglas Sleeper Transport, more than 10,650 will be built between 1935 and 1947.

22 December
Brigadier General Oscar "Tubby" Westover assumes command of the Air Corps replacing a disgruntled Major General Benjamin D. Foulois. In the wake of the Air Corps "Air Mail Fiasco," Foulois suffers though congressional hearings and interrogations that leave his reputation in shambles. Westover is promoted to major general shortly thereafter.

27 December
Aircraft from the Fifth Composite Group divert a deadly lava flow around Hilo, Hawaii, by precision bombing of the Mauna Loa volcano.

1936

19 February
Airpower advocate Billy Mitchell dies at Doctor's Hospital in New York City at age 57. He is buried in Milwaukee, Wisconsin. Mitchell enlisted in the Army as a private but soon rose to officer rank and led massive air raids against enemy positions during World War I.

12 May
The largest high-speed wind tunnel facility in the world, directed by the National Advisory Committee for Aeronautics, becomes operational at Langley Field. The test section measures eight feet across and the tunnel is capable of producing winds from 85 to 500 miles per hour.

Left: *The Boeing YB-17 in flight during testing near its Seattle plant. (NASM)*

6 June
100-octane aviation gasoline is mass-produced by the Saucony-Vacuum Oil Company, Inc., by the catalytic cracking method.

7 June
Major Ira C. Eaker accomplishes the first-ever blind transcontinental flight when he flies from New York to Los Angeles completely reliant upon cockpit instruments.

16 June
The Air Corps issues its first contract for a single-seat, closed-cockpit, retractable-gear fighter plane. Seversky Aircraft Company wins the contract for what will become the P-35 fighter.

29 June
Major General Frank M. Andrews and Major John Whitely set a world airline distance record for amphibious aircraft when they fly a Douglas YOA5-2 Wright 800 from San Juan, Puerto Rico, to Langley Field, Virginia, a distance of 1,430 miles.

18 July
The Spanish Civil War begins, during which air units from several countries participate.

2 August
Aviation pioneer Louis Blériot dies near Paris. He is 54 years old. In 1909, Blériot became the first man to cross the English Channel in a heavier-than-air vehicle—the Type XI.

1937

11 February
Major J. McDuffie commands eight Martin bombers during a 4,000-mile flight from Langley Field, Virginia, to Airbrook Field, Panama. The bombers accomplish this flight over water without water landing equipment—the first time a major body of water has been crossed by land-configured Army aircraft.

1 March
The first YB-17A arrives at Langley Field, Virginia, assigned to the 2d Bombardment Group.

26 March
Dr. Robert H. Goddard launches a liquid-fuel rocket which features moveable air vanes powered by direct contact with the rocket's exhaust and linked to the gyrostabilizer in the nose. This rocket climbs to nearly 9,000 feet.

12 April
The first practical jet engine undergoes lab testing at Cambridge University, England. Royal Air Force officer Frank Whittle, the engine designer and builder, conducts the tests.

7 May
The Lockheed XC-35 high-altitude research aircraft flies for the first time. The aircraft tests methods for full-cabin pressurization and wins the Collier Trophy for achievement as the first pressurized cabin flown.

Right: *Amelia Earhart piloted this Lockheed Electra in which she and Fred Noonan were lost over the Pacific Ocean. (NASM)*

30 June
Chief of the Air Corps Major General Oscar Westover is forced to terminate the Army's balloon program due to lack of congressional funding. The Army transfers what equipment it has to the Navy.

1 July
The Signal Corps Weather Service is transferred to the Army Air Corps.

2 July
American aviatrix Amelia Earhart and Fred Noonan are lost over the Pacific Ocean. The duo was attempting to fly around the world in a Lockheed Electra and had taken off from San Francisco on May 21.

7 July
Japanese forces invade China, beginning the Second Sino-Japanese War and the first battles of World War II in the Pacific. The Marco Polo Bridge Incident results in the first occupation of Chinese territory (proper) by Japanese military forces.

20 July
The first uniform insignia authorized for an independent American air unit, the General Headquarters Air Force, is authorized for wear.

26 July
Jacqueline Cochran sets a women's American speed record of 203.9 miles per hour piloting a Beechcraft airplane over a 1,000-kilometer circuit.

23 August
The first completely automatic landing, a Fokker C-14B flown by Captain George V. Holloman, occurs at Wright Field, Ohio. The automatic pilot equipment is turned on in flight and Holloman monitors the event. The inventor, Captain Carl J. Crane, and Holloman receive the Mackay Trophy and Distinguished Flying Crosses for the achievement.

1 September
Lieutenant Benjamin Kelsey pilots the first flight of the Bell XFM-1 Airacuda at Buffalo, New York. The Airacuda, a multiplace fighter, results in failure.

21 September
Aviatrix Jacqueline Cochran establishes a new international women's three-kilometer flight speed record of 293 miles per hour flying a civilian version of the Seversky P-35 pursuit plane.

15 October
The Boeing XB-15 flies for the first time under the hand of test pilot Eddie Allen at Boeing's Seattle, Washington, plant. Only one is ever built.

Left: *Six Army YB-17s in formation over New York City during their flight to South America. Leading the flight is Lieutenant Colonel Robert Olds. A few months later, Olds would lead three of these YB-17s during the intercept of the ocean liner* Rex, *more than 600 miles out to sea. (USAF)*

3 December
Major Alexander P. Seversky flies from New York City to Havana, Cuba, in a record setting time of five hours and three minutes.

1938

17 February
A flight of six Boeing B-17s under the command of Lieutenant Colonel Robert D. Olds takes off from Miami, Florida, enroute to Buenos Aires, Argentina. The B-17s are to participate in the inaugural for President Roberto Ortiz.

27 February
The 10,000 mile goodwill flight to Argentina that had begun 10 days earlier ends successfully when the six B-17s involved land at Langley Field, Virginia. All the flight crew are awarded the Mackay Trophy for the mission. The trip to Argentina is the longest nonstop mass flight in Air Corps history, taking 33 hours and 30 minutes.

6 April
The Bell XP-39 Airacobra flies for the first time at Wright Field, Ohio. Bell test pilot James Taylor makes the flight. P-39s will be produced in large numbers as part of the Lend-Lease program between the U.S. and the Russians. Some 4,800 P-39s will be used in the ground attack role on the Eastern Front.

22 April
World War I flying ace Eddie Rickenbacker buys Eastern Airlines from North American Aviation, Inc., for $3.5 million.

12 May
Attempting to validate the Air Corps' mission of coastal defense, three B-17s intercept an ocean liner, the *Rex*, at a range of 700 miles from U.S. territory. The U.S. Navy, miffed at the Air Corps' success, react by demanding that a 100-mile coastal limit be enacted to limit Army bombers.

3–12 August
Major Vincent J. Meloy leads a flight of three B-17s from Langley Field, Virginia, to Bogota, Columbia.

19 August
The B-18 bomber flies its first transcontinental flight from Hamilton Field, California, to Mitchel Field, New York, logging 15 hours and 18 minutes flight time.

12 September
A high-pressure wind tunnel capable of simulating flight up to 35,000 feet begins operation at the Guggenheim School of Aeronautics at the Massachusetts Institute of Technology (MIT). It is the first tunnel of this type in the United States. Named after the Wright Brothers, the eight-foot test section is capable of creating wind speeds of 400 miles per hour.

Right: *The Douglas A-20 was built in large numbers and flew combat around the globe during World War II. (NASM)*

21 September
Major General Oscar Westover is killed when the Northrop A-17AS Staff Special aircraft he was flying crashes in Burbank, California. The accident board results show that the crash was caused by gusty and unpredictable wind conditions coupled with rising heat currents near the ground. Westover's mechanic, Staff Sergeant Hymes, is also killed in the crash.

29 September
Although Brigadier General Hap Arnold was immediately named as Chief of the Air Corps and promoted to major general the day after Westover's untimely death, he is not confirmed for the position until this date. A source close to President Roosevelt had claimed that Arnold had a drinking problem in an effort to insert a different officer as chief. The attempt to tarnish Arnold's reputation failed.

The Munich conference ends with Prime Minister Neville Chamberlain of Britain proclaiming "Peace in our time," after agreeing to cede the Sudetenland to Germany. He will resign in May 1940 and Winston Churchill will take his place.

14 October
The Curtiss XP-40, the prototype for the P-40 Warhawk, flies for the first time near Buffalo, New York. Test pilot Edward Elliot conducts the test. Almost 14,000 P-40s will be built before the end of the construction run in 1944.

26 October
The Douglas Model 7B, later the A-20 Havoc, flies for the first time near El Segundo, California. The A-20 Havoc would be the most produced Army surface attack aircraft and would see service in all combat theaters.

14 November
During a top-level White House conference, President Roosevelt suggests that the size and scope of the Army Air Corps be increased. Major General Hap Arnold calls the meeting the "Magna Carta" of American airpower.

10 December
James H. Wyld of the American Rocket Society makes a static test of his regeneratively cooled rocket motor design at the ARS test facility near New Rochelle, New York. Wyld's coolant jacket concept is a significant breakthrough in rocket technology.

1939

5 January
Amelia Earhart Putnam is declared legally dead.

21 January
Dr. George W. Lewis is elected to head the Institute of Aeronautical Sciences.

Left: *Major General Hap Arnold escorts President Roosevelt on a tour at Bolling Field just after taking command of the Army Air Corps in 1938. (Robert and Kathleen Arnold Collection)*

Below: *Ju-87 Stuka dive bombers led the German thrust into Poland. (NASM)*

27 January
Experienced test pilot 1st Lieutenant Benjamin Kelsey flies the XP-38, later the P-38 Lightning, at March Field, California, on its maiden flight. In February, Kelsey attempts to break the transcontinental speed record in the XP-38 but crashes just before landing at Mitchel Field. Nonetheless, the performance until that point still convinces the Air Corps to purchase the plane in large numbers.

10 February
The North American NA-40, later the B-25 Mitchell, flies for the first time piloted by test pilot Paul Balfour. The B-25 will become famous when 16 Mitchell bombers are launched from the USS *Hornet* in a daring raid against Tokyo in April 1942.

14 February
Major Caleb V. Haynes and his crew of four officers and six enlisted men fly more than 3,200 pounds of supplies and medicine to Chile for the relief of earthquake victims. Taking off from Langley Field in the Air Corps' only XB-15, the mission demonstrates potential uses of the aircraft for humanitarian relief as well as demonstrating the range and payload potential of heavy bombers. The flight takes 29 hours and 53 minutes flight time to complete. The crew receives the Mackay Trophy for the mission.

21 March
A board of officers is selected to evaluate permanent and auxiliary airfields in Puerto Rico. The members of the board are Colonel Hugo E. Pitz, Lieutenant Colonel Joseph T. McNarney, Karl S. Axtator, and Major George Kenney. McNarney and Kenney would later rise to four-star rank during World War II.

24 March
Jacqueline Cochran sets a women's national altitude record flying a Beechcraft over Palm Springs, California, of 30,052 feet.

3 April
President Roosevelt signs the National Defense Act that will direct military spending for the following year. Air Corps strength is expanded to 48,000 personnel, 6,000 aircraft are authorized, and the budget explodes to $300 million for 1940. The act also authorizes training for African American pilots and crew.

18 April
Major General Hap Arnold recalls Colonel Charles A. Lindbergh to active duty and assigns him the task of evaluating the weaknesses in military airpower expansion underway at that time.

1 June
Civilian flying schools around the country are tasked to train Air Corps flying cadets in an effort to rapidly produce a large trained flying force. This is part of Hap Arnold's balanced air plan that he had put into effect after taking command of the Air Corps.

15 June
The Heinkel He-176 rocket aircraft makes its first flight near Peenemünde-West, Germany. Flights are made until July 3, when the risks are determined to be too great to continue the flight testing program.

30 July
Major Caleb V. Haynes and Captain W. D. Olds fly a B-15 loaded with a payload of 15.5 tons to an altitude of 8,200 feet, reestablishing the payload-carrying altitude world record.

24 August
The first jet-powered aircraft, the Heinkel He-178, makes its first flight, piloted by Erich Warsitz. The second flight, a bit longer in duration, occurs three days later.

26 August
Major Charles M. Cummings and Major Stanley Umstead fly a B-17A from Miami to the Panama Canal Zone in six hours. The 1,200-mile flight demonstrates the speed at which the Panama Canal Zone can be reinforced if needed.

1 September
Germany invades Poland after aerial bombardment of the Polish lines by Junkers Ju-87 Stuka dive bombers. World War II in Europe begins.

2 September
At the annual Cleveland Air Races, songwriter Robert Crawford sings "Nothing Will Stop the Air Corps Now," designated the official Army Air Corps song. His was selected by a group of Air Corps wives from over 700 entries.

14 September
Igor Sikorsky's VS-300 becomes the first practical helicopter to fly. Sikorsky pilots the craft to a height of three feet for a duration of 10 seconds.

Above: *More Consolidated B-24 Liberators were built than any other U.S. fighter or bomber aircraft—more than 18,400 by the war's end. (NASM)*

15 September
Jacqueline Cochran sets a new international speed record for a 1,000-kilometer course of 305.9 miles per hour while piloting a Seversky AP-9 monoplane near Burbank, California.

19 October
Vannevar Bush is elected chairman of the National Advisory Committee for Aeronautics, succeeding Joseph Ames. Bush, president of the Carnegie Institute, is frequently at odds with the military over control of aeronautical research in the U.S.

16 December
Major General Delos Emmons assumes command of GHQ Air Force, replacing Major General Frank Andrews.

23 December
Anthony Fokker dies at the age of 49 from meningitis. Fokker had become one of the most famous and successful aircraft designers of all time.

29 December
The Consolidated XB-24 Liberator makes its maiden flight, piloted by company test pilot Bill Wheatley. More than 18,000 B-24s will be built, a total exceeding any other U.S. military aircraft in history.

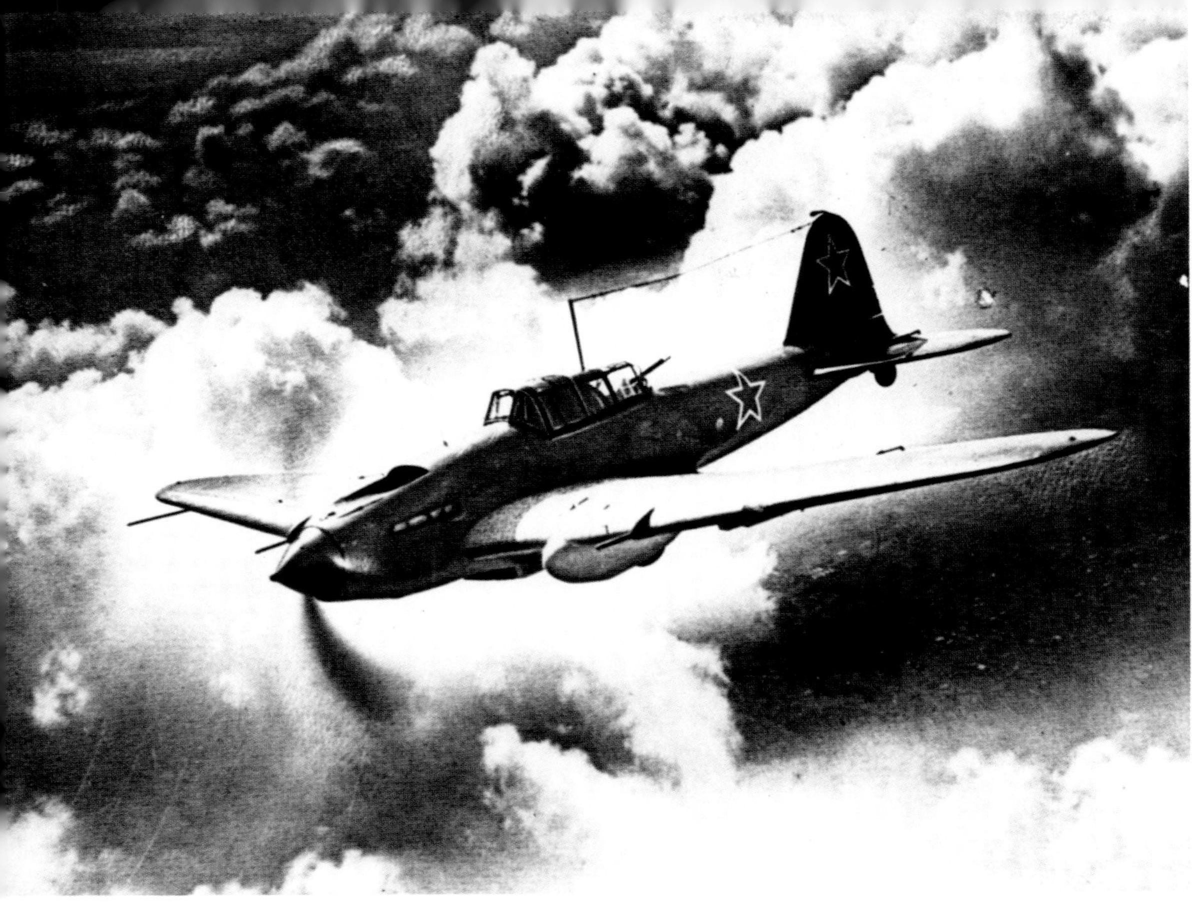

Left: *The Soviet Union built more than 36,000 Il-2 Sturmovik fighters—more than any other combat plane in history. (NASM)*

Below: *The Northrop N-1M was America's first successful flying wing. Today, it is part of the National Air and Space Museum's collection and is on display at the Steven F. Udvar-Hazy Center near Dulles International Airport. (NASM)*

Opposite: *Curtiss C-46 Commando poster. (NASM Poster Collection)*

30 December
The prototype for the Russian IL-2 Sturmovik makes its first flight. The aircraft will become an effective ground attack platform and will be built in greater numbers than any other plane in history; more than 38,000 are constructed.

1940

19 January
Major Jimmy Doolittle (USAR) is elected to be president of the Institute of Aeronautical Sciences.

23 January
Bombers assigned to the 7th Bombardment Group at Hamilton Field, California, move a complete field unit by air, testing the ability of air forces to airlift combat forces. An entire battalion of the 65th Coast Artillery is moved 500 miles by 38 aircraft. It is the first American large-scale experiment with air mobility of ground forces.

2 February
George DeBothezat dies in Boston at age 58. In 1923, DeBothezat built the Army Air Service's first helicopter and, although the Army delayed acquisition of helicopters until many years later, the DeBothezat machine was a significant step forward in the development of early helicopters.

21 February
A resonant-cavity microwave generator—the magnetron—is created by two scientists working at the University of Birmingham, England. The magnetron is essential to further development of airborne radar.

26 February
Air Defense Command is established to integrate the defenses of the U.S. against air attacks. Brigadier General James E. Chaney will assume command on March 15, 1940.

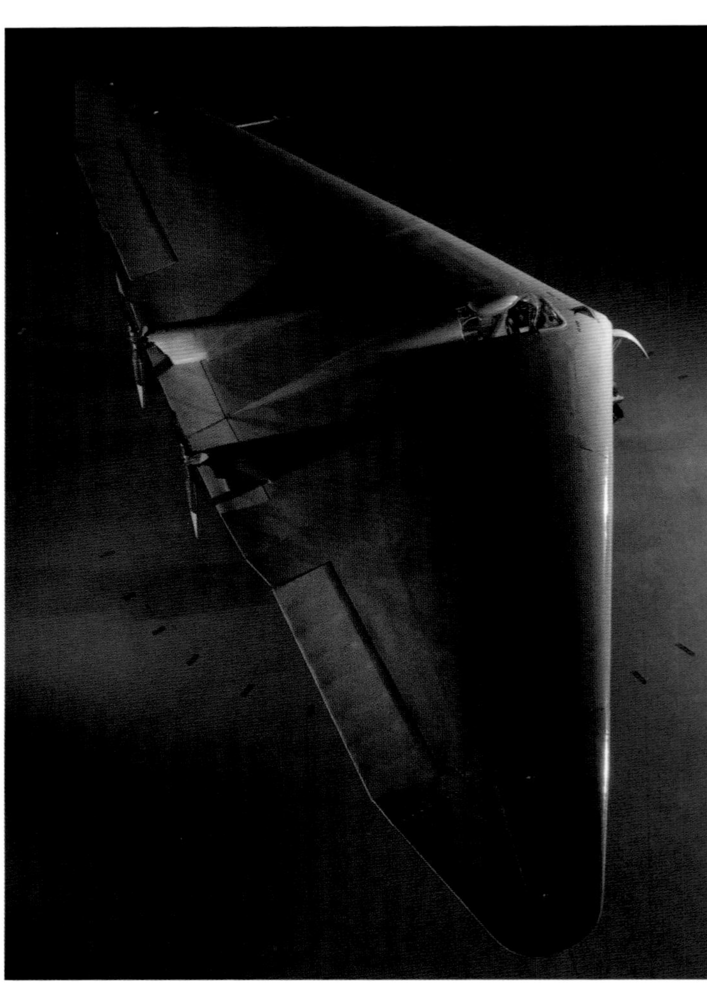

25 March
Under the liberalized release policy, Air Corps aircraft contractors are authorized to sell modern Army aircraft to anti-Axis nations. This establishes a production base for future Air Corps expansion.

26 March
The Curtiss CW-20T, the prototype for the C-46 Commando, makes its first flight at St. Louis, Missouri. The Commando is one of the largest and heaviest aircraft in the Air Corps inventory and will survive to see service in Korea and even the early years of Vietnam.

9 April
Germany invades Denmark and Norway. For the first time in the history of war, troops dropped from aircraft are used to secure a landing field to stage operations by air.

15 April
Field Manual 1-5 is issued by the War Department. It concerns the employment of air power and advocates centralized control of aerial assets. A more detailed regulation, FM 100-20, will not be published until 1943.

18 April
Ames Aeronautical Laboratory located at Moffett Field, California, is formed. Later this facility will be renamed NASA Ames Research Center.

16 May
After weighing advice given him by Chief of the Air Corps, Major General Hap Arnold, President Roosevelt calls for increased aircraft production—up to 50,000 planes per year.

23–25 May
Military maneuvers simulating European combat conditions take place near Barksdale Field, Louisiana. More than 300 aircraft participate with the Third Army, which conducts ground operations.

28 May
Dr. Robert H. Goddard meets with commander of the Army Air Corps Major General Henry H. Arnold to discuss rockets and their potential for military uses. Although Arnold shows limited interest in Goddard's work, he had already directed Caltech professor Theodore von Kármán to begin work on rocket-propelled takeoff devices to increase flight range of heavy bomber-type aircraft.

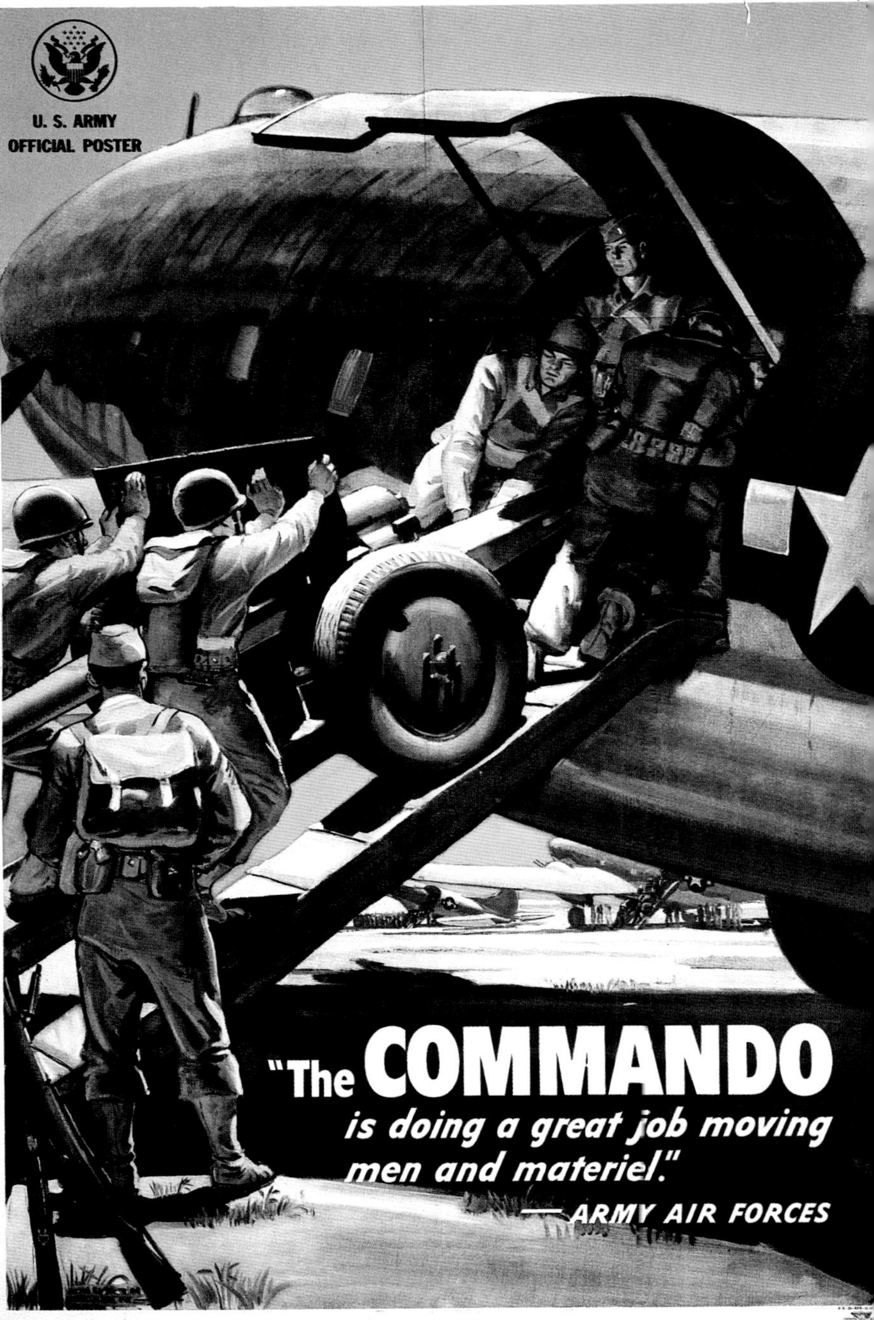
U. S. ARMY
OFFICIAL POSTER
"The COMMANDO
is doing a great job moving
men and materiel."
—ARMY AIR FORCES

11 June
A Goddard rocket fires for a duration of 43.5 seconds, becoming the longest duration burn of a rocket engine during Dr. Goddard's research.

27 June
The Council of National Defense creates the National Defense Research Committee, which will work on special scientific projects during the war. Among these is the development of the AZON guided bomb project, a controlled trajectory weapon.

3 July
The N-1M Flying Wing is introduced by the Northrop Aircraft Company. The aircraft is a single airfoil design and the pilot sits inside the wing itself.

Left: *Top view of a North American B-25 Mitchell bomber. (NASM)*

Below: *Nose section of the Martin B-26 Marauder* Flak Bait. *The aircraft survived more than 200 combat missions and is preserved in the National Air and Space Museum in Washington, D.C. (NASM)*

8 August
Eagle Day. The Luftwaffe begins aerial attacks over Great Britain in an attempt to neutralize the Royal Air Force so that a cross-channel invasion might be accomplished in the near future. This day marks the first in the struggle that will be known as the Battle of Britain.

16 August
The first bombardier training school opens at Lowry Field, Colorado. The first class consists of a cadre who will become the instructors for the school.

19 August
Test Pilot Vance Breese makes the first flight of the North American B-25 Mitchell bomber. The original design, the NA-40, was significantly modified and renamed the NA-62. Larger engines and redesigned wings and tail were part of the modifications that became the famous B-25.

16 September
The Selective Service Act is passed by Congress. The act institutes the peacetime draft and requires all military services to enlist African Americans.

17 September
The Battle of Britain ends with the RAF victorious. The Luftwaffe loses more than 1,700 aircraft and crew. Germany's invasion of Great Britain is postponed indefinitely. Fighting continues into October.

8 October
The first Eagle Squadron is formed in Great Britain. Although under Royal Air Force command, American volunteers participate in support operations in these units.

26 October
The North American NA-73, later the P-51 Mustang, makes its first flight near Inglewood, California. Company test pilot Vance Breese makes the flight. Once the original Allison V-1710 V-12 liquid-cooled engine is replaced with a Packard-built Rolls Royce Merlin V-1650 engine and additional fuel tanks added, the Mustang becomes the premier escort fighter of the war.

1 November
The Hawaiian Air Force is activated at Fort Shafter, Hawaii.

19 November
GHQ Air Force is removed from the Chief of the Air Corps chain of command and assigned to the general commanding the field forces. The step forward achieved with the establishment of the GHQ was neutralized by the realignment.

25 November
The Martin B-26 Marauder flown by company test pilot William "Ken" Ebel makes its first flight near the Martin Middle River plant in Maryland.

The prototype for the de Havilland DH-98 Mosquito flies for the first time at Hatfield, England. Mostly made of wood, the Mosquito is fast, maneuverable, and is capable of long-range missions. The Mosquito also serves with Army Air Forces units later in the war.

Above: *German bombers struck deadly blows against London after August 8—Eagle Day. (NASM)*

MUCH MORE THAN A SCHOOLHOUSE

Dennis M. Drew

Left: *"B-17s over Europe" by David Ray. (USAF Art Collection)*

Opposite: *The visual version of AWPD-1, crafted by members of the Air Corps Tactical School. It served as the basic plan for the air war over Europe during World War II. (USAF)*

Historical coincidence has seldom been so appropriate as in 1931 when the Air Corps Tactical School (ACTS), which traced its lineage directly to William "Billy" Mitchell, the firebrand prophet of airpower, relocated to Maxwell Field in Alabama, once a flying school operated by Wilbur and Orville Wright. But even this historical confluence did not foreshadow the impact the ACTS and its successor, Air University, would have on the development of airpower into the twenty-first century.

ACTS had always been much more than just a schoolhouse for airmen. In its previous home at Langley Field, Virginia, it became a "think tank" focused on the future of American military aviation. In the balmy, air-condition-less climes of south Alabama, the ACTS "think tank" became a "hothouse" where new ideas quickly germinated, grew, and matured into the concepts that shaped American airpower in World War II. The key to the process was the faculty, which during the 1930s read like a "who's who" of future senior air leadership. Often enough they disagreed about the future of airpower and its ultimate role in modern warfare. The debates were particularly heated when those who championed pursuit or attack aviation pitted their ideas against those who advocated strategic bombardment as the key to victory in modern, mechanized warfare.

The so-called bomber barons prevailed; thus during the 1930s, faculty and students alike argued, debated, challenged, taught, and learned their ideas. As the decade progressed, their notions grew and matured into the doctrine of precision daylight bombardment of an adversary's "industrial web" by large formations of self-defending, high-flying, unescorted, heavy bombers. Their doctrine put muscle on the skeleton of ideas developed by Billy Mitchell in the previous decade, and set the course for the development and use of American airpower as World War II approached.

In the sweltering days of August 1941, Chief of the Army Air Forces "Hap" Arnold called four former ACTS faculty members to Washington to quickly develop the air annex for the war production plan that might be needed should America be drawn into the Second World War. The result, nine days later, was a document labeled AWPD-1 (Air War Plans Division-1) based primarily on the strategic bombing concepts developed at ACTS. Although somewhat revised in 1942, it remained as the basic American air strategy document throughout World War II and it stands as tribute to the intellectual ferment at the ACTS. But as the war clouds darkened, the attention of the Army Air Force turned to training for current combat rather than to educating for future conflicts, and the ACTS became a casualty of the war.

After the war, Air University, an organization with a much larger and broader educational mission, opened its doors at Maxwell Field. During the ensuing decades—now with air conditioning—the university has grown to include four fully accredited, graduate-level, degree-granting institutions; the world's largest community college; two company-grade military education schools; the capstone school for noncommissioned officers; two commissioning sources—the Reserve Officer Training Corps (ROTC), the Officer Training School (OTS); a myriad of professional courses for various career specialties; and one of the world's largest distance-learning programs. At the heart of it, in the center of Academic Circle, is the Muir Fairchild Library, the finest library focused on airpower in the world.

But like the old ACTS, Air University has always been much more than a schoolhouse. Just as superior teaching and classroom debate continue to advance understanding, so also the large contingent of international officers leavens the entire university, resulting in broader and deeper understandings of many global issues. To broaden its reach, the university gave birth to the professional journal of the Air Force, which influences and stirs the debate of serious issues worldwide and is printed in five languages. Serious research and writing, which naturally flows from academic activity, spawned Air University Press, which edits, publishes, and distributes airpower-related books, monographs, theses, and pamphlets worldwide.

The dramatic rise of airpower to a position of critical importance in modern conflict has been a product of critical thinking about the art of war and innovative ideas about aerial warfare conceived at the Air Corps Tactical School and its successor, the Air University. Much more than schoolhouses, ACTS was, and Air University remains, an Alabama "hothouse" germinating ideas for the future of military aviation.

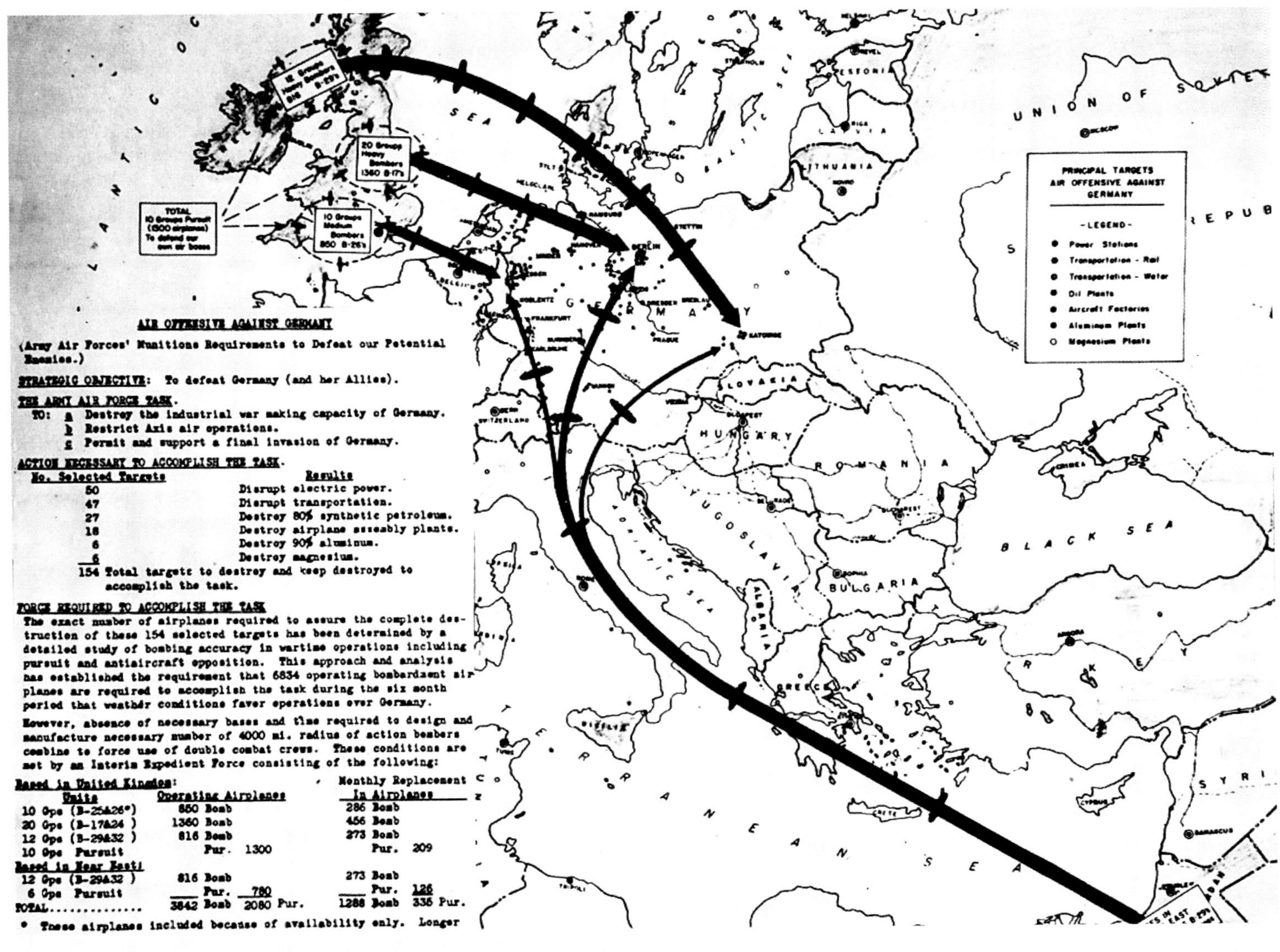

AIR OFFENSIVE AGAINST GERMANY

(Army Air Forces' Munitions Requirements to Defeat our Potential Enemies.)

STRATEGIC OBJECTIVE: To defeat Germany (and her Allies).

THE ARMY AIR FORCE TASK.

TO: a Destroy the industrial war making capacity of Germany.
b Restrict Axis air operations.
c Permit and support a final invasion of Germany.

ACTION NECESSARY TO ACCOMPLISH THE TASK.

No. Selected Targets	Results
50	Disrupt electric power.
47	Disrupt transportation.
27	Destroy 80% synthetic petroleum.
18	Destroy airplane assembly plants.
6	Destroy 90% aluminum.
6	Destroy magnesium.
154	Total targets to destroy and keep destroyed to accomplish the task.

FORCE REQUIRED TO ACCOMPLISH THE TASK

The exact number of airplanes required to assure the complete destruction of these 154 selected targets has been determined by a detailed study of bombing accuracy in wartime operations including pursuit and antiaircraft opposition. This approach and analysis has established the requirement that 6834 operating bombardment airplanes are required to accomplish the task during the six month period that weather conditions favor operations over Germany.

However, absence of necessary bases and time required to design and manufacture necessary number of 4000 mi. radius of action bombers combine to force use of double combat crews. These conditions are met by an Interim Expedient Force consisting of the following:

Units	Operating Airplanes		Monthly Replacement In Airplanes	
Based in United Kingdom:				
10 Gps (B-25&26*)	850 Bomb		286 Bomb	
20 Gps (B-17&24)	1360 Bomb		456 Bomb	
12 Gps (B-29&32)	816 Bomb		273 Bomb	
10 Gps Pursuit	Pur.	1300	Pur.	209
Based in Near East:				
12 Gps (B-29&32)	816 Bomb		273 Bomb	
6 Gps Pursuit	Pur.	780	Pur.	126
TOTAL..............	3842 Bomb	2080 Pur.	1288 Bomb	335 Pur.

* These airplanes included because of availability only. Longer

WORLD WAR II: GLOBAL MISSIONS

1941–1943

WORLD WAR II: GLOBAL MISSIONS

1941–1943

As in the Great War, American politicians were unwilling to leap headlong into a major European conflict. Unlike that war, America's military had already begun preparing, industrially and organizationally, within each branch of service. Those who had served during the First World War made efforts to ensure that the dismal state of America's armed forces in 1917 would not be repeated. Those in the Air Corps had the farthest to go. Transforming the automobile industry into an aircraft industry had been the result of the urgencies of war in 1917. The expansion of America's aircraft industry had proceeded gradually during the interwar years, establishing the foundations for the massive expansion necessitated by Lend-Lease agreements and the explosion of production to support America's military buildup in 1940–44. Leading the effort to build a modest force of aircraft during the 1920s and 1930s were men such as Billy Mitchell, Mason Patrick, Oscar Westover, Frank Andrews, and Henry "Hap" Arnold.

The Japanese had already invaded China in 1937 and were working their way down the east coast occupying town after town, slaughtering soldiers and civilians alike along the way. The prospect of global war prompted a growing concern for military readiness in the United States, and in January 1939 Congress authorized the purchase of more than 3,000 military aircraft. American aircraft production accelerated, but Army Air Corps leaders realized they would need every airplane that American factories could produce just to equip U.S. fighting forces with aircraft of existing designs.

After Germany invaded Poland in September 1939, much of that production was diverted to Great Britain and France. In 1940, President Franklin Delano Roosevelt called for a five-fold increase in aircraft production. His challenge to produce 50,000 planes each year seemed a tall order to everyone but General Hap Arnold, who had recommended double that number to the President. The buildup of American air forces had begun.

While American industry organized to increase the production of airplanes, Arnold and his commanders forged the administrative structure to handle a massive personnel expansion. His days in Washington were consumed arranging training courses for new pilots, negotiating locations for new bases, or haggling over congressional legislation. In the months prior to Pearl Harbor and in the early years of American involvement in World War II, Arnold continued touring factories and encouraging factory workers because he believed they were vital to the creation of real American air supremacy. He also feared that aircraft production was behind schedule.

Before Pearl Harbor, Lend-Lease was more than just a transfer of supplies to the embattled British and others—it was a political demonstration of support to those under attack by the Axis powers. Lend-Lease was also more than just a political program. Strategically, supplies sent to Russia helped in keeping open a third combat front, one that tied up vital Axis resources while the Allies' war machines built up production momentum. American technicians who assisted in the maintenance of Lend-Lease aircraft also provided intelligence reports on the air war. By 1942, and throughout the remainder of the war, reverse Lend-Lease between local British industry and deployed Army Air Forces units worked to ease American supply requirements while lessening the total Lend-Lease bill.

Concurrently with the beginning of Lend-Lease, Allied war planners feverishly devised, revised, and improvised. The first strategic discussions between Britain and America took place in March 1941. The necessity of a coordinated air campaign against Germany and other Axis-controlled zones as top priority was never in doubt.

As the air arm expanded, reorganization and restructuring of the War Department became an obvious necessity. With the establishment of the Army Air Forces in June 1941, General Arnold was named overall commander, and Robert Lovett was appointed Assistant Secretary of War for Air. As Chief of the Army Air Forces, Arnold also acted as General Marshall's Deputy Chief of Staff for Air. Under the War Department General Staff reorganization in March 1942, Army Air Forces were recognized as co-equal with Army Ground Forces and Army Services of Supply.

By the spring of 1942, Arnold was also a member of the Joint Chiefs of Staff. In that capacity, he was directly responsible to the Secretary of War and to the Army Chief of Staff for all Army air operations. Unity of command was finally achieved within Army aviation. Thereafter, Arnold, now the Commanding General of the Army Air Force, ensured continued preparations for war while also encouraging scientific and technological advances in his blossoming command.

By August 1942, war plans had been adapted to deal with the global nature of World War II in the air. Emphasis was placed upon gaining and maintaining complete air ascendancy over Axis forces. Allied planners recognized that freedom to operate in the air, in large measure, permitted freedom to operate on the ground and over the seas.

In January 1943, the Allies met at Casablanca to solidify plans to defeat the Axis. This conference was the most important for American airpower during the entire war. Averell Harriman and Harry Hopkins, Roosevelt's assistants, informed Arnold that the British were planning to push hard for American night bombing. Arnold summoned Major General Ira C. Eaker, Eighth Air Force Commander, to plead the American case directly to Prime Minister Winston Churchill. Arnold and his commanders advocated the "continuous application of massed airpower against critical objectives," arguing that both day and night attacks would challenge the Luftwaffe and stretch German resources.

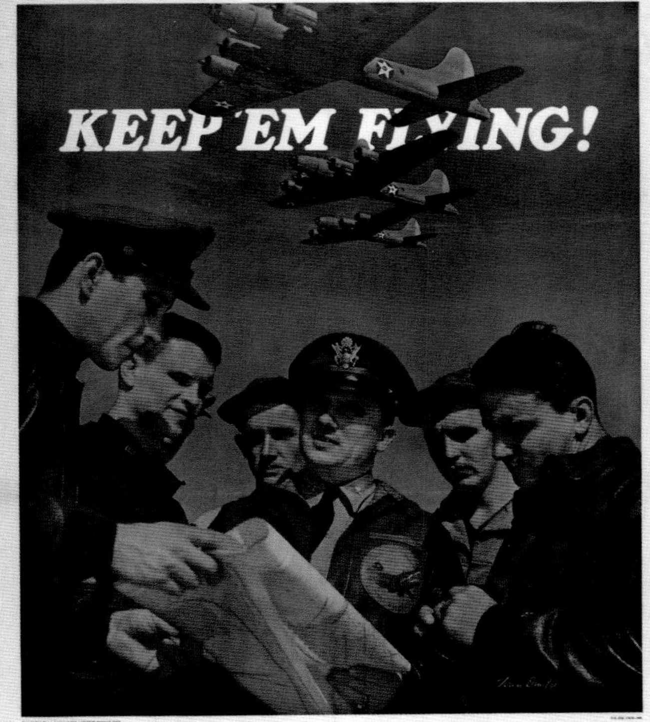

Although Churchill may have already been leaning toward a combination of day and night operations, Eaker drove the American argument home. Churchill agreed to an Allied air campaign subsequently called the Combined Bomber Offensive (CBO). The Casablanca Directive established the air objectives for the remainder of the war in Europe: the progressive destruction and dislocation of the German military, industrial, and economic system and the undermining of the morale of the German people to a point where their capacity for armed resistance was fatally weakened. After plans were finalized, the combination of Royal Air Force night attacks and Army Air Forces daylight attacks began the systematic destruction of Germany's war machine.

Pages 138–139: *"Tora, Tora, Tora," by Robert T. McCall. (NASM Art Collection)*

Above: *Recruiting and the national draft began before the Japanese attack upon Pearl Harbor. Posters like this emphasized teamwork and the glamour of military flying. (Army Art Collection)*

WORLD WAR II: GLOBAL MISSIONS

1941–1943

"Of all the Air Force's faults, its greatest has always been the fact that it has made its work seem too easy."
—Henry H. "Hap" Arnold, Commanding General, U.S. Army Air Forces

1941

January
During this month, RCA makes a proposal to the National Defense Research Committee (NDRC) to design and develop a radio-controlled, rocket-powered missile that will be guided using a television in the nose. The National Bureau of Standards (NBS) is assigned to build the device.

11 January
The Army Air Corps successfully tests robot planes that are controlled by radio signals sent either from ground stations or from another airplane. Further development of guided glide bombs and torpedoes is initiated in February.

February
The Air Corps experiments with and adopts a special flash technique for taking night reconnaissance photographs. Major George W. Goddard works with technicians at Eastman Kodak to develop heavy cylindrical flares that, when dropped from a plane and allowed to fall without a parachute, explode in mid-air to create a burst of light of several million candle power. A camera in the photo aircraft activates when the flare bursts, using a photoelectric cell which senses the light. Up to 20 square miles may be photographed using this technique.

5 February
A photoelectric sensor that is capable of measuring the height of clouds above the ground is developed by the National Bureau of Standards (NBS).

11 March
President Franklin Delano Roosevelt signs the Lend-Lease Act into law. The act authorized industry to lend war materiel to several Allied nations. The majority of goods went to the British and the Russians. More than 43,000 aircraft were eventually shipped to these nations to fight the war against Nazi Germany.

Left: *A bombardier guides a radio-guided B-17 bomber from the mothership. Used for only a short time during the war, they were deemed too dangerous after several crews were lost while arming or bailing out of the "orphan" aircraft. Later, drone flight became much more routine. ("Controlling a Drone Plane in Flight," by Milton Marx, USAF Art Collection)*

Right: *The Tuskegee Airmen made their mark in the skies over Europe. Their squadrons broke the barriers of race while making tremendous contributions flying escort missions from bases in Italy. (USAF)*

22 March
Captain Harold R. Maddux takes command of the 99th Pursuit Squadron at Chanute Field, Illinois. The 99th is the Army's first African American flying unit and is one of three units in the 332d Fighter Group.

26 March
The Air Corps Technical Training Command is established.

27 March
Authorities for the United States and Great Britain sign a base-lease agreement which stipulates that the U.S. may use eight air and naval bases located in the British Atlantic and Caribbean possessions in exchange for a number of U.S. naval destroyers. The agreement is implemented in September.

28 March
The American "Eagle Squadron" becomes fully operational. This unit, under British command, is made up of American volunteers. In 18 months, these units will be integrated into the American 4th Fighter Group in England.

1 April
The U.S. and Mexico sign an agreement allowing reciprocal transit of military aircraft through each other's territory.

9 April
A United States-Danish agreement gives the U.S. the right to build and use airfields in Greenland.

11 April
Anticipating the eventuality of world war, and fearing that Europe might fall under complete Nazi control, the Army Air Corps requests both Consolidated and Boeing Aircraft companies to submit concept designs for a very-long-range bomber. These early conceptualizations are later realized in the Convair B-36 Peacemaker.

Left: *During the Battle of Britain, the 303rd "Kosciusko" Squadron, RAF (Polish), shot down more Germans than any other—126 confirmed victories. ("Witold Urbanowicz's Hurricane, September 1940," by Max Crace, USAF Art Collection)*

Left: *"The Flying Tigers," by Robert Taylor. (Courtesy of Robert Taylor)*

Below: *Production posters reminded Americans that raw materials were important to the war effort. Rationing, scrap drives, and victory gardens emphasized the critical shortage of certain wartime materials. (NASM Poster Collection)*

15 April
President Roosevelt authorizes Reservists who are on active duty to resign from the Air Corps and sign up for duty with a new P-40 volunteer group under the command of Claire L. Chennault. The American Volunteer Group (AVG), also known as the Flying Tigers, created by FDR's secret and unpublished executive order, is tasked to help China fight the Japanese in the air.

Igor Sikorsky, piloting a Vought-Sikorsky, accomplishes the first officially recorded rotor helicopter flight in the Western Hemisphere of duration greater than one hour. The flight lasts just over one hour and five minutes and is flown near Stratford, Connecticut.

18 April
Construction on the new $50 million Consolidated Aircraft plant begins near Fort Worth, Texas. The factory will become a major builder of B-24 and B-32 bombers during World War II.

6 May
Republic test pilot Lowery Brabham flies the XP-47B Thunderbolt for the first time. This durable aircraft flies in every combat theater during the war as both an escort fighter and ground attack aircraft.

13–14 May
The first mass flight of bomber aircraft across the Pacific Ocean takes place when 21 Boeing B-17D Flying Fortresses take off from Hamilton Field, California, and land at Hickam Field, Hawaii. The flight takes 13 hours and 10 minutes.

15 May
Royal Air Force Lieutenant Sayer makes the first official flight of the Gloster E28/39, the first British turbojet airplane, at Cranwell, England. The hop lasts approximately 17 minutes.

YOU GIVE HIM WINGS!
U. S. ARMY
The Army needs
LUMBER for Training Planes

Right: *A Consolidated B-24D Liberator on final approach. (USAF)*

Below: *A call to enlist in the Army Air Forces that uses the AAF symbol above the easily recognizable P-47 Thunderbolt fighter. (NASM Poster Collection)*

21 May
The Army Corps Ferrying Command is created. By the end of the war in Europe, the command (renamed as the AAF's Air Transport Command) possesses nearly 2,500 transport aircraft—one-fourth of those are four-engine heavy lifters.

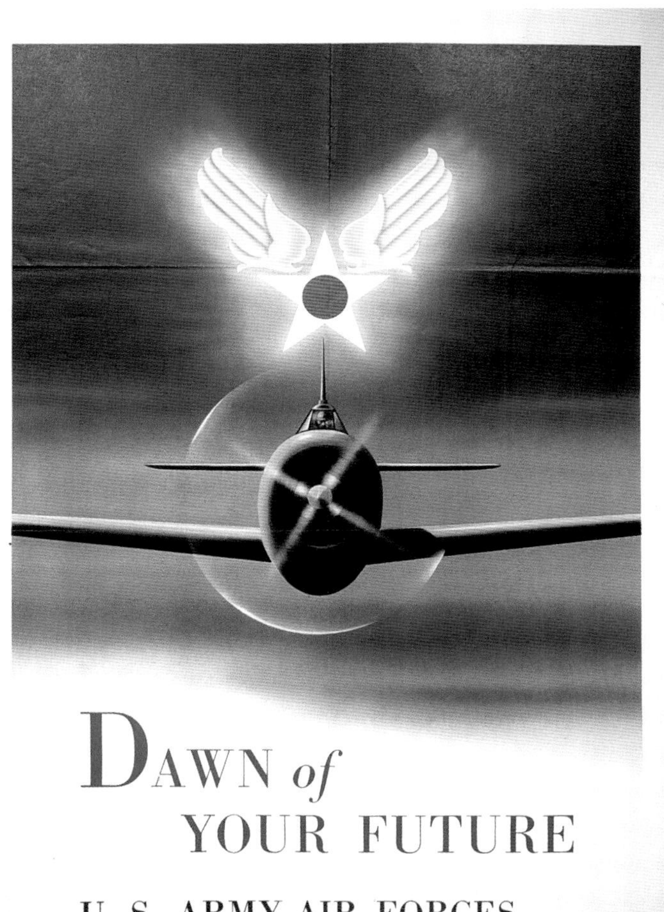

DAWN of
YOUR FUTURE
U. S. ARMY AIR FORCES
—Enlist Today!

31 May
Major General George H. Brett is appointed Chief of the Air Corps.

4 June
The grade of "Aviation Cadet" is created replacing the "Flying Cadet" designation. Once a cadet earned his wings, he was commissioned as an Army officer.

16 June
The Consolidated B-24 Liberator bomber began Army service. More Liberators were built during the war years than any other type of American plane.

20 June
The Army Air Forces is established when the War Department consolidates the Office of the Chief of the Air Corps and the Air Force Combat Command, formerly GHQ Air Force. This is part of a major reorganization of the U.S. Army. Major General Hap Arnold assumes command as Commanding General, Army Air Forces (CGAAF)—a position he holds until 1946.

22 June
BARBAROSSA, the German invasion of Russia, is launched by more than 100 Nazi surface divisions and overwhelming Luftwaffe strikes. These claim to have destroyed 900 Russian aircraft on the ground and 400 in the skies.

27 June
Lieutenant Colonel Stanley Umstead, Major Howard G. Bunker, and a flight crew of five fly the Douglas XB-19 bomber on its maiden flight.

Left: *Famed leader of the Tuskegee Airmen, Lieutenant Colonel Benjamin O. Davis, in the cockpit of his P-51 Mustang. (USAF)*

28 June
President Roosevelt creates the Office of Scientific Research and Development (OSRD) when he signs Executive Order 8807.

30 June
A joint Army-Navy contract is awarded to Northrop Aircraft to design an aircraft gas turbine engine rated to 2,500 horsepower and weighing less than 3,215 pounds. Named the Turbodyne, it is the first turboprop power plant to operate in North America.

1 July
Lieutenant Colonel C. V. Haynes flies a B-24 from Bolling Field, D.C., to Scotland. His route takes him and his crew through Montreal and Newfoundland. This is the first AAF overseas transport service flight, made even before direct U.S. involvement in the global conflict.

8 July
A force of 20 B-17Cs given to the Royal Air Force is used to attack the German city of Wilhelmshaven. Largely ineffective because of the RAF's bombing doctrine, it is the first operational use of the B-17 in combat. The British name for the B-17 is Fortress I.

16 July
At Langley Field, full-scale wind tunnel tests of the A-1 weapon—a power-driven controllable bomb—are conducted.

19 July
The first class of African American pilots to be trained by the military arrives at the Tuskegee Institute in Alabama. Among these twelve aviation cadets is Captain Benjamin O. Davis, a West Point graduate. Davis will command the Tuskegee Airmen for much of the war.

Captain Claire Chennault, a retired Air Corps pilot, assumes duties as chief instructor of Marshall Chiang Kai-shek's Chinese Air Force. Chennault is called back to active duty in 1942 and leads the American Volunteer Group (AVG)—the Flying Tigers—into combat against the Japanese.

24 July
Dr. Jerome C. Hunsaker is elected as Chairman of the National Advisory Committee for Aeronautics as well as the Chairman of the NACA's Executive Committee.

1 August
In an attempt to impact Japanese aviation, President Roosevelt prohibits the export of aviation fuel outside the Western Hemisphere. The exception to the prohibition is Great Britain or other countries engaged in resisting Axis aggression.

Three successful tests of the J. Wyld liquid fuel rocket motor are accomplished. The company, Reaction Motors, Inc., is founded one year later to continue work on this design.

Right: *An AAF Ercoupe is propelled over a test ribbon during early Jet-assisted Takeoff (JATO) experiments. The added thrust came from small rocket motors affixed to the underside of the wings. (USAF)*

Below: *Captain Homer Boushey, seated in the cockpit of the AAF Ercoupe, was the JATO test pilot. (USAF)*

4 August
The Air War Plans Division (AWPD) begins to formulate the Army Air Forces' strategic bombardment campaign plan that will eventually be executed in Europe and then Japan. The plan—AWPD-1—is completed on August 12 after around-the-clock War Department sessions.

12 August
Captain Homer Boushey makes the first takeoff in an Army Ercoupe aircraft specially modified with three small solid-propellant rockets fastened beneath each wing. Incorrectly dubbed Jet Assisted Takeoff (JATO), the system is designed to save fuel on takeoff and thus extend range and increase payload capability of heavy combat aircraft.

23 August
In a follow-on JATO test, Captain Boushey controls his Army Ercoupe from which the propeller has been removed. This time, six rocket motors are mounted under each wing. The project scientist, Professor Theodore von Kármán, calculated that with 12 rockets, the Ercoupe could become airborne, provided it was moving at 25 knots ground speed. A one-half-ton pick-up truck pulled in front of the propellerless plane and a sturdy rope is tied to the truck trailer hitch. Boushey grabs the other end and holds on tight until the truck has pulled the Ercoupe to a speed of about 30 miles per hour. The daring pilot releases the rope, hits the switch that ignites the rockets and accomplishes the first rocket-powered flight attaining an altitude of 10 feet for a short, straight-ahead distance.

27 August
Pilot Officer William R. "Wild Bill" Dunn, flying a Hawker Hurricane as a volunteer with 71 Squadron of the Royal Air Force, is the first U.S. citizen to become an aerial ace during World War II when he shoots down his fifth and sixth enemy aircraft over Belgium. Three RAF Eagle Squadrons are composed primarily of American volunteers under RAF command.

28 August
Under the command of Lieutenant Colonel Eugene L. Eubank, 35 B-17s of the 19th Heavy Bombardment Group begin deployment to Clark Field, Philippines. By early November, all Flying Fortresses have arrived to assist in the reinforcement of the Philippines.

September
During this month, Dr. Robert H. Goddard begins work on development of liquid propellant JATO engines for the AAF and the U.S. Navy. He delivers useable units one year later.

THEODORE VON KÁRMÁN

Left: *Professor Theodore von Kármán not only was a gifted aerodynamicist, but was equally talented in the classroom environment. Many of America's finest aeronautical scientists studied under Kármán at Caltech. (USAF)*

Of all the civilians who had worked for the Army air arm during the 1930s and 1940s, one had such sweeping influence and long-term impact that his ideas continue to shape today's modern Air Force—Theodore von Kármán.

Near Budapest, Hungary, in the spring of 1881 von Sköllöskislaki Kármán Todor was born to Helen and Maurice von Kármán. At age six, Todor showed off for visitors by multiplying six-digit numbers in his head with the speed of a present-day calculator. At sixteen, Todor was awarded the Eötvös Prize as the finest mathematics and science student in all of Hungary. This was only the beginning of an academic, scientific, and engineering career that would have few equals in the first half of the twentieth century. Theodore von Kármán had always admired Sir Isaac Newton, the quintessential theoretical scientist as well as practical engineer. Not only had Newton postulated the Universal Law of Gravity but had also, for example, designed a footbridge over a river near his alma mater, King's College, in Cambridge, England. Kármán emulated Newton and throughout his career in aerodynamic theory maintained a strong interest in the actual engineering of his own formulations as well as applications of his work to worldly problems.

Kármán's early studies with European scientists built a strong foundation for his lifelong career in aeronautics. Facing rising nationalism, he was eventually convinced to move to California and head the Caltech Aeronautics Laboratory by university president Robert Millikan. It was his application of fluid mechanics coupled with his theoretical understanding of science which impressed Millikan so much. Yet Kármán never believed that theoretical aerodynamics should be sacrificed at the expense of practical applications of the science. It was in this position that his influence reached the leadership of the Army Air Corps.

Kármán was encouraging the study of unconventional ideas. Although solving theoretical problems was Kármán's strength, he was not afraid to challenge accepted theory or, when necessary, get his hands dirty during experimental evaluation. On at least one occasion, the professor climbed into his wind tunnel with a handful of modeling clay and modified an airplane wing root that he suspected of causing high-speed turbulence. The modification became known throughout the world as "Kármáns," small wing fillets which minimized turbulence at high speeds.

Kármán was also a gifted teacher. He was well prepared and organized for each lecture. The countless equations that covered his chalkboards were a thing of beauty and his lines were straight, properly aligned, and legible. More importantly, the professor possessed the gift of creative scientific conception at its highest level, expert at clarifying and reducing those concepts to clear and understandable form. Finally, Kármán understood the essential physical elements involved in complicated engineering problems and was able to solve them by successive approximation.

His teaching ability was undoubtedly one reason why Henry "Hap" Arnold, a man of somewhat rudimentary academic ability, came to like and respect Kármán. Kármán's wind tunnel lessons were at the heart of this respect. Driven by the belief that high-speed aircraft were a definite possibility, better wind tunnels became one of Kármán's greatest obsessions.

In 1939, while Caltech scientists were building small rocket engines to assist heavy bombers to get off the ground faster, Chief of the Air Corps, General Hap Arnold told Kármán of his belief that experimental research was the only way to get and keep American aeronautics on top in the world. Arnold wanted to know what type of equipment the Air Corps needed to begin the process. Kármán suggested a high-speed wind tunnel be built at Wright Field, one which required 40,000 horsepower to operate. At that time it was a revolutionary piece of equipment, and General Arnold immediately found funding for the project.

Despite his reputation as a gifted research scientist, aeronautical engineer, and teacher, Kármán remained a man of humility. During his lifetime, he shared the company of Einstein, Henry Ford, Daniel Guggenheim, Gandhi, Jane Mansfield, Orville Wright, Pope Pius XII, Joseph Stalin, and President John F. Kennedy. The ease with which he moved through these impressive social and political circles demonstrated that, although he was aware of his personal standing as a renowned scientist, he remained comfortable in any social situation.

In reality, Kármán was interested in everything surrounding him. The professor once said, "The greatest progress in my lifetime has consisted of the elimination of what I call the scientific prejudices." While at Caltech, Kármán developed a similar vision to that held by General Hap Arnold; both realized the advantages of a cooperative aeronautics establishment between civilian scientists and military men working on the same team. During his life he successfully breached barriers between engineers and scientists, work and leisure, art and science, students and teachers, home and classroom, thinkers and laborers, as well as "long hairs" and military men.

Right: *Kármán had calculated the number of JATO rocket engines that would be needed to lift the test Ercoupe into the air without assistance of its propeller. He explains his solution to Clark Millikan, Martin Summerfield, Frank Malina, and Captain Boushey. (USAF)*

6 September
The Boeing B-17E makes its first flight. The B-17E, larger and more heavily armored than previous models, includes a stabilizing fin which extends forward of the tail to provide high-altitude stability.

11 September
Under the direction of Brigadier General Brehon B. Somerville, construction begins on a new building to house the offices of the War Department. The Pentagon project is completed at a cost of $83 million in less than 18 months. The Pentagon is the largest single office building in the world, home to more than 25,000 civil and military employees.

17 September
Paratroopers are dropped in a tactical combat exercise for the first time. Thirteen DC-3 aircraft are used during the Louisiana operation to drop a parachute company.

20 September
Philippine Department Air Force is activated at Nichols Field, Luzon. This organization is the parent to the Far East Air Force and, later, the Fifth Air Force.

9–16 October
A week-long test of the U.S. air defense network takes place. More than 40,000 civilian aircraft spotters of the Aircraft Warning Service search the skies for "enemy" bombers using searchlights and anti-aircraft artillery sighting mechanisms. The communications network is evaluated when more than 1,800 observations stations are tested.

30 October
Major Alva L. Harvey pilots a B-24 Liberator during a record-setting around-the-world flight. During the mission, a 3,150-mile nonstop segment between Great Britain and Moscow is flown carrying the Harriman Mission. The total distance covered during the around-the-world portion of the mission is 24,700 miles, in 17 days.

7 November
The AAF launches its first guided glide bomb—the GB-1.

12 November
The AAF launches its first radio-controlled glide bomb—the GB-8. The GB series of weapons are part of Project APHRODITE, a program centered on developing standoff weapon capability and more precise guidance for bombs.

30 November
The Army's oldest pilot, Brigadier General Frank P. Lahm, retires. Lahm was the Army's first military aviator.

1 December
The Civil Air Patrol is established by executive order. Their primary mission is coastal patrol and detection of enemy submarine activity.

7 December
Japanese carrier-based naval aircraft (torpedo bombers, dive bombers and fighters) launch a surprise attack upon American and British possessions in the Pacific. Nearly 200 Japanese aircraft hit Pearl Harbor and nearby Hickam Field at 7:55 a.m. Shortly thereafter, Bellows and Wheeler Field are also attacked in the same strike. Of the nearly 3,000 American casualties that day, 163 Army Air Forces personnel are killed and 390 others are wounded or missing. Of the 231 aircraft assigned there, 64 are destroyed and many others are damaged. The U.S. Navy loses more than 2,000 sailors in these attacks. Six Army pilots shoot down 10 Japanese aircraft in air-to-air combat. Lieutenant George Welch, flying a P-40, is credited with four of the 10.

8 December
Japanese forces attack U.S. military installations in the Philippine Islands. More than 100 aircraft, including 18 B-17s, 56 P-40s and P35s, and 26 other types of aircraft, are destroyed on the ground at Clark and Iba airfields. Although the AAF is dealt a severe blow, five pilots shoot down seven Japanese during the attack. Lieutenant Randall B. Keator is credited with shooting down the first Japanese aircraft in his P-40 over the Philippines. About 80 more American servicemen are killed in these attacks. The few B-17s that survive the raid fly night combat attacks against nearby Japanese forces two days later. One of these B-17s, later known as *The Swoose*, is preserved in the collection of the National Air and Space Museum.

8 December
Congress declares war on Japan.

Bell test pilot Robert Stanley makes the first flight of the XP-63 Kingcobra near Buffalo, New York. This powerful fighter is similar to but larger than the P-39 Airacobra.

10 December
In the first U.S. offensive operations against Japan, five B-17s attack Japanese targets off Aparri, near Luzon, Philippines. These 93d Bomb Squadron planes are the first U.S. aircraft to sink an enemy vessel by aerial bombardment. In one such attack, Captain Colin P. Kelly Jr., piloting a B-17, attacks a Japanese landing force and is hit by enemy fire. Kelly orders the crew to bail out of the bomber but is killed when he fails to escape the crippled plane himself. For his actions he is awarded the Distinguished Service Cross (posthumously) for his attack upon the heavy cruiser *Ashigara*.

11 December
Pilot Officer John Gillespie Magee Jr. is killed during a midair collision over Great Britain. An American serving with the Royal Canadian Air Force, and only 19 years old at the time of his death, Magee had authored the poem "High Flight" during the previous summer and had mailed a copy of the lyric to his parents. The poem has since been put to music, made into a patriotic video tribute, and is often seen as a sign-off piece when television network programming stops in the middle of the night.

Opposite: *"Battleship Row," by Robert T. McCall. (NASM Art Collection)*

Right: *"Colin Kelly's Last Flight," by James Dietz. (USAF Art Collection)*

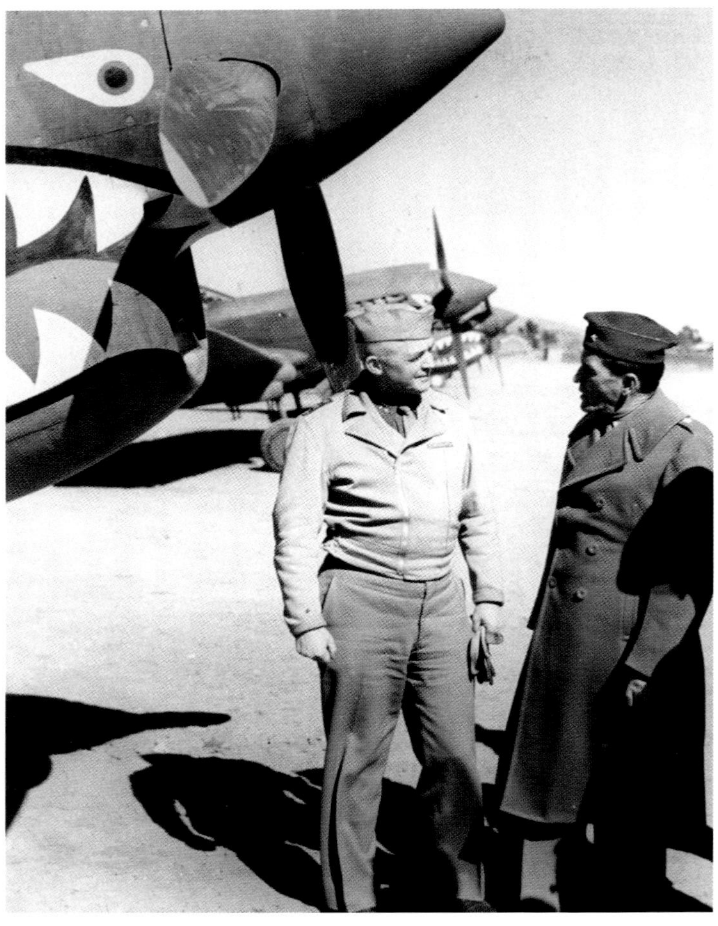

12 December
Captain Jesus Villamor shoots down a Japanese bomber over the Philippines. Villamor is flying the Army's first monoplane fighter, the obsolete P-26 Peashooter. This aerial victory is the only one credited to a Peashooter in its history.

Major General H. A. Dargue, a close personal friend of General Hap Arnold, is killed when his plane crashes into a mountain near Bishop, California.

16 December
Lieutenant Boyd D. "Buzz" Wagner becomes the first American ace of World War II. A member of the 17th Pursuit Squadron, flying a P-40, he shoots down his fifth enemy aircraft in air-to-air fighting in the Philippines.

18 December
Reaction Motors Inc. (RMI) holds its incorporation meeting on this date. RMI is the first commercial American rocket company that concentrates on liquid-propellant engines based on Wyld's rocket designs. The company will produce engines used in the Bell X-1 and the X-15 rocket planes.

20 December
The China-based American Volunteer Group flies its first missions against the Japanese. Led by the dynamic Claire Chennault, the Flying Tigers, initially all volunteers, fly the P-40 Warhawk painted with easily distinguishable shark teeth on the engine intake.

30 December
The AAF formally requests that the National Defense Research Committee (NDRC) begin development of "controlled trajectory bombs." Eventually, programs like Azon and Razon guided weapons result from this request.

1942

6 January
President Roosevelt calls for a greatly expanded air force. He challenges America and the aircraft industry to produce 100,000 combat planes during the coming year.

14 January
The Sikorsky XR-4 makes its first flight. The helicopter is a single-rotary wing, two-man craft.

15 January
Alaskan Air Forces is activated at Elmendorf Field. In February it will be redesignated as Eleventh Air Force.

28 January
Eighth Air Force is activated at Savannah Army Air Base, Georgia. During the war, the Mighty Eighth will become America's strategic bombardment powerhouse and participate with the Royal Air Force during the Combined Bomber Offensive (CBO).

February
Billy Mitchell is posthumously promoted to the rank of Major General.

Above: *General Arnold meets with Brigadier General Claire Chennault at his headquarters in China. The shark-mouthed P-40 Warhawks were the hallmark of the American Volunteer Group (AVG) and the units that followed American entry into the war. (Robert and Kathleen Arnold Collection)*

Opposite, above: *"Defense in the Air Begins on the Ground," by Ralph Iligan. (NASM Poster Collection)*

Opposite, below: *The AAF's first combat helicopter, the Sikorsky XR-4. (NASM)*

5 February
Far East, Caribbean, Hawaiian, and Alaskan Air Forces are redesignated Fifth, Sixth, Seventh, and Eleventh Air Forces respectively.

12 February
Tenth Air Force is activated at Patterson Field, Ohio. During the war they will fight the Japanese in China, launching attacks from bases in South Asia.

20 February
The War Production Board announces that aircraft production has been elevated to the same priority as that of tanks and ships. Government allocations of materials for wartime production are modified accordingly.

Major General Ira C. Eaker and six staff officers arrive in the United Kingdom.

23 February
Major General Ira C. Eaker takes charge of VIII Bomber Command. Eaker, long-time friend of General Hap Arnold, will lead his B-17 bombers on the first American raid over Nazi-occupied territory. Eaker's staff continues to arrive in England to establish his headquarters.

A Japanese submarine (I-17) that has infiltrated U.S. Pacific Coast waters fires artillery shells at an oil refinery near Santa Barbara, California. Although aircraft are sent to intercept and destroy the sub, it escapes.

27 February
Allied air and naval units—including all available AAF B-17s, A-24s, P-40s, and LB-30s—try to intercept a Japanese convoy of approximately 80 ships approaching Java. An Allied Naval force is decisively defeated when they meet these enemory forces near Surabaya.

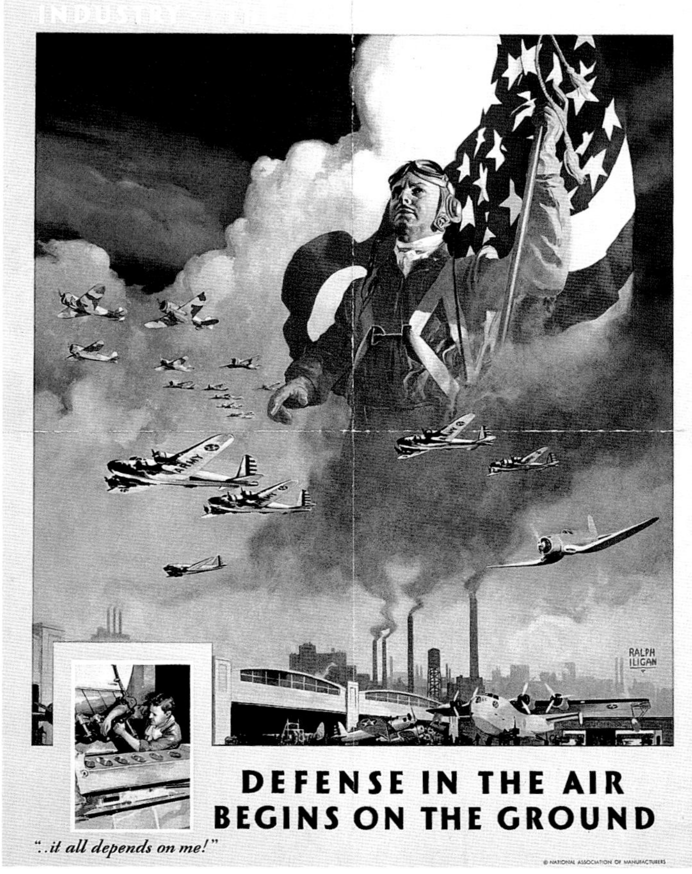

RALPH ILIGAN
DEFENSE IN THE AIR
BEGINS ON THE GROUND
"..it all depends on me!"

2 March
By Presidential order, the War Department is reorganized into three sections: Air Force, Ground Forces, and Service Forces. Within the autonomous Army Air Forces branch remains the Air Corps, which operates as the combatant arm of the division.

6 March
The first class of African American pilots graduates from the aviation school established at the Tuskegee Institute, Alabama. Captain Benjamin O. Davis is among them.

19 March
A group of five scientists from Caltech form the Aerojet Engineering Company. Dr. Theodore von Kármán is the president. Aerojet will produce both liquid- and solid-fuel rockets for the Army and the Navy. They become one of the largest rocket engine producers in the world.

25 March
Flying a British Spitfire with the RAF, Major Cecil P. Lessing becomes the first Eighth Air Force pilot to fly a combat mission over France. His 36-plane formation is recalled when it is determined that they are facing a force of 50 German planes.

Left: *The Douglas C-54 Skymaster, one of the AAF's airlift workhorses. (NASM)*

26 March
The Douglas C-54 flies for the first time at Clover Field near Santa Monica, California, flown by company pilot John F. Martin.

27 March
The War Department and the Department of the Navy jointly announce that the U.S. Navy will take command of anti-submarine operations with jurisdiction over both American coastlines. In this function, the Navy will hold authority over AAF aircraft in accomplishing this mission.

2 April
Tenth Air Force launches its first combat mission. Three heavy bombers, led by Major General Lewis H. Brereton, strike shipping targets near Port Blair.

6 April
In preparation for the arrival of 16 B-25s in China, 10 Pan American Airways DC-3s begin to transfer 30,000 gallons of fuel and 500 gallons of oil from Calcutta to Asansol. Subsequently, this fuel is moved again to China. This airlift is the first resupply mission over the "Hump." The B-25s that were expected to land at these Chinese airfields never make it. They launch a daring raid against Tokyo from the aircraft carrier USS *Hornet* but are unable to locate the fields when the aircraft carrying the radio homing beacon crashes in poor weather.

7–24 April
Liquid-propellant rockets developed by Dr. Frank S. Malina, a member of the Caltech "Suicide Club," are tested on an AAF Douglas A-20A aircraft. A total of 44 Jet Assisted Takeoffs (JATO) are made during the test program.

9 April
Attempting to defend Luzon against the Japanese, Army air and ground forces had moved to Bataan. Badly outnumbered, these forces held the line until finally overwhelmed on this date when they are forced to surrender Bataan. Approximately 75,000 American soldiers and Filipino soldiers and civilians are forcibly marched to northern Luzon as captives and slave labor. Approximately 10,000 die or are killed by their captors during the march in the sweltering heat. Those who survive the march are kept in prisons near Manila, where an additional 1,600 Americans and nearly 16,000 Filipinos die from a variety of causes. The Bataan Death March demonstrates the ruthlessness of the Japanese Army and incenses American military and political leaders.

12 April
Famed World War I ace, Eddie Rickenbacker, at the request of the pilots of the 94th Pursuit Squadron, asks General Hap Arnold to reinstate the "Hat in the Ring" emblem to the 94th. In 1924, the emblem had been changed to the Indian Head of the 103d Aero Squadron.

18 April
Lieutenant Colonel Jimmy Doolittle leads 16 North American B-25 Mitchell bombers on a raid against the Japanese. The aircraft take off from the aircraft carrier USS *Hornet* (CV-8) almost 800 miles from their targets in Japan. The Tokyo Raid is the first attack upon the Japanese homeland by U.S. forces during

World War II. Fifteen of the 16 aircraft crash or ditch near the southeast Chinese coast and one flies to a Russian airfield, where the crew and the plane are interred. That crew escapes to friendly hands in 1943. Sagging American morale gets a tremendous boost when headlines announce the strike against the Japanese capital city.

2 May
Carl A. "Tooey" Spaatz is designated Eighth Air Force commander. He assumes command three days later at Bolling Field, D.C.

4–8 May
For the first time in maritime history, a major naval engagement is fought only by employment of aircraft. During the Battle of the Coral Sea, the main battle fleets never come within visual range of the opposing fleet forces. Losses are heavy as the Japanese lose two carriers, 100 aircraft, and 3,500 men. The U.S. lose one carrier, 65 planes, and 540 men.

6 May
All organized American resistance to Japanese forces in the Philippines ends when forces defending Corregidor surrender.

12 May
Eighth Air Force advance echelon arrives at High Wycombe, England. A group numbering 39 officers and 348 enlisted troops begin preparations for the arrival of the flying units soon to follow.

17 May
The Army Air Forces first helicopter, the Sikorsky XR-4 Hoverfly, is delivered to Wright Field, Ohio. Igor Sikorsky and C. L. "Les" Morris deliver the craft personally. It is accepted by the AAF on May 30. The XR-4 will be the only U.S. helicopter to see action during World War II.

18 May
Seventh Air Force is placed on alert anticipating an attack at Midway Island. Immediately, older model B-18s, although still used, are rapidly replaced with newer B-17s.

Above: *Igor Sikorsky (right) delivers one of his XR-4 Hoverfly helicopters to Wright Field, Ohio. (NASM)*

Right: *Doolittle's Raiders gather on the deck of the aircraft carrier* Hornet *with ship captain Marc Mitscher before the historic strike against Tokyo, Japan. (NASM)*

THE TOKYO RAID, APRIL 18, 1942

After Japanese airpower dealt a stunning tactical blow to U.S. military forces in Hawaii, Franklin D. Roosevelt became a wartime president. One of his most immediate concerns was to initiate a retaliatory strike against the Japanese. FDR had challenged his general staff to come up with a way to attack the heart of Japan.

By mid-January 1942, the concept of a carrier-based strike had been accepted as the most plausible solution to FDR's earlier request. Captain Low, a submariner, had approached Admiral Ernest J. King, Chief of Naval Operations, to evaluate the possibilities. King passed the idea to General Arnold, who tasked his staff problem-solver, Lieutenant Colonel Jimmy Doolittle, to work out the details with the Navy.

After preliminary test flights, the North American B-25 Mitchell bomber was selected for the mission. Eighteen B-25s flew from their home base in Oregon to Minnesota and then Indiana, where they were all fitted with additional fuel tanks. The range of an unmodified Mitchell was only 1,300 statute miles on a good day.

Additional modifications were also made. The weight of the guns was too much for a mission of such duration, so they were removed. The highly secret Norden bombsight was removed. It was not very accurate at low altitude and, if captured by the Japanese, could have revealed classified bombsight technology. In its place was installed a simple metal aiming sight. Ten five-gallon cans of gas were stowed in the radio operator's seat. The aircraft radios had also been removed since the mission was going to be executed under strict radio silence. When all was added together, each aircraft was capable of carrying just over 1,100 gallons of useable fuel. Under typical flight conditions, the calculated range for this amount of fuel was 2,400 statute miles. After all of the modifications were made, four 500-pound bombs barely fit into the bomb bay.

The crews and the airplanes met up with each other at Eglin Field, Florida, where they began training. Only mission-essential elements were covered during those few weeks at Eglin; cross-country techniques, night flying procedures, low-altitude navigation and bombing, rapid escape, and target

identification. The most intense training during March took place on a small auxiliary runway, about five miles away from the main landing field (aux field #9), where each pilot learned the finer points of carrier take-off procedures. By April Fools Day the crews were ready and had flown to Alameda, California, to board the newly built aircraft carrier USS *Hornet*. The ship unmoored and anchored in the middle of San Francisco Bay that evening. Doolittle reminded his men that the mission was top secret, and then released his crews to go ashore for dinner and a night on the town. The *Hornet* and support ships set sail the next morning.

Doolittle's original concept for the raid consisted of a daylight attack after a night carrier launch. Naval planners, however, considered the problems associated with night operations too risky and that plan was scrapped. The second plan consisted of a dawn departure, daylight raid, and then landing at dusk at Chinese airfields. This idea was shelved as fears over daylight detection mounted among the planners.

The plan finally agreed upon between the Army and the Navy included a near-dusk take off and night raid on Tokyo. This option, it was thought, stood the best chance of achieving complete surprise. The plan depended upon a fast carrier run-in at night to get as close to the mainland as possible just prior to launch. After the planes were away, the fleet would make an immediate turn back toward Hawaii and a run for waters beyond the range of Japanese land-based aircraft to preserve the limited fleet that remained in the Pacific. On April 13, Task Force 16.2 had met up with Task Force 16.1, Halsey's force, near Hawaii, and proceeded toward the Japanese mainland with a total of 16 ships that included his flagship, the aircraft carrier USS *Enterprise*.

Doolittle was to lead these 16 planes and 80 men and had planned to take off two hours before the rest of the aircraft, attack Tokyo with incendiary bombs, setting fires that the others could follow to the city. Such a plan, however, would have exposed the 15-ship main force to a prepared defense and blacked-out target cities. In retrospect, this plan was as tactically flawed as the daylight attack. In the end it did not matter because the B-25 crews were forced to launch early. The night attack plan was disrupted when Japanese picket boats spotted Task Force 16 early on the morning of the 18th. There were no other acceptable options; the mission had to be launched immediately.

Fuel considerations for this mission being what they were, close formation flying was never considered viable. But a loose formation technique, trailing slightly behind and to the side of another by three to five miles, was practical. Throttle settings between the airplanes were quite evenly matched, and inflight visibility during the day allowed such visual contact. This technique was used at times during the raid by several of Doolittle's crew, but was not part of the original plan.

Opposite: *B-25s pack the deck of the aircraft carrier* Hornet *along with a few naval fighters as Task Force 16, under the overall command of Admiral Halsey, speeds toward their launch point. (USAF)*

Above: *The Raiders practiced low-level navigation over the Gulf of Mexico before flying to Alameda, California, to board the* Hornet. *(Photo by Fred Bamberger, courtesy of the Air Force Association)*

Essentially, owing to the added distance at the takeoff point, there was no certain plan for how or where to land these aircraft when Doolittle took off at 8:20 a.m. ship's time. When Doolittle's B-25 rose from *Hornet's* undulating deck, he already knew the mission was in jeopardy and might end under a billowing parachute or in a sinking plane somewhere at sea. Halsey and Doolittle shared in the responsibility for the launch decision, and they acted with the clear intention of completing the mission as FDR had demanded.

Hornet steered into the wind while the deck pitched in heavy seas. Engines roared to life and Doolittle taxied his plane forward a few feet onto three cork pads that were laid to provide enough friction for the tires to hold the B-25 as the engines were pushed to full throttle. Takeoff procedures worked as advertised. Doolittle's original mission objective was to ignite fires for the rest to follow, but that element of the plan was unnecessary.

After traveling more than 700 statute miles, minuscule errors in heading control were amplified to many miles off course. Several of the B-25 crews were totally lost when they finally made landfall around noon. Doolittle himself flew well north of his planned route, but quick work by his navigator, Henry Potter, steered him back on course—much to the relief of those following him in trail. The sun was shining brightly about half-past noon when Doolittle became the first pilot to bomb the Japanese homeland in fulfillment of FDR's orders.

Unknown to the Raiders, the aircraft that had carried the homing radio beacons for the landing fields in China had crashed, and with it any chance of finding the strips at night and in bad weather. Lastly, but fortunately, the original targets, planned for night recognition and attack, were large industrial zones and hitting at least some part of the complex would be much easier during broad daylight.

But what of American strategic bombing doctrine? The final plans for this raid centered upon a night attack. The doctrine with which the AAF entered the air war was one of precision, high-altitude, daylight bombing. None of these precepts was followed in the actual planning or in the subsequent execution of the Tokyo Raid. The attack was not intended to do maximum damage; rather, it was intended to make a spectacle. The attack was designed so that the Japanese people would clearly know that a foreign enemy had bombed Tokyo. The fires Doolittle had planned to set were to serve not only as beacons to the other 15 B-25s, but also to dramatically, and undeniably, announce that the capital city had been bombed. The plan to spread the attack force over a 50-mile front, at night, would have created the effect of many, perhaps hundreds of aircraft, attacking the island. Further, the order forbidding the bombardment of the radio towers near Tokyo revealed that immediate dissemination of the news by Japanese radio was expected. This was exactly what FDR had in mind when he ordered the attack.

It was not doctrine, but flexible, creative thinking that generated such an audacious plan. In reality, the Doolittle Raid violated almost every accepted doctrinal idea for bombardment openly held by the AAF. The reality was, however, that the raid occurred in daylight. Post-mission reports revealed that several of the B-25 wingmen actually flew a three-mile trail-type formation behind their flight leader all the way to the target area. By practicing this basic form of mutual combat support, it is likely that those following Doolittle, some admittedly lost when they reached Japan, were steered to the target area by Doolittle's excellent lead navigator.

In almost every case primary targets were bombed. The damage done far exceeded expectations, largely as a result of the highly inflammable Japanese construction, the low altitude attack, the clear weather over Tokyo, and the careful target study that the crews had done while at sea. Weapons were released anywhere between 600 and 2,500 feet above the ground. All 16 planes had descended to extremely low altitude and egressed the target area at high speed. All 16 crews began the task of calculating how much fuel they had remaining and how far they could fly with that reserve. Initial calculations were not encouraging. Davey Joneses' navigator, Lieutenant Eugene F. McGurl, half-heartedly joked, "Hey, I don't think we're gonna have to swim more than one hundred miles."

Doolittle's Raiders got a lucky break that evening. A stiff 30-knot tailwind had developed between Japan and China and, much to the surprise of the navigators, several of the planes appeared to be getting pretty

good gas mileage and making good time. Only one bomber had insufficient fuel to make the Chinese mainland. It diverted to Russia rather than risk having to ditch in the middle of the North China Sea. Ski York's plane was the only aircraft to land on its wheels, and the five crewmen were interned in Russia until they managed to escape into Iran in May 1943.

Once the raiders made landfall over China, luck ran out. Fearing air raids by the Japanese, and not knowing of the timing of Doolittle's raid on the Japanese capital, when the B-25 engines were heard, all lights were extinguished on the ground. Additionally, unfavorable flight weather over the China coast made safe landings impossible. As a result all planes either landed in the water near the coast or the crews bailed out with their parachutes.

The attack on Tokyo blurred the rules that defined mission success and failure. General Arnold cringed at the loss of all 16 B-25s during the mission and, in that regard, considered the attack a failure. The loss of men and equipment, however, was necessarily weighed against political and strategic impact before the mission was actually launched. In the weeks following the raid, American morale appeared to soar.

The Japanese vented their anger upon those Chinese who provided aid to the Raiders during their escape. Some 250,000 Chinese soldiers and peasants were massacred by invading Japanese air and ground forces during the next four months as the Japanese pushed their invasion deeper into China.

The Tokyo Raid was the first and, at that time, the only combat mission flown by these 80 men. For the planning, execution, and leadership during the Raid, Doolittle received the nation's highest military award. On May 19, 1942, President Roosevelt, the man who had ordered the mission, personally decorated the newly promoted Brigadier General, James H. Doolittle, with the Medal of Honor in a private White House ceremony. During a time when America was desperately seeking heroes, FDR gave America Doolittle and his Raiders.

Opposite: *B-25 bomber #297 was flown by the No. 14 crew—piloted by Major John A. Hilger—during the raid. Here, the bomber makes a practice flight from Auxiliary Field #9 in Florida. (Photo by Fred Bamberger, courtesy of the Air Force Association)*

Above: *Lieutenant Donald G. Smith, the No. 15 aircraft during the Tokyo Raid, dives for the water after delivering his bombs to his target in Kobe, Japan, on Osaka Bay. ("Doolittle Raiders Bomb Osaka," by George Guzzi, USAF Art Collection)*

Left: *The Northrop P-61 Black Widow was the first American aircraft specifically designed for night fighting. (NASM)*

Below: *In a private White House ceremony, FDR presented the Medal of Honor to newly promoted Brigadier General Jimmy Doolittle. In attendance are (left to right) General Arnold, Joe Doolittle, Doolittle, General Marshall, and FDR (seated). (Robert and Kathleen Arnold Collection)*

Base loan agreements are signed between the U.S. and Panamanian officials that provide for the use of a number of air bases to defend the Panama Canal.

26 May
The Northrop P-61 Black Widow piloted by Vance Breese flies for the first time near Hawthorne, California. The P-61 is designed specifically to fight at night using radar.

30–31 May
The first 1,000-plane raid of the war is accomplished by the Royal Air Force Bomber Command when they attack Cologne, Germany.

3 June
Army Air Forces units participate in the operations in preparation for the upcoming Battle of Midway when B-17s from the Seventh Air Force attack the Japanese fleet consisting of 45 ships, nearly 600 miles southwest of Midway. The next day, other B-17s from Kauai attack a different attacking force only 150 miles north of Midway.

4–6 June
The Japanese launch a major offensive in an attempt to capture Midway Island, expanding their sphere of influence in the Pacific. Alerted to their plans, U.S. forces, primarily Navy but also some Army Air Forces bombers, defend the important strategic island. Four Japanese carriers go to the bottom of the Pacific to the loss of one American carrier, *Yorktown*. More importantly, more than 300 Japanese aircraft are on board the carriers as they sink. The loss of these aircraft and their crews debilitate the Japanese fleet. After the Battle of Midway, the Japanese are forced to a defensive posture for the remainder of the war.

6–7 June
Seventh Air Force commander Major General Clarence L. Tinker is lost while leading a flight of LB-30s from Midway on a pre-dawn bombing raid on Wake Island. Tinker is the first AAF general officer to be killed in action during the war.

9 June
President Roosevelt awards newly promoted Brigadier General James "Jimmy" Doolittle, then assigned to the First Special Aviation Project, the Medal of Honor for his role in planning and leading the Tokyo Raid.

Right: *General Hap Arnold and Air Vice Marshal Sir Charles "Peter" Portal, Royal Air Force, discuss European air matters. (Robert and Kathleen Arnold Collection)*

11 June
Eleventh Air Force strikes Kiska for the first time. Five B-24s and five B-17s attack recently occupied Kiska harbor installations and attack Japanese ships. For the next year, whenever the Eleventh Air Force had spare airpower, the island of Kiska is bombed.

12 June
Colonel Harry A. Halverson leads the first strategic air raid in the European-African-Middle Eastern Theater. He leads 13 Consolidated B-24 Liberator bombers from Fayid, Egypt, against oil fields near Ploesti, Romania, with little effect.

18 June
Major General Carl A. "Tooey" Spaatz assumes command of the Eighth Air Force in London, England, where approximately 85 airfields are prepared and handed over for operations to the AAF.

21 June
The Arnold-Portal-Towers Agreement is signed in London. It addresses U.S. air commitments in Europe and provides support for Operation BOLERO, the buildup of USAAF in Europe. The agreement is approved by the Joint Staff on June 25, and by the Combined Staff on July 2.

23 June
Even before the formal approval of BOLERO flights, the first aircraft deployed in support of the operation leave Presque Isle, Maine, for bases in Great Britain.

4 July
Members of the Fifteenth Bomb Squadron accomplish the first Army Air Forces bomber mission over Western Europe. Flying six American-built Royal Air Force Boston III bombers (these RAF versions of the Douglas A-20 Havoc were also built in the U.S.) these airmen are part of an overall 12-plane British low-level attack against enemy airfields in the Netherlands. The American commander of the 15th Bomb Squadron, Captain Charles C. Kegelman, suffers severe damage from flak but manages to coax his aircraft back to Swanton Morley. A week later, Kegelman is decorated by General Spaatz with the Distinguished Service Cross (DSC) for extraordinary gallantry and heroism during the raid. He is the first in the Eighth Air Force to be awarded the DSC.

The American Volunteer Group ("Flying Tigers") is replaced when the AAF Chinese Air Task Force (CATF) is activated. Major General Chennault is named commander of the unit which is incorporated into the AAF as the 23d Pursuit Group. The AVG had destroyed 300 Japanese aircraft both in the air and on the ground, losing 50 aircraft of their own and only nine pilots. Only five pilots join the AAF after the merger. The rest of the volunteers return home.

6 July
An AAF P-40 Warhawk test fires 4.5-inch, M8-type rockets for the first time while in flight.

Left: *"The Return from the Raid Over Rouen," by Peter Hurd. (Army Art Collection)*

Below: *The Douglas A-26 Invader was a highly effective medium bomber. (NASM)*

7 July
A German submarine (U-701) is sunk by an Army Air Forces Lockheed Vega A-29 assigned to the Zone of the Interior (ZI) off the coast of North Carolina. It is the first time that an AAF aircraft successfully sends a German sub to the bottom during World War II. The victorious crew is from the 396th Bombardment Squadron.

10 July
The Douglas XA-26 Invader prototype flies for the first time when company pilot Ben O. Howard takes it into the air near El Segundo, California. Initially plagued by problems with control during conditions of asymmetric thrust, the aircraft was eventually made safe and was used with great effect during the war.

21 July
General Eisenhower assigns Eighth Air Force the mission, with the Royal Air Force, of providing air dominance over western France by April 1, 1943.

6–7 August
Captain Harl Pease Jr., Seventh Air Force, voluntarily repairs an unserviceable B-17 and then flies it himself the next day in an attack near Rabaul, New Britain. He and his crew destroy several Japanese Zeros and drop bombs on target before his damaged plane is shot down by the enemy. Pease and his crew are lost.

12 August
British Prime Minister Winston Churchill and American ambassador to Moscow Averill Harriman meet with Stalin in Moscow. The discussions last for three days. Stalin is informed about Operation TORCH and a possible second front in the European War.

14 August
Lieutenant Joseph D. Shaffer and Lieutenant Elza E. Shahan team up to shoot down a German FW 200 off the coast of Iceland. This is the first aerial victory for the AAF in the European Theater.

17 August
Colonel Frank A. Armstrong leads 12 97th Bomb Group B-17 bombers in an attack against Rouen-Sotteville rail yards in occupied France. This is Eighth Air Force's first bombing raid conducted over Europe during the war. During the raid, Sergeant Kent R. West shoots down a German fighter—the first Eighth Air Force gunner to receive credit for a combat kill.

19 August
More than 5,000 Allied troops, most of them Canadian, execute a raid on Dieppe. In an effort to decoy German fighters away from the location of the raid, 22 Eighth Air Force B-17s drop more than 30 tons of bombs on airfields near Abbeville and Drucat. Not only is it successful at deceiving enemy fighters, the raid causes extensive damage to the airfield targets.

Right: *General George Kenney was "MacArthur's Airman" in the Pacific Theater. (USAF)*

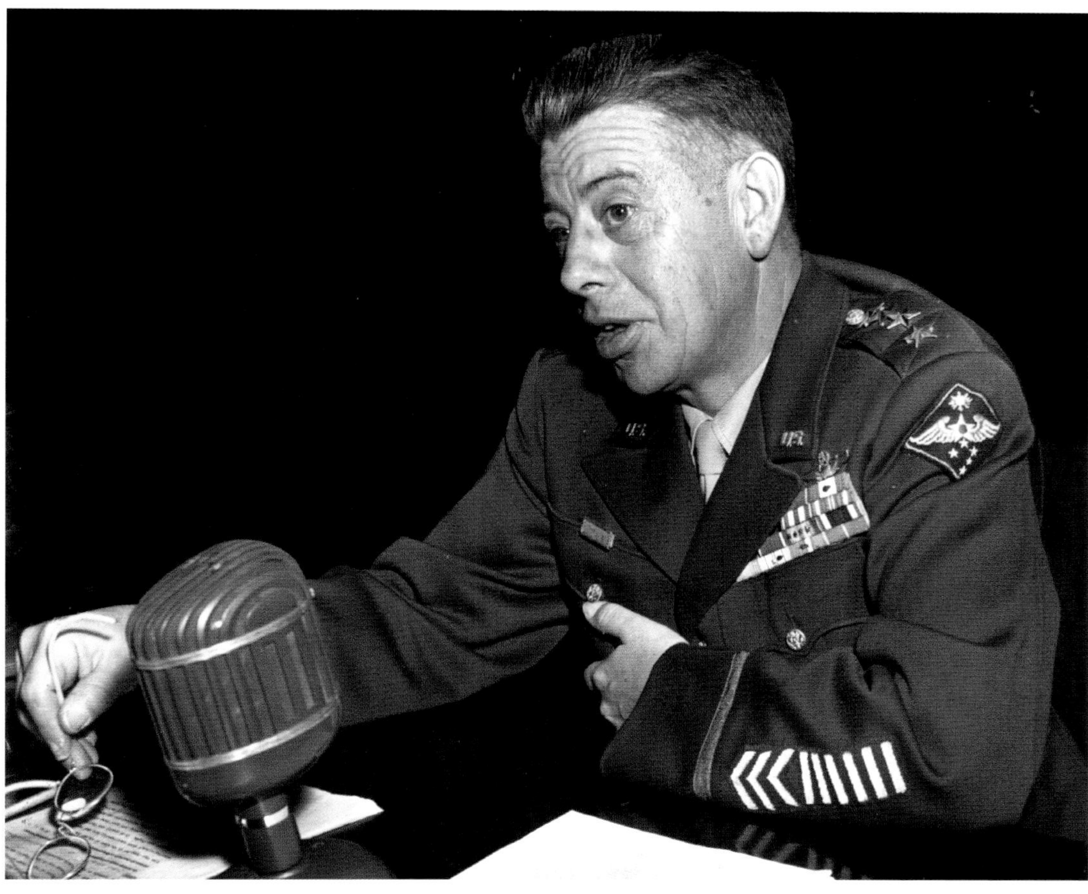

20 August
Twelfth Air Force is activated at Bolling Field, D.C. Three months later, the Twelfth moves to the Mediterranean Theater, providing airpower support for the invasion campaigns in North Africa and Italy.

21 August
General Eisenhower assigns General Spaatz additional responsibilities as Air Officer ETOUSA and head of the air section of its staff. This assignment ensures that the theater air arm is fully represented in operations planning meetings.

22 August
Sixth Air Force aircraft from the 45th Bomb Squadron sink German submarine U-654 off the Panamanian coastline.

3 September
Lieutenant General George C. Kenney takes command of Fifth Air Force in Brisbane, Australia. Kenney replaces Lieutenant General George Brett who had been recalled to the U.S. one month before.

In the first U.S. raid over the city of Hanoi, a B-25 Mitchell medium bomber drops both bombs and pamphlets on the city and an airfield, where munitions and nine aircraft were destroyed or damaged.

5 September
General Spaatz convinces General Eisenhower to rescind his previous order suspending bomber operations of the Eighth Air Force from Britain to totally support Twelfth Air Force preparations for Operation TORCH.

6 September
Two B-17s from the Eighth Bomber Command fail to return from a combat mission over Meaulte, France, becoming the first aircraft lost in combat by that unit.

9 September
A Japanese Yokosuka E14Y-1 floatplane is launched from a submarine near the U.S. West Coast. The aircraft makes two flights near the northern California border, drops four bombs, and starts a forest fire. This attack is the only direct attack upon the continental United States by enemy aircraft during the war.

General Hap Arnold submits AWPD-42, the AAF plan to achieve air ascendancy over the enemy, to General Marshall. The plan, approved through the President's office by November, contains the basic tenets of both American bombing doctrine and the Combined Bomber Offensive, which will be placed into action during the following year.

Left: *This Boeing B-29 production poster illustrates the technological complexity of the most advanced aircraft ever built at that time. (NASM Poster Collection)*

Opposite: *During the invasion of North Africa, Twelfth Air Force and Royal Air Force fighters operated together to drive back enemy forces. (USAF)*

12 September
The 4th Fighter Group is activated at Bushey Hall, England. Anchored by former Eagle Squadron pilots, the 4th takes up the task of providing fighter escort for heavy bombers during raids over Europe. The official transfer of personnel from the RAF to the VIII Fighter Command begins on September 29.

17 September
Army Engineer, Brigadier General Leslie R. Groves, takes command of the most secret military project of the war. The Manhattan Project objective is to build an atomic weapon for wartime use.

21 September
The Boeing XB-29 Superfortress flies for the first time. Boeing test pilot Eddie Allen is at the controls at the Boeing factory near Seattle, Washington.

23 September
Brigadier General Jimmy Doolittle assumes command of the Twelfth Air Force in Great Britain—known during its build-up as Eighth Air Force "Junior."

25 September
In a rare day of acceptable flying weather, Eleventh Air Force aircraft team up with Canadian P-40 Kittyhawk fighters in attacks against Kiska and Little Kiska. This operation is the first combined mission of the Eleventh and the Canadians.

1 October
The Bell XP-59, America's first turbojet-powered airplane, accidentally makes its first flight at Muroc Field, California, piloted by Bell test pilot Robert Stanley. During high-speed taxi tests, Stanley takes the plane airborne but lands immediately. Officially, the XP-59A, powered by two GE I-16 engines modeled after the Whittle engine prototype, is flown for the Army Air Forces brass the next day—October 2d. Stanley flies two sorties and then the Army project officer, Colonel Lawrence C. "Bill" Cragie, becomes the first military pilot to fly a jet when he takes the plane on its third mission of the day.

2 October
At the Aeromedical Laboratory located at Wright Field, Ohio, Major J. G. Kearby ascends to a simulated altitude of 60,200 feet in the altitude chamber as part of a series of full-pressure-suit tests.

3 October
The liquid fueled V-2 "vengeance" rocket is fired successfully for the first time at the Peenemünde test facility in northern Germany, near the Polish border. The five-and-one-half-ton German rocket travels 120 miles.

21 October
Tenth Air Force B-24 Liberator bombers of the Indian Air Task Force (IATF) attack the Lin-his coal mines near Kuyeh. The attack marks the first use of heavy bombers in China and the AAF's first attacks north of the Yangtze and Yellow rivers.

22 October
Twelfth Air Force begins moving its headquarters from England to North Africa.

The first practical U.S. jet engine, completely American in design, begins its fabrication at the Westinghouse Electric company. They are authorized to build two 19A axial-flow turbojet power plants. The X-19A engine will take five months to complete. It is the precursor of the J30, J34, J40, J46, and J54 engines.

23 October
British General Montgomery leads his Eighth Army in a massive ground offensive at El Alamein. Heavy fighting continues into the next day. American B-25s provide close support operations west of El Alamein to assist the British advance.

24 October
Putting an end to Samuel Pierpont Langley's claims to manned powered flight, Smithsonian Secretary Charles G. Walcott admits that the Langley Aerodrome A that had crashed off the end of houseboat in 1903 had no chance of actually flying. This opens the way for the return of the Wright Flyer to the Smithsonian Institution.

8 November
The invasion of North Africa begins. Operation TORCH, commanded by Major General Dwight D. Eisenhower, is supported by Twelfth Air Force airlift and fighter aircraft. Royal Air Force Supermarine Spitfires also participate by providing air cover. C-47 aircraft attempting to land troops find French fighter planes hostile as they shoot down several transports. Spitfires flown by the 31st Fighter Group respond by shooting down three French fighters.

Colonel Demas T. Craw, Army Air Forces, dies near Port Lyautey, French Morocco, after volunteering to accompany assault boats through enemy territory in order to secretly secure an armistice from the Vichy French commander there to avoid a full-scale attack by General George S. Patton's army. Upon landing, Craw is killed by a burst of machine-gun fire at point-blank range, but his intelligence officer and driver make it to Casablanca, secure the truce, and radio back to stop the assault. Craw is posthumously awarded the Medal of Honor.

Major Pierpont M. Hamilton, Army Air Forces, volunteers to execute a dangerous mission into Port Lyautey, French Morocco to locate a French commander in order to bring about a cessation to hostilities. After undergoing heavy machine-gun fire and capture, Hamilton secures the truce, returns to the beach, and completes the mission by communicating by radio back to Patton's army. For his heroic actions he is awarded the Medal of Honor.

9 November
The Piper L-4 flies into combat for the first time when three of these Army observation aircraft are launched from a Navy carrier to assist ground forces going ashore during Operation TORCH. The planes are piloted by Lieutenant William Butler, with Observer Brenton Deval, Lieutenant John R. Shell, and Captain Ford Allcorn.

10 November
In support of Operation TORCH, P-40s from the 33d Fighter Group are transported by Navy aircraft carriers to Port Lyautey, Morocco. More than 100 planes launch from the carriers *Chenango* and *Archer* and then land at their deployed location over a period of three days.

11 November
All French resistance in Northwest Africa ends after French Admiral Darlan broadcasts the orders to cease resistance the previous evening.

12 November
U.S. Middle East Air Force is dissolved and Ninth Air Force is established, commanded by Lieutenant General Lewis Brereton. They begin providing tactical air support to the British Army as they press their attacks to the west against the enemy.

12–15 November
The Battle of Guadalcanal begins. P-38s of the 339th Fighter Squadron move to Henderson Field to support the defense of Guadalcanal. On the night of the 13th, one of the P-38s is destroyed by Japanese naval artillery fire. During the three-day battle, airplanes from the 11th Bomb Group, 69th, 70th, and 72d Bomb Squadron and the 39th and the 339th Fighter Squadrons participate.

13 November
After being forced to ditch in the Pacific Ocean some 600 miles north of Samoa, World War I ace Eddie Rickenbacker, Colonel Hans C. Adamson, and Private John F. Bartek are rescued by a Navy Vought-Sikorsky OS2U Kingfisher seaplane. The trio had been afloat on a raft for 21 days.

15 November
Lieutenant Harold Comstock and Lieutenant Roger Dyer set a speed record for airplanes when they dive their P-47 Thunderbolt from 35,000 feet and accelerate to a remarkable 725 miles per hour.

Left: *Army L-4s were launched from modified LSTs by the 3d Infantry Division for observation missions during the invasion of Italy. These mini-aircraft-carriers were used in the Mediterranean during subsequent invasions. (National Archives)*

Opposite, above: *"Thunderbolts Blast 'em!" from the AAF Poster Series. (NASM Poster Collection)*

16 November
Supporting the British First Army's march into Tunisia, Twelfth Air Force C-47s drop British paratroopers near Souk el Arba. Additionally, six B-17s of the 97th Bomb group in Algiers fly a raid on Sidi Ahmed airfield at Bizerte. The 97th had flown the first heavy bomber mission against enemy targets from the United Kingdom on August 17; now, having been transferred to the Twelfth, they fly their first combat mission in Africa as well.

1 December
Eighth Air Force commander Major General Tooey Spaatz is transferred to the Mediterranean Theater. He flies to Algeria to act as General Eisenhower's air advisor. Major General Ira C. Eaker is named as his replacement to command the Mighty Eighth.

The first issue of the AAF's *Air Force Magazine* is published, taking the place of the AAF Newsletter.

2 December
Scientists at the University of Chicago successfully accomplish the first nuclear chain reaction in history.

4 December
B-24s of the Ninth Air Force flying from bases in Egypt attack targets in Italy for the first time. The targets—harbor installations, a rail yard, and several ships, including a battleship—are near Naples.

5 December
It is announced that General Carl A. Spaatz will become the Deputy Commander in Chief for Air, Allied Forces in Northwest Africa.

8 December
A VIII Bomber Command study of attacks on German submarine pens reveals that current weaponry is not capable of penetrating the structure of the facilities from any safe bombing altitude. Programs such as APHRODITE, unmanned aerial bombs, are developed to penetrate such defenses.

22–23 December
Twenty-six 307th Bombardment Group B-24 Liberators from Seventh Air Force carry out the first full-scale attack upon enemy airfields when they bomb Wake Island after staging through Midway Island.

24 December
President Roosevelt announces that he has selected General Dwight D. Eisenhower to lead the invasion to liberate France. Ike's title will be Supreme Allied Commander, Allied Expeditionary Force.

South Pacific AAF P-39s working with Marine Corps F4Fs and SBDs attack Munda with devastating results. Twenty-four enemy aircraft are destroyed in the air and on the ground, with no friendly losses.

Below: *The first version of* Air Force Magazine—*December 1942—took the place of the Army Air Forces newsletter. (Air Force Association)*

THE AIR FORCE AND THE JET ENGINE

Jeremy R. Kinney

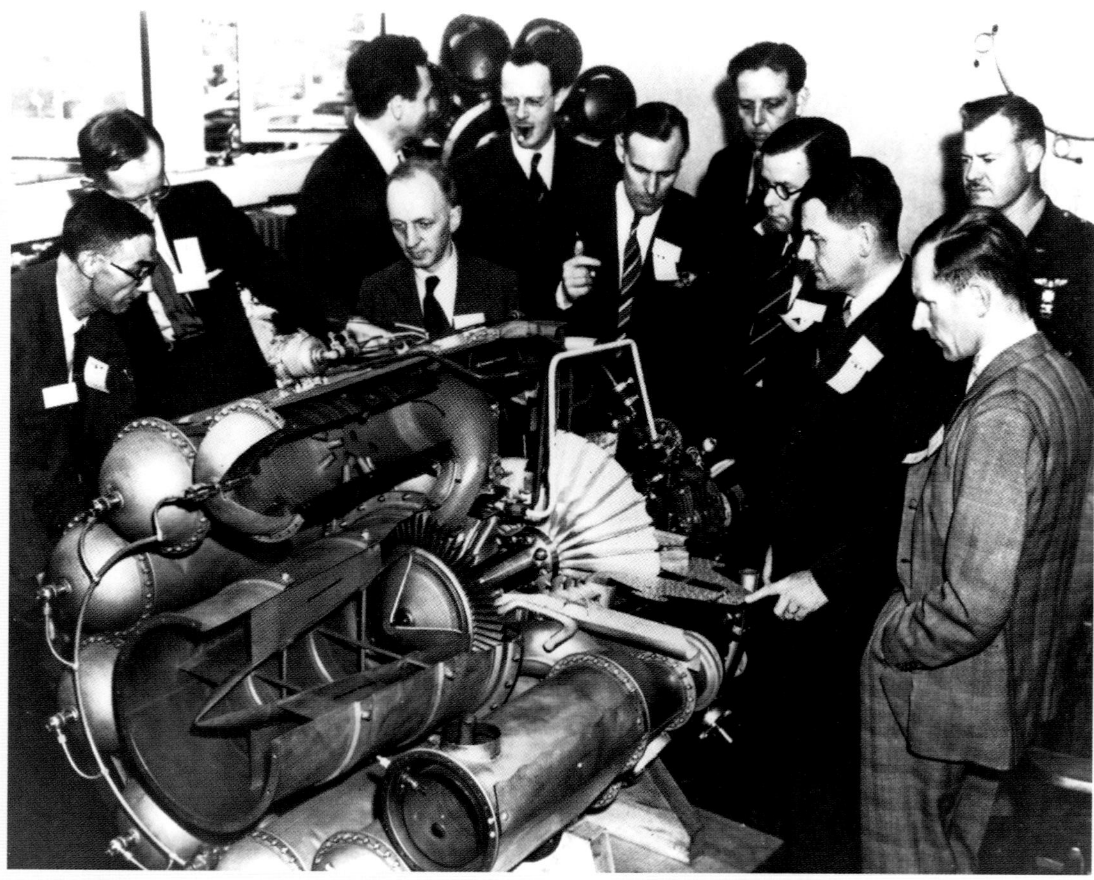

Left: *Visiting British scientists get their first look at the improvements made to the original Whittle engine as they view the General Electric I-A cut-away model. (AFMC/HO, from the GE Collection)*

Test pilot Robert M. Stanley took off from Muroc Dry Lake in America's first jet airplane, the Bell XP-59A Airacomet, on October 1, 1942. The Airacomet, powered by two turbojet engines of British origin, reached a speed of 390 miles per hour in the skies over the high desert of California. The world's first practical jet airplane, the Messerschmitt Me 262, first had flown in Germany just a few months earlier in July. When it entered combat, the Me 262 was faster than anything in the sky. As America's first foray into jet aircraft, the Airacomet proved to be a humble start. It appeared that the United States was behind in the development of a new technology, pioneered independently in Great Britain and Germany, which revolutionized aerial warfare.

The aeronautical community realized as early as the 1920s that the speed of conventional propeller-driven airplanes would reach an ultimate limit and sought an alternative to the piston-engine-propeller combination. One viable option was an aeronautical gas turbine engine, which takes in air, compresses it, mixes it with vaporized fuel, and ignites it to create thrust. In Great Britain, Frank Whittle (1907–1996), a young Royal Air Force officer, patented a jet engine concept that went unnoticed by the British air ministry as well as the international aeronautical community in 1930. Whittle generated private support for his invention and founded Power Jets, Ltd., in March 1936 and successfully operated a test engine a year later. Power Jets built a new engine for the Gloster E.28/39, which flew on May 15, 1941, at speeds up to 340 miles per hour. The British government built upon that success with the Gloster Meteor, the first and only jet-powered airplane to serve with the Allies during World War II.

Unaware of Whittle's pioneering work, Hans von Ohain (1911–1998), a young physicist in Germany, patented an aeronautical gas turbine engine in November 1935 and constructed a small working model. Von Ohain met with aircraft manufacturer

Ernst Heinkel, known widely for his obsession with high-speed aircraft, who quickly gave him a job developing Germany's first jet engine. Even though von Ohain started after Whittle, the Heinkel He 178 flew on August 27, 1939, making it the world's first gas turbine-powered, jet-propelled airplane. Powered by von Ohain's engine, the He 178 flew at speeds up to 360 miles per hour.

The He 178 took to the air five days before the Nazi invasion of Poland started World War II. Its success encouraged the German air ministry to develop further jet-propelled aircraft during the war, especially as Nazi fortunes worsened over the course of the war. The Luftwaffe introduced the first practical jet airplane, the Messerschmitt Me 262, in July 1942, and it entered combat two years later. The single-seat, all-metal fighter featured a swept wing, tricycle landing gear, four heavy cannons in its nose, and was capable of speeds approaching 540 miles per hour. Two Junker Jumo 004B turbojets, considered the first practical jet engine, powered the Me 262. Developed by Dr. Anselm Franz (1900–1994), the Jumo 004B was an axial-flow design where an alternating series of rotating blades and stationary blades moved the overall flow path along the axis of the engine, the configuration used by all modern jet engines.

The success of the British and the Germans sent the United States, the dominant aerial power during World War II, scrambling for its own jet technology. The American military and the aviation industry relied upon conventional propeller and reciprocating piston engine propulsion technology to fight the war. Even though many believed jet propulsion was simply impractical, Air Corps Chief, General Henry H. Arnold, saw the flight of the E.28/39 and received assistance from the British government in starting an American gas turbine program. A complete Whittle engine and drawings arrived at the General Electric factory in Lynn, Massachusetts, in October 1941. With two new I-A engines, the Bell XP-59A exhibited performance roughly equal to the best piston-engine propeller-driven fighters.

The limited success of the Airacomet led to the first practical American jet, the Lockheed P/F-80 Shooting Star, which first flew in January 1944 and gave the American military and the industry valuable experience in designing, manufacturing, and operating military jet fighter aircraft. The success of the Shooting Star led to a new generation of American military jet aircraft, primarily the Boeing B-47 Stratojet bomber and the North American F-86 Sabre fighter. These aircraft benefited from the merging of the German swept wing with American innovations such as engine pods and the country's first practical axial-flow turbojet, the General Electric J47. The United States quickly took the lead in jet aircraft technology as the Cold War unfolded in the late 1940s.

Right: *The Bell XP-59A Airacomet was America's first turbine-powered plane. The dual jet engines had been developed by General Electric with the help of a British-built Whittle engine shipped to the U.S. at the request of General Hap Arnold. (USAF)*

27 December
2nd Lieutant Richard I. Bong shoots down two Japanese aircraft while flying a Lockheed P-38 Lightning. Bong will go on to score 40 victories, all while flying P-38s, and ends the war as America's leading ace.

29 December
The Collier Trophy for 1941 is jointly awarded to the U.S. Army Air Forces and the private airline companies of the nation.

1943

3 January
In the first attack against German submarine pens at Saint-Nazaire, 68 VIII Bomber Command aircraft strike using precision formation bombing techniques. Although considerable damage to the target is achieved, opposition over the target is stiff. Seven aircraft are lost, 47 are damaged, and 70 aircrew are missing and another five are reported killed in the raid. This attack is the first in which formation precision bombing is attempted.

5 January
Brigadier General Kenneth N. Walker, commander, V Bomber Command, Fifth Air Force, leads a bombing attack on a shipping harbor in Rabaul, New Britain, with direct hits on nine enemy vessels, for which he receives the Medal of Honor. During the mission, Walker's B-17 is brought down by enemy fighters. Walker is one of the architects of AWPD-1, the first air plan for fighting the war in the air.

As a result of the successful invasion of North Africa, General Dwight David Eisenhower activates Allied Air Forces in the Northwest African Theater. Major General Carl Spaatz assumes command as Ike's chief air advisor—the Air CinC. The command is activated one week later, consisting of Twelfth Air Force plus Royal Air Force and potentially some French units, when available for combat duty.

9 January
The Lockheed C-69 transport, the military version of the Lockheed Constellation, makes its first flight in Burbank, California. Boeing test pilot Eddie Allen and Lockheed test pilot Milo Burcham pilot the aircraft during the flight.

13 January
Thirteenth Air Force is activated in Espiritu Santo and New Caledonia. From there, fighters and bombers will operate throughout the Southwest Pacific under overall command of Major General Nathan F. Twining. They fly combat missions near Munda.

14–24 January
The Casablanca Conference opens in Morocco. The U.S. and British discuss general objectives for the conduct and outcome of the war, including an offensive in the Mediterranean. Roosevelt and Churchill agree to defeat Germany first while accepting only unconditional surrender of enemy forces. On the 21st, Lieutenant General Ira C. Eaker, at the insistence of General Hap Arnold, presents the Army Air Forces' daylight bombing strategy to Churchill, which eventually leads to the Allies 'round-the-clock bombing strategy—the Combined Bomber Offensive (CBO).

Above: *Lieutenant Richard I. Bong in his P-38 Lightning. Bong will rise to be America's leading ace in World War II with a total of 40 victories. (USAF)*

Below: *General Ira C. Eaker held several vital command positions in Europe during World War II. (NASM)*

15 January
The Pentagon opens its doors as the new headquarters for the War Department.

27 January
Heavy bombers from the Eighth Air Force attack U-boat construction pens, power-generating facilities, and docks at Wilhelmshaven and Emden. These raids, accomplished by Eighth Air Force, 1st Bombardment Wing, are the first American bombing raids of the war against enemy targets in Germany.

Thirteenth Air Force commander, Major General Nathan Twining, and his crew of 14 are reported down at sea somewhere between Guadalcanal and Espiritu Santo. All 15 men are rescued by a Navy PBY-5 Catalina flying boat on February 1. The rescue raft was not equipped with a radio for signaling, thereby complicating the search. After this rescue, dinghy radio sets become standard equipment on rescue rafts.

28 January
Hugh Dryden of the Bureau of Standards is elected president of the Institute of the Aeronautical Sciences.

4 February
Lieutenant General Frank Andrews assumes command of European Theater of Operations, U.S. Army (ETOUSA). General Eisenhower assumes command of the North African Theater of Operations, U.S. Army (NATOUSA).

15 February
Major General Ira C. Eaker will replace Major General Carl A. Spaatz as commander of the Eighth Air Force. Spaatz will move to the Mediterranean to command the air operations for TORCH, the invasion of North Africa.

17 February
The Mediterranean Air Command is activated. Air Chief Marshal Sir Arthur Tedder assumes overall command of three subordinate operational commands. One of these, the Northwest African Air Forces, is commanded by Major General Tooey Spaatz.

18 February
During a test flight of the XB-29, all aboard are lost when it crashes—including Edmund T. Allen, "the greatest test pilot of them all."

The first class of flight nurses graduate from the AAF School of Air Evacuation located at Bowman Field, Kentucky.

Above: *General Frank M. Andrews. (USAF)*

Below: *Portrait of Dwight D. Eisenhower, by Nicodemus Hufford. (Army Art Collection)*

21 February
The 93d Bombardment Group is relieved from duty in the Middle East under Ninth Air Force and is reassigned to Eighth Air Force in England. The 93d will jump back and forth between England and North Africa and by war's end will log the highest total number of combat missions of any Eighth Air Force Bomb Group. The 93d flies 396 missions as a group, including 43 from North Africa. The 93d was one of the units that bombs Ploesti, Romania, on August 1, 1943.

26 February
Major General Jimmy Doolittle assumes command of XII Bomber Command.

1 March
In Algeria, Lieutenant General Spaatz assumes command of Twelfth Air Force.

2–4 March
Fifth Air Force heavy bombers, both B-17s and B-24s, attack a convoy of 16 Japanese ships just north of New Britain. Attacking in coordination with the Royal Australian Air Forces, four of the transport ships are sunk on the first day. By the second day of the battle, B-25 bombers, fighters, and PT boats from the U.S. Seventh Fleet have engaged the enemy convoy, sinking all but four escort destroyers. By the third day, the combined American force has finished off the rest of the Japanese ships, ending the battle. More than 40,000 tons of Japanese shipping are sunk, more than 50 enemy aircraft are destroyed, and approximately 75 percent of the Japanese troops bound for Lae are killed at sea. General Douglas McArthur calls the Battle of the Bismarck Sea the "decisive aerial engagement" in his theater of operations.

10 March
The Army Air Forces activate Fourteenth Air Force at Kunming, China. Commanded by Major General Claire Chennault, former Flying Tigers leader, the Fourteenth includes a flying wing composed of both American and Chinese pilots. Flying P-40s, these pilots conduct armed reconnaissance missions into Burma from Kunming.

18 March
1st Lieutenant Jack W. Mathis, Eighth Air Force, is assigned as the lead bombardier for the 359th Bomb Squadron of 22 B-17Fs attacking submarine yards at Vegesack, Germany. His aircraft is hit with anti-aircraft fire and he is mortally wounded, knocked from the bomb sight to the rear of the bombardier compartment. 1st Lieutenant Mathis drags himself back to the sight, releases his bombs, and then dies at his post. Mathis becomes the first Eighth Air Force Medal of Honor recipient. Tragically, his brother Mark replaces him as a bombardier in the 359th and is lost over the North Sea during a bombing mission on May 14, 1943.

19 March
Lieutenant General Henry H. Arnold is promoted to the rank of general. His four stars are the first for an Army Air Forces commander.

26 March
Lieutenant Elsie S. Ott, an Army nurse, is awarded the first Air Medal ever issued to a woman. She earns the medal after an evacuation of five patients flown from India to Washington, D.C.

2 April
The research building of the AAF School of Aviation Medicine, which houses four altitude decompression chambers, opens. Twenty-seven officers and 35 civilians man the facility.

4 April
Eighth Air Force heavy bombers strike industrial targets near Paris, France. These 85 bombers hit the Renault armaments factory and motor works inflicting heavy damage. Four U.S. bombers are lost when enemy fighters attack the formation.

5–22 April
Operation FLAX begins. This North African Air Force campaign is designed to destroy the enemy's system of ferrying personnel and supplies to Tunisia by attacking air transports and their escorts. NAAF aircraft claim more than 60 enemy planes are destroyed during operations.

8 April
Eighth Fighter Command bolsters its capability by adding the 56th and 78th Fighter Groups to the already operational 4th FG. Increased activity is now possible that will eventually allow deep fighter penetration with long-range heavy bombers.

11 April
The first hybrid rocket design in the U.S. is tested by the California Rocket Society. The rocket is powered by liquid oxygen and a carbon rod.

Opposite: *Consolidated B-24 Liberators at treetop level in North Africa. The Ploesti Raid was executed by using such tactics. (USAF)*

Below: *Army Air Forces aircraft were critical in disrupting Japan's sea-based supply lines. ("B-25s Skip Bombing Wewak," by Tony Fachet)*

GENERAL OF THE AIR FORCE HENRY HARLEY "HAP" ARNOLD

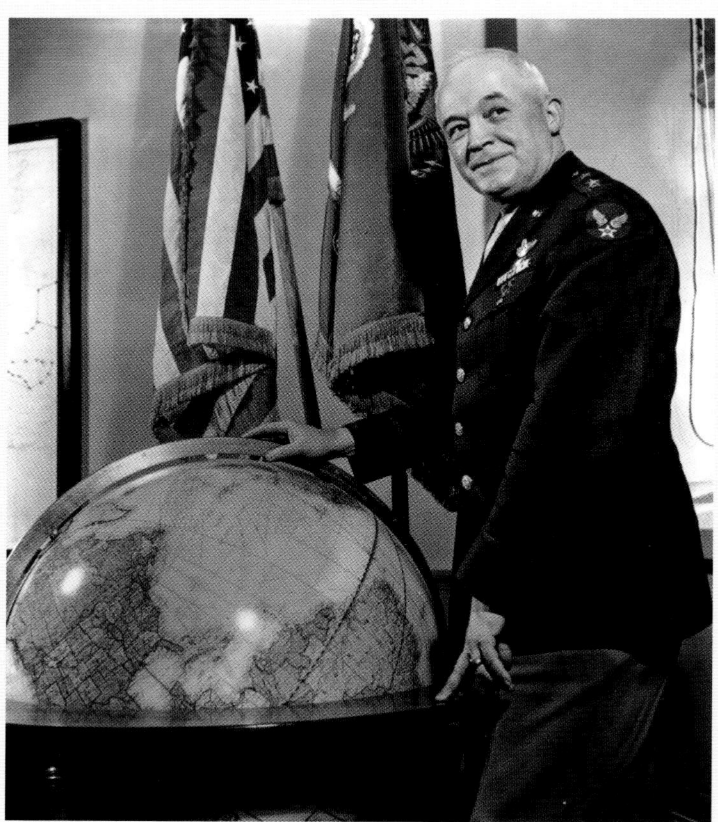

America's least known five-star general, Henry Harley "Hap" Arnold commanded the largest air armada ever assembled in history. As Commanding General of the U.S. Army Air Forces during World War II, he held sway over three active wartime theaters: Europe, the Pacific, and the Mediterranean.

Raised near Philadelphia, Harley Arnold began his military career after graduating from West Point in 1907. Military Academy classmates called him "Pewt," after a mischievous fictional character popular at that time—Pewter Purinton. After a few years of infantry service mapping remote locations in the Philippines, Arnold volunteered to join the newly established ranks of army aviators in the Signal Corps. In 1911, he earned his pilot wings at the Wright Brothers flying school in Dayton, Ohio, although he never actually flew with either of the inventors of the airplane. From that day forward, his military career entwined with the development of military aviation and aeronautical technology.

As a young lieutenant, Arnold became one of the Army's first instructor pilots at the newly established College Park aviation school. He was involved in early experiments involving expanded uses for military aircraft, set early altitude records, practiced using radios to communicate from airplanes to ground forces, and spearheaded the first deployment of an American aerial unit when the aviation school moved south to Georgia in search of better flying weather. After a near fatal flying accident, Arnold grounded himself in 1912 and did not fly again until late in 1916.

During America's involvement in World War I, Arnold served on the Army General Staff in Washington, D.C., as assistant chief of the aeronautical division. In that capacity he oversaw the building up of the infant American aircraft industry as well as the creation of a system for training both pilots and aircraft mechanics for war. He never saw combat but was involved with secret weapons projects like the "Flying Bomb," an unmanned, biplane-like missile, which he revitalized during World War II. Arnold was also directly involved in overseeing the massive demobilization of the army following World War I.

During the court-martial of General Billy Mitchell in 1926, Arnold acted as his military escort and made heartfelt attempts to convince Mitchell to argue his case for airpower from his convictions rather than reading from his latest book. During the late 1920s, Arnold attended Army Command and General Staff school and stayed on at Fort Leavenworth, Kansas, as an instructor, where he introduced the concepts of military airpower to traditional Army ground officers.

It was not until 1934 that Arnold began using the nickname "Hap" following the sudden death of his mother. She had always called him "Sunny," but her death hit Arnold hard and he could not bear the constant reminder of her that his old nickname generated. He assumed command of the 1st Wing that was based at March Field, California, in 1931, where he honed his skills as a large unit commander. While there, he made friends with Theodore von Kármán, head of the Caltech wind tunnel laboratory. From that time forward, Arnold had access to the most gifted

Opposite: *Henry Harley Arnold rose to the rank of five-star general during World War II, the only airman so honored. In 1949, President Truman redesignated him as General of the Air Force. (Robert and Kathleen Arnold Collection)*

Right: *General Hap Arnold always looked to the future. His belief in aeronautical science and the technology that developed from it was steadfast. (Robert and Kathleen Arnold Collection)*

aeronautical scientist in America. Their relationship continued until Arnold retired from service in 1946.

During 1934, Hap Arnold commanded the Western Division during the Air Corps assignment to carry the U.S. mail in the face of civil contract disputes. Seen as a failed exercise by most, Arnold realized that the airmen had performed with distinction, considering their outdated equipment and inadequate instrument flight training. Once Congress realized the same, military aviation funding began to improve, allowing progress in aviation technology once again. The first real indication of such advancement was the 1934 flight of 10 Martin B-10 bombers to the Alaskan frontier and back during August and September. Hap Arnold commanded the round-trip mission, demonstrating the potential of airpower over long distances and harsh terrain.

Arnold continued his advocacy of strategic bombardment throughout and even following his career. He once called the B-17 Flying Fortress airpower that "you could put your hands on." He insisted upon the production of the mammoth B-29 Stratofortress despite early engine problems and other technological difficulties. He spent much of 1939–1942 touring aircraft factories, encouraging workers, pressuring bosses, and demanding success. Hap Arnold's World War II Army Air Forces grew from a tiny force of about 50,000 officers and men in 1938, to 2.4 million by war's end. Arnold directed the expansion and the reorganization that supported it with dedication and selflessness. Arnold believed in true unification of military airpower—that all military airplanes belonged under one command. His view, politically untenable in the face of the U.S. Navy and the development of the aircraft carrier, was not adopted until 1947 when Congress passed the National Defense Act, although a separate U.S. Air Force was created.

Hap Arnold lived only a few years after World War II ended. He died of a massive heart attack on January 15, 1950, at his retirement ranch, "El Rancho Feliz," in Sonoma, California. He was buried at Arlington National Cemetery on January 19 during a terrible sleet storm that grounded hundreds of aircraft that were slated to fly over the cemetery during the services. Arnold was 63. He had served during two world wars, yet had never fired a shot or dropped a bomb during combat.

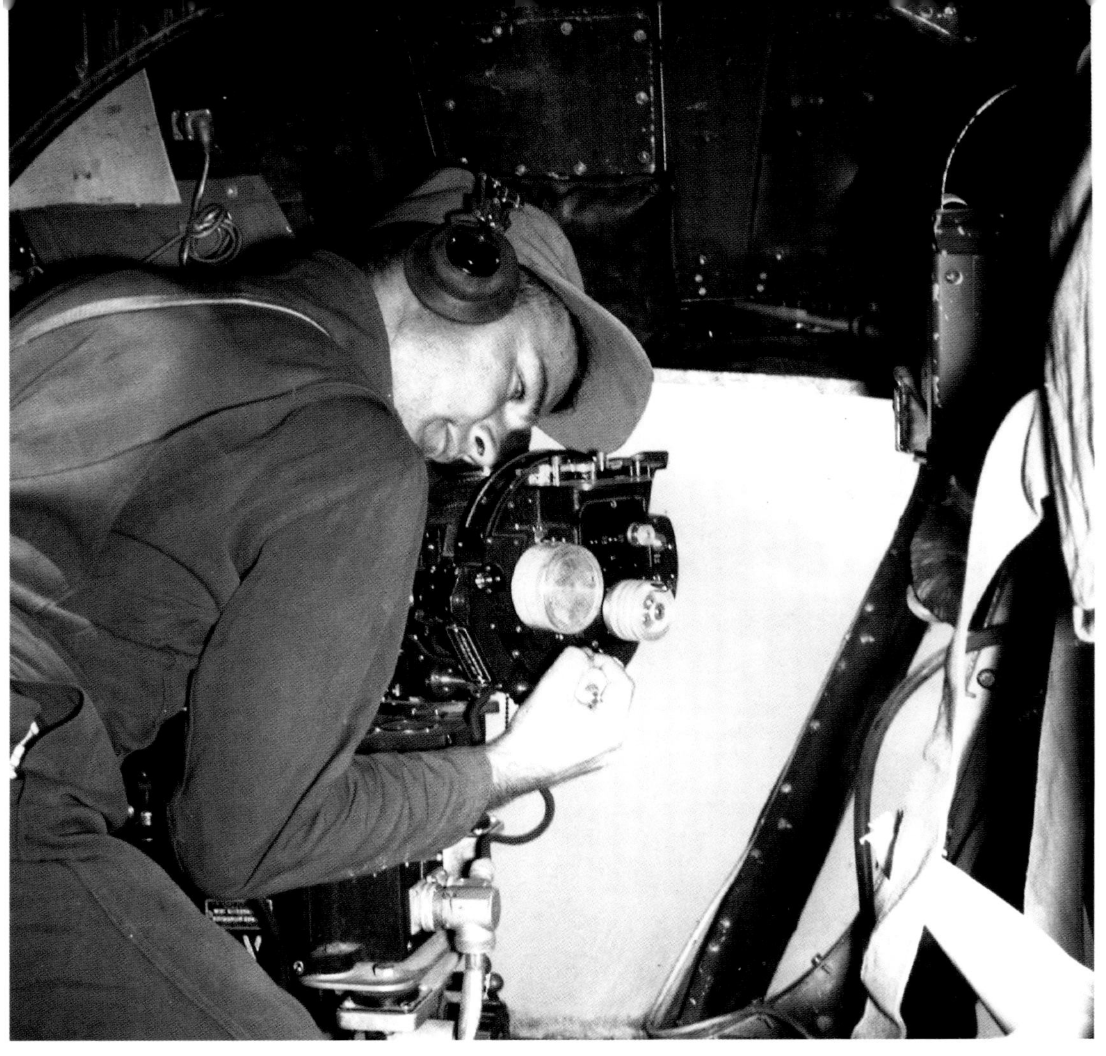

Left: *The Norden bombsight enabled the AAF's doctrine of high-altitude daylight precision bombardment envisioned during the 1930s. (USAF)*

12 April
Details of the highly secret Norden bombsight are released by the War Department. The sight is said to remain locked to a target irrespective of aircraft speed or other movement.

14 April
Fleet Radio Unit, Pacific Fleet, intercepts and decodes a "hot" Japanese naval communiqué. Admiral Isoroku Yamamoto was planning an inspection visit to three bases of operations near Bougainville Island. A fairly precise itinerary, escort aircraft force, and the date for the inspection are deciphered. The decoded message is immediately transmitted to Admiral Chester Nimitz, Commander in Chief Pacific Fleet, who assigns the planning of the attack to Admiral William F. Halsey. Army Air Forces P-38 Lightning fighters stationed on Guadalcanal are selected for the intercept mission.

Under the Zone of the Interior, the Weather Wing is activated, assuming responsibility for supervision of the AAF Weather Service from Headquarters AAF.

17 April
Eighth Air Force launches its first 100 planes raid against one single target. They attack the Focke-Wulf factory in Bremen, Germany. Fifteen bombers are shot down and another is brought down by AAA fire. German fighter opposition is so strong that fighter commanders call for 20 additional Fighter Groups to balance the odds over Europe.

18 April
Major John W. Mitchell leads a flight of 16 P-38s on a daring surprise attack to intercept and shoot down Admiral Yamamoto, who was inspecting bases near Bougainville. The P-38s fly at very low altitude in a circuitous route to a predetermined location 35 miles from Kahili. At about 9:35 a.m., two Mitsubishi Betty bombers and six Zeros are identified by the fighters. The Betty carrying Yamamoto takes evasive action but is attacked by Captain Thomas G. Lanphier Jr. and 1st Lieutenant Rex T. Barber. The pilots rake the bomber with machine-gun fire, and it crashes into the jungle. The following day, Yamamoto's body is recovered from the wreckage, flung from the aircraft but still strapped into his pilot seat, clutching the hilt of his samurai sword.

1 May
Staff Sargeant Maynard H. Smith, Eighth Air Force, is on his first combat mission as a gunner in the 423d Bomb Squadron when his aircraft is subjected to anti-aircraft fire and fighter attacks near the Saint Nazaire U-boat base and shipyard. Smith singlehandedly administers first aid to a wounded crew member, mans the waist guns, and fights intense flames while others of the crew bail out. Smith is one of four Army Air Forces enlisted men to receive the Medal of Honor during the war.

3 May
Lieutenant General Frank M. Andrews is killed when his aircraft crashes into the side of a mountain in Iceland during an approach to land. Andrews, the Commanding General, European Theater of Operations, U.S. Army, had been instrumental in educating Army senior staff officers in the use of American airpower.

4 May
Sixty-five Eighth Air Force heavy bombers strike the Ford and General Motors factories at Antwerp. As part of the mission, a group of 30 B-17s and B-24s launch a diversion toward the French coast. More than 100 German fighter planes sent to intercept the feint are distracted long enough to allow the main strike force to attack with limited fighter interruption. P-47s are able to provide escort to the bombers for 175 miles.

6 May
Captain H. Franklin Gregory makes the first helicopter landing on a ship when he pilots his Sikorsky XR-4 Hoverfly to a touchdown on the deck of USS *Bunker Hill*, at anchor in Long Island Sound, New York.

8 May
Aircraft of the North African Air Force and the Ninth Air Force—B-26s, B-25s, and P-40s—begin bombardment of the Pantelleria landing ground. RAF Wellington bombers also participate. Additionally, P-40s strafe and attack other targets in the Gulf of Tunis. The aerial attacks at Pantelleria, an island located between Italy and Tunisia, will continue through June.

14 May
A "Maximum Force" bombing effort—just over 200 VIII Bomber Command bombers—is launched simultaneously on four separate objectives. Primary targets are submarine yards and naval facilities at Kiel and factories near Antwerp. This is the first time that more than 200 AAF bombers have been dispatched on a single mission.

17 May
The crew of the B-17F Flying Fortress *Memphis Belle* completes 25 combat missions in the European Theater. Led by Captain Robert Morgan, the crew returns to the U.S. with their bomber and fly it on a nationwide tour selling war bonds.

After 10 of 11 322d Bomb Group B-26s are lost on a single low-level bombardment mission, Eighth Air Force suspends such operations for medium bombers indefinitely.

18 May
The plan for the Combined Bomber Offensive is approved by the Combined Chiefs of Staff. The CBO—around-the-clock bombing of the enemy—lists the destruction of German fighters as an immediate priority. Primary targets of CBO: German submarine bases, aircraft industry, ball bearings, and oil production and refining. Secondary objectives are synthetic rubber, tires, and military vehicles.

Right: *From North Africa to Tokyo, the North American B-25 Mitchell bomber was used in every theater of operation during World War II. (NASM Poster Collection)*

Left: *"When Prayers are Answered," by William Phillips. (© William Phillips)*

Opposite: *Lieutenant Charles Hall became the first African American to score a victory over an enemy fighter. He celebrated with a bottle of cola. (NASM)*

2 June
Lieutenant William B. Campbell and Lieutenant Charles B. Hall fly the first combat mission of the 99th Fighter Squadron out of Fardjouna, Tunisia.

10 June
The Combined Bomber Offensive against Germany begins. The Royal Air Force attacks German targets by night while Eighth Air Force attacks industrial targets during the day. The around-the-clock bombardment was intended to destroy Germany's wartime capability and morale. The Combined Operational Planning Committee is established as the agency for coordinating efforts between Allied air forces.

10-11 June
North African Air Force Wellington bombers, naval aircraft, and Ninth Air Force B-25s continue to bomb the island of Pantelleria while escorted by P-40 fighters. After more than one month of aerial bombardment, the British 1st Division lands on Pantelleria unopposed. The enemy forces there surrender unconditionally to the British forces without a shot being fired. This marks the first time that a major military objective surrenders solely because of the impact of airpower.

15 June
The first AAF unit to fly the B-29 operationally is formed at Marietta, Georgia—the 58th Bombardment Wing.

16 June
Captain Jay Zeamer Jr., Fifth Air Force, pilots a B-17 on a photographic mission in Buka, Solomon Islands. During the mission, Zeamer's bomber is attacked, and he receives wounds in both arms and legs. Zeamer continues to pilot the plane and destroys at least five enemy aircraft.

2nd Lieutenant Joseph R. Sarnoski, Fifth Air Force, makes possible the completion of the photographic mapping mission over the Buka Area of the Solomon Islands. He volunteers as bombardier of the B-17 crew and fights off 20 enemy fighters. Even after he is knocked under the catwalk by a 20mm shell, he crawls back and keeps post until he collapses dead on his guns.

22 June
Conducting the first large-scale daytime raid on targets in the Ruhr, 182d Eighth Air Force bombers strike chemical works and synthetic rubber plants. Another formation bombs the Ford and General Motors plants at Antwerp.

24 June
Lieutenant Colonel William R. Lovelace of the AAF Aeromedical Laboratory makes a record-setting parachute jump near Ephrata, Washington. He leaps from an altitude of 40,200 feet.

Physicist Joseph S. Ames dies at the age of 78. Ames chaired the NACA from 1927 to 1939. The NACA research facility at Moffett Field, California, is named in his honor.

30 June
The invasion of New Georgia begins with landings by Army and Marine Corps forces on nearby Rendova Island. Thirteenth Air Force and other allied fighters cover these landings. During the landings, some 60 Japanese aircraft attack the task force and suffer approximately 50 percent casualties.

1 July
A classified memo from Major General B. Giles to General Hap Arnold stresses the need for greater numbers of fighter escort airplanes in numerical relation to bombers. The memo suggests that for every two heavy bomber groups, one fighter group is required.

4 July
After a staged flight totaling 28 hours, the first transatlantic Skytrain lands in Great Britain. A fully loaded Waco CG A-4 glider was towed for more than 3,500 miles carrying medical supplies for Russia and a variety of other mechanical parts.

7 July
Adolph Hitler establishes the German V-2 program as the highest military priority.

In the Zone of the Interior, AAF Training Command is established and takes up the responsibility formerly assigned to Technical Training Command and Flying Training Command.

8 July
Colonel Malcolm G. Grow, a surgeon in the Eighth Air Force, develops the flak vest and steel pot helmet for use by aircrew in aerial combat. For his ingenuity, Grow receives the Legion of Merit. Untold lives are saved by these pieces of equipment, which are capable of deflecting shrapnel that hits and penetrates aircraft during combat.

9–10 July
The first major airborne operation of the war takes place when British and U.S. contingents are dropped on the island of Sicily to facilitate the coming amphibious assault. On the 12th, C-47s drop another contingent of troops but suffer terrible losses, as 20 of the transports are lost.

13 July
The Office of Strategic Services (OSS) is activated under the command of Brigadier General "Wild Bill" Donovan. This unit, the forerunner of the CIA, is tasked with special missions behind enemy lines.

19 July
Flying from bases in Benghazi, Libya, NASAF B-17s carry out the first AAF bombing raids on Rome, attacking the rail yards there. B-25 and B-26 medium bombers strike nearby airfields.

21 July
While flying his eighth mission, 99th Fighter Squadron pilot, Lieutenant Charles B. Hall, shoots down a German FW-190 over Sicily. Hall is the first African American Army Air Forces pilot to shoot down an enemy aircraft.

22 July
A British intelligence report indicates that the CBO is having good effect against Germany. According to the report, the Luftwaffe has been forced into a defensive posture employing more than 50 percent of its forces on the Western Front. Additionally, damage to industrial areas seems extensive, particularly in transportation, rubber, fuel, iron, and coal in the Ruhr.

Left: *"Hamburg Raid, July 1943," by Floyd Davis. (Army Art Collection)*

24 July
In its first attack in Norway, 167 Eighth Air Force heavy bombers attack nitrate works at Horoya. Work at the plant is disrupted for more than three months. Aluminum and magnesium plants are also attacked and subsequently abandoned by the Germans. On this raid, new techniques are used by the bombers to join above overcast weather conditions using "splasher beacons." Future poor weather missions are made possible by such rejoin methods.

24 July–3 August
The RAF and AAF begin Combined Bomber Offensive raids against the German port city of Hamburg. On the first night, 750 RAF heavy bombers do considerable damage to the target. The following day, more than 200 Eighth Air Force heavy bombers attack the Hamburg shipyard and the U-boat base at Kiel. Effective German fighter tactics bring down 19 American planes. These attacks are costly, as escort fighters are unable to accompany the bombers all the way to targets in Germany.

26 July
Eight Seventh Air Force B-24 Liberators fly the last mission against Wake Island from Midway. The formation is intercepted by enemy fighters, and the bombers claim that 11 of those are shot down during the engagement.

28 July
More than 300 Eighth Air Force heavy bombers are launched for attacks into Germany. Because of horrible weather, only about 80 reach their targets, the aircraft factory at Kassel and the major Focke-Wulf factory at Oschersleben. These raids mark the deepest penetration into Germany yet by American bombers from England. Although the targets suffer significant damage, 22 of the bombers are lost when the Luftwaffe fighters attack their formations with rockets. P-47s with jettisonable auxiliary fuel tanks for the first time on a mission escort the bombers into Germany, but not all the way to their targets. Another formation of P-47s meet them after the attack and engage about 60 enemy fighters already engaging the bomber force. One P-47 is shot down but the Germans lose nine aircraft.

Flight Officer John C. Morgan, Eighth Air Force, serving as co-pilot on a 326th Bomb Squadron mission, earns a Medal of Honor for his bravery. Prior to reaching the German coast his B-17 is attacked by a large force of enemy fighters, and the pilot's skull is split open by enemy fire, leaving him in a crazed condition. Fighting off the pilot, Morgan flies the aircraft over the target and back to a friendly base wholly unassisted. Morgan is later shot down over Germany and spends more than one year as a prisoner of war.

29 July
More than 200 Ninth Air Force P-40s attack a number of targets around Messina. This is the single largest number of fighters operating during a single day in the Sicilian Campaign.

1 August
Consolidated B-24 Liberators from the Ninth Air Force staging out of Libya, including some on loan from Eighth Air Force, conduct a treetop raid on the oil refineries at Ploesti, Romania. Although 40 percent of the refineries' capacity is destroyed, 54 planes of the 177 launched are lost along with 532 aircrew. Operation TIDALWAVE demonstrates without doubt the dire need for long-range fighter escorts for heavy bombers. More than 230 Ninth Air Force P-40s attack multiple targets around Messina. This is Ninth Air Force's single largest attack to date. Five Medals of Honor are awarded for this single mission:

Colonel Leon W. Johnson, Ninth Air Force (detached to Eighth), leads a mass low-level bombing attack against Ploesti oil refineries in Romania as part of the Ploesti raid. Though his flight becomes separated from the rest and loses the element of surprise that was vital, he carries out the planned attack and completely destroys the important refining plants and installations that were the object of his mission.

Colonel John R. Kane, Ninth Air Force, leads a group of B-24 heavy bombardment aircraft during the Ploesti raid in Romania. Kane leads his element through dangerous cumulus clouds, intensive anti-aircraft fire, enemy fighter planes, and oil fires in order to complete the mission successfully.

Lieutenant Colonel Addison E. Baker, Ninth Air Force (detached to Eighth), leads an aerial attack on enemy oil refineries and installations in the Ploesti raid, Romania. His plane is brought down by a large-caliber anti-aircraft shell.

Major John L. Jerstad, Ninth Air Force (detached to Eighth), dies during the Ploesti raid in Romania while leading a formation in a low-level attack against an enemy oil refinery. His aircraft is hit and set on fire three miles from the target. He has the option to land in a field before reaching the refinery, but choses to complete the mission. He is able to drop his bombs on target, but his aircraft is no longer capable of functioning and crashes into the target zone.

2nd Lieutenant Lloyd H. Hughes, Ninth Air Force (detached to Eighth Air Force), is tasked with bombing oil refineries in Ploesti, Romania. Before reaching the target, Hughes's aircraft is severely damaged by enemy fire. He has the option to land in a grain field before reaching the refineries. However, he choses not to jeopardize the mission and drops the bombs on target. Unfortunately, his aircraft is no longer capable of landing and bursts into flames.

Below: *B-24 Liberators attacked the refineries at Ploesti, Romania, from smokestack heights. Losses were horrific and heroism marked the day. Five airmen were awarded the Medal of Honor for their combat actions. (USAF)*

Above: *Perhaps the most famous image of a quartet of WASPs after delivering the B-17* Pistol Packin' Mama *to its destination. (NASM)*

5 August
Merging the Women's Flying Training Detachment with the Women's Auxiliary Ferrying Squadron, the AAF forms the Women's Airforce Service Pilots (WASP). Famed aviatrix Jacqueline Cochran assumes duties as the director, while Nancy Harkness Love becomes the WASP executive with the Ferrying Division of Air Transport Command.

7 August
In the Zone of the Interior, the AAF Redistribution Center is created to process and reassign military personnel as they return from overseas tours of duty.

11 August
The Quebec Conference begins this date. FDR and Churchill hold discussions covering all phases of military operations around the globe and decide upon future actions for the Allies.

13 August
Bombers of the Northwest African Strategic Air Forces, flying from bases in Italy, attack targets in Germany for the first time.

15 August
After months of reconnaissance and relentless bombardment, U.S. and Canadian troops invade Kiska. They discover that the Japanese have evacuated their garrison during heavy fog and poor weather predominant there.

16 August
Fifth Air Force aircraft attack oil tanks and transport vessels at various locations. P-38s and P-47s intercept a force of 25 enemy fighters preparing to attack the Allied bomber force, shooting down 12 Japanese fighters. This marks the first use of the P-47 Thunderbolt in the Southwest Pacific Theater.

17 August
Exactly one year after the first heavy bombers began attack from bases in the United Kingdom, Eighth Air Force launches 315 B-17s on a mission that will send them farther into Germany than any raid to date. The primary targets are the Messerschmitt aircraft factory at Regensburg and the anti-friction-bearing plant at Schweinfurt. During the unescorted, two-pronged attack, B-17s unload 724 tons of bombs over their targets with mixed results. German fighters attack the bomber formations at will. Those aircraft that had attacked the bearing factory at Regensburg land at bases in North Africa. By the end of the day it is determined that 60 of the original 315 have been lost during the air battle.

17–18 August
Royal Air Force heavy bombers attack German V-weapon sites at Peenemünde during a massive night raid. More than 500 RAF aircraft drop 2,000 tons of bombs on the target. This raid marks the beginning of Operation CROSSBOW—the attacks to destroy German V-1 and V-2 capability—ordered by British Prime Minister Winston Churchill. The raid is effective and delays the V-weapons program for a short period of time during the rebuilding of the facility. Many structures are leveled, and between 600 and 800 people are killed, including the man in charge of rocket development, Walter Thiel.

More than 200 airplanes launch from a secret airfield near Lau, New Guinea. The attack force makes a surprise attack on the enemy airfields near Lae-Salamaua, destroying nearly all of the Japanese aircraft located on those fields.

The Sicily Campaign officially ends at 1000 hours (10 a.m.) when the 3d Division of Lieutenant General George S. Patton's Seventh Army enters Messina. The fall of Sicily allows increased operations against Italy.

18 August
During a mid-morning attack at locations around Wewak, New Guinea, more than 70 heavy and medium bombers covered by 100 fighters cause heavy damage to Japanese aircraft on the ground in those locations. Additionally, more than 30 enemy fighters are claimed to have been shot down.

Major Ralph Cheli, Fifth Air Force, dies near Wewak, New Guinea, while leading his squadron in an attack on the Dagua Airdrome. Cheli's B-25 is severely damaged by the enemy, and although he has the opportunity to abandon the plane at the necessary altitude, he choses to complete the attack. Upon ending the mission, it is impossible for Cheli to gain the necessary altitude to parachute, thus he instructs his wingman to lead the formation and crashes his plane into the sea. He is captured by the Japanese and held prisoner and does not survive the war. He is awarded a Medal of Honor for the mission.

20 August
At New Delhi, India, the AAF, India-Burma Sector, CBI Theater, is activated with Major Genernal George E. Stratemeyer in command.

21 August
During the Aleutian Campaign, Eleventh Air Force records show that since June 3, 1942, 69 enemy aircraft have been destroyed, 21 ships sunk, 29 ships damaged, and 29 of its own aircraft have been lost during the struggle. Preparations are made to redeploy Eleventh Air Force combat units to the U.S.

25 August
Operation STARKEY planning begins when Eighth Air Force heavy bombers are assigned to attack the German Air Force to keep their forces contained in Western Europe rather than be transferred to the Eastern Front. The Allies hope to prod the Luftwaffe into an air battle of attrition before the actual invasion of France in 1944.

27 August
In the first Eighth Air Force attacks on German rocket-launching sites, more than 180 heavy bombers are used to attack construction sites at Watten. These missions against V-weapon sites are later designated NOBALL targets.

Equipped with special radar equipment, 10 Thirteenth Air Force SB-24 "Snooper" bombers begin all-weather bombing operations from Carney Field, Guadalcanal.

1 September
Since May 1940 through this date, the United States has produced 123,000 airplanes and nearly 350,000 aircraft engines.

More than 70 Fifth Air Force B-24s and B-25s strike Alexishafen-Madang with 201 tons of bombs. This is the heaviest single mission total dropped by the Fifth in one day.

Right: *In factories such as this one, race and gender boundaries were broken as more than 16 million working-age men entered military service during World War II. (USAF)*

JACQUELINE COCHRAN: PIONEER AVIATOR

Alfred Beck

Among the foremost women fliers of her time, Jacqueline Cochran began a poverty-stricken existence as Bessie Mae Pittman in Florida on May 11, 1906. Largely inventing the story of her early life as an orphan, Cochran kept the surname of her deceased first husband and later styled herself Jacqueline while pursuing a beautician's career in New York. By sheer will and native ability, she overcame her gritty beginnings and proved to be an intuitive, aggressive manager and a natural pilot.

On a wager, millionaire Floyd Odlum underwrote Cochran's first flying lesson at Roosevelt Field, New York. In 1932, she soloed after two sessions and had a pilot's license in three weeks. Cochran's marriage to Odlum in 1936 gave her the means to continue both an aerial career and a lucrative cosmetics enterprise. After gaining a commercial license in 1934, she entered a 12,000-mile England to Australia race, a precursor to her entry into the daunting Bendix Transcontinental American Race in 1935. She won the women's division of the contest in 1937 and beat all her male competitors in 1938, flying the hazardous Seversky P-35 fighter. This capped her record-breaking New York-to-Miami run in just over four hours. Cochran ran up a succession of altitude and speed records that gave her the Harmon Trophy three years running after 1939, the first of her 15 Harmons.

She joined the beleaguered British in June 1942 with 25 American women pilots she recruited. Flying from Canada to Britain in a Lockheed Hudson, she became the first woman to take a bomber across the Atlantic. Under 18-month contracts with the British Air Transport Auxiliary, her pilots teamed with British female aviators to ferry aircraft from factories to airfields in Britain, releasing male pilots for combat. General Henry "Hap" Arnold, Commanding General of the Army Air Forces, early doubts overcome, pressed Cochran in September 1942 to return home to found an identical service for the AAF.

On September 11, 1942, Cochran launched the Women's Flying Training Detachment under the Air Corps. The program's 25,000 applicants were culled down to fewer than 1,200, and only those with required physical characteristics and 200 flying hours were admitted. Later that requirement was dropped to only 35 flying hours. The detachment was housed at Howard Hughes Field near Houston, Texas, but moved to Avenger Field, Sweetwater, Texas, in May 1943, for the duration of its existence. The women endured the same military discipline and training as male pilots and eventually graduated 1,074 fliers to ferry the Army's aircraft where needed in the continental United States. Women piloted every airframe type in the American inventory during the war, including the B-29 Superfortress. Repeated attempts to commission these women failed, however, and their civil service status left them paying most of their own expenses. Congress thus denied them military insurance and subsidy—even for funerals.

Cochran also discovered a rival organization in 1942 in the Women's Auxiliary Ferrying Squadron under Nancy Harkness Love, a pilot with 1,200 hours experience. An understated contrast to Cochran's forceful personality, Love owned an aviation

company in Boston with her husband, a staff officer in the AAF's Air Transport Command. She recruited 23 female acquaintances, each with over 1,000 flying hours, into the ferrying business in 1942 at New Castle, Delaware. Cochran prevailed on the Army to merge the two women's organizations in July 1943 under her sole command as the Women Airforce Service Pilots (WASP), eventually with uniforms, but still no military status, though Cochran herself achieved Air Staff membership.

By November 1944, the demand for combat pilots declined and trained males were being released to the ground army or other flight duty. Threatened civilian flight instructors also factored in to help end the WASP program on December 20, 1944. WASPs had flown over 60 million miles and delivered 12,650 aircraft; 38 died. Only in 1979 did Congress retroactively grant these women the military status and awards due them.

Jackie Cochran received a Distinguished Service Medal for her wartime work and continued setting aerial marks, especially in a professional association with legendary Brigadier General Charles E. Yeager. In June 1953, she became the first woman to exceed the sound barrier, flying an F-86 Sabre jet at Edwards AFB. After additional milestones in intercity and distance flying in the early 1960s, she pushed an F-104G Starfighter to 1,429.297 miles an hour on May 11, 1964. She gained international recognition for her aerial feats and repeated formal awards for her business contributions and acumen. The Air Force awarded her the Distinguished Flying Cross in 1969 and subsequently a Legion of Merit. She died in Indio, California, on August 9, 1980.

Opposite: *Jacqueline Cochran commanded the Women Airforce Service Pilots during World War II. The WASPs flew more than 60 million miles, delivering military aircraft to a variety of points of departure. (NASM)*

Above: *Nancy Harkness Love led the Women's Auxiliary Ferry Squadron until it merged with Jackie Cochran's WASPs. (NASM)*

Right: *WASPs talk shop near their B-17 Flying Fortress at Buckingham Field, Florida, during World War II. Women flew every type of AAF aircraft during the war as part of their stateside duties. ("Ladies Soaring Circle," by Dottie Knight, USAF Art Collection)*

Left: *A high-flying formation of B-17s leaves contrails over targets in Germany. (USAF)*

Opposite: *"Fortress Engaged in Commemoration of the B-17," July 1995, by Keith Ferris. (USAF Art Collection)*

5 September
Paratroopers are dropped near Nadzab airfield in the first such operation in the southwest Pacific. Fifth Air Force uses 82 C-47s to insert the forces after the drop zone was bombed by 52 medium bombers. Nadzab airfield is quickly restored to operational status and becomes a major operating location for the Allies.

6 September
In the first raid of more than 400 bombers, Eighth Air Force launches a multi-pronged attack on targets near Stuttgart, Germany. Poor weather wrecks the primary plan but secondary strikes are accomplished under heavy enemy fighter attacks. Forty-five American bombers are lost.

9 September
D-day for Operation STARKEY, intended to bring the Luftwaffe into the sky to fight, is a relative failure. More than 300 heavy bombers hit targets around Paris and on the French coast, but enemy fighters stay away for much of the raid.

The invasion of Italy begins as Operations AVALANCHE and SLAPSTICK. Twelfth Air Force support for these operations is significant and continues in intensity through the rest of the year.

13 September
The 82d Airborne Division drops into the beachhead at Salerno. The 1,200 paratroopers are delivered to the combat zone by 80 aircraft from the 52d Troop Carrier Wing. There are no losses of men or planes during this airborne operation, one of the most successful of the war.

One of America's leading glider experts and special assistant to CGAAF Hap Arnold, Richard DuPont, is killed in a glider accident at March Field, California. DuPont had been working as an advisor to the AAF glider program at the time of his untimely death.

18 September
In Rome, all Italian forces not under German command surrender to the Allies. Those forces will turn to fight against the Axis from this point forward.

20 September
1st Lieutenant Henry Meigs II, flying a 6th Night Fighter Squadron P-38 against Japanese night attackers over Bougainville, shoots down two aircraft within one minute.

22–23 September
For the first time, American B-17s fly a night bombing mission with the Royal Air Force. The few aircraft that participated were experimenting with possible alternatives to the daylight raids, which had resulted in tremendous losses the month before.

Ninth Air Force flies its last bombing mission from Africa when its B-24s hit two enemy airfields. The Bomb Groups of IX Bomber Command are transferred to Twelfth Air Force.

27 September
For the first time in Europe, American bombers fly daylight bombing raids led by pathfinder aircraft equipped with H2S direction-finding equipment. More than 1,000 tons of bombs are dropped from above complete overcast. For the group of 244 bombers that attack the port of Emden, P-47s equipped with belly tanks are able to escort the formation all the way to the target in Germany and return with the bombers to their home bases, setting a new distance record of more than 600 miles on the mission. This marks the beginning of long-range fighter escort for American bombers.

29 September
General Eisenhower and Marshal Badoglio sign Italian surrender documents on board HMS *Nelson* off the coast of Malta.

1 October
Intelligence reports issued by the Eighth Air Force indicate that despite attempts to slow German aircraft production, German fighter production has continued to increase and enemy fighter strength on the Western Front remains strong.

2 October
The Rocket Ram, America's first military rocket-powered plane, is glide tested by John Myers. The Ram is equipped with an Aerojet XCAL-200, monoethylaniline-fueled engine.

3 October
The first American turbojet engine afterburner is built at the NACA Lewis Flight Propulsion Laboratory.

5 October
In General Spaatz's absence, General Jimmy Doolittle takes command of Twelfth Air Force.

7 October
The first night propaganda leaflet drop in a series of such missions is accomplished by aircraft of the Eighth Air Force.

8 October
More than 350 heavy bombers attack industrial areas and the city of Bremen as well as the U-boat facilities at Vegesack. At Bremen, about 30 U.S. aircraft are lost. This is the first bombing mission to use airborne jammers (Carpet Equipment) against German radar.

10 October
The AAF demonstrates control of a drone aircraft using television to provide visual feedback.

11 October
Colonel Neel E. Kearby, Fifth Air Force, leads a flight of four B-47s to reconnoiter the enemy base in Wewak, New Guinea. Kearby single-handedly shoots down six enemy aircraft after observing enemy installations at four airfields and securing important tactical information. Colonel Kearby is awarded the Medal of Honor for his accomplishment.

12 October
Allied Air Forces, including aircraft of the Fifth Air Force, assault Rabaul from the air. Approximately 350 B-24s, B-25s, P-38s, and Royal Australian Air Force aircraft attack the town, harbor, and local airfields. Three enemy ships and 50 enemy aircraft are destroyed.

13 October
Italy declares war against Germany.

14 October
American heavy bombers return to attack the anti-friction-bearing plants at Schweinfurt. Of the 230 bombers launched on the raid, 60 are shot down, many by German fighters firing long-range rockets, and 138 more are so badly damaged that Eighth Air Force is forced to cancel operations deep into Germany without escort fighters.

15 October
Eighth Air Force commander, Lieutenant General Ira C. Eaker, assumes additional responsibility when HQ U.S. Army Air Forces, United Kingdom, is activated to coordinate operations between Eighth and Ninth Air Forces in Britain with Major General L. H. Brereton in command. The establishment of HQ USAAFUK is essential as the forces of Ninth Air Force arrive in England after being redeployed from the Mediterranean Theater. The Ninth begins to arrive the next day, filling the role of the tactical air arm for the AAF in Europe.

16 October
General Hap Arnold proposes that the Fifteenth Air Force be established in Italy. The new command would be used as needed to supplement Combined Bomber Offensive operations against Germany. Arnold makes this recommendation based upon discussion with experienced weather forecasters who believe that better weather in Italy will allow a greater number of effective combat missions to be flown during the harsh winter months.

25 October
More than 60 Fifth Air Force B-24s bomb airfields near Rabaul, destroying some 20 enemy aircraft on the ground. The B-24s claim 30 aerial kills in addition to those caught on the ground. These attacks continue through the Allied landings on Bougainville on November 1.

November
During this month, General Hap Arnold, Commanding General, AAF, directs an expanded guided missile program based upon German advances in the field. His scientific advisor, Theodore von Kármán, submits a proposal to develop long-range surface-to-surface missiles.

1 November
Fifteenth Air Force is activated at Tunis, Tunisia. Major General James H. "Jimmy" Doolittle assumes command. His forces will move rapidly from Africa to Italy to accomplish bombing missions in preparation for the invasion of France.

2 November
Fifteenth Air Force flies its first combat missions of the war. Seventy-four B-17s and 38 B-24s attack an aircraft factory and industrial complex at Wiener-Neustadt. They are escorted by P-38 fighters. B-25s and B-26s bomb targets near Amelia.

In support of Allied operations at Bougainville, Fifth Air Force B-25s and P-38s attack Rabaul airfield and harbor. During the intense air battle, 11 ships are sunk and dozens of enemy aircraft are destroyed in the air and on the ground. Anti-aircraft fire and enemy fighters take a deadly toll, and by the end of the day, 21 U.S. aircraft are lost.

Major Raymond H. Wilkins, Fifth Air Force, leads the Eighth Bomb squadron in an attack over Simpson Harbor near Rabaul, New Britain. Before reaching the target on his 87th combat mission, Wilkins' B-25 is hit by intense anti-aircraft fire. Nevertheless, he strafes a group of harbor vessels and makes a low-level attack of an enemy destroyer. Wilkins destroys two enemy vessels, but his aircraft is uncontrollable and crashes into the sea. He is awarded the Medal of Honor for his heroism.

3 November
A new blind bombing tool, the H2X, is used in combat for the first time. More than 530 Eighth Air Force B-17s and B-24s, nine of them using the new device, attack the German port of Wilhelmshaven. P-38s escort the bombers nearly the entire trip in their first European Theater operation.

6–7 November
Tenth Air Force B-24s begin a series of night mining missions when they drop mines into the Rangoon River. Similar missions continue through November.

11 November
In a joint attack, Fifth and Thirteenth Air Forces along with U.S. Navy carrier aircraft attack targets on Rabaul. This is the first time that the Thirteenth Air Force has attacked Rabaul.

13 November
In one of the heaviest Allied raids in New Guinea, Fifth Air Force 57 B-24s and 62 B-25s attack Alexishafen and Madang airfields while P-40s strafe targets in the same area. A second flight of B-24s attack targets at Gasmata, Kaukenau, and Timoeka. A B-25 and Royal Australian Air Forces Beaufighters sink a small freighter.

In the longest escort mission by American fighters to date, 115 Eighth Air Force heavy bombers attack the port of Bremen. Another 100 bombers abort due to poor weather. Of the 47 P-38s escorting the bombers, seven do not return after facing an overwhelming number of German fighters over the target.

14 November
In the first ever AAF raids on Bulgaria, 90 Twelfth Air Force B-25 Mitchell medium bombers drop 135 tons of bombs on targets in Sofia.

20 November
The Army Air Forces sponsored play, *Winged Victory*, opens on Broadway. Written by playwright Moss Hart, it is a story of the struggles made by flying cadets to earn their wings. The cast of 300 actors are nearly all military service members. Later, the play will be made into a movie.

XX (Twentieth) Bomber Command is activated at Smokey Hill Army Air Field, Salina, Kansas. Major General Kenneth B. Wolfe assumes command one week later. Wolfe had assisted in the development of the B-29 Superfortress, which is assigned to the XX Bomber Command.

Opposite, above: *Lieutenant General Ira C. Eaker. (NASM)*

Opposite, below: *Ships burn after medium bombers attack near Rabaul. (USAF)*

Above: *In one of Fifth Air Force's most famous missions, Major John Henebry, piloting* Notre Dame de Victories, *and Captain Richard Ellis, piloting* Seabiscuit, *attack Japanese cargo ships near Rabaul. ("Simpson Harbor," by Michael Hagel, USAF Art Collection)*

22–26 November
During talks at the Cairo Conference, FDR, Churchill and Chiang Kai-Shek discuss the use of the B-29 Superfortress against Japan—called Operation TWILIGHT. It is agreed that the long-range bomber will initially be based in the China-Burma-India Theater (CBI) to initiate these attacks.

26 November
Eighth Air Force launches 440 bombers against Bremen. Weather is poor and 29 U.S. aircraft are shot down.

28 November
At the Tehran Conference, FDR, Churchill and Stalin give OVERLORD, the invasion of Northern France, and ANVIL, the invasion of Southern France, priority over all other operations. Stalin agrees to enter the war against Japan after Germany is defeated.

29 November
The AAF accomplishes its first raid against targets in Sarajevo, Yugoslavia. Twelfth Air Force sends 25 B-25 Mitchell medium bombers to attack rail yards and military bases.

3 December
A memo from Air Chief Marshal Portal to the Combined Chiefs of Staff suggests that Operation POINTBLANK is well behind schedule if air superiority is to be attained before OVERLORD begins (tentatively set for May 1). Pressure on Eighth Air Force mounts to destroy the Luftwaffe.

5 December
SEXTANT, the Second Cairo Conference, begins. During the three-day talks, Roosevelt and Churchill discuss global operations, set a timetable for offensive operations in the Pacific Theater, and establish a Unified Command for the Mediterranean Theater effective on December 10. FDR decides that General Eisenhower will hold Supreme Allied Command for the invasion of France—Operation OVERLORD.

Of 250 B-26 bombers launched to attack enemy targets in France, 200 turn back due to poor weather. Ninth Air Force P-51s of the 354th Fighter Group, equipped with long-range fuel tanks, escort Eighth Air Force heavy bombers nearly 500 miles to targets in northern Germany. This is the first escort mission flown by American P-51 Mustangs during World War II.

8 December
General Hap Arnold notifies General Tooey Spaatz that he will become the overall air commander in Europe for the invasion scheduled for summer 1944.

13 December
A massive attack is launched against port areas in Bremen and Hamburg as well as the U-boat yards at Kiel by 649 B-17s and B-24s. This is the first Eighth Air Force raid in which more than 600 bombers attack targets. Initially 710 were dispatched for the mission.

17 December
Orville Wright presents the 1942 Collier Trophy to General Hap Arnold for his leadership and organization of the U.S. Army Air Forces.

18 December
General Hap Arnold sends Air Chief Marshal Portal his commander list for 1944. The list includes: General Eaker, Allied Air Forces in the Mediterranean Theater of Operations; General Cannon, Twelfth Air Force; General Twining, Fifteenth Air Force; General Spaatz, U.S. Strategic Air Forces; General Doolittle, Eighth Air Force; and General Brereton, Ninth Air Force.

20 December
The Mighty Eighth launches a massive attack against the port of Bremen. Enemy fighters knock down 27 bombers, while P-51s and P-38s engage the Luftwaffe in a fierce air battle. The Germans use twin-engine,

Opposite: *General Carl A. "Tooey" Spaatz. (USAF)*

Right: *A flight of four North American P-51 Mustangs patrol the skies over Europe. Long-range Mustangs helped to turn the tide of the air war as escorts for heavy bombers on strikes deep into Germany. (NASM)*

rocket-firing fighters, that are protected by single-engine fighters. Anti-radar strips of metal foil, called window, are used in combat by Eighth Air Force for the first time.

Tech Sergeant Forrest L. Vosler, Army Air Corps, Eighth Air Force, is a B-17 radio operator-air gunner on a bombing mission over Bremen, Germany. Vosler is hit in the chest and face after a 20mm enemy shell explodes into the aircraft. Although pieces of metal penetrate both of his eyes, Vosler refuses first aid and helps the crew to safety. He is one of four enlisted men to receive the Medal of Honor during the war.

Mediterranean Allied Air Forces is established under the overall command of Air Chief Marshal Tedder. General Spaatz commands the U.S. element, USAAF North African Theater of Operations.

22 December
Orders are issued implementing General Arnold's command plan of December 18.

24 December
In the largest single raid against enemy targets in Europe to this date, Eighth Air Force sends 670 B-17s and B-24s on their first CROSSBOW mission to destroy German V-1 and V-2 sites in the Pas de Calais area of northern France. Significantly, no American aircraft are lost during the attacks. Twenty-six heavy bomber groups are now active in the European Theater.

Fifth Air Force bombardment of Cape Gloucester reaches its peak as 190 B-24 Liberators, B-25 Mitchells, and A-20 Havoc medium bombers pound the area during extensive daylight attacks.

26 December
Fifth Air Force accomplishes highly successful pre-invasion bombardment operations at Cape Gloucester, New Britain. Enemy targets are completely destroyed. The U.S. 1st Marine Division lands at Cape Gloucester that morning, while B-25s and B-24s continue bombing enemy positions. P-40s, P-38s and P-47s provide air defense and claim 60 victories in the enemy filled skies. The term "Gloucesterizing" is coined to describe such missions of total destruction as were flown by Fifth Air Force during this day.

31 December
A 500 plane raid against targets near the coast of France is launched by the Eighth Air Force. Twenty-five planes do not return. After this raid, the total tonnage of bombs dropped by the Eighth exceeds that dropped by RAF Bomber Command for the first time.

WORLD WAR II: AIR WAR IN EUROPE

1944–1945

WORLD WAR II: AIR WAR IN EUROPE

1944–1945

Until the spring of 1944, the Luftwaffe remained a powerful adversary. During the fall of 1943, American Combined Bomber Offensive operations were halted temporarily following catastrophic losses on daylight raids against targets in Ploesti, Romania, and Schweinfurt, Germany. From August to October, Army Air Forces in Europe lost nearly 10 percent of the attack force aircraft and approximately 2,000 airmen. The bomber, it appeared, would not always get through. The American attacks deep into occupied territory could not resume until the arrival of the P-47 Thunderbolt, the long-range P-38 Lightning, and the Rolls Royce Merlin–powered P-51 Mustang—effective fighters with external drop tanks for extended range. It was only after the arrival of long-range fighter escorts in the form of P-38s with drop tanks and finally the P-51 Mustang that allowed serious attacks in support of POINTBLANK directives. POINTBLANK, frequently associated only with the massive Allied effort to destroy and disrupt the German air forces, actually referred to the entirety of CBO missions which, at times, seemed completely centered on the counterair offensive in Europe. Even with escort fighters, large numbers of bombers—more than 1,000 on every raid—could not be mustered until April and May of 1945.

The role of the Mediterranean Allied Air Forces (MAAF) was an important, yet complicated one. Lieutenant General Ira C. Eaker was moved from command of the Eighth Air Force to command the MAAF in January 1944. He was a skilled administrator and had long experience with the Eighth, but now he would command not only one large unit with one major mission, but three with three distinct missions: strategic bombardment, tactical air support for the ground armies in Italy, and coastal defense and protection of the sealanes. A long list of secondary tasks also kept him busy during his tenure in Italy. Unlike his command in England, which was one American air force, in Italy he held sway over participants from six Allied nations. The tasks were monumental when taken in their entirety, and Eaker has never received the credit he deserved for his accomplishments in that command.

One week after his arrival, the MAAF spearheaded the Allied invasion of Italy at Anzio. In their most active day of the war, Twelfth and Fifteenth Air Forces paved the way for the ground forces that grabbed a toehold at the beachhead and never let go. Eaker answered to two different command structures, one strategic and one tactical, and had to balance his resources to manage and please both masters. Additionally, he had to manage the dismemberment of Twelfth Air Force to fulfill other needs on the continent while Fifteenth Air Force swelled with strategic assets at the same time. Perhaps the most difficult of his problems was the winter weather in Italy, which kept his forces on the ground too often to allow effective operations until March.

When American operations against Germany resumed in 1944, the tale was different. Late in February, the Royal Air Force and the Army Air Force began a devastating series of around-the-clock raids against German aviation industry targets. Together, these air forces flew more than 6,000 sorties and dropped 20,000 tons of bombs in one week. The

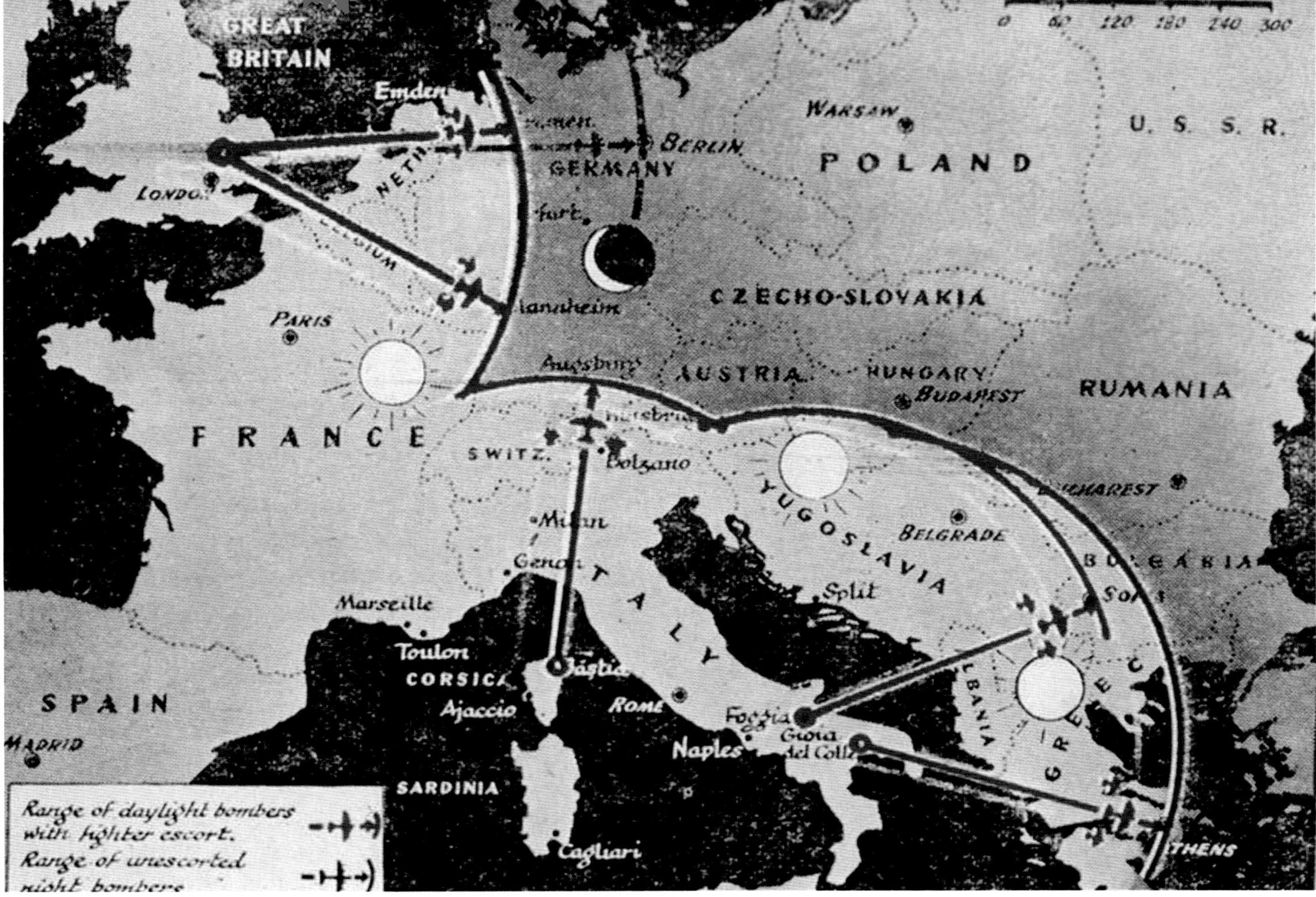

Allied air forces lost more than 400 aircraft—nearly 7 percent. But the rate of German fighter losses to the American escorts was even higher. Big Week, as it came to be known, initiated the final collapse of the Luftwaffe. By April the rate of German fighter pilot attrition soared to 25 percent. The German Air Force had been dealt a serious blow. During May and June, Allied air forces were busy preparing the landing zones for the cross-channel effort, including an ongoing effort to disrupt and destroy German V-weapon sites across France. German U-boat facilities also remained high on the list of priorities. It was against these targets that American bombers dropped a series of remotely guided weapons designed to keep airmen away from deadly flak.

OVERLORD took place on June 6, 1944, under an umbrella of complete air supremacy, the result of months of bombing efforts and sacrifice by Allied aviators. But losses of airmen and airplanes had been high. Once a beachhead had been established in France, the Allied strategic air forces—the Eighth and the Fifteenth—were relieved from many of their tactical support obligations and returned to a punishing campaign, striking deep into the heart of Germany. From July 1944 through May 1945, Allied bombers dropped over 70 percent of the total tonnage of bombs employed against Germany during the entire war. As a comparison, in 1943 the Allies had dropped over 220,000 tons of bombs in the European Theater. In 1944 that number more than quadrupled to 1.2 million tons.

In August, Eaker's MAAF reacted with flexibility as preparations for Operation ANVIL/DRAGOON began while Allied armies in Italy paused to regroup, resupply, and prepare for the invasion of southern France. The operation was a political nightmare for the planners, as it had been scheduled and cancelled several times over a six-month period. The final plan was one favored by American commanders as it had the potential to end the war quickly. The British opposed it because it disregarded political considerations of the postwar situation in the Balkans. In the end, it was the XII Tactical Air Forces that did most of the heavy lifting during the operation, with some support from the RAF.

By the last year of the war, the Army Air Forces had grown from 26,500 men and 2,200 planes to over 2.4 million men and women and nearly 200,000 combat and training aircraft. America's Eighth Air Force flew its last combat mission against Germany on April 25, 1945. Many of the aircrew soon shipped out for Pacific bases to fly more strategic bombing missions.

Pages 192–193:
"D-Day—The Airborne Assault," by Robert Taylor. (NASM Art Collection)

Above: *Daylight and night bomber ranges in Europe before long-range escort operations became routine in late 1943.*

WORLD WAR II: AIR WAR IN EUROPE

1944–1945

"Destroy the Enemy Air Force wherever you find them, in the air, on the ground, and in the factories."

—General of the Army Hap Arnold to his Air Commanders, December 27, 1943

1944

January
A Romanian pilot defects with a brand new Ju-88 and delivers it to a British airfield on Cyprus. The plane ends up being flown into U.S. custody and closely examined at Wright Field in Dayton, Ohio.

1 January
Caltech's rocket laboratory begins research and development of a long-range missile at the request of the Army. Project ORDCIT eventually results in the development of the Private "A" and Corporal missiles.

4–5 January
During the day, more than 500 Eighth Air Force heavy bombers attack German ports. That evening, U.S. aircraft begin flying supplies supporting the underground resistance in Western Europe. Lieutenant Colonel Clifford Heflin flies the first of these nighttime Operation CARPETBAGGER missions from Tempsford, England, to locations in France.

6 January
AAF command changes begin to go into effect when Lieutenant General Carl A. Spaatz assumes command of the U.S. Strategic Air Forces in Europe (USSAFE, which will be changed to USSTAF on February 4). Lieutenant General Jimmy Doolittle takes over Eighth Air Force, and Lieutenant General Ira C. Eaker moves to Italy as commander of Mediterranean Allied Air Forces. Additionally, to better coordinate Combined Bomber Offensive (CBO) attacks across Europe, control of Fifteenth Air Force is placed under General Spaatz as CG, USSAFE.

8 January
The Lockheed XP-80 flies for the first time at the Muroc test center. Piloted by Milo Burcham, the *Lulu Belle*, named after a popular cartoon character, blazes the trail for future production aircraft that will exceed 500 miles per hour in level flight. Development of the prototype took only 143 days and was designed by Clarence L. "Kelly" Johnson.

11 January
Some 600 Eighth Air Force B-17s and B-24s strike enemy industrial targets in Germany. Enemy fighter opposition is fierce, with some estimates ranging as high as 500 enemy fighters in the fight. Another 60 bombers are lost during the raid. For the first time, B-24 Liberators are used as Pathfinder aircraft to accomplish bombing through overcast skies.

Right: *This Corporal Missile, a result of Project ORDCIT, is being positioned for firing at White Sands Proving Ground in 1956—more than a decade after the commencement of the program. (NASM)*

Right: *The Lockheed XP-80 and some of the project personnel at Edwards AFB. (NASM)*

Below: *General Carl A. Spaatz deplanes in Europe. (NASM)*

Major James H. Howard, Ninth Air Force, was the leader of a group of P-51s over Halberstadt, Germany, tasked with providing support for a heavy bomber formation. Howard single-handedly destroys a German FW-190 and two Bf-110s along with another probable Me-109. He fights off a formation of more than 30 enemy airplanes, protecting the attacking bomber formation with extreme risk to his life and personal safety. Howard is the only P-51 pilot to receive the Medal of Honor during the war.

Fifteenth Air Force B-17s, escorted by P-38 Lightning fighters, attack the harbor near Piraeus, Italy. Although the attackers claim eight enemy kills, flying in extremely poor weather results in midair collisions that claim six B-17s.

13 January
During a raid against NOBALL targets (German rocket sites) in France consisting of nearly 200 Ninth Air Force B-26 Marauders, several aircrew report being fired upon by anti-aircraft rockets.

14 January
While more than 500 Eighth Air Force heavy bombers attack 20 V-weapon sites in the Pas de Calais, France, Twelfth Air Force aircraft blast targets throughout Italy in support of the Fifth Army's efforts in the Monte Trocchio area. B-25s, A-20s, A-36s, and P-40s are all in the air in coordinated close support missions.

15 January
Lieutenant General Ira C. Eaker assumes overall command of Mediterranean Allied Air Forces and also takes responsibility for the AAF forces serving in that combined command.

16 January
General Eisenhower assumes his post as the Supreme Allied Commander, Allied Expeditionary Force. In this position, Ike holds overall command of OVERLORD, the invasion of France.

21 January
More than 500 Eighth and Ninth Air Force bombers strike V-weapon sites across northern France. An additional 400 bombers are forced to turn back to England without attacking 19 other targets because of low clouds.

22 January
Mediterranean Allied Air Forces (MAAF) support Operation SHINGLE, the Allied invasion of Anzio in western Italy which begins at 2:00 a.m. Both Twelfth and Fifteenth Air Forces launch a maximum effort totaling more than 1,200 sorties in support of the Fifth Army's VI Corps. Missions are flown to isolate the battlefield and to maintain air superiority during the landings at the Anzio beachhead. Support missions continue into February.

JAMES H. "JIMMY" DOOLITTLE

Carroll V. Glines

Few aviators in history have achieved as much international renown as James H. "Jimmy" Doolittle. He is the only individual ever awarded the Medal of Honor for valor in war and the Presidential Medal of Freedom for lifetime achievements.

Doolittle was commissioned in 1917, at the time when Army pilots were encouraged to demonstrate the capabilities of military aircraft through air shows and aviation record-setting. He was rated as a pilot in the Army Air Service and served during and after World War I as a flying instructor, test pilot, and engineering officer.

In 1922, Doolittle achieved an "aviation first" by flying from the East Coast to the West Coast in less than 24 hours. He was also first to cross the country in less than 12 hours. But this was only the beginning of more than two decades during which he often pushed aircraft beyond commonly accepted limits of their capabilities, thereby keeping aviation in the public eye. In 1923, Doolittle enrolled for a master's degree at Massachusetts Institute of Technology (MIT), where he examined the stresses—G forces—placed on an aircraft during various flight maneuvers. This area of study had previously been determined by loading an aircraft with sandbags on the wings and stabilizers until they gave way under the weight. However, this theory had never been tested in flight. Doolittle put an aircraft through maneuvers to the point of near failure and proved that the predetermined point of failure had been reached. His research ended when the wings on his test plane began to fracture.

Doolittle continued his work toward a doctorate at MIT in 1924 and chose for his dissertation a study of the wind velocity gradient at various altitudes and the effect of wind on aircraft at sea or ground level. Thus, this Army test pilot became a qualified aeronautical engineer, the first in the growing Air Service, and the recipient of the first such degree ever awarded by MIT. During the next decade, he used his knowledge to win numerous air racing trophies and establish many point-to-point speed records.

In the history of heavier-than-air flying, hundreds of pilots had lost their lives while trying to fly through clouds and fog. They relied on their senses but crashed when they became disoriented, losing their relationship with the ground. Aviation could not progress if pilots were unable to fly safely in adverse weather.

In 1929, Doolittle was chosen to participate in blind flying experiments sponsored by the Daniel Guggenheim Fund for the Promotion of Aeronautics and headed the Full Flight Laboratory at Mitchel Field, Long Island. He soon found that the primary need was for accurate, reliable, and easy-to-read instruments that showed pilots the exact heading and precise altitude and attitude of their aircraft, especially during the landing phase of flight. Doolittle worked with radio and instrument manufacturers for seven months helping to design flight instruments and radio navigation devices. He made more than 100 practice

Opposite: *Jimmy Doolittle was a true "Renaissance Man" of the twentieth century. (NASM)*

Right: *"Lieutenant General James H. Doolittle," by Howard Chandler Christie. (NASM Art Collection)*

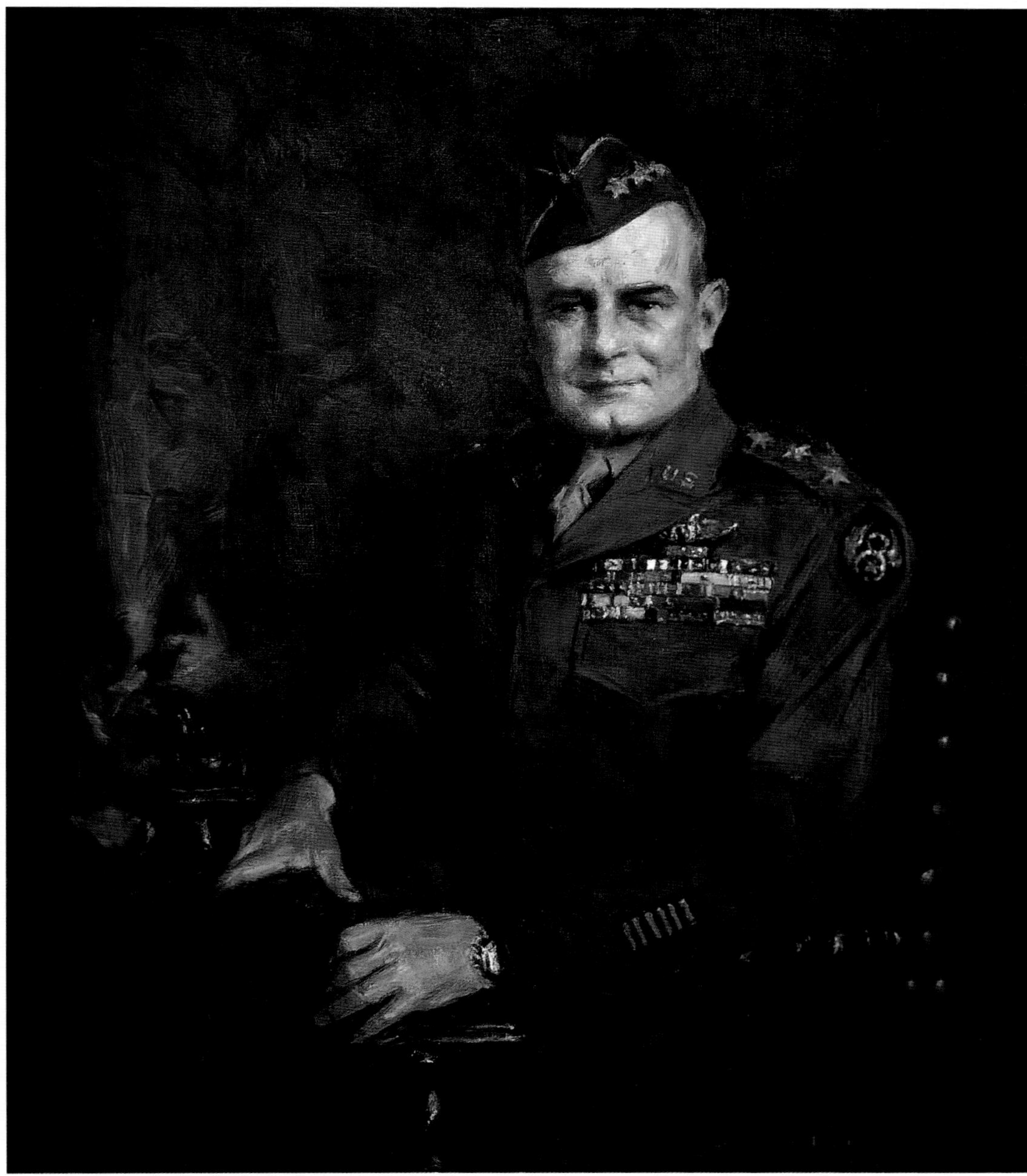

flights during these months, and on September 24, 1929, made a solo takeoff in early morning fog, flew a rectangular course, and landed without seeing the ground. He repeated the flight with a safety observer in the front cockpit while he was enclosed beneath a canvas cockpit hood.

This technological breakthrough marked the point in aviation history when a pilot's worst adversary, the weather, could finally be defeated. Military and civilian pilots could fly through fog and clouds with assurance and safety. Aviation, able to advance beyond its primitive state, entered a new era of great progress. Doolittle always believed that this achievement was his greatest contribution to aviation.

Returning to civilian life in 1930, Doolittle used his piloting experience and scientific credentials to encourage the development of 100-octane fuel that sparked the development of more powerful and efficient engines for commercial and military aircraft. This advance helped the Allies to prevail in the skies during World War II.

Although Doolittle made his mark in aviation history by garnering awards and international recognition for his flying capability, the military mission that forever placed him in the history books, after his return to active duty in 1940, was the surprise raid against Japan in April 1942. He organized and led a force of 16 medium bombers launched from the U.S. Navy aircraft carrier, USS *Hornet* that attacked five cities in Japan in retaliation for the raid against American forces at Pearl Harbor.

After this heroic mission, Doolittle's dedication, knowledge, and leadership were rewarded as he rose to command the Eighth Air Force, the mightiest bombing force the world has ever known.

27 January
In a very active day, Twelfth Air Force P-40s and Fifteenth Air Force P-38s and P-47s provide close air support and air cover to the Fifth Army while shooting down dozens of enemy fighters over Rome and Florence.

28 January
Eighth Air Force B-24s equipped with the Gee-H blind-bombing device strike V-weapon sites near Bonnieres, France. This is the first time Gee-H is used in combat. It is more accurate than previous devices but is range-limited, only effective up to 200 miles.

29 January
Eighth Air Force Pathfinder aircraft lead a force of 763 bombers to strike targets near Frankfurt, Germany. Although almost 1,900 tons of bombs are dropped on the city, nearly 30 American bombers go down in the face of fierce enemy fighter attacks. This is the first Eighth Air Force mission in which more than 700 bombers attack enemy targets.

30 January
During the second 700-plane raid launched by Eighth Air Force, blind bombing devices allow the formation to attack industrial targets near Brunswick, Germany. A smaller force of airplanes bombs targets near Hanover. During the raid, 20 bombers are lost to enemy fighter attacks. Meanwhile, Twelfth Air Force fighters find no enemy air opposition over Anzio during their daily fighter sweeps over the beachhead. The 451st Bomb Group flying B-24 Liberators is activated in Fifteenth Air Force, bringing the total number of heavy bomb groups in that command to nine.

February
During this month, the Army Ordnance Division and the AAF initiate the development of a surface launched, supersonic, guided, high-altitude missile designed to intercept aircraft. This project develops the Nike I surface-to-air missile.

The first VB-1/2 AZON controllable, vertical glide bombs were sent to Europe. These weapons are controllable in "azimuth only" (from side to side) and are guided through a bombsight by means of radio remote control. A total of 15,000 AZONs are produced through November 1944.

1 February
Major General Elwood Quesada takes charge of IX Air Support Command, which controls all of the fighter and reconnaissance units of the IX Fighter Command.

3 February
The 358th Fighter Group is added to the 354th FG already flying missions for the Ninth Air Force. Between this date and May 1, 1944, 16 more Fighter Groups will be added to the Ninth for duty during OVERLORD.

8 February
Fifteenth Air Force activates the 454th Bomb Group flying B-24 Liberator bombers. This brings the total of heavy bomb groups in the Fifteenth to 10. B-17s continue to bomb targets near Orvieto, Piombino, and Prato under escort by P-47 and P-38 fighters.

9 February
Twelfth Air Force aircraft strike hard in support of Fifth Army operations around Cassino, Italy.

13 February
The Combined Chiefs of Staff alter the Combined Bomber Offensive plan to refine and reduce the number of targets essential to the campaign prior to D-day. The modifications reflect adaptations based upon Germany's industrial revitalization and dispersal plans. Disruption of lines of communications and destruction of the Luftwaffe remain critical priorities of the CBO.

15 February
In an effort to soften up the enemy for the U.S. Fifth Army and the British Eighth Army as they push toward Rome, Twelfth Air Force B-25s and B-26s attack the Monte Cassino Benedictine Abbey, suspected of being home to enemy forces. The abbey is destroyed by the force of 254 American B-17s, B-25s, and B-26s attacking in two waves. Another 60 B-24 Liberators strike other targets surrounding Cassino. Follow-on attacks occur over the following three days.

18 February
HQ Eighth Air Force establishes the 8th Reconnaissance Wing (provisional) in Cheddington, England, to provide independent capability as the D-day invasion date approaches. Colonel Elliot Roosevelt, FDR's son, is given command of the unit.

18–19 February
Twelfth Air Force A-20s, P-40s, and A-36 aircraft struggle to hold off the advancing German counter-attack around the Anzio beachhead, which began on the 16th. The deepest German advance occurs on the 18th, but they are pushed back on the 19th by more than 200 close support sorties and a hard-fought Fifth Army counterattack.

Opposite, above: *"Shatzi Over Frankfurt", by Michael Resmussen. (USAF Art Collection)*

Opposite, below: *"Bob Baseler over the Adriatic," by Dan Witkoff. (USAF Art Collection)*

Above: *Major General Elwood R. "Pete" Quesada. (NASM)*

Left: *With two failed engines, crewmen finally are forced to bail out of their disabled B-17 at low altitude over Europe. Note the opening parachute in the upper left corner of this combat image, as well as the crewman just aft of the right wing's trailing edge. (NASM)*

Below: *Consolidated B-24 Liberators fly through a dense flak belt over their target area. (USAF)*

20–26 February
Lieutenant General Jimmy Doolittle's Eighth Air Force launches more than 1,000 heavy bombers for the first time, initiating Big Week. Nearly 900 of these make it to their targets in Germany. Priorities during the coming week are aircraft factories and airfields as well as attempting to draw German fighters into the skies where massive numbers of AAF escort fighters are waiting. Losses are heavy but less than expected. From this week forward, the German air forces continue to lose punch. Although the Luftwaffe is still formidable at times, this campaign marks the turn in the tide of the air war over Europe.

1st Lieutenant William R. Lawley Jr., Eighth Air Force, B-17 pilot, manages to save his crew after his co-pilot is killed and he and two others wounded during a German fighter attack over the target area in Germany. When it is determined that two of the crew are not able to bail out after coming off target, their bombs frozen to the racks of their B-17, Lawley overcomes severe blood loss and flies the severely crippled B-17 back to an emergency gear-up landing in England. He is awarded the Medal of Honor for his bravery

2nd Lieutenant Walter E. Truemper, Eighth Air Force, navigator on a B-17 bombing mission over Germany, is hit by an enemy fighter shell in the cockpit, killing the co-pilot and severely wounding the pilot. Truemper and the crew manage to fly the plan back to the home station. While he attempts to land the plane by listening to instructions from the control tower, he eventually stalls and crashes in the traffic pattern during his third attempt. Truemper, his engineer, and the co-pilot are killed in the crash, but Truemper is awarded the Medal of Honor for the mission.

Sergeant Archibald Mathies, Eighth Air Force, B-17 mechanic and gunner during an attack over Germany, assists the bombardier in flying the plane after the pilot is severely wounded and the co-pilot is killed by a German fighter attack. Sergeant Mathies and the crew take control of the plane and return to the airfield. After the rest of the crew jumps to safety, Mathies, assisted by Lieutenant Truemper, heroically tries to land the plane. After their second attempt, the plane crashes after stalling near the airfield and Mathies, the navigator, and the wounded pilot perish. Mathies is one of four enlisted men to receive the Medal of Honor during the war.

Right: *"Aerial Gunner," by Peter Hurd. (Army Art Collection)*

21 February
In another massive air raid, 764 B-17s and B-24s attack aircraft factories and storage facilities near Brunswick and Diepholz. Weather prevents further attack by other forces. Results are difficult to determine, as weather is a factor.

22 February
In a coordinated attack between Eighth and Fifteenth Air Force heavy bombers, 101 B-17 strike aircraft factories around Halberstadt. Another 154 bombers attack a variety of aircraft production and airfield targets across Germany, including the aircraft works around Regensburg. During the attacks, more than 50 heavy bombers are lost to enemy fighters.

23 February
Poor weather grounds the majority of Eighth Air Force air operations over Europe this day while Fifteenth Air Force B-24s attack the industrial complex at Steyr. Escort fighters claim 30 victories during the raid.

24 February
When the weather breaks for the day, Eighth Air Force launches several 200-plus plane bombardment missions. First, 231 B-17s strike ball bearing works at Schweinfurt; second, 238 B-24s hit the aircraft factory and airfield at Gotha, losing 33 bombers; third, 236 bombers attack targets near Rostow after poor weather forces a shift to their secondary target. Overnight, Royal Air Force Bomber Command targets Schweinfurt again as part of the CBO. Fifteenth Air Force B-17s once again attack the factories in Steyr and also the oil refinery at Fiume, where they lose 19 bombers to enemy defenses.

25 February
In a major combined effort, Allied heavy bombers strike targets at Regensburg, Augsburg, Furth, Stuttgart, Zara Harbor, and Fiume. The strikes are flown both day and night, and German fighters take punishing losses against Allied fighter escort and Fortress and Liberator gunners.

26 February
Big Week comes to an end when poor weather grounds most aircraft during the day. Big Week succeeds in halting the momentum of German airpower. From this point forward, Germany becomes more and more defensive in the air. They are forced to withdraw into Germany as larger and more frequent air raids demand more and more airplanes to defend the German homeland.

March
It is reported by the U.S. Office of War Information that as of the end of 1943, more than 7,800 aircraft—mostly combat planes—have been shipped to the Soviet Union in an effort to equip the Red Air Force through Lend-Lease.

NACA Lewis Laboratory performs high-altitude testing of the P-59 propulsion system. The program that follows these initial tests makes major contributions to the development of turbojet engines.

BIG WEEK

Left: *"P-51 Mustang and Me 109," by J.B. Deneen. (NASM Art Collection)*

The Army Air Forces entered the Second World War touting the doctrine of high-altitude daylight precision bombardment. The realities of poor weather and the dispersal of enemy industrial and military targets forced a modification in the doctrine developed in the laboratory environment of Maxwell Field classrooms. From the fall of 1943 until the end of the war the AAF conveniently flowed back and forth between attempts at daylight precision and poor weather and night area bombardment, not only in Europe but also over Japan. Adaptability to the realities of the European air war and its challenges opened the floodgates for massive air raids over vast expanses of German territory. Criticism over ethical and moral issues raised by nonprecision area bombardment resulted.

It was with this reality-driven bombing philosophy that Eighth Air Force began operations in 1944. Having replaced Lieutenant General Ira Eaker as Eighth Air Force commander in January 1944, Lieutenant General Jimmy Doolittle's forces would fracture the Luftwaffe's concrete hold in the air and begin the shift in momentum over Germany. Doolittle enjoyed an ever-increasing arrival of bombers and long-range fighters and had few real restrictions to operations such as Eaker had in the previous months.

Overall European air commander General Carl "Tooey" Spaatz and his deputy, Major General Frederick L. Anderson, Doolittle from the Eighth, Nathan Twining from the Fifteenth, and Eaker now commanding the Mediterranean Allied Air Forces planned and executed the missions that were flown in February against Germany under the code name ARGUMENT—massive attacks against the Luftwaffe and the German aircraft industry. This operation became known as Big Week.

Allied troops were expected to hit the beaches at Normandy during the summer of 1944. By mid-February, there were enough bombers and escort fighters to begin the massive raids against Germany that General Arnold had requested, but the weather remained untenable. Army meteorologists finally delivered a favorable weather forecast for February 20—an extended period of weather suitable for visual bombardment.

Spaatz, after weighing the concerns expressed by his fighter pilots about icing conditions at altitude, personally issued the order to begin the raids. Doolittle's Eighth Air Force contributed the majority of the attacking forces on this first 1,000-plus-bomber mission, accompanied by more than 900 fighters over Axis territory. Technically, only 971 of the bombers received mission credit, but more than 1,000 were launched. Sixteen bomber wings, 17 fighter groups, and 16 RAF fighter squadrons filled the skies over Europe. Six of the bomber wings flew unescorted to Poland on a northern route while the rest attacked aircraft industry targets around Leipzig and Brunswick in central Germany. Damage was heavy in many target areas, but machine tools used in aircraft construction escaped significant destruction. Only 21 bombers were lost, while damage to the targets was moderate—although the actual extent of the damage was not known in detail until after the war. This raid had been preceded by an RAF night attack on Leipzig and would be followed each night by coordinated RAF area bombing raids upon targets that complemented the combined forces bombing scheme.

Similar raids were flown on February 21 and 22 with less success. On the 23rd, the entire Eighth Air Force was grounded due to unforecasted low clouds and icing conditions, while the Fifteenth flew only 102 bombers against ball-bearing factories in Austria. On Thursday and Friday, the clouds broke and once again massive attacks upon Germany were launched. More than 800 bombers of the Eighth and Fifteenth Air Forces attacked Augsburg, Stuttgart, Schweinfurt, and Regensburg under clear skies, suffering modest losses. Damage inflicted during these raids was severe and largely the result of a precision visual attack. When the Friday missions had landed, Big Week officially came to an end.

During ARGUMENT, of the 3,800 bombers that flew during the week, the Eighth lost 158, the Fifteenth lost 89, and more than 2,600 airmen were either killed, wounded, or captured. Roughly 10,000 tons of bombs were dropped by American forces, which equaled the total tonnage dropped by the Eighth Air Force during its entire first year of operations. The RAF attacks against five German cities were equally massive. More than 2,300 bombers dropped nearly 10,000 tons of bombs while losing 157 aircraft during that week.

Despite the remarkable tonnage unleashed, the effects upon German industry were (and still are) difficult to assess precisely. Of greater impact than the material attacks upon industrial targets was the fact that the Luftwaffe came up to meet the bombers and were shot out of the sky in large numbers. Their losses of fighter aircraft and pilots during this week continued into March and by April resulted in the Luftwaffe's inability to combat the Allied air offensive. By the end of February, the Eighth had lost a total of 300 bombers. Luftwaffe casualties, however, numbered one-third of their single-engine fighters and nearly 20 percent of their fighter pilots.

Big Week signaled a transformation of the air campaign against Germany. It was the first time that the AAF had launched 1,000 bombers on any single raid—a force size approaching what Air Chief Marshal Sir Arthur "Bomber" Harris, AOC-in-C of RAF Bomber Command, had advocated for many months. It was also during this week that the overall strength of Eighth Air Force surpassed that of RAF Bomber Command for the first time during the war. For all of the squabbling that had occurred over air doctrine and command of air forces during the first two years of the air campaign, American airmen, in large measure, assumed greater control of the air war after Big Week's conclusion.

Right: *"Berlin, December 5, 1944" by Alfred Vetromile. (USAF Art Collection)*

Left: *P-38 Lightning fighters of the 20th Fighter Group head for targets in France. ("Dawn Chorus," by Nicolas Trudgian, USAF Art Collection)*

Below: *"First American Daylight Bombing of Berlin, 4th March 1944," by John Rayson. (USAF Art Collection)*

2 March
Fifteenth Air Force gains another group of B-24 Liberators, the 459th Bomb Group, and employs 300 B-17s and B-24s, escorted by 150 P-47s and P-38s, to provide support for Fifth Army's operations at Anzio.

4 March
Thirty-one Eighth Air Force B-17s bomb Kleinmachnow in southwest Berlin. Originally, more than 200 B-17s are launched, but many abort the mission when German radio transmissions "recall" the bombers, citing bad weather. The 95th Bombardment Group presses ahead anyway and bombs the target from 28,000 feet. This is the first time that U.S. bombers have attacked the German capital city. Although General Doolittle asks General Spaatz to allow him to lead the raid, Spaatz refuses to grant him permission. Had Doolittle flown on any raid over Berlin, he would have been the only American airman to have bombed all three Axis capitals during the war.

6 March
Eighth Air Force launches a major offensive strike against Berlin; 658 heavy bombers attack the city and surrounding areas. Although more than 1,600 tons of bombs are dropped on the capital, more than 10 percent of the attacking bombers, 69 aircraft, are lost on the raid, the highest number of bombers lost on any single day of the war. Medium bombers of the Ninth Air Force continue to strike NOBALL targets in France while Twelfth Air Force hammers targets around the Anzio beachhead.

8 March
Once again targeting ball bearings in Germany, Eighth Air Force sends 460 bombers to strike the factory at Erkner. Seventy-five additional aircraft strike Wildau and Berlin. During the raids, resistance is stiff, and 36 bombers are shot down.

9 March
More than 450 Eighth Air Force bombers continue to attack deep into Germany, including strikes on Berlin, Brunswick, Hannover, and Nienburg.

15 March
Ninth Air Force P-51 fighters are released from direct commitment to Eighth Air Force bombers unless needed for specific missions.

Right: *In close formation, North American B-25 Mitchell medium bombers deliver the blow that closed the Brenner Pass on March 25. (NASM)*

In an all-out effort, AAF aircraft destroy the Monastery at Monte Cassino, while the Third Battle of Cassino is underway on the ground. The battle for Monte Cassino goes on for four months.

16 March
The National Advisory Committee for Aeronautics (NACA) releases a proposal to develop a jet-propelled transonic research aircraft at a conference held at the NACA Langley Laboratory. Although the aircraft ultimately flies with a rocket engine, this proposal leads to the development of the Bell X-1, the first plane to break the sound barrier in level flight.

18 March
In continued attempts to draw German fighters into the skies, Eighth Air Force sends 679 bombers against aircraft factories across Germany. The Luftwaffe comes out in force. Forty-three American bombers and 13 escort fighters are lost, but the enemy suffers heavy losses during their attacks. The loss of German pilots begins to impact the effectiveness of defensive responses over Germany. Fifteenth Air Force continues to pound targets in Italy, destroying a sizeable number of enemy planes during their raids.

19 March
Lieutenant General Eaker's Mediterranean Allied Air Force launches Operation STRANGLE in support of Allied ground forces in Italy. The objective of the seven-week campaign is to interdict enemy supplies across Italy by disrupting rail service, and marshalling yards and ports. By mid-May, the Allies will have dropped more than 26,000 tons of bombs while flying more than 50,000 sorties during the campaign.

20 March
The 67th Tactical Reconnaissance Group completes a series of 83 missions to map the French coast in preparation for the D-day invasion. More than 9,500 prints are made of the beaches, and no aircraft are lost during this one-month project.

22 March
Mount Vesuvius, near Naples, Italy, erupts during combat operations. B-25s from the Twelfth Air Force fly through the ash, which burned through control surfaces made of fabric and pitted and chipped Plexiglas windshields.

25 March
Fifteenth Air Force bombers successfully close the Brenner Pass between Italy and Austria, disrupting enemy transportation of supplies and communications.

26 March
Eighth Air Force sends more than 500 B-17s and B-24s to attack V-weapon sites in Pas de Calais and Cherbourg areas. Ninth Air Force sends 338 B-26s and A-20s to strike torpedo-boat pens near Ijmuiden. Another 140 P-47s and P-51s dive-bomb marshalling yards at Creil and other military installations in France.

Left: *The B-24* Paper Doll *returns from a difficult mission to its base in Italy. Later, this B-24 would be lost during a raid over Romania. ("Maximum Effort," by James Dietz, USAF Art Collection)*

Below: *When fitted to the P-51 Mustang, the Rolls Royce Merlin engine turned a good fighter into a great one. This Merlin is part of the propulsion collection at the National Air and Space Museum in Washington, D.C. (NASM)*

27 March
In a massive effort against airfields and aircraft works, the Mighty Eighth launches 700 heavy bombers to strike targets across France.

28 March–2 April
Rapidly expanding its forces, Fifteenth Air Force launches a series of massive strikes on March 28 in support of Operation STRANGLE. Nearly 400 B-17s and B-24s are launched to strike marshalling yards and railroads around Verona and Cesano in their first "thousand-ton raid." P-38s and P-40s provide excellent cover for the bombers, and no bombers are lost in the first 1,000-ton bombing raid flown by the Fifteenth. Three American fighters are shot down by the enemy during the mission. The next day, more than 400 heavy bombers are launched, topping the record number for attacks over Turin, Milan, and Bolzano, Italy. Six aircraft are lost during these raids. On April 2, now with 16 heavy bomber groups at the ready, more than 530 bombers are launched against the ball bearing factory at Steyr, among other targets near Mostar. Escort fighters claim more than 100 enemy fighters are destroyed, while 19 bombers go down.

April
During this month, the 100,000th Rolls-Royce Merlin engine is produced. The first Merlin came off the assembly line in July 1937 and has evolved through several models from 1,000 to 1,600 horsepower. Used on American P-51 Mustangs, the Merlin also powers 13 other operational aircraft, from RAF Halifax bombers to the Spitfire.

3 April
IX Bomber Command establishes a new operational leave policy as aircrew shortages develop. Bomber crews are allowed a maximum of one week of leave between their 25th and 30th missions. They are allowed a maximum of two weeks between the 40th and 50th missions.

3–4 April
Fifteenth Air Force launches two days of strong attacks against aircraft industrial targets around Budapest, Hungary. More than 450 bombers strike on the 3rd, and escorts and bombers claim two dozen kills during the raid. More than 300 heavy bombers strike the next day, and a wild fighter-to-fighter battle ensues. Ten bombers, including the B-24 *Paper Doll*, go down, and dozens of enemy aircraft are shot down.

Right: *The B-24*E-Z Duzit *depicted in the upper right corner of this painting was shot down on this raid. The crew was captured and held prisoner in Stalag Luft I until the end of the war. ("Brunswick, 8 April 1944," by Paul Jones, USAF Art Collection)*

4 April
Twentieth Air Force is secretly activated in Washington, D.C. Initially the Joint Chiefs of Staff, through General Hap Arnold, will control the B-29 assets which will strike mainland Japan.

5 April
Fifteenth Air Force bombers return to Ploesti and target the refineries and marshalling yards near the town. Enemy fighters and anti-aircraft fire claim 13 bombers during the struggle.

8 April
In a major assault upon German airfields and aircraft factories, 13 combat wings from the Eighth Air Force launch in three separate forces. The largest single force of 192 B-17s attacks factories in Brunswick. The Eighth loses 34 bombers during this ambitious offensive. Ninth Air Force is also hard at work launching one of the largest tactical raids of the war. More than 200 B-26s and P-47s attack targets near Hasselt, Belgium.

9 April
Eighth Air Force B-24s and B-17s continue to pound away at airfields and aircraft factories in Germany and Poland. Of 399 bombers launched, 32 are shot down. No diversionary raids are flown, allowing the enemy to concentrate on the attacking formations. Raids of this type continue through D-day.

11 April
More than 800 Eighth Air Force B-17s and B-24s attack fighter production factories and airfields across Germany. Enemy fighters shoot down 64 bombers in the second bloodiest day of the war in the air for the Eighth Air Force. Ninth Air Force launches more than 300 aircraft against a variety of targets in France. B-26s, A-20s, and P-47s are all active.

1st Lieutenant Edward S. Michael, Eighth Air Force, pilot of a B-17 aircraft, comes under heavy attack over Germany and his plane is severely damaged. Lieutenant Michael gives the order for the crew to bail out and attempts a crash landing, despite his own wounds. After passing out at the controls for a short time, he evades enemy attacks from the air and the ground for nearly an hour to save the life of a bombardier whose parachute had been damaged. Michael regains consciousness in time to land the aircraft despite his injuries and the damage to the aircraft. Michael is awarded the Medal of Honor for the mission.

Left: *Bristling with machine guns, this YB-40 was designed for aerial gunnery practice, firing ceramic bullets at heavily armored RP-63A Kingcobras. The rounds shattered on contact, leaving a powdery mark that could be scored after the target plane landed. (USAF)*

13 April
General Eisenhower, the Supreme Commander, assumes direction of most air operations from his headquarters in the United Kingdom. He also assumes command of the majority of ground and naval forces that will participate in the invasion of France. More frequent attacks will be made in northern France by Eighth and Ninth Air Forces as D-day approaches. Continued attention is given to attacking V-weapon sites and lines of communication. Twelfth Air Force continues to hit hard in Italy, supporting ground forces engaged there.

Fifteenth Air Force continues to grow and launches its largest raid of the war to date. More than 530 heavy bombers attack multiple aircraft factories and airfields across Germany and Hungary. More than 200 fighter sorties are flown to support the raids, the largest to date by the Fifteenth. More than 120 enemy aircraft are destroyed on the ground while fighters claim an additional 40 victories in the air. Raids of this magnitude continue through D-day.

5 May
Swelling in size, Fifteenth Air Force now reaches 20 heavy bombardment groups in strength, allowing them to launch more than 640 bombers against targets near Ploesti, Romania. Fighters fly 240 support sorties for the largest bomber force ever dispatched by the Fifteenth.

7 May
Eighth Air Force sends more than 900 heavy bombers to attack industrial centers near Munster and Osnabruck, Germany. Additional sorties are flown later that day against marshalling yards near Liege. It is the first time 900 airplanes attack targets in one day for Eighth Air Force during the war.

8 May
General Eisenhower sets the date for the invasion of France—June 5.

9 May
An Allied offensive against airfields in France begins. Eighth Air Force sends a total of 797 heavy bombers to attack more than a dozen airfields in France and other targets in Luxembourg. The objective is to keep Germany from rebuilding these fields for use prior to D-day. Ninth Air Force joins in the pre-invasion airfield offensive two days later.

Lieutenant Colonel R. E. Horner makes the first flight of an aircraft modified to demonstrate high-lift boundary layer control. The project was initiated by the AAF in May 1942.

11 May
Allied ground forces attack the Gustav Line under an umbrella of artillery fire. The Fifth Army spearheads the attack on the enemy stronghold just south of Rome.

Operation STRANGLE, which began on March 19, comes to a close. In total, Mediterranean Allied Air Force drops 26,000 tons of bombs during 50,000 sorties flown during the campaign.

12 May
The Mighty Eighth sends 800 bombers to attack oil production targets at Merseburg, Chemnitz, and Brux. Facing massive enemy fighter concentrations, 46 heavy bombers are shot down. Ninth Air Force executes a dress rehearsal for the airborne invasion of Normandy. Operation EAGLE tests the tactics and techniques of all specific missions related to the pre-dawn plan to land paratroopers and supplies inland to help establish the beachhead. Weather is a mitigating factor, but the exercise goes on as planned.

Fifteenth Air Force, now expanded to its full combat strength of 21 heavy bombardment groups, launches its largest raid against German headquarters at Massa d'Albe and Monte Soratte. The 730 bombers launched also strike airfields and transportation targets north of Rome. More than 250 fighters escort the bomber formations during the day.

13–14 May
Twelfth Air Force aircraft put pressure on enemy forces as the Fifth Army and the French Expeditionary Force make a hard-fought breakthrough against the Gustav Line. Country-wide attacks north of Rome continue throughout the day. Fifteenth Air Force sends approximately 700 heavy bombers on interdiction strikes supporting the ground forces engaged at the Gustav Line. These support operations continue through May.

19 May
As Fifteenth Air Force pushes its attack to the north of Rome, enemy fighter opposition disappears.

22 May
Eighth Air Force occupies its last station in Britain, bringing the total to 77, including 66 airfields, which are home to 82 operational or headquarters units.

25 May
In one of its busiest days of the war, Twelfth Air Force strikes targets throughout Italy while the German Army retreats from Anzio. Support operations for the Fifth Army are critical as all surface forces—from Anzio and from the west coast—finally meet, forming a solid Allied front in the move to the north.

27 May
Enjoying excellent weather throughout the European Continent, more than 2,000 aircraft attack targets across France and Germany to Italy. Losses are heaviest over Germany as 24 Eighth Air Force bombers are shot down. Raids by the Mighty Eighth routinely number more than 800 bombers plus escort aircraft. Ninth Air Force missions number between 300 and 600 escorts. Fifteenth Air Force is capable of launching more than 800 bombers on one mission.

29 May
In a daring new experiment, Captain Charles T. Everett flies a test A-20 aircraft nicknamed *Alclad Nag*, which is fired upon by the top turret gunner of a YB-40. They are testing the "frangible" bullet. This ceramic bullet is made to disintegrate on contact with the target plane for aerial gunnery training. In September, a highly armored version of the P-63—an RP-63A Kingcobra—is flown by company pilot Robert Stanley, the same pilot to fly America's first jet plane. The RP-63 is designed to be the target aircraft for the frangible bullets tested by Everett.

30 May
In England, assault forces for OVERLORD begin loading equipment and supplies for the upcoming invasion of France. Theater-wide aerial assaults continue to ensure that the Luftwaffe has no operational capability near the landing zones.

Right: *An RP-63A aerial target aircraft. Today, aerial gunnery is accomplished by firing training rounds against targets towed by specially modified aircraft. (USAF)*

Left: *"Marauder Mission," by Robert Taylor. (Courtesy of Robert Taylor)*

Opposite, below: *Allied forces establish a beachhead in France as the massive supply operation continues after the initial D-day amphibious assault. (USAF)*

31 May
A Vertical Bomb—the VB-7—is tested for the first time. It is one of the earliest attempts to guide a weapon by remote radio signals, moving control fins for steering.

2 June
In preparation for OVERLORD, heavy bombers continue to attack transportation and airfield targets in northern France. Heavy attacks upon the coastal defenses at Pas de Calais are made as part of Operation COVER—the deception campaign designed to disguise the actual landing zones for the invasion—that continues until June 4. Two waves of bombers totaling more than 1,000 aircraft accomplish these strikes with minimal enemy fighter contact. Only eight bombers are lost to anti-aircraft fire. American tactical forces meet with ground liaison officers to finalize targets to be struck before and during the landing. V-weapon sites, fuel depots, bridges, and railroads are high on the list.

In Italy, Twelfth Air Force continues to fly sorties in support of the Allied push north of Rome.

Lieutenant General Ira C. Eaker commands the first raids flown during Operation FRANTIC, the shuttle-bombing between Italy and the Soviet Union. Eaker, taking off from bases in Italy, leads a flight of 130 B-17s escorted by 70 P-51 Mustangs to bomb marshalling yards at Debreczen, Hungary, then go on to land at an airfield in Poltava, Soviet Union. Later, more than 400 heavy bombers taking off from airfields in both Italy and the United Kingdom attack deep targets and then land in the Soviet Union.

4 June
General Eisenhower postpones OVERLORD for 24 hours due to poor weather for the amphibious landings. Aerial preparation of the landing zone continues as more than 500 tactical strikes are made against bridges and coastal gun batteries.

Twelfth and Fifteenth Air Forces are hard at work in Italy supporting advancing ground forces and hitting targets in the northwest near the French border in preparation for Operation ANVIL/DRAGOON, the invasion of southern France.

Rome falls to Allied forces. The first unit to enter the city is the 88th Reconnaissance Troop, 88th Infantry Division. The Fifth Army converges on the city, and 3d Infantry Division is tasked to garrison the area.

5 June
Eighth Air Force sends 629 heavy bombers to strike coastal targets in France. A handful of P-51s bomb and strafe coastal targets as well. Six bombers are downed by anti-aircraft fire. Relentless aerial attacks continue throughout Italy and France in an effort to soften up Axis defenses.

Lieutenant Colonel Leon R. Vance Jr., Eighth Air Force, leads a B-24 Heavy Bombardment Group as the Command Pilot in an attack against the enemy near Wimereaux, France. Suffering a direct hit from German anti-aircraft fire, Vance's right foot is nearly severed from his leg. He takes over for his dead pilot and recovers the plane to the English Channel, where he ditches the Liberator. Miraculously, he is rescued hours later by a British ship. Vance disappears on return to the U.S. when his transport plane vanishes between Iceland and Greenland. In addition to receiving the Medal of Honor, Vance Air Force Base, Enid Oklahoma, is named in his honor.

Right: *During the early morning hours on June 6, paratroopers from the 82d and 101st Airborne Divisions drop beyond the beachhead to attack enemy rear areas. ("Hour of Liberation," by Larry Selman. © Larry Selman)*

5–6 June
D-day invasion begins in the middle of the night as more than 1,400 Ninth Air Force transport planes and gliders drop airborne troops and land in the rear of German-occupied France. These three full airborne divisions are to secure the inland approaches to the Normandy beachhead. At dawn, in the largest amphibious attack in history, Allied forces staging from England land on the beaches of Normandy in northern France—the American zones are Omaha and Utah beaches. Three bombers are lost; two in a mid-air collision and one in a ground fire, but none to enemy attacks. American and British air forces fly more than 15,000 sorties in support of the invasion force during this 24-hour period.

Eighth Air Force reaches its peak strength of 40 heavy bomb groups. Four missions are launched during the day. In the first mission, more than 1,000 bombers attack enemy positions near the landing beaches. As the day progresses, targets move farther inland against communications centers and transportation hubs. Approximately 3,600 tons of bombs fall from Eighth Air Force planes during the day. More than 1,800 Eighth Air Force fighter sorties are flown, including escort, transportation target attacks, and strafing of enemy convoys. In all, 25 fighters are lost, most to anti-aircraft fire at low altitude. The Eighth will continue to support ground forces when needed but will also attack strategic targets, particularly oil production, deep into Germany for the remainder of the European war.

Ninth Air Force sends 800 medium bombers to northern France along with more than 2,000 fighters flying sweeps over the beaches and inland in support of the bombers. Including the losses from the nighttime air drop, 30 aircraft are lost. The Ninth will continue to support the ground armies as they prepare for the breakout from the beachhead, and then continue to blast lines of communications and V-weapon sites during the coming months.

Twelfth Air Forces continue to strike tactical and strategic targets throughout Italy, and will continue to do so throughout the rest of the Italian campaign.

Fifteenth Air Force continues shuttle-bombing missions against targets in Ploesti, Brasov, and Turin. Nearly 700 bombing sorties are flown against these targets. During these missions, primary targets are oil facilities in Hungary and Yugoslavia and will remain high priority for the duration of the war.

Left: *The German V-1 "Buzz Bomb" terrorized British cities and was responsible for many deaths during the last months of the war. (NASM)*

Below: *American military commanders visited the beaches at Normandy observing operations first hand. General Eisenhower (second from left) remarked, "If I didn't have air supremacy, I wouldn't be here." (Robert and Kathleen Arnold Collection)*

7 June
Eighth and Ninth Air Forces continue to interdict incoming German reinforcements while also resupplying airborne troops dropped into the enemy rear the night before. Again, more than 2,000 sorties are flown in support of the invasion forces now established on five beaches in northern France. Fifteenth Air Force reaches its peak strength of 21 heavy bomber groups and seven fighter groups and continues strikes against targets in northwestern Italy. As long as weather cooperates, thousands of AAF aircraft are airborne each day across Europe, accomplishing close support, interdiction, and strategic bombardment missions.

8 June
General Carl Spaatz places oil targets as the highest priority target for aircraft of the USSTAF.

9 June
Allied units begin operating from bases in northern France.

11 June
The first Operation FRANTIC mission is completed when Fifteenth Air Force B-17s and P-51s take off from bases in the Soviet Union, bomb several oil distribution and refinery targets, and return to their home bases in Italy.

12 June
General Hap Arnold joins the Joint Chiefs of Staff as they cross the English Channel in a landing craft and inspect the six-day-old lodgments.

13 June
The Germans launch V-1 weapons against England for the first time. The first one explodes at Swanscombe, Kent, at 4:18 a.m. Eleven are detected, and four of these strike random locations in London. General Hap Arnold, in England to monitor the air war, insists on being driven to one of the V-1 impact locations to inspect the damage. Attacks continue throughout the month of June.

21–22 June
Eighth Air Force continues Operation FRANTIC shuttle-bombing missions when 144 heavy bombers strike oil targets enroute to two landing fields in the Soviet Union. This raid is flown in conjunction with a massive 900-plane attack against targets in and around Berlin. More than 900 fighters are involved in escorting these bombers to and from the targets. Additional B-24s strike CROSSBOW targets and rocket sites at Siracourt. During this night, German bombers attack the 73 B-17s that landed at Poltava, Soviet Union. Flares are dropped to illuminate the

airfield as the Germans damage or destroy almost every American bomber parked on the ground. Ninth Air Force provides 700 escort fighters for the Eighth Air Force bombers over Berlin.

22 June
The GI Bill is signed into law. The Servicemen's Readjustment Act of 1944 makes higher education attainable for millions of returning veterans of World War II and all future conflicts.

23 June
Across northern France, Allied efforts are made to strike at V-1 sites in an effort to curtail continued attacks against southern England. Weather keeps many aircraft on the ground, but Fifteenth Air Force launches a raid against oil targets near Ploesti, Romania. More than 400 bombers and 300 escort fighters are tasked for this mission. The results are particularly bloody as more than 100 bombers and fighters are shot down in the effort.

2nd Lieutenant David R. Kingsley, Fifteenth Air Force, flies as a bombardier during a raid against oil targets in Giurgiu, near Ploesti, Romania. During the mission, his aircraft is severely damaged and the crew are wounded. After the pilot gives the order to bail out, Kingsley assists the gunners with their parachutes and discovers that the tail-gunner's parachute harness is missing. Kingsley gives his parachute to his tail-gunner and assists him to bail out of the doomed B-17. He dies when the plane crashes but is awarded the Medal of Honor.

3 July
The Northrop P-61 Black Widow, America's only aircraft built specifically as a night fighter, flies its first operational mission in Europe.

5 July
The Northrop MX-324, America's first rocket-powered plane, flies for the first time. Pilot Harry H. Crosby takes the plane aloft at Harper Dry Lake near Barstow, California.

7 July
Eighth, Twelfth and Fifteenth Air Forces hit petroleum, oil, and lubricant (POL) targets throughout the theater. Of the approximately 3,000 sorties flown during the day, around 60 planes are shot down.

8 July
Lieutenant Colonel Clifford Heflin flies a C-47 into France to rescue Allied airmen forced to parachute from damaged aircraft behind enemy lines. This is the first time such a mission has been attempted.

9 July
Fifteenth Air Force launches its first Pathfinder-led mission against oil targets at Ploesti, Romania. Escort is provided by P-38 and P-51 fighters who meet between 40 and 50 enemy fighters, shooting down 14 of them.

1st Lieutenant Donald D. Pucket, Fifteenth Air Force, flying a B-24 on a "restrike" of oil targets near Ploesti, completes his mission but shortly afterward his plane is hit with anti-aircraft fire. Assessing the damage and casualties, he orders his crew to bail out but three refuse. Pucket remains on board and makes one final attempt to control the aircraft, but the bomber crashes into a mountainside, killing the three men remaining on board. Pucket is awarded the Medal of Honor for his heroic actions.

Pieces of a wrecked V-1 "buzz bomb" (Fiesler Fi 103) are delivered to Wright Field, Ohio, for evaluation. In 17 days, the Ford Motor Company builds a copy of the pulse-jet motor. By October, Republic Aircraft Company has been able to copy the weapon's airframe design. U.S.-built duplicates are called the JB-1 "Loon."

Above: *Colonel Benjamin Davis leads a flight of P-51s during a Fifteenth Air Force escort mission. Also known as the Tuskegee Airmen, Davis's all-black 352d Fighter Group compiled a superb combat record. ("Red Tails," by Melvin Brown, USAF Art Collection)*

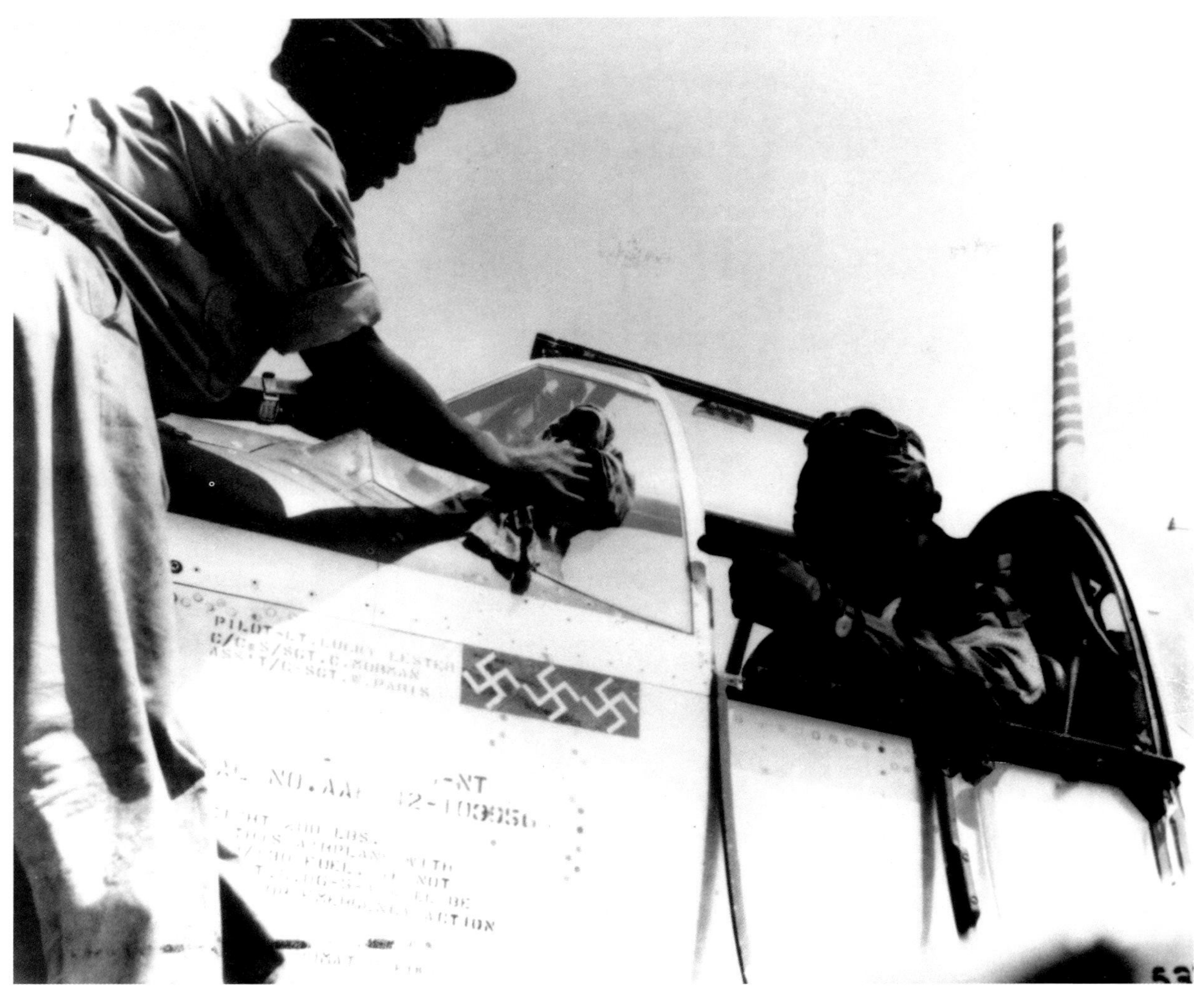

Left: *Lieutenant "Lucky" Lester shot down three German aircraft in one engagement on July 18, 1944. (NASM)*

11–16 July
Eighth Air Force launches a massive raid against targets in Munich. Nearly 800 fighters fly escort support for the strike. The next day, more than 1,000 bombers and 750 fighters attack Munich, again striking marshalling yards and aircraft engine factories. On the 13th, 888 bombers strike again with 548 fighters as escort. Fifteenth Air Force begins to strike targets in southeast France while continuing strikes against oil production in Romania, as ground forces continue to press north through Italy.

17 July
Napalm, a jellied form of gasoline, is dropped by a P-38 during a close support mission against a fuel depot at Coutances, France. This is the first use of napalm in combat by U.S. forces.

18 July
Lieutenant C. D. "Lucky" Lester, flying a 100th Fighter Squadron P-51 out of Ramitelli, Italy, is part of an escort mission for a group of B-17s bombing targets over southern Germany. His flight engages a formation of Bf 109s, and in the ensuing six minutes of combat, he shoots down three enemy aircraft. Through his many aerial engagements, he never returned with a bullet hole in his aircraft. The 100th was one of four black fighter squadrons that made up the 332d Fighter Group. Black military pilots flew more than 15,000 missions while destroying 261 enemy aircraft.

19 July
Eighth and Fifteenth Air Forces team up when the Mighty Eighth launches 1,100 bombers against targets in Germany, including Munich, followed one hour later by the Fifteenth, which launches 400 bombers against targets in the same area. During the raids, more than 1,000 fighter sorties are flown in support of the bombers. For the rest of the month, Munich will be targeted regularly by near-1,000-plane raids.

22 July
As part of Operation FRANTIC, the first shuttle bombing mission flown exclusively by fighter aircraft occurs this date when P-38s and P-51s based in Italy attack an airfield near Ploesti, Romania, and then go on to land at Russian bases.

25 July
Operation COBRA begins this date. After being postponed for 24 hours due to poor weather, 1,495 Eighth Air Force heavy bombers are tasked with saturation bombardment targets in front of Allied-held lines in preparation for the breakout attempt in the VII Corps area. Unfortunately, 35 aircraft drop their bombs on the American side of the forward line of troops, resulting in the deaths of 102 soldiers, including Lieutenant General Leslie McNair, commander of the U.S. Army Ground Forces. Additionally, 380 troops are wounded in the fratricide incident. Ninth Air Force medium bombers are also tasked to provide

saturation attacks for the ground troops and also drop some of their bombs short, killing and wounding more American soldiers. Lieutenant General Doolittle, distraught by the mistakes, is reassured by General Eisenhower that the positive effects of the bombardment far outweighed its consequences.

27 July
The National Advisory Committee for Aeronautics Executive Committee discusses possible uses for robots in the military as well as in other areas.

16 August
A flight of five German Me 163 Komet rocket planes attack a formation of B-17s near Merseburg, Germany. The aircraft could sustain flight for 8–10 minutes and obtain speeds in excess of 590 miles per hour. The rocket-plane attack, the first of its kind, is ineffective. Only 279 Komets are built by the end of the war.

4 August
The first APHRODITE launches occur against V-weapon sites in the Pas de Calais, France. Also known as "Weary Willie," or "Willie Orphan," war-torn heavy bombers—B-17s and B-24s—were modified with an open cockpit arrangement so that the crew—a pilot and weaponeer—could fly the bomber near the coast of England, arm the 20,000 pounds of nitrostarch that had been packed into the airframe, and then parachute to safety as the control ship started to direct the weary bomber from a safe distance behind. A dozen such attempts are made before the heavy bomber part of APHRODITE is cancelled. Glide bombs fitted with radio and TV guidance devices are still used during the project.

4–6 August
Eighth Air Force launches 1,250 heavy bombers in strikes against four oil refineries, four aircraft factories, torpedo factories in Germany, coastal batteries in Calais, two V-weapon sites, two airfields, and the Peenemünde rocket research facility. Raids of this scope demonstrate the capability of the bomber force to carry out a variety of missions concurrently, and similar raids are accomplished for the next two days as well. Shuttle-bombing begins again in addition to the main attack forces that are launched and return to home base. These FRANTIC missions continue until August 12. More than 70 Fifteenth Air Force P-38s and P-51s respond to a Soviet request for AAF air strikes. This is the first such request made by the Soviet leadership.

6 August
Major George E. Preddy Jr., Eighth Air Force P-51 pilot, shoots down six enemy fighters in one mission, within the time span of only five minutes. Preddy was on escort duty for a formation of B-17s enroute to bomb factories near Brandenburg, Germany. Preddy does not survive the war, as he is killed by anti-aircraft fire on Christmas Day 1944. He becomes the highest scoring P-51 pilot of the war, with 24 kills.

Right: *Major George Preddy's P-51 depicted during his six-victory day. ("Cripes A Mighty," by Willie Jones, USAF Art Collection)*

PROJECT APHRODITE

Left: *War-weary B-17s and B-24s were modified as massive bombs. Filled with dynamite, the pilot and the bomb specialist were to parachute to safety once remote control of the aircraft was passed to a trailing mother ship. (USAF/HRA)*

Below: *Glide bombs were also part of Project APHRODITE. (USAF/HRA)*

Although the U.S. Army Air Forces went to war with high-altitude daylight precision bombardment as their primary doctrine, other top secret programs were being developed by Army scientists to produce remotely piloted weapons that might keep American airmen out of deadly flak belts over heavily defended targets.

The most experimental of these was a project known as APHRODITE. Two different types of weapons, a series of glide bombs, and remote-controlled, nitrostarch-filled, war-weary bombers were used during this operation. The bombs, capable of gliding one mile for each thousand feet of altitude, were initially unguided and not very accurate with a computed error of up to one mile from the target. On later models, bombardiers used radio-controlled steering and aimed the weapon by using a television camera mounted on the bomb. This method, of course, was dependent upon clear visual conditions. Standard one- or two-thousand-pound bombs affixed with a small set of wings were employed by the Eighth Air Force from mid-1943 until May 1944. The development of the largely inaccurate glide bomb series of weapons (GB-1 thru GB-8, which included radio-controlled steering, television cameras for aiming, and even a torpedo modification) demonstrated one thing very clearly—General Hap Arnold, Army Air Forces commander, was not completely sold on precision bombing doctrine as it stood, even after the Casablanca Conference in January 1943, where the Combined Bomber Offensive (CBO) became the plan of attack against Germany.

Immediately after the GB series was shelved due to its inaccuracy, radio-controlled "Weary Willy" modified aircraft took to the skies. These surplus bombers occupied valuable space and, more critically, valuable maintenance time. By late 1943, General Arnold had directed Brigadier General Grandison Gardner's Eglin Field engineers to outfit these "Weary

Willies" with automatic pilots. These airplanes, both B-17s and B-24s, could then be remotely flown into enemy targets, such as U-boat construction facilities on the French Coast.

Each orphan B-17 or PB4Y (a naval B-24) was packed with 20,000 pounds of TNT (trinitrotoluene). The pilot took off, flew the plane to the English Channel, the bombardier armed the explosives, and then the two crewmen parachuted out of a hole cut in the fuselage of the bomber. These monstrous bombs were originally to be used against German submarine pens and missile sites. A few blew themselves out of the sky when the explosives went off prematurely. About a dozen actually were guided across the Channel and detonated near targets in France, inflicting little damage but creating huge holes in the ground. APHRODITE, a concept well ahead of the technology of the day, was eventually shelved when the British began to fear reprisals by V-2 rockets as a result of APHRODITE attacks.

These programs demonstrated that the AAF had accepted the fact that precision daylight bombing was not the only way to attack enemy targets from the air. Successful, accurate employment of such "stand-off" weapons would have to wait for decades.

APHRODITE was clearly a nonprecision system of weapons. Yet, Arnold staunchly supported its development even before Germany launched V-1 attacks that began in the early morning hours of June 13, 1944. After these attacks began, Eisenhower ordered Air Marshal Arthur Tedder, his air deputy, to get the upper hand on Crossbow targets at the expense of everything but urgent requirements of the ongoing cross-channel attack. An overreaction considering that Eisenhower himself described German missile attacks only as "very much of a nuisance." Successful APHRODITE missions would have allowed the AAF to claim that they were executing Ike's orders, but at the same time, would not have required the transfer of any additional bomber assets away from attacks into Germany.

Not only were "Weary Willies" capable of carrying large amounts of explosives, using them as guided missiles assured that none would remain in American stockpiles. Arnold remembered the painful Liberty engine lessons from World War I production days.

Project APHRODITE demonstrated Arnold's willingness to supplement precision bombing doctrine in an effort to save the lives of American aircrew. It deflected political pressure away from his air forces. Arnold was confident that the war in Europe was, essentially, under control by late spring 1944.

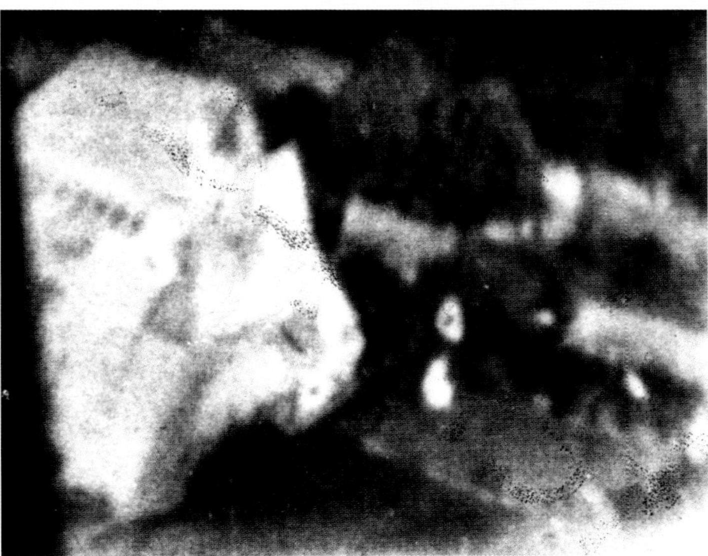

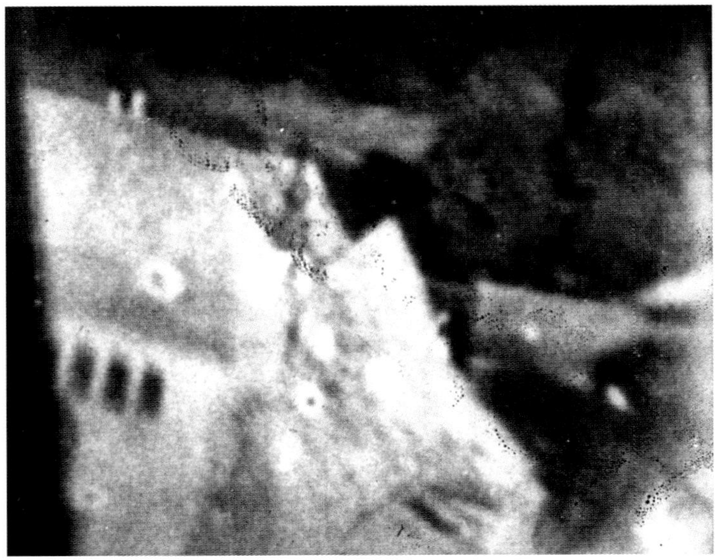

Above (top to bottom): *Televisions mounted on some glide bombs allowed the bombardier to steer the weapon to its target by radio remote control. This sequence, taken by a Polaroid camera in the mother ship, shows such an attack. (USAF/HRA)*

8 August
Lieutenant General Hoyt S. Vandenberg assumes command of Ninth Air Force.

9 August
Captain Darrell R. Lindsey, Ninth Air Force, leads a formation of 30 B-26 medium bombers to destroy the L'Isle Adam railroad bridge over the Seine River in France which the enemy was using to move troops, supplies, and equipment. Despite being hit by heavy flak, and having an engine on fire, he remains at the head of the formation. After completing the mission, he goes down with his plane in order to give his crew the chance to escape the burning plane. Captain Lindsey is awarded the Medal of Honor.

10–14 August
Despite being hampered by poor weather, Twelfth Air Force B-25s, B-26s, and P-47s strike targets on the French and Italian coast west of Genoa. On the 12th, heavy bombers from Fifteenth Air Force join in the strikes in preparation for the invasion of southern France. Eighth and Ninth Air Forces continue to cover operations around Paris.

13 August
As part of Project APHRODITE, two GB-4 television-guided, radio-controlled glide bombs are launched against E-boat pens in LeHavre, France. Additional strikes are launched through September 13.

14 August
Captain Robin Olds, son of famed aviator General Robert Olds who led six B-17s on a long-distance flight to South America and intercepted the ocean liner *Rex* more than 700 miles out at sea in 1938, shoots down his first enemy fighter. By the 4th of July 1945, Olds accumulates 11 more victories. Almost 22 years later, on January 2, 1967, Olds shoots down an enemy MiG in Southeast Asia. This victory makes Olds the only American ace ever to shoot down enemy aircraft in nonconsecutive wars.

14–15 August
Operation ANVIL/DRAGOON is launched in the south of France. As the amphibious convoy approaches the coastline, Twelfth and Fifteenth Air Forces pound defenses in the Toulon-Nice-Genoa area. Hundreds of heavy and medium bombers escorted by approximately 200 fighter escorts operate with impunity along the French Mediterranean Coast. As the invasion begins, Fifteenth carpet bombs the landing beaches and the Twelfth provides deadly close support for the landing forces. By the 17th, Fifteenth Air Force has returned to hammering away at POL targets while the Twelfth continues to provide tactical air support to the ANVIL/DRAGOON invasion forces. This is the greatest one-day effort by MAAF in the war thus far.

18 August
Ninth Air Force B-26s and A-20s, having begun their forward deployment to continental Europe from England, strike fuel and ammo dumps and road and rail chokepoints in an effort to disrupt retreating German ground troops. More than 1,000 fighters fly cover during these operations and provide close support and interdiction for the ground armies as they return to their bases. The noose is tightening around the neck of the German military machine.

25 August
Paris is liberated from Nazi occupation.

28 August
Two Eighth Air Force P-47 pilots, Major Joseph Myers and 2nd Lieutenant Manford Croy Jr., shoot down a German Me-262 in the first aerial victory over a jet aircraft.

8 September
The world's first ballistic missile—the German V-2—is launched in combat. The first explodes in a Paris suburb, and the second in a London suburb a few hours later. Scientist Wernher von Braun developed the missile at the secret research facility at Peenemunde. Von Braun comes to America after the war as part of Operation OVERCAST (renamed Project PAPERCLIP in 1946), where he will continue his work developing rockets for the U.S. More than 100 German scientists and engineers come to Fort Bliss, Texas, in December 1945, under agreements reached through PAPERCLIP.

9 September
While Ninth Air Force bombers fly leaflet drop missions over coastal France and Belgium, more than 700 transport planes deliver supplies, evacuate wounded, and pick up interned Allied personnel across the rapidly expanding theater of operations.

10 September
While more than 1,000 Eighth Air Force heavy bombers attack aircraft, tank, and jet propulsion plants in south central Germany, Ninth Air Force assigns railroad targets in an effort to cut them both west and east of the Rhine. More than 800 transports complete supply and evacuation missions across France as the Normandy invasion force meets the DRAGOON force. Luxembourg is liberated as the advance toward Berlin continues.

The Fairchild C-82, the first World War II aircraft exclusively designed to carry cargo, flies for the first time near Hagerstown, Maryland.

Opposite, above: *General Hoyt S. Vandenberg would become the second Chief of Staff of the independent Air Force. (NASM)*

Opposite, below: *C-47s drop paratroops during Operation DRAGOON, the invasion of Southern France on August 15, 1944. ("Skytrains on Track," by Craig Kodera, USAF Art Collection)*

Above: *A German Me-262 jet fighter passes through an American pilot's gun sight. Moments later the aircraft explodes. (NASM)*

Previous Spread: *Two German Me-262s zoom past a P-38 Lightning over Europe. ("Jagdfieber," by Heinz Krebs, USAFAP)*

Left: *Troops dropped during Operation MARKET GARDEN prepare to secure the Maas River Bridge in the distance while additional troops from the 82d Airborne Division are delivered by AAF C-47s. ("Making It Happen," by James Dietz)*

11 September
FDR and Winston Churchill confer at the second Quebec Conference. Plans for the completion of the European war are finalized and Pacific war plans are discussed.

Troops from the V Corps are first to cross into Germany near Thionville.

12 September
In a rare demonstration of defensive airpower, the Luftwaffe launches approximately 400 fighters against an Eighth Air Force bomber package numbering more than 800 bombers and hundreds of fighters. The bombers suffer 45 losses and about 12 P-51s go down.

14 September
A Douglas A-20 intentionally penetrates a hurricane to collect scientific data. The first "Hurricane Hunters" are Colonel Floyd B. Wood, Major Harry Wexler, and Lieutenant Frank Reckord. They return safely to their home base.

17–30 September
Operation MARKET GARDEN begins. More than 1,500 Allied transports and nearly 500 gliders carry 20,000 troops of the First Allied Airborne Army to the Netherlands. Their initial objective is to secure the Rhine River bridges at Arnhem to secure the axis of advance for the British Second Army. A second wave drops the next day, and the air battle is intense around Arnhem as 16 B-24s and 21 fighters go down during operations. For the next two weeks, Allied airpower provides support to the First Allied Airborne Army by bombing targets throughout the Netherlands.

20 September
Republic rolls out the 10,000th P-47 Thunderbolt at Farmingdale, New York. It would take another 10 months to build the next 5,000 of these durable, multipurpose fighter aircraft.

21 September
As the offensive continues, many Eighth Air Force B-24s become gasoline tankers and begin delivering fuel to ground and air forces in France. More than 80 such sorties are flown on this date. More than 100 are flown on the 22nd. By the 28th of September, nearly 200 B-24s will be delivering fuel to advancing forces in France while B-17s and other B-24s continue to attack German fuel resources.

1 October
With the assignment of the 5th Photo Group, Reconnaissance, Fifteenth Air Force, reaches its full wartime authorization of 21 heavy bomber groups, seven fighter groups, and one reconnaissance group.

2 October–2 November
During the next 30 days, when not stopped by poor weather, Eighth Air Force will launch 12 raids of more than 1,000 bombers and another six raids of more than 450 bombers. Fighter escorts accompany each mission, and between five and 17 groups of "little friends"

participate. Priority targets remain airfields, oil production and refineries, motor works, and munitions plants. Targets surrounding the city of Cologne are hit particularly hard and often during the month. German defenses are inconsistent, but are still formidable when launched in mass. On October 7, 52 bombers and 15 fighters are shot down. Some of the bombers fall victim to new Me 262 jet fighters. Not a superweapon, four of the enemy Me 262 jets are shot down on that same day.

12 October
In support of Operation PANCAKE, Twelfth Air Force sends 700 heavy bombers to strike ammo and fuel dumps, barracks, vehicle repair facilities, and munitions factories. Another 160 P-51 Mustangs strafe rail, airfield, and river targets supporting the U.S. Fifth Army offensive near Bologna.

14 October
Allies liberate Athens, Greece.

Field Marshal Edwin Rommel commits suicide.

1 November
At Caltech, the nation's first rocket research and development center is reorganized and renamed the Jet Propulsion Laboratory (JPL). The JPL will become the center of American rocket development during the early years of the Cold War.

2 November
During a 1,100-plane raid at the synthetic oil plant at Merseburg/Leuna, approximately 500 enemy fighters attack wave after wave of American bombers. Although 17 fighter groups provide escort for the mission, 40 bombers and 28 fighters are lost. American fighter pilots claim more than 150 victories during the massive air battle.

2nd Lieutenant Robert E. Femoyer, Eighth Air Force, is tasked with navigating a bomber over Merseburg, Germany. The bomber is hit by three enemy anti-aircraft shells, and Femoyer is severely injured. However, he refuses an injection of morphine in order to keep his mind clear enough to navigate through enemy flak belts. He dies of blood loss and shock shortly after guiding his crew to safety. Femoyer is one of two navigators to receive the Medal of Honor during the war.

4 November
Twelfth Air Force sends more than 300 medium bombers to bomb rail lines and roads in Brenner Pass and west of the Po Valley. Four P-47 Thunderbolt fighter/bombers attack a Milan hotel where it is believed that Hitler is staying.

Right: *Two Eighth Air Force P-47s provide cover for the 381st Bomb Group over Europe. ("Big Friends—Little Friends," by Paul Jones, USAF Art Collection)*

Left: *The Boeing XC-97, derived from the basic B-29 airframe, served as both a cargo plane and aerial tanker until after the Korean War. (NASM)*

Opposite: *Defended by more than 700 guns, the last operating oil plant in Germany is attacked by more than 1,000 B-17s. Fifty-six B-17s were lost. ("Leuna Oil Refinery, Merseburg, November 30, 1944," by Alfred Vetromile, USAF Art Collection)*

5 November
In their largest operation against one single target during the war, Fifteenth Air Force sends 500 B-24s and B-17s to bomb the Vienna/Florisdorf oil refinery. Nearly 140 P-38s and 200 P-51s provide escort and support for the mission.

7 November
General Hap Arnold tasks Professor Theodore von Kármán to investigate the potential of airpower for the future. The report and recommendations of Kármán's report, *Toward New Horizons*, will form the basis of the scientific foundation of the future Air Force.

9 November
As Third Army launches a full-scale ground attack on Metz, Eighth Air Force sends more than 1,100 heavy bombers to attack targets near Metz, Thionville, and Saarbrucken. Eleven fighter groups escort, but still 40 bombers and fighters are lost during the operation.

1st Lieutenant Donald J. Gott, Eighth Air Force, is pilot of a B-17 on a bombing run at the marshalling yards at Saarbrucken, Germany. During the mission, the aircraft is severely damaged by anti-aircraft fire—one engine is completely blown off the aircraft—and the radio operator's arm is nearly severed. In order to save the radio operator, Gott attempts a crash landing in Allied-occupied territory. Despite Gott's attempt to land successfully, the damage is too severe, and the aircraft explodes, killing those who remain on board and the tail gunner, whose parachute becomes tangled in the aircraft empennage. Gott is awarded the Medal of Honor.

2nd Lieutenant William E. Metzger Jr., Eighth Air Force, is B-17 co-pilot on a bombing run at the marshalling yards at Saarbrucken, Germany. He remains with the pilot and a wounded radio operator after completing their mission over Saarbrucken, Germany, in order to attempt a crash-landing on friendly territory. Their aircraft had been severely damaged by flak and was leaking fuel and finally explodes before a landing attempt can be made. Metzger and his pilot, radio operator, and tail gunner are killed in the fireball. Metzger receives the Medal of Honor for his heroic actions.

12 November
Fighter pilot combat tour length is set at 270 flight hours.

15 November
Army Ordnance initiates a program of research and development of ballistic missiles. The project is called *Hermes* and begins with a prime contract issued to

the General Electric Company. Plans are also made under the *Hermes* program to study captured German V-2 rockets.

The Boeing XC-97 Stratofreighter makes its first flight.

16 November
More than 4,000 Allied aircraft drop over 10,000 tons of bombs in front of the First and Ninth Armies in preparation for a major ground offensive.

21 November
During another 1,000-plus plane raid over oil targets at Merseburg/Leuna, Eighth Air Force bombers are hit hard despite 16 groups of fighters as escort. Approximately 35 bombers and fighters go down during the fighting. In a similar raid four days later, Pathfinder bombers will lead a force of 900 bombers over these same targets. Weather is extremely poor and more than 65 aircraft fail to return to their home airfields. It is later determined that many had diverted to emergency landing fields in Allied-occupied territory.

27 November
During an Eighth Air Force raid against targets in the Magdeburg-Munster-Hannover area, approximately 750 German fighters are observed by American aircrew. This is the single largest aerial defense mounted by the Luftwaffe during the war, demonstrating the ability to concentrate forces when needed.

30 November
The Mighty Eighth sends 1,200 heavy bombers against four synthetic oil plants at Bohlen, Zeitz, Meresburg/Leuna, and Lutzkendorf. A total of 19 fighter groups (16 from the Eighth and three from the Ninth) escort the package. Flak is intense and brings down 29 bombers. Another dozen are shot down by enemy fighters.

1–16 December
Only 11 months after the establishment of Project ORDCIT, two-dozen Private A rockets are launched by JPL at Camp Irwin, California.

5 December
Launching more than 500 bombers against targets over Berlin and Munster, Eighth Air Force escort fighters meet 300 Luftwaffe fighters in heated aerial combat and claim dozens are destroyed.

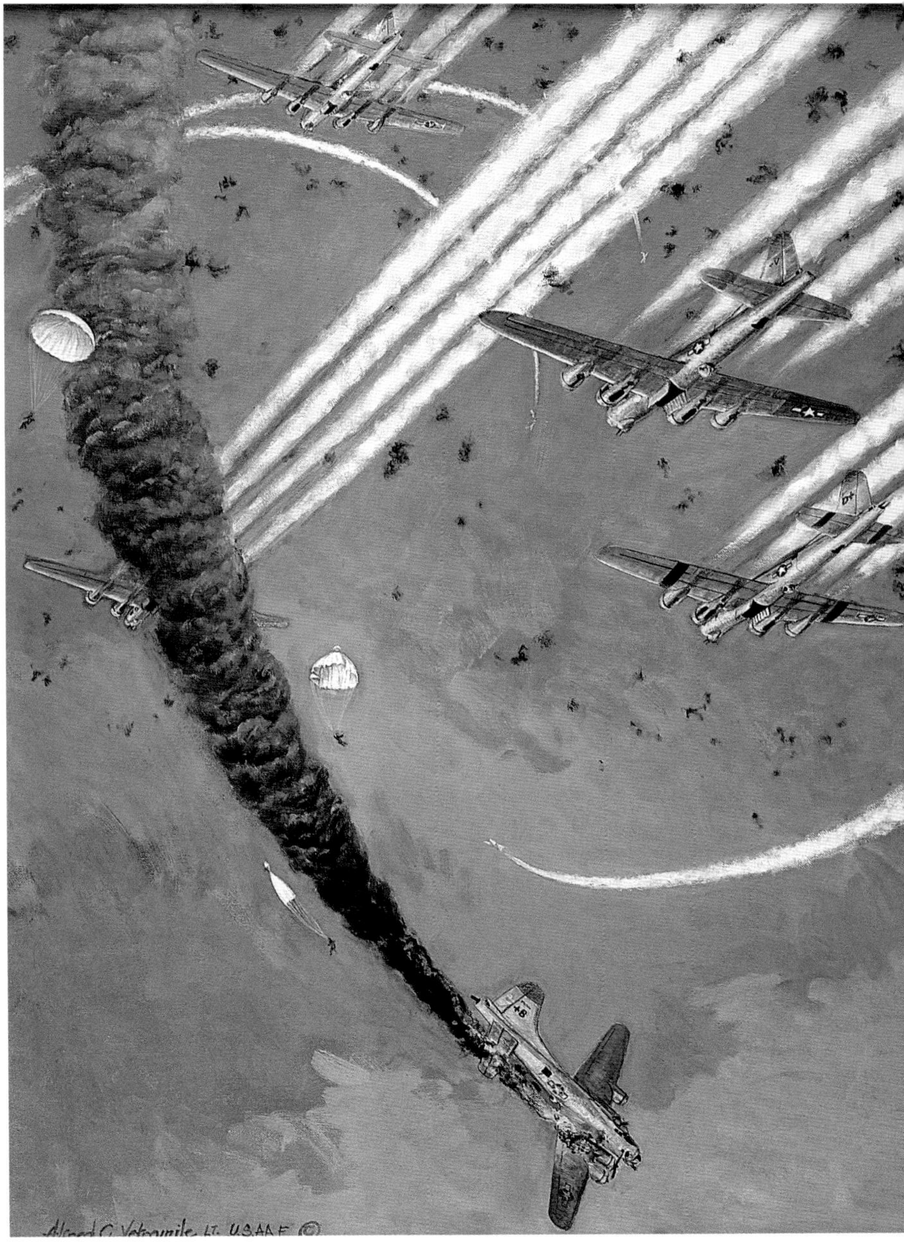

15 December
Army Major Glenn Miller, famous band leader, departs from France bound for England as a passenger aboard a Noorduyn C-64 Norseman aircraft. His aircraft disappears and no wreckage is ever found.

FDR signs legislation authorizing the five-star ranks of General of the Army and Admiral of the Fleet.

16 December–28 January
The Battle of the Bulge—the Ardennes Offensive. In a surprise night attack, two German Panzer armies strike U.S. First Army front line units. Unprepared for the attack, one division is virtually destroyed and others retreat. The ensuing battle is one of the most intensely fought in the history of the U.S. Army.

17–18 December
Although no Ninth Air Force bombers fly, more than 1,000 fighters launch against the German counter offensive in the Ardennes Forest. Ninth Tactical Air Forces attack the leading edge of the enemy troops by strafing and bombing Field Marshal von Rundstedt's Panzer tanks. After the 18th, poor weather grounds most planes until December 23.

21 December
General Hap Arnold is promoted to five-star rank—General of the Army.

23 December
Weather finally allows Ninth Air Force to launch some 500 B-26 and A-20 sorties in support of the engaged Allied forces at the Bulge. Although 31 bombers are shot down, many more German fighters are destroyed by the escort forces.

24 December
In the largest Eighth Air Force operation to date, more than 2,000 heavy bombers are dispatched to targets across Europe. Eleven airfields, 14 communications centers, five cities, and dozens of targets of opportunity are bombed by more than 1,900 of the original force. Thirteen Allied fighter groups in an escort role—more than 1,000 fighters—meet 200 enemy fighters, shooting down about 25 percent of the enemy planes.

Brigadier General Frederick W. Castle, Commander, 4th Bomb Wing, Eighth Air Force, dies in the skies over Belgium after his crippled B-17 Flying Fortress is shot down by enemy fighters. Brigadier General Castle, leader of more than 2,000 heavy bombers, initially lost an engine but in attempting to keep from endangering friendly ground

Opposite: *This Army five-star flag belonged to General of the Army Henry "Hap" Arnold and is now on display at the National Air and Space Museum, Steven F. Udvar-Hazy Center, near Dulles International Airport. (NASM)*

Right: *"Bastogne," by Olin Dows. (Army Art Collection)*

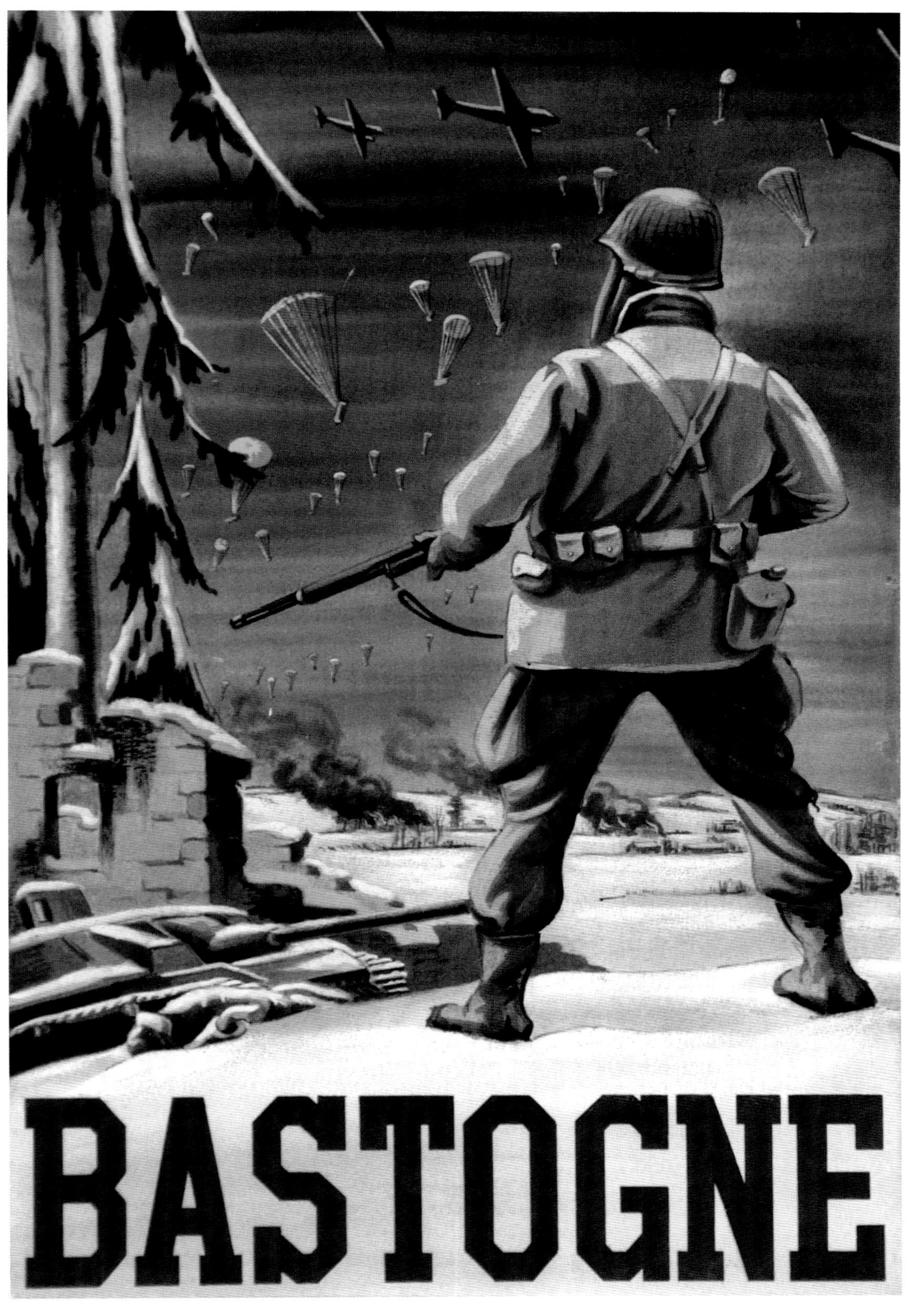

troops, lost maneuverability. Left vulnerable to enemy fighters, he remained at the controls of his aircraft to afford four of his crewmembers an opportunity to escape. Castle had been with the Mighty Eighth since its first days in England and was flying his 30th combat mission. In addition to being awarded the Medal of Honor, Castle Air Force Base in Merced, California, is named in his honor.

Ninth Air Force continues to fly support sorties for the III, VIII, and XII Corps along the south flank at the Bulge. These missions continue as weather permits for the remainder of the month, including direct support for the U.S. 4th Armored Division as it breaks through to the besieged 101st Airborne Division at Bastogne—surrounded since December 22.

1945

1 January
Eighth Air Force's three Bomb Divisions are redesignated as Air Divisions. During the month, weather allows, hundreds of bombers and fighters are sent to attack targets throughout Europe. Transportation targets (bridges over the Rhine River, rail junctions, and tank and troop concentrations in the western sections of Germany) are added to the target list with increased priority. Hundreds of escort fighters accompany the bombers on every mission, striking lines of communications as they return to base. Weather throughout the continent limits bombing operations throughout the winter months.

The German Air Force attacks AAF and Allied airfields in Brussels, Eindhoven, and Metz. More than 120 Allied aircraft are destroyed on the ground. It is estimated that of the 700–800 attacking aircraft, some 460 are shot down by Allied fighters and anti-aircraft fire.

31 January
After nine days grounded by winter weather, Fifteenth Air Force finally gets more than 670 B-17s and B-24s in the air to bomb oil refinery targets at Moosbierbaum and marshalling yards at Graz and Maribor. Escort duties are performed by P-51 and P-38 fighters while more B-24s drop supplies to ground forces in northern Italy. Another 300 bombers make it over the same oil plant the next day.

3 February
In a typical good-weather, daylight raid for Eighth Air Force, more than 1,200 B-17s and B-24s launch to attack marshalling yards in Berlin as well as a synthetic oil plant and transportation targets in Magdeburg. More than 900 escort fighters provide top cover. The mission against Berlin, which includes 959 of the dispatched bombers, is the largest of the war thus far. The Mighty Eighth will deliver a similar blow to the same target area three days later.

5 February
Twelfth Air Force is busy striking a variety of tactical targets in the Po Valley and fly more than 270 medium bombers against targets in northern Italy. Fifteenth Air Force sends more than 730 heavy bombers to take out petroleum storage tanks at Regensburg as well as communications targets. Huge numbers of P-38s and P-51s escort the bombers and also cover B-24s delivering supplies to Yugoslavia. Two days later, Fifteenth will use a similar number of bombers and fighters to attack eight different oil production facilities.

4–11 February
The United States, Great Britain, and the Soviet Union issue the Yalta Declaration as a result of the Crimea Conference—more commonly referred to as the Yalta Conference. This declaration establishes the methods for world organization following the end of the war, to include the "dismemberment of Germany." It is also agreed that a few months after Germany surrenders and hostilities end in Europe, the Soviet Union shall enter the war against Japan while at the same time concluding a pact of friendship and alliance with China.

7 February
The Consolidated Vultee XP-81 prototype makes its first flight. This aircraft includes a conventional turboprop in the nose and a turbojet in the aft fuselage.

13–15 February
Allied bombers bombard targets in Dresden, Germany. Late on the 13th, RAF aircraft carefully mark the

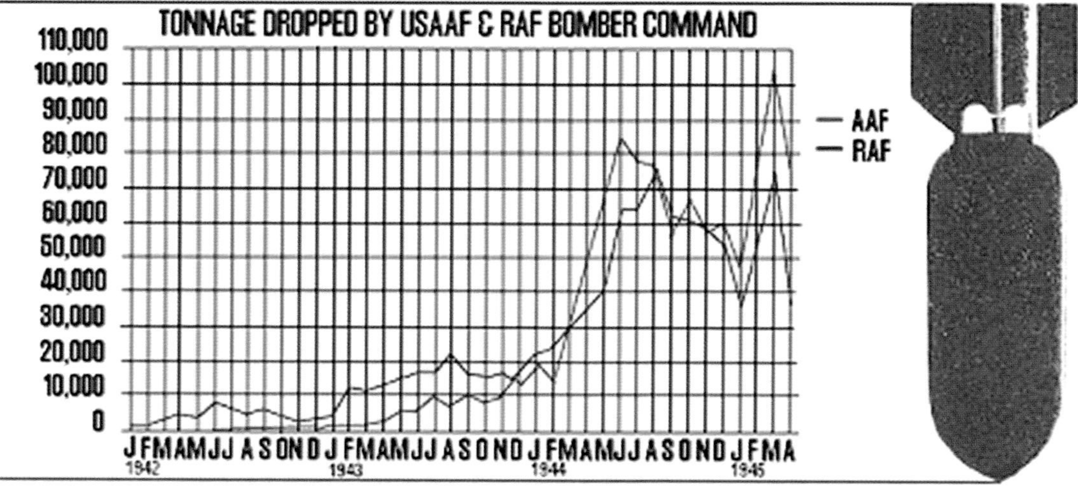

Left: *Bombing tonnage increased dramatically as D-day approached, and continued for some time after the invasion began.*

Opposite: *P-51s and Fw 190s battle over Germany in February 1945. ("Aces Meet," by Loren Blackburn, USAF Art Collection)*

target for the attacking bombers who strike just after 1:00 a.m. on the morning of the 14th. American B-17s continue the bombardment during the day after launching more than 1,300 heavies to attack several marshalling yards and oil depots. Sixteen groups of escort fighters fill the skies over Germany during these raids, which continue again that night and the following day. The bombing starts a firestorm in the city, killing 25,000.

20 February
Henry Stimson, the Secretary of War, approves plans to create a rocket test area near White Sands, New Mexico.

22–23 February
Eighth, Ninth, Twelfth, and Fifteenth Air Forces participate in Operation CLARION—an Allied attempt to systematically destroy all forms of transportation available to Germany throughout occupied territory in Europe in 24 hours. Approximately 9,000 aircraft launch strikes from bases in England, France, Holland, Belgium, and Italy over an area covering 250,000 square miles. Eighth launches more than 1,350 bombers against more than 40 different targets; Ninth launches some 1,500 medium bombers and more than 800 fighter/bomber sorties; Twelfth sends aircraft although still committed to supporting the U.S. Fifth Army; Fifteenth sends more than 350 bombers and more than 300 fighters to more than 50 specific targets in southern Germany and Italy. Additional missions against similar targets are flown on the 23rd in anticipation of the crossing of the Rhine River into Germany by Allied ground forces.

25 February
The Bell XP-83 flies for the first time. This pressurized turbojet aircraft evolves from Bell's earlier P-59 Airacomet—America's first jet aircraft.

March
The NACA publishes *Summary of Airfoil Data*, by Ira H. Abbott, A.E. von Doenhoff, and Louis Stievers from the Langley Laboratory. It becomes the classic reference for airfoil data for the NACA.

Plans for Project PAPERCLIP, the recruitment of German missile scientists, begin in the Pentagon.

11 March
In an AAF raid on Essen, the AAF drops the greatest tonnage on one raid during the war. More than 1,000 bombers release 4,738 tons of bombs on the town.

THE AIR ATTACK ON DRESDEN, FEBRUARY 13–15, 1945

Tami Davis Biddle

The Anglo-American air attack on the city of Dresden, carried out on February 13-15, 1945, has made its way powerfully into the popular memory of the Second World War. It is, paradoxically, one of the most frequently mentioned but least understood events of that war.

Late on the night of February 13, British Lancaster and Mosquito marker aircraft began dropping target indicator bombs across Dresden to guide subsequent bombers to their aim points. The target marking was so tight and effective that the Lancaster bombers of Bomber Command's No. 5 Group could readily locate the glow of the red beacons. The 244 bombers met with little German resistance, and the uncontested air space allowed pilots to come down low and make careful, concentrated bomb runs. High explosive bombs blasted structures into combustible bits that were ignited by accompanying incendiary bombs. Though the first wave of bombers stayed over the city only briefly, their precise work seeded intense fires that were fanned by steady westerly winds.

Just as the first target markers began to fall over Dresden, a second group of 550 heavy bombers was taking off from Britain, headed for the beleaguered German city. They too arrived largely unmolested. When the second wave's master bomber arrived over Dresden at about 1:20 A.M. on the 14th, he discovered a city so consumed by fire that smoke obscured his aim points. Crews mostly bombed blind through the fire and smoke, thus extending the area of the fire. Just before 2:00 A.M. the last of the bombers departed, leaving behind fires that could be spotted from 100 miles away. So intense were the flames that they pulled oxygen in from all directions, creating the fierce winds that characterize a firestorm. Material around the center of the conflagration was subject to spontaneous combustion: the heat was so intense that actual contact with flame was unnecessary. Carbon monoxide seeped into shelters to poison those who sought refuge beneath the ground.

Shortly after noon on the 14th, B-17 bombers of the U.S. Eighth Air Force's 1st Air Division approached the city. This raid on Dresden originally had been scheduled to precede the British attack, but had been postponed due to weather. It was one of three American air attacks in the region that day; targets in Magdeburg and Chemnitz were also hit. Nine of 12 1st Air Division Groups reached Dresden; the other three bombed the city of Prague by mistake. Some crews were able to bomb the Dresden railway marshalling yards, while most others, again inhibited by smoke, bombed on instrument and thus scattered their bombs widely across the city. All told, 311 B-17s of the 1st Air Division dropped 771 tons of bombs, including 294 tons of incendiaries, on Dresden. A day later (February 15), 210 B-17 bombers that had failed to reach their designated target—a synthetic oil plant—bombed Dresden as a "secondary" target. They dropped another 461 tons of bombs on the ailing city.

In their execution, these attacks were no different from other Anglo-American air attacks carried out in the same period. Dresden contained military targets, and it met the fate that had befallen many other German cities. In terms of lives lost and damage done, the Dresden raid was less destructive than the American air attack on Tokyo waged on the night of March 9–10, 1945—an attack that most Americans living today barely know. And it was less destructive than the devastating firestorm imposed by Bomber Command on the city of Hamburg in late July 1943. Recent, careful accounting has revealed that the death toll at Dresden was about 25,000.

But while the operational and tactical aspects of the Dresden raid resembled other Anglo-American raids carried out at the time, it was an unusual air attack in two important respects. First, it was intended—along with attacks on several other cities in eastern Germany—to aid the advance of the Red Army at a time when the advance of that army appeared to be of critical importance for a timely conclusion to the war in Europe. And second, it saw the falling away of the last remaining constraints around targeting for aerial bombing in Europe during the Second World War.

The catalyst for the Dresden raid can be traced to the autumn and winter of 1944, when a series of setbacks unnerved an Allied High Command that had been confident and optimistic following the Normandy breakout earlier in the year. The vague instructions that guided the Dresden raid stated, ". . . where heavy attack will cause great confusion in civilian evacuation from the east and hamper movement of reinforcements from other fronts." The blurring effect of euphemistic phrases created a space in which moral dilemmas could be avoided. But it meant that the Allies would block movement and reduce the resources available to the Germans for warfighting; it meant that the Allies were prepared to use the presence of large numbers of refugees on the Eastern Front as a lever against the Wehrmacht—as a kind of human wall to hinder the Wehrmacht's ability to conduct efficient maneuver warfare.

Left: *B-17s dropped tons of incendiary bombs on the German city of Dresden during a round-the-clock operation with Royal Air Force night attacks. (USAF)*

Opposite: *A firestorm that consumed much of the city also left approximately 25,000 dead. (National Archives)*

Left: *Lieutenant John Kirk bags a jet-powered Me-262 over Germany. ("Small Boy Here," by John Amendola, USAF Art Collection)*

Below: *"Glider Assembly Area," by Olin Dows. (Army Art Collection)*

14 March
In Italy, AAF bombers in cooperation with Russian war planners strike targets in Austria, Hungary, and Yugoslavia in support of the Red Army.

21 March
Lieutenant John Kirk, at 25,000 feet on an escort mission over Rutland, Germany, gives chase after a diving Me 262 that had just fired at the B-17 formation. Accelerating past 500 miles per hour, Kirk engages and disables the German jet. The pilot bails out at 10,000 feet. The P-51, *Small Boy Here*, was never hit in combat nor did it ever abort any mission of the 60 flown by the aircraft during the war. Ironically, it was the last American aircraft to crash in Europe when a replacement pilot misjudged a landing after a mission flown on V-E Day.

24 March
After one month of generally good weather which allowed routine bombing missions numbering more than 1,000 heavy bombers, Allied ground forces cross the Rhine River into Germany. In support of this operation, approximately 7,000 sorties are flown by Eighth and Ninth Air Forces. The attack strikes rail yards and bridges, flak positions, communications centers, and many other targets in support of the Allied offensive—also known as PLUNDER-VARSITY. Additionally, more than 2,000 transports and gliders drop two Allied airborne divisions on the other side of the Rhine, near Wesel, Germany. Although one of the most successful airborne drops in history, some 50 aircraft and 11 gliders are shot down during the offensive. Ninth Air Force continues support missions for the Allied ground forces as they push forward into Germany.

Bombers from the Fifteenth Air Force strike Berlin for the first time, sending 150 B-17s to drop more than 350 tons of bombs on industrial targets in the city.

April
Supersonic wind tunnel tests of sweptback wing sections are accomplished at the Aberdeen Proving Ground. Theodore von Kármán suggests such tests be run at Mach numbers approaching 1.75.

1–13 April
Seventeen Private F rockets are fired by JPL at the Hueco Range at Fort Bliss, Texas.

Right: *A 4th Fighter Group P-51 shoots down an Me 262 during an escort sortie. ("Mission to Wesendorf, Germany—4 April 1945," by John McCoy, USAF Art Collection)*

4–11 April
During the coming week, Eighth Air Force bombers relentlessly pummel targets throughout Germany. More than 1,000 bomber raids are launched each day against ordnance depots, armament factories, airfields, aircraft plants, rail yards, industrial complexes, oil storage facilities, marshalling yards, headquarters buildings, and jet aircraft operating bases. Enemy jet fighters attack formations regularly but only in small numbers, which limits their effectiveness as the bombers are escorted by hundreds of P-51 escort fighters. During operations on the 10th, 10 U.S. bombers are shot down by jet fighters, the highest total in one day brought down by jets, but 20 Me 262s go down in the sights of American pilots. Hundreds of German aircraft are destroyed on the ground during several of these raids.

12 April
President Franklin Delano Roosevelt dies. Harry S. Truman takes the oath of office as the new President and Commander-in-Chief.

15 April
In the largest operational effort of the war for this command, Fifteenth Air Force sends 830 B-17s and B-24s to attack gun positions, supply dumps, troop concentrations, and headquarters areas along the highway that leads from Bologna. P-38s escort the formations. Another 300-plus strike bridges and ammo dumps in northern Italy. Hundreds of fighters fill the skies, providing close support of the U.S. Fifth Army as they march north toward Germany. During this 24-hour period, 1,142 bombers strike targets. This day marks the greatest number of fighters and bombers launched for attack and the heaviest bomb tonnage dropped by the Fifteenth during the war. Concentrated raids on transportation and POL targets continue during the next six days.

Eighth Air Force bombers drop napalm weapons on German ground forces in pillboxes, tank trenches, and artillery batteries. Even though more than 850 bombers drop these weapons, they are considered ineffective. This is the sole operational employment of napalm by Eighth Air Force bombers during the war.

21 April
Soviet forces reach Berlin.

21–26 April
Twelfth Air Force A-20s and B-26s press the attack upon retreating enemy forces. By the 25th, all combat operations are aimed at plugging retreat routes and disrupting transportation in the north Po Valley. During the attacks, more than 1,000 vehicles are destroyed, bridge spans are dropped, and airfields are strafed.

Left: *Red-nosed Mustangs at their home field in England. ("Air Force Field at Debden, England—April 1945," by John McCoy, USAF Art Collection)*

24–25 April
1st Lieutenant Raymond L. Knight, Twelfth Air Force, shows remarkable courage and dedication while leading low-level strafing missions over a two-day period in his P-47 Thunderbolt fighter-bomber aircraft. Knight's flight leadership results in the destruction of 14 grounded enemy aircraft and the wreckage of 10 others. During the second day's attacks, his plane already crippled by anti-aircraft fire, he continues to press the attack and refuses to abandon his plane or his wingmen. Instead he attempts to guide his fighter back to a friendly area. His P-47 finally becomes uncontrollable when it hits severe mountain turbulence and breaks apart. Knight is killed in the ensuing crash. Knight's was the last aviation Medal of Honor awarded during World War II.

25 April
The Mighty Eighth flies its last combat mission against industrial targets in the war. Around 275 B-17s strike the armament works at Plzen-Skoda, Czechoslovakia, while a similar number of B-24s hit a transformer near Traunstein and other nearby targets.

27 April
By this date, P-51s, B-17s, and B-24s used as replacement aircraft in Europe stop flowing into combat units. Aircraft strength authorization for bomber groups is reduced from 68 planes to 48 planes, while fighter group strength is reduced from 96 planes to 75 planes. This is the first step in the massive demobilization that is about to occur in the European Theater.

28 April
Allied ground forces take the town of Venice, Italy. Italian partisans capture and hang Benito Mussolini and his family.

30 April
Adolph Hitler commits suicide in his chancellery bunker.

1 May
In the face of poor weather, Fifteenth Air Force B-17s bomb the main marshalling yard at Salzburg with P-38 and P-51 escort. This mission is the final Fifteenth Air Force bombing raid flown during World War II.

1–7 May
A massive humanitarian mission—Operation CHOWHOUND—is launched as Allied bombers drop nearly 8,000 tons of food to starving civilians in the Netherlands. The drops are made in open areas and airfields by agreement with the Germans. Eighth Air Force sends about 400 B-17s each day to a variety of locations. British Avro Lancaster bombers had begun delivering food on April 28.

Right: *Lieutenant General Bedell Smith countersigns the formal German surrender at Rheims, France, on May 7, 1945. (National Archives)*

Below: *A combat veteran P-51D on patrol in the skies over Europe. (NASM)*

2 May
German forces surrender in Italy.

Wernher von Braun and the rest of his V-2 technical group surrender to American forces in Germany near the Austrian border. Many are transferred to Fort Bliss, Texas, to continue their work on V-2 technology. Later, some are relocated to Huntsville, Alabama, where they assist in the development of America's missile forces. The Redstone and the Saturn V launch vehicles are developed under von Braun's guidance.

5 May
Ground forces of the Red Army occupy the German rocket test center at Peenemünde. Since June 1944, some 20,000 V-1 and V-2 weapons had been fired by the Germans. It is estimated that about 1,100 V-2 rockets had been successfully fired against England and around 1,600 had been fired against targets in continental Europe.

7–8 May
The German High Command surrenders unconditionally to Allied forces in a small school building at Reims, France. The surrender is to be effective on May 9. President Truman declares Victory in Europe (V-E Day) as May 8.

8 May
V-E Day
All Fifteenth Air Force combat operations cease on V-E Day. All operations after today are related to transport, supply, and training missions. Twelfth Air Force continues to fly evacuation and supply missions throughout the theater. Ninth Air Force aircraft are tasked to fly "demonstration missions" over a variety of once hostile target areas and also over liberated concentration camps. By the end of the month, the redeployment of most air forces in Europe is proceeding at full speed.

10 May
Lieutenant General Jimmy Doolittle is relieved of his Eighth Air Force command and reassigned to HQ AAF in Washington, D.C. Major General William E. Kepner replaces him.

Left: *The first atomic detonation in history—the Trinity Test—occurred in the New Mexico desert on July 16, 1945. (USAF)*

Opposite: *General George Catlett Marshall, here as a lieutenant general. (NASM)*

18 May
Within the organization of Eighth Air Force, Troop Movement Section is organized under the Director of Operations. They will control all redeployment movements.

19 June
Dr. Frank L. Wattendorf, an engineer at Wright Field and a member of the AAF Scientific Advisory Group (later Scientific Advisory Board), recommends to the Chief of the Engineering Division that a major Air Force Development Center be built far away from facilities in Dayton, Ohio, and located near a large, cheap source of power. This center would include facilities for the development of supersonic aircraft and missiles. The center becomes a reality, and in 1950 is dedicated as the Arnold Engineering Development Center in Tullahoma, Tennessee.

25 June
Ground is broken and construction begins at the White Sands Proving Grounds in New Mexico. Dozens of captured V-2 rockets as well as the first American large-scale, liquid-fuel rockets will be tested there in the coming years. The center officially opens on July 13.

26 June
The United Nations Charter, the constitution of the United Nations, is signed by the 50 original member nations in San Francisco. The charter will enter into force on October 24, 1945. The five founding members are the United States, China, the Soviet Union, the United Kingdom, and France.

30 June
A summary of production for the "Arsenal of Democracy" shows that between July 1940 and July 1945, U.S. industry produced 297,000 aircraft, 86,338 tanks, and 17.4 million rifles, carbines and pistols.

16 July
A nuclear device is tested deep in the desert of New Mexico. It is a "Fat Man"–type of weapon rather than the less complicated version that is eventually dropped over Hiroshima.

16 July–2 August
The Big Three leaders meet in Potsdam, Germany. President Truman, Soviet Premier Stalin, and British Prime Ministers Churchill and Atlee disagree about postwar Europe arrangements. The war in the Pacific is also discussed during the conference held near Berlin.

28 July
A B-25 Mitchell bomber flying in dense fog crashes into the 79th floor of the Empire State Building in New York. Nineteen are killed and 29 more are hurt.

10 August
Dr. Robert H. Goddard, American rocket pioneer, dies at age 63 in Baltimore, Maryland. In 1926, Goddard launched the world's first liquid-fuel rocket.

September
The first U.S. jet aircraft, the Bell XP-59 Airacomet, goes on exhibit at the Smithsonian Institution. Today, the XP-59 hangs in the Milestones of Flight Gallery in the National Air and Space Museum.

5 September
The Douglas C-74 prototype, the Globemaster, flies for the first time at the Douglas Plant located in Santa Monica, California.

8 September
William F. Durand, one of the original 1915 NACA members, retires.

26 September
The Army WAC Corporal is launched on its first development flight. The missile establishes a U.S. record of 43.5 miles height, and is the first liquid propellant rocket developed with government funding. It is fired at the White Sands range in New Mexico.

11 October
The first U.S.-built ballistic missile is launched at the White Sands Proving Ground near Alamogordo, New Mexico. A Tiny Tim booster with a WAAC Corporal rocket reaches an altitude of 43 miles.

20 October
Lieutenant General Nathan F. Twining is commander of a flight of three B-29s which fly from Guam to Washington D.C. via India and Germany—a new route to the American capital. The 13,000 mile journey takes just under 60 hours.

24 October
The United Nations' Charter goes into effect.

7 November
A remote controlled version of the P-59 jet fighter is flown by the Bell Aircraft Corporation. A television was affixed to the inside of the cockpit to read the flight instruments while in the air.

18 November
General of the Army, George Catlett Marshall, a graduate of the Virginia Military Institute, retires from active service. General of the Army Dwight D. Eisenhower replaces him the next day.

20 November
The Nuremberg Trials begin. The Trial of Major War Criminals Before the International Military Tribunal (IMT) tried 24 Nazi leaders as well as a host of lesser war criminals during the second phase of the legal proceedings. The trials lasted until 1949. Original indictments were made earlier in October.

A B-29 sets a world nonstop, non-refueling distance record of 8,198 miles during a flight from Guam to Washington, D.C. The flight takes just over 35 hours.

29 November
The Army Air Forces School moves from Orlando, Florida, to Maxwell Field in Montgomery, Alabama. The school was assigned to the AAF as a major command, later becoming the Air University.

3 December
The 412th Fighter Group at March Field, California, becomes the first AAF unit to be equipped with jet fighters—the Lockheed P-80.

14 December
Bell Aircraft is awarded a contract for the development of a supersonic, rocket-powered, flight research aircraft. Three Bell X-2s are eventually built to fulfill the contract requirements.

17 December
General Carl "Tooey" Spaatz accepts the Collier Trophy from President Truman for "demonstrating the airpower concept" in the air war over Europe.

19 December
President Truman submits a plan to Congress for the unification of the armed forces.

WORLD WAR II: AIR WAR IN THE PACIFIC

1944–1945

R
27

WORLD WAR II: AIR WAR IN THE PACIFIC

1944–1945

The Boeing B-29 Superfortress became the chosen instrument for strategic bombing in the Pacific Theater. The B-29 was originally intended for use in Europe when Army Air Corps leaders feared a German victory over England and anticipated that operating bases for air attacks against Germany might be difficult to come by. When the Royal Air Force successfully defended Great Britain against the German onslaught in 1940, the B-29 was redesignated for the Pacific Theater. The aircraft's tremendous battle range would be required if the Allies were to attack the Japanese homeland.

For much of 1942 and 1943, however, air operations in the Pacific were predominantly tactical in nature—close air support, interdiction, airlift, and reconnaissance—many at night or in poor weather. Theater operations stretched from the Aleutian Island chain to Australia and from China to Hawaii—covering approximately half the globe. Operations were often planned and flown as a joint force, particularly Thirteenth Air Force operations in the central Pacific. It was essential for all air assets, Army, Marine Corps, and Navy to know what the others were doing. The Island Hopping Campaign made such cooperation essential. Often, AAF fighters flew top cover for Navy dive bombers while AAF aircraft prepared the landing beaches for Allied amphibious forces. Deep in China, Air Transport Command made a viable defense against the invading Japanese possible by flying tons of supplies over the Himalayas—an extremely dangerous mission. Operations in the Pacific were far more "joint" than in Europe. To compound matters, air operations often flown with Allies—Australia, New Zealand, India, and China—played significant roles in fighting both the air and the ground wars in the Pacific Theater.

Strategic air operations against Japan started relatively late in the war and differed significantly from air operations over Europe. While tactical air operations generally were tied to forces taking "one damn island after another," strategic missions depended upon obtaining airfields close enough to Japan to launch effective missions with the new, but untried B-29 Superfortress. Not until the Mariana Islands were taken as operating bases could the strategic bombardment of the Japanese home islands begin in earnest.

To take advantage of the closer staging bases, General Arnold had selected Major General Curtis E. LeMay to direct the final air assault against Japan, moving him from the European Theater to succeed Brigadier General Haywood S. "Possum" Hansell Jr. as commander of XXI Bomber Command in the Marianas. Initial B-29 raids flown from bases in China and India had been disappointing. Arnold directed LeMay to make the best use of the new B-29 Superfortress and its significant advances in aircraft technology. These included pressurized crew space, remotely aimed machine guns, and massive engines that could carry the bomber on long-range missions lasting more than 12 hours, often reaching targets 1,500 miles or more from their home base.

Poor weather over Japan frequently prevented visual acquisition of targets, and strong high-altitude winds resulted in less-than-accurate radar bombing. LeMay, understanding that General Arnold was a man of results and not excuses, was driven to action. He ordered low-level attacks—his only option. After shifting from high-altitude to low-altitude bombardment, LeMay wrote Arnold and explained that weather rather than any preconception about low-altitude bombing was the reason he changed. LeMay's decision, almost immediately, was fateful for the Japanese.

Bombing of the Japanese home islands was anything but precise. Nearly three-quarters of the raids against Japan were flown at night. Area bombing and the use of incendiary weapons became the hallmark of the strategic campaign against the Japanese. Fully two-thirds of the air attacks against Japan were area-bombing raids on industrial centers and cities. During the final six months of the war, nearly 60 percent of the bombs dropped on Japan were incendiary weapons. More than 60 Japanese cities—home to 20 million—were attacked during the war. Compared to the war in Europe, far more square miles of Japanese cities were devastated, with most destruction attributed to massive fires resulting from incendiary attacks.

Early on the morning of August 6, 1945, B-29s from the 509th Bomb Group lifted off a coral runway on Tinian Island. Colonel Paul Tibbets and the crew of the *Enola Gay* set course for Japan, along with two other bombers that would measure the effects of the weapon as it detonated over its target—both visually and electronically. A little more than six hours later, over the city of Hiroshima, Tibbets dropped *Little Boy*, the first atomic bomb, which destroyed a significant portion of the downtown area, and killed between 70,000 and 80,000 occupants instantly.

After the war, the Japanese reported over one-half million casualties and the destruction of 2.5 million buildings resulting from the totality of the American aerial assault. The United States lost 400 B-29s and more than 1,000 airmen, a majority of these in non-combat-related accidents, such as engine fires. The bombing effort against Japan cost one-tenth that of the strategic bombing campaign against Germany—and no land invasion was required to finish the job.

In the end, airpower in the Pacific played a vital role in the defeat of Japan, but not the decisive one. Naval aircraft and U.S. ships destroyed the Japanese navy, allowing surface forces to capture vital staging bases closer to Japan. Marines valiantly stormed the beaches into hostile enemy gunfire, enabling the establishment of forward airfields.

Airpower advocates were both correct and mistaken in their predictions concerning the effectiveness of the air weapon in World War II. On one hand, air forces denied the enemy free use of the third dimension of the battlefield—the air. On the other hand, airpower was not singularly decisive —but it was a vital contributor to Allied victories around the globe. The battle for "complete air ascendancy" called for in AWPD-42 proved costly and protracted.

Clearly Allied airpower punished enemies by destroying cities and factories. Production was more or less impeded, transportation of supplies disrupted, and morale strengthened in some cases, weakened in others. But in the end, strategic airpower ensured the existence of theater air supremacy.

In the Pacific, the Japanese attack on Pearl Harbor had been a tactical masterpiece. Surprise, mass, unity of command, and deception led to a stunning tactical victory. Strategically, the attack on Pearl Harbor led inevitably to the devastation of Japan's industrial infrastructure and the destruction of more than 60 Japanese cities. The final blow, unleashing the atom against those who had "awakened the sleeping giant" only four years before, put the world on notice that the United States would use its awesome power if the cause was perceived as just. The bombs dropped on Hiroshima and Nagasaki, for good or ill, set an ominous precedent for the years to come.

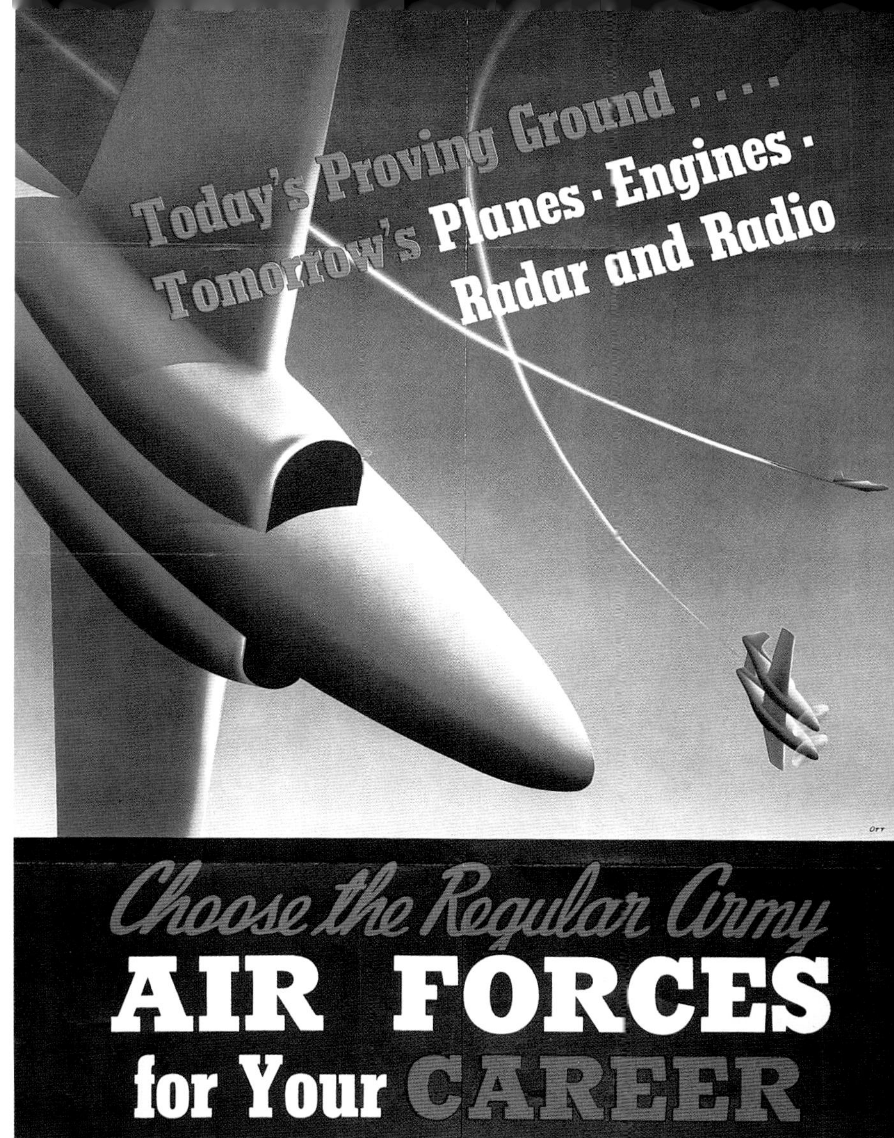

Pages 240–241: *B-29s did most of their firebombing at night. This painting depicts the first low-level, night fire bombing raid over Tokyo. ("Spring Morning (Lucky Strike)," by William Phillips, NASM Art Collection)*

Above: *One of the earliest recruiting posters based upon the lure of high technology and the opportunity to make the Air Force a lifelong career. (NASM Poster Collection)*

WORLD WAR II: AIR WAR IN THE PACIFIC

1944–1945

"It was the B-29s and the B-29s only that could put tons and tons of bombs on Japan. The fleet couldn't do it; Naval Air couldn't do it; the Army couldn't do it. The B-29s could."
—General of the Air Force Henry H. "Hap" Arnold, 1949

1944

1 January
In Burma, Tenth Air Force B-25s and P-38s attack a bridge over the Mu River. Bombing from extremely low altitude, Major Robert A. Erdlin drops his bombs while pulling up to avoid a ground obstacle. The low-angle loft successfully drops two spans of the bridge. The "Burma Bridge Busters" are born.

Fifth Air Force launches more than 120 B-24s, B-25s, and A-20s in attacks against targets in the Saidor area. More medium bombers strike troop concentrations and supply targets in preparation for the Allied invasion of New Guinea.

2 January
Tenth Air Force medium and heavy bombers attack targets at Yenangyaung, Burma, setting fire to oil storage facilities and damaging a power plant.

7 January
Lieutenant General Hubert R. Harmon assumes command of Thirteenth Air Force. In June, he will assume command of the Sixth Air Force (later the Caribbean Air Command). Harmon, a 1915 West Point graduate and classmate of General Eisenhower, will be recalled from retirement in 1953 to be appointed as the first superintendent of the newly established U.S. Air Force Academy, effective August 14, 1954.

10 January
More than 100 Fifth Air Force medium and heavy bombers hit targets near Madang, Alexishafen, and Bogadjim. P-39s strafe villages and barges in New Britain.

Lieutenant General Harmon's Thirteenth Air Force begins a night bombardment campaign. B-24s strike airfields at Lakunai and Vunakanau, as well as supply depots near Buka. Operating jointly on a regular basis, Thirteenth Air Force and U.S. Navy fighters provide air cover while Navy dive bombers hit targets at Cape St. George. Similar missions are flown the next day.

13 January
Lieutenant General Kenneth B. Wolfe, XX Bomber Command, arrives at New Delhi, India, with his advanced staff. His arrival marks the beginning of personnel deployment for Operation MATTERHORN—the B-29 offensive against Japan.

Left: *Army Air Force medium bombers were very effective disrupting in Japanese oil production and storage. (NASM)*

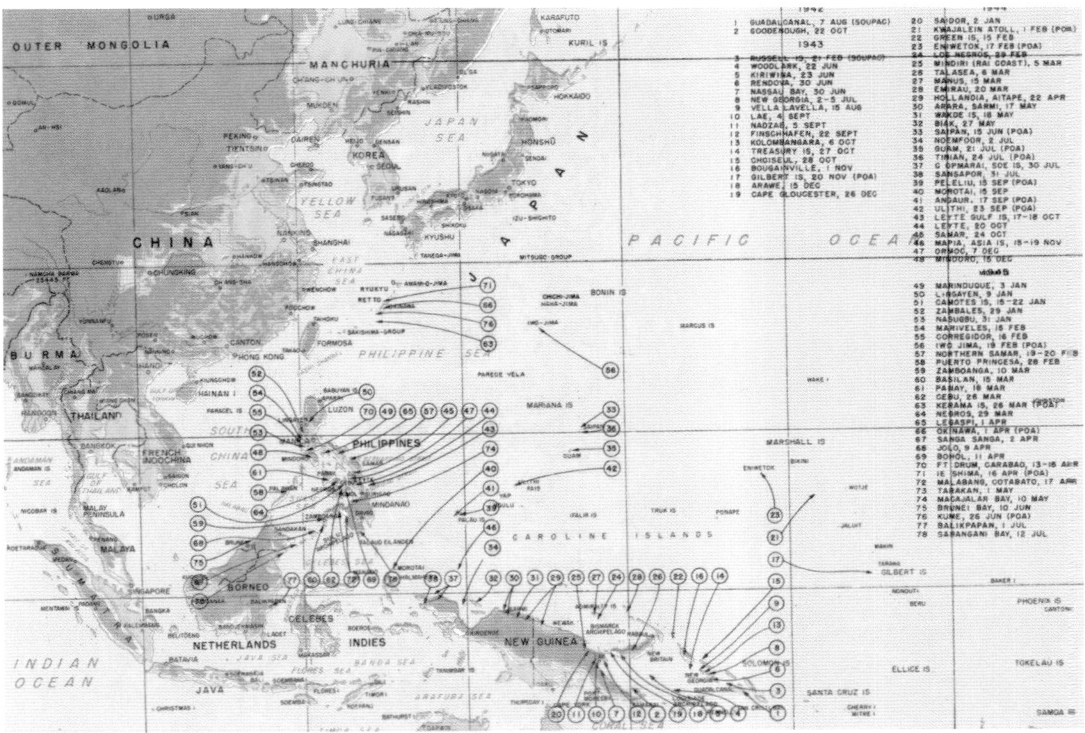

Right: *This map depicts each amphibious landing made in the Pacific during World War II, and airpower involved with every one.*

Below: *P-38 Lightning ground crew load the machine guns in preparation for another mission. (USAF)*

13–14 January
During a midnight raid, two dozen Thirteenth Air Force B-24 Liberators bomb the airfields at Vunakanau and Malaguna. At dawn, a dozen B-25s strike targets near Buna. AAF P-39s join Navy SBDs in attacking Wakuni. During the day, a large joint operation is carried out when 70 Allied fighters support 50 Navy dive bombers in an attack against shipping in Simpson Harbor.

18 January
In a very active day for Fifth Air Force, 40 Liberators bomb Hansa Bay while another group of B-24s attack Laha. More than 70 B-25s attack Madang and Bogadjim and other positions near Shaggy Ridge. More than 50 AAF fighters engage 50 Japanese fighters over Wewak. Three P-38s are lost. Similar attacks continue for the next three days.

22 January
Continuing to bomb the town of Rabaul and airfield targets, Thirteenth Air Force medium and heavy bombers do serious damage to the runway at Lakunai and will continue to attack its airfield until the end of the month. On this day, more than 90 fighter planes provide cover for the attacking bombers.

29 January
Seventh Air Force B-24 Liberators open a 'round-the-clock offensive against enemy bases in the Gilbert Islands as the Allied invasion forces approach. Single-plane night attacks and formation day strikes keep the Japanese on the defensive. Formations of P-39s flying in flights of four patrol the airfield at Mille, denying its use to the enemy. Liberators bomb Kwajalein during the next 48 hours to soften the landing zone fcr the invasion forces.

31 January
U.S. troops invade Kwajalein as part of the Marshall Islands Campaign. During the next week, Allies will fight to take control of both Kwajalein and Majura Atoll.

1 February
Seventh Air Force fighters rapidly advance to occupy newly established bases in the Gilberts. Operation CATCHPOLE begins with the goal of taking Eniwetok from the Japanese and then using its airfields to strike important airfields in the Marianas.

3 February
Colonel Philip Cochran leads five P-51 Mustangs on the first air-commando combat mission. The mission is flown against Japanese targets in the China-Burma-India (CBI) Theater of operations.

Fifth Air Force fighters and bombers pound Wewak. During the attack, approximately 80 enemy aircraft are destroyed on the ground. P-39s and B-25s continue to strike shipping targets in the Bismarck Sea.

16 February
The Allied attack on Truk Island begins and will continue for one full month.

19 February
In an effort to keep the pressure on the Japanese, more than 60 Tenth Air Force A-36s, P-51s, and B-25s strike a variety of targets in Burma. Fuel supply depots, rail cars, and river traffic are attacked over a wide area.

Fourteenth Air Force continues flying "sea sweeps" in search of targets of opportunity from the Formosa Straights to Indochina. On this date, B-24s, B-25s, and P-40s claim three ships sunk and bridges, trains, and depots also attacked.

U.S. forces land on Eniwetok while Seventh Air Force bombers pound targets near Ponape and Wotje. The island is under complete control of U.S. forces in 48 hours.

22 February
Fifth Air Force sends B-24s and B-25s to attack Iboku Plantation as Marines advance.

One Seventh Air Force P-40 launches rockets against an airfield target. This is the first rocket attack made by any Seventh Air Force plane.

28 February
Fifth Air Force B-24s continue to pound landing areas for advancing Allied troops. Airfields in Nubia and Awar as well as targets in Hansa Bay area are struck in anticipation of the landings.

Thirteenth Air Force launches a multi-wave attack upon Rabaul. First, 22 B-25s with fighter escort bomb targets; next, P-38s dropping glide bombs strike the same targets; then, only two minutes later, 11 B-24s pound the same target again. P-39s and Navy dive bombers hit other targets as well.

29 February
Fifth Air Force bombers destroy two Japanese air bases at Alexishafen, New Guinea. These strikes are in support of Allied landings in the Admiralty Islands and complete the isolation of Rabaul, New Britain.

1 March
Supporting invasion forces, Fifth Air Force strikes targets on Los Negros Island and near Lorengau in the Admiralty Islands. More than 100 bombers and fighters attack targets near Wewak, New Guinea, including enemy positions at Madang and Saiba. These attacks continue until the April 22nd landings at Hollandia. Direct support continues as U.S. troops begin the process of taking the island.

Opposite, above: *Cartoon depicting a B-24 Liberator inflicting damage on Japanese forces amassed at Wewak. (Army Art Collection)*

Opposite, below: *View from inside an Army Air Forces glider in tow. (USAF)*

Right: *As soon as Japanese islands were captured, AAF aircraft deployed to crude airfields and continued to support Allied amphibious forces during the Pacific campaign. (USAF)*

The XI Strategic Air Force is activated at Shemya, Alaska. It is composed of the forces of XI Bomber and XI Fighter Command and becomes a tactical operating agency along the Aleutian Islands, Alaska.

3 March
Operation FORAGER begins this date to capture the Mariana Islands. Airfields there will allow B-29 bombers to strike the Japanese mainland with efficiency. Seventh Air Force is tasked to neutralize enemy air activity in the Carolines and continue to bomb Wake Island while bypassing the Marshall Islands during the operation.

5 March
Under the cover of darkness, Special Forces under the command of British Brigadier General Orde C. Wingate land at locations in central Burma. They arrive in Army gliders flown by Colonel Philip G. Cochran's Air Commandos (not officially formed until later in the month). Both U.S. and British troops are inserted 50 miles northeast of Indaw. Almost half of the gliders land successfully, but the Japanese had cut trees across the second landing zone and those troops could not deploy. In total, 539 men, three mules, and more than 30 tons of equipment are delivered. This mission successfully halts the Japanese invasion of India.

11 March
Seventh Air Force B-24 Liberators operate from runways on Kwajalein Island for the first time. Their first attack against Wake Island from bases in the Marshalls is supported by P-40s and B-25s operating from the Gilbert Island chain.

15 March
Seventh Air Force B-24 Liberators make the first land-based air attack on Truk Atoll. Eleven bombers attack from their base on Kwajalein. B-25s from Tarawa strike Maloelap, while many of the fighters that were used during Operations FLINTLOCK—the occupation of Kwajalein and Majuro Atolls—and CATCHPOLE—the occupation of Eniwetok Atoll—are back in the Hawaiian Islands for refitting, rest, and recreation leave.

General Hubert Harmon becomes the Commander, Air, Solomons (COMAIRSOLS), and his forces continue to provide support throughout the region.

20 March
U.S. Marines land on Emirau Island, in the Admiralty chain, meeting little resistance. Construction of air bases begins immediately. Thirteenth Air Force provides support by attacking surrounding airfields. Radar-equipped SB-24s cover the Marine's approach to the island.

29 March
In India, the AAF activates the 1st Air Commando Group. Its mission is to provide air support and airlift to Allied soldiers in Burma. This unit, commanded by Lieutenant Colonel Philip G. Cochran and John R. Alison, is responsible for rear area harassment and rescue operations.

Left: *The P-39 Airacobra was powered by a non-supercharged engine mounted behind the pilot. The Airacobra saw extensive combat duty in the southwest Pacific. Of the nearly 10,000 built, almost half were used by Soviet pilots as part of Lend-Lease. (USAF)*

Below: *American troops displaying captured Japanese flags and "good luck" banners after an island victory. (USAF)*

Thirteenth Air Force bombers carry out the first daylight raids on Truk Atoll. Many Japanese aircraft are destroyed, but two B-24s are also lost.

30 March
Fifth Air Force carries out its first major daylight raid against Hollandia, New Guinea. Targets include airfields, fuel dumps, and troop concentrations along the north coast from Wewak to Madang. A mix of A-20s, B-25s, P-39s, P-40s, and P-47s strike at will in these areas.

2 April
Colonel Leonard F. Harman lands the first operational XX Bomber Command B-29 at Chakulia, India.

4 April
Twentieth Air Force is activated in Washington, D.C., under General Hap Arnold, who assumes direct command two days later. This command is established to ensure that the B-29s attacking Japan from India and China, then later from the Mariana Islands, come under direct supervision of the Joint Chiefs of Staff.

16 April
More than 170 Fifth Air Force planes are sent to attack targets near Hollandia. After the mission, severe weather interferes with their return to landing fields. Unable to find suitable runways, 37 A-20s and B-25s are lost.

22 April
After six weeks of constant bombardment, Allied amphibious landings begin west of Hollandia and Aitape, New Guinea. Fifth Air Force now shifts to close support and continues to attack Wewak and Hansa Bay with B-24s, B-25s, and A-20s. Close support and interdiction attacks continue for several months after the landings.

24 April
The first two Twentieth Air Force B-29s to fly the Hump to China land at Kwanghan. The XX Bomber Command commander, Lieutenant General Kenneth B. Wolfe, pilots one of the Superfortresses.

25 April
Flying a Sikorsky YR-4 helicopter, Lieutenant Carter Harman, 1st Commando Group, rescues four men from the Burmese jungle. It is the first time that a helicopter has been used by the AAF in a combat rescue operation.

Right: *B-29s of the 468th Bomb Group release their massive bomb load over Rangoon, Burma. (NASM)*

Seventh Air Force B-24s flying from Eniwetok attack enemy targets on Guam. This is the first time land-based bombers have attacked the island. The bombers are accompanied by a Navy PBY reconnaissance aircraft.

1 May
The first four very heavy bomber (VHB) bases open in Chengtu, China. B-29s are arriving daily.

10 May
After five months of intense labor by more than 400,000 Chinese laborers, five B-29 bases and six fighter bases are completed as part of Project CHENGTU—the establishment of B-29 operations against Japan from these locations deep in mainland China near the city of Chengtu.

17–18 May
The combination of Allied aerial bombardment and naval artillery results in an unopposed landing at Arare and Sarmi, New Guinea. Fifth Air Force sends more than 100 B-24s to hammer away at enemy defensive positions, while more than 100 medium bombers continue to provide support near Wewak.

27 May
Twenty-four rocket-firing Fourteenth Air Force P-40s strike military installations and enemy troops at Nanchang, China.

Fifth Air Force B-24s and B-25, attack Babo airfield and Biak Island in preparation for the amphibious landing planned at Biak this date. Additionally, Fifth Air Force medium bombers continue to provide close support to ground forces around Wewak.

5 June
In the first B-29 raid of the war, Superfortresses assigned to XX Bomber Command, Twentieth Air Force, strike rail targets in Japanese-occupied Bangkok, Thailand. More than 75 bombers participate in the attack. Originally, 98 B-29s were launched, and five crashed due to non-combat-related problems.

During the evening hours on Wakde, Japanese fighters attack the Allied airfield there, destroying several aircraft on the ground.

6 June
The D-day invasion begins in Europe.

U.S. STRATEGIC BOMBING OF JAPAN DURING WORLD WAR II

Kenneth P. Werrell

Left: *High altitude daylight raids were ineffective because of poor weather and strong jet-stream winds that sometimes exceeded the Norden's capability to compensate. This image is taken looking down from open bomb bay doors at the top of the formation. (USAF)*

The strategic bombing campaign against Japan did not mirror the one waged against Germany. In contrast, it was shorter and less costly than the European effort, with different problems and solutions. Paradoxically, the results were more impressive, but less significant.

The AAF's bomber workhorses of the German war were unable to bring the war to the Japanese homeland because of the vast distances of the Pacific Theater. Thus a new bomber, the Boeing B-29 Superfortress, would be the Army Air Forces' (AAF) weapon. Designed for the war against Germany, it featured a number of innovations (cabin pressurization, remotely controlled defensive armament, and a flight engineer) and was the most sophisticated and best performing heavy bomber of the war. Ironically it would operate from the most primitive base areas of the war.

The Twentieth Air Force launched its first attack on Japan on June 14, 1944, from forward bases in China. Despite heroic efforts to build these bases, largely by hand, and supply them over the Himalayas, the infamous "Hump," the B-29s could barely reach southern Japan. Thus these early attacks were few and ineffective. In June 1944, the U.S. attacked Japanese positions in the Marianas, where it established bomber bases. On November 24, 1944, B-29s began daylight strikes on Japan from these bases.

Japanese air defenses were considerably weaker than those the Americans encountered over Europe; however, other obstacles were more formidable. Weather, poor visibility, and high winds aloft (the jet stream) made high-altitude, precision-bombing very difficult. In addition, high-altitude, long-range operations with heavy bomb loads intensified the bombers' mechanical and fuel management issues.

This is the situation General Curtis LeMay found when he took command of the Marianas B-29s in January 1945. Although things were improving, including increasing numbers of bombers, a new maintenance system, and shortly the seizure of Iwo Jima that gave the aviators an emergency airfield between their bases and targets, the bombing campaign was not achieving results. Facing extremely poor weather for visual bombing, LeMay put his command and career on the line when he introduced new and daring tactics that radically deviated from established Army Air Forces doctrine. He launched his bombers with a maximum bomb load of incendiaries, ordered them flown at night at low altitude. Each B-29 was to fly one behind the other rather than in complex formations. This change from precision bombing marked the progression from the "desired" to the "doable," from the theoretical to the practical.

On the first of these raids against Tokyo, March 9–10, 1945, the B-29s burned out 16 square miles of Japan's capital, killed more people than any other bombing raid in history—including the two atomic bomb attacks—and lost only 14 of 334 bombers. The airmen then proceeded to burn out all of Japan's major cities (except Kyoto) as well as many smaller cities whose population was greater than 60,000 people. Compared to the strategic air war against Germany, the B-29 campaign was relatively quick, efficient, and cheap. In less than half the time of the European campaign, the B-29s destroyed twice the urban area in Japan as in Germany with far fewer sorties and less bomb tonnage. Perhaps more remarkable was the dramatic difference in American casualties; the European strategic bombing effort cost 35 times the bombers and 47 times the casualties as did the Japanese campaign.

The bombing of Japan, while more efficient than that of Germany, was at the same time less effective and less important. Although it devastated large urban areas and the factories therein, industrial output had already declined because American submarines, along with the very successful B-29 mining of Japan's coastal waters, reduced imports of raw materials. The major result of the bombing campaign was to demonstrate to all Japanese (civilians and decision makers alike) the superiority of the American military and the impotency of the Japanese defenders.

The bombing campaign against Japan was a great success. American industry and airmen put a superior weapon into operation in record time and sustained intense operations despite having to build bases and infrastructure, and supply and fight a war literally halfway around the world. The AAF had fewer problems with air defenses than against Germany, but had to overcome more difficult logistical, weather, and infrastructure problems, along with the bomber's mechanical difficulties. The performance of the AAF's strategic bombers in World War II, coupled with the promise of long-range bombers armed with nuclear weapons in the future, were major factors in the creation of an independent air force (USAF) in 1947.

Below: *Although daylight precision bombardment was AAF doctrine, devastating effects were achieved by night firebombing of Japanese cities. This is another view from atop a bomb bay as the weapons are released. (USAF)*

Left: *Aircraft maintenance crews await the return of their aircraft from a raid. ("Return from Kiska Raid," by Edward Laning, Army Art Collection)*

Opposite: *The Soviets produced the Tupolev Tu-4 Bull bomber, a copy of the Boeing B-29 Superfortress. Markings are enhanced on this image. (NASM)*

15 June
Far East Air Force (FEAF) is created with command authority over Fifth and Thirteenth Air Forces. General George Kenney assumes command with his headquarters at Brisbane, Australia. Fifth Air Force strikes north at Truk Atoll, while Thirteenth strikes south at airfields at Bougainville and continues the mission of neutralizing Rabaul and attacking Bougainville. These attacks continue through August.

In the first raid on Japanese home islands since the carrier-launched strike on targets in and around Tokyo on April 18, 1942, 68 B-29s from Chengtu, China, attack a steel factory on the island of Kyushu. The target is more than 1,500 miles away from their bases in China. Results are poor as bombardiers are not familiar with high-altitude winds, which ruin aiming points.

The Mariana Island Campaign begins as U.S. forces begin attacks to capture the heavily fortified island of Saipan. The hard-fought battle rages for three weeks.

19–20 June
The Battle of the Philippine Sea, also known as "The Great Marianas Turkey Shoot," takes place during these 48 hours. American carrier-based aircraft shoot down more than 370 Japanese carrier-based planes, effectively neutralizing Japanese naval airpower for the remainder of the war.

24 June
Seventh Air Force P-47s now flying from newly captured airfields on Saipan provide close support to Allied ground forces by strafing remaining enemy positions on Saipan and Tinian.

25 June
Continuing their campaign of reconnaissance and harassment in the Aleutians, two Eleventh Air Force B-24s bomb the suspected enemy airfield at Kurabu Cape. Even though the major campaign had ended when Kiska was reoccupied in August 1943, the Eleventh continued to fly bombing and reconnaissance missions against the Japanese in the Kurile Islands for the remainder of the war. This was the only campaign in World War II fought on and over North American soil. The Territory of Alaska was not an American state during World War II.

26 June
The P-61 Black Widow begins flying night patrols over Saipan. Along with P-47 sweeps during the day, these missions continue well into July.

2 July
Along with a battery of naval guns, aircraft of the Far East Air Forces hit the Kamiri area of Noemfoor Island, in preparation for an amphibious landing there. The Allies have little trouble establishing a beachhead.

6 July
A Northrop P-61 Black Widow piloted by 1st Lieutenant Francis Eaton intercepts and shoots down a Japanese Mitsubishi G4M3 Betty bomber. This is the first victory scored by the P-61 during the war. Second Lieutenant James Ketchum is the radar operator and the gunner is Staff Sergeant Gary Anderson.

Fourteenth Air Force B-25s, P-40s, and P-51s support Chinese ground forces by striking Japanese targets along the Yangtze River. B-25s drop supplies to fielded armies and attack the Tien Ho airfield at Canton overnight. The Fourteenth provides air superiority, close support, and supply over the Hump for ground forces from Tsinan in the north to Indochina in the south; and from Chengtu in the west to Formosa in the east. At war's end, they will officially be credited with destruction of more than 2,300 enemy aircraft, 350 bridges, 1,200 locomotives, and 700 railroad cars.

11 July
Seventh Air Force P-47s relentlessly attack in the Marianas. Tinian and Pagan are today's targets but the P-47s out of Saipan continue to soften up the beaches for the landings scheduled for less than two weeks from this date. In the coming days, B-24s stage out of Eniwetok and strike Tinian twice a day until the landings begin.

19 July
FEAF B-24 Liberators attack the airfield on Yap. They attack in two waves, while others bomb Ngulu and Sorol Islands. Fighter-bombers lend close support to ground forces on the Sarmi-Sawar sector on the north coast of New Guinea.

21 July–10 August
After success on Saipan, American forces had turned their attention toward Guam, the largest island in the Marianas. After three tough weeks of bombardment and fighting, the island is secured by Marines and Army forces and becomes a valuable staging base for operations against Japan. Seventh Air Force aircraft support the operation by attacking Japanese forces on Truk and Tinian.

24 July–1 August
During the Battle of Tinian, U.S. Marines secure the island with tactical support from AAF bombers and fighters. Seventh Air Force P-47s drop napalm canisters to clear heavy vegetation concealing enemy positions.

29 July
Chengtu-based B-29s attack the Showa Steel Works at Anshan, China. The first combat loss of a B-29 occurs when a crippled bomber proceeds to attack a secondary target and is intercepted by five enemy fighters. A flak-damaged B-29 is forced to make an emergency landing on a small Soviet field at Tarrichanka. The crew is interned, and the aircraft, along with two other B-29s which fall into Soviet hands in August and November, are given to the Tupolev Design Bureau to be copied. The Tu-4 is the result, looking remarkably like an American B-29. The Tu-4, the Soviet's first strategic bomber, does not become operational until 1947.

GEORGE C. KENNEY

Tom Griffith

Few commanders in the Second World War depended more on airpower and his air commander than General Douglas MacArthur. The rugged mountains, thick jungles, and vast distances of the Southwest Pacific put a premium on using aircraft to bypass and then isolate the Japanese forces. As a result, MacArthur's Airman, General George C. Kenney, played a critically important role in planning and winning the southwest Pacific campaigns.

Kenney landed in Australia in late July 1942 at a low point in the Allied fortunes. Japanese attacks at Pearl Harbor and in the Philippines had effectively wiped out the American air units in the Pacific. When Kenney joined MacArthur, it appeared that the emperor's forces might invade Australia.

Although maps of the southwest Pacific at the time showed Japanese control over the whole territory, in reality their defensive perimeter depended on holding a few key areas. Rather than eliminate each outpost piecemeal, MacArthur's "leap-frog" strategy avoided the most strongly held, but widely separated, Japanese garrisons. In carrying out these plans he depended on airpower and Kenney. Kenney believed gaining air superiority was an essential prerequisite for any operation, and he attacked Japanese aircraft on the ground and in the air to eliminate them. Simultaneously, he isolated Japanese garrisons by bombing supply ships at sea or attacking enemy supply lines on land. He argued forcefully for using aircraft in this interdiction role, rather than bombing enemy positions at the front lines. Problems with locating and attacking targets on the ground through the thick jungle canopy endangered friendly forces. Kenney's forces also flew Allied soldiers and their equipment to the battle areas and brought out wounded by air.

Kenney demonstrated great flexibility in carrying out these missions. Early in the war he realized that airmen had been overly optimistic about the ability of bombers to defend themselves from enemy attacks. Kenney installed external fuel tanks on fighter aircraft to increase their range, allowing them to accompany the bombers to their targets. He also shifted from the prewar doctrine of high-altitude bombing to low-altitude attacks of Japanese shipping. His most dramatic success occurred during the Battle of the Bismarck Sea in early March 1943, when Kenney's airmen devastated a 16-ship convoy. The attack not only wiped out the reinforcements and their supplies, but also shocked the Japanese high command, who abandoned further attempts at reinforcing eastern New Guinea.

In June 1943, MacArthur began a series of amphibious assaults along the northern coast of New Guinea. Through the rest of the year MacArthur's forces moved westward to outflank the Japanese stronghold at Rabaul in a series of well-integrated air, land, and sea operations. With New Guinea secure, MacArthur returned to the Philippines, landing on the island of Leyte on October 20, 1944, and captured Manila in March 1945. Kenney now turned his attention northward—first, in support of the landings at Okinawa with attacks on the airfields on Formosa, and, later, in preparation for an American invasion of the Japanese homeland. In the end, atomic bombs and the Soviet Union's entry into the war made the planned invasion unnecessary.

Kenney's efforts as the air commander were enhanced by several advantages. He served with a theater commander who learned to appreciate the benefits offered by air power. In addition, the ability to intercept and decode Japanese radio transmissions gave Kenney an extremely accurate picture of his foe, allowing him to concentrate forces on high-payoff attacks.

Knowledge about the enemy's capabilities and intentions explain only part of Kenney's success; equally important was having enough aircraft available to carry out the missions. While aviators received much of the publicity, Kenney realized how important the hard-working mechanics were to winning the war and raised their morale by awarding military decorations and improving living conditions. Although the number of aircraft in his command increased slowly, the total number of sorties flown grew rapidly, an indication of the strenuous efforts expended on flying the same few aircraft more often.

Throughout the war, Kenney proved capable of meeting a wide range of problems with innovative solutions. He built and maintained a strong and effective working relationship with MacArthur and focused on using airpower to avoid large concentrations of enemy forces and outflank the enemy through the air. Although the basic strategy of bypassing the enemy is as old as war itself, MacArthur's campaigns in the Southwest Pacific would not have been possible without intelligently applied air power. In this regard, General George C. Kenney, MacArthur's Airman, proved vital to the Allied victory and secured his place as one of the Air Force's premier combat commanders.

Below: *General Carl Spaatz, General George Kenney, and General Douglas MacArthur meet somewhere in the Pacific. Kenney was "MacArthur's Airman." (USAF)*

Left: *A pilot concentrates on an accurate attack. Note the lucky charm—a four leaf clover. ("Fighter Pilot at Work," by Tom Lea, Army Art Collection)*

Below: *A pilot's view as he takes aim at his objective. ("Target Practice #3," by Henry Billings, Army Art Collection)*

30 July
After months of constant attacks, Tenth Air Force fighter-bombers force a Japanese withdraw from Myitkyina, Burma (present-day Myanmar). After the Japanese commander orders the pullout, he commits suicide. By August 1, the town will be retaken by the Allies.

FEAF creates a diversion for an unsupported surprise landing at the far western tip of New Guinea, near Mar. No opposition is seen as Task Force TYPHOON lands. AAF aircraft keep the Japanese occupied by bombing targets near Wewak and Aitape.

1 August
General Millard F. Harmon assumes command of the Army Air Forces Pacific Ocean Areas (AAFPOA), activated to consolidate logistics and administrative functions as well as all tactical operations throughout the Pacific Theater. Harmon also assumes the position of Deputy Commander of Twentieth Air Force and is responsible for strategic operations and reports directly to General Hap Arnold.

Seventh Air Force control is transferred to AAFPOA. As organized resistance on Tinian ends, Saipan-based P-47 and P-61 aircraft continue day and night combat patrols over Guam, Rota, and Pagan. These patrols are in direct support of Marine and Army forces fighting on Saipan, Guam, and Tinian.

8 August
Fourteenth Air Force aircraft hit targets in Hengshan, China, where they destroy several trucks, and at Hamoy and Swatow, where they destroy radio stations and munitions storage facilities.

Right: *A military bulldozer clears island vegetation to make room for hundreds of incoming B-29 bombers in the Mariana Islands. (USAF)*

10 August

Control of the Marianas Islands is secured. From bases constructed on Guam, Tinian, and Saipan, American B-29s begin an aerial campaign against Japan that will last until mid-August 1945.

Seventh Air Force begins the AAF's campaign to neutralize enemy defenses on Iwo Jima. B-24s flying from Saipan pound Japanese forces there. For the next six months, he Seventh will attack Iwo Jima, Chichi Jima, and other enemy locations in the Carolines and Marianas in preparation for the amphibious landings on Iwo Jima.

10–11 August

In one of two missions flown during this period, 24 Twentieth Air Force B-29s flying out of Chengtu bomb targets in Nagasaki, Japan. In the second mission, staged through China Bay, Sri Lanka (Ceylon), more than 30 B-29s bomb the oil refinery at Palembang, Sumatra, Indonesia, while eight others mine the nearby Moesi River. The mission from Ceylon to Sumatra—some 3,900 miles—is the longest single-staged combat flight made by a B-29 during the war.

14 August

Seventh Air Force is reorganized. The new "mobile tac airforce" retains only functional combat units. In a typical day of combat missions, B-24s based in Saipan bomb Iwo Jima, B-25s strike Pagan, P-47s strike Rota, and B-25s flying from the Marshalls hit Ponape, while B-24s attack the Wotje Islands.

16–17 August

As Japanese resistance in India is defeated, Tenth Air Force presses attacks against multiple targets in northern Burma.

20 August

Seventh Air Force B-24s based on Saipan strike targets on Yap for the first time. Marshall Island-based Liberators continue to bomb Truk.

23 August

Thirty-two Tenth Air Force P-47s provide close support to advancing British ground forces by attacking troop concentrations, gun batteries, and headquarters buildings.

Left: *The Douglas C-46 Commando was one of the workhorses in the CBI. (USAF)*

Below: *Maintenance men were the backbone of successful operations. ("Portrait of Sgt. R. H. Hulse, Crew Chief," by Tom Lea, Army Art Collection)*

Opposite: *A kamikaze strikes the carrier* Belleau Wood, *October 30, 1944. (NASM)*

24 August
Brigadier General "Rosie" O'Donnell lands his 73d Bomb Wing HQ advanced echelon in the Marianas. This is the first contingent of Twentieth Air Force to arrive at these island bases.

28 August
Brigadier General Haywood "Possum" Hansell assumes command of XXI Bomber Command, and Brigadier General Laurence Norstad becomes Chief of Staff for the Twentieth Air Force.

29 August
Major General Curtis E. LeMay assumes command of the XX Bomber Command.

4 September
Tenth Air Force B-24s begin to haul thousands of gallons of fuel to staging areas in Kunming for use by advancing forces. By this time, Brigadier General William H. Tunner has taken charge of the Air Transport Command and is responsible for the Hump operations.

8 September
Nearly 100 Chengtu B-29s bomb the Showa Steel Works again in Anshan, China. During the night, Japanese bombers attack one of the Chengtu bases, damaging a B-29, a C-46, and wounding two soldiers.

10 September
Troop Carrier missions and cargo-hauling continue throughout the China-Burma-India Theater. B-24s haul fuel over the Hump. Similar missions continue throughout September. During November and December, Airlifters fly the Hump an average of 300 times each day to resupply forces operating in China.

15 September–25 November
The Palau Islands Campaign begins to secure staging locations for the invasion of Japan. The island of Peleliu is eventually taken by U.S. Marines, but only after the loss of more than 2,300 lives and 8,400 casualties. The assault on nearby Angaur begins two days later and the island is quickly captured with far fewer casualties.

23 September
Thirteenth Air Force moves its base of operations from Hollandia to Noemfoor Island, New Guinea.

Seventh Air Force B-24s continue to attack Chichi, Haha, and Ani Jima.

24 September
Eleventh Air Force B-24s bombing the airfield at Kurabu Cape are attacked by a dozen Japanese fighters. One damaged B-24 is forced to land in the Soviet Union.

28 September
More than 100 Fourteenth Air Force fighters perform armed reconnaissance missions throughout southeastern China and to a lesser degree, in southwestern China and Indochina.

2 October
Tenth Air Force troop carrier sorties continue, more than 250 each day, to locations throughout the CBI. These missions continue to provide the lifeblood of supplies and troops during the month of October. By the 16th, more than 300 airlift sorties will be flown during one day.

12 October
The first Twentieth Air Force B-29, *Joltin' Josie, The Pacific Pioneer*, lands on Saipan to establish temporary headquarters. Brigadier General Haywood Hansell, Commanding General of XXI Bomber Command, personally pilots the aircraft.
Later this day, regular elements of the 73d Bomb Wing also arrive.

14 October
For the first time, more than 100 Twentieth Air Force B-29s from Chengtu mass to attack targets. This raid against the aircraft plant at Okayama is the first in a series of attacks against Formosa flown in conjunctions with U.S. landings at Leyte, Philippines.

20 October–10 December
The Philippines Campaign begins as General Douglas MacArthur leads an Allied force at Leyte. Four days later, the important airstrip at Tacloban is secured. The fight continues until mid-December and becomes the bloodiest Pacific campaign until losses are surpassed at Okinawa the following April. AAF aircraft make limited contributions as the distance from New Guinea to Luzon is too great. Once airfields closer to the action become available, AAF participation increases.

21 October
The first Japanese kamikaze mission—Divine Wind—is launched against Allied naval vessels during the opening phases of the Battle of Leyte Gulf (October 23–26). Generally considered the largest naval battle in history, U.S. and Australian naval forces soundly defeat the Japanese.

Two Seventh Air Force B-24s fly the first combat missions from Guam. They attack the small island of Yap.

26 October
Fourteenth Air Force B-24s and B-25s attack Japanese shipping off the east Luichow Peninsula. Additionally, B-25s strike rail yards at Hsuchang.

Major Horace S. Carswell Jr., Fourteenth Air Force, dies while co-piloting a B-24 Liberator against a Japanese shipping convoy in the South China Sea. The plane is crippled and his pilot is killed by flak during the attack. Upon reaching land, only two of his four engines working, he orders the crew to bail out—his own parachute is damaged beyond use. Carswell remains at the controls in an attempt to reach a base. Unfortunately, a third engine fails, and the aircraft crashes into a mountainside. Carswell dies in a valiant attempt to save himself and an injured crewman. He is awarded the Medal of Honor, and Carswell Air Force Base is named in his honor.

27 October
Pilots from the 9th Fighter Squadron are the first to fly missions from bases in the Philippine Islands since 1942. They are based at Tacloban airfield flying P-38 Lightning fighters. Major Richard Bong is among the first pilots to score victories from this base in the Philippines.

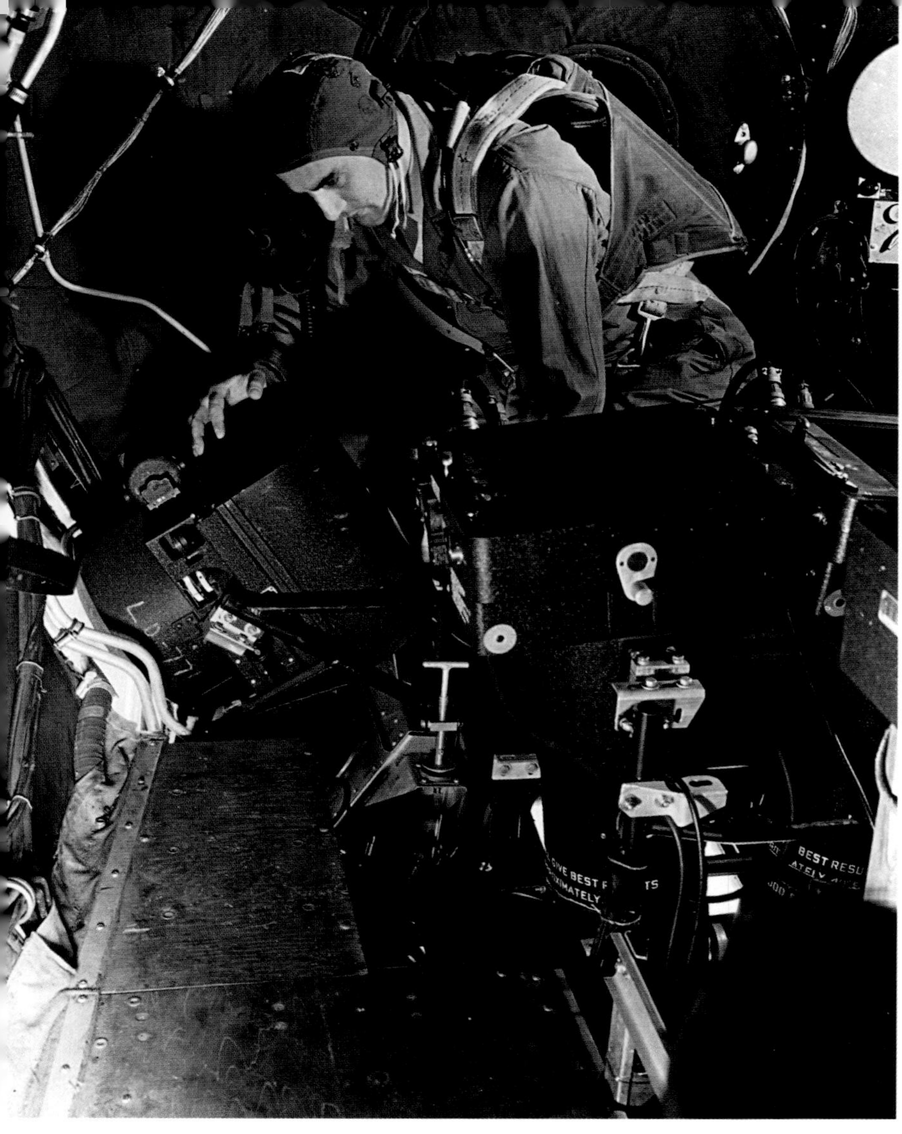

By this date, FEAF fighters, many Fifth Air Force P-38s, are striking targets throughout the Philippine Islands—primarily airfields in the central Philippines, on the northeast peninsula of Celebes, and on Halmhera. Direct support and interdiction will continue until Japanese resistance is finally crushed.

28 October
The XXI Bomber Command flies its first combat mission from the Marianas when 14 B-29s hit submarine pens on Dublin Island. The mission is less than a success, and Brigadier General Hansell is forced to abort due to aircraft problems.

1 November
A photo reconnaissance version of the B-29, the F-13, makes a flight over Tokyo, Japan. It is the first flight by an American aircraft over the city since April 18, 1942—the Doolittle Raid.

3 November
The Japanese launch Fu-go balloons by the thousands against the United States. These weaponized balloons are intended to ascend into the jet stream, which at times reaches upwards of 150 miles per hour, and detonate after reaching the American mainland. These paper or rubberized silk balloons are launched over a five-month period. Several hundred actually reach American soil, and 285 cause explosions from the West Coast as far east as Michigan. These balloons are 32 feet in diameter and hold nearly 20,000 cubic feet of hydrogen. Most are launched from the island of Honshu, Japan. The first report of such a balloon reaching the U.S. is received by officials of the Zone of the Interior the next day, November 4.

5 November
Twentieth Air Force sends 24 B-29s from the Marianas to attack Iwo Jima airfields. This attack begins tactical operations to prepare for the invasion, which will take place in February. From Calcutta, 53 B-29s are sent to bomb targets in Singapore. These bombers seriously damage the King George VI Graving Dock, making it unusable for several months.

6 November
Beginning this date, Seventh Air Force B-24s fly a series of missions through December 24, during which they will lay mines in a number of anchorages throughout the Bonin Islands.

8 November
During a strike against Iwo Jima by 17 Twentieth Air Force B-29s, the XXI Bomber Command loses its first bomber during combat operations when it is forced to ditch after being attacked by Japanese aircraft that drop phosphorus bombs into the B-29 formations.

10 November
Fifth Air Force sends 36 B-25 Mitchell medium bombers to attack a Japanese convoy near Ormoc Bay, Philippines. Three ships are destroyed.

Opposite, above: *Interior of the F-13 (B-29 reconnaissance version) with mapping cameras and operator. (NASM)*

Opposite, below: *Thousands of Japanese Fu-go bomb-carrying balloons were launched against America. (NASM)*

Right: *B-29 bombardiers often used Mount Fuji as an initial point before attacking Tokyo and surrounding areas. (USAF)*

24 November
Flying their first mission against Japan from the Marianas after their combat training sorties are completed on November 11, 88 B-29s of the XXI Bomber Command bomb the Mushino Aircraft Factory in Tokyo. It is the first bombardment of the capital since Doolittle's Raiders attacked on April 18, 1942. Brigadier General O'Donnell leads the mission, and his co-pilot is Major Robert K. Morgan, the pilot of the famous B-17 *Memphis Belle.* The mission is less than perfect; of the original 111 bombers launched, 17 abort, 50 bomb secondary targets, and a handful cannot drop their bombs at all due to mechanical problems. One B-29 is rammed by a Japanese fighter, the first XXI Bomber Command loss to direct enemy action, and one is forced to ditch when it runs out of fuel on the way back to the Marianas.

8 December
Major Richard I. Bong, Fifth Air Force, though technically assigned as a gunnery instructor, volunteers for repeated P-38 combat missions over Balikpapan, Borneo, and Leyte area of the Philippines. His aggressiveness results in his shooting down eight enemy airplanes during the period of October 10 to November 15. General Kenney had also submitted the recommendation for the Medal of Honor to recognize Bong as America's "Ace of Aces."

After a down period of more than one week, more than 60 Marianas based Twentieth Air Force B-29s combine with Seventh Air Force P-38s, B-24s, and Navy cruisers to attack Japanese airfields and aircraft on Iwo Jima. The targets are home to enemy forces launching limited attacks against U.S. bases in the Marianas. These attacks continue until early 1945, during which 11 B-29s are destroyed on the ground and 43 more are damaged.

13 December
Twentieth Air Force B-29s from the Marianas strike the Mitsubishi aircraft engine factory at Nagoya. Bombing accuracy seems improved, as significant damage is done during the raid.

15 December
The Battle of Luzon begins this date and will continue until July 1945. This massive campaign provides a major base of operations for the eventual invasion of Japan planned for the coming summer. FEAF aircraft support the invasion by striking airfields and shipping, and providing close support during the battle.

17 December
Colonel Paul W. Tibbets takes command of the newly activated 509th Composite Group at Wendover Field, Utah. Under Tibbets' command, they will train to deliver the first nuclear weapons in combat.

Major Richard I. Bong shoots down his 40th enemy aircraft in the Pacific. Bong, who had shot down his first Japanese aircraft on December 27, 1942, is grounded by General George Kenney after achieving his 40th victory. Bong ends the war as America's leading ace.

Left: *"Major Tom McGuire Flying in* Pudgy V *with the 431st FS," by Francis McGinley. (USAF Art Collection)*

Below: *Five stars for airpower! General of the Army, Hap Arnold. (Robert and Kathleen Arnold Collection)*

18 December
Flying from Chengtu, 84 B-29s carry out their first firebombing raid of the war when they attack the docks at Hankow, China, along the Yangtze River. Joining the attack are 200 Fourteenth Air Force aircraft.

19 December
Fourth Air Force fighters are launched to search for a reported Japanese balloon spotted over Santa Monica, California. No balloon is found.

21 December
General Henry H. "Hap" Arnold is promoted to the rank of General of the Army. He remains the only airman to hold five-star rank.

25–26 December
Major Thomas B. "Mickey" McGuire Jr., Fifth Air Force, volunteers to lead a P-38 fighter escort mission for bombers targeting Mabalacat Airdrome—a base for kamikaze operations on Luzon, Philippine Islands. Outnumbered 3 to 1 by the Japanese, he continues to fight after his guns jam. He risks his own life to save a crippled bomber while downing several enemy aircraft. McGuire is killed in action on January 7, 1945, over Los Negros Island. He scored 38 victories, second only to Major Richard I. Bong, and is awarded the Medal of Honor.

31 December
Brigadier General Hansell closes his headquarters on Saipan and moves the XXI Bomber Command Forward Echelon Staff to Guam to join the ground echelon already in place.

1945

2 January
Tenth Air Force troop carriers launch an amazing 546 sorties to frontline areas and resupply bases in China. During this year, they will average more than 500 sorties each day that they fly.

Far East Air Forces (FEAF) continues to attack multiple targets in the Philippines. P-38s and A-20s strike San Fernando Harbor. B-24s bomb Clark Field (still occupied by the Japanese), while B-25s bomb Batangas and enemy airfields in the central Philippines. Other FEAF aircraft conduct armed reconnaissance missions across the Philippines. These missions will continue for the remainder of the war.

5 January
Operation GRUBWORM is completed. One month after its initiation, two entire Chinese divisions, Chinese Sixth Army Headquarters, a heavy mortar company, 249 American soldiers, and two portable surgical hospitals have been airlifted from Burma to China into the combat zone. In all, more than 1,300 transport sorties are required to complete the operation; only three aircraft go down while completing the mission.

7 January
FEAF operates jointly with Third Fleet in attacking enemy airfields in northern Luzon. In the largest joint mission of the war in the southwest Pacific, more than 130 light and medium bombers are launched against five airfields.

9 January
AAF participates in the opening strikes of the Luzon and Philippines Campaign as amphibious forces of the U.S. Army land at Lingayen Gulf at 9:30 a.m. FEAF aircraft strike multiple targets and provide close support to the invasion forces.

B-24s from the Eleventh Air Force bomb targets using H2X equipment when they attack Suribachi Bay airfield. This is the first use of such equipment by the Eleventh.

10 January
Thirteenth Air Force moves its operating base again, from New Guinea to Leyte as part of the FEAF offensive in the Philippines.

11 January
Captain William A. Shomo, Fifth Air Force, flying an F-6D (a reconnaissance version of the P-51 Mustang) over Luzon, Philippines, spots enemy aircraft and orders an attack against unknown enemy odds. Shomo destroys seven enemy aircraft and his wingman destroys three, all in one action—unparalleled success in the southwest Pacific. His achievement, earning him the Medal of Honor, is remarkable as tactical reconnaissance pilots rarely engage enemy planes. Two other pilots, one AAF and one Navy, had seven-victory days during the war, but neither received the Medal of Honor.

17 January
Ninety-one B-29s take off from Chengtu, China, on a bombing mission over Shinchiku, Formosa. About a dozen abort before reaching the target. It is the last mission launched from Chinese bases against Japanese targets. After this mission, the XX Bomber Command B-29 will move to India where Brigadier General Roger Ramey will take command until they join the rest of the B-29s in the Marianas in March. The 58th Bomb Wing, the last remaining in the XX Bomber Command, provides support for British and Indian ground forces in Burma by interdicting targets in Indochina, Thailand, and Burma. Oil refineries in distant Singapore and the East Indies are also attacked until redeployment in March. The capture of several of the Mariana Islands will allow a closer base of operations for B-29s against Japanese home islands.

20 January
Brigadier General Haywood S. "Possum" Hansell is replaced as commander of the XXI Bomber Command by Major General Curtis E. LeMay. Hansell, an excellent administrator, has suffered from inadequate numbers of aircraft, continuing mechanical deficiencies, and extremely strong high-altitude wind conditions that have negatively impacted bombing results. Brigadier General Roger M. Ramey assumes command of the XX Bomber Command.

22 January
Fifth Air Force B-24s with P-38 escort make their first strikes on Formosa. Other FEAF aircraft continue to support ground operations in south and central Luzon.

Above: *Major General Curtis E. LeMay replaced Brigadier General Haywood Hansell as commander of XXI Bomber Command on January 20, 1945. LeMay would begin the firebombing campaign a few weeks after taking command. (USAF)*

GREAT AIR FORCE NOVELS

Phil Meilinger

Left: *"Lieutenant Fred J. Christensen Bags Two Fw 190s in Combat Near Dummer Lake, 15 March 1944 (56th FG)," by George Guzzi. (USAF Art Collection)*

There are hundreds of novels dealing with the U.S. Air Force, Air Corps, and Army Air Forces. The best of these are entertaining, but more importantly, they also address overarching themes regarding air warfare. War is an intensely human act, and its outcomes are often total—people and nations die—so men and women are forced to confront the basics of existence. Life and death, duty, the morality of force, friendship, integrity, courage, leadership, and the needs of one versus the needs of the many are all worthy topics for an airwar novel.

In addition, there are clear themes in novels on air warfare. The first reflects the attitude Americans have felt toward war. In World War I, the first to witness extensive use of aircraft, men volunteered with a nobility of purpose that would strike many today as naïve. By war's end such feelings had been supplanted by pessimism and cynicism. One of America's top aces, Elliott White Springs, wrote in *War Birds: Diary of an Unknown Aviator:* "War is a horrible thing, a grotesque comedy. And it is so useless. This war won't prove anything. All we'll do when we win is to substitute one sort of Dictator for another. In the meantime we have destroyed our best resources. Human life, the most precious thing in the world, has become the cheapest."

World War II was so massive and total there was little room for such introspection, and the characters in these air novels are primarily concerned with survival. This would again change in the wars of Korea and Vietnam. Especially regarding the latter, the novels are almost universally bitter—bitter regarding restrictive Rules of Engagement, an inane strategy dictated from Washington, and the burdens of a corrupt South Vietnamese ally.

There are other trends in airwar novels. Regarding World War II, more books have been written about air warfare over Europe than about the Pacific Theater, and the former tends to focus on the B-17s of the Eighth Air Force, usually narrated by a pilot. Far fewer are B-24 accounts, or those told by

Right: *"B-17s of the 91st bomb Group Over Germany," by George Guzzi. (USAF Art Collection)*

navigators, bombardiers, or enlisted personnel.

In contrast, Korean War air novels focus on fighter pilots and their duels in "MiG Alley." The literature of the Vietnam War is similarly biased toward fighter pilot accounts—usually those who flew over North Vietnam rather than over the south.

A dominant theme in all these novels is death. Aviators and their families constantly face the fact that each mission flown may be the last; every dinner in the mess may reveal empty chairs; every knock at the door may bring a telegram. Death was a constant companion, and it was often a solitary experience. Aircraft took off, but sometimes they simply disappeared. In James Salter's *The Hunters*, one pilot reflects:

> *You lived and died alone, especially the fighters. Fighters. Somehow, despite everything, that word had not become sterile. You slipped into the hollow cockpit and strapped and plugged yourself into the machine. The canopy ground shut and sealed you off. . . You were as isolated as a deep-sea diver, only you went up, into nothing, instead of down. You were accompanied. They flew with you in heraldic patterns and fought alongside you, sometimes skillfully, always at least two ships together, but they were really of no help. You were alone. At the end, there was no one you could touch. You could call out to them, as he had heard someone call out one day going down, a pitiful, pleading "Oh, Jesus!" but they could touch you not.*

Another enduring topic is that of friendship. The voices here speak of the unshakeable bond that exists between those in war, a bond forged by shared experiences, danger, fear and a sense of loss. Sometimes that loss involves those who have already fallen, but it also infers a loss of innocence, of a life that once was but that can never be again.

Here are some suggestions for the Top Ten Air Force Novels, arranged in a roughly chronological fashion:

The War Lover, by John Hersey (the crew of a B-17 in England)

Goodbye Mickey Mouse, by Len Deighton (a P-51 squadron in England)

Island in the Sky, by Ernest K. Gann (a ferry pilot goes down in the Canadian wilderness)

Guard of Honor, by James Gould Cozzens (racism and fatigue at a peacetime training base)

Captain Newman, M.D., by Leo Rosten (survivors of air combat in a psychiatric ward)

The Hunters, by James Salter (F-86 fighter pilots in the Korean War)

Cassada, by James Salter (leadership and followership in a Cold War fighter squadron)

The Laotian Fragments, by John Clark Pratt (the clandestine war in Laos)

Termite Hill, by Tom Wilson (electronic warfare in the air over North Vietnam)

Cadillac Flight, by Marshall Harrison (F-105s against North Vietnam, and Washington)

Left: *Strong surface winds, demonstrated by the smoke in the distance, were always a factor in accurate bombing. (USAF)*

Below: *A C-46 Commando traverses the dangerous Himalayas delivering supplies to China. (USAF)*

24 January
Twentieth Air Force B-29s join Seventh Air Force B-24s in bombing Iwo Jima. Additional B-24s act as spotters for naval bombardment.

Eleventh Air Force fighters shoot down a Japanese Fu-go balloon southeast of Attu.

FEAF B-24s begin attacks on Corregidor. These strikes continue for three weeks in anticipation of Allied amphibious landings there.

25–26 January
More than 70 Twentieth Air Force (XX Bomber Command) B-29s mine Singapore Harbor, Cam Ranh Bay, Pakchan River, Phan Rang Bay, and other land approaches.

27 January
Twentieth Air Force completes the redeployment of all B-29s from Chengtu to bases in India. The move is largely a result of untenable logistics dependent upon supplies flown into the theater over the Hump. Maintaining the XX Bomber Command in China takes approximately 15 percent of total supplies flown in by the Air Transport Command.

Marianas-based B-29s attack targets near Tokyo and meet heavy enemy fighter attacks that shoot down five B-29s of the original 130 launched on the mission. Four more Superfortresses ditch or crash-land on their return to base. B-29 gunners claim dozens of enemy fighters are shot down during the engagement.

28 January
After several weeks of aerial attacks, Clark Field, one of the most important air bases in the Philippines, is retaken by U.S. ground forces. Clark had fallen to the Japanese in January 1942.

The Burma Road is reopened, allowing supplies to flow over land from Ledo to Kunming. Although this takes some pressure off of the CBI Airlifters who have been flying the Hump since 1942, they continue to launch approximately 500 sorties during each flying day. Relocation of B-29s previously based at Chengtu allows for more supplies for Tenth and Fourteenth Air Force aircraft and advancing ground troops.

4 February
Twentieth Air Force launches two B-29 wings against the Japanese mainland for the first time. Nearly 100 bombers attack Kobe and Natsusaka. Japanese fighter defenses are strong, and although only one B-29 is shot down, 35 others are damaged.

5 February
In the heaviest bombardment of Corregidor to date, 60 FEAF B-24 Liberators pound the island. B-25s, and additional B-24s continue support missions for Allied ground troops.

10 February
Twentieth Air Force B-29s strike the Nakajima aircraft plant at Ota, Japan. Of the 84 XXI Bomber Command airplanes that reach the target, 12 are lost to fighters. Fighter escort for the B-29s is not yet available.

12–14 February
Marianas-based B-29s fly reconnaissance missions for U.S. naval forces approaching Iwo Jima. Two dozen B-29s bomb targets there to suppress anti-aircraft batteries. Seventh Air Force relentlessly attacks defensive targets on Iwo Jima, anticipating the arrival of landing forces that will take the island.

16 February
AAF C-47s drop more than 2,000 paratroopers, "The Rock Force," on Corregidor in support of the amphibious operations to secure Manila Bay. Hundreds of FEAF aircraft bomb a complex target set around the island for most of the day. FEAF will continually pound targets in support of the ground troops. At one point, Marine Corps Corsairs are placed under the tactical control of the FEAF commander to carry out napalm strikes against airfields and other tactical targets in the central Philippines.

Tenth Air Force transports fly more than 600 resupply sorties over the Hump. This pace continues though the end of the war, weather permitting. By war's end, ATC will deliver more than 700,000 tons of material to China at a loss of 910 aircrew.

17 February
The final paratrooper action of the Pacific campaign occurs. A company of troops is dropped near a prison camp in Manila, where interned civilians and military personnel have been kept since 1942.

19 February
Marines land on the Japanese-occupied island of Iwo Jima. The strategic location of the island will provide an emergency landing field for B-29s, will allow escort fighters to accompany the bombers on long-range missions over Japan, and will eliminate an important Japanese staging location for Pacific operations. Twentieth Air Force launches more than 150 B-29s in an effort to draw Japanese aerial reinforcements away from Iwo Jima. More than 110 of these bomb targets in Tokyo and six bombers are lost to enemy fighters. The island is finally secured by the Marines Corps on March 26, at the cost of 6,520 American lives.

Right: *"B-24s Bombing Corregidor—February 1945." by Robert Laessig. (USAF Art Collection)*

Tenth Air Force engages in combined operations with British and Chinese ground forces as they push forward into China. B-25s cut roads and bridges, while fighter-bombers attack enemy troop concentrations at the front line of the battle.

25–26 February
Lieutenant General Millard F. "Miff" Harmon, the commander of the AAF, Pacific Ocean Areas, is lost when his B-24 Liberator disappears during a flight between Kwajalein Island and Hawaii. No wreckage or survivors are ever found; he is declared dead one year after the accident.

27 February
The last B-29 wing remaining in Calcutta begins deployment to the Marianas. The move is completed by June 6.

Left: *Marines raising the flag on Mount Suribachi, February 23, 1945. (Tom Lovell, Marine Corps Art Collection).*

Below: *A Fu-go balloon bomb is detected over the Atlantic Ocean by Eleventh Air Force aircraft. Only one fatal detonation occurred in the U.S. (NASM)*

23 February
Owing much to air support from the AAF, the Marines, and the Navy, brave Marines and one Navy medic raise the American flag on Mount Suribachi on Iwo Jima.

24 February
Using only incendiary bombs, 105 XX Bomber Command B-29s set fire to the Empire Dock area in Singapore. More than 40 percent of the warehouse space is destroyed. This is the last time that India-based XX Bomber Command will launch more than 100 bombers on any one raid.

25 February
Twentieth Air Force B-29s bomb urban Tokyo using incendiary bombs from high altitude. This raid is the largest flown by XXI Bomber Command and the first time that three wings of B-29s have attacked the same target.

Right: *The city of Tokyo burns after low-level incendiary attacks are directed by Major General Curtis LeMay. One-quarter of a million buildings were incinerated, and approximately 100,000 people died in the flames. (USAF)*

Below: *The devastation caused by the firebombing of Tokyo spread for 16 square miles. (USAF)*

4 March
A B-29 makes the first emergency landing on a captured airfield on Iwo Jima. Through the end of the war, more than 2,400 such landings will be made on the small but important island.

6 March
Seventh Air Force deploys 28 P-51s and 12 P-61 Black Widow night fighters to Iwo Jima. The bases on Iwo Jima allow fighter escort to accompany B-29s to their targets over Japan.

9–10 March
Departing from three bases in the Marianas, 325 B-29s from XXI Bomber Command under Major General Curtis LeMay, launch to strike Tokyo, Japan, on an overnight, low-level firebombing raid. Although many abort after takeoff, 279 B-29s reach the target. The attack burns approximately 16 square miles of the city. It is estimated that between 80,000 and 100,000 are killed and more than 250,000 buildings are destroyed in the raid. Fourteen bombers, each carrying a crew of 10 men, are lost to anti-aircraft fire during the raid. This raid signals the shift from high-altitude precision bombardment to low-level night area bombing. Bombing altitudes for this raid varied between 4,900 feet and 9,200 feet. This aerial attack is generally considered to be the most devastating single air raid in history.

11–12 March
In a similar raid to that flown by Twentieth Air Force B-29s over Tokyo two days before, 285 aircraft firebomb the city of Nagoya, Japan, causing similar results.

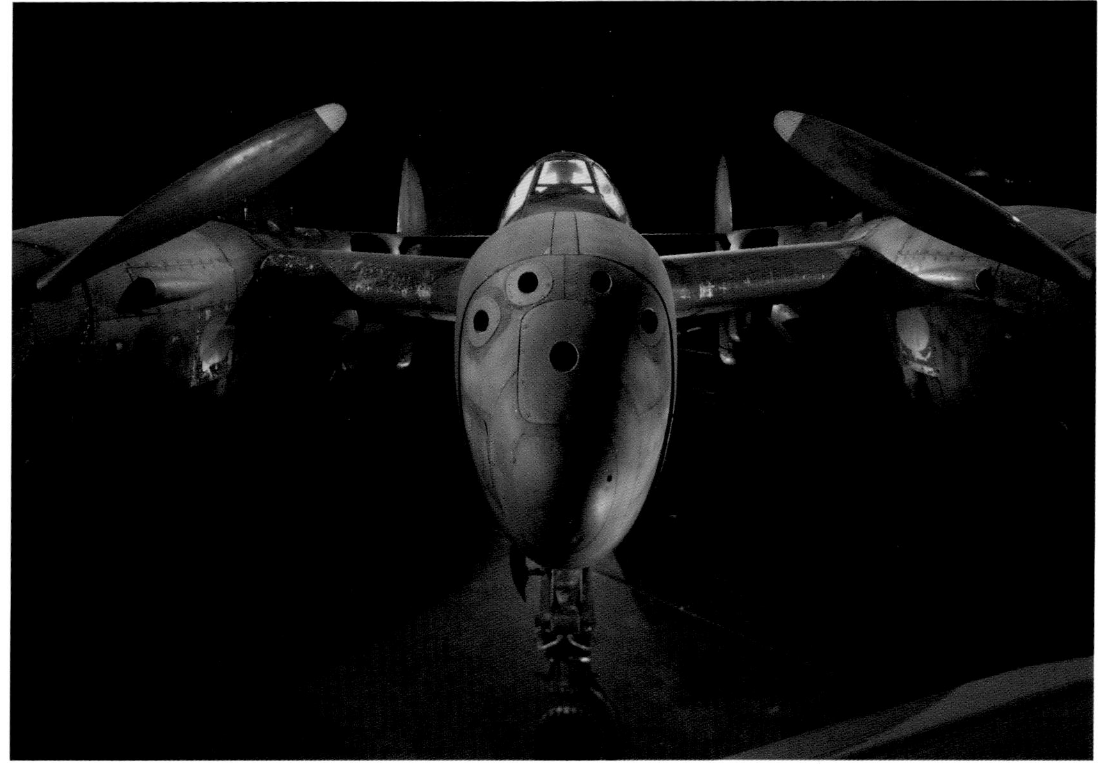

Left: *The Lockheed P-38 flew in every combat theater of war. This particular Lightning was flown by Major Richard I. Bong on April 16, 1945, at Wright Field during fuel system tests and is now on display at the National Air and Space Museum, Steven F. Udvar-Hazy Center, near Dulles International Airport. (NASM)*

13 March
Tenth Air Force P-47s provide support for the Chinese 50th Division by striking Japanese forces along the Namtu River. P-38s continue to hit targets in central Burma.

13–14 March
Twentieth Air Force launches its third firebombing raid. This time, 274 bombers strike Osaka, destroying eight square miles of the city center.

16–17 March
Twentieth Air Force launches its largest raid to date when more than 300 XXI Bomber Command aircraft firebomb Kobe, Japan. One-fifth of the city is destroyed by the 2,300 tons of incendiary bombs dropped on the city. Kobe was home to a large aeronautical research facility, including a wind tunnel designed in the late 1920s for the Japanese by Professor Theodore von Kármán, then a young fluid dynamics teacher at the University of Aachen, Germany.

18–19 March
The city of Nagoya is once again targeted for firebombing by the XXI Bomber Command. Nearly 300 B-29s strike the city from low altitudes during the night. This is the last firebombing raid flown in March.

21 March
In the Zone of the Interior, a P-63 from Walla Walla Airfield intercepts, chases, and then shoots down a Japanese Fu-go balloon. The pilot must refuel twice along the way but eventually shoots the balloon down over a desolate area near Reno, Nevada.

27–28 March
In the first mine-laying mission flown by B-29s flying from the Marianas, about 100 bombers drop mines in the Shimonoseki Strait between Honshu and Kyushu. Many other mine laying operations are flown during the rest of the war. More than 150 other B-29s begin to strike targets in support of the upcoming invasion of Okinawa.

29 March
In the last mission flown by XX Bomber Command aircraft operating from India, 24 B-29s attack oil fields on Bukum Island late at night.

30 March
XX Bomber Command launches its final B-29 mission from bases in India when 26 bombers attack Japanese targets on Bukum Island near Singapore.

31 March
As a diversion for the invasion of Okinawa scheduled for the following day, 137 B-29s hit the airfield at Omura and the Tachiahari machine plant.

1 April
The Ryukyu Islands Campaign begins with the invasion of Okinawa. By the time operations have ended, this will be the largest sea-land-air battle in history and the most deadly for U.S. troops during the

Right: *The PT-19 Cornell was used as an observation plane above the Burmese jungle. (NASM)*

Below: *President Franklin D. Roosevelt's funeral procession. (Naval Historical Center)*

war. Almost 19,000 die during the struggle to subdue Okinawa. Japanese casualties are far worse. It is estimated that more than 200,000 soldiers and civilians died or killed themselves during the fighting, which will last until June 21.

2 April
In the early morning hours, Twentieth Air Force B-29s mine the harbors around Hiroshima and Kure and also bomb the Nakajima aircraft factory in Tokyo.

4 April
A Tenth Air Force YR-4 helicopter assigned to the Air Jungle Rescue Detachment rescues a downed PT-19 pilot in enemy territory in Burma.

5 April
The Soviet Union notifies Japan that it wishes to abolish a 1941 document that assures five years of Russo-Japanese neutrality.

7 April
Twentieth Air Force is now able to launch escort fighters from the newly captured island of Iwo Jima. The first of these escort missions is flown this date when 280 B-29s, accompanied by 91 P-51s, attack three separate targets in Japan.

12 April
President Franklin Delano Roosevelt dies. Harry S. Truman is sworn in as the new President.

Staff Sergeant Henry E. "Red" Erwin, Twentieth Air Force, was tasked with dropping phosphorus smoke markers out of the mission lead B-29 over Koriyama, Japan. One of the phosphorus markers explodes in the ejector chute and shoots back into the aircraft, blinding him. Erwin realizes that if the burning bomb remains in the plane, the entire crew will be lost. Now blind and aflame, Erwin takes the smoke marker between his forearm and his body, struggles to the cockpit, and throws it out of the co-pilot's window, saving the aircraft and the crew. Remarkably, Erwin survives his extensive third-degree burns and other injuries. In a personal letter, General of the Army Hap Arnold writes that he regards Erwin's action "one of the bravest in the records of this war." Erwin is one of only four enlisted men to receive the Medal of Honor during World War II.

Left: *The 509th lead echelon arrived on Tinian in May 1945. The B-29* Enola Gay *arrived on July 6. This color photo was taken after the mission had been successfully completed. (NASM)*

13–14 April
More than 330 Twentieth Air Force B-29s bomb the Tokyo arsenal area.

Eleventh Air Force P-38s and P-40s scramble to intercept unknown radar returns. They turn out to be 11 Fu-go balloons. The pilots shoot down nine of them over the Aleutians.

15–16 April
Nearly 300 B-29s strike the cities of Kawasaki and Tokyo during the night.

16 April
For the first time, P-51s from Iwo Jima attack targets in Japan. A B-29 provides navigational aid for the fighters to and from their targets at Kanoya. North American B-25s will pick up the navigational duties on all subsequent P-51 missions, which will continue until August 14.

17 April
From this date until May 11, priorities for the Twentieth Air Force shift from strategic bombardment of Japan to direct support and interdiction in the campaign to take Okinawa. The XXI Bomber Command will assign three-quarters of its missions to destroying aircraft and enemy operations on 17 airfields on Kyushu and Shikoku. From these airfields, the Japanese launch attacks upon the invasion forces at sea and on the ground at Okinawa—including kamikaze attacks.

23 April
Navy PB4Y Liberator bombers (Consolidated B-24-type) launch two Bat missiles against Japanese ships. This is the first combat use of automatic homing missiles in history.

1 May
As the Japanese begin to withdraw from southern China, Fourteenth Air Force concentrates on attacking transportation targets to impede the enemy's mobility in retreat. Lieutenant General George E. Stratemeyer assumes command of the AAF in the China Theater with his headquarters at Chunking. Both Tenth and Fourteenth Air Forces fall under his control.

3 May
Rangoon, Burma, falls to the Allies. Forces of the Indian Army occupy the city. Apart from cleanup operations, the war against the Japanese in Burma is over. For the rest of the month, AAF operations are drastically reduced, and Tenth Air Force is relieved from combat duty and redeployed to Piardoba, India. A single squadron of P-38s remain in Burma to patrol roads into China.

5 May
Suffering the only U.S. casualties on American soil inflicted by enemy action during World War II, a woman and five children are killed when a Japanese submarine-launched Fu-go balloon bomb explodes near Lake View, Oregon.

8 May
V-E Day (Victory in Europe). World War II ends in Europe with Germany's unconditional surrender.

10 May
A high-priority missile program is initiated to defend against Japanese suicide Baka bombs. Using standard Jet Assisted Takeoff (JATO) units developed at Caltech by the "suicide club" of rocket scientists, the Naval Aircraft Modification Unit develops the Little Joe ship-to-air weapon. The missile is tested in July for the first time.

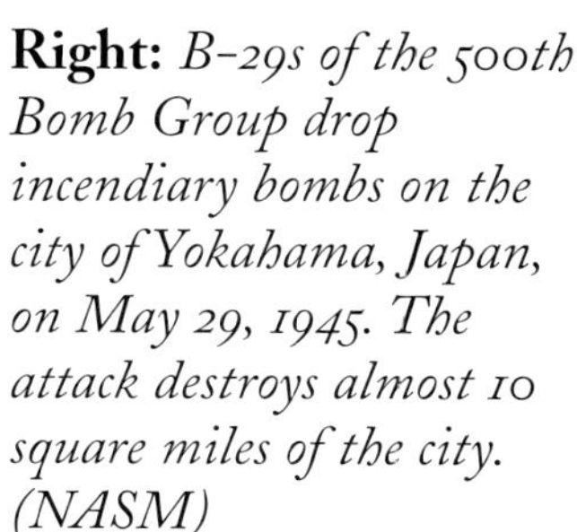

Right: *B-29s of the 500th Bomb Group drop incendiary bombs on the city of Yokahama, Japan, on May 29, 1945. The attack destroys almost 10 square miles of the city. (NASM)*

Eleventh Air Force and Fleet Air Wing Four complete their most successful and largest operation to date. A dozen B-24s bomb enemy ships at Kataoka naval base and follow up with photo reconnaissance of Paramushiru. Later, 16 B-25s based on Attu hit shipping targets as well.

14 May
Twentieth Air Force launches its first raid in which four B-29 wings attack the same target. North Nagoya is attacked by 472 bombers. During the raid, 11 B-29s are lost. The XXI Bomber Command now consists of the 58th, 73d, 313th, and 314th Bomb Wings.

16 May
Far East Air Force P-38s attack the Ipo Dam area of Luzon, Philippines. The attack is made by nearly 100 aircraft that drop napalm canisters in the largest single use of that weapon in the war.

17 May
During early morning darkness, 457 B-29s strike Nagoya once again, this time in the southern part of the city.

18 May
The Advanced Air Echelon of the 509th Composite group arrives on Tinian and sets up shop on North Field. The 509th has been tasked to drop the atomic bomb on Japanese targets if needed.

19 May
Twentieth Air Force sends 272 B-29s to bomb the city of Hamamatsu, Japan.

23–24 May
Of the 562 B-29s launched this night, 520 reach the designated industrial area target on the west side of Tokyo Harbor. This is the largest number of B-29s to fly during a single mission during World War II. Seventeen B-29s go down during the operation.

25 May
Operation OLYMPIC, the invasion of Japan, is authorized by the Joint Chiefs of Staff. The invasion date is set for November 1.

Twentieth Air Force sends 464 B-29s to strike targets in Tokyo once again. The price is high on this raid as 26 bombers do not return home. This is the greatest loss of B-29s in one day during World War II. The VII Fighter Command based on Iwo Jima is assigned to Twentieth Air Force both operationally and administratively.

29 May
Firebombing raids begin again as 454 B-29s, escorted by 101 P-51s from Iwo Jima, attack Yokohama and destroy the main business district. About nine square miles of the city are in ruins. Japanese offer stiff fighter defense and shoot down seven bombers and three fighters.

1 June
During a raid flown in marginal weather, 458 B-29s attack the city of Osaka. Of the 148 fighters assigned to meet the formation, 27 collide when severe turbulence ruins their formation. Only a few dozen fighters meet the bombers as escorts. Ten B-29s do not return from the raid.

Left: *The contingent of SILVERPLATE B-29s—those modified to drop atomic bombs—fly over the runway at Tinian. (NASM)*

5 June
Kobe, Japan, is once again the target of an incendiary attack. Four square miles of the town are burned by the 473 attacking bombers. Eleven B-29s are lost on the raid.

7 June
Twentieth Air Force strikes Osaka with high explosive bombs and incendiary weapons. The mission is accomplished by radar bombing techniques and destroys approximately 55,000 buildings in the city. Another small group of B-29s mine the Shimonoseki Strait.

9 June
Japanese Premier Suzuki states that Japan will fight to the bitter end rather than accept unconditional surrender.

A combined force totaling 110 B-29s attack aircraft factories in Akashi, Nagoya, and Narao, Japan.

During a mission flown by Eleventh Air Force B-25s, one of the bombers is shot down over the Kamchatka Peninsula by Soviet anti-aircraft gunners. Another is damaged and crash-lands in Petropavlovsk. This is the first instance of Soviet fratricide of an American aircraft.

11 June
Combat aircrew of the 509th Composite Group begin to arrive on Tinian with their specially modified SILVERPLATE B-29s. All gun turrets have been removed, extra fuel tanks have been installed, new propellers have replaced factory models, and special radar and radio monitoring equipment has been installed for the special mission planned for this unit.

14 June
The Joint Chiefs of Staff are advised to prepare for the immediate occupation of Japan should a sudden end to the war occur.

15 June
Forty-four B-29s fly the final firebombing raid against large Japanese cities as Osaka is once again targeted.

17–18 June
Incendiary attacks against a series of smaller Japanese cities begin when more than 450 B-29s target Omuta, Hamamatsu, Yokkaichi, and Kagoshima. Additional B-29s mine the waters around Kobe and the Shimonoseki Strait.

Night and day intruder missions begin this date, flown by Seventh Air Force P-47s and P-61s over Kyushu and the Ryukyu Islands. A dozen such missions are accomplished during June.

19–20 June
Twentieth Air Force targets Toyohashi, Fukuoka, and Shizuoka with 480 B-29s. Mining operations also continue throughout Japanese waters.

22 June
After a bloody struggle to take Okinawa, the island is declared "captured" by Allied forces. Nearly 50,000 casualties are suffered by American forces during the four-month campaign; more than 12,000 of these are killed.

Right: *Since January 1945, AAF medium bombers had been crucial to success in the Pacific. Here, A-20s attack Japanese air forces on Luzon just before Clark Field was recaptured by the Allies. "Roarin 20s Over Clark Field," by Steve Ferguson. (USAF Art Collection)*

FEAF bombers continue to pound Balikpapan in preparation for Allied landings there scheduled for early July.

Twentieth Air Force B-29s target several aircraft plants and the naval arsenal at Kure using nearly 300 Superfortresses.

26 June
Widespread attacks are flown by more than 450 Twentieth Air Force B-29s against aircraft plants, light metal works, weapons arsenals, oil refineries, and the city of Tsu.

28–29 June
The towns of Okayama, Sasebo, Moji, and Nobeoka are targets of incendiary bombardment by 487 B-29s.

30 June
The 509th Composite Group begins training missions operating out of Tinian in the Marianas. Each crew is expected to fly five or six practice flights to familiarize themselves with the Pacific Theater. Many will drop large practice bombs, called pumpkins, to simulate the effect of a 10,000-pound bomb dropped from their aircraft.

1 July
FEAF B-24s continue to attack Balikpapan as the Australians carry out amphibious landings there. B-25s, P-38s, and B-24s support the landings by hitting nearby enemy airfields. These attacks continue into the next day as Australian forces capture the island and its oil facilities.

Twentieth Air Force launches more than 530 B-29s with incendiary bombs against the cities of Ube, Kure, Shimonoseki, and Kumamoto. Mines are also dropped in Japanese waters.

In the Zone of the Interior, Headquarters, Weather Wing is redesignated Headquarters, Army Air Forces Weather Service.

3 July
Fighters from the Fifth Air Force fly their first missions over Japan.

Twentieth Air Force B-29s launch incendiary attacks with 560 bombers against the cities of Kochi, Himeji, Takamatsu, and Tokushima.

6 July
Lieutenant General Stratemeyer officially assumes command of the AAF, China Theater.

THE NORDEN BOMBSIGHT

Stephen L. McFarland

Left: *The bombardier sat in the nose of the B-29. Note how the mission "crush" hat actually got that way. (USAF)*

Opposite: *The Norden Bomb Sight—the most accurate mechanical high-altitude, horizontal bombsight ever built. (USAF)*

Product of a secret Navy project in the 1930s to develop a synchronous bombsight capable of hitting maneuvering ships from high altitudes, the Norden bombsight was the analog computer at the heart of the Army Air Forces' strategy of daylight precision strategic bombing. Targeting enemy industries, the concept was to fly over defending armies and navies to strike at interconnected industrial networks. This strategy, developed at the Air Corps Tactical School between the world wars, would destroy the machines of war rather than the human beings who supported an enemy's war economy and avoid the trench warfare of World War I. The United States produced 90,000 Norden bombsights during World War II at a cost two-thirds that of the atomic bomb project. Norden sights aimed the first bombs dropped by the United States Army Air Forces against Germany on August 17, 1942, and the last bombs dropped on Japan on August 10, 1945. Mounted on B-17s, B-24s, B-26s, and B-29s, the Norden was the brain behind this revolution in warfare—one that also served in the Korean and Vietnamese conflicts, delivering acoustic sensors along the Ho Chi Minh Trail. It was the most accurate mechanical high-altitude, horizontal bombsight ever built.

Carl Norden, a Dutch citizen living in Brooklyn, designed his 50-pound, bomb-aiming device in the early 1930s to track the movement of an aircraft over the ground, automatically compensating for changes in altitude, wind direction, barometric pressure, speed, and dozens of other variables. It was synchronous; meaning the turning of gears in the bombsight matched the movement of the bomber aircraft over the ground, thereby fixing the device's optical telescope on the intended target. Linked to a C-1 Minneapolis-Honeywell automatic pilot, once locked on a target the Norden bombsight flew a bomber to that precise point in the sky where bombs could be dropped with a reasonable expectation of hitting a target. Wartime patriots claimed it capable of dropping a bomb into a pickle barrel from four miles up. With the exception of the Manhattan Project, the Norden was America's most closely guarded secret until combat losses delivered it into Axis hands.

The Norden bombsight, and the doctrine that gave it life, provided America with a moral high ground and launched a long-standing imperative to rely on technology to defeat America's enemies. Beyond the hyperbole that surrounded the device was a reality in war that belied the doctrine of daylight precision strategic bombing. Against Germany in World War II, Norden bombsights aimed more than seven million bombs at the German industrial infrastructure and transportation grid, but only 31 percent landed within one thousand feet of their aiming points. The strength of enemy defenses, bad weather, dust, smoke, poor intelligence information, and the enemy ability to repair bombing damage combined to limit the Norden's effectiveness that was so dependent on making visual contact with the target. Postwar studies indicated that as bombers flew higher to escape anti-aircraft artillery fire, each one-hundred feet of altitude added 6.1 feet in circular error. Each additional flak gun caused 4.5 feet of error. When defenders used smoke screens, almost 300 feet of error resulted. Wartime mass-production techniques caused inaccuracies more than five times greater than specifications allowed.

The Navy gave up on high-altitude bombing in favor of dive bombing, while the Army Air Forces organized 60 or so bombers into combat formations that dropped all bombs on a signal from a lead bombardier—carpeting the target zone with several hundred bombs with the hopes that one or more would hit something valuable rather than relying on Norden-guided precision. Of 123,586 tons of bombs dropped on German targets during the oil offensive of 1944–1945, only 4,326 tons hit anything significant. The greatest example of the failure of precision bombing was Curtis LeMay's firebombing campaign against 68 Japanese cities using Norden-equipped B-29s in the last nine months of World War II, causing more than 800,000 civilian and military casualties.

In war time, prewar development and training accuracies were not achieved. Norden-aimed bombs diverted critical quantities of enemy resources to defending Axis skies and to repairing the damage they caused. Offensive forces diverted to defense and weapons-not-produced justified the effort. America's pursuit of precision in aerial bombing began in World War I with British-supplied Wimperis and French-supplied Michelin bombsights, was accelerated by the Norden bombsight in World War II, and reached maturity with the laser-, television-, and GPS-guided weapons of today.

6–7 July
Twentieth Air Force B-29s firebomb the cities of Chiba, Akashi, Shimizu, and Kofu, sending 517 aircraft over the targets.

9–10 July
Twentieth Air Force firebombing targets for the day are Sendai, Sakai, Gifu, and Wakayama. An additional 60 bombers strike the oil refinery at Yokkaichi.

12–13 July
Continuing their deadly routine, Twentieth Air Force sends 453 B-29s to firebomb the cities of Utsonomiya, Ichinomiya, Tsuruga, and Uwajima. Another 53 bombers attack the petroleum center at Kawasaki.

14 July
Seventh Air Force is officially assigned to the Far East Air Forces (FEAF). The aircraft will be completely moved to Okinawa by July 28.

16 July
General Carl Spaatz takes command of the U.S. Army Strategic Air Force in the Pacific while Major General Curtis LeMay assumes command of Twentieth Air Force, in actuality the XXI Bomber Command since the inactivation of the XX Bomber Command. Until this date the Twentieth had been under the direct control of the Joint Chiefs of Staff through General of the Army Hap Arnold.

At the Trinity site in New Mexico, the first atomic bomb, sometimes called "the gadget," is successfully tested. The weapon produces a yield of 19 kilotons of TNT.

16–17 July
Four more Japanese cities are firebombed by 466 B-29 bombers—Numazu, Oita, Kuwana, and Hiratsuka.

19–20 July
Along with mine laying operations and attacks on oil facilities at Amagasaki, 470 B-29s firebomb the cities of Fukui, Hitachi, Chosi, and Okazaki.

20 July
The 509th Composite Group begins precision bombing practice over previously bombed Japanese cities. The practice missions, flown on this date and the 24th, 26th, and 29th of July, are intended to familiarize the crews with tactics that will be used during the actual atomic missions. An additional reason for these practice missions is to allow the Japanese to see small B-29 formations overhead during daylight hours—as if they were reconnaissance missions.

24 July
Using high-explosive weapons, 570 B-29s strike aircraft factories at Hando, Nagoya, and Takarazuka. Metal works at Osaka and the cities of Tsu and Kawana are also attacked.

Left: *The "Little Boy" atomic bomb as it is being loaded into the bomb bay of the B-29* Enola Gay *on Tinian. Dropped on August 6, it was the first atomic bomb used in combat in history. (NASM)*

Right: *A conglomeration of AAF aircraft get their final checks before being made ready for combat in this huge depot hangar. (USAF)*

26 July
The components of the first atomic weapon—"Little Boy"—arrive and are unloaded at Tinian. This small island in the Marianas is the home base for the 509th Composite Group tasked to deliver the bomb over a Japanese target.

The Potsdam ultimatum is issued by the Big Three, Winston Churchill, Franklin D. Roosevelt, and Joseph Stalin. Japan is told to surrender or face "utter destruction."

26–27 July
Firebombing continues when 350 B-29s attack the cities of Matsuyama, Tokuyama, and Omuta.

28–29 July
With no word from the Japanese, the firebombing continues as 471 B-29s attack the cities of Tsu, Aomori, Ichinomiya, Ujiyamada, Ogaki, and Uwajima. Seventy-six additional B-29s hit the oil refinery at Shimotsu.

29 July
FEAF aircraft strike targets throughout the theater. A-26 medium bombers, B-24s, B-25s operating out of Okinawa, and P-47s from Ie Shima strike a multitude of targets on the Japanese home islands. The A-26 target is the naval base and engine works at Nagasaki.

1 August
In the single largest operational day for Twentieth Air Force B-29s during World War II, 836 Superfortresses are launched; 627 firebomb the cities of Hachioji, Toyama, Nagaoka, and Mito; 120 bomb the Kawasaki oil plant; 37 drop mines in the Shimonoseki Strait.

2 August
Major General Nathan F. Twining assumes command of Twentieth Air Force. Major General LeMay moves to be Chief of Staff for USASTAF. Twining will rise to the rank of full general and become Chief of Staff of the U.S. Air Force in 1953. In 1957, Twining becomes the first USAF officer to hold the Chairmanship of the Joint Chiefs of Staff.

3 August
Fighter aircraft of the VII Fighter Command stationed on Iwo Jima fly 100 sorties over Tokyo, striking airfields and rail equipment.

5 August
FEAF aircraft strike targets from Luzon to Kyushu with impunity. More than 330 aircraft participate in this far-reaching aerial offensive. Sortie numbers will increase each day for the next 10 days.

5–6 August
Incendiary raids continue as 470 B-29s attack the cities of Saga, Mae Bashi, Imabari, and Nishinomiya-Mikage. Another 100 B-29s bomb the coal processing plant at Ube.

The VII Fighter Command is officially assigned to Twentieth Air Force for the duration of the war.

6 August
Colonel Paul W. Tibbets Jr., Commander of the 509th Composite Group, pilots the B-29 *Enola Gay*, named after his mother, from Tinian to Hiroshima, Japan, where he and his crew drop the "Little Boy" atomic bomb over the center of the city. After a flawlessly executed mission, the weapon detonates approximately 2,000 feet above the ground, almost precisely over the aiming point. Within moments, the bomb explodes in a colossal, churning, purplish fireball and destroys much of Hiroshima, immediately killing between 70,000 and 80,000 of its inhabitants. Two additional B-29s, the *Great Artiste* and aircraft number 91, accompany the *Enola Gay* to observe the explosion, drop measuring equipment, and take photographs.

About 100 Twentieth Air Force P-51s attack Tokyo from their bases on Iwo Jima.

FEAF aircraft continue to attack targets in Kyushu and the coast of Korea. Until the end of the war, these attacks will continue throughout southern Japan, the Ryukyus, and Korea.

America's "Ace of Aces," Major Richard I. Bong, dies when the F-80 Shooting Star he is flying stalls on takeoff. He ejects but is too low for his parachute to fully deploy.

Left, above: *The* Enola Gay *flight crew. (NASM)*

Left, middle: Enola Gay, *now restored, is on display at the Steven F. Udvar-Hazy Center near Dulles International Airport. (NASM)*

Left, below: *The cockpit. Colonel Tibbets sat in the left-hand seat as pilot in command. (NASM)*

Opposite, below: *The "Little Boy" detonated over Hiroshima just after noon on August 6, 1945. (USAF)*

Right: *Major Richard I. Bong, America's "Ace of Aces," died when he ejected out of the envelope during a flight test of an F-80 Shooting Star the same day that the B-29* Enola Gay *bombed Hiroshima. (NASM)*

Below: *"Atomic Landscape (Japanese Burial Detail)," by Robert M. Graham. (Army Art Collection)*

7 August
B-29s begin targeting Kyushu and are accompanied by FEAF P-47 fighters during the raid.

8 August
The Soviet Union declares war on Japan and then invades Manchuria the following day.

FEAF and Twentieth Air Force aircraft relentlessly pound away at targets on Kyushu, while additional B-29s attack Yawata with incendiary bombs. Later that day, 60 B-29s attack targets in Tokyo, while during the night another incendiary raid is accomplished over Fukuyama.

9 August
Major Charles W. Sweeney pilots the B-29 *Bockscar* from Tinian to Kokura, the intended target, but poor weather prevents visual bombing. Sweeney proceeds to the secondary target, Nagasaki, where weather is also poor, but a cloud break allows a visual attack. The "Fat Man," an implosion-type, plutonium weapon, is dropped over the city and detonates approximately 2,000 feet above the target area, creating nearly double the energy of the "Little Boy" bomb, a gun-type, uranium weapon. Approximately 35,000 occupants of Nagasaki are killed instantly, a number lower than might have been the case had Sweeney's bomb fallen more accurately on the aiming point.

9–10 August
Twentieth Air Force sends 95 B-29s to attack the Nippon Oil Refinery located near Amagasaki.

10 August
During a 95-plane B-29 strike on Amagasaki, Japan, a record average bomb load per plane of 20,648 pounds is delivered to the target. The ability to carry more than 10 tons of weapons in each airplane made the B-29 a ruthlessly efficient delivery system.

12 August
Soviet troops advance into Korea from the north.

13 August
Eleventh Air Force flies its final combat mission of the war. Six B-24s attack the Kashiwabara staging area by radar bombing and leave enormous smoke plumes behind.

Above: *"Last Aerial Combat of WW II," by William Reynolds. (USAF Art Collection)*

Right: *Repatriated Pacific Theater POWs are readied for their return home. More than 130,000 Americans were held as POWs around the globe during World War II. (USAF)*

14 August
The final B-29 missions are flown against targets throughout Japan. Some 400 bombers accompanied by P-51s drop mines and strike six cities with conventional bombs. A record total number of effective aircraft sorties—754 bombers and 169 fighters—are in the air. As the P-51s return to Iwo Jima, they strike airfields near Nagoya. These are the last U.S. fighter attacks flown against the Japanese.

14–15 August
The final night incendiary raid is flown by more than 160 B-29s which attack Kumagaya and Isezaki. A mission launched from the Marianas to attack the Nippon Oil Company in Tsuchizakiminato is the longest round-trip, unstaged mission flown during the war, covering 3,650 miles.

15 August
For the first time in history, the Japanese people hear the voice of their emperor over the radio as Hirohito personally broadcasts the Imperial Rescript of Surrender to a stunned nation.

All offensive actions against Japan end. General of the Army Douglas MacArthur is made Supreme Commander for Allied Powers. In this role, he will officiate at the formal surrender ceremony.

18 August
Two Consolidated B-32 Dominators fly a reconnaissance mission over Tokyo, Japan. They are attacked by more than a dozen Japanese fighters. During the attack, one U.S. crewman is killed and two more wounded by the fighters' guns. The Dominator crews shoot down two of the attackers and damage others. The aircraft are able to land at a base in Okinawa. This is the final combat action against Japan by American forces.

27 August
Twentieth Air Force B-29s drop supplies to Allied prisoners of war (POWs) held in Japan, Korea, and China. The first drop is made this date into Weihsien Camp near Peking, China. Over the next month, more than 900 sorties are flown and 4,470 tons of supplies are dropped to more than 150 POW camps. About 63,000 POWs are still alive throughout the theater.

28 August
Advanced forces arrive in the Japanese home islands, marking the beginning of the postwar occupation.

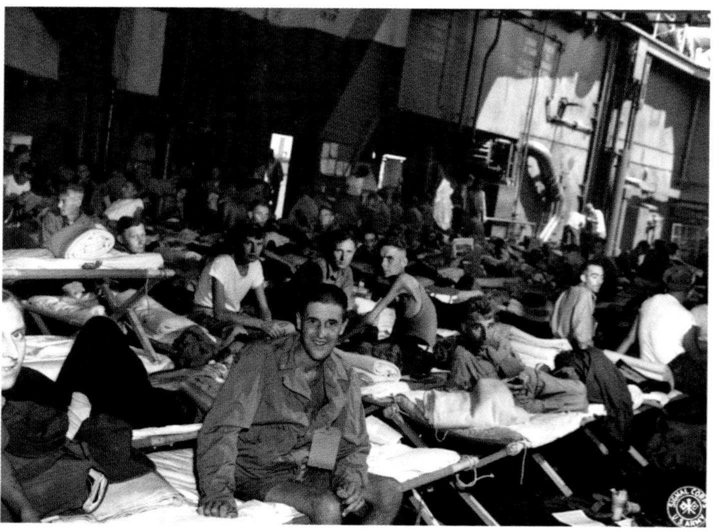

American ground forces begin coming ashore near Tokyo the next day. The occupation of Japan will continue through 1952.

29 August
While a B-29 is dropping supplies to POWs in Korea, a Soviet fighter shoots it down.

2 September
Aboard the USS *Missouri*, anchored in Tokyo Bay, a delegation of Japanese officials led by Japanese Foreign Minister Mamoru Shigemitsu sign the instruments of surrender under the watchful eye of General of the Army Douglas MacArthur and dozens of attending Allied officers. Massive formations of Navy and Marine Corps fighters and Army Air Forces bombers fly over the ship in celebration of victory.

After the surrender ceremony concludes, the films taken of the event are transferred to a waiting Douglas C-54 Skymaster, which then makes a record flight from Tokyo to Washington, D.C., to make the delivery. The flight takes 31 hours and 25 minutes, but because of the peculiarities inherent in crossing the international dateline, the flight begins and ends on the same calendar day.

4 September
Two Eleventh Air Force B-24s are intercepted by Soviet fighter planes during a high-altitude reconnaissance mission. This is the first instance of an event which will continue during the Cold War over the next five decades. All Eleventh Air Force missions are cancelled two days later.

13 October
The Army Air Forces hold a large open house at Wright Field near Dayton, Ohio. On display for public viewing are many pieces of captured enemy aeronautical technology. The AAF uses the opportunity to tell the public, more than one million during one week, about the AAF's contribution to victory in World War II.

24 October
The United Nations is officially established. The Charter, which was drafted over the summer, is ratified by the United States, the United Kingdom, the People's Republic of China, France, and the Soviet Union. These are the five permanent members of the Security Council. A majority of the remaining member states also approve and the Charter takes effect.

Above: *Mass formations of B-29s fly over the deck of the USS* Missouri *in celebration of the Japanese surrender earlier during the day. World War II had ended. (National Archives)*

Left: *The Boeing bombers—B-17 (left) and B-29 (right)—along with other AAF aircraft, made victory tours across the United States immediately after the war was over. (NASM)*

INDEPENDENCE!

1946–1949

AIR FORCE
Shealy

INDEPENDENCE!

1946–1949

America's air forces would never again be as large as they had been during the Second World War. The 2.4 million airmen and some 300,000 aircraft had been the product of global necessity. The combination of weapons technology, aircraft capability, and the need for large numbers of aircrew on each aircraft resulted in these huge numbers. But these conditions would not last long. Immediately after the Japanese surrendered, the Army Air Forces began to demobilize, and in less than one year released 1.7 million airmen.

Life had changed at home as well. GIs flooded the job market, which had been filled admirably by women and minorities. Education was a prime concern for many who had gone off to war, and the passage of the GI Bill assured that many of those returning would receive the opportunity to earn an education. Technologies developed during the war soon found uses in the civilian world. One of the most significant was broadcast television—similar to the technology used on Project APHRODITE television glide bombs.

The establishment of the United States Air Force seemed inevitable after the war, but the Navy was not satisfied with losing a strategic attack mission that now clearly belonged to the Army Air Forces' long-range bombers. Arguments between the air arm and naval advocates reached a peak when monies were cut from the Navy's carrier budget and transferred to the Air Forces' heavy B-36 Peacemaker program. The "Revolt of the Admirals" reflected the rancor that existed between airmen and seamen concerning budget and mission that Brigadier General Billy Mitchell had initiated when he sunk the *Ostfriesland* in 1921. In some ways, this rivalry continues to the present.

Despite shrinking budgets, the immediate postwar years were a time of tremendous technological change. The sound barrier was broken by a United States Air Force test pilot, Captain Charles "Chuck" Yeager, in a rocket-propelled ship that was little more than a tube housing a rocket motor and a set of straight, thin wings. But the barrier had been broken. The XS-1 (later X-1) was only the first in a series of X-planes that tested all manner of aerodynamic principles, from the sweptback wing to different kinds of fuselage shapes.

Jets and rockets were seen as the future of the independent Air Force. Groups of advisors and scientists published reports and made recommendations to the Air Force Chief and his generals that aircraft speed and the ability to reach into space would be essential to any future military victory. The U.S. Air Force Scientific Advisory Board was influential. After their first forecast report, *Toward New Horizons*, was published in 1945, programs and policies were dictated by scientists rather than generals or doctrinaires.

Less than one year after the Japanese surrender, it was becoming apparent that a new conflict was brewing. Britain's wartime prime minister, Winston Churchill, explained to Americans that an "iron curtain" had fallen over Europe. The stage was set for the beginning of a series of confrontations between the democratic West and the Communist Soviet Union and her satellite nations. The Cold War would continue for more than five decades, and against this global backdrop of an ideological war between

Pages 284–285: *The Consolidated B-36 Peacemaker provided the means for the Air Force to carry out its strategic bombing mission anywhere on earth. ("B-36," by Charles Shealy, USAF Art Collection)*

Right: *The United States Air Force Seal.*

democracy and communism, regional military struggles would erupt from Central Europe to Asia.

Air Force leadership was also changing. The United States Air Force witnessed a changing of the guard. In the years between the Second World War and the Korean War, almost all of the Army Air Forces' senior leadership retired. Pioneer aviators like Arnold, Spaatz, and Doolittle were gone. Those young airmen who had risen rapidly during wartime remained to carry the airpower torch. Vandenberg and LeMay were among those who had been trusted by the "old guard" of air force with the future.

The birth of any new organization is painful and challenging, none more so than the creation of U.S. Air Force. Needed were new administrative structures, new staff organization, new methods of promotion, and new budget management mechanisms. To highlight the break from the Army, the USAF led the other services in racial integration in 1947, conducted a major, long-duration airlift operation in 1948–49 (The Berlin Airlift), and also established the Women's Air Force (WAF) in 1948. A large part of the Air Force's budget was channeled toward weapons and aircraft development—primarily jets, guided missiles and the weapons that would arm them—both nuclear and conventional.

By 1950, the Army Air Forces of World War II seemed a distant memory. The next war would serve as the transition between the piston-driven air forces of the past and the turbojet-powered air forces of the future.

INDEPENDENCE!

1946–1949

"From Stettin in the Baltic to Trieste in the Adriatic an iron curtain has descended across the Continent."
—Former British Prime Minister Winston S. Churchill, March 5, 1946

1946

During the year, Wright Field engineer J. W. McGee begins to investigate high-temperature aluminum alloys. By mid-1947, McGee will develop the "ML" Alloy as a useful material for aircraft construction. It is named for the Materials Laboratory where the work was accomplished.

10 January
C. A. Moeller and D. D. Viner soar in an Army R-5 to an unofficial world helicopter altitude record of 21,000 feet. The sortie is flown at Stratford, Connecticut.

16 January
The U.S. Upper Atmospheric Research Panel is initiated. A panel of experts is selected to conduct tests on more than 60 captured German V-2 rockets. As a result of this project, other rocket programs spring up at Johns Hopkins University and the Naval Research Laboratory.

19 January
Bell Aircraft test pilot Jack Woolams makes the first glide test of the XS-1 rocket research aircraft. The test is flown at Pinecastle Army Air Base, Florida.

26 January
The First Experimental Guided Missile Group is formed at Eglin Field, Florida, to develop and test rocket-powered missiles.

Colonel William H. Councill, piloting a Lockheed F-80 Shooting Star, flies coast-to-coast in four hours and 13 minutes. The average speed of 584 miles per hour sets a new transcontinental record.

3 February
The Army Air Forces announces that it has been developing a plane that includes a totally automatic flight profile system. This system will require the onboard pilot to do nothing during a mission except monitor the controls for safety.

4 February
The Air Force Association is incorporated, and Lieutenant General Jimmy Doolittle (retired) becomes the first AFA national president. The AFA has evolved into an independent, nonprofit, civilian organization promoting public understanding of aerospace power and the pivotal role it plays in the security of the nation.

Left: *"Lieutenant General James H. Doolittle," by Leopold Seyffert. (USAF Art Collection)*

Right: *The XP-84 Thunderjet set speed records during its testing. The cockpit section of this aircraft is part of the National Air and Space Museum collection. (NASM)*

Below: *The first three Air Chiefs: (left to right) Generals Arnold (1939–1946), Spaatz (1946–1948), and Vandenberg (1948–1953) render a salute during a pass in review. (USAF)*

9 February
General of the Army Hap Arnold relinquishes command of the Army Air Forces (AAF) to General Tooey Spaatz. Spaatz will officially take command on March 1.

26 February
The last of 112 former AAF airbases in England closes, marked by a special ceremony at Honington Air Station, Suffolk, England. The last Eighth Air Force B-17 must delay its departure, however, due to a terrible snowstorm.

28 February
Major William Lien flies the Republic XP-84 Thunderjet for the first time. Later, as the F-84, the Thunderjet will fly extensively over the skies of Korea and will also become the first fighter equipped to carry a tactical nuclear weapon. The F-84 is the AAF's first post–World War II fighter.

March
During this month, the AAF establishes Project RAND as a division of the Douglas Aircraft Company. Project RAND will begin studies and investigate the potential of missiles, earth satellites, and supersonic flight.

In an effort to develop a weapon capable of intercepting a V-2-type weapon, the AAF establishes a program for ballistic missile defense.

5 March
Former British Prime Minister Winston Churchill presents a speech at Westminster College, Fulton, Missouri. During his talk he uses the term "iron curtain" to describe the developing situation in Europe.

8 March
The Civil Aeronautic Agency certifies the Bell Model 47 helicopter for flight, the first rotary wing aircraft to receive such a certification. The military designation of the Model 47 becomes the UH-13.

11 March
At the NACA Lewis Altitude Wind Tunnel, an afterburner is operated in conditions simulating high altitude for the first time.

12 March
At Maxwell Field, Alabama, the AAF School is redesignated as Air University, having jurisdiction over the Air Command and Staff School, the Air War College, and other bases for support and training.

15 March
The first static test firing of a captured German V-2 rocket is accomplished at White Sands Proving Ground, New Mexico.

21 March
On this date, Air Defense Command (ADC), Strategic Air Command (SAC, redesignated from Continental Air Forces), and Tactical Air Command (TAC) are activated.

22 March
The WAC becomes the first American rocket to leave the earth's atmosphere when it reaches an altitude of 50 miles. It is the result of a joint Jet Propulsion Laboratory (JPL)-Army Ordnance project.

1 April
In one of the few remaining missile contracts surviving postwar budget cuts, Bell Aircraft signs a contract with the AAF to build a guided missile capable of striking targets 100 miles distant. The Rascal missile will eventually result from this contract, also known as Project MX-776.

16 April
Military scientists launch a captured German V-2 rocket in the United States for the first time at White Sands Proving Ground, New Mexico.

19 April
Consolidated-Vultee contracts with the AAF to develop a true intercontinental ballistic missile (ICBM). This effort is called Project MX-774 (Missile Experimental).

22 April
The Glenn L. Martin Company contracts with the AAF to develop and build a surface-to-surface guided missile. The Matador missile with a 600-mile range results from this arrangement, officially the MX-771.

In an effort to better understand weather phenomena, the U.S. Weather Bureau, in coordination with the AAF, Navy, NACA, and universities, begins test flights using unpiloted Northrop P-61 Black Widow aircraft and piloted gliders to gather scientific data.

Opposite: *A captured German V-2 rocket is launched at the White Sands Proving Ground, New Mexico. Missile performance was measured by electronic equipment developed by the General Electric Company. (Courtesy of General Electric)*

Right: *The XB-43, America's first jet-powered bomber, was a stepping stone to future bomber development. (NASM)*

16 May
The AAF Institute of Technology is established at Wright Field. This technically oriented advanced studies school is intended to graduate around 350 officers annually.

17 May
The Douglas XB-43 flies for the first time. This experimental twin-engine platform will later result in production of the four-engine B-45 Tornado.

28 May
The AAF initiates a study to investigate the potential of using atomic energy for aircraft propulsion. Project NEPA will continue for a decade.

29 May
The War Department Equipment Board issues a report that emphasizes the importance of missiles in future warfare. The seven systems recommended for further consideration include ground-to-ground missiles with ranges from 150 to several thousand miles.

3 June
Lieutenant Henry A. Johnson establishes a world record for a 1,000-kilometer course flying a P-80 to an average speed of 462 miles per hour. His record time is 1 hour 20 minutes and 31 seconds from start to finish.

5 June
The Army Air Forces announces it has ordered two prototype multi-engine, jet-powered bombers—the XB-47 is the result.

17 June
Professor Theodore von Kármán chairs the first meeting of the newly established Scientific Advisory Board (SAB) at the Pentagon. The group sprang from the original group of 33 scientists and engineers that participated in Operation LUSTY, a search for German secret technology, at the end of World War II. The SAB is still active as a sounding board and think tank for current and future Air Force technologies.

22 June
The first jet-powered airmail delivery is made when two AAF P-80s take off from Schenectady, New York. One flies to Washington, D.C., while the other flies to Chicago, Illinois.

26 June
The Knot (one nautical mile per hour) and Nautical Mile (1.15 the length of a statute mile) are made the standard unit of measure for speed and distance by the AAF and the U.S. Navy.

THE USAF SCIENTIFIC ADVISORY BOARD

Left: *General of the Army Hap Arnold congratulates Professor Theodore von Kármán after the completion of the Scientific Advisory Group's technology study,* Toward New Horizons. *(USAF)*

When General Hap Arnold convinced Dr. Theodore von Kármán to assist the Army Air Forces in the exploitation of German technology at the end of World War II, he solidified a long-term relationship with the leading aerodynamic mind in America. Kármán's theoretical and applied aerodynamics ideas formed the cornerstone for future American air forces technology. This foundation was based largely upon Kármán's first official scientific report titled, *Where We Stand,* and the major technology forecast that followed in 1945, titled *Toward New Horizons.* These documents established a blueprint for achieving and maintaining American air supremacy for the future.

America's air forces witnessed three distinct periods of scientific development relating to the institutionalization of these Kármán reports. The first began in mid-1945 and lasted through General Arnold's retirement in February 1946. The second period was characterized by the professor's active involvement in the follow-up and implementation of major recommendations made in the *Toward New Horizons* report. The third and most extensive period began with Kármán's resignation as SAB Chairman and may be characterized by the maturation of the SAB as an institution.

Arnold's final contribution to the Air Forces involved convincing incoming AAF Chief General Carl A. Spaatz to activate a permanent version of the Scientific Advisory Group (SAG) as part of his staff—which he did. With his health rapidly deteriorating, Arnold retired from active service in February 1946, the same month that the original charter for the 1944 Scientific Advisory Group expired. His dream of preparing a blueprint for future American Air Forces had been completed.

The last formal meeting of the AAF Scientific Advisory Group was held on February 6, 1946. Arnold thanked them for their efforts during the past year and decorated the SAG members with the Meritorious Civilian Service Award. He asked that they continue to support the scientific aspects of the Army Air Forces in

Right: *Theodore von Kármán (head of table) organized and led the original AAF Scientific Advisory Group (SAG). Today's Air Force Scientific Advisory Board (USAFSAB) is the direct descendent of Kármán's group. (USAF)*

times of peace. On March 1, 1946, Kármán resigned his government position as AAF Scientific Advisor, ending one of the most intellectually active chapters ever in U.S. Army history.

The second period of implementation of Kármán's studies began without the dynamism of General Arnold or the bottomless funding that the Army Air Forces had enjoyed during the previous five years. The newly formed Scientific Advisory Board (SAB) forwarded its first set of program recommendations to General Spaatz on August 29, 1946. Spaatz approved implementation of the proposals, but made it clear that any additional funding for the recommendations was nonexistent.

Questions originated in the AAF about the function of the newly renamed Scientific Advisory Board. Would it be relegated to immediate issues? Would it have a forecasting function? Some officers, in the face of budget constraints and Air Force reorganization, saw no useful purpose for the SAB. Others, although realizing its possible importance in providing advice on scientific matters, were at a loss when deciding just how this might be accomplished. The fledgling SAB was going through growing pains because there existed no organizational precedent for such a permanent group. Confusion at all levels of the Army chain of command concerning the possible uses of the SAB hindered its effectiveness.

An important change occurred when Lieutenant General Laurence Craigie took command of the newly established Directorate of Research and Development. Craigie had been involved with Arnold's XP-59A project from the beginning and was absolutely aware of the importance of R&D to the future success of the Air Force. Kármán, realizing this, immediately convinced Craigie that the SAB needed to report directly to the Chief of Staff, as he had recommended in *Toward New Horizons*, rather than through other Rearch and Development channels.

In December 1945, Kármán had emphasized that the successful accomplishment of the SAG's mission was assured because of Arnold's insistence that they disregard current projects, and report directly to the Commanding General. The professor had reemphasized this recommendation in 1946 to LeMay, who disregarded the advice. Spaatz, following Craigie's careful explanation of the problem, not only supported the recommendations, but also enacted them immediately. On May 14, 1948, the Spaatz-Kármán agreement went into effect as Air Force Regulation (AFR) 20-30, the SAB charter. The process was finalized in the summer and enacted September 18, 1947, when the Air Force became an independent service branch. By April 1948, with Craigie's help, initial administrative obstacles were removed and with Kármán still acting as the Chairman, the SAB and the independent Air Force began the real work of attaining and maintaining military air supremacy.

Left: *The Able Test was the first in a series of nuclear weapons tests carried out at Bikini Atoll in the Pacific. Valuable data was gathered by airborne drones during these blasts. (USAF)*

Below: *The early B-36 Peacemakers had only one large wheel on the main landing gear and did not have jet engines mounted on the wings. ("Texas Giant," by K. Price Randel, USAF Art Collection)*

28 June
Boeing contracts with the AAF to design a new long-range heavy bomber—it will become the B-52.

1 July
Operation CROSSROADS testing begins when 509th Composite Group pilot Major Woodrow Swancutt drops a Fat Man atomic weapon over Bikini Atoll from a Boeing B-29 Superfortress named *Dave's Dream*. The test is designed to measure specific effects of the bomb on a variety of targets, from ships at sea to unmanned aircraft (drones) in flight. The 23-kiloton-yield bomb misses its target by a significant margin, but measurements are still collected and damages assessed after the detonation. This test explosion is generally referred to as the Able Test. Even though the bomb is off target, five ships are sunk and nine more are heavily damaged of the 73 naval vessels in the test zone.

25 July
As part of Operation CROSSROADS, a second nuclear device is detonated at Bikini Atoll. The bomb is detonated 90 feet beneath the ocean surface, sinking eight ships and causing radioactive contamination.

6 August
Two unmanned B-17 bombers are flown nonstop from Hilo, Hawaii, to Muroc Dry Lake, California, controlled entirely by radio.

8 August
The Convair XB-36 prototype flies for the first time at the Fort Worth, Texas, factory. Company pilots Beryl A. Erickson, "Gus" Green, and a crew of seven are on board. The massive bomber had been under development since 1941.

12 August
President Harry S. Truman signs an appropriations bill that includes $50,000 to establish the National Air Museum within the Smithsonian Institution, Washington, D.C. The museum's stated mission is to "memorialize the development of aviation." Originally housed in a variety of Smithsonian buildings, the name is changed to National Air and Space Museum in 1966, and a decade later the museum on the National Mall opens in 1976. The most visited museum in the world, 100 million visitors had walked through the doors by December 1986.

Right: *A modified B-29 carried the X-1 rocket research plane to high altitude, where it was dropped. The XS-1's (later changed to X-1) rocket motor then propelled the plane through the sound barrier. (NASM)*

17 August
Sergeant Larry Lambert becomes the first person in the U.S. to use an ejection seat to escape from an aircraft. He "punched out" of a P-61 over Ohio at an altitude of 7,800 feet, flying 302 miles per hour.

19 August
Major General Laurence C. Craigie, the first military man to pilot a jet aircraft, presents a plaque to Orville Wright on behalf of the Army Air Forces, praising his contributions to aeronautics. The presentation takes place at Orville's laboratory on his 75th birthday.

31 August
In the first postwar Bendix Cup Race, Colonel Leon Gray wins the Jet Division race when he flies from Los Angeles, California, to Cleveland, Ohio, in four hours and eight minutes. Gray pilots a P-80 Shooting Star to victory with an average speed of 495 miles per hour.

30 September
The NACA Muroc Flight Test Unit is deployed to Muroc, California, to assist in the X-1 program. Thirteen engineers, technicians and observers had been sent from the Langley Laboratory under the direction of Walter Williams. This unit is the original unit that provides the impetus for future development of the NASA Flight Research Center at Edwards, California. It would also be a source of friction between the Air Force and NACA as to which organization would actually fly the first supersonic flight.

4–6 October
Colonel C. S. Irvine pilots the B-29 *Pacusan Dreamboat* on the first nonstop unrefueled flight over the North Pole. Irvine's route takes him from Hawaii, over the magnetic North Pole, and to Egypt, a distance of approximately 10,000 miles, taking nearly 40 hours to complete.

7 October
The first XS-1 is shipped from the Bell Niagara Falls Plant to Muroc Dry Lake, California. Two more will follow, and eventually the name will be changed to X-1.

10 October
When a captured V-2 rocket (No. 12) is launched at White Sands Proving Ground to collect spectroscopic readings during its flight into space, the discipline of space science technically begins.

24 October
A V-2 rocket carries a movie camera—a DeVry 35-mm commercial version—to an altitude of 65 miles and records film of the earth. Rocket No. 13's camera covers an area of 40,000 square miles.

10 November
Sanford Moss, pioneer of turbines that made turbo-supercharging possible on AAF aircraft during World War II, dies at the age of 74.

8 December
Bell pilot Chalmers Goodlin makes the first powered flight in the XS-1 research aircraft. After the XS-1 drops from the B-29 carrier aircraft, the RMI XLR-1 rocket engine fires, lifting the aircraft to 35,000 feet and a top speed of Mach .75.

THE AIR FORCE AND THE SUPERSONIC REVOLUTION: TOO MUCH OF A GOOD THING?

Richard P. Hallion

Left: *The Bell X-1 rocket research plane zooms past the FP-80 chase aircraft. This photo was taken by the FP-80 pilot, Robert A. Hoover, during a 1951 flight at Muroc. (NASM)*

On October 14, 1947, Air Force test pilot Captain Charles E. "Chuck" Yeager became the first pilot to exceed the speed of sound, attaining Mach 1.06 (700 miles per hour) at an altitude of 43,000 feet. Yeager accomplished the feat as pilot of the Bell XS-1 (later X-1) rocket-propelled research airplane, air launched from a modified Boeing B-29A Superfortress bomber. With this flight, fears of a "sound barrier" forever limiting the progress of aviation ended, and the supersonic era was born.

Concern about supersonic flight dated to the early 1920s, when wind tunnel researchers had first discovered that air moving over a wing section moving at Mach 1 generates a strong shock wave that robs the wing of lift and dramatically increases its drag, the retarding force tending to hold an airplane back. By the late-1930s, when the first manned aircraft began encountering so-called compressibility effects as they dove at speeds over 500 miles per hour, the need for reliable supersonic design information assumed critical importance. Aircraft of all the major aeronautical powers mysteriously broke up at "transonic" speeds, those velocities where an airplane was moving at speeds below that of sound, while the accelerated airflow over the wing was moving at the speed of sound, or even faster. Because of the unreliability of contemporary wind tunnel testing, researchers in England, Germany, and America opted instead to develop instrumented supersonic research airplanes. Eventually, only the United States possessed the resources to embark upon such an effort, resulting in the so-called X-series research aircraft program, of which the XS-1 was the first in a series that continues to the present day.

Discoveries in the rubble of Nazi Germany revealed the tremendous extent of German research in high-speed aerodynamics, and turbojet and rocket propulsion. Army Air Forces Chief of Staff General Hap Arnold had already arranged for Dr. Theodore von Kármán, director of the Guggenheim Aeronautical Laboratory, California Institute of Technology, to undertake a comprehensive assessment of the state of contemporary aeronautics, and forecast what future trends and needs might be. Out of this came a landmark report, *Toward New Horizons*, which,

Right: *Near the end of World War II, American scientists evaluated German aeronautical programs at first hand. This enormous high-speed wind tunnel was under construction as the Allies advanced toward Berlin. It was disassembled and shipped back to the U.S. (USAF/HRA)*

together with the results of the X-series, dramatically reshaped future Air Force research, development, and acquisition. The Air Force embarked on a future based upon supersonic fighters and bombers, "robot" cruise and guided missiles, and exotic orbital rocket-boosted hypersonic gliders and nuclear-powered airplanes.

In 1950, the Air Force went to war in Korea, a war that highlighted the serious decline in American combat capabilities that had occurred in the hasty and unfocused drawdown of forces after V-J day. In that conflict, Air Force fighter aircraft under development at the end of the war, particularly the F-86 Sabre (which had been redesigned as a sweptwing airplane following exposure to German technology), generally performed very well, fulfilling traditional "swing role" air-to-air and air-to-surface missions. But after the war, Air Force doctrine increasingly emphasized meeting the perceived needs of strategic nuclear warfare, and thus research, development, and acquisition efforts increasingly emphasized supersonic dash, missile armament, and the ability to carry nuclear weapons, or intercept and destroy enemy nuclear-equipped airplanes. This resulted in a tremendous expenditure of effort on so-called Century series fighters (F-100, F-101, F-102, F-103, F-104, F-105, F-106, F-107, F-108, and F-109), of which only the F-100, F-101, F-102, F-104, F-105, and F-106 actually entered service. (The F-110 and F-111 were later developments and not strictly within the original intent of the Century series; the F-110 was the initial designation for the Air Force F-4, and the F-111 was the result of the McNamara-era TFX strike aircraft competition.) These aircraft were less fighters than interceptors or nuclear strike airplanes; only the F-100 represented the kind of balanced design approach that had characterized earlier successful fighters. The quest for sustained supersonics drove high-speed bomber programs such as the B-58 and XB-70 as well. Yet again, the utility of these proved far less than originally contemplated, so that neither outlasted the Boeing B-52, which preceded them. Only the Mach 3+ Lockheed SR-71 Blackbird, designed to meet strategic reconnaissance requirements, justified sustained supersonic performance.

The 1960s and, particularly, the Vietnam War, revealed very different needs, forcing the Air Force to hastily acquire the Navy's F-4 fighter, and the A-1 and A-7 attack airplanes. Post-Vietnam Air Force acquisition showed far greater realism in tying performance and capabilities to perceived needs, resulting in outstanding balanced systems such as the F-15 Eagle.

17 December
President Harry S. Truman sends a congratulatory telegram to Orville Wright on the occasion of the 43rd anniversary of the first flight at Kitty Hawk, North Carolina. To celebrate the occasion, more than 200 fighter planes fly past the Wright Memorial in salute. Lieutenant General Nathan F. Twining lays a wreath at the grave of Wilbur Wright as part of the remembrance.

The National Institute of Health initiates a space biological research program at Holloman AFB, New Mexico. Part of the program will be the high-speed rocket test sled used with subjects to study their reaction to high-G situations.

A V-2 launched at White Sands Proving Ground sets a velocity and altitude record for a single-stage rocket—3,600 miles per hour and 116 miles altitude. The rocket is carrying a payload of fungus spores as an experiment.

1947

13 January
Milton Caniff creates the comic strip *Steve Canyon*, which highlights the contributions of airpower to America. The main character is a strapping, tall, blond pilot whose adventurous life is followed by many readers, particularly energetic adolescent boys.

23 January
For the first time in the U.S., telemetry is transmitted from a V-2 in flight to a ground receiving station that records its performance. This capability is part of Project HERMES, ongoing at White Sands.

28 January
During a special ceremony at the Mission Inn, Riverside, California, retired General of the Army Hap Arnold affixes a pair of wings autographed by Orville Wright to the highest point on the legendary "Fliers' Wall."

5 February
The Atomic Energy Commission and the Secretaries of War and Navy recommend that nuclear weapons production should continue. President Truman agrees.

10 February
Major E. M. Cassell pilots a Sikorsky R-5A helicopter to an unofficial altitude record of 19,167 feet in the skies over Dayton, Ohio.

17 February
A WAC Corporal missile launched at White Sands Proving Ground attains an altitude of 240,000 feet.

20 February
The Blossom Project, a test where canisters are ejected from V-2 rockets at their apex, begins when the No. 20 rocket is fired.

27 February
A North American P-82B Twin Mustang sets the record for the longest nonstop, unrefueled flight by a propeller-driven aircraft. Pilot Lieutenant Colonel Robert Thacker and co-pilot John M. Ard fly *Betty Jo* from Hickam Field, Hawaii, to LaGuardia Airport, New York. The flight covers more than 5,000 miles and takes 14 hours and 33 minutes.

12 March
President Harry S. Truman declares support for Greece and Turkey in their struggle to fight

Opposite: *Retired General Hap Arnold affixes Orville Wright's wings to the pioneer aviators wall at the Mission Inn. (Robert and Kathleen Arnold Collection)*

Right: *The addition of multiple external fuel tanks helped this P-82* Betty Jo *fly nonstop from Hawaii to New York. (NASM)*

Below: *Henry Ford (left) confers with General Arnold during World War II. (Robert and Kathleen Arnold Collection)*

Communist insurgency. The Truman Doctrine shapes American policy for the next 50 years.

16 March
The Convair 240 airliner flies for the first time near San Diego, California. The 240 will be adopted by the Air Force as the T-29 navigator training aircraft and the C-131 MedEvac aircraft. The T-29 will fly for the first time on September 22.

17 March
North American test pilot George Krebs flies the XB-45 Tornado for the first time at Muroc Army Airfield, California. The B-45 is the first production jet bomber used by the Air Force and the first American four-engine jet bomber built. Four Allison J-35 engines make up the propulsion system for the aircraft.

7 April
Automobile pioneer Henry Ford dies. During his life, the Ford Company participated in a number of aviation projects such as the Tri-Motor, the B-24, and the Liberty Eagle (Flying Bomb). In eulogy the following day, Orville Wright states that Ford "did more to promote the welfare of the American people, and particularly the working class, than any man who ever lived in this country."

15 April
Captain William P. Odom pilots a modified A-26 *Reynolds Bombshell* to a new round-the-world flight record of 78 hours and 56 minutes. The mission covered approximately 20,000 miles.

30 April
The Army and Navy adopt a standardized set of nomenclature for guided missiles. The basic designations will contain combinations of A (Air), S (Surface), and U (Underwater), with the first letter indicating the missile's origin and the second indicating its target.

21 May
Engineers at NACA Langley demonstrate a near-silent aircraft. A special propeller with five blades is coupled with an engine with muffled exhausts.

27 May
The first Army Corporal E, a guided surface-to-surface missile, exceeds all expectations after its initial launch this date.

Left: *W. Stuart Symington is sworn in as the first Secretary of the Air Force, marking the official "birthday" of the newly independent military service. (USAF)*

5 June
The European Recovery Act—better known as the Marshall Plan—is announced.

A New York University team under contract with Air Materiel Command launches the Army Air Forces' first research balloon. The cluster of rubber spheres is released at Holloman, New Mexico.

19 June
Colonel Albert Boyd, the Air Force's chief test pilot, flies a Lockheed P-80R Shooting Star to an absolute speed record of 623.8 miles per hour in the skies over Muroc Dry Lake, California.

25 June
The Boeing B-50 makes its first flight. More powerful than the B-29, the B-50 will serve until 1955 as a bomber and until 1965 as a tanker.

30 June
Army Air Forces and NACA representatives meeting at Wright-Patterson agree to divide responsibilities for the X-1 testing program. The AAF attains the responsibility for breaking the sound barrier as soon as possible, while NACA will acquire details of research information during the program.

1 July
With budget money drying up, the AAF cancels the MX-774 program. The program will be revived, and the Atlas missile will be the result.

3 July
The AAF sends aloft a 10-balloon cluster carrying a 50-pound set of atmospheric instruments provided by New York University scientists. The balloon cluster reaches 18,550 feet altitude.

18 July
President Harry S. Truman assigns Thomas K. Finletter as chair of a five-man working group whose task is to provide a broad plan that will give the United States the "greatest possible benefits from aviation." The report is due by the first of the year.

26 July
Following months of congressional debate and inter-service argument and compromise, President Harry S. Truman signs the National Defense Act of 1947 onboard his presidential aircraft, the Douglas VC-54C *Sacred Cow*. His signature sets legislation into motion that will result in the foundation of the independent United States Air Force in September. The legislation creates a Department of the Air Force equal to the Departments of the Army and Navy. Additionally, an Air National Guard is created as a reserve component of the Air Force.

22 August
Dr. Hugh L. Dryden becomes the director of Aeronautical Research at NACA. Dryden had served as Theodore von Kármán's deputy during the post–World War II study, *Toward New Horizons*.

28 August
The first of 22 production B-36A Peacemaker aircraft flies for the first time. This model, an unarmed version, is designed for training B-36 aircrew at Carswell AFB, Texas.

18 September
The United States Air Force becomes a separate, independent military service. W. Stuart Symington is sworn in as the first Secretary of the Air Force.

22 September
A robot-controlled USAF C-54 becomes the first aircraft to cross the Atlantic in such a configuration. The plane takes off from Stephenville, Newfoundland, and lands at Brize Norton, England, after a 2,400-mile flight.

25 September
The liquid-propellant Aerobee sounding rocket is successfully launched for the first time at White Sands

Left: *The XJR2F-1 Albatross was a highly successful rescue aircraft and saw extensive service during the Korean War. (NASM)*

Proving Ground, New Mexico. Variations of the Aerobee will be used to launch experiments until 1985.

26 September
General Carl A. Spaatz takes command of the United States Air Force as its first Chief of Staff. With Spaatz now in command, the official transfer of personnel, facilities, and aircraft from the Department of the Army to the Department of the Air Force is ordered by the Secretary of Defense, James W. Forrestal.

Major General William E. Kepner is named chief of a new USAF unit, the Atomic Energy Division. Kepner led the 8th Fighter Command in the skies over Europe during World War II.

1 October
The North American XP-86 flies for the first time at Muroc Dry Lake, California. The Sabre, the USAF's first swept-wing fighter plane, is piloted by legendary test pilot George "Wheaties" Welch. Welch was one of a few AAF pilots who was able to climb into an airplane during the Japanese attack at Pearl Harbor and fight back that day.

The Grumman XJR2F-1 amphibian makes its first flight at Bethpage, New York. The Albatross will be produced in large numbers and flown by the USAF as the SA-16 and the HU-16. The prototype SA-16 flies for the first time on October 24th. The Albatross will amass a stunning save record during the Korean and Vietnam Wars by rescuing some 1,000 downed airmen.

6 October
The Air Force's first modern guided air-to-air missile, the Firebird XAAM-A-1, is launched successfully.

10 October
After a 17-year wait, the U.S. Patent Office finally issues a patent to Carl L. Norden for his bombsight.

Orville Wright suffers a heart attack while traveling to a meeting at the National Cash Register Company. He is taken to Dayton's Miami Valley Hospital where he is treated and released four days later.

14 October
Captain Charles "Chuck" Yeager breaks the sound barrier and earns the Mackay Trophy. Piloting the XS-1, Yeager fires the rocket motor as the experimental plane is dropped from the B-29 mother ship. He accelerates to Mach 1.06 over the dry lake bed at Muroc, California. This is the first time that supersonic speed is reached in level flight by a powered aircraft. This is a joint USAF-NACA program.

Below: *Captain Charles "Chuck" Yeager seated in the Bell X-1* Glamorous Glennis *in which he became the first man to fly faster than the speed of sound in level, powered flight. This image was taken in May 1948. (NASM)*

CARL ANDREW SPAATZ (1891–1974)

David R. Mets

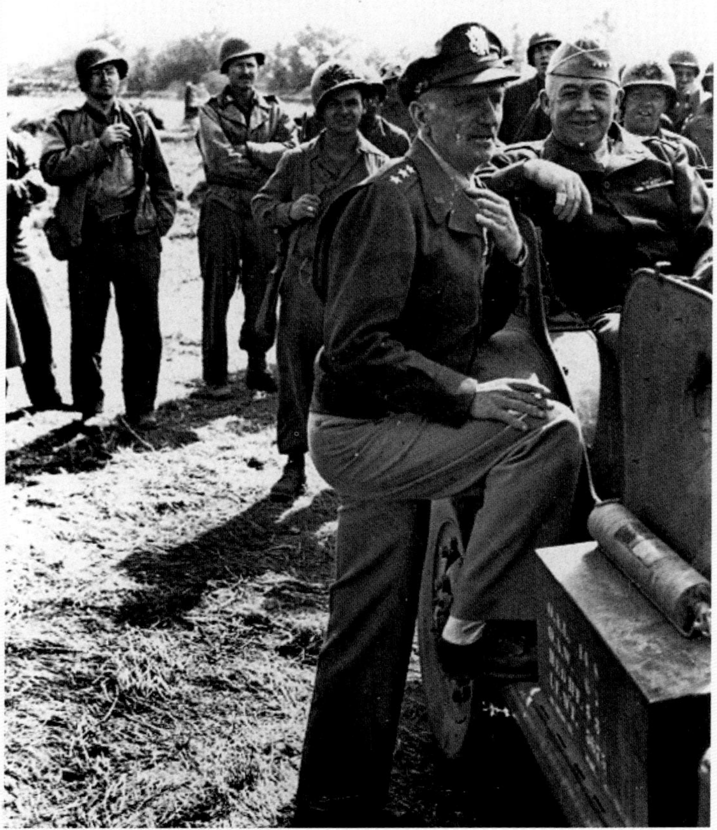

Carl A. Spaatz was laid to rest in the USAF Academy cemetery in the summer of 1974. Spaatz was 12 years old when the Wright brothers first flew, and he lived to meet the first humans to land on the moon. His inherently pragmatic intelligence enabled him to cope with the changes of a dynamic century and to influence events in important ways. He ended his military career as the first Chief of Staff of the independent United States Air Force.

Carl A. Spaatz was in the third generation of a German family that had immigrated in the mid-nineteenth century. He was one of five siblings in a Pennsylvania newspaperman's family. He graduated from West Point in 1914 with a middling academic record, a very poor conduct record, and as a "cleansleeve"—holding no cadet rank at all. Spaatz served with a black infantry unit in Hawaii for a year before going to pilot training in San Diego, and thence to Mexico with Benjamin Foulois's First Aero Squadron on Pershing's 1916 Punitive expedition.

Three years after graduation, as a major, he commanded America's largest air training center in France, leading a unit of about 5,000 people and up to 1,000 aircraft. But before he came home, he insisted on some time at the front and managed to shoot down three Fokkers during the St. Mihiel Offensive. He came away with the Distinguished Service Medal and high regard from Brigadier General William "Billy" Mitchell, the overall air commander of that battle.

Spaatz was one of only about 90 majors in an Air Service where promotions were slow. That led to a long string of important assignments, first as the commander of the only pursuit group—which was then deemed the elite unit. He attended the Air Service Tactical School, testified at the Billy Mitchell trial (1925), and commanded the famous 1929 *Question Mark* endurance flight to win the Distinguished Flying Cross. Just as bombers were rising to preeminence, Spaatz took command of the 7th Bomb Group and then a bomber wing. The rest of the 1930s was filled with assignments to the air staff and unimpressive attendance at the Command and General Staff School.

Early in World War II, Spaatz went to England as an observer of the Battle of Britain, gaining some valuable lessons for building air forces and making important connections with RAF leaders. He returned to England in June 1942 to command Eighth Air Force in the initial strategic bombing of Nazi-occupied Europe. By the end of the year, when Eisenhower went south for the invasion of North Africa, he reached back for Spaatz to come serve as his air commander. In the course of the battle, Spaatz helped resolve the organizational problems associated with offensive ground campaigns and especially helped sell the airmen's traditional tactical air preferences to the soldiers. That resulted in the famous *Field Manual 100-20*, the principles of which are still pillars of USAF tactical doctrine. Meanwhile, Eighth Air Force in England had run into deep trouble.

Opposite: *General Carl A. Spaatz was General Arnold's operational commander in Europe for much of the war. Here, "Hap" pays "Tooey" a visit in Europe. (USAF)*

Right: *"General Carl Spaatz," by Thomas Stephens. (USAF Art Collection)*

At the end of 1943, Eisenhower returned to England to plan OVERLORD, and Spaatz again came along to run the strategic air campaign—and less directly to influence the tactical air campaign. The strategic offensive of the fall of 1943 had failed to establish air superiority over the Luftwaffe; without that, there could be no landings. It looked as though Eighth Air Force would run out of crews before command of the air could be achieved.

There were tough decisions to be made. One was on the length of crew tour length—Spaatz agonized but extended it from 25 to 35 missions. Some RAF leaders wanted to transfer control of Eighth Air Force to the tactical leaders for the interdiction campaign—still short of having command of the air. Eisenhower's decision went against Spaatz in the end, but it was hedged and delayed enough that the Luftwaffe could not stop the invasion by air. Had the D-day invasion been attempted without command of the air, certain disaster would have followed. Spaatz barely got out to the Pacific to lead the closing air campaign against Japan when the nuclear weapons were dropped, and the war ended. He was the only individual to attend the surrender ceremonies of both Germany and Japan.

What difference did Carl Spaatz make to America? Without overestimating the impact of individuals on history, he was one of the principals who conceived and built the Air Force we have today. A protégé of both General Billy Mitchell and General Hap Arnold, he lived to take the helm of the newly independent Air Force in its infancy. Not only was he a builder of the theory and doctrine that guided our airpower through World War II, he was the operational commander of what remains the most massive and decisive application of airpower in history. He was not alone,of course, but the painful choices he made helped win command of the air, and did so sooner than it otherwise would have. Without that, the victory in North Africa would have been longer in coming; without that, the Normandy landings might have failed; without that many more would have died.

Left: *The Northrop YB-49 was the jet-powered follow-on to the piston-powered XB-35 Flying Wing. (NASM)*

Below: *The McDonnell XH-20 "Little Henry" ramjet-powered helicopter. (NASM)*

21 October
The Northrop YB-49 jet bomber flies for the first time. The flying wing design is developed from the propeller-driven version, the XB-35, which first flew on June 25, 1946.

30 October
Long-time engineer-in-charge of the Langley Aeronautical Laboratory, Dr. H. J. E. Reid, receives the Medal of Merit from President Truman for his wartime contributions to American airpower.

4 November
The first complete Aerobee rocket is launched, reaching an altitude of 190,000 feet above White Sands Proving Ground.

15 November
The USAF reveals that it has been flight testing a ramjet helicopter for the past six months. The *Little Henry* is the world's first ramjet-powered helicopter and first flies a tethered flight on May 5, 1947, and its first free flight on August 26. The first XH-20 is now on display at the National Museum of the USAF in Dayton, Ohio.

23 November
The Convair XC-99 makes its first flight at Lindbergh Field in San Diego, California. The XC-99, the cargo version of the B-36 Peacemaker, is the world's largest landplane. The flight is piloted by Convair pilots Russell R. Rogers and Beryl A. Erickson.

26 November
Langley scientists successfully operate the first hypersonic-flow wind tunnel. which includes an 11-inch test section.

10 December
Lieutenant Colonel John P. Stapp, a member of the Air Force Medical Corps, straps in and takes his first rocket-propelled research test ride on a 2,000-foot-long test track. The purpose of the experimental rides is to gain accurate information on rapid deceleration from a disabled jet aircraft.

17 December
The Boeing XB-47 Stratojet flies for the first time between Moses Lake AFB and the Boeing factory airfield in Seattle, Washington, flown by Boeing test pilots Robert Robbins and Scott Osler. Originally designed with straight wings, the plans were changed when Boeing engineers were able to review German wind tunnel test results at the end of World War II. A simple telegram to the Boeing Company was sent which instructed the designers to "sweep back the wings of the B-47. . . data follows." It is the first bomber to be designed with six engines mounted on wing pylons. The XB-47 spans 116 feet and has a gross weight of 125,000 pounds.

Right: *The Convair XC-99 was the largest land plane ever built, capable ofcarrying 400 troops. The flight crew stands beneath the nose of the plane as Convair employees line-up to tour their creation. (NASM)*

Below: *A Strategic Air Command Boeing B-50A Superfortress crew receives their final briefing before takeoff. (NASM)*

1948

2 January
The Air Force Technical Museum is established at Patterson Field, Ohio.

4 January
Scientists at the University of California announce the completion of a pilot model for the world's first low-pressure supersonic wind tunnel.

27 January
Orville Wright suffers a heart attack, his second, and is taken to Miami Valley Hospital in Dayton, Ohio.

30 January
Orville Wright dies of a heart attack at Miami Valley Hospital. He is 76 years old.

6 February
V-2 Rocket No. 36 is launched at White Sands Proving Ground and controlled from the ground during the entire flight, using the Hermes A-1 flight control system. This test marks the dawn of a truly guided missile.

16 February
Strategic Air Command deploys B-29 Superfortresses to Germany as part of a long-distance operations exercise. The Royal Air Force (RAF) practices intercepting the bombers as they fly over southern England.

18 February
The co-executors of Orville Wright's will announce that the original Wright Flyer will be returned to the United States from the Science Museum in London and turned over to the National Museum in Washington, D.C.

20 February
Strategic Air Command accepts its first Boeing B-50 Superfortress. The B-50, while similar to the B-29, has more powerful engines, a larger tail section, and in-flight refueling capability.

10 March
The USAF reports that a B-29 has dropped a non-explosive monstrous bomb weighing 42,000 pounds at Muroc Dry Lake, California.

11–14 March
Military service chiefs sign the Key West Agreements to delineate service roles and missions. Military aeronautical and rocket research is not exclusively assigned to one service for development.

22 March
The Lockheed TP-80C, the prototype of the two-seat T-33 jet trainer, flies for the first time near Van Nuys, California. Lockheed pilot Tony LeVier flies the mission.

28 March
Strategic Air Command completes a series of aerial refueling tests with specially converted B-29s, designated KB-29M. These tankers carry a 2,300-gallon fuel tank in the bomb bay and are equipped with the hose and reel system for refueling.

21 April
The United States Air Force is assigned the mission of protecting the United States by Defense Secretary James V. Forrestal.

26 April
The Air Force announces a policy of racial integration in the ranks. Seen as wasteful and inefficient, segregation is to be eliminated by using "Negro personnel in free competition for any duty within the Air Force for which they may qualify." This policy predates President Truman's mandate to integrate the military, which is not issued until July.

30 April
General Hoyt S. Vandenberg becomes the second Chief of Staff of the Air Force, succeeding General Carl A. Spaatz.

20 May
The first production model F-86 Sabre Jet flies for the first time at Inglewood, California. It is powered by a General Electric J47 turbojet that produces 5,000 pounds of thrust. More than 6,000 Sabres will be built during years of production.

24 May
Jacqueline Cochran sets a world speed record of 432 miles per hour for propeller-driven aircraft over a 1,000-kilometer course.

26 May
The Civil Air Patrol is established as a civilian auxiliary of the USAF when President Harry S. Truman signs the legislation.

1 June
The Military Air Transport Service (MATS) is established by consolidating both Air Force and Navy transport commands into one Air Force–directed command.

10 June
The USAF confirms that the X-1 has been able to attain supersonic speed several times since Captain Chuck Yeager's first successful penetration of Mach I.

11 June
With the publication of Air Force Regulation 65-60, designations for USAF aircraft are changed to reflect advances in roles and missions. The antiquated "P" for Pursuit is changed to "F" for Fighter; "R" for Reconnaissance takes the place of "F," which in the 1930s represented Fotographic; "H" takes the place of "R" for Rotary wing craft in designation for helicopters.

The Office of Air Force Chaplains is established.

12 June
The Women's Armed Service Integration Act is passed by Congress, establishing Women in the Air Force (WAF). Passage allows women to serve as permanent, regular members of all the services but limits the number of women in each branch to 2 percent of the total force.

16 June
Colonel Geraldine P. May, the first woman in the Air Force to achieve the rank of colonel, is named as the first director of Women in the Air Force.

18 June
The first two aerial refueling squadrons in the USAF are established at Davis-Monthan AFB, Tucson, Arizona, and Roswell AFB, New Mexico. Both are equipped with the Boeing KB-29M.

26 June
The Berlin Airlift begins. Douglas C-47s transport 80 tons of supplies to the blockaded city of Berlin. USAFE commander, General Curtis E. LeMay, gathers the required aircraft to begin the airlift mission.

The USAF accepts its first operational B-36 Peacemaker bomber. By the end of the year, 35 B-36s will be assigned to the 7th Bomb Wing in Fort Worth, Texas. Its 230-foot wingspan and 160-foot length make the B-36 the world's largest airplane.

12 July
George W. Lewis dies at Lake Winola, Pennsylvania. Lewis had served as the director of aeronautical research for NACA from 1924 until 1947 and executive officer of NACA from 1919 to 1924. In 1936, he received the Guggenheim Medal for directing aeronautical research. Under his leadership, NACA contributions to the evolution of American airpower were significant.

13 July
The Convair MX-774 rocket is successfully launched for the first time. This rocket incorporates gimballed engines (moveable, directional nozzles) and other design features that will eventually be incorporated into the Atlas ICBM.

17 July
American B-29s arrive in England to undergo training at British bases. These are the first U.S. bombers based in the United Kingdom since the end of World War II.

Opposite, top: *General Hoyt S. Vandenberg is sworn in as the second USAF Chief of Staff as General Spaatz looks on. (USAF)*

Opposite, bottom: *Chaplains play a significant role in maintaining morale in the field. (USAF)*

Above: *C-47s line up and load up during the Berlin Airlift. (USAF)*

Left: *Not since the end of World War II had American bombers visited the United Kingdom until these B-29s arrived to participate in combined exercises. Such deployments occur throughout the Cold War. (USAF)*

WILLIAM H. TUNNER: FATHER OF MODERN AIRLIFT

Alfred Beck

Over a 42-year career, William Tunner became the acknowledged expert on American military airlift. He directed military air transport efforts during three world crises and commanded the foremost American military airlift command at his retirement.

Born July 14, 1906, in Elizabeth, New Jersey, Tunner graduated in the West Point class of 1928 and completed advanced flight training at Kelly Field in 1929. A four-year stint as instructor pilot at Randolph Field was followed by field service in Panama and the air defense of the canal. He spent another three years absorbing the techniques of air-ground support. By early 1939, Tunner commanded a reserve unit recruiting local fliers around Memphis, Tennessee, for the U.S. Army Air Corps. With war engulfing Europe, he joined and eventually commanded the fledgling Ferrying Command, which later merged into the wartime Air Transport Command. The ATC first moved aircraft to Hitler's enemies through Lend-Lease and grew to rival commercial airlines in its far-flung routes and number of pilots employed. Under Tunner, Ferrying Command, renamed the Ferrying Division of the ATC, delivered more than 14,000 aircraft to the Allies and circled the earth with its services.

Efforts to keep China in the war against Japan prompted the start of an airlift operation in April 1942 to resupply Chinese and American forces there from bases in India. The command suffered management inefficiencies and a serious loss of pilots on the treacherous flights through Himalayan mountain passes—known as "the Hump." Statistically, for each thousand tons delivered, three American pilots died. Though C–47 and C–46 haulers were hard pressed to carry the load, freight deliveries rose against these odds and freakish weather to a sustained 10,000 tons a month. Brigadier General Tunner assumed command of the India-China Division of the ATC on September 3, 1944. With a systems approach derived from managing transport functions at home, he recast control of the entire enterprise. Using officers commissioned directly from American air and ground transport companies, he adopted an industry-standard production-line maintenance system that consolidated all work on a specific aircraft type in centralized, dedicated facilities. Aircraft in-commission rates rose by 50 percent.

Pilot fatigue was rampant in a "cowboy culture" in which men flew maximum hours in the shortest time to reach a ceiling at which they rotated out of the theater. Tunner revised this with new ceilings, mandatory rest periods, and more comfortable living conditions. He replaced the older transports with four-engine C–54 Skymasters, and delivery rates reached 45,000 tons a month in 1945. By war's end Tunner's managerial system decreased the accident rate to well below 1 percent while moving 650,000 tons from 13 bases in northeast India to six fields around Kunming, China. His methods, still in use in the Air Force today, were tested again within three years.

Amid Cold War tensions gathering in Europe in 1948, Russian authorities, attempting to force the Allies from the city, closed off road, rail, and canal traffic into Berlin on June 29, stopping the life blood of 2.5 million

Opposite: *Major General William H. Tunner—Master of Airlift. (USAF)*

Right: *Tunner commanded a combined and joint force that included airlift forces from Britain, U.S. Navy Douglas RD5s, and Air Force C-47s and C-54s during the Berlin Airlift. He would go on to establish the airlift system used during the Korean War as well. (NASM)*

souls. The American occupation command immediately began an airlift employing three flight paths into two inadequate Berlin commercial fields at Tempelhof and Gatow. Initially, this process carried only 1,750 of the 4,500-ton daily requirement for the metropolis. Major General Tunner, summoned in October to unify the effort, set about making the airlift as "systematic as a metronome." With British and American assets in a Combined Air Task Force, he tightened the schedules to the city to the desired three-minute intervals, expanded the city's existing fields and built another, and sent planes to their destination by two of the existing corridors and brought them out by the third, central one. Aircraft missing a landing proceeded back west instead of stacking up over the city. Averages rose to 8,000 tons a day. On Sunday, April 16, 1949, Tunner staged a round-the-clock "Easter Parade" that delivered 12,941 tons in planes landing every 63 seconds. The Russians lifted the blockade in May, though Operation VITTLES continued through September. Tunner made airlift a flexible diplomatic instrument and allowed the West to prevail in the first major confrontation of the Cold War.

In Korea two years later, Tunner's Combat Cargo Command supported Marine forces retreating from the Chosin Reservoir after Chinese armies intervened there in late 1950. Transports evacuated casualties and even dropped a bridge span that permitted the safe withdrawal of American forces.

From 1953 to 1957, Tunner commanded U.S. Air Forces, Europe, receiving a third star. He served briefly on the Air Staff before his last assignment in 1958 commanding the Military Air Transport Service. He retired in 1960 and died at Ware Neck, Virginia, in 1983.

Left: *The B-45 Tornado was America's first operational jet-powered bomber, capable of conventional and nuclear weapons delivery. (NASM)*

Below: *The McDonnell XF-85 Goblin during testing extended below a modified EB-29B Superfortress nicknamed "Monstro." The Goblin was designed to fit inside the bomb bay of the B-36 Peacemaker to provide short-range fighter cover during missions. (NASM)*

20 July
The first west-to-east transatlantic jet deployment occurs when 16 F-80 Shooting Stars take off from Selfridge Field, Michigan, and fly to Scotland, en route to Fürstenfeldbruck, West Germany. The flight takes nine hours and 20 minutes and is led by Colonel David Schilling.

30 July
The first operational North American B-45A Tornado is delivered to the Air Force. It will be the first USAF aircraft to carry a tactical nuclear weapon.

6 August
Two 43d Bomb Group B-29s, *Gas Gobbler* and *Lucky Lady*, complete a 20,000-mile around-the-world flight in 15 days.

8 August
A Consolidated B-36B Peacemaker completes a nonstop, unrefueled, round trip flight between Fort Worth, Texas, and Hawaii. The distance covered during the flight is more than 9,400 miles.

16 August
The Northrop XF-89 Scorpion flies for the first time at Muroc Dry Lake, California. The Scorpion is the first all-weather, electronic-intercept-capability jet designed for the Air Force.

23 August
In a near disastrous test flight, the McDonnell XF-85 Goblin parasite fighter flies for the first time. Designed to be dropped from the bomb bay of a B-36 Peacemaker, the Goblin would provide fighter cover for the bomber over the target area and then hook up to the bomber and be retracted back into the bomb bay. Test pilot Ed Schoch is unable to complete the hookup, then collides with the trapeze and shatters his canopy. Dazed, Schoch successfully lands the Goblin on the dry lake bed below. Schoch will successfully complete the hookup on October 14.

3 September
Operation DAGGER, a major joint air defense exercise involving RAF aircraft and American B-29s, takes place.

15 September
Major Richard L. Johnson sets a new world speed record of 671 miles per hour while flying a F-86A Sabre in the skies over Muroc Dry Lake, California.

12–16 September
The diary of Orville Wright that describes the first flight in some detail is displayed publicly for the first time at the Library of Congress on the occasion of the first anniversary of United States Air Force independence.

18 September
On the first anniversary of the establishment of the USAF, Convair pilot Sam Shannon flies the Model 7-002 for the first time at Edwards AFB, California. The 7-002 will be accepted by the Air Force and redesignated as the XF-92, a valuable delta-wing test aircraft based upon a radical design pioneered by Germany's Alexander Lippisch.

28 September
The NACA Flight Propulsion Research Laboratory in Cleveland, Ohio, is renamed in memory of Dr. George W. Lewis. The Lewis Flight Propulsion Laboratory will continue to make significant contributions to aircraft propulsion development.

15 October
Major General William H. Tunner assumes command of the American and British Combined Berlin Airlift Task Force. His experience in command of flying "The Hump" during World War II pays valuable dividends during the Berlin Airlift.

19 October
General Curtis E. LeMay is appointed Strategic Air Command commander.

20 October
The McDonnell XF-88 flies for the first time. Although cancelled in 1950, experience gained during the program led to the development of the very successful F-101 Voodoo.

The president of the Franklin Institute presents Theodore von Kármán with the Franklin Medal. The citation for the award praises his outstanding engineering and mathematical achievements and his "unusual leadership whereby some measure of his own genius is constantly instilled in those who work with him."

31 October
The USAF reveals that a modified F-80 has been flown at altitude powered only by two wingtip ramjet engines. This marks the first such use of a ramjet on a piloted aircraft.

4 November
The RAND Corporation is formed as a result of the joint USAF-Douglas Project RAND. RAND will serve as a direct advisor to the USAF by bringing together scientific, industrial, and military brainpower to provide advice to Air Force decision-makers.

5 November
The marking "USAF" is officially approved for all Air Force war planes except those operated by MATS.

Right: *The Convair XF-92A provided much data for future delta-wing aircraft development. (NASM)*

HAP ARNOLD, RAND, AND A SCIENCE OF WAR

Martin Collins

In November 1945, Army Air Forces Commanding General Hap Arnold published an essay entitled "The 36-Hour War: Arnold Report Hints at the Catastrophe of the Next Great Conflict" in *Life*, then the premier mass-circulation magazine of the time. Arnold's article did more than hint; with the aid of artist's renderings showing rockets streaking earthward and massive explosions, he forewarned of the possibility and harrowing effects of a nuclear strike on Washington, D.C.

As with many other leaders in the military and civil society, he saw a new age in the aftermath of World War II. Ocean-crossing missiles and bombers and nuclear weapons left the U.S. open, from border to border, to the possibility of crippling attack, with scant warning. War in the high-technology age loomed as total war—not as army versus army, but as society versus society.

In response to this readily visualized threat, Arnold pushed for a national commitment to ongoing military preparedness, especially through a systematic program to apply the latest science and technology to improve and invent new weapons. But this view of preparedness stood in direct contrast to a long history of American suspicion of large, standing forces. For Arnold and the military the challenge was this: how to work with the great sources of scientific and technical knowledge in American civil society and assure military preparedness?

Although Arnold retired in early 1946, he left a rich legacy, and a deep effect on the future Air Force, in his response to this question. One part was the creation of a Scientific Advisory Group, under the leadership of famed aerodynamicist Theodore von Kármán. Another was the founding, in collaboration with Donald Douglas, president of Douglas Aircraft Company, of RAND—the first "think tank."

RAND's founding dramatized the new circumstances and preoccupations of the postwar period. Established in March 1945 via a "letter" contract from the Army Air Forces to the Douglas Company, RAND (allegedly an acronym for Research and No Development) had two primary purposes: to provide an institutional home for a range of academic

disciplines—from engineering to the social sciences and humanities—and to focus that collective expertise in assisting the Air Staff with nearly all aspects of planning for war. RAND represented a key idea of the time: that modern war, total in extent, required a comprehensive science of war. The only way to achieve this new "science" was by drawing together elements of American society previously distinct— the military, aircraft industry, and academia—an exercise unimaginable in prewar years.

This spirit of experimentation and overturning of tradition lent a certain mystery to RAND as it caught the public eye in the early 1950s. By then RAND had outlived its initial home at Douglas and become a separate nonprofit corporation, in Santa Monica, California. Much of its work was classified, adding secrecy to mystery—a 1951 article in *Fortune* magazine dubbed RAND "a secular monastery."

RAND's "science of war" research, too, fueled this view. Its very first study, capturing Arnold's view of the importance of future-oriented technology, was entitled "Preliminary Design of an Experimental World-Circling Spaceship." Inaugurated in 1945, this research

Opposite: *General Hap Arnold (left) and Donald Douglas during a World War II visit to the factory. (Robert and Kathleen Arnold Collection)*

Right: *Donald Douglas at his office in Southern California. The Douglas-Arnold relationship went beyond just business. Arnold's family often went sailing on Douglas's beautiful boat. Arnold's son, Bruce, eventually married Douglas's daughter Barbara. (Robert and Kathleen Arnold Collection)*

extended for nearly a decade, helping to establish the engineering and social basis for satellites of varying applications—communications, meteorology, and, especially, military reconnaissance—all well before Sputnik started the Space Age in 1957.

But RAND's place in the national psyche grew more from its intimate role in dealing with the central and overriding fact of the Cold War: the ability of both the U.S. and the Soviet Union to fight a nuclear war. Arnold's 1945 speculation on ready and able nuclear forces was soon a reality. Over the 1950s, atomic and then hydrogen bombs were perfected and produced by the hundreds, and long-range bombers, including the still-used B-52, and long-range missiles made those bombs formidable. The "Strategic Bombing Systems Analysis," published in 1950 after three years of exhaustive research, presented a detailed breakdown of technologies and plans for mounting an Air Force nuclear attack on the Soviet Union. This was RAND's first attempt to study war scientifically—and "systems analysis" thereby became a catchphrase for researching and defining options for complex problems. Over the 1950s and into the 1960s, RAND studies were at the center of military and political discussion of how to fight and defend against nuclear war—to live with the "delicate balance of terror." In 1964, director Stanley Kubrick explored the prominence of RAND-type thinking in the film *Dr. Strangelove or: How I Stopped Worrying and Learned to Love the Bomb*. By then RAND's "science of war" began to turn to a "science of society" as systems analysis found new applications in the Kennedy and Johnson administrations as urban and other social challenges received greater attention.

Left: *A family portrait of Air Force Flight Test Center "X-planes." The planes are (left front then clockwise) Bell X-1A, Douglas D-558 jet (Navy), Convair XF-92, Bell X-5, Douglas D-558 rocket (Navy), Northrop X-4, and X-3 Stiletto. (NASM)*

10–12 November
The School of Aviation Medicine holds the first-ever symposium discussing problems associated with space travel.

22 November
The original Wright Flyer arrives in Washington, D.C., after being on display in the Science Museum in London. It had been on display there for two decades.

30 November
A Douglas C-54 Skymaster makes a remarkable descent from 15,000 feet to 1,000 feet in one minute and 22 seconds. This extreme descent rate is possible by the installation of Curtiss-Wright reversible-pitch propellers.

1 December
Continental Air Command is activated this date.

2 December
The Beech Model 45 demonstrator flies for the first time. The Model 45 will begin Air Force service as the T-34A Mentor, the first primary trainer to be added to the inventory since the end of World War II. The T-34 will be used by the USAF until 1961.

9 December
In a tremendous demonstration of capability, a B-36 and a B-50 each complete a nonstop, round-trip flight from Carswell AFB, Texas, to Hawaii. The massive B-36 flies the entire trip without refueling and the B-50 refuels three times from a KB-29M to make the 35-plus hour trip.

16 December
The Northrop X-4 Bantam flies for the first time at Muroc Dry Lake, California, piloted by Charles Tucker of Northrop. The Bantam program, which consists of two X-4 aircraft, will experiment with the characteristics of semi-tailless, swept-wing aerial vehicles. It is a joint NACA-USAF test program.

17 December
The original Wright Flyer is donated to the Smithsonian Institution, bringing a 45-year-old dispute between Wrights and the institution to a close. President Harry S. Truman sends a message to the crowd while members of the Wright family, the Smithsonian staff, and many foreign dignitaries attend the ceremony.

The presentation of the first annual Wright Memorial Trophy is made that evening at a formal dinner. The National Aeronautical Association presents the trophy for "significant public service of enduring value to aviation in the United States." Dr. William F. Durand, the first recipient of the award, is unable to attend.

Right: *A Douglas C-47 outfitted with skis makes a landing on a snow-packed runway. (NASM)*

Below: *The original Wright Flyer is on display at the National Air and Space Museum in Washington, D.C. (NASM)*

28 December
Lieutenant Colonel Emil Beaudry lands a ski-equipped C-47 Skytrain on a remote Greenland icecap to rescue a dozen airmen who had been stranded there in two separate crashes. On December 9, an Arctic storm had forced a C-47 to land on the icecap, stranding them there. In a failed B-17 and towed glider rescue attempt, five more airmen were stranded with the original C-47 crew. His successful rescue of these 12 airmen earns Beaudry the Mackay Trophy.

29 December
Defense Secretary Forrestal announces that the U.S. is developing an "earth satellite vehicle program" to study the feasibility of placing objects into earth orbit.

31 December
On this date, the 100,000th Operation VITTLES flight is made during the Berlin Airlift.

1949

3 January
Severe blizzards strike several western states. The USAF begins Operation HAYRIDE and airlifts more than 4,700 tons of livestock and supplies in more than 200 sorties through March 15.

5 January
Major Chuck Yeager sets an unofficial climbing speed record greater than 13,000 feet per minute in the Bell X-1 during the first standard takeoff made by a rocket-propelled research aircraft. This is the first and only takeoff made from the ground during the rocket research plane program.

19 January
The Martin XB-61 Matador tactical missile is successfully test fired at Holloman Air Force Base, New Mexico. The Matador is a short range, mobile, surface-to-surface missile.

Left: *The Boeing B-50A* Lucky Lady II *takes fuel from a KB-29 during a practice flight prior to its Mackay Trophy-winning around-the-world flight. (NASM)*

25 January
Air Force "slate" blue uniforms are mandated for the service for the first time. These replace the Army "olive buff" uniforms previously worn.

8 February
Shattering all existing transcontinental records, a Boeing B-47 Stratojet flies from Moses Lake airfield, Washington, to Andrews AFB, Maryland, in three hours and 45 minutes. The jet bomber averages more than 600 miles per hour during the near-2,300-mile trip, cutting the previous record nearly in half.

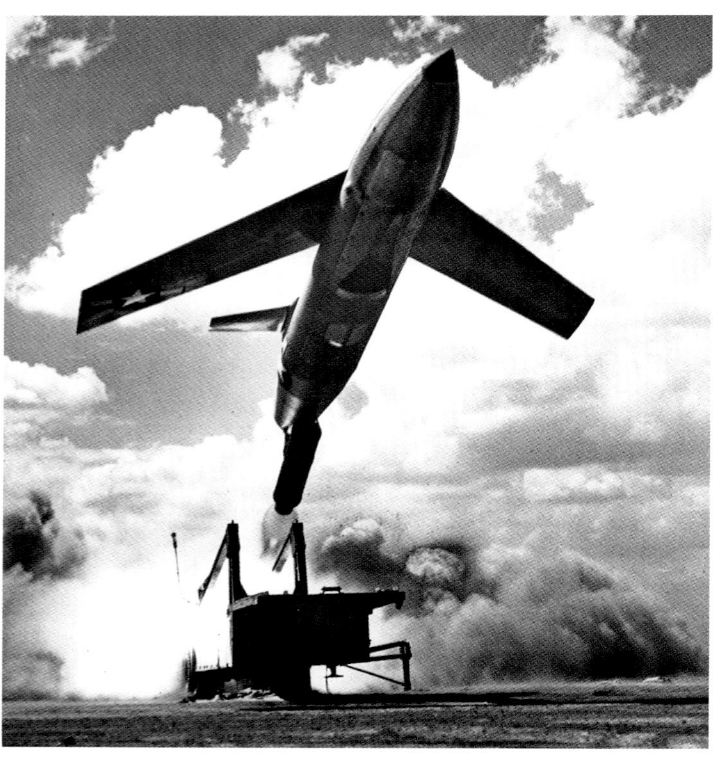

Below: *A Matador missile is launched from a tactical platform. (NASM)*

9 February
The Air Force establishes a Department of Space Medicine at its School of Aviation Medicine located at Randolph AFB, Texas.

24 February
In the first rocket launch from Cape Canaveral, Florida, a two-stage rocket made of a captured German V-2 and a WAC Corporal second stage is launched this date. The rocket reaches an unprecedented altitude of 244 miles and a speed of 5,150 miles per hour. Part of Project BUMPER, this launch demonstrates the potential of two-stage rockets.

26 February–2 March
The Boeing B-50 *Lucky Lady II* makes the first nonstop flight around the world when the aircraft, piloted by Captain James Gallagher, lands at Carswell AFB, Texas. The flight covers more than 23,400 miles and takes just over 94 hours to complete. Four in-flight refuelings are accomplished between the B-50 and six KB-29M tanker aircraft over the Azores, Arabia, the Philippines, and Hawaii. A probe-and-drogue technique developed by the British and practiced by the AAF during World War II is used to grab the refueling hose and transfer fuel between aircraft. The crew earns the Mackay Trophy for the mission.

4 March
Berlin Airlift aircraft have transported more than one million total tons of coal, food, and supplies since the effort began in June.

Right: *"Staying Power—Berlin," by Gil Cohen. (USAF Art Collection)*

Below: *In a June 1, 1949, ceremony, President Truman decrees that retired Army five-star General Hap Arnold will become General of the Air Force. (USAF)*

15 March
Global Weather Central is established by the Military Air Transport Service (MATS) to support Strategic Air Command (SAC) operations.

26 March
A B-36D Peacemaker outfitted with an additional four J47-GE-19 jet engines housed in two wing pods makes its first flight. The increased performance demonstrated by this configuration allows the cancellation of the B-54 project and stokes the fire of contention burning between the USAF and the Navy over strategic nuclear missions. The B-36D has a new top speed of 440 miles per hour and can carry approximately 85,000 pounds.

30 March
A bill which provides for the first permanent radar defense network for the U.S. is signed into law by President Truman.

4 April
The North Atlantic Treaty Organization (NATO) is created.

6 April
During a high point of operations during the Berlin Airlift, one aircraft lands at Tempelhof Airport every four minutes or less for six consecutive hours. One particular ground approach crew directs more than 100 landings in just over six hours.

16 April
The Lockheed YF-94 Starfire flies for the first time at Van Nuys, California. The mission is flown by test pilot Tony LeVier and test engineer Glenn Fulkerson. The YF-94 is modified from the basic TP-80 design to act as an interim all-weather interceptor.

Airlift to Berlin hits its high point when 12,940 tons of supplies are airlifted to Berlin by 1,398 aircraft sorties.

7 May
Hap Arnold, retired five-star general, is given the permanent rank of General of the Air Force by a special act of Congress. A formal pinning-on ceremony occurs later in June.

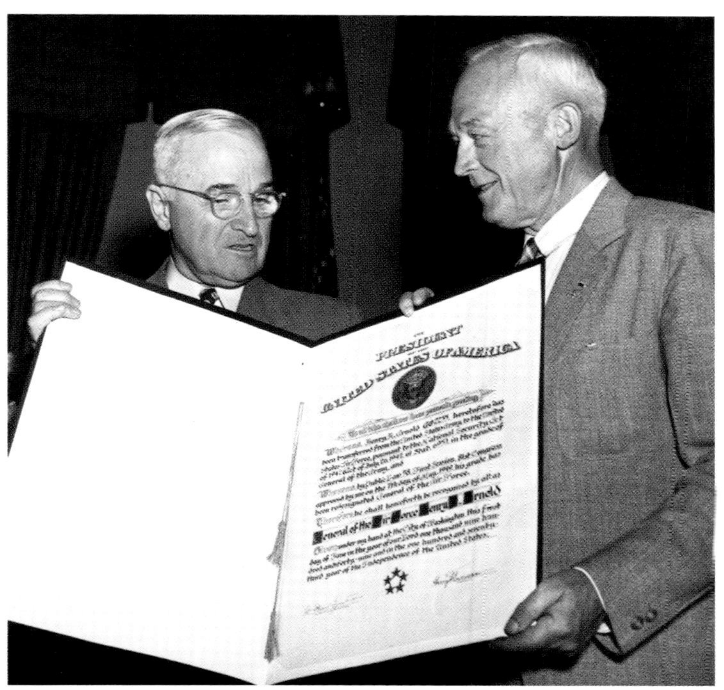

THE INCOMPARABLE GOONEY BIRD: THE DOUGLAS C-47 SKYTRAIN

David R. Mets

Left: *Not only a cargo carrier, the C-47 also delivered paratroopers and their equipment behind enemy lines. (USAF)*

During the Roaring Twenties, many Americans thought commercial aviation to be the wave of the future and an essential part of the nation's airpower. Attempts were made to promote aviation as a safe, efficient, and dependable mode of transportation. Events such as the Transcontinental Air Race of 1919 and the seven-day air refueled reliability test in an Air Corps Fokker Tri-Motor were arranged for just such a purpose. But fatal crashes, sometimes involving famous Americans like Notre Dame football coach Knute Rockne, who died in a fiery Fokker aircraft disaster, undermined these efforts. The infant airlines were having a tough time turning a profit because of poor aircraft reliability and safety. Such unreliability caused the airlines to search for new equipment. One such request from Trans World Airlines (TWA) arrived on the desk of Donald Douglas in Santa Monica, California.

Donald Douglas, who had helped design the Martin bomber that Billy Mitchell's airmen used to sink several ships in 1921, including the captured German battleship *Ostfriesland*, soon left Martin to start his own company. Douglas manufactured the Air Service World Cruisers, which were commissioned for the round-the-world flight of 1924. Until 1932, though, all his work had been for the military. TWA's search for a new commercial airliner stimulated him to branch out into civilian airplane design.

Douglas's first attempt was the DC-1 purchased by TWA's Howard Hughes. Only one of these was built. It was the prototype for the slightly larger DC-2, which was bought in numbers by both U.S. and foreign airlines, becoming the foundation for airline profitability. The DC-2 served as the parent of the famous DC-3 Skytrain—known by all as the Gooney Bird.

As Douglas's designs evolved, the all-metal, low-wing, twin-engine aircraft had grown from about 18,000 pounds takeoff weight to about 27,000—though in many

Right: *Workers at Douglas Aircraft Company, Long Beach, California, assemble fuselage sections for more than a dozen Skytrains on the production line. (NASM)*

wartime situations it flew at much heavier weights. Douglas built more than 10,000 Gooney Birds, and by World War II their Pratt & Whitney power plants were generating 1,300 horsepower. Unlike early Fokkers, they had controllable pitch propellers, and by the mid-1930s the props had a full-feather capability enabling relatively easy flight on one engine. In the passenger configuration, 21 seats were provided. The standard load in the cargo version was around 6,000 pounds, though during wartime they were known to have taken off with loads more than twice that.

Thus, the Gooney Bird was available in numbers and had been thoroughly tested before Pearl Harbor. Countless pilots confess to having a love affair with the C-47 as being a kind and gentle bird that was a pleasure to fly—safe, and of high reliability and maintainability under austere conditions that surpassed most other aircraft. The DC-3 was fast for its time and with temporary tanks could span the oceans. It could operate in and out of fields too short and rough for almost all other transport planes.

The C-47 filled multiple roles. It was at first the standard equipment for the long haul routes of the Air Transport Command and the Naval Air Transport Service. It was the main equipment that birthed the American airborne capability and served the paratroopers even into the Korean War. The C-47 was there for the disastrous drops on Sicily. It pulled the gliders to Normandy. It was there for the Bridge Too Far at Arnhem. It was the main instrument for the landing of troops for the very successful recapture of Corregidor. The Goon was used as a bomber dropping 55-gallon drums of napalm on enemy ground forces. Enterprising crews in the China-Burma-India Theater (CBI) built gun mounts for strafing ground targets. The C-47 did yeoman service everywhere helping wounded soldiers survive through medical evacuation. It sprayed for mosquitoes and provided executive transportation for many VIPs.

But the story did not end when World War II ended. The C-47 saw the garrison through the grim early weeks of the Berlin Blockade when the European Command had only 100 of them plus two C-54s. They held the door open until hundreds of C-54s could be redeployed thence from all over the world. USAF C-47s helped succor the Marines and carry away their wounded during the grim retreat in the first winter in Korea.

The Douglas B-26s had done important work in the Korean War, too, and even in the early days of Vietnam. But their more ancient brethren, the Gooneys, outlasted them to soldier on longer as transports, side-firing gunships, and electronic countermeasure platforms throughout Vietnam and afterwards. Even at the dawn of the twenty-first century they go on in similar roles in Latin America, Africa, and even in North America—an incomparable airplane if there ever was one.

9 May
The XF-91 flies for the first time under the control of Republic Company test pilot Carl Bellinger. The XF-91 Thunderceptor has a jet-and-rocket hybrid propulsion system and a unique inverse-taper, variable incidence design wing.

11 May
President Truman signs a bill authorizing a guided missile test range for USAF use. Cape Canaveral will be established as its main facility.

11–12 May
The Soviets lift the surface blockade of Berlin. Airlift sorties continue.

21 May
Captain H. D. Gaddis pilots a Sikorsky S-52-1 helicopter to a new international altitude record of 21,220 feet in the skies over Bridgeport, Connecticut.

24–26 May
The Institute of Aeronautical Sciences and the Royal Aeronautical Society hold the Second International Conference on Aeronautics in New York.

4 June
The Lockheed XF-90 flies for the first time. The XF-90 never went into production, as it lost the competition against the McDonnell XF-88, eventually developed as the F-101 Voodoo.

1 July
The USAF Medical Service is established, and Major General Malcolm C. Grow becomes the first surgeon general of the Air Force.

Lockheed test pilot Tony Lavier flies the F-94A all-weather interceptor for the first time. A derivative of the Lockheed T-33, it includes a nose radome, four .50-caliber machine guns, and an afterburner-equipped J-33 engine.

3 July
The Boeing B-29 *Enola Gay* is presented to the Smithsonian Institution at Park Ridge, Illinois, where it will be stored temporarily.

8 August
Major Frank K. "Pete" Everest pilots the X-1 to an unofficial altitude record of 71,902 feet, the highest altitude attained by the first generation of X-1 research aircraft.

10 August
The Department of Defense is established when President Truman amends the National Security Act of 1947, changing the name from the National Military Establishment.

25 August
Major Pete Everest is saved by his T-1 partial pressure flight suit when the Bell X-1 he is piloting suffers a rapid decompression while flying at 69,000 feet. This is the first emergency use of the T-1 pressure suit. Everest is able to land the aircraft safely.

Opposite, top: *Republic XF-91 Thunderceptor. (NASM)*

Opposite, bottom: *Joe I, the Soviet's first atomic bomb test. (Courtesy Air Force Association)*

Right: *North American T-28A Trojan. (NASM)*

Below: *This C-54 flew the final Berlin Airlift sortie into Berlin. More than 1.7 million tons were delivered during the operation. (NASM)*

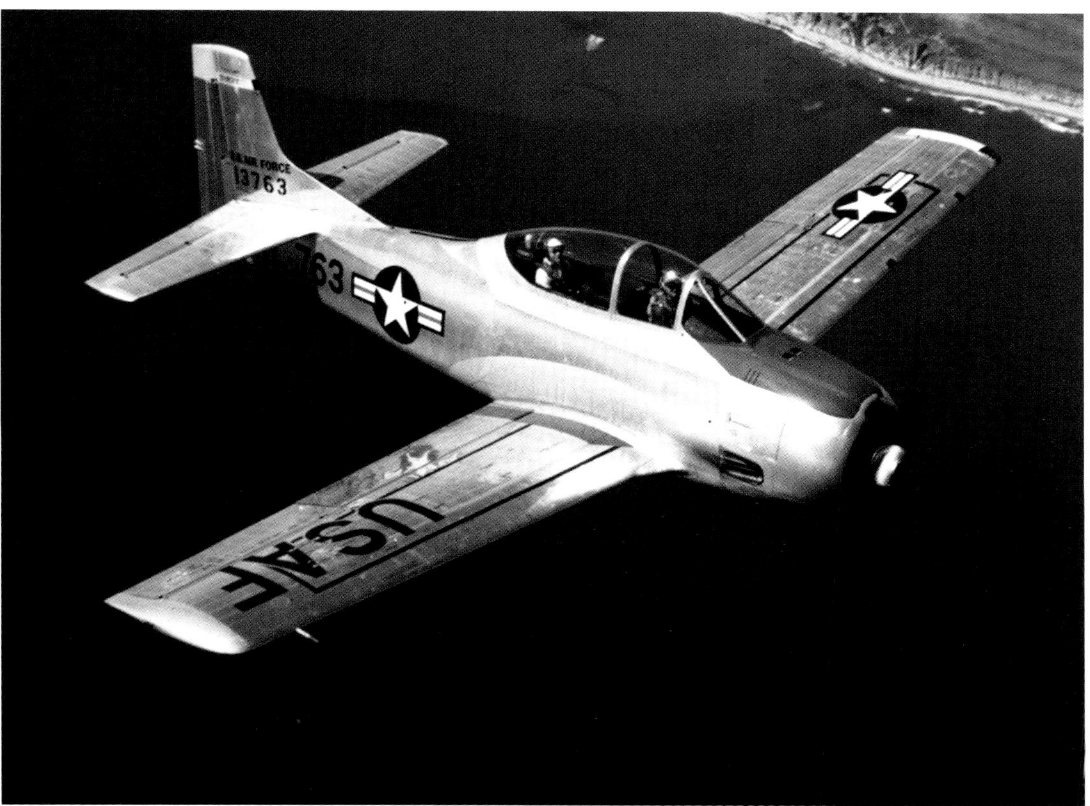

9 August
The Soviet Union detonates its first atomic bomb at Semipalatinsk, a nuclear test range. The bomb is similar to the Fat Man weapon detonated over Nagasaki, Japan. The Soviets call it *First Lightning*.

24 September
The North American T-28 Trojan flies for the first time at Inglewood, California. The T-28 is used as a trainer for a time and also as an attack aircraft during the early stages of the Vietnam War.

30 September
The Berlin Airlift ends. Total tonnage carried exceeds 2.34 million tons, carried by more than 277,000 flights. In the end, USAF, U.S. Navy, RAF, Commonwealth air forces, and British independent airlines participate in the operation. USAF aircraft carry 1.78 million of the total tonnage.

1 October
With Mao Zedong as leader, the Peoples Republic of China is established.

The Long-Range Proving Ground at Cape Canaveral is activated. Major General W. L. Richardson takes command.

14 October
Chase Aircraft Company pilots make the first flight of the XC-123 assault transport. The USAF acquires the aircraft (C-123 Provider), which will see extensive use during the Vietnam War.

27 October
The Unitary Wind Tunnel Act authorizes the construction of several important wind tunnel testing facilities across the country. More than $100 million is earmarked to build the Arnold Engineering Development Center (AEDC) in Tullahoma, Tennessee.

Left: *A Sikorsky YH-19 Chickasaw during a transition flight. (NASM)*

Below: *The YF-84D "Sabre Dog" was designed to fire only rockets. (NASM)*

10 November
The Sikorsky YH-19A flies for the first time. This influential helicopter design is the first to have a cabin unrestricted by the engine, which was mounted in the nose.

18 November
A Douglas C-74 Globemaster I *The Champ*, flies from Mobile, Alabama, to RAF Marham, United Kingdom, a 23-hour-long trip. The mission sets a transatlantic record by carrying 103 passengers and crew. It is the first airplane to cross the Atlantic with more than 100 people aboard.

29 November
The Douglas YC-124 Globemaster II, known as "Old Shakey" to many who fly it, flies for the first time. Operationally it will be able to carry 50,000 pounds of cargo for 850 miles. This type serves in almost every command in the Air Force from MATS to SAC.

December
During this month, reaching a major milestone in wind tunnel technique, the first-ever continuous-flow, transonic flow is demonstrated at NACA Langley's eight foot high-speed tunnel using the "slotted-throat" technique.

2 December
The USAF fires its first Aerobee research (RTV-A-1a) rocket at Holloman Air Development Center, New Mexico.

5 December
The USAF diverts $50 million funding to begin building radar sites in Alaska and other northern tier locations to protect against any Soviet bomber incursions. Specified detection range is to be 300 miles.

17 December
A commemorative six cent airmail stamp is issued by the Post Office at Kitty Hawk, North Carolina. The etched stamp shows the Wright Flyer in flight and small portraits of Orville and Wilbur in the upper right corner of the stamp.

22 December
The North American YF-84D Sabre Dog all-weather interceptor flies for the first time. The modified F-86 has a redesigned nose to accommodate large radar and a retractable rocket tray.

Above: *Jacqueline Cochran smiles as she prepares to deplane. Cochran will set a number of speed records during her flying career. (USAF)*

25 December
Stupalith, a special ceramic product that contracts when heated and expands when cooled, is revealed by the USAF. Stupalith can withstand temperatures in excess of 2,000 degrees and is used to insulate and protect certain parts of jet and rocket engines.

28 December
After a two-year investigation, the USAF determines that "flying saucers" do not exist and disbands Project SAUCER at Wright-Patterson AFB, Ohio.

29 December
Flying a F-51 without payload, famed aviatrix Jacqueline Cochran sets an international speed record of 436.995 miles per hour over a 500-kilometer course and lands at the Mount Wilson course, Desert Center, California.

US AIR FORCE
12733

PROPPING UP DOMINOES

1950–1953

PROPPING UP DOMINOES

1950–1953

By 1950, the United States Air Force had established itself as the "get-things-done" service. Fresh from the success of the Berlin Airlift, the USAF had much to be proud of, even if they had relatively little to work with in terms of personnel or capabilities. By June, the USAF consisted of 48 combat flying wings and only 400,000 officers and airmen—two million fewer than at its peak strength during World War II.

It was already clear that the Soviet Union and the United States were embarked upon conflicting ideological paths. Inevitably, the Cold War was slowly heating up. In June it also became clear that the means by which the Cold War would be fought was through brushfire wars initiated by Soviet satellite nations around the world.

The Korean War, a surprise to both the United States and the United Nations, made it possible to begin rebuilding America's military to necessary and acceptable levels. More than simply enlarging the tactical forces required to stem the hostile aggression in Korea, the Air Force also began expansion to counter the growing Soviet nuclear threat and to ensure the national security of the United States. Procuring long-range heavy bombers like the B-36 Peacemaker and perfecting in-flight refueling were essential to accomplishing long-range nuclear attack options. Thus, the Air Force had two priorities, fighting a tactical war in a faraway place while also providing the capability to respond to a significant Soviet nuclear threat.

With these responsibilities, the Air Force did not use most of its top-of-the-line equipment during the Korean War. F-82 Twin Mustangs, F-80 Shooting Stars, and F-51 Mustang fighters were involved throughout the air battle and were eventually joined by the F-84. Except for the F-86 Sabre, deployed to the theater only after the appearance of the deadly MiG-15, new equipment like the B-36 Peacemaker and the B-47 Stratojet were held in reserve to carry out the mission of strategic deterrence.

To augment the immediate needs of the ground war, the Air Force mobilized 22 wings of the Air National Guard and 10 wings of the Air Force Reserve, plus an additional 100,000 Air Force Reservists. Many of these units operated outdated aircraft, like the B-26 bomber. By the end of the Korean War, in response to the Joint Chiefs of Staff authorization to expand the USAF, 93 combat-ready wings and more than 975,000 personnel were actively serving.

Development of technologically advanced weapons systems continued during these years. Guided missiles and bombs were at the forefront of weapons development. The Tarzon radio-guided bomb actually took out a few bridge spans with one blow during strikes in Korea. By the end of the war, the last piston-driven fighters had been retired and replaced with turbine-powered, air-refuelable aircraft.

Stateside, speed records fell in rapid succession as the F-86 and the new F-100 Super Sabre took to the skies over Edwards AFB, California. A series of "X-planes" were built to test everything from swept-wing designs to a variety of fuselage shapes. Rocket development also continued, and the first long-range missiles were actually tested during these years.

Living creatures were sent aloft as subjects in a number of missiles.

Although aging B-29 and B-50s accomplished the majority of strategic bombardment during the Korean War, there were few decidedly "strategic" targets that were not quickly destroyed in the first few months of the air war. By 1955, the B-47, powered by six turbojet engines, would replace the vast majority of propeller-driven bombers, and the B-52 Stratofortress, powered by four pairs of turbojets, began to replace the 10-engine B-36 as the long-range component of Strategic Air Command.

The development of aerial refueling was also influenced by the transition to jet power. KC-97 piston-driven tankers could not match the operational speeds necessary for safe refueling of jet fighters and bombers. The KC-135 eventually filled the gap and provided the refueling capability that the Air Force would depend upon for the following four decades.

Tactical aircraft like the B-26, which saw significant use in Korea, were soon replaced by the jet-powered B-57 Night Intruder (originally the British Canberra), and the B-66 and RB-66 came shortly after that.

Airlift missions had been vitally important to the success of UN ground forces during the war. Major General William Tunner impressed many ground force commanders who found themselves in tough defensive struggles during the ebb and flow of the first 18 months of the war. Soon, the C-130 Hercules would begin to replace aging C-119 and C-123 cargo planes, beginning a remarkable aircraft career that continues to the present day.

During the Korean War, the Air Force flew more than 720,000 sorties from Japan and from bases in Korea. Approximately 1,500 Air Force aircraft were destroyed during the war, while some 520 USAF personnel were killed. Of the nearly 500,000 tons of weapons dropped on the enemy during the war, 85 percent of them were delivered by USAF aircraft, which earned the respect and admiration of ground commanders like Army Lieutenant General Walton Walker, who freely admitted that without airpower, it would have been impossible for his forces to stay in Korea.

In the end, although a clear UN victory remained out of reach, support for the democratic South Korean people was strong; thanks to American involvement, their nation remained free from communism. More significantly, the Korean War demonstrated to the American people that peace might be easier to maintain through strength rather than through political displays of ambivalence or military weakness.

Pages 324–325: *The North American F-86 Sabre, an Air Force mainstay in the Korean War, gained and maintained air superiority over the Korean Peninsula in the finest of airpower tradition. ("Gabby Scores Again," by Robert Cunningham, courtesy of Lockheed Martin)*

Right: *"Beauteous Butch," by Mark Waki. (USAF Art Collection)*

PROPPING UP DOMINOES

1950–1953

"You have a row of dominoes set up, you knock over the first one, and what will happen to the last one is the certainty that it will go over very quickly. So you could have a beginning of a disintegration that would have the most profound influences."
—Dwight D. Eisenhower, Presidential News Conference, April 7, 1954

1950

15 January
General of the Air Force Hap Arnold dies of heart failure at his Sonoma, California, ranch—El Rancho Feliz. Arnold's life spanned the development of military aviation from the Wright brothers to the advent of jet and rocket propulsion.

18 January
The YF-94C Starfire flies for the first time. The USAF's first rocket-armed interceptor, it entered service in 1953.

23 January
The USAF Research and Development Command is established. The creation of this command separates basic research functions from logistics and procurement functions, which remain with the Air Materiel Command. On September 16, it is renamed as Air Research and Development Command (ARDC).

31 January
President Harry S. Truman orders the development of the hydrogen bomb—a thermonuclear weapon—also known as "the Super Bomb."

14 February
The Soviet Union and China sign a treaty of alliance and mutual assistance.

1 March
The Boeing B-47A Stratojet rolls out of the factory and is turned over to the USAF for engineering inspection prior to flight.

15 March
In a statement delineating basic roles and missions for the armed services, the USAF receives formal and exclusive responsibility for strategic guided missiles.

22 March
Under the Atlantic Pact arms aid program, the first four of 70 Boeing B-29 Superfortresses are delivered to the Royal Air Force at Marham, England.

18 April
The USAF announces the planned procurement of 1,250 new aircraft at the cost of slightly more than $1.2 billion.

24 April
Thomas K. Finletter is appointed Secretary of the Air Force.

5 May
The Northrop YRB-49A flies for the first time near Hawthorne, California. The experimental flying-wing design is powered by six Allison J-35 turbojet engines housed within the 172-foot wingspan.

Left: *General Hap Arnold's closest friends felt that this image best represented Arnold's retirement life as well as his spirit. He finished his memoir,* Global Mission, *only months before he died. (Robert and Kathleen Arnold Collection)*

Right: *This North American F-82G Twin Mustang, credited with the first aerial victory of the war, is outfitted for night interceptor operations, including a black paint scheme and radar pod mounted on the wing between cockpits. (NASM)*

Below: *Republic F-84F Thunderstreak, the swept-wing version of the original F-84. (NASM)*

10 May
President Harry S. Truman signs legislation that creates the National Science Foundation.

12 May
The Bell X-1 No.1 rocket research plane makes its final flight. After the mission it is donated to the Smithsonian Institution.

3 June
The Republic YF-96A Thunderstreak flies for the first time, piloted by test pilot O. P. Bud Haas. The swept-wing variant of the F-84 Thunderjet, its designation was changed to YF-84F in September.

25 June
The Korean War begins when North Korean forces invade South Korea. North Korean fighter planes, flown by Russian pilots, strike Kimpo Air Base.

United Nations resolutions demand that North Korea immediately withdraw their forces from the south. Major General Earle E. Partridge, Fifth Air Force Commander, puts his forces on alert and increases surveillance activities between Korea and Japan.

26 June
During evacuation operations at ports near Inchon, South Korea, Air Force F-82 Twin Mustangs and SB-17 aircraft provide top cover for foreign ships rescuing U.S. citizens and then transporting them to bases in Japan.

27 June
President Harry S. Truman orders the U.S. Air Force into combat over Korea. Airlift aircraft begin evacuation operations from airfields near Seoul and Suwon while fighters and medium bombers provide cover.

In an F-82, 1st Lieutenant William "Skeeter" Hudson and his radar observer, Lieutenant Carl S. Fraser, shoot down a North Korean Yak-11 aircraft near Seoul, earning the first credited kill of the war. Five other USAF pilots are credited with a total of seven kills that day, the highest number of kills recorded in one day during 1950.

Fifth Air Force B-26s attacks enemy targets in South Korea staging from their bases in Japan. Bad weather makes their raids ineffective. Fifth Air Force establishes an advance echelon at Itazuke AB, Japan, and RF-80s deploy to Itazuke AB for missions in Korea.

THE AIR WAR OVER KOREA

Conrad Crane

USAF leaders entered their first air war as an independent service hoping to achieve air superiority quickly and then gain victory with a bombing campaign to destroy the enemy's capacity and will to wage war. Instead they found themselves with a nagging problem: neutralizing effective MiG-15 interceptors. Unable to strike the true source of enemy war-making resources, they were forced to look for alternative methods to achieve decisive results in a limited, non-nuclear war.

When North Korea invaded the South in June 1950, America retaliated with air strikes by B-26 bombers of the Far East Air Forces. However, FEAF's main component, the Fifth Air Force, was trained and configured primarily for the air defense of Japan. F-80 Shooting Stars, F-51 Mustangs with D-day markings, and B-29s from Strategic Air Command were rushed to the Far East. The Superfortresses arrived with a plan to recreate the incendiary campaign that had played a major role in defeating Japan in World War II, but General Douglas MacArthur, the theater commander, would not approve it. Instead FEAF, having quickly destroyed the small North Korean Air Force, concentrated initially on unfamiliar ground support operations that helped turn the tide of battle. Once the battle front stabilized in August, FEAF Bomber Command began a systematic destruction of North Korean industry that would virtually eliminate that capacity by October. The major exceptions were hydroelectric facilities and complexes around Rashin along the Soviet border, placed off limits because of political restrictions.

FEAF and Naval aircraft from the carriers of Task Force 77 also provided close air support, a key enabler for MacArthur's drive north. However, just as orders were issued to redeploy the SAC bombers back home, a new threat appeared. On November 1, MiG-15s from China flown by Soviet pilots began attacking American and United Nations aircraft over North Korea. There were also reports of a great buildup of Chinese forces in Manchuria. MacArthur responded by asking for permission from the Joint Chiefs of Staff for FEAF to conduct "hot pursuit" into China and bomb bridges over the Yalu River. When these requests were refused, he unleashed Bomber Command to create a wasteland across North Korea to

Opposite: *"Glory in the Sun," by Harley Copic. (USAF Art Collection)*

Right: *Two F-86 pilots walk through the gate on the way to another mission over MiG Alley. (USAF)*

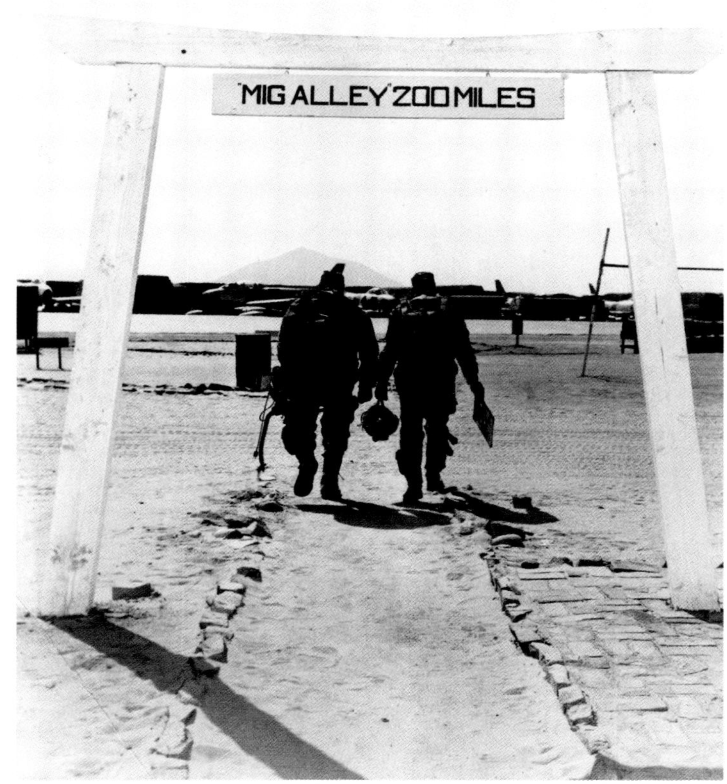

prevent any Chinese incursion. Faith in airpower was a key factor in MacArthur's decision to continue the drive north to the Yalu, which was repulsed by surprise Chinese attacks in late November.

In the long retreat that followed, FEAF, Navy, and Marine aircraft again provided invaluable aid to ground forces with bombing strikes and supply drops. Eventually the tide turned again, and these same air forces supported the return of UN forces to the 38th Parallel. By the fall of 1951, the ground war was stalemated while negotiators at Panmunjom tried to agree on armistice terms. The air war continued unabated. General Matthew Ridgway, who had replaced MacArthur, knew that airpower was now his primary tool to coerce his enemy, and he pursued a number of interdiction campaigns to achieve that end. However, the limited needs of a stagnant front, Communist countermeasures, and the growing number of MiG-15s frustrated his efforts. The United States had rushed new F-86 Sabres to Korea to counter the MiGs, but they were always heavily outnumbered, and by October 1951 the slow B-29s had been driven from the daylight skies by enemy jets. The brunt of day bombing efforts would now be borne by jet fighter-bombers, a significant change in air warfare.

The exploits of F-86 aces were front page news in American papers, and there were exorbitant claims of kills on both sides of the Yalu. Best estimates give American pilots about a 2-1 edge over their Communist foes in air-to-air combat. As the Soviets left, less-experienced Chinese and North Korean pilots took over, who proved more reluctant to cross over the Yalu into "MiG Alley." By 1953 unauthorized Chinese incursions by F-86 pilots hunting MiGs, and headlines, were common, despite determined efforts by higher headquarters to stop the practice.

These incursions were just one signal to the Communists that UN forces might be willing to escalate the war. When General Mark Clark replaced Ridgway as UN Forces Commander in mid-1952, he authorized a new air campaign. General O. P. Weyland, FEAF commander, and his deputy for operations, Brigadier General Jacob Smart, who had led the low level raid against the oil fields at Ploesti, Romania, during World War II, proposed an "air pressure" approach to force Communist compliance at the peace table. Instead of interdiction, destruction of key targets was the goal. Initially Air Force and Navy jets hit the now unrestricted hydroelectric complexes. Then they moved to destroy "supply centers," basically every town and city in North Korea. In May 1953, the last target was struck and irrigation dams sustaining the North Korean rice crop were damaged.

When the armistice was signed in July 1953, USAF leaders were convinced their air pressure attacks had been the decisive force in Korea that coerced the Communists to concede. In reality internal factors were probably more important in leading to Communist concessions, though the massive destruction imposed by airpower, especially on Chinese troops, was definitely important. President Eisenhower was convinced that his threat to expand the war, including the use of atomic bombs, had been decisive. After the war, his emphasis on nuclear weapons, the USAF's perception of the success of air pressure, and continued focus on the Soviet threat distracted American airmen from the lessons learned over Korea. They would have to be relearned, at considerable cost, in the skies over Vietnam.

The Royal Australian Air Force (RAAF) No. 77 Squadron arrives in Korea to support the Fifth Air Force.

North Korean forces reach Samchock on the east coast and in the west cross the Han River, threatening Suwon airfield.

Far East Air Forces (FEAF) begin evacuation of Suwon Airfield and authorize improvement of Kumhae Airfield, 11 miles northwest of Pusan, to compensate for the loss of Kimpo and Suwon.

The first Fifth Air Force Tactical Air Control Parties (TACPs) arrive at Suwon.

1 July
USAF airlift forces deliver the first U.S. ground forces from Japan to airfields near Pusan, South Korea, when the 374th Troop Carrier Wing (TCW) begins airlifting the U.S. Army 24th Infantry Division from Itazuke, Japan, to Pusan, South Korea.

Fifth Air Force gains operational control of the Royal Australian Air Force No. 77 Squadron.

3 July
FEAF continues to airlift Army troops to Korea using smaller C-46s and C-47s instead of heavier C-54s, which damage the Pusan runways.

6 July
The first strategic bombing raid of the war is executed when nine B-29 Superfortresses attack the Rising Sun Oil Refinery at Wonsan and chemical plants near Hungnam in North Korea.

28 June
North Korean ground forces occupy Kimpo Air Base and the city of Seoul while Douglas B-26 Invaders and Boeing B-29 Superfortresses of the Far East Air Forces (FEAF), commanded by General George E. Stratemeyer, launch the first offensive attacks of the war, striking targets north of Seoul to the North Korean border. Lieutenant Bryce Poe pilots an RF-80A on the first jet reconnaissance sortie of the war.

29 June
In the first USAF attacks against targets in North Korea, B-26 medium bombers based in Japan strike airfields near Pyongyang, the capital.

General MacArthur directs General Stratemeyer to attack the Han River bridges and the North Korean troops massed north of the river that divides North and South Korea. F-82s use napalm for the first time in the war. Five North Korean planes that were attacking the airfield at Suwon are shot down. Eight B-29s attack enemy-held Kimpo airfield and the Seoul railroad station. Enemy aircraft attack a formation of B-29 bombers as they return to Japan, and their gunners shoot down one of the opponent's airplanes. This marks the first victory by a B-29 gunner during the war.

The 8th Tactical Reconnaissance Squadron (TRS) begins photographic reconnaissance of North Korean airfields. RB-29s also start operations over Korea, taking off from Yokota AB, Japan.

30 June
President Truman orders the use of U.S. ground troops in Korea, a naval blockade of North Korea, and aerial attacks across the North Korean border.

Above: *Strategic Air Command B-29s delivered the opening strategic blows of the Korean War. (NASM)*

Right: *Major General George E. Stratemeyer. (NASM)*

Right: *The Stinson L-5G was utilized as a forward air controller (FAC) during the Korean War, directing F-80 air strikes. (NASM)*

Below: *The Boeing SB-17G was modified to drop lifesaving equipment to downed airmen in the waters surrounding Korea. (NASM)*

James H. "Jimmy" Doolittle is named "Aviator of the Decade" and Jacqueline Cochran is named "Outstanding Aviatrix" by the Harmon International Aviation Awards Committee.

8 July
President Truman designates General Douglas MacArthur as Commander in Chief of United Nations forces in the Korean Theater.

Major General Emmett "Rosie" O'Donnell Jr., a former advisor to General Hap Arnold during World War II, assumes command of Bomber Command (Provisional) at Yokota AB, Japan, a FEAF organization.

9 July
Forward air controllers (FAC) begin using L-5G and L-17 liaison airplanes to direct F-80 air strikes.

10 July
The Fifth Air Force employs the North American T-6 Texan trainer aircraft for FAC missions. The first "Mosquito" mission is responsible for calling in a flight of F-80s, which target and destroy a column of enemy tanks.

F-80s catch an enemy convoy stopped at a bombed-out bridge near Pyongtaek. Along with B-26s and F-82s, they attack the convoy and destroy dozens of trucks, tanks, and half-tracks.

12 July
Four Military Air Transport Service (MATS) airplanes arrive in Japan from the United States carrying 58 large 3.5-inch rocket launchers (bazookas) and shaped charges desperately needed to destroy North Korean tanks.

Enemy fighters shoot down one B-29, one B-26, and one L-4, the first enemy aerial victories of the war.

13 July
Forty-nine FEAF Bomber Command B-29s attack marshaling yards and an oil refinery near Wonsan, North Korea.

The 3d Air Rescue Squadron (ARS) flies an SB-17 aircraft off the Korean coast, dropping rescue boats to downed B-29 aircrew.

Lieutenant General Walton H. Walker, Commander, Eighth Army in Korea, assumes command of all U.S. ground forces in Korea.

14 July
The first USAF units deploy to bases in South Korea—the 35th Fighter-Interceptor Group to Pohang and the 6132d Tactical Air Control Squadron to Taegu.

Left: *Following a tactical mission, the pilot debriefs the crew chief and describes any problems with the plane or offers congratulations for a "code 1" jet. ("After a Mission—Korea 1953" by David Hall, USAF Art Collection)*

Below: *The Grumman SA-16 Albatross (later HU-16) was able to accomplish rescue missions on land or sea. One version was even fitted with skis for landing in snow and ice. The Albatross rescued about 1,000 victims during the Korean War. (USAF)*

15 July
The 51st Fighter Squadron (FS) (Provisional) at Taegu flies the first F-51 Mustang combat missions in Korea.

A Fifth Air Force operations order assigns the "Mosquito" call sign to all airborne controllers flying T-6 airplanes.

19 July
In an aerial engagement near Taejon, Fifth Air Force F-80s shoot down three enemy Yaks, the highest daily number of aerial victories for the month.

In the campaign to establish air superiority in the theater, seven F-80s of the 8th Fighter-Bomber Group (FBG) destroy 15 enemy airplanes on the ground near Pyongyang.

20 July
Despite FEAF close air support, the North Korean Army takes Taejon and forces the 24th Infantry Division to withdraw to the southeast.

Major General Otto P. Weyland arrives in the Far East to assume the position of FEAF vice commander for operations.

Fifth Air Force pilots in F-80s shoot down two more enemy aircraft, the last aerial victories in Korea until November. Enemy air opposition by this time has virtually disappeared, a sign of UN air superiority.

22 July
The U.S. Navy aircraft carrier USS *Boxer* arrives in Japan with 145 USAF F-51s aboard.

The 3d ARS deploys the first H-5 helicopter in Korea to the airbase at Taegu.

24 July
Fifth Air Force in Korea is established at Taegu near Eighth Army headquarters for ease of communication and coordination.

General MacArthur takes command of the UN Command formally established in Tokyo. He assigns responsibility for ground action in Korea to Eighth Army Commander General Walton Walker; naval action to Vice Admiral C. Turner Joy, Commander, Naval Forces, Far East; and air action to General George Stratemeyer, Commander, FEAF.

28 July
The first amphibious SA-16 Albatross aircraft arrives in Japan for air rescue service off the Korean coast.

30 July
Forty-seven B-29s bomb the Chosin nitrate explosives factory at Hungnam on the east coast of North Korea.

Right: *This North American F-51D is ready for takeoff on a ground attack mission loaded with auxiliary furl tanks and rockets. (NASM)*

1 August
Major General Mason M. Patrick, the first Chief of the U.S. Army Air Service, is memorialized when the Air Force renames its long-range proving ground, Patrick AFB, after him.

2–3 August
The 374th Troop Carrier Group (TCG) airlifts 150 tons of equipment and supplies from Ashiya, Japan, to the Eighth Army in Korea in 24 hours, establishing an airlift record for the war.

3 August
SA-16 amphibious rescue aircraft began flying sorties along the Korean coast to retrieve pilots forced down during operations.

4 August
B-29s attack key bridges north of the 38th parallel, initiating FEAF Interdiction Campaign No. 1.

5 August
F-51D Mustang pilot, Major Louis Joseph Sebille, is killed when he intentionally crashes his damaged plane into an enemy ground position near H'amchang, South Korea. For his actions he receives the first Medal of Honor awarded to a member of the United States Air Force. An Air Force medal is not designed until the 1960s, so the Army Medal of Honor is awarded until then.

10 August
The Air Force calls up two stateside Reserve Air Force units for service. They are the first of 25 USAF Reserve flying units activated for service during the Korean War.

11 August
The Fairchild XC-120 Packplane flies for the first time. It was derived from the C-119B Flying Boxcar wings and tail assemblies.

C-119 Flying Boxcars begin airlifting trucks from Tachikawa AB in Japan, to Taegu AB, Korea.

16 August
In an effort to disrupt an impending attack by the North Koreans against the Pusan Perimeter, nearly 100 B-29 bombers attack a 30-square-mile troop concentration zone near Waegwan, dropping 800 tons of bombs on suspected troop concentrations. Not since the weeks following the D-day invasion at Normandy had such massive ground support operations been accomplished.

19 August
American airpower helps drive North Korean forces at the Yongsan bridgehead near Seoul back across the Naktong River, ending the Battle of the Naktong Bulge.

22 August
Chinese anti-aircraft gunners fire at RB-29s from across the Yalu River. This marks the first hostile Chinese action against UN aircraft.

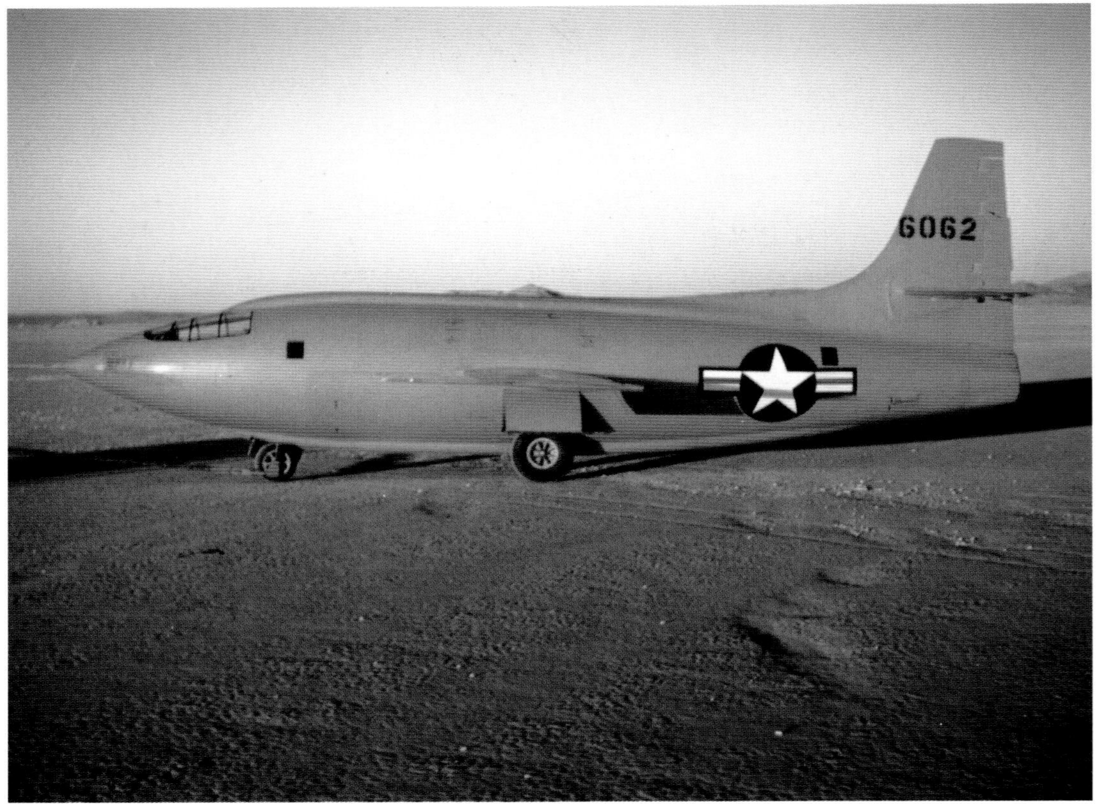

Left: *The Bell X-1 rocket research plane is towed across the dry lake bed at Muroc. The X-1 is on display at the National Air and Space Museum in Washington, D.C. (NASM)*

23 August
General MacArthur sets September 15th as the date for the UN amphibious assault at Inchon.

The first Razon guided-bomb mission of the war is flown but enjoys limited success. Only one bomb strikes the railroad bridge west of Pyongyang, its target.

25 August
FEAF directs Fifth Air Force to maintain round-the-clock armed surveillance of enemy airfields in an effort to prevent enemy buildup of air strength before the Inchon invasion.

26 August
Fifth Air Force gathers C-46s from all over the Far East to augment FEAF airlift in preparation for UN offensives planned for September.

At Ashiya, FEAF organizes the 1st Troop Carrier Task Force (Provisional) as the nucleus of the new Combat Cargo Command (Provisional). Major General William H. Tunner, the architect of American airlift, assumes command.

General Hoyt Vandenberg, Air Force Chief of Staff, presents Alexander Wetmore, Secretary of the Smithsonian Institution, the Bell X-1 No.1, rocket-propelled, supersonic research aircraft.

31 August
North Korean forces launch a coordinated offensive against the Pusan Perimeter. Fifth Air Force provides close air support for the embattled UN troops there.

Seventy-four B-29s bomb mining facilities, metal industries, and marshaling yards at Chinnampo in the largest strategic bombing mission of the month.

4 September
Flying a Sikorsky H-5 helicopter, Lieutenant Paul W. van Boven accomplishes the first combat rescue of a pilot shot down behind enemy lines during the Korean War.

9 September
FEAF Bomber Command begins a rail interdiction campaign north of Seoul to slow enemy reinforcements to Inchon. Air Force medium bombers combine attacks on marshaling yards with rail-cutting raids at multiple points along key resupply routes.

15 September
Striking deep behind the invading North Korean ground forces, UN forces under the command of General Douglas MacArthur complete an amphibious landing at the port of Inchon, South Korea. Ground forces are supported by Navy and Marine Corps aviation while FEAF assets bomb targets near Pusan in preparation for the upcoming breakout attempt.

Right: *General MacArthur's amphibious landing at Inchon was strategically brilliant, but operationally risky. (Naval Historical Center)*

15 September–28 October
Flying F-84E Thunderjets, the 27th Fighter Escort Wing begins a deployment to West Germany. Departing their home station at Bergstrom AFB, Texas, two waves of fighters make the journey; the first on September 15–18, the second on October 15–28.

16 September
In coordination with the Inchon Landing the previous day, Eighth Army ground forces supported by the USAF begin the advance from the Pusan Perimeter. These combined operations result in the retreat of North Korean ground forces, fearing total encirclement by the pincer offensive from the west and the south.

17 September
U.S. Marines capture Kimpo airfield near Seoul.

Supporting the Eighth Army, Fifth Air Force F-51s and F-80s carry out napalm attacks, killing hundreds of enemy soldiers near the Naktong River.

FEAF initiates the drop of four million psychological warfare leaflets. The operation will take one full week.

18 September
Forty-two B-29s carpet-bomb two areas near Waegwan, effectively destroying enemy troop concentrations blocking the Eighth Army offensive.

This day marks the third anniversary of the establishment of the USAF as an independent service.

19 September
Combat Cargo Command initiates airlift to Kimpo AB in Seoul as 32 C-54s land with equipment and supplies for engaged ground troops.

Fifth Air Force provides close air support for advancing American ground forces.

20 September
Combat Cargo Command begins round-the-clock airlift into Kimpo airfields. Night operations are made possible by using night-lighting equipment transported the previous day.

22 September
The first nonstop jet fighter flight over the Atlantic Ocean is completed by Colonel David C. Schilling in an F-84 Thunderjet. Taking off from RAF Manston near Kent, England, he refuels three times enroute and, after a journey of 10 hours and 3,300 miles, lands in Limestone, Maine (the future Loring AFB). His wingman, Lieutenant Colonel William D. Ritchie, runs out of fuel and ejects safely over Labrador. Schilling is awarded the Harmon Trophy for completing the hazardous crossing.

North Korean resistance evaporates along the Pusan Perimeter.

Lieutenant George W. Nelson, a USAF pilot in a "Mosquito" aircraft, drops a note to 200 enemy troops near Kunsan demanding their surrender. They comply and move to a designated hill, where they are formally captured by nearby UN ground troops.

AIRLIFT IN THE UNITED STATES AIR FORCE

Robert C. Owen

Left: *The C-47 set the early standard for airlift reliability and durability and, with the C-46 and C-54, hauled the bulk of Berlin Airlift supplies. ("Operation Vittles, 1948," by Nicholas Mosura, USAF Art Collection)*

Airlift was the inevitable result of military aviation in World War I. Exploiting the capabilities of the combat aircraft at hand, aviators in that war also used them to move important individuals, mail, and supplies to isolated battalions, or themselves to Paris for the weekend. By war's end, Allied plans for 1919 included an airdrop of 10,000 paratroops behind German lines.

In the interwar years, Army airmen experimented with every facet of modern airlift. Early on, the Army Air Service modified light bombers and purchased civilian aircraft to lift the injured from the field and to carry engines, weapons, and dispatches out to squadrons deployed for exercises. After 1926, major military air exercises included transport squadrons to link air units to their bases of supply.

Airlift came of age in World War II. As war raged in Europe and Asia, American airmen established troop carrier aviation to support the Army's new paratroop units. This new arm quickly grew into a force of over 10,000 twin-engine C-47s and C-46s assigned to Troop Carrier commands in every theater. In 1941 the Ferrying Command began to deliver aircraft to allied countries and to fly important diplomatic missions, first over the Atlantic and then worldwide. Formed after Pearl Harbor, the Air Transport Command (ATC) absorbed Ferrying Command, and then built a global air transport system of over 3,000 C-87s, C-54s, other aircraft, and 300,000 personnel. By 1944, Troop Carrier forces in Europe could launch a stream of planes 200 miles long to drop an airborne corps during Operation Market Garden—the largest airborne drop in military history. ATC's contributions to victory included the Hump airlift over the Himalayas and the feat of flying more planes over the Atlantic during a typical day in 1945 than had made the trip in total between 1919 and 1940.

During the Cold War the Air Force sized and equipped the Military Air Transport Service (which replaced ATC in 1948) with C-118 and C-124 aircraft to support Strategic Air Command unit movements and the post-strike "reconstitution" of SAC bombers recovering at bases around the world. To organize the peacetime and wartime contributions of the airlines,

Right: *The four-engine Douglas C-54 Skymaster provided improved capabilities for military airlift. More than 1,200 were built. (USAF)*

the President activated the Civil Reserve Airlift Fleet (CRAF) in 1952. Meanwhile, the Air Force Reserve and National Guard took over most of the troop carrier mission, augmenting a handful of active-duty units with over 20 wings of C-119 and C-123 aircraft.

The Vietnam War provided the impetus to organize all airlift forces into a single global system. Airlift emerged during this conflict as a core element of American combat capabilities, rather than simply a valuable adjunct to them. Virtually every American soldier traveled into and out of the theater by air, along with about 10 percent of all cargo flowing into the war zone. Moreover, the new fleet of turbine-powered C-130s, C-133s, C-141s, and C-5s (their efficiency enhanced by the new 463L cargo handling system) allowed the Air Force to deploy entire combat wings and Army brigades directly from the United States to the battle zone by air in a matter of days. In 1966 the Air Force activated the Military Airlift Command (MAC) to emphasize that global airlift was a combat, as well as a logistic, element of America's defense.

After Vietnam, the Air Force moved to make MAC truly "global." First, it consolidated virtually all theater and long-range airlift forces under MAC's control. Then, it replaced all of the piston and turboprop aircraft remaining in its long-range fleet with turbofan-powered jumbo jets; C-5Bs, KC-10s, and after a long political fight, the remarkable C-17. These aircraft made airlifts of entire joint combat forces over global distances a core Air Force capability.

The First Gulf War validated this new global capability. MAC "T-tails" and mobilized CRAF airliners lifted all of the initial air and ground combat echelons into the theater and virtually all subsequent personnel, whose equipment was transported by ship. The scale of the lift—500,000 passengers and 526 thousand tons of cargo—was equivalent to moving the population of Kansas City and its vehicles almost 8,000 miles by air. Meanwhile, C-130s and C-21s carried the in-country load, including major supply lifts in support of the final offensive.

Since the Gulf War, military airlift has become a predominant feature of American military operations and diplomacy. In 1992, the Air Force consolidated virtually all of its air refueling and airlift units under the new Air Mobility Command, an action that improved the agility and capacity of its global reach.

The finest example of operational air mobility during the last decade was the intervention into Afghanistan following the terrorist attacks of September 11, 2001. The entire theater was opened by airlift, including the transport of three Predator UAVs to the theater which were the first aircraft to fly over hostile territory after the attacks, since there were no reliable surface lines of communication into that landlocked country, nor time to wait to establish them. There and in dozens of greater and lesser conflicts and contingencies, the Air Force's airlift community continues to fulfill their motto—Anything-Anywhere-Anytime.

Above: *America's first operational combat jet, the F-80 Shooting Star, was called to duty in the skies over Korea. ("F-80s Over the Target in Korea" by Francis H. Geaurgureau, USAF Art Collection)*

23 September
Headquarters Fifth Air Force in Korea deploys inland from Pusan to Taegu.

In the first recorded special operations mission of the war, SB-17 aircraft make a classified flight over Korea.

26 September
Continuing the UN counterattack to the north, U.S. military forces from Inchon and Pusan link up near Osan. South Korean troops, supported by Fifth Air Force aircraft, advance northward along the Korean east coast toward the 38th parallel.

Twenty B-29s of the 22d BG bomb a munitions factory at Haeju, destroying the power plant and five related buildings. Another group of B-29s attack the Pujon hydroelectric plant near Hungnam. These attacks mark the end of the first strategic bombing campaign against North Korea.

27 September
After destroying or disabling nearly all strategic targets in northern North Korea, the Joint Chiefs of Staff halt all strategic bombardment there.

While U.S. Marines drive enemy forces from Seoul, more than 100 Communist troops waving "safe conduct" leaflets dropped earlier by B-29s, surrender to U.S. forces.

28 September
Launched to an altitude of 97,000 feet in a balloon, eight white mice return to earth unharmed at Holloman Air Force Base, New Mexico.

South Korean troops advance into North Korea for the first time while General MacArthur officially restores Seoul to South Korean president Syngman Rhee.

The first jet fighter squadron to operate from a base in Korea arrives from Itazuke, Japan.

Three RB-45 Tornadoes, the first jet reconnaissance aircraft in the USAF inventory, arrive for duty in the Far East.

29 September
Jumping from an altitude of 42,449 feet, Captain Richard V. Wheeler sets a record for a parachute jump in the skies over Holloman Air Force Base, New Mexico.

2 October
FEAF B-29s attack a North Korean military training area at Nanam, destroying 75 percent of the buildings and slowing North Korean efforts at reinforcement.

The 8th Tactical Reconnaissance Squadron (TRS) deploys forward to Taegu (K-2 Airbase), becoming the first USAF day reconnaissance squadron stationed in Korea.

4 October
FEAF assumes operational control of all land-based aircraft in Korea, including USMC squadrons at Kimpo.

The South African Air Force No. 2 Squadron arrives in the theater and is attached to the FEAF.

6 October
The U.S. Air Force takes charge of Kimpo airfield, held by the Marines since its capture.

Eighteen B-29s attack an enemy arsenal at Kan-ni, North Korea, while FEAF issues a new interdiction plan canceling attacks on bridges south of Pyongyang and Wonsan.

7 October
The UN General Assembly overwhelmingly approves a resolution authorizing General MacArthur to strike north into North Korea. Almost immediately, U.S. troops cross the 38th parallel for the first time.

USAF airplanes drop food to a group of 150 former POWs who had escaped during the North Korean retreat.

8 October
Razon bomb missions resume after more reliable radio-guided weapons from the U.S. arrive in Korea.

10 October
The first of 66 Air National Guard units is mobilized for service during the Korean War. Approximately 45,000 members of the Air Guard will be called upon during the conflict.

Lieutenant General Lauris Norstad takes command of Unites States Air Forces in Europe.

20 October
Over a three-day period, a combined force of more than 100 C-119 Flying Boxcars and C-47 Skytrain transport planes drops more than 4,000 U.S. Army paratroopers and hundreds of tons of supplies in locations 30 miles north of Pyongyang, the North Korean capital.

24 October
General MacArthur removes restrictions on U.S. troop movement into North Korea. He gives orders to pursue the enemy all the way to the Chinese border.

25 October
Communist China enters the Korean War.

FEAF Bomber Command temporarily suspends B-29 combat missions for lack of legitimate strategic targets in Korea. Additionally, FEAF removes all restrictions on close air support missions near the Yalu River, allowing fighter operations supporting General MacArthur's newest offensive all the way to the Chinese border.

Combat Cargo Command establishes a new daily record by airlifting 1,767 tons of equipment within Korea.

26 October
UN forces reach the Yalu River along the Chinese border at Chosin in northwest Korea.

Combat Cargo Command C-119s drop supplies to friendly ground troops cut off in North Korea. More than 28 tons of ammunition, fuel, and oil are dropped.

Below: *C-119s drop paratroopers over Korea in 1951. Such operations were rare during the Korean War. (U.S. Army Military History Institute, Carlisle Barracks, Pennsylvania)*

AERIAL REFUELING

In July 1923 at Rockwell Field, California, the Army Air Service, America's early air arm, conducted the first successful air-to-air refueling that used a hose to refuel one aircraft from another. The mission, sanctioned by Major Henry H. Arnold, was flown in a de Havilland DH-4 by two lieutenants—Frank Seifert and Virgil Hine. Arnold wrote:

> *While the great benefits to be derived from refueling in the air are probably unappreciated at this time by many people in aviation circles, it can only be a matter of a few years until the pioneering work done at this station will be the basis for operating airplanes on long cross-country flights whenever it is needed to carry great loads or carry materiel or personnel to greater distances than the capacity of gas and oil tanks will permit.*

His predictions today seem remarkably accurate.

In service lore, the most famous of the early refueling experiments was flown on New Year's Day 1929 in a Fokker C-2 Tri-Motor nicknamed *Question Mark*. Since the duration of the mission was unknown, the nickname seemed appropriate. While Georgia Tech edged California 8-7 in a hard-fought Rose Bowl victory, the first aerial hook-ups were made directly over the Pasadena stadium. Two Douglas biplanes served as tankers and kept *Question Mark* aloft for seven days before a mechanical failure forced the aircraft back to the ground. During that flight, the aircrews transferred 5,660 gallons of fuel and 245 gallons of oil, as well as food and other supplies.

During the Second World War, the Army Air Forces conducted experiments with B-17s and B-24s fitted with air refueling apparatus, but the process was never mastered. After the Second World War, long-range requirements of the Strategic Air Command, based upon their new nuclear delivery mission, renewed interest in practical aerial refueling. It was not until the Korean War that a probe-and-drogue system was developed that is still used today by the U.S. Navy and some special operations aircraft. During these operations, the tanker extends a refueling hose that terminates in a funnel-shaped drogue that looks like a badminton shuttlecock. The drogue serves to stabilize the hose and houses the refueling contact mechanism. Although a major advance in refueling technology, there remained many problems with stability and transfer rates. These limitations became more problematic with the development of high-speed jet fighters.

Further experimentation finally resulted in the development of the refueling process most familiar to military pilots—the flying boom. The boom operator literally flies the boom into place using small airfoils, called ruddervators, located near the tip of the refueling boom. Pilots are guided by a system of colored lights located on the underside of the tanker aircraft. The refueling pilot must maintain a position

Left: *The KB-47 Stratotanker simplified refueling problems between turbine- and piston-powered aircraft. (NASM)*

Opposite: *The KC-135 Stratotanker jet-refueling aircraft has been the backbone of Air Force operations since 1957. (USAF)*

behind and below the tanker and must remain within established parameters so that the mechanical limits of boom equipment are not exceeded.

The first boom-equipped tankers were KB-29s modified in 1950. B-50 bomber aircraft were equipped to receive fuel from that tanker at the same time. But again there were problems. Propeller-driven KB-29s were too slow for faster turbojet aircraft, so B-50 aircraft were modified as tankers; a jet was added to each wing to increase its speed during operations. The B-50 dragged three hoses, one from the fuselage and one each from a pod mounted under each wing, allowing three aircraft to refuel at the same time. It made for a remarkable aerial picture. At the same time, boom-equipped aircraft advances were being made as well.

The Boeing KC-97 Stratotanker had entered service and booms had been added. With turrets removed, the boom section moved forward on the fuselage. The boom operator lay face down and looked through a window at the receiving aircraft.

As the Air Force converted to an all-jet fleet, a new tanker was needed to match the capabilities of the Boeing B-52 Stratofortress and newer, faster, fuel-hungry fighters. Boeing developed the KC-135, and the prototype first flew in July 1954. The first KC-135A entered service in 1957.

Ever-expanding involvement around the globe indicated that a more capable military tanker fleet was needed. Civil aircraft were evaluated for possible use as refueling platforms, and the Lockheed DC-10 was finally selected as the supplemental tanker aircraft. Nearly five years after selection, the first KC-10 Extender was put into service in 1981. The KC-10 system includes both drogue and boom capability and is itself air refuelable, allowing it to serve as a flying gas station for aircraft virtually anywhere around the world.

One airman summarized the mission of the modern tanker force when he said, "We provide the legs to carry the muscle, to deliver the punch." Few missions can be accomplished in the modern global world without aerial refueling.

28 October
F-84E Thunderjets of the 27th Fighter Escort Wing complete the first-ever jet aircraft unit deployment from the U.S. to Europe. For successful completion of this transatlantic mass jet fighter deployment, the unit receives the Mackay Trophy.

2 November
A FEAF RB-45 Tornado flies its first reconnaissance mission of the war.

4 November
B-26s providing close support for Eighth Army strike enemy troops near Chongju, killing hundreds of enemy soldiers and providing hard-pressed UN troops some needed relief.

5 November
Bomber Command begins incendiary raids on North Korean cities and towns. Twenty-one B-29s of the 19th BG drop 170 tons of fire bombs on Kanggye, 20 miles south of the Chinese border. The attack destroys 65 percent of the city center.

8 November
In the largest incendiary raid of the Korean War, 70 B-29 Superfortresses drop 580 tons of fire bombs on Sinuiju on the Chinese border. Other B-29s attack bridges over the Yalu River for the first time.

Lieutenant Russell J. Brown, flying a jet-powered F-80 Shooting Star, engages and shoots down a jet-powered, Soviet-piloted MiG-15. It is the first successful jet-to-jet aerial combat engagement in history.

9 November
Airman Harry J. LeVene scores the first B-29 jet victory of the Korean War when he shoots down an attacking MiG-15. His damaged RB-29 limps back to Japan, where it crash lands. Five crewmen perish during the crash-landing.

10 November
The first air-to-air loss of an American B-29 occurs when a MiG-15 engages and shoots one down near the Yalu River. The 307th Bombardment Group crew parachute behind enemy lines and become POWs.

18 November
The 35th Fighter Interceptor Group moves to a base located in North Korea. The 35th FIG, also the first fighter group based in South Korea, staged out of Yonpo airfield, near Hungnam.

19 November
In the first mass light bomber attack of the Korean War, 50 B-26s staging from Japan drop incendiary bombs on Musan, North Korea, on the Tumen River bordering with China. The attack destroys 75 percent of the town's barracks area.

23 November
FEAF B-29s attack North Korean communications, supply centers, and Yalu River bridges; Fifth Air Force fighters intensify close air support missions; and Combat Cargo Command airdrops ammunition to frontline troops in support of UN offensive operations that begin this day.

25 November
Chinese Communist Forces launch a ferocious counteroffensive and, with almost double the number of General MacArthur's U.S. troops, stop the newly begun UN offensive in its tracks.

Opposite: *"Russell Brown Downs a MiG," by Keith Ferris. (Courtesy of the artist)*

Below: *Airman Harry L. LaVene shot down a MiG from his post as an RB-29 tail gunner—the first of the war. Fortunate to survive a crash landing after that mission, LaVene later posed with his machine guns. (USAF)*

The Royal Hellenic Air Force detachment, a C-47 transport unit representing Greece's airpower contribution to the war, arrives in the Far East and is immediately attached to FEAF.

26 November
B-26s fly their first close air support night missions under TACP direction while the Chinese continue to drive the Eighth Army in northwest Korea and X Corps in northeast Korea southward.

28 November
For the first time, B-26s, using advanced radar equipment, bomb targets within 1,000 yards of friendly front lines.

Airlift of supplies and troops and evacuation of sick and wounded Marines from North Korea begin. Nearly 1,600 tons of supplies and equipment are delivered to the 1st Marine Division surrounded by the enemy at the Chosin Reservoir. These operations continue until mid-December.

4 December
MiG-15s shoot down one of the three USAF RF-45 Tornado reconnaissance aircraft in the theater.

5 December
UN forces abandon Pyongyang, North Korea, which they had held since October 19.

Greek C-47s join the Combat Cargo Command airlift to supply UN troops surrounded in northeastern Korea. The command flies 131 flights and evacuates 3,925 patients from Korea to Japan during the most active day of the war for aeromedical airlift. Transports fly most of these from a frozen airstrip at Hagaru-ri.

6 December
The 27th Fighter Escort Wing, flying F-84 Thunderjets, accomplish combat missions over Korea and return to Itazuke, Japan.

7 December
FEAF B-29s bomb North Korean towns in the Changjin Reservoir area to relieve enemy pressure on U.S. Marine and Army units attempting to break out. Troops finally link up and build crude airstrips allowing Combat Cargo Command airplanes to land carrying food and ammunition and to evacuate casualties.

Eight C-119s drop bridge spans to surrounded U.S. troops so that they might cross a 1,500-foot-deep gorge and break the enemy encirclement. This is the first air-dropped bridge in the history of warfare.

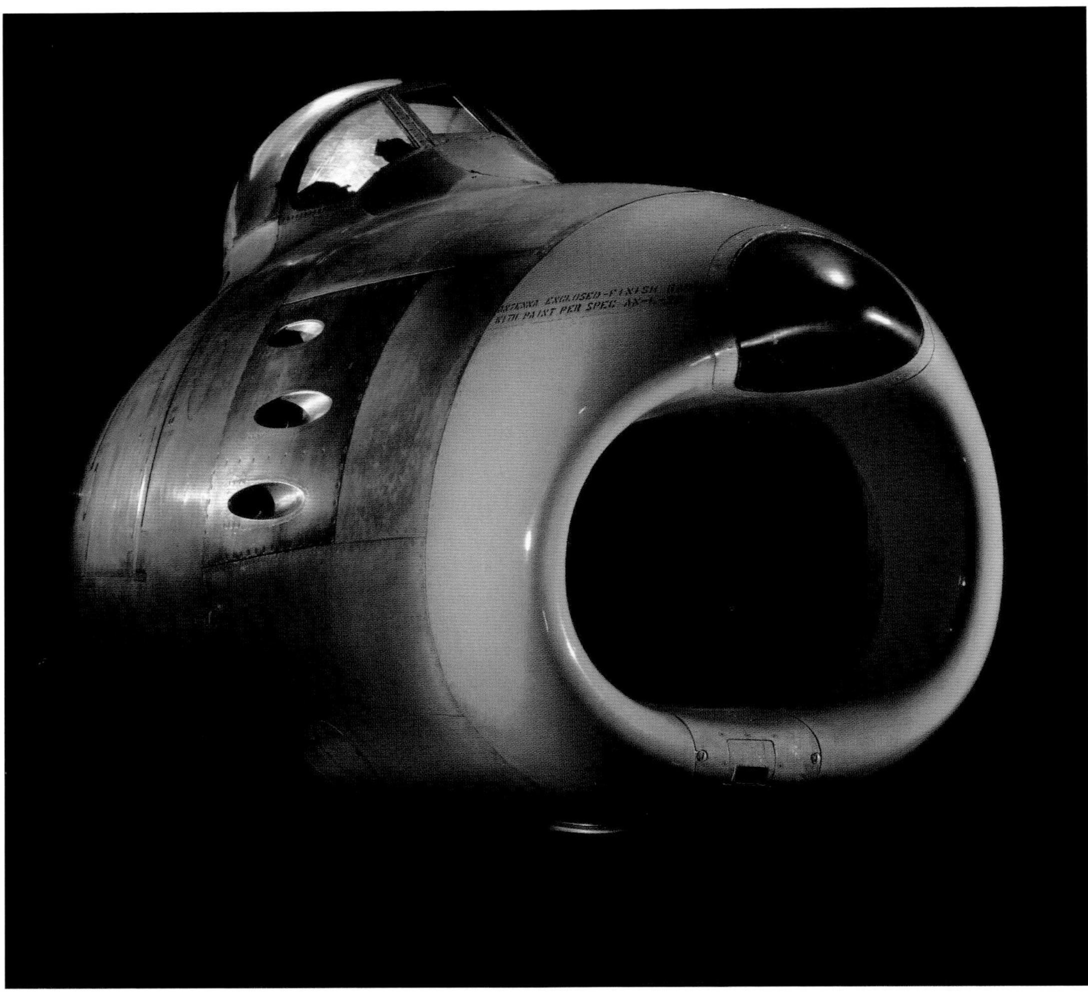

Left: *The North American F-86 Sabre was America's premier fighter during the Korean War. This F-86 is on display at the National Air and Space Museum, Steven F. Udvar-Hazy Center, near Dulles International Airport. (NASM)*

Opposite: *"First Swept Wing Encounter," by Keith Ferris. (Courtesy of the artist)*

14 December
The first Tarzon radio-guided bomb used in Korea is dropped on a tunnel near Huichon, with limited effectiveness. The Tarzon bomb is a six-ton version of the Razon bomb, but despite its size does not live up to expectations.

14–17 December
As Chinese forces approach, Combat Cargo Command begins a three-day aerial evacuation from Yonpo airfield near Hamhung. During these 72 hours, 228 patients, 3,891 passengers, and 20,088 tons of cargo will be transported from the airfield.

15 December
The 4th Fighter Interceptor Group begins F-86 Sabre operations in Korea.

Bomber Command launches its first mission in a new zone interdiction plan.

17 December
Lieutenant Colonel Bruce H. Hinton, flying a swept-wing F-86A Sabre, shoots down a MiG-15, achieving the first ever swept-wing versus swept-wing jet fighter victory in history.

20 December
Twelve 61st Troop Carrier Group C-54 Skymasters airlift hundreds of South Korean orphans from Kimpo Air Base to an island located to the south of Pusan during Operation CHRISTMAS KIDLIFT.

22 December
One Navy and five Air Force pilots shoot down six MiG-15s, the highest daily total since June.

A MiG-15 shoots down an F-86 for the first time.

Headquarters Fifth Air Force, Eighth Army in Korea headquarters, and the Joint Operations Center move from Seoul to Taegu as enemy ground forces threaten in the north.

23 December
Three H-5 helicopter crews with fighter cover rescue 11 U.S. and 24 South Korean soldiers from a field eight miles behind enemy lines.

Eighth Army commander Lieutenant General Walton Walker is killed in a vehicle accident north of Seoul. General Matthew B. Ridgway replaces him three days later.

24 December
The evacuation of the U.S. Army X Corps is completed at Hungnam. More than 105,000 troops and 91,000 civilians had escaped since the exodus began two weeks earlier. Air Force B-26s and U.S. Navy gunfire pin the enemy down as the last ships depart.

25 December
Chinese forces recross the 38th parallel into South Korea.

26 December
Lieutenant General Matthew B. Ridgway takes command of Eighth Army in Korea, now bolstered by the transfer of the X Corps.

29 December
RF-51 reconnaissance aircraft begin flying tactical reconnaissance missions in Korea for the first time. Flying from bases in Taegu, the efficient F-51 is capable of flying longer and to greater ranges than the turbine-powered RF-80s.

31 December
Chinese Communist Forces in Korea launch an offensive against UN troops south of the 38th parallel. General Ridgway orders Eighth Army troops to fall back to a new defensive line 70 miles farther south.

1951

1 January
Half a million Chinese and North Korean troops launch a massive ground offensive against UN forces. Fifth Air Force begins counter-force raids on enemy troop columns.

Air Defense Command, once abolished, is restored to full status as a major command. Lieutenant General Ennis C. Whitehead takes command one week later.

2 January
A C-47 drops aerial flares illuminating B-26 and F-82 night attacks. The flares also deter enemy night surface attacks upon U.S. troops.

3 January
Chinese troops cross the frozen Han River east and west of Seoul, while Eighth Army evacuates the South Korean capital.

In one of the largest Bomber Command air raids, more than 60 B-29s drop 650 tons of incendiary bombs on Pyongyang. A similar raid is carried out two days later. In all, FEAF fly 958 combat sorties, a one-day record.

4 January
For the third time in six months, Seoul changes hands as Chinese forces move in. The last USAF aircraft evacuates Kimpo airfield in the face of the Chinese onslaught. After the last aircraft departs, American bombers crater the runway, making it unusable for the enemy's aircraft.

6 January
Combat Cargo Command concludes the airlift of supplies to the U.S. 2d Infantry Division, which was fighting to prevent a break in the UN defensive line across South Korea.

10 January
Brigadier General James E. Briggs replaces General O'Donnell as commander of Bomber Command. In stark contrast to policy during World War II, Air Force Strategic Air Command decides to change commanders of Bomber Command every four months to provide wartime experience to as many officers as possible.

12 January
FEAF B-29s drop 500-pound general purpose bombs fused for airburst, showering enemy troops with thousands of steel fragments. The new use of the weapon stalls the enemy advance.

13 January
FEAF employs a six-ton Tarzon radio-controlled bomb against an enemy-held bridge at Kanggye. The bomb is guided to the center span for a direct hit, destroying nearly 60 feet of the structure.

14 January
Chinese Communist Forces advance into South Korea and capture the town of Wonju. This marks the farthest penetration that the CCF will make during the war.

16 January
The first six Strategic Air Command B-36 Peacemaker intercontinental bombers land at RAF Lakenheath, United Kingdom, after deploying 7,000 miles from Bergstrom AFB, Texas.

The Air Force initiates Project MS-1593, which will become the Atlas missile program. Convair wins the contract for two design studies: a pure ballistic missile and a boost-glide vehicle.

17 January
A small detachment of F-86 Sabres begin operating from Taegu again, restoring F-86 operations in Korea. For the first time, the F-86s take the air-to-ground role as fighter-bombers, conducting armed reconnaissance and close air support missions.

17–18 January
Combat Cargo Command launches 109 C-119 sorties, dropping more than 550 tons of supplies to frontline UN troops.

19 January
FEAF launches an intensive air interdiction campaign to curtail enemy supplies and reinforcements.

21 January
Large numbers of MiG-15s attack USAF jets, shooting down one F-80 and one F-84. Lieutenant Colonel William E. Bertram shoots down a MiG-15 to score the first USAF aerial victory by an F-84 Thunderjet.

23 January
Thirty-three F-84s staging from Taegu attack Sinuiju, provoking a furious half-hour air battle with MiG-15s from beyond the Yalu River in China. The Thunderjets shoot down three of the MiGs.

A force of 46 F-80s strike Pyongyang's anti-aircraft artillery batteries. An additional force of 21 B-29s crater airfields located there.

25 January–9 February
Operation THUNDERBOLT, the first UN offensive of the year, begins. Eighth Army objectives are to recapture the port of Inchon and the airfield at Suwon. Approximately 70 C-119s deliver 1,162 tons of supplies, including fuel, oil, sleeping bags, C rations, and signal wire to help sustain the offensive.

26 January
A C-47 control aircraft, is modified to hold enough communications equipment to maintain radio contact with all airborne T-6 Mosquitoes, TACP, and the Tactical Air Control Center—a Korean War–era ABCCC.

31 January
A special operations unit of the 21st TCS drops a UN agent behind enemy lines near Yonan, on the Korean west coast just south of the 38th parallel. This marks the first such spy mission of the war.

8 February
FEAF aircraft launch an all-out attack on rail lines in northeastern Korea.

Brigadier General John P. Henebry replaces Major General Tunner as commander of the 315th Air Division in charge of all airlift operations in the Korean Theater.

13 February
The 315th Air Division airlifts more than 800 sick and wounded troops from forward airstrips to Taegu and Pusan. So many C-47s are used that they could not fulfill other airlift demands.

13–16 February
Three Chinese divisions surround UN troops, including the U.S. 23d Infantry Regiment and a French battalion, at Chipyong-ni in central Korea. Despite heavy enemy ground fire, nearly 100 transports drop 420 tons of food and ammunition to the encircled troops. The Fifth Air Force provides close air support for the surrounded troops until the arrival of a friendly armored column. During the fighting, an H-5 helicopter crew braves snow and strong winds to deliver medical supplies to the embattled ground forces.

16 February
The U.S. Army begins using the L-19 Bird Dog for forward air control, artillery spotting, and other front-line duties, relieving Fifth Air Force of demands for these types of missions.

Opposite: *F-86 maintenance personnel work diligently to prepare these F-86 Sabres for combat operations in Korea. (NASM)*

Below: *Flying in close formation with other transports near Chungju, Korea, a FEAF C-119 Flying Boxcar from Combat Cargo Command drops rations and gasoline to troops on a snow-covered battlefield. More than 300 tons are delivered. ("Buck Nineteen" by Nicholas Mosura, USAF Art Collection)*

Left: *Paratroopers jump from a formation of C-46 Commando aircraft somewhere over Korea. (USAF)*

Below: *A jet-powered B-47 refuels behind a piston-driven KC-97 tanker. Note the difference in the angle of attack of each aircraft while refueling. (NASM)*

17–18 February
B-26s fly a night bombing mission using Shoran, a short-range navigation system employing an airborne radar device and two ground beacon stations for precision bombing. It is the first such mission of the war.

23 February
Bomber Command begins flying missions using the accurate MPQ-2 radar, bombing a highway bridge seven miles northeast of Seoul.

24 February
Airlifters drop a record 333 tons of cargo to frontline UN troops. The theater workhorse C-119 provides the bulk of the 70 aircraft used in the mission.

1 March
Twenty-two F-80s, sent to escort 18 B-29s over North Korea, miss their rendezvous time and arrive on station ahead of the Superfortresses. Low on fuel, they return to base. The unescorted B-29s are then attacked by a force of enemy MiGs, damaging 10, three of which had to land in South Korea.

Thule Airbase, Greenland, is established by the USAF this date. Located 690 miles north of the Arctic Circle, it is the USAF's northernmost operational base.

4 March
Fifty-one C-119s drop 260 tons of supplies to the 1st Marine Division in the largest airdrop of the month.

6 March
F-86 Sabres begin patrols over the Yalu River after months away from the area.

14 March
Communist forces quietly withdraw from Seoul after General Mathew B. Ridgway's troops capture the high ground surrounding the city.

B-26s drop tetrahedral tacks on highways to puncture the tires of enemy vehicles.

15 March
UN forces enter Seoul, the fourth time the city has changed hands since the war began.

A Boeing B-47 Stratojet is refueled by a KC-97A tanker for the first time.

16 March
FEAF flies 1,123 effective sorties, a new daily record for the war.

20 March
Strategic Air Command's 7th Air Division is activated with its headquarters at South Ruislip, near London.

Right: *The SH-19A Chickasaw served with distinction during wartime in Korea. (NASM)*

23 March
In the largest single-day airborne operation of the war—Operation TOMAHAWK—120 C-119s and C-46s, escorted by 16 F-51s, deliver the 187th Airborne Regimental Combat Team and two Ranger companies deep behind enemy lines northwest of Seoul. The airlifters drop more than 3,400 men and 220 tons of equipment and supplies.

On the same day, 22 B-29s of the 19th and 307th BGs, protected from MiGs by 45 F-86s, destroy two bridges in northwestern Korea.

24 March
An H-19 is used for the first time in Korea for the air evacuation of wounded troops. The H-19 is considerably larger, more powerful, and has greater range than the H-5s.

29 March
FEAF B-29s attack the Yalu River bridges, as the ice over the river has begun to thaw.

2 April
The Air Research and Development Command (ARDC) becomes operational this date with Major General David M. Schlatter in command.

3 April
The service-test YH-19 helicopter flying missions with the 3d Air Rescue Squadron (ARS) picks up a downed F-51 pilot southeast of Pyongyang, while under enemy fire.

6 April
A Labor Department report reveals that aircraft plant employment has risen by 100,000 people during the first six months of the Korean War.

9 April
Jacqueline Cochran establishes a women's world speed record of 469.5 miles per hour over a 16-mile course while flying a North American F-51 Mustang at Indio, California. This is her fifth speed record in piston-powered aircraft.

12 April
Forty-six B-29s attacking the Yalu River bridge at Sinuiju, accompanied by 100 fighters, encounter more than 100 enemy fighters. Three bombers are shot down and seven damaged. Fighting back, B-29 gunners destroy seven MiGs, and F-86 pilots score victories over four more. The bridge is not destroyed.

At President Truman's direction, Eighth Army commander General Matthew Ridgway replaces General Douglas MacArthur. MacArthur had publicly criticized Truman's wartime and foreign policies.

17 April
President Truman signs an executive order extending U.S. military enlistments involuntarily by nine months to mitigate manpower shortages existing in the military services during the war.

Warrant Officer Donald Nichols leads a special operations team to recover MiG-15 wreckage and technical information. For his efforts, Nichols is awarded the Distinguished Service Cross.

18 April
A monkey is launched into space atop an Aerobee rocket at Holloman AFB. Unfortunately, the monkey does not survive the return to earth.

Left: *This F-86 Sabre has just launched a volley of missiles during a ground attack training mission at Nellis AFB, Nevada. (NASM)*

Opposite, top: *Major General Earle E. Partridge took command of the FEAF in May 1951. (NASM)*

Opposite, bottom: *Captain James Jabara became the world's first jet ace on May 20, 1951. (NASM)*

19 April
After a fleet-wide upgrade project, the first modified and reconditioned C-119 Flying Boxcar returns to combat service.

21 April
The Fairchild XC-123 Provider flies for the first time, powered by four J47 turbojet engines. This prototype is not produced, but it remains the first U.S.-developed jet transport plane.

23 April
FEAF flies 340 close air support sorties, one of the highest daily totals prior to 1953. F-86 aircraft begin operating from Suwon airfield, located in the north of South Korea, improving their time-on-station in MiG Alley near the Yalu River.

23–26 April
FEAF flies more than 1,000 combat sorties, inflicting enemy casualties and destroying supplies needed to sustain the offensive.

30 April
Fifth Air Force launches 960 effective sorties—a new daily record.

9 May
In one of the largest joint counterair efforts of the war to date, Fifth Air Force and 1st Marine Air Wing fighter-bombers fly more than 300 sorties against Sinuiju airfield in extreme northwestern Korea.

16–26 May
Airlift aircraft fly an average of 1,000 tons of supplies for 10 days straight from Japan to Korea in support of UN ground forces in contact with enemy forces.

17–22 May
Bomber Command B-29s fly 94 close air support sorties against enemy ground forces—most of them at night.

20 May
Captain James Jabara earns his fifth and sixth victories in the skies over Korea and becomes the world's first jet ace.

General Stratemeyer, FEAF commander, suffers a severe heart attack.

21 May
Major General Earle E. Partridge assumes command of FEAF. Major General Edward J. Timberlake replaces him as Fifth Air Force commander.

27–28 May
Specially equipped C-47s flying leaflet-drop and voice-broadcast sorties over enemy troops encourage them to surrender to the U.S. Army's IX Corps. Some 4,000 enemy soldiers, many carrying surrender leaflets, do surrender. The captives report that morale is low, largely due to UN air attacks.

31 May
Fifth Air Force begins Operation STRANGLE, an interdiction campaign against enemy supply lines in North Korea.

June
During this month, the Air Force Flight Test Center (AFFTC) is formally established at Edwards AFB, California.

1 June
Major General Frank F. Everest assumes command of Fifth Air Force, replacing Major General Timberlake.

Major John P. Stapp, an Air Force aeromedical researcher, begins a series of rocket sled experiments that test the onset of "Gs" and how the human body handles those forces. For an instant, Stapp endures the force of 48 "Gs" which occur at the rate of 500"Gs" per second, far more than previous medical science suggested was survivable.

3 June
Two C-119 Flying Boxcars are shot down by friendly artillery during a drop operation. This fratricide incident leads to the establishment of new procedures to identify friendly aircraft during airdrops.

10 June
In Tokyo, Lieutenant General Otto P. Weyland assumes command of FEAF, replacing General Earle Partridge.

20 June
Bell pilot Jean L. "Skip" Ziegler makes the first test flight of the Bell X-5 research plane. The X-5 is the first practical variable-sweep wing aircraft flown.

A Martin B-61 Matador (later TM-61) tactical guided missile is launched for the first time.

25 June
The Arnold Engineering Development Center (AEDC) is dedicated at Tullahoma, Tennessee. President Harry S. Truman and Mrs. Henry "Hap" Arnold are present for the ceremony.

1 July
Kim Il Sung, North Korean premier, and Paeng Te-huai, commander of Chinese forces, agree to participate in truce negotiations.

Colonel Karl L. Polifka, commander, 67th Tactical Reconnaissance Wing (TRW) and aerial reconnaissance expert, is shot down and killed while flying an RF-51 near the front lines.

6 July
An Air Materiel Command KB-29M tanker, flown by a Strategic Air Command crew assigned to the 43d Air Refueling Squadron (ARS) conducts the first in-flight refueling over enemy territory under combat conditions. Four RF-80 Shooting Stars tank up prior to flying reconnaissance missions over North Korea.

Left: *A B-26 Invader crew conducts a night interdiction briefing. B-26s flew 60 percent of the night sorties of the war, logging more combat time than any other aircraft type—224,000 combat flying hours. (USAF)*

14 July
The Ground Observer Corps begins round-the-clock skywatch across the United States.

30 July
Ninety-one F-80s suppress enemy air defenses while 354 USMC and USAF fighter-bombers attack their targets. The Joint Chiefs of Staff do not release information on the mission to the news media so as not to generate negative publicity in the face of ongoing peace negotiations.

17 August
Colonel Fred J. Ascani, flying an F-86E Sabre at the National Air Races in Detroit, sets a new world speed record of 635.6 miles per hour over a 100-kilometer closed course. Ascani receives the Mackay Trophy for the achievement.

18 August
FEAF expands Operation STRANGLE against North Korean railroads.

Colonel Keith Compton wins the first USAF jets-only Bendix Trophy transcontinental race. He flies an F-86A Sabre from Muroc, California, to Detroit, Michigan, in three hours and 27 minutes at an average speed of 553.8 miles per hour.

24 August
General Hoyt S. Vandenberg, USAF Chief of Staff, announces that tests of theater tactical nuclear weapons had been accomplished during the month of February.

24–25 August
B-26s claim more than 800 trucks are destroyed during night anti-truck operations throughout Korea.

25 August
As part of a challenging joint mission, 35 B-29s, escorted by USN fighters, drop 300 tons of bombs on marshaling yards at Rashin in northeastern Korea. The target, a major supply depot, lies less than 20 miles from the Soviet border. The target is destroyed and no aircraft are lost.

28 August
The Lockheed XC-130 wins a USAF competition for a turboprop-powered transport aircraft. The design will evolve into the C-130 Hercules, one of the most durable and widely used aircraft in aviation history.

5 September
The Air Force awards a contract to Convair to modify a B-36 to use a nuclear reactor as its power source. The NB-36H flies in 1955, but a host of technical and environmental problems result in the cancellation of

the program before a practical atomic-powered aircraft can be developed. General Electric was contracted to build the atomic-powered engine.

9 September
Seventy MiGs jump 28 F-86 Sabres between Sinanju and Pyongyang. In the face of three to one odds, F-86 pilots, Captain Richard S. Becker and Captain Ralph D. Gibson, each destroy a MiG, increasing the number of jet aces from one to three.

13 September
The USAF establishes its first pilotless bomber squadron located at the Missile Test Center near Cocoa, Florida.

14 September
Captain John S. Walmsley Jr., flying a night B-26C interdiction sortie, attacks an enemy train until all his weapons are gone. Now unarmed, he uses an experimental searchlight that is mounted on his aircraft's wing, illuminating the target for another B-26. Walmsley and his bombardier and photographer are killed when his aircraft is downed by ground fire. His gunner miraculously survives, is captured, and lives through the war.

20 September
Air Force scientists make the first successful recovery of animals from a rocket flight—a monkey and 11 mice—after being launched to an altitude of 236,000 feet in an Aerobee research rocket.

23 September
Eight B-29s knock out the center span of the Sunchon rail bridge despite nine-tenths cloud cover. They use Shoran to accomplish their mission.

25 September
An estimated 100 MiG-15s attack 36 F-86 Sabres flying a fighter sweep mission over the Sinanju area. Sabre pilots destroy five MiGs in aerial combat.

27 September
During Operation PELICAN, a service-test C-124A Globemaster II delivers its first payload to Korea, airlifting 30,000 pounds of aircraft parts to Kimpo airfield near Seoul.

28 September
A Yokota-based RF-80 logs a 14 hour and 15 minute Korean combat sortie. During the mission the pilot refuels multiple times from two KB-29M tankers.

Above: *The B-26C accomplished a variety of interdiction missions in the skies over Korea. (NASM)*

Left: *Even with a KB-29 boom attachment, a late modification to the World War II bomber, the KB-29 was too slow to refuel jet aircraft effectively. (NASM)*

Left: *The six-engine Boeing B-47 Stratofortress soars near the Boeing factory at Seattle, Washington. (USAF)*

Opposite: *Upon his death in combat, Major George A. Davis Jr. had accumulated a total of 14 victories. (USAF)*

30 September
Brigadier General Joe W. Kelly assumes command of Bomber Command.

16 October
Fifth Air Force Sabre pilots destroy nine MiG-15s in aerial combat establishing a single-day record.

22 October
Two SA-16 Albatross rescue aircraft save the 12-man crew of a downed B-29, the highest number rescued by SA-16s on any single day of the war.

23 October
MiG-15s destroy three B-29s and one F-84 while damaging another five bombers during one of the most deadly, and through this date, the largest aerial engagements of the war. Fighter and B-29 gunners shoot down five MiGs during the battle.

The first production Boeing B-47 Stratojet enters service with the 306th Bombardment Wing at MacDill AFB, Florida.

4 November
Thirty-four F-86s encounter an estimated 60 MiG-15s in the Sinanju area. The F-86 pilots destroy two and damage three others.

9 November
Landing on the beaches of Paengnyong-do Island, off the southwest coast of North Korea, a C-47 crew rescues 11 crewmen of a downed B-29.

12 November
Peace negotiations move to Panmunjom, a village less than five miles east of Kaesong, in a newly established demilitarized zone on the 38th parallel.

16 November
Fifth Air Force fighter-bombers successfully cut rail lines throughout North Korea. Additional targets hit are bridges, gun positions, supply buildings, fuel dumps, and enemy freight cars.

30 November
F-86 pilots engage 44 enemy aircraft flying south to bomb UN targets. The Sabre pilots destroy 12 and damage three others. Major George A. Davis Jr. is the first man to become an ace in two wars—World War II and Korea.

13 December
Outnumbered two to one, 29 F-86s engage 75 MiG-15s over Sinanju. In a hard-fought battle, the F-86 pilots shoot down nine MiGs. Combined with other actions that day, Air Force pilots score a total of 13 aerial victories.

27 December
FEAF aircraft accomplish 900 sorties, damaging or destroying locomotives, railcars, buildings, vehicles, and gun positions. This marks the greatest sortie number for the month.

1952

7 January
The Air Force announces expansion plans for a 50 percent increase in combat strength. This will result in growth to 143 operational flying wings and 1.27 million Airmen. This will be revised downward to 120 wings in May 1953.

8–13 January
During Exercise Snowfall, 100 516th Troop Carrier Wing aircraft airlift more than 8,600 troops from Fort Campbell, Kentucky, to Wheeler-Sack Airfield, New York. This marks the largest airlift of troops to date.

12 January
F-84s trap three supply trains racing for the shelter of a tunnel by bombing the tunnel entrance shut. The pilots then systematically destroy the boxcars and two locomotives.

1 February
The USAF acquires its first high-speed digital computer. The Univac I is a vacuum-tube-based computer.

9 February
Ten medium bombers, dropping their weapons by using radar-aiming methods, deliver 100 tons of 500-pound bombs and render the north bypass to the Chongju rail bridge unserviceable.

10 February
Major George A. Davis Jr. leads a flight of 18 F-86E Sabres on patrol near the Manchurian border where they engage 12 MiG-15s in aerial combat. Davis shoots down two enemy aircraft becoming the first USAF ace in two different wars, but the surviving MiGs catch him off guard and shoot down his aircraft. For valor demonstrated against superior numbers while successfully protecting the fighter-bombers in his flight, Davis posthumously receives the Medal of Honor.

20 February
President Harry S. Truman appoints retired Lieutenant General Jimmy Doolittle to head a presidential commission to study how to relieve airport congestion near large American cities.

March
During this month, the Air Force initiates its Rocket Engine Advancement Program (REAP) to study

liquid oxygen and hydrocarbon fuel combinations. As a result of the study, standard liquid rocket fuels are changed for large-scale liquid-fuel missiles, including the Atlas Intercontinental Ballistic Missile (ICBM).

3 March
Operation SATURATE begins this date. The round-the-clock rail interdiction campaign replaces Operation STRANGLE against North Korean targets.

11 March
In one of the most effective napalm attacks of the war, fighter-bombers drop 150 tons of bombs and approximately 33,000 gallons of napalm on a four-square-mile supply storage and troop training area.

19 March
The newly improved F-86F-25 Sabre flies for the first time. This Sabre has modified leading edges to improve maneuverability at high altitudes where previously it had been at a disadvantage against the MiG-15. By June, these aircraft will begin arriving in Korea to battle enemy MiG-15s.

25 March
Fifth Air Force flies 959 interdiction sorties from Sinanju to Chongju striking rail networks and highways.

GABBY GABRESKI

Conrad Crane

Left: *"The Shooter and The Wingman," by Philip Alexander, MD. (Courtesy of the artist)*

Francis "Gabby" Gabreski shot down more enemy aircraft than any other American fighter pilot in the European Theater during World War II. He also served in the Korean conflict, one of those few individuals who became an ace in two wars. He retired from the Air Force in 1967 after more than a decade working on a series of new jet aircraft and was inducted into the Aviation Hall of Fame in 1978.

Gabreski was born to Polish immigrant parents in Oil City, Pennsylvania, in 1919. His interest in airplanes began in 1932 when he saw Jimmy Doolittle win the Thompson Trophy in a GeeBee R-1 at the Cleveland Air Races. After graduating from high school, Gabreski was admitted to Notre Dame in 1938, but he left after two years to enter Army pilot training. Second Lieutenant Gabreski was learning to fly P-40s in Hawaii on December 7, 1941, and spent most of that day flying combat patrols in aircraft salvaged from the Japanese attacks. He then managed an assignment with the RAF in late 1942, and flew Supermarine Spitfires with the Polish 315th Squadron. Though he downed no German planes, he did gain a great deal of valuable combat experience and honed his flying skills.

By early 1943, the Eighth Air Force was building up its own fighter force, and in February he reported in to Hubert "Hub" Zemke's soon-to-be-famous 56th Fighter Group. Gabreski was initially treated as an outsider, but he soon earned everyone's respect with his superior flying ability and aggressive fighting spirit. He rose to be the flying executive officer for the group, which included many fighter aces including Bob Johnson and Robin Olds. Flying a P-47 Thunderbolt, Gabreski shot down his first German on August 24, 1943. At that time he was commanding the 61st Fighter Squadron—one of "Zemke's Wolfpack." Three months later he became an ace, shooting down two Me-110s near Bremen. On July 5, 1944, he achieved his 28th kill, becoming the top American ace in the European Theater.

Gabreski was scheduled to fly back to the United States on July 20, but he couldn't resist flying on one last mission as an escort for a bombing raid near Frankfurt, his 166th. On the way back to base after the raid, in accordance with approved Army Air Forces doctrine, he decided to strafe a German airfield. He came in so low on his run that he clipped the ground

Right: *Francis "Gabby" Gabreski in the cockpit of his P-47 Thunderbolt. Gabreski tallied more victories than any other American pilot in the skies over Europe. He achieved "ace" status during the Korean War as well. (USAF)*

with his propeller, forcing him to land in a nearby wheat field. After eluding his pursuers for five days, he was captured and spent the rest of the war as a prisoner. He tried to get assigned to a prison camp in Poland, thinking it would then be easier to blend into the population if he escaped, but instead ended up at Stalag Luft I, along with Colonel Zemke, until they were liberated by the Russians in May.

After World War II he flew F-51s and F-80s, and when the Korean War began in June 1950 he was commanding his old 56th Fighter Group. That month they began transitioning to the new F-86. Gabreski became very skilled with the Sabre and clamored for a return to combat. In May 1951 he reported to the 4th Fighter Wing, and the next month he flew his first combat mission. The 32-year-old pilot downed his first MiG on July 8 while escorting fighter-bombers near Pyongyang. In December he took command of the 51st Fighter Wing to lead them through their transition to Sabres. By the time he was reassigned to the United States in June 1952, he was credited with six and a half kills, again achieving ace status.

Upon his return home, he accepted the personal thanks of President Truman in the Oval Office, and then began a career working with new jets, including the F-100 and F-101. He helped pioneer air-to-air refueling for transatlantic flights before retiring as a colonel to take a job with Grumman Aerospace. He spent some time as president of the Long Island Railroad, but eventually returned to work on air force projects for Grumman. He retired as assistant to Grumman's vice president for marketing in 1987. He lived in retirement on Long Island, New York, for many years as "America's Greatest Living Ace," before passing away in January 2002.

Left: *The Boeing YB-52 Stratofortress provided the capability to strike strategic targets around the globe without refueling. About 100 B-52s, updated to accommodate modern precision-guided weapons, will continue flying well into the twenty-first century. (NASM)*

1 April
Fifth Air Force Sabre pilots destroy 10 MiGs while losing one F-86. Colonel Francis "Gabby" Gabreski shoots down a MiG to become the eighth jet ace of the war.

The USAF redesignates its former Army enlisted grades from private first class, corporal, and buck sergeant to airman third, second, and first class respectively.

10 April
Brigadier General Chester E. McCarty assumes command of the 315th for the remainder of the war.

15 April
The Boeing YB-52 Stratofortress makes its first flight. Boeing test pilot Alvin M. "Tex" Johnson is at the controls near Seattle, Washington.

18 April
The Convair YB-60 flies for the first time at Carswell AFB, Texas. The YB-60 unsuccessfully competes with the Boeing YB-52 for the Air Force production contract.

29–30 April
A C-47, a C-119, and a C-46 crash during these 48 hours claiming the lives of 16 people. This is the greatest collective loss for the 315th Air Division in the first half of 1952.

3 May
Lieutenant Colonels William Benedict and Joseph Fletcher fly an Air Force C-47 equipped with ski-and-wheel landing gear, making the world's first successful landing at the North Pole.

7 May
The Lockheed X-7 air-launched ramjet test vehicle is launched from a test B-29 and flies for the first time. This ramjet design anticipates the Bomarc missile program.

8 May
As part of a major interdiction campaign, Fifth Air Force fighter-bombers fly 465 sorties against a large enemy supply depot 40 miles southeast of Pyongyang, the largest one-day attack since the war began. UN pilots damage or destroy more than 200 supply buildings, personnel shelters, revetments, vehicles, and gun positions. An F-86 on a dive-bombing strike is lost when hit by ground fire. It marks the first loss of a Sabre on a fighter-bomber sortie.

16–17 May
C-119, C-54, and C-46 aircraft transport 2,361 members of the 187th Airborne Regimental Combat Team and their combat equipment from Japan to Pusan. The team quells rioting enemy POWs at Koje-do, the UN's prisoner compound.

22 May
Fifth Air Force launches nearly 500 sorties against industrial areas southwest of Pyongyang. The site, which produces hand grenades, small arms, and ammunition, is almost completely destroyed.

26 May
The first C-124 Globemasters are received by the 315th Air Division as two squadrons begin the conversion from C-54s.

29 May
The first combat aerial refueling takes place when 12 Republic F-84 Thunderjets "top off" after a strike mission on their way back to their base at Itazuke AB, Japan. Three more similar missions will be flown as part of Operation RIGHTSIDE.

30 May
Lieutenant General Glenn O. Barcus takes command of Fifth Air Force.

7 June
During Operation HIGHTIDE, a test of aerial refueling capabilities, 35 F-84 Thunderjets depart from Japan, refuel using KB-29M aircraft over Korea, and then attack targets in North Korea.

10–11 June
During a night attack by B-29s against the rail bridge at Kwaksan, enemy MiGs, operating in conjunction with radar-controlled searchlights and flak, destroy two of the bombers and badly damage a third. A startling revelation, this new night intercept capability prompts FEAF to improve electronic countermeasures to confuse enemy radar.

15 June
2nd Lieutenant James F. Low becomes an ace just six months after completing flight training.

23 June
A strike force of fighter-bombers, with F-86 Sabres flying cover, attack North Korean hydroelectric power plants. The Sui-ho complex sustains severe damage and is rendered non-functional.

24 June
FEAF flies more than 1,000 sorties, the highest daily total for the month. Fifth Air Force fighter-bombers reattack hydroelectric power plants from the previous day. This two-day campaign is the largest single air effort since World War II.

3 July
The first operational C-124 arrives in Korea.

C-47s drop more than 22 million leaflets over targets in North Korea. This number represents more than one-sixth of all such drops during the month.

4 July
During a major air battle over the North Korean military academy, MiGs, some with Soviet pilots, engage the UN attack force of 50 F-86s and 70 F-84s. Fifth Air Force pilots down 13 MiG-15s while losing two Sabres. Four MiGs break through the protective fighter screen but fail to destroy any fighter-bombers. Unfortunately, the attack is only marginally effective.

4–17 July
Led by Colonel David C. Schilling, 58 F-84Gs make the first large-scale Pacific crossing of jet fighters supported by in-flight refueling. The flight covers 10,895 miles and is accomplished with seven ground stops and two aerial refuelings.

10 July
Over the next three weeks, the 315th AD will airlift the 474th fighter/bomber wing from Misawa to Kunsan, the largest complete unit movement by air to date.

Right: *The Douglas C-124 Globemaster II entered service in 1950 and flew strategic airlift missions for two decades. (USAF)*

Left: *"F-86 Sabre Dance," by Harley Copic. (USAF Art Collection)*

Below: *The North American RB-45 was equipped with five camera stations and designed to accomplish both day and night reconnaissance missions. (NASM)*

11 July
In the first raid of Operation PRESSURE PUMP, most operational air units in the Far East attack 30 major targets in and around Pyongyang. This marks the largest single strike of the war. The North Korean Ministry of Industry is completely destroyed. Most other targets sustain heavy damage. FEAF flies a month-high 1,329 sorties.

13–31 July
Two Air Force Sikorsky H-19 helicopters, named *Hopalong* and *Whirl O Way*, complete the first transatlantic helicopter flight when they fly from Westover Field, Massachusetts, to Prestwick, Scotland, in a five-stage effort.

29 July
Major Louis H. Carrington, Major Frederick W. Shook, and Captain Wallace D. Yancey crew a 91st Strategic Reconnaissance Wing RB-45 from Elmendorf, Alaska, to Yokota, Japan. This marks the first nonstop, transpacific flight by a jet aircraft, covering 3,640 miles and taking nine hours and 50 minutes to complete. For the mission, the crew is awarded the Mackay Trophy.

30–31 July
Sixty B-29s destroy 90 percent of the Oriental Light Metals Co. facility located only four miles from the Yalu River. Despite tough night counterair efforts, the B-29s are extremely effective and target damage is severe. The attacking bombers suffer no losses.

6 August
In what will be the major air-to-air battle of the month, 34 F-86s destroy six of 52 attacking MiG-15s.

8 August
Fifth Air Force fighters fly 285 close air support sorties, the highest daily total for the month. At night, B-26s fly three voice-broadcast sorties totaling almost four hours over enemy-held positions near the eastern Korean coast.

22–23 August
Three C-47s fly 60-minute voice-broadcast sorties near the front lines as UN Command begins to place more emphasis on psychological war.

29 August
As a show of force requested by the U.S. Department of State, FEAF aircraft attack Pyongyang in the largest air attack of the war to date. This strike was to coincide with a visit by China's foreign minister, Chou En-lai, to the Soviet Union. FEAF aircraft, protected by Air Force F-86 Sabres and RAAF Meteors, launch 1,400 air-to-ground sorties, damaging 31 targets, while three UN aircraft are lost to ground fire.

Right: *A Collection of Korean War aces: (left to right) Captain Lonnie R. Moore, 10 MiGs; Colonel Vermont Garrison, 10 MiGs and 7.33 victories in World War II; Colonel James K. Johnson, 10 MiGs and one victory in World War II; Captain Ralph S. Parr, 10 victories in Korea; and Major James Jabara, 15 MiGs and 1.5 victories in Europe in World War II. ("Five Men, 55 MiGs," by Nixon Galloway, USAF Art Collection)*

3–4 September
B-29s fly 52 effective sorties, and all but two target the Chosin hydroelectric power plant complex.

4 September
Seventy-five fighter-bombers attack targets well north of the Chongchon River, flushing out an estimated 90 MiGs from their Manchurian bases. The 39 F-86 Sabres screening the attacking F-84s engage the MiGs and shoot down 13 of them, equaling the single-day record. The mission is not without loss as four F-86s are shot down by MiG pilots. Major Frederick "Boots" Blesse destroys his fifth enemy aircraft to become an ace during the fighting.

9 September
Forty-five F-84s attack the North Korean military academy at Sakchu, and three are shot down by attacking MiGs. The F-86s flying top cover suffer no losses during the aerial battle and destroy five MiGs.

16 September
B-26s flying predominantly at night carry out 110 armed reconnaissance and interdiction sorties. Using new roadblock tactics, the light bombers damage or destroy more than 100 enemy vehicles.

19 September
Thirty-two B-29s with F-86 escorts attack an enemy barracks and two supply areas southwest of Hamhung during daylight. This marks the first day raid flown by B-29s in nearly 11 months. A jet-powered RB-45 flies pre-strike reconnaissance before the B-29 formation attacks, while an RB-29 orbits in the assembly area to provide target weather information.

21 September
F-86 Sabre pilot, Captain Robinson "Robby" Risner, shoots down two MiG-15s to become an ace as the enemy responds to an attack on the Pukchong munitions plant by 41 F-84s.

30 September
The Bell GAM-63 Rascal is launched for the first time. It is an air-launched strategic missile.

3 October
Great Britain tests its first atomic bomb in the Monte Bello Islands off the northwest coast of Australia.

4 October
Brigadier General William P. Fisher assumes command of Bomber Command.

8 October
Ten B-29s execute a rare daylight visual bombing mission in eastern Korea, coordinating with Navy fighter-bomber attacks.

Truce talks at Panmunjom recess over the issue of forced repatriation of POWs. UN delegates propose to allow enemy POWs to choose repatriation or not; the Communist delegates insist upon the forced repatriation of all POWs at the end of the war.

13 October
After nearly a year, the enemy, using small fabric-covered biplanes, conduct harassment raids near Seoul. These are known as "Bedcheck Charlie" raids.

20 October
Douglas test pilot William Bridgeman flies the sleek X-3 Stiletto for the first time. The test program proves useful in experimental titanium machining and construction. It also provides a large collection of useful data on short-span, low aspect-ratio wings, as well as inertial coupling in flight.

31 October
The U.S. tests its first thermonuclear device at Eniwetok in the Marshall Islands. The "Mike Shot" equals more than 10 million tons of TNT, more than 1,000 times more powerful than the Little Boy bomb dropped on Hiroshima at the end of World War II.

10 November
The 315th AD passes a major milestone when it air evacuates the 250,000th patient from Korea to Japan.

12–13 November
Six 98th Bomb Wing B-29s knock out four spans of Pyongyang's recently restored railway bridges.

15 November
A C-119 Flying Boxcar, returning 40 travelers to Korea from rest leave in Japan, crashes, killing all on board.

19 November
Captain J. Slade Nash flies an F-86D to a new world record over a three-kilometer course. He streaks over the Salton Sea, California, at 698.5 miles per hour.

North American Aviation's XLR-43-NA-3 rocket engine is test fired. It is the first American rocket to produce more than 100,000 pounds of thrust.

22 November
Major Charles J. Loring Jr., leading a flight of four F-80s, is hit near Sniper Ridge by enemy ground fire. He deliberately crashes his aircraft into the midst of enemy gun emplacements and destroys them, ending the threat from enemy troops at the cost of his own life. Loring is posthumously awarded the Medal of Honor for his sacrifice.

26 November
The turbojet-powered, subsonic, long-range Northrop N-25 Snark (later B-62) missile is launched from a zero-length launcher for the first time. The Snark is America's first intercontinental guided missile.

2–5 December
President-elect Dwight D. Eisenhower tours the front lines in Korea and also meets with South Korean President Syngman Rhee.

9 December
The United Nations adopts strategy 14/1, which bases a strategy of defense of Europe upon the use of American nuclear weapons.

11 December
A fully loaded B-26 catches fire at Kunsan airfield and explodes, destroying three additional B-26s and causing damage to six F-84s parked nearby.

16 December
The first Air Force helicopter squadron is activated by the Tactical Air Command this date. The squadron is equipped with Sikorsky H-19A helicopters.

17 December
Two F-86 Sabre pilots sight an IL-28 twin-jet bomber escorted by two MiG-15s a few miles south of the Sui-ho reservoir. One Sabre chases the trio across the Yalu River.

Opposite, top: *A Snark Missile (MX-775) takes off from its mobile launcher at Cape Canaveral, Florida, in May 1960 (NASM)*

Opposite, bottom: *"Loring's Final Flight," by Gerald Asher. (USAF Art Collection)*

Right: *A SAC B-52 and a B-47 are depicted as they appeared during the 1960s. ("Cold War Deterrent" by Karl Neumann, USAF Art Collection)*

1953

17–18 January
Eleven B-29s attack the underground radio station at Pyongyang with 2,000-pound bombs. The target is only 1,000 feet from a suspected POW camp. Despite 10 direct hits, the weapons do not penetrate deeply enough to destroy the radio station.

22 January
The last F-51 Mustangs are withdrawn from combat in preparation for the transition to F-86 Sabres. This move ends the use of Air Force single-engine, propeller-driven aircraft in offensive combat during the Korean War.

24 January
Captain Dolphin D. Overton III sets a record for becoming a Korean War jet ace in the shortest time. He achieves five victories in four days.

28–29 January
A B-29 explodes over its target southwest of Sariwon. This is the fourth B-29 lost since December and the last one lost in the war.

30 January
An F-86 pilot intercepts and downs a Russian-built Tu-2 twin-engine bomber over the Yellow Sea, northeast of Pyongyang. This is the first reported Tu-2 kill since November 1951.

The Boeing B-47E Stratojet flies for the first time. The major production model, more than 1,300 B-47Es will be built, along with 255 RB-47E reconnaissance versions.

30–31 January
Approximately 10 enemy fighters so badly damage a 307th BW B-29 that it barely makes an emergency landing in South Korea.

4 February
Harold E. Talbott is appointed Secretary of the Air Force.

8 February
Aviation Medicine is recognized by the American Medical Association as a medical specialty, the first medical discipline that has evolved from military practice and research.

14 February
The Bell X-1A takes its first glide test under with pilot Jean "Skip" Ziegler at the controls. The X-1A is longer, has a greater fuel capacity, and has a revised cockpit from the original X-1. It makes its first powered flight on February 21.

15 February
Twenty-two F-84 Thunderjets strike the Sui-ho hydroelectric power plant and suffer no losses. Eighty-two escorting F-86 Sabres attack 30 MiGs, while the Thunderjets deliver their 1,000-pound bombs. The attack disrupts power production at Sui-ho for several months.

Left: *A flight of four F-86 Sabres pitch out from echelon formation over Korea. (USAF)*

16 February
Captain Joseph C. McConnell Jr. becomes an ace this date.

18 February
Four F-86s attack a formation of 48 MiG-15s just south of the Sui-ho reservoir, shooting down two enemy aircraft. Two other MiGs, attempting to follow an F-86 through evasive maneuvers, enter uncontrolled spins and crash. In this battle, Captain Manuel J. Fernandez becomes an ace by downing his fifth and sixth MiGs.

18–19 February
More than 500 jet aircraft deliver high-explosive bombs on a tank and infantry school southwest of Pyongyang, destroying 243 buildings.

5 March
Fifth Air Force completes 700 sorties, including a flight of sixteen F-84 Thunderjets attacking an industrial area just 60 miles from the Siberian border. Fighter-bombers flying ground support missions report damage to 56 bunkers and gun positions, 14 personnel shelters, and 10 supply shacks.

10 March
Two Czechoslovak Air Force MiG-15s attack two Air Force F-84 Thunderjets in Western European airspace. One F-84 is shot down, but the pilot ejects safely. This is the first such attack during the Cold War.

14 March
In an attempt to provoke aerial engagements with Communist fighters, Air Force combat crews drop leaflets asking, "Where is the Communist air force?" over ground targets after they attack them.

21–22 March
Attacks upon bridges become important once again during Operation SPRING THAW. Medium bombers knock out spans of two principal bridges at Yongmi-dong and render the third unserviceable.

27 March
MiG-15s equipped with external fuel tanks attack two RF-80s and two RAAF Meteors only 38 miles north of the front lines.

31 March
The last F-80C Shooting Star is retired from front-line Korean War service.

12 April
An H-19 helicopter rescues Captain Joseph C. McConnell Jr., an F-86 pilot with eight victory credits to date, from the Yellow Sea after he had ejected from his battle-damaged fighter.

7 April
The Atomic Energy Commission announces that it is using Lockheed QF-80 drones to study radioactive clouds during atomic tests. The drones are controlled from other airborne aircraft using Sperry control equipment.

Right: *The YF-100, a sleek fighter, was used to establish several speed records. The operational F-100 Super Sabre played an active role in the early years of the Vietnam War. (NASM)*

Below: *Captain Joseph C. McConnell Jr., seated in his F-86 Sabre* Beauteous Butch II. *McConnell became the first triple ace of the Korean War and the leading ace of the war, with 16 total victories. (USAF)*

13 April
The F-86F Sabre flies its first air-to-ground combat mission.

20 April–3 May
During Operation LITTLE SWITCH, Communist and UN forces exchange sick and injured prisoners.

26 April
Armistice negotiations between Communist and UN forces reconvene after a six-month break.

26–27 April
A B-29 medium bomber drops leaflets over North Korea, initiating the search for an enemy MiG-15. The operation is called Project MOOLA.

13 May
Four waves of 59 F-84G Thunderjets carry out the first attack against previously prohibited irrigation dams 20 miles north of Pyongyang. Floodwaters destroy six miles of embankment, five bridges, two miles of the major north-south highway, render Sunan airfield inoperable, and, most critically, ruin five square miles of prime rice crop.

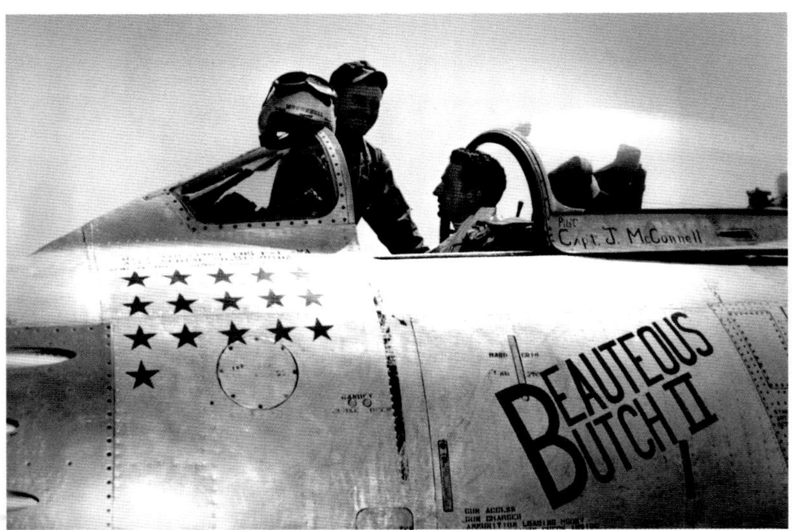

16 May
Ninety dedicated F-84G attack sorties breach the Chasan irrigation dam, destroying three railroad bridges and spoiling rice ripening in surrounding fields.

18 May
Captain Joseph C. McConnell Jr. downs three more MiG-15s to become the first triple jet ace and, with 16 victories, the highest scoring ace of the Korean War.

Jacqueline Cochran sets a new world speed record over a 100-kilometer course in a Canadian-built F-86E Sabre. In the skies over Edwards AFB, California, Cochran flies 652.3 miles per hour. Earlier in the day, Cochran became the first woman to exceed the speed of sound. Major Chuck Yeager flew the chase aircraft during that flight.

25 May
North American test pilot, George "Wheaties" Welsh, flies the YF-100 Super Sabre for the first time and easily breaks Mach 1 during the one-hour sortie. The flight occurs at the Air Force Flight Test Center at Edwards AFB, California. Nearly 2,300 F-100s will be built during its service life.

27 May
The Air Force Historical Foundation (AFHF) is officially established. It is dedicated to the preservation, perpetuation, and appropriate publication of the history and heritage of American aviation.

31 May
Lieutenant General Samuel E. Anderson assumes command of Fifth Air Force.

THUNDERBIRDS — THE LEGACY

Debby Baer Becker with General Lloyd W. "Fig" Newton

Left: *The Thunderbirds flying T-38 Talons over the Hoover Dam near their home—Nellis AFB, Nevada. (USAF)*

Opposite: *"Thunder in the Canyon," by William Phillips. (NASM Art Collection)*

When the United States Air Force Thunderbirds take flight, they become part of more than 50 years of aerial demonstration history. Crowds gather in anticipation as the sun glints from the perfectly polished paint and shining metal of the machines aligned on the airfield. Cheers erupt when they see the flash, hear the roar, and feel the thunder of blazing afterburners. America's colors wave, trail, and blur. After more than five decades of inspiring air show crowds around the globe, the Thunderbirds have earned their place as the ultimate expression of American idealism. They are dreamed perfection. They are the public face of the United States Air Force and represent all those who have ever worn Air Force blue.

Without exception, the Ambassadors in Blue represent the very best of our Air Force and our nation. They bring skill and professionalism as polished as the jets they fly. As a team they strive for perfection day after day—year after year—just like the thousands of Air Force professionals that they represent. The American public deserves nothing less. Yet this team, which has touched so many hearts over the years, started with so few.

In 1953, a fledgling Thunderbird team took to the skies above Luke Air Force Base, Phoenix, Arizona. The Air Force wanted to showcase its front-line fighter, the F-84G Thunderjet. The flight leader was Major Richard "Smokey" Catledge. With final approval on all matters concerning the unit, Catledge handpicked his wingmen, Captains Buck and Bill Patillo (identical twins), and slot pilot Captain Bob Kanaga to shape the essential Thunderbird diamond formation. Spare pilot Captain Robert McCormick, maintenance officer 1st Lieutenant A. D. "Brownie" Brown, narrator Captain Bill Brock, and 23 top-notch maintenance personnel rounded out the first team.

Within three weeks, this distinctively red, white, and blue diamond formation graced the skies of Nellis Air Force Base, Las Vegas, Nevada, and chose the name "Stardusters." Shortly thereafter, they flew a demonstration at Williams AFB near Phoenix, Arizona. By the time they flew their first official air show on July 19, 1953, in Cheyenne, Wyoming, the Air

Force had renamed them "Thunderbirds." In 1954, these highly skilled pilots and state-of-the-art aircraft performed their first official overseas tour in South America for more than three million spectators. This challenging State Department–commissioned goodwill tour required creative solutions to tough challenges. Flying high performance jets into several small South American airfields had previously been considered too difficult, while team logistics to support such deployments initially seemed impossible. By overcoming such obstacles the Thunderbirds performed aerial demonstrations for many people who had never seen a jet airplane before. The creativity, ingenuity, professionalism and pride established by the first team remains the legacy of all Thunderbird teams.

That legacy is today shouldered by eight pilots, three support officers, and 122 enlisted members. They travel approximately 200 days each year to showcase the skills and dedication of the more than 690,000 active duty, Air Force Reserve, and National Guard men and women. Each Thunderbird represents these airmen who work diligently every day around the globe to keep our homeland safe and free. This honorable task demands excellence both in the air and on the ground. Perfection is their objective, and the Thunderbirds accept nothing less.

After a two- or three-year tour, Thunderbirds go on to complete Air Force careers or to work in our communities. The tradition of "Once a Thunderbird, Always a Thunderbird" demands that the traits of the legacy carry on in its members. Hours of sacrifice to answer the call of duty and fulfill the responsibilities of the Thunderbird team build significant character in our members—a reflection of the character of our Air Force. Only 2,200 men and women have worn the Thunderbird patch since 1953.

Many Thunderbirds have gone on to great service both in the USAF and the American community. Richard "Smokey" Catledge retired from the Air Force as a Major General. Bill Creech (solo and wingman, 1954–1955) became a general and commander of Tactical Air Command. The first Thunderbird to become an astronaut was Bill Pogue (solo, 1955–1957). Sam Johnson (slot, 1957–1958), seven-year POW, has been serving Texas District 3 as a U.S. congressman since 1992. Lacy Veach (solo, 1975–1978) flew space shuttle missions STS-39 and STS-52. Vickie Graham (air show coordinator, 1976–1978) served as managing editor for *Airman Magazine* and is a photojournalist and communications consultant. Brock McMahon (egress technician, 1974–1982) is the director of programs and resources for the largest maintenance group in Air Mobility Command. Dee Holloway Pfeiffer (crew chief, 1976–1977) heads up her own nationwide executive recruitment agency and organized Thunderbird stories for the book, *We Rode the Thunder*.

The Thunderbirds are an American symbol of strength and freedom. Countless hours of dedication and discipline have earned them their place, not only in America's culture, but also in America's heart. The Thunderbirds embody freedom, courage, dignity, loyalty, and the traditions of the United States Air Force. Their thunderous presence awakens the American spirit.

Left: *The Martin B-57 Canberra, a British-designed aircraft, was used as both a fighter-bomber and a reconnaissance platform. This model is configured with four pairs of five-inch high-velocity aircraft rockets (HVAR) and wingtip fuel tanks. (NASM)*

8 June
The 3600th Air Demonstration Flight—the USAF Thunderbirds—perform an unofficial show at their home station, Luke AFB, Arizona, for the very first time. The team flies F-84G Thunderjets with special red, white, and blue highlights.

11 June
Thirteen F-84s attack an airfield located midway on the North Korean-Manchurian border. This strike is the deepest flown into enemy territory to date. Pilots report that the raid has rendered the runway unserviceable.

13–18 June
F-84s, B-29s, and Marine F4U Corsair fighter-bombers strike irrigation dams at Toksan and Kusong in an effort to flood nearby airfields. The raids fail because the Communists had lowered water levels to decrease water pressure on the dam.

15 June
Brigadier General Richard H. Carmichael assumes command of Bomber Command.

16 June
Fifth Air Force flies 1,834 sorties. More than half are close support missions against enemy troops in the Pukhan Valley area.

18 June
A Douglas C-124 Globemaster crashes after takeoff near Tokyo, Japan. All 129 on board are killed, making this the worst air disaster in history.

21 June
The Thunderbirds perform for the American public for the first time at Cheyenne, Wyoming.

30 June
F-86 Sabres score 16 MiG victories in a single day, a new daily record.

General Nathan F. Twining becomes the Air Force Chief of Staff.

11 July
South Korean president Syngman Rhee agrees to accept a cease-fire agreement in return for promises of a mutual security pact with the United States.

Major John Bolt, a Marine flying with the USAF 39th Fighter Interceptor Squadron (FIS), shoots down his fifth and sixth MiGs, becoming the Marines' only Korean War ace.

15 July
Major James Jabara earns his 15th aerial victory to become the world's second triple jet ace.

16 July
Commander Guy Bordelon, flying with Fifth Air Force, becomes the sole Navy ace of the Korean War.

Lieutenant William Barnes breaks the recognized absolute speed record to 715.7 miles per hour. Flying an F-86D, he takes the record from an earlier F-86D flight, a rare occurrence in the setting of such records.

19 July
The final session of armistice negotiations at Panmunjom convene. After meeting only one day, the top negotiators adjourn while technical experts work out the cease-fire details.

20 July
The Martin B-57A Canberra (Night Intruder), the American production version of the British original, flies for the first time at the Martin Middle River, Maryland, plant. The Canberra is the first aircraft of non-American design accepted by the USAF.

21–22 July
Eighteen B-29s fly the final mission of the war for Bomber Command when they attack Uiju airfield.

22 July
The final air-to-air battle of the Korean War occurs between three Air Force F-86F Sabres and four Communist MiGs. During this engagement, Lieutenant Sam P. Young records the last MiG kill of the Korean War.

27 July
Lieutenant General William K. Harrison, U.S. Army, the senior delegate for the UN Command, and General Nam Il, the senior delegate for the North Korean Army and the Chinese Volunteers, sign the armistice agreement establishing the cease-fire in the Korean War.

Captain Ralph S. Parr Jr. downs an enemy transport plane, becoming a double ace with the last air-to-air victory of the Korean War. He downs an Il-2 nearly two and one-half hours after the armistice goes into effect.

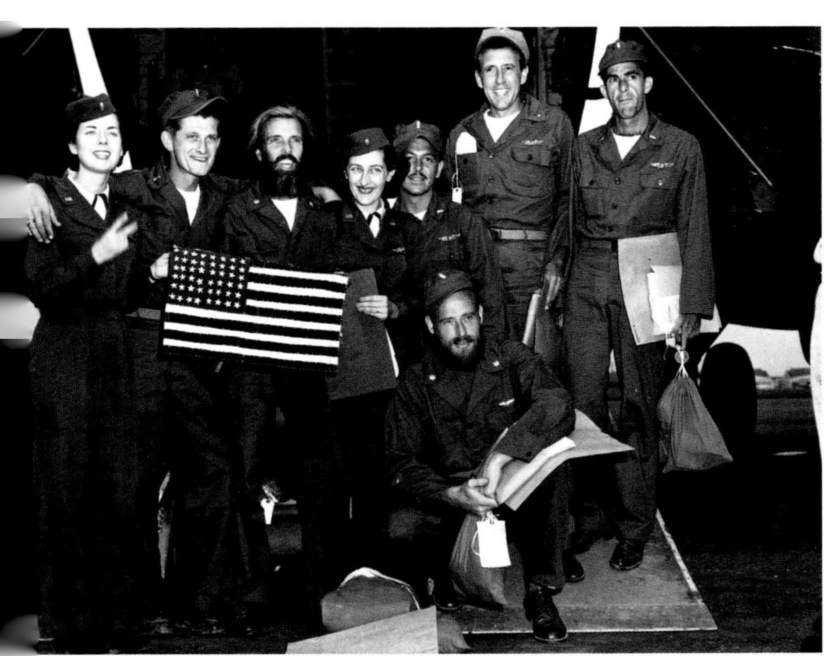

Twenty-four minutes before the cease-fire becomes effective, a B-26 crew drops the last bombs of the Korean War in a night, radar-directed, close support mission. Aircraft from the same squadron had flown the first combat strike into North Korea three years before.

An RB-26 of the 67th Tactical Reconnaissance Wing (TRW) flies the last combat sortie of the war over North Korea.

The Korean War formally ends at 10:01 p.m. All FEAF's aircraft are south of the front line or more than three miles from North Korea's coastline.

In accordance with the armistice agreement, in August POWs will be exchanged in Operation BIG SWITCH. More than 77,000 Communist prisoners are released and 12,700 UN prisoners are returned, including 3,597 Americans. The operation begins on August 6.

28 July
A Strategic Air Command B-47 Stratojet establishes a transatlantic speed record when it flies from Limestone, Maine, to Fairford, England, in four hours and 43 minutes, averaging 618 miles per hour.

Above: *This RB-26 crew flew the last combat sortie of the Korean War: (left to right) Airman 3rd Class Dennis Judd, Lieutenant Donald Mansfield, and Lieutenant Bill Ralston. (NASM)*

Left: *Repatriated prisoners and their medical staff prepare to board their transport home. (USAF)*

RICHARD WHITCOMB AND THE AREA RULE

John D. Anderson, Jr.

Left: *The redesigned Convair F-102 featured effective use of Whitcomb's "area rule." Notice how the aircraft fuselage narrows as the delta wings sweep outward in an effort to keep a uniform total cross-sectional area. (NASM)*

Opposite: *Richard Whitcomb examines a wind tunnel model built to test his area rule theories. Note the hourglass shape of the model's fuselage between the wings. (NASM)*

In 1943, a young engineer passed through the front gate of the National Advisory Committee for Aeronautics (NACA) Langley Aeronautical Laboratory at the outskirts of the sleepy tidewater town of Hampton, Virginia. Richard T. Whitcomb, an honors graduate from Worcester Polytechnic with a degree in mechanical engineering, had just been hired by the NACA to work as a wind-tunnel engineer. He came with a fertile background. Born in Evanston, Illinois, on February 21, 1921, Whitcomb was influenced by his grandfather, who had known Thomas A. Edison. Much later, in an interview with the *Washington Post* on August 31, 1969, Whitcomb stated, "I used to sit around and hear stories about Edison. He sort of developed into my idol." Whitcomb, with an already intense interest in aeronautical engineering, had read a *Fortune* magazine article on the research facilities at the NACA Langley Laboratory and decided that was where he belonged. Little did he know, as he began his first work day at the laboratory, that he was destined to become one of America's most innovative aerodynamicists.

In the late 1940s, the Langley wind tunnels were busy with research on high-speed flight. Chuck Yeager had broken the sound barrier in the Bell X-1 in 1947, and supersonic flight was, by far, the dominant aspect of NACA's research programs. Whitcomb was assigned to the Langley eight-foot transonic wind tunnel, where in a few years he developed a reputation as an "idea person" for his combination of knowledge and intuition about aerodynamic flows. By 1950 he was deeply involved with a study of the large drag increase that an airplane experiences approaching the speed of sound. He realized that the physics of the airflow changed dramatically as it expands from subsonic to supersonic speeds. Insightfully he hypothesized the shape of the fuselage in the vicinity of the wing should have a reduced area to help smooth this expanding flow and to mitigate the formation of shock waves that served to increase the drag. Whitcomb also knew that ballistics engineers had been designing artillery shells and machine gun bullets for years with a smooth distribution of cross-sectional area in order to reduce the wave drag at supersonic speeds. To better understand this concept, imagine taking a tube of sausage and cutting it vertically into a number of thin round slices. The area of each slice is essentially the same. Similarly, if you slice an artillery shell the same way, the area of the slices will smoothly vary with distance along the shell, good for reducing drag. In contrast, if you slice a conventional airplane in the same way, you will find a smooth distribution of cross-sectional area along the fuselage until you get to the wing. There the cross-sectional area of the wing suddenly adds to that of the fuselage, creating a sudden and discontinuous change in the area. Whitcomb reasoned that to preserve a smooth distribution of area in the vicinity of the wing, and thus reduce drag, the area of the fuselage would have to be reduced accordingly, giving the fuselage a type of "coke-bottle" shape. This concept later became known as the area rule.

Although Whitcomb encountered some skepticism from his colleagues, Adolph Busemann, a famous and respected German aerodynamicist then working for the NACA, called Whitcomb's area rule concept "brilliant." Whitcomb was allowed to pursue the idea, carrying out a number of wind tunnel experiments in early 1952 that proved its validity. Indeed, the data showed that the large drag rise near Mach 1 was reduced almost 60 percent when the fuselage was sufficiently indented in the vicinity of the wings. These dramatic results, however, were not immediately appreciated by the aircraft industry.

Serendipity stepped in. At the same time that Whitcomb was spending long days and nights in the wind tunnel conducting his research, Convair was designing one of the new "century series" fighters intended to fly at supersonic speeds. Designated the YF-102, the aircraft had a delta-wing configuration and was powered by the Pratt & Whitney J-57 turbojet, the most powerful engine in the United States at that time. Aeronautical engineers at Convair expected the YF-102 to fly supersonically. On October 24, 1953, flight tests of the YF-102 began at Muroc Air Force Base (now Edwards), and a production line was being set up at Convair's San Diego plant. As the fight tests progressed, however, it became painfully clear that the YF-102 could not fly faster than sound—the transonic drag rise was simply too great for even the powerful J-57 to overcome. Then the Convair engineers began to pay attention to Whitcomb's research results. After consultation with Whitcomb and other NACA aerodynamicists at Langley, and inspecting the area-rule findings that had been obtained in the Langley eight-foot tunnel, the Convair engineers modified the airplane to become the YF-102A, with an area-ruled fuselage. On December 20, 1954, the prototype YF-102A left the ground at Lindbergh Field, San Diego, and exceeded the speed of sound while still climbing. Whitcomb's area rule had increased the top speed of the airplane by 25 percent. The production line rolled, and 870 F-102As were built for the U.S. Air Force. The area rule made its debut in dramatic style.

Whitcomb's area rule remained classified until September 1955. Two months after his work became public, Whitcomb was awarded the Collier Trophy, the most prestigious award in aeronautics. The Collier Trophy citation read, "For discovery and experimental verification of the area rule, a contribution to base knowledge yielding significantly higher airplane speed and greater range with the same power." Richard Whitcomb had made aeronautical engineering history.

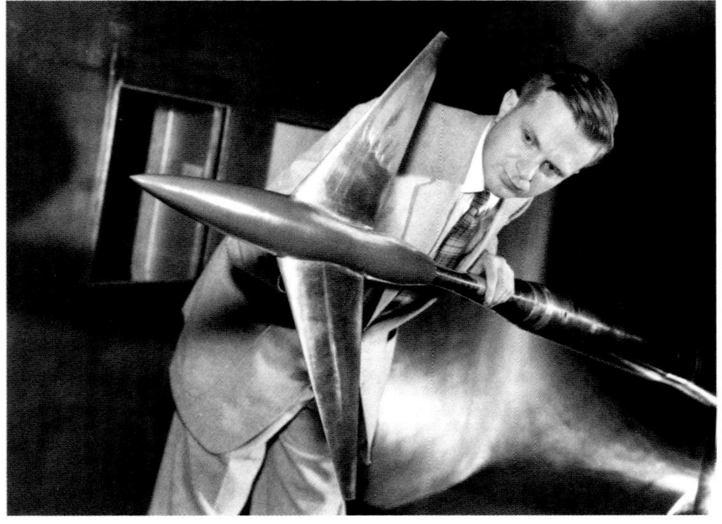

Left: *This B-36 was specially modified to tow three RF-84 fighters to increase range during reconnaissance missions. Aerial refueling eventually eliminated the need for such creative solutions. (NASM)*

Below: *This MiG-15 was flown by defector Noh Kum Suk into American hands at Kimpo Airfield. (NASM)*

29 July
An RB-50 operating out of Yokota AB, Japan, is shot down by a Soviet MiG-15 when it flies over Soviet waters near Vladivostok. Only one of the 17 crew survives the incident.

The Air Force announces the final combat numbers for the Far East Air Force during the Korean War. FEAF aircraft shot down a total of 839 MiG-15 jet fighters, are credited with 154 more as probable, and damaged 919 more in the 37 months of the air war. UN air forces lose 110 aircraft in dogfights, 677 to ground fire, and 213 to other causes.

3 August
A Redstone missile is test fired for the first time at the Air Force Missile Test Center, Cape Canaveral, Florida.

12 August
The Soviet Union explodes its first thermonuclear device.

20 August
During Operation LONGSTRIDE, a 40th Air Division deployment of F-84 Thunderjet aircraft from U.S. bases to Europe, the first nonstop, transatlantic crossings by fighter aircraft are made by two separate Thunderjet units. This demonstration of Strategic Air Command's rapid, long-range deployment capability earns the 40th Air Division the Mackay Trophy.

25 August
A B-36 that has been modified to launch and recover an RF-84F in flight is revealed by the USAF and called the "flying aircraft carrier." The first successful test of the Fighter Conveyor (FICON) system is made this date.

September
A B-47 Stratojet is refueled in flight by a KB-47B Stratojet tanker aircraft. This marks the first jet-to-jet aerial refueling in history. The refueling is accomplished using the hose and drogue system.

11 September
A Sidewinder infrared-guided, air-to-air missile intercepts and destroys an airborne target for the first time during a test firing at China Lake, California. The AIM-9 series of missiles will be produced into the twenty-first century.

21 September
North Korean pilot Lieutenant Noh Kum Suk flies his MiG-15 fighter to Kimpo Airfield and defects. He is given $100,000 and asylum in the United States. The MiG is disassembled and flown to Okinawa, Japan, where it undergoes test flights. After the North Koreans refuse an American offer to return the plane,

it finds its way to the USAF Museum (now National Museum of the United States Air Force).

October
During this month, the first of 10 RC-121C Super Constellations is delivered to the Air Defense Command.

1 October
The Air Force activates its first airborne early warning and control squadron. The 4701st is stationed at McClellan AFB, California, and operates the RC-121 aircraft.

14 October
The prototype of the B-64 Navaho guided missile, called the X-10, is fired for the first time. The Navaho is a ramjet-powered, surface-to-surface weapon.

23 October
The Piasecki YH-16 Transporter twin-rotor helicopter flies for the first time.

24 October
Convair test pilot Richard L. Johnson flies the YF-102 at Edwards AFB, California, for the first time. Performance is poor and not until a serious redesign, including the use of the "area rule" in the fuselage design, does aircraft performance improve to acceptable levels.

29 October
Lieutenant Colonel Frank K. "Pete" Everest pilots a North American YF-100 to a new world speed record of 755.15 miles per hour. Four runs are flown over a nine-mile course near the Salton Sea in California. This speed record, which approaches the speed of sound, is the last absolute speed record achieved at "low altitude," defined as lower than 330 feet above the ground.

31 October
The Strategic Missiles Evaluation Committee is formed under the guidance of special assistant to the Secretary of the Air Force, Trevor Gardner. Noted mathematician John von Neumann subsequently heads the committee.

6 November
A Boeing B-47 sets a new transatlantic speed record when the crew takes off from Limestone AFB, Maine, and lands at RAF Brize Norton, United Kingdom. The flight takes four hours and 57 minutes.

Above: *This Navaho X-10 (B-64) Missile is ready for launch. (NASM)*

20 November
A. Scott Crossfield, a NACA test pilot, flies the Douglas D-558-2 Skyrocket twice the speed of sound—the first to accomplish this feat. The aircraft is powered by a Reaction Motors XLR-8-RM-6 four-chamber, liquid-propellant rocket engine. Retired in 1957, the Skyrocket now hangs in the National Air and Space Museum.

12 December
Major Chuck Yeager attains the speed of 1,650 miles per hour in a Bell X-1A rocket-propelled research aircraft launched from a modified B-36 Peacemaker. That speed equals approximately Mach 2.4 at the test altitude of 70,000 feet. The flight takes place over Edwards AFB, California. Losing complete control of the aircraft at high altitude, he recovers to level flight at 25,000 feet and lands successfully. Yeager receives the Harmon Trophy for the flight.

FUEL VENT
© Machat '83

BEYOND THE ATMOSPHERE

1954–1963

BEYOND THE ATMOSPHERE

1954–1963

Following the cessation of hostilities on the Korean Peninsula, the Air Force began to reassess its warfighting capabilities. The air war in Korea did not resemble the air battles fought during World War II. The strategic war had been relatively short and was fought using outdated piston-driven bombers. The jet bomber force was deployed to deter the threat of war with the Soviet Union. It was this Cold War Air Force that would see a massive expansion during the coming decade.

Guided missile programs stepped to the forefront of American strategic planning and policy. Scientists and their recommendations continued as important tools for decision-makers in the government and the Pentagon. Dr. John von Neumann's Teapot Committee envisioned the potential for developing nuclear weapons and the capability of powerful rocket technology. No longer would heavy bombers provide the sole capability for delivering devastating nuclear strikes around the globe. Guided intercontinental ballistic missiles (ICBMs) would soon provide a delivery method which, at the time, was perceived as almost impossible to stop. These ICBMs, designed for launch from underground or undersea platforms, made their detection and destruction unlikely.

The combination of turbojet development, improving missile technology, and extremely powerful light-weight nuclear weapons and warheads resulted in a race to catch up with, and then keep one step ahead of the Soviet Union in space. In 1961, to confront a series of Soviet space-related firsts, a frustrated President John F. Kennedy challenged the nation to land an American on the moon before 1970—a challenge that energized the aerospace industry with a virtually bottomless federal budget. Former aircraft companies became "aerospace" companies overnight.

Although manned space flight received the most attention across the country, developments in military space flight were closely interrelated both by project and by agency. The establishment of the National Aeronautics and Space Administration (NASA), a civilian government agency, drew heavily on America's military for its operational personnel. Seven of the first 16 NASA astronauts were USAF pilots.

Many significant orbital projects were also managed by the Air Force, some in cooperation with NASA and the Navy. The X-15 hypersonic rocket-powered research plane program was one such cooperative effort that provided much of the data needed to explore beyond the atmosphere in manned vehicles. But other programs also made vital contributions to understanding aerospace science—with the emphasis on space.

Extreme high-altitude balloon experimentation exposed daring aeronauts to brutal atmospheric conditions, while testing the equipment needed to protect them from such harsh environments. Sub-orbital missile test programs not only validated the technology needed to enter earth's orbit, but also gathered life science information by sending a variety of animal subjects skyward in experimental capsules. Often, more information was learned from a failed launch or capsule recovery than might have been learned from a totally successful one.

By the late 1950s, military missiles, both intercontinental and intermediate-range versions, became operational around the globe. Rocket

Pages 376–377: *The X-15 hypersonic research plane was considered one of the most successful in history. The experimental program proved that manned aircraft could return from orbit and land at a predetermined location. ("Return from Mach 6," by Michael Machat, USAF Art Collection)*

Right: *Depiction of the first flight of the XB-58 on November 11, 1956. ("World's First," by K. Price Randel, USAF Art Collection)*

propellant development began to emphasize solid fuels that were easier to maintain and had a longer "silo life," simplifying maintenance and shortening operational response times when compared to older liquid-fueled missiles.

Perhaps some of the most significant accomplishments during this decade were related to satellite technology. The world got a first look at itself from space during these years, thanks to early weather satellites. Communications satellites demonstrated that it was possible for one person to speak with another halfway around the globe in real time. Initially, communications from space were something of a novelty. President Dwight Eisenhower sent the globe a Christmas message from an orbiting transmitter, but it was the haunting "beep, beep, beep" returned from the Soviet *Sputnik* satellite that initiated the space age as well as the space race in October 1957.

Satellites and specialized high-altitude aircraft provided the ability to secretly watch military movements and deployments that previously had been executed completely unobserved. The surprise attack across the 38th Parallel into South Korea was carried out with such operational freedom. But with high-altitude reconnaissance aircraft, orbital observation stations, and spy satellites came a corresponding loss in an adversary's ability to carry out surprise attacks. Of course, this ability also resulted in dangerous conflicts such as the Cuban Missile Crisis.

By the early 1960s, significant numbers of American military advisors had been sent to South Vietnam—a nation under siege by the forces of communism. At home, the aircraft that would be used in the skies over Vietnam were under development during these years—everything from the KC-135 Stratotanker to the venerable C-130 Hercules and C-141 Starlifter, to the F-105 Thunderchief and the F-4 Phantom II, and the UH-1 Huey and the amazing SR-71 Blackbird. Aircraft technology did not stagnate despite the massive budget for space activities.

By the time the U.S. officially entered the Vietnam War in 1964, Strategic Air Command had become an all-jet force, Air Force astronauts had ventured into space, and the strategic nuclear triad consisting of global-range bombers, ICBMs, and submarine launched intercontinental ballistic missiles (SLBMs) was a reality even if not completely operational. Aviation records continued to be set and broken by newer and faster jet and rocket aircraft by both men and women flyers. Conflict with the Soviet Union had heated up on occasion, and Americans had lost their lives during routine reconnaissance missions when fired upon by Soviet interceptors. The Soviets finally succeeded in shooting down a high-flying U-2 spyplane on May 1, 1960, resulting in an embarrassing international crisis for President Dwight Eisenhower.

President Kennedy's election in 1960 resulted in the race to the moon, a greater emphasis on conventional military forces over nuclear ones, and greater direct involvement in military training in Vietnam. America's eventual official entry into the conflict in Vietnam had far-reaching political, social, and military effects that still resonate more than three decades after the evacuation of the American Embassy in Saigon in 1975.

BEYOND THE ATMOSPHERE

1954–1963

"Now it is time to take longer strides—time for a great new American enterprise—time for this nation to take a clearly leading role in space achievement, which in many ways may hold the key to our future on earth."
—President John F. Kennedy, Special Message to Congress on urgent national needs, May 25, 1961

1954

10 February
Under the guidance of Dr. John von Neumann, the Air Force Strategic Missile Evaluation Committee reports a major technological breakthrough in the size of nuclear warheads. The Teapot Committee, as it is known, recommends that a special group be formed to accelerate the development of Intercontinental Ballistic Missiles.

1 March
The United States detonates a 15-megaton fusion bomb in the Marshall Islands. This first hydrogen bomb is far more powerful than earlier atomic weapons.

5 March
Lockheed test pilot Tony LeVier pilots the XF-104 Starfighter on its first official test flight at Edwards AFB, California. It had hopped off the ground during high-speed taxi tests the week before. The airplane was designed by Clarence "Kelly" Johnson of the Lockheed "Skunk Works."

1 April
President Dwight D. Eisenhower establishes the United States Air Force Academy by legislative action. The first class will begin training at Lowry AFB, in Denver, Colorado, while construction of the Academy is underway. The permanent site, selected over the next few months, is just north of Colorado Springs—15,000 acres of land nestled in the foothills of the Rocky Mountains.

The Military Air Transport Service receives the first of 26 Convair C-131A transport planes—a military version of the Convair 240 airliner.

8 April
The office of the Air Force Assistant Chief of Staff for Guided Missiles is established at the Pentagon.

8 May
The French Army is defeated at Dien Bien Phu, ending their struggle in Indochina.

18 June
The Martin B-57B, the interdiction variant of earlier Canberras, flies for the first time.

4 June
Major Arthur Murray pilots the X-1A research aircraft to a world record altitude of over 90,000 feet.

Left: *The Lockheed XF-104 Starfighter was designed by Clarence "Kelly" Johnson, who would later design the U-2 and SR-71. (NASM)*

Right: *The Douglas RB-66A Destroyer in flight. Large aft speedbrakes are deployed and natural metal engine nacelles are prominent in this photo. (NASM)*

Below: *The Boeing 367-80 "Dash 80" is towed into position at the National Air and Space Museum, Steven F. Udvar-Hazy Center, near Dulles International Airport. (NASM)*

21 June
Under the leadership of Major General Walter C. Sweeney Jr., Fifteenth Air Force commander, three 22d Bombardment Wing B-47 Stratojets complete a nonstop flight from March AFB, California, to Yakota AB, Japan. The 6,700-mile journey takes just under 15 hours and requires two aerial refuelings with KC-97 tankers.

26 June–17 July
In the aftermath of the French defeat in Indochina, the 315th Air Division and the Military Air Transport Service (MATS) airlift more than 500 wounded French soldiers from Vietnam. During Operation WOUNDED WARRIOR, the injured are moved via Japan and the United States on their way back to France.

28 June
The Douglas RB-66 Destroyer reconnaissance aircraft makes its first flight.

1 July
In accord with instructions issued in June, the Air Force establishes the Western Development Division. Under the command of Brigadier General Bernard A. Schriever, this division will be responsible for the development of the USAF Ballistic Missile Program. Of critical importance is the acceleration of the Atlas ICBM program.

15 July
The prototype for the Boeing KC-135 Stratotanker flies for the first time in Seattle, Washington. Legendary Boeing test pilot Tex Johnson later flies a demonstration of the "Dash-80" for Air Force evaluators. During the flight, Johnson accomplishes a barrel roll at near tree-top level. The Air Force orders hundreds of the plane, later redesignated as the KC-135 Stratotanker.

26 July
Lieutenant General Hubert R. Harmon is appointed as the first superintendent of the USAF Academy.

5 August
The first production B-52 flies for the first time. This operational variation of the earlier B-52 prototype seats the pilot and copilot side-by-side instead of in tandem like the YB-52 or the B-47.

Left: *In this formation of two YC-130 Hercules transport aircraft, the original cockpit configuration and flat nose section will undergo several modifications during its service life. (NASM)*

Below: *A Cessna T-37A with natural metal finish. Most "Tweety Birds" were painted white during operational service. (NASM)*

6–7 August
In two separate demonstrations of long-range strategic potential, two 308th Bombardment Wing B-47 Stratojets take off from Hunter AFB, Georgia, and fly a 10,000-mile nonstop round-trip mission to French Morocco and return. At the same time, two B-47s from the 38th Air Division depart from Hunter AFB and fly a simulated bombing mission over Morocco. For these achievements, the 308th Bombardment Wing and the 38th Air Division are awarded the Mackay Trophy.

23 August
The YC-130 prototype of the turboprop-powered Hercules tactical airlift aircraft flies for the first time.

24 August
The Communist Control Act is signed into law by President Eisenhower. The Communist Party is outlawed in the United States.

26 August
Major Arthur "Kit" Murray pilots the Bell X-1A to a record altitude of 90,440 feet, where he reports that he is able to see the curvature of the earth. The mission is flown at the Air Force Flight Test Center in California.

1 September
The Fairchild C-123B Provider makes its first flight.

The Air Force establishes the Continental Air Defense Command with its headquarters at Colorado Springs, Colorado. General Benjamin W. Chidlaw is named commander.

27 September
The F-100A Super Sabre achieves operational status at George AFB, California.

29 September
The F-101A Voodoo flies for the first time. Evolved from the XF-88 all-weather interceptor, the Voodoo was designed as a twin-engine interceptor and bomber escort. Later, a reconnaissance version, the RF-101 would participate in the Cuban Missile Crisis, flying low-altitude tactical reconnaissance sorties over Cuba.

8 October
Major Kit Murray pilots the Bell X-1B rocket research aircraft during its maiden flight over Edwards AFB, California.

9 October
As part of the attempt to speed up Atlas missile development, $500 million is added to the current year's budget for guided missile programs.

12 October
The Cessna XT-37 side-by-side jet trainer makes its first flight at Wichita, Kansas. The first production model will fly in September 1955.

18–19 October
An Ad Hoc committee of the Air Force Scientific Advisory Board meets under the guidance of Theodore von Kármán. The committee evaluates the possibility of applying nuclear energy to missile propulsion. In the final report, continuous study of nuclear propulsion is recommended so that the Air Force will not fall behind in this field.

27 October
Benjamin O. Davis Jr. becomes the first African American general officer in the U.S. Air Force.

1 November
The Boeing B-29 Superfortress is withdrawn from Air Force service.

2 November
While testing the Convair XFY-1 Pogo, pilot James F. "Skeets" Coleman accomplishes a vertical takeoff, transitions to horizontal flight and flies for 21 minutes, and then lands vertically to complete the mission flown at Lindbergh Field in San Diego, California. For his work in this program, Coleman receives the Harmon Trophy.

7 November
An RB-29 reconnaissance aircraft operating near Hokkaido, Japan, is shot down by two Soviet MiGs.

17–19 November
Colonel David A. Burchinal remains aloft while piloting a Boeing B-47 Stratojet for 47 hours and 35 minutes. This endurance record is the result of extremely poor weather over England and France, forcing Burchinal to refuel nine times while waiting for the weather to clear.

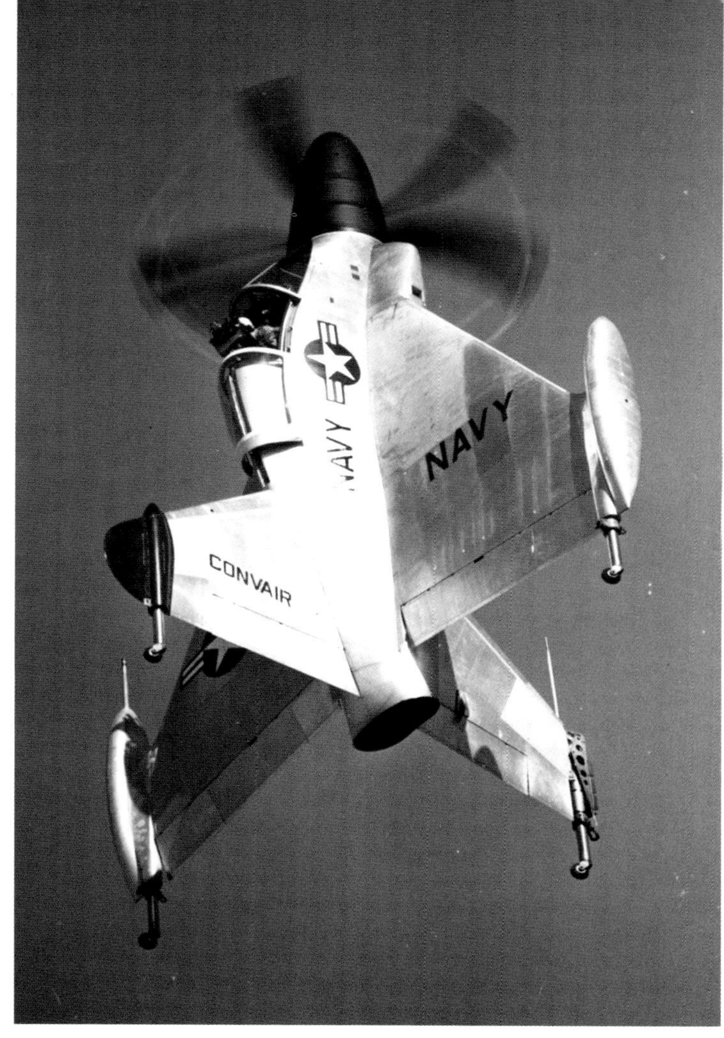

7 December
An Air Force Navaho X-10 missile is recovered by a fully automatic approach and landing system at Edwards AFB, California.

10 December
Colonel John P. Stapp tests the limits of human endurance during a rocket-propelled sled run. Stapp is accelerated to more than 630 miles per hour and then experiences the equivalent G-force of ejecting from an aircraft at 35,000 feet at Mach 1.7. He proves that humans can withstand the forces of such hostile ejection environments.

20 December
The revised Convair F-102A flies for the first time. The F-102A is the first Air Force fighter armed only with guided missiles and rockets.

23 December
A joint Air Force-Navy-NACA memorandum is signed, thus beginning the development of a new hypersonic research aircraft. The X-15 will result ftom this agreement.

Above: *The Convair XFY-1 in vertical flight and also with Navy markings. (NASM)*

Left: *Colonel John Paul Stapp reaches 632 miles per hour during this rocket sled run in March 1953, easily beating the chase plane to the finish line. (NASM)*

BENJAMIN O. DAVIS JR.

Von Hardesty

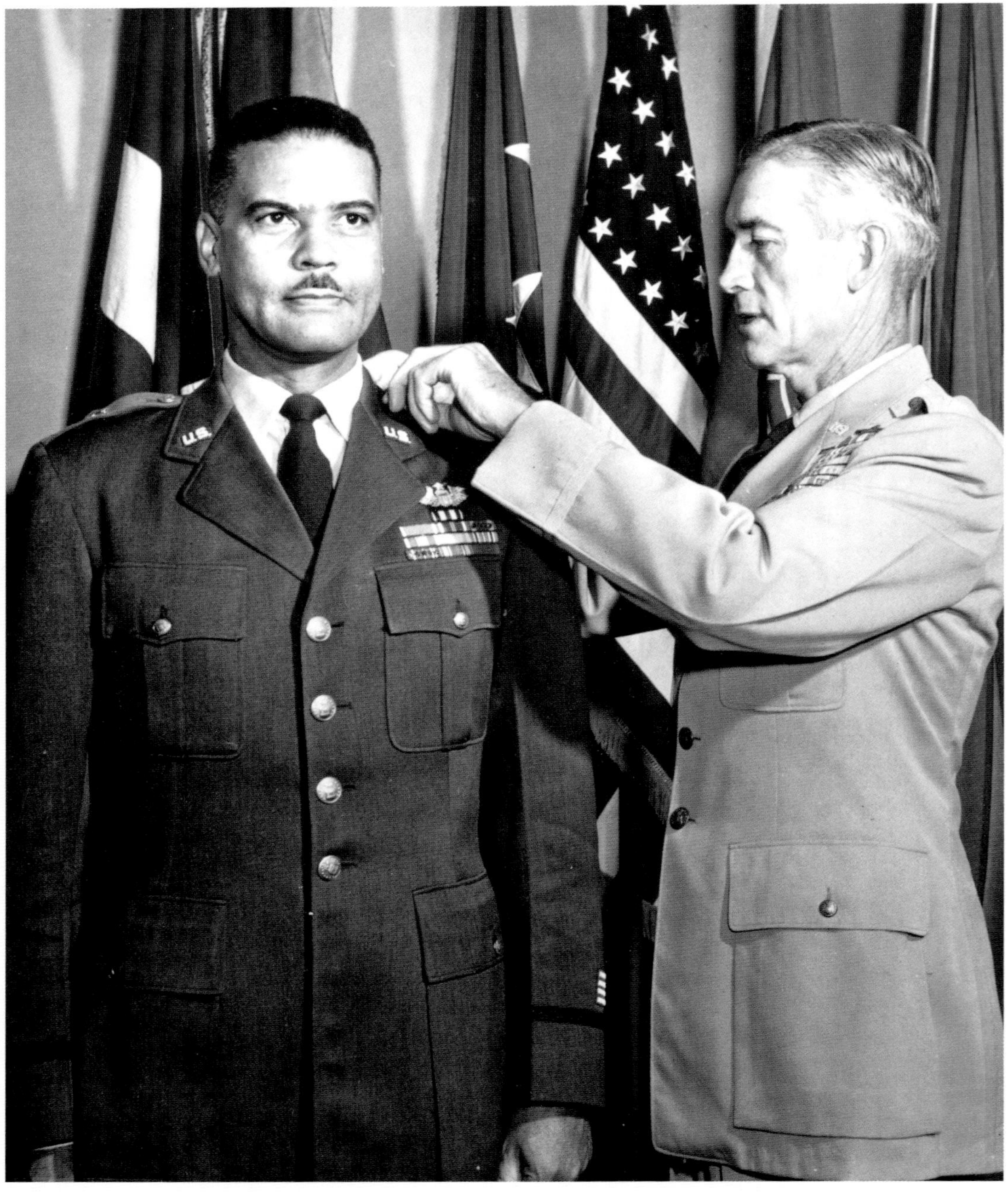

Left: *General Earle E. Partridge, FEAF commander, pins brigadier general stars on the shoulders of Benjamin O. Davis Jr. in October 1954. (NASM)*

Opposite: *Major Benjamin O. Davis Jr. poses in front of his fighter plane during World War II, his West Point graduation ring prominent on his left hand. (USAF)*

On the eve of World War II, the United States Army Air Corps approved the recruitment of African Americans for training as pilots and ground crew specialists. Benjamin Oliver Davis Jr., a career army officer and West Point graduate, quickly moved to the epicenter of this controversial experiment, winning his silver wings at the Tuskegee Army Airfield on March 7, 1942. That same year Davis assumed leadership of the fledgling 99th Pursuit (later Fighter) Squadron, which became the nucleus for the Tuskegee airmen.

Davis went on to lead the 99th and then the 332d Fighter Group in the Mediterranean Theater of Operations during World War II. The war years became an arena for Davis to challenge the racial discrimination then prevailing in the American military—in particular, the widely held notion that African Americans lacked the requisite aptitude and courage to become effective military pilots. Davis worked diligently to demonstrate that his pilots possessed the same skills and patriotism as their white counterparts in the war effort.

Deployed to North Africa in 1943, Davis led the 99th Fighter Squadron into combat, flying mostly tactical support missions. From the start, the 99th faced an uphill fight in winning acceptance from some theater air commanders, who complained that the all-black air unit lacked competence and aggressiveness. In the face of such hostility, Davis displayed moral courage in his spirited defense of his pilots. His forceful stand against racial bigotry ultimately proved effective, setting the stage for an enlarged role for black airmen in the war. The creation of the 332d Fighter Group in September 1943 offered Davis personal vindication and a clear signal that the Tuskegee experiment would continue, notwithstanding its many critics.

Early in 1944, Davis assumed command of the newly organized 332d Fighter Group, now deployed to Italy in the aftermath of the Anzio campaign. In the final months of the war, the 332d Fighter Group flew 200 escort missions to enemy targets in Germany and Nazi-occupied Europe. Pilots of the 332d Fighter Group downed 111 enemy aircraft in air-to-air combat, including some of the earliest aerial victories over German jets. These same pilots won the affection of bomber crews for their sustained and reassuring presence as escorts on the many arduous bombing missions to Nazi Germany. Davis himself flew 60 combat missions in World War II, ending the war as a full colonel. He was the recipient of the Distinguished Flying Cross and the Silver Star for personal bravery.

The postwar years soon brought dramatic new opportunities for Davis and all African American airmen. The creation of an independent United States Air Force in 1947 coincided with the benchmark decision to end official segregation in the American armed forces. For Davis, there were a series of new assignments as the most senior African American officer in the Air Force. He graduated from the Air War College in 1950, and then accepted a staff position at the Pentagon. During the Korean War, Davis assumed command of the 51st Fighter-Interceptor Wing in South Korea. As commander of an integrated air unit, Davis quickly earned the respect of all personnel under his command. He was subsequently transferred to Japan where he was appointed director of operations and training in the Far East Air Forces. Given his manifest skills, Davis was promoted to brigadier general in 1954. Davis became the first African American to achieve the rank of major general in 1959. His third star as a lieutenant general followed in 1965.

During these active years Davis held a sequence of key posts in the Air Force. As vice commander of Thirteenth Air Force, he was instrumental in shaping the air defenses of Taiwan. Later he joined the Twelfth Air Force in Europe, serving as deputy chief of staff for operations. After a brief stint in the Pentagon as director of manpower and organization, General Davis moved on to several assignments in Korea, including command of the Thirteenth Air Force in 1967. One of his final posts was deputy commander of United States Strike Command. He retired from the Air Force in 1970.

In retirement, General Davis held a number of civilian positions, most notably as director of civil aviation security for the Department of Transportation—a key position offered to Davis in response to a series of aircraft hijackings. His tenure at the Department of Transportation ultimately led to appointment as assistant secretary of transportation in 1971.

In a special ceremony on December 9, 1998, President Clinton awarded Davis his fourth star as full general, being the first African American ever to receive such an honor in retirement.

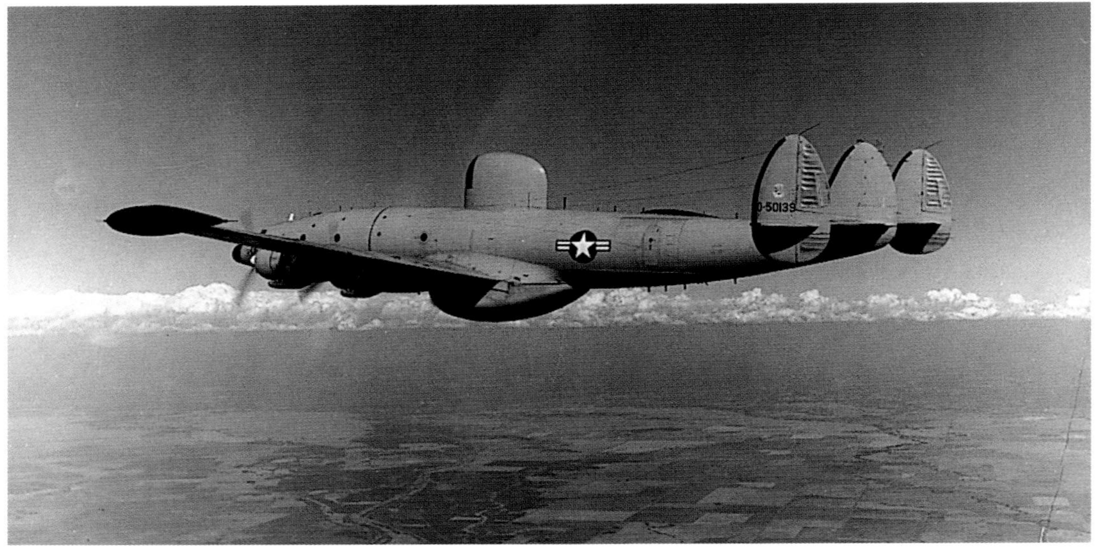

Left: *The Lockheed RC-121 Constellation provided early warning of enemy fighters to American aircraft operating in the skies over Vietnam. (USAF)*

Below: *"Trevor Gardner," by Jay Ashurst, Air Force Space Command. (USAF)*

1955

20 January
The U.S., Great Britain, and France reach an agreement with the South Vietnamese government to assist with the modernization of their military equipment.

26 February
North American test pilot George Smith ejects from his F-100 Super Sabre while traveling at Mach 1.05. Smith is the first man to survive a supersonic ejection from an aircraft. He sustains an instantaneous 64 "Gs" upon exit from the jet and then .3 seconds at 20 "Gs." He is knocked unconscious by the forces, and his parachute blows nearly one-third of its panels. He falls into the Pacific Ocean, where he is picked up by a fishing boat. After a long recovery period, he returns to flight testing.

1 March
The Lockheed RC-121D Constellation airborne early warning platform begins operating from Otis AFB, Maine, providing coverage to the continental east coast of the United States.

Trevor Gardner becomes the first Secretary of the Air Force for Research and Development.

2 March
Boeing demonstrates the 367-80 (future KC-135) with a modified in-flight refueling boom.

6 March
General Nathan F. Twining, Air Force Chief of Staff, reports that Atlas, Navaho, and Snark missile programs are being accelerated due to Soviet progress in missile technology.

6 April
A B-36 Peacemaker bomber launches a nuclear-tipped missile from 42,000 feet which detonates six miles above Yucca Flat, Nevada. This is the highest altitude that a nuclear device is detonated during testing.

21 April
An Aerobee-Hi sounding rocket is launched by the Air Force for the first time. The rocket reaches an altitude of 123 miles with a near-200-pound payload.

2 May
The Air Force approves the Western Development Division's proposal to build a second type of ICBM—the Titan.

10 May
The last of 448 Douglas C-124 Globemaster II transport aircraft are delivered to the Air Force. The C-124 was referred to by most of those who flew it as Old Shaky.

Right: *A U.S. Air Force Lockheed U-2. USAF U-2s were generally natural metal, while the CIA painted their secret spy aircraft black. (NASM)*

A GE XJ-79 turbojet engine is test flown in a B-45 jet bomber. The J-79 will power many future Air Force fighter and bomber planes.

15 May
Final details are reached concerning the construction of the Defense Early Warning (DEW) line. The American-Canadian agreement is aimed at protecting the North American continent from a surprise Soviet attack.

11 June
An experimental all-magnesium F-80C aircraft is delivered to Wright-Patterson AFB, Ohio. Its purpose is to test strength and weight of magnesium alloys in aircraft construction.

The first Atlas rocket is test fired this date.

29 June
The first operational Boeing B-52 Stratofortress (an RB-52B) arrives at Castle AFB, California, and enters service with the 93d Bombardment Wing.

1 July
The Air Force reactivates its research on weightlessness. Dr. S. J. Gerathewohl directs the program, which will later include airborne parabolic flight profiles to produce zero-gravity conditions.

11 July
The first USAF Academy class is admitted at its temporary location at Lowry AFB, Colorado. The Class of 1959 consists of 306 male students. The Air Force Academy is dedicated at that same location on the same day.

20 July
An Air Force NB-36H Peacemaker housing a nuclear reactor flies for the first time. The reactor is not activated.

1 August
The first zero-gravity research flights are carried out by Lockheed F-80 and T-33 aerial simulation platforms.

4 August
The Lockheed U-2 flies for the first time at Groom Lake, a secured airfield in central Nevada.

Crew from the B-29 special operations plane *Stardust 40* are released by the Chinese after being captured during the Korean War on January 13, 1953. These men were held as prisoners longer than any other captives in Korea.

8 August
In a bizarre in-flight accident, an X-1A explodes just before being dropped from its B-29 mother ship. Pilot Joseph A. Walker is saved, and the experimental plane is jettisoned from the B-29.

Left: *The Republic F-105 Thunderchief. ("Rolling Thunder," by Keith Ferris, USAF Art Collection)*

Below: *The Ryan X-13 Vertijet during one of its tethered test flights. (NASM)*

20 August
Colonel Horace A. Hanes, director of flight testing at the Air Force Flight Test Center, Edwards AFB, California, pilots a North American F-100 Super Sabre to a new speed record of 822.1 miles per hour over the Mohave Desert. This is the first absolute speed record set at high altitude. His record-setting flight earns him the Mackay Trophy.

5 October
The Air Force signs the initial production order for 29 Boeing KC-135 Stratotankers. More than 700 will eventually be built, with an additional 88 as different variants of the original model.

20 October
The last Strategic Air Command B-50 Superfortress is withdrawn from service. It is a B-50D from the 97th Bombardment Wing at Biggs AFB, Texas.

22 October
The Republic F-105 Thunderchief, piloted by Republic test pilot Rusty Roth, flies for the first time and easily breaks the sound barrier during the mission. The "Thud" will become one of the USAF's frontline fighters during the Vietnam War.

8–14 November
The Air Force adopts special procedures to streamline the approval process for the ICBM and IRBM programs that have been named by the President as the nation's highest priority. These are commonly called "Gillette Procedures," like the brand name of the most popular U.S. razor of the day.

18 November
The Bell X-2 research airplane makes its first powered flight when it is dropped from a specially modified Boeing EB-50A. The rocket plane reaches 627 miles per hour, the equivalent of Mach .95, during the sortie. Lieutenant Colonel Frank Everest pilots the X-2 during the mission.

4 December
Pioneer aviator Glenn L. Curtiss dies in Baltimore. He is 69 years old.

10 December
The Ryan X-13 Vertijet flies for the first time. Although designed to take off and land vertically, this first flight is flown using a set of temporary landing gear so that a traditional profile can be flown.

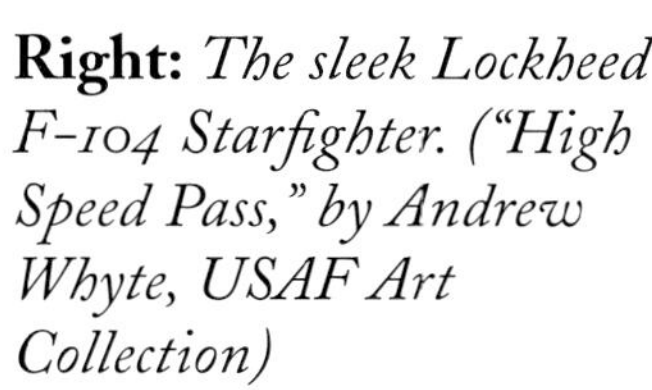

Right: *The sleek Lockheed F-104 Starfighter. ("High Speed Pass," by Andrew Whyte, USAF Art Collection)*

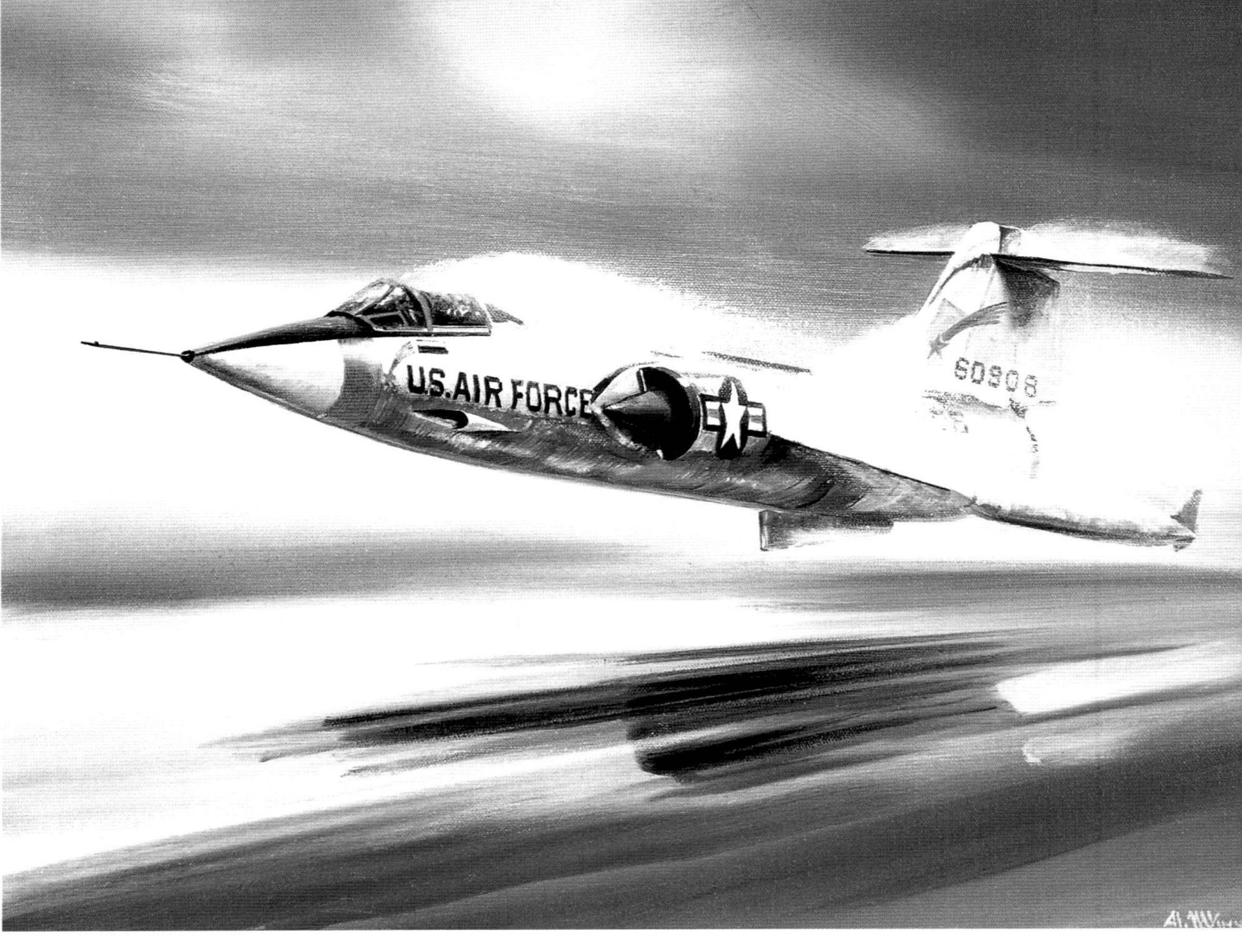

1956

1 January
The 1,000th Boeing B-47 Stratojet enters service with the Strategic Air Command.

17 January
It is revealed by Defense Department authorities that a semi-automatic ground environment (SAGE), an automated, electronic air defense system has been created. The system communicates via phone lines to combat centers where important information can be processed by large computers.

10 February
Marshal of the Royal Air Force, Lord Hugh Trenchard, dies at his home in London. Trenchard, "The Father of the Royal Air Force," was 83 years old.

17 February
The first production model F-104 Starfighter flies its first flight.

9 March
The Boeing B-52C Stratofortress flies for the first time. The B-52C, heavier than previous models, carries larger underwing tanks.

28 March
Airman D. F. Smith remains sealed inside the USAF's space cabin simulator for 24 hours.

23 April
The Douglas C-133A Cargomaster flies for the first time. Fifty of this type will be delivered to the USAF over the next two years.

7 May
The Air Force begins operating its "Texas Tower" early warning radar on Georges Bank, located 100 miles east of Cape Cod, Massachusetts.

21 May
Major David Crichlow, piloting a 93d Bombardment Wing B-52B Stratofortress at an altitude of 50,000 feet, delivers the first airborne hydrogen bomb, detonating in the test area on Bikini Atoll in the Pacific.

31 May
The first RB-57D reconnaissance aircraft is delivered to the 4080th Strategic Reconnaissance Wing located at Turner AFB, Georgia.

4 June
The Boeing B-52D flies for the first time near Wichita, Kansas. Although similar to the B-52C, the "D" is equipped only as a long-range bomber.

22 June
The USAF Reserve begins Operation SIXTEEN TON, a cargo airlift to the Caribbean. This operation marks the first sustained operation carried out by the USAF Reserve.

MISSILE EVOLUTION: ATLAS, THOR, TITAN, AND MINUTEMAN LAUNCHERS

Roger D. Launius

During the early 1950s all the armed services of the United States worked toward the fielding of intercontinental ballistic missiles (ICBM) that could deliver warheads to targets half a world away. Competition was keen among the services for a mission in the new "high ground" of space, whose military importance was not lost on the leaders of the world.

In April 1946 the Army Air Forces gave Consolidated Vultee Aircraft (Convair) Division a study contract for an ICBM. This led directly to the development of a succession of ballistic missiles in the 1950s. This was a revolutionary effort that gave humanity for the first time in its history the ability to attack one continent from another. And there was no effective defense. It shrank the size of the globe, and the United States—which had previously been protected from outside attack by two massive oceans—could no longer rely on natural defensive boundaries or distance from its enemies. During Dwight D. Eisenhower's two terms as President, the United States moved from a position of having essentially no space access to possessing ICBMs with a significant capability. The Air Force entered the race to build long-range ballistic missiles with dramatic flair during 1954, receiving formal approval for the rapid development of three major liquid-fueled weapons—Atlas, Titan, and Thor.

The Atlas was the first of these. It received high priority from the White House and hard-driving management from Brigadier General Bernard A. Schriever (1910–2005), a flamboyant and intense Air Force leader. Known as the SM-65 Atlas program, it initially went by the name "Weapon System 107A." This work officially began in February 1954. The first Atlas rocket was test fired on June 11, 1955, and a later generation rocket became operational in 1959.

At first many engineers believed Atlas to be a high-risk proposition. To limit its weight, Convair engineers under the direction of Karel J. Bossart (1904–1975), a pre–World War II immigrant from Belgium, designed the booster with a very thin, internally pressurized fuselage instead of massive struts and a thick metal skin. The "steel balloon," as it was sometimes called, employed engineering techniques that ran counter to the conservative engineering approach used by other designers.

A second ballistic missile, Thor, dates from January 1956 when the Air Force began developing it as a 1,500 mile intermediate-range ballistic missile (IRBM). The Thor program progressed rapidly under contract to the Douglas Aircraft Corp., and the first flight of the Thor IRBM took place on January 25, 1957. In August 1958 the USAF's first operational squadron entered service in Europe. The Thor was a stopgap measure, however, and once the first generation of ICBMs based in the United States became operational, the Thor missiles were quickly retired. The last of the missiles was withdrawn from operational alert in 1963. With modification and the addition of an upper stage, it remained a workhorse in America's fleet of launchers under the name Thor-Delta (later as Delta), propelling moderate payloads into orbit until the twenty-first century. It launched, among other payloads, the first CORONA satellite reconnaissance spacecraft into orbit beginning in 1960.

The Titan ICBM program emerged not long after Atlas, and proved to be an enormously important ICBM effort and later a civil and military space launch asset. To consolidate ballistic missile development efforts, Secretary of Defense Charles E. Wilson (1886–1972) issued a decision on November 26, 1956, that effectively took the Army out of the ICBM business and assigned responsibility for land-based systems to the Air Force and sea-launched missiles to the Navy. The Air Force began what became the Titan family of boosters in October 1955 by awarding the Glenn L. Martin Company a contract to build an intercontinental ballistic missile (ICBM). This missile had been ordered as a backup to the Atlas ICBM then in development—the Air Force had some nagging doubts about some of the technology being incorporated into Atlas. The new missile became known as the Titan I, the nation's first two-stage ICBM. Designed to be based in underground silos, the USAF deployed 54 Titan Is followed by 54

Right: *"Minuteman II Launch, Vandenberg AFB, California," by Arthur Kane. (USAF Art Collection)*

improved Titan IIs. The first Titan II ICBMs were activated in 1962, and modified Titan IIs were selected to launch NASA's Gemini spacecraft into orbit during the mid-1960s. As a result of arms and nuclear reduction treaties, the Titan II weapon system was deactivated during the mid-1980s, with the missiles being removed and the silos destroyed. As a result of this decommissioning, several of the old Titan II ICBMs found use as space launchers in the post-*Challenger* era of commercial space operations in the 1990s.

While Atlas, Thor, and Titan were important steps forward in ballistic missile capability, their liquid-fuel propulsion system took too much time to prepare for launch. A solution was a solid-propellant system that obviated the need for constant servicing. Hence, the silo-based, three-stage Minuteman became America's workhorse ICBM. Approved for development in September 1958, the USAF put on alert its first 10 LGM-30A/B Minuteman missiles at Malmstrom Air Force Base, Montana, in October 1962. This Minuteman I, which had a single nuclear warhead, was superseded by Minuteman II and III missiles that could carry up to three independently targeted nuclear warheads. In all, 550 Minuteman IIIs were deployed in the United States beginning in 1970. This ICBM remains on alert, a total of 500 single reentry vehicle Minuteman IIIs serving as the nation's ICBM deterrent through 2020.

4 July
The first U-2 overflight of Soviet territory occurs this date. The spy plane overflies Minsk, Leningrad, and the Baltic in search of Soviet bomber aircraft and then returns to Wiesbaden AB, Germany.

15 July
Sixteenth Air Force is activated in Torrejon, Spain.

18 July
The last Boeing KC-97G is completed at the Boeing factory in Renton, Washington.

23 July
Lieutenant Colonel Frank K. Everest pilots the Bell X-2 rocket-powered research plane to set a speed record just over 1,900 miles per hour. He climbs to 75,000 feet during the sortie flown over Edwards AFB, California.

27 August
The first static firing of the Thor rocket engine takes place at the Air Force Flight Test Center at Edwards AFB, California.

31 August
The Boeing KC-135 Stratotanker production model flies for the first time near Renton, Washington.

7 September
Captain Iven C. Kincheloe pilots the Bell X-2 rocket research plane to a new altitude record for manned flight—126,200 feet. Kincheloe is the first man to fly an aircraft above 100,000 feet. For this flight over Edwards AFB, Kincheloe receives the Mackay Trophy.

15 September
The Air Force's first Tactical Missile Wing is activated as part of Twelfth Air Force and is located at Hahn AB, Germany. The 701st TMW is to be equipped with the Matador missile.

20 September
The first Jupiter C three-stage missile is launched at Cape Canaveral, Florida. It reaches an altitude of 680 miles and travels 3,300 miles downrange.

27 September
Captain Milburn G. Apt is killed after setting a new speed record of 2,094 miles per hour. Launched from a B-50 bomber, Apt quickly reached Mach 3.2 as he descended through 65,000 feet, becoming the first man to fly in excess of three times the speed of sound. Apt lost control of the last surviving X-2 during a turn back toward Edwards AFB and could not recover. He attempted to eject but was killed when the escape capsule impacted the desert floor.

20 October
Lawrence Bell, developer of America's first jet aircraft, the Bell XP-59, and the first aircraft to break the sound barrier, the Bell X-1, dies in Buffalo, New York.

26 October
Bell pilot Floyd Carson pilots the first flight of the XH-40 helicopter at Fort Worth, Texas. The helicopter will later be designated at the UH-1 Iroquois—more commonly known as the Huey.

31 October
British, French, and Israeli aircraft attack Egypt as the Suez Crisis erupts into open warfare.

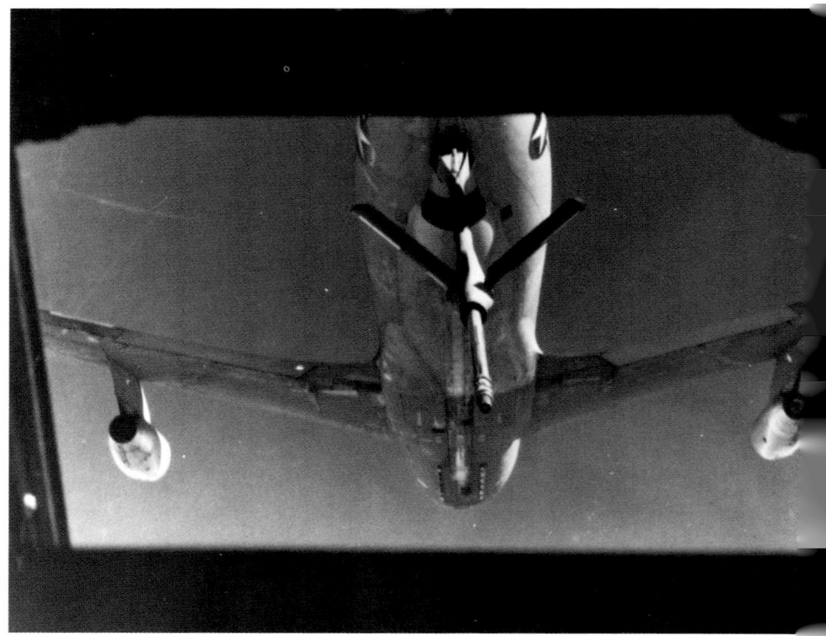

Opposite, top: *The Bell UH-1 Iroquois helicopter would be flown by all services and several foreign countries during its operational life. This UH-1H "Huey" is on display at the National Air and Space Museum, Steven F. Udvar-Hazy Center. (NASM)*

Opposite, bottom: *Pre-contact position behind a KC-135 Stratotanker. (NASM)*

Right: *The Convair B-58 Hustler configured without its massive centerline pod. Early flight tests were accomplished without it. (NASM)*

November
During this month, in support of British and French operations against Egypt over the Suez Canal, the first operational U-2 missions are flown monitoring military activity in Britain, France, and Israel. USAF aircraft are deployed to respond quickly should hostilities worsen.

6 November
The Air Force's first full-scale Navaho ramjet ICBM is launched from Cape Canaveral, Florida. The missile breaks apart after nearly 30 seconds in flight.

11 November
After several redesigns to incorporate the fuselage-narrowing "area rule" into the plans, the Convair B-58 Hustler makes its first flight at the factory airfield in Fort Worth, Texas. Piloted by B. A. Erickson with John D. McEachern as systems specialist and Charles P. Harrison as flight test engineer, the Hustler flies to a speed of Mach .9. It was not fitted with the massive centerline pod that became a recognizable fixture on most B-58s.

16 November
The Department of Defense transfers a portion of Camp Cooke, California, to the Air Force. This parcel of land becomes Vandenberg AFB and will become the first USAF ICBM base.

26 November
The USAF is given operational jurisdiction over long-range missiles by Secretary of Defense Charles E. Wilson.

30 November
The Martin Matador (TM-61) is certified operational after its final test flight. The medium-range Matador cruises at 35,000 feet at a speed of 650 miles per hour. The Matador is the Air Force's first operational tactical missile—an early form of the now familiar cruise missile.

9 December
The first C-130 Hercules is delivered to the 463d Troop Carrier Wing. The C-130 is capable of flights of 2,500 miles, while carrying either 25 tons of cargo or 92 troops and their equipment. It can take off and land on a 4,000-foot unprepared runaway.

11 December
Operation SAFE HAVEN begins. During the next seven months, Military Air Transport Service aircraft will airlift more than 10,000 Hungarian refugees from West Germany to the United States, where they are granted asylum. The crisis had resulted when Communist troops crushed an anti-Communist revolution in their country.

21 December
Major Arnold I. Beck ascends to a simulated altitude of 198,770 feet, the highest ever, inside an Air Research and Development Command altitude chamber located at Dayton, Ohio.

26 December
The Convair YF-106 Delta Dart flies for the first time at Edwards AFB, California.

1957

16–18 January
Lucky Lady III, a B-52 assigned to the 93d Bombardment Wing, and two additional B-52s make the world's first nonstop, around-the-world, jet flight in history. Major General Archie J. Old Jr. acts as mission commander aboard the lead aircraft, which is piloted by Lieutenant Colonel James H. Morris. The three mission aircraft and two spares depart from Castle AFB, California, on January 16, and after the second refueling, the spares are released. The three B-52s continue the journey with an additional three aerial refuelings. The flight covers 24,325 miles and takes 45 hours and 19 minutes to complete. For demonstrating SAC's ability "to strike any target on the face of the earth," the 93d Bombardment Wing is awarded the Mackay Trophy and each member of the flight crew receives the Distinguished Flying Cross.

25 January
The first attempt made by the Air Force to launch its Thor intermediate-range ballistic missile (IRBM) fails. The Thor missile took thirteen months from production contract to finished missile. The first successful launch will occur in August.

18 February
Continuing its tradition of philanthropy, the Guggenheim Foundation awards $250,000 to the Harvard University Aviation Health and Safety Center.

10 March
The NACA Lewis Laboratory begins research on the ion engine that electrically charges an inert gas and accelerates it to a speed of approximately 30 kilometers per second. When these ions are emitted as exhaust from a spacecraft at this speed, the craft is propelled in the opposite direction.

27 March
The McDonnell F-101B Voodoo, the two-seat version of the original Voodoo, flies for the first time.

1 April
Strategic Air Command begins to transfer its remaining fighter wings to the control of Tactical Air Command. By July, all seven wings will make the migration to Tactical Air Command.

11 April
The Ryan X-13 Vertijet VTOL research aircraft becomes the first jet aircraft to take off vertically, transition to conventional level flight, and land vertically.

19 April
The Douglas Thor (XSM-75) is successfully launched for the first time at Cape Canaveral, Florida, but is destroyed by the range safety officer during the flight.

6 May
William M. Holaday is named Department of Defense Special Assistant for Guided Missiles.

2 June
Captain Joseph W. Kittinger Jr. remains aloft in a lighter-than-air craft—*Man High I*—for six hours and 34 minutes in the skies over Minnesota. Two hours are spent at altitudes above 96,000 feet. Kittinger establishes a new altitude/endurance record for a manned lighter-than-air aircraft during this mission. This is the first solo balloon flight into the stratosphere.

The first operational Lockheed U-2 spy plane arrives at Laughlin AFB, Texas, where it is assigned to the 4080th Strategic Reconnaissance Wing.

11 June
The 4080th Strategic Reconnaissance Wing located at Laughlin AFB, Texas, accepts its first Lockheed U-2. During these early years, the CIA painted their U-2s black while USAF U-2s retained their factory aluminum finish.

28 June
The first operational Boeing KC-135 Stratotanker arrives at Castle AFB, California, and is assigned to the 93d Air Refueling Squadron.

1 July
The first USAF intercontinental ballistic missile (ICBM) wing is activated at Cooke (later Vandenberg) AFB, California. The unit is designated as the 704th Strategic Missile Wing (SMW) and is armed with the Northrop Snark.

Opposite: Lucky Lady III *and her two B-52 wingmen rest after more than 24,000 nonstop miles in flight. (NASM)*

Above: *A Genie air-to-air nuclear rocket ready for loading on the Convair F-106 Delta Dart in the background. (NASM)*

Far East Air Forces, best known as FEAF to those Korean War veterans who served in the theater, is renamed Pacific Air Command (PACAF) and its headquarters moved to Hickam AFB, Hawaii.

General Thomas D. White becomes Chief of Staff, United States Air Force.

10 July
The Convair B-58 Hustler is unveiled to the American public for the first time.

13 July
President Dwight D. Eisenhower becomes the first President to fly in a helicopter, a USAF Bell UH-13J, from the White House to a secret meeting held at an undisclosed location.

19 July
A Northrop F-89J Scorpion fires a Douglas MB-1 Genie missile. The Genie is a nuclear-armed, air-to-air missile and this marks the first launch of such a weapon by an aircraft.

24 July
The Distant Early Warning (DEW) line, a series of early radar warning stations across the Canadian Arctic region, becomes operational.

15 August
General Nathan F. Twining is appointed to serve as the Chairman of the Joint Chiefs of Staff. The first Air Force officer to hold that post, Twining is the USAF Chief of Staff from 1954–1956.

19–20 August
Major David G. Simmons sets a new world altitude record for a balloon ascent in *Man High II*. The balloon is launched from Crosby, Minnesota, and reaches a peak altitude of 101, 516 feet, landing after 32 hours aloft in Elm Lake, South Dakota.

3 September
The NACA submits a *Study of the Feasibility of a Hypersonic Research Plane* to the Air Force.

4 September
The Lockheed CL-329 Jetstar flies for the first time. This utility transport plane will enter service with MATS as the C-140 and would be used to test and evaluate navigation aids and communications.

20 September
The first completely successful launch of a Thor IRBM takes place at Cape Canaveral, Florida.

1 October
A Snark intercontinental ballistic missile is launched from Cape Canaveral by an Air Force crew for the first time.

4 October
Sputnik, the world's first artificial space satellite, is launched into orbit by the Soviet Union. This event marks the beginning of the Space Age. *Sputnik I* ceases transmission of its radio signal on October 26 and remains in orbit until January 4, 1958.

20 October
The Air Force launches a three-stage, solid-fuel rocket from a balloon that is 19 miles above the earth. Project Far Side, sponsored by the Air Force Office of Scientific Research, is an attempt to gather data on cosmic rays and other space phenomena to an altitude of 4,000 miles.

24 October
Air Force Air Research and Development Command (ARDC) issues a proposal calling for the development of a hypersonic glide rocket weapon system (WS 464L). This will become the Dyna-Soar project (Dynamic Soaring).

7 November
President Eisenhower announces the creation of a Special Assistant to the President for Science and Technology and appoints Massachusetts Institute of Technology (MIT) president James R. Killian to the post.

13 November
Air Force Vice Chief of Staff, General Curtis E. LeMay pilots a Boeing KC-135 Stratotanker, establishing a new nonstop distance record of 6,350 miles when he travels from Westover AFB, Massachusetts, to Buenos Aires, Argentina. On the return trip, LeMay sets a Buenos-Aires-Washington, D.C., speed record flying the 5,200 miles in 11 hours and five minutes. He receives the Distinguished Flying Cross upon landing.

21 November
Francis E. Warren AFB, Wyoming, is selected as the location for the construction of the first dedicated Air Force ICBM base.

The NACA, under the leadership of retired General Jimmy Doolittle, forms a special committee on space technology.

27 November
Operation SUN RUN is executed by RF-101C Voodoos of the 363d Tactical Reconnaissance Wing. Over the course of the day, three transcontinental speed records are shattered by careful planning and the use of KC-135 Stratotankers to refuel at high altitudes and at Mach .8.

Both the Thor and Jupiter IRBMs are ordered to production. The missiles will eventually be assigned to the Air Force.

29 November
Both intercontinental and intermediate ballistic missile programs are assigned to Strategic Air Command by the USAF Chief of Staff, General Thomas D. White.

12 December
Major Adrian Drew sets a new absolute speed record flying a McDonnell F-101A Voodoo to an average speed of 1,207.6 miles per hour. The flight takes place at Edwards AFB, California, during Operation FIREWALL.

15 December
The first USAF operational Snark (SM-62) squadron is activated at Patrick AFB, Florida. The unit is designated as the 556th Strategic Missile Squadron (SMS).

17 December
The Air Force launches its first Atlas intercontinental ballistic missile (ICBM), which successfully delivers its reentry cone into the designated target area some 500 miles downrange.

This date marks the fifty-fourth anniversary of the Wright brothers' first flight.

19 December
A Thor intermediate-range missile is test launched for the fourth time. The test is the first fully guided missile flight that uses an all-inertial guidance system.

23 December
The USAF awards a contract for the production of an intercontinental, Mach 3 strategic bomber. North American Aviation will begin to design and develop the B-70 Valkyrie.

Opposite: *Portrait of General Nathan F. Twining by Robert Brackman. (USAF Art Collection)*

Below: *The McDonnell F-101 Voodoo, extremely fast at low altitude, developed into a nuclear strike aircraft. A reconnaissance version, the RF-101, provided tactical support and proved effective at locating missile sites in Cuba. (USAF)*

Left: *A Jupiter-C launch vehicle lifts* Explorer 1 *into orbit. (NASM)*

Opposite: *An F-100 Super Sabre blasts off from a tactical launcher using a rocket booster. (NASM)*

1958

1–15 January
The first Air Force unit assigned to develop and train to use the Bomarc unmanned supersonic interceptor missile is activated at Cooke AFB, California. The 672d Strategic Missile Squadron (SMS) along with the 4751st Air Defense Missile Wing and the 864th SMS, equipped with the Jupiter IRBM, are all activated during this period.

29 January
The Department of Defense announces plans to create the National Pacific Missile Range as part of the Naval Air Missile Test Range at Point Magu, California, to be used for long-range missile tests.

31 January
The first orbital American satellite, *Explorer I*, is launched atop a Jupiter-C rocket from Cape Canaveral, Florida. It carries an experiment devised by James A. Van Allen that, when completed, reveals the radiation belt around the earth.

1 February
The Strategic Air Command (SAC) activates its first Atlas Missile Wing, the 706th SMW, located at Francis E. Warren AFB, Wyoming.

7 February
The Advanced Research Projects Agency (ARPA) is established by the Department of Defense to assume leadership of the nation's outer space program.

18 February
At the Arnold Research Development Center in Tullahoma, Tennessee, USAF scientists using wind tunnels create an airflow speed of approximately 32,400 miles per hour for one-tenth of a second.

17 March
America's second satellite, *Vanguard I*, is launched into orbit from Cape Canaveral. It operates on solar-powered batteries and has a 1,000-year life expectancy. Data returned from the satellite proves that the earth has a slight pear shape.

21 March
On the Holloman AFB sled test track, a two-stage rocket propels the sled to speeds in excess of 2,700 miles per hour.

26 March
A North American F-100D Super Sabre fitted with an Astrodyne rocket motor is launched from a rail launcher for the first time.

27 March
The Air Force Ballistic Missile Division is tasked by ARPA to carry out three lunar probes using a Thor-Vanguard system.

2 April
President Dwight D. Eisenhower proposes the formation of a National Aeronautics and Space Agency that would incorporate the NACA as well as conduct civilian space programs and assist military space initiatives.

5 April
The Air Force successfully launches an Atlas ICBM from Cape Canaveral, Florida, to an impact area located 600 miles downrange.

8 April
An Air Force KC-135 Stratotanker establishes a nonstop, unrefueled jet flight record during a mission from Tokyo, Japan, to Lajes Field, Azores. The flight covers 10,288 miles.

1 May
Unexpected data gathered from the orbital *Explorer* satellites reveals a high-intensity band of radiation extending from 600 miles to 8,000 miles above the earth. Dr. James Van Allen describes the radiation as "1,000 times as intense as could be attributed to cosmic rays." This comes to be known as the Van Allen Radiation Belt.

7 May
Major Howard C. Johnson establishes a new altitude record when he pilots an air-breathing F-104A Starfighter to an amazing 91,243 feet.

12 May
North American Air Defense Command (NORAD) is formally established to protect the North American continent from short notice enemy attacks. This international American-Canadian Command is headquartered in Colorado Springs, Colorado.

16 May
Captain Walter W. Irwin sets a new absolute speed record, piloting a Lockheed F-104A Starfighter to an average speed of 1,404.2 miles per hour.

BERNARD A. SCHRIEVER (1910–2005)

Walter J. Boyne

General Bernard A. "Bennie" Schriever ranks with the legendary figures of Hap Arnold and Curtis LeMay in terms of his contributions to the United States Air Force. Schriever is most appreciated for his development and acquisition of a reliable force of intercontinental ballistic missiles, an almost unimaginably difficult achievement. Yet he also recognized that the missile technology would have direct application to space. He is the father of both the American ICBM and the American space programs.

Schriever was born in Bremen, Germany, on September 14, 1910, and as a young boy watched with fascination as German Zeppelins passed overhead to bomb England. His father, an engineering officer on a German ship line, had been interned in the United States, and his mother brought Bernard and his brother to America in January 1917.

His family moved to Texas, where his father was killed in an industrial accident in 1918. His mother worked to bring up the two children, and Schriever became a naturalized citizen in 1923. Graduating near the top of his class at Texas A&M in 1931, he earned a commission as a 2nd Lieutenant in the Field Artillery. In 1932 he reverted from officer to aviation cadet status and graduated in June 1933 from Kelly Field. His first assignment was at March Field, flying Keystone B-4 and Martin B-10 bombers under the command of Lieutenant Colonel Henry H. Arnold.

Schriever flew the B-4 and the Douglas O-38 on the hazardous Salt Lake City to Cheyenne air mail route in 1934. Temporarily forced to reserve status in 1935, he ran a Civilian Conservation Corps camp in New Mexico. Restored to active duty in 1936, he flew a Boeing P-12 at Albrook Field in Panama. Budget cuts bumped him out of the service again, and he became a pilot with Northwest Airlines in 1937. In October, 1938, he won a hotly contested competition for 20 regular officer slots, and was sworn in as a second lieutenant—for the third time.

After a brief stint at Hamilton Field, he took on test pilot duties at Wright Field and graduated from the Air Corps Engineer School in June 1941. He received one of the rare assignments to a civilian graduate school, earning a master's degree in mechanical engineering (aeronautics) from Stanford University, in June 1942.

Schriever flew 38 combat missions in B-17s, B-25s, and C-47s in the South Pacific, becoming an engineering officer for General George C. Kenney, commander of the Fifth Air Force. A colonel by 1943, he kept his engineering headquarters in the forefront of the fighting, once landing at the Manila airport while the fighting was still going on.

After the war, General Arnold made use of Schriever's well-proven engineering and management skills by appointing him Chief, Scientific Liaison Section, Deputy Chief of Staff, Materiel. Here he could mix with the brilliant scientists who made up Arnold's revolutionary and decisively important Scientific Advisory Board.

Working with such legendary scientists and leaders as Simon Ramo, John von Neumann, and Trevor Gardner, Schriever was assigned the task of

creating an intercontinental ballistic missile force. On August 2, 1954, Schriever took command of the Western Development Division in Inglewood California. Here he mustered the military/academic/scientific/industrial team whose task eventually surpassed the Manhattan project in size, funding, and importance. If the Manhattan project had failed, Japan would still have been defeated. But with the Soviet Union's advances in nuclear and missile technology, the development of a U.S. ICBM was an absolute necessity.

The successful October 4, 1957, launch of *Sputnik* dealt a tremendous blow to American prestige and morale. However, it was a stroke of good fortune for Schriever, for it began an era of almost unlimited funding allowing him and his team to meet their challenges.

Schriever elicited almost miraculous achievements from his organization, fielding no less than three operational ICBM systems (Atlas, Titan, and Minuteman) and the Thor intermediate-range ballistic missile within a span of eight years.

Despite the end of the Cold War, there still remains a requirement for ballistic missile systems. More importantly, Schriever's achievements paved the way to the exploitation of space. Our modern war-fighting capability, so dependent upon the modern network of intelligence, meteorological, communication and navigation satellites, owes its existence to the brilliant work of Schriever and his team. He was given the signal honor in 1998 of having an Air Force base named after him while still living.

General Schriever died on June 20, 2005, in Washington, D.C. He was 94 years old.

Opposite: *"First Silo Shot, Vandenberg AFB, May 1961," by Nixon Galloway. (USAF Art Collection)*

Above: *Lieutenant General Bernard A. Schriever was the man most responsible for the success of the American ICBM program. (USAF)*

Left: *The Bell X-14 vertical flight research plane. (NASM)*

Opposite: *A North American T-39 Sabreliner—the Air Force used these as an executive transport, cargo carrier, and radar training aircraft. (NASM)*

24 May
The Bell X-14 makes its first transition from hover to horizontal flight, taking approximately 30 seconds to do so. The X-14 was built from the fuselage of a Beech T-34 and the wings of a Beech Bonanza. It is the only open cockpit X-plane. The Air Force flew this vertical flight testbed until 1960, when it was transferred to NASA.

Captain E. L. Breeding experiences up to 83 "Gs" for a fraction of a second during rocket sled testing at Holloman AFB, New Mexico.

27 May
The 335th Tactical Fighter Squadron accepts the first group of F-105B Thunderchiefs at Eglin AFB, Florida.

The YF4H-1, the prototype for the F-4 Phantom II, flies for the first time, piloted by Robert C. Little.

3 June
The Air Force and NACA jointly announce the details of an inertial guidance system that will be used on X-15 hypersonic research aircraft to control correct pitch attitude for atmospheric reentry during high-altitude testing.

4 June
The Air Force launches a Thor missile from a "tactical-type" launcher at Cape Canaveral, Florida.

16 June
The Phase I Dyna-Soar boost-glide orbital spacecraft development contract is awarded to the Martin Company, the lead contractor, and the Boeing Company as a major subcontractor.

27 June
At Cape Canaveral, the first military launch of a Snark intercontinental missile is carried out by members of the 556th Strategic Missile Squadron.

30 June
The NACA reports that a full 50 percent of its research is now related to solving the problems of missiles and other types of space vehicles.

14–15 July
Composite Air Strike Force Bravo is sent to Lebanon following a request from the Lebanese president, who feared a coup. During Operation BLUE BAT, U.S. airlift moves some 2,000 troops from Germany to the Middle East in support of this Tactical Air Command operation.

26 July
Captain Iven C. Kincheloe is killed in the crash of an F-104 Starfighter at Edwards AFB, California.

1 August
High over Johnson Island in the Pacific, a nuclear anti-ICBM missile is detonated in an effort to assess whether such a weapon might be able to destroy or damage an incoming enemy ICBM.

2 August
The Air Force launches an Atlas missile that uses a full-power profile for both the sustainer and booster engines.

6 August
The Air Force contracts with Rocketdyne Division of North America to develop a one-million-pound thrust rocket engine.

19 August
Dr. T. Keith Glennan is sworn in as the Administrator of the National Aeronautics and Space Administration (NASA), which will begin activities on October 1. Dr. Hugh L. Dryden is sworn in as his Deputy Administrator. They will hold the first NASA senior staff meeting on September 24.

21 August
The National Advisory Committee for Aeronautics (NACA) holds its final meeting under the gavel of chairman, Jimmy Doolittle, as its replacement, NASA, gears up for operations.

23 August
Congress creates the Federal Aviation Administration (FAA). The FAA will assume responsibility for military and civilian air traffic control and establishing the locations of airports and missile sites. In September, retired General Elwood "Pete" Quesada will be appointed by President Eisenhower to head the organization.

2 September
An Air Force C-130 on an ELINT (electronic intelligence) mission along the Turkish-Soviet border is attacked and shot down by five Soviet MiGs.

3–9 September
Beginning the last week in August, a Tactical Air Command force deploys to the Far East, responding when Red China threatens the Taiwan Straits. During Operation XRAY TANGO, the composite strike force consisting of F-100 Super Sabres, F-101 Voodoos, B-57 Canberras, and C-130 Hercules aircraft earns the Mackay Trophy for its swift and effective accomplishment.

9 September
A Lockheed X-7 pilotless ramjet test platform reaches Mach 4 after being air-launched from a Boeing EB-50.

16 September
The North American NA-246 jet transport plane flies for the first time. This six-seat carrier will fly for the USAF as the T-39 Sabreliner.

19 September
The first Kaman H-43A Huskie flies for the first time. Eighteen Huskies will join the USAF inventory as a firefighting and crash recovery helicopter at Tactical Air Command bases.

Left: *The B-52G was capable of carrying two GAM-77/AGM-28 Hound Dog missiles—one under each wing. The Hound Dog was a long-range, air-to-ground, low-yield nuclear weapon with an effective range between 200 and 700 miles, depending upon its launch altitude. (USAF)*

24 September
Launched by semiautomatic controls in Kingston, New York, a Bomarc missile located at Cape Canaveral destroys an incoming target drone flying at 48,000 feet and 1,000 miles per hour. The successful intercept occurs 75 miles from the launch point in Florida.

1 October
The National Aeronautics and Space Administration (NASA) officially begins work, replacing the National Advisory Committee for Aeronautics (NACA), which had limited authority or expertise in issues pertaining to space exploration.

26 October
The Boeing B-52G flies for the first time. This variant is capable of carrying two AGM-28 Hound Dog missiles, one under each wing. The first B-52G is delivered to the 5th Bombardment Wing located at Travis AFB, California, in February 1959.

1 November
The USAF Kaman H-43B flies for the first time. This helicopter is powered by a Lycoming T53-L-1A shaft-turbine engine. More than 200 will be used around the globe; in 1962 they will be redesignated as the HH-43B.

8 November
The third and last attempt by the Air Force to launch a lunar probe is made from Cape Canaveral, Florida. The third stage fails to ignite and, after reaching an altitude of 1,000 miles, the *Pioneer 2* falls back to earth and burns up over East Central Africa. This marks the last authorized moon shot made by the USAF.

28 November
The Air Force makes its first operational Atlas missile test flight that covers more than 6,300 miles and lands near its intended target.

3 December
The Jet Propulsion Laboratory, since World War II under the guidance of Caltech, is transferred to NASA by President Eisenhower.

16 December
A Thor missile is launched in the new Pacific Missile Test Range, the first ballistic missile fired over the waters of the Pacific Ocean. Another Thor is fired from Cape Canaveral at the same time.

A Military Air Transport Service (MATS) C-133 Cargomaster lifts 117,900 pounds of cargo to an altitude of 10,000 feet, setting a new record for weight carried to that altitude.

18 December
The first artificial communications satellite is launched into orbit by a four-ton Air Force Project SCORE Atlas rocket. The next day, broadcasts begin of a taped Christmas message from the President of the United States, Dwight D. Eisenhower—the first human voice heard from space.

23 December
The first Atlas-C missile is successfully test fired at Cape Canaveral, Florida.

Right: *The Consolidated B-36 Peacemaker was retired from service in February 1959. (USAF)*

Below: *A two-stage Titan I leaves the launch pad. The Titan was capable of striking targets more than 5,500 miles away. (NASM)*

1959

4 January
The Pacific Missile Range and Vandenberg AFB, California, are declared operational for missile test firings.

6 January
Fidel Castro becomes the Cuban premier after an uprising overthrows the government.

21 January
Although an Army weapons system at this time, a Jupiter intermediate-range missile strikes its target some 1,700 miles away from the launching point.

1 February
The Royal Canadian Air Force assumes control of the DEW line (Distant Early Warning) from the United States Air Force.

6 February
The first Titan I ICBM is successfully launched by the USAF. The Titan is a two-stage, liquid-fueled rocket capable of reaching targets some 5,500 miles away.

12 February
The last B-36 Peacemaker is retired from active service. Strategic Air Command becomes an all-jet bomber force.

17 February
Dr. J. Allen Hynek, Associate Director of the Smithsonian Astrophysical Observatory at Cambridge, Massachusetts, recommends to the Air Force that they should take an active and positive approach to investigation of all UFO sightings, use scientific means to determine what caused the sightings, and keep the public informed of existing policy concerning the phenomena.

19 February
A two-stage rocket research sled fired at Holloman AFB, New Mexico, attains a top speed of 3,090 miles per hour, roughly four times the speed of sound.

28 February
The *Discoverer I* satellite is successfully launched from Vandenberg (formerly Cooke) AFB, California, into polar orbit using a Thor-Hustler launch system. This is the first satellite launched into polar orbit and the first launched into orbit from Vandenberg.

10 March
The first X-15 hypersonic research plane captive flight is made with test pilot A. Scott Crossfield in the

L. GORDON COOPER JR., VIRGIL "GUS" GRISSOM, AND DONALD K. "DEKE" SLAYTON: THE FIRST THREE USAF ASTRONAUTS

Roger D. Launius

Three of the first seven of America's astronauts—the Mercury Seven selected in April 1959—came from the ranks of the United States Air Force, L. Gordon Cooper Jr., Virgil I. "Gus" Grissom, and Donald K. "Deke" Slayton. The selection had been grueling, for NASA had pursued a rigorous process of winnowing candidates that involved record reviews, biomedical tests, psychological profiles, and a host of interviews. The initial candidates included five Marines, 47 Navy aviators, and 58 Air Force pilots, and the final selectees included the three Air Force officers, three naval aviators, and one Marine. The three Air Force selectees went on to make a significant impact on the history of human spaceflight.

The first to fly was Captain Virgil I. Grissom, known to all as Gus. The second American into space, Grissom repeated Alan Shepard's 15-minute sub-orbital flight on July 21, 1961. Unfortunately, this mission ended with Grissom being rescued from drowning after the hatch to his *Liberty Bell 7* spacecraft unexpectedly blew off after splashdown. The two suborbital Mercury flights proved valuable for NASA technicians, who found ways to solve or work around literally thousands of obstacles to successful spaceflight. Grissom also commanded the maiden flight of the Gemini program, *Gemini III,* on March 23, 1965, with John W. Young, a naval aviator chosen as an astronaut in 1962, accompanying him. Grissom died in the *Apollo I* capsule fire on January 27, 1967, during a ground test where he and his two crewmates, Edward White and Roger Chaffee, were practicing for what would have been the first piloted flight of the Apollo spacecraft in Earth orbit. At 6:31 p.m., after several hours of work, a fire broke out in the spacecraft, and the pure oxygen atmosphere intended for the flight helped it burn with intensity. In a flash, flames engulfed the capsule and the astronauts died of asphyxiation. It took the ground crew five minutes to open the hatch. As the nation mourned the loss of the three astronauts, NASA appointed an eight-member investigation board, and it set out to discover the details of the tragedy: what happened, why it happened, could it happen again, what was at fault, and how could NASA recover? The members of the board learned that the fire had been caused by a short circuit in the electrical system that ignited combustible materials in the spacecraft fed by the oxygen atmosphere. They also found that it could have been prevented and called for several modifications to the spacecraft, including a move to a less oxygen-rich environment. Changes to the Apollo capsule followed quickly, thanks to the efforts of a dedicated team of engineers. Within a little more than a year it was ready for flight.

Gordon Cooper, better known as "Gordo" to his colleagues, flew the last mission of the Mercury program, a 22-orbit flight on May 15–16, 1963, setting the stage for the Gemini program. Cooper also commanded the *Gemini V* mission on August 21–29, 1965, where he and Pete Conrad set a spaceflight endurance record with an eight-day orbital mission. Cooper left NASA in 1970 and worked in a variety of aerospace enterprises. He died at age 77 at his home in Ventura, California, on October 4, 2004.

Deke Slayton was originally scheduled to pilot the final Mercury mission but had to be removed from flight status because of a heart condition in August 1959. Early on, he took the lead for the Mercury Seven and later officially headed the astronaut office. Accordingly, he assigned crews for each mission and oversaw the full range of astronaut activities. Slayton returned to flight status in March 1972, and flew as a member of the crew of the Apollo-Soyuz Test Project (ASTP) in 1975. This mission was the first human spaceflight mission managed jointly by two nations and served an important Cold War objective of demonstrating détente between the U.S. and the Soviet Union. It tested the compatibility of rendezvous and docking systems for American and Soviet spacecraft and suggested that future international ventures in space had potential. Thereafter, Slayton directed the space shuttle flight research effort, retiring from NASA in 1982. He died on June 13, 1993, from brain cancer.

Collectively, these three USAF astronauts played exceptionally significant roles in the development of the nation's space program. Slayton oversaw the entire effort for its first 23 years, Grissom was the quintessential test pilot who flew shakedown missions for both the Mercury and Gemini programs, and Cooper flew the first long-duration space missions, demonstrating the ability of humans to survive in the exceptionally harsh environment of space.

Opposite: *Of the Mercury Seven astronauts, three were Air Force pilots: Virgil I. "Gus" Grissom (center, back row), Donald K. "Deke" Slayton (second from left, front row), and L. Gordon Cooper (far right, back row). (NASM courtesy of NASA)*

Above: *The original Mercury Seven astronauts with a USAF F-106B jet aircraft. From left to right: M. Scott Carpenter, Leroy Gordon Cooper, John H. Glenn Jr., Virgil I. Grissom Jr., Walter M. Schirra Jr., Alan B. Shepard Jr., and Donald K. Deke Slayton. (NASA)*

Left: *A B-52 Stratofortress fires an AGM-28 Hound Dog air-to-ground missile. B-52s carried one missile under each wing and could use the missile's J-52 engine to boost aircraft thrust when carrying extreme loads. (NASM)*

cockpit. Several additional captive flights are made attached to the B-29 mother ship before the first test drop is made.

2 April
The Mercury Seven astronauts are selected by NASA from a field of 110 applicants.

9 April
The Mercury Seven are announced by NASA. This core group of military pilots is to be trained as the first men in space. Among the seven selected are three Air Force pilots—Captains L. Gordon Cooper, Virgil I. Grissom, and Donald K. Slayton.

10 April
The Northrop YT-38 Talon supersonic jet trainer flies for the first time. The aircraft is derived from the N-156 fighter design and will go on to be one of the most durable training platforms in Air Force history.

23 April
The Hound Dog air-to-ground supersonic, nuclear-capable missile is launched from a B-52 test plane. The test firing takes place over the Atlantic Missile Range.

27 April
The 1958 Annual report of the NACA is submitted by Congress to the President. It is the 44th and last report of the NACA, which was established in 1915. Most significantly, historical sections of the report are written by Jimmy Doolittle and Jerome C. Hunsaker, previous NACA heads.

28 April
Douglas Aircraft Company signs a $24 million contract with NASA to build the three-stage Thor-Vanguard launching rocket called Delta.

1 May
The Smithsonian Optical Tracking Station located at Woomera, Australia, photographs the *Vanguard I* satellite at its farthest distance from earth: 2,500 miles. Scientists compare the achievement to successfully photographing a golf ball from 600 miles away.

6 May
The Air Force's Jupiter IRBM is declared operational after the successful launch of one of the missiles covers 1,500 miles from Cape Canaveral, Florida.

12 May
The first of three VC-137A executive transport planes, redesignated from the Boeing 707 series, is delivered to the 1298th Air Transport Squadron (later the 89th Military Airlift Wing at Andrews AFB, Maryland).

The Air Force launches a Thor missile that carries a GE Mark 2 nose cone to an altitude of 300 miles and 1,500 miles down range. The data capsule is recovered containing a photograph of the earth taken from that altitude.

15 May
General Bernard A. Schriever, Air Research and Development Center commander, publicly displays the first re-entry vehicle ever recovered after an intercontinental-range missile flight.

25 May
Air Defense Command accepts its first Convair F-106 Delta Dart. This interceptor will eventually replace the F-102 Delta Dagger.

28 May
Able, a rhesus monkey, and Miss Baker, a South American squirrel monkey, are launched to an altitude of 300 miles in the nose cone of a Jupiter rocket. They are recovered in the ocean near Antigua Island after the successful mission. The aerospace medicine branches of all three services cooperate in gathering and evaluating the medical data collected during the mission. Able died the next day during surgery to remove electronic measuring devices implanted for the previous day's mission. Ill effects from the mission were not the cause of death.

3 June
The first USAF Academy class graduates 207 of the original 306 cadets. The cadet wing had moved from its temporary location at Lowry AFB to the new location in August 1958.

8 June
The X-15 hypersonic rocket plane, under the control of Scott Crossfield, accomplishes its first nonpowered drop test when it is released from the B-52 carrier aircraft at 38,000 feet over the Mohave Desert.

23 June
The Arnold Engineering Development Center (AEDC) is tasked to prepare operational and design requirements for a "large-scale" space test facility for testing military space weapons.

1 July
As part of the nuclear space rocket program, the first experimental reactor—Kiwi-1—operates at full temperature and duration at a test facility in Jackass Flats, Nevada.

24 July
An Air Force Thor data capsule, which contains film taken of the nose cone separation sequence, is recovered near Antigua.

30 July
The Northrop N-156F flies for the first time and exceeds Mach 1 during the sortie. Later, the aircraft will be redesignated at the F-5 Freedom Fighter.

7 August
Two F-100 Super Sabres make the first jet fighter flight over the North Pole.

A U.S. satellite, *Explorer 6*, transmits crude television pictures from space for the first time. Additionally, an intercontinental relay of a voice message via satellite is made when Major Robert G. Mathis transmits the first message.

24 August
An Air Force Atlas-C is launched to an altitude of 700 miles and flies 5,000 miles downrange. During the flight, the nose cone camera photographs approximately one-sixth of the earth's surface. The nose cone is successfully recovered.

Right: *The Northrop YF-5A breaks the sound barrier on its maiden flight. It will be developed as the F-5 Freedom Fighter. (NASM)*

THE USAF ACADEMY: A PERSONAL REFLECTION

Dik A. Daso

Left: *The Class of 1981 during graduation ceremonies at Falcon Stadium. (Courtesy of author)*

Opposite, bottom: *Cadet Lieutenant Colonel Dik Daso leads his squadron in the march to Mitchell Hall for the noon meal. (Courtesy of author)*

It was the summer of 1977 when my family and I made the long drive from Bay Village, a Cleveland suburb, to Colorado Springs. I was to enter the U.S. Air Force Academy as a member of the Class of 1981 ("Second to None"). I, like most in the entering class, was not really sure what I had gotten into.

It has been three decades since our class walked up the "Bring Me Men" ramp in groups of thirty-or-so anxious young men and women (we were the second class to include women). What I remember now are largely impressions and feelings rather than specifics—although I do remember a few details as if they were yesterday. One of those memories was that first haircut. We collectively had more hair than a herd of sheepdogs; after all, it was 1977. But that haircut was significant because it was our first step toward becoming the "Class of '81" and leaving our individualism behind—a major step for most of us high school over-achievers. Everything that we had accomplished in high school might just as well have fallen to the floor of the cadet barber shop with our curly locks that day.

It took a few years, but by the time we were "Firsties," we had learned the real meaning of "cooperate and graduate." We were a real class and those who made it to June Week, particularly those of us in the lower middle of the Class Order of Merit, knew that we had not made it alone. The pride within each of us as we crossed that stage in the football stadium was indescribable. The joy that erupted from us as the roar of the Thunderbirds overhead signaled the beginning of our Air Force careers was only heard by God—the afterburners muffled the shouting. As I look back on those years, I am thankful for the rigors of cadet life, the exceptional quality of the Academy education, the Honor Code, the wonderful extra-curricular opportunities, but most of all I am thankful for my classmates, who were at once, brilliant, athletic, hilarious, and thoughtful. Even though today I only see a few of them, and only occasionally, I think of them from time to time and wonder how they are, what they are doing, and realize how each of them made a positive impact on my life.

For many of us in the Class of '81, graduation meant pilot training. I'd be lying if I told you that I initially went to the Academy for any other reason than to become an Air Force pilot. After graduation, I went to Laughlin AFB, Texas, where I earned my wings. But I got quite a different look at the Academy when I returned as a history instructor in 1992 as a young major.

Right: *A cadet helps unveil the sculptured Air Force core values during a ceremony on September 29, 2004. The core values replaced the poetic reference "Bring me men..." after Air Force leaders decided the words were out of step with the current culture. (USAF)*

Following spread: *A panoramic view of the Air Force Academy in Colorado Springs, Colorado. ("Thunderbirds Flyby at the Air Force Academy," by Alfred Johnson, USAF Art Collection)*

Integrity First
Service Before Self
Excellence in All We Do

After a decade of exciting Air Force flying as a T-38 instructor pilot, an RF-4C pilot, and an F-15 "Eagle Driver," I ended up back in the Department of History (a long story that revolved around closing squadrons and too many pilots), where I had majored in the subject as a cadet. Much was the same at the Academy as it had been, but there was one significant difference: Banked Pilots. In a class of 1,000 cadets, only 250 were to be awarded pilot slots when the Class of 1992 graduated. The only problem was, more than 500 were pilot eligible. What I witnessed during my time as an instructor was that the "cooperate and graduate" philosophy was basically gone. Morale was low, particularly for the 250 or so cadets that were not going to attend pilot training immediately after graduation, and the concept of teamwork was left behind, particularly by those in the number 200–300 positions on the Graduation Order of Merit. It was truly the dark ages that February and March.

But the Academy reflected the ebb and flow of the real Air Force. There were too many pilots for the available flying positions throughout the Air Force and many of the young cadets would enter the banked pilot holding pattern for the exact same reason. Fortunately, those days have passed.

The good news was, and remains, that cadets are still bright, funny, energetic, and dedicated. The classroom was a tremendous place to interact with them because of the academic freedom of the environment. Teaching at the Academy ranks as one of the most personally rewarding assignments of my 20-year career. As the Air Force Academy struggled (and continues to struggle) with the social pressures that arise when 4,000 young adults are placed in a closed environment, the foundation of the Academy remains the same—to train the finest officers in the Air Force. I've met outstanding officers throughout my life, but there is just something about having graduated from the "Zoo" that grads carry with them for the rest of their lives—for each grad it is something just a little different, but it transcends "The Ring" and today makes the entire Academy experience well worth the effort.

JV
482
ALFRED "CHIEF" JOHNSON
1988

Left: *"Mach/ing Bird 66671," by Ricardo Scozzari. (USAF Art Collection)*

29 August
Project OXCART, the program to develop a high-speed, high-altitude reconnaissance aircraft, awards the development contract to Lockheed. The project will be managed by the Air Force but provide aircraft for the CIA.

1 September
The USAF Atlas ICBM operations are taken over by the Strategic Air Command at Vandenberg AFB, California.

2 September
Famed aeronautical scientist Theodore von Kármán is appointed chairman of an international committee that will establish an International Academy on Astronautics.

9 September
An Air Force Atlas missile is launched for the first time from Vandenberg AFB, California. The ICBM flies 4,300 miles and attains a top speed of 16,000 miles per hour. Strategic Air Command declares the Atlas weapon system fully operational after the successful test. Concurrently, a second Atlas is fired at Cape Canaveral on the same day.

17 September
Test pilot Scott Crossfield pilots the X-15 hypersonic research aircraft during its first powered drop from a modified B-52 bomber. Crossfield climbs above 53,000 feet and accelerates to Mach 2.11 during the sortie.

1 October
The USAF Aerospace Aeromedical Center is activated at Brooks AFB, Texas, consolidating a number of related aerospace medical functions.

2 October
Air Force Missile Test Center commander Major General Donald N. Yates is appointed as the Department of Defense representative for Project Mercury support operations.

6 October
The Air Force launches an Atlas ICBM and a Thor IRBM to their full flight range at Cape Canaveral, Florida.

13 October
The Air Force launches Bold Orion, an air-launched ballistic missile, from a B-47 bomber. The missile passes within four miles of the orbiting *Explorer 6* satellite and reaches an altitude of 160 miles.

28 October–19 December
The 4520th Aerial Demonstration Squadron—the Thunderbirds—complete a demonstration tour throughout the Far East. For the successful accomplishment of this goodwill tour, the team is awarded the Mackay Trophy.

31 October
An Atlas ICBM (Series D) goes on alert status equipped with a nuclear warhead. This is the first American ICBM placed on alert status capable of delivering a nuclear payload.

3 November
A C-133 Cargomaster delivers an Atlas missile to its operational destination. The C-133 is the first transport designed for just such a missile-delivery mission.

7 November
Although *Discoverer VII*, a USAF satellite, is successfully placed into a polar orbit, its capsule cannot be recovered after its return to earth. The same fate awaits *Discoverer VIII,* launched two weeks later.

14 November
A new Aerospace Medical Center is dedicated at Brooks AFB, Texas.

16 November
Captain Joseph W. Kittinger makes a record parachute jump when he leaps from an altitude of 76,400 feet carried aloft by the balloon *Excelsior I.*

17 November
A Department of Defense decision assigns all satellite and space vehicle programs to the military service with "primary interest" in each project. The Air Force is assigned the Discoverer, Midas, and Samos projects when they are transferred from Advanced Research Projects Agency (ARPA).

8 December
Major General Don R. Ostrander leaves ARPA to become the Director of NASA's Office of Launch Vehicle Programs, responsible for development and operations.

9 December
A Kaman H-43B rescue helicopter climbs to 29,846 feet setting a new rotary-wing altitude record.

At Akron, Ohio, a USAF Goodyear unmanned balloon is launched to an altitude of 100,000 feet. A radar "photograph" of the earth's surface is taken from the payload gondola of the balloon.

11 December
Captain Joseph Kittinger ascends in the balloon *Excelsior II* to an altitude of 74,400 feet, where he jumps from the gondola and freefalls for 55,000 feet—a freefall record.

Above: *The Convair F-106 was designed as a high-altitude, high-speed interceptor. ("High Contact," by William Phillips, USAF Art Collection)*

Setting a new speed record for a 100-kilometer course, Brigadier General J. H. Moore averages 1,216.5 miles per hour in an Air Force F-105B Thunderchief.

14 December
Captain Joseph B. Jordan pilots a Lockheed F-104C Starfighter to an altitude of 103,389 feet. This marks the first time that an air-breathing, jet-powered aircraft climbs above 100,000 feet altitude. The flight takes place at Edwards AFB, California.

15 December
Major Joseph W. Rogers establishes a new official world absolute speed record when he flies a Convair F-106A Delta Dart to an average speed of 1,525.95 miles per hour. The flight takes place at Edwards AFB, California.

Left: *CIA pilot Francis Gary Powers and the Lockheed U-2. Powers is wearing red and white high-visibility overalls over his high-altitude partial pressure suit. (NASM)*

Below: *The TIROS I was designed to test experimental television techniques to monitor weather from orbit. (NASM)*

1960

30 January
The Central Intelligence Agency (CIA) orders twelve Lockheed A-12 high-altitude, high-speed reconnaissance aircraft. Funding for the purchase is approved.

9 February
The Air Force dedicates the National Space Surveillance Control Center (SPACETRACK) at Bedford, Massachusetts.

24 February
On its longest flight to date, a Titan ICBM flies 5,000 miles downrange. It was launched from Cape Canaveral.

1 April
The first U.S. weather satellite, TIROS 1 (television infrared observation satellite), is launched into earth orbit atop a Thor-Able from Cape Canaveral, Florida. The satellite took pictures of earth's cloud formations for two months, including a tornadic system near Wichita, Kansas, on May 19. During its operational life, TIROS will transmit 22,952 picture frames of earth's cloud formations and complete more than 1,300 orbits.

13 April
Transit 1B becomes the first U.S. navigation satellite in space.

1 May
CIA pilot Francis Gary Powers is shot down while flying a U-2C during an overflight mission of the Soviet Union. The Soviets fire a salvo of several SA-2 Guideline missiles and one brings Powers down. Another actually destroys a Soviet fighter that had been in pursuit of the U-2C. Powers is captured near Sverdlovsk, Russia, put on trial and convicted of spying. He is exchanged for a Soviet spy in 1961.

19 May
Major Robert M. White flies the X-15 hypersonic research plane to an altitude of 107,000 feet—its highest yet achieved.

20 May
In the longest flight of an Atlas missile to date, the missile attains an apogee of 1,000 miles and travels more than 9,000 miles from the Atlantic Missile Range to the Indian Ocean.

23 May
During Operation AMIGOS, the Air Force carries out a massive humanitarian airlift of supplies and relief equipment to the victims of earthquakes in Chile. During the next month, more than 1,000 tons of aid will be lifted more than 4,500 miles from bases across the U.S.

24 May
The first anti-missile early-warning satellite, MIDAS II, is launched into earth orbit.

25 June
The Air Force establishes Aerospace Corporation, a nonprofit civilian group to manage engineering, research, and development of missiles and military space programs. Dr. Ivan A. Getting of Raytheon will be named the first president of Aerospace Corporation in July.

28 June
The Smithsonian Institution awards the Langley Medal, the Institution's highest award, to rocket pioneer Robert H. Goddard (posthumously).

1 July
A 55th Strategic Reconnaissance Wing RB-57H operating in international waters over the Barents Sea is attacked and shot down by a Soviet MiG-17 Fresco. The co-pilot and navigator are the only survivors of the six-man crew and are held by the Soviet Union as spies until they are released the following January.

8 July
Operation NEW TAPE begins when unrest threatens the newly established Democratic Republic of the Congo. U.S. citizens are evacuated and UN troops are airlifted into the country. The operation continues for the next four years.

As part of the Project ROVER nuclear rocket program, the second experimental reactor (Kiwi-A Prime) is tested at full power and for its full duration at a test facility at Jackass Flats, Nevada.

14 July
Operation SAFARI, the airlift of some 38,000 UN troops across the African continent, begins. More than 100 C-130 Hercules and C124 Globemaster II transports participate.

17 July
USAF balloons carry aloft the first of three NASA experiments that contain 12 mice. The balloon ascends to 130,000 feet and remains there for nearly 12 hours. The purpose of the test is to study the effect of cosmic ray particles on the subject animals.

1 August
The first Convair B-58 Hustler arrives at Carswell AFB and becomes a member of the 43d Bombardment Wing. The B-58 is designed to fly twice the speed of sound, refuel in flight, and carry a nuclear warhead deep into the Soviet Union.

10–11 August
A 300-pound capsule that had been ejected from an orbiting satellite, the Air Force's *Discovery XIII*, is the first man-made object retrieved from space. Although originally planned as an aerial retrieval, the capsule falls so far out of the intended recovery zone that Navy divers are required to locate and salvage the capsule. The silk 50-star American flag housed in the capsule will be presented to the President on August 15.

12 August
Major Robert M. White establishes a new altitude record for manned vehicles when he pilots the X-15 to 136,500 feet.

Below: *Afterburners blaze as this Convair B-58 Hustler heads off on a mission. (NASM)*

JOSEPH W. KITTINGER JR.

Tom Crouch

Born on July 27, 1928, Joe Kittinger grew up near Orlando, Florida. He came to aviation via a familiar path that began with a flight aboard a Ford Trimotor with his father, followed by a few years of intensive model aircraft building, culminating in hours spent learning to fly a Piper Cub owned by a returning veteran. After two years at the University of Florida, he enlisted in the USAF and received his wings and second lieutenant's bars at Las Vegas, Nevada, in March 1950.

During a flying assignment in Germany, Kittinger was accepted as a test pilot flying F-84G fighters for NATO. That experience led to his next assignment to a team led by Colonel John Paul Stapp at the Aero Medical Field Laboratory, Holloman AFB, New Mexico, investigating new techniques and equipment designed for use in high altitude, high speed aircraft. Kittinger tested a variety of partial pressure suits, "flew" to altitudes of over 100,000 feet in pressure chambers, and served as a test subject for experiments involving his reaction to everything from claustrophobia to extreme temperatures.

Ironically, the balloon, the oldest type of flying craft, was also the natural choice to carry instruments and test pilots to extreme altitudes. During the first flight of Project Man High, on June 2, 1957, Kittinger reached an altitude of 96,000 feet in a three-by-seven-foot sealed gondola dangling beneath an enormous helium-filled balloon made of plastic film only two-thousandths of an inch thick.

His next assignment was as test director for Project Excelsior at the Aerospace Medical Research Laboratory at Wright-Patterson AFB. The new series of tests was designed to develop equipment and escape procedures that might save the lives of aviators and astronauts. He made his first high altitude parachute jump from an Excelsior balloon at an altitude of 76,400 feet in November 1959. It nearly became his last when the shroud lines of a small stabilizing parachute wrapped around his neck. He was able to free himself when his emergency parachute opened at 12,000 feet. Three weeks later he made a jump from 74,000 feet without incident. For his contributions to the Excelsior program, President Eisenhower awarded Kittinger the Harmon International Trophy for ballooning achievements.

Kittinger made his ultimate jump on August 16, 1960, when he stepped out of an open balloon gondola at an altitude of 102, 800 feet over New Mexico and began his "long, lonely leap." During the hour and a half ascent, temperatures dipped to 94 degrees below zero. "There is a hostile sky above me," he told ground controllers. "Man will never conquer space. He may live in it, but he will never conquer it. The sky above is void and very black, and very hostile." If his pressure suit failed at this altitude, he would lose consciousness within 12 seconds and be dead in two minutes.

He fell from the top of the atmosphere for four minutes thirty-seven seconds, until his main parachute opened at 17,500 feet. In the thin upper atmosphere near the beginning of his free fall, he was traveling very close to the speed of sound. As he dropped into the denser atmosphere near 50,000 feet, his speed

slowed to 250 miles per hour. His hand swelled painfully when one of his gloves had failed during the jump. Eight minutes after his main parachute opened, he was back on solid ground in the White Sands Missile Range, having set three world records: the highest ascent in an open gondola, the longest free fall, the longest parachute descent.

Kittinger spent another two years conducting high altitude balloon research with the Stargazer project, which carried astronomers to high altitudes, before embarking on the first of his three combat tours in Vietnam. His first two tours were with the Air Commandos, flying Douglas A-26 attack aircraft. During his final tour as vice-commander of a fighter wing operating the F-4D Phantom II, he scored a victory over a MiG-21. On May 11, 1972, just four days before the end of his third tour, Kittinger was shot down and spent the next 11 months as a prisoner of war.

Retiring as a colonel in 1978, Joe Kittinger most certainly did not retire from flying. Returning to his native Florida, he flew balloons and antique biplanes for an air show, Rosie O'Grady's Flying Circus. He began entering, and winning, gas balloon races and events in 1982, and quickly emerged as a major international competitor. He set a world's record for the longest distance flown in a 1,000-cubic meter balloon, traveling 2,001 miles from Las Vegas, Nevada to Franklinville, New York in 72 hours. He combined another world's record for the longest distance flown with a 3,000-cubic foot balloon with the first solo balloon flight across the Atlantic, traveling 3,543 miles from Caribou, Maine, to Montenotte, Italy, in 86 hours.

Colonel Joseph Kittinger has enjoyed an extraordinary career. His decorations include the Silver Star with Oak Leaf Cluster, Legion of Merit with Oak Leaf Cluster, the Distinguished Flying Cross with five Oak Leaf Clusters, a Bronze Star with "V" device and two Oak Leaf Clusters, the Harmon International Trophy, and two Montgolfier Diplomas for achievement in the air. He has logged over 11,000 hours of flight time in 62 different aircraft types. Test pilot, combat aviator, POW, and record-holding sport pilot, Joseph W. Kittinger Jr. is a genuine aerospace pioneer and hero.

Opposite: *Captain Joseph W. Kittinger Jr. (USAF)*

Above: *Captain Kittinger takes "The Long Lonely Leap" from 102,800 feet from an open gondola carried aloft by a balloon. (USAF)*

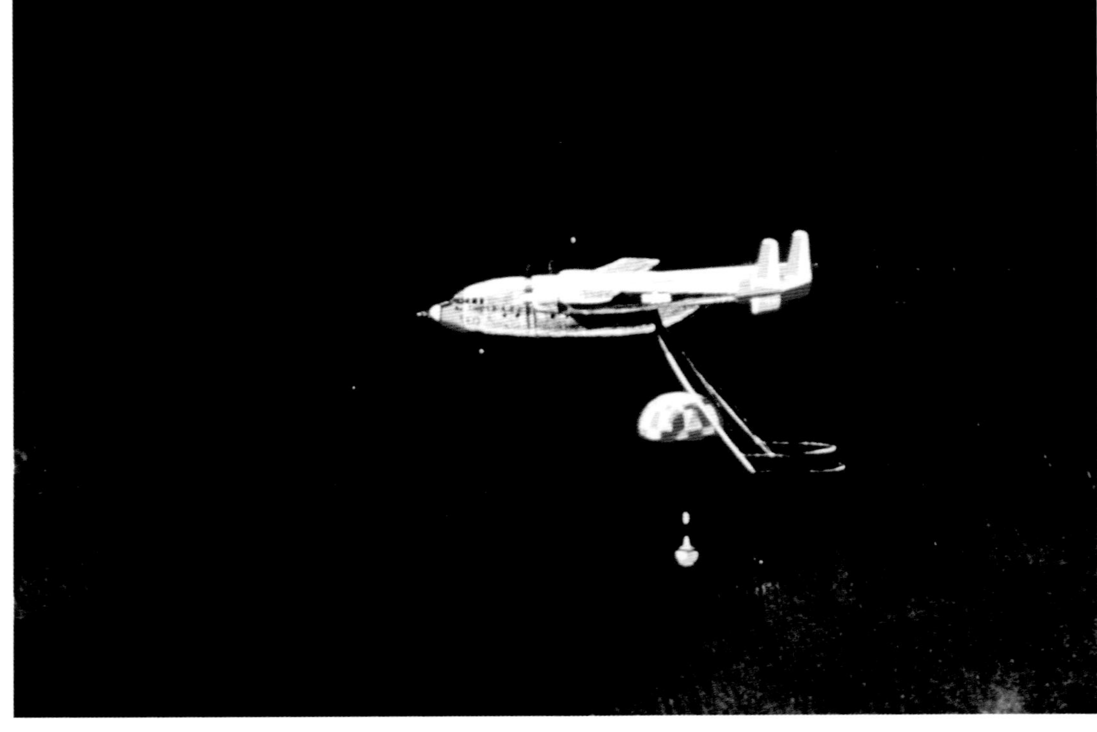

Left: *"Mid-air Capture of* Discoverer 14 *Satellite," by Ren Wicks. (USAF Art Collection)*

Below: *The BMEWS station at Clear Air Force Station, Alaska, was designed to provide 30 minutes warning of any incoming ICBM attack. (USAF)*

16 August
Captain Joseph W. Kittinger Jr. ascends to 102,800 feet in the *Excelsior III* balloon and then makes a parachute jump from that altitude, setting several records in the process. This is the highest parachute jump on record and Kittinger achieves the longest freefall jump (17 miles) of four minutes and 37 seconds during his descent. He also becomes the first man to approach the speed of sound without an aircraft when he reaches 614 miles per hour during the freefall from altitude.

18 August
The Air Force launches *Discoverer XIV* into polar orbit from Vandenberg AFB, California.

19 August
Captain Harold F. Mitchell piloting a C-119 Flying Boxcar retrieves the *Discoverer XIV* reentry capsule in midair on his third attempt, snagging it at 8,000 feet near Honolulu, Hawaii. This is the first time that such a feat has successfully been accomplished. The 6593d Test Squadron (Special) receives the Mackay Trophy for their development of techniques required for aerial recovery of returning space capsules.

26 August
Under the guidance of the USAF and ARPA, construction begins on the world's largest radar located in Arecibo, Puerto Rico. When completed, the radar will be capable of bouncing signals off of Venus, Mars and Jupiter.

30 August
With six Atlas ICBMs now on alert, the 564th Strategic Missile Squadron located at F. E. Warren AFB, Wyoming, becomes the first fully operational Air Force ICBM unit.

10 September
During SKY SHIELD, a NORAD exercise, civil aviation operations cease for six hours while military aircraft participate in defense maneuvers across the United States.

15 September
Air Force pilots Captain W. D. Habluetzel and Lieutenant J. S. Hargreaves complete a simulated journey to the moon by remaining in a space cabin simulator at the School of Aviation Medicine located at Brooks AFB, Texas, for more than 30 days.

21 September
At Nellis AFB, Nevada, the Tactical Air Command accepts its first Republic F-105 Thunderchief all-weather, nuclear-capable fighter.

1 October
The first Ballistic Missile Early Warning System (BMEWS) station becomes operational at Thule, Greenland. BMEWS is designed to provide Strategic Air Command with adequate warning of an

impending missile attack, allowing the launch of strategic alert forces in the United States.

12 October
An Air Force C-130 establishes a heavy-equipment parachute drop record when it delivers 41,740 pounds over El Centro, California.

12 November
A restartable rocket motor is tested for the first time during the launch of *Discoverer XVII* from Vandenberg AFB, California.

14 November
An Air Force C-119 successfully retrieves the *Discoverer XVII* capsule in mid-air after the capsule flies 31 earth orbits. This is the second aerial retrieval of a returning satellite made by an Air Force C-119. In the capsule is the first letter carried by orbital satellite. The letter is from the Air Force Chief of Staff, General Thomas D. White, and addressed to the Secretary of Defense.

23 November
TIROS II weather satellite is launched by a Thor-Delta rocket, becomes the 14th successful U.S. satellite launch in 1960.

1 December
The Air Force delivers to Jet Propulsion Laboratory (JPL) the first 1:1,000,000 scale map of the lunar landing site selected by NASA.

3 December
A Titan ICBM explodes in its silo at Vandenberg AFB, California. Night refueling of the missile is underway when the disaster occurs.

10 December
Concluding the most successful Discoverer mission thus far, a C-119 piloted by Captain Gene Jones, snags the *Discoverer XVIII* capsule at 14,000 feet. The capsule had flown a total of 48 earth orbits and carried human tissue to study the effects of radiation in space. Earlier tissue experiments indicate that human tissue survives the effects of solar radiation.

14 December
Setting a new world and jet distance records for unrefueled flight over a closed course, a B-52G cruises an astonishing 10,079 miles in 19 hours and 44 minutes. The Stratofortress is part of the 5th Strategic Bombardment Wing stationed at Travis AFB, California.

Above: *A Hawaii-based C-119 crew practices techniques needed for aerial recovery of space capsules. (NASM)*

16 December
Strategic Air Command launches its first missile from Vandenberg AFB, California. The Atlas-D with Mark 3 nose cone travels 4,384 miles to Eniwetok Atoll.

19 December
NASA uses a Redstone rocket booster to carry out the first unmanned test of the Project Mercury space capsule. The missile is launched from Cape Canaveral and travels 235 miles downrange, attains a speed of 4,200 miles per hour, and reaches an altitude of 135 miles. The capsule safely parachutes back to earth, splashing down in the Atlantic Ocean, and is recovered in about 30 minutes time.

20 December
Ho Chi Minh, leader of the Republic of Vietnam, organizes the National Liberation Front of South Vietnam (NLF). The NLF is committed to the unification of Vietnam, the ousting of U.S. advisors, and the overthrow of the acting South Vietnamese government.

1961

13 January
A Convair B-58 Hustler piloted by Major H. J. Deutschendorf breaks six world speed records. The B-58 averages 1,200.2 miles per hour over the closed course, and then averages 1,061.8 miles per hour over two runs, carrying a payload of 4,408 pounds and a three-man crew.

22 January
The Titan II launch vehicle is selected to boost the USAF Dyna-Soar into space.

31 January
HAM (Holloman Aero Medical), a chimpanzee, is launched in a Mercury space capsule atop a Redstone booster rocket to test the effects of space flight on primates. HAM is aloft for 18 minutes and suffers no ill effects. This successfully tests the Mercury life support system.

A USAF Samos II, a 4,100-pound photographic test satellite, is placed into orbit atop an Atlas-Agena from Point Arguello, California.

The first LGM-130 Minuteman ICBM, a three-stage sold-propellant missile, is launched from Cape Canaveral and under full guidance strikes its target some 4,600 miles downrange.

3 February
Strategic Air Command initiates the *Looking Glass* program. Converted KC-135 Stratotankers are modified to enable direct communication with the Joint Chiefs of Staff, any SAC base, or any airborne SAC aircraft. The airborne command post operates 24 hours a day, every day.

6 March
The first Boeing B-52H Stratofortress flies at Wichita, Kansas. This version of the B-52 is designed to carry the GAM-87A Skybolt air-to-surface missile. The B-52H is equipped with turbo-fan engines.

7 March
Major Robert M. White pilots the North American X-15 hypersonic research airplane to a speed of 2,905 miles per hour. In accomplishing this, White becomes the first man to exceed Mach 4 during this mission.

Strategic Air Command authorizes the use of the Quail (GAM-72A) on B-52s. The Quail is designed as a diversionary missile.

Opposite, top: *"Ham, Able, and Baker—The Space Chimpanzee and Monkeys That Were Instrumental in the Space Program," by Susan Jackel. (USAF Art Collection)*

Opposite, bottom: *The T-38 Talon became the Air Force's advanced primary jet trainer. ("T-38 Randolph AFB," by Robert McCall, USAF Art Collection)*

Right: *"Firefly Across the Atlantic," by Steven Bricker. (USAF Art Collection)*

17 March
The first Northrop T-38A Talon supersonic jet trainer is delivered to Randolph AFB, Texas, where it enters operational service with Air Training Command.

1 April
Two Air Force commands are redesignated this date. Air Materiel Command becomes Air Force Logistics Command (AFLC) and Air Research and Development Command becomes Air Force Systems Command (AFSC).

12 April
Cosmonaut Yuri Gagarin becomes the first man into space and the first to orbit the earth.

17 April
A USAF Cambridge Research Center project, a constant-altitude balloon, is launched from Vernalis, California, and remains aloft at 70,000 feet for nine days while supporting a 40-pound payload.

More than 1,000 CIA-trained Cuban refugees fail to "liberate" Cuba. The incident will become know as the "Bay of Pigs."

19 April
Four Air National Guard B-26 crew members lose their lives on a special operations mission during the failed Bay of Pigs invasion in Cuba.

21 April
Captain Robert M. White pilots the X-15 to an altitude of 105,000 feet. During the climb, White accelerates to 3,074 miles per hour at 70,000 feet and becomes the pilot of the first aircraft to exceed 3,000 miles per hour.

2 May
A B-58 Hustler crew sets a transatlantic record when they fly *Fire Fly* from New York to Paris in three hours and 56 minutes. On June 3, while participating in the Paris Air Show, the aircraft engines stall at Mach 1.3 and the aircraft crashes, killing all three crewmen.

3 May
An Air Force Titan I ICBM is launched for the first time from a hard "silo lift" launcher. The Air Force Systems Command (AFSC) crew is stationed at Vandenberg AFB, California.

5 May
Astronaut Alan B. Shepard Jr., a U.S. Navy commander, becomes the first American in space during his suborbital flight in the *Freedom 7* capsule. The flight lasts 15 minutes and 28 seconds, achieving a peak altitude of 116.5 miles.

11 May
President Kennedy authorizes American advisors to assist the South Vietnamese in their struggle against the forces of North Vietnam.

26 May
A Convair B-58 Hustler crew from the 43d Bombardment Wing sets a transatlantic speed record when they fly from New York to Paris in just under three hours and 20 minutes. Major William R. Payne, Captain William L. Polhemus, and Captain Raymond Wegener average just over 1,300 miles per hour during the flight. They are awarded the Mackay Trophy for their achievement, which is intended to commemorate the 34th anniversary of Charles Lindbergh's first transatlantic crossing.

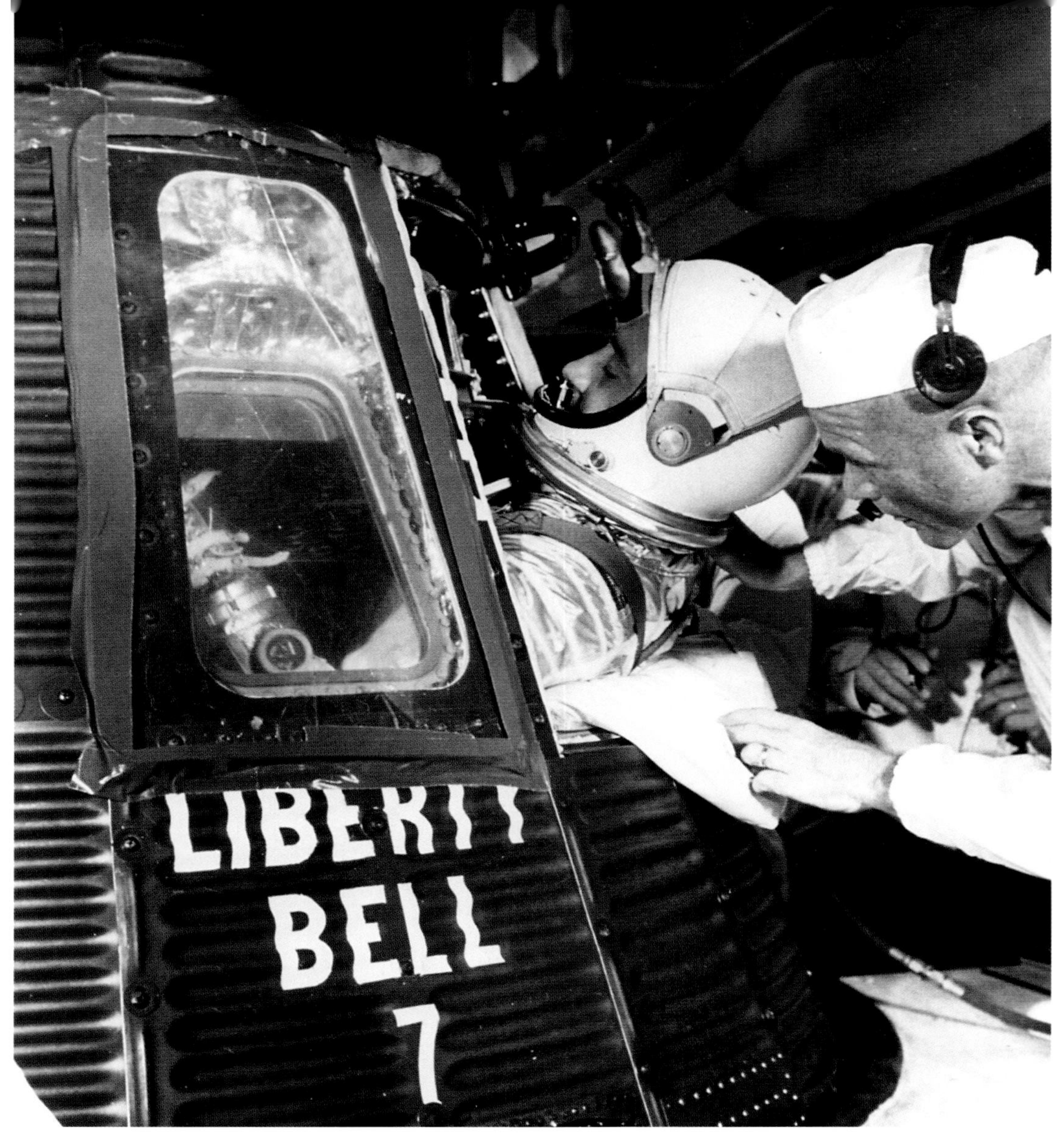

Left: *Virgil I. "Gus" Grissom, an Air Force test pilot, climbs into Liberty Bell 7 on his way to becoming America's second man in space. (NASA)*

1 June
The first Bomarc-B site is declared operational at Kincheloe AFB, Michigan.

9 June
The first C-135A Stratolifter is delivered to the Military Air Transport Service (MATS). This delivery marks the beginning of the modernization of the piston-driven transport fleet.

23 June
Major Robert M. White pilots the X-15 to a speed of 3,603 miles per hour and becomes the first man to exceed Mach 5 during this mission.

30 June
General Curtis E. LeMay becomes the Chief of Staff, United States Air Force.

July
The Strategic Air Command places a full 50 percent of its airborne assets on 15-minute ground launch alert.

1 July
The North American Air Defense Command begins to catalog man-made space objects using a specially designed detection and tracking system.

The Air Force Communications Service is activated. The AFCS is responsible for communications and air traffic control at Air Force installations around the world.

12 July
The Air Force launches Midas 3 into orbit using the "kick in the apogee" technique. The Agena-B second stage is reignited in flight, boosting the satellite to a near-circular polar orbit approximately 1,850 miles above the earth.

21 July
Captain Virgil I. "Gus" Grissom becomes the second American in space—the first USAF astronaut in space—when he is launched in a Mercury capsule atop a Mercury-Redstone 4 booster. Grissom's flight in the *Liberty Bell 7* attains an altitude of 118 miles, a speed of 5,310 miles per hour, and flies 300 miles downrange. Grissom receives the first-ever General Thomas D. White USAF Space Trophy for the flight. The award will be presented in June 1962.

8 August
The USAF tests its first Atlas F missile from Cape Canaveral, Florida. The Atlas F ICBM, designed for long-term storage of liquid-fuel propellants, is the only

Right: *"Jacqueline Cochran Breaking Another Record," by Si Mezerow. (USAF Art Collection)*

Below: *Jacqueline Cochran next to her record setting T-38 Talon. This T-38 aircraft now resides in the collection of the National Air and Space Museum. (USAF)*

Atlas version that be emplaced in hardened underground silos, and will have a significantly shorter launch sequence than other Atlas missiles.

12–13 August
The Berlin Wall is completed and most entry points to the city are closed by the Soviets.

24 August
Jacqueline Cochran establishes a new women's absolute speed record when she flies her Northrop T-38 Talon to a speed of 844.2 miles per hour.

25 August
The Lockheed C-130E flies for the first time.

8 September
Jacqueline Cochran pilots a Northrop T-38 Talon to a new women's 1,000-kilometer closed course speed record of 639.4 miles per hour.

15 September
Breaking the women's closed course distance record, Jacqueline Cochran pilots a T-38 a total distance of 1,346.4 miles.

19 September
A Bomarc B pilotless interceptor launched from Eglin AFB, Florida, is controlled from Gunter AFB, Alabama, and intercepts a supersonic Regulus II supersonic drone at seven-mile altitude and 250 miles from the launch point. The Bomarc successfully executes a 180-degree turn to complete the intercept.

1 October
In response to the Berlin Wall crisis during Operation STAIRSTEP, 18,500 members of the Air National Guard report for duty, while several ANG units are deployed to Europe in support of USAF in Europe (USAFE) wings already in place.

11 October
Major Robert M. White soars to an altitude of 217,000 feet while piloting the X-15. This is the first manned aircraft to top 200,000 feet above the earth.

12 October
Jacqueline Cochran establishes a new women's altitude record when she pilots a Northrop T-38 Talon to 56,071 feet.

13 October
Eugene Bullard, the only black U.S. aviator to serve during World War I, dies in New York at the age of 67.

Left: *Minuteman ICBMs provided the backbone of strategic deterrence since the early 1960s. (USAF)*

18 October
An Air Force Kaman H-43B climbs to a new record altitude of 32,840 feet.

20 October
In response to increased activity by the North Vietnamese Army, McDonnell RF-101C Voodoo reconnaissance aircraft begin flying missions over Vietnam.

9 November
Major Robert M. White pilots an X-15 hypersonic rocket plane to a top speed of over 4,000 miles per hour. White pushes the rocket throttle wide open and reaches an altitude of 101,600 feet. This is the first time that Mach 6 is exceeded by a manned aircraft on the 45th X-15 flight in the program.

Flying AT-28 Trojans, Douglas SC-47s, and B-26 Invaders, the *Farm Gate* Detachment (Air Commandos) in Vietnam becomes operational. Beginning in January, these advisors will instruct Vietnamese pilots in the T-28 ground-attack aircraft.

15 November
When the 2d Advanced Echelon, Thirteenth Air Force, is activated in Saigon, South Vietnam, the Air Force officially enters the Vietnam War.

17 November
The first silo launch of a Minuteman ICBM occurs at Cape Canaveral, Florida. The Minuteman soars more than 3,000 miles down the Atlantic Missile Range.

21 November
An Air Force Titan ICBM is launched at Cape Canaveral by a military crew of the 6555th Aerospace Test Wing. The Titan carries a nose cone that will be used in the Nike-Zeus anti-missile tests.

22 November
A highly secret Samos reconnaissance satellite is boosted into orbit atop a Atlas-Agena B launch system.

29 November
Two chimpanzees are successfully launched into orbit and recovered. The primates are launched from Cape Canaveral and complete the two orbits in one of NASA's Mercury capsules.

1 December
The first operational Minuteman unit is activated at Malmstrom AFB, Montana—the 10th Strategic Missile Squadron (SMS). The unit will become operational on February 28, 1963.

15 December
North American Air Defense Command's SAGE system is declared fully operational when the 21st and final control center is completed at Sioux City, Iowa.

Right: *The Skyblazers were a superb flight demonstration team, thrilling millions across Europe during the 1940s–1960s. ("Skyblazers," by Bill Dillard, USAF Art Collection)*

Below: *This depiction of C-123s spraying defoliant (Agent Orange) over a dense jungle in the Central Highlands of Vietnam circa 1965. ("Operation Ranch Hand," by Richard Weaver, USAF Art Collection)*

1962

January
The *Skyblazers*, USAFE's official jet acrobatic team and the Air Force's first jet display team, is disbanded. The Skyblazers were formed in 1948 and flew F-80s, F-84s, F-86s, and F-100s during their existence.

7 January
A detachment of Air Force C-123 Providers that had been specially modified to dispense herbicide and insecticide is tasked to participate in Operation TRAIL DUST, at Tan Son Nhut AB, near Saigon, South Vietnam. This deforestation project is designed to reveal Communist positions and road locations in Vietnam. The Air Force part of this project, called *Ranch Hand*, would continue for exactly nine years in the skies over Vietnam, the last mission flown on January 7, 1971.

10–11 January
Major Clyde P. Evely pilots a Boeing B-52H to a new world record for unrefueled flight distance. He and his crew fly from Okinawa, Japan, to Madrid, Spain, in 22 hours and 10 minutes—a distance of 12,532 miles.

13 January
The first full-scale *Ranch Hand* defoliation mission is carried out this date.

29 January
The last of 47 test firings of a Titan I missile occurs at Cape Canaveral, Florida. In total, 34 are totally successful, and only three are complete failures.

2 February
A C-123 Provider flying a defoliant training mission crashes in South Vietnam. Captain Fergus C. Groves II, Captain Robert D. Larsen, and Staff Sergeant Milo B. Coghill become the first official USAF fatalities in Vietnam.

11 February
CIA U-2 pilot Francis Gary Powers returns to the U.S. after serving one and one-half years of a 10-year sentence. He is exchanged for a Soviet spy held by the U.S.

20 February
Marine Lieutenant Colonel John H. Glenn Jr. becomes the first American astronaut to orbit the earth. He remains in orbit in his capsule, *Friendship 7*, for nearly five hours.

5 March
A 43d Bombardment Wing B-58 Hustler crewed by Captains Robert G. Sowers, Robert MacDonald, and John T. Walton flies a round-trip sortie from New York to Los Angeles and back, establishing three speed records during the mission. The round-trip portion takes four hours, 41 minutes, 11 seconds with an average speed of 1,044.5 miles per hour. Not only does this crew win the Mackay Trophy for the mission but they also win the Bendix Trophy.

16 March
The first Titan II missile is successfully launched at Cape Canaveral, Florida. The missile stands more than 100 feet tall.

21 March
To test the ejection capsule in the B-58 Hustler, a bear is taken aloft in the aircraft and then ejected in the pod at 35,000 feet at a supersonic speed of 870 miles per hour. The bear lands safely after more than seven minutes under parachute.

22 March
Following reports that unidentified aircraft had been spotted over South Vietnam, four Convair F-102 Delta Daggers are deployed to Tan Son Nhut AB in South Vietnam from their home base at Clark AFB, Philippines.

18 April
Strategic Air Command's 724th Strategic Missile Squadron (SMS) becomes operational at Lowry AFB, Colorado. Nine Titan I missiles are located in hardened underground silos—the USAF's first missile base of this kind.

20 April
The first Titan I ICBM begins operational alert at Lowry AFB, Colorado.

22 April
Famed aviatrix Jacqueline Cochran pilots a Lockheed Jetstar named *Scarlet O'Hara* across the Atlantic Ocean. She is not only the first woman to fly a jet across the Atlantic, but she also sets 68 different flight records in the process. She flew this leg of a longer journey between Gander, Newfoundland, and Shannon, Scotland.

26 April
Lockheed test pilot Lou Schalk makes the first flight of the prototype of the A-12 aircraft designed and built at the Lockheed Skunk works. The A-12 is the predecessor of the SR-71 Blackbird.

27 April
The Special Air Warfare Center is established at Eglin AFB, Florida.

4 May
The Lockheed A-12 achieves supersonic speed for the first time—Mach 1.1.

19 June
The Dyna-Soar winged re-entry vehicle is classified as the X-20.

Opposite, top: *An ejection pod carrying a bear is test-fired from a B-58 Hustler. (NASM)*

Opposite, bottom: *This is an artist's conception of a Dyna-Soar launch. The Air Force space plane was to be launched into space atop an Atlas IIIC missile. The project, cancelled before it flew, still contributed vital data to future manned space plane projects. (USAF)*

Right: *A U-2 discovers missiles in Cuba on October 14, 1962. ("Videmus Omnia," by Douglas Nielson, USAF Art Collection)*

29 June
The first Minuteman missile launched by a military crew occurs at Cape Canaveral, Florida. The solid-fueled missile is launched from its underground silo to a target area 2,300 miles downrange.

9 July
After a few failed attempts, a hydrogen bomb of one megaton-plus yield is detonated at 248 miles altitude over the Pacific Missile Range after being launched from Johnston Island, the highest thermonuclear blast yet made during Operation DOMINIC. The electromagnetic pulse generated by the blast is felt as far away as the island of Oahu, Hawaii, some 800 miles from the test site.

17 July
Major Robert M. White pilots the X-15-1 hypersonic research aircraft to a world record altitude of 58.7 miles, attaining a speed of 3,784 miles per hour. On this mission White, the first man to exceed 300,000 feet altitude, becomes the first Edwards test pilot to earn astronaut's wings by flying an airplane in space.

19 July
A Nike-Zeus anti-missile missile fired from Kwajalein Island intercepts an Atlas missile nose cone after it is launched from Vandenberg AFB, California. This marks the first interception of an ICBM by another missile.

1 August
The first Atlas F test launch from an underground silo occurs at Vandenberg AFB, California. The USAF-launched missile travels 5,000 miles to its target area in the Pacific Test Range.

9 August
Multiple countdown capability is demonstrated at Cape Canaveral when the USAF launches two Atlas D missiles in rapid succession.

14 September
NASA announces the names of the next nine astronauts selected for the space program. Four of these are Air Force officers: Major Frank Borman, Captain James A. McDivitt, Captain Edward H. White, and Captain Thomas P. Stafford. Among the others are civilian NASA pilot Neil A. Armstrong, who will become the first man to walk on the surface of the moon in July 1969.

18 September
Edwards AFB test pilot Major Fitzhugh L. Fulton pilots a Convair B-58 Hustler carrying an 11,000-pound payload to a record altitude of 85,360 feet. This category record still stands today.

14 October
An Air Force U-2 reconnaissance flight, piloted by Major Steve Heyser of the 4080th SRW, uncovers indisputable evidence of Soviet medium-range ballistic missiles located in Cuba. This marks the beginning of the Cuban Missile Crisis.

17 October
A Vela Hotel satellite performs the first space-based detection of a nuclear explosion in space.

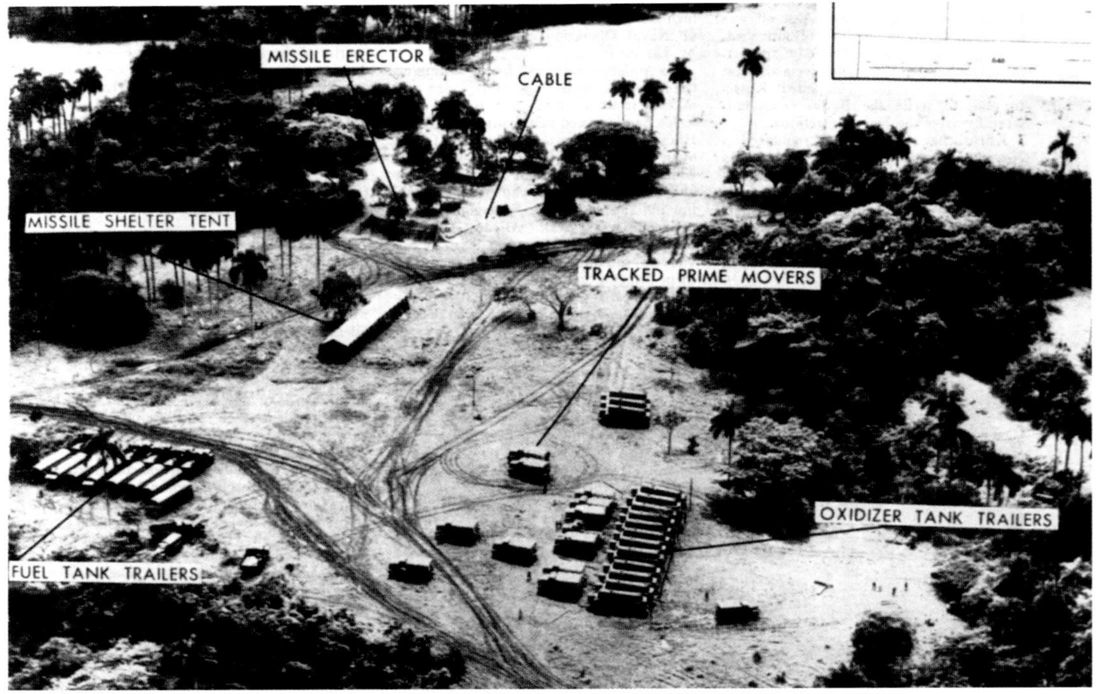

Left: *One of a number of tactical reconnaissance photos taken by a low-flying RF-101 Voodoo confirming the presence of Soviet missile support equipment. (USAF)*

Below: *"Cuban Sightseeing Tour—1962," by Michael Machat. (USAF Art Collection)*

17–22 October
During this tense period, seven U-2 missions are flown over Cuba daily to monitor Soviet missile movements. RF-101C and Navy RF-8 aircraft carry out low-level reconnaissance missions, which uncover Soviet bombers located at Cuban airfields.

22 October
President John F. Kennedy announces a blockade of Cuba and demands the removal of missiles located there.

All Strategic Air Command forces are placed on 24-hour alert. The B-47 force is dispersed for protection and B-52s, armed with nuclear weapons, take up orbital alert just outside of Soviet airspace. All missile sites are placed on high alert as well.

25 October
Strategic Air Command RB-47 Stratojets and KC-97 tankers join the Navy in a massive aerial search for incoming Soviet shipping to Cuba.

27 October
Major Rudolph Anderson Jr., from the 4080th Strategic Reconnaissance Wing, is shot down in his U-2 over Cuba. Anderson is posthumously awarded the first Air Force Cross for his sacrifice during the mission.

The first 10 Minuteman I missiles are placed on alert at Malmstrom AFB, under the operational control of the 10th Strategic Missile Squadron. Two more flights of Minuteman Is become operational in December.

28 October
The Soviet Union agrees to remove their missiles from Cuba. The U.S. will also move missiles that have been stationed in Turkey as part of the agreement.

29 October
Air Force RF-101 Voodoos confirm that the Soviet missiles are being dismantled in Cuba.

2 November
President John F. Kennedy authorizes Operation LONG SKIP to airlift military equipment to Calcutta, India, after China invaded that country. In only two weeks, MATS will transport more than 1,000 tons of cargo, mostly by relying on the new C-135 jet transports.

20 November
President Kennedy terminates the Cuban blockade, as all missiles have been removed from Cuba.

24 November
The Department of Defense awards General Dynamics and Grumman the prime contract for the

Right: *"Jacqueline Cochran" by Chet Engle. (NASM Art Collection)*

Tactical Fighter Experimental (TFX). The variable-swept-wing twin-engine jet fighter is to carry 20,000 pounds of weapons and travel at Mach 2.5. This will become the F-111 "Aardvark" fighter for the Air Force and will never be procured by the Navy.

5 December
The Atlas missile flight-test program ends with the last of 151 launches at Cape Canaveral, Florida. Of these, 108 are successful.

13–14 December
Project Stargazer, a balloon flight designed to provide the clearest view of the stars to this date, is launched over New Mexico. Captain Joseph A. Kittinger and a U.S. Navy civilian astronomer, William C. White, ascend to 82,000 feet altitude and remain there for more than 18 hours. White uses a telescope mounted on the crew gondola to view the heavens and make observations.

27 December
The first six Lockheed SR-71 reconnaissance aircraft are ordered. The SR-71 is the follow-on of the A-12 already under development.

1963

2 January
All Allied forces at Bien Hoa AB are committed to execute Operation BURNING ARROW, a concentrated one-hour strike against Viet Cong targets near Ap Bac. After the main strike is completed, Piasecki H-21 helicopters drop supplies and lift troops into the area.

1 February
The American Institute of Aeronautics and Astronautics is officially formed. The AIAA combines the American Rocket Society (1930) and the Institute of the Aerospace Sciences (1932) into one organization.

6 February
A Titan II missile is fired by an Air Force crew for the first time when members of the 655th Aerospace Test Wing launch the missile at Cape Canaveral.

12 April
While flying an F-104 Starfighter at 30,000 feet, Jacqueline Cochran establishes a women's world speed record of 1,273.2 miles per hour over a straight course.

1 May
Jacqueline Cochran sets another women's speed record when she pilots a TF-104 Starfighter over a 100-kilometer course to a mark of 1,203.7 miles per hour. The flight is made at Edwards AFB, California.

7 May
Theodore von Kármán, long-time scientific advisor to the Air Force general staff and aerodynamics wizard, dies in Aachen, West Germany. He was 81 years old. The previous February, the Hungarian immigrant

Left: *"Major Cooper's Spacecraft," by Robert McCall. (USAF Art Collection)*

Below: *Initially designed for the Navy, the F-4 became the workhorse of the USAF for two decades. (NASM)*

Opposite: *The four-engine Lockheed C-141 Starlifter served as the USAF's primary strategic airlifters until the turn of the century. (NASM)*

received the first National Medal of Science for leadership aeronautics, astronautics, and defense from President John F. Kennedy.

An RB-57E flew an operational reconnaissance mission in support of Project PATRICIA LYNN in Vietnam.

15–16 May
Major L. Gordon Cooper Jr. launches into orbit in the *Faith 7* Project Mercury capsule. Cooper completes 22 earth orbits and remains aloft for 34 hours and 19 minutes, becoming the first American to remain in orbit for more than one day before returning to earth. The *Faith 7* capsule successfully splashes down 60 miles southeast of Midway Island. This is the final Project Mercury flight.

24 May
The first Lockheed A-12 crash occurs this date when tail number 06926 goes down near Wendover, Utah.

27 May
The McDonnell Douglas F-4C Phantom II flies for the first time near St. Louis, Missouri. More than 580 F-4C models will be built for the Tactical Air Command.

8 June
The first operational Titan II ICBM squadron is activated at Davis Monthan AFB, Arizona. Strategic Air Command's 570th Strategic Missile Squadron assumes the launch duties.

16 June
Junior Lieutenant Valentina Tereshkova, a Soviet cosmonaut, becomes the first woman in space when she spends three days in orbit in her *Vostok 6* capsule.

17 June
The Sikorsky CH-3C heavy transport helicopter flies for the first time. It features a hydraulically operated rear ramp.

20 July
The Lockheed A-12 makes its first flight over Mach 3.

20–21 July
The crew of an Air Force C-47, using the call sign "Extol Pink," take off on a flare-drop mission over a Vietnamese village under attack by Communist Viet Cong forces. During the mission, the pilot, Captain Warren P. Tomsett, is tasked to fly the Gooney Bird to a remote, unlighted airfield near the Cambodian border to extract six wounded Vietnamese soldiers. Accepting the mission, his crew quickly goes to work making fuel and navigation calculations, preparing to receive the patients, and keeping an eye out for enemy activity. The co-pilot, Captain John Ordemann, the navigator, Captain Donald Mack, the Crew Chief, Technical Sergeant Edsol Inlow, and the load masters, Staff Sergeant Jack

Morgan and Staff Sergeant Frank Barrett, prepare for the dangerous landing at Loc Ninh. Landing under fire, Tomsett and his crew successfully recover the wounded and return them to Bien Hoa AB for treatment. For their gallantry, the crew receive the Mackay Trophy.

25 July
A partial Test Ban Treaty that limits atmospheric, underwater, or outer space testing of nuclear weapons is initialed by Britain, the U.S., and the Soviet Union, and formally signed by the parties in Moscow on August 5. President Kennedy signs the treaty on October 7.

26 July
The first satellite to be placed in a geosynchronous orbit, *Syncon 2*, is launched this date. This type of orbit allows the satellite to remain over the same location on the earth at all times.

1 August
The Air Force launches NASA's Mariner II satellite, which is designed to make the first orbit around the sun—a journey of 540 million miles.

7 August
The YF-12A, designed as an advanced interceptor, is test flown at Groom Lake, Nevada, for the first time. James Eastham is the test pilot.

22 August
NASA test pilot Joe Walker pilots the X-15 to a remarkable 67 miles above the surface of the earth. This is the highest altitude reached during the X-15 test program.

16 October
The Air Force launches twin satellites supporting Project Vela Hotel. These 20-sided, 475-pound satellites—assume a 7,000-mile high, circular orbit on opposite ends of the earth.

A Convair B-58 Hustler flies from Tokyo, Japan, to RAF Greenham Common, Berkshire, United Kingdom, in eight hours and 35 minutes. This flight, part of Operation GREASED LIGHTNING, establishes a new world record for the 8,028-mile-long journey.

22 October
The YAT-37D flies for the first time. This aircraft is the armed version of the Cessna T-37 jet training aircraft.

1 November
South Vietnamese president Ngo Dihn Diem is assassinated.

22 November
President John Fitzgerald Kennedy is assassinated in Dallas, Texas.

29 November
A week after the assassination of President Kennedy, Cape Canaveral is renamed Cape Kennedy by executive order signed by President Lyndon B. Johnson.

10 December
Robert S. McNamara, Secretary of Defense, assigns development of the Manned Orbiting Laboratory to the Air Force.

The Air Force's X-20, Dyna-Soar program, is cancelled without ever flying.

Colonel Chuck Yeager suffers severe burns while egressing from a rocket-augmented NF-104A. The aircraft falls from an altitude of 90,000 feet to 10,000 feet, stuck in a flat spin, before he is able to eject from the plane. He recovers from his injuries.

17 December
The Lockheed C-141A Starlifter flies for the first time at Dobbins AFB, Georgia. This date also marks the 60th anniversary of the first manned, powered flight made by the Wright brothers at Kitty Hawk in 1903.

31 December
President Lyndon B. Johnson authorizes the deployment of U-2s to Vietnam. The 4028th Strategic Reconnaissance Squadron will begin flying operational missions in mid-January 1964. They will join the more than 100 USAF aircraft already deployed to South Vietnam.

TRAGIC LIMITS

1964–1975

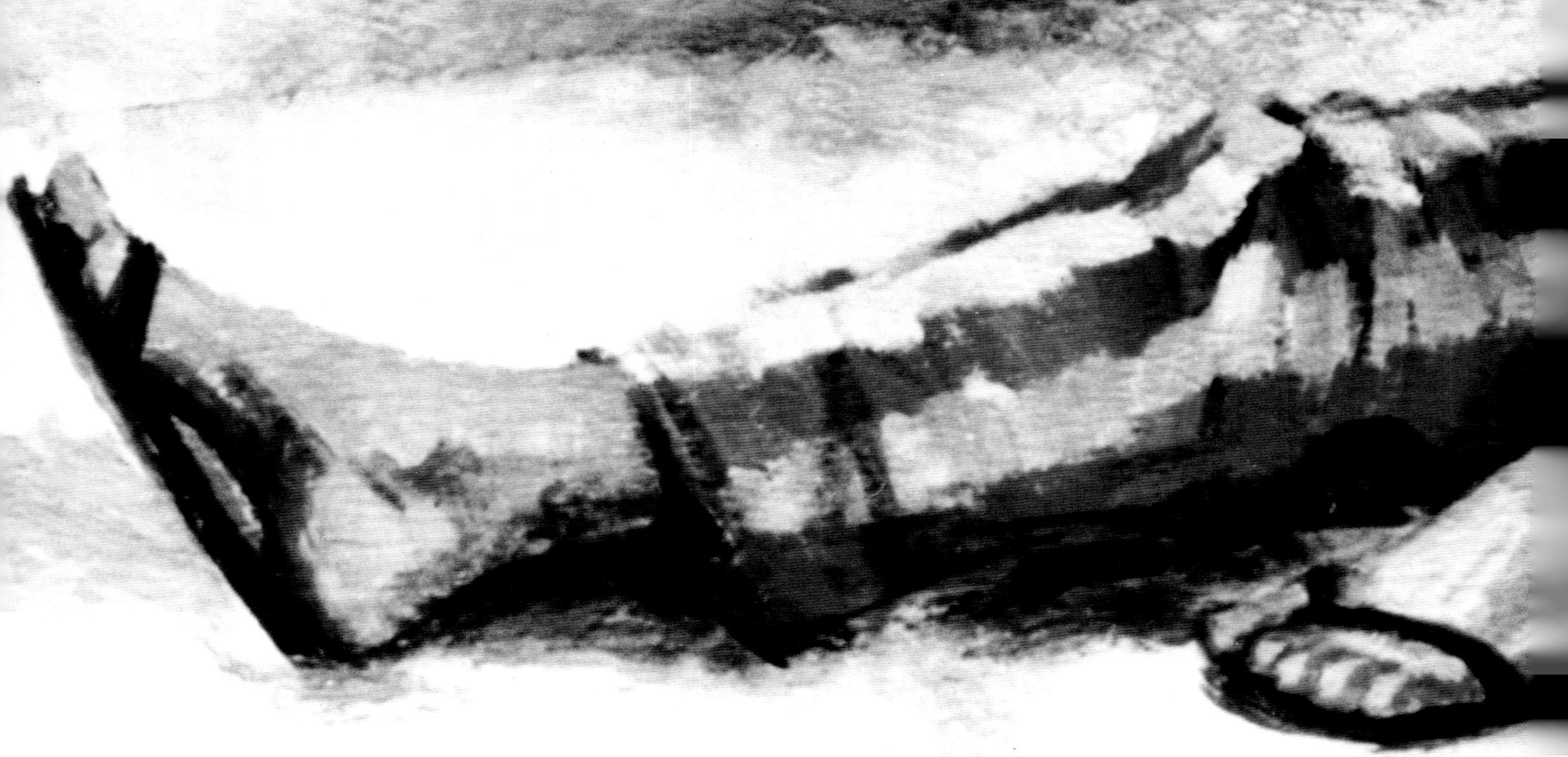

TRAGIC LIMITS

1964–1975

In August 1964, President Lyndon B. Johnson sent U.S. forces to Southeast Asia to fight a war that, over the next decade, would divide the nation and debilitate the military. Restrictions placed on war fighters resulted in ineffective campaigns against an elusive enemy that fought with tenacity and had apparent unlimited support from other major Communist nations. Yet, entry into this conflict, a Vietnamese civil war, was more gradual than a simple response to a single attack upon a few naval vessels in the Gulf of Tonkin. Americans had been providing military support in Southeast Asia for many years as "advisors" and "instructors" to the South Vietnamese military.

There may be no other decade in American history where such dichotomy between foreign and domestic events takes place concurrently. While the world had narrowly escaped cataclysmic destruction during October 1962, the race into space that followed on the heels of the Cuban Missile Crisis eventually resulted in one of the single greatest technological achievements in history—landing a man on the moon. This accomplishment was a great source of national pride at a time when the repercussions of the Tet Offensive still echoed across the country. Some questioned the wisdom of continuing the struggle to save South Vietnam from defeat at the hands of the Communists, while spending staggering amounts of money to beat the Soviets to the moon. Yet, the conflict continued, the nation's wounds grew deeper, and NASA landed on the moon six times.

Events during the Vietnam era were the result of complex political and social policies, but the technological advances in rockets and payload capacity had direct application to strategic military operations. Many of these developments fell to the Air Force for management and operational expertise. After all, it had been the Western Development Division, under General Bernard Schriever, that had forged the path to NASA's rocket program. In fact, the Air Force continued a close relationship with NASA throughout the 1960s and 1970s, both by supplying astronauts for the manned flight programs and also providing support for space flight operations such as radar tracking and communications.

It seems tremendously significant that the nadir of American involvement in Southeast Asia transpired within roughly the same time period (1968–1969) as astronaut Neil A. Armstrong's first steps on the moon.

During the Vietnam War, the first practical precision-guided weapons were used to drop bridge spans where conventional bombs could not deliver the knockout blow. At the tactical level, the Air Force proved highly effective at destroying targets as long as the targets could be located. Gaining and maintaining air superiority proved a greater struggle, particularly since aerial attacks into enemy sanctuaries in China were forbidden. Not wishing to involve such powerhouse enemy nations, overwhelming airpower was not contemplated until December 1972 during Linebacker II, and then only upon the North Vietnamese capital.

Building upon the developments of the previous decade, a wide variety of satellites were placed into orbit, enhancing communications and weather observation, and reconnoitering potential enemy nations. Basic rules for the use of space were established and emergency recovery plans were made in case any astronauts became stranded in orbit. The X-15 hypersonic research program continued to gather data that would be vital to the creation of NASA's space shuttle program.

Aeronautical development also continued during the decade. The massive Lockheed C-5 Galaxy, the largest aircraft built at the time, would make significant contributions to the Air Force's ability to carry out humanitarian missions around the world—from California to Pakistan. Although technical problems initially plagued the program, the C-5 grew into its heavy lift role and eventually became a superb complement to the capabilities of the Lockheed C-130 Hercules and C-141 Starlifter. Of course, beyond the many humanitarian missions flown by Air Force airlift aircraft, their primary responsibility

Pages 434–435: *Brutal treatment, torture, and death awaited captured military men in Southeast Asia. This depiction is intended to capture the plight of fellow pilots as seen by those who remained safe. ("For Our Comrades Up North," by Maxine McCaffery, USAF Art Collection)*

Right: *A group of Republic F-105 Thunderchiefs descends along Thud Ridge toward targets in North Vietnam. ("The Song Begins," by William Phillips, USAF Art Collection)*

remained the resupply and transport of troops throughout Southeast Asia.

Fighter development was not dormant either. The burly McDonnell F-4 Phantom II replaced the North American F-100 Super Sabre and served as an excellent complement to the Republic F-105 Thunderchief ("Thud") during the early years of the war in Southeast Asia. But, by the end of the war, more efficient and more powerful jet fighters were being designed—the twin-engined McDonnell Douglas F-15 Eagle and the single-engine General Dynamics F-16 Fighting Falcon took wing even before South Vietnam fell to the North in 1975. The durable F-4 would eventually be phased out in the Air Force inventory during the 1990s.

If it were not enough to be facing a war during which tragic limits were imposed upon America's military, there still loomed the Soviet Union—America's Cold War nemesis. To assure that any strategic nuclear attack could be adequately responded to, the Department of Defense built and then strengthened the strategic triad—long-range B-52 bombers, land-based ICBMs, and submarine-launched ballistic missiles (SLBMs). Forays to build faster high-flying bombers, like the XB-70 Valkyrie, were eventually abandoned. Aerial refueling assets were procured and then expanded, forming the early foundations of global attack and deployment capability. During the following decade, mass deployments from the U.S. to bases in Europe were a part of the war plan to stop any potential Soviet ground thrust through the Fulda Gap. At the same time, agreements between the U.S. and the U.S.S.R. to reduce nuclear arsenals were signed and enforced.

With the end of the war, many who had fought in it returned to the United States with burdensome memories and experiences that they would carry with them for a generation. Some pressed hard for changes in training. Major Moody Suter's brainchild resulted in the establishment of RED FLAG, a realistic combat exercise that sought to provide "combat" experience to young, inexperienced aircrew.

None suffered more than those held captive by the Communists. The brutal conditions tested the inner strength and resolve of each prisoner, each in his own way. Yet, most continually resisted their captors by communicating through the simply designed "tap code," a five-by-five matrix of the alphabet used to spell words and phrases. By establishing a chain of command and enforcing it, POWs were able to gain strength from those who together endured the unthinkable.

With the loss of Vietnam to Communism, the American military began to reassess itself, but the Cold War did not fade with the memories of Vietnam. For nearly 15 years, the Air Force would continue to train for all-out war with the Soviets. The triad remained on constant alert. Satellites and reconnaissance planes kept constant watch on potential trouble spots around the world. The Cold War Air Force was a vigilant force, a force ready at a moment's notice to strike a devastating strategic nuclear attack against the Soviet Bear.

TRAGIC LIMITS

1964–1975

"Houston, Tranquility Base here. The Eagle has landed."
—Astronaut Neil A. Armstrong, July 20, 1969, from the Lunar Excursion Module on the Moon

1964

29 February
President Lyndon B. Johnson reveals the existence of the Lockheed A-12 reconnaissance plane. He identifies the jet as the "A-11," an earlier name for the aircraft design, and also displays a photograph of the YF-12A that is supposedly still classified.

28 March
Operation HELPING HAND, a humanitarian relief mission to an earthquake-devastated region near Anchorage, Alaska, commences. During the next three weeks, more than 1,800 tons of supplies are delivered to those in need. All available cargo planes are utilized during the operation.

21 April
A strategic benchmark is reached when the number of Strategic Air Command intercontinental ballistic missiles (ICBMs) equals the number of ground alert SAC bombers for the first time.

11 May
Jacqueline Cochran establishes a new women's speed record when she flies a Lockheed F-104G Starfighter to the speed of 1,429.3 miles per hour over a 15-25-kilometer course.

18 May
The first production RF-4C Phantom II reconnaissance aircraft flies for the first time. The RF-4C will provide a significant amount of the tactical reconnaissance information gathered by the USAF during the Vietnam War.

9 June
Air Force KC-135 Stratotankers are used in refueling operations in Southeast Asia when eight F-100s refuel before striking Pathet Lao gun emplacements in Laos.

24 July
President Johnson announces the existence of the high-flying RS-71, accidentally calling it the "SR-71." The name sticks.

2 August
North Vietnamese patrol boats attack the Navy destroyer USS *Maddox* in the Gulf of Tonkin. Additional attacks allegedly carried out two days later by North Vietnamese ships on the *Maddox* and the USS *Turner Joy* out at sea are disputed.

Left: *The B-52 Stratofortress and the ICBM force formed two legs of the strategic nuclear triad during the Cold War. The third leg was the strategic submarine force capable of launching SLBMs. (NASM)*

Opposite, top: *RF-4s normally flew their low-level reconnaissance missions alone, and often at night. It was RF-4 crews that lived by the phrase—"Alone, Unarmed, and Unafraid." ("Night Sky Over Vietnam," by Frank Wootton, NASM Art Collection)*

Right: *Douglas A-1Es of the 1st Special Operations Squadron escort an HH-53 from the 40th Aerospace Rescue and Recovery Squadron over Vietnam. (USAF)*

Below: *The XB-70 was designed as a Mach 3 strategic nuclear bomber. High cost, improvements in ICBMs, and a deadly accident, which destroyed one of the two prototypes, resulted in the cancellation of the program. (NASA)*

5 August
President Johnson orders air strikes against North Vietnam in response to the events of August 2–4—the first overt steps that involve the U.S. in the Vietnamese civil war.

USAF assets begin to deploy to Southeast Asia in mass. B-57s, F-100s, F-102s, RF-101s, and F-105s all arrive in theater to prepare to begin operations.

7 August
The Gulf of Tonkin Resolution that authorizes "all necessary steps, including the use of armed force" to repel any attacks and assist allies, is passed by Congress.

14 August
The first Atlas/Agena D standard launch vehicle is successfully fired at Vandenberg AFB, California.

21 September
Piloted by company pilot Alvin White and Colonel Joseph Cotton, the North American XB-70A Valkyrie flies for the first time at Palmdale, California. This massive six-engine, delta-wing bomber will fly three times the speed of sound at altitudes up to 70,000 feet. Several problems occur during the flight. A landing gear retraction failure light illuminates, so the gear must be left in the "down and locked" position. One engine overspeeds and must be shut down. And upon landing, two tires blow when the wheels lock during touchdown. No production models are ever built, as surface-to-air missile technology poses too great a threat for such an aircraft.

1 November
Four Americans are killed and 72 wounded when the Viet Cong launch a mortar attack against Bien Hoa Air Base. Five B-57s are destroyed and 15 more are damaged. Additionally, four Vietnamese Air Force A-1s are destroyed in the attack.

10 November
Although the CIA denies it, the first operational sortie flown by an A-12 takes place over Cuba.

17 November
Lockheed C-130s from the 464th Troop Carrier Wing participate in Operation DRAGON ROUGE, a rescue of hostages being held in Zaire. After delivering the paratroopers, the 464th TCW transports refugees from Zaire to France. For their achievement, the 464th TCW is awarded the Mackay Trophy.

9–10 December
Air Force A-1H and A-1E Skyraiders inflict hundreds of casualties upon Viet Cong forces in Quang Tin and Binh Din Provinces.

10 December
The Air Force lofts a 3,700-pound satellite into orbit using a Titan III booster with "Transtage" technology. This new method places the entire third stage of the launch vehicle into orbit and then launches the satellite into a separate orbit.

UNSUCCESSFUL AIR WAR: VIETNAM

Mark Clodfelter

For the United States Air Force, the eight-year span of America's combat involvement in Vietnam was a frustrating mix of tactical successes and strategic disappointments. Air power's destruction of North Vietnamese divisions at Khe Sanh in 1968 and An Loc in 1972, complemented by airlift to beleaguered troops in both locations, contrasted starkly with the dismal failure of the 1965–1968 Rolling Thunder air campaign. In the end, America's strategic mistakes proved decisive, and air power could not salvage a war that did not conform to previous American experience and expectations.

Air Force aircraft, in concert with Navy and Marine Corps counterparts, dropped roughly eight million tons of bombs on Southeast Asia from 1964 to 1973. Four million tons fell on South Vietnam, three million tons on Laos, and a million on North Vietnam. For the air commanders, those bombs falling on North Vietnam counted the most, and Rolling Thunder aimed to sever North Vietnamese support from the Viet Cong insurgency in the South. Yet severe political restrictions, stemming from President Lyndon Johnson's fear of active Chinese or Soviet intervention, and from his desire to keep the American public focused on his "Great Society" programs, stymied the effort. Moreover, the type of war faced during much of the Johnson presidency—an infrequently waged guerrilla struggle in which the enemy fought an average of only one day a month, and hence needed minimal resupply—significantly limited bombing effectiveness. Onerous military controls, such as the notorious "Route Package" system that delineated Air Force from Navy bombing, further disrupted Rolling Thunder.

In South Vietnam, Air Force efforts to provide close air support and disrupt enemy positions achieved mixed results. When air power caught isolated enemy forces, such as in the Ia Drang Valley in November 1965, it proved devastatingly effective. When employed near populated areas, however, its results were less certain. The Viet Cong tactic of firing from inhabited villages could lead to air strikes that in turn produced

Opposite: *The first internal gun-equipped F-4E squadron in Southeast Asia stationed at Korat, Thailand, continued to deliver bad news to the North Vietnamese after replacing a squadron of aging F-105s. ("Bad News for Uncle Ho," by Keith Ferris, USAF Art Collection)*

Right: *B-52s delivered carpet bombing during ARC LIGHT operations in South Vietnam. ("Diamond Lil Under Fire (B-52D Over North Vietnam, December 1972)," by Richard Allison, USAF Art Collection)*

collateral damage. In a war fought to achieve a stable, independent, noncommunist South Vietnam, the winning of "hearts and minds" was vital to achieving success. Villagers killed by American bombing undermined that goal.

The air wars in Southeast Asia continued during Richard Nixon's presidency. The bombing of Communist forces and supply lines in Laos, started during the Johnson era as Operations BARREL ROLL and STEEL TIGER, gave way to Operation COMMANDO HUNT. Nixon also secretly began attacking suspected enemy positions in Cambodia with B-52s. Yet he initially refrained from bombing North Vietnam. In the aftermath of the public dismay spawned by the 1968 Tet Offensive, America now sought to end its military involvement, secure the release of its prisoners, and provide the South Vietnamese with a chance to control their own destiny. After Tet until spring 1972, the war continued as a predominantly guerrilla conflict waged largely by the NVA. American firepower, including great amounts of air power, had wrecked much of the Viet Cong when it chose to reveal itself during Tet's conventional battles.

At the end of March 1972, the North Vietnamese launched a large-scale invasion of South Vietnam dubbed the Easter Offensive. Nixon responded by bombing the Northern heartland and mining Northern ports. He could do so because his détente policy had effectively severed the North Vietnamese from their Chinese and Soviet benefactors. His use of air power in Operation LINEBACKER proved far more effective than Johnson's ROLLING THUNDER because of Hanoi's decision to wage a fast-paced, conventional war of movement that required significant amounts of logistical resupply. Stiffening South Vietnamese Army resistance also helped, and by July 1972 the Easter Offensive had stalled. In early October, North Vietnamese negotiators in Paris abandoned their demand for South Vietnamese President Nguyen Van Thieu's removal, but Thieu balked at an agreement that would leave North Vietnamese troops in South Vietnam. In December, Nixon turned to a second LINEBACKER air campaign, aimed at persuading the North Vietnamese to sign the agreement and to demonstrate to Thieu that bombing would resume if Hanoi violated the agreement's terms. Thieu reluctantly endorsed the settlement at the final hour, and the January 1973 Paris Peace Accords ended America's active involvement in the conflict.

Debate began almost immediately over whether Nixon-style use of air power would have achieved success in spring 1965. The conventional war in 1972 differed vastly from that of the Johnson presidency. More importantly, had bombing persuaded the North Vietnamese to stop fighting and had the Chinese and Soviets remained pacified, victory still could not follow unless the U.S. addressed the key reason that most Viet Cong fought—the corrupt, out-of-touch Saigon regime. As long as such a government endured, the application of American military force—and air power in particular—would have likely yielded dismal results.

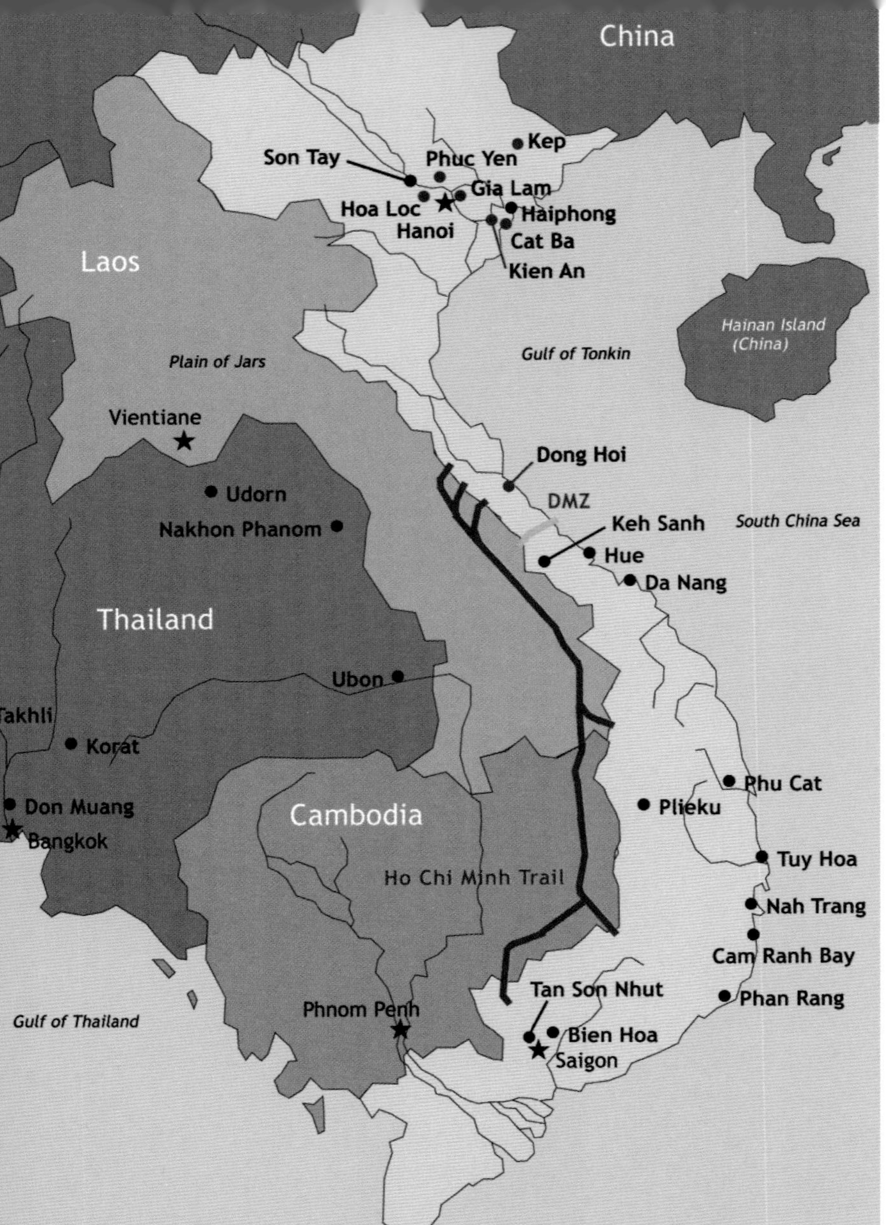

14 December
Operation BARREL ROLL, USAF armed reconnaissance and ground support missions over northern Laos, begin. BARREL ROLL missions will continue until the following April.

15 December
Captain Jack Harvey and crew carry out the first aerial gunship mission of the Vietnam War. Flying an FC-47 modified with side-firing Gatling guns, the crew strafes ground targets.

21 December
The YF-111A flies for the first time at Fort Worth, Texas.

22 December
Operation BIGLIFT, a humanitarian mission to deliver supplies to flood victims in California and Oregon, begins. During the following month, 1,500 tons of basic necessities are delivered to the region.

The prototype SR-71A Blackbird flies for the first time, accelerating to 1,000 miles per hour during a one-hour sortie.

1965

6 January
General Dynamics test pilots flying the F-111A fighter-bomber change the geometry of the swing-wing design in flight for the first time. The test is accomplished at 460 miles per hour at 27,000 feet, and no ill effects are noted during the transition.

21 January
The Air Force launches a 100-pound satellite from a pod on an Atlas ICBM. This Aerospace Research Satellite is designed to measure radiation and micrometeorites in orbit. The satellite is placed into a westward orbit, the first known orbit in such an orientation.

24 January
Former British Prime Minister Winston Churchill, dies at Hyde Park Gate, London. He was 90 years old. His funeral on January 30 is broadcast by television to the U.S. via the *Telstar 2* satellite.

1 February
General John P. McConnell becomes the USAF Chief of Staff.

4 February
The Air Force successfully test fires a Titan IIIC solid-fuel rocket booster. During the two-minute test, the engine generates 1.25 million pounds of thrust—25 percent more than anticipated.

7 February
Eight Americans are killed and 100 wounded when the Viet Cong attack Pleiku and other airbases in South Vietnam.

7–11 February
During Operation FLAMING DART, retaliatory strikes for the February 7 raids, Allied aircraft strike targets in North Vietnam. On February 8, an Air Force F-100 Super Sabre flies air cover for attacking South Vietnamese fighters. This marks the first air strike over North Vietnam by an Air Force aircraft.

18 February
The first USAF jet attacks against enemy troops in South Vietnam are flown by B-57 Canberras and F-100 Super Sabres against targets near An Khe.

Opposite: *This map of Southeast Asia highlights the route of the Ho Chi Minh Trail.*

Right: *During the early years of the Vietnam War, the F-100 Super Sabre, commonly known as the Hun, was the USAF ground attack workhorse. ("Bombs Away—Nam '67," by Charles Shealy, USAF Art Collection)*

2 March
Operation ROLLING THUNDER, a sustained air campaign against the North Vietnamese, begins and will continue for the next three and one-half years, until October 31, 1968.

On the first day of ROLLING THUNDER, Lieutenant Hayden J. Lockhart, flying an F-100 over North Vietnam, is shot down. Although he evades capture for one week, he is finally caught and held prisoner for the next eight years.

8 March
The first U.S. ground forces arrive in Vietnam; Marines are deployed to defend the air base at Da Nang.

23 March
Air Force Major Virgil "Gus" Grissom becomes the first two-time astronaut in space when he and Navy Lieutenant Commander John W. Young are launched in the first manned *Gemini III* space flight.

30 March–23 May
Air Force C-124s fly nearly 3,000 Danish United Nations Peace Keepers and 76 tons of cargo to Cypress to quell violence there.

3 April
Operation STEEL TIGER, the interdiction of the enemy's main supply route, begins. The campaign to disrupt the Ho Chi Minh Trail continues for the rest of the war.

3–4 April
F-105s make the first of many attacks against bridges in Vietnam. Although the attacking Thuds damage the Thanh Hoa Bridge, they cannot take out any of the spans. Two F-105s are shot down—the first USAF losses in aerial combat during the war.

20 April
The last production Atlas missile is placed into storage for use as a research and development launch vehicle. This marks the final phase-out of America's first-generation liquid-fueled ICBMs.

Left: *The Ho Chi Minh Trail, barely wide enough for a truck, was normally busy at night, filled with small motored vehicles and heavily laden bicycles and foot traffic. (USAF)*

AMERICAN POWS IN THE VIETNAM WAR

Colonel Fred Kiley, USAF (Ret)

More than 800 Americans fell into captivity in Southeast Asia between 1961 and 1973, about 770 of them military and 50 civilian; 130 died in captivity. Those totals are small for 12 years of conflict—for example, there were 7,140 U.S. POWs in the Korean War (1950–53), 2,701 of whom died in captivity. Fully half of the Vietnamese War POWs were incarcerated much longer than prisoners during the Korean War or World War II. The Vietnam War POWs were treated brutally and lived in miserable conditions as the enemy tried to exploit them for propaganda. That so many resisted so well for so long and survived the enemy's savagery is remarkable.

Before the first U.S. pilot was shot down over North Vietnam in 1964, about 15 POWs—advisors and reconnaissance pilots—had been taken in Laos, and almost 25—four Army officers, 12 Army NCOs, and eight civilians—in South Vietnam. Of these 40 captives, 14 died in captivity, six escaped, and 18 were eventually released—one of them, Army Captain Jim Thompson, after nine years.

These prisoners soon established communication—the life blood of resistance—by using a "tap code" learned at survival school by Air Force Captain "Smitty" Harris, one of the first POWs. This code was based on a 5 by 5 matrix of the alphabet, using C for both C and K. A great virtue of this code was that it could be tapped, coughed, blinked, flashed by hand, swept by broom, and flapped while drying laundry. Two of the early resistance leaders—Navy Commander Jim Stockdale and Air Force Lieutenant Colonel Robby Risner—who were captured in September 1965, immediately recognized its value and ordered all prisoners to learn it and to teach it to other prisoners.

The movements of prisoners between camps, along with continued torture and constant indoctrination, were aimed at intimidating, confusing, and demoralizing the POWs. A particularly low point came in July 1966, when 52 POWs were taken into central Hanoi, handcuffed together in pairs and paraded through the streets, exposed to crowds estimated at 70–100,000 people, and filmed for television. The guards assigned to the march did little to contain the brutality of the mobs. Every POW was brutalized.

However, conditions in the North Vietnamese prisons moderated in 1969 after the death of Ho Chi Minh. The POWs faced less torture, but they had new enemies born of introspection—boredom, loneliness, guilt, and despair. Elemental day-to-day coping became a supreme test. But on Thanksgiving Day 1970, a watershed event occurred. U.S. Special Forces raided an outlying prison near Son Tay, 22 miles northwest of Hanoi near the Red River. Unfortunately, the POWs had recently been moved from that camp, but the raiders killed a large group of Chinese advisors in a nearby building and escaped with no casualties. The effect of this daring raid was immediate and widespread. The next morning, the Vietnamese began emptying all the outlying prisons and concentrating the POWs in the big camps in central Hanoi. The result was that hundreds of men living in tiny cells found themselves in large barrack-style rooms of 30 or more. Morale and physical welfare improved at once. The men generally remained in those large camps for the rest of the war, with the exception of the movement in May 1972 of 210 POWs from the Hilton to Dogpatch, a new camp near the Chinese border.

Back in Hanoi, under the leadership of Colonel John Flynn, now the overall senior officer, new programs were in place to improve health, create an accurate list of all POWs who had ever been imprisoned in and around Hanoi, and pass time productively by holding classes in everything from aerodynamics to languages. Flynn created a new organization—the 4th Allied POW Wing. He called it the 4th because Vietnam was the 4th U.S. war of the century. He chose "Allied" because of the presence of South Vietnamese Captain Nguyen Quoc Dat, a long-held and much respected prisoner, and "Wing" because most of the POWs were aviators.

In December 1972, B-52 strikes shook the prisons and rained dust upon hundreds of cheering Americans. Word of the signing of the Peace Accords spread and rumors of repatriation began to look like more than rumors. The Dogpatch group returned in January 1973, and Operation HOMECOMING soon began. By late March the last of the Northern POWs had left Hanoi and returned to tumultuous receptions at home.

The saga of the POWs in South Vietnam is much more difficult to summarize. Their numbers were smaller—for example, about 50 in mid-1966, and 100 by late 1967, including about 20 civilians. These POWs were scattered in small groups up and down South Vietnam. Most were young soldiers and Marines. All were brutalized.

The story of the POWs in Southeast Asia—including two pilots and two CIA operatives who survived prison in China—is an epic tale of survival, endurance, resistance, and escape, a saga of tragedy and humor and death and, ultimately, of victory over a cruel enemy during dark and unforgiving times. No one has summed up their story better than Stuart Rochester in the conclusion of *Honor Bound*:

> *The company of men who walked the Hanoi March, trekked the Ho Chi Minh Trail at the point of a bayonet, and battled the enemy behind the lines from Briarpatch to Dogpatch survived the jails and jungles of Southeast Asia at great odds. Blessed in many instances with advantages of skills and seasoning when compared with U.S. POWs of previous generations, they were also endowed with qualities of heart and soul, grace and courage that in the end mattered more than their relatively high ranks. Those who made it back gave their countrymen an occasion to celebrate patriotism and heroism unencumbered by the vexing moral and political issues that beclouded so much of the war effort. In helping to achieve a healing, uplifting closure to the bitterly divisive conflict, the POWs, even when they were no longer incarcerated, continued to wield a symbolic power out of proportion to their small numbers. Their proud return to a grateful nation remains one of the few truly shining memories of that troubled era.*

Opposite: *"In His Country's Service," by Maxine McCaffrey. (USAF Art Collection)*

Below: *Colonel Robinson "Robbie" Risner makes a brief address to the crowd upon his return to freedom after seven and one-half years as a captive in North Vietnam. (USAF)*

Left: *Depicted in this painting is the first Air Force attack upon a SAM site. Several aircraft were shot down and, as it turned out, the target was a dummy built to deceive the attackers. After the strike, the Thud drivers, including a young Charles "Chuck" Horner, lit the afterburners and escaped over Thud Ridge. ("Thud Ridge," by James Laurier, USAF Art Collection)*

23 April
The first operational Lockheed C-141 Starlifter arrives at Travis AFB, California. The C-141 is a swept-wing, four-engine jet transport aircraft.

29 April–5 May
Operation POWER PACK, a peacekeeping and stabilization mission in the Dominican Republic, begins. Air Force C-130s and C-124s transport 12,000 troops and more than 17,000 tons of supplies to the region from Pope AFB, North Carolina. Air Force participants come from the Air Force Reserve, the Air National Guard, and from regular Air Force fighters and reconnaissance aircraft.

1 May
Colonel Robert L. Stephens sets a new world speed record when he pushes his Lockheed YF-12A to 2,070 miles per hour. Eight other speed records are established as well as a sustained level flight altitude record. For these accomplishments, the YF-12A/SR-71 Test Force consisting of Lieutenant Colonel Daniel Andre, Lieutenant Colonel Walter F. Daniel, Major Noel T. Warner, and Major James P. Cooney earn the Mackay Trophy.

12–18 May
Bombing over North Vietnam is halted during this period in an effort to test Hanoi's willingness to negotiate an end to the fighting. They do not respond.

22 May
F-105 Thunderchiefs attack North Vietnamese Army barracks for the first time, striking targets above the 20th parallel.

Below: *On June 3, 1965, Edward H. White II (USAF) became the first American to step outside his spacecraft (*Gemini IV*) and let go, effectively setting himself adrift in the zero gravity of space. For 23 minutes, White floated and maneuvered himself around the Gemini spacecraft, logging 6,500 miles during his space walk. In his right hand, White carries a Hand Held Self Maneuvering Unit (HHSMU), which is used to move about the weightless environment of space. The visor of his helmet is gold plated to protect him from the unfiltered rays of the sun. (NASA)*

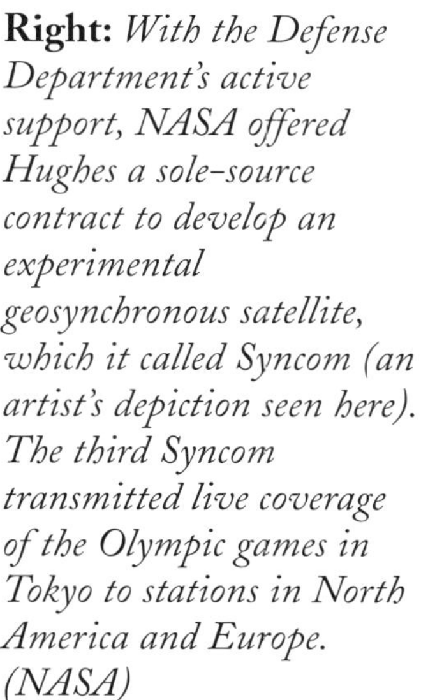

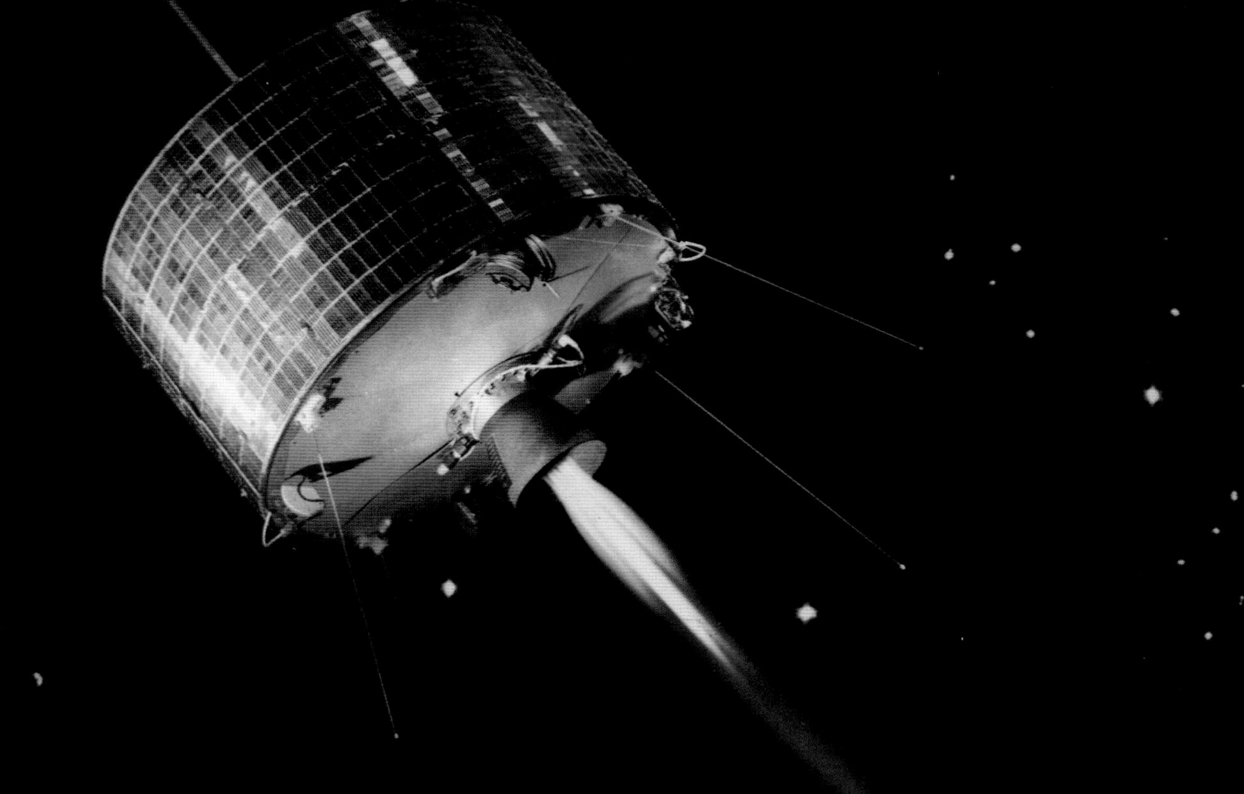

Right: *With the Defense Department's active support, NASA offered Hughes a sole-source contract to develop an experimental geosynchronous satellite, which it called Syncom (an artist's depiction seen here). The third Syncom transmitted live coverage of the Olympic games in Tokyo to stations in North America and Europe. (NASA)*

3 June–7 June
Air Force astronauts Major Edward H. White and Major James A. McDivitt establish a U.S. space endurance record, just over 97 hours during 63 earth orbits. Major White becomes the first U.S. astronaut to walk in space when he leaves the capsule tethered by life support lines and carries a hand-held gas propulsion unit for mobility on June 4. He floats freely in space for 23 minutes before returning to the security of the capsule.

18 June
The first B-52 Stratofortress raids of the Vietnam War—called ARC LIGHT—are carried out by 28 Guam-based bombers. Viet Cong targets near Saigon are attacked. This is the first time that the mammoth B-52 is used in combat. ARC LIGHT missions will be flown until August 1973.

An Air Force Titan III lifts a 10.5-ton payload into orbit. This is the most powerful launch vehicle built at this time and generates 2.5 million pounds of thrust when the central liquid three-stage rocket and the two strap-on solid boosters are fired.

30 June
The 800th Strategic Air Command Minuteman I ICBM becomes operational at F. E. Warren AFB, Wyoming. This is the last Minuteman I to join the operational ranks.

8 July
The Air Force Satellite Control Facility assumes control of former NASA satellites *Syncom II* and *III* on behalf of the Department of Defense. These satellites were placed in geosynchronous orbits to relay communications and weather information.

10 July
Two 45th Tactical Fighter Squadron (TFS) F-4C Phantom II aircraft shoot down two MiG-17 Fresco jet fighters. These are the first USAF aerial victories of the Vietnam War. The aircrews flying the mission are: Captains Thomas S. Roberts (pilot) and Ronald C. Anderson (WSO); and Captains Kenneth E. Holcombe (pilot) and Arthur C. Clark (WSO).

16 July
The YOV-10A Bronco flies for the first time. The Bronco will begin operations in Vietnam in 1968 and will serve as an excellent Forward Air Controller (FAC) platform.

23 July
A Soviet-built SA-2 surface-to-air missile (SAM) shoots down a U.S. aircraft, an F-4 Phantom II, for the first time.

5 August
During a full-duration static test of the *Saturn V* first-stage booster, the five engines generate approximately 7.5 million pounds of thrust during the 2.5-minute burn. This test stage never flies in space and is turned over to the Smithsonian Institution, National Air and Space Museum. The *Saturn V* is lent to the Kennedy Space Center, where it is on display.

21–29 August
Air Force astronaut L. Gordon Cooper Jr. and Charles Conrad Jr. (USN) launch onboard *Gemini V* for a week-long mission in space. During the week, they orbit the earth 120 times.

Right: *The UH-1 Iroquois, best known as the Huey, provided mobility, rescue, and observation capability and was used by all services. (NASM)*

Below: *Hurricane Betsy, which struck the Gulf Coast in September 1965, as photographed by the TIROS VII satellite. Betsy took 75 lives and became the first storm to cause more than $1 billion in damages. The storm became known as "Billion Dollar Betsy." (NOAA)*

8 September
Photographed by four different *Tiros* satellites, Hurricane Betsy is the first significant weather event monitored by satellite.

15–21 September
Air Force C-130s evacuate more than 1,000 Americans during Operation NICE WAY after India and Pakistan go to war.

23 October
The first Northrop F-5E fighters arrive in Vietnam for duty with the 4503d Tactical Fighter Squadron.

31 October
The first 10 Minuteman II ICBMs are accepted by the Air Force and assigned to the 447th Strategic Missile Squadron at Grand Forks AFB, North Dakota. The second-generation Minuteman is larger and more accurate than the first but still fits into the same silo as the Minuteman I.

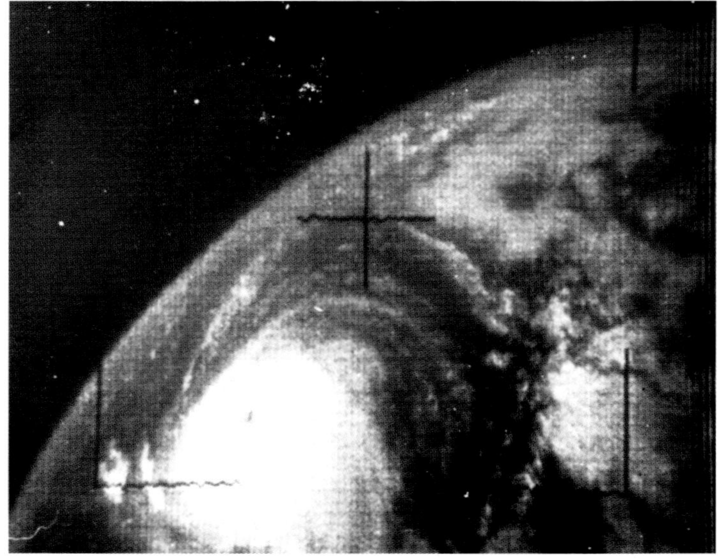

1 November
Colonel Jeanne M. Holm is appointed director of the Women of the Air Force. Holm will rise to the rank of Major General, the first woman in American military history to achieve two-star rank.

14–16 November
The Battle of Ia Drang Valley marks the first major battle between U.S. troops and North Vietnamese Army regulars (NVA) inside South Vietnam. American Army troops of the 1st Cavalry Division (Airmobile) under the command of Lieutenant Colonel Hal Moore respond to the NVA threat by using helicopters to fly directly into the battle zone. Upon landing, the troops quickly disembark then engage in fierce fire fights, supported by heavy artillery and B-52 air strikes, marking the first use of B-52s to assist combat troops. The two-day battle ends with NVA retreating into the jungle. Seventy-nine Americans are killed and 121 wounded. NVA losses are estimated at 2,000.

27 November
Approximately 35,000 war protesters encircle the White House and then proceed to the Washington Monument to continue their protest.

15 December
The crews of *Gemini VI* and *Gemini VII* rendezvous in space. Captain Walter M. Schirra (USN) and USAF Major Thomas P. Stafford in *Gemini VI* maneuver

Right: *During early Wild Weasel missions, an F-100 would attack and destroy the enemy radar site and the F-105s would then find the surface-to-air missiles and attack them. (NASM)*

Below: *The Lockheed SR-71 has a smooth chine from the tip of the nose to the wing root, a modification from earlier A-12 models flown by the CIA. ("Edge of Space," by Harold McCormick, USAF Art Collection)*

within feet of the *Gemini VII*. The *Gemini VII* crew, Air Force astronaut Frank Borman and James A. Lovell (USN), establish a 14-day endurance record that lasts for the next five years.

10 December
U.S. Pacific Command divides North Vietnam into six "route packages"; Route Pack 1, 5, and 6B were covered by the USAF, while RP 2, 3, 4, and 6A belonged to the Navy.

22 December
In the first successful Wild Weasel attack, Air Force F-100F Wild Weasels attack and destroy a Fan-Song radar north of Hanoi, while F-105 Thunderchiefs attack and destroy the associated SA-2 SAM site.

23 December–23 January
Operation BLUE LIGHT, the transport of the U.S. Army's 3d Infantry Division from Hickam AFB, Hawaii, to Pleiku, South Vietnam, begins. The airlift operations will take nearly one month and will end up as the largest such combat operation in history. More than 4,600 tons of equipment and 3,000 soldiers are delivered.

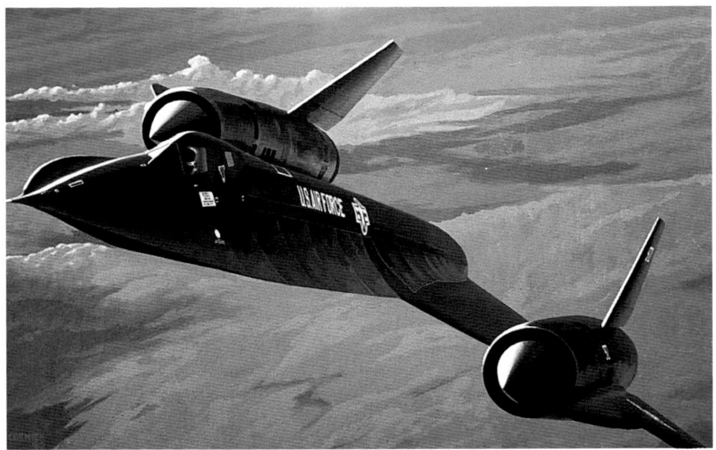

1966

1 January
Air Force airlift units are redesignated. Military Air Transport Service (MATS) becomes Military Airlift Command (MAC), Eastern Air Transport Force becomes Twenty-first Air Force and Western Air Transport Force becomes Twenty-second Air Force.

Augmenting the Military Airlift Command (MAC), the Air National Guard (ANG) begins flying 75 airlift sorties each month from the U.S. to Southeast Asia. The ANG already covers about 100 overseas sorties each month in a variety of roles.

7 January
The first operational SR-71 is delivered to the 4200th Strategic Reconnaissance Wing at Beale AFB, California.

17 January
A B-52 carrying four nuclear weapons collides with a KC-135 Stratotanker in the skies over Spain. Seven of the eleven flight crew members are killed in the collision. Thankfully, all of the nuclear weapons are recovered from the crash site; the last one is recovered at a depth of 2,500 feet in the Atlantic Ocean off the Spanish coast.

4 March
Three enemy MiG-17s attack a flight of F-4C Phantom II fighters over North Vietnam. After an unsuccessful attack the MiGs disappear to their base near the Communist capital.

JEANNE M. HOLM

Rebecca Grant

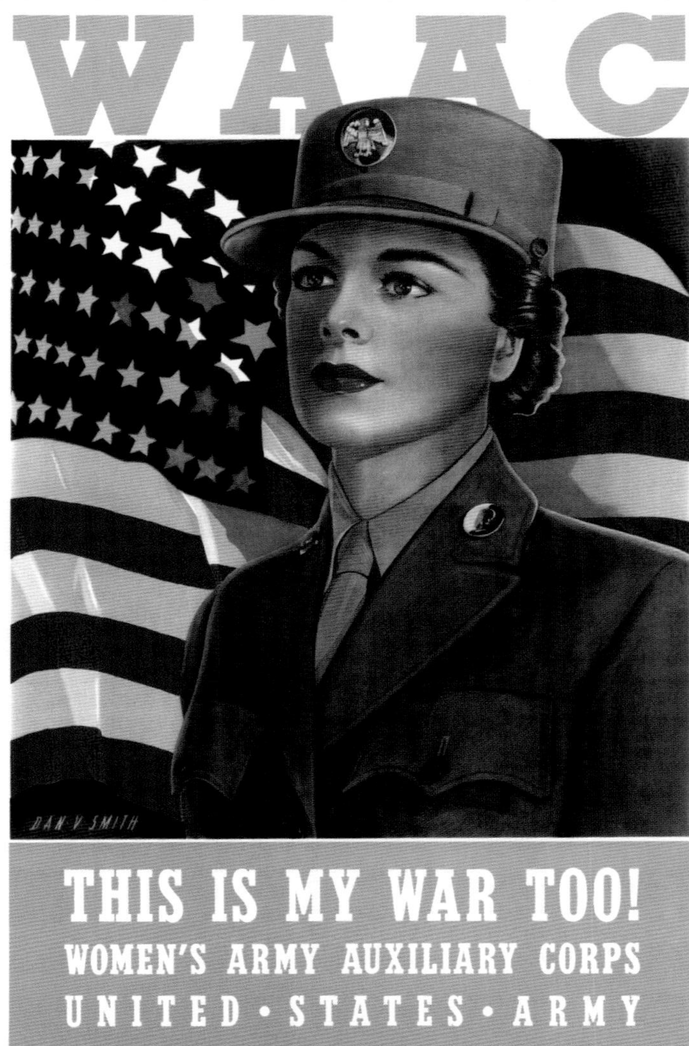

WAAC
DAN V. SMITH
THIS IS MY WAR TOO!
WOMEN'S ARMY AUXILIARY CORPS
UNITED • STATES • ARMY

She was a professional silversmith from Portland, Oregon, when America entered World War II. Enlisting in the Women's Army Auxiliary Corps in July 1942 was step one for young Jeanne Marjorie Holm—a step that led her to a distinguished career, when she became the Air Force's first female brigadier general and the first American woman to pin on the two stars of a major general.

Holm's Army service in World War II took her to command of a basic training regiment in Georgia, then to command of the 106th WAC Hospital Company in West Virginia. With the war over, she returned to civilian life as a college student. But the outbreak of the Berlin crisis in 1948 saw Holm return to active duty, and in 1949 she transferred to the Air Force. Her next assignment was Germany, and here Holm began to make her mark. In a three-year overseas tour she worked in plans and operations for a depot wing and as a War Plans officer for the 85th Air Depot Wing.

Holm was the first woman officer to enroll in the Air Command and Staff College at Maxwell Air Force Base, Alabama—the first of her notable "firsts." After graduation, Holm took an assignment to the Pentagon in personnel, plans, and programs. It was the beginning of the main focus of her Air Force career. Personnel and manpower assignments at NATO and back at USAF headquarters followed. In 1965, Holm was named as Director, Women in the Air Force, a position under the Deputy Chief of Staff for Personnel. Holm retained the position for eight years. As the top officer directing women's careers, Holm had far-reaching responsibilities at a time of change for women in society at large as well as in the Air Force.

She brought a can-do attitude to the task and turned it into a basis for enlightened personnel policy. Her philosophy was that the Air Force needed its women members, and those of the "old school" should learn to incorporate women for the good of national defense. That philosophy was never clearer than in a 1981 statement by Holm defending the ability of women to serve.

"It's a bum rap," Holm said of the stereotype criticisms of women in uniform. "Men get VD, men go AWOL, men get into fights. The military has learned to deal with these as personnel problems that can be solved. But military leaders aren't used to women's problems and there is a tendency to see them as insoluble. The military must realize that it must deal as effectively with women's problems as it has with men's problems if it wants to have an effective force," said Holm.

While Holm was Director for Women in the Air Force she more than doubled the number of women officers on active duty. She set policies to open more job opportunities to women and took time to update the uniforms they wore. Later, as the services moved

Opposite: *Jeanne Holm signed up for the Women's Army Auxiliary Corps (WAAC) during World War II. (U.S. Army Women's Museum)*

Right: *General Jeanne Holm, born in 1921, in Portland, Oregon, enlisted in the Army in July 1942, soon after the Women's Army Auxiliary Corps was established by Congress. She attended Officer Candidate School at Fort Des Moines, Iowa, and in January 1943, received a commission as a "Third Officer," the WAAC equivalent to second lieutenant. (USAF)*

into the era of the all-volunteer force, Holm consistently made the point that opening military jobs to women was an essential component of the policy—and of avoiding a return to the peacetime draft.

Holm herself was a major role model. Effective July 16, 1971, she was promoted to brigadier general. Wearing the star, Holm made history. She did it again with her promotion in 1973 to the rank of major general (with a 1970 date of rank). Holm was the first woman ever to serve in that grade in the U.S. Armed Forces.

Although Holm retired from active duty in 1975, her public service did not end. She served as special assistant to the President for Women from 1974–1977. In later years, Holm continued to champion the military service of women. She authored two books on women in the military, and was a longtime member of DACOWITS, the authoritative Defense Advisory Committee on Women in the Services.

In 2003, the Air Force Association honored Major General Holm with a lifetime achievement award for her long service and singular accomplishments in expanding career and duty assignments for women.

Holm stands as a role model, but her 1998 written description of women who served in World War II also sums up the foundation of her own success: "They loved their country and enriched its traditions with their bravery." So did she.

Left: *A Boeing B-47 crew runs to the jet during an alert exercise. (USAF)*

Below: *Airman First Class William H. Pitsenbarger volunteered to remain behind with Charlie Company to treat its wounded. Rather than endanger his crewmates and his helicopter during a pickup, he stayed with the soldiers, fighting and treating wounded until he himself was killed in action. ("A1C William H. Pitsenbarger," by John Witt, USAF Art Collection)*

7 March
French president Charles DeGaulle announces that he is taking France out of the United Nations and requests that all foreign military forces within French borders be removed. Several USAFE units are forced to relocate to honor this request.

10 March
Major Bernard F. Fisher, 1st Air Commando Squadron, successfully lands his A-1E aircraft in Vietnam on the dangerous A Shau Valley airstrip, which had been overrun by the enemy and was covered with battle debris and parts of an exploded aircraft. Fisher rescues downed airman Major Dafford Myers and is later awarded the Medal of Honor. This is the first USAF Medal of Honor awarded during the Vietnam War and the first to be awarded as the "Air Force Medal of Honor" having been redesigned in 1963.

16–17 March
Gemini VIII is forced to return to earth earlier than expected due to a faulty spacecraft maneuvering thruster, which induced a disorienting roll of one revolution per second in the *Gemini* capsule. Astronauts Neil A. Armstrong and Air Force astronaut David Scott are forced to splashdown in the secondary return zone. Air Force rescue forces recover the capsule within 20 minutes after splash down in the Pacific Ocean. This marks the first time that Air Force has participated in a *Gemini* recovery operation.

31 March
The Boeing B-47 Stratojet stands down as the Strategic Air Command completes the phase-out of the first jet, swept-wing bomber.

1 April
Seventh Air Force, headquartered at Saigon, is made a subcommand of Pacific Air Forces (PACAF), with Lieutenant General Joseph H. Moore maintaining his command.

6 April
The U.S. Army transfers its intratheater tactical airlift to the Air Force. The USAF redesignates these as the C-7A Caribou and the C-8A Buffalo.

11 April
Airman 1st Class William H. Pitsenbarger, 38th Aerospace Rescue and Recovery Squadron, gives his life 35 miles east of Saigon while providing aid for the wounded during field evacuations. He remains with those awaiting pickup, despite mortal wounds, and cares for those wounded who remained in the field. For his heroism, Pitsenbarger is posthumously awarded the Medal of Honor.

Right: *An American POW is paraded through the streets of North Vietnam subjected to jeering crowds and brutality. (USAF)*

12 April
B-52 Stratofortresses strike targets in North Vietnam for the first time, striking supply lines at the Mu Gia Pass 85 miles north of the border.

25 April
The 447th Strategic Missile Squadron becomes fully operational—the first SMS equipped with Minuteman II ICBMs.

26 April
An F-4C piloted by Major Paul J. Gilmore with weapon system officer, 1st Lieutenant William T. Smith, becomes the first USAF crew to score an aerial victory over a MiG-21 fighter. The crew was providing escort for a flight of F-105 Thunderchiefs on a ground attack mission near Hanoi.

3 May
An HC-130H successfully completes the first "air snatch" of a man from the ground at Edwards AFB, California. The Fulton Recovery System—a scissors device attached to the nose of the aircraft—is used to accomplish this daring and dangerous extraction.

2 June
Surveyor 1 soft-lands on the surface of the moon—the first U.S. spacecraft to do so. Live television pictures of the lunar surface are relayed via the *Early Bird* satellite to televisions around the world.

3 June
Air Force astronaut Thomas Stafford and Eugene Cernan (USN) launch into orbit aboard *Gemini IX*. Orbital problems with the Agena Target Docking Adapter result in an early return to earth on June 6.

8 June
In a tragic accident, the second XB-70A prototype is destroyed when an F-104N Starfighter piloted by NASA Chief test pilot Joe Walker collides with the right wingtip and vertical tail of the supersonic bomber. Several aircraft powered by General Electric–built engines were flying in formation for a "family portrait" when the F-104N was enveloped by the Valkyrie's strong wingtip vortices. Walker's F-104 struck the XB-70's right wing and then was hurled in a corkscrew-like maneuver into the vertical tails and then through the left wing. His F-104 exploded almost immediately and the aft tip and vertical tail of the XB-70 were torn from the jet. Air Force Major Carl Cross, on his first flight in the Valkyrie, is unable to activate his escape capsule and dies in the ensuing crash.

1 July
General William W. Momyer assumes command of the Seventh Air Force at Tan Son Nhut AB, Vietnam.

6 July
Radio Hanoi reports that American POWs have been paraded through the streets and forced to endure the jeering of North Vietnamese crowds.

Left: *The MiG-21 was difficult to see in the air, particularly head-on. It was small compared to the F-4, but did not have the same radar targeting capability of American jet fighters. (USAF)*

Below: *Depicted here is a Lockheed C-141B on the ice at McMurdo Sound, Antarctica. The resident emperor penguins show little fear of anything except polar predators. ("Inspection Party," by Keith Ferris, USAF Art Collection)*

18–21 July
Gemini X crew, astronaut John Young and Air Force astronaut Michael Collins, make two rendezvous and docking maneuvers with the Agena target vehicle and complete a tricky EVA before returning to earth.

20–29 August
U.S. forces react swiftly after a massive earthquake hits four Turkish provinces. With initial estimates of more than 2,000 deaths, more than 50 tons of medical supplies, water filtration units, and medical personnel are transported to the region by USAFE aircraft.

3 September
North Vietnamese MiG-21s launch in large numbers for the first time. These new fighters are based at airfields around Hanoi which is off limits for attack by U.S. forces.

20 September
Lieutenant Colonel Donald M. Sorlie pilots a NASA M2-F1 "lifting body" to a successful landing at the Air Force Flight Test Center at Edwards AFB, California. This marks the first lifting body flight piloted by an Air Force pilot.

11 November
The last Gemini flight, *Gemini XII*, carries James Lovell (USN) and Air Force astronaut Buzz Aldrin into orbit where they complete three EVAs and dock with the Agena target vehicle.

14 November
Captain Howard Geddes pilots a C-141 Starlifter to a safe landing in the Antarctic. The 2,200-mile flight originated in Christchurch, New Zealand, and lands on the ice at McMurdo Sound.

14 December
While participating in a night combat strike in Southeast Asia, Colonel Albert R. Howarth demonstrates exemplary airmanship and courage under fire. For his actions his is awarded the Mackay Trophy.

24 December–31 January 1967
President Lyndon B. Johnson calls for a Christmas cease-fire to press for a peace initiative with Communist leaders in Hanoi who do not respond to the overture.

Right: *The Lockheed C-130 was the tactical airlift workhorse during the war in Southeast Asia. This aircraft has just redeployed ground troops from the highlands to the coastal plain. (USAF)*

Below: *Astronauts (left to right) Gus Grissom, Ed White, and Roger Chaffee, pose in front of Launch Complex 34, home to their Saturn 1 launch vehicle. The astronauts later died in a fire in the capsule on the pad. (NASA)*

1967

2 January
Demonstrating flexibility and cunning, the 8th Tactical Fighter Wing, commanded by Colonel Robin Olds, conducts Operation BOLO, an attempt to trick enemy MiGs to engage in combat with superior American F-4 Phantom IIs. Flying the normal routes used by F-105s, three waves of F-4s are engaged by unknowing MiG-21s. Seven different F-4 crews shoot down seven MiGs in 12 minutes during the mission over the Red River Valley—a single-day aerial victory record. Colonel Olds adds one victory to his total and becomes the only USAF ace to score victories in World War II and Vietnam.

18 January
A Titan IIIC launch vehicle sends a Defense Department package of eight Defense Communications Satellites into orbit simultaneously.

27 January
During a mission rehearsal and emergency procedures exercise, three *Apollo* astronauts, Lieutenant Colonel Gus Grissom (USAF), Lieutenant Colonel Ed White (USAF), and Lieutenant Commander Robert Chaffee (USN), are killed when a fire ignites inside their sealed *Apollo* capsule at Cape Kennedy, Florida.

8–13 February
A cease-fire is held for the Tet religious holiday in hopes that North Vietnam might be willing to open negotiations. Instead, the Communists move 25,000 tons of supplies farther south to support operations there.

22 February
The first paratroop drop of the Vietnam War is made when 23 C-130 Hercules aircraft insert the 173d Airborne Brigade in support of Operation JUNCTION CITY, a massive search and destroy operation spearheaded by the Army's Big Red One along the Cambodian border.

24 February
Captain Hilliard A. Wilbanks, 21st Tactical Air Support Squadron, flying an unarmed Cessna O-1 Bird Dog as a forward air controller near Dalat, South Vietnam, gives his life to act as close air support for a forward element of South Vietnamese rangers. He fires his rifle and smoke rockets from the side window of his aircraft to distract the advancing enemy ground forces. During his last attack, he is mortally wounded and his bullet-riddled aircraft crashes between the opposing forces. For his bravery, Wilbanks is awarded the Medal of Honor.

Above: *An F-105 pilot guns down an enemy MiG-17 Fresco over Vietnam. (NASM)*

10 March
Captain Merlyn H. Dethlefsen, 354th Tactical Fighter Squadron, flying F-105s over North Vietnam, is hit while attacking a surface-to-air missile site. Ignoring the enemy's overwhelming air and ground firepower, he repeatedly attacks and renders ineffective many of the enemy defenses, allowing the accompanying fighter-bombers to complete their mission. For his bravery, Dethlefsen is awarded the Medal of Honor.

Captain Max Brestel, flying an F-105 Thunderchief, shoots down two MiG-17s in one mission.

10–11 March
The Thai Nguyen steel factory is attacked by Air Force F-105s and F-4s stationed at Ubon, Thailand, for the first time.

11 March
Air Force aircraft attack the Canal des Rapides bridges four miles outside of Hanoi.

15 March
The Sikorsky HH-53B makes it first flight. The HH-53 is the largest and fastest helicopter in the USAF inventory.

22 March
A B-52 base is established at U-Tapao, Thailand, to absorb the overflow from Anderson AB, Guam.

25 March
The 6th Strategic Wing at Eielson AFB, Alaska, assumes the primary role of strategic reconnaissance. The 6th flies a variety of RC-135 aircraft including the RC-135S *Cobra Ball* and the RC-135X *Cobra Eye*.

3 April
Paul W. Airey becomes the first Chief Master Sergeant of the Air Force. In this position, Airey will advise senior leadership on enlisted force issues.

10 April
B-52s flying out of U-Tapao Air Base, Thailand, fly their first missions of the war. Staging from Thailand, the bombers do not need aerial refueling during their missions.

19 April
Major Leo K. Thorsness, 357th Tactical Fighter Squadron, successfully completes his mission of destroying a surface-to-air missile site and then, despite low fuel and heavy anti-aircraft defenses, remains in the target area to ward off multiple MiG-17 aircraft. For his actions, Thorsness is awarded the Medal of Honor. Thorsness is later shot down on April 30 and spends the rest of the war as a POW.

Opposite, bottom: *Chief Master Sergeant of the Air Force Paul Wesley Airey was adviser to Secretary of the Air Force Harold Brown and Chief of Staff of the Air Force General John P. McConnell. He was the first chief master sergeant appointed to this position. (USAF)*

Right: *Four Missouri Air National Guard F-4s fly past the "Phantom Phactory." McDonnell and Douglas merged in April 1967. (NASM)*

25 April
Retired Major General Benjamin D. Foulois dies at age 87 at Andrews AFB, Maryland. Foulois had taught himself to fly a Wright aircraft, had accompanied General John J. Pershing into Mexico in search of Pancho Villa, and as Chief of the Air Corps had taken on the task of delivering the U.S. mail when civilian contracts had been cancelled in 1934.

26 April
MiG airfields at Kep and Hoa Lac are attacked for the first time. Airfields closer to Hanoi remain off limits.

28 April
Operation CREEK PARTY begins when KC-97L Air National Guard crews initiate refueling operations from Ramstein Air Base, Germany. This all-volunteer operation will continue for the next decade and marks the first time that the ANG has conducted sustained overseas operations supporting real contingencies.

The McDonnell Douglas Corporation is formed when the two parent companies merge.

Five *Vela* satellites are launched into orbit aboard one Titan IIC launch vehicle. These satellites are designed to detect nuclear tests on earth, particularly in China.

31 May
Major John H. Casteel and his crew—Captain Dean L. Hoar, Captain Richard L. Trail, and Master Sergeant Nathan C. Campbell—fly a 902d Air Refueling Squadron KC-135 Stratotanker over the Gulf of Tonkin, Vietnam. During their station time, the KC-135 crew executes several emergency refueling rendezvous with fuel-starved Navy fighters. Casteel's crew manages to save six of these Navy planes. For this remarkable achievement, Casteel and his crew receive the Mackay Trophy.

31 May
The first A-12 mission flown over North Vietnam is piloted by the CIA's Mel Vojvodich. The mission lasts three hours and 40 minutes.

1 June
Air Force flight crews fly two Sikorsky HH-53E helicopters from New York to Paris, retracing Lindbergh's first solo transatlantic flight taking nearly 31 hours. Nine refuelings are required to accomplish this feat—the first nonstop transatlantic helicopter flight.

5–11 June
When war breaks out between Israel and the Arab states, U.S. citizens are moved to Wheelus Air Base, Libya, to protect them from elevated hostilities in the region. During this week, more than 8,000 people are evacuated to Europe from Libya and Jordan during these operations.

1 July
The Air Force launches a Titan IIIC launch vehicle that places six additional satellites into orbit, supporting the Initial Defense Communications Satellite Program (IDCSP). These satellites join the 18 already in orbit in support of IDCSP.

30 July
The General Dynamics FB-111A makes its first flight at Forth Worth, Texas.

THE SIKORSKY H-53

John F. Guilmartin Jr.

It is unlikely that the Air Force, ever ambivalent about helicopters, would have procured a large, high-performance helicopter were it not for the demands of the air war against North Vietnam. The stimulus was the need to rescue airmen downed deep within enemy territory. When the ROLLING THUNDER campaign began in March of 1965, the primary Air Force rescue helicopter was the Kaman HH-43B, (the first H for rescue and the second H for helicopter), with limited speed, range and payload that rendered it useless for long range combat rescue. Something better was needed, and the Sikorsky H-53, being developed for the Marine Corps as a heavy-lift helicopter, was the obvious choice. The H-53 was not yet operational, and the Marines had first call on production. In the interim, the Air Force procured a modified version of the smaller Sikorsky H-3. The result was the HH-3E, fitted with an external hydraulic rescue hoist; jettisonable auxiliary fuel tanks; armor protection for crew members, engines, gearboxes, and critical flight control components; a full radio communications suite; a Doppler radar navigation system; and, later, a retractable probe for air-to-air refueling. The HH-3E entered combat in October 1965 and proved highly successful, but was power-limited to a crew of four—pilot, co-pilot, flight mechanic, and pararescueman—and a meager defensive armament of two 7.62mm M-60 machine guns.

The H-53, designed in response to a 1961 Marine Corps request, evolved from the piston-engine HR2S, the product of 1946 requirement for a helicopter with a

5,000-pound payload—enormous for the time—and the H-53's exceptional performance was a result of the far-sighted requirement. The H-53 inherited the HR2S's drive train and tail rotor and a main rotor with the same 72-foot diameter, but with six blades instead of five. The key difference was the replacement of the HR2S's R-2800 piston engines with General Electric T-64 turbines, yielding enormous dividends in speed, payload, and reliability. The H-53 beat the tandem-rotor Vertol CH-47 in a 1962 Navy/Marine design competition. Secretary of Defense Robert McNamara delayed production by ordering the Vertol proposal reconsidered in the interests of interservice commonality, forcing re-design and delaying operational deployment until September 1966.

The HH-53 (rescue version of H-53) received the same equipment as the HH-3, plus radar homing and warning gear as well as extra armor. In addition, the HH-53's power margin permitted an armament of three Gatling-type, electrically driven, six-barreled 7.62mm mini-guns mounted in the crew door, (the left forward cabin window and on the cargo ramp. Additionally, a second pararescueman was added to the crew. The firepower and extra pararescueman added significantly to the HH-53's combat rescue capabilities. The first two HH-53Bs, an interim design with the external fuel tanks supported by external struts, reached Udorn, Thailand, in August 1967, and six more were delivered prior to the arrival of the first HH-53C in September 1969. Tactical Air Command subsequently procured the CH-53 for special operations, but without aerial refueling or the ramp mini-gun. The Air Force ultimately bought 44 H-53s of all types.

The H-53 was an extraordinarily capable machine, making some of the most spectacular combat rescues of the Vietnam War and serving with distinction in the special operations role over Northern Laos and along the Ho Chi Minh Trail. Perhaps the most dramatic HH-53 mission of the war was the November 1970 Son Tay Raid. In a largely overlooked commitment, 10 Air Force CH-53s and two HH-53s flying from the USS *Midway* provided some 20 percent of helicopter airlift for the April 1975 Saigon evacuation. Air Force CH and HH-53s took Marines to and from their objectives in the May 1975 SS *Mayaguez*/Koh Tang operation.

By 1972, HH-53s were being fitted with a limited night recovery system that was expanded into a full night/adverse weather capability with the Pave Low modification, incorporating a forward-looking infrared system and terrain-avoidance radar. Following the April 1980 Desert One fiasco all Air Force H-53s were upgraded to Pave Low III/MH-53Js and transferred to Special Operations Command, where they served with distinction in the 1991 Gulf War, over Haiti, in the Balkans, over Afghanistan, and in the Second Gulf War. Upgraded to Pave Low IV/MH-53Ms with infra-red absorptive paint, titanium/composite main rotor blades in place of the aluminum originals, an elastomeric main rotor system, integrated defensive avionics, and a .50-caliber machine gun on the ramp, the Air Force H-53 remains the most capable special operations and combat rescue helicopter in the world.

Opposite: *An HH-53B: the first operational Air Force version of the H-53. The H-53 had significantly greater range and lift capacity than the H-43 and was procured to replace its H-3 precursor. It proved to be an extraordinary tough and capable helicopter. (USAF)*

Above: *An HH-53 returning to its base at Dan Nang illustrates the stark contrast between the Vietnamese lifestyle and the technology available to American forces. ("Two Worlds," by Bill Edwards, USAF Art Collection)*

10 August
The Senate Appropriations Committee cuts $172 million from the Navy's F-111B carrier aircraft program after it is determined that it may be too heavy to land on current carriers. The cuts apply until further carrier trials are conducted. The program is effectively killed the following March when no improvements in the compatibility of the airplane and carriers seem possible. The Navy will never field the F-111B.

11 August
F-105s of the 355th and 388th Tactical Fighter Wings based in Thailand drop two of the Paul Doumer bridge's spans, a heavily defended vital supply line. Future attacks take place in October and December.

21 August
U.S. pilots report that 80 SAMs are launched at their formations during the day, the most of any single day of the war.

24 August–4 September
A temporary halt to bombardment around Hanoi as a demonstration of goodwill yields no response from authorities in Hanoi.

26 August
Major George E. Day, 4th Allied POW Wing, is hit by ground fire while flying his F-100F "Fast FAC" over North Vietnam. Badly injured after ejecting, he is captured and tortured by the enemy. Later, Day escapes and evades enemy forces for two weeks, despite his injuries, when he is detected, shot, and recaptured by the Viet Cong. Day continues to provide maximum resistance despite his severe injuries. For his bravery, Day is awarded the Medal of Honor.

28 August
The improved Lockheed U-2R flies for the first time. Twelve will be built during the production run—six for the USAF, and six for the CIA.

9 September
Sergeant Duane D. Hackney demonstrates remarkable bravery in the face of the enemy during a combat rescue of a downed Air Force pilot. Hackney receives the Air Force Cross for his actions, the first living enlisted man so awarded.

3 October
The North American X-15 hypersonic research aircraft sets an unofficial absolute world speed record for a non-orbiting manned aircraft. After launch from a B-52, the X-15, piloted by Air Force test pilot William Knight, reaches a speed of Mach 6.7 (4,543 miles per hour) at an altitude of 102,100 feet.

The first McDonnell Douglas F-4E Phantom II with an internal 20-millimeter Vulcan gun in the aircraft nose is delivered to the USAF.

10 October
President Johnson signs the Outer Space Treaty that prohibits the use of nuclear weapons in space, provides for the rescue of stranded astronauts, and prohibits any nation from claiming rights to any planet or the moon.

Left: *This photo illustrates how the X-15 rocket-powered aircraft was taken aloft under the wing of a B-52. Because of the large fuel consumption, the X-15 was air-launched from a B-52 aircraft at 45,000 feet at about 500 miles per hour. The X-15s made a total of 199 flights and were manufactured by North American Aviation. (NASA)*

Right: *The F-111 was created from controversy. Originally part of the TFX (tactical fighter experimental) project, the USAF ended up with the fighter originally intended for both the Navy and the Air Force. The F-111 had a variable geometry wing that allowed extremely high speeds when swept back, and acceptable landing speeds when in the forward position. (USAF)*

16 October

The first operational F-111A supersonic tactical fighter (called the Aardvark by most) is delivered to Nellis AFB, Nevada. The variable-geometry-winged fighter is capable of night terrain-following, hands-off operations.

24 October

A joint force of Air Force, Navy, and Marine Corps aircraft attack North Vietnam's largest air base—Phuc Yen.

30 October

During an A-12 sortie over North Vietnam, six SA-2 missiles are fired at the plane. Upon inspection after the mission, a small missile fragment is found lodged in the lower wing fillet area.

9 November

Captain Gerald O. Young, 37th Aerospace Rescue and Recovery Squadron, piloting an HH-3E helicopter, flies a mission to extract an Army ground reconnaissance team inside Laos. Despite the loss of other rescue helicopters to heavy enemy fire, he continues to attempt the rescue until his helicopter is also shot down. He manages to crawl out of the window of his aircraft and help others who are wounded in the crash. Young, having refused any rescue attempts because of the proximity of the enemy, evades capture for 17 hours until he is able to call for his own evacuation with his emergency radio. For his heroic actions, Young is awarded the Medal of Honor.

Captain Lance P. Sijan, 4th Allied POW Wing, evades capture for six weeks after ejecting from his disabled F-4 Phantom II over North Vietnam. During his capture, he is interrogated and severely tortured. Yet again, he escapes but is recaptured, contracts pneumonia, and dies on January 21, 1968, as a prisoner at age 25. Sijan is the first USAF Academy graduate to receive the Medal of Honor.

15 November

Air Force Major Michael J. Adams is killed when his X-15 hypersonic research plane enters a Mach 5 spin over the test range. The airplane eventually breaks apart at 65,000 feet during oscillations that exceed 15 Gs. Adams receives his astronaut wings posthumously.

17 November–29 December

Operation EAGLE THRUST, a long-range deployment of soldiers from the U.S. to Southeast Asia, is accomplished. C-133 and C-141 aircraft airlift more than 10,000 paratroops and 5,000 tons of equipment, including 37 helicopters, to the combat zone. This is the largest and farthest deployment of troops (not a paradrop) and equipment during the war. The forces deploy from Fort Campbell, Kentucky, to Bien Hoa AB, South Vietnam.

Left: *A Lockheed C-130 delivers pallets of cargo to the Khe Sanh airstrip in 1967. The technique used here, low-altitude parachute extraction (LAPES), allowed the aircraft to deliver cargo to "hot" or dangerous fields, minimizing exposure to enemy fire. (USAF)*

Below: *The Lockheed AC-130 Spectre gunship delivers precise firepower with its 40 and 105 mm guns and targeting equipment. (USAF)*

8 December
Air Force Major Robert H. Lawrence Jr., the first African American chosen by NASA to be an astronaut, dies when the test pilot he is instructing crashes their F-104 at Edwards AFB, California. Lawrence is killed before he travels into space.

29 December
The final operational flight made by a Boeing B-47 takes place when the last RB-47H retires from the 55th Strategic Reconnaissance Wing at Offutt AFB, Nebraska.

1968

1 January
Air Defense Command is redesignated as Aerospace Defense Command (ADC).

21 January
The U.S. Marine base at Khe Sanh comes under attack by regular forces of the People's Army of North Vietnam, also called the North Vietnamese Army (NVA). Outnumbered by more than three to one, the Marines stage a hard-fought defense bolstered by airpower. Air Force airlift delivers an average of 165 tons of equipment, weapons, and supplies each day of the 77-day-long siege. Throughout the siege, B-52s carpet bomb the hills and trench works surrounding the base, disrupting PAVN attempts to strike.

22 January
A B-52G carrying four nuclear weapons crashes while attempting an emergency landing at Thule AB, Greenland. Radioactive waste spread across sea ice at North Star Bay takes months of cleanup before acceptable radioactive readings return.

23 January
While on an electronic intelligence collection mission in international waters off Wonsan, North Korea, the vessel USS *Pueblo* (AGER-2) is attacked and seized by North Korean local forces. One crewman, Seaman Duane Hodges, is killed and the other 82 are taken prisoner and held in captivity for one year. The ship is never returned to the U.S. and is held as a trophy in a museum in North Korea.

Right: *The first combat deployment of the F-111A occurred when Detachment 1 of the 428th TFS deployed to Thailand from Nellis AFB. The name of this deployment was COMBAT LANCER, a six-month test of the low-level, terrain-following, all-weather, deep-interdiction capability expected from the F-111. ("COMBAT LANCER—F-111 in Combat over Southeast Asia," by Jack Fellows, USAF Art Collection)*

26 January
The CIA flies an A-12 mission over North Korea while events surrounding the USS *Pueblo* unfold.

29 January
The U.S. and South Vietnamese begin a 36-hour cease-fire for the Tet religious holiday.

30–31 January
In a series of coordinated attacks throughout South Vietnam, forces of both the PAVN and the Viet Cong strike in battalion-sized units, surprising Americans and South Vietnamese forces across the country. The offensive will continue into 1969. Although Tet operations turn out to be a military disaster for the North, suffering between 60,000 and 100,000 casualties, U.S. public opinion will become more critical of President Johnson's policies overseas.

27 February
The AC-130A Hercules gunship is used to attack targets along the Ho Chi Minh Trail in Vietnam for the first time.

29 February
Public law takes effect that eliminates previous restrictions regarding promotions for women to ranks higher than full colonel. As a result, Colonel Jeanne Holm and Colonel Helen O'Day become the first two Air Force women promoted to the permanent rank of colonel. Holm continues as the Director of Women in the Air Force while O'Day is assigned to the Office of the Air Force Chief of Staff.

25 March
The F-111 flies combat missions in Vietnam for the first time, striking military targets north of the border during Operation COMBAT LANCER. The F-111's terrain-following radar (TFR) allowed strikes on previously unreachable targets in more mountainous regions in North Vietnam. Largely an experiment, the F-111s do not remain in Vietnam very long, returning to their home bases after the initial deployment.

27 March
Cosmonaut Yuri Gagarin, the first man in space, is killed in a MiG-15 crash near Moscow. He was 34 years old.

31 March
President Johnson orders a partial halt to the bombardment of North Vietnam above the 20th parallel; political pressure moves the line farther south to the 19th parallel. President Johnson proposes that peace talks should begin with North Vietnam. He surprises the nation by announcing that he will not seek reelection.

4 April
The Reverend Dr. Martin Luther King Jr. is assassinated in Memphis. Racial unrest erupts in about 100 American cities.

10 April
SR-71A operations over Vietnam begin as sorties are flown by aircraft of the 9th Strategic Reconnaissance Wing from Kadena AB, Japan.

Left: *This C-130, hit by an enemy mortar, and destroyed, killing 150 people at Kham Duc, South Vietnam. (USAF)*

Opposite, top: *The Apollo 7 Saturn IB space vehicle is launched from the Kennedy Space Center's Launch Complex 34 at 11:03 a.m. October 11, 1968. A tracking antenna is on the left and a pad service structure on the right. (NASA)*

Opposite, bottom: *The Apollo 8 crew is photographed on a Kennedy Space Center (KSC) simulator in their simulator suits. From left to right: James A. Lovell Jr., William A. Anders (USAFR), and Frank Borman (USAF). (NASA)*

3 May
The 120th Tactical Fighter Squadron stationed at Buckley Air National Guard Base, Colorado, arrives in South Vietnam and flies its first combat missions two days later. This is the first ANG unit called to active duty during the Vietnam War.

12 May
Air Force C-130s and C-123s with Army and Marine helicopters evacuate forces from Kham Duc in the face of superior enemy forces. One of the C-130s is hit by mortar fire and explodes, killing 150 people. In all, 1,500 are evacuated, while 259 lives are lost along with eight aircraft. Four airmen earn the Air Force Cross during these operations, one the Medal of Honor, and Lieutenant Colonel Daryl D. Cole receives the Mackay Trophy for heroic actions during this evacuation.

Lieutenant Colonel Joe M. Jackson, 311th Air Commando Squadron, subjects himself to intense hostile fire by landing his C-123 Provider aircraft on a dangerous runway at Kham Duc, South Vietnam, in order to rescue a three-man Combat Control Team. His actions earn him the Medal of Honor.

5 June
Robert F. Kennedy is shot and mortally wounded in Los Angeles after winning the California democratic presidential primary.

13 June
Eight orbital communications satellites are placed into orbit atop a Titan IIIC launch vehicle. This constellation of satellites augments the Defense Satellite Communications System. The launch takes place at Cape Kennedy, Florida.

17 June
The first McDonnell Douglas C-9 Nightingale rolls out of the factory in Long Beach, California. The C-9 will be used as an aeromedical evacuation aircraft within the borders of the United States.

30 June
The first Lockheed C-5A Galaxy (first of five in the development phase) flies for the first time. Military Airlift Command (MAC) will eventually acquire 81 of these monster airlift aircraft.

1 August
General George S. Brown assumes command of the Seventh Air Force, replacing General Momyer.

8 August
Two Air Force medics die when the Army helicopter they are riding aboard crashes after picking up a critically wounded Turkish official.

The first USAF C-9A Nightingale makes its first flight.

25 August
The North American OV-10 Bronco begins its 90-day combat evaluation in South Vietnam.

1 September
Lieutenant Colonel William A. Jones III, 602d Special Operations Squadron, flies his heavily damaged A-1H Skyraider near Quang Binh Province, North Vietnam, to rescue a downed pilot. Jones endures burns to his arms, hands, neck, shoulders, and face as a result of the enemy anti-aircraft fire during the rescue. For his heroism, Jones receives the Medal of Honor.

26 September
The Air Force LTV A-7D Corsair II makes its first flight.

11 October
Apollo 7, the first manned Apollo mission, is launched from Cape Kennedy, Florida. USAF units support the launch, and Air Force Major Donn F. Eisele is one of the astronauts on board. The capsule is mounted atop the powerful Saturn IB launch vehicle.

31 October
Operation ROLLING THUNDER, the bombardment of North Vietnam, ends.

1 November
All bombing of North Vietnam is suspended while reconnaissance missions continue. This cessation of bombing continues until April 1972. However, bombing operations in Laos intensify during Operation COMMANDO HUNT which begins this date and continues until March 30, 1972.

26 November
1st Lieutenant James P. Fleming, 20th Special Operations Squadron, completes a successful rescue of a six-man Studies and Observation Group (SOG) near Duc Co, South Vietnam, despite an extreme low fuel state and hostile fire in the area. Flying his UH-1F Huey, Fleming makes one of the most dramatic combat rescues in history, earning the Medal of Honor.

21–27 December
Apollo 8, a NASA mission designed to orbit the moon for the first time, is launched from Cape Kennedy, Florida. On board are Air Force astronauts Colonel Frank Borman, Colonel William Anders, and Navy Captain James Lovell Jr. This is the first time in history that a manned spacecraft leaves earth's orbit.

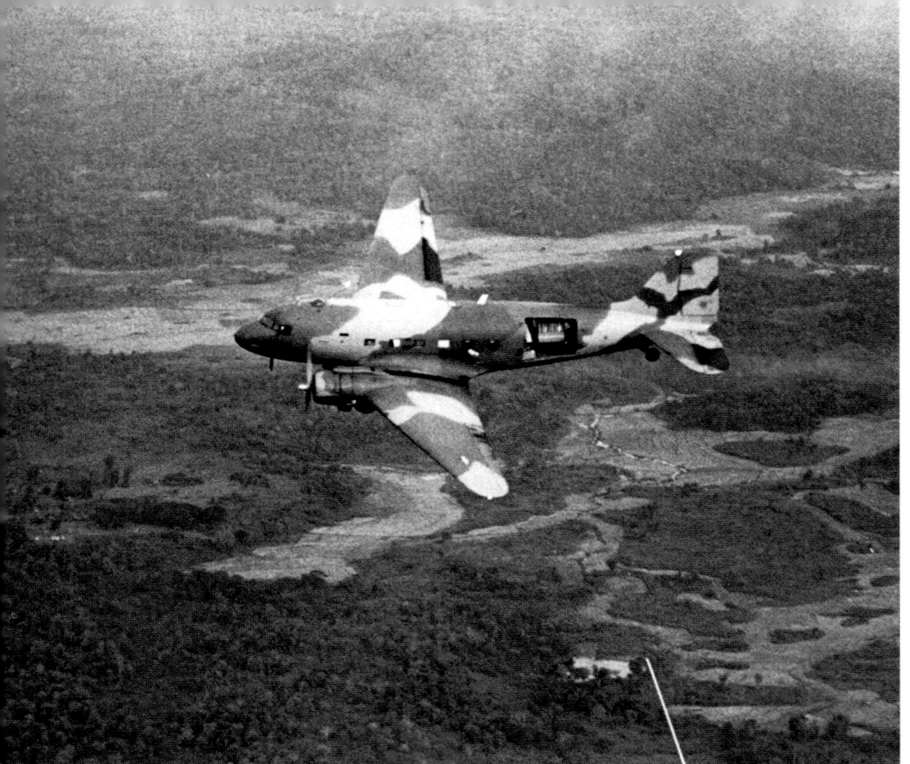

9 February

TACSAT 1, the world's first tactical communications satellite, is launched into geosynchronous orbit atop a Titan IIIC. The satellite is designed to relay communications between mobile land, sea, or airborne tactical stations.

24 February

Airman First Class John L. Levitow, 3d Special Operations Squadron, is loadmaster on an AC-47 gunship during a night mission over Bien Hoa Province, South Vietnam, when the aircraft is struck by a mortar shell. Now heavily damaged, a magnesium flare begins to smoke while inside the plane. Although Levitow is severely wounded, he realizes that the aircraft and entire crew is in danger and selflessly throws himself on the burning flare and then tosses it out the cargo door just before it fully ignites. For his heroism, Levitow receives the Medal of Honor on November 8, 2000, when a previous award is upgraded.

1969

1 January

The first AC-119 Shadow gunship mission is flown over South Vietnam by an Air Force Reserve unit—the 71st Special Operations Squadron.

7 January

The 1,000th Northrop T-38 Talon supersonic jet training aircraft is delivered for Air Force service.

9 January

NASA announces the selection of the crew that will be first to land on the moon. Former naval pilot turned civilian Neil A. Armstrong, Air Force Colonel Edwin E. "Buzz" Aldrin, and Air Force Lieutenant Colonel Michael Collins.

4 February

The remaining XB-70 Valkyrie flies from Edwards AFB, California, to Wright-Patterson AFB, Ohio, where it will be preserved at the National Museum of the USAF.

3–13 March

During *Apollo 9*, Air Force astronauts Colonel James A. McDivitt and Colonel David R. Scott, joined by civilian Russell L. Schweickart, carry out the first in-space tests of the lunar module while orbiting the earth. During the mission, a crew transfers between space vehicles using an internal connection for the first time.

18 March

Air Force B-52s begin covert bombardment of enemy sanctuaries located in Cambodia. Called "The Menus," operations over Cambodia code named BREAKFAST, LUNCH, DINNER, and SNACK will continue until May 1970. The bombers will fly more than 4,300 missions against Communist targets there.

Above: *In an AC-47 gunship like this one that Airman First Class John Levitow saved his crewmates and earned the Medal of Honor. (NASM)*

Right: *This oblique view featuring International Astronomical Union (IAU) Crater 302 on the moon surface was photographed by the Apollo 10 astronauts in May of 1969. Note the terraced walls of the crater and central cone. One of the Apollo 10 astronauts aimed a handheld 70mm camera at the surface from lunar orbit for a series of pictures in this area. (NASA)*

4–10 April
During a redeployment of 72 F-4D Phantoms from Spangdahlem AB, Germany, to Holloman, AFB, New Mexico, the 49th Tactical Fighter Wing conducts 504 aerial refuelings without incident. For this feat, the 49th TFW is awarded the Mackay Trophy.

17 April
Air Force Major Jerauld Gentry pilots the Martin X-24A lifting body on its first glide test. The program will evaluate the possibility of reusable and maneuverable future spacecraft.

14 May
Operation COMBAT MOSQUITO, an effort to control an encephalitis outbreak in Ecuador, begins when two C-141s deliver 50 tons of insecticide to that country. Two UC-123s immediately begin to spray the mosquito breeding grounds from the air, killing almost all of the population of pests by the end of the month.

18–26 May
Apollo 10 is the second spacecraft to orbit the moon. It is the first time that a complete Apollo spacecraft—Command and Service Modules plus the Lunar Module—travels to the moon. This mission is a dry run for the *Apollo 11* moon landing that will be launched in July. Air Force Colonel Thomas Stafford joins Eugene Cernan (USN) in the Lunar Module as they establish an orbit only 5.5 miles above the lunar surface, taking photographs of potential landing sites.

21 May
A Lockheed C-5A Galaxy takes off with a gross weight of 728,100 pounds, setting a new world record.

June
The USAF Aerial Demonstration Squadron—Thunderbirds—begins its season at the USAF Academy graduation ceremony flying the F-4 Phantom II.

5 June
Air Force bombers renew attacks on North Vietnam, the first since last November's cease-fire.

10 June
The No. 1 X-15 hypersonic rocket research plane is presented to the Smithsonian Institution by Air Force Systems Command staff. Secretary of the Air Force R. C. Seamans Jr. makes the official presentation. The craft now hangs in the National Air and Space Museum Milestones of Flight Gallery in downtown Washington, D.C.

1 July
The USAF Air Rescue and Recovery Service (ARRS) completes its 2,500th rescue in Southeast Asia.

8 July
The first troops withdrawn from Vietnam under President Richard M. Nixon's policy of Vietnamization are flown to McChord AFB, Washington. C-141 Starlifters transport 25,000 troops during this operation.

Above: *The Thunderbirds flew the F-4 Phantom II from 1969–1973. During those years, they were truly the "thunder" birds, as the F-4 was one of the loudest jet fighters ever built. (NASM)*

Left: *The X-15-1 now hangs in the National Air and Space Museum in Washington, D.C. (NASM)*

16 July
At 9:32 a.m. the Saturn V launch vehicle carrying the *Apollo 11* capsule and crew lifts off from pad 39A at Cape Kennedy, Florida. The objective of the mission is to land the first human on the surface of the moon.

20 July
The *Apollo 11* Lunar Module *Eagle* touches down on the surface of the moon at 4:17 p.m. EDT. At 10:56 p.m. EDT, astronaut Neil A. Armstrong becomes the first human to set foot on an extraterrestrial body. About an hour later, Air Force astronaut Edwin E. "Buzz" Aldrin Jr. joins Armstrong on the moon, and

Above: *The Apollo 11 Saturn V space vehicle lifts off with astronauts Neil A. Armstrong, Michael Collins (USAF) and Edwin E. Aldrin Jr. (USAF), at 9:32 a.m. EDT July 16, 1969, from Kennedy Space Center's Launch Complex 39A.*

Top: *Portrait of the prime crew of the Apollo 11 lunar landing mission. From left to right: Commander Neil A. Armstrong, Command Module Pilot Michael Collins, and Lunar Module Pilot Edwin E. Aldrin Jr.*

Center: *Astronaut Edwin E. "Buzz" Aldrin Jr., Lunar Module pilot, is photographed during the Apollo 11 extravehicular activity on the moon.*

Bottom: *Astronauts Neil Armstrong, Michael Collins, and Buzz Aldrin Jr. await helicopter pickup from their life raft. They splashed down at 12:50 pm EDT July 24, 1969, 900 miles southwest of Hawaii after a successful lunar landing mission.*

Opposite: *The four principal HL-10 pilots are seen here with the lifting body aircraft. Left to right:* *Air Force Major Jerauld R. Gentry, Air Force test pilot Peter Hoag, and NASA pilots John A. Manke and Bill Dana. The HL-10 was one of five lifting body designs flown at NASA's Dryden Flight Research Center, Edwards, California, from July 1966 to November 1975 to study and validate the concept of safely maneuvering and landing a low lift-over-drag vehicle designed for reentry from space. (NASA)*

the astronauts set up experiments and collect samples to be returned to earth. The total lunar excursion time is two hours and 32 minutes. Air Force astronaut Michael Collins remains in lunar orbit during the mission. Following his NASA career, Collins will become the first director of the National Air and Space Museum in Washington, D.C.

24 July
After eight days, three hours, 18 minutes, and 35 seconds, *Apollo 11* returns to earth carrying its three-man crew and approximately 50 pounds of lunar rocks and soil samples. The crew remains under quarantine for three weeks to ensure that no contaminants have been returned to earth from the lunar surface.

1 August
General John D. Ryan becomes the USAF Chief of Staff.

Donald L. Harlow becomes the new Chief Master Sergeant of the Air Force

19 August
Air Force aircraft begin airlifting supplies to southern Mississippi in support of humanitarian relief efforts in the wake of Hurricane Camille—one of the most powerful hurricanes ever to hit the United States. During the next month, nearly 6,000 tons of supplies will be delivered using several different kinds of aircraft.

8–14 October
Three Air Force HH-53 helicopters from Wheelus AB, Libya, deploy to Tunisia to assist with rescue efforts for flood victims. During their deployment, they rescue more than 400 people. The unit returns the following week to deliver needed supplies and then continues rescue operations during which 2,000 more are saved.

6 November
The largest balloon ever launched lifts a 13,000-pound payload into the skies over Holloman AFB, New Mexico. The 34-million-cubic-foot balloon is 1,000 feet tall when ready for launch.

14–24 November
An all-Navy crew of astronauts goes to the moon aboard *Apollo XII*. Astronauts Conrad and Bean gather samples and explore the surface for seven and a half hours before returning to earth.

18 December
The Maverick (AGM-65) air-to-surface television-guided missile is test fired against moving targets by crews of the Air Force Missile Development Center.

1970

16 January
The last operational B-58 Hustlers are flown from Grissom AFB, Indiana, to Davis Monthan AFB, Arizona, where they are decommissioned.

17 February
Air Force B-52 bombers strike targets in Northern Laos for the first time. President Nixon makes these raids public on March 6.

Left: *The Lockheed C-5 Galaxy carries a cargo load of 270,000 pounds, can fly 2,150 nautical miles, offload, and fly to a second base 500 nautical miles from the original destination without aerial refueling. With aerial refueling, the aircraft's range is limited only by crew endurance. (NASM)*

18 February
Air Force test pilot Major Peter C. Hoag flies the HL-10 experimental lifting body to a speed of Mach 1.86 at 65,000 feet. Air Force Major Jerauld Gentry pilots the first powered flight the following month.

27 February
Pratt and Whitney is selected as the prime engine development and production company for both the McDonnell Douglas F-15 Eagle and the Grumman F-14 Tomcat by the Department of Defense.

15 March
All U.S. military bases are connected by phone when the overseas part of the automatic voice network is completed. This telephone network will be known to all as the "Autovon."

18 March
In coordination with a ground offensive into Cambodia to engage North Vietnamese troops, B-52s begin nighttime operations against targets in Cambodia. During the next two months they will fly more than 4,300 sorties and drop more than 120,000 tons of bombs during this campaign.

11–17 April
The Air Force provides mission support for the ill-fated *Apollo 13* spacecraft. One of the three astronauts, John L. Swigert Jr., was a former Air Force pilot and also flew with the Massachusetts and Connecticut Air National Guard.

14 April
A Minuteman III ICBM is loaded aboard a C-141 Starlifter and transported from Hill AFB, Utah, to Minot AFB, North Dakota. This is the first airlift of an operational Minuteman III missile.

17 April
An important Minuteman III milestone is reached when the first missile is placed in a silo assigned to the 741st Strategic Missile Squadron, Minot AFB, North Dakota. At the end of December, the 741st SMS becomes the first SAC Minuteman III squadron to achieve operational status.

2 May
College campuses erupt in violence over the invasion of Cambodia.

4 May
Four students are killed by Ohio National Guardsmen attempting to maintain order at Kent State.

5 May
After successful trial programs are completed at four major universities, the Air Force Reserve Officer Training Corps (AFROTC) is opened to women nationwide.

8 May
During an attack on one of the most heavily defended roads in Southeast Asia, the crew of an AC-119K Shadow gunship is hit by ground fire, losing about 15 feet of the left wing and one of its ailerons. The crew had destroyed several trucks and successfully return to base. For their extraordinary airmanship, Captain Alan D. Milacek and his crew receive the Mackay Trophy.

2 June
Following a disastrous earthquake in Peru, USAF Southern Command begin humanitarian operations to the devastated region. In one month, airlift assets throughout the command deliver more than 750 tons of supplies and nearly 3,000 passengers while also evacuating 500 medical patients.

6 June
Military Airlift Command accepts its first operational Lockheed C-5 Galaxy. The C-5 is the largest operational aircraft in the world at this time.

24 June
The U.S. Senate repeals the 1964 Gulf of Tonkin Resolution.

13 July
Retired Army Lieutenant General Leslie Groves, military head of the Manhattan Project that developed the atom bomb during World War II, dies at the age of 73.

31 July
Under the program of Vietnamization, the first class of Vietnamese pilots graduate from training at Keesler AFB, Mississippi.

24 August
Flying from Eglin AFB, Florida, to Da Nang Airport, Vietnam, two HH-53 Sea Stallions accomplish the first nonstop, transpacific helicopter crossing.

1 September
General Lucius D. Clay Jr. assumes command of the Seventh Air Force, replacing General Brown.

6 September
After Palestinian terrorists hijack three airliners over Europe and fly them to the Middle East, USAFE planners make preparations for possible rescue and evacuation of the passengers onboard. C-130s and F-4s are deployed to Incirlik AB, Turkey, in case Operation FLAT PASS is called to execution. The hijackers release their captives after landing but then blow up all three airliners; the operation is never carried out.

28 September–31 October
Providing medical support to Jordan in the Middle East during Operation FIG HILL, USAF airlifters transport 200 personnel and more than 180 tons of supplies and equipment. Jordan had been battling guerrillas from the Popular Front for the Liberation of Palestine (PFLP) in the bloody conflict. Air Force and Army medical teams perform more than 1,200 medical procedures before they return to their home bases in Europe.

2 October
The new Bell UH-1N Twin Huey arrives at Hurlburt Field, Florida. The Special Operations Center there will acquire 79 of these durable helicopters—the first USAF unit to fly this helicopter.

21 November
Brigadier General Leroy J. Manor commands the Son Tay Raid, an attempt to rescue POWs from a camp 23 miles west of Hanoi. Army Colonel Arthur D. "Bull" Simons leads the search and rescue forces—inserted by helicopter and escorted by C-130E Combat Talon aircraft—on the ground but none are recovered, as the prisoners had already been moved to a different location. Practice runs for the raid are made at Eglin AFB, Florida, in a replica of the camp. During the raid, F-4s fly combat air patrol (CAP) for the rescue force. F-105s provide a diversionary attack to draw enemy air defense away from the intended target. Five A-1 Skyraiders provide close air support around the camp. The Navy also participates and launches a massive air strike against Haiphong.

Right: *Ground crew at Da Nang carries an AIM-9 Sidewinder heat-seeking missile to be loaded onto the wing station of an F-4. ("Sidewinder," by Henry Lozano, USAF Art Collection)*

Left: *C-123 crew members return from a defoliation mission over Vietnam. ("Ranch Hand," by Henry Lozano, USAF Collection)*

Opposite, top: *Apollo 15 Lunar Module pilot James B. Irwin (USAF) loads the "Rover," Lunar Roving Vehicle (LRV), with tools and equipment in preparation for the first lunar extravehicular activity (EVA-1) at the Hadley-Apennine landing site. (NASA)*

1971

28 January
The last Ranch Hand defoliation mission is carried out by Fairchild UC-123B sprayers.

31 January–9 February
Apollo 14 is the third successful launch and moon landing. The Command Module pilot is Air Force astronaut Stuart A. Roosa. The lunar excursions, totaling more than nine hours, are made memorable when astronaut Alan Shepard hits a golf ball about 400 yards on the lunar surface with a makeshift club.

1 March
The Capitol building in Washington is damaged by a bomb apparently planted in protest of the invasion of Laos.

17 March
2d Lieutenant Jane Leslie Holley becomes the first woman commissioned through a USAF ROTC program. She graduates from Auburn University, Alabama.

11 April
Former Air Force astronaut Michael Collins, *Apollo 11* Command Module Pilot, becomes the first director of the National Air and Space Museum. During his tenure, the Air and Space Museum building is completed on the National Mall—on time and under budget.

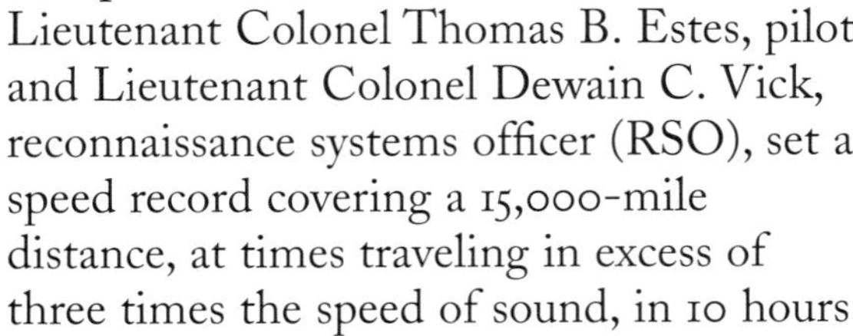

26 April
Lieutenant Colonel Thomas B. Estes, pilot, and Lieutenant Colonel Dewain C. Vick, reconnaissance systems officer (RSO), set a speed record covering a 15,000-mile distance, at times traveling in excess of three times the speed of sound, in 10 hours and 30 minutes nonstop. For this flight, the SR-71 crew receives the Mackay Trophy and the Harmon International Aviator Award.

16 July
Colonel Jeanne M. Holm, director of Women of the Air Force, is promoted to Brigadier General making her the first woman general officer in the Air Force.

16 June–18 July
During Operation BONNY JACK, the relocation of refugees in India, some 23,000 are airlifted from Tripura to Gauhati, India, having fled civil war in East Pakistan. Both C-130s and C-141s deliver more than 2,000 tons of food, medicine, and supplies during this month to the displaced refugees.

26 June
The F-100 Super Sabre is relieved from front-line operational duty as the 35th Tactical Fighter Wing redeploys from Phan Rang AB, Vietnam.

12 July
Benjamin O. Davis is nominated by President Nixon to be Assistant Secretary of Transportation. Davis was the first African American USAF general officer.

26 July–7 August
Apollo 15 launches from Cape Kennedy for the moon with an all Air Force flight crew. Loaded on board for the first time is a dune buggy–like vehicle—the lunar rover—to increase their exploration ranges on the lunar surface.

29 July
Flight testing of the X-24A lifting body is concluded. The program yields a tremendous amount of data useful in NASA's space shuttle program.

1 August
General John D. Lavelle assumes command of the Seventh Air Force replacing General Clay.

21 September
Using LORAN (long-range aid to navigation) to carry out the first all-instrument air strike of the war, nearly 200 tactical fighters hit oil and fuel storage areas near Dong Hoi and destroy some 350,000 gallons of fuel.

1 October
Richard D. Kisling becomes the new Chief Master Sergeant of the Air Force.

28 October
Director of the National Air and Space Museum, Michael Collins, announces that a new $40 million building plan will replace the original $70 million plan due to cuts necessitated by the Vietnam War.

7–8 November
Air Force aircraft attack North Vietnamese airfields at Dong Hoi, Vinh, and Quan Lang.

26–30 December
Striking targets south of the 20th parallel, U.S. aircraft launch the largest raid of the war since 1968 when more than 1,000 sorties are launched during this period.

1972

5 January
President Nixon announces that $5.5 billion has been earmarked in the budget to fund the space shuttle program.

17 February
An 89th Military Airlift Wing VC-137 Stratoliner departs from Andrews AFB, Maryland, carrying President Richard M. Nixon to China. Nixon becomes the first president to visit China, where he will meet with Mao Tse-tung and Chinese Premier Chou En-lai.

20 February
An Air Force HC-130H sets a new world record for unrefueled flight by a turboprop aircraft when it flies from Taiwan to Illinois.

23 March
The U.S. terminates their participation in the Paris peace talks, as no progress to end the fighting in Vietnam is achieved.

30 March
U.S. airpower halts a large-scale North Vietnamese offensive across the demilitarized zone (DMZ) separating North and South Vietnam. The "Easter Offensive" consists of approximately 40,000 troops and 400 armored vehicles and includes thrusts into the Central Highlands and areas north of Saigon.

Left: *The VC-137 was a highly modified 707 flown by the USAF in the service of the President of the United States. (USAF)*

The Joint Chiefs of Staff authorize up to 1,800 B-52 sorties may be flown each month in Southeast Asia. This is an increase of 600 sorties per month from a February 8 JCS directive.

6 April
Bombardment of North Vietnam, on hold since November 1968, resumes.

7 April
General John W. Vogt Jr. assumes command of the Seventh Air Force, replacing General Lavelle.

7 April–13 May
As a result of the Easter Offensive, the USAF deploys 200 stateside aircraft to Southeast Asia to stop an all-out invasion of South Vietnam during Operation CONSTANT GUARD.

16–27 April
Apollo 16 accomplishes the fifth landing on the surface of the moon. Air Force astronaut Charles Duke Jr. accompanies astronaut John Young (USN) during a record stay of nearly 71 hours on the lunar surface. Each astronaut totaled 20 hours and 15 minutes of lunar exploration time. Astronaut Thomas Mattingly orbits in the command module during the mission.

27 April
USAF fighters drop 2,000-pound *Paveway I* laser-guided bombs on the Thanh Hoa Bridge, closing it. More than 850 previous sorties had failed to make a significant dent in the structure.

5 May
The Pave PAWS system, designed to detect sea-launched ballistic missiles, achieves initial operational status.

10 May
Sustained bombing of North Vietnam begins once again. Originally known as Operation FREEDOM TRAIN, the name will be changed to LINEBACKER I and operations will continue until October 23. In one of the early LINEBACKER I missions, 8th Tactical Fighter Wing F-4 Phantoms use precision-guided bombs to damage and close the Paul Doumer Bridge in Hanoi.

The Fairchild YA-10 prototype makes its first flight. It is designed to carry large bomb loads and hold up well under difficult combat conditions. The YA-10 will be adopted for production in January 1973.

13 May
In a follow-on raid against the Thanh Hoa Bridge, 14 Air Force F-4 Phantom IIs employ a variety of guided weapons and conventional iron bombs and succeed in dropping one of the bridge spans. The structure is unusable for rail freight for the rest of the year.

30 May
Three Japanese terrorists attack passengers in the Tel Aviv Airport.

11 June
Boeing B-52 Stratofortresses drop laser-guided bombs on a hydroelectric plant near Hanoi, North Vietnam, destroying it.

Above: *Apollo 16 astronaut Charles M. Duke Jr. (USAF), pilot of the Lunar Module "Orion," stands near the Rover at Station No. 4, near Stone Mountain at the Descartes landing site. (NASA)*

Below: *A ground crew loads a laser-guided bomb onto an F-4. These LGBs were used effectively against bridges where conventional bombs had failed. (USAF)*

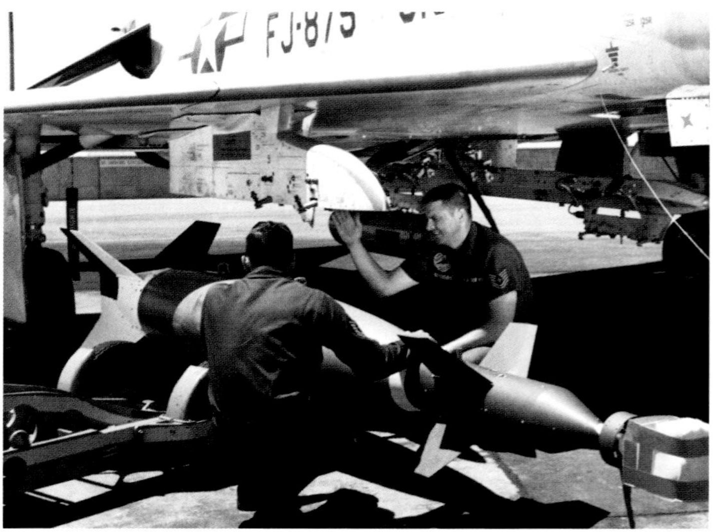

20 June
Airline pilots around the world walk off the job. The pilots insist on tighter security measures for air travel.

29 June
Captain Steven L. Bennett, 20th Tactical Air Support Squadron, notices a friendly unit in danger of ambush while flying his OV-10A Bronco in Quang Tri Province, South Vietnam. Though there is no air or artillery support available, Bennett strafes the enemy forces to protect the unit. His plane is hit and the parachute of his observer is destroyed, so he attempts to make a water landing but dies during the ditching attempt. This action saves the life of his observer. For his sacrifice, Bennett is posthumously awarded the Medal of Honor.

26 July
NASA announces that Rockwell International will be the prime contractor for the space shuttle.

27 July
The prototype McDonnell Douglas YF-15A Eagle flies for the first time.

11 August
The Northrop F-5E Tiger II flies for the first time.

28 August
Captain Richard S. "Steve" Ritchie becomes the first ace of the Vietnam War when he, along with Captain Charles DeBellevue (weapon system officer), shoots down his fifth MiG-21.

9 September
Captain Charles B. DeBellevue, weapon system officer in an F-4, shoots down his fifth and sixth enemy aircraft, becoming the leading ace of the Vietnam War.

11 September
Air Force aircraft attack and destroy the Long Bien Bridge over the Red River in downtown Hanoi. Precision-guided bombs are once again used.

2 October
The Air Force launches two satellites into orbit atop an Atlas-Burner two-stage rocket from Vandenberg AFB, California. One satellite, *Space Test Program 72-1*, will measure radiation effects upon spacecraft; the other, *Radcat*, is a passive radar and optical calibration target.

13 October
Captain Jeffery S. Feinstein, an F-4 weapon system officer, shoots down his fifth MiG and becomes the third and final Air Force ace of the Vietnam War.

The three Air Force aces of the Vietnam War—Captain Richard S. Ritchie, Captain Charles B. DeBellevue, and Captain Jeffery S. Feinstein—are awarded the Mackay Trophy for their airmanship.

Top left: *Captain Richard "Steve" Ritchie, the first Air Force ace in Vietnam. (USAF)*

Top center: *Captain Charles B. DeBellevue, leading USAF ace with six victories. (USAF)*

Top right: *Captain Jeffery S. Feinstein, last USAF ace of the Vietnam War. (USAF)*

"THE SHOOTDOWN OF TRIGGER 4"

Todd P. Harmer and C. R. Anderegg

Left: *A flight of F-4s laden with bombs and AIM-7 Sparrow missiles refuels before a combat mission over Southeast Asia. (USAF)*

Of the many axioms that rule the swirling world of aerial combat, none is truer than "get the first shot." Fighter pilots will go to almost any length to beat their adversaries to the draw because that puts the opponent on the defensive. It is a very uncomfortable feeling to be defensive, when one is operating a machine that is by its nature offensive. Granted, the fighter may perform defensive missions, but even the newest fighter pilot learns that the best defense is to attack before the target knows it is under attack.

More than three decades ago, fighter crews received a weapon system designed to let them throw the first punch—the AIM-7 Sparrow. The first versions of the AIM-7 seemed impressive; it was a large missile, 425 pounds, with a devastating warhead of expanding steel rods propelled outward in a circle by nearly 70 pounds of explosives. Since the missile was radar-guided, the F-4 Phantom II crews could fire it under any conditions—at night, in the weather, at targets more than 10 miles away. In fact, they could fire it beyond visual range without ever seeing the target. The Sparrow seemed to be an ideal weapon to fire from long range at Soviet nuclear bombers attacking America and her military forces.

Then came the air war over Vietnam, and the fighter crews soon realized that AIM-7 shots beyond visual range were often more dangerous to other Americans than to the enemy because there was no way to identify the target as friend or foe. Air Force and Navy crews alike quickly reverted to the eye as the only sure way to discriminate. They abandoned the capability of the AIM-7 to reach beyond visual range, as they came to distrust shots that might lead to a fratricide—the inadvertent downing of a friendly fighter.

By 1972 a new system was in place in a handful of Air Force F-4s. Nicknamed Combat Tree, the system allowed these few fighters to interrogate the beacons North Vietnamese MiGs carried to show their position to their radar controllers. If an F-4 identified a MiG with Combat Tree, and the crew was certain that no friendlies were in front of them, the crew could shoot beyond visual range. Nonetheless, it was risky for several reasons. Navy fighters entered from one direction as Air Force fighters entered from another. Groups were on different radio frequencies. Sometimes targets changed at the last minute. The weather was almost always a factor. It was common for strike packages to become confused as flights got separated. Combat leaders often had to improvise in a thick fog of war.

On July 29, 1972, Air Force F-4s and North Vietnamese MiG-21s engaged 50 miles northeast of

Hanoi. An F-4 shoots down two MiGs, and the MiGs shoot down an F-4 whose call sign was Trigger Four (the fourth F-4 in a flight of four). Such is the historical record, but some pilots thought the record was wrong. They suspected that only one MiG fell, and that the second kill was really Trigger 4, a victim of fratricide perpetrated by Cadillac 1, the lead F-4 of another flight of four on a MiG sweep out of Udorn, Thailand. Fortunately, the two-man crew of Trigger 4 survived the shootdown and several months of imprisonment in Hanoi.

In the mid-1990s, an Air Force Reserve colonel conducted research on Trigger 4's loss. His skill and persistence permitted him to gather an impressive body of data on this one engagement. He reached a preliminary conclusion that it was "highly probable" Trigger 4 had been a victim of fratricide. Yet, he knew his conclusion was based on circumstantial evidence; he had no smoking gun. In 2000, the pilot of Trigger 4, who was aware of the Air University researcher's findings, wrote to the Air Force Chief of Staff and asked that the matter be investigated fully. In response, the Chief of Staff directed a team of experts to reconstruct the air battle of July 29, 1972, to determine the facts surrounding the shootdown. He also appointed a nonpartisan senior mentor to monitor the team's thoroughness and methodology.

The Project Trigger team began work in the Checkmate division of the operations directorate. They built upon the foundation provided by the Air University researcher. Even though this air battle occurred nearly 29 years earlier, the team found and interviewed 27 participants and other interested observers of the battle. The team was fortunate to be able to synchronize five cockpit tapes, including three from the F-4 flight leaders in the battle, and two others from the flight whose kill was in question. Using the tapes, eyewitness accounts, intelligence records, and the team's considerable experience, they reconstructed the engagement and determined that a MiG-21 shot down Trigger 4. A short time later, the team briefed the pilot of Trigger 4 and his backseater on their findings, thus putting to rest, once and for all, the question of who had shot down Trigger 4. The details of their study may be found at the Air Force Historical Research Agency, Maxwell Air Force Base, Alabama under the title "The Shootdown of Trigger 4."

Below: *Aerial combat often occurred in the blink of an eye—much like the shootdown of Trigger 4. For example, on July 8, 1972, near Banana Valley, North Vietnam, 40 miles southwest of Hanoi, Captain Steve Ritchie is 2.3 seconds from destroying his second MiG-21 in less than two minutes. This double MiG kill raised his score to four victories. Ritchie became the only USAF pilot ace of the Vietnam War on August 28, 1972, when he downed his fifth MIG-21. ("Splash Two," by Mark Waki, USAF Art Collection)*

22 November
A B-52 struck by a surface-to-air missile over North Vietnam becomes the first bomber lost to enemy action. The crew safely ejects over Thailand and is rescued.

1 December
Astronaut Thomas P. Stafford (USAF) becomes the youngest brigadier general in the Air Force at age 42. Stafford flew on *Gemini VI, Gemini IX*, and *Apollo 10*.

7–19 December
The final Apollo mission is launched from Cape Kennedy, Florida. *Apollo 17* is the last manned effort to land on the moon.

18–29 December
President Richard M. Nixon orders that the full-scale bombardment and mining of North Vietnam resume, called Operation LINEBACKER II. During the operation, 741 B-52 sorties are launched (12 aborted) accompanied by 796 suppression of enemy air defense (SEAD), combat air patrol (CAP), and Chaff sorties. Fifteen B-52s are lost during the operation along with two F-111s, three F-4s, two A-7s, two A-6s, one EB-66, one HH-53, and one RA-5C.

On the 18th, Staff Sergeant Samuel O. Turner, a B-52 tail gunner, shoots down a MiG-21 as it converges on his bomber. Six days later, Airman First Class Albert E. Moore, also a B-52 tail gunner, scores a victory over another MiG-21. These are the only aerial gunner victories of the war.

1973

8 January
The last USAF victory of the Vietnam War is scored by Captain Paul D. Howman and 1st Lieutenant Lawrence W. Kullman, firing an AIM-7 radar missile against a MiG near Hanoi.

15 January
As the Paris peace talks approach, all offensive operations against North Vietnam—mining and bombardment—cease.

18 January
The Department of Defense awards the A-X contract to Fairchild Republic. The winning company will produce the durable and deadly A-10 "tank-buster" close support aircraft.

22 January
Former President Lyndon B. Johnson dies of a heart attack at age 64. Johnson was a behind-the-scenes force that influenced John F. Kennedy to press for landing a man on the moon by the end of the 1960s.

27 January
A Vietnam cease-fire date of January 28 is established when U.S., South Vietnamese, Viet Cong, and North Vietnamese representatives meeting in Paris sign an "Agreement on Ending the War and Restoring Peace to Vietnam."

28 January
The final Operation ARC LIGHT mission, the B-52 campaign in Southeast Asia, is flown against enemy targets in South Vietnam.

The Vietnam War officially ends as the cease-fire takes effect. Operations continue in neighboring countries.

12 February
Operation HOMECOMING, the airlift of released prisoners from Hanoi to Clark Air Base, begins when the first of 591 former POWs board USAF C-141s to make the trip. From Clark AB, the former POWs are flown to U.S. hospitals, then are reunited with their families. The Military Airlift Command (MAC) aircrews that participate in this special airlift receive the Mackay Trophy.

21 February
Bombing operations in Laos are halted when a cease-fire is signed by authorities. Violations of the agreement result in B-52 strikes, which will continue until April.

28 March
The last USAF aircraft departs South Vietnam. Other USAF aircraft remain on duty in surrounding countries for the time being. Seventh Air Force moves to Nakhon Phanom AB, Thailand, assuming the dual role of U.S. Support Activities Group and Seventh Air Force.

10 April
The Boeing T-43A navigation trainer, modified from the 737-200 series, flies for the first time.

17 April
Guam-based B-52s carry out the final bombing mission over Laos in retaliation for cease-fire violations.

18 April
Staff Sergeant Roy W. Hooe dies at age 78 in Martinsville, West Virginia. Hooe was the mechanic who flew on board the 1929 *Question Mark* endurance flight over Pasadena, California. The craft remained airborne for 150 hours.

14 May–22 June
The unmanned *Skylab 1* is launched into orbit, where a critical meteoroid shield rips off and takes one of the lab's two solar panels with it. Internal temperatures reach 126 degrees Fahrenheit. Delays result in the manned program until a plan for safely operating the station can be established and practiced. *Skylab 2* will be launched on May 25, and the crew will accomplish the necessary repairs to begin working in the station.

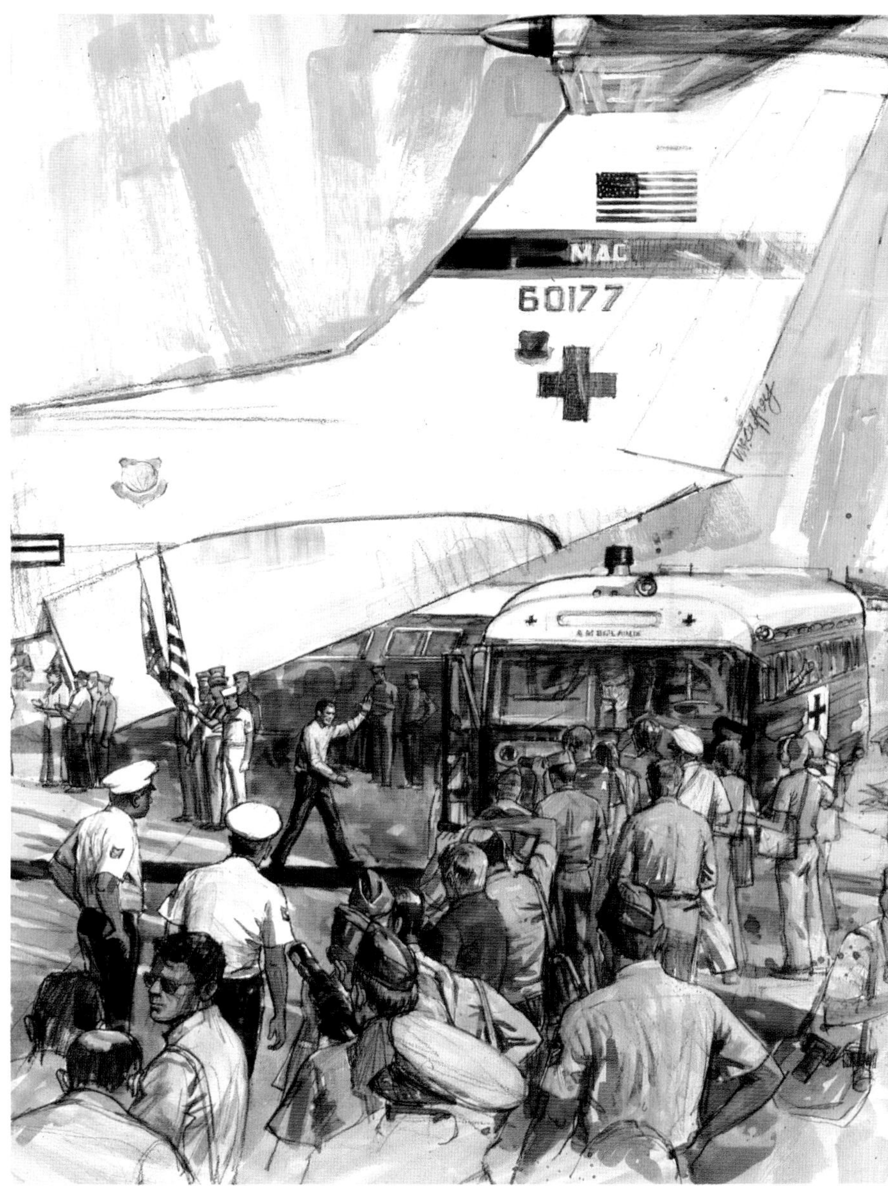

Opposite: *The SA-2 Guideline surface-to-air missile was a deadly complement to anti-aircraft fire near enemy targets. AAA claimed the vast majority of aircraft that were shot down during the war. (USAF)*

Right: *"Operation Homecoming—Return of POWs—Clark AFB, 1973," by Maxine McCaffery. (USAF Art Collection)*

LINEBACKER II

Mark Clodfelter

Left: *A B-52 refuels as its crew prepares to attack targets in North Vietnam. KC-135 Stratotankers orbited in secure areas in Southeast Asia and could easily be "tapped" by pre- and post-strike fighters and bombers. (USAF)*

At 1945 local time on December 18, 1972, the first of three waves totaling 129 B-52 bombers attacked military targets surrounding Hanoi, launching the air campaign known as Linebacker II. President Richard Nixon ordered the attacks, aimed at compelling North Vietnamese negotiators to sign an accord ending America's military involvement in the war, and, at the same time, persuading South Vietnamese President Nguyen Van Thieu that American air power would enforce the agreement's terms. Twelve days and 20,000 tons of bombs later, Nixon ended the offensive, convinced that it had forced the North Vietnamese to return to Paris for serious negotiations.

The President had long desired to attack the Northern capital with heavy bombers. In response to North Vietnam's Easter Offensive, in April 1972 Nixon sent B-52s on five separate raids against targets in the Northern heartland, including installations in Haiphong. National Security Advisor and chief American negotiator Henry Kissinger dissuaded the President from attacking Hanoi with B-52s as part of the initial Linebacker campaign in May, but Eighth Air Force Headquarters in Guam began planning for such attacks in the summer. In late November, after the Paris negotiations appeared deadlocked, Nixon told the Joint Chiefs of Staff to prepare for B-52 raids to strike Hanoi. On December 14, the President ordered the attacks to begin on the night of the 18th.

Nixon aimed to demonstrate the seriousness of America's efforts to end the Vietnam War, and thus he intended for B-52s to provide the center of gravity for a campaign that also included fighter-bombers. After the Easter Offensive, he had placed more than 200 B-52s—

Right: *Waves of B-52s attacked Hanoi during the opening hours of Linebacker II. AAA, SAMs, and enemy fighters filled the skies as bombers and fighters defended themselves with flares. F-4 Wild Weasels attacked SAM sites to protect the B-52s. ("High Road to Hanoi," by Jack Fellows)*

half the bomber strength of Strategic Air Command—in Guam and Thailand. With the exception of the five raids in April, the bulk of the B-52 effort had gone towards wrecking the advance of the North Vietnamese Army in South Vietnam. By sending the bomber force against the heavily defended Northern heartland, Nixon knew that he jeopardized a key component of America's nuclear triad. Indeed, North Vietnamese surface-to-air missiles (SAMs) had downed their first B-52 of the war during a November 22, 1972, raid in the North Vietnamese panhandle.

To succeed, Linebacker II had to achieve significant results with minimum losses. On the first two nights, the losses appeared acceptable. Three B-52s were shot down on the 18th, while two were severely damaged, and though two bombers out of 93 were damaged on the 19th, none were shot down. On December 20, however, the North Vietnamese scored their greatest triumph of the campaign, destroying six B-52s and damaging a seventh out of 99 attackers. Two key factors produced the losses: repetitive routing to and from target, and the lack of the most sophisticated SAM-jammers for 41 of the 98 G-Model B-52s. As a result, for the remainder of the campaign only D-Model B-52s attacked the most heavily defended targets, and only 30 B-52s flew each of the next four nights. After a raid on the 21st resulted in the loss of two more bombers, the next three attacks occurred away from the immediate vicinity of Hanoi.

When the bombing resumed after a 24-hour lull on Christmas Day, Linebacker's direction had transferred from SAC Headquarters at Offutt AFB, Nebraska, to Eighth Air Force at Guam. On the 26th, 120 B-52s attacked 10 different targets in Hanoi and Haiphong in 10 minutes from seven different directions. Two aircraft were downed—a loss rate of 1.66 percent—and the imaginative routing continued for the duration of the campaign. Sixty bombers attacked each of the next three nights, with two B-52s downed on the 27th, the campaign's final losses. For the eleven-day effort, North Vietnamese gunners fired more than 1,000 SAMs and shot down 15 B-52s.

Nixon stopped the offensive after receiving word on the 28th that the North Vietnamese would return to Paris and negotiate seriously. He believed that the campaign had made the decisive difference in the North Vietnamese signing the Paris Peace Accords on January 27, 1973. His view was not shared by many Democrat leaders in Congress, who referred to the offensive as an "outrage" and "a Stone Age Tactic," or by much of the world press, which was outraged by Nixon's "Christmas bombings." Despite the outcry, the bombing caused minimal civilian casualties, and Hanoi's mayor claimed that 1,318 civilians had died in his city, while Haiphong reported a death toll of 305.

The true impact of Linebacker II remains more difficult to discern. The North Vietnamese commander in South Vietnam contended that the damage from the continued bombing there caused Hanoi's leaders to sign the agreement, and South Vietnamese President Thieu waited until almost literally the final hour before endorsing the accord.

Left: *Air Force Security Police guard the perimeter of this air base with the help of canines trained to smell out and stop infiltrators. (NASM)*

15 May
In support of Operation AUTHENTIC ASSISTANCE, a C-130 flies the first of 541 missions delivering supplies and food to combat drought in Chad, Mali, and Mauritania. In all, 19 C-130s deliver more than 9,200 tons during the operation that lasts until October.

13 June
The first of four E-4A advanced airborne national command posts flies for the first time.

1 July
The draft is terminated in the U.S., but American men are still required to register for Selective Service as directed by law.

23 July
Captain Edward V. Rickenbacker dies at age 82. Rickenbacker was America's leading World War I ace.

28 July–25 September
The crew of *Skylab 3* carries out extensive scientific and medical research concerning long-duration space flight. Maintenance of the station is ongoing, including the installation of a twin-pole solar shield during an EVA.

15 August
The final B-52 strike against Cambodia is flown by B-52 bombers. This concludes almost eight continuous years of bombardment by B-52s and almost nine years by fighter/bombers.

A Thailand-based A-7D Corsair II flies the final Southeast Asian War combat mission. An EC-121 based in Korat, Thailand, is credited with flying the last U.S. mission of the war. More than 5.25 million sorties had been flown by USAF aircraft in Southeast Asia since February 1962 and more than 1,700 Air Force aircraft had been lost due to combat actions.

20 August
Aircraft from the Military Airlift Command (MAC), Tactical Air Command (TAC), and the Air Force Reserve (USAFR) begin airlifting more than 2,400 tons of humanitarian relief equipment and supplies to flood-ravaged victims in Pakistan.

1 October
Thomas N. Barnes becomes the new Chief Master Sergeant of the Air Force.

12 October–6 April
During Operation GIANT REACH, nine SR-71 missions are flown over the Middle East from the U.S. These missions are flown to monitor the Arab-Israeli Yom Kippur War.

14 October–14 November
Operation NICKEL GRASS, the delivery of war materiel to Israel during the Yom Kippur War, is carried out by Air Force C-5 Galaxy and C-141 Starlifter aircraft. Sometimes called "the airlift that saved Israel," 22,400 tons of supplies offset Soviet airlift to Egypt and Syria, helping Israel to overcome shortages in critical military equipment and strengthen their overall military posture in the region. C-5s fly 145 sorties to deliver more than 10,000 tons, while C-141s fly 421 sorties to deliver 11,600 tons.

Right: *The first preproduction F-16 touches down at Edwards AFB, California, and joins the Combined Test Force there. The use of fly-by-wire controls was a remarkable development and allowed the relocation of the "stick" from between the pilot's legs to the right-hand instrument panel just outboard of the pilot's right knee. A slightly reclining seat was installed to increase "G" tolerance and pilot comfort in flight. (USAF)*

6–24 October
After the Yom Kippur War ends, Air Force airlift assets assist in transporting United Nations peacekeeping forces to Egypt during Operation NIGHT REACH. Additional humanitarian aid is rendered in the form of blankets delivered to Israel and medicine to Jordan.

16 November–8 February
The crew of *Skylab 4* will spend 84 days in orbit around the earth. One of their primary missions will be to observe and photograph the comet Kohoutek, which passes near the earth.

5 December
Pioneer radar developer Sir Robert Watson-Watt dies at age 81. He invented the first electronic device to detect airborne aircraft and was critical to the outcome of the Battle of Britain.

13 December
The YF-16 is rolled out of the General Dynamics factory at Fort Worth, Texas.

1974

20 January
The YF-16 makes an inadvertent flight during high-speed taxi tests. The first "official" flight will take place on February 2.

2 February
The General Dynamics YF-16 officially flies for the first time at Edwards AFB, California.

10 April
As part of the Camp David peace agreement between Egypt and Israel, minesweeping operations at the Suez Canal are conducted by a combined force of U.S., French, and British specialists. This includes C-130 communications flights that supported Operation NIMBUS STAR.

9 May
Congress begins impeachment hearings against President Nixon related to the Watergate break-in.

1 July
General David C. Jones becomes the USAF Chief of Staff.

14 July
General Tooey Spaatz dies in Washington, D.C., at age 83. Spaatz was the first Chief of Staff of the independent United States Air Force.

25 July
Following a Greek-led coup against Cyprus President Makarios, followed closely by a Turkish invasion of the island, the United Nations arranges for a truce. USAFE C-130s deliver 10,000 blankets, 7,500 cots, 600 tents, and two water trailers for use by refugees of the violence.

9 August
President Nixon resigns under the scrutiny of the Watergate hearings. Gerald Ford is sworn in as President of the United States.

17 August
As part of Operation COMPASS COPE, the Air Force completes its first test of the Teledyne-built remotely piloted vehicle (RPV).

26 August
Charles A. Lindbergh dies of cancer at his Maui, Hawaii, home at the age of 74.

1 September
The 9th SRW SR-71A, crewed by Major James V. Sullivan and Major Noel Widdifield, makes a record-breaking flight from New York to London in 1 hour 54 minutes and 56 seconds, averaging more than 1,800 miles per hour.

3 September
The last alert Minuteman I ICBM is removed during replacement by Minuteman III missiles at F. E. Warren AFB, Wyoming.

13 September
An SR-71 crewed by Captain Buck Adams and William Machorek establishes a world record from London to Los Angeles in three hours 47 minutes and 39 seconds averaging 1,436 miles per hour.

24 October
A Minuteman I ICBM is successfully launched after being dropped from the hold of a C-5 Galaxy flying at 19,500 feet above the Pacific Ocean.

24 November
The Vladivostok Strategic Arms Limitation Accord is signed by President Gerald R. Ford and General Secretary Leonid I. Brezhnev. The accord limits the deployment of strategic delivery vehicles and multiple, independently targetable reentry vehicles (MIRVs).

2 December
The Air Force announces that the NAVSTAR global positioning satellite system has been approved by the Department of Defense. This joint USAF-Navy

Above: *With wings swept forward, a Rockwell B-1 Spirit moves into the contact position behind a Boeing KC-135R Stratotanker. ("Team Mates," by Keith Ferris, USAF Art Collection)*

Below: *"Portrait of Major Jim Sullivan and Major Noel Widdifield—The Record Setting SR-71 Crew, 1974," by Maxine McCaffery. (USAF Art Collection)*

Opposite: *Operation BABY LIFT ended when a C-5, carrying hundreds of orphans, crashed in Hanoi. Half of the passengers perished. (USAF)*

project will revolutionize global navigation and assist in weapons accuracy.

23 December
The prototype variable-geometry Rockwell B-1A flies for the first time.

30 December
John Victory, the first employee of the National Advisory Committee for Aeronautics (NACA), dies at the age of 82. As executive secretary, Victory played a crucial role in the rise of the NACA to international renown.

1975

13 January
John L. McLucas, Secretary of the Air Force, authorizes the purchase of the General Dynamics F-16 fighter.

16 January–1 February
Flying the "Streak Eagle," a pre-production F-15 aircraft, Majors Roger J. Smith, David W. Peterson, and Willard R. MacFarlane set eight time-to-climb records. The final record-breaking flight reached an altitude of 98,425 feet in three minutes, 27.8 seconds and then continued climbing to 103,000 feet. The plane is flown in a natural metal finish to minimize weight. For their airmanship, these pilots are awarded the Mackay Trophy for the year 1974.

7 February
The Air Force begins testing the DIGITAC system on an LTV A-7 Corsair II. The digital flight control system will eventually allow aircraft such as the F-117 to remain controllable even though unstable in their design.

25 March
The U.S. organizes a number of airlift operations to evacuate refugees from invading Communist forces throughout Southeast Asia. As part of the effort, Military Airlift Command (MAC) becomes the single airlift manager for all Air Force military airlift—strategic and tactical. The most intense operations will occur during the month of April.

4 April
During Operation BABY LIFT, the evacuation of orphans from Saigon, South Vietnam, begins tragically when a fully loaded C-5 Galaxy crashes in a rice field near the city. More than half of the passengers survive the disaster.

Airlift operations also begin in Cambodia when C-130s rescue nearly 900 Cambodians who had been surrounded by the Khmer Rouge in Phnom Penh.

12 April
Operation EAGLE PULL, the final evacuation of Phnom Penh, is executed by USAF and Marine Corps helicopters escorted by Air Force fighters and gunships. More than 280 are evacuated before the city falls to the Communist forces of the Khmer Rouge on April 17.

29–30 April
Operation FREQUENT WIND, the final evacuation of Saigon, begins when more than 6,000 people board Air Force helicopters staging off the deck of the Navy aircraft carrier USS *Midway*. Additionally, more than 150 aircraft will be evacuated from South Vietnam to Thailand—among these are C-130s, F-5s, and AT-37s.

Operation NEW LIFE, the aerial evacuation of some 45,000 from Saigon, continues. More than 5,600 U.S. citizens are among those who board MAC C-141 and C-130 aircraft in Saigon.

The last Americans, 10 Marines from the embassy, depart Saigon at 8:35 a.m., ending U.S. presence in Vietnam.

29 April–16 September
During Operation NEW ARRIVALS, more than 120,000 Indochinese refugees are airlifted from staging areas in the Pacific to the United States. Commercial airliners, C-141s, and C-130s accomplish the mission.

Left: *An HH-53 on the ground at Kho Tang, Cambodia, during the attempt to rescue the captured crew of the* Mayaguez. *The fierce fighting that continued through the night marked the last major military action in Vietnam. (USAF)*

30 April
North Vietnamese forces capture Saigon and unify the country under Communist rule. The complicated and divisive conflict in Southeast Asia finally ends.

12 May
The Cambodian Navy seizes the American merchant ship SS *Mayaguez* in international waters off the Cambodian coast.

14 May
During the early morning hours, 175 Marines load into helicopters of the 3d Aerospace Rescue and Recovery Group and the 21st Special Operations Squadron, and reach their landing zone at Kho Tang, Cambodia. Although expecting only light resistance, the force is met by 150–200 heavily armed Communist Khmer Rouge troops, who shoot down three of the first eight helicopters and damage two others. About 100 Marines are put ashore. Air Force A-7s, F-4s, OV-10s, and AC-130s support the Marines while the Navy strikes inland targets with carrier aircraft. Soon, a fishing boat carrying the captured crew of the *Mayaguez* approaches the Navy destroyer *Wilson*. A withdrawal under fire is accomplished through the night until all 230 Marines are retrieved. Several Air Force personnel receive the Air Force Cross for heroism that day, and Major Robert W. Undorf receives the Mackay Trophy for conspicuous gallantry, initiative, and resourcefulness during the operation—the last major American military action in Southeast Asia.

15 July–24 July
In a rare display of cooperation, two Soviet cosmonauts and three American astronauts rendezvous space craft in orbit after launching on the same day from sites in each country. Fifty-two hours after the Soviet Soyuz space craft launches, a rendezvous and docking maneuver is accomplished successfully. For two days the international crew carries out a variety of experiments. The mission commander, Brigadier General Thomas P. Stafford, goes on to command the Air Force Flight Test Center and retire from the USAF as a lieutenant general. Astronaut Deke Slayton had flown combat missions during World War II and been recalled to Air National Guard duty during the Korean War.

31 July
The last Lockheed F-104 is retired from Air National Guard service after 18 years of service.

8–15 August
Air Force Reserve and Air National Guard C-130s deliver 1,400 tons of fire retardant over blazing forest fires in California.

1 September
Daniel "Chappie" James Jr. is elevated to the rank of general, the first African American four-star general in American military history.

7 October
President Gerald R. Ford signs legislation that will permit women to enter military service academies. The academy classes that will graduate in 1980 enter the following June. All service academy classes will include women from this point forward.

31 October
The Boeing E-3A Sentry (AWACS) flies for the first time.

November
The USAF and DARPA reveal that detailed study has been underway to develop aircraft that are nearly invisible to radar detection. This stealth technology is part of a program called HAVE BLUE.

29 November
The first ever "Red Flag" begins at Nellis AFB, Nevada. Recognizing that combat experience increases the chances of survival in wartime, this realistic training exercise provides an initial margin of "combat" for American aviators. Today, "Red Flag" offers both Combined and Joint training.

6 December
The F-4G Wild Weasel flies for the first time. A total of 116 will be modified from F-4E Phantom IIs for SEAD and SAM suppression missions.

Above: *General Daniel James Jr. was the first African American promoted to the rank of Air Force four-star general. He was a Tuskegee Airman, but did not see action until the Korean War. His career spanned three wars and 30 years. James was a recognized civil rights pioneer. (USAF)*

Left: *An F-5 Aggressor returns from another RED FLAG sortie at Nellis AFB, Nevada. (USAF)*

RED FLAG: MOODY SUTER'S BRAINCHILD

James "Snake" Clark

Left: *A four-ship formation of Royal Australian Air Force (RAAF) F-111s participates in RED FLAG exercises. A typical exercise involves more than 85 aircraft flying to the Nevada Test and Training Range. Along with the USAF, units from the Army, Navy, Marine Corps, United Kingdom, and Australia participate. (USAF)*

Today there is a war going on over the desert of Nevada. As the strike package of F-16s and F-15s approach the target supported by AWACS, KC-135 tankers, and RC-135 Rivet Joint aircraft, they are jumped by camouflaged F-16 aggressor aircraft. The formation breaks and goes defensive, but it is too late. The shots have already been taken—the "kills" made. The survivors hit the deck and try to dodge the challenging integrated air defenses and avoid the occasional B-52, C-130, and HH-60 leaving the range on their return to Nellis AFB. No one was actually killed today, but there was lots of gun camera film reviewed and then validated by computer data collected during the mission. After the debriefing and some determined grumblings like, "we'll get' em next time," there are the traditional drinks bought at the O-Club.

Today, the dogfight was not over Berlin, Hanoi, Belgrade, or Baghdad but in the skies of Nevada. Mistakes were made, but everyone made it home alive. This war is called "RED FLAG," the most intense and realistic combat training in the world.

The concept of RED FLAG was born deep in the basement of the Pentagon, not far from the famed purple water fountain, with a group known as the "fighter mafia." These brash majors would all grow up to be generals—Bill Kirk, Chuck Horner, and John Corder. But the chief visionary and great salesman was Major Moody Suter. Like their predecessors in the Lafayette Escadrille, the Flying Tigers, Sabres over MiG Alley, and the Phantoms in the "Triple Nickel," all combat pilots know that the first 10 combat missions are the most dangerous and deadly. Air-to-air training during the Vietnam era consisted of canned scenarios—scripted engagements with F-4s training against other F-4s, using U.S. tactics. This somewhat-less-than-realistic training led to the development of bad habits that were then carried into combat. Over North Vietnam poor habits resulted in an appalling and unacceptable one-to-one kill ratio. The "fighter mafia" conceived of an idea for realistic combat training. The exercise concept included the gathering of operational pilots from around the nation to fly against a dedicated "aggressor" force flying dissimilar aircraft and using enemy tactics. The entire exercise would take place in a SAM/AAA-rich environment. The Air Force had to give its young fighter pilots something similar to the experience of their first 10 combat missions without actually shooting at them. This concept was refined at the club and documented on bar napkins. Eventually, it was transformed into the RED FLAG Concept Brief in early 1975. (The original viewgraphs for this briefing sit on my desk today.)

Right: *An aggressor F-16 Fighting Falcon heads out to the war zone during RED FLAG 06-1—held during January and February 2006. RED FLAG tests aircrews' war-fighting skills in realistic combat situations simulating an air war. Aircraft fly missions day and night at the Nevada Test and Training Range. (USAF)*

The first RED FLAG exercise began on November 29, 1975, with 561 people and 37 aircraft flying 552 sorties. This revolutionary change in fighter training initially involved greater than normal risk of aircraft accidents. General Robert J. Dixon, then Commander of Tactical Air Command, was willing to take that risk to improve combat capabilities. In over 30 years of RED FLAGs, almost half a million warriors from all four services and 23 nations have flown more than 400,000 sorties, totaling some 700,000 flying hours.

The 414th Combat Training Squadron (RED FLAG) continues to provide this realistic training, which has expanded to include all aspects of air and space power. RED FLAG has changed with the times, now incorporating bombers, tankers, airlift, special ops, recce, UAVs, SAM and EW threats, Information Warfare, and space operations. It was, and still is, the closest thing to war that a pilot can experience. An aircrewman who flew in Desert Storm said after his first combat mission, "It was almost as tough as RED FLAG." The same is true today as it was 12 years ago for combat pilots returning from missions during Operation IRAQI FREEDOM.

Each year more than 800 aircraft, 4,000 aircrew members, and 11,000 maintenance and support personnel fly more than 10,000 sorties in the Nevada skies. Since that first RED FLAG, no U.S. aircraft has been lost in air-to-air combat. U.S. and Allied pilots have dominated in the skies over Iraq, Bosnia, Kosovo, and Afghanistan.

From my first RED FLAG in April 1978 to today, I see the same enthusiasm and classic learning curve from the first terrifying RED FLAG missions to the graduation missions, where the concentrated two weeks of training comes together. To honor the tradition of those innovators who made RED FLAG a reality, on July 11, 1996, the Air Force named the RED FLAG building after its founder. Suter Hall now stands as a monument to their vision and dedication. I think that Moody would be proud.

Left: *In a conventional conflict, the B-52 can perform air interdiction, and offensive counter-air and maritime operations. During DESERT STORM, B-52s delivered 40 percent of all the weapons dropped by coalition forces. It is highly effective when used for ocean surveillance, and can assist the U.S. Navy in anti-ship and mine-laying operations. Two B-52s, in two hours, can monitor 140,000 square miles of ocean surface. (USAF)*

ON ALERT

1976–1989

Chris Kenyon '75

ON ALERT

1976–1989

While the Air Force strategic forces maintained a posture of readiness after the war in Southeast Asia was over, the tactical forces were drastically changing. The next generation of high performance fighter planes began entering the USAF inventory. The F-15 Eagle, the newest air superiority fighter, arrived at Langley AFB, Virginia, in 1976. Operational F-16 Fighting Falcons were not too far behind. The tank-busting A-10 Thunderbolt II, designed to provide the firepower needed to stop a massive Soviet tank invasion in Europe, came on line at about the same time. Although F-4 Phantom IIs remained operational, they soon were relegated to second-class status behind the next generation fighters. Special mission F-4s—F-4G Wild Weasels and RF-4C reconnaissance platforms—survived a bit longer in the inventory than the fighter-attack models—but not much longer.

Perhaps the most significant development during these years mated low-observable technology to fighter and bomber design. The remarkable F-117 Nighthawk, developed in complete secrecy during the early 1980s, incorporated computer-assisted flight controls, without which the unstable plane would become nearly impossible to fly. The aircraft built during these years would remain the backbone of the USAF fighter force into the twenty-first century.

Improvements were made to the airlift force as well. The C-141, "stretched" and fitted with a refueling receptacle, became a global rather than a theater airlift asset. The C-5 fleet underwent an important modification to its wings, which were suffering from stress cracks caused by under-engineering and heavy loads. Lift and deployment capability was also enhanced by the acquisition of the KC-10 Extender. Not only did that aircraft provide refueling advantages, but it was also capable of transporting personnel and their equipment to the deployment location. These improvements to the airlift fleet were essential as humanitarian relief efforts at home and abroad seemed continual. Air Force airlifters delivered supplies around the world. Among their missions were the delivery of supplies to Antarctica, hurricane relief to the Marshall Islands, peacekeeping troops to Honduras, hay to snowbound livestock in the Midwestern States, drug enforcement officials to the Bahamas, and a host of other humanitarian and relief missions around the globe.

It was not until 1985 that the strategic bomber force also acquired of new aircraft to bolster the aging B-52 fleet. The Rockwell B-1B Lancer, reactivated by the Reagan administration, was followed in 1989 by the futuristic Northrop B-2 Spirit stealth bomber, a design reminiscent of flying wings first flown in the 1930s.

Development of new and highly accurate stand-off weapons reached an unprecedented level of effectiveness. When coupled with stealthy delivery systems, the Air Force's ability to deliver debilitating attacks against enemy targets was unmatched in the world. As weapons technology developed to levels of accuracy only dreamed of two decades before, the need for tremendous numbers of massive heavy bombers began to decline. Accuracy coupled with more powerful and better designed warheads effectively neutralized targets with one or two precision weapons. In the not-so-distant past these targets would have required tens or even hundreds of "dumb bomb" sorties to achieve the desired results. More and more, the lines which had once delineated strategic from tactical aircraft began to blur.

Traditional social barriers were also being recast during these years. Laws preventing women from entering military academies, pilot training, and navigator school were abolished. By 1982, an all-female crew had flown a SAC KC-135 on a refueling mission from Mather AFB. Barriers to women in combat would also eventually be changed, allowing women to serve as bomber, fighter, and combat airlift crew.

African Americans also continued to fracture social barriers on their way to significant achievements in the air and in space. Lieutenant Colonel Guy Bluford became the first African American in space, and Captain Lloyd "Fig" Newton was selected as the first African American Thunderbird pilot. Newton rose to the rank of four-star general as Commander of Air Education and Training Command.

The Air Force continues to develop space systems to better support the warfighter. One of the most significant developments, one that today impacts every mission and nearly every combat weapons drop, is the establishment of the Global Positioning System constellation of satellites. The ability to navigate and deliver weapons and cargo without reliance upon radio navigation has revolutionized both military and civilian aviation. The accuracy which comes from GPS enhances the capabilities of modern weapons to such a point that it is no longer the aircraft that is "strategic" or "tactical," but the weapon delivered and the effect provided that determine the nature of the strike.

In addition to launch activities, the USAF has remained a crucial participant in NASA's continuing exploration of space. Air Force astronauts have participated in the space shuttle program since its first test flight. Air Force Major Susan Helms became the first military woman in space. Colonel Eileen Collins was the first female shuttle pilot in orbit and also the first to command a shuttle mission—a most difficult return to flight mission, the first into orbit after the space shuttle *Columbia* disintegrated during reentry in January 2003.

Flight records are continually broken during this period. Several that fall are long-distance flights such as the Rutan *Voyager* unrefueled around-the-world journey in December 1986. Others are speed records and many are shattered by the Rockwell B-1B Lancer while more fall to the Mach-3-plus Lockheed SR-71 Blackbird.

During these years, the Air Force is called upon to enforce the political will of the nation to enforce sanctions or deter aggression. Operation ELDORADO CANYON, strikes by F-111s based in England upon targets in Libya, and Operation JUST CAUSE, the capture of Manuel Noriega in Panama, signal a significant dependence upon airpower in preserving freedom and human rights around the globe. America's ability to engage targets unreachable by other nations will continue to improve during these years and will be put to the test shortly after the collapse of the Soviet Union in 1989.

The Air Force played a major role during these years of constant vigilance in applying pressure upon the Soviet Union, mired in their own Vietnam-like war in Afghanistan, to tear down the walls of oppression. A combination of political upheaval, religious enlightenment, and economic collapse finally resulted in the demise of what had been, for more than half a century, America's Cold War adversary. Even while the Soviet Union was fracturing into several independent states, a different threat to freedom was just beginning to gain momentum—the forces of Islamic extremism and the spread of the ideology of global terrorism.

Pages 490–491: *Training, training, and more training was essential to preparedness during the Cold War. The B-52 in this painting is on a nine-hour training mission from Mather AFB, California. A typical mission included one or two aerial refuelings, a low-level navigation segment, and transition practice in the traffic pattern. ("Put Out My Hand, and Touched the Face of God," by Chris Kenyon, USAF Art Collection)*

Above: *The vigilance of an American eagle creates the background for another watchful warrior—an E-3 Sentry. ("Vigilant Warriors," by Roderick Lees, USAF Art Collection)*

ON ALERT

1976–1989

"General Secretary Gorbachev, if you seek peace, if you seek prosperity for the Soviet Union and Eastern Europe, if you seek liberalization: Come here to this gate! Mr. Gorbachev, open this gate! Mr. Gorbachev, tear down this wall!"
—President Ronald Wilson Reagan, Remarks at the Brandenburg Gate, West Berlin, June 12, 1987

1976

9 January
The first operational McDonnell Douglas F-15 Eagle air superiority fighter arrives at Langley AFB, Virginia, where it is assigned to the 1st Tactical Fighter Wing.

31 January
The Air Force transfers control of Udorn AB to the Royal Thai Air Force. The USAF has operated the base since the beginning of the Vietnam War. American presence at Korat will end a month later.

5 February–3 March
During Operation EARTHQUAKE, the shipment of relief supplies and equipment to Guatemala, Air Force airlifters deliver nearly 1,000 tons and 700 personnel to the region. Medical, communication, and engineering specialists help during the crisis. Some assistance continues through June.

1 March
The USAF ceases operations at Taipei Air Station, Taiwan, after two decades of service there.

15 March
NASA launches two nuclear-powered Air Force communications satellites—*Les-8* and *Les-9*—atop an Atlas IIIC launch vehicle.

21 March–9 June
In the wake of a series of typhoons in the Pacific, Air Force airlift aircraft deliver needed relief supplies to bases from Guam to the Philippines. Air Rescue and Recovery helicopters are busy after the storms, saving more than 700 flood victims in the Philippine Islands.

22 March
The last Strategic Air Command aircraft in Southeast Asia, a U-2, departs U-Tapao Airfield, Thailand, ending SAC's presence in the region.

A-10 Operational Test and Evaluation (OT&E) begins at Davis Monthan AFB, Arizona, with the arrival of the first version of the heavily armored, close-air support aircraft.

26 March
The NASA Flight Research Center at Edwards AFB, California, is renamed for Hugh L. Dryden, the agency's former deputy administrator.

Left: *The Fairchild Republic A-10 Thunderbolt II—better known as the Warthog—will become one of the deadliest tank-killing weapons in the history of warfare. The A-10, armed with the General Electric GAU-8/A Avenger 30 mm, seven-barrel Gatling gun, can carry a large complement of ground attack weapons. (USAF)*

Right: *Viking 1 obtained this color picture of the martian surface and sky on July 24, 1976. The sky has a reddish cast, probably due to scattering and reflection from reddish sediment suspended in the lower atmosphere. (NASA)*

Below: *The Smithsonian Institution, National Air and Space Museum opened in 1976 on the National Mall in Washington, D.C. (NASM)*

2 April
The last C-118A Liftmaster is retired from the active inventory and flies its final sortie to land at Davis Monthan AFB for storage.

6 May–5 June
Air Force personnel stationed at Aviano AB, Italy, provide medical assistance and manpower to assist in relief for earthquake victims in northeastern Italy.

7–24 June
An exercise designed to test and evaluate the Korean tactical air control system is held and includes several South Korean bases. Operation TEAM SPIRIT will become one of the largest field exercises in the world.

28 June
The first women enter the USAF Academy as members of the class of 1980.

1 July
The Smithsonian Institution, National Air and Space Museum opens its doors to the public. In only two years, museum visits top the 20 million mark, making it the most visited museum in the world.

15 July
Naval and Marine Corps navigators begin training at Mather AFB, California, consolidating all navigator training at one location.

20 July
On the seventh anniversary of astronaut Neil A. Armstrong's first step on the moon, the *Viking I* probe successfully soft-lands on Mars. The probe transmits photographs back to earth of the martian surface viewed on televisions around the world.

27–28 July
Three Lockheed SR-71s set three flight records. The first sets three records over a 1,000-kilometer closed-circuit course—a world absolute speed record of 2,092 miles per hour, a world jet speed record with 2,200 pound payload, and world jet speed record without payload. The second establishes two new records—an absolute world and jet speed record over a 15/25-kilometer course of 2,193 miles per hour. The third sets two new records—new absolute and jet records for sustained altitude in level flight at 85,069 feet.

THE MODERN CRUISE MISSILE AND THE USAF

Clayton Chun

Left: *A weapons technician controls the loading of an air-launched cruise missile into the belly of a B-52 Stratofortress. (USAF)*

Opposite: *The cruise missile combined advanced navigation and propulsion technology to deliver accurate strikes against distant targets. ("Air Launched Cruise Missile," by Charles Shealy, USAF Art Collection)*

The Air Force has used cruise missiles since the early 1950s. A cruise missile is a long-range, low-flying guided missile. Cruise missiles are akin to unmanned aircraft, but are much slower than ballistic missiles. USAF crews operate air and ground launched cruise missiles that can be nuclear or conventionally armed. Using these missiles extended the Air Force's ability to attack targets without the use of aircraft or forcing an aircrew to penetrate enemy air defenses to deliver ordnance. Today, the Air Force operates several types of air launched cruise missiles.

The first operational cruise missile used in combat was the German V-1 during World War II. After the war, the USAF developed several ground-launched cruise missiles. These cruise missiles gave the service the ability to deliver nuclear or conventional warheads hundreds to thousands of miles away. This capability gave the nation a limited nuclear retaliatory capability before the country could deploy intercontinental ballistic missiles. The nation built the ground-launched Matador, Snark, and Mace to attack surface targets. USAF crews also operated the Hound Dog, its first cruise missile launched from a bomber. The Hound Dog allowed B-52 crews to avoid sophisticated air defenses that included supersonic interceptors and surface-to-air missiles. Unfortunately, these cruise missiles were relatively large and many of the ground launched cruise missiles were inaccurate and had readiness concerns. Air Force interest in cruise missiles increased as technology resulted in improved accuracy and lowered costs in a world where Soviet air defenses had improved greatly.

Cruise missiles became popular in the late 1970s. When President Jimmy Carter came to office, he cancelled many defense programs he believed wasteful that included the B-1 bomber. The Carter administration sought an alternative and expanded the use of cruise missiles in 1977. As a result of a joint Air Force and Navy development effort, the USAF adopted two cruise missiles, the AGM-86B Air Launched Cruise Missile (ALCM) and a Navy-developed AGM-109 Tomahawk as the Ground Launched Cruise Missile (GLCM).

A cruise missile must navigate its flight path and guide itself to its target. Early guidance systems were unreliable and inaccurate. Missile designers compensated for the lack of accuracy by using larger payloads, which required large carrier missiles. During

ALCM and Tomahawk development, a new guidance system was designed that replaced earlier programmed flight activities. Computer and guidance component improvements matched images of terrain taken from the missile to those stored internally on a computer. Terrain contour matching (TERCOM) increased accuracy and thus allowed both missiles to become smaller. Smaller missiles allowed aircraft to carry more ALCMs, made them harder to detect, and made the GLCM more mobile.

Cruise missiles gave the USAF several innovative deployment options. The ALCM rejuvenated the B-52. B-52s could launch nuclear-armed ALCMs against the thickest of air defenses and strike a target within 1,550 miles with a 200-kiloton nuclear device. Mobile GLCMs confounded the Soviets and could hit objectives 1,500 miles away with an 80-kiloton nuclear warhead. Its guidance system could place the missile within 100 feet of a planned impact site. However, the end of the Cold War and arms control agreements reduced nuclear weapons. The GLCM was deployed in Europe, but was retired through the Intermediate-Range Nuclear Force Treaty in 1987. The last GLCM served in Europe until 1991.

The ALCM found new life as a conventional weapon. During the 1991 Persian Gulf War, seven B-52Gs took off from Barksdale AFB, Louisiana, and launched 35 conventionally armed ALCMs (AGM-86C) on January 17 against Iraqi targets. The AGM-86C used a Global Positioning System (GPS), satellite navigation constellation guidance system. GPS guidance gave the AGM-86C even greater precision than TERCOM. Since its operational debut in 1991, the USAF has used the AGM-86C in the Balkans and during Operation IRAQI FREEDOM.

Cruise missiles expand the Air Force's ability to strike targets over very long ranges, with exacting precision, and without risking an aircraft loss to enemy air defenses. These weapons evolved from simple, highly inaccurate unmanned aircraft to sophisticated stealth weapons that allow the USAF to strike targets worldwide. Instead of building a larger fleet of penetrating bombers, the nation was able to provide a potent nuclear deterrent by using cruise missiles. Today, the USAF and Navy operate cruise missiles that can strike any known target anywhere in the world.

1–2 August
Two Air Force Hueys (UH-1 Iroquois) respond to a flash flood in Colorado's Big Thompson Canyon. While on station, they rescue 81 stranded persons.

6 September
A Soviet MiG-25 Foxbat pilot, Lieutenant Victor Belenko, defects and flies his aircraft to Hokkaido, Japan, where he asks for asylum in the United States. After a detailed examination, the MiG is returned to the Soviets dismantled.

9 September
The first fully guided air-launched cruise missile (ALCM) flight takes place at the White Sands Missile Range, New Mexico. The missile guides itself during the flight based upon preset coordinates. The first unguided flight is made on March 5.

29 September
Ten women enter undergraduate pilot training at Williams AFB, Arizona. This is the first time since WASPs flew during World War II that women will be taught to pilot U.S. military aircraft.

10 December
Air Force search and rescue teams discover the whereabouts of downed U.S. balloonist Ed Yost. He is afloat in the Atlantic Ocean after his balloon loses helium and is forced to float on the surface in his catamaran-shaped gondola until a nearby West German tanker can pick him up.

29 November–December
Military Airlift Command C-141s begin delivery of humanitarian relief supplies to Incirlik AB following a devastating earthquake in Turkey. C-130s pick up the tents and heaters and take them to Van Airport, more austere but closer to the affected region.

1977

1 January
The 479th Tactical Training Wing is formed at Holloman AFB, New Mexico. Three squadrons of AT-38B Talon aircraft modified with gunsights and practice bomb dispensers will form the backbone of Fighter Lead-in School.

8 January
The first "stretched" version of the Lockheed C-141 Starlifter rolls off of the production line at their Marietta, Georgia, factory. The C-141B, more than 23 feet longer than the original model, is now equipped with an aerial refueling receptacle, allowing virtually unlimited range and increased cargo capacity.

31 January–11 February
After a massive blizzard covers New York and Pennsylvania, MAC airlifts some 1,160 tons of snow removal equipment and transports 430 passengers to Buffalo, Pittsburgh, and Niagara Falls.

23 March
Tactical Air Command receives its first Boeing E-3A Sentry aircraft at Tinker AFB, Oklahoma. The distinctive rotating disk is a new feature atop the Air Force airborne warning and control aircraft.

24 March
The Lockheed YC-141B flies for the first time. The stretched version of the C-141 is 23 feet longer and includes a refueling port.

27–30 March
Two Boeing 747 airliners carrying 643 passengers collide at Tenerife, Canary Islands, in the worst civil aviation disaster in history. Air Force C-130 Hercules aircraft take

medical support personnel to Tenerife and then airlift 56 survivors of the disaster to a staging area at Las Palmas, Canary Islands. There, C-141s pick up the injured for the journey to medical facilities in the United States.

14 April
A Soviet Tu-95 Bear suddenly appears off the coast of Charleston, South Carolina. Radar controllers lose contact when the aircraft descends below radar coverage. The Bear makes its way along the eastern coast and then reappears near Jacksonville, Florida, much to the embarrassment of defensive radar experts.

May
The de Havilland Canada DHC-6 Twin Otter is selected for use by the USAF Academy parachute jump program. The commercial aircraft is redesignated as the UV-18B and stationed at nearby Peterson AFB, Colorado.

2 May
First Lieutenant Christine E. Schott becomes the first woman undergraduate pilot training student to solo in the T-38 Talon supersonic jet training aircraft. Famed pilot Jacqueline Cochran accumulated many solo hours in the aircraft and set several performance and time-to-climb records in the early 1960s.

19 May
During a severe aircraft emergency, B-52 pilot Captain James A. Yule demonstrates remarkable presence of mind and gallantry in recovering and safely landing his plane. For his actions, Yule is awarded the Mackay Trophy for 1976.

16 June
An Air Force Lockheed C-5 Galaxy lands in Moscow. This is the first C-5 landing in the Soviet Union. Flying from Chicago to Moscow nonstop, the C-5 required two aerial refuelings to complete the trip.

Captain David M. Sprinkel and crew of C-5 Aircrew Mission AAM 1962-01 composed of members from the 436th Military Airlift Wing and the 512th Military Airlift Wing airlift a large superconducting electromagnet, support equipment, and personnel in support of joint U.S.-USSR energy research program. For this mission, the crew receives the 1977 Mackay Trophy.

Dr. Wernher von Braun dies at age 65. Former director of the German secret rocket program, he came to the U.S. after World War II as part of Operation PAPERCLIP, and was part of the team that developed the Redstone and Saturn V rockets.

Opposite: *This Northrop AT-38B was used as a fighter lead-in trainer at Holloman AFB, New Mexico. It can be viewed at the National Museum of the USAF in Dayton, Ohio. (USAF)*

Right: *A Lockheed C-141A Starlifter (in background) parks next to its younger big brother, the C-141B. The stretched version is 23 feet longer and also includes a refueling port. (USAF)*

30 June
President James E. Carter cancels the Rockwell B-1 program after four prototypes are built.

3 August
Cadet Colonel Edward A. Rice Jr. is named Cadet Wing Commander at the USAF Academy. He is the first African American to hold this, the highest cadet leadership position within the Cadet Wing.

4 August
The last operational T-33 Shooting Star is retired from service and flies to its final landing at Davis Monthan AFB, Arizona, for storage.

12 August
The Space Shuttle Enterprise makes its first glide test after being released from atop a specially modified Boeing 747 Shuttle Carrier Aircraft (SCA) at the Dryden Flight Research Center at Edwards AFB, California. Former Air Force fighter pilot Fred Haise and Air Force Colonel C. Gordon Fullerton crew the early glide tests. The first glide begins at 22,800 feet and ends with a smooth landing on the Rogers Dry Lake bed to cheers from a crowd of approximately 70,000 people.

2 September
The first 10 female Air Force pilots receive their wings.

30 September
Using a Delco inertial navigation system (INS), a Charleston-based C-141 Starlifter flies a transatlantic mission without a navigator. The mission terminates at Rota Naval Station, Spain. Navigators are gradually phased out as crew members as navigational technology continues to improve.

1 October
Air Force Reserve and Air National Guard C-130 aircraft and crew begin quarterly deployments to Howard AFB in the Panama Canal Zone during Operation VOLANT OAK.

12 October
Five women graduate from undergraduate navigator training (UNT), the first to do so.

Left: *The shuttle* Enterprise *is commanded by former* Apollo 13 *Lunar Module pilot Fred Haise (left) with C. Gordon Fullerton as pilot. The shuttle orbiter* Enterprise *was named after the fictional Starship* Enterprise *from the popular 1960s television series,* Star Trek. *(NASA)*

Below: *Captains Mary E. Donahue, Connie J. Engel, Kathy Lasauce, Susan D. Rogers, Lieutenants Victoria K. Crawford, Mary M. Livingston, Carol Ann Scherer, Christine E. Schott, Sandra May Scott, and Kathleen A. Bambo received their wings and became USAF jet-qualified pilots. In a class of 46 cadets, Captain Engel took top honors and Captain Donahue finished first scholastically. ("Air Force Wings to First Women Jet Pilots," by Nathalee Mode, USAF Art Collection)*

Right: *The Lockheed C-5 Galaxy suffered through numerous mechanical problems and a fleetwide wing-box modification. Nonetheless, the massive cargo plane will overcome these troubles to serve as the USAF's heavy strategic lift aircraft for the next three decades. (USAF)*

1978

8–17 February
During Operation SNOW BLOW II, Air Force C-5s, C-141s, and C-130s deliver more than 2,300 tons of snow-removal equipment and other supplies to snowbound New England. More than 1,000 passengers are airlifted to safety during the operation as well.

22 February
The first Global Positioning System (GPS) satellite—*NAVSTAR 1* (Navigation System with Timing and Range)—is launched into orbit atop an Atlas F booster rocket. A "constellation" of 24 GPS satellites will revolutionize not only navigation, but weapons accuracy as well. The final satellite will launch to orbit in 1994.

16–27 May
During Operation ZAIRE I, airlift support for Belgian and French troops in Zaire, two C-5 Galaxy crew, one under the command of Lieutenant Colonel Robert F. Shultz, and the other piloted by Captain Todd H. Hohberger, overcome a number of challenging obstacles, including flight fatigue and major mechanical problems, to deliver their cargo and passengers to the forces engaged in rescuing European workers threatened by Katanga rebels from Angola. For their actions during May 20–23, the aircrews receive the Mackay Trophy.

31 May–16 June
During the second phase of operations—ZAIRE II—Air Force airlifters transport African peacekeeping troops into Zaire and evacuate beleaguered Belgian and French troops. During 72 sorties, they transport more than 1,600 tons cargo and 1,225 passengers.

12 July
After a quarter-century of service, the last KC-97L Stratofreighter is retired from the Air Force inventory.

27 July
The 81st Tactical Fighter Wing arrives at Bentwaters/Woodbridge Royal Air Force bases flying the A-10 Thunderbolt II. This marks the first overseas deployment for this aircraft.

14–16 August
An Air Force C-141 Starlifter delivers 26 tons of relief supplies to flood victims in Khartoum, Sudan.

17 August
The Air Force accepts the first production model General Dynamics F-16 Fighting Falcon at the factory in Fort Worth, Texas.

22–29 November
After a mass suicide in Jonestown, Guyana, HH-53 Jolly Greens of the 55th Aerospace Rescue and Recovery Squadron (ARRS) transport more than 900 bodies to makeshift morgues in Georgetown, Guyana. HC-130s refuel the helicopters as they ferry back and forth between locations. C-141s will retrieve the bodies and return them to the United States on November 28.

9 December
After the collapse of the Shah of Iran's government, Air Force C-5 and C-141 aircraft evacuate some 900 people from Tehran.

GLOBAL POSITIONING SYSTEM: A NAVIGATIONAL MIRACLE

Clayton Chun

Precise, easily accessible means to determine exact locations have been sought since man ventured outside a cave or dwelling. Navigation on air, land, and sea was made easy, accurate, and inexpensive through the use of satellites. In the 1960s, the Air Force started development of a constellation of navigation satellites that evolved into the Global Positioning System (GPS). Today, GPS serves the nation and the world by providing precise three-dimensional position, time, and velocity information for any point on the earth, around-the-clock, for anyone with a GPS receiver. By using the GPS constellation, military and civilian users can replace traditional maps or celestial navigation.

The U.S. Navy first developed a satellite navigation system—Transit. Transit was created to support the operation of the Polaris submarine-launched ballistic missiles. This capability allowed submarines to gather key location data to support guidance and launch operations. Transit was available to submarine and surface ship captains. The first Transit satellite was launched for the Navy by an Air Force Thor-Ablestar from Cape Canaveral, Florida, on April 13, 1960. From its circular polar orbit, Transit provided radio transmissions that used Doppler shift measurements to determine positions. Unfortunately, this early program proved too time-consuming, lacked precision, and did not provide information on velocity that aircraft, cruise missiles, or land-based ballistic missiles required in flight.

These added demands for time, location, and velocity required a new satellite system that would provide three-dimensional information. A joint USAF/USN program was established in 1973. This program, with the Air Force as executive agent, was named Navstar GPS on May 2, 1974. The GPS constellation was conceived to provide a worldwide range of services to anyone who could receive its radio transmissions—military or civilian. The only constraint to using GPS is possession of a receiver unit. A GPS user receives signals from a minimum of four orbiting satellites. A receiver unit has a device that can determine the difference in time from the transmitting satellite to the receiver unit. This difference is the basis for calculating range information that helps plot location and velocity. GPS offers an almost unlimited application for military and civilian navigation uses. GPS design engineers incorporated two cesium and two rubidium atomic clocks for extreme accuracy. The satellite consists of the clocks and a communications system to broadcast its radio signals.

The GPS system was designed to have an initial 16-satellite constellation that would expand to 21 satellites with at least three on-orbit spare satellites. In 2004, there were 28 satellites in the constellation. Each satellite would orbit the earth every 12 hours and have an original service life of about seven and a half years. These satellites were placed at regular intervals in orbit at about 12,600 miles. The satellites transmit their signals on two L-band frequencies.

Aside from the GPS satellites, the system includes a series of ground stations. These five ground stations allow controllers to update computer software, monitor satellite health, and adjust the atomic clocks. They are located at Schriever Air Force Base (master control station) in Colorado, Hawaii, Diego Garcia, Kwajalein Island, and Ascension Island.

The Air Force first launched a GPS satellite, a research and development model, on February 22, 1978, from Cape Canaveral. Since then, it has improved satellites with greater lives, better clocks, and faster processors. GPS critics were concerned about the system's vulnerability to jamming or its signal being used by a foe. However, many of these complaints were muted during the 1991 Persian Gulf War when GPS was used to guide aircraft strikes, ease land navigation over deserts without landmarks, and substitute for inadequate maps. Demand was so great for GPS services that military units faced severe shortages of hand-held receiver units.

GPS has made a profound change to the USAF and the world. Military units no longer have to fear being lost with GPS. GPS signals have given all branches of the military and our allies the ability to navigate accurately, greatly improve weapons delivery, map areas, coordinate military operations like aerial refueling, and achieve successful search and rescue missions. These functions have significantly multiplied the USAF's combat power without adding to expensive weapons systems.

Opposite: *The NAVSTAR Global Positioning System (GPS), a constellation of orbiting satellites, provides 24-hour navigation services and three-dimensional position information to an unlimited number of users. (Artist's rendition, USAF)*

Below: *An artist's rendition of the NAVSTAR II satellite. Twenty-four NAVSTAR satellites make up the constellation needed to provide global positioning to anyone with a receiver. (USAF)*

Left: *More than a decade after the arrival of the first operational F-16 at Hill AFB, a ground crewman guides a 388th Tactical Fighter Wing F-16C Fighting Falcon aircraft onto the taxiway in Saudi Arabia during Operation DESERT SHIELD. (USAF)*

Opposite, below: *As* Voyager 1 *flew by Jupiter, it captured this photo of the Great Red Spot. The Great Red Spot is an anti-cyclonic (high-pressure) storm on Jupiter that can be likened to a massive hurricane on earth. It is so large that three earths could fit inside it. (NASA)*

1979

6 January
The first operational General Dynamics F-16 Falcon single-engine fighter aircraft is delivered to Hill AFB, Utah, where it will be assigned to the 388th Tactical Fighter Wing.

27 February
The improved F-15C flies for the first time at St. Louis, Missouri.

9 March
Responding to perceived threats upon the northern border of Saudi Arabia, two E-3A AWACS aircraft deploy to Riyadh during Operation FLYING STAR.

31 March
While piloting his H-3 helicopter, Major James E. McArdle rescues 28 shipwrecked Taiwanese seamen. Hovering above the endangered ship that ran aground in the Yellow Sea, McArdle and his crew hoist the survivors to safety. For his airmanship, McArdle is awarded the Mackay Trophy.

Following the March 28 nuclear plant accident at Three Mile Island, Pennsylvania, MAC aircraft begin flying the first of 15 missions in support of activities after the mishap.

3–5 April
After Typhoon Meli strikes Fiji, two C-141s deliver more than 20 tons of relief supplies to the people of the island.

13 April
A Military Airlift Command (MAC) C-141 delivers 20 tons of vegetable seeds to be planted by the people of Zaire. It is hoped that this delivery will assist in the long-term solution to undernourishment in that region.

19–20 April
MAC aircraft deliver more than 130 tons of humanitarian relief supplies and equipment to Titograd International Airport, Yugoslavia, after an earthquake strikes the Adriatic coastal regions.

May
As part of the tactical planning for military conflict in Europe, the first Forward Operating Base (FOB) for A-10s based in England is activated at Sembach AB, Germany. The FOB concept would allow close-air support sorties to originate near the forward edge of the battle area should the Soviets invade Europe.

1 June
The Community College of the Air Force relocates from Lackland AFB, Texas, to Maxwell AFB, Alabama, becoming part of Air University.

Right: *Since the first GLOBAL SHIELD exercise in 1979, USAF heavy bombers have continued to practice weapons delivery at ranges around the globe. This Boeing B-52H Stratofortress drops a load of M-117 750-pound bombs while on a training run. During Operation DESERT STORM, B-52s delivered 40 percent of all the weapons dropped by coalition forces. (USAF)*

5 June
President Jimmy Carter authorizes full-scale development of the MX Missile.

8–16 July
During Operation GLOBAL SHIELD, SAC's first test of the Single Integrated Operational Plan (SIOP), hundreds of bombers, tankers, and strategic missiles are called to alert. Others are dispersed to locations from which mock attacks upon targets located in various bombing ranges are launched. GLOBAL SHIELD will continue to be SAC's major annual operational exercise throughout the Cold War. Often, more than 100,000 SAC personnel are involved during exercise operations.

9 July
The space probe *Voyager 2* reaches Jupiter and begins sending photographic images back to earth. The craft had been launched in 1977. *Voyager 1* had transmitted images to earth earlier in March, but the images from *Voyager 2* result in significant scientific discoveries related to orbiting Jovian Moons, like Europa and Io, and images from Saturn and Neptune as well.

26 July
The 400th Minuteman ICBM is launched from Vandenberg AFB, California.

31 August
As Hurricanes David and Frederic approach the U.S., Air Force cargo planes begin delivering relief materiel to devastated Caribbean Islands. Nearly 3,000 tons of supplies will be delivered by November 21.

15–22 September
Eight Air Force Reserve and Air National Guard (California and Wyoming) C-130s drop fire suppressant over forest fires blazing in southern California. More than 250 sorties deliver 732,000 gallons of the liquid during one of the largest aerial firefighting operations in history. These missions are flown over the same areas where the Air Service had once accomplished fire-spotting missions during the early 1920s.

Left: *The Cold War cat-and-mouse game between American interceptors and Soviet long-range reconnaissance aircraft began soon after World War II and continued for more than five decades. Here, Air Force F-15s intercept a Soviet Tu-95 Bear. (USAF)*

Opposite: *This painting depicts a gathering of eight RH-53 helicopters over the carrier* Nimitz *in the Persian Gulf, en route to a desert rendezvous with six Air Force C-130s in Iran. ("The Iranian Rescue Mission Begins," by Ren Wicks, USAF Art Collection)*

1 October
The Air Defense Command begins to deactivate by transferring its responsibilities between Strategic and Tactical Air Command. For the moment, NORAD maintains operational control of former ADC resources.

4 November
The U.S. Embassy in Tehran, Iran, is stormed by 3,000 Iranian militants. They take 66 U.S. citizens captive and hold them hostage. President Carter seeks a diplomatic solution.

2–21 December
MAC aircraft deliver 650 tons of relief supplies and transport 250 personnel to Majuro Atoll, Marshall Islands, after Typhoon Abby sweeps through the region leaving widespread destruction.

20 December
The Air Force launches a Minuteman I ICBM in a test of the Advanced Maneuvering Reentry Vehicle (AMARV). AMARV has a fully autonomous navigation system for evading enemy anti-missile missiles.

27 December
The Soviet Union sparks a coup in Afghanistan establishing a new puppet regime to govern, which provokes a civil war. Thousands of Soviet troops are airlifted into the country.

1980

2–4 January
MAC airlifters deliver relief to earthquake victims on Terceira Island, Azores, after a Portuguese government request for assistance. A week later, MAC aircraft deliver humanitarian aid to victims of Cyclone Claudette on Mauritius Island in the Indian Ocean. More than 17 tons of supplies are delivered to assist in relief efforts.

25–28 February
Clark AB-based F-15s intercept Soviet Bear D and F aircraft after they penetrate the Philippine air defense zone. Probes such as these are routine during the Cold War.

12–14 March
Two 644th Bombardment Squadron B-52 Stratofortresses fly around the world in 43 and one-half hours as they are tasked to locate Soviet naval vessels in the Arabian Sea. Flying more than 22,000 miles at an average speed of 488 miles per hour, the crew is awarded the Mackay Trophy. This marks the third time that SAC aircraft have flown around the world nonstop; once in 1949 and again in 1957.

6 April
The first C-141B operational mission is flown from Beale AFB, California, to RAF Mildenhall, United Kingdom. One refueling is required to complete the 11 hour 12 minute flight.

7 April
After an Iranian takeover of the U.S. embassy in Tehran, Iran, all Iranian military trainees in the United States are ordered to leave the country within four days. Many are enrolled in Air Training Command courses.

24 April
Operation EAGLE CLAW, an attempt to rescue 66 hostages being held by Iranian militants, turns into a disaster as poor weather and mechanical problems ground the forces in the desert. During the withdrawal operations, one of the RH-53 helicopters collides with an EC-130, exploding and killing eight members of the contingent.

May
The first woman to attend Air Force undergraduate helicopter training, 2d Lieutenant Mary L. Wittick, begins her ground training.

18 May–5 June
After Mount St. Helens erupts, aircraft from the Aerospace Rescue Service (ARRS), Military Airlift Command, and the 9th Strategic Reconnaissance Wing begin humanitarian efforts throughout the affected regions in northwest Washington State. ARRS helicopters rescue 61 stranded individuals and transport them to safety while SR-71s fly reconnaissance sorties to photograph affected areas to assist ground rescue personnel.

28 May
The first Air Force Academy class that includes women graduates at ceremonies held at Falcon Stadium. Ninety-seven women are commissioned out of the original 157 that entered the academy Class of 1980 during traditional June Week activities.

8 July
The McDonnell Douglas FSD F-15B flies for the first time. This modified F-15 demonstrates ground attack capabilities and has a redesigned front cockpit and a rear cockpit designed for a Weapons Systems Officer (WSO). This modification will become the F-15E Strike Eagle.

14 August
The first C-5A modified with new wings flies at Dobbins AFB, Georgia. Lockheed will modify 77 of the heavy-lift aircraft, significantly extending service life.

Left: *An RC-135U Combat Sent aircraft flies a training mission from Offutt Air Force Base, Nebraska. The RC-135U Combat Sent is the modern descendant of the original RC-135 and provides strategic electronic reconnaissance information to the President, Secretary of Defense, Department of Defense leaders, and theater commanders. (USAF)*

10 July–3 October
Air Force F-4 Phantom II aircraft deploy to Egypt for the first time during Operation PROUD PHANTOM. Twelve Moody AFB, Georgia, F-4Es make the deployment as part of a tactical exercise designed to train the Egyptians in the use of the F-4. Egypt had just purchased the front-line fighter from the U.S.

2 September
At Johnson Space Center, Houston, Texas, an Air Weather Service detachment is activated to support NASA operations in the area.

16 September
An Air Force RC-135 electronic surveillance aircraft is forced to use drastic measures to evade an attack by a Libyan MiG-23 fighter. The Boeing RC-135 is conducting an ELINT mission along the Libyan coast.

18 September
An explosion at a Titan II launch complex at Little Rock AFB, Arkansas, kills one USAF maintenance person and destroys the missile silo.

1 October
Operation ELF ONE, the deployment of E-3A AWACS and KC-135 Stratotankers to Saudi Arabia, begins in an effort to better monitor Saudi airspace during the Iran-Iraq War. The operation will continue for the next eight years during the Iran-Iraq War.

3 October
Captain John J. Walters and his HH-3 Jolly Green Giant helicopter crew hoist and airlift 61 passengers and crew of the Dutch cruise ship *Prinsendam,* afloat and on fire in the Pacific Ocean, to a nearby oil tanker. For the mission, Captain Walters receives the Mackay Trophy.

12–23 October
After a devastating earthquake hits El Asnam, Algeria, MAC aircraft deliver more than 400 tons of relief supplies to assist the more than 100,000 who are homeless after the massive quake.

20–23 October
Responding to floods in Nicaragua, 40 tons of medical supplies and food is transported by Southern Air Division aircraft to the troubled country.

20 November
The first Pave Tack–modified F-111 arrives at RAF Lakenheath, where it is assigned to the 48th Tactical Fighter Wing.

21 November
Nellis-based helicopters and aircraft respond to a massive fire in the MGM Grand Hotel, where they rescue more than 300 hotel guests from the burning 26-story structure.

10 December
With Poland in crisis and the Iran-Iraq War fully engaged, four USAFE E-3A aircraft are moved to Ramstein AB to participate in Operations ELF, the monitoring of airspace throughout the Middle East. Additional E-3s are deployed to Europe to support these ongoing operations.

1981

January–June
In an effort to assist the Salvadorian government in their struggle against leftist guerillas, Air Force C-130s deliver more than 500 tons of arms, helicopters, and other war materiel to that country. The shipments are valued near $5 million.

Left: *The C-9 is a twin-engine, T-tailed, medium-range, sweptwing jet aircraft used primarily for Air Mobility Command's aeromedical evacuation mission. The C-9A is capable of carrying 40 litter patients, 40 ambulatory and four litter patients, or various combinations thereof, providing the flexibility for Air Mobility Command's worldwide aeromedical evacuation role. (USAF)*

11 January
Boeing delivers the first and second air-launched cruise missile (ALCM) to the 416th Bombardment Wing at Griffiss AFB, New York. The ALCM (AGM-86B) is a nuclear-capable weapon with a range after launch of approximately 1,500 miles. Once dropped, the missile navigates to its target by using a terrain-matching system that allows extremely low-altitude ingress to the target.

12 January
Nine A-7D Corsair II aircraft assigned to the ANG in Puerto Rico are destroyed by terrorists.

18–25 January
After 444 days as hostages in Iran, 52 Americans are released and transported from Tehran to Rhein-Main AB, Germany, where they receive medical treatment at the USAF hospital in Wiesbaden. They return to the U.S. four days later aboard a VC-137.

1 February
Donald W. Douglas, founder of the Douglas Aircraft Company in 1920, dies at age 88. His company was responsible for aircraft such as the Douglas World Cruisers, the C-47 Skytrain, and the DC series of commercial airliners.

17 March
The first McDonnell Douglas KC-10 Extender is delivered to Strategic Air Command. Larger than the KC-135, the Extender carries more fuel and cargo and is capable of refueling a greater variety of aircraft.

1 April
Flying the Northrop F-5E Tiger II, the 527th Tactical Fighter Training Aggressor Squadron is activated at RAF Aclonbury, United Kingdom.

12–14 April
Aerospace Defense Command and Air Force Communications Command facilitate the space shuttle *Columbia*'s first flight with space tracking and communications systems.

2 May
The Airborne Laser Laboratory (ALL) successfully shoots down an aerial target drone over White Sands Missile Range, New Mexico.

Right: *The Boeing NKC-135A Airborne Laser Lab can be viewed at the National Museum of the USAF in Dayton, Ohio. (USAF)*

THE NATIONAL MUSEUM OF THE UNITED STATES AIR FORCE: THE WORLD'S LARGEST AND OLDEST MILITARY AVIATION MUSEUM

Left: *A C-141 Starlifter aircraft, known as the* Hanoi Taxi, *flies over its soon-to-be new home at the National Museum of the U.S. Air Force adjacent to Wright-Patterson AFB, Ohio. This particular aircraft was used to return American POWs home at the end of the war in Southeast Asia. (USAF)*

Everyone who has ever been part of the Air Force family should, at some point in their lives, take a trip to Dayton, Ohio. There, located adjacent to Wright-Patterson AFB in a complex of hangar-like buildings, is housed the single largest and most complete collection of U.S. Air Force aircraft and artifacts in the world. More than 75,000 small items, including flight suits, G-suits, and space suits; trophies and memorabilia; technical equipment and more than 300 aircraft and missiles are displayed and stored throughout the open Quonset hut-shaped hangars. It is a treasure trove of Air Force history—heaven for those seeking to understand the evolution of American military airpower.

Since 1918, artifacts related to Army aviation have been collected and displayed in a variety of locations. The original museum opened at McCook Field, near Dayton, Ohio, in the spring of 1923 and its name has changed four times since. The most recent change occurred in 2004 when it was redesignated as the National Museum of the United States Air Force.

On display at the museum are balloons and aircraft made of wire, wood, and fabric dating from the early 1900s. Nearby are displayed many of today's most modern aircraft, molded of composite materials formed to reflect radar energy—some piloted, some not. Creative displays invite visitors to imagine the challenges faced by pilots and support troops throughout Air Force history.

For example, a whimsical, yet familiar, life-size exhibit of an early military training aircraft transports anyone who has ever flown or maintained training aircraft back to a moment immediately following a minor "ground looping" accident. A mannequin represents the angry instructor pilot; the dejected student, the irritated maintenance chief and the dumbfounded apprentice all attest to the consequences arising from the crash of the nosed-over propeller-driven plane. For every piece of hardware in the museum there are personal stories related to its creation, construction, and utilization. Diorama displays help to expose the personality of the artifact instead of simply glorifying the technology itself. Of course, this type of display cannot be created for each

Right: *The famous B-17F* Memphis Belle *is readied for restoration (USAF).*

Below left: *Museum officials formally induct a Northrop B-2 Spirit stealth bomber into the institution's aircraft collection. The Air Force's national museum is the first place to permanently exhibit the stealth bomber to the public. (USAF)*

Below right: *The Boeing B-29 Superfortress* Bockscar *is on display in the Air Power Gallery. Bockscar dropped the second atomic bomb on Nagasaki, Japan, on August 9, 1945. (USAF)*

airplane and missile in the museum's inventory. There is still ample opportunity to wander through acres and acres of some of the most rare and iconic aircraft ever flown by America's air arm.

The museum, the world's largest and oldest military aviation museum, today is charged with portraying the heritage and traditions of the Air Force through specialized exhibits. But the NMUSAF is far more than just one museum in Dayton. As a large part of the Office of Air Force History, the staff provides technical and professional guidance to the U.S. Air Force Heritage Program. This extensive program includes 12 field museums and 260 domestic and international heritage sites. In addition, the staff is required to maintain accountability for more than 6,000 historical artifacts and aircraft and spacecraft on loan to 450 civilian museums, cities, municipalities and veterans' organizations throughout the world. It is a daunting task, but the reward for those who make the journey will be enjoyed for a lifetime.

18 June
The Lockheed F-117A Nighthawk flies for the first time. Until 1988, all flights are accomplished at night and under extreme secrecy.

3 August
When civilian air traffic controllers stage an illegal strike across the country, USAF controllers man their empty chairs, allowing the commercial airline industry to continue operations and service.

15 August
The first B-52G Stratofortress specifically configured to deploy air-launched cruise missiles (ALCM) arrives at Griffiss AFB, New York.

7 September
Edward A. Link, inventor of the Link Trainer, dies at age 77. His first patented simulator was built in 1929 and later adopted for use by the Army Air Corps in 1934.

15 September
The first Lockheed TR-1A strategic reconnaissance aircraft is accepted by the Strategic Air Command. The aircraft first flew on August 1.

The first USAF squadron to be outfitted with the Pave Tack laser weapons system becomes operational. The 494th Tactical Fighter Squadron flying F-111F aircraft are assigned to RAF Lakenheath, United Kingdom.

October
Euro-NATO Joint Jet Pilot Training begins at Sheppard AFB, Texas.

2 October
President Ronald W. Reagan announces that 100 B-1B Lancers will be built for the USAF. This reverses former president Jimmy Carter's previous cancellation of the program. Additionally, Reagan announces that the MX missile will continue and be deployed in existing missile silos.

5 November
Eventually replacing the EB-66 and EB-57 aircraft, the first EF-111A is delivered to the 388th Electronic Combat Squadron (ECS) at Mountain Home, Idaho. The EF-111 Raven, first flown in March 1977, will unofficially become known as the "Spark-vark," used primarily as an electronic warfare defense-suppression asset.

23 November
Eight Boeing B-52 Stratofortresses establish a record for the longest B-52 bombing mission during exercise BRIGHT STAR '82. The bombers take off from North Dakota and fly 15,000 miles in 31 hours, dropping training bombs on a practice target in Egypt.

Above: *The Lockheed F-117 was designed to reflect radar energy making it virtually invisible to enemy radar detection. (USAF)*

Below: *The General Dynamics EF-111 Raven provided electronic warfare support for strike packages until the late 1990s. (USAF)*

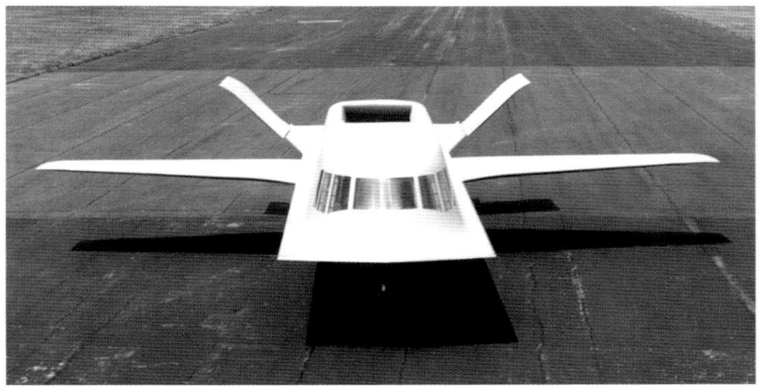

1982

18 January
During a practice demonstration sortie flown at Indian Springs, Nevada, four T-38 Thunderbird aircraft crash into the desert floor, killing all four pilots. This event caps a terrible flight season; during the previous May and September aircraft accidents cost the lives of two other team members as well.

26 January
Former Air Force astronaut Major General Michael Collins (USAFR) flies his "fini flight" in an Edwards AFB F-16 Fighting Falcon. General Collins orbited the moon while astronauts Neil Armstrong and Buzz Aldrin landed on the surface on July 20, 1969.

5 February
Northrop's Tacit Blue stealth technology demonstrator flies for the first time in complete secrecy. The "Whale" will fly an additional 135 sorties that are designed to evaluate radar cross-section reduction techniques.

24 February
The first NATO AWACS arrives at Geilenkirchen NATO AB, Germany. Eventually, 18 E-3A Sentry aircraft will be stationed there.

3 March
The first six A-10 Thunderbolt II aircraft arrive at Suwon, South Korea, for eventual beddown. The project is known as Commando Vulcan.

Above: *Tacit Blue, also known as the Blue Whale, was used to test low-observable designs later incorporated into the Northrop B-2 Spirit bomber. (USAF)*

Right: *CFM-56 turbofan engines fitted to the KC-135 made it more economical and will extend the aircraft's life well into the twenty-first century. (USAF)*

2 April–14 June
Argentinean military forces invade the Falkland Islands, a British colony. During the battle that follows, airpower plays an important role in the conduct of operations. British forces will defeat the invaders with support from carrier-based Harrier aircraft.

10 June
For the first time, an all-female crew flies a Strategic Air Command KC-135 Stratotanker on an aerial refueling training mission. The women are assigned to Castle AFB, California.

1 July
The first Gryphon ground-launched cruise missile wing in USAFE, the 501st Tactical Missile Wing, is stationed at RAF Greenham Common, United Kingdom. Five additional GLCM wings will eventually be activated.

15 July
Strategic Air Command launches its 1,500th missile from Vandenberg AFB, California. Launches began at Vandenberg in December 1958.

4 August
The first re-engined KC-135 Stratotanker makes its maiden flight. Marking the beginning of a major fleet overhaul, the aircraft is fitted with the CFM-56 turbofan engine—much more efficient and powerful than older turbojets.

September
Air Force Space Command (officially redesignated on November 15, 1985) is activated at Peterson AFB, Colorado.

Left: *Modified to carry air- launched cruise missiles, B-52s expanded their role and improved their capability as global bombers. (USAF)*

Below: *Chimpanzee Ham is greeted by the recovery ship commander after his flight on the Mercury Redstone rocket in 1961. Ham died in 1983 at the age of 26. (NASA)*

16 September
B-52 Crew E-21, Strategic Air Command successfully lands their crippled B-52 under almost impossible conditions, thereby saving their lives and a valuable Air Force aircraft. The pilot Captain Ronald L. Cavendish and his crew, 2d Lieutenant Frank A. Boyle, Captain Ronald D. Nass, 1st Lieutenant James D. Gray, 1st Lieutenant Michael J. Connor, Technical Sergeant Ronald B. Wright, and 1st Lieutenant Gerald E. Valentini, receive the Mackay Trophy for their superior airmanship during the in-flight emergency.

21 September
A 416th Bombardment Wing B-52 launches a cruise missile for the first time during operational testing. By December, 16 Griffiss-based B-52Gs are modified to carry six ALCMs under each wing.

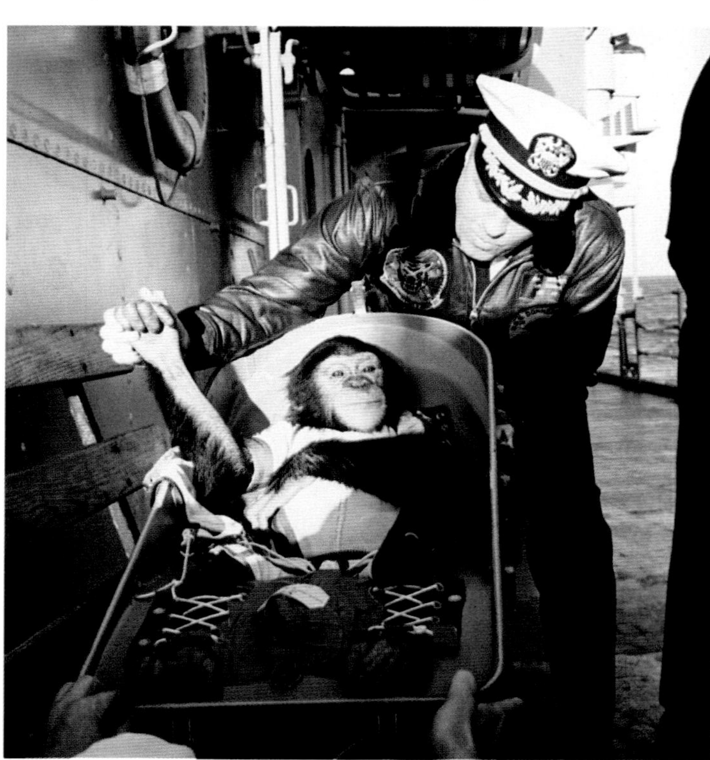

16 November
Space shuttle *Columbia* with its crew of four astronauts lands at Edwards AFB, California, after its first operational flight. This marks the first time that a spacecraft carries more than three people into orbit.

16 December
The first air-launched cruise missile is placed on alert at Griffiss AFB, New York.

24–30 December
Ten days after an earthquake strikes Yemen, MAC aircraft deliver more than 80 tons of relief supplies, including tents, medicine, and generators, to help those most affected by the event.

1983

1 January
United States Central Command (USCENTCOM), a unified command, is activated at MacDill AFB, Florida.

19 January
HAM (Holloman Aero Med), the chimpanzee that flew into space in January 1961, dies in the North Carolina Zoological Park at the age of 26.

1 February
A joint, combined training exercise involving many Air Force units begins in Honduras. *Ahuas Tara I* is designed to demonstrate American support for noncommunist regimes in Central America.

3 February
Strategic Air Command completes the retrofit of 300 Minuteman III ICBMs with new reentry systems. These modifications will help to enhance retaliatory capability of the Minuteman III force.

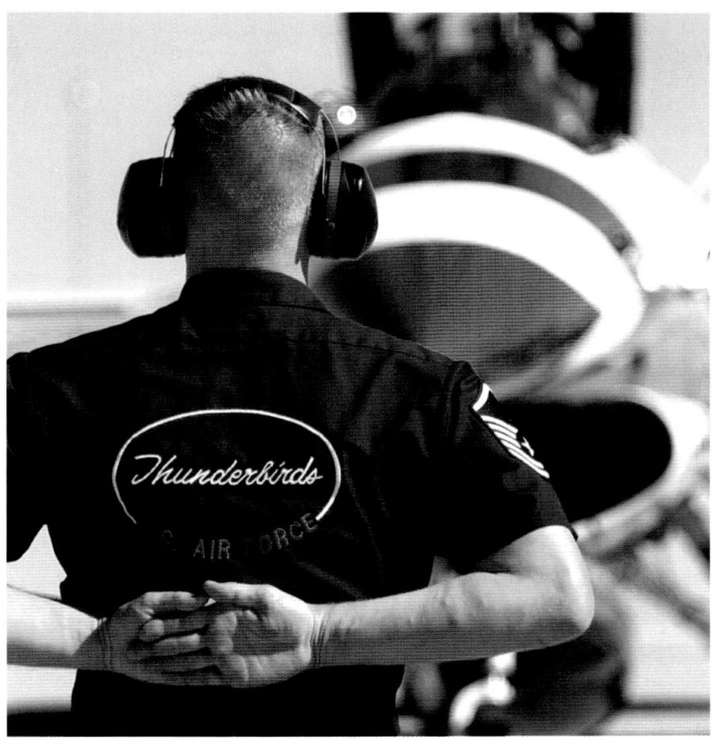

March
The Air Force Demonstration Squadron—Thunderbirds—begin performances in their newest aircraft, the F-16A. The first demonstration is flown at Langley AFB, Virginia.

1 March
Twenty-Third Air Force is activated at Scott AFB, Illinois. A diverse command, the organization will be responsible for providing combat rescue, weather reconnaissance, security for missile sites, and training for special operations aircrew.

1 March
Hurlburt Field, Florida, is reassigned to Military Airlift Command as is the 1st Special Operations Wing assigned there.

7 March
The largest peacetime aerial sea mining exercise—TEAM SPIRIT '83—includes three wings of B-52D and B-52G Stratofortresses. The bombers simulate laying mines off the coast of South Korea.

15–28 March
B-52s successfully launch three AGM-84 Harpoon missiles in the Pacific Missile test Range, near Kwajalein Atoll, as SAC explores methods to accomplish sea interdiction missions.

1 April–1 May
Space Command inherits four installations and 31 operational units from the Strategic Air Command. Most of these are related to missile warning and space surveillance missions.

5–10 April
After torrential rains flood southeastern Louisiana, Air Force C-141s deliver 83 tons of relief equipment for the victims of the massive flooding. More than 40,000 homes are underwater and the state penitentiary is also flooded.

18 April
Terrorists bomb the U.S. embassy in Beirut, Lebanon.

1 May
Air Force aircraft begin flying surveillance missions of the Bahamas helping law enforcement to apprehend drug smugglers.

13 June
Pioneer 10, a deep space probe, becomes the first space craft to leave the solar system.

17 June
The Peacekeeper missile is launched for the first time from Vandenberg AFB, California. In this test, multiple dummy warheads reenter earth's atmosphere and land in the Kwajalein range in the Pacific Ocean.

Above and below: *Thunderbird aerial demonstrations always begin with an enthusiastic ground performance by the aircraft crew chiefs. Strict precision, total dedication, and perfect military bearing lead up to the moment when six F-16 jet engines roar to life and the aerial demonstration begins. (USAF)*

"ON ITS WAY TO A SPACE AND AIR FORCE"

Sheila Widnall

General Tony McPeak has described Desert Storm as the first space war. Beyond the expected use of space assets for communications, surveillance and weather, the truly revolutionary use by ground troops of commercially available hand-held GPS units—many from Radio Shack—as well as the innovative use of Digital Signal Processing received in Colorado Springs to cue Patriot batteries in Iraq about scud launches made it clear that the integration of space into combat operations would provide revolutionary capabilities.

During the 1990s, the Air Force entered a period of sustained experimentation and demonstration of the integration of air and space capabilities. GPS-guided weapons offered revolutionary capability to combat aircraft. Remote location of the Combined Air Operations Center (CAOC) demonstrated that space and air operations could be conducted away from the front. Integration of surveillance information from U-2, Global Hawk, and satellites into real-time intelligence, making possible retargeting of combat aircraft en route, demonstrated these capabilities. Space capabilities changed the concept of operations. These experiments also helped to integrate the people of our air and space components, setting the stage for combined space-air-ground operations in Afghanistan and Iraq, and transition the Air Force to new capabilities in the integration of air and space.

In 1994, in recognition of the importance of space to the Air Force, General Thomas Moorman, the former Vice Commander of Air Force Space Command, became Vice Chief of Staff. General Moorman, who must hold the record for blue ribbon space commissions, has served on every important space study during his active duty career and beyond. He brought extraordinary experience and capability to the Air Force Leadership Team during the creation of the 1997 document *Global Engagement*.

Opposite: *The joint direct attack munition (JDAM), guided by satellite information, is accurate to within six feet of the programmed aim point. (USAF)*

Right: *Global Hawk, an unmanned, high-altitude reconnaissance platform, cruises at extremely high altitudes and can survey large geographic areas with pinpoint accuracy, giving military decision-makers the most current information about enemy location, resources, and personnel. The aircraft can operate autonomously from takeoff to landing or can accept commands via satellite communications. (USAF)*

Global Engagement reflected that the USAF was transitioning from an Air Force into an Air and Space Force on an evolutionary path to a Space and Air Force—an attention-getting phrase. The challenge became the integration of these space capabilities into one seamless Air Force: the people, their skills and commitment, the equipment, and operations. The focus on integrating space into combat operations was enhanced by the selection of Generals Charles A. Horner, Joseph W. Ashy and Howell M. Estes III, in turn, to serve as CINC Space. They had served as Commander of Air Operations in Desert Storm, in Bosnia and as Director of Operations in the Pentagon.

The Air Force also made a commitment to cost-effective access to space by fielding the Evolved Expendable Launch Vehicles (EELVs). These vehicles serve as a basis for future assured access, and will perhaps prove crucial to the new manned exploration initiative. They also mark the first time that the U.S. has developed a large liquid rocket engine since the space shuttle engine of the 1970s.

As it strengthened its capabilities in space, the Air Force drew on its rich history of accomplishment in missiles, launch vehicles, and satellites. It stood on the shoulders of giants such as the late General Bernard A. Schriever and many others. Schriever organized the Air Force's ballistic and systems divisions developing such ballistic missiles as Thor, Atlas, Titan, and Minuteman, and all the aerospace systems that were launched into orbit in those years, including support for NASA during Project Mercury.

Both Air Force and related National Reconnaissance Office (NRO) accomplishments in space have been recognized as truly extraordinary by significant national awards. The earliest surveillance satellite was the CORONA program. Long classified, it was made public in 1995, and in 2005 its developers Minoru Araki, Francis Madden, Edward Miller, James Plummer, and Don Schoessler were awarded the National Academy of Engineering Draper Prize for Engineering Achievement. This follows the award in 2003 of the Draper Prize to Colonel Brad Parkinson (USAF ret.) and Dr. Ivan Getting for the development of GPS.

But the promise of space will be hard won. Space is a challenging and hostile environment for the development of new systems. Ground testing and simulation must replace the test flights of aircraft development. Every component must be qualified beyond typical flight conditions before being assembled into subsystems, which again must be tested. When this rule has been violated or shortcuts taken in program development, problems occur. Anomalies that occur in flight must be understood and resolved to ensure future success. The potential for space operations is great and will provide our Air Force access to the new high ground of the future.

Left: *On* Challenger's *middeck, Mission Specialist (MS) Guion Bluford, restrained by a harness and wearing a blood pressure cuff on his left arm, exercises on the treadmill. Although not the first African American astronaut, Bluford is the first African American to fly in space. (NASA)*

7 July
The first F-16A Falcons are delivered to the Air National Guard. The first unit to receive them is the 169th Tactical Fighter Group based at McEntire ANGB, South Carolina.

General Dynamics rolls out its 1,000th F-16 (including exports). A total of 2,165 have been ordered by the USAF.

7 August
A bomb explodes at the officer's club at Hahn AB, Germany. Supposition is that the explosive was planted by anti-nuclear activists.

30 August
Air Force astronaut lieutenant Colonel Guion S. Bluford becomes the first African American astronaut in space. Bluford graduated from Penn State University in 1964 as a distinguished Air Force ROTC graduate. He attended pilot training at Williams Air Force Base, Arizona, where he received his pilot wings. He attended F-4C combat crew training in Arizona and Florida and was then assigned to the 557th Tactical Fighter Squadron, Cam Ranh Bay, Vietnam. He flew 144 combat missions, 65 of them over North Vietnam.

1 September
After a Soviet Su-15 shoots down a Korean 747 jetliner north of Japan, Kadena-based HC-130s conduct search operations for survivors supported by KC-135 Stratotankers. None of the 269 aboard the plane survived.

3–25 September
During Operation RUBBER WALL, Military Airlift Command transports deliver 4,000 tons of supplies to American Marines in Lebanon. More than 100 heavy lift sorties are flown to accomplish this mission.

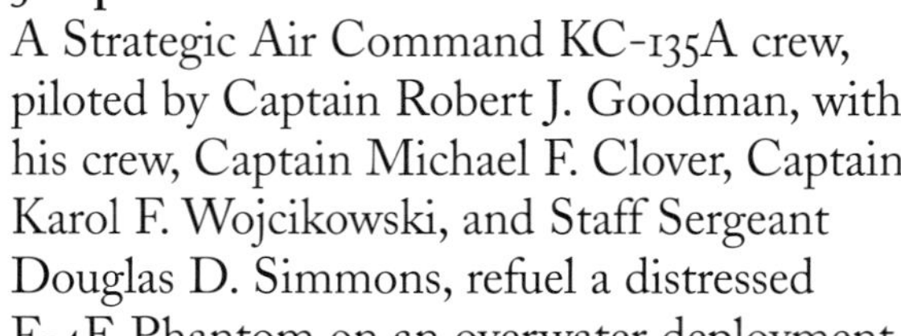

5 September
A Strategic Air Command KC-135A crew, piloted by Captain Robert J. Goodman, with his crew, Captain Michael F. Clover, Captain Karol F. Wojcikowski, and Staff Sergeant Douglas D. Simmons, refuel a distressed F-4E Phantom on an overwater deployment four times and effectively tow the F-4 with the refueling boom until the aircraft is able to land safely. For outstanding achievement while on a routine refueling mission involving F-4E aircraft, saving a valuable aircraft from destruction and its crew from possible death, the crew are awarded the Mackay Trophy..

Right: *The Nighthawk is aptly named, as the vast majority of F-117 missions are flown under the cover of darkness. Though operational in 1982, the aircraft will fly for six years before it is revealed to the American public. (USAF)*

28 September
The EF-111A tactical electronic jamming aircraft is officially designated "Raven," but will become known throughout the Air Force as the "Spark-Vark," the electronic version of the "Aardvark."

1 October
The last B-52D is retired from duty and flies to the "Boneyard" at Davis Monthan for displosal.

6 October
Three B-52G Stratofortresses are modified to carry the AGM-84 Harpoon anti-ship missile.

13 October
The Smithsonian Institution, National Air and Space Museum announces that a large addition will be built near the Dulles International Airport. Planned to house such aircraft as the Concorde and the space shuttle *Enterprise*, the huge hangar-like museum will open to the public during the Centennial of Flight celebration—on December 15, 2003.

23 October–9 December
After terrorists attack a Marine Corps barracks in Beirut, Lebanon, Military Airlift Command and Air Force Reserve forces are responsible for transporting 239 dead and 95 wounded Americans to the U.S. and European hospitals for care and burial.

25 October–3 November
A joint American military force stages a raid into Grenada, a Caribbean island, to accomplish three objectives; restore democracy, evacuate U.S. citizens (mostly medical students attending college), and eliminate a hostile Cuban/Soviet base. During Operation URGENT FURY, Military Airlift Command and Air Force Reserve transports fly approximately 500 missions and transport more than 11,000 passengers and 7,700 tons of cargo to Grenada. Air Force tankers, fighters, and Air National Guard EC-130s support ground operations. A flight of MC-130 Hercules aircraft drop paratroops during the assault on Point Salinas while under hostile fire. Lieutenant Colonel James L. Hobson Jr., aircraft commander of the lead MC-130E during the Grenada rescue mission, earns the Mackay Trophy for his actions as flight lead carried out from his disabled aircraft during the assault.

26 October
Lockheed F-117s stationed at the Tonopah Test Range achieve Initial Operational Capability (IOC). Five stealth fighters are on station and 18 A-7 Corsairs are flown for proficiency in single seat aircraft and also to provide a cover story for Tonopah operations.

1–5 November
Following a severe earthquake in northern Turkey on October 30, four C-141s and six C-130s deliver 234 tons of humanitarian aid to the victims.

6 December
The National Transonic Wind Tunnel is dedicated at Langley AFB, Virginia. The tunnel is designed specifically for testing the newer, extremely fast, jet aircraft currently being designed and built.

Left: *The sleek Lear C-21 is used as executive transport by the Air Force. Here, maintenance is performed before flight. (USAF)*

Opposite: *The B-1B Lancer is an improved variant of the original B-1A, construction of which was initiated by the Reagan administration in 1981. Major modifications include increased payload capability of 74,000 pounds, improved radar, and a significant reduction of the aircraft's radar cross section (RCS). (USAF)*

1984

1 January
Space Command assumes total resource management of the Global Positioning System (GPS).

28 January
The 419th Tactical Fighter Wing at Hill AFB, Utah, is the first USAF Reserve Unit to receive the F-16A Fighting Falcon. They replace the last F-105s still flying in the Reserves.

4 February
The HH-60A Night Hawk flies for the first time. It is a combat-rescue version of the H-60 series. Eventually, funding is not received for production and the HH-60A Night Hawk project is abandoned.

24 February
The F-15E Strike Eagle is selected as the USAF dual-role fighter over the F-16XL. Nearly 400 aircraft will be procured.

19 March–9 April
When Libya threatens Egypt and the Sudan, the Air Force flies 45 airlift sorties and deploys one E-3A Sentry to support the threatened nations.

6 April
The first of 80 Learjet C-21A aircraft are delivered to the Air Force. These sleek business jets will eventually replace the CT-39 Sabreliner.

11 April
The first Beech Aircraft Corporation C-12F is accepted by the 375th Aeromedical Airlift Wing for use as an operational support aircraft.

16 May
Air Force C-141s deliver 22 tons of medical supplies to Peshawar, Pakistan, to assist Afghan refugees who had fled their country in the face of war.

25 May
The Unknown Soldier of the Vietnam War is transported aboard an Air Force C-141 on the journey to internment at Arlington National Cemetery. Later identified by DNA testing as Air Force 1st Lieutenant Michael J. Blassie, his remains were returned to his family and he was buried in his home town of St. Louis, Missouri, on July 11, 1998.

June
The USAF Academy accepts its first Schweizer TG-7A motorglider. Used in airmanship programs, the TG-7A is assigned to the 94th Air Training Squadron.

16 June
The improved F-16C flies for the first time at Fort Worth, Texas. The airplane features an improved Head Up Display (HUD) and improved multi-mode radar. General Dynamics will deliver the first aircraft to the USAF on July 19.

21 June
A 22d Air Refueling Wing KC-10 Extender operating out of Christchurch International Airport, New Zealand, refuels a MAC C-141B Starlifter, which is responsible for the midwinter airdrop of supplies to the U.S. Antarctic bases and McMurdo Sound.

July
The 69th Bombardment Squadron at Loring AFB, Maine, receives the first Harpoon (AGM-84) missile. The Harpoon is an anti-ship missile.

7 August–2 October
During Operation INTENSE LOOK, Air Force airlift aircraft transport nearly 1,000 passengers and more than 1,300 tons of cargo to locations near the Red Sea in support of U.S. minesweeping operations. Both Egypt and Saudi Arabia had requested such support after ships mysteriously blew up while operating there.

8 August
The first C-23 Sherpa aircraft begin service in USAFE. The Sherpa is designated to facilitate the European distribution system by flying short hops between operational locations and depot distribution nodes.

29 August
The last OV-10 Broncos depart Sembach AB, Germany after a decade of service in USAFE.

14–18 September
On the 37th anniversary of the establishment of the USAF, famed aeronaut and high-altitude balloon operator Colonel Joe Kittinger Jr. (USAF, ret.), completes the first solo crossing of the Atlantic Ocean in a balloon. His journey begins in Caribou, Maine, and ends 3,550 miles and 84 hours later when he lands in Savona, Italy. This flight also establishes a new balloon distance record.

18 October
The *Star of Abilene*, the first Rockwell B-1B Lancer, flies at Palmdale, California.

1 December
The Air Force Reserve at Kelly AFB, Texas, receives their first C-5A Galaxy.

22 December–March 1985
In an effort to combat famine in the Sahel region of Africa, eight C-141 Starlifters deliver more than 200 tons of food and relief supplies from Italy to Kassala, Sudan. More than 100,000 refugees are in dire need of these supplies.

THE ROLE OF AIR FORCE HISTORIANS

George M. Watson Jr.

Since its founding as an official office in March 1942 in response to President Franklin Delano Roosevelt's letter requesting that the wartime agencies keep a record of their activities, the Air Force (then Army Air Forces) historical function has been to serve as the air arm's institutional memory, preserve its heritage, promote a better understanding of the present, and help to plan for the future. Following World War II, when Headquarters Army Air Forces (AAF) called for historical programs to be maintained at all major commands and independent activities, AAF historians continued their tradition of gathering historical documents.

After the war, archives were established at Maxwell Air Force Base with an accompanying group of civilian historians and archivists hired to maintain the growing number of documents as well as to write studies. In the field, historians were aligned under the Intelligence branch until the late 1950s, when history became an additional duty of the Public Affairs function. In addition to the history offices that were established at major commands, a history liaison office was established in Washington, D.C., that served as an operating location of the larger office at Maxwell AFB, Alabama. Eventually, the responsibilities of the Washington office expanded to include direct response to the Office of the Secretary of the Air Force and Air Staff, a logical task as the records of both those offices were stored at the National Archives in Washington, D.C.

As time progressed, some thinkers and planners believed that the Air Force history function ought to have a command presence in the nation's capital in much the same manner as the Army and the Navy's historical programs. To consider this issue, the Chief of Staff and Secretary of the Air Force established the Advisory Committee of the Air Force Historical Program. Meeting several times both in Washington and at Maxwell AFB, the committee presented its findings in 1968. As a result, the chief and the secretary established the Office of Air Force History. The office was assigned under the Chief of Staff and headed

Above: *Air Force Flight Test Center chief historian James Young (far right) participates in the retirement of NASA's B-52B at Edwards AFB, California. One of the historian's jobs is to document base history and significant events. (USAF)*

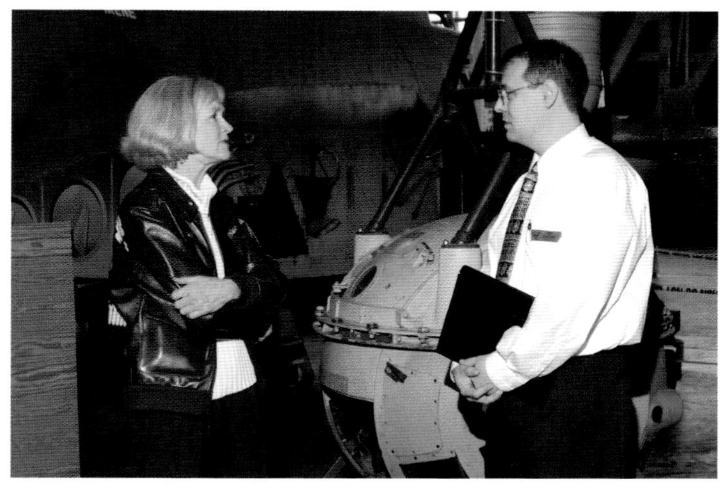

by a general officer with a staff that included a distinguished scholar to serve as Chief Historian. The nucleus of the new organization was absorbed from the former USAF Historical Division Liaison Office. More importantly, when the new office opened on January 15, 1969, it had gained its independence from the Office of Information and answered directly to the Air Force Chief of Staff.

Worldwide field history offices also blossomed in the command sections and on the staff of major command headquarters, centers, and wings where they prepared official histories and created a permanent repository of historical documentation, perspective, and advice. As these field documents were published, they were transferred to the Maxwell AFB archives.

The new Office of Air Force History provided policy guidance and direction to the field and eventually to the Air Force Museum. At headquarters, priority shifted from providing short studies, called "Blue Book" monographs, to book-length coverage of topics. The new headquarters office initiated a special series on the war in Southeast Asia that consisted of two general volumes plus several detailed studies that covered a broad range of important operational subjects. Other books were written that required extensive research at the National Archives and the Suitland Record Center in the greater Washington, D.C., area and presidential libraries as well.

During the 1980s, a survey of Air Staff historical requirements resulted in the publication of an excellent series of case studies on such topics as close air support, strategic bombardment, and air superiority that added to the office's growing list of published works. Coverage and documentation of wartime missions remained a fundamental part of the Air Force History Program's mission. To accomplish this, historians deployed to the theater of operations to furnish historical perspective, record events, and collect combat data. An Air Force Historical Research Agency documentation team usually followed up this effort in the specific theater of operations shortly after the termination of hostilities. Almost immediately Air Force historians began producing chronologies, briefings, analyses, and finally books that interpreted the wartime experience. This effort has continued through all conflicts since World War II.

Above: *The* Memphis Belle *pilot's widow, Linda Morgan, speaks with museum historian Jeff Duford beside the dismantled* Memphis Belle *at the National Museum of the U.S. Air Force. Collecting oral history interviews related to Air Force personnel preserves the rich personal history of the service. (USAF)*

Right: *Major General Robert Elder discusses operations with deployed historians at the Combined Air Operations Center (CAOC). Historians use interviews to collect and archive information about air operations around the globe. (USAF)*

Left: *A tremendous view of the shuttle* Discovery *as it approaches the Mir Russian Space Station. It was aboard* Discovery *that the first all-military shuttle mission was flown in 1985 under the command of USAF Colonel Loren Shriver. (NASA)*

1985

1 January
Exhausting all possible avenues to extend the nose gear during 13 hours of flight time in his KC-135 Stratotanker, Lieutenant Colonel David E. Faught makes a successful nose gear-up landing. His heroism and outstanding airmanship save the lives of eight crewmembers and prevent the loss of an Air Force aircraft. For his actions, he is awarded the Mackay Trophy.

24–27 January
The first all-military space shuttle mission is flown aboard *Discovery*. Air Force Colonel Loren J. Shriver pilots STS-51C, launched from Kennedy Space Center, Florida. STS-51C performs a Department of Defense mission, which includes deployment of a modified Inertial Upper Stage (IUS) vehicle from the space shuttle's payload bay. One of the shortest shuttle missions, *Discovery* returns to earth after 73 hours, 33 minutes.

8 March
In the Bahamas, Military Airlift Command helicopters assist police and U.S. Drug Enforcement Agency officials in a bust that results in the confiscation of $320 million in cocaine. Counter-drug support continues in the region throughout April during Operation BAHAMAS AND TURKS, a joint drug interception campaign.

25 March
The combat exclusion policy for women is modified to allow women to serve as forward air controllers (FAC) and crew various C-130s. Combat exclusions are gradually eliminated over the next decade.

4 April
James H. Doolittle is promoted in retirement to four-star rank. Doolittle is the first Air Force Reserve officer to hold the rank of full general.

29 June
The 159th Tactical Fighter Group stationed at NAS New Orleans becomes the first Air National Guard unit to receive the F-15 Eagle.

Right: *The first operational B-1B Lancer* Star of Abilene *sits on static display at Dyess AFB, Texas—the aircraft's first assigned home base. (USAF)*

Below: *In the spring of 1989, the last of 50 Lockheed C-5B Galaxy aircraft was added to the 76 C-5As in the Air Force's airlift force structure. The C-5B includes all C-5A improvements as well as more than 100 additional system modifications to improve reliability and maintainability. (USAF)*

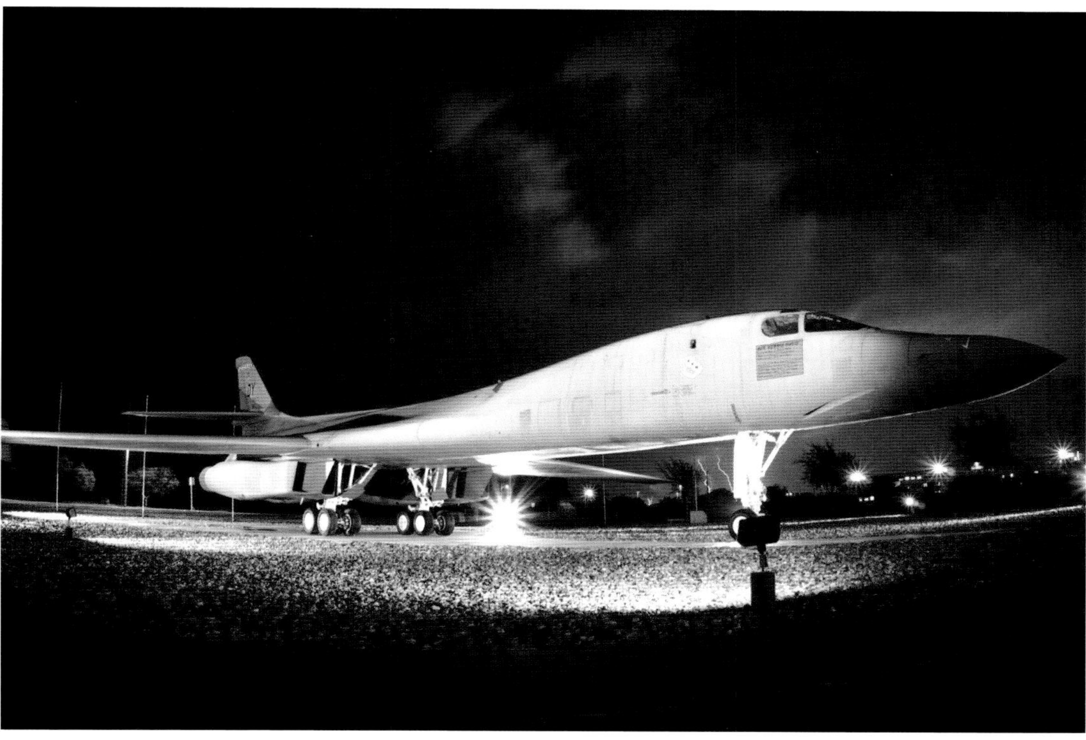

21 June–25 July
To stop a grasshopper infestation in Idaho, C-123K Providers spray insecticide over more than 735,000 acres of land during 73 sorties.

30 June
The final above ground test launch of the Peacekeeper missile is accomplished at Vandenberg AFB, California.

2–10 July
While forest fires burn in Idaho and California, Air Force C-141s transport 285 fire fighters and more than 450 tons of flame retardant to staging areas near the fires. C-130s fly 200 sorties to deliver the retardant over nearly 1.5 million acres of burning timberland.

7 July
The first operational B-1B Lancer arrives at Dyess AFB, Texas, where it is assigned to the 96th Bombardment Wing. The Wing achieves Initial Operating Capability (IOC) in October 1986.

23 August
The first Minuteman "cold launch" is accomplished at Vandenberg AFB, California. By ejecting the missile from the silo using compressed gas and then igniting the rocket motor, less damage occurs to the silo and shorter reload times are the result.

10 September
The Lockheed C-5B flies for the first time. Fifty of these monstrous transports will be built for the USAF. The last one will be delivered in April 1989.

13 September
An Air Force F-15 Eagle launches an anti-satellite missile—the Vought ASM-135—which destroys an orbiting satellite—Defense Department P78-1. This test is the first time that a successful anti-satellite intercept takes place.

21–30 September
In the wake of devastating earthquakes in Mexico City, Air Force airlift units deliver more than 360 tons of humanitarian relief supplies to the region.

18 October
Fitted with the mission adaptive wing (MAW), the AFTI General Dynamics F-111A flies for the first time.

THE USAF STRIKES INTO SPACE: ANTI-SATELLITE CAPABILITY AND SPACE CONTROL

Clayton Chun

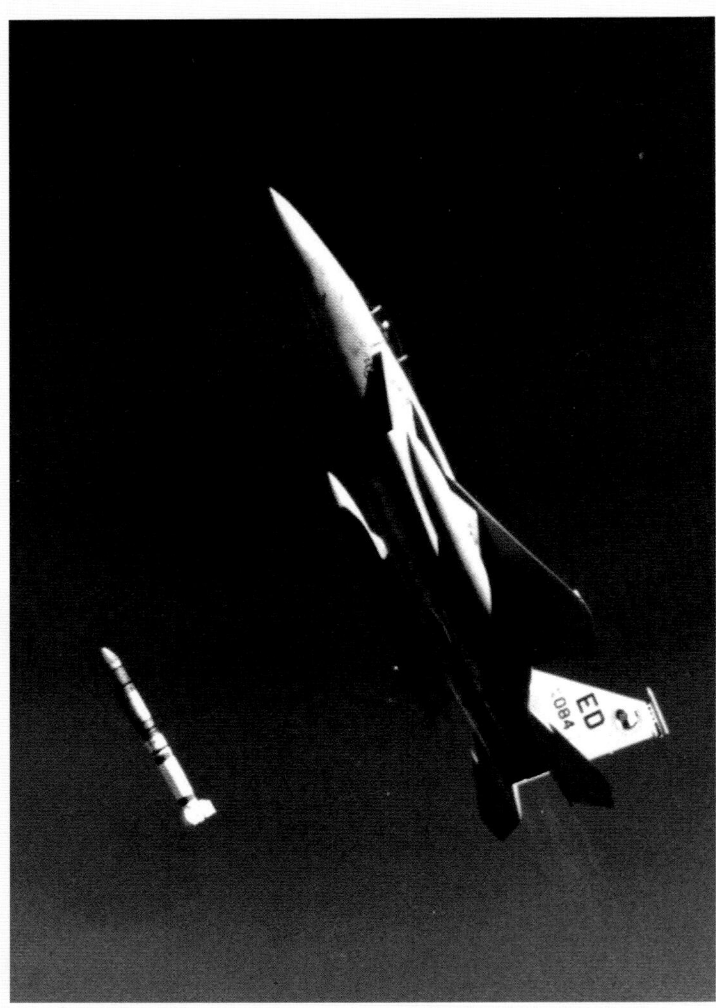

Left: *The Air Force experimented with an F-15-delivered missile designed to strike a satellite in low earth orbit. This missile was composed of existing rocket propulsion systems and a new kinetic-kill vehicle that the Air Force tested for operational use. The anti-satellite weapon was never made operational. (USAF)*

Space satellites have become a key part of military operations and have aided global economies. Military activities have come to rely on instant, accurate, and widely available communications, navigation, early warning, intelligence, and weather functions from satellites. Today, campaign planning, flight activities, and precision-guided weapons delivery would be imperiled without space access. In the early days of the space age, the Air Force recognized that military space operations would have a profound impact. This resource required protection for satellite resources and capability to take action against hostile ones. Anti-satellite (ASAT) capability became a key topic in the early 1960s. These activities would result in development of several ASAT projects that produced an operational ASAT missile system using a ballistic missile in 1963 and an F-15 fighter-delivered missile in the early 1980s.

A nation that does not want a satellite to operate may either destroy the satellite's capability or simply disable it. Other considerations can also involve targeting the satellite itself, the communications link between the satellite and a ground station, or the ground station itself. ASAT weapons may be surface-launched, air-launched, or orbital. Although most nations might believe that ASAT activities are built for a defensive strategy, a country could use this capability in an offensive role. Unlike anti-ballistic missile actions, ASAT capability could conceivably be used in a preemptive manner.

The United States Air Force developed an ASAT weapon based on a series of high-altitude nuclear tests in the Pacific. In the Starfish Prime test on July 9, 1962, over Johnston Island in the Pacific, a nuclear device was detonated at an altitude of 248 miles. The test produced a large electromagnetic pulse (EMP) that seriously crippled three orbiting satellites by damaging their solar panels.

Fears of the Soviet Union's growing space and nuclear capability and their propaganda claims of orbiting nuclear weapons helped motivate the United States to build a countermeasure. The Air Force's Program 437 operated two nuclear-armed Thor intermediate range ballistic missiles based on Johnston Island tests. Program 437 replicated the same effect on Soviet satellites as it did on Starfish Prime. Nuclear devices provided a relatively inexpensive means to produce large bursts of EMP. Due to a lack of precision guidance, large nuclear devices had to be used. USAF crews operated the system until 1970. EMP was effective, but there was no way to limit damage to other satellites or its long-term effects. Additionally, the USAF operated only two missiles, could only attack certain targets in a particular orbital inclination, and was costly.

The United States believed that the Soviet Union was testing and building an orbital ASAT device. The fragmentation device would intercept a target and explode, destroying the delicate satellite. In 1980, development contracts were awarded to Vought and Boeing to begin work on an aircraft-delivered miniature homing vehicle (MHV) that could destroy a target by hitting it at great speeds. The F-15 would act as an initial booster for this kinetic kill vehicle. This airborne missile was composed of a first stage from the Boeing Short-Range Attack Missile, second stage Vought Altair III (used as a fourth stage on the Scout space launch vehicle), and the infrared guided MHV. The Air Force conducted five tests on the system from 1984 to 1986. One test, on September 13, 1985, was conducted against a satellite.

The F-15 MHV program offered much flexibility over Program 437. Air Force F-15s could operate from many airfields, collateral damage from the MHV impact was minimal, and F-15s could conduct missions several times a day. Air Force planners selected bases at Langley AFB, Virginia, and McChord AFB, Washington, for the ASAT-armed F-15s. Despite its successful development, Congress cancelled the program when it restricted further ASAT testing in the late 1980s.

The USAF pioneers development in all aspects of military space operations. ASAT development was a natural part of extending its space mission. By using existing programs, Air Force scientific, engineering, and operational crews adopted technologies and systems to meet a new challenge from space. ASAT operations offered a challenge that was met with innovation and vision. Protecting national and Air Force space assets continues as a key mission for the military.

Below: *During the early 1960s, the Air Force's Aerospace Defense Command operated two nuclear-armed Thor intermediate range ballistic missiles as anti-satellite weapons under Program 437. Air Force crews operated the weapons on Johnston Island, south of Hawaii, and used electromagnetic pulse to destroy a satellite. The system operated until 1970. (USAF)*

Jarvis, Ronald E. McNair, Air Force Lieutenant Colonel Ellison S. Onizuka, Judith A. Resnik, Michael J. Smith and Christa McAuliffe, a New Hampshire school teacher. It was later discovered by NASA engineers and a separate panel of scientists commissioned by President Ronald Reagan that the vehicle broke up during the launch due to the failure of rubber seals in the booster engines called "O rings." Subsequently, NASA adopts much stricter safety standards for shuttle missions, which will not resume until September 1988.

18–22 February
During severe floods in northern California, Air Force H-3 Jolly Greens, HH-53s, and C-130s begin evacuation of more than 500 stranded civilians, deliver more than 3,000 sandbags to Army troops on the ground, and rescue 33 people from life-threatening situations.

4 March
Air Force tankers refuel foreign aircraft for the first time during Operation BRIGHT STAR. This exercise is conducted with a combined force of U.S. and Egyptian forces.

15–18 November
After a volcanic eruption in Columbia, Air Force aircraft transport 50 tons of supplies to those affected by the blast.

12 December–20 January
After the fatal crash near Gander Newfoundland of a DC-8 airliner carrying members of the 101st Airborne Division, USAF C-130s and C-141s are called upon to carry the victims' remains back to U.S. soil. During the period, 125 tons of cargo are transported with 770 passengers to assist in the accident cleanup and investigation. In all, 248 soldiers and eight civilians perished in the tragedy.

31 December
McDonnell Douglas is awarded a contract to build the C-17A long-range, heavy-lift cargo transport. The design requirement stipulates that the aircraft be capable of carrying C-5 size payloads into austere fields formerly accessible only to a C-130.

1986

8 January
Military Airlift Command accepts its first C-5B Galaxy. The plane is delivered to Altus AFB, Oklahoma.

28 January
The space shuttle *Challenger* explodes 73 seconds after liftoff from Kennedy Space Center, Florida. On board are former Air Force pilot Francis R. Scobee (shuttle commander), former Air Force officer Gregory B.

Opposite, top: *STS-51L Challenger crew members. (NASA)*

Opposite, bottom: *Hurtling out of the conflagration at 78 seconds are the* Challenger*'s left wing, main engines, and the forward fuselage. (NASA)*

Right: *Ground crew members arm the weapons on this F-111, bound for targets in Libya during Operation ELDORADO CANYON in 1986. (USAF)*

5 March
Following a precipitous and hazardous launch in near zero-zero weather during a CORONET EAST deployment, the crew of a KC-10 assigned to the Strategic Air Command's 68th Air Refueling Wing provides emergency refueling to a KC-10 and three Navy A-4Ms over the Atlantic Ocean. The pilot of the KC-10, Captain Marc D. Felman, and his crew—Captain Thomas M. Ferguson, Master Sergeant Clarence Bridges Jr., Master Sergeant Patrick S. Kennedy, Master Sergeant Gerald G. Treadwell, Technical Sergeant Lester G. Bouler, Technical Sergeant Gerald M. Lewis, Staff Sergeant Samuel S. Flores, Staff Sergeant Scott A. Helms, and Staff Sergeant Gary L. Smith—receive the Mackay Trophy for their actions.

25 March
An all-woman Minuteman missile crew serves on alert for the first time at Whiteman AFB, Missouri. They are members of the 351st Strategic Missile Wing.

14–15 April
Operation ELDORADO CANYON, the bombing of terrorist targets in Libya, is executed under the cover of darkness by 18 48th Tactical Fighter Wing F-111Fs and five EF-11As stationed in England. Forced to fly a circuitous 5,500-mile round-trip route around continental Europe when France and Spain do not grant overflight permission, the F-111s are refueled by more than two dozen KC-10 and KC-135 tankers. Air Force EF-111s and Navy aircraft jam radar sites and attack air defense networks while the Aardvarks execute the attack. One F-111 is believed shot down by a surface-to-air missile and the crew killed.

28 April–7 May
Two days after a nuclear reactor accident at Chernobyl, Soviet Union, Air Weather Service units are mobilized to track the path of the airborne nuclear fallout. Air Force WC-130s conduct air-sampling missions during this period. A variety of Air Force C-135 aircraft operating out of Mildenhall are involved in gathering details surrounding the accident.

17 June
The last UC-123K Provider is retired from active service as a spraying aircraft.

July
During this month, both the Air National Guard and the USAF Reserve take delivery of their first C-141B Starlifters.

19–28 July
During Operation SOUTHERN HAYLIFT, 24 C-141s and eight C-130s deliver 19,000 bales of hay to drought-stricken farmers across the southeastern U.S. The delivery saves hundreds of cattle and allows many ranchers to save enough livestock to survive until next season.

1 September
The last operational Cessna O-2 is retired and sent into storage at Davis Monthan AFB, Arizona.

1 October
The Goldwater-Nichols Department of Defense Reorganization Act is the most significant restructuring of DoD since the creation of the Air Force in 1947. Operational authority is centralized through the Chairman of the Joint Chiefs, who also is designated as the principal military advisor to the President, National Security Council, and Secretary of Defense. The Act emphasizes jointness as the path to future military success.

10 October
The Air Force MX missile—known as the Peacekeeper—goes on alert for the first time. The Peacekeeper (LGM-118A) is capable of attacking 10 targets with its multiple independent reentry vehicle (MIRV) capability.

11–16 October
After a powerful earthquake destroys El Salvador's capital city, Air Force airlift delivers medicine and relief supplies to the survivors of the disaster.

14–23 December
Flying experimental aircraft *Voyager*, Richard G. Rutan and Jeana L. Yeager fly nonstop and unrefueled around the world. During the journey, the duo covers just less than 25,000 miles. A unique aircraft, constructed of lightweight graphite-honeycomb composite materials, lifted off from Edwards AFB, California, on the morning of December 14 and returns nine days later. For their record-breaking flight, the pilots, the designer, Burt Rutan, and the crew chief, Bruce Evans, earn the Collier Trophy. The *Voyager* is on display at the National Air and Space Museum in Washington, D.C.

1987

16 January
A B-1B Lancer fires a short-range attack missile (SRAM) while in the Tonopah Test Range, Nevada. This is the first SRAM test accomplished.

3 February
The first Sikorsky UH-60A helicopter (modified to *Credible Hawk* standards) is delivered to the 55th Aerospace Rescue and Recovery Squadron stationed at Eglin AFB, Florida. These helicopters will be refitted with the Pave Low III forward-looking infrared systems to assist in accomplishing their mission of long-range search and rescue.

16 February
Staffed by both Army and Air Force medical personnel, the Joint Military Medical Command is activated in San Antonio, Texas.

Above: *The deployment of the Peacekeeper (also known as the MX) missile increases the strength and credibility of the ground-based leg of the U.S. strategic triad. (USAF)*

Below: The Rutan Voyager *flown by Dick Rutan and Jeana Yeager made the first non-stop, unrefueled flight around the world. The aircraft is on display in the National Air and Space Museum in Washington, D.C. (NASM)*

Left: *The MH-53J Pave Low III heavy-lift special operations helicopter is the largest, most powerful, and technologically advanced helicopter in the Air Force inventory. The terrain-following and terrain-avoidance radar, forward-looking infrared sensor, inertial navigation system with global positioning system, along with a projected map display enable the crew to follow terrain contours and avoid obstacles, making low-level penetration of hostile territory possible. (USAF)*

14 April
Making its longest flight to date, a Dyess-based B-1B flies more than 9,400 miles in 21 hours and 40 minutes. The flight requires five in-flight refuelings.

5 May
The last Titan II ICBM is removed from service at Little Rock AFB, Arkansas. The Titan II, the largest of American ICBMs, was liquid-fueled.

1 June
Special Operations Command is established.

10 June
The Rockwell B-1B makes its first appearance at the Paris Air Show.

4 July–17 September
Lieutenant Robert Chamberlain and his crew set 12 new records for speed and payload combinations while flying the Rockwell B-1B Lancer. In September, Major Brent A. Hedgpeth and his B-1 crew set an additional nine speed records. For this exceptional performance, Detachment 15, Air Force Plant Representative Office and B-1B System Program Office, Air Force Systems Command, receive the Mackay Trophy.

17 July
The first Sikorsky MH-53J Pave Low III is delivered to Hurlburt Field, Florida, and will begin operational service in 1988.

22 July–21 December
During Operation EARNEST WILL, the protection of Kuwaiti oil tankers during the Iran-Iraq War, Air Force E-3s are deployed to patrol the Persian Gulf. USAF C-141 and C-5 aircraft transport minesweeping

Below: *A B-1B Lancer flies overhead during the Edwards Air Force Base Open House and Air Show. The B-1B unofficially set or broke almost 50 world speed records during the October air show. Record-breaking courses included three 15-, 25-, 100-, 500- and 1,000-kilometer speed dashes. (USAF)*

Left: *The CV-22 Osprey fires countermeasures out of one of the rear buckets—storage areas for countermeasures—during a safe-separation test over the precision impact range area at Edwards AFB, California. (USAF)*

equipment and personnel to Southwest Asia and KC-10s and KC-135s provide aerial refueling to Navy escort fighters.

17 September
A Rockwell B-1B breaks nine flight records during a five-hour sortie. The bomber carries a payload of 66,140 pounds over a distance of 3,107 miles, averaging 655 miles per hour during the mission.

24 September
The USAF Demonstration Squadron—Thunderbirds—perform in Beijing, China, for an enthusiastic crowd of some 20,000.

28 September
The first B-1B Lancer is lost during a low-level training mission after striking a large bird.

1 October
Onizuka Air Force Station, California, and the Air Force Satellite Control Network are transferred from Systems Command to Space Command.

19 November
The Department of Defense awards a $2 billion contract to Northrop for the production of the B-2 stealth bomber.

8 December
Representatives from the United States and the Soviet Union sign an agreement that will result in the removal of six Air Force tactical missile wings throughout Europe. The Intermediate-Range Nuclear Forces Treaty eliminates the deployment of missiles with effective ranges between 620 and 3,415 statute miles.

1988

1 January
For the first time, SAC strategic missile crews are permitted to serve alert in mixed male/female teams.

19–22 February
An 86th Military Airlift Wing C-141 delivers 50 tons of construction material to typhoon-wrecked Marshall Island housing areas. Typhoon Roy had damaged much of the inhabited area.

17–18 March
Air Force aircraft transport 3,200 U.S. soldiers to Honduras during exercise GOLDEN PHEASANT. The deployment is the result of threatening moves by Nicaraguan Sandinista forces there.

17 April–23 July
Captain Michael Eastman and his C-5 Galaxy crew, assigned to the 436th Military Airlift Wing (MAC), fly the first of the missions that deliver nuclear test monitoring equipment to the Soviet Republic of Kazakhstan for joint verification experiments. Eastman and his crew are the first U.S. airmen to land at the site where the Soviets exploded their first nuclear weapon in 1949. For successful mission accomplishment, the crew receives the Mackay Trophy.

30 April
Rockwell delivers the 100th-production B-1B Lancer to the 384th Bombardment Wing at McConnell AFB, Kansas. This, the last production model built, rolled out of the factory on January 20.

23 May
The first of six Bell/Boeing V-22A Osprey prototype aircraft rolls out of the factory in Arlington, Texas. The program is cancelled by the Reagan administration in January 1989 because of budget cuts.

7 July
The last F-106 Delta Dart is taken off of alert and will be retired from on-line service on August 1. This aircraft had served as an interceptor since 1959. Many are converted to drones used as missile targets over the Gulf of Mexico.

15–28 August
To monitor a cease-fire between Iraq and Turkey, United Nations peacekeepers are airlifted to the region. These 500 UN troops are flown from Canada to the region on Air Force C-5 Galaxies.

22 August –6 October
In an effort to fight more than a dozen fires burning in Yellowstone National Park, MAC aircraft, both active and reserve, transport 4,000 firefighters to the scene as well as 2,500 tons of equipment to the disaster zone.

September
Air Force E-3A Sentry aircraft fly missions over the Seoul Olympics to deter any North Korean aggression. The Sentry is escorted by U.S. fighter planes based in South Korea.

10–15 September
In the aftermath of a catastrophic flood in Bangladesh that leaves more than 20 million people homeless, four Airlift Wings deliver more than 100 tons of relief equipment and supplies as well as a field hospital.

29 September
NASA launches the space shuttle *Discovery* into orbit. This is the first shuttle launch since the *Challenger* exploded shortly after launch on January 28, 1986. Air Force Colonel Richard O. Covey is the shuttle pilot during the mission (STS-26).

10 November
The existence of the Lockheed F-117 Nighthawk stealth fighter is revealed to the American public for the first time. The F-117 is designed with the lowest observable radar cross-section of any aircraft then flying.

15 November
The Soviet Union launches *Buran*, its version of a space shuttle. *Buran* looks remarkably like the U. S. space shuttle but is unmanned, has no main rocket engines, and is completely automated. After being launched into orbit by an Energia booster, the *Buran* completes two orbits and lands uneventfully. It never flies again.

Above: *The Return to Flight launch—the first shuttle launch since the* Challenger *exploded during ascent in January 1986—of the space shuttle* Discovery *and its five-man crew took place on Pad 39-B on September 29, 1988. (NASA)*

Left: *The Northrop B-2A was the culmination of more than five decades of flying-wing research. Data gathered from the Tacit Blue experimental aircraft were essential to the B-2s low radar cross-section. (USAF)*

STEALTH: ATTACKING THE FUNDAMENTAL UNDERPINNINGS OF AN ENEMY'S AIR DEFENSES

Richard P. Hallion

Left: *On March 3, 2006, the first F/A-22A Raptors assigned to the 94th Fighter Squadron arrive at Langley AFB, Virginia. The 94th is the second squadron to receive the F-22 (redesignated F/A-22), which is capable of supersonic cruising speeds without afterburner thrust—known as "Supercruise." (USAF)*

Opposite: *The ability to penetrate enemy radar defenses using stealth technology creates air superiority in a new way. An enemy cannot hit what they cannot see. (USAF)*

The Gulf War of 1991 is probably best remembered for the image of precision bombs dropping on opening night on key Iraqi air defense and strategic targets—bombs dropped by F-117 stealth fighters. The war highlighted a new kind of aerial attacker, an aircraft (or later a cruise missile) that could enter enemy airspace in the face of radar-based detection and defensive systems, evade those systems, proceed to a target, and destroy it. "Stealth," a word circulating among aerospace circles since the late 1970s, became a common expression among the media and public, thanks to its success over Iraq.

The concept of aerial stealth predated the Gulf War by fully 75 years. During the First World War, various combatants had experimented with cladding wood-and-wire biplanes with transparent or semi-transparent coverings to reduce their optical signatures. The introduction of radar for aircraft detection in the late 1930s led to some early studies for means to reduce radar return, but, with the exception of coatings developed for submarine schnorkels in the late-Second World War period, little else was done. American aircraft loss rates in the Second World War, averaging approximately one percent of all sorties, but, on some specific raids, sometimes exceeding 25 percent of the force dispatched, highlighted the vulnerability of aircraft to radar-based defenses. However, the true magnitude of the radar threat only became apparent in the 1950s, as the first efforts were made in America and the Soviet Union to construct "integrated" air defense networks consisting of search radars, fire control radars, communications, airborne radar-equipped interceptors with infrared-guided or radar-guided missiles and cannon, and radar-guided surface-to-air missiles. The Vietnam experience showed the United States the difficulties of operating in an environment where an enemy possesses strong radar-and-missile-based defenses, and resulted in development of counter-SAM "Wild Weasel" suppression of enemy air defense efforts. But it was the 1973 Arab-Israeli war that really provided the impetus to attempt the first "stealth" aircraft development. In that war, over 19 days, the Israel air

forces lost 109 aircraft, representing fully 35 percent of its prewar combat aircraft strength.

Even before the loss of Francis Gary Powers's U-2 on May 1, 1960, Lockheed had studied various ways to reduce the radar signature of the high-altitude reconnaissance airplane, resulting in so-called iron ball paint coatings. For the subsequent Lockheed SR-71 Blackbird, the company made use of materials, shaping, and coatings to significantly reduce its signature. Even so, true stealth remained an elusive dream until the advent of advanced computer modeling and simulation enabled predictive radar cross-section (RCS) reduction. (RCS is the apparent size of an aircraft as it is "seen" by a radar, and bears no relationship to the actual physical cross-section of the aircraft; thus a small airplane can have an RCS far greater than a large airplane.) In the mid-1970s, the Defense Advanced Research Projects Agency (DARPA) launched a study effort to generate a genuine "low observable" technology demonstrator. In response, Lockheed developed a radical "faceted" testbed, the XST, which made use of shaping, coating, and radar-absorbent structure and materials to reduce its signature. Both static "pole tests" and dynamic in-flight tests of this aircraft demonstrated that the facets worked, minimizing the testbed's vulnerability to search and fire control radars. Encouraged, in November 1978 the Air Force contracted with Lockheed for development of a single-seat twin-engine stealth aircraft. The result was the F-117, which first flew on June 18, 1981, and, after extensive test and development, entered operational service with the Tactical Air Command in October 1983. A total of 59 were produced between 1981 and 1990, at a flyaway cost of $42.6 million per aircraft. The F-117 incorporated the same faceted concept as the XST, but differed greatly in design details and size. It had a so-called 2-D exhaust for reduced infrared and radar signature, special inlet grids to reduce radar return, an electronic flight control system (without which it could not fly as it is both statically and dynamically unstable in flight), and a distinctive all-moving V-tail.

In addition to the F-117, the potential of stealth led to development of the AGM-129 stealth cruise missile, the B-2 strategic bomber, the F-22 (now F/A-22) fighter, and the contemporary Joint Strike Fighter, the F-35. The success of the F-117 in the Gulf and in subsequent conflicts, has led to the recognition that stealth is a genuinely revolutionary military technology comparable in impact to the advent of the turbojet or the swept wing.

22 November
The prototype for the Northrop B-2A is rolled out of the factory at Palmdale, California.

29 November
The 60th and last KC-10 Extender, the first modified with wing-mounted refueling pods in addition to a centerline boom, is delivered to the USAF.

9 December
After a particularly destructive earthquake in Yerevan, Armenia, Mikhail Gorbachev accepts U.S. aid, which is delivered over the next two months. MAC delivers more than 300 tons of relief and, for the first time, is permitted to enter Russian airspace without Soviet observers onboard each incoming flight.

22 December
The second E-8A JSTARS airframe is delivered to Grumman for modification. The first airframe flies for the first time during this month. These two E-8s will deploy to Southwest Asia during Operation DESERT STORM and be used in combat while still under development.

Above: *A Northrop B-2 Spirit approaches a tanker for aerial refueling. (USAF)*

1989

10 January
The first AGM-136 Tacit Rainbow is launched from a B-52 Stratofortress. The unmanned missile is designed to fly to predetermined coordinates and then loiter until radar energy is detected and identified. Once enemy radar is located, the missile homes in and destroys the site.

16 February
Northrop closes the T-38 Talon production line after building 3,806 of the supersonic jet training aircraft.

27 March
After the oil tanker *Exxon Valdez* spills 10 million gallons along the Alaskan coastline, MAC transports begin delivering more than 1,000 tons of equipment and supplies to assist in the cleanup efforts.

17 April
The *Elf One* Boeing E-3 detachment, assigned to Riyadh, Saudi Arabia, for more than eight years, returns to Tinker AFB, Oklahoma.

11–18 May
During Operation NIMROD DANCER, U.S. forces in Panama are reinforced with more than 2,600

Right: *The 16th Tactical Reconnaissance Squadron flying RF-4Cs participated in Hurricane Hugo relief efforts both on the ground and in the air. Regional National Guard commanders were provided with aerial images of the damage caused by numerous tornadoes along the path of the storm. (USAF)*

soldiers, Marines, and nearly 3,000 tons of equipment. Air Force airlift takes 85 sorties to accomplish the mission.

16 May–29 June
During Operation BLADE JEWEL, nearly 6,000 nonessential U.S. personnel are evacuated from Panama in the face of unrest.

14 June
Launched for the first time, a Titan IV heavy-lift booster rocket takes a Defense Department satellite into orbit.

6 July
President George H. W. Bush presents famed American aviator James H. Doolittle the Presidential Medal of Freedom. Doolittle is the only American awarded the nation's two highest awards—the Medal of Honor and the Medal of Freedom.

17 July
The Northrop B-2A makes its first flight, taking off at Palmdale and flying to Edwards AFB, California.

21 September–15 November
In the aftermath of Hurricane Hugo, Air Force aircraft transport more than 4,300 tons of humanitarian relief equipment throughout the Caribbean Sea and to South Carolina. Shaw AFB, South Carolina, RF-4C reconnaissance aircraft conduct low-altitude missions along the route of the storm from Charleston to the northern parts of the state. The images are used by National Guard troops conducting rescue and recovery missions on the ground.

Left: *A Titan IV launch vehicle lifts a military satellite into orbit. The USAF had decided to trust its heavy space-lift to the expendable Titan IV booster, which could launch 10,000-pound payloads into geosynchronous orbit. (USAF)*

AWACS

George Williams

Left: *An E-3 Airborne Warning and Control System aircraft from Tinker AFB, Oklahoma, flies a mission. The E-3 Sentry is modified Boeing 707/320 commercial airframe with a rotating radar dome. (USAF)*

The military need filled by AWACS (Airborne Warning and Control System) has been clearly identifiable since men have waged war. Unimpeded surveillance lays bare the enemy's strength so that his intentions can be deduced, partially or totally. Aggressors have to fight uphill, a marked drawback. The fortunate defenders are not only separated by stand-off distance from their attackers but also are able to survey the battle area with a panoramic view of the other side of the hill. These direct advantages provide warning time to organize a defense or a counterattack.

Technology has enhanced and also complicated this quest for ever higher ground. Over time, airplanes supplanted earthbound observation assets. Aerial platforms, whether tethered balloons (American Civil War), aircraft (World War I), or airborne radar (World War II), were able to provide increased stand-off distances and achieve greater surveillance volumes in order to extend critical warning times. Inevitably, as computers, sensor systems, and communications capabilities were designed or adapted for aerial use, aircraft became larger and heavier. A number of intermediate airborne early warning systems had been fielded since the Korean War, such as the Air Force's EC-121 Warning Star, the U.S. Navy's E-2C Hawkeye and the RAF's Avro Shackleton. However, they all suffered from one significant drawback: their radars could provide surveillance coverage over water but lacked the ability to detect and track airborne targets amid ground clutter. This persistent difficulty would not be remedied until the advent of the E-3 AWACS with its pulse Doppler radar and state-of-the-art integrated computer system.

As early as 1963 the U.S. Air Force had identified its fundamental operational requirements for a proposed AWACS: a reliable, survivable air platform capable of

Right: *A NATO AWACS aircraft flies with three F-16 Fighting Falcons during a recent test mission. The AWACS was at Edwards AFB, California, to conduct airborne interoperability testing on the Link-16, part of the F-16 Mid-Life Upgrade test program. (USAF)*

detecting and tracking airborne targets over land and water and providing surveillance information—as well as command, control, and relay data—to other ground and air facilities. However, the specific technological means to satisfy the USAF requirements did not yet exist. From the field of candidates to develop the system, Boeing was eventually selected for its 707-size aerial vehicle, while Westinghouse won the competition to build suitable radar for the new AWACS. In 1975 the first Westinghouse AN/APY-1 radar was built onto and into a Boeing 707 airframe and the first E-3A emerged. Its most striking external feature, a disk-shaped rotodome, six feet thick and 30 feet in diameter, is mounted above the aft fuselage on two titanium struts and houses the main radar antenna and the identification-friend-or-foe (IFF) antenna. When in surveillance orbit, the AWACS multi-mode radar has an effective range in excess of 250 miles.

Successful operation of this dependable airplane and its versatile radar relies almost entirely upon the processing power of an onboard IBM System 4Pi computer. On the AWACS, this wall-locker-sized computer dominates the center of the information web, correlating sensor and data link inputs with geographical position data from the E-3 aircraft to provide a coherent, integrated, real-time situational display for the mission crewmembers.

The composition of each AWACS crew is easily adjusted to meet the specific demands of a variety of mission requirements. The flight crew can be augmented for long-duration missions; the mission crew, consisting of technicians (for communications, computer and radar), weapons and surveillance sections, and the battle staff, can likewise be tailored to fit the specific sortie for that day. Common elements of each mission normally stipulate that the E-3 arrive on time, set up communications links (data link as well as voice channels) with other airborne or ground-based stations, conduct effective surveillance in its assigned volume of airspace, control other friendly assets (such as interceptors and aerial tankers), and survive to fly another day.

Despite initial misgivings about its expense and untested capabilities, the E-3 AWACS has subsequently proven its technical, tactical, and strategic worth in a variety of roles. While the United States possesses the largest fleet, NATO and several other nations have purchased their own AWACS. All have periodically upgraded their capabilities, keeping their aircraft up to date with major improvement and modification initiatives. Globally, typical scenarios have included long-range early warning in conjunction with an established area defense system of ground radars and control centers, deployments to integrate with other tactical air assets already in theater to fight a fluid, dynamic air war, and rapid responses to contingencies anywhere in the world.

1 October
General Hansford T. Johnson becomes the first USAF Academy graduate to make his fourth star.

3 October
The last production model U-2R is delivered to the USAF. This completes the run of nine U-2Rs, 26 TR-1As, and two TR-1Bs in the USAF inventory.

4 October
A 96th Bombardment Wing B-1B Crew lands its aircraft with a retracted nose gear. For superior airmanship, the crew—Captain Jeffrey K. Beene, Captain Vernon B. Benton, Lieutenant Colonel Joseph G. Day, Captain Robert H. Hendricks—receive the Mackay Trophy.

A C-5B lands in Antarctica for the first time when a 60th Military Airlift Wing crew sets down in a normal landing at McMurdo Station. The Galaxy carries two UH-1 Huey helicopters, 72 passengers, and 84 tons of cargo to the remote area.

17 October
Military Airlift Command assets respond after a massive earthquake in the San Francisco Bay area of California. More than 250 tons of relief equipment flow into the region.

14 December
Women are permitted to serve as combat crew on C-130 and C-141 airdrop missions for the first time, marking the entry of women into combat crew roles.

20 December–4 January 1990
Operation JUST CAUSE, an effort to restore democracy in Panama, begins when six F-117 stealth fighters, used in combat for the first time, and special operations AC-130H aircraft commence operations there. The stated goals of the invasion were to safeguard American lives, defend democracy and human rights, combat drug trafficking, and protect the integrity of treaties governing the canal. During the operation, MAC aircraft transport 9,500 troops from

Above: *The Lockheed U-2 provided overhead coverage of Soviet targets until Gary Powers was shot down on May 1, 1960. Project CORONA satellites picked up the mission soon thereafter. (USAF)*

Left: *The AC-130 gunship's primary missions are close air support, air interdiction, and force protection. AC-130s played a key role during Operation JUST CAUSE in Panama. (USAF)*

Left: *A C-130 Hercules drops U.S. Army 82d Airborne Division soldiers over the Normandy Drop Zone at Fort Bragg, North Carolina, during a joint forcible entry exercise (JFEX). JFEX is a joint airdrop exercise designed to enhance service cohesiveness between the U.S. Army and Air Force. Lessons learned during JUST CAUSE still influence this training today. (USAF)*

Pope AFB, North Carolina, to Panama in less than 36 hours—part of this includes the largest night combat airdrop since World War II. USAF Reserve forces fly 455 sorties carrying 3,700 tons of cargo and nearly 6,000 passengers. Reserve refuelers support all forces during operations and also contribute 29 AC-130 gunship sorties during the operation.

20 December

During the initial invasion of Panama, after having flown a difficult eight-hour night flight in poor weather, and accomplishing several aerial refuelings, the crew of *Air Papa 06* spearheads the opening aerial barrage on the headquarters of the Panamanian Defense Forces (PDF). La Comandancia (PDF headquarters) is the critical target in the early hours of the invasion. Using surgical precision, the AC-130H Spectre gunners destroy the target and then attack and destroy several barracks and anti-aircraft artillery guns. Their precise application of firepower results in virtually no collateral damage to the civilian buildings and homes surrounding La Comandancia. For their airmanship and outstanding professionalism during aerial flight over the Republic of Panama during Operation JUST CAUSE, the AC-130H crew of *Air Papa 06*, 16th Special Operations Squadron, is awarded the Mackay Trophy.

29–31 December

Following a bloody anti-Communist revolution in Bucharest, Romania, Air Force transports deliver more than 30 tons of medical supplies to treat the wounded.

Below: *F-117 stealth fighters saw combat for the first time during Operation JUST CAUSE. Six aircraft were launched during the initial invasion. Two of the six attacked a barracks complex while the others dropped diversionary bombs to distract Panamanian Defense Forces. (USAF)*

RS
AF
89
137

GLOBAL REACH, GLOBAL POWER

1990–2006

GLOBAL REACH, GLOBAL POWER

1990–2006

The end of the Cold War in 1989 obviated the need for large numbers of around-the-clock alert forces. It also seemed to suggest that a smaller military could still provide the necessary forces required to meet any threat to the nation, at home or abroad. Since 1989, American military forces have been actively engaged in combat operations in Southwest Asia or some other part of the world almost continually.

During the 1991 Persian Gulf War, the E-3 and the E-8 Joint STARS performed magnificently. The first true test of the efficacy of the F-117 occurred over Baghdad during the air campaign. Making up only a fraction of the attacking force, the F-117 struck a disproportionate number of critical targets during the first several nights of the battle with tremendous effect—and suffered no casualties—not even one bullet hole. Stealth worked and so did precision guided munitions. Yet, it is interesting to note that only about ten percent of the bombs dropped during the 1991 Persian Gulf War were precision-guided munitions (PGMs). The percentage of PGMs dropped from the skies over Bosnia and Afghanistan a decade or so later would approach 70 percent.

During the mid-to late 1990s, in support of America's commitment to the North Atlantic Treaty Organization (NATO), the Air Force participated in NATO's first-ever combat operations over Bosnia-Herzegovina during Operations DENY FLIGHT and DELIBERATE FORCE, the latter of these because a campaign of genocide against non-Serb Bosnian men was taking place. Predominantly an air campaign, DELIBERATE FORCE attacked targets with razor-like precision over a three week period in 1999. These NATO air strikes, almost all of them using precision-guided munitions, paralyzed the Bosnian Serbs and brought and end to the slaughter in the region.

Then, on the morning of September 11, 2001, everything changed. Within three days, three Air Force RQ-1 Predator UAVs were flying sorties over Afghanistan, the home of the Taliban government, well known as supporters of global terrorist activities. Soon, Special Forces and Air Force combat air controllers had descended upon Afghanistan searching for Bin Laden and systematically eliminating any Taliban resistance that could be found. The battle was joined on many fronts, but on the ground the war was being fought by less than 500 CIA operatives and Special Forces troops, all supported overhead by an armada of airpower assets. Precision strikes were called in by ground troops in pursuit of enemy forces. Within minutes, targets were destroyed as GPS-guided weapons dropped from orbiting B-52s, B-1Bs, and for the first time, from UAVs armed with Hellfire missiles.

After successful completion of operations in Afghanistan, President George W. Bush, the son of the 1991 Persian Gulf War president, turned his attention to Saddam Hussein's Iraq. For many years Iraq had violated UN Resolution after Resolution. By 2003, intelligence reports seemed to indicate that Iraq had been stockpiling, or was attempting to manufacture, weapons of mass destruction (WMD)—either nuclear or chemical/biological—which violated UN Resolution 1441. Although the U.S. failed to garner the same support as it had during the 1991 Persian Gulf War, or locate any WMDs, Britain and many other nations contributed troops and other forms of support when a predominantly U.S. and British force invaded Iraq in March 2003.

Coalition Special Forces had infiltrated Iraqi territory well before the first bombs fell on Baghdad, presumably to disable critical communication and radar nodes so as to blind and cut off leadership from their armies. Although airpower initiated the battle, the plan approved by General Tommy Franks (US

Army) was a simultaneous thrust consisting of airpower and ground assaults. The "rapid dominance," approach, called "shock and awe" by some, was intended to bypass cities and towns along the road to Baghdad where the ground forces were to capture or eliminate Iraqi leadership and remove Saddam Hussein from office.

Hussein's Ba'ath Party toppled, and Hussein himself fled into the countryside where he was captured on December 13th, while hiding in an underground dugout little larger than a full-sized bathroom with a very low ceiling. Operations against insurgent groups are ongoing. As of March 16, 2006, 2,310 Americans had perished during Operation IRAQI FREEDOM. In Afghanistan, 278 had died during Operation ENDURING FREEDOM.

As the new century dawns, a wave of technological aviation wonders are just beginning their operational careers. The stealthy, F/A-22 Raptor is capable of supersonic flight without the use of afterburners. The F-35 Joint Strike Fighter is on the near horizon and may also prove to be the last large-scale purchase of manned fighter aircraft that the Air Force ever makes. The stout but remarkably capable C-17 Globemaster III combines the wide-body capacity of a C-5 with the ability to operate in austere locations like a C-130. Upgrades and modifications continue on the half-century-old B-52H. It appears that about half of the remaining fleet of heavy bombers will be grounded for good, leaving about 50 of the most highly modified for service during the next few decades. Early in this century the Air Force will retire the now quarter-century-old F-117 Nighthawk—the original stealth fighter. The follow-on may actually be a stealthy unmanned bomber—perhaps the B-3.

Undoubtedly, the decisive technology for the future is the remotely piloted unmanned aerial vehicle (UAV). The idea of being able to strike an enemy from the air, without risking the life of a pilot, and yet maintaining the same level of situational awareness and flexibility in targeting, is not far from becoming a reality. From tiny "bumble-bee" sized aircraft to the massive Global Hawk and the lightning-fast X-45 Combat UAV, the future of combat operations and weapons deliver will certainly evolve toward UAVs. Where people are part of the mission, it is unreasonable to expect that an airplane will ever fly without the personal attention of a pilot on board.

Regardless of the development of future technologies, in the end it will be the people who wear the many colors of Air Force uniforms that will continue to ensure that America's Air Force is not only the best in the world, but best by an unfair margin. The future of the Air Force and our nation rests in the hands of those who serve and those who will one day serve as members of the finest fighting force the world has ever known.

Pages 542-543: *On February 28, 1994, Captain Bob "Wilbur" Wright engages and shoots down three Serb attack aircraft in the first NATO combat action in history. ("Sting of the Black Viper," by Rick Herter, USAFAP)*

Above: *F-117s of the 8th FS prepare to deploy to the Balkans from their home base at Holloman AFB, New Mexico. The "Black Sheep" participated in operations over Kosovo in 1999. ("Countdown to Kosovo," by Herb Mott, USAFAP)*

GLOBAL REACH, GLOBAL POWER

1990–2006

"The battle is now joined on many fronts. We will not waver, we will not tire; we will not falter; and we will not fail. Peace and freedom will prevail."

—President George W. Bush addressing the nation, October 7, 2001

1990

3 January
Operation JUST CAUSE comes to a close when Manuel Noriega, in the face of a massive manhunt including a $1 million dollar reward for his capture, seeks refuge in the Vatican diplomatic mission in Panama City. Surrounded by American military forces which employ psychological pressure on him and diplomatic pressure on the Vatican mission, Noriega surrenders. He is taken into custody and immediately placed aboard an Air Force C-130 and extradited to the United States. By mid-January, American combat forces have begun to withdraw. During Operation PROMOTE LIBERTY, U.S. forces remain in Panama to support the reconstruction of the newly installed Panamanian government.

31 January
Operation CORONET COVE ceases. For more than a decade, the Air National Guard has rotated units to the region to defend the Panama Canal. During the operation, the ANG has flown nearly 17,000 flight hours during 13,000 sorties, including missions during Operation JUST CAUSE.

26 February
The Lockheed SR-71 Blackbird is officially retired from active service. High operational costs and improvements in satellite technology impact the decision.

6 March
When an SR-71A (64-17972) flies from coast to coast on its final flight, it sets four transcontinental speed records. After landing at Dulles International Airport, the crew taxis the aircraft into the custody of the National Air and Space Museum. This SR-71 is now on permanent display in the Stephen F. Udvar-Hazy Center, just a few miles from where it landed after its final flight.

4 April
The last McDonnell Douglas KC-10 Extender is accepted by the USAF for duty.

11 April
The first ground-launched cruise missile to be destroyed in accordance with the INF Treaty is loaded onto a C-5 Galaxy for transport to the United States.

21 April
The Lockheed F-117 Nighthawk is revealed to the public at Nellis AFB, Nevada. More than 100,000 visitors come to see the arrowhead-shaped formerly secret plane.

24 April
NASA launches the Hubble Space Telescope into orbit aboard the space shuttle *Discovery*. Hubble will provide the most detailed images of the universe ever collected once it becomes operational.

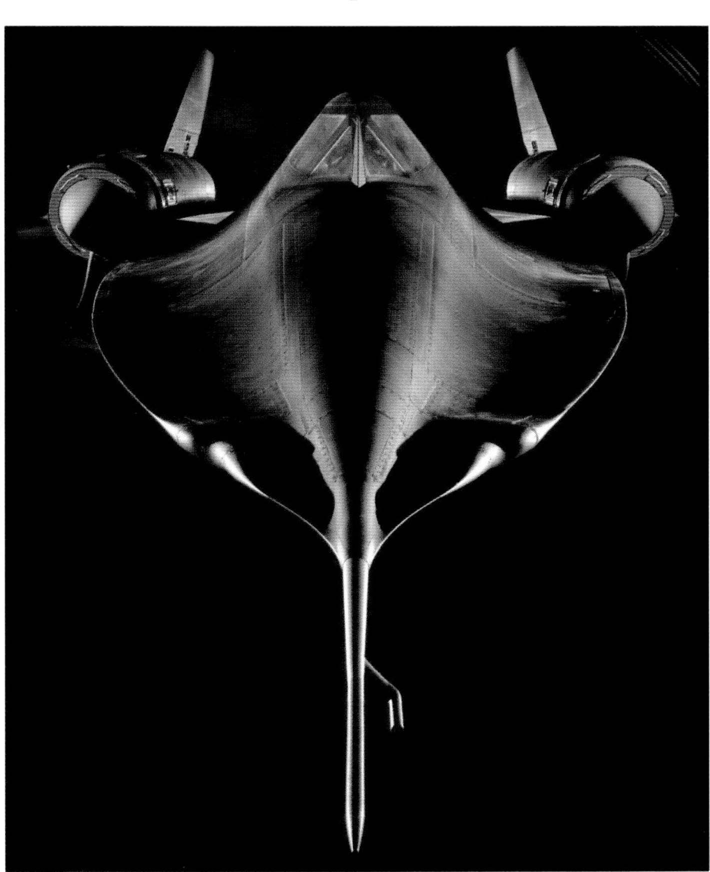

4 May
The AIM-120 advanced medium-range air-to-air missile (AMRAAM) is approved for use by Air Force fighters.

22 May
Twenty-Third Air Force is redesignated as Special Operations Command (SOC).

15 June
Global Reach-Global Power, a post-Cold War planning document, is released by Secretary of the Air Force Donald B. Rice.

1 July
General Michael J. Dugan becomes Chief of Staff, USAF.

Opposite: *At Howard AFB, Panama, Manuel Antonio Noriega Moreno is turned over to DEA officials for transport to the U.S. to stand trial in civil court. ("The Capture of Noriega," by Francis McGinley, USAFAP)*

Above: *The Lockheed KC-10 Extender is an Air Mobility Command advanced tanker and cargo aircraft designed to provide increased global mobility for U.S. armed forces. Although the KC-10's primary mission is aerial refueling, it can combine the tasks of a tanker and cargo aircraft by refueling fighters while simultaneously carrying the fighter support personnel and equipment to overseas deployments. (USAF)*

Left: *The Lockheed SR-71 was finally grounded in 1999. This record-setting aircraft can be viewed at the Steven F. Udvar-Hazy Center located near Dulles Airport 15 miles outside of Washington, D.C. (NASM)*

Left: *F-15C Eagle aircraft of the 58th Tactical Fighter Squadron take off on deployment to Saudi Arabia during Operation DESERT SHIELD. (USAF)*

12 July
The last production F-117 Nighthawk is delivered to the USAF by Lockheed.

17 July
In a televised speech, Saddam Hussein warns that he will attack Kuwait if Iraqi demands are not met. His demands include border disputes, decreasing Kuwaiti oil production, and reducing Kuwait's share of oil from the Rumaila oil field, which extends under Iraqi territory.

17 July–1 August
After a massive earthquake destroys Baguio in the Philippines, Air Force airlifts deliver nearly 600 tons of relief equipment and supplies and some 2,500 passengers to nearby Clark Air Base to assist the homeless and provide medical care for the injured.

24 July
EC-135C Looking Glass, airborne nuclear command and control posts, continuous flight operations cease. In the three decades of operations, the airborne control center has never had an accident during hundreds of thousands of flight hours.

30 July
The CIA reports that 100,000 Iraqi troops and 300 tanks are massed on the Kuwait border. Iraqi, Kuwaiti, and Saudi representatives meet in Jeddah, Saudi Arabia, to attempt reconciliation but the talks fail.

2 August
At 1 a.m., local time, Iraq invades Kuwait, using land, air, and naval forces. President George H. W. Bush issues Executive Orders 12722 and 12723, declaring a national emergency. The Joint Staff reviews options, including CENTCOM Operations Plan 1002-90, a top-secret contingency plan to move ground troops and supporting air and naval forces to the region over three to four months. Central Command (CENTCOM) staff starts formulating the air campaign for the defense of Saudi Arabia. The UN Security Council passes Resolution 660 calling for the unconditional withdrawal of Iraqi troops from Kuwait.

7 August
Operation DESERT SHIELD, the protection of Saudi Arabia and the build-up of American forces in the region, begins after Iraq's armed invasion of Kuwait.

The 71st Tactical Fighter Squadron (TFS) deploys from Langley AFB, Virginia. Twenty-four F-15C Eagles make the 8,000-mile trip to Dhahran, Saudi Arabia, in 15 hours with the help of 12 in-flight refuelings.

8 August
The first USAF Reserve crew into the Gulf lands their C-141 at Dhahran, Saudi Arabia. This is the first USAF aircraft in crisis zone. F-15s from 1st TFW and the elements of 82d Airborne Division also arrive in Saudi Arabia. U.S. Airborne Warning and Control System (AWACS) aircraft augment Saudi AWACS orbiting over the Kingdom of Saudi Arabia.

Iraq annexes Kuwait.

Headquarters USAF activates contingency support staff. General H. Norman Schwarzkopf asks Vice Chief of Staff Lieutenant General Mike Loh to create the plans for the strategic air campaign. Checkmate, the Air Staff planning group under Colonel John Warden, starts developing the basic plan for strategic air war.

9 August
Alaskan Air Command is redesignated Eleventh Air Force and assigned to Pacific Air Forces (PACAF).

10 August
At CENTCOM at MacDill AFB, Florida, Colonel Warden and his staff brief the air concept plan to General Schwarzkopf, and it is approved. F-16s from Shaw AFB, South Carolina and C-130s from Pope AFB, North Carolina, arrive in the Gulf. General Charles "Chuck" Horner and his staff draw up contingency plans for coalition forces in the event that Iraq attacks Saudi Arabia before sufficient defensive forces are in place.

12 August
Thirty-two KC-135 tankers deploy to Saudi Arabia, the first of more than 300 KC-10s and KC-135s that will arrive in the theater. MH-53J Pave Low helicopters of 1st Special Operations Wing arrive in Dhahran.

14 August
The Soviet Union joins the coalition in a naval quarantine of Iraq. The Department of Defense announces the presence of E-3 AWACS, KC-10s, KC-135s, and RC-135s in the Gulf region.

15 August
Highly classified F-117 stealth fighters from Tonopah, Nevada, and F-4G Wild Weasels from George AFB, California, deploy.

16 August
A-10 attack aircraft from Myrtle Beach AFB, South Carolina, deploy to the Gulf.

17 August
The Civil Reserve Air Fleet (CRAF) is activated for the first time since it was authorized in 1952. President George H. W. Bush authorizes the call-up to speed the build-up of forces in the Middle East.

Iraqi forces in Kuwait, heavily reinforced, build defensive positions along the Saudi border. Colonel Warden briefs a revised concept plan to General Schwarzkopf and the CENTCOM staff and is then directed to go to Saudi Arabia and brief it to General Horner.

Air Force Space Command establishes Defense Satellite Communications Systems (DSCS) links for DESERT SHIELD. Pre-positioned ships begin off-loading in Saudi Arabia.

19 August
The initial contingent of F-117 stealth fighters departs their home base at Tonopah, Nevada, for Saudi Arabia. After a brief layover at Langley AFB, Virginia, 18 F-117s from the 415th TFS continue to Khamis Mushait Air Base, Saudi Arabia, for Operation DESERT SHIELD. The next contingent of 18 jets will follow in early December.

20 August
General Horner concludes that air and ground strength in the region is now sufficient to defend Saudi Arabia against any potential Iraqi invasion. Colonel Warden and his group brief the concept plan to General Horner in Riyadh.

Right: *Ground crews service the F-117A aircraft of the 37th Tactical Fighter Wing (TFW) on the flight line. The 37th is preparing to deploy to Saudi Arabia for Operation DESERT SHIELD. (USAF)*

PLANNING THE DESERT STORM AIR CAMPAIGN

Phil Meilinger

Left: *Without the USAF aerial refueling force, global engagement would simply not be possible. Here, a KC-135R refuels an F-15. (USAF)*

When Saddam Hussein invaded Kuwait on August 2, 1990, President George Bush stated that this aggression "would not stand." He then began the difficult job of building a global coalition to oust Iraq from Kuwait.

General Norman Schwarzkopf had no less daunting a task. Iraq's army, the world's fourth largest, had recently waged a lengthy war against Iran. It was battle-hardened, fighting on its own turf, and enjoying interior supply lines. It also had a sizeable air force of over 700 aircraft. Finally, Saddam had two terror weapons: ballistic missiles, called Scuds, and chemical weapons that could slaughter thousands. Indeed, Saddam had already employed such deadly weapons against Iran and his own Kurdish minority.

Schwarzkopf turned to his planning staff for options, but was unhappy with their suggestions. His planners, largely ground officers, presented him with a traditional scheme calling for a massive ground assault through the middle of the Iraqi lines. It would be a bloodbath, incurring as many as 20,000 American casualties.

Disappointed, Schwarzkopf looked elsewhere for ideas, asking the acting Air Force Chief of Staff, General John "Mike" Loh, if his staff could prepare an air option.

Coincidentally, one of Loh's planners was already in the midst of doing just that. Colonel John Warden was a fighter pilot who had flown in Vietnam, commanded a fighter wing, and was known as a creative and iconoclastic thinker. He was on a Caribbean cruise with his wife when Saddam invaded Kuwait. Jumping ship at the next port, Warden grabbed a plane back to Washington, gathered together a number of officers, and began devising an air campaign plan to drive the Iraqis from Kuwait. When Loh called on August 8, Warden and his team doubled their efforts, bringing in experts from the other services, as well as intelligence agencies.

Warden envisioned a strategic air campaign that bypassed Iraqi fielded forces and instead attacked the enemy's vital centers. Warden saw an enemy nation as a series of key centers of gravity: its fielded forces, population, infrastructure, key industries, and political leadership. He portrayed these centers schematically as concentric rings. The outermost ring of the bull's-eye was an enemy's fielded forces; the center was its leadership. Warden argued that the old method of

Right: *The E-3 Sentry is an airborne warning and control system aircraft that provides all-weather surveillance, command, control, and communications needed by commanders of U.S. and NATO air defense forces. During Operation DESERT STORM, the E-3 proved itself the premier air battle command and control aircraft in the world. (USAF)*

making war—engaging in a bloody battle of attrition against an enemy's army—was the most costly and time-consuming of military strategies. Instead, he advocated "inside-out" warfare—airpower would be used to hit the center and inner rings of the Iraqi state. Attacks on Saddam's leadership, and communications and transportation networks would disrupt the enemy and have a cascading effect outwards toward the fielded forces, which would be left directionless.

It was vital that such air strikes occur quickly and with overwhelming force. Warden dubbed his plan INSTANT THUNDER—to distinguish it from the disastrous policy of gradual escalation followed in Vietnam. That air campaign, termed ROLLING THUNDER, was a ghost the airmen wanted to expunge.

On August 10, Warden briefed INSTANT THUNDER to Schwarzkopf, who exclaimed, "You have given me a plan that won't get my command destroyed." He then directed Warden to brief Lieutenant General Charles "Chuck" Horner, his air component commander who had already deployed forward to Saudi Arabia.

Horner was a crusty, savvy and no-nonsense fighter pilot who had served two combat tours in Vietnam and had been an operational commander most of his career. Frankly, he was a bit annoyed that a young colonel from the Pentagon was coming out to tell him how to do his job. It smacked too much of Vietnam when armchair generals in Washington dictated the targets, tactics and weapons used in an air war thousands of miles distant. Nonetheless, he listened attentively to Warden's briefing and saw its merit. Still, it wasn't complete. Not so much a campaign plan as a theoretical construct, he gave INSTANT THUNDER to his own planner, Brigadier General Buster C. Glosson, and told him to convert it into something usable. At the same time, Horner realized, as did Schwarzkopf, that Warden's emphasis on strategic targets was misguided. At some point, ground operations would be necessary, and when they occurred, the Iraqi Army must already be broken. Horner directed a major reemphasis on attacking the 50-plus Iraqi divisions arrayed against coalition forces. A goal of this revised air campaign plan, which would be used in DESERT STORM, was to hammer the Iraqi forces by air until they were below 50 percent of their strength—*before* a coalition ground assault took place.

Overall, only 2 percent of the coalition air effort was directed at the leadership targets that Warden thought so crucial, but these strikes were disproportionably effective. Although Saddam survived, he was cut off from his military forces, resulting in a disjointed and ineffective military response.

Warden's strategic air campaign plan reframed the debate on war strategy. Instead of a plan that focused on a potentially bloody ground assault, it postulated an air campaign that set the conditions for a concluding ground phase. This indeed occurred. Intelligence agencies concluded after the war that all 52 front-line Iraqi divisions had suffered attrition greater than 50 percent of their strength by the air campaign before the ground phase even began.

During this conflict, airpower had changed the face of war.

21 August
USAF Gulf presence includes A-10s, C-130s, E-3 AWACS, F-4Gs, F-15s, F-15Es, F-16s, F-117s, KC-135s, KC-10s, and RC-135s. Needing 6,000 Air Force reserve volunteers, 15,000 step up and are ready to go in 72 hours.

22 August
By this date, more than 20 percent of the total USAF Reserve force has volunteered for service during Operation DESERT STORM—more than 15,000 in all. They have flown more than 4,300 hours and transported more than 7 million tons of cargo and more than 8,000 people to the region.

23 August
The first of two highly modified Boeing 747-200B airliners—the VC-25A—arrives at Andrews AFB and is assigned to the 89th Military Airlift Wing. When occupied by the President of the United States, the aircraft call sign is *Air Force One.*

Secretary of Defense Richard B. Cheney authorizes the recall to active duty of necessary reserves for the Gulf crisis. Eventually, more than 20,000 reservists will serve on active duty during the Gulf War period.

24 August
117th Tactical Reconnaissance Wing, Birmingham, Alabama, deploys six RF-4C aircraft to the Gulf, joining RF-4Cs deployed by 67th TRW, Bergstrom AFB, Texas.

25 August
Air Force F-111 fighters from RAF Lakenheath, United Kingdom, deploy.

28 August
Iraq declares Kuwait to be its 19th province. President Bush, in a meeting with 170 members of Congress, defines U.S. objectives in the Gulf—"Immediate, complete, and unconditional withdrawal of all Iraqi forces from Kuwait, the restoration of Kuwait's legitimate government, security and stability of Saudi Arabia and the Persian Gulf, and the protection of American citizens abroad."

Air Force F-16 fighters from Torrejon AB, Spain, deploy to Qatar.

5 September
Five Air National Guard units begin deployment of C-130H aircraft.

6 September
A 40-cent postage stamp featuring the portrait of Claire Chennault, famed leader of the Flying Tigers during World War II, is released.

8 September
The USAF promotes Colonel Marcelite Jordan Harris to the rank of brigadier general. She is the first African American woman to achieve this rank in the Air Force.

Left: Air Force One, *a modified Boeing 747, flies over Mount Rushmore. The principal differences between the VC-25A and the standard Boeing 747 are the electronic and communications equipment, its interior configuration and furnishings, a self-contained baggage loader, front and aft air-stairs, and in-flight refueling capability. These aircraft are flown by the presidential aircrew, maintained by the presidential maintenance branch, and are assigned to Air Mobility Command's 89th Airlift Wing, Andrews Air Force Base, Maryland. (USAF)*

Right: *The YF-22 performs an early test-mission profile. ("High Alfa," by K. Price Randall, USAF Art Collection)*

8–12 September
First AC-130H gunships from 16th Special Operations Squadron arrive in the Gulf.

13 September
In Riyadh, Brigadier General Buster Glosson (deputy commander, Joint Task Force Middle East) briefs Generals Schwarzkopf and Powell on the complete operational air war plan. General Colin Powell, Chairman of the Joint Chiefs of Staff, asks when USAF could execute the plan. Glosson says, "Within 24 hours."

17 September
The Chief of Staff, USAF, General Michael J. Dugan, is relieved of his command by the Secretary of the Air Force, Richard B. Cheney, for making unauthorized comments to media personnel while in Saudi Arabia concerning DESERT SHIELD. General Jon Michael Loh serves as CSAF (acting) until a replacement can be appointed.

18–28 September
Air Force airlift wings deliver tents, cots, and blankets to Jordan for use by the hundreds of thousands of refugees fleeing there from Kuwait after the Iraqi invasion.

20 September
Iraq's Revolutionary Command Council declares that there will be no retreat and that the "mother of all battles" is inevitable.

29 September
The Lockheed/General Dynamics/Boeing YF-22A prototype flies for the first time. It makes a ferry flight to Edwards AFB, California.

1 October
Patrick AFB, Florida, is assigned to Air Force Space Command along with space-launch responsibilities formerly held by Air Force Systems Command. Vandenberg AFB, formerly a SAC asset, will also be reassigned to Space Command in January 1991.

3 October
General Curtis LeMay, legendary World War II and Cold War commander, dies in Riverside, California, at the age of 83.

10 October
Air Force fighter units arriving in theater fly training sorties to prepare for desert warfare. F-15C combat air patrol (CAP) is becoming routine.

30 October
General Merrill A. McPeak is appointed as Chief of Staff, USAF.

Operation DESERT EXPRESS, overnight airlift of crticial items to the Gulf, begins.

3 November
With Lockheed test pilot Dave Ferguson at the controls, the YF-22A Advanced Technology Fighter (ATF) prototype, configured with General Electric YF120 prototype turbofans, becomes the first fighter aircraft in history to achieve sustained supersonic flight without employing afterburners. The aircraft attains a "supercruise" speed of Mach 1.58 at 40,000 feet.

Left: *The E-8C Joint Surveillance Target Attack Radar System (Joint STARS) is a joint Air Force-Army program. The Joint STARS uses a multi-mode side-looking radar to detect, track, and classify moving ground vehicles in all conditions deep behind enemy lines. The aircraft is the only airborne platform in operation that can maintain real-time surveillance over a sizeable portion of the battlefield. This system was used with tremendous effect during the Battle of Khafji in late January 1991. (USAF)*

17 November
Air Force Space Command repositions a DSCS II satellite over the Indian Ocean to improve communications for DESERT SHIELD operations.

21 November
OA-10s from Davis Monthan AFB, Arizona, deploy.

29 November
The UN Security Council passes Resolution 678, authorizing use of force to expel Iraq from Kuwait. The resolution allows a grace period, giving Iraq one final opportunity to comply with previous resolutions. Iraq does not comply.

5 December
152d TRG, with RF-4Cs, deploys to Saudi Arabia to replace 117th TRW aircraft and personnel.

21 December
Air Force EF-111s deploy to the Gulf.

22 December
Clarence "Kelly" Johnson, legendary aircraft designer and founder of the "Skunk Works," dies at the age of 80.

29 December
The first deployed Air National Guard unit—the 169th Tactical Fighter Group—arrives in the Persian Gulf for operational duty.

1991

2 January
The Air National Guard's 174th TFW deploys 18 F-16s to Saudi Arabia and, along with 169th TFG, is incorporated into 4th TFW (Provisional).

CENTCOM announces U.S. strength in the Gulf exceeds 325,000 troops.

11 January
Two pre-production E-8A JSTARS aircraft are flown to Riyadh, Saudi Arabia, for possible use in the Persian Gulf.

12 January
Congress, after intense debate, clears U.S. forces for war against Iraq. The House votes 250–183 to authorize the President to use military force to implement UN Resolution 678 to force Iraq to withdraw from Kuwait. The Senate votes 52–47 in favor of the same authorization.

15 January
The deadline for Iraq's withdrawal from Kuwait passes.

16 January
CENTCOM announces that 425,000 U.S. troops are deployed in theater, supported by ground forces of 19 nations and naval efforts of 14 nations. The first

Right: *An RF-4C from the Reno Air National Guard overflies Kuwait City after liberation by coalition forces in 1991. ("RF-4 Over Kuwait City," by Frances Caraker, USAF Art Collection)*

elements of the USAFE Joint Task Force Headquarters deploy from Ramstein AB, Germany, to Incirlik AB, Turkey, and prepare to establish USAF's first wartime composite wing. Seven B-52Gs, launching from Barksdale AFB, Louisiana, become the first aircraft to take off on a DESERT STORM combat mission and are carrying the never-before-used AGM-86C Conventional Air Launched Cruise Missiles.

16–17 January
Operation DESERT STORM begins during the early morning hours as Air Force MH-53 Special Operations helicopters lead an Army helicopter force to destroy enemy radar sites. B-52Gs flying from Barksdale AFB, Louisiana, fire a volley of 35 cruise missiles against targets in Iraq. This attack marks the longest combat mission in history—more than 35 hours—as the bombers return to Barksdale after the missiles are launched.

F-117s strike strategic targets throughout Iraq, most in Baghdad. Remarkably, this small force of stealth aircraft attack 30 percent of the strategic targets during the first day of the Gulf War. Despite facing a massive array of Iraqi anti-aircraft fire during the attacks, only one F-117 suffers minor incident flak damage during the entire conflict.

Air Force C-130s begin airlifting the Army's XVIII Airborne Corps from the rear area to the Saudi-Iraqi border. Landing and unloading in 10-minute intervals, the C-130s deliver nearly 14,000 troops and more than 9,000 tons of cargo. This operation is the forward deployment; General Norman Schwarzkopf later calls it the "Hail Mary maneuver."

An Air Force F-15C, piloted by Captain Jon K. "JB" Kelk of the 33d TFW, shoots down a MiG-29 early in the morning hours for the first aerial victory of the operation.

Coalition forces fly more than 750 attack sorties from land bases during the opening day of the war. The U.S. Navy launches 228 combat sorties from six carriers in the Red Sea and the Arabian Gulf. Turkey approves Air Force use of Incirlik AB and Turkish airspace to open a northern front against Iraq. USAFE immediately deploys aircraft to Turkey.

18 January
Air Force aircraft based in Incirlik, Turkey, begin attacks against targets in northern Iraq. This "second front" strategy is designed to keep the Iraqis from reinforcing their deployed forces in Kuwait.

19 January
Two F-16Cs from the 614th Tactical Fighter Squadron (Torrejon AB) are shot down by surface-to-air missiles and the pilots taken prisoner. Three Iraqi Scud surface-to-surface missiles hit Israeli territory, injuring 10.

Iraq parades seven captured coalition airmen on television.

Left: *The McDonnell F-15E Strike Eagle remains a versatile ground attack aircraft. During DESERT STORM, F-15Es often were tasked to search for and then destroy mobile SCUD launchers in the deserts of Iraq. (USAF)*

21 January
While attempting to rescue a downed Navy F-14 pilot in Iraqi territory, Captain Paul T. Johnson destroys a threatening enemy vehicle with his A-10 Thunderbolt II. His actions allow an Air Force MH-53J Pave Low helicopter to ingress and rescue the pilot. Johnson receives the Air Force Cross. The Pave Low crew from the 20th Special Operations Squadron, Captain Thomas J. Trask, Major Michael Homan, Master Sergeant Timothy B. Hadrych, Technical Sergeant Gregory Vanhyning, Technical Sergeant James A. Peterson Jr., Staff Sergeant Craig Dock, and Sergeant Thomas W. Bedard receive the Mackay Trophy for extraordinary heroism and self-sacrifice of the crew during the rescue of a downed Navy F-14 pilot in Iraq.

22 January
Corvette 03, an F-15E deployed to the Middle East from Seymour-Johnson AFB, is shot down over Iraq. Colonel David W. Eberly, director of operations for the 4th Fighter Wing (Provisional), and his weapon systems officer, Lieutenant Colonel Tom Griffith, evade Iraqi soldiers for three days but are captured near the Syrian border and held until March.

22–27 January
F-111F "Aardvarks" attack Al Asad AB, Iraq, with laser guided bombs. These precision-guided munitions successfully penetrate hardened aircraft shelters, destroying the aircraft and equipment inside.

Additional strikes are made to prevent the flow of crude oil into the waters of the Persian Gulf. By destroying the pressure manifold, the flow is greatly reduced.

23 January
Only five Iraqi air bases remain functional after a week of bombing. Iraqi sorties fall from 235 to 40 per day. Iraq begins dumping Kuwaiti oil into the Persian Gulf and torching Kuwaiti oil wells.

24 January
A Saudi F-15C pilot shoots down two Iraqi F-1 Mirages attempting to attack coalition ships with Exocet missiles. The coalition flies 2,570 attack sorties, for a total of 14,750 during the first eight days of war.

25 January
Coalition forces destroy three Iraqi bombers on the ground. Major attacks on Iraqi hardened aircraft shelters begin. USAF, using the new I-2000 bomb, has remarkable success.

26 January
Air emphasis shifts to strikes against the Iraqi fielded forces in Kuwait. Iraq sends aircraft to Iran for sanctuary (122 in all by the end of the war).

In Washington, marchers protest the war in the Persian Gulf. Anti-war protesters march in Bonn, Berlin, Switzerland, and France. Demonstrations also occur in support of war in several U.S. cities, among them Boston and Chicago.

27 January
Air Supremacy is declared over Iraq. Success in this 10-day campaign allows Coalition Forces in the region freedom to attack at will and also provides freedom from attack during operations.

F-111s, using GBU-15 guided bombs, destroy oil-pumping manifolds at the Kuwaiti terminal, drastically reducing the flow of oil into the Gulf. F-16 KILLER SCOUT operations begin.

29–31 January
Iraqi forces cross into Saudi Arabia and reach the town of Khafji. Although an AC-130 Spectre is shot down and all 14 crew members are lost, USAF tactical airpower helps to fix the force in place while coalition ground troops force the Iraqis to retreat within three days. E8-C Joint STARS plays a vital role during the flight.

2 February
An Air Force B-52 bomber goes down in the Indian Ocean while attempting to return to Diego Garcia after a mission over Kuwait. Three crew members are rescued; three are lost.

3 February
Iraq withdraws troops from Khafji area.

6 February
Captain Robert Swain (USAFR) shoots down an Iraqi helicopter in his A-10 Thunderbolt II. This is the first air-to-air victory by an A-10 in history.

9 February
Iraqi Scud missiles hit Israel, injuring 26.

"Tank plinking," picking off individual tanks with smart weapons, begins and is accomplished with effect by A-10 Thunderbolt II aircraft. Coalition sources reveal that 15 percent of Iraq's armor, about 600 tanks, and between 15 percent and 20 percent of overall fighting ability, have been destroyed.

11 February
Coalition aircraft fly 2,900 attack sorties, bringing the 26-day total to 61,862.

12 February
Air attack destroys three downtown Baghdad bridges: Martyr's Bridge, Republic Bridge, and July 14 Bridge.

13 February
When F-117s attack the Al Firdos bunker in downtown Baghdad, several hundred civilians who are using the building as a bomb shelter are killed by the deadly accuracy of the bombing. Unfortunately, prior intelligence reports on the building, a communications center, did not reveal its civilian use. Coalition authorities take tighter control of missions over Baghdad after this event.

14 February
Two USAF crewmen are killed when their EF-111A is lost in Saudi Arabia after an electronic jamming mission over Iraq.

Anti-war demonstrators splash blood and oil on the Pentagon doorway.

Right: *Captain Bob Swain, an Air Force Reserve A-10 pilot, makes history when he shoots down an Iraqi Gazelle helicopter northeast of Kuwait City using his 30mm cannon. It is the first air-to-air victory for an A-10 ("Captain Swain and his A-10 Make History," by William Lacy, USAF Art Collection)*

THE TALE OF CORVETTE 03

Tom Griffith

In American military history Operation DESERT STORM stands out as a unique conflict for many reasons, not the least of which is the small number of Americans killed, wounded, or captured. The numbers, while small, should not blind us to the very real cost of that war. On the fourth night of Operation DESERT STORM, I launched on my third combat mission and learned the cost of war firsthand.

That night, I was flying with Colonel David Eberly, the 4th Fighter Wing Director of Operations, and our target was the Scud missile storage areas in western Iraq near the town of Al Qaim. We took off as the number three aircraft in the first group of twelve. As we crossed into Iraq, the tension began to build. We turned off our external lights and flew in a line of aircraft at 20,000 feet. About 15 miles from the target, explosions from the anti-aircraft guns lit up the sky. It was like being on top of a deadly fireworks display during the 4th of July. As we began our bombing run, the Iraqis fired a dozen missiles at the formation. Two of the missiles on the right side of the plane appeared to be guiding on us. I put out some chaff to decoy them and we began an evasive maneuver, but as we banked up, a missile hit us on the left side. We ejected immediately and I landed in the middle of the target area where I could see and feel the bombs being dropped by other aircraft and the missiles being shot by the defenders on the ground. After getting my bearings, I scrambled out of there. A couple of hours later, I joined up with Colonel Eberly. We moved slowly in the darkness and as dawn approached, we found a small hill where we hid and awaited rescue.

We waited for two days on that hill with no food and only a little water. When we ran out of water, we started walking toward the Syrian border about

15 miles away. We walked all night and had almost made it to the border when the Iraqis grabbed us. After being taken to various posts and paraded through a jeering crowd on the streets of an Iraqi town, we were handcuffed, blindfolded, and driven to Baghdad, where we stayed for the remainder of the war.

During my captivity, I was held in four prisons. They all differed in some ways, but from a prisoner's perspective they also had certain similarities. The Iraqis kept us in solitary confinement the entire time. We got one meal a day, usually a piece of pita bread and a small bowl of broth. There was no heat in our cells and it was cold that winter in Baghdad. They dressed us in thin yellow prison suits and we slept on the concrete floor with just one blanket. We all had dysentery most of the time.

Of course, we were also interrogated for information. We got asked about our planes, our tactics, and the war in general. Naturally, no one wanted to answer those questions, but that didn't matter very much. The Iraqis weren't terribly sophisticated interrogators, but they did know how to get the information they needed. Generally, they would use beatings or death threats as a means of coercion. In the videos they released, what you couldn't see was the guy just out of camera range with the gun pointed at our heads. Sometimes there would be beatings just for their amusement.

Getting through those days was hard. We went from flying a high performance fighter engaged in a high-tech war to waging a personal war of wits and will. What got us through those times? Our training was a big help. Every Air Force aviator goes through a mock prisoner of war experience as part of survival training. The training removed some of the fear, but at the same time it can't prepare you for everything. When you're in training you know it will only be a few days; in the real situation you don't know how long it's going to be. In training you know that they aren't going to kill you, but that was a constant worry in Baghdad. Saddam Hussein's regime had no regard for its own citizens, so I didn't think they would care much about killing us. And, of course, in training they can't create the same intensity of pain that you receive as a prisoner.

What also helped us was knowing that we would not be forgotten. Knowing that the American public would ensure that we were accounted for, no matter how long it took.

Opposite: *The F-15E is capable of carrying a mixed weapons load as this refueling Strike Eagle displays. Precision-guided bombs, air-to-air and air-to ground missiles, and additional fuel tanks allow long-range strikes and self-defense options during combat missions. (USAF)*

Above: *Depicted here is a 4th Tactical Fighter Wing F-15E in action over Iraq during the 1991 Persian Gulf War. The F-15E Strike Eagle deployed to Iraq just one week after the Iraqi invasion of Kuwait. The 335th Tactical Fighter Squadron was the first interdiction unit to deploy to the theater. ("Eagle of the Persian Gulf," by Nixon Galloway, USAF Art Collection)*

Left: *An F-15C, two F-15Es, and two F-16s fly over burning oil fields in Kuwait. (USAF)*

Opposite, top: *The 84th Aerial Port Squadron airmen work with a C-141 Starlifter crew to load cargo in Haiti, while supporting Operation SECURE TOMORROW. (USAF)*

Opposite, bottom: *Two USAF Reserve C-5s on the ramp at Westover ARB, Massachusetts. (USAF)*

15 February
Saddam's five-man Revolutionary Command Council announces that Iraq is ready "to deal" with the UN resolution that requires withdrawal from Kuwait.

17 February
Heavy bombing of the Iraqi army in Kuwait has destroyed 1,300 of Iraq's 4,280 tanks and 1,100 of its 3,110 artillery pieces. Iraq's foreign minister arrives in Moscow for talks with Soviet president Mikhail Gorbachev.

19 February
A mixed force of F-4Gs and F-16s from the composite wing in Turkey launch daylight attacks on Baghdad from the north. Coalition air forces fly a record 3,000 attack sorties—a total of 83,000 so far during the war.

21 February
An Air Force C-141 delivers 55 tons of relief cargo to Sierra Leone to help suffering people through economic hardships there.

23 February
B-52s relentlessly bomb Iraqi troop positions. Iraqi troops set 100 more Kuwaiti oil wells on fire. General Schwarzkopf determines that attrition of Iraqi combat effectiveness is sufficient for a successful ground offensive with limited casualties.

24–28 February
The "Hail Mary" ground offensive into Iraq begins. As Coalition ground forces overwhelm the entrenched Iraqi army, more than 3,000 combat reconnaissance, close air support, and interdiction sorties are carried out over the 100 hour invasion. Approximately 80,000 Iraqis are captured and held captive until after the war ends.

25 February
Air Force F-16s attack Iraqi forces attempting to surround an Army Special Forces team. Their defensive actions allow an Army UH-60 Black Hawk helicopter to accomplish the rescue.

Iraqi Scuds hit the Dhahran barracks used by U.S. Army Reservists, killing 28 and wounding more than 100. Baghdad Radio airs Saddam's order for Iraqi forces to withdraw from Kuwait. At least 517 oil wells are burning in Kuwait.

26 February
The "mother of all retreats" features Iraqi soldiers attempting to escape envelopment in Kuwait. Thousands of military and civilian vehicles, loaded with looted goods, clog the four-lane highway out of Kuwait City. Repeated air attacks destroy much of the army's equipment. Coalition forces engage Republican Guards between the Kuwait-Iraq border and Basra. Other coalition forces seize Kuwait City and Al Jahrah.

27 February
Coalition forces liberate Kuwait City and envelope Iraqi forces. Two specially made 4,700-pound GBU-28 bombs destroy the "impregnable" Iraqi command bunker at Al Taji. Attack sorties reach a one-day record of 3,500. President Bush announces that coalition forces will suspend offensive operations the next day at 8 a.m. local time. Iraq must end military action, free all POWs, third country nationals, and Kuwaiti hostages, release remains of coalition forces killed in action, agree to comply with all UN resolutions, and reveal the location of land and sea mines.

28 February
The Gulf War ends when a Coalition-declared cease-fire takes effect at 8 a.m. During DESERT STORM, the USAF flew a total of 59 percent of the Coalition's combat sorties. Nearly 2,000 USAF aircraft comprised 75 percent of the total numbers of planes involved in combat operations during the war. Less than 10 percent of the bombs dropped during the operations are precision-guided munitions. Space satellite technology plays a vital role in communications and navigation for coalition forces in the air and on the ground, highlighting the critical importance of space operations to future conflicts.

March–December
During these months, a massive global airlift effort continues. With the war over, Military Airlift Command aircraft begin transporting relief supplies to Kuwait after the seven-month-long Iraqi occupation of that country. For the next six months, C-5s and C-141s deliver more than 1,000 tons of fire-fighting equipment and 100 professional firefighters to Kuwait, where the process of extinguishing some 500 oil fires set by retreating Iraqi soldiers begins. During the same period, C-5s airlift 150 tons of relief supplies to Bucharest, Romania, providing assistance to the government during a critical economic situation that has ignited street violence. C-5s deliver more than 200 tons of medical supplies to Lima, Peru, to assist in controlling the spread of cholera that afflicts 150,000 people. While Iraqi Kurds flee the country to escape from Saddam Hussein's forces, MAC aircraft deliver more than 7,000 tons of humanitarian aid to southeastern Turkey and surrounding areas where Kurdish refugees are gathering. MAC aircraft, along with Special Operations airlifters, deliver 3,000 tons of relief to flood victims in Bangladesh during May and June. To assist in drought-suffering Ethiopia, 19 MAC missions are flown to deliver more than 1,000 tons of relief cargo from June through September. Additional humanitarian missions are flown to Chad, Mongolia, China, and Albania during the summer months.

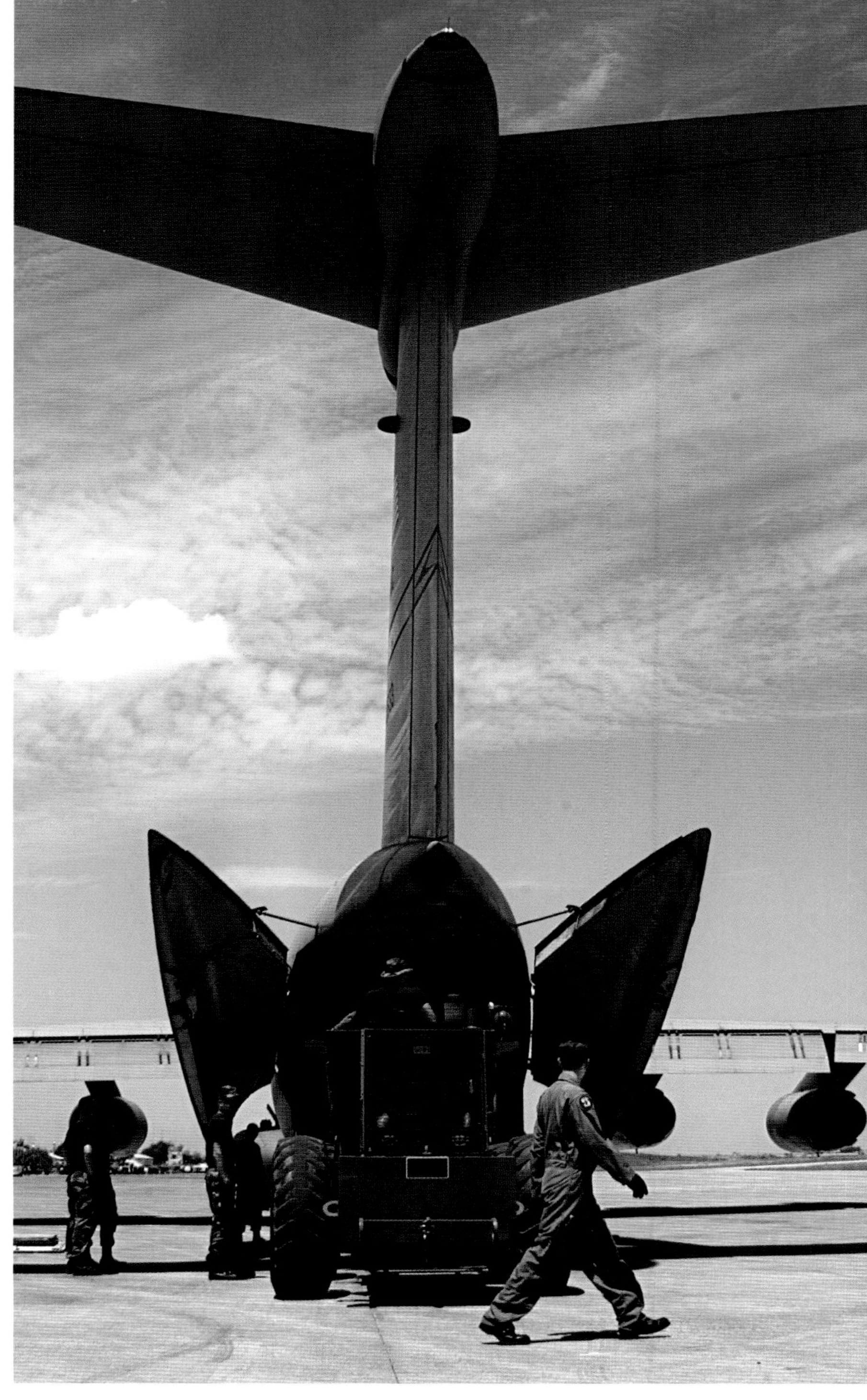

3 March
At Safwan in Iraq, General Schwarzkopf and Lieutenant General Khalid ibn Sultan, Saudi commander of Joint Arab Islamic Force, and their associates from other coalition countries, meet eight Iraqi officers, led by Lieutenant General Sultan Hashim Ahmad, commander of Iraqi III Corps in Iraq. Military leaders discuss cease-fire arrangements and the formal cease-fire is signed.

CENTCOM reports the total number of Iraqi equipment destroyed or captured increased to 3,300 tanks, 2,100 armored vehicles, and 2,200 artillery pieces. The number of POWs is estimated at 80,000.

5 March
Iraq releases 35 POWs including 15 U.S. military personnel. One of these is Army Major Rhonda Cornum, the second female held captive.

Left: *The Boeing C-17 Globemaster III made its maiden flight on September 15, 1991. The first production model was delivered to Charleston AFB, South Carolina, on June 14, 1993. The first squadron of C-17s, the 17th Airlift Squadron, was declared operationally ready January 17, 1995. (USAF)*

10 March
Iraq releases 21 U.S. POWs including eight members of USAF. Former POWs return to Andrews AFB, where they are met by Secretary Cheney, General Powell, and several thousand spectators.

20 March
An Air Force F-15C on patrol over Iraq shoots down a Su-22 Fitter that is flying in violation of the cease-fire agreement.

11 April
Iraq accepts all terms of the UN cease-fire resolution. The Persian Gulf War officially ends at 10 a.m. EST.

12 April
A Soviet AN-74 Coaler transport aircraft is intercepted by F-15 Eagles stationed at Galena Airport, Alaska. This is the first time that an AN-74 has been intercepted by USAF aircraft.

18 April
A Martin Marietta/Boeing MGM-134A (small ICBM) is launched for the first time from Vandenberg AFB, California. The missile impacts 4,000 miles downrange in the Kwajalein Missile Range.

31 May
The 501st Tactical Missile Wing at RAF Greenham Common, United Kingdom, is inactivated. The 501st TMW was the first ground-launched cruise missile wing to activate in Europe and is the last one to close down.

8 June–2 July
After the eruption of Mount Pinatubo near Clark AB, Philippines, covers the base and the surrounding area with tons of volcanic ash, the Air Force begins evacuation of personnel stationed there. During Operation FIERY VIGIL, more than 15,000 people are evacuated from the disaster zone and 2,000 tons of humanitarian cargo are delivered to assist survivors there—the largest evacuation operation since the fall of South Vietnam in 1975.

10 July
The final FB-111A flight is made from Plattsburgh, New York, to Davis Monthan AFB, Arizona, where the aircraft will be placed in storage.

22 August
The Air Force charters the Gulf War Air Power Survey (GWAPS) to evaluate the role played by airpower during the Gulf War.

15 September
The Boeing C-17A Globemaster III makes its first flight. The C-17 is designed to carry oversized cargo but deliver it to austere landing zones.

The Beech T-1A Jayhawk flies for the first time. The T-1 will become the backbone for the airlift and tanker tracks for specialized undergraduate pilot training. The trainer will become operational in February 1993.

Right: *An aerial gunner assigned to the Air Force Special Operations Detachment at a forward-deployed location loads a 40 mm cannon aboard an AC-130 gunship. The aircraft is armed with a 25 mm gun, 40 mm cannon and a 105 mm cannon. Its primary missions include close-air support for convoys, air interdiction of targets and force protection for base defense. This aircraft and crew are deployed from Hurlburt Field, Florida, the home of Air Force Special Operations. (USAF)*

27 September
President Bush orders that SAC alert will cease the following day. SAC has been at operational alert since October 1957. This event symbolically represents the end of the Cold War for the U.S. Air Force.

October–December
As governments and economies in the former Soviet Union begin to collapse, Air Force airlift delivers aid to many in need of medical supplies from Mongolia to the Ukraine. Food and relief supplies are delivered to Moscow and St. Petersburg, Russia, Yerevan, Armenia, and Minsk, Byelorussia.

26 November
Clark AB, Philippines, closes. This event marks the end of more than 90 years of U.S. military presence in the region.

21 December
The Rockwell AC-130U flies for the first time. This improved gunship includes increased firepower, better accuracy, and improved ability to locate targets on the ground.

1992

30 January
Air Force Space Command (AFSPACECOM) assumes control of Defense Department satellites as well as the operation and management of the Air Force Satellite Control Network.

10–29 February
During Operation PROVIDE HOPE I and II, Air Force C-5s and C-141s deliver food and medical supplies to newly independent states within the former Soviet Union.

4 March
Two Air Force B-52 Stratofortresses land in Russia on a mission of friendship. This is the first landing of U.S. bombers in Russia since World War II.

19 March
Alaska-based F-15s intercept two Russian Tu-95 Bear aircraft off the Alaska coast. This is the first such intercept since the collapse of the Soviet Union.

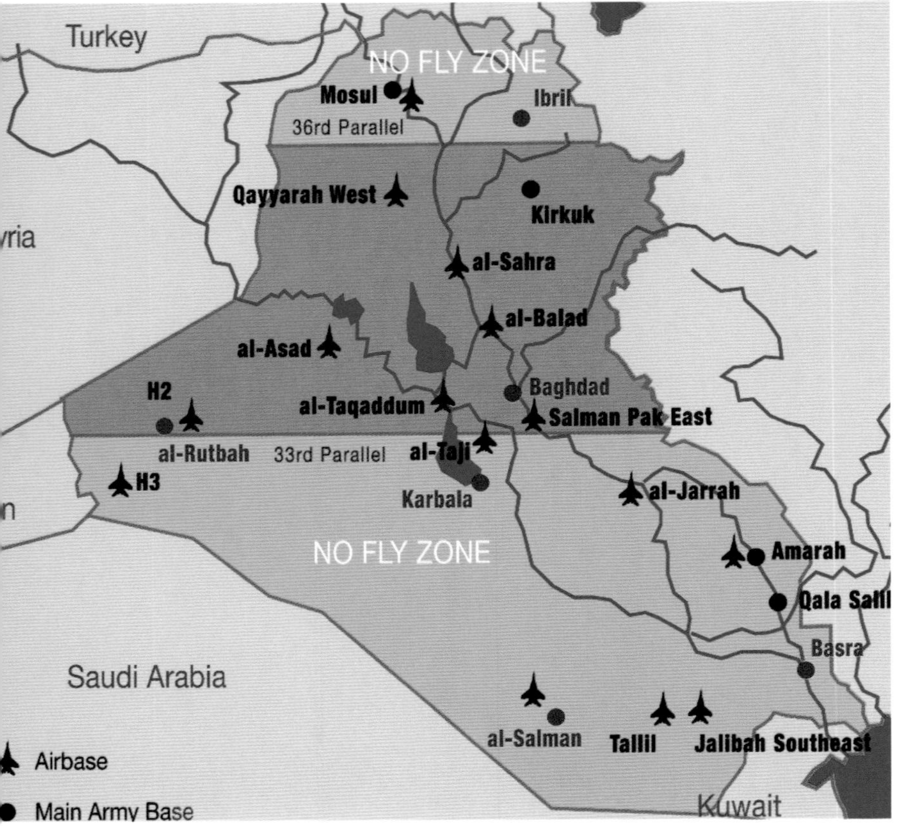

Left: *Map of the Iraqi Theater of Operations.*

Opposite: *Astronaut Susan J. Helms (USAF) was the first American military woman in space when she flew on* Endeavour *(STS-54). (NASA)*

24 March
Twenty-four nations sign the Open Skies Treaty, which allows unarmed reconnaissance flights over the territory of any signatory nation.

27 March
Former NASA administrator James E. Webb dies at age 85. He led the NASA program from 1961 to 1968 and was instrumental in establishing the environment needed to land men on the moon.

7 April
The U.S. recognizes Bosnia-Herzegovina, Croatia, and Slovenia as independent nations. Serb rebellions in the region will eventually generate a civil war. Ten days later, Air Force C-141s deliver humanitarian relief cargo to Sarajevo, capital of Bosnia-Herzegovina, as regional economies collapse.

24 April
An unarmed Air Force C-130 Hercules, flying a counter-narcotics mission, is attacked by two Peruvian Su-22 Fitter fighter planes over international waters. Heavily damaged with one crewman killed and four others wounded, the C-130 crew, assigned to the 310th Airlift Squadron, Howard AFB, Panama, is able to make an emergency landing. For extraordinary resourcefulness and unusual presence of mind during an unprovoked attack in international airspace, the crew receives the Mackay Trophy.

7–8 May
For the first time, a U.S. military band visits the Russian capital when the Command Band of the Air Force Reserves marches in a Moscow parade.

1 June
Strategic, Tactical, and Military Airlift Commands are inactivated as part of a significant Air Force reorganization. The assets of the three former commands are reconstituted by function and assigned to newly activated Air Combat Command (ACC) and Air Mobility Command (AMC).

The Department of Defense activates the United States Strategic Command (USSTRATCOM), a new unified command. The former SAC commander assumes command of the newly established USSTRATCOM.

1 July
The Air Force inactivates Logistics Command (AFLC) and Systems Command (AFSC) and consolidates the two into the Air Force Materiel Command (AFMC).

2 July
Operation PROVIDE PROMISE, regular deliveries of humanitarian relief to Bosnia-Herzegovina, begins. All types of Air Force airlift assets will deliver medicine and food to Sarajevo until 1996.

18 August
Operation SOUTHERN WATCH begins. From this point forward, Iraqi flights are not permitted south of the 32-degree north latitude line.

14 August–28 February 1993
During Operation PROVIDE RELIEF, a humanitarian supply mission to Somalia, Air Force aircraft deliver more than 23,000 tons of food and supplies while flying 3,000 sorties to the war-torn and drought-ridden country.

25 August–28 October
Hurricane Andrew devastates southern Florida. Damage to Homestead AFB near Miami is so severe that the base is permanently closed. Air Mobility Command (AMC) aircraft transport 13,500 relief workers and deliver 21,000 tons of equipment to the region. A total of 724 missions are flown.

26 August
Operation SOUTHERN WATCH, the enforcement of UN sanctions against Iraq, begins. Air Force aircraft take to the skies over southern Iraq, providing protection to Shiite Muslims living there and to prevent any further Iraqi activity against Kuwait or Saudi Arabia.

28 August
The 67th Reconnaissance Wing (RW) at Bergstrom AFB, Texas, is inactivated. This marks the end of regular Air Force RF-4C Phantom II operations.

September–December
Air Mobility Command aircraft fly missions around the globe delivering humanitarian relief to typhoon victims in the Pacific, transporting sick children from Chernobyl to medical treatment facilities, transporting UN peacekeepers from Pakistan to Somalia, evacuating Americans from civil disturbances in Tajikistan, and delivering flood relief to Pakistan.

4 December–4 May 1993
Operation RESTORE HOPE, support for UN peacekeeping efforts in Somalia, begins. During the Operation, AMC performs more than 1,000 airlift missions—Air Force Reserve crews fly an additional 190 sorties, delivering 1,500 tons of cargo. By the end of the operation, more than 50,000 passengers (most of these by commercial carrier) and 40,000 short tons of cargo will be delivered to the region.

16 December
During a night flight, a B-52 suffers a series of catastrophic emergencies which result in the loss of all four engines on the left side of the aircraft. The crew is able to start two of the engines and eventually makes a successful emergency landing. For quick thinking, immediate reaction, and astute situational awareness enabling them to return a crippled B-52 aircraft to stable flight and safe landing, the 668th Bomb Squadron crew, Captain Jeffrey R. Swegel, Major Peter B. Mapes, Captain Charles W. Patnaude, Lieutenant Glen J. Caneel, and Captain Joseph D. Rosmarin, receive the Mackay Trophy.

27 December
An Air Force F-16 shoots down an Iraqi MiG-25 that is violating the United Nations no-fly zone near the Iraqi border. This is the first aerial victory achieved with an AIM-120 AMRAAM air-to-air missile. The "Slammer" is a launch-and-leave missile that extends the F-16's effective engagement range.

1993

3 January
President Bush and President Boris Yeltsin of Russia sign the second Strategic Arms Reduction Treaty (START II) that will eliminate MIRV-capable ICBMs and reduce the number of nuclear weapons carried by bombers.

13 January
When Iraqi surface-to-air missiles are discovered south of the 32-degree north latitude line, and Iraqi troops cross over newly established lines bordering Kuwait, President Bush orders the destruction of Iraqi missile sites and command centers as a protective measure for those patrolling the no-fly zone. F-117 stealth fighters participate in the attacks.

Air Force Major Susan J. Helms flies as a mission specialist aboard the space shuttle *Endeavour* (STS-54). Helms becomes the first U.S. military woman in space.

17 January
During a suppression of enemy air defenses mission with an F-4G Wild Weasel, an Air Force F-16 is vectored to intercept an Iraqi MiG-23 and engages the aircraft with an AIM-120 AMRAAM (nicknamed "Slammer") missile and destroys it.

BRIGADIER GENERAL SUSAN J. HELMS: FIRST MILITARY WOMAN IN SPACE

Left: *Astronaut Susan J. Helms, Expedition-2 flight engineer, works at the Human Research Facility's (HFRF) Ultrasound Flat Screen Display and Keyboard Module in the Destiny/U.S. Laboratory. While on board the ISS, Helms also set a record for the longest space walk in history: 8 hours and 56 minutes. (NASA)*

When the space shuttle *Endeavour*, (STS-54) rocketed into orbit on January 13, 1993, Air Force Major Susan Helms became the first American military woman in space. A remarkable accomplishment, but not the only "first" during her military career. Helms began at the United States Air Force Academy in 1976 as a member of the first graduating class (1980) that included women.

Helms received a Bachelor of Science degree in aeronautical engineering from the U.S. Air Force Academy in 1980, and after commissioning, was assigned to Eglin AFB, Florida, as an F-16 weapons separation engineer with the Air Force Armament Laboratory. In 1982, she became the lead engineer for F-15 weapons separation and by 1984, she was selected to attend graduate school. She received her degree from Stanford University in 1985 and was assigned as an assistant professor of aeronautics at the U.S. Air Force Academy. In 1987, she attended the Air Force Test Pilot School at Edwards Air Force Base, California. After completing one year of training as a flight test engineer, Helms was assigned as a USAF Exchange Officer to the Aerospace Engineering Test Establishment, Canadian Forces Base, Cold Lake, Alberta, Canada, where she worked as a flight test engineer and project officer on the CF-18 aircraft. She was managing the development of a CF-18 Flight Control System Simulation for the Canadian Forces when selected for the astronaut program.

Selected by NASA in January 1990, Helms became an astronaut in July 1991. She flew on STS-54 (1993), STS-64 (1994), STS-78 (1996), STS-101 (2000) and served aboard the International Space Station as a member of the *Expedition-2* crew (2001). A veteran of five space flights, Helms has logged 5,064 hours in space, including an EVA of 8 hours and 56 minutes (world record).

Helms shuttle flights have accomplished a wide variety of objectives. Her first mission, STS-54 *Endeavour*, January 13–19, 1993, deployed a $200-million NASA Tracking and Data Relay Satellite (TDRS-F) and collected x-ray data that will enable investigators to answer questions about the origin of the Milky Way galaxy. As part of the flight, the crew demonstrated the physics principles of everyday toys to an interactive audience of elementary school students across the United States. The STS-54 mission duration was almost six days.

Helms second shuttle mission, STS-64 *Discovery*, September 9–20, 1994, validated the design and operating characteristics of Lidar in Space Technology Experiment (LITE) by gathering data about the earth's

troposphere and stratosphere. Additionally, the crew deployed and retrieved SPARTAN-201, a free-flying satellite that investigated the physics of the solar corona, and the testing of a new Extra Vehicular Activity (EVA) maneuvering device. During the flight, Helms served as the flight engineer for orbiter operations and the primary RMS operator aboard the space shuttle. The STS-64 mission lasted nearly 11 days.

During STS-78 *Columbia*, June 20 to July 7, 1996, Helms was the payload commander and flight engineer. This became the longest space shuttle mission yet flown. The mission included studies sponsored by 10 nations and five space agencies, and was the first mission to combine both a full microgravity studies agenda and a comprehensive life science investigation. The Life and Microgravity Spacelab mission served as a model for future studies on board the International Space Station. The STS-78 mission lasted nearly 17 days.

Helms final dedicated shuttle flight (STS-101) was aboard *Atlantis,* May 19–29, 2000. The mission was dedicated to the delivery and repair of critical hardware for the International Space Station (ISS). Helms prime responsibilities during this mission were to perform critical repairs to extend the life of the Functional Cargo Block (FGB). In addition, she had prime responsibility for the onboard computer network and served as the mission specialist for rendezvous with the ISS. After landing, Helms had amassed another 10 days in space.

The culmination of her astronaut career occurred when Helms lived and worked onboard the International Space Station as a member of the second crew to inhabit the International Space Station Alpha. The *Expedition-2* crew (two American astronauts and one Russian cosmonaut) launched from Cape Canaveral on March 8, 2001, onboard STS-102 *Discovery* and successfully docked with the ISS on March 9, 2001. The *Expedition-2* crew installed and conducted tests on the Canadian made Space Station Robotic arm (SSRMS). During her stay onboard the space station, Helms installed the airlock that had been delivered by STS-104 mission using the robot arm. She and her crewmates also performed a "fly around" of the Russian Soyuz spacecraft and welcomed the visiting Soyuz crew which included the first space tourist, Dennis Tito.

On March 11, she performed a world record 8-hour and 56 minute space walk to install hardware to the external body of the laboratory module. Helms spent a total of 163 days aboard the space station. She returned to earth with the STS-105 crew aboard *Discovery* on August 22, 2001.

After a 12-year NASA career that included 211 days in space, Helms returned to service in the U.S. Air Force in July 2002 to take a position at HQ USAF Space Command. Since that time Colonel Helms has been the 45th Space Wing vice wing commander at Patrick AFB, Florida, and in June 2006 took command of the 45th Space Wing as a newly promoted brigadier general.

Above: *The International Space Station as seen from an approaching spacecraft. (NASA)*

Opposite: *Astronaut Story Musgrave and Jeffery Hoffman complete servicing tasks on the Hubble Space Telescope (HST). (NASA)*

Left: *"Operation PROVIDE HOPE—Somalia, 1993" by Gil Cohen, USAF Art Collection*

Below: *The Delta Clipper Experimental (DC-X) lifts off for the first time from White Sands Missile Range. (White Sands Missile Range, Public Affairs)*

18 January
After being fired upon, F-4G Wild Weasels strike surface-to-air missile sites in northern Iraq. F-16s bomb Iraqi airfields after being fired upon by anti-aircraft guns.

2 February
After having established a joint airlift operations cell in Zagreb, Croatia, U.S. forces begin transporting injured noncombatants from the Bosnian war zone. By the end of the month, Operation PROVIDE PROMISE airdrop missions will deliver aid to refugees who are fleeing advancing Serbian troops.

31 March
The United Nations Security Council establishes a no-fly zone over Bosnia. The zone will become active on April 5.

12 April
Operation DENY FLIGHT, enforcement of a UN no-fly zone over Bosnia, begins. The U.S. provides most of the forces to accomplish this mission.

28 April
Secretary of Defense Les Aspin announces that women will be permitted to fly combat aircraft.

11 June
An Air Force AC-130 Spectre gunship takes part in a retaliatory raid in Somalia called Operation CONTINUE HOPE. Somali warlord attacks upon UN forces precipitated the strike.

14 June
The first operational C-17A Globemaster III is accepted by the USAF upon its delivery to Charleston AFB, South Carolina. The C-17 will be assigned to the 437th Airlift Wing.

29 June
In support of the Open Skies Treaty, a specially modified OC-135B aircraft is tested at Wright-Patterson AFB, Ohio. This aircraft will be used under the treaty rules to overfly treaty nations.

1 July
Air Training Command (ATC) is redesignated. The new Air Education and Training Command (AETC) absorbs former ATC units and the Air University (AU).

Twentieth Air Force is transferred from Air Combat Command to Air Force Space Command. The Twentieth is responsible for the daily operations of the ICBM force.

Fourteenth Air Force is activated to perform missile warning and space surveillance at Vandenberg AFB, California. The base is assigned to AFSPACECOM.

19 July
The "don't ask, don't tell" policy, which lifts the direct ban on homosexuality in the military, is announced by the Department of Defense. Homosexual conduct is still forbidden by the policy.

6 August
Dr. Sheila E. Widnall becomes the first woman to serve as an armed service secretary when she takes the oath of office as Secretary of the Air Force.

18 August
Air Force Space and Missile Center (SMC) personnel provide field test operations support throughout Delta Clipper Experimental (DC-X) testing at the White Sands Missile Range (WSMR), including its first flight on this date. Test support personnel provide assistance with range liasion for WSMR and supporting USAF test assets, collecting firsthand information regarding the benefits and gaining first-hand experience in the test program. The DC-X and the DC-XA developed key technologies for a vertical take-off and landing (VTOL) rocket while reintroducing America's aerospace industry to the vision of routine, reliable "aircraft-like" access to and from space. Among the many achievements of the DC-X program are; the first rocket-powered, computer-piloted vertical takeoff and landing; first demonstratation of a gaseous H_2/O_2 Reaction Control System in flight; first soft rocket landing on an unprepared ground; and the first successful rocket powered in-flight abort with safe landing on the desert floor.

27 September
General James H. "Jimmy" Doolittle, aerospace visionary, combat commander, aeronautical scientist, racing pilot, and daredevil, dies at his home in southern California. He was 96 years old.

1 October
For the first time, the Air Force Reserve activates a B-52 unit—the 93d Bomb Squadron at Barksdale AFB, Louisiana.

3–4 October
Air Force pararescueman Technical Sergeant Tim Wilkerson runs in and out of enemy fire to retrieve five wounded Army Rangers and then treat their injuries during a battle in Mogadishu, Somalia. For

his actions under fire and while wounded, he receives the Air Force Cross.

5–13 October
During Operation RESTORE HOPE II, AMC delivers reinforcements to Somalia. USAF C-5 and C-141 aircraft deliver 18 Abrams tanks, 44 Bradley fighting vehicles, and 1,300 troops from the U.S. to Somalia in nine days.

8 October
The sustained airlift of humanitarian supplies to Bosnia, Operation PROVIDE PROMISE, becomes the longest sustained airlift operation in USAF history, surpassing in duration, but not tonnage or sorties, the Berlin Airlift.

2–13 December
Air Force astronaut Colonel Richard O. Covey is the pilot on the space shuttle *Endeavour* during the first mission flown to repair the orbiting Hubble telescope. The $2 billion Hubble will need one additional repair mission before it will begin to provide scientists with some of the most spectacular images of the universe ever seen.

17 December
Marking the 90th anniversary of the Wright Brothers' first flight, the first B-2 Spirit stealth bomber, *The Spirit of Missouri*, arrives at Whiteman AFB, Missouri, where it is assigned to the 393d Bomb Squadron.

Left: *Lieutenant Bob "Wilbur" Wright, flying an F-16C, downed his third low-flying Bosnian Serb J-1 Jastreb with an AIM-9 Sidewinder air-to-air missile while enforcing a UN-sanctioned no-fly zone during Operation DENY FLIGHT. ("Triple Over the Balkans," by K. Price Randel, USAF Art Collection)*

1994

10 January
Two Air Force rescue HH-60G Pave Hawk helicopter crews rescue six sailors from their damaged tugboat off the coast of Iceland. The crew of Air Force Rescue 206 and the crew of Air Force Rescue 208 receive the Mackay Trophy for extraordinary heroism and self-sacrifice during rescue operations in heavy seas and strong winds.

13 January
The Air Force closes operations in the Netherlands after four decades when the last F-15A leaves Soesterberg Air Base.

17–25 January
Following a powerful earthquake in southern California, AMC C-5s and C-141s airlift trucks, generators, and disaster relief specialists to the region. When completed, more than 150 tons of material and 270 personnel will be delivered.

25 January
Clementine I, an unmanned lunar probe, is launched into space by an Air Force Titan II booster rocket. This marks the first lunar mission since *Apollo 17* left the moon in 1972.

5 February
A mortar attack in Sarajevo's central marketplace leaves 68 dead and 200 injured. Four C-130s and a medical team are sent to evacuate the wounded to hospitals in Germany.

7 February
The first Military Strategic and Tactical Relay Satellite is launched into orbit by a Titan IV/Centaur rocket, also being used for the first time. The satellite is part of a system designed to provide secure, survivable communications during any conflict.

10 February
1st Lieutenant Jeannie Flynn graduates from F-15E RTU and becomes the Air Force's first woman fighter pilot.

28 February
During Operation DENY FLIGHT over Bosnia-Herzegovina, two F-16s from the 86th Fighter Wing shoot down four Serb Jastreb-Galeb attack aircraft. This event marks the first NATO combat operation in history. Lieutenant Bob "Wilbur" Wright shoots down three, two with AIM-9 Sidewinders and one with an AIM-120 "Slammer." His accomplishment is the first "hat trick" since the Korean War. After another F-16 shoots down a fourth Jastreb, the F-16 aerial record reaches 69 victories and zero losses.

Right: *Since Lieutenant Jeannie Flynn graduated from F-15E RTU as the first woman fighter pilot, women have filled many different combat aircrew positions—from A-10 pilot to C-130 loadmaster. In 2006, a female fighter pilot, Captain Nicole Malachowski, will take the Number 3 (right wing) position with the USAF Thunderbirds. (USAF)*

13 March
Two military satellites are launched into orbit atop a Taurus booster rocket. This is the first use of the Taurus booster to launch a payload into orbit.

25 March
A C-5 Galaxy departs Mogadishu carrying the last American military personnel and ending American military involvement in Somalia.

6–12 April
During Operation DISTANT RUNNER, Air Force aircraft help to evacuate and transport Americans and other foreign citizens from Burundi to Nairobi, Kenya, when civil unrest forces them to leave Rwanda.

10 April
Air Force F-16s attack a Bosnian Serb command post after the Bosnian Serbs attack UN personnel. This attack marks the first NATO air-to-ground attack in history.

14 April
Two Army UH-60 Black Hawk helicopters are shot down when misidentified by two Air Force F-15s while patrolling over Northern Iraq. Twenty-six people are killed in the fratricide incident, including 15 Americans.

3 May
The last B-52G is flown to Davis Monthan AFB, Arizona, where it will be stored in the "Boneyard." About 100 remaining B-52Hs will continue to fly well into the twenty-first century.

6 May
1st Lieutenant Leslie DeAnn Crosby becomes the first Air Force Reserve woman fighter pilot when she graduates from Air National Guard F-16 RTU in Tucson, Arizona.

24 June
The F-117A is officially named Nighthawk. Over its operational life, it has been called "Stink Bug," "Wobblin' Goblin," and "Black Jet."

30 June
Forty-six years after the Berlin Airlift, the USAF in Europe (USAFE) leaves Berlin with the inactivation of Detachment 1 of the 435th Airlift Wing.

July
The final production F-15 Eagle is delivered to the USAF.

1 July
Air Combat Command transfers responsibility for ICBM readiness to the Air Force Space Command. This consolidates space operations within AFSC.

The 184th Bomb Group (Kansas ANG) receives the B-1B Lancer bomber. They are the first ANG unit so equipped.

16–22 July
Fragments of the Shoemaker-Levy Comet impact Jupiter. The Hubble space telescope, *Voyager*, and other space cameras are focused on the event—the first time that such a cosmic collision has been viewed.

21 July
A C-141 damaged by small arms fire returns from a PROVIDE PROMISE mission over Bosnia. Operations there are temporarily suspended.

22 July–6 October
Air Force aircraft fly humanitarian relief to Rwandan refugees in Zaire during Operation SUPPORT HOPE.

2 August
During a global-power mission to Kuwait, two B-52s from the 2d Bomb Wing fly for 47 continuous hours, needing five aerial refuelings to complete the record-setting circumnavigation of the earth.

9 September
The space shuttle *Discovery* (STS-64) launches into orbit carrying a cew of six: Commander Richard N. Richards, Pilot L. Blaine Hammond Jr. (USAF), and Mission Specialists Mark C. Lee (USAF), Carl J. Meade (USAF), Susan J. Helms (USAF), and Dr. J. M. Linenger. Mission Specialists Lee and Meade also are scheduled to perform an extravehicular activity (EVA) during the mission.

19 September
During Operation UPHOLD DEMOCRACY, a non-combat operation designed to restore Haiti's democratically elected president and curtail the flow of refugees into the U.S., the Air Force provides the airlift support needed to accomplish the mission, which will continue until 1996.

10 October
As Iraqi troops mass near the Kuwaiti border, USAF aircraft begin to move closer to the Persian Gulf region during Operation VIGILANT WARRIOR. During the month, the USAF triples its forces there.

14–16 October
Two C-17 Globemaster III cargo planes fly their first operational strategic mission. The newest AMC heavy lifter carries supplies from Langley AFB, Virginia, to Saudi Arabia.

26 October
General Ronald R. Fogleman is appointed Chief of Staff, USAF. General Fogleman is the first U.S. Air Force Academy graduate (1963) to assume command of the USAF.

Above: *A B-1B Lancer refuels. In July 1994, B-1s were assigned to the Air National Guard for the first time. (USAF)*

Below: *The versatile C-17 is capable of rapid strategic delivery and can also perform tactical airlift and airdrop missions when required. (USAF)*

31 October–1 November
Two B-1B Lancers fly round-trip from Ellsworth AFB, North Dakota, to a bombing range in Kuwait, and back again. This 25-hour flight marks the first time that the Lancer appears in the Persian Gulf.

21–23 November
During Project SAPPHIRE, two C-5s transport 1,300 pounds of highly enriched uranium to the U.S. from the former Soviet Republic of Kazakhstan, where it will be safeguarded from terrorists and other threats.

22 December
The first of three SR-71 Blackbirds reactivated for NASA service flies at Edwards AFB, California.

1995

5 January
Ben Rich, Lockheed Skunk Works designer and "Father of the F-117," succumbs to cancer at age 69.

17 January
The 17th Airlift Squadron becomes the first operational C-17 Globemaster III squadron in the Air Force. Concurrently, the AMC commander authorizes the C-17 for routine uses.

3 February
Air Force astronaut Lieutenant Colonel Eileen Collins becomes the first woman space shuttle pilot during her mission into orbit aboard *Discovery*.

3–10 February
Supporting UN peacekeeping missions, eight C-141s transport more than 400 Nepalese troops from Katmandu to Haiti for service there.

5 March
Russian weapons inspectors arrive at Malmstrom AFB to monitor the destruction of American Minuteman II missiles. The visit is part of Strategic Arms Reduction Treaty (START) enforcement.

24 March
The last 1960s-era Atlas E booster rocket launches a satellite into polar orbit from Vandenberg AFB, California.

8 April
An Air Force C-130 is hit by small arms fire during a PROVIDE PROMISE mission into Sarajevo. The crew returns without incident to its home base.

Above: *STS-93 Commander, Eileen M. Collins, an Air Force colonel, was the first woman to command a space shuttle mission. (NASA)*

Right: *This SR-71 was one of two retired from USAF service but loaned to NASA for high-speed research testing at Dryden FRC, Edwards, California. (NASA)*

Left: *On May 25, 1995, a composite NATO strike package attacks an ammunition storage complex in Bosnia-Herzegovina in support of Operation DENY FLIGHT. This block 40 F-16C drops two laser-guided bombs using its LANTIRN targeting pod. Four Aviano-based F-16s struck their targets during this raid with pinpoint accuracy marking the first combat use of LGBs by F-16s. ("Through the Eye of the Needle," by K. Price Randel, USAF Art Collection)*

19 April
After a domestic terrorist detonates a powerful car bomb at an Oklahoma City federal building, Air Force aircraft transport firefighters, urban search and rescue teams, investigators, and medical personnel from across the country to the disaster zone. Personnel from Tinker AFB provide support to ongoing operations there.

27 April
The Air Force Space Command declares the GPS constellation complete and fully operational.

8–11 May
After 22 inches of rain falls on sections of Louisiana, the Air National Guard rescues thousands of trapped flood victims over a two-day period.

25–26 May
Air Force aircraft are part of NATO attacks on Serb military bunkers in Bosnia in an attempt to stop artillery attacks against Sarajevo. Air Force F-16Cs drop laser-guided bombs on their targets, the first use of that system by an F-16 in combat.

2–3 June
Two Air Force B-1B Lancer bombers fly around the world in 36 hours, 13 minutes, and 36 seconds. During the mission, the Lancers refuel six times and accomplish live bombing activity in three bombing ranges on three continents in two hemispheres. For this demonstration of global reach and global power, the crew, Lieutenant Colonel Doug Raaberg, Captain Gerald Goodfellow, Captain Kevin Clotfelter, Captain Rick Carver, Captain Chris Stewart, Captain Steve Adams, Captain Kevin Houdek, and Captain Steve Reeves, receive the Mackay Trophy.

2–8 June
After being shot down over Bosnia in his F-16, Captain Scott O'Grady evades capture for six days until he is rescued by Marine Corps helicopters.

27 June–7 July
The space shuttle *Atlantis* docks with the Russian space station *Mir* for the first time.

Opposite, bottom: *A forward observer for the 5th Infantry Regiment consults his GPS locator as he recites his position to pilots during emergency close-air-support training. (US Army)*

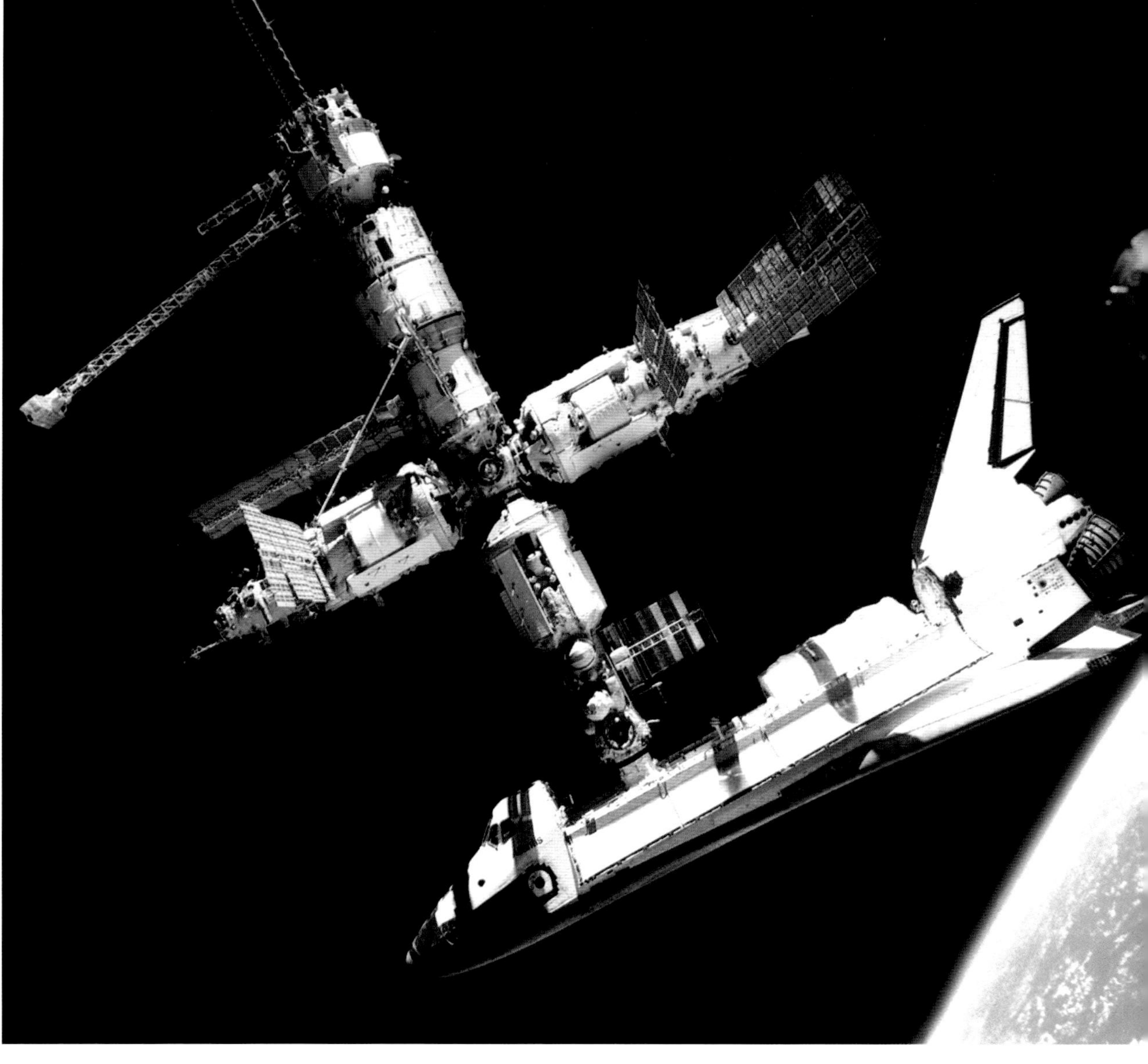

Right: *This view of the space shuttle* Atlantis, *connected to Russia's* Mir *space station, was photographed by the* Mir-19 *crew on July 4, 1995. Cosmonauts temporarily undocked the Soyuz spacecraft from the cluster of* Mir *elements to perform a brief fly-around. (NASA)*

17 August
Flight testing begins on the E-8C joint surveillance target attack radar system (JSTARS), the aircraft that will eventually replace the experimental versions that were deployed to the Persian Gulf during the 1991 war.

25–29 August
Eleven C-17 Globemaster IIIs move 300 tons of troops and equipment to Kuwait, participating in their first major exercise as an operational unit.

30 August
Operation DELIBERATE FORCE, air strikes on Serb positions in Bosnia by NATO forces, begins and USAF aircraft participate in the strike force. The first bombs of the surgical attack are delivered just after 2 a.m.

31 August
Three NATO strike packages attack targets in the Sarajevo area. The majority of targets are Integrated Air Defense Systems (IADS) nodes, ammo depots, and equipment storage and maintenance facilities. A 24-hour suspension of air strikes is requested for the next morning in support of NATO negotiation efforts.

1 September
German assets are tasked to accomplish reconnaissance while all direct attack missions are placed on ground alert.

5 September
Negotiations are unsuccessful and at 10 a.m., NATO air strikes resume with re-attacks of some earlier targets at the top of the list.

6 September
Five strike packages and one re-strike package are launched. Targets are similar to those already attacked with the addition of key bridges and chokepoints as required to meet theater commander objectives. Italian Tornados fly air strike missions for the first time. An Air Force F-16C (block 50) from the 23d FS detects a Serb Sa-6 radar site and uses the HARM Targeting System Pod to launch an AGM-88 missile against the site and destroys it. This is the first combat use of the newest HARM system.

7 September
Six strike and two re-strike packages are flown and missions are added late in the day against six bridges and one choke point.

DELIBERATE FORCE

Rob Owen

Left: *Air Force airlift plays a major role in every modern military or humanitarian operation. Here, a KC-10 refuels a C-141B while a C-5 waits in line for fuel during Operation DELIBERATE FORCE. (USAF)*

The limited duration and scope of DELIBERATE FORCE, the air campaign to bring peace to war-torn Bosnia-Herzegovina, belied its significance as a milestone in the evolution of modern warfare. The campaign ran from August 30 through September 21, 1995, but weather and bombing pauses limited actual bombing to only 12 days during that period. During those days, NATO flew a total of 3,535 attack and support sorties against the Bosnian Serbs and dropped a total of 1,026 bombs. These small numbers reveal the character of DELIBERATE FORCE as the first "post-industrial" war, in which air forces achieved strategic effect through the use of precision weapons, integrated intelligence systems, and targeting aimed at achieving effects rather than mere destruction of enemy assets. DELIBERATE FORCE left a clear message that there were now two kinds of air forces in the world—precision air forces and non-precision air forces.

When the first strike package entered Bosnian airspace in the early hours of August 30, NATO had been in confrontation with the Serbian faction of the Bosnian Civil War for over three years. In the fall of 1992, the United Nations Protective Force asked NATO to enforce its ban on unauthorized flights over the region. When such flights continued, largely by Serb factions, the UN and NATO agreed in April 1993 to activate Operation DENY FLIGHT, which included provisions for shooting down offending aircraft. The UN also declared demilitarized safe areas around six major Bosnian cities and asked NATO to be ready to strike any forces launching attacks against these cities—strikes likely to be made only by the Serbs.

Despite the steady expansion of the DENY FLIGHT mission, the UN and NATO confronted the Bosnian Serbs with excruciating restraint. The memberships of both organizations were divided between governments advocating strong punitive actions against the Serbs and those who doubted that an intervention, based on air power, would do any good in a region where violence was endemic and no warring faction had clean hands. Division led to indecision and, when they were authorized at all, half-hearted pinprick attacks. In April 1994, for example, NATO jets began striking Serb positions near the Gorazde safe area. But when the Serbs took 150 UN

Right: *For the first time in its history, NATO forces participated in combat operations. Here, a German MiG-29 leads two Air Force F-16s on a mission. (USAF)*

peacekeepers hostage, the UN called off further strikes and brokered a cease-fire. Later, when NATO jets attacked the Serb airfield at Udbina, the UN ambassador, Yasushi Akashi, allowed them to crater the runway and attack anti-aircraft weapons actually firing at them, but not to strike any other weapon emplacements, parked aircraft, or buildings for fear of inflicting casualties and, thereby, prompting more UN hostage takings. These events only served to emphasize the indecision of the intervention, while leaving the UN humiliated and NATO air commanders fuming.

The Bosnian Serb Army soon forced the issue. In June of 1995 it took the safe area city of Srebrenica, drove out the UN peacekeepers, and a few days later murdered more than 6,000 Bosnian men in cold blood. Despite the warnings of a conference of NATO ministers in July, the Serbs continued to advance and, on August 28, dropped a mortar shell into a Sarajevo market place that killed 37 people and forced both the UN and NATO military commanders to authorize air strikes. Following a pause to allow UN peacekeepers to concentrate in defensible positions, the attack began.

And they hit hard. On the first night, NATO aircraft struck enemy air defenses—radar sites, missile launchers, and radio relay towers—throughout southern Bosnia, and they hit Serb gun positions around Sarajevo. Strikes continued for three days, supported by hundreds of tanker, electronic warfare, surveillance, and other sorties. For the first time, medium-range Predator unmanned aerial vehicles (UAVs) flew daily reconnaissance sorties that provided extremely valuable imagery of Serb positions and movements. Active bombing ceased for the first four days of September while the UN and a group of American, British, French, Russian, and German ambassadors tried to negotiate with the Serbs. When the Serbs balked, NATO renewed the bombing.

What ensued was a whirlwind of destruction and bewilderment for the Bosnian Serbs. Strike sorties went in at the rate of about 70 per day, supported by around 200 other sorties. NATO smart bombs picked off target after target—bridges, radar sites, barracks, and munitions dumps—with profound accuracy and in rapid order, while causing fewer than 30 civilian deaths. By September 14, all but a handful of the 357 targets originally on NATO's target list were damaged or destroyed. The pace of destructiveness was so rapid that NATO commanders feared that their Serb counterparts no longer had the ability to know what was happening to their forces.

Precision Guided Munitions (PGMs) enabled all of these characteristics of DELIBERATE FORCE—rapid rate of destruction, minuscule collateral damage rates, and micro-management of individual strikes. PGMs had been around since World War II, and their use during the first Gulf War (1990–1991) had demonstrated their potential to revolutionize warfare. During DELIBERATE FORCE, that potential became reality.

Left: *AF fighters carried high-speed anti-radiation missiles (HARM) during Operation DELIBERATE FORCE. This missile homes in on energy emitted from enemy tracking radar, and destroys it. (USAF)*

Below: *Air Force OV-10 Broncos provided Forward Air Control (FAC) during Operation DELIBERATE FORCE. (USAF)*

8 September
Planning for attacks on integrated air defense (IADS) targets in northwest Bosnia-Herzegovina are refined and finalized. Final planning is accomplished for the use of stand-off weapons, allowing targets in well-defended areas to be attacked from outside the range of enemy air defenses. Four strike packages are tasked against 15 re-strike targets and 19 CAS aircraft are re-tasked against eight fixed targets. Reconnaissance assets continue to gather battle damage assessment data.

9 September
Five strike packages are planned but two are aborted because of poor weather. Three are delayed but successfully executed later in the day. Stand-off weapons including High Speed Anti-Radiation Missiles (HARM) and GBU-15, 2,000-pound glide bombs are employed against IADS targets in well-defended areas.

10 September
Tomahawk Land Attack Missiles (TLAM), HARM, and other stand-off weapons including the Stand-Off Land Attack Missile (SLAM) are employed during attacks on key IADS nodes. At 2:25 p.m., UN requests CAS support following BSA shelling of UN positions near the Tuzla airport. Three flights of fighters support the CAS request. Two command bunkers and an artillery position are identified, targeted, and successfully engaged.

11 September
Four strike packages planned against 10 targets are launched, taking advantage of favorable weather conditions. Additional attacks using stand-off weapons are conducted in northwest Bosnia-Herzegovina. Additional reconnaissance missions are tasked to support ongoing efforts to refine target damage assessments and develop re-strike requirements.

12 September
Ammo storage depots in the Doboj area northwest of Tuzla are attacked. These targets were approved on September 11, following BSA shelling of the Tuzla airport two days before. Strike packages are tasked with re-strike missions and placed on alert. They are launched after validated re-strike targets can be assigned and strike plans coordinated and briefed.

14 September
Offensive operations are suspended at 10:00 p.m. in response to a Force Commander, UN Protection Force letter to Commander in Chief SOUTH; representatives of the warring factions had agreed to the UN-brokered Framework Agreement. The initial suspension is scheduled to last 72 hours. Compliance with initial conditions would result in an additional 72-hour suspension after which UN/NATO would review progress toward full compliance with the agreement.

20 September
UN and NATO leadership agree that DELIBERATE FORCE objectives have been met, the military mission accomplished, and end state objectives achieved. The operation comes to a close.

15–21 September
After Hurricane Marilyn strikes the eastern Caribbean Sea, Air Force airlift—Regular, Guard, and Reserve—deliver relief supplies to the region. During this mission, the C-17 is used for humanitarian relief efforts for the first time.

1 October
Chief Master Sergeant Carol Smits assumes the duties of senior enlisted advisor of the USAF Reserve. She is the first woman to hold this position.

28 Otober–18 December
Operation VIGILANT SENTINEL, the first test of the expeditionary force concept, is held when F-16s of the 20th and 357th Fighter Wings deploy to Bahrain.

1 November
Peace negotiations between the presidents of Bosnia, Croatia, and Serbia begin in an effort to end fighting in the former Yugoslavia. The talks are held at Wright-Patterson AFB, Ohio.

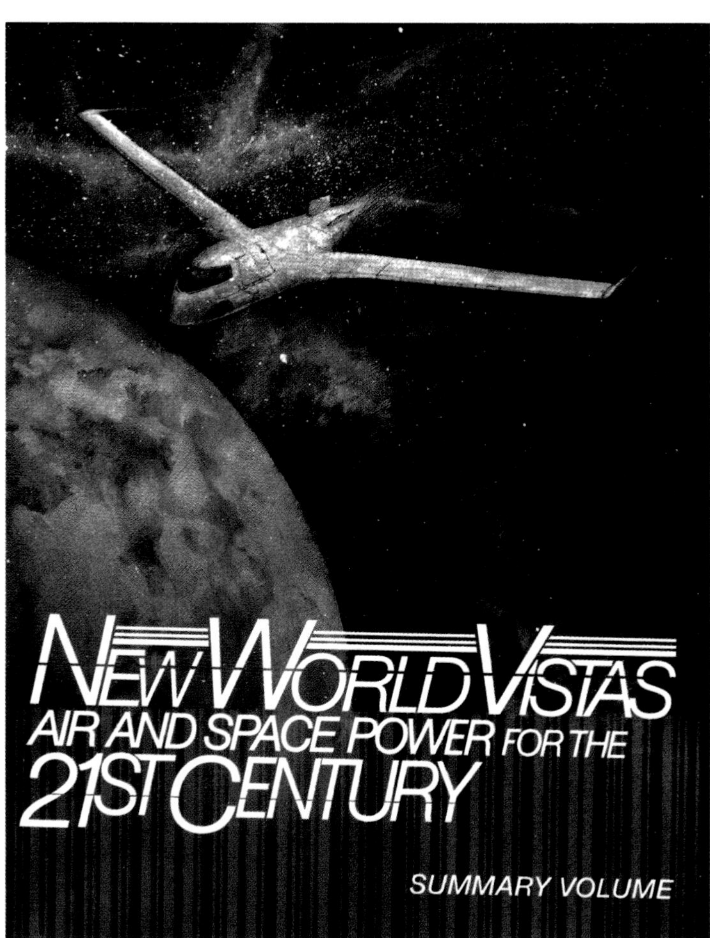

December
Dr. Gene McCall, head of the Air Force Scientific Advisory Board (SAB), releases *New World Vistas*, an air and space science and technology forecast that was ordered by the Secretary of the Air Force, Dr. Sheila Widnall. Air Force Chief of Staff General Ronald R. Fogleman.

6 December
In support of NATO peacekeeping operations, AMC aircraft begin transporting American troops and equipment to Bosnia to implement the peace agreement arranged at Wright-Patterson AFB. The document will be signed in Paris later in the month.

Top: *An A-10 crew chief tightens a few fasteners before the airplane is ready for its next mission over Bosnia. (USAF)*

Center: *Secretary of the Air Force, Dr. Sheila Widnall, and General Ronald R. Fogleman, Air Force Chief of Staff, watch as SAB Chair Dr. Gene McCall delivers* New World Vistas, *a 13-volume science and technology study. (USAF)*

Left: *The* New World Vistas *Summary Volume was the first of 13 volumes written to look at possible futures for USAF technology and operations. (USAF)*

Left: Endeavour *touches down after a successful 10-day mission. On board the shuttle is mission commander John H. Casper (USAF), pilot Curtis L. Brown (USAF), and four mission specialists. ("Endeavour Touchdown," by Paul Kreutziger, USAF Art Collection)*

Below: *The Lockheed Martin/Boeing Tier III-(minus) unpiloted aerial vehicle is inspected by NASA September 14, 1995, following its arrival at the Dryden Flight Research Center, Edwards, California. The Dark Star prototype made its first flight on March 29, 1996. (NASA)*

1996

4 January
Operation PROVIDE PROMISE, the transport of humanitarian aid to Bosnia-Herzegovina, comes to a close. It is the longest sustained relief operation in history. In support of the operation, the USAF has flown more than 4,500 sorties of the nearly 13,000 accomplished by NATO Allies and delivered more than 62,000 metric tons of cargo. The final relief missions into the region are flown on the 9th.

19–29 March
The STS-77 crew flies a 10-day mission aboard space shuttle *Endeavour*. The crew performs a record number of rendezvous sequences (one with a SPARTAN satellite and three with a deployed Satellite Test Unit) and approximately 21 hours of formation flying in close proximity of the satellites. Air Force Colonel John R. Casper brings *Endeavour* back to earth at Cape Canaveral after 160 earth orbits and 4.1 million miles. The mission duration is 240 hours and 39 minutes.

29 March
The Dark Star (Tier Three Minus) unmanned aerial vehicle (UAV) flies for the first time at Edwards AFB, California. Dark Star is a jet-powered, high-altitude, long-loitering reconnaissance platform that incorporates stealth technology. Only five will be built and the program will eventually be cancelled in favor of other UAV systems.

3 April
An Air Force CT-43 transport plane crashes near Dubrovnik, Croatia. Thirty-five people are killed including Commerce Secretary Ronald Brown, who was on the way to an economic development meeting in the former Yugoslavia.

9–25 April
During Operation ASSURED RESPONSE, the evacuation of noncombatants from Liberia, Air Force AC-130s, MC-130s, C-130s, and MH-53J Pave Low helicopters, supported by KC-135 Stratotankers, fly 94 missions.

Right: *Boeing C-17 Globemaster III aircraft eventually replaced the C-141 fleet, the last of which retired in 2006. (USAF)*

17 April
Operation UPHOLD DEMOCRACY, the effort to support a democratic government in Haiti, comes to a close after 18 months.

18 April
Two Air Force C-17 Globemaster III transport aircraft airlift two MH-53J Pave Low helicopters from duty in Sierra Leone to a base in England. This is the first time such a massive payload is transported in this fashion, saving deployment time and fuel costs.

30 April
Tacit Blue, the test vehicle for early stealth technology, is made public for the first time.

1 May
For the first time, a foreign military officer assumes command of a unit on an Air Force base in the United States when a German pilot takes command of the German Tactical Training Center at Holloman AFB, New Mexico.

31 May
The Air Force orders 80 additional Boeing C-17 Globemaster III aircraft to be built over the next seven years. The C-17 will gradually replace the C-141 in the Air Force inventory. The $16.2 billion order is the largest military contract ever awarded.

11 June
Air Combat Command accepts its first production E-8 JSTARS aircraft delivered for evaluation in March. Test models flew during DESERT STORM and JOINT ENDEAVOR.

21 June
Commander David Cheslak (USN) becomes the first Naval flight officer to command an Air Force squadron. He assumes command of the 562d Flying Training Squadron at Randolph AFB, Texas. The unit began operations in April and is responsible for training all navigators for the Air Force and the Navy.

25 June
The Khobar Towers residential building near King Abdul-Aziz Air Base in Saudi Arabia is bombed by terrorists. Nineteen Air Force personnel are killed and another 300 Americans are wounded. The Americans there are deployed in support of Operation SOUTHERN WATCH.

1 July
An Air Force EA-6B crew makes a carrier takeoff from the USS *Constellation,* deployed to the Pacific. This is the first time that an Air Force crew flies a Prowler off the deck of a carrier.

Left: *The E-3 Sentry (AWACS) is essential to successful operations around the globe. The E-3 Sentry has served in all major Air Force operations since DESERT STORM. (USAF)*

27 July
The General Dynamics F-111 is retired from active Air Force service.

2–3 September
During Operation DESERT STRIKE, a response to Iraqi seizure of the city of Irbil, Air Force B-52s staging out of Guam from Barksdale AFB, launch 13 cruise missiles against Iraqi military targets. This is part of a coordinated cruise missile attack launched against the Iraqi air defense infrastructure, including surface-to-air missile sites and command and control nodes in southern Iraq. During this mission, the B-52 and Conventional Air Launched Cruise Missile (CALCM) weapon systems demonstrate their capability for rapid en route retargeting, providing the joint force with additional target coverage and strike flexibility. Two bombers and 29 tankers made this Global Engagement mission possible. The next day, a second strike of 17 Tomahawk sea-launched cruise missiles is executed and the operation comes to a close. For performing the first combat employment of the B-52H in history, Duke 01 flight is awarded the Mackay Trophy.

3 September
The first Air Force unit dedicated to operating an unmanned aerial vehicle (UAV), the 11th Reconnaissance Squadron, begins operating the RQ-1 Predator over Bosnia-Herzegovina to help ensure the terms of the peace agreement are upheld. The unit was activated in July 1995.

14 September
Air Force Security Forces assist in providing security during the first free elections in Bosnia since the end of their civil war.

15 September
During Operation PACIFIC HAVEN, the Air Force transports thousands of northern Iraqi refugees to Anderson AFB, Guam, where they are processed for settlement in the U.S.

30 September
Seventeenth Air Force inactivates after more than four decades of service in Europe.

21 November
Dr. Sheila Widnall, Secretary of the Air Force, and General Ronald R. Fogleman, CSAF, release *Global Reach, Global Power*, a position paper describing concepts for Air Force power projection in the twenty-first century.

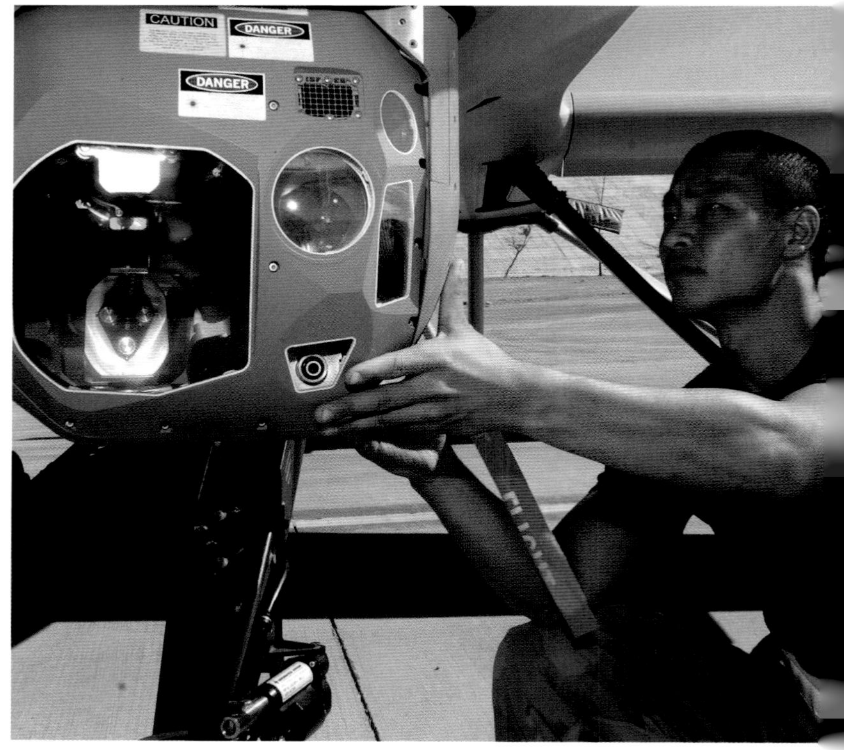

Right: *The F-22 Raptor will assure air dominance well into the twenty-first century.*

Opposite, bottom: *An RQ-1 Predator is prepared for another combat mission. The sensors provide real-time intelligence to combat commanders. (USAF)*

1997

1 January
Operation NORTHERN WATCH supplants Operation PROVIDE COMFORT designed to enforce the no-fly zone in Iraq above the 36th north latitude line.

6 January
Abdullah Hamza Al-Mubarek is commissioned by the Air Force and selected as its first Muslim chaplain candidate.

31 January
The 31st Air Expeditionary Wing (AEW) is activated. It is the first AEW in the Air Force and is available for rotational worldwide deployment.

17 February
The Air Force Reserve becomes a major command. This elevated status reflects the increasing importance of the Reserves in America's global military commitments.

18 February–3 March
During Operation ASSURED LIFT, the movement of African peacekeeping forces to Liberia, five C-130s airlift 1,160 peacekeepers and more than 450 tons of cargo from several African nations to Liberia.

17 March
During Operation GUARDIAN RETRIEVAL, a joint task force evacuation of individuals threatened by civil unrest in Zaire, Special Operations Command and Air Mobility Command provide assets to achieve the mission. During the next two weeks, a combined 57 missions will evacuate 532 passengers.

21 March
Air Force Lieutenant Colonel Marcelyn A. Atwood assumes command of a Navy training squadron at Pensacola, Florida. Atwood is the first Air Force officer to command a Navy squadron.

1 April
At Whiteman AFB, Missouri, the first six B-2 Spirit bombers achieve operational status with the 509th Bomb Wing.

All stateside Air Combat Command C-130 Hercules aircraft are assigned to Air Mobility Command.

9 April
The first production model Lockheed-Martin-Boeing F-22 Raptor rolls out of the factory during a well-attended ceremony. About 3,000 attend the event in Marietta, Georgia, including Speaker of the House Newt Gingrich, SECAF Sheila Widnall, and CSAF Ronald Fogleman. The first flight will not occur until September 7.

Left: *The MC-130E Combat Talon I and MC-130H Combat Talon II provide infiltration, exfiltration, and resupply of special operations forces in hostile territory. Secondary missions include psychological operations and helicopter air refueling. Both aircraft feature terrain-following and terrain-avoidance radars capable of operations as low as 250 feet in adverse weather. (USAF)*

10 June
An Air Force Special Operations Command MC-130H Combat Talon II crew rescues 56 people from destruction and civil war in the Republic of the Congo. The crew, assigned to the 352d Special Operations Group at RAF Mildenhall, England, also delivered 12 Army and Navy Special Forces personnel to survey the situation in the capital city of Brazzaville. The Talon II mission took more than 21 hours and included three in-flight refuelings to retrieve the 30 Americans, 26 foreign nationals, and a young boy's dog. For overcoming hostile gunfire, three heavyweight air refuelings, and over 13 hours flying 3,179 nautical miles to their objective to insert a European survey and assessment team and extract 56 people from Brazzaville, Republic of Congo, and achieving this goal while on the ground for less than 23 minutes, the crew of *Whick* 05, under the command of Lieutenant Colonel Frank J. Kisner, is awarded the Mackay Trophy.

1 August
McDonnell Douglas and Boeing merge and begin operating as one company with more than 220,000 employees.

1 September
General Ronald R. Fogleman retires one year earlier than expected. A collection of politically charged events are at the root of his decision to step aside. General Ralph "Ed" Eberhart becomes CSAF (acting) until a replacement is appointed.

6 October
General Michael E. Ryan becomes Chief of Staff, USAF.

1998

11 February
A Joint Direct Attack Munition (JDAM) is dropped for the first time in a test at China Lake, California. A B-1B Lancer makes the drop.

23 February
B-2 Spirit bombers from Whiteman AFB, Missouri, deploy to an overseas location for the first time when they travel to Anderson AFB, Guam.

28 February
The Teledyne Ryan Aeronautical Company's RQ-4 Global Hawk unmanned aerial vehicle flies for the first time at Edwards AFB, California. The jet-powered UAV is designed to conduct reconnaissance operations at 65,000 feet for long periods at a time.

27 May
Air National Guard HH-60 Pave Hawks from the 210th Rescue Squadron rescue six survivors trapped inside an airplane that had crashed on a glacier on Mount Torbert, Alaska. The survivors, including two small children, were in grave danger because of the extreme weather and temperature. The crew makes a limited approach to land at the 10,500-foot level on the side of Mount Torbert with extremely unpredictable winds. For their daring and intrepidity, the crew receive the Mackay Trophy.

12 June
The space shuttle *Discovery* completes the ninth and final shuttle flight to the Russian space station *Mir*. The shuttle-*Mir* missions began in 1995.

Right: *The B-1B was first used in combat in support of operations against Iraq during Operation DESERT FOX in December 1998. (USAF)*

Below: *STS-95 mission commander Curtis L. Brown (USAF) and payload specialist John Glenn (right) are photographed on the aft flight deck of* Discovery *during a press conference from space. Glenn became the oldest human in space during this shuttle mission. (NASA)*

20 August
After attacks upon the U.S. embassies in Kenya and Tanzania on August 7, U.S. Naval forces launch 75 cruise missiles against targets in the Sudan and Afghanistan in retaliation.

22 September
Air Force airlifters begin delivering hurricane relief supplies to Puerto Rico, the Dominican Republic, and Mississippi after Hurricane George leaves behind a path of destruction.

29 October
John H. Glenn Jr., former senator from Ohio and the first American to orbit the earth, returns to space in the space shuttle *Discovery*. He will participate in experiments concerning microgravity and the elderly. Glenn, who is 76 years old, becomes the oldest person in space.

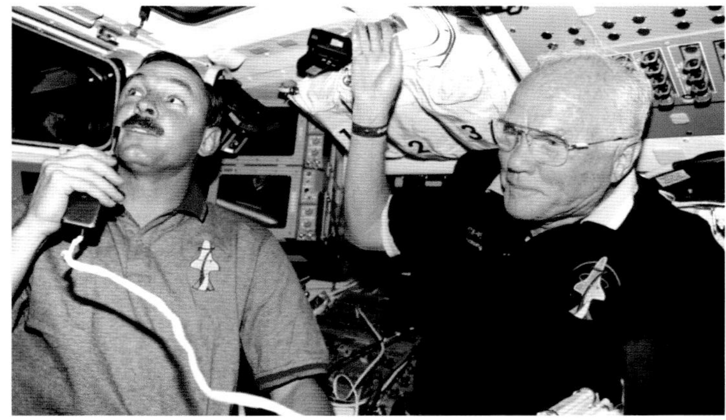

6 November
After Hurricane Mitch sweeps through Honduras, Nicaragua, Guatemala, and El Salvador, leaving 10,000 dead in its wake, Air Force transports begin to deliver more than 3,500 tons of relief materiel during more than 200 missions. The missions will continue until March 1999.

4–15 December
NASA launches the space shuttle *Endeavour* on a mission that begins the assembly of the International Space Station (ISS). *Endeavour* astronauts join the *Unity* module to the *Zarya* module during the 11-day flight.

9 December
President Bill Clinton presents an honorary fourth star to Benjamin O. Davis Jr., the first African American Air Force general officer and the leader of the "Tuskegee Airmen" during World War II.

16–20 December
Operation DESERT FOX, a punitive strike against Iraqi targets after their refusal to allow weapons inspectors access to specific sites, begins. During the four-day operation, the largest in the region since 1991, approximately 100 enemy targets are attacked and weapons production facilities are destroyed. B-1B Lancers are used in combat during this operation for the first time.

GLOBAL REACH, GLOBAL POWER: TRANSFORMATIONAL DOCTRINE FOR THE 1990S

Richard P. Hallion

Left: *The importance of aerial refueling capability cannot be over emphasized. Without it, there is no sustained global reach. Here, a stealthy B-2 bomber approaches the contact position during refueling operations. (USAF)*

The last years of the Cold War spawned a renaissance of military thought. The Air Force's *Global Reach, Global Power* strategic planning framework, undertaken by Air Force Secretary Donald B. Rice, was key to the success American arms enjoyed in the Gulf and afterwards.

The origins of *Global Reach, Global Power* stem from the inspiration of Colonel John A. Warden III, who had written a highly influential National War College thesis, *The Air Campaign: Planning for Combat.* This work emphasized thinking of an enemy in terms of its vital centers, seeking to achieve nodal effects, not merely generating large numbers of sorties against lists of targets. As such, it was the first comprehensive argument for what would eventually become the normative approach to warfare the Department of Defense would pursue in the post-Cold War era. Warden's thinking, particularly the notion that "territory is a dangerous enchantress in war... [it] may well be the political objective of a campaign, but it rarely should be the military objective," influenced a number of creative Air Staff officers, particularly then-Lieutenant Colonel David A. Deptula. Deptula, on the staff of the Secretary's Staff Group (SAF/OSX), subsequently became one of the two principle authors (the second being The RAND Corporation's Dr. Christopher Bowie, on detail to the Secretary's Staff Group) of *Global Reach, Global Power*.

Global Reach, Global Power was written within SAF/OSX over the spring of 1990, at a time when the Air Force was in the midst of a great internal debate over its future following the end of the Cold War. Secretary Rice, a former president of the RAND Corporation, had called for a survey of the Air Force senior leadership and found that no clear consensus existed on what the Air Force brought to war-fighting, or even why the United States needed an independent Air Force in the post-Cold War era. The service had lived with organizations such as the Tactical Air Command and Strategic Air Command for so long that thinking about use of aircraft had lost all flexibility. Bombers were "strategic" and fighters were "tactical." The notion that aircraft were weapons platforms and could accomplish either strategic or tactical missions depending on what weapon was dropped upon what target was generally unrecognized. Troubled by this widespread lack of unity and understanding, Rice actively sought to generate a strategic planning framework that would enunciate what the Air Force's air and space power brought to the Department of Defense. Deptula and Bowie, working with other officers throughout the Air Staff, prepared point papers and drafts that were circulated and reviewed at the highest levels of the service.

Global Reach, Global Power emphasized the transformational nature of modern aerospace power, particularly its ability to reach rapidly across global distances to intervene in crisis regions, either for humanitarian or combat purposes. Deptula and Bowie identified five key characteristics inherent in aerospace power: speed, range, flexibility, precision, and lethality. Aerospace power could sustain nuclear deterrence; provide versatile combat forces for power projection and combat operations; supply rapid global mobility via airlifters supported by a "tanker bridge"; ensure control of the high ground via surveillance, communications, and navigations systems, thereby furnishing global knowledge and situational awareness; and, finally, build U.S. influence via airlift, crisis response, shows of force, and similar operations.

Global Reach, Global Power was issued as an Air Force White Paper in June 1990 and immediately generated controversy for its bold advocacy of aerospace. Indeed, it was the first time that the United States Air Force had issued a senior-level enunciation of how the service contributed to national security, reflecting Rice's unusually strong role in shaping the service's future—arguably the most direct intervention of an Air Force Secretary in the history of the service. When the United States went to war in the Gulf, *Global Reach, Global Power* became fact, not simply assertion. The reorganization and merger of commands that followed the Gulf War reflected the core thinking of *Global Reach, Global Power*, as has every subsequent doctrinal and organizational restructuring of the service since its release in 1990.

Right: *The U-2 provides continuous day or night, hight-altitude, all-weather, stand-off surveillance in direct support of U.S. and allied ground and air forces. It provides critical intelligence to decision makers through all phases of conflict. The U-2 is capable of collecting multi-sensor photo, electro-optic, infrared and radar imagery, as well as performing other types of reconnaissance functions. (USAF)*

Left: *The C-130J incorporates state-of-the-art technology to reduce manpower requirements, lower operating and support costs, and provide life-cycle cost savings over earlier models. The C-130J-30 is a stretch version, adding 15 feet to the fuselage. (USAF)*

1999

24 January
An AGM-154A joint stand-off weapon (JSOW) is launched for the first time in combat by a Navy F-18 against an Iraqi radar site.

17 February
The first Lockheed C-130J, a new transport modified with six-bladed propellers, arrives at Keesler AFB, Mississippi, and is assigned to the 403d Wing.

24 March
Operation ALLIED FORCE, a NATO air campaign to cripple the Serbian war machine in Kosovo, begins. The Air Force part of the operation—NOBLE ANVIL—includes air superiority sorties and the first use of the B-2 bomber in combat. On this day, two Air Force F-15C aircraft attack enemy MiG-29s and Captain Jeffrey G. J. Hwang shoots down two during one engagement. In recognition of an exceptionally meritorious F-15C flight during combat operations in support of Operation ALLIED FORCE using AIM-120 AMRAAM air-to-air missiles, Hwang receives the Mackay Trophy.

26 March
An Air Force F-16 pilot shoots down two MiG-29s in combat over Yugoslavia.

27 March
An F-117 Nighthawk stealth fighter is lost due to ground fire near Belgrade. This is the first combat casualty during the F-117's history. A-10 pilot, Captain John A. Cherrey, locates the downed pilot who is rescued that same day. Cherrey receives the Silver Star for his actions.

"Melissa," an extremely disruptive computer virus, is held in check by an anti-virus program developed by ACC personnel.

4 April
Air Force C-17s begin delivering relief supplies to Tirana, Albania, to assist Kosovo refugees. Operation SUSTAIN HOPE is the humanitarian counterpart of Operation ALLIED FORCE. Over the next month, more than 3,000 tons of food and shelters will be airlifted to the region.

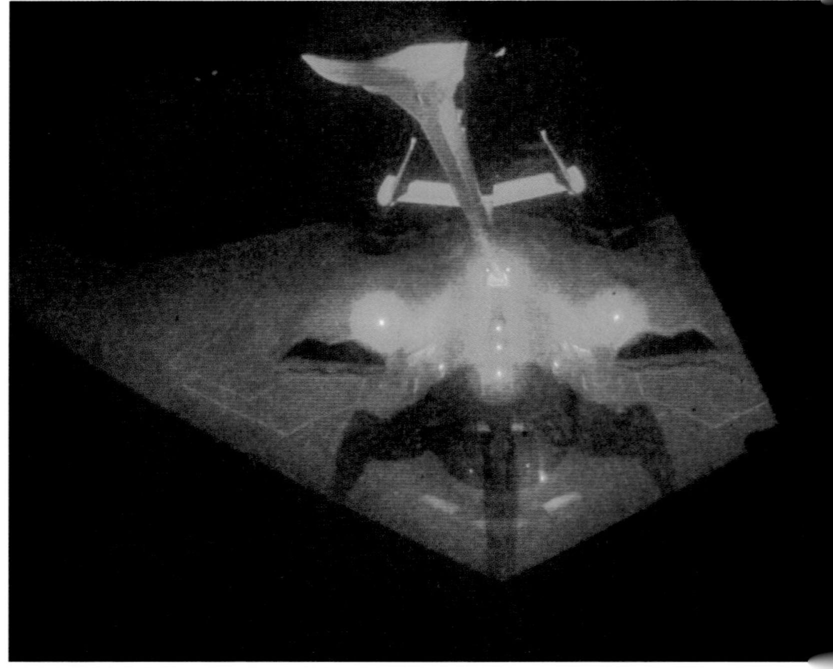

Right: *A C-17 Globemaster III transfers its cargo to a mobile ramp loader during Operation SUSTAIN HOPE. ("Kosovo Here We Come," by Nilo Santiago, USAF Art Collection).*

Opposite, below: *The combat effectiveness of the B-2 was remarkable during Operation ALLIED FORCE. B-2s were responsible for destroying 33 percent of all Serbian targets in the first eight weeks of operations. (USAF)*

17 April
An RQ-1 Predator flies over Serbia during Operation ALLIED FORCE. This marks Predator's first combat sortie in its history.

21 April
Two NATO missiles damage the headquarters of Yugoslavia's ruling Socialist Party.

23 April
NATO bombs the headquarters of Serbian state television. NATO leaders meeting in Washington reject an offer by Slobodan Milosevic that suggests the creation of an "international presence" in Kosovo.

1 May
The first Air Force Reserve Command tanker wing mobilized for duty enters active service. Five more, and one rescue wing, will enter active duty service during the coming months.

Forty-seven bus passengers are killed when NATO bombs a bridge in Kosovo.

2 May
Serbian ground forces shoot down an F-16 over Yugoslavia. The pilot is rescued. This is the second and final loss during the operation for the Air Force.

Three captured U.S. soldiers are released from Kosovo into the custody of U.S. civil rights leader Jesse Jackson.

4 May
The final Air Force victory of ALLIED FORCE occurs when an F-16CJ pilot shoots down a MiG-29 over Kosovo.

6 May
Foreign ministers from the Group of Eight (G8) agree on a framework for a peace plan which calls for the return of all refugees and the deployment of an international security force in Kosovo.

8 May
The Chinese embassy in Belgrade is mistakenly hit by NATO missiles, which kill three people.

23 May
Fighting flares on the border between Serb forces and Albanian police. President Bill Clinton says he no longer rules out "other military options."

26 May
NATO agrees to boost the number of troops in a future Kosovo peacekeeping mission from 28,000 to 45,000.

Left: *An Air Force B-1 Lancer flies past the Egyptian Pyramids during exercise BRIGHT STAR '99. Aside from operational deployments in Southwest Asia, this is CENTCOM's largest deployment exercise to date. (USAF)*

Opposite, bottom: *An Air Force Titan IVB space launch vehicle lifts off carrying a Defense Support Program (DSP) satellite that will add to a constellation of similar satellites the Air Force uses to provide early warning of missile launches worldwide. (USAF)*

27 May
Slobodan Milosevic and four other top officials are indicted for war crimes by the International Criminal Tribunal in The Hague.

30 May
NATO says it wants a clear, personal statement from Milosevic that he accepts alliance conditions before it will halt air raids.

3 June
Talks in Belgrade resume for a second session. A Russian spokesman in Moscow says Yugoslavia viewed the peace plan as a "realistic" way out of the Kosovo crisis.

9 June
NATO and Yugoslav military authorities sign an agreement on the withdrawal of Yugoslav security forces from Kosovo.

10 June
NATO suspends air strikes when the Yugoslavian president agrees to the withdrawal of Serb forces in Kosovo. This is the first war in history won by airpower alone.

23 July
Air Force astronaut Colonel Eileen Collins becomes the first woman commander of a space shuttle mission when *Columbia* launches into orbit.

1 October
Aerospace Expeditionary Force 1 deploys to Southwest Asia. This is the first deployment of 10 such forces that will rotate availability for global deployments. This new system is designed to respond to worldwide contingencies more effectively while also making deployments more predictable. This lack of predictability is cited by Air Force members as one of the causes of low morale.

6 October
The destruction of the first of 150 Minuteman III silos in eastern North Dakota begins in accordance with the Strategic Arms Reduction Treaty with Russia.

10 October–2 November
Exercise Bright Star 99/00 is the most significant coalition military exercise conducted by USCENTCOM to date, the tenth in a series, and will be the largest-scale deployment and employment exercise conducted in the USCENTCOM area of responsibility outside of the Arabian Gulf.

2 November
Howard Air Base is officially turned over to Panama.

20–28 December
Air Force aircraft delivers humanitarian aid to Venezuela in the aftermath of massive flooding that leaves 200,000 homeless.

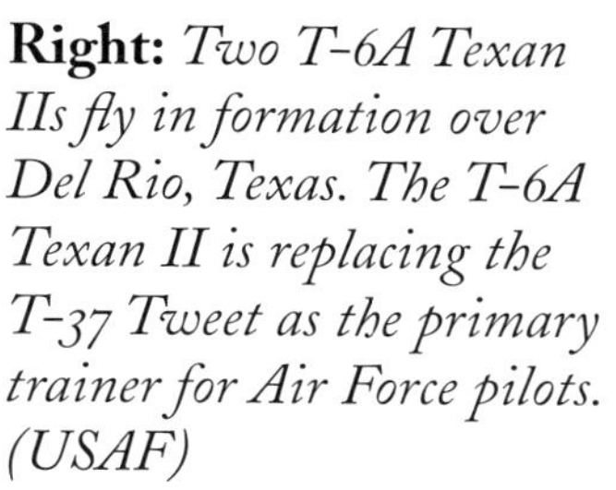

Right: *Two T-6A Texan IIs fly in formation over Del Rio, Texas. The T-6A Texan II is replacing the T-37 Tweet as the primary trainer for Air Force pilots. (USAF)*

2000

2 March
Floods in Mozambique leave an estimated one million people homeless. Operation ATLAS REPONSE begins when Air Force heavy-lifters deliver humanitarian relief from Europe to southern Africa. C-130s and helicopters distribute the supplies to austere locations.

3 May
For the first time in nearly four decades, an Air Force officer is appointed to serve as the supreme allied commander in Europe. General Joseph W. Ralston heads all NATO forces.

8 May
An Air Force Titan IVB launch vehicle lifts a Defense Support Program satellite into orbit from Cape Canaveral, Florida. The DSP satellites are designed to provide early warning of missile launches worldwide. The Titan IVB is the nation's largest and most powerful expendable launch vehicle.

23 May
The first production T-6A Texan II arrives at Randolph AFB, Texas. The Texan will eventually replace both the T-37 and the Navy T-34 for use as the primary pilot training aircraft.

18 September
The first Air Force CV-22 arrives at Edwards AFB, California. The Osprey takes off like a helicopter and flies like an airplane when the propulsion system rotates during flight.

12 October
NASA launches the 100th space shuttle mission. *Discovery* carries several parts for the international space station into orbit and four space walks are required to assemble them during the mission. The first permanent crew will arrive via Russian space craft on November 2.

15 October
After the October 12 terrorist attack on the USS *Cole*, which is docked in Yemen, aircrew from the 75th Airlift Squadron and the 86th Aeromedical Evacuation Squadron transport 28 victims across the ocean to Norfolk, Virginia, for treatment. For these heroic rescue efforts for victims of the USS *Cole* tragedy during the 6,000-mile round-trip journey between Aden, Yemen, Djibouti, Africa and Ramstein, Germany, the aircrew is awarded the Mackay Trophy.

21 October
Lockheed Martin X-35A test piolot Tom Morgenfeld completes medium-speed taxi tests on Runway 7 at Palmdale. Maximum speed during test is 94 knots.

24 October
Tom Morgenfeld verifies the airworthiness of the X-35A during its 30-minute first flight from Palmdale to Edwards AFB. The X-35 climbed to an altitude of 10,000 feet, maintained an airspeed of 250 knot and accomplished a series of figure-eight maneuvers to demonstrate key handling qualities and to validate design predictions.

3 November
The lead government test pilot, Air Force Lieutenant Colonel Paul Smith, flies the X-35A on its fifth flight.

5 November
Lieutenant Colonel Paul Smith flies a 0.8-hour mission in the morning followed by a 0.6-hour mission by Tom Morgenfeld in the afternoon. This marks the first time that the X-35A flies two sorties in one day.

7 November
Lieutenant Colonel Smith completes several aerial refuelings in the X-35A with a KC-135 stratotanker, transferring a total of 14,000 pounds of fuel. This 2.9-hour tenth flight is the longest of the program thus far.

9 November
Tom Morgenfeld approaches the supersonic threshold in the X-35A during the eleventh and twelfth flights, reaching speeds of 0.98 Mach at 25,000 feet.

10 November
USMC test pilot Major Art "Turbo" Tomassetti gets his turn in the X-35A cockpit. The test flight coincides with the 225th birthday of the Marine Corps.

18 November
Hot pit refueling (ground refueling without shutting down the engine) allows the flight test team to complete four flights in one day. The first two of four are also the first X-35A flights for BAE test pilot Simon Hargreaves. The last two of the day are the first X-35A flights for Royal Air Force test pilot Squadron Leader Justin Paines.

21 November
Tom Morgenfeld completes several field carrier landing practice approaches (FCLPs), air-to-air tracking, and formation maneuvering and reaches an altitude of 34,000 feet on a 2.5-hour flight before taking the X-35A to supersonic speeds for the first time.

22 November
Lieutenant Colonel Paul Smith completes aerial refueling qualification with a KC-10 in a 2.5-hour flight. Later in the day, Morgenfeld flies the aircraft from Edwards back to Palmdale after collecting performance data and completing serveral touch and goes. The X-35A acculumlates 27.4 hours in 27 flights, achieves Mach 1.05, and reaches a maximum altitude of 34,000 feet during the aircraft's final flight as the X-35A. The aircraft heads back into the hangar to its transformation into the X-35B STOVL version.

Opposite, top: *The Lockheed Martin X-35A conventional takeoff and landing version of the Joint Strike Fighter (JSF) will eventually replace aging F-16s and A-10s as the fighter for the twenty-first century. (USAF)*

Opposite, bottom: *The X-35A refuels for the first time with a KC-135 Stratotanker over Edwards AFB. (USAF)*

Right: *Chief Test Pilot Simon Hargreaves hover-tests the X-35B short takeoff and vertical landing (STOVL) variant. (USAF)*

2001

11 January
The X-35B lift fan installation, completed in less than three hours, includes the fan, a vectored nozzle, clutch, and all actuation and service systems in an integrated unit.

23 January
Lieutenant Colonel Paul Smith flies the X-35C (U.S. Navy version) for the first time on an aerial refueling qualification mission. Tom Morgenfeld flies the X-35C on the second and third flights of the day. X-35C reaches 20 total flights.

3 February
In an operation to deliver relief supplies to earthquake victims in India, AMC C-5s deliver cargo to Guam where four C-17s pick up the load and deliver it to the devastated regions. KC-135s refuel the Globemasters in the journey across the Pacific and Indian Oceans. More than 30,000 are suspected dead and hundreds of thousands are homeless.

21 February
A modified RQ-1 Predator fires a Hellfire missile on the range at Nellis AFB, Nevada. The Predator, tail number 3034, is the first unmanned vehicle to fire an offensive missile.

24 February
Lieutenant Colonel Stayce D. Harris becomes the first African American woman to command an Air Force flying squadron when she takes charge of the 729th Airlift Squadron at March AFB, California.

16 March
The hover pit tests of the X-35B are conducted with a special landing gear that allows load cells to measure STOVL lift forces and moments directly, while keeping the airplane from lifting into hover at higher power settings. Over a two-week period, the team conducts more than 100 tests with all control functions commanded by the pilot.

1 April
After a Chinese fighter plane collides with a Navy EP-3 reconnaissance aircraft over international waters, the pilot is forced to make an emergency landing on Hainan Island, China. After being held for 11 days, the crew of 24 (which includes one Air Force member) is released.

23 April
The RQ-4A Global hawk completes the first nonstop crossing of the Pacific Ocean by an unmanned aerial vehicle (UAV). The trip originates at Edwards AFB, California, and terminates in Edinburgh, Australia. The 7,500-mile flight takes 23 hours.

Left: *To demonstrate interoperability between U.S. and Australian military systems, Global Hawk flew 7,500 miles nonstop across the Pacific to Australia on April 22–23, 2001, setting new world records. (USAF)*

Below *The Pentagon after the September 11, 2001 attack. (USAF)*

8 May
The Secretary of Defense designates the USAF as the executive agent for the Pentagon's space activities.

5 June
The commander of the Air Force Reserve, James E. Sherrard, III, is promoted to three-star rank. This reflects the growing importance of the contribution of Lieutenant General Sherrard's reserve forces to the Air Force "total force" concept.

24 June
Chief test pilot Simon Hargreaves completes a vertical "press up" to 20 feet above the ground and maintains a controlled hover for 30 seconds before completing a soft vertical landing. This marks the X-35Bs first vertical takeoff to a sustained altitude.

27 June
Simon Hargreaves hovers for two minutes on two separate flights during the day. Hot pit refueling is used between flights.

3 July
Simon Hargreaves performs a functional check flight and takes the X-35B from Palmdale to Edwards AFB in a 30-minute flight, the twelfth mission for the STOVL demonstrator. The doors of the lift fan bay remain open during the flight.

20 July
Major Art "Turbo" Tomassetti completes short takeoff, takes the X-35B to Mach 1.05, and lands vertically in the same flight in his second flight of the day—Mission X.

6 August
Tom Morgenfeld flies the X-35B from Edwards to Palmdale, bringing the 10-month JSF X-35 flight test program to a close. The 3.7-hour flight, which includes a maximum power climb to 34,000 feet and accelerations to Mach 1.2, is the longest flight of the X-35 flight test program. The X-35B accumulates 21.5 hours of flying time, reaches a speed of Mach 1.2, lands vertically 17 times, and completes 14 short takeoffs and six short landings in 39 flights.

24 August
The last Minuteman III missile silo is destroyed at Minot AFB, North Dakota, in accordance with the terms of the Strategic Arms Reduction Treaty.

6 September
General John P. Jumper becomes the Chief of Staff, USAF. General Jumper is a distinguished graduate (1966) of the Virginia Military Institute, Lexington, Virginia.

11 September
Terrorists with ties to the Middle East hijack four U.S. airliners in flight, crashing two of them into the World Trade Center complex in New York City. A third is flown into the Pentagon in Washington, D.C. The fourth aircraft crashes in rural Pennsylvania after heroic passengers, aware of the fate of the other planes, attempt to regain control of their plane. The FAA grounds civilian air traffic for several days, a crushing economic blow to the airline industry, and military flights over key U.S. cities are initiated. Operation NOBLE EAGLE, which includes a wide variety of new domestic security measures, will continue indefinitely.

14 September
President George W. Bush authorizes the call-up of 50,000 reservists to active duty to fight the war against terrorism.

19 September
A contract is issued for the initial production of 10 Lockheed Martin F-22 Raptor fighter planes.

20 September
President Bush establishes the Office of Homeland Security with Pennsylvania governor Thomas Ridge as its head.

29 September
The U.S. launches satellites into orbit from the Kodiak Launch Complex in Alaska. This is the first time that an American spacecraft has been launched from a location other than California or Florida.

1 October
Air Force Space Command assumes control of the AMC Space and Missile System Center. All space systems within the USAF are now under AFSC control.

General Richard B. Myers is appointed the Chairman of the Joint Chiefs of Staff. It has been nearly two decades since an Air Force officer has held that position.

Above: *Two Air-Defense Fighter (ADF) F-16A Fighting Falcons from the North Dakota Air National Guard's 178th Fighter Squadron lead an F-15C Eagle from the 27th Fighter Squadron at Langley Air Force Base, Virginia, in formation during a combat air patrol mission in support of Operation NOBLE EAGLE. More than 11,000 airmen—most of them Air National Guard and Air Force Reserve—continue to protect America's skies. The 27th began flying NOBLE EAGLE sorties in the F-22 Raptor in January 2006. (USAF)*

Left: *Air Force General Richard B. Myers, Chairman of the Joint Chiefs of Staff. (USAF)*

THE AIR FORCE IN AFGHANISTAN

Rebecca Grant

Left: *Crew chiefs with the 40th Expeditionary Maintenance Squadron prepare to launch a B-52 Stratofortress at a forward-deployed location during Operation ENDURING FREEDOM. They are deployed from Minot Air Force Base, North Dakota. (USAF)*

Opposite: *A deployed weapons loader straps a JDAM to a cradle for transport to be loaded on a B-52 Stratofortress during ENDURING FREEDOM. (USAF)*

Afghanistan was target number one in the war on terrorism that began after September 11, 2001. But going after al Qaeda's main base was no easy task. It demanded a new level of effort from the Air Force, and big breakthroughs in air warfare.

On September 16, USAF Predator Unmanned Aerial Vehicles took to the skies over Afghanistan gathering vital intelligence. Operation ENDURING FREEDOM officially began on October 7, 2001. B-2 bombers flying from the United States, B-1s and B-52s from theater bases, and fighters from two aircraft carriers launched strike sorties. C-17s flew the 6500-mile mission from Germany on the first night of the war to airdrop supplies as part of a humanitarian mission.

Securing air superiority was the first task. Then, U.S. Central Command (CENTCOM), headed by Army General Tommy Franks, planned for Special Operations Forces (SOF) controllers on the ground to call in airstrikes and clear the way for Afghan opposition forces to take back Taliban-controlled cities. It was airpower's ability to kill emerging targets that created the payoff on the ground. Along the way the USAF forged a new standard for precision, persistence, rapid response, time-critical targeting, and sustainment.

Guaranteed precision was essential to the campaign. Newer weapons such as the satellite-guided

Joint Direct Attack Munition (JDAM) plus tried and true laser-guided bombs delivered impressive results. Generating sorties for persistent battlespace coverage was also a challenge. The Combined Air Operations Center (CAOC), responsible for directing the air war, organized aerial refueling tracks to bring fighters and bombers into Afghanistan for designated periods of time. These soon got the nickname "vul" periods—meaning the aircrews were "vulnerable" or on-call for retasking during the cycles. B-1s and B-52s from in-theater bases could extract four or five hours of time on station from their 10-hour missions. Navy carriers in the North Arabian Sea split duties as "night" or "day" carriers to keep up a steady supply of sorties. USAF F-15Es and F-16s flying from bases in the Persian Gulf flew long sorties to deliver a few combat-ready hours over Afghanistan.

Next came the task of responding when controllers on the ground called for precision strikes. During October, CENTCOM sent in teams and supplies and built working relations with the warlords of the Northern Alliance and other Afghan opposition forces. At the same time, aircrews on station were often diverted to strike time-critical targets, such as al Qaeda leadership. Rapid coordination enabled CENTCOM leadership to communicate with and task individual aircraft.

Joining the fight were unmanned aerial vehicles—UAVs. Predator supplied streaming video for target identification and also became an offensive weapon when Hellfire-C missiles were successfully used to attack high-value targets. Global Hawk made its combat reconnaissance debut with long hours of imaging enemy targets.

For months, supplies for SOF, regular forces, and Afghan allies all came by air. Airlifters at remote locations from Uzbekistan to the Persian Gulf made expeditionary warfare possible.

By November, the pieces were in place for rapid success. Key cities fell as the ability to keep bombs on target raised the confidence of the Afghan opposition forces. "Every day the targeting and effectiveness has improved, and that has clearly played a critical role in killing Taliban and al Qaeda troops," Secretary of Defense Donald Rumsfeld said on November 13. Two days later, Afghanistan's capital city of Kabul was in Northern Alliance hands and by December 20, 2001, Afghanistan had a new interim government. The tactics of persistent air coverage and precisely controlled strikes had scored the first major victories of the global war against terrorism.

As Afghanistan's new government took hold, on-call airpower helped secure the country. Coalition forces

continued to hunt down remaining concentrations of terrorists. March of 2002 saw an immediate airpower response when Operation ANACONDA, a major ground assault by U.S. forces, encountered unexpected al Qaeda resistance. The quick reactions of on-call airpower from carriers and land bases saved the day.

For two weeks, air commanders funneled fighters and bombers into tight airspace of the mountainous battle area of Tora Bora. These airmen delivered more bombs per square mile than they had during the peak strikes of Operation DESERT STORM. A new level of air integration—and a new level of joint military power—was now at hand.

The Air Force role in Afghanistan continued during stability operations after the fall of the Taliban. A-10s, B-1s, and other aircraft established top cover and delivered close support for Coalition ground operations when needed. Airmen flew "a dozen show of force missions on election day," in the fall of 2004, said Deputy CFACC Major General Norm Seip. "Airpower played a complementary role to what was being done on the ground—to provide security over Afghanistan's 4,000 polling places," he said.

Shortly after elections, Army Lieutenant General David Barno, the top commander in Afghanistan, praised, "Air power from all the services—intelligence, surveillance and reconnaissance assets, mobility aircraft, close air support and space systems—gives ground forces in Afghanistan the ability to operate in smaller units and respond quicker with more accurate weaponry than at any other point in history."

7 October
Operation ENDURING FREEDOM, attacks against the Taliban and associated terrorist targets in Afghanistan, begins. Lieutenant General Charles W. Wald is the Joint Force Air Component Commander (JFACC) for the operation. A significant force of land-based Air Force aircraft and sea-based Navy aircraft surgically strike targets throughout the country. Special Operations forces from all services play a significant role in the fight against the Taliban and al Qaeda.

8 October
C-17 Globemaster III aircraft make their first airdrops in combat zones when they deliver pallets of Humanitarian Daily Rations (HDRs) over Afghanistan territory under friendly control.

9 October
NATO deploys AWACS aircraft to the United States during Operation EAGLE ASSIST to help monitor American skies. This is the first real-world NATO deployment to the U.S. During the next seven months, NATO aircraft will fly 360 sorties.

26 October
Lockheed Martin gets the nod over Boeing to build the next-generation fighter—a single-engine fighter that will be built in three variants—conventional landing, short takeoff and vertical landing, and carrier landing—and purchased by the Air Force, Navy, Marine Corps, and the Royal Air Force.

2 November
After an MH-53 Pave Low Special Operations helicopter crashes in the mountains of Afghanistan, another MH-53 goes out in search of survivors. For rescuing the crew of a sister ship under extremely hazardous weather conditions behind enemy lines in Afghanistan, the crew of Knife 04 is awarded the Mackay Trophy.

28 November
U.S. ground forces are transported by C-17 to a base near Kandahar as part of Operation SWIFT FREEDOM. Navy Seabees and other Special Operations forces will arrive in small numbers throughout the operation.

12 December
A B-1B Lancer crashes some 10 miles north of its base at Diego Garcia. The crew ejects and survives. This marks the first USAF aircraft loss during ENDURING FREEDOM and the first B-1 lost in combat.

2002

11 January
A C-141 makes the first transfer of Taliban and al Qaeda detainees from Afghanistan to the U.S. Naval base at Guantanamo, Cuba. They will be held there until a decision to close the prison is made in March 2006.

31 January
U.S. military forces join Philippine forces in a campaign against Abu Sayyaf, a terrorist group seeking the establishment of an independent Islamic state in the southern Philippines.

4 February
An MQ-1B Predator fires a Hellfire missile in combat. The missile accurately finds its target, killing a group of senior al Qaeda leaders in southeastern Afghanistan. This marks the first offensive use of an unmanned aerial vehicle in combat.

1 March
As Operation ENDURING FREEDOM continues, the Coalition forces launch Operation ANACONDA in eastern Afghanistan. On station B-52s, B-1Bs, AC-130s, A-10s, and F-15s support ground forces that are attempting to surround and kill enemy forces regrouping near Gardez. Laser-guided thermobaric bombs are used in attacks on the caves where suspected terrorists are hiding. The weapon detonates and consumes the oxygen in the enclosed space, killing the inhabitants.

2 March
While flying a close air support mission over the Shah-e-Kot Valley on the second day of Operation ANACONDA, the 14-man crew of an Air Force AC-130 Spectre gunship engage enemy forces who had surrounded elements of the 10th Mountain Division in rough terrain. While overhead during a two-hour, nighttime operation, the crew continually attacked enemy positions until two Black Hawk helicopters were able to land and pick up the battered troops. For their actions, the crew of GRIM 31 receives the Mackay Trophy.

4 March
Two airmen are killed during helicopter assault operations near Gardez in support of ANACONDA.

13 May
The U.S. and Russia agree to reduce their nuclear arsenals by two-thirds.

22 May
The X-45A unmanned combat aerial vehicle (UCAV) flies for the first time at Edwards AFB, California. This stealthy vehicle will be capable of accomplishing ground attack missions under extremely hostile conditions, alone or in formations.

18 July
Benjamin O. Davis Jr., the Air Force's first African American general officer, dies at age 89. Davis led the Tuskegee Airmen during World War II.

22 July
The YAL-1A, the "Airborne Laser" platform, completes its first flight. This weapon is designed to destroy enemy ICBMs in the boost phase rather than the end game.

Opposite, top: *U.S. Marines wait to board an Air Force C-17 as their equipment is placed aboard during Operation ENDURING FREEDOM. (USAF)*

Opposite, bottom: *The X-45 Unmanned Combat Aerial Vehicle (UCAV) undergoes testing at Edwards, AFB, California. The program will terminate in 2006.*

Above: *Aerial refueling is a challenging maneuver for any pilot. Here, an F-16 pilot has contact with the boom and is taking on fuel. (USAF)*

PREDATOR: A PERSONAL HISTORY

James "Snake" Clark

Left: *Predator #034 was the first RQ-1 modified to fire the Hellfire-C missile. The same aircraft was the first UAV to fire an offensive missile in combat during the hunt for al Qaeda operatives in Afghanistan after the terrorist attacks of September 11. Having flown more than 260 combat sorties, #034 was retired and is now part of the National Air and Space Museum modern military aircraft collection. (USAF)*

Opposite: *Ground crew load a Hellfire-C missile onto Predator's pylon. (USAF)*

The history of the Unmanned Aerial Vehicle (UAV) can be traced back to early World War I. In World War II, unmanned flying bombs, such as the JU-88/FW-180 and B-24, were used and in Vietnam, reconnaissance drones flew over Hanoi and Haiphong. But it was not until the late 1990s that the UAV came of age in the skies of Bosnia and Kosovo and became known as the Predator.

In an age of Supersonic Fighters, Stealth Bombers, and precision-guided weapons, the Predator was the most unusual hero. It was a 48-foot glider with an Austrian snow mobile engine and a television camera; it raced through the skies at a blistering 70 knots. What the Predator did was solve one of the most basic commanders problems; identify threats that lay on the other side of the hill. Predator solved that problem by electronically downloading a television picture over enemy territory to the battlefield commander miles away, but what was even more remarkable was that the aircraft could be controlled from hundreds and now thousands of miles away through the use of some very creative communication satellite technology. Now a commander on the ground and the President in Washington could see the battle in real time. What is even more remarkable is how quickly this "transformation" technology was developed, fielded in combat, and quickly modified to meet the emerging enemy threat. In a time when weapons systems are usually developed over a series of years the Predator's wartime capabilities were modified in days and weeks—a great example of "Yankee technology" in the twenty-first century.

The Predator was born in 1994 as an advanced concept technology demonstration (ACTD). Rather than test it in the safe skies over Nellis AFB, Nevada, the Predator was raced into action in Albania in 1996 and was deployed to Operation JOINT ENDEAVOR in Bosnia. It was there that my path crossed with the Predator. General Ronald Fogleman, Chief of Staff, United States Air Force, wanted someone with no

UAV experience to take an objective outsider's look into Predator operations. As an Air Force Colonel assigned to the Pentagon, I was "volunteered" to do the study. What I found was remarkable; this little drone could fly hundreds of miles away and provide color television and infrared video surveillance of enemy activity, without risking the life of a pilot. In a control van, which was a converted NASCAR transporter trailer, I watched pilots and sensor operations sitting in front of computer screens actually flying this thing—simply remarkable. After my report was done, I was convinced that the Predator and I would go our own separate ways—I was wrong!

On April 2, 1999, while sitting at my desk at the Pentagon, the phone rang. It was General Michael Ryan, Chief of Staff of the Air Force, and he said, "Snake, I just got done talking to Lieutenant General Mike Short, 16th Air Force Commander, and he needs the ability to get precision geographic locations of where the Predator TV is looking." I said, "It doesn't exist." He said, "I know. . . invent it!"

Forty-eight hours later, I was on a plane to NATO Combined Air Operations Center (CAOC) at Vicenza, Italy, to invent "precision Predator." Operational Allied Force was to ensure the safety of the people of Kosovo. Predators were flying surveillance mission's 24 hours a day, identifying hostile/illegal activities. The rules of engagement were that moving targets had to be identified visually by a forward air controller before any air strike. In Kosovo, most villages and buildings looked the same so "talking" the eyes of the strike pilot on to the target was a challenge. To solve this problem, we turned to a legendary Air Force group called "Big Safari," who made the routine possible. They recommended installing a laser designator on to the Predator to pinpoint accurate target locations from the Predator to the F-16s. Within 18 hours, Big Safari had "borrowed" a laser designator from the U.S. Navy and begun testing. They deployed it in 38 days for its first combat mission. On the 39th day, the Serbs surrendered.

As part of the "lessons learned" from Operation ALLIED FORCE, it was determined that if the Predator had a weapon on it, we could cut the time between identifying a target and then destroying it. This is called the "Kill Chain." General John P. Jumper, Commander, Air Combat Command, and later, Chief of Staff of the United States Air Force, directed the weaponization of Predator with the Army Hellfire missiles. The original plan was a multi-year, multi-million-dollar program. General Jumper thought that this was too long and too much, and told them to try again. In September 2000, real world events clearly

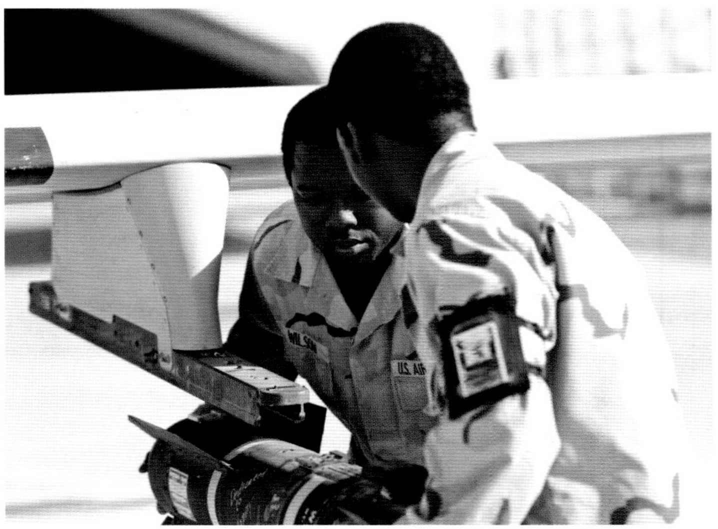

demonstrated the immediate need for the weaponized Predator. General Ryan, who was the Air Force Chief of Staff, directed us to do this quick and dirty. Thanks to the magicians at Big Safari, 61 days later Predator #3034, chained to a test stand at the Navy Test Center, China Lake, fired our 1st Hellfire. The wings rocked and tail fluttered but did not fall off (parts of that Hellfire sit on my desk today).

Twenty-four days later, on February 16, 2000, Predator #3034 took its first successful Hellfire shot from the air, and to all of our surprise, it worked! Through the spring and summer, we aggressively tested our new Predator capability.

The morning of September 11, 2001, changed our lives and the future of the Predator. From my office in the Pentagon, we watched the attack on the World Trade Center until the explosion on television, until we were shocked by Flight #77 as it crashed into the Pentagon. Late on the evening of September 12, a lone C-17 took off from an airfield on the west cost with its cargo of Predators and Hellfire missiles. Days later, one of America's first responses to the terrorist attacks on 9/11 was in place and ready for combat.

During Operation ENDURING FREEDOM combat operations in Afghanistan, the Hellfire Predator combination proved to be lethal against the Taliban. Today in Afghanistan and Iraq, armed Predators are patrolling the skies providing support and protection to U.S. and coalition forces being flow from 8,000 miles away at a base in the United States.

Predator #3034, the first UAV to test the Hellfire, and the first to shoot in combat on October 7, 2001, has returned safely to the United States and now awaits its place of honor at the Smithsonian National Air and Space Museum in Washington, D.C. Our "unmanned hero"—not bad for a glider with a snow mobile engine and a team of outstanding Americans who made it possible.

30 July
For the first time, supersonic combustion occurs in a scramjet engine during flight in the atmosphere. Scramjets (air-breathing supersonic combustion engines) have the potential to revolutionize air transport and the launching of small payloads.

21 August
The first launch of the Lockheed-Martin Atlas V, part of the Air Force's Evolved Expendable Launch Vehicle program, is made from Cape Canaveral, Florida. Each launch vehicle consists of a standard booster and then additional boosters are added as needed depending upon the payload being delivered.

1 October
U.S. Northern Command is activated as a new unified command. General Ed Eberhart assumes command and the responsibility for the military protection of North America. General Eberhart also retains command of North American Aerospace Defense Command.

General John P. Jumper, CSAF, authorizes the deactivation of the Peacekeeper ICBM system. The end of the Cold War has mitigated the need for weapons with this kind of destructive power.

9 December
After Typhoon Pongsona devastates the island of Guam, AMC forces begin to deliver relief aid to the region. During the next 10 days, 58 C-5 sorties deliver 1,200 tons of supplies to the people of Guam.

2003

16 January–1 February
The space shuttle *Columbia*, STS-107, breaks up 15 minutes before the scheduled landing killing all seven astronauts aboard. The mission was the first in recent years that was not related to International Space Station (ISS) activities. The seven crew members helped oversee 80 microgravity experiments on board. It was the 28th mission for the space shuttle *Columbia*.

8 February
The Department of Defense enlists commercial airlines to transport troops and equipment as part of the buildup for possible war with Iraq with the activation of the Civil Reserve Air Fleet (CRAF).

17 February
The Federal Aviation Administration determines that the 270-foot-tall Air Force Memorial, which will be built on part of the old Naval Annex, is not a danger to air navigation. Construction may now proceed.

7 March
UN chief weapons inspector Hans Blix reports to the UN Security Council that Iraqi disarmament will take several months. The U.S. and Britain present a revised draft resolution to the Security Council giving Saddam Hussein an ultimatum to disarm by March 17 or face war.

17 March
President George W. Bush delivers an ultimatum that requires Saddam Hussein and his sons to leave Iraq in 48 hours.

18–19 March
Nearly 3 million informational leaflets are scattered over western and southern Iraq near military and civilian sites in 20 locations during this two-day period. EC-130 Commando Solo aircraft broadcast messages into the country to inform Iraqi citizens. Meanwhile, coalition aircraft bomb a series of Iraqi targets in response to Iraqi anti-aircraft artillery in violation of the southern no-fly zone.

19 March
Operation IRAQI FREEDOM begins. In a "decapitation" attack, a suspected gathering of Iraqi national leaders in Baghdad is struck by a salvo of cruise missiles and precision-guided bombs dropped by F-117 Nighthawk stealth aircraft. Coalition air forces also strike long-range artillery, air defense sites, and

Opposite, left: *Space shuttle* Columbia *AF astronauts Lieutenant Colonel Michael Anderson (left) and Colonel Rick Husband were eulogized in a joint message by the Secretary and the Chief of Staff of the AF. (USAF).*

Right: *F-15Cs secured the air during the first three days of combat over Iraq. (USAF)*

Below: *Weapons loaders affix a Joint Direct Attack Munition to a fighter during IRAQI FREEDOM. (USAF)*

surface-to-surface missile sites. President George W. Bush addresses the nation from the Oval Office as Operation IRAQI FREEDOM begins: "On my orders, coalition forces have begun striking selected targets of military importance to undermine Saddam Hussein's ability to wage war...I want Americans and all the world to know that coalition forces will make every effort to spare innocent civilians from harm. A campaign on the harsh terrain of a nation as large as California could be longer and more difficult than some predict. And helping Iraqis achieve a united, stable, and free country will require our sustained commitment."

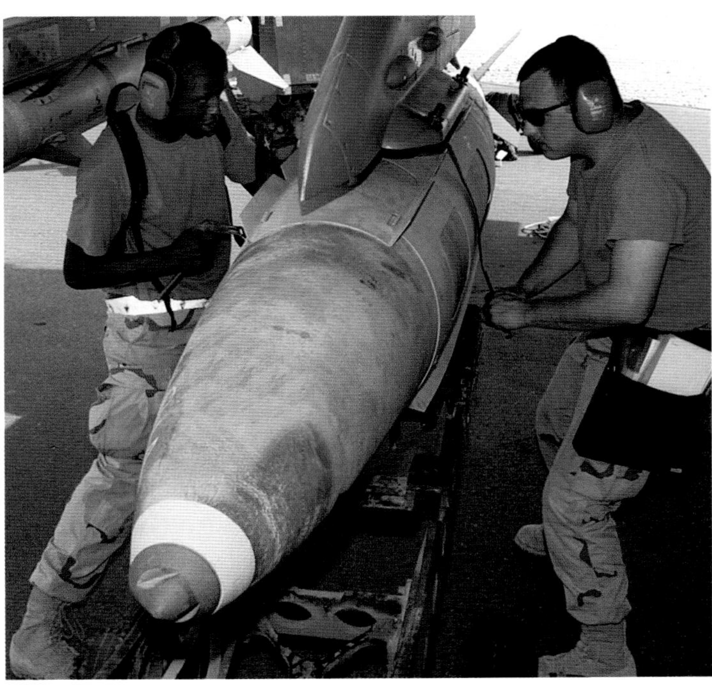

21 March
Coalition forces launch a massive missile assault on Baghdad. In the next 24 hours, coalition forces strike hundreds of Iraqi targets.

22 March
More than 1,000 air sorties and an equal number of cruise missiles attack Iraqi targets.

24 March
Coalition forces conduct 24 hours of nonstop bombardment against Iraq's northern oil capital of Kirkuk.

26 March
The C-17 Globemaster III crew of VIJAY 10 leads the insertion of 15 C-17A strategic transport aircraft during Operation NORTHERN DELAY. In so doing, the crew directs the largest formation airdrop since World War II. The mission results in the safe and successful airdrop of 990 paratroopers and 20 heavy platforms into Bashur Airfield. The crew work in support of Operation IRAQI FREEDOM and effectively establishes the United States' second combat front in Iraq. Air Force officials call the mission the most demanding C-17 mission in the aircraft's history, and it marks the first time paratroops drop from the C-17 during combat. The superb execution of this mission earns the crew of VIJAY 10 the Mackay Trophy.

Left: *This B-52 crew is carrying a load of joint direct attack munitions (JDAM) during a mission in support of Operation IRAQI FREEDOM (USAF).*

Below: *IRAQI FREEDOM is a combined operation. Here, an AF KC-135 leads a formation of F-15Es, F-16s, a Navy F-18, an F-117, and a RAF Tornado. (USAF)*

2 April
For the first time in combat history, Air Force aircraft drop CBU-105 wind-corrected munitions dispensers. B-52 Stratofortresses drop the armor-piercing, sensor-fused weapons in central Iraq to stop an Iraqi tank column on its route toward coalition troops.

7 April
A B-1 Lancer from the 34th Bomb Squadron, Ellsworth Air Force Base, South Dakota, drops four GBU-31 satellite-guided joint direct attack munitions (JDAM) against a suspected meeting of senior Iraqi regime leaders in Baghdad. Hussein and other officials are believed to be at the meeting when the bombs strike.

The first coalition aircraft, a C-130 Hercules, lands at Baghdad International Airport, bringing in troops under the cover of darkness.

8 April
An A-10 Thunderbolt II goes down near Baghdad International Airport. The pilot ejects safely from the aircraft and is recovered by coalition ground forces near the airport. Meanwhile, two airmen are missing after their F-15E Strike Eagle goes down in Iraq. They are later confirmed dead.

11 April
For the first time in combat history, a B-52 uses a Litening II advanced airborne targeting and navigation pod to strike facilities at an airfield in northern Iraq.

16 April
President Bush calls for the UN Security Council to end Iraq's economic sanctions. General Tommy Franks, commander of U.S. Central Command, enters Baghdad for the first time since the war began.

May
The Air Force announces that it will lease 100 Boeing KC-767 tankers to replace the oldest of its KC-135 tankers, subject to congressional approval. The proposed lease would be for six years, starting in 2006. An option to buy at the end of the lease is to be included in the deal. A Defense Science Board review of the USAF's proposed lease concludes that further studies were required before a decision could be made. In November 2004, it is announced that a study to clarify service needs and an analysis of alternative strategies are required, to be followed by a request for competitive bids. In January 2006, the lease program is cancelled by the Secretary of Defense.

Right: *An F-15 Eagle and an Indian air force Mirage 2000 fly together over the Himalayas during Cope India '04, the first dissimilar air combat training exercise between the two air forces in more than 40 years. (USAF)*

1 May
President Bush lands on the aircraft carrier *Abraham Lincoln* and declares an end to major combat operations in the Iraq war.

29 August
The Air Force launches the last of 14 Defense Satellite Communications System (DSCS III) satellites atop a Delta IV launch vehicle. The first DSCS III was launched into orbit in 1981.

9 September
The "Father of the H-bomb," physicist Edward Teller, dies at age 95. Teller fled from Nazi Germany and then became a pivotal member of the Manhattan Project. He also served as an adjunct member of the Air Force Scientific Advisory Board for four decades.

11 October
Ivan Getting, originator of the concept which eventually led to the Global Positioning System (GPS), dies at age 91. Getting was a contributing member of the Air Force scientific Advisory Board since its inception in 1945.

The original 1903 Wright Flyer is placed on public display at ground level in its own gallery at the National Air and Space Museum. The exhibition is part of the Centennial of Flight celebration.

15 December
The Smithsonian Institution, National Air and Space Museum opens the Stephen F. Udvar-Hazy Center located near Dulles International Airport just outside of Washington, D.C. When fully outfitted, the center will house nearly 80 percent of the national aircraft and spacecraft collection—all of which will be on public display.

2004

February
At Edwards AFB, California, a modified KC-135 Stratotanker sprays water on an F/A-22 during an airborne ice test. The Raptor is the first aircraft to use the "rain and ice tanker" for testing.

14–25 February
Elmendorf AFB F-15s and approximately 150 personnel deploy to Gwalior Air Force Station, India, to participate in COPE INDIA 04. F-15 Eagles and an Indian Air Force Mirage 2000s fly together over the Himalayas in the first dissimilar air combat training exercise between the two air forces in more than 40 years.

16 April
The crews of JOLLY 11 and 12 undertake a rescue operation near Kharbut, Iraq. The two crews are sent to the location of a U.S. Army CH-47 Chinook helicopter that had crashed in a sandstorm in near-zero visibility. The conditions render infrared and night vision goggles ineffective and make navigation extremely difficult. All five survivors of the crash are located and rescued. During egress, JOLLY 11 and JOLLY 12 avoid surface-to-air missiles and evacuate the combat zone unharmed. For their actions, both crews are awarded the Mackay Trophy.

5 June
Ronald Wilson Reagan, the 40th president of the Unites States, dies peacefully at his Los Angeles home. He was 93 years old. He served as the military's commander-in-chief from 1981 to 1989.

AIRBORNE LASER

Rebecca Grant

Left: *The YAL-1A Airborne Laser aircraft flies over the Edwards AFB test range while officials recertified the aircraft's airworthiness. The YAL-1A soon resumed testing of the system's laser beam control system. (USAF)*

It was 1977 when Phillips Lab tested a unique new weapon: the chemical oxygen iodine laser, or COIL. Demonstration of this high-energy laser beam opened up the possibility for a new application of aerospace power. After further research and development, a contract was awarded in 1996 to field the Airborne Laser (ABL) aboard a modified 747 jet. Its mission: locate, track, and destroy enemy missiles during the initial launch, or boost phase.

The post-Cold War era refocused attention on the proliferation of ballistic missiles around the world. Iraqi Scud missile attacks on Saudi Arabia and Israel during Operation DESERT STORM demonstrated the disruptive potential of ballistic missiles. By 1998, a total of over 10,000 ballistic missiles existed worldwide, scattered among the arsenals of more than 30 nations. Experts agreed that the best time to intercept a missile was in the boost phase, just after its launch. Targeting and attacking launchers and missiles in boost phase became a dominant concept for countering the ballistic missile threat, especially in regional conflicts. This concept required a system that could cover multiple areas. Enter the Airborne Laser, with its global reach and ability to reposition to any hotspot rapidly.

The Airborne Laser engagement sequence begins with passive detection of the hot missile plume from off-board and on-board systems. Then, ABL refines the target with use of a special laser. Next, adaptive optics—developed by the Air Force under the Starfire Optical program, based at Kirtland Air Force Base, New Mexico—rapidly compensate for atmospheric distortion and optical turbulence. The effect is like finding a path through clouds and other disturbances to keep the COIL power beam at peak strength. The COIL beam then engages the missile target by firing a three-to-five second burst from the nosecone of the ABL aircraft. It kills the enemy missile by burning a hole through the missile body and causing its engine to burn out.

COIL technology was the prime choice for the first working speed-of-light laser weapon because chemical lasers generate megawatts of power—enough to take out an enemy missile. However, the challenge was packing the large, chemical system into an aircraft.

It took a special aircraft to house the weapon

Right: *An artist's conception of the YAL-1 attacking and disabling an enemy ballistic missile during the boost phase of the launch. Eventually, the Airborne Laser platform will be fully deployable to any global location. (USAF)*

system. In December 1999, a 747-400 fresh off the Boeing production line became the first ABL aircraft. Engineers reconfigured this jumbo jet to house a 14,000-pound nosecone section as the housing for the laser weapon. Next came installation of a mid-section bulkhead to separate the crew from the chemicals. The aircraft's underbelly was lined with titanium. New support beams compensated for the unusual weight distributions. A refueling receptacle, infrared sensors, and miles of electric wiring completed the refit.

Airborne Laser moved from USAF management to the Missile Defense Agency in 2001. Now ABL is part of national as well as regional missile defense concepts.

The ABL concept became a flyable reality between 2002 and 2004. "We've spent 25 to 30 years developing the technology," the Air Force's program director for the Airborne Laser said in December 2002. "Now is the time for the engineers to take what those smart physicists and scientists have done and put it in the field."

One test milestone was passed in 2002 with flights of the extensively modified host aircraft to verify airworthiness. Installation of the laser in the special 747 came next, creating a true prototype dubbed the YAL-1. In November 2004, the ABL team fired the laser successfully in a test at Edwards AFB. Then in December 2004, the ABL team logged "first flight" with the integrated battle management and beam control/fire control systems.

By 2010, plans call for a fleet of seven operational ABL aircraft. Four crewmembers will man each plane. The concept of operations calls for a pair of ABL aircraft to set up an orbit at high altitude, over secure territory, and maintain watch for missile plumes. ABL is also demonstrating capability to be an information node in the network of layered missile defenses. In addition to its attack options, ABL is built to determine and pass on information on the missile launch site, target track, and predicted impact point of the enemy missile.

The ABL has faced many challenges over the years, from technical hurdles to changes in funding. Yet, ABL remains at the leading edge of development of laser weapons for twenty-first-century operations. Although hurdles remain, ABL is on track to put "photons on target" and revolutionize counter ballistic missile attack.

Left: *One of two pararescuemen from 38th Rescue Squadron at Moody AFB, Georgia, follows an inflatable boat out the back of an HC-130 during a rescue mission in the Caribbean on July 23, 2004. (USAF)*

Below: *Captain Nicole Malachowski was selected for the 2006 Air Force Demonstration Squadron—Thunderbirds. She is the first female demonstration pilot on any U.S. military high performance jet team. (USAF)*

7–16 June
More than 9,000 Airmen, Sailors, Marines, Soldiers, and Coast Guardsmen from active duty, reserve and National Guard units participate in Operation NORTHERN EDGE, Alaska's premier annual military training exercise. The joint training exercise is designed to enhance interoperability among the various branches by sharpening and honing joint service techniques and procedures.

23 June
A Boeing Delta II rocket lifts off from Cape Canaveral carrying a replacement satellite for the Air Force's Global Positioning System (GPS) into orbit. The Delta II is a three-stage launch vehicle.

23 July
Two pararescuemen from the 38th Rescue Squadron at Moody Air Force Base, Georgia, deploy with a rubber boat out the back of an HC-130 during a rescue mission 350 miles northeast of the Caribbean island of St. Maarten. They provide medical support to a Chinese fisherman who sustained a life-threatening chest injury the day before.

4–8 September
Members of the 45th Space Wing's hurricane response team begin to assess damage caused by Hurricane Frances surrounding Patrick AFB, Florida.

14 September
At Mountain Home AFB, Idaho, the Thunderbirds No. 6 aircraft bursts into flames, and the pilot safely ejects less than a second before the plane impacts the ground during an air show. The pilot guided the jet away from the crowd of more than 60,000 people. This is the second crash since the Air Force began using F-16 Falcons for its demonstration team in 1982.

15 September
Formal ground breaking ceremonies are held at the site of the Air Force Memorial, just across the Potomac River from the Pentagon. Dedication ceremonies are scheduled for October 2006.

Right: *Lieutenant Colonel James Hecker, 27th FS commander, flies over Fort Monroe before delivering the first operational F-22 Raptor to its permanent home at Langley Air Force Base, Virginia on May 12, 2005. This is the first of 26 Raptors to be delivered to the 27th Fighter Squadron. (USAF)*

Below: *President George W. Bush (from left), Air Force Chief of Staff General John P. Jumper and Vice President Dick Cheney salute as the American flag passes during the 2005 Presidential Inaugural Parade, January 20, 2005. (USAF)*

3 December
The Boeing Airborne Laser (ABL) team flies an aircraft equipped with the integrated battle management and Beam Control/Fire Control (BC/FC) systems for the first time at Edwards Air Force Base, California. The ABL places a megawatt-class, high-energy Chemical Oxygen Iodine Laser (COIL) on a Boeing 747-400F aircraft to detect, track, and destroy ballistic missiles in the boost phase of flight. ABL is also able to pass vital enemy missile launch information to other global ballistic missile defense systems.

12 December
The Airborne Laser team announces that they have successfully completed a series of tests involving its high-energy laser at the Systems Integration Lab at Edwards AFB, California. During these tests, lasing duration and power are demonstrated at levels suitable for the destruction of multiple classes of ballistic missiles. The Northrop Grumman-built laser has been operated at simulated altitude, and achieved steady state operations under full optical control.

2005

3–8 January
HH-60G Pave Hawk helicopters are deployed to Sri Lanka in C-17 Globemaster III aircraft to support tsunami relief efforts after a large region is devastated by the massive wave on December 29. More than 145 tons of relief supplies will be sent to the region. C-130s from Yokota AB, Japan, also continue to deliver relief supplies and equipment to tsunami-devastated areas in Thailand, Indonesia, and Sri Lanka.

20 January
President George W. Bush is inaugurated for his second term. As a pilot in the Texas Air National Guard, George W. Bush is the first American president to have served as a member of the United States Air Force.

12 May
The first operational F-22A Raptor arrives at its permanent home at Langley Air Force Base, Virginia. This is the first of 26 Raptors to be delivered to the 27th Fighter Squadron. The 1st Fighter Wing will achieve initial operational capability on December 15.

13 May
Sixteen C-17 Globemaster IIIs fly in a mass formation from Charleston AFB, South Carolina, to Biggs Army Air Field, Texas. The event marks the largest C-17 formation ever to fly across the country.

16 June
Spokesmen for the U.S. Air Force Thunderbirds announce their new pilots for the 2006 demonstration season, which includes the first female demonstration pilot in the 52-year history of the Thunderbirds. Captain Nicole Malachowski, of the 494th Fighter Squadron at RAF Lakenheath, England, will join the team as the first female demonstration pilot on any U.S. military high-performance jet team.

Left: *Technical Sergeant L. Torres and a young boy are lifted to safety from the roof of the child's flooded home. The pararescueman, from the 38th Rescue Squadron at Moody AFB, Georgia, is deployed to New Orleans for Hurricane Katrina search-and-rescue operations. (USAF)*

Below: *Students with the 332nd and 335th Training Squadrons at Keesler AFB, Mississippi, prepare to be evacuated to Sheppard AFB, Texas, aboard a C-17 Globemaster III from the 58th Airlift Squadron, Altus AFB, Oklahoma. More than 2,400 students and non-essential personnel were evacuated from Keesler because of Hurricane Katrina's aftermath. (USAF)*

22 June
A catastrophic, cascading sequence of events causes a U-2S surveillance and reconnaissance aircraft to crash in Southwest Asia, killing the pilot. The 9th Reconnaissance Wing aircraft was returning to a forward-deployed location from a high-altitude intelligence, surveillance, and reconnaissance mission at the time of the accident.

29 July
The United States provides airlift for 1,200 Rwandan troops to Sudan in support of the African Union Mission there.

23–29 August
Hurricane Katrina forms over the Bahamas on August 23rd and crosses southern Florida as a minimal hurricane before strengthening rapidly in the Gulf of Mexico. Katrina quickly becomes the strongest hurricane ever recorded in the Gulf. Although the storm weakens considerably before making its second landfall August 29 in southeast Louisiana, it is possible that Katrina is the largest hurricane of its strength to approach the United States in recorded history. The storm surge causes catastrophic damage along the coastlines of Louisiana, Mississippi, and Alabama, including the cities of Mobile, Alabama, Biloxi, and Gulfport, Mississippi, and Slidell, Louisiana. Levees separating Lake Pontchartrain from New Orleans, Louisiana, are breached by the surge, ultimately flooding roughly 80 percent of the city and many areas of neighboring parishes. Many Air Force personnel are affected by the storm, bases are evacuated, and relief efforts begin as soon as conditions permit.

Right: *Langley AFB F/A-22s deployed for the first time to test their combat sortie generation capability. The Raptor unit is operating out of Hill AFB, Utah, during this historic event. (USAF)*

Below: *This is the first of 24 TH-1H Hueys that will be modified to train helicopter student pilots. The TH-1H is the newest aircraft to join the Air Force inventory. (USAF)*

2 September
General T. Michael "Buzz" Moseley is sworn in during a ceremony at Andrews AFB, Maryland, and becomes the Air Force Chief of Staff.

10 September
Over the skies of Southwest Asia, an all-female C-130 crew flies a combat mission for the first time.

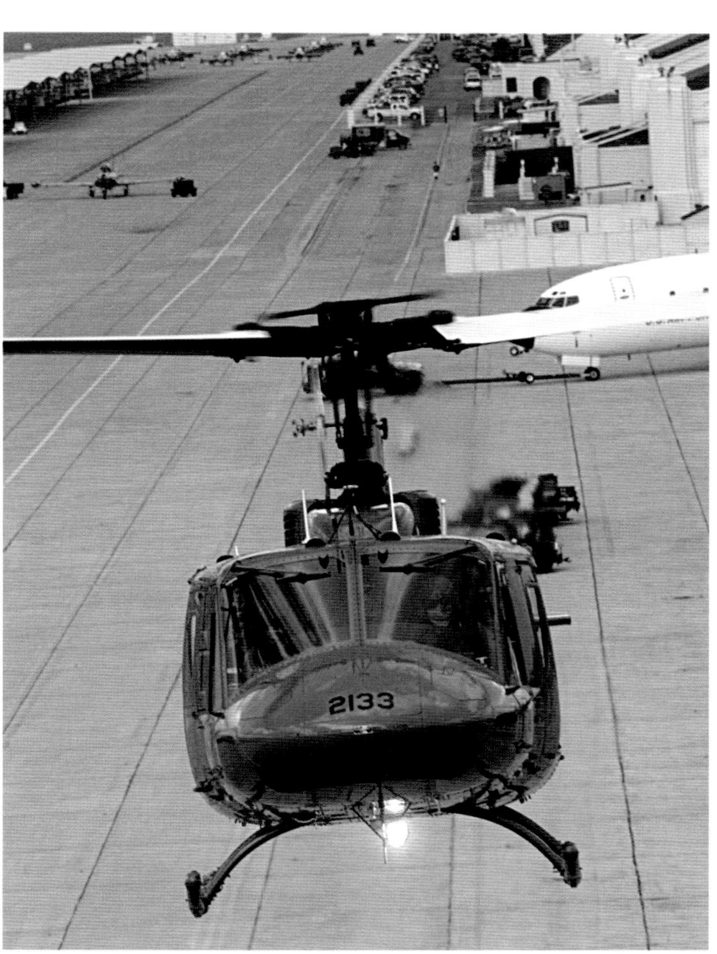

24 September
Hurricane Rita makes landfall in southwestern Louisiana as a Category 3 storm, causing extensive damage along the Louisiana and extreme southeastern Texas coasts and completely destroying some coastal communities. Relief efforts are extensive and are now coordinated with Katrina responses as well. The Civil Air Patrol is active in damage assessment in the Houston area during relief efforts flying their new Gippsland Aeronautics GA-8 Airvan aircraft.

15 October
In the first-ever operational F-22 Raptor squadron deployment, the 27th Fighter Squadron from Langley AFB, Virginia, flies to Hill AFB, Utah, to participate in Operation COMBAT HAMMER. The deployment will test whether the unit can generate a combat-effective sortie rate away from their home base. During the deployment, JDAMs are dropped by the F-22s for the first time on October 18. This event demonstrates the aircraft's potential in the ground attack role.

5 November
The first of 24 TH-1H Huey IIs is delivered to the Air Force at Randolph AFB, Texas, where training will begin. The helicopter is the newest aircraft to join the Air Force inventory.

November–January 2006
Supporting the Combined Joint Task Force—Horn of Africa, Air Force C-130s participate in exercises at Camp Lemonier, Africa, as part of the Expeditionary Rescue Squadrons established there.

Left: *As part of an ongoing effort to train Iraq's military forces, Iraqi trainees and a coalition soldier receive instruction on proper weapons use at Al Kasik Training Base firing range. (USAF)*

Below: *Airmen unload a UH-60L Black Hawk from a C-5 Galaxy. The C-5 is from the 137th Airlift Squadron, New York Air National Guard, in support of Operation NEW HORIZONS—Honduras. (USAF)*

2006

14 January
In an effort to bolster recruiting, and already actively sponsoring a NASCAR vehicle, the USAF enters the "Monster Truck" circuit. *Afterburner*, the Air Force's new monster truck, flies over cars at the San Antonio Monster Jam, making the finals, but is edged out by *Grave Digger* in the final match.

29 January
The Iraqi Air Force makes history when it flies its first C-130E flight with an all-Iraqi-aircrew outside of Iraq. On board is Bayan Jabr, Minister of Interior, and his staff. They attend the 23rd Annual Minister of Interior Summit in Tunisia. The summit is sponsored by the Arab League, and it is the first time that Iraq has attended the event. The Iraqi Air Force 23d Transport Squadron flies three C-130E model aircraft. The squadron will move to Tallil Air Base in February, where U.S. Air Force operators and maintainers work side-by-side training Iraqi airmen.

Right: *The first Hawaii-based C-17 Globemaster III flies past Diamond Head volcanic crater on its way to Hickam for the official arrival ceremony. (USAF)*

Bottom: A "wall" of four F-15 Eagles target and fire radar missiles at a drone as part of a joint service training exercise. (USAF)

3 February
At Al Kasik Training Station, Iraq, Captain LeeAnn Roberts instructs Iraqi trainees and a coalition soldier on proper weapons firing techniques at the firing range. She is the first female coalition military assistance training team instructor assigned to the base, and the only female out of 8,000 assigned personnel.

January–April
Operation NEW HORIZONS 2006—Honduras, a joint training exercise between the U.S. military and Honduran government, is accomplished. The Joint Task Force Asegurar el Futuro, or "Securing the Future," is charged with providing engineering, medical and support training for the U.S. and Honduran forces. When the mission is completed, the JTF will have built four schoolhouses and a maternity clinic, and provided free humanitarian medical assistance to 14 different locations in and around the coastal city of La Ceiba.

8 February
The first C-17 Globemaster III flies to Hickam AFB, Hawaii, to become a permanent resident there. This is the first operational C-17 unit to be based on the island.

22 February
Four F-15s target and then fire simultaneously at an aerial decoy during a joint service training exercise. The three-day exercise, conducted near Okinawa, Japan, tests the capabilities of the 18th Wing and Pacific Command's assets.

22 March
A-10 Thunderbolt IIs begin to arrive at Davis Monthan AFB, Arizona, to kick off "Hawgsmoke 2006."

7 June
Two USAF F-16s, employing precision weapons, kill al Qaeda terrorist leader Abu Musab al-Zarqawi and one of his key lieutenants in a raid near Baqubah, Iraq.

ENDURING HERITAGE: MEMORIALS TO AIRMEN

David A. Lande

Memorials to the United States Air Force honor the courage and commitment of many generations of men and women who served their country. Ranging from small bronze plaques to static B-52s, the memorials are symbols that remind us of hard-won freedom and air power's legacy in protecting it.

The Air Force Memorial, dedicated in 2006, salutes American air power since its earliest beginnings, and the contributions of millions of men and women who served in the U.S. Air Force and its predecessor organizations. On high ground overlooking the Pentagon from the southwest, the memorial features three stainless steel spires soaring some 270 feet into the sky, creating the sense of bold and graceful flight. The majesty of the promontory setting, with its panoramic view of the nation's capital, symbolizes the Air Force's protective vigil over America—consistent with the pledge of the Air Force mission "to defend the United States and protect its interests through air and space power."

Most Air Force bases have their own memorials, often in the form of vintage aircraft mounted near entry gates or bronze busts of legendary air leaders for which the base or its buildings are named. An entire collection of aircraft is displayed on a grassy common in front of the Air University at Maxwell Air Force Base, celebrating the evolution of air power technology from the Wright Flyer to modern-day fighters. Some on-base memorials are artistic expressions, such as the stained glass arches of the All Veterans' Memorial located at Eglin Air Force Base; and the Airmen's Memorial, a three-bladed propeller embedded in a Georgia granite obelisk dedicated "In memory of all airmen who gave their lives in service to their country," at Patrick Air Force Base. Also at Patrick is a large plaque with the names of all Air Force recipients of the Medal of Honor.

The U.S. Air Force Academy, home to many memorials and monuments, includes a statue of Brigadier General Robinson Risner, senior ranking POW in Vietnam after his capture in 1965. In front of Arnold Hall is a larger-than-life statue of General Henry "Hap" Arnold standing alongside a globe, his index finger pointing to some distant trouble spot in the world, perhaps subtly embodying the USAF vision statement of "Global Vigilance, Reach and Power." The Academy's Cemetery Memorial Wall is a place where "air-related units remember comrades who died as a result of combat," and its Honor Garden features, among numerous other famous World War II aircraft, a 1/6-scale bronze B-24 Liberator mounted on a three-ton polished blue granite base bearing the message "To honor all the brave airmen lost, the valiant ones who survived, and to all who designed, produced, maintained, and flew this stalwart aircraft."

Above: *The stained glass memorial to the World War II B-24 crew of the* Lady Be Good, *once displayed at the Wheelus AFB chapel in Libya, is now on display at the National Museum of the USAF. (USAF)*

The National Museum of the United States Air Force at Wright-Patterson Air Force Base has a Memorial Park with more than 400 plaques and statuary monuments that commemorate fighter wings, bomber wings, aircraft maintenance units, engineer aviation battalions, air refueling squadrons, troop carrier squadrons, and hundreds of other commands of all sizes and types.

The majority of memorials are dedicated to specific units in specific wars, and probably none can claim more memorials than the U.S. Eighth Air Force in World War II. The Eighth had among the highest casualties ever suffered by a military unit, accounting for more than 26,000 men killed in action, and 6,000 bombers and fighters lost between August 1942 and May 1945. The number of memorials is commensurate with the staggering losses. Cambridge American Cemetery, located near Madingly, England, is the final resting place for 3,812 Americans, many from the Eighth Air Force. On Cambridge's Wall of the Missing are the names of 5,126 more who died, but whose remains were never recovered or identified. Scattered across the gently undulating landscape of East Anglia where the Eighth's 120 bomber and fighter stations once stood, poignant memorials now mark the last place heroic American airmen set foot on earth. Many of the memorials were placed with the help of a grateful local populace. Near Savannah, Georgia, where the Eighth was activated in January 1942, the Chapel of the Fallen Eagles and Memorial Gardens were created on the grounds of the Mighty Eighth Air Force Museum. The chapel is designed as a composite of the many Gothic chapels dotting the rural English countryside, where Eighth Air Force personnel often worshipped during the war.

Other monuments remember groundbreaking accomplishments like those of the Tuskegee Airmen, commemorated at the former Walterboro Army Airfield, now a regional airport outside Walterboro, South Carolina, and the Women Airforce Service Pilots (WASPs) at the former Avenger Field in Sweetwater, Texas—wartime training sites for each respectively. Both also have memorials at Wright-Patterson, the Air Force Academy, and other locations.

Across the United States, the Air Force is recognized along with America's other armed forces in the war memorials of countless town squares and parks, and in national memorials in Washington, D.C. Our nation's most hallowed ground at Arlington National Cemetery is the eternal resting place for thousands of air veterans, including Hap Arnold, Claire Chennault, Jimmy Doolittle, Thomas E. Selfridge, and Daniel "Chappie" James Jr. On the grounds are numerous memorials to air and space

Above: *The Cambridge American Cemetery, Madingly, England. (Courtesy of David A. Lande)*

Left: *A concept drawing of the United States Air Force Memorial scheduled for completion in the fall of 2006. (Pei Cobb Freed & Partners Architects LLP)*

Below: *This stained glass window is one of many that are part of the Mighty Eighth Air Force Museum's Chapel of Fallen Eagles in Savannah, Georgia. The 100th Bomb Group—also known as the "Bloody Hundreth" because of their casualty rate—is memorialized in this beautiful glass work. (Courtesy of Michelle Gagner)*

power, such as the space shuttle *Challenger* monument, the "Second Schweinfurt Memorial" tree plaque commemorating the costly bombing raid on Schweinfurt, Germany, on October 14, 1943, and living memorials in the form of trees dedicated to many bombardment groups and the Glenn Miller/AAF Orchestra. Near the entrance of Arlington Cemetery is the Women in Military Service for America Memorial, which prominently recognizes women of the Air Force. Also in Washington, the Vietnam Veterans Memorial has the names of 2,581 Air Force KIAs on "the Wall," and the Korean War Veterans Memorial has images of airmen and F-86 Sabres among its 2,400 ghostly photographic etchings on polished black granite. And the World War II Memorial remembers the valor of the U.S. Army Air Forces in the form of bas-relief panels, inscriptions, the Freedom Wall honoring America's war dead, and the USAAF seal in bronze alongside the sister services' seals.

Memorials look forward in time as well as backward. They remind us of what is best about our nation and why we defend it—and give us a legacy to uphold in our future. Perhaps the legacy of memorials to airmen is best expressed by Hap Arnold, whose words are inscribed at the entrance of the Air Force Memorial: "It is the American people who will decide whether this Nation will continue to hold its air supremacy... In the final analysis, our striking force belongs to those who come from the ranks of labor, management, the farms, the stores, the professions and colleges and the legislative halls... Air Power will always be the business of every American citizen."

A NOTE ABOUT SOURCES

A number of important studies were essential to the creation of this work. Many are official Army and Air Force works that have been published by service historical research agencies. For the Great War, Mauer Maurer's, *The U.S. Air Service in WW I* (Office of Air Force History, 4 vols.) was essential. For World War II, those desiring an exhaustive chronological recounting of air operations should consult Carter and Mueller's, *The Army Air Forces in World War II: Combat Chronology, 1941-1945* (Office of Air Force History, 1973), the companion volume to the *History of the Army Air Forces in WW II* (commonly known as Craven and Cate). From the Army "Green Books," Mary H. Williams' *United States Army in World War II: Chronology, 1941-1945* (Center of Military History, 1989) provided the details of ongoing surface engagements that were supported by the Army Air Forces. Dr. A. Timothy Warnock's *Air Force Operations During the Korean War* is definitive. This study helped to shape the foundation of the chronological entries in Chapter VIII. Those wishing to review the detailed official narrative history of the Air Force in the Korean War should refer to Robert F. Futrell's work, *The United States Air Force in Korea, 1950-1953* (Duell, Sloan, and Pierce, 1961). Eliot A. Cohen's *Gulf War Air Power Survey (GWAPS), Vol. V: A Statistical Compendium and Chronology* goes into significant details of air operations during DESERT SHIELD/DESERT STORM. For Medal of Honor entries, see Barrett Tillman's *Above and Beyond: The Aviation Medals of Honor* (Smithsonian Institution Press, 2002). Additional sources:

AFP-210-1-1, *Historical Data: A Chronology of American Aviation Events* (Department of the Air Force, 1955)

Meghan Cunningham, *Logbook of Signal Corps No. 1, The U.S. Army's First Airplane* (Air Force History and Museums Program, 2004)

Eugene Emme, *Aeronautics and Astronautics, 1915-1960* (NASA, USGPO, 1961)

Daniel L. Haulman, *One Hundred Years of Flight: USAF Chronology of Significant Air and Space Events, 1903-2002* (AF History and Museums Program, 2003)

Peter R. March, *Sabre to Stealth: 50 Years of the United States Air Force* (Royal Air Force Benevolent Fund Enterprises, 1997)

Arthur George Renstrom, *Wilbur and Orville Wright: A Reissue of a Chronology Commemorating the Hundredth Anniversary of the Birth of Orville Wright—August 19, 1871* (NASA Publication SP-2003-4532)

Frederick J. Shaw Jr. and A. Timothy Warnock, *The Cold War and Beyond: Chronology of the United States Air Force, 1947-1997* (Air Force History and Museums Program, 1997)

Frank H. Winter and F. Robert Van Der Linden, *100 Years of Flight: A Chronology of Aerospace History, 1903-2003* (American Institute of Aeronautics and Astronautics, 2003)

A significant number of Internet resources are available and almost every military air unit operates a web page. Many of these contain unit chronologies detailing combat tours, deployments, and significant unit events. I found the following sites useful as general resources.

Air Force History Support Office Web Site
https://www.airforcehistory.hq.af.mil/index.htm

Air Force Magazine Online, Up from Kitty Hawk: A Chronology of Aerospace Power since 1903.
http://www.afa.org/magazine/kittyhawknew/kittyhawk.asp

American Institute of Aeronautics and Astronautics (AIAA) History of Flight Chronology Web Page
http://www.aiaa.org/content.cfm

History of Flight, U.S. Centennial of Flight Commission
http://www.centennialofflight.gov/timeline/search_timeline.cfm

National Aeronautical Association, Mackay Trophy
http://www.naa.aero/html/awards/index.cfm?cmsid=70

National Aeronautics and Space Administration, NASA History Division, History Timelines
http://history.nasa.gov/timeline.html
http://www.hq.nasa.gov/office/pao/History/Defining-chron.htm

This list is not exhaustive, but highlights the most useful resources. The difficulty in assembling any chronological history is that, at times, dates do not agree between sources. When this occurred, the official sources took precedence unless primary documents revealed more convincing evidence to the contrary.

CREDITS

Photographs
(USAF)
All USAF photos are from the collection of the Air Force Association (*Air Force Magazine*) or from the Official Air Force web page—Air Force Link—(*http://www.af.mil/photos/index.asp*) unless otherwise noted.

(NASM)
The following images are used with permission from the National Air and Space Museum archives (NASM images used with Boeing special permission are in BOLD type):

CH I
11; 2002-16646
14; 2002-15559
16; 2003-30820
17; 74-11523
18; 2003-30818, 2003-12985
20; 73-1859
24; A-3615
26; 80-13477, 2002-16644
27; 95-8963
28; 72-9661, 75-11635
30; 2B31051
31; 97-15334
32; A-43519-G
33; 2003-12491, 85-18299
36; 85-8305
37; 78-15867, 78-10241
39; 72-10099
41; 77-711
42; 2003-12979
43; 2003-33191, 93-16253
45; 73-4311
47; 86-4711

CH II
53; 98-20430
54; 96-16162
55; 90-4832, 75-6979
59; 95-2855
61; 75-16334
65; 2005-22942
67; 91-3993
70; 2004-11948
71; 78-2721
75; 97-16507
76; 85-12301

CH III
86; 86-13274
88; 2002-2568
93; 77-6903
96; 81-876
97; 77-6876
98; 040309-F-9999G-003
99; 99-15615
100; 80-10443
101; 78-4647
102; 88-7415
104; 80-2080
105; 84-3662
107; 94-7993
110; 79-9405
113; 75-5419
114; 00013491
120; A-45869-D
121; 2002-17266
122; USAF-30193AC
124; 87-751
126; 80-16287
127; 71-1060
129; **87-6048**
130; 74-3210
131; **82-1729**
132; 82-8315, 9A02864
133; 20680
134; 82-14280, 97-15363
135; 92-4102

CH IV
143; 1980.049
153; 91-3946, 98-20685
154; 87-9655
155; 76-17318, 81-880
160; 90-2374
162; 83-16107
170; 92-9387
177; 98-20678
179; 99-15448
182; 91-1471
184; A-4567-F
185; 96-15604
188; 86-5932
191; 2000-1601

CH V
196; 2005-8169
197; 77-12895
198; A-672
201; 85-7751
207; 86-12533
208; A-1108
214; 76-3443
216; 99-15470
220; 84-2978
221; 97-16973
226; 83-8718
233; USAF-57193AC
237; 98-15407
239; 86-1003

CH VI
244; 9A03546
249; 2006-6753
253; 76-13911
259; 98-15681
260; 87-13689, 2003-6574
268; 86-12160
270; A01595
271; 75-5394
272; 99-42697
273; 2004-41000
274; 76-17743
278; 87-9675
280; 2003-18753, 9A03248, 98-15873, 2003-18754
281; 89-22018
283; 88-10680

CH VII
289; 87-5932
290; 77-1012
291; 89-21534
295; 76-2221
296; 97-17485
299; **86-323**
300; USAF-33548AC
301; 94-1904, 98-15403
304; 97-15291, 95-3378, **89-1886**, **88-12570**
308; 84-3724
310; **89-2985**, 93-4045
311; 76-17593
314; 98-15404
315; 92-1254, 79-759
316; 2003-4844, 2005-8166
319; 90-13787
320; 87-5952
321; 91-19685, 76-3183
322; 92-704, **74-3346**

CH VIII
329; 81-10617, 93-11858
332; 85-11309, 92-4095
333; 73-8710, 82-8321
335; 90-5962
336; 79-14866
342; 88-11876
346; 9A01571
348; 97-1357
350; 88-87
351; 80-14001
352; 80-5653
353; 85-12339, 88-10683
355; 72-7759, **90-12956**
360; **83-8904**
362; 89-5914
364; 95-8684
367; 75-3050
370; 89-9935
371; 99-15491
373; 75-4846
374; 81-5068, 83-9139
375; 93-12553

CH IX
380; 82-5416
381; 99-41458, 2003-44751
382; 78-15952, 71-1352
383; 75-14812, **92-4094**
384; 91-7002
387; 80-8646
388; 2003-35053
392; 2004-17767, 83-5355
393; 92-6569
394; 92-7029
395; **2005-8165**
399; **87-741**
402; 89-5898
403; 94-4517
405; 96-15333
406; 2004-28035
408; 93-4160
409; 99-40835
416; 93-16025, 2003-4990
417; 93-4176
421; 84-9594
428; 92-6566
432; **97-17041**
433; 92-4053

CH X
438; 95-8693
448; 2006-622
449; 90-7046
456; 97-15343
457; 7A32766
466; 2003-40702
467; 79-833
470; 75-15198
473; 94-10743
482; 83-5366

CH XI
495; 2000-753
530; 2001-7671

CH XII
547; 2000-9346-640

(NASA)
All credited NASA images can be located on the NASA photo web site: *http://grin.hq.nasa.gov/subject-space.html*

Art
(U.S. Army Art Collection)
All U.S. Army Art Collection works are used with permission of the Center of Military History, Army Art Collection.

(NASM Art Collection)
All NASM Art Collection works are used with permission of the National Air and Space Museum, the artist, or the artist's estate. Special thanks to:
Robert T. McCall, William S. Phillips, Robert Taylor, Mrs. Chet Engle, Ms. Christie Woodfin, Phyllis S. Corbett McKenzie, The Leach International Corporation, and the Flight Safety Foundation, Inc.

(U.S. Air Force Art Collection)
All Air Force art is used with the permission of the USAF Art Program. Additional permission to publish USAF art was received from these artists:

John F. Amendola, Richard Allison, Gerald W. Asher, Steven B. Bricker, Melvin S. Brown, Frances Caraker, Gil Cohen, Harley F. Copic, James J. Dietz, Keith Ferris, Steve W. Ferguson, George Guzzi, Rick Herter, Paul R. Jones, Arthur C. Kane, Paul Kreutziger, Michael J. Machat, Robert T. McCall, Nicholas Mosura, William S. Phillips (copyright retained by artist), Kenneth Price Randel (with special thanks to Lockheed Martin, who sponsored his works), Mark Waki, Andrew C. Whyte, and John Witt.

Special permission for additional artwork was given by: James J. Dietz, Tony Fachet, Jack Fellows, Keith Ferris, William S. Phillips, and Larry Selman. (Copyrights retained by artists).

Maps
Maps used in this volume are compiled from a variety of government sources and created by Mr. Guy Aceto for use in this book.

ABOUT THE AUTHOR

Dik Alan Daso (BS, USAF Academy; MA, PhD, University of South Carolina) is curator of Modern Military Aircraft at the Smithsonian Institution, National Air and Space Museum, Washington, D.C. A retired Air Force lieutenant colonel and command pilot, he has logged more than 2,700 flying hours in RF-4C Phantom II and F-15A Eagle fighters and T-38 Talon supersonic training aircraft. During his Air Force career, he also served as a history instructor at the USAF Academy, an executive officer on the Air Force Scientific Advisory Board, and chief of Air Force doctrine in the Pentagon.

Aside from his Air and Space Museum duties at the Smithsonian, he also served as co-curator for a new permanent exhibition at the National Museum of American History—*The Price of Freedom: Americans at War*. This 18,000-square-foot exhibition examines America's wars as pivotal events in American history and spans the years 1755–2005. He edited and compiled the exhibition books for the National Air and Space Museum, Steven F. Udvar-Hazy Center, and for *The Price of Freedom* exhibit.

He has written chapters that were included in *West Point: Two Centuries and Beyond* (McWhiney Foundation Press, 2004) and *The Air Force* (Hugh Lauter Levin Associates, Inc., 2002) and was the editor of the "Ancillary Volume" of *New World Vistas: Air and Space Power for the 21st Century* (USAF Scientific Advisory Board, 1995). His books include *Architects of American Air Supremacy* (Air University Press, 1997); *Hap Arnold and the Evolution of American Airpower* (Smithsonian Institution Press, 2000), which won the American Institute of Aeronautics and Astronautics (AIAA) History Manuscript Award; and *Doolittle: Aerospace Visionary* (Potomac Books, 2003).

CONTRIBUTORS (ALPHABETICAL)

C. R. Anderegg (BA, Hobart College; MS, Troy University) is director of Air Force History and Museums Programs and Policies, Headquarters USAF, Washington, D.C. A retired Air Force colonel and command pilot with more than 3,800 hours in fighters, he also flew 170 combat missions in Vietnam. His publications include *The Ash Warriors* (PACAF History Office, 2000); *Sierra Hotel: Flying Air Force Fighters in the Decade after Vietnam* (Office of Air Force History, 2001); and *The Shootdown of Trigger Four* (HQ USAF, 2001).

John D. Anderson Jr. (BAE, University of Florida; PhD, The Ohio State University) is the curator for aerodynamics at the Smithsonian's National Air and Space Museum. He served as an Air Force lieutenant and scientist at Wright Field in Dayton, Ohio from 1959-1962. He chaired the Department of Aerospace Engineering at the University of Maryland in 1973. He has published ten books including, *A History of Aerodynamics* (Cambridge University Press, 1997) and *The Airplane: A History of Its Technology* (American Institute of Aeronautics and Astronautics, 2002). He is an Honorary Fellow of the American Institute of Aeronautics and Astronautics and a Fellow of the Royal Aeronautical Society.

Debra Baer Becker (BA, University of Calgary) is a freelance writer and former member of the Thunderbirds team. She is a director of the Thunderbirds Alumni Association and a feature writer for their magazine, *Thunder Rolls*. Debra lives in Portland, Maine.

Alfred M. Beck (BA, St. Francis College, Pennsylvania; MA, PhD, Georgetown University) is a former chief editor of the Air Force History program. Retired since 1994, he is sole owner of Habemus Duam, Ink, a firm concentrating on historical research, editing, and publication consulting. His publications include *Hitler's Ambivalent Attaché: Lt. Gen. Friedrich von Boetticher in America* (Potomac Books, 2005); *With Courage: U.S. Army Air Forces in World War II* (Air Force History, 1995); and *The Corps of Engineers in the War Against Germany* (U.S. Army Center of Military History, 1985).

Walter J. Boyne (BSBA, University of California, Berkeley; MBA, University of Pittsburgh; Honorary Doctorate, Salem University) is a retired Air Force colonel with 5,000 hours flying time. A former director of the National Air and Space Museum, he has an extensive consulting clientele in industry, publishing, and television. His publications include some 50 books, including seven novels. Their titles include *The Influence of Air Power on History*; *Beyond the Wild Blue*; and *Dawn over Kitty Hawk*. He is one of the few authors to have had best-sellers on both the fiction and non-fiction lists of *The New York Times*.

Clayton K. S. Chun (BS, University of California, Berkeley; MS, University of Southern California; MA, University of California, Santa Barbara; PhD, RAND Graduate School) is chairman of the Department of Distance Education at the U.S. Army War College, Carlisle Barracks, Pennsylvania. He is a retired Air Force officer with assignments in missile, space, education, and strategy development. He has published widely in the areas of national security, space and missile history, military history, and economics.

James G. "Snake" Clark (BA, Catholic University; MA, Troy State University) is a senior USAF civilian executive and serves as director, Air Force Combat Support at HQ USAF Pentagon. A retired Air Force colonel, he served as a pilot and weapons controller in a number of operational and staff positions, including the 527th USAF Aggressor Squadron, in his 28-year career. Currently, he is the USAF's expert in Unmanned Aerial Vehicle (UAV) combat operations.

Mark Clodfelter (BS, USAF Academy; MA, University of Nebraska; PhD, University of North Carolina) is professor of military history at the National War College, Washington, D.C. A retired Air Force lieutenant colonel and radar officer, he has 20 years of teaching experience, including service at the Air Force Academy, Air Force School of Advanced Air and Space Studies (SAASS), University of North Carolina, and National War College. He is the author of *The Limits of Air Power: The American Bombing of North Vietnam* (Free Press, 1989; Bison Books, 2006).

Martin Collins (PhD, University of Maryland) is a curator in the Space History Division, Smithsonian Institution, National Air and Space Museum, Washington, D.C. He has examined the history of RAND and the Air Force in two related studies: The RAND History Project, an oral history of dozens of RAND and AF personnel, and *Cold War Laboratory: RAND, the Air Force, and the American State, 1945–1950* (Smithsonian Institution Press, 2002).

Conrad C. Crane (BS, USMA; MA, PhD, Stanford) is currently director of the U.S. Army Military History Institute in Carlisle, Pennsylvania. A retired Army lieutenant colonel, he spent 10 years as professor of history at the U.S. Military Academy and held the Douglas MacArthur Chair of Research at the U.S. Army War College. His publications include *Bombs, Cities, and Civilians: American Airpower Strategy in World War II* (University Press of Kansas, 1993) and *American Airpower Strategy in Korea, 1950–1953* (University Press of Kansas, 2000).

Tom D. Crouch (BA, Ohio University; MA, Miami University; PhD, The Ohio State University) is senior curator of aeronautics at the Smithsonian's National Air and Space Museum. He has written many books on the history of flight technology, including *A Dream of Wings* (1981); *Eagle Aloft: Two Centuries of the Balloon in America* (1983); *The Bishop's Boys: A Life of Wilbur and Orville Wright* (1989); *Wings: A History of Aviation from Kites to the Space Age* (2003); and *Rocketeers and Gentlemen Engineers: A History of the American Institute of Aeronautics and Astronautics and What Came Before* (2005).

Tami Davis Biddle (BA, Lehigh University; M.Phil., University of Cambridge; PhD, Yale University) is the George C. Marshall Chair of Military Studies in the Department of National Security and Strategy at the U.S. Army War College, where she teaches national security strategy and military history. Her research focus is warfare in the twentieth century, in particular the history of air warfare and the history of the Cold War. Her book *Rhetoric and Reality in Air Warfare: The Evolution of British and American Ideas about Strategic Bombing, 1914–1945* (Princeton University Press, 2002) was a *Choice* outstanding academic book, 2002, and was recently added to the Royal Air Force Chief of Air Staff's Reading List.

Dennis M. Drew (BA, Willamette University; MS, University of Wyoming; MA, University of Alabama) is a professor and associate dean at the School of Advanced Air and Space Studies, Air University, Maxwell AFB, Alabama. A retired Air Force colonel, he has published and lectured extensively on military and airpower history and theory in books and leading academic journals in the United States and overseas. His recent books include *Lexington to Desert Storm and Beyond* (M.E. Sharpe, 2000) and *Making 21st Century Strategy* (Air University Press, 2006)

Carroll V. Glines (BBA, MBA, University of Oklahoma; MA, American University) is curator of the Doolittle Library, Special Collections, University of Texas, Dallas. A retired Air Force colonel, he has logged 5,600 hours in more than 35 types of aircraft. He has been editor of *Air Cargo*, *Air Line Pilot*, and *Professional Pilot* magazines and has written 36 books including *Doolittle's Tokyo Raiders; Chennault's Forgotten Warriors; Around the World in 175 Days;* and *Those Legendary Cubs*, biographies of Roscoe Turner and Bernt Balchen, and has co-authored the autobiographies of Generals James H. Doolittle and Benjamin D. Foulois.

Rebecca Grant (BA, Wellesley College; PhD, London School of Economics, University of London) is founder and president of IRIS Independent Research. Formerly with RAND in Santa Monica, California, she worked on missile proliferation issues and analyzing Operation Just Cause in Panama. She has lectured at Air University and Air Command and Staff College, Maxwell AFB, and is the author of *The First 600 Days of Combat*, an official Air Force report on the war on terrorism; *The B-2 Goes to War* (2001); and *The Radar Game* (1999), as well as numerous magazine articles on aspects of air and space power.

Thomas E. Griffith Jr. (BS, USAF Academy; MA, University of Alabama; PhD, University of North Carolina) is commandant of the School of Advanced Air and Space Studies at Maxwell AFB, Alabama. He has published a variety of articles on targeting strategies, military innovation, and is the author of *MacArthur's Airman: General George C. Kenney and the War in the Southwest Pacific* (University of Kansas Press, 1998). Colonel Griffith has over 2,000 hours in the F-4 and the F-15E and has had a variety of operational, command, and staff assignments. He was a prisoner of war during Operation Desert Storm.

John F. Guilmartin Jr. (BA, USAF Academy; MA, PhD, Princeton University) is a professor of history at The Ohio State University, where he teaches military history, early modern European history, and the history of the Vietnam War. A retired Air Force lieutenant colonel and senior pilot, he completed two Southeast Asia combat tours, flying HH-3E "Jolly Greens" in 1965–1966 and HH-53C "Super Jolly Greens" in 1975, including participation in the Saigon evacuation. His publications include *Gunpowder and Galleys: Changing Technology and Mediterranean Warfare at Sea in the Sixteenth Century* (1974, 2nd ed. 2003) and *America in Vietnam: The Fifteen-Year War* (1992).

Richard P. Hallion (BA, PhD, University of Maryland) is senior adviser for Air and Space Issues, Headquarters U.S. Air Force, responsible for analysis and insight regarding sensitive national aerospace programs and related subject areas. He teaches and lectures widely, has been a visiting professor at the U.S. Army War College and the National Air and Space Museum of the Smithsonian Institution, and is the author and editor of numerous books, articles, and essays relating to aerospace technology and operations, most recently *Taking Flight: Inventing the Aerial Age* (Oxford University Press, 2003).

Von Hardesty (PhD, The Ohio State University) is a curator at the Smithsonian National Air and Space Museum. He has published widely on aviation topics, including *Red Phoenix: The Rise of Soviet Air Power, 1941–1945; Air Force One: The Aircraft That Shaped the Modern Presidency; Great Aviators and Epic Flights; Lindbergh: Flight's Enigmatic Hero*; and *Black Aviator: The Story of William J. Powell*, among other publications. He served as a curator for the exhibition *Black Wings: The American Black in Aviation*.

Todd P. Harmer (BS, USAF Academy; MS, Embry-Riddle; MAS, Naval War College; MAAS, School of Advanced Airpower Studies; MAS, National War College) is vice commander, 388 Fighter Wing, Hill Air Force Base, Utah. He is an Air Force colonel and command pilot with over 3,000 hours in fighter aircraft, including combat experience in Southwest Asia. In addition to numerous articles on fighter tactics, his publications include *Enhancing Operational Art: The Command and Control of Airpower* (Air University, 2000) and *The Shootdown of Trigger Four* (HQ USAF, 2001).

Peter L. Jakab (BA, MA, PhD, Rutgers University) is chairman of the aeronautics division at the Smithsonian Institution, National Air and Space Museum, Washington, D.C. He is curator of early aviation at NASM and has produced numerous exhibitions and publications on the Wright brothers and the invention of the airplane. His books include *Visions of a Flying Machine: The Wright Brothers and the Process of Invention* (Smithsonian Institution Press, 1990); *The Published Writings of Wilbur and Orville Wright* (Smithsonian Institution Press, 2000); and *The Wright Brothers and the Invention of the Aerial Age* (National Geographic, 2003).

Frederick Kiley (BS, University of Massachusetts; MA, Trinity University; PhD, University of Denver) is a professor in the Department of English at the Air Force Academy. A retired Air Force colonel and Vietnam veteran, he was formerly director of National Defense University Press and the NDU Senior Fellows Program. His publications include *Honor Bound* (Naval Institute Press, 1999)—a history of American POWs during the Vietnam War and Pulitzer Prize runner-up—*A Catch-22 Casebook* (T.Y. Crowell, 1973); and *Satire from Aesop to Buchwald* (Odyssey Press, 1971).

Jeremy R. Kinney (BA, Greensboro College; MA, PhD, Auburn University) is curator of aeronautical propulsion systems and interwar military aircraft at the Smithsonian Institution, National Air and Space Museum, Washington, D.C. In 2003, he served as the Centennial of Flight lecturer at the University of Maryland at College Park. His publications include *Airplanes* (Greenwood Press, 2006), and he co-authored *The Wind and Beyond: A Documentary Journey Through the History of Aerodynamics in America* (Government Printing Office, 2003).

David A. Lande (BA, Concordia College; MA, University of Wisconsin Oshkosh) served on the staff of the World War II Memorial in Washington, D.C., and contributed to *The World War II Memorial: A Grateful Nation Remembers* (Smithsonian Institution Press, 2004). He is the author of seven books about World War II and aviation, including *From Somewhere in England* about the U.S. Eighth Air Force. His work has appeared in *MHQ; Aeroplane; World War II; Aviation History*; and *Sport Aviation*, among others. He served as a commissioned officer in the air wing of the U.S. Marine Corps.

Roger D. Launius (BA, Graceland College; MA, PhD, Louisiana State University) is chair of the Division of Space History at the Smithsonian Institution's National Air and Space Museum in Washington, D.C. He served as NASA chief historian, 1990–2002. He is the author of many books on aerospace history, including *Space Stations: Base Camps to the Stars* (Smithsonian Books, 2003), which received the AIAA's history prize; *To Reach the High Frontier: A History of U.S. Launch Vehicles* (University Press of Kentucky, 2002), with Dennis R. Jenkins; and *Imagining Space: Achievements, Possibilities, Projections, 1950–2050* (Chronicle Books, 2001), with Howard E. McCurdy.

Stephen L. McFarland (BA, University of Kansas; MA, PhD, University of Texas) is associate provost for academic affairs, dean of the graduate school, and professor of history at Auburn University, Auburn, Alabama. His publications include *America's Pursuit of Precision Bombing, 1910–1945* (Smithsonian Institution Press, 1995 & 1997); *A Concise History of the U.S. Air Force* (Government Printing Office, 1997); and *To Command the Sky: The Struggle for Air Superiority over Europe, 1942–1944* (Smithsonian Institution Press, 1991 & 1993 & 2002; University of Alabama Press, 2006) with Wesley P. Newton.

Phillip S. Meilinger (BS, USAF Academy; MA, University of Colorado; PhD, University of Michigan) is a senior analyst in Northrop Grumman's Analysis Center. A retired Air Force colonel, pilot, staff officer, and educator, he has flown all over the world, served in the Pentagon, and taught at the Air Force Academy, Air University, and the Naval War College. His publications include *Hoyt S. Vandenberg: The Life of a General* (Indiana University Press, 1989); *Airwar: Theory and Practice* (Frank Cass, 2003); and *The Paths of Heaven: The Evolution of Airpower Theory* (Air University Press, 1997).

David R. Mets (BS, U.S. Naval Academy; MA, Columbia University; PhD, University of Denver) is military defense analyst at Air University, Maxwell AFB, Alabama. His principal book is *Master of Airpower: General Carl A. Spaatz* (Presidio Press, 1988), and he was a command pilot in the Air Force with a total of 5,600 flying hours, including 300 in the Gooney Bird.

Lloyd W. "Fig" Newton (BS, Tennessee State University; MA, George Washington University; Honorary Doctorate, Embry-Riddle Aeronautical University and Benedict College) is vice president for military international programs and business development for Pratt & Whitney, East Hartford, Connecticut. A retired Air Force general and command pilot, he has logged more than 4,000 flying hours in the T-37, T-38, F-4, F-15, F-16, C-12, and F-117 Nighthawk. He flew 269 combat missions from DaNang Air Base, South Vietnam. He was selected to join the U.S. Air Force Aerial Demonstration Squadron—Thunderbirds—in November 1974. He held several positions including narrator, slot pilot, and right wingman.

Robert C. Owen (BA, MA, University of California, Los Angeles; MA, PhD, Duke University) is a professor of aeronautical science, Embry-Riddle Aeronautical University, Daytona Beach, Florida. A retired Air Force colonel, Professor Owen holds both command pilot and commercial pilot ratings and has flown over 4,000 hours in a variety of aircraft. His publications include the *Chronology* volume of the *Gulf War Air Power Survey* (USGPO, 1993) and *Deliberate Force: A Case Study in Effective Air Campaigning* (Air University, 2000).

Jeffery S. Underwood (BA, MA, University of Oklahoma; PhD, Louisiana State University) is the National Museum of the United States Air Force historian at Wright-Patterson AFB, Ohio. He has worked as a historian for the U.S. Air Force since 1988. His publications include *The Wings of Democracy: The Influence of Air Power on the Roosevelt Administration, 1933–1941* (Texas A&M University Press, 1991) and articles and book reviews for professional journals and magazines. He has given television and radio interviews for the History Channel, PBS, National Geographic, BBC, C-SPAN, NPR, CNN, MSNBC, CBS, and others.

George M. Watson Jr. (BA, University of Maine; MA, Niagara University; PhD, The Catholic University of America) is a veteran of the Vietnam War serving with the 101st Airborne Division from 1969 to 1970. A long-time historian for the Air Force, he is the author of *The Office of the Secretary of the Air Force, 1947–1965* (Center for AF History, 1993) and co-author of *With Courage: The Army Air Forces in World War II* (AF History and Museums Program, 1995). His latest book is *Biographical Sketches: Chiefs of Staff and the Secretaries of the Air Force* (AF History and Museums Program, 2002).

Kenneth P. Werrell (BS, USAF Academy; MA, PhD, Duke University) served briefly in the USAF piloting weather reconnaissance aircraft before retiring from Radford University after 26 years of teaching. He has lectured at Canadian, Army, and USAF service schools, and authored seven books on military aviation, the most recent of which are *Sabres over MiG Alley: The F-86 and the Battle for Air Superiority in Korea* (Naval Institute, 2005); *Archie to SAM: A Short Operational History of Ground-Based Air Defense* (Air University, 2005); and *Chasing the Silver Bullet: U.S. Air Force Weapons Development from Vietnam to Desert Storm* (Smithsonian Institution Press, 2003).

Robert P. White (MA, Georgetown University; PhD, The Ohio State University) is the historian for the Air Force Office of Scientific Research in Arlington, Virginia. He retired from the U.S. Air Force after a 21-year career. He initially served as a signals intelligence officer in positions that included squadron operations officer and squadron commander. He served as an Air Force historian and chief of air staff history at the Pentagon. His biography on Major General Mason Patrick was published by the Smithsonian Institution Press in 2001.

Sheila Widnall (B.Sc., MS, and Sc.D., Massachusetts Institute of Technology) served as Secretary of the Air Force from 1993 to 1997. Under her watch, together with CSAF General Ronald Fogleman, the Air Force issued its long-range vision, *Global Engagement*, which laid out the path for the integration of air and space and technology-enabled core capabilities. Secretary Widnall articulated the Air Force's core values—integrity, service above self, and excellence—integrating these into every facet of Air Force life. Dr. Widnall is an Institute Professor at MIT and has been a member of the faculty since 1964. She was born in Tacoma, Washington, near McChord Field.

George K. Williams (BS, U.S. Military Academy; MA, Cornell; D.Phil., Oxford) was executive director of the Air Force Historical Foundation during the genesis of this book. A retired Air Force colonel with 3,000 flying hours in the AWACS aircraft, he served initially with the Army for 10 years, including a tour in Vietnam as a platoon leader and company commander in the First Squadron, First Cavalry. He has published *Biplanes and Bombsights: British Bombing in World War I* (Air University Press, 1999).

ACKNOWLEDGMENTS

The completion of this book was made possible by the contributions and dedication of a sizeable cast of players—a team that provided essays, images, editing skills, design expertise, artistic talents, historical research, and inspiration. To all of these contributors I owe a tremendous amount of thanks.

At the Smithsonian's National Air and Space Museum, the archives and library staff provided superhuman support in gathering images used throughout the book. Kristine Kaske Martin and Kate Igoe responded to endless photo requests with a smile (most of the time) and were able to quickly address issues of permission and photo quality that are reflected in these pages. Allan Janus, Melissa Keiser, Brian Nicklas, Dan Hagedorn, and Dana Bell all assisted me in finding the most appropriate photo for each page. Barbara Weitbrecht was also instrumental in developing a way to accomplish photo research electronically outside of the library, a tremendous time-saver.

The museum's photo staff provided the artifact images and accomplished much of the scanning and processing of art transparencies used throughout the book. Mark Avino, Eric Long, Dane Penland, and Carolyn Russo are the best in the business and I am thankful for the time and effort made by each during this project.

Trish Graboske handled early contracting issues with grace. Her work has allowed this collaboration—unique in this military series—between the National Air and Space Museum and the Air Force Historical Foundation.

I offer my thanks to George Schnitzer, who is a volunteer in the Division of Aeronautics. George accomplished the monumental task of tracking down the permission granting authorities for a number of paintings in the Air and Space Art Collection. His diligence has allowed us to publish several pieces of art that have never been seen in print before.

I am also grateful for the efforts of three U.S. Air Force Academy cadets who served as interns during the research phase of this book. C1C Tiffany Scheivert, C1C Erin Frazier, and C1C Suzie Crespo crafted the chronological entries for the Medal of Honor and did the tedious work of researching and validating a complete list of commanders of the air arm from 1907 to 2006. It was refreshing to see the enthusiasm that these cadets brought to Washington during their assignment here.

The USAF Art Collection staff provided massive support during this project. Leading the division, Rusty Kirk has been gracious and patient with repeated requests for transparencies and slides. Nick Mosura, a talented artist in his own right, Greg Thompson, and Milt Lawler have gathered the required materials and researched dozens of questions about artists and captions with care and consideration.

The Air Force was originally a branch of the Army. Ms. Renee Klish, curator of the U.S. Army Art Collection, was kind enough to offer me and three of my summer interns a personal tour of the Army's stored paintings. Her complete knowledge of the collection allowed me to add several pieces of aviation-related Army art to the early chapters. I am indebted to her and the U.S. Army for their participation.

The Office of Air Force History in the Pentagon under the direction of Mr. Dick Anderegg has supported this project from the beginning. Without open access to previous research projects and the expertise of several Air Force historians, completion of this book would have taken much longer than it did. The Air Force Association and *Air Force Magazine* staffs opened their archives for use during this project. For the past five decades, the *Air Force Magazine* has run articles on just about every major figure, event, or technological development related to the evolution of the USAF. Their archives are a treasure trove of Air Force images, stories, and facts that were often helpful in clearing discrepancies in dates or details of events. Leading the charge in the process of filtering through the thousands of images in the *Air Force Magazine* archives was Guy Aceto. A gifted photojournalist, Guy helped me locate appropriate images for the text and also created maps and graphics for the book. He provided never-before-published images and located unique photos to support my sometimes impossible requests. Without Guy's help, this book would be far less visually exciting. Thanks Guy, you did a great job.

I offer warm thanks to my friend Robert Arnold and his new bride, Kathy, who has always been supportive of my Air Force books. The grandson of General Hap Arnold, Robert's keen eye for history and his gracious permission to publish several photographs from his family collection enrich this book and make it possible to document early Air Force history in a way that no other collection can.

There are not enough words or amount of praise that could possibly express my gratitude to those at Hugh Lauter Levin Associates, Inc., that have been involved in this project: series editor, James O. Muschett; designer, Charles Ziga; copy editor, Deborah T. Zindell; and design production, Chris Berlingo. Most of all, I am indebted to the project editor, Melissa C. Payne, who became the best "wingman" that a fighter pilot could ever have. She kept the book on schedule, organized the massive computer image files into a useable mass of data, read and provided technical editing for the entire manuscript, provided motivation when I needed it, and a swift kick when appropriate. Her infectious giggle was my signal that, despite lack of sleep from pulling an all-nighter, everything was on-track. Melissa's efforts made this book come to life and it has been my greatest pleasure to work with such a tremendous professional.

This book would not have been published without the sponsorship of the Air Force Historical Foundation. Not only are they the co-publishers for the volume, the executive director, George Williams—a retired Air Force officer and historian—read each chapter. George not only offered an important outside look at the project, but also contributed to the work as one of more than 30 vignette authors. His keen eye and red pen saved me from several errors and offered improvements to style that only a writer would discern. George's contributions have made this a more readable volume for all of you. George, I salute you and the Air Force Historical Foundation.

As important as those directly involved in this book are those who supported me at home. I thank my Mom and Dad, Dick and Pat Daso, who taught me that any obstacle could be overcome by hard work and diligence. My wife Patty gave me time, space, and encouragement when I needed it. Her understanding and tolerance in the face of looming deadlines were essential to the completion of this project. I owe her thanks and a long vacation. To Lindsey, Natalie, Taylor, and Tyler, I thank you for providing me the inspiration that comes from the joy of watching you thrive at college on the difficult journey to adulthood. I am proud of all that you have achieved and anticipate the future with excitement.

—Dik Daso

INDEX

*Page numbers in italic indicate illustrations.